3rd International Conference on Materials Science and Intelligent Manufacturing (MSIM 2025)

Journal of Physics: Conference Series Volume 3009

Guangzhou, China
10-12 January 2025

ISBN: 979-8-3313-2046-1
ISSN: 1742-6588

Printed from e-media with permission by:

Curran Associates, Inc.
57 Morehouse Lane
Red Hook, NY 12571

Some format issues inherent in the e-media version may also appear in this print version.

Institute of Physics
Dirac House, Temple Back
Bristol BS1 6BE UK

Phone: 44 1 17 929 7481
Fax: 44 1 17 920 0979

techtracking@iop.org

Additional copies of this publication are available from:

Curran Associates, Inc.
57 Morehouse Lane
Red Hook, NY 12571 USA
Phone: 845-758-0400
Fax: 845-758-2633
Email: curran@proceedings.com
Web: www.proceedings.com

TABLE OF CONTENTS

Journal of Physics: Conference Series **3009** (2025) 011001 doi:10.1088/1742-6596/3009/1/011001

Preface

The 2025 3rd International Conference on Materials Science and Intelligent Manufacturing (MSIM 2025) was successfully held in Guangzhou, China, from January 10 to 12, 2025. Building upon the achievements of the previous editions, MSIM 2025 provided an international forum for researchers, engineers, and industry professionals to share their latest advancements and explore emerging trends in materials science and intelligent manufacturing.

This year, MSIM 2025 attracted about 160 participants from various countries and regions, highlighting the growing significance of these interdisciplinary fields. The conference received over 300 high-quality submissions, out of which 115 papers were accepted after a rigorous peer-review process, ensuring a high academic standard and fostering meaningful discussions.

The technical program covered a broad spectrum of topics in materials science and intelligent manufacturing, including advancements in electronic and magnetic materials, computational materials science, nanotechnology, phase transition mechanisms, intelligent production processes, digital manufacturing, engineering optimization, and manufacturing process control. These discussions facilitated knowledge exchange and fostered new research collaborations among attendees.

The success of MSIM 2025 was made possible by the dedication of our esteemed authors, reviewers, and keynote speakers, whose contributions were instrumental in maintaining the high academic quality of the conference. We extend our sincere gratitude to all participants and supporting institutions for their invaluable efforts. We hope that the knowledge shared during MSIM 2025 will inspire further research and technological advancements in materials science and intelligent manufacturing.

MSIM 2025 Organizing Committee

Journal of Physics: Conference Series **3009** (2025) 011001 doi:10.1088/1742-6596/3009/1/011001

Committee Member

Conference General Chair
Prof. Dr. Meharan Emadi Andani, University of Verona, Verona, Italy

Publication Chair
Assoc. Prof. Ata Jahangir Moshayedi, Jiangxi University of Science and Technology

Technical Program Committee Chair
Prof. Dr. Zeashan Khan, Interdisciplinary Research Center for Intelligent Manufacturing & Robotics, KFUPM, Saudi Arabia

Technical Program Committee
Prof. Vikas Gupta, Technocrats Institute of Technology, Bhopal, M.P., India
Prof. Tirumala Ramashri, SVUniversity College of Engineering, India
Prof. Chandra Singh, Department of ECE, India
Prof. Miadreza Shafie-khah, University of Vaasa, Finland
Prof. PRIOU, University Paris Nanterre, France
Prof. Ian R. McAndrew, Capitol Technology University, USA
Prof. Badrul Hisham bin Ahmad, Universiti Teknikal Malaysia Melaka, Malaysia
Prof. Yipin Wan, Chang'an University, China
Prof. Junxi Bi, Inner Mongolia University of Technology, China
Prof. Ning Sun, Nankai University, China
Prof. Nianbing Zhong, Chongqing University of Technology, China
Prof. Liangliang Tian, Chongqing University of Arts and Sciences, China
Prof. Bin Cao, Hebei University of Technology, China
Prof. Shaolin Liao, Sun Yat-sen University, China
Prof. Ning Sun, Nankai University, China
Assoc. Prof. Mingwei Qin, Southwest University of Science and Technology, China
Assoc. Prof. Mingwei, Qin, Southwest University of Science and Technology, China
Assoc. Prof. Chuanjun Zhao, Shanxi University of Finance and Economics, China
Dr. Tian Su, School of Civil Engineering and Geomatics, China

Peer Review Statement

All papers published in this volume have been reviewed through processes administered by the Editors. Reviews were conducted by expert referees to the professional and scientific standards expected of a proceedings journal published by IOP Publishing.

- **Type of peer review:** Single Anonymous
- **Conference submission management system:** Morressier
- **Number of submissions received:** 189
- **Number of submissions sent for review:** 153
- **Number of submissions accepted:** 112
- **Acceptance Rate (Submissions Accepted / Submissions Received × 100):** 59.3
- **Average number of reviews per paper:** 3
- **Total number of reviewers involved:** 25
- **Contact person for queries:**
  Name: Xuexia
  Email: xx.ye@keoaeic.org
  Affiliation: AEIC Academic Exchange Information Centre

Journal of Physics: Conference Series **3009** (2025) 012001 doi:10.1088/1742-6596/3009/1/012001

Luminescence of a new mineral beryl ($Be_3AI_2Si_6O_{18}$) for anti-counterfeiting

Xiaojuan Peng, Tuerxun Aidilibike, Bahetiguli Asilibike*

School of Electronics and Engineering, Laboratory of Optoelectronic Materials and Devices, Yili Normal University, Yining 835000, Xinjiang, China

*Corresponding author's e-mail: bahet58@163.com

Abstract In this paper, a series of phosphors of $Be_3Al_2Si_6O_{18}$: x% Eu^{3+} (x = 0.07, 0.1, 1, 2, 4) were prepared by high-temperature solid-state method. The structure of the samples was characterized by XRD, and the microstructure was further studied by X-ray photoelectron spectroscopy (XPS). The excitation spectrum and emission spectrum were measured by a spectrometer, and the color coordinates were analyzed. Studies have shown that the natural mineral beryl ($Be_3Al_2Si_6O_{18}$) in Xinjiang can be effectively excited in the ultraviolet region to form blue-green or red luminescence. On the basis of this luminescence, a series of phosphor samples doped with Eu^{3+} ions were prepared. The experimental results show that the phosphor exhibits red luminescence from matrix impurity ions and Eu^{3+} ions at the excitation wavelengths of 300 nm and 394 nm, respectively. At the excitation wavelength of 365 nm, when the doping concentration of Eu^{3+} ion is 0.1 %, the balanced red, blue, and green tricolor luminescence is formed, and the white luminescence is obtained. The color coordinates in CIE1931 are(0.3567, 0.3375), and the color temperature is 4492 K (positive white light). The research on the luminescent properties of natural mineral beryl as a matrix material has not been reported. Therefore, based on the research in this paper, it is proved that this natural mineral can be used as a new resource to promote the development of new tracks, new markets, and emerging industries in the field of luminescence, such as multiple anti-counterfeiting and LED.

1. Introduction

Currently, research on the use of beryl as a phosphor material is still in its infancy at home and abroad. Owing to its beautiful color and rare luminescent properties, beryl has always been loved by people. Therefore, the study of beryl as a phosphor matrix has a broad source of raw materials and potential applications. The crystal structure of beryl is silicate, and its structure is stable. Therefore, phosphors prepared by doping with rare earth ions have excellent luminescent properties [1-7]. The Eu^{3+} ion has a rich energy-level structure, especially its $^5D_0$-$^7F_2$ energy level transition can produce a strong red emission near 615 nm. This red emission is particularly important in the preparation of white LEDs because the red component plays a key role in improving the color rendering index and reducing the correlated color temperature [8-13]. The luminescence properties of CaF_2: Eu^{2+} nanopowders have also been studied [14]. Meng X et al. prepared red-emitting Ba_2GdVO_6: Eu^{3+} phosphor for blue-converted warm white LED [15]. Guo L et al. studied the tunable luminescence chromaticity of CaZnOS: Bi^{3+}, Eu^{3+} based on energy transfer [16].

Used as the matrix, natural minerals have good physical and chemical stability and can withstand high temperatures and pressure without losing the structure or luminescence properties. Studies have shown that natural beryl has blue-green light or red light under ultraviolet excitation. This feature can be effectively applied to luminescence and contribute to green mining, new energy, and new material industries. In this paper, the luminescence properties of rare earth ions doped in natural samples have

Journal of Physics: Conference Series **3009** (2025) 012001 doi:10.1088/1742-6596/3009/1/012001

been studied. A series of phosphor materials of $Be_3Al_2Si_6O_{18}$: x % Eu^{3+} (x = 0.07, 0.1, 1, 2, 4) were prepared by the high-temperature solid phase method. The luminescence mechanism of Eu^{3+} ions in beryl was discussed. It has been proven that the natural mineral beryl has great potential value for LED and anti-counterfeiting, further promoting the development of new industries.

2. Experiment

2.1 Material preparation

A series of phosphor materials of $Be_3Al_2Si_6O_{18}$: x% Eu^{3+} (x = 0.07, 0.1, 1, 2, 4) were prepared by the high-temperature solid-phase method. The raw materials used were natural beryl powder and Eu_2O_3 (99.99%). The material was weighed according to the stoichiometric ratio, and the weighed material was put into the agate mortar and ground for 30 minutes. The ground powder was put into the ceramic crucible and then put into the high-temperature energy-saving box for ignition at high temperature. After 3 hours of ignition to 1150°C, the sample was kept for 2 hours, then removed and placed in a sample container after cooling. During the experiment, it is necessary to ensure that the mortar, crucible, and container remain free of impurity pollution, to ensure that the experimental results are accurate and not affected by other impurities.

2.2 Instruments

The phase identification and structure analysis of natural mineral and synthetic material powders were carried out using D8 A powder diffraction system (voltage 40 kV, current 40 mA, wavelength 1.5418 Ω, scanning step 0.01°) (Brook, Germany). The specific data of the test were compared with the standard card PDF # 84-1139. The X-ray photoelectron energy meter ESCALAB 250 (0.46 eV) was used to detect the samples by XPS, and the elements of the samples were qualitatively and quantitatively analyzed. The excitation and emission spectra of the samples in the range of 300 nm-800 nm were measured using a 150 W xenon lamp as the excitation light source and a steady-state spectrometer (FSL1000, British Edinburgh company). During the experiment, according to the actual needs, the corresponding filters were correctly placed at the entrance of the observation grating to achieve the effect of eliminating stray light.

3. Results and discussion

3.1 Analysis of the structural properties of the samples

XPS plays a key role in the study of the composition analysis of natural samples, which can help reveal the chemical elements and chemical bonds in the samples. The experiment was carried out at room temperature with a binding energy of 1350-0 eV. In the spectral analysis, the elements Cr (Cr 2p, 585.98 eV), Mn (Mn 2s, 782.47 eV) , Mn (Mn 2p1, 645 eV), O (O 1s, 532 eV), C (C 1s, 284.8 eV), Be (Be 1s, 118.6 eV), Si (Si 2p, 102.66 eV), Al (Al 2p, 74.46 eV) and Fe (Fe 2p1, 717.93eV) were detected in natural minerals. Figure 1 (a) shows the XPS survey spectra of elements in natural beryl samples, and (b) shows the SEM image. A detailed analysis of the microscopic surface state of the sample can be obtained from the SEM image. In the analysis of the charge effect, charge correction was carried out with the reference standard of C 1S. Natural beryl belongs to silicate with good chemical stability and thermal stability. This sample is a mineral discovered by our research group in the Altay region of Xinjiang.

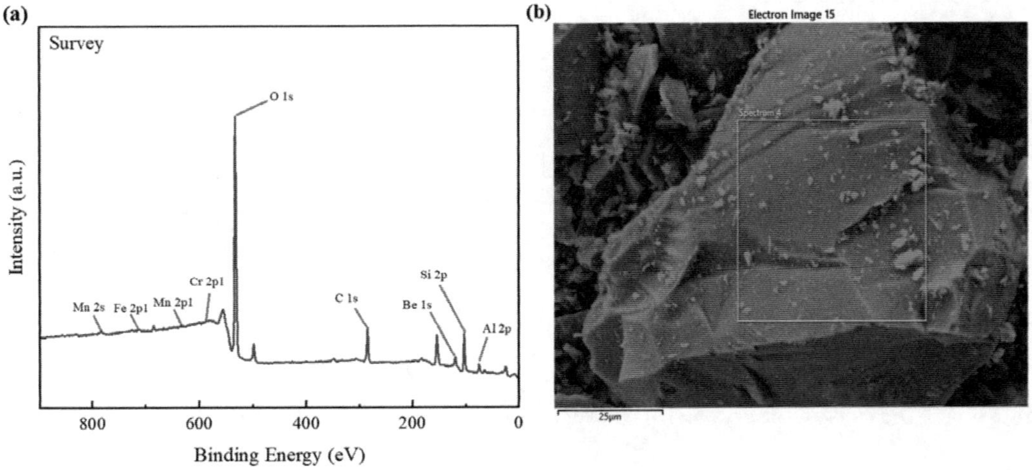

Figure 1 (a) XPS survey spectra of natural beryl; (b) SEM image.

Figure 2 (a) shows a comparison of the XRD patterns of $Be_3Al_2Si_6O_{18}$: Eu^{3+} phosphors and the natural mineral beryl with the standard card PDF # 84-1139 of $Be_3Al_2Si_6O_{18}$. By comparing the X-ray diffraction pattern of natural beryl with the standard card data of beryl, it is found that the important diffraction peak positions of natural beryl are consistent with those in the standard card, which proves that the impurities in natural minerals have no significant effect on the crystal structure of beryl. In the XRD pattern of rare earth ion Eu^{3+} doped, there is no diffraction peak of the Eu_2O_3 impurity phase, and the crystal structure of the matrix is not destroyed. A small number of Eu^{3+} ions successfully entered the crystal structure of the $Be_3Al_2Si_6O_{18}$ matrix, indicating that $Be_3Al_2Si_6O_{18}$: x% Eu^{3+} (x = 0.07, 0.1, 1, 2, 4) phosphors were successfully synthesized.

Figure 2 (b) shows that the crystal structure of beryl is hexagonal. The crystal of beryl is composed of basic structural units composed of $[SiO_4]$ tetrahedrons, which are connected to form a typical hexagonal ring structure. These rings are perpendicular to the c-axis and are arranged in parallel along the c-axis. In this structure, the aluminum ion (Al^{3+}) usually exhibits a six-fold coordination to form an aluminum-oxygen coordination octahedron, while the beryllium ion (Be^{2+}) exhibits a four-fold coordination to form a distorted beryllium-oxygen coordination tetrahedron. Both ions are located outside the ring. Typical parameters of the unit cell are as follows: $a = b \approx 0.921$ nm, $c \approx 0.917$ nm, space group P6 / mcc, Z = 2 (indicating that there are two chemical formula units in the unit cell), and volume of the unit cell of 673.626611 $Å^3$. The crystal morphology of beryl is mostly hexagonal columnar, and longitudinal stripes are common on the cylindrical surface. According to the different impurity elements, the beryl can show different colors. The ion radii of Be^{2+}, Al^{3+}, and Si^{4+} in beryl $Be_3Al_2Si_6O_{18}$ are 0.045 nm, 0.0535 nm, and 0.040 nm, respectively. The ion radii of Be^{2+}, Al^{3+}, and Si^{4+} in beryl $Be_3Al_2Si_6O_{18}$ are 0.045 nm, 0.0535 nm and 0.040 nm, respectively. The percentage difference of ion radii (Dr) was calculated by substitution relationship [17]:

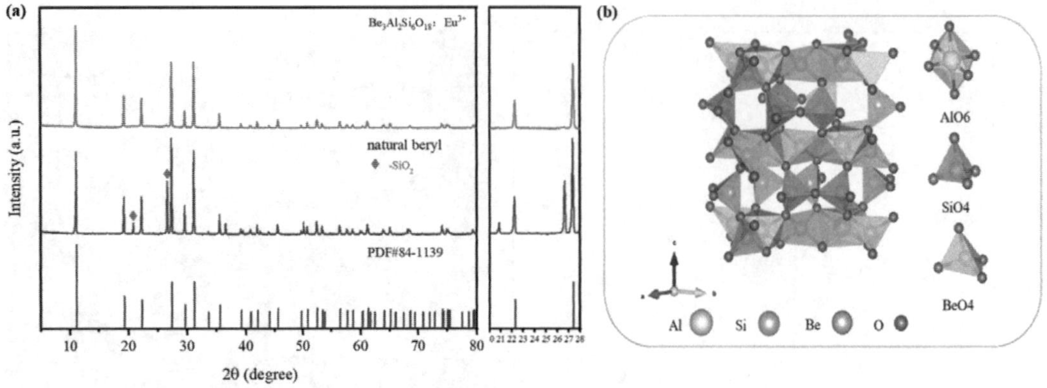

Figure 2 (a) XRD pattern of the crystal; (b) Crystal structure diagram of $Be_3Al_2Si_6O_{18}$.

$$D_r = \left(\frac{R_h = R_i}{R_h}\right) \times 100\% \qquad (1)$$

where *Rh* represents the radius of the cation in the matrix, and *Ri* represents the radius of the incorporated cation. After calculation, based on the principle of similar ion radius, it is concluded that a small amount of Eu^{3+} ions is more likely to occupy the six-coordinated Al^{3+} ion site.

3.2 Spectral analysis of matrix samples

Figure 3 shows the excitation spectra, emission spectra, and color coordinate analysis of the natural beryl $Be_3Al_2Si_6O_{18}$ matrix. A spectrometer was used to continuously change the excitation light to explore the luminescence of the matrix material. It is found that the material has good absorption of ultraviolet light and can produce emission light that can be used. The research data can prove that the emission peaks of the matrix itself originate from the trace elements Cr, Mn, and Fe in natural minerals [18]. As shown in Figure 3 (a), when the excitation wavelength is 300 nm, there are wide emission spectra in the range of 600-800 nm, and the broadband at 720 nm is related to Fe^{3+} ions. When the excitation wavelength is 360 nm, the spectra of the matrix in the range of 400-700 nm are much stronger than those after 700 nm, and the main emission peaks are at 419, 452, 468, 512, and 650 nm. Among them, the emission spectra at 400-550 nm are attributed to the tetrahedral and octahedral coordination of the Mn^{2+} ions. The presence of Cr^{3+} ions affects the spectra in the range of 600-700 nm. The above phenomena are caused by the difference in the optimal excitation wavelength affecting each ion.

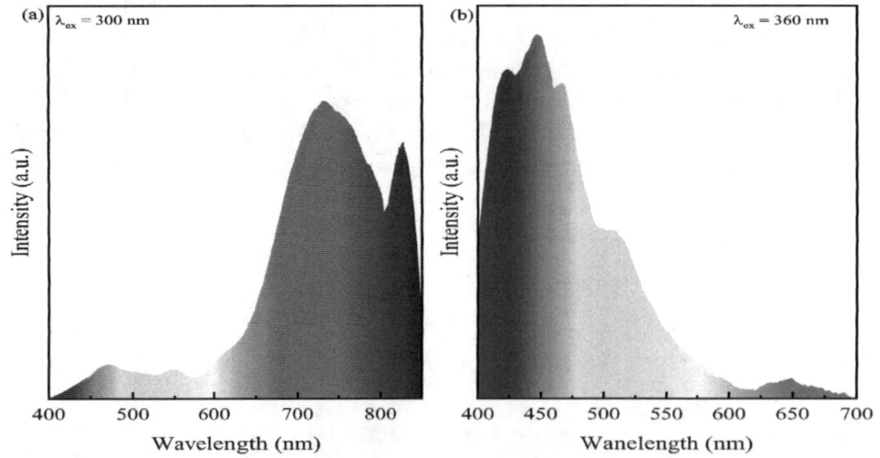

Figure 3 Emission spectra of natural beryl ($Be_3Al_2Si_6O_{18}$). (a) The excitation spectra at 300 nm; (b) The excitation spectra at 360 nm.

As shown in Figure 4 (a), the excitation spectra are measured in the range of 350-470 nm at 419 nm, 452 nm, 468 nm, 512 nm, and 650 nm as the monitoring wavelengths. Owing to the influence of different ions in each emission band, the excitation spectra at different monitoring wavelengths are slightly different. By comparing the data in the diagram, it is proven that this beryl material can indeed be effectively excited under ultraviolet and produce luminescence. Figure 4 (b) shows the chromaticity coordinate diagram. The data detected by the spectrometer are quantitatively described in the CIE 1931 to accurately characterize the color. At different excitation wavelengths, the two-color coordinates are (0.2032, 0.1975) and (0.4034, 0.3366), respectively. Compared with the standard green coordinates (0.21, 0.71), standard blue coordinates (0.14, 0.08), and standard red coordinates (0.683, 0.317) specified by NTSC, the matrix itself has blue-green or red luminescence.

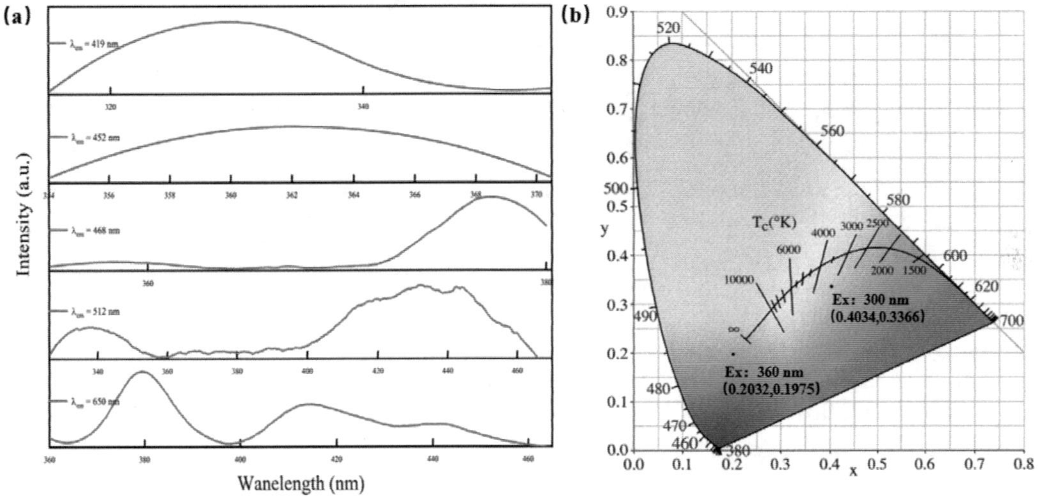

Figure 4 (a) Excitation spectra of natural beryl at 419 nm, 452 nm, 468 nm, 512 nm, and 650 nm; (b) CIE 1931 chromaticity graph.

3.3 Spectral analysis of phosphor samples

A series of $Be_3Al_2Si_6O_{18}$: x% Eu^{3+} (x = 0.07, 0.1, 1, 2, 4) phosphors were prepared at high temperatures. Under ultraviolet excitation, the phosphor increases the emission spectra of the rare earth Eu^{3+} ion, in addition to its emission light. It is found that there are obvious differences in the effects at different excitation wavelengths and concentrations. Therefore, the following two aspects were studied in depth.

Study 1: Due to the good absorption of the samples in the ultraviolet region, the luminescence effect of the samples was observed by changing the excitation wavelength.

As shown in Figure 5, under ultraviolet excitation, the emission peaks of the $^5D_0 \rightarrow {}^7F_M$ (M = 1, 2, 3, 4) transitions of Eu^{3+} ions increased, and the peaks are located at 585 nm, 615 nm, 656 nm, and 700 nm, respectively. In Figure 5 (b), corresponding to the transition from the $^5D_0$ level to the $^7F_1$,$^7F_2$,$^7F_3$,$^7F_4$ levels of Eu^{3+} ions, the strongest emission peak is located at 615 nm ($^5D_0 \rightarrow {}^7F_2$) red emission peak. Figure 5 (a) shows the normalized data diagram. The results show that the emission peaks of Eu^{3+} ions are continuously enhanced when the excitation wavelength is continuously close to the optimal excitation wavelength of Eu^{3+} ions. The fluorescent powder also achieves the overall regulation of color from red to white and then red. At the excitation wavelength of 365 nm, the emission peak of Eu^{3+} ions compensates for the lack of white luminescence of the matrix. The spectra are lost, and the three primary colors of red, green, and blue are formed. Therefore, white luminescence is realized under these conditions.

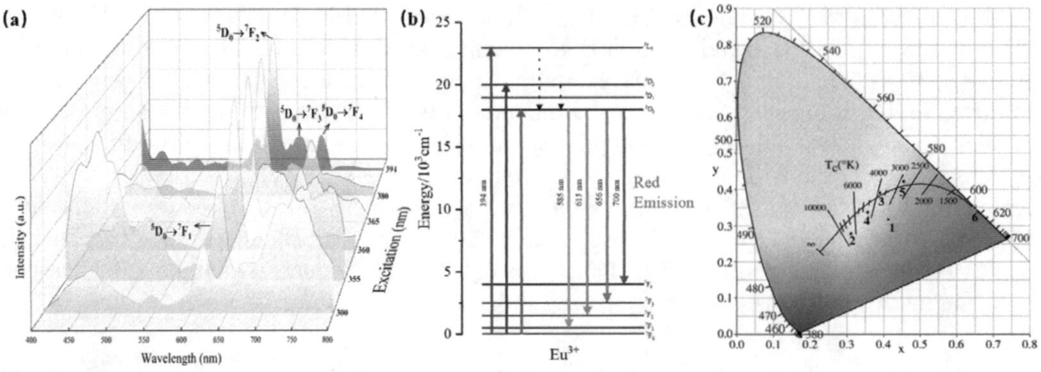

Figure 5 (a) Emission spectra of $Be_3Al_2Si_6O_{18}$: 0.1 % Eu^{3+} at 400 nm-800 nm (λ_{em}=615 nm): the excitation spectra at 300 nm, 355 nm, 360 nm, 365 nm, 380 nm, and 394 nm; (b) Energy level scheme of Eu^{3+}; (c) CIE 1931 chromaticity graph of $Be_3Al_2Si_6O_{18}$: 0.1% Eu^{3+}.

The color coordinates (0.3567, 0.3375) obtained according to CIE 1931 are close to the standard white coordinate (0.33, 0.33) specified by NTSC. According to formula (2), the color temperature value was calculated to be 4492 K (positive white light), indicating that the phosphor emits white light. When excited at 394 nm, the strongest emission of Eu^{3+} ions and strong red luminescence can be achieved. The color coordinates are (0.6509, 0.3487). The changing trend of the luminescence effect is shown in the chromaticity coordinate diagram in Figure 5 (c), The color coordinates of the $Be_3Al_2Si_6O_{18}$: 1% Eu^{3+} phosphor at different excitation wavelengths are listed in Table 1.

The formula is as follows [19]:

$$n = (x-0.3320)/(y-0.1858),$$
$$CCT = -437*n^3 + 3601*n^2 - 6861*n + 5514.3 \quad (2)$$

where the values of (x, y) are the CIE color coordinates, and CCT represents the related color temperature.

Table 1 CIE diagram at different excitation wavelengths.

Excitation wavelength (nm)	Coordinates (x, y)
300	(0.4138, 0.3177)
355	(0.3115, 0.2782)
360	(0.39, 0.3904)
365	(0.3567, 0.3375)
380	(0.4558, 0.4223)
394	(0.6509, 0.3487)

Study 2: When the monitoring wavelength is determined, the doping concentration of Eu^{3+} ions is changed to achieve the effect of adjusting the light intensity. The following is the emission spectrum analysis and color coordinate analysis of a series of phosphors of $Be_3Al_2Si_6O_{18}$: x% Eu^{3+} (x = 0.07, 0.1, 1, 2, 4).

The emission spectra, emission intensity with Eu^{3+} doping concentration, and color coordinates of the matrix doped with different Eu^{3+} ion concentrations at the excitation wavelength of 394 nm are shown in Figure 6. From the emission spectra shown in Figure 6 (a) and (b), it is found that with the change of concentration, the emission peaks do not change, while the emission intensity changes. When the concentration is 0.5%, the concentration quenching is achieved. The luminescence mechanism of rare earth ions arises from the radiative transition between their energy levels. At the same time, the lanthanide contraction effect provides an excellent shielding for the 4f layer orbit of the rare earth ion, so that it is not disturbed by the external crystal field environment. Therefore, adjusting the doping concentration of Eu^{3+} ions mainly affects the intensity of the emission peaks rather than their positions. When the doping concentration of Eu^{3+} ions further increases, especially more than 0.5%, the distance between ions is significantly shortened, which promotes the rapid transfer of energy between luminescent centers. However, this rapid energy transfer may cause energy to be transferred to the surface quenching center, resulting in energy loss, triggering fluorescence quenching, and ultimately weakening the upconversion luminescence intensity. The concentration quenching effect imposes a strict limit on the choice of doping concentration, which not only affects the number of luminescent ions, but also hinders the enhancement of emission intensity. To further understand the concentration quenching mechanism between Eu^{3+} ions, formula (3) proposed by Blasse [20] is used to calculate the critical distance Rc of energy transfer between Eu^{3+} ions, namely

$$Rc = 2\left(\frac{3v}{4\pi x_e N}\right)^{\frac{1}{3}} \tag{3}$$

Where, X_e is the critical doping amount for fluorescence quenching, N is the number of cations in the unit cell, and v is the cell volume. For $Be_3Al_2Si_6O_{18}$: Eu^{3+}, the crystallographic parameters are as follows: v = 673.626611 nm^3, N= 3, and X_e= 0.005. The critical distance is estimated to be 0.2046 nm. Blasse pointed out that when the distance between activator ions is less than 0.5 nm, the exchange interaction is the main method of energy transfer.

In Table 2 and Figure 6 (c), the color coordinate changes of the doped Eu^{3+} ions at different concentrations are analyzed in detail. The results show that red luminescence can be realized when the host material is doped with Eu^{3+} ions.

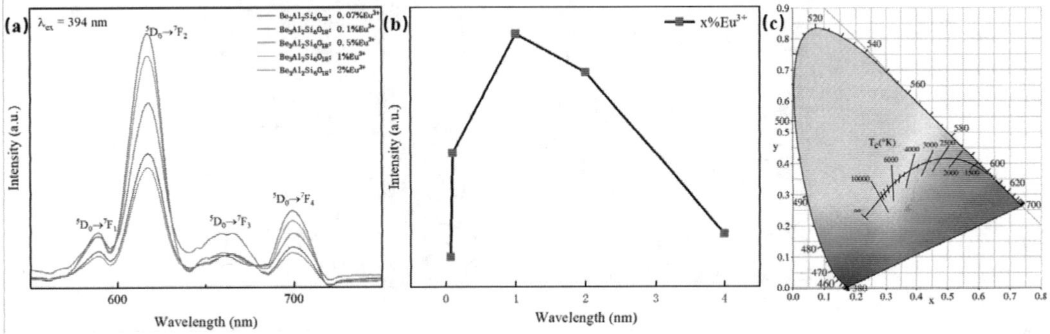

Figure 6 (a) Emission spectra of $Be_3Al_2Si_6O_{18}$: x% Eu^{3+} (x = 0.07, 0.1, 1, 2, 4) at 550 nm-750 nm (λ_{ex}=394 nm); (b) Variation of emission intensity with Eu^{3+} doping concentrations; (c) CIE 1931 chromaticity graph of phosphor.

Table 2 CIE diagram at different concentrations.

Concentration	Coordinates (x, y)
0.07%	(0.6371, 0.3624)
0.1%	(0.6509, 0.3487)
0.5%	(0.6499, 0.3497)
1%	(0.6425, 0.3571)
2%	(0.6424, 0.3572)

3.4 Practicability analysis of phosphor samples

Figure 7 (a) shows the sample obtained by mixing the phosphor with the resin under natural light. Figure (b-d) show the luminescence spectra of the samples at the excitation wavelengths of 300 nm, 365 nm, and 394 nm in the spectrometer. This luminescent material exhibits different colors and luminescence intensities under excitation light of different wavelengths. Such a change process is difficult to replicate and can be used as a unique anti-counterfeiting feature. For example, some counterfeiting labels for

Journal of Physics: Conference Series **3009** (2025) 012001 doi:10.1088/1742-6596/3009/1/012001

valuable documents, certificates, or artwork can also be used in the packaging or labels of products. Consumers can verify the authenticity of the product through a specific wavelength of light source (such as a small ultraviolet lamp). Under normal light, the label may not look special, but under specific excitation light, it will show a preset color or pattern. For example, in the packaging of some high-end cosmetics, liquor, and electronic products, this multi-wavelength excitation anti-counterfeiting technology can be used to increase the anti-counterfeiting level of products. Such a change process is difficult to replicate and can be used as a unique anti-counterfeiting feature.

In addition, it can be applied to combined encryption anti-counterfeiting: the luminescence characteristics of different excitation samples are combined with other anti-counterfeiting technologies, such as reactions with chemical substances, temperature changes, etc., to form a multi-encrypted anti-counterfeiting system. For example, the luminescence properties of a luminescent material will change in the presence of a specific chemical substance, and the color or intensity of the luminescence after excitation will also be different under specific temperature conditions. Applying this material to the anti-counterfeiting marking of products, only when specific chemical and temperature conditions are met at the same time, can the correct luminescence effect be observed, thereby improving the safety and reliability of anti-counterfeiting.

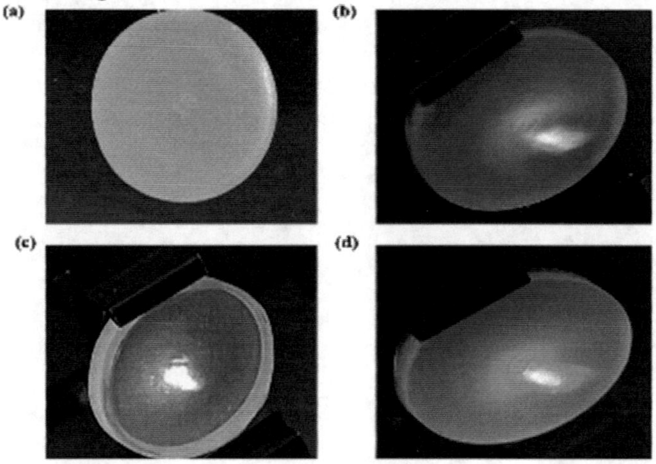

Figure 7 (a) $Be_3Al_2Si_6O_{18}$: Eu^3 phosphor under natural light; (b-d) Luminescence effect diagrams of the sample at the excitation wavelengths of 300 nm, 365 nm, and 394 nm respectively.

4. Conclusions

In this study, a series of $Be_3Al_2Si_6O_{18}$: x% Eu^{3+} (x = 0.07, 0.1, 1, 2, 4) phosphor samples were successfully prepared using a high-temperature solid-phase method. To fully explore the structural characteristics and luminescent properties of these phosphors, X-ray diffraction (XRD), X-ray photoelectron spectroscopy (XPS), spectral analysis, the 1931 International Commission on Illumination (CIE) chromaticity diagram, and other technical means were used to conduct a systematic study. The results show that the host material exhibits blue-green or red luminescence under ultraviolet light excitation. By doping different concentrations of rare earth ions Eu^{3+} into the matrix material, the phosphors prepared in this study exhibit the characteristic red luminescence of Eu^{3+} ions under ultraviolet light excitation. Further in-depth research has shown that these samples can achieve red and white luminescence under different conditions, and the prepared phosphors show great potential application value in the fields of light-emitting diodes (LEDs) and multiple anti-counterfeiting.

This study proves that the natural mineral beryl has unique advantages in the field of luminescence. Tunable luminescence has been realized through a preliminary study doped with rare earth ions Eu3+ and by comparing the luminescence effect of the samples at different excitation wavelengths and different concentrations. In the future, we will continue to study and develop the application of natural beryl in phosphor materials, such as studying the luminescence characteristics of beryl doped with more diverse rare earth ions or transition ions, and gradually expand its application.

5References

[1] Ma Y, Gao X, Zhang W, et al. Enhanced red luminescence of $Ca_3Si_{2-x}M_xO_7$:Eu^{3+}(M=Al, P)phosphors via partial substitution of Si^{4+} for applications in white light-emitting diodes [J]. Rare Metals,2024,43(02):736-748.
[2] Kong Li, Qiu Lu, Wei Qiye, et al. Effect of flux on the spectral performance of red phosphor $SrMoO_4$:Eu^{3+} for white LEDs [J]. Journal of Artificial Crystals, 2019, 48(04): 616-620+626.
[3] Liu Jingxiang. Preparation of $BaY_4Si_3O_{13}$:Eu^{3+} red phosphor for white LED and doping modification with Ce^{3+} and Tb^{3+} [J]. Harbin Institute of Technology, 2022.
[4] Li Zhao, Wu Hongqiang, Shi Bingyao, et al. Preparation and performance of $GdVO_4$:Eu^{3+} red phosphor for white LED [J]. Functional Materials, 2020, 51(06): 6186-6189.
[5] Wang Lei, Aierken Sidike. Improvement of the luminescent performance of white LED devices using red phosphor Mg_2SiO_4:Eu^{3+}, Tb^{3+} [J]. Paper Equipment and Materials, 2022, 51(01): 97-99.
[6] Cai Xiaoyong, Jiang Hongxi. Preparation and optical properties of red phosphor $CaWO_4$:Eu^{3+},Bi^{3+}[J/OL]. Journal of Artificial Crystals: 1-11 [May 13, 2024].
[7] Wang B, Sun S, Li Z, et al. A novel Tb^{3+} and Eu^{3+} co-doped dual-emitting phosphate $K_3SrBi(P_2O_7)_2$ phosphor for application in FIR thermometers [J]. Journal of Rare Earths,2024,42(03):455-463.
[8] Liu Zhe. Analysis of Application of Rare Earth Luminescent Materials in Solid White Light LED Lighting [J]. Contemporary Chemical Industry Research, 2024(01): 116-118.
[9] Zhao Jianguo, Suo Borian, Xu Ru, et al. Effect of Fluorescent Pigment Proportion and Excitation Wavelength on High-Quality White Light LED [J]. Acta Optica Sinica, 2024, 45(01): 103-110.
[10] Dong G, Li X, Pan H, et al. Crystal structure, luminescence enhancement and white LEDs application of NaZnPO4: Eu^{3+} red phosphors [J]. Journal of Luminescence, 2019,206: 260-266.
[11] Chen Hao, Hui Zengzhe, Guo Feifei, et al. Study on photochromic and photochromic control characteristics of Ca_2SnO_4:Eu^{3+} ceramics [J]. Functional Materials, 2024,55(01):1117-1123.
[12] Lu Chendong, Xu Yingchao, Zhou Qiong, et al. Study on the properties of Eu^{3+}-doped $CaSnSiO_5$ orange-red phosphor for white LED [J]. New Chemical Materials,2023,51(11):128-132+137.
[13] Zhang Wenna, Li Zhongyin, Ma Jitao, et al. Preparation and tunable luminescence of Bi^{3+}, Eu^{3+} co-doped La_4GeO_8 phosphor [J]. Acta Luminosa Sinica,2024,45(03):450-457.
[14] Zhang Yue, Guo Yakun, Jing Zekun, et al. Hydrothermal synthesis and luminescence properties of CaF_2: Eu^{2+} nanopowders [J]. Functional Materials,2023,54(11):11152-11156.
[15] Meng X, Huang S, Shang M. Red emitting Ba_2GdVO_6: Eu^{3+} phosphors for blue light converted warm white LEDs [J]. Inorganic Chemistry Communications, 2020, 113: 107768.
[16] Guo L, Li T, Zhu C, et al. Tunable luminescent chromaticity of CaZnOS: Bi^{3+}, Eu^{3+} with white emitting based on energy transfer [J]. Journal of Alloys and Compounds, 2022, 905: 164262.
[17] Blasse G. Phys. Lett. A,1968,28:444.
[18] Michael Gaft, Renata Reisfeld, Gerard Panczer. Modern Luminescence Spectroscopy of Mineralsand Materials [M]. Springer-Verlag Berlin Heidelbery 2005:97-99.
[19] Hu Liulian, Erken Shidike, Wan Ying, et al. Luminescent Characteristics of Dy^{3+}, Tm^{3+} Co-doped $Ca_2MgSi_2O_7$ [J]. Chinese Journal of Lasers, 2018, 39(07): 948-954.
[20] BLASSE G. Energy transfer in oxidic phosphors [J]. Physics Letters A, 1968,28： 444-445.

Journal of Physics: Conference Series **3009** (2025) 012002 doi:10.1088/1742-6596/3009/1/012002

Study on fatigue crack initiation behavior of two types of lattice structures

Yinghan Sui[1,2] and Jia Liu[1,2,*]

[1]School of Mechanical and Electric Engineering, Changchun University of Science and Technology, Changchun 130022, PR China
[2]National Base of International Science and Technology Cooperation for Optics, Changchun 130022, PR China

[*]Corresponding author: liujia@cust.edu.cn

Abstract: As a new type of multi-functional structure with lightweight and high strength, the lattice structure can meet the lightweight of its structure while taking into account the overall mechanical properties. The performance of the lattice structure is not only related to the material itself but also to its topological structure. Lattice structures are subjected to cyclic loading in practical applications. Fatigue fracture is the main failure mode. In this study, we selected the body-centered cubic and the octahedron structures for finite element analysis. Firstly, their uniaxial compression statics are simulated, and the stress distribution programs of the two structures are obtained. Secondly, their fatigue life distributions are simulated. The simulation results show that the lowest fatigue life of the lattice structure is located at the node location. Finally, the fatigue crack initiation simulation of the two structures is carried out. The results show that multiple short fatigue cracks are generated at the nodes of the lattice structure. The prediction of the crack initiation position is in agreement with the results of fatigue life distribution.

1. Introduction

A lattice structure is a porous structure composed of rod elements and nodes arranged periodically according to certain rules [1]. The lattice structure has attracted wide attention due to its excellent mechanical properties. Lattice structures are different from traditional materials; they have high strength, are lightweight, and have strong designability while meeting load-bearing performance. Through structural design and manufacturing, the structural parts can be lightweight and meet the functional requirements of the structure. The study of lattice structure has attracted more and more scholars' interest. However, due to the complex structure of lattice, manufacturability limits its development [2]. Traditional manufacturing methods are difficult to process and cannot meet their application and development. With the rapid development of additive manufacturing technology, because it can transform multi-dimensional manufacturing into two-dimensional manufacturing [3], it solves the constraints and limitations encountered by traditional manufacturing methods and can realize the flexible manufacturing of complex lattice structures.

Lattice structures have great adjustable design space. By designing its topological structure, the cell configuration with excellent load-bearing and multi-functional characteristics can be realized [4]. Researchers have constructed a variety of three-dimensional lattice cell configurations [5]. The truss lattice structure composed of beams has attracted wide attention because of its simple geometric form and simple design method. The basic structure of the lattice structure is body-centered cubic (BCC), octahedral structure (Octa), and their derivatives.

However, lattice structures are subjected to cyclic loads in practical applications. Fatigue damage has become a common phenomenon. The research on the fatigue performance of lattice structures is still in its infancy. Structural design and molding quality optimization still need to be studied. Additive manufacturing technology can manufacture lattice structures with geometric complexity, but it is not possible to completely ignore process constraints. The printing process does not adapt to the rapid development of technology. It is difficult to accurately control various process parameters during the preparation process, which will inevitably produce inherent defects [6]. Defects lead to unstable quality of lattice structure, which seriously reduces the service life of lattice structure and affects the reliability and safety of structural parts. Therefore, it is very important to evaluate the safety of additive manufacturing lattice structures.

In this paper, the body-centered cubic lattice structure and octahedral lattice structure are studied. We simulate the stress distribution of the lattice structure under unidirectional compression and simulate the fatigue life distribution by using the finite element method to guide the fatigue crack initiation position of the lattice structure. Finally, the fatigue crack initiation of lattice structure unit cells is simulated based on the extended finite element method built in Abaqus software. The simulation results show that the prediction of crack initiation position is in good agreement with the fatigue life distribution map.

2. Finite element analysis

2.1. Unit cell design

The lattice structure has been serving in a complex environment for a long time. It is subjected to alternating loads such as high amplitude wind and airflow, resulting in fatigue fracture. Among all kinds of lattice cells, the envelope space of cubic lattice cells is a standard cube. The shape of cells is regular, which is convenient for cell arrangement and theoretical calculation. The basic structure of lattice structure is body-centered cubic (BCC), octahedral structure (Octa), and its derivatives. This paper mainly studies BCC structure and Octa structure as the basic units.

The cell side length a is 10mm. The weight loss ratio β is 80 %. The lattice cell size parameters are shown in Table 1.

Table 1. Lattice cell size parameters.

	BCC	Octa
Relative density (r)	0.2	0.2
Cell side length (a/mm)	10	10
Rod diameter (t/mm)	2.13	1.94

The 3D model of lattice cells is established according to the size parameters shown in Table 1. Lattice cell models are shown in Figure 1.

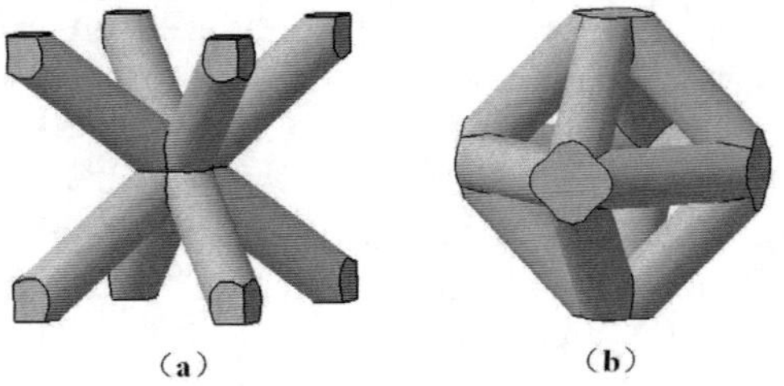

Figure 1. Schematic diagram of cubic lattice structure: (a) BCC; (b) Octa.

2.2. Material properties

In this study, structural steel was selected as the material for finite element analysis. It is widely used in construction, bridges, machinery, and other industries. Structural steel is often used in finite element analysis. The density of the material is 7850 Kg/m^3, the material's elastic modulus is set as 210GPa, and the Poisson's ratio is 0.3.

2.3. Finite element model

The static simulation model of the lattice cell is established in the Standard module of Abaqus software, as shown in Figure 2. The finite element simulation should try to reproduce the experimental process as much as possible. To simulate the pressure head and base of the electronic universal testing machine, two rigid panels are created at the upper and lower ends of the lattice structure model. The upper and lower surfaces of the lattice structure are bound to the rigid contact surface. The rigid plate at the bottom of the cell is completely fixed, and a static load of 3000 N along the vertical direction is applied at the reference point of the top rigid plate. The simulation material is set as structural steel. The uniaxial compression static simulation of BCC and Octa lattice structures is carried out to obtain the stress distribution cloud diagram of the structure under compressive static load. Statics results can be imported into FE-SAFE software for subsequent calculation of fatigue life distribution. Using sine wave load and stress ratio R=-1, the fatigue life distribution cloud diagram of the lattice structure is calculated.

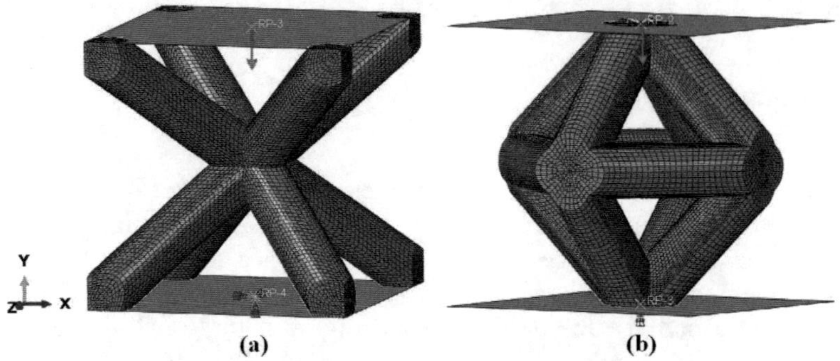

Figure 2. Uniaxial compression statics model schematic diagram: (a) BCC; (b) Octa.

In this study, the fatigue damage model of BCC and Octa structures is established by using the extended finite element method (XFEM) built in ABAQUS software to analyze their fatigue crack initiation position. The finite element method needs to restore the experimental process as much as possible. To apply the concentrated force, the upper surface of the cell is coupled to the node RP1, the lower surface is coupled to the node RP2, and fixed constraints are applied. The upper reference point RP1 is subjected to a vertical upward (Y-axis direction) 0.5mm cyclic displacement load. The rotational degrees of freedom in three directions are limited. The stress ratio R = -1 is used to simulate the cyclic loading of lattice structure under actual working conditions. Due to the complex structure of the lattice, the external software Hypermesh is used for meshing. The three-dimensional eight-node hexahedral mesh (C3D8R) is used for analysis. The mesh size is set to 0.2 mm. Using the XFEM method to simulate the fatigue cracks propagation behavior needs to preset a crack source in the lattice structure. A circular crack model with a diameter of 0.2 mm is preset. The crack model does not need to be meshed and assigned to the material. The simulation material model is structural steel. The density of the material is 7850 Kg/m^3, the elastic modulus is 210GPa, and the Poisson's ratio is 0.3. The maximum principal stress criterion is selected as the material's fatigue damage failure mode. The maximum principal stress is 100MPa. The critical strain energy release rate G_{IC} is 40J/m^2. The fatigue crack initiation position and fatigue crack growth propagation of the lattice cells are obtained by simulation. The solution results can be viewed through the visualization module. The fatigue damage model based on XFEM in this paper is shown in Figure 3:

Journal of Physics: Conference Series **3009** (2025) 012002 doi:10.1088/1742-6596/3009/1/012002

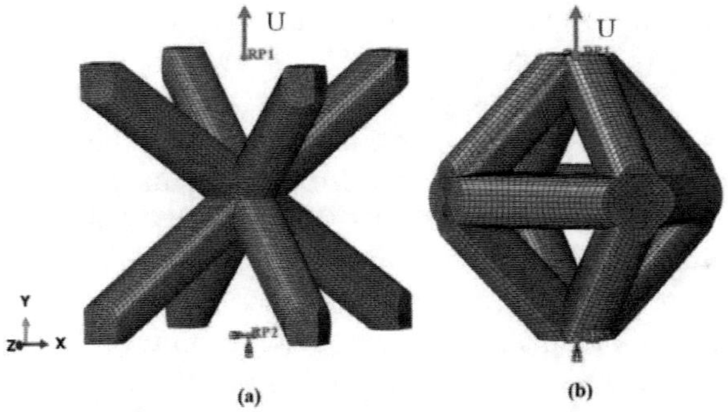

Figure 3. Fatigue damage model schematic diagram: (a) BCC; (b) Octa.

3. Results and analysis

To study the stress distribution of BCC structure, the stress distribution of BCC lattice structure under static compression is obtained by numerical simulation, as shown in Figure 4(a). The stress gradually increases from blue to red. It can be seen from the figure that the stress of the BCC structure is mainly concentrated on the inclined rod. A relatively symmetrical stress distribution can be observed. There is evident stress concentration at the node position. It indicates that the BCC structure can easily fail first at the node position. The fatigue life distribution of the BCC structure is shown in Figure 4(b). The red area is a dangerous area with low fatigue life, where it is the most easily fatigued fracture of the lattice structure. The fatigue life results show that the behavior is consistent with the stress distribution. The fatigue life of the stress concentration position is short, which means that the fatigue crack is most likely to initiate at the node position of the BCC structure under cyclic loading. When the preset crack propagation leads to the complete fracture of the bar, the subsequent cracks initiate at the joint position of the BCC structure, which is consistent with the stress distribution results obtained by statics and the fatigue life distribution obtained by FE-SAFE. The fatigue fracture morphology of the BCC lattice structure is shown in Figure 5. Under the cyclic load, the bar is completely broken. The fracture section is rough.

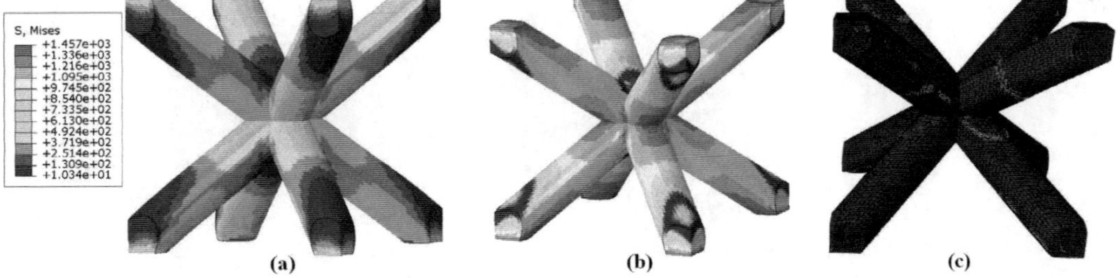

Figure 4. Finite element simulation results of BCC lattice structure: (a) stress distribution; (b) fatigue life; (c) fatigue crack propagation.

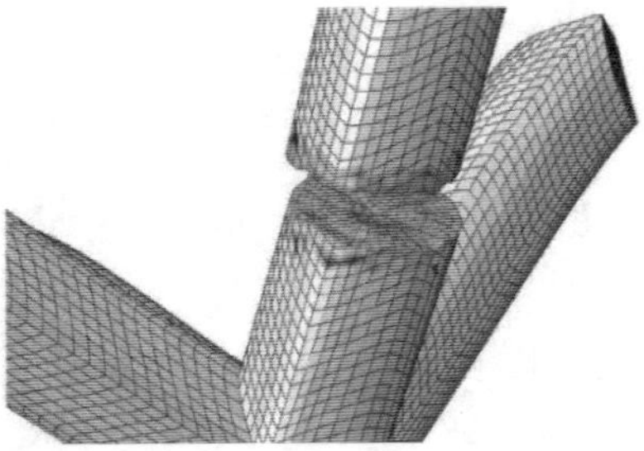

Figure 5. The schematic diagram of fatigue fracture morphology of BCC lattice structure.

Journal of Physics: Conference Series **3009** (2025) 012002 doi:10.1088/1742-6596/3009/1/012002

The stress distribution of the Octa structure is mainly distributed on the inclined rod. There is evident stress concentration at the node position. As shown in Figure 6(a), the stress gradually increases from blue to red. The whole structure shows a symmetrical stress distribution. The fatigue life distribution of the Octa structure is shown in Figure 6(b). The fatigue life increases from blue to red. The red position is the most easily affected by fatigue damage. The fatigue life of the stress concentration position is short, which means that the fatigue crack is most likely to initiate at the node position of the Octa structure under cyclic loading. The fatigue crack propagation model of the Octa lattice structure shown in Figure 6(c) proves this. It is observed that when the preset crack ultimately propagates, the rod fracture failure occurs, the structure continues to suffer fatigue damage, and multiple new cracks are initiated at the node of the lattice structure. In the absence of preset cracks, the fatigue crack initiation of the Octa structure is shown in Figure 7. The crack initiates at the inner side of the node of the lattice structure. Due to the compression of the upper layer, fatigue cracks appear on the outer side of the upper node of the lattice structure. The location of fatigue crack initiation is consistent with the stress distribution and the fatigue life distribution.

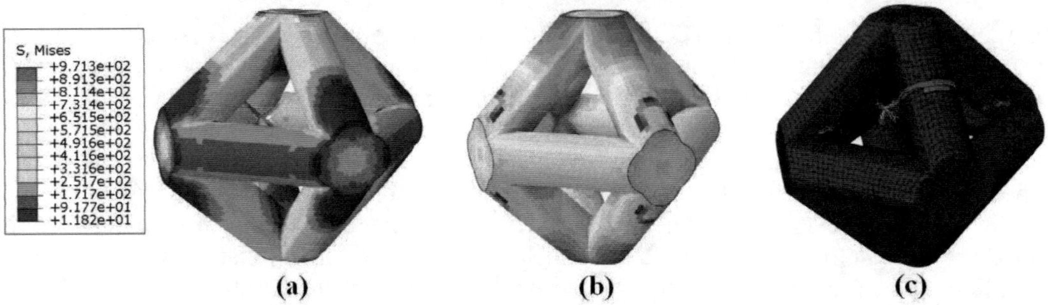

Figure 6. Finite element simulation results of Octa: (a) stress distribution; (b) fatigue life; (c) fatigue crack propagation.

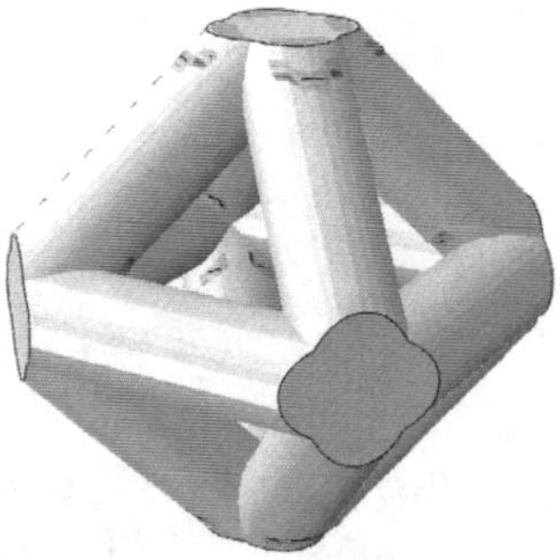

Figure 7. The fatigue crack initiation diagram of Octa structure.

4. Conclusion

With the development of additive manufacturing technology, the safety of metal 3D-printed lattice structures has gradually become the research focus. In this paper, the mechanical properties and fatigue properties of the BCC structure and the Octa structure are studied. The main findings of this study are:

(1) Both structures show a relatively symmetrical stress distribution and stress concentration at the node position. The unit cell type dramatically affects the overall performance of the lattice structure.

(2) The fatigue life distribution diagram shows that the positions where the two types of lattice structures are most prone to fatigue damage are located at the node stress concentration position.

(3) The fatigue crack initiation position predicted by the lattice structure model based on XFEM

agrees with the fatigue life diagram based on FE-SAFE.

In general, the simulation of fatigue crack initiation of lattice structure by XFEM can lay a foundation for the subsequent research on fatigue crack propagation and fatigue life prediction of lattice structure.

References

[1] Liu W, Song H, Wang Z, et al. (2019) Improving mechanical performance of fused deposition modeling lattice structures by a snap-fitting method. Materials & Design, 181. DOI: 10.1016/j.matdes.2019.108065.
[2] Najmon J C, Raeisi S and Tovar A. (2019) Review of additive manufacturing technologies and applications in the aerospace industry. Additive Manufacturing for the Aerospace Industry, pp7-31. DOI: 10.1016/B978-0-12-814062-8.00002-9.
[3] ZHANG Long, LI Ang, et al. (2021) Optimal Design and Experimental Verification of an Enclosed Skin Lattice Support Structure. Journal of Mechanical Engineering, 57(22): 35-42. DOI: 10.3901/JME.2021.22.035.
[4] BAI Long, XIONG Fei, et al. (2018) Multi-objective Structural Optimization Design of Ti6Al4V Lattice Structure Formed by SLM. Journal of Mechanical Engineering, 54(5): 156-165. DOI: 10.3901/JME.2018.05.156.
[5] DING Yi, WU Weitao, et al. (2024) Topology Review and Convective Heat Transfer Comparison of 3D Lattice Structures. Applied Mathematics and Mechanics, 45(8): 1001-1023. DOI: 10.21656/1000-0887.450184.
[6] Lozanovski B, Downing D, et al. (2020) Non-destructive simulation of node defects in additively manufactured lattice structures. Additive Manufacturing, 36: 101593. DOI: 10.1016/j.addma.2020.101593.

"Green" S-nZVI loaded alginate/chitosan double-gel interpenetrated porous beads for methylene blue removal from water: preparation, adsorption behavior and pH stability

Xiaohong Wang [1*], Fengyan Fu [1], Tianjia Wang [2], Jie Zhang [1], Zhihua Gao [1]

[1] Department of Chemistry, Hengshui University, Hengshui 053000, China
[2]AEW Technology Group Co., Ltd, Tianjin 301739, China

[*]Corresponding author's e-mail: hsxywang@qq.com

Abstract. In this study, sulfurized nano iron(S-nZVI) loaded sodium alginate/chitosan composite gel spheres (S-nZVI/SA/CTS) were prepared from natural biodegradable sodium alginate (SA), chitosan (CTS) and S-nZVI, which was"green" synthesized from pomegranate leaf extract. The removal of methylene blue (MB) from water by S-nZVI/SA/CTS was investigated using an interpenetrating network constructed with sodium alginate and chitosan as the skeleton, and sulfide nano iron as the main material for adsorbing methylene blue. The wet composite gel spheres of S-nZVI/SA/CTS prepared under certain conditions could remove MB up to 98.52 %, and which were used 6 times, the removal rate of MB can still reach more than 95%. When the pH of the solution is between 3-11, the removal rate of MB by gel ball is above 93.5%, which indicates that the gel spheres of S-nZVI/SA/CTS had good acid and alkali resistance. The adsorption kinetic data of S-nZVI/SA/CTS on MB was more in line with the pseudo second-order kinetics model (R^2=0.9998). The synthesis of a green, cost-effective, and highly efficient S-nZVI/SA/CTS gel-sphere adsorbent holds significant importance for treating dye wastewater.

1. Introduction

The textile, food, printing, cosmetic, and paper industries have widely used organic dyes in their manufacturing processes in recent years, which have improved the value and appearance of the products [1], but the dye wastewater is characterized by high discharge, complex composition and biohazardousness, which can cause immeasurable harm to human life and health and the natural environment, if it enters the water environment without thorough treatment [2]. Therefore, the removal of dyes from wastewater is of great significance. Adsorption has been widely used in the field of environmental remediation and wastewater treatment due to its low cost, ease of operation, and good removal of pollutants [3].

Chitosan (CTS) and sodium alginate (SA) belong to natural polysaccharide polymers, which are widely characterized by their own good biocompatibility and ability to biodegrade [4]. However, sodium alginate is intolerant to alkali, and chitosan is soluble in acid. Both of them have a narrow pH range for use and poor mechanical strength.CTS-SA forms an interpenetrating bi-gel network, which improves the mechanical strength of hydrogel. Nano-zero-valent iron (nZVI) is widely used in environmental treatment because of its large specific surface area and high reactivity, which can adsorb, degrade, and even mineralize certain organic substances [5]. However, nZVI also has the disadvantages of short lifetime and easy agglomeration [6]. nZVI is traditionally prepared using the physical ball milling method and chemical reagent reduction method, which has disadvantages such as secondary pollution. Therefore, adopting green and environmentally friendly methods to synthesize nano-iron has become the development trend of nano-iron.

This study extracted a common plant to synthesize nZVI and sulfurized the surface to improve the agglomeration phenomenon. This approach prolongs the lifetime of nZVI and effectively avoids the secondary pollution formed in the traditional chemical synthesis method. Meanwhile, CTS-SA double hydrogel was utilized as a skeleton to support and disperse S-nZVI to reduce its agglomeration and increase the adsorption effect of the composite hydrogel, which is of great significance for the treatment and remediation of organic wastewater in the water environment.

2. Materials and Methods

2.1 Materials

Ferrous sulfate, methylene blue, sodium alginate (SA), acetic acid (30%), and the above drugs are analytically pure, produced by Tianjin Damao Chemical Reagent Factory; chitosan (CTS) is BR grade deacetylated 90% viscosity 50-200mpa-s, produced by Tianjin Damao Chemical Reagent Factory; sodium thiosulphate barium chloride is analytically pure, produced by Tianjin Guangfu Science and Technology Development Co. The pomegranate leaves were taken from the pomegranate tree at Hengshui University.

Visible spectrophotometer (722G, Shanghai Yidian, China), ultrasonic cleaner (BL6-180A, Shanghai Bilang, China), nano-laser particle-size distributor (BT-90, Dandong Baxter, China), FT-IR spectrometer (WQF-510A, Riley, China), Thermogravimetric Analyzer (TGA4000, PerkinElmer Amethyst, America), SEM-EDS(NanoSEM230-EDSXMAX50, FEI, America), and XRD(Bruker D8 Advance, Germany) are used in this study.

2.2 Synthesis of S-nZVI and S-nZVI/SA/CTS

2.2.1 Synthesis of sulfurized nano-iron (S-nZVI). Preparation of plant extract: pomegranate leaves were washed, dried, and cut into pieces, added with deionized water, kept at a constant temperature of 353K for 2h, and cooled down, i.e., the plant extract was obtained. Preparation of nano zero-valent iron: drop the deoxygenated deionized water prepared $FeCl_3$ solution into the plant extract under stirring, after 10min to get the nano iron suspension solution, vacuum filtration, and drying to get the nano iron powder.Preparation of iron nanosulfide: take the above iron nanosulfide solution, add sodium sulfide and sodium thiosulfate solution drop by drop under stirring, stir for 10min, get iron nanosulfide solution, vacuum filtration, dry to get iron nanosulfide powder (S-nZVI).

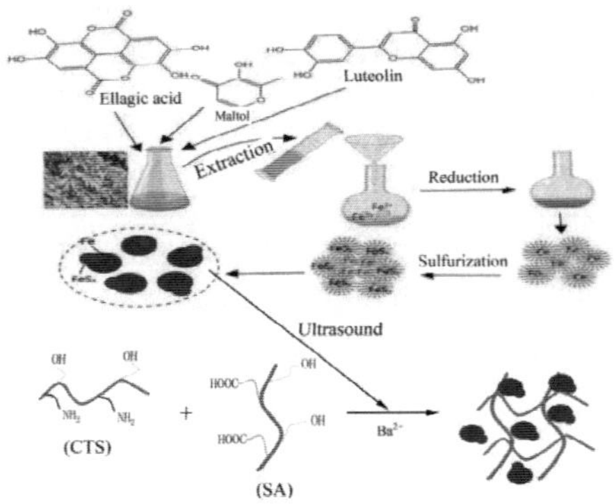

Figure 1. A Green route to synthetic S-nZVI/SA/CTS.

2.2.2 Synthesis of S-nZVI/SA/CTS. We weighed 2.0 g of SA and 1.0 g of CTS, respectively, to 98 mL of water and 99 mL of acetic acid (2%, v/v) and ultrasonized for 1 h. Then, we dissolved them fully, standing for 30 min to obtain SA (2%, w/v) solution and CTS (1%, w/v) solution. The prepared SA solution was poured into the CTS solution, and a small amount of NaOH solution was added dropwise for sufficient stirring to obtain the SA/CTS mixture. The S-nZVI particles were added to a mortar, and a small amount of gluconolactone was added for thorough grinding to obtain S-nZVI powder. It was added to the SA/CTS mixture prepared above and sonicated for 10 min to obtain the S-nZVI/SA/CTS mixture. The S-nZVI/SA/CTS mixture was injected drop by drop into $BaCl_2$ (1%, w/v) solution using a syringe at a low rotational speed to obtain S-nZVI/SA/CTS. After soaking in $BaCl_2$ solution for some time, the S-nZVI/SA/CTS were removed and washed with distilled water 5-6 times and stored in distilled water for spare use.The preparation process of S-nZVI/SA/CTS is shown in the Figure 1.

2.3 Characterization

2.3.1 Scanning electron microscopy (SEM). A field emission scanning electron microscope (SEM) (Tecnai G2 F30, FEI, USA) equipped with an x-ray energy spectrometer (EDS) was used to study the surface and cross-sectional morphology and elemental distribution of the samples.

2.3.2 Particle Size Analysis. The S-nZVI was finely ground, and a certain amount of the sample was weighed into a beaker containing water with a small amount of dispersant and ultrasonicated for 5 min.

2.3.3 Infrared analysis (FTIR). We weighed a certain amount of dry and ground SA samples in agate, added a certain amount of dry potassium bromide powder under the infrared light to be fully ground, and pressed it into a transparent sheet under the pressure of 10-15 Mpa, and then carried out the infrared test. The above steps are repeated for CTS, SA/CTS, and S-nZVI/SA/CTS for the IR test.

2.3.4 Thermogravimetric analysis (TGA). The experiments were carried out by heating the SA, CTS, and S-nZVI/SA/CTS samples to 700 ℃, 900 ℃ and 900 ℃ under the conditions of heating rate and nitrogen flow rate of 10 ℃/min and 20 mL/min, respectively.

2.4 Batch experiments

The adsorption properties of the materials were investigated through a series of batch experiments. The concentration of residual methylene blue was determined by standard curve method using a visible spectrophotometer (λ_{max} =664 nm). The above experimental parameters were kept constant in the following experiments unless otherwise specified. The removal efficiency and adsorption were calculated by Equations (1) and (2), respectively.

$$Removal\ efficiency = (C_0\text{-}C_e)/C_0 \times 100\% \quad (1)$$

$$q_t = (C_0 - C_e)V)/m \quad (2)$$

C_e in the above equation is the mass concentration of MB at time t (mg/L); C_0 is the initial mass concentration of MB (mg/L); q_t is the adsorption capacity after a certain contact time (mg/ g); m is the dose of wet adsorbent (g), and V is the initial volume of the solution (L), in the above equation.

2.5 Adsorption kinetics

The experimental data for MB removal by S-nZVI/SA/CTS were fitted using the proposed primary and proposed secondary kinetic models in this experiment. Equations (3) and (4) are the proposed primary and proposed secondary kinetic equations, respectively.

$$ln(q_e\text{-}q_t)=lnq_e\text{-}k_1t \quad (3)$$

$$t/q_t=1/k_2q_e^2+t/q_e \quad (4)$$

where q_t and q_e are the adsorption amount and the equilibrium adsorption amount in mg/g at time t, respectively;k_1 is the proposed primary adsorption rate constant (/min); k_2 is the proposed secondary adsorption rate constant ($g \cdot mg^{-1} \cdot min^{-1}$).

3. Results and Discussion

3.1 Scanning Electron Microscopy

The SEM images of nZVI, and S-nZVI are displayed in the Figure 2(a,b), showing that some particles gathered together, which were nearly spherical particles with rough surfaces. Because nZVI and S-nZVI have small particle sizes, large surface energy and magnetic properties, they are easy to agglomerate.

nZVI and S-nZVI do not have much difference in SEM morphology, and both of them are approximately spherical. s-nZVI has a slightly increased particle size, and the degree of agglomeration is somewhat improved compared with that of the unsulfurized nZVI. To determine the elemental composition of the synthesized S-nZVI particles, we did EDS spectroscopic characterization (Figures2c-f), which showed that the elements of C, O, Fe, and S were uniformly distributed in S-nZVI, with C (57.3 wt%), O (39.3 wt%), Fe (12.5 wt%), and S (0.9 wt%), which suggests that the nano iron has been successfully sulfidized. The carbon and oxygen elements were mainly derived from pomegranate leaf biomass, and probably, the biomolecules in the pomegranate extract covered the surface of the S-nZVI particles, which could further improve the stability of the nanoparticles.

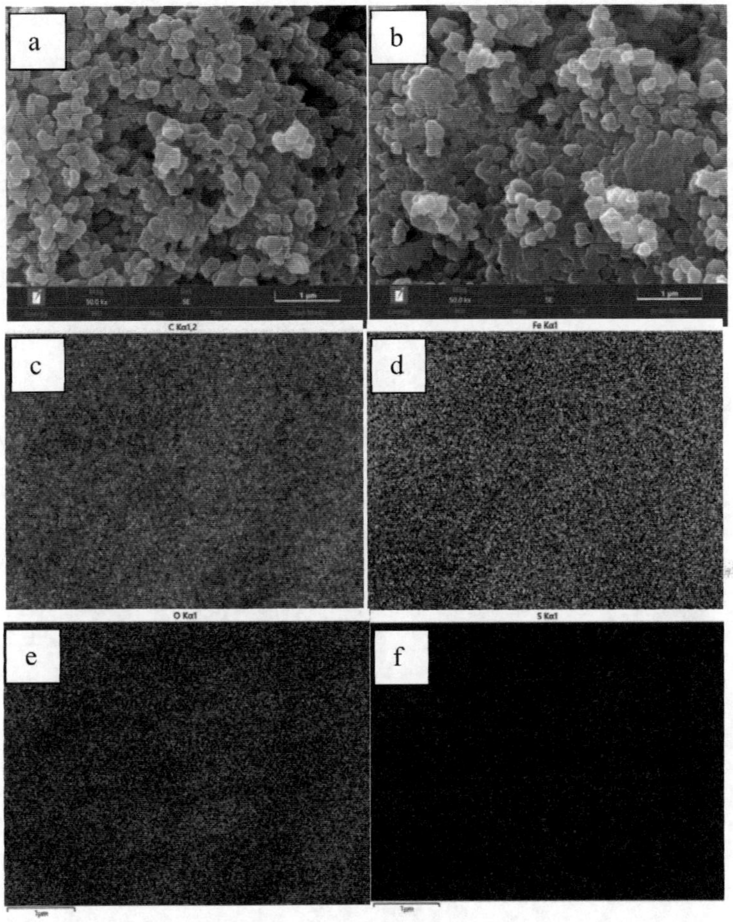

Figure 2. SEM and EDS electron spectra of particles: SEM images of nZVI(a) and S-nZVI(b); EDS surface scan images of C(c),Fe(d),O(e),S(f) in the S-nZVI.

3.2 Particle size analysis

To investigate the particle size distribution of nZVI and S-nZVI, it was measured using a nano-laser particle size distributor. The results are shown in the Figure 3. nZVI has a median diameter of 45.5 nm and a particle size distribution between 28.1-79.2 nm, with nearly 50% of the particles in the range of 40-50 nm. S-nZVI has a median diameter of 179 nm and a particle size distribution between 111-281 nm, with nearly 50% of the particles in the range of 158-199 nm.

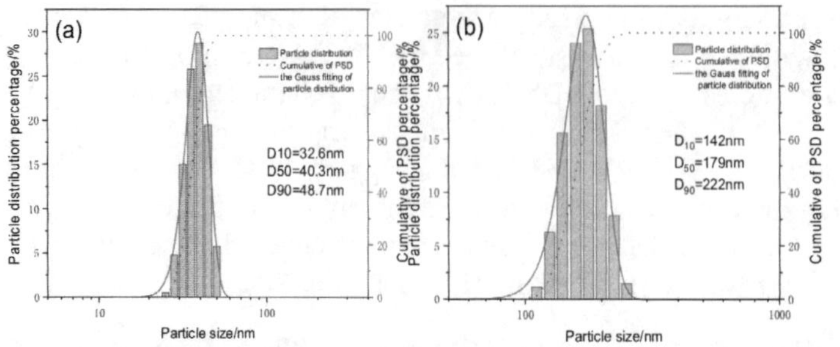

Figure 3. Particle size distribution of S-nZVI(a); and nZVI (b).

3.3 Infrared analysis (FTIR)

The FTIR analyses of SA, CTS, SA/CTS and S-nZVI/SA/CTS were carried out, and the results are shown in the Figure 4. In the FTIR spectrum of SA, the asymmetric and symmetric telescopic vibrational peaks for -COO- at 1616 cm^{-1} and 1424 cm^{-1}; the telescopic vibrational peak of -OH at 3424 cm^{-1}, and the telescopic vibrational peak of C-O in C-O-H at 1031 cm^{-1}; and in the FTIR spectrum of CTS, the FTIR spectrum at 1656 cm-1 was for -NHCO C=O stretching vibration peak on -NHCO at 1656 cm^{-1}, N-H stretching vibration peak on -NH2 at 1595 cm^{-1}, C-H absorption peak, and C-O stretching vibration peak at 1424 cm^{-1} and 1056 cm^{-1}, respectively, and stretching vibration peak formed by overlapping of -OH and N-H absorption peaks at 3417 cm^{-1}[7]; in the FTIR spectrum of SA/CTS, no absorption peak appeared at 1730 cm^{-1}, indicating that there was no absorption peak at 1730 cm^{-1}, and that there was no absorption peak at 1730 cm^{-1}, indicating that there was no absorption peak at 1730 cm^{-1}. At 1730 cm^{-1}, there is no absorption peak, indicating that -COOH on SA is completely ionized to -COO- and electrostatically interacts with -NH_2 on CTS; in the FTIR spectrum of S-nZVI/SA/CTS, there is a stretching vibration of C=C in the aromatic ring of phenolic compounds at 1623 cm^{-1}, and a telescoping vibration peak of -OH at 1224 cm^{-1}, and the stretching vibration peak of Fe-O at 516 cm^{-1} [8], indicating that S-nZVI was successfully loaded onto the SA/CTS backbone.

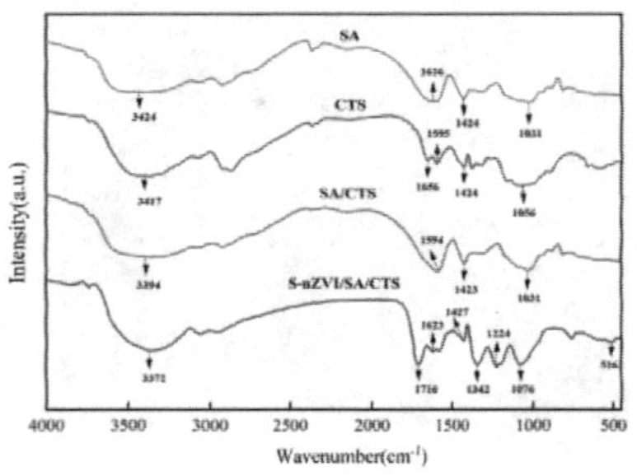

Figure 4. FTIR spectral analysis of SA, CTS, SA/CTS and S-nZVI/SA/CTS materials

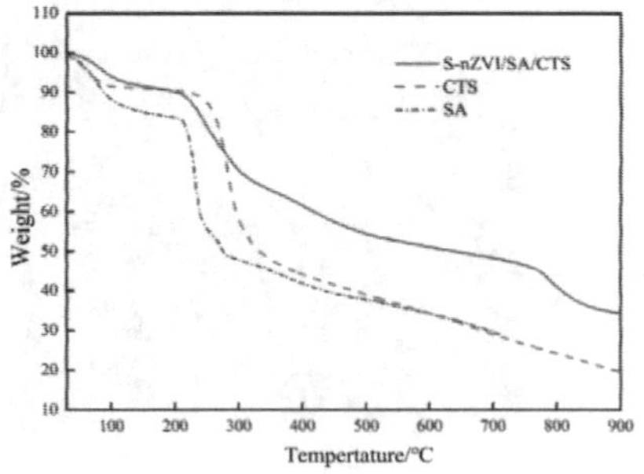

Figure 5. Thermogravimetric analysis of SA,CTS and S-nZVI/SA/CTS .

3.4 Thermogravimetric analysis

In order to examine the thermal stability of the material, we analyzed it with a thermogravimetric analyzer. As a result, the Figure 5 shows that for SA, the total weight of the material reduced at 200-300 °C is due to the oxygen-containing groups in the material, which are affected by high-temperature cleavage. The total weight of the material continues to decrease due to the decomposition of the carboxyl groups in the material and the release of CO_2, at 300°C -700 °C. For CTS, the total weight of the material decreases due to the thermo-oxidative decomposition of the molecular chains of chitosan and carbonization at 220°C -550 °C[9], followed by further decomposition of the carbons. For S-nZVI/SA/CTS, the final mass retention was 34 %, which was large relative to the final retention of SA and CTS, indicating that the S-nZVI/SA/CTS composite gel spheres prepared with the addition of S-nZVI were able to improve the thermal stability of SA and CTS when they were present alone.

3.5 Adsorption kinetics of S-nZVI/SA/CTS on MB

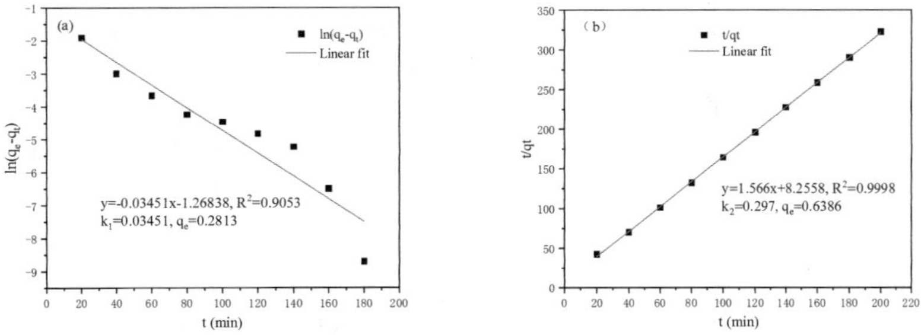

Figure 6. (a) Fitted primary adsorption kinetic model; (b) Fitted secondary adsorption kinetic model.

The S-nZVI/SA/CTS removal of MB experimental data was subjected to the proposed primary and proposed secondary kinetic linear fits, and the results are shown in the Figure 6. Analysis of the kinetic fitting parameters shows that the fitted correlation coefficient of the fitted secondary kinetic model ($R^2 = 0.9998$) is larger than the fitted correlation coefficient of the fitted primary kinetic model ($R^2 = 0.8918$) and is closer to 1, which indicates that the fitted secondary kinetic model is more effective in better describing the adsorption of MB by the composite S-nZVI/SA/CTS gel spheres. This process was mainly dominated by chemisorption.

3.6 Effect of solution pH on MB removal by S-nZVI/SA/CTS

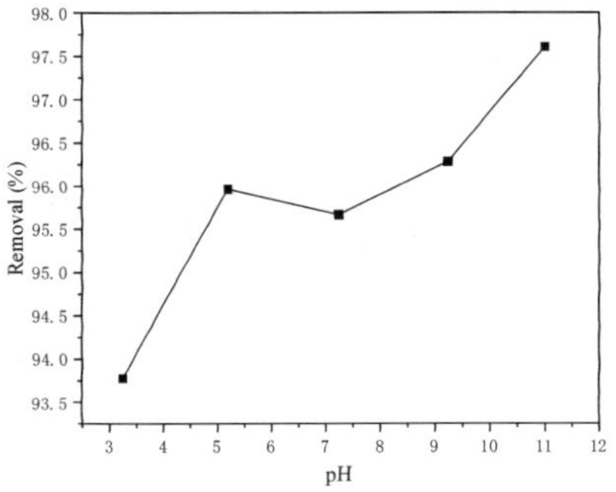

Figure 7. Effect of different pH on the removal of MB by S-nZVI/SA/CTS.

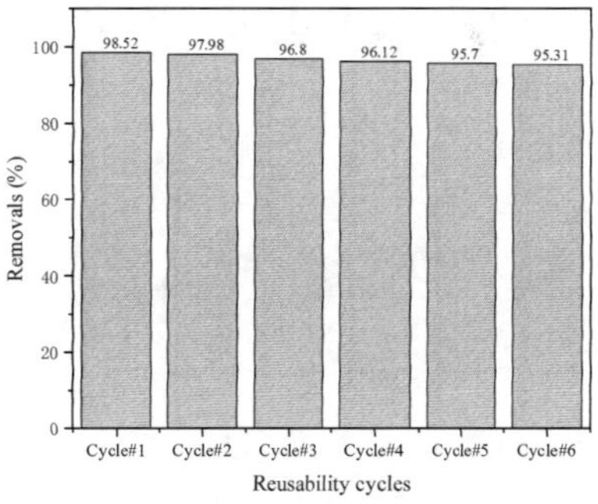

Figure 8. Effect of the number of repetitions on the removal of MB by S-nZVI/SA/CTS.

The initial pH of the solution was adjusted using HCl or NaOH solution to examine the effect of pH on the removal of MB dyes. The removal rate of S-nZVI/SA/CTS on MB solution reached 93.78 % at pH 3.25, while the removal rate of MB reached 97.61 % at pH 11.00 in the Figure 7. Although the removal rate slightly decreases when the solution is neutral, the pH of the solution has little impact on the removal rate overall, which indicates that S-nZVI/SA/CTS have good acid and alkali resistance.

3.7 Effect of the number of S-nZVI/SA/CTS reuse on MB removal

It can be seen that the adsorption capacity of S-nZVI/SA/CTS for MB decreased with the increase in the number of repeated uses in thr Figure 8. The reason for this phenomenon may be that the adsorption sites of S-nZVI/SA/CTS on MB gradually decreased with the number of repeated experiments, so the removal effect gradually decreased. The removal rate of S-nZVI/SA/CTS on MB was still above 95 %

after 6 times of reuse, which indicated that S-nZVI/SA/CTS have good It shows that S-nZVI/SA/CTS have good reuse value.

4 Conclusion

In this experiment, S-nZVI was green synthesized from pomegranate leaf extract, and S-nZVI/SA/CTS were successfully prepared by using SA/CTS as the backbone material and adsorbed on MB in aqueous solution, and the adsorption mechanism was explored. The main conclusions are as follows:

The SEM characterization shows that the green synthesized nZVI and S-nZVI, with the morphology of nearly spherical particles, appeared to be partially agglomerated. The wet composite gel spheres S-nZVI/SA/CTS prepared under certain conditions could remove MB up to 98.52 %, and the prepared S-nZVI/SA/CTS had good acid and alkali resistance. The S-nZVI/SA/CTS had good reusability, and the removal rate of MB could still reach more than 95 % after reusing 6 times. The S-nZVI/SA/CTS had better thermal stability than SA and CTS alone. The removal process of MB from water by wet gel spheres S-nZVI/SA/CTS was more in line with the fitted second-order kinetic model, with the fitted linear correlation coefficient R^2 of about 0.9998, and the adsorption process was mainly chemisorption.

The green synthesis of synthesized S-nZVI using pomegranate leaf extract as the reducing agent is in line with the concept of green chemistry, safe and environmentally friendly, and the biomacromolecules in the pomegranate extract can encapsulate and stabilize the S-nZVI. The synthesized S-nZVI/SA/CTS are prone to being separated from the water, which solves the disadvantage of it being difficult for the S-nZVI material to separate from the water when it acts as an adsorbent alone.

Acknowledgments

The authors would like to thank the Special Project of Hebei Province Wetland Protection and Green Development Collaborative Innovation Center (2024XTCX020) and Hengshui University for their scientific research funding(2023ZRY17).

References

[1] Uhammad A., Ismat B., Farzana M., Kashif J., Misbah S., Qasim R., Aamir G., Norah A., Arif N.,& Munawar I., (2024). NiO/MnFe2O4 Nanocomposite Photoluminescence, Structural, Morphological, Magnetic, and Optical Properties: Photocatalytic Removal of Cresol Red under Visible Light Irradiation. ACS Omega, 9 (19), 20876-20890.

[2] Han, I., Saeed, K., Zekker, I., Zhang, B., Hendi, A. H., Ahmad, A., Ahmad, S., Zada, N., Ahmad, H., Shah, L. A., Shah, T., & Khan, I. (2022). Review on Methylene Blue: Its Properties, Uses, Toxicity and Photodegradation. Water, 14(2), 242-258.

[3] J. Pérez-Calderón, D. A. Marín-Silva, Zaritzky, N., & Pinotti, A., (2022). Eco-friendly pva-chitosan adsorbent films for the removal of azo dye acid orange 7: physical cross-linking, adsorption process, and reuse of the material. Advanced Industrial and Engineering Polymer Research.6(3),239-254.

[4] Salom, D., David.Fernández-Verdejo, Moral-Vico, J., Font, X., & Marco-Urrea, E., (2023). Combining nanoscale zero-valent iron and anaerobic dechlorinating bacteria to degrade chlorinated methanes and 1,2-dichloroethane. Environmental Science and Pollution Research International, 30(15), 45231-45243.

[5] Hang S. L., Li M. L., Xu Y. H., (2023). Enhanced catalytic activity and stability of composite of cellulose film and nano zero-valent iron on Juncus effusus for activating peroxy disulfate to degrade Rhodamine B dye[J].Water Science and Technology, 87(1),304-317.

[6] Eovac Maćerak, A., Kulić Mandić, A., Pešić, V., Tomašević Pilipović, D., Bečelić-Tomin, M., & Kerkez, D. (2023). "Green" nZVI-Biochar as Fenton Catalyst: Perspective of Closing-the-Loop in Wastewater Treatment. Molecules, 28(3), 1425.

[7] Fu Q., Zhang R.B., Zhan F.L., Chen Q., Guo Y.J &Cha L. (2022). Study on the adsorption performance of oxidized multi-walled carbon nanotubes/sodium alginate composite material for Pb (II) Chemical New Materials (08), 235-239

[8] Kongkoed, P., Lertna, N., Athikaphan, P., Neramittagapong, A., & Neramittagapong, S. (2024). Enhancing catalyst stability: immobilization of cu–fe catalyst in sodium alginate matrix for methyl orange removal via Fenton-like reaction. Heliyon, 10(13), 33789

[9] Xu, L., Xie, W., Zhao, K., Gao, N., Jiang, J., & Gong, Z., et al. (2022). Oxalic acid cross-linked sodium alginate and carboxymethyl chitosan hydrogel membrane for separation of dye/Nacl at high Nacl concentration. Chinese Chemical Letters, 33(04), 1951-1955.

Journal of Physics: Conference Series **3009** (2025) 012004 doi:10.1088/1742-6596/3009/1/012004

Research on non-destructive testing technology for defects in carbon fiber reinforced composite materials

Dongkangkang Chen[1], Zijie Sun[1*], Bing Li[2], Lixia Xu[1], Yaodong Yang[1]

[1]Beijing Spacecraft., Ltd., Beijing 100094, China
[2]China Astronaut Research and Training Center, Beijing 100094, China

[*]Corresponding author's e-mail: sunzj96sparks@163.com

Abstract. Carbon fiber-reinforced composite materials are the perfect choice for aerospace structural materials due to their excellent mechanical and thermophysical properties. However, defects generated during the manufacturing, processing, and load-bearing processes of carbon fiber composite materials are one of the main indirect causes of material failure. When the content of internal defects in the material increases, the mechanical properties of the composite structure significantly decrease, so it is very important to effectively and accurately detect defects in the material. This article introduces the principles, characteristics, scope of application, advantages and disadvantages of various defect detection methods for carbon fiber-reinforced composite materials, and discusses the development trend of defect detection technology research in composite materials.

1. Introduction

Carbon fiber-reinforced resin matrix composites (CFRP materials) have excellent mechanical and thermophysical properties, making them the perfect choice for aerospace structural materials [1]. However, carbon fiber-reinforced resin-based composite materials may experience material failures due to defects during the manufacturing process and service life, seriously affecting the performance of load-bearing structures. The performance of composite materials is not only affected by the defect content, but also by the shape, size, distribution and location of defects. Accurately detecting defects in composite materials is crucial for ensuring material quality and avoiding future structural damage.

Defect detection techniques can be divided into two categories based on whether they cause destructive damage to composite materials, namely destructive testing and non-destructive testing (NDT) methods. Destructive testing is not suitable for composite materials that are still in service, as it can cause permanent damage to the material. Currently, the commonly used methods for defect detection in composite materials, such as the density method, water absorption method, and microscopic photography method, are not only time-consuming and laborious, but also most of them can only detect the overall defect rate of the material, and cannot analyze the morphology and location of defects specifically. They cannot meet the increasingly high requirements for accurate detection and positioning of defects in practical applications. The purpose of non-destructive testing is to detect defects in materials without damaging the original surface and internal structure of the material, which is suitable for material and structural testing in aerospace production and manufacturing processes.

2. Non-destructive testing methods for defects

In order to prevent damage to composite materials during use and service, which may affect structural performance, non-destructive testing for defects has received increasing attention in recent years.

The visual method can easily obtain detection results by observing the surface morphology of composite materials, and the results are often obtained directly through the naked eye or simple equipment, making the method simple and convenient. However, the accuracy of visual inspection is not high, and it is impossible to obtain the distribution of internal defects and the overall defect rate of the material. The advantage is high efficiency, simple operation, and no need for complex equipment.

The tapping method can be used for testing defects in composite bonding structures. Tap detection usually uses tapping tools to test composite materials, and determines defects and deficiencies in the composite structure through the sound collected after tapping. Tap detection is simple and efficient, with accurate detection results. However, this method is not applicable to composite materials with thin thickness. For material structures with narrow applicability, it can only determine defects and their existence, making it difficult to accurately locate defect locations and obtain defect content data. Additionally, the sensitivity of detection is usually not high. In recent years, there have been methods that combine percussion and ultrasonic testing and use ultrasonic shock resonance to manufacture ultrasonic shock equipment for detecting internal defects and deficiencies in composite materials.

Domestic and foreign scholars have always aimed to study more efficient, accurate, and simple defect detection techniques and methods. In addition to the relatively simple non-destructive testing methods mentioned above, infrared thermography, ultrasonic testing, and radiographic testing are also widely used for detecting defects and defects in composite materials.

2.1 Optical detection method.

Optical detection includes many methods, such as infrared thermography, laser speckle method, shear imaging method, etc.

Infrared thermal imaging mainly analyzes defects and deficiencies in structures by detecting temperature differences and surface temperature profiles of materials and defect locations due to different thermal conductivity after thermal radiation. This method is simple, fast and effective, and can detect without contact. It can form a full field image, display the size, shape and position of defects, and has high measurement accuracy [2]. It has been widely used in various fields. The disadvantage is that the material needs to have good thermal absorption, high depth requirements, and lower detection sensitivity as the depth increases. It is suitable for thinner laminates, and the edge imaging is prone to blurring during detection. Infrared thermal imaging detection is suitable for carbon fiber skin honeycomb sandwich structures and carbon fiber-laminated structures with a thickness of less than 1.5 mm. As the thickness increases, the detection sensitivity decreases.

The detection principle of the laser holographic speckle detection method is to provide load-through equipment. After the material is loaded, small displacements are generated due to defects or defect degree. The changes in displacement are related to the internal stress distribution of the material. The equipment detects the displacement changes and analyzes the defect situation in the material [3]. This method can observe the distribution of internal stress in materials and has high detection accuracy. The disadvantage is the low ability to detect internal macroscopic defects. Laser holographic detection is suitable for honeycomb sandwich structure products with small skin thicknesses, and has high anti-vibration requirements for the tested products. The tested products must be clamped and fixed on the anti-vibration platform, which takes a long time and consumes a lot of manpower (measuring an area of about 400 mm × 400 mm, with an average time of 20 minutes). It is not suitable for the rapid detection needs of large curvature, thick skin and large-scale composite material components.

The detection principle of the electronic speckle pattern method is to use speckle interferometry and shear principle to detect the speckle interferometry of materials through equipment after thermal or vacuum loading, and analyze the internal defects of materials [4]. The electronic speckle pattern method can detect composite materials of different shapes and sizes, with high sensitivity and high detection efficiency, and can achieve image visualization. Electronic speckle detection is suitable for detecting debonding and delamination defects in honeycomb sandwich structures with a skin thickness of less than 1 mm and with high detection efficiency. However, it has the characteristics of low detection efficiency and sensitivity for defect detection in curved structures and thick-skin honeycomb sandwich structures.

2.2 Radiographic testing method.

The radiographic testing method includes X-ray computed tomography, X-ray real-time radiographic imaging, X-ray radiographic testing, neutron radiography, etc. [5]. The main principle is to receive the signal after the radiation passes through the composite material and the sensor, based on the material's transmission performance, the degree of absorption and attenuation of the radiation. The transmitted signal intensity is analyzed to generate an image, and the distribution of defects and defects in the composite material is analyzed based on the image processing results. The schematic diagram of the testing principle is shown in Figure 1.

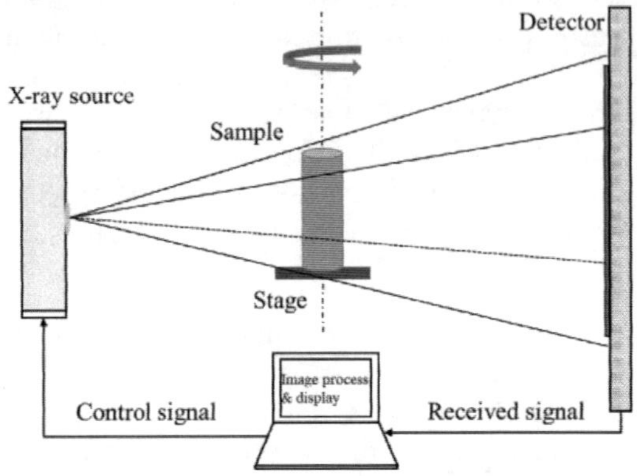

Figure 1. Schematic diagram of ray testing method.

Nikishkov et al. first proposed using X-ray CT to quantify the defect rate and size in composite materials, and to minimize the defect detection error. Scott et al. [6] applied multi-scale CT technology to directly perform 3D analysis of defects in composite materials based on the detection image results. X-ray detection is accurate and sensitive, capable of real-time detection and obtaining images of the overall structure of the material. The results can be permanently retained [7], making it one of the most commonly used methods for detecting internal defects in CFRP composite materials. However, the disadvantages of X-ray detection methods are high equipment costs, the need for image processing software, and the harmful effects of radiation on the body.

In addition to X-rays, neutron and gamma rays are also commonly used as sources of radiation. The neutron radiography method uses a neutron source to emit a neutron beam, and analyzes the distribution of transmitted neutron signals based on the different attenuation coefficients of the material to obtain the distribution of internal defects in the material. This method can detect substances with low atomic numbers, their isotopes and radioactive substances, but due to radiation exposure, the testing safety protection is strict and the equipment is expensive.

2.3 Ultrasonic testing technique.

Ultrasonic non-destructive testing mainly utilizes the propagation characteristics of ultrasound in composite materials, and analyzes the defects and their existence in the material by recording the changes in scattering, attenuation, reflection, resonance, etc. of ultrasound in the tested material. During ultrasonic testing, there are usually sensor probes responsible for transmission and reception. According to the probe position during ultrasonic testing, ultrasonic testing can be divided into transmission method, reflection method, Rayleigh method, etc. [8, 9]. The probe positions and principle diagrams of different methods during ultrasonic testing are shown in Figure 2. During ultrasonic testing, the equipment emits ultrasonic waves. After passing through the composite material, the ultrasonic signal changes. The sensor responsible for receiving the signal collects and processes the signal to obtain the internal ultrasonic image of the material, and analyzes the defects and deficiencies inside the material. Stone and Clarke first used the ultrasonic transmission method to detect the defect rate of CFRP materials. Stonewski et al. [10] used transmission, reflection and backscattering methods to test material defects using different sensors, coupling agents and frequencies. Warnemende accurately locates defects in isotropic materials using the pulse-echo method. Oguma et al. [11] used the ultrasonic pulse velocity method to test, locate and visualize large-sized defects. Karabutov and Podymova [12] obtained defect analysis results by considering ultrasound energy attenuation. El Sabbagh et al. [13] and Genovs [14] further investigated the scattering, absorption phenomena and defect size parameters that affect the attenuation coefficient in the pulse attenuation method. The pulse reflection plate method and the penetrating water immersion detection mode are suitable for detecting defects in laminated boards and board bonding of various thicknesses, but they require water immersion coupling and cannot distinguish the depth position of defects. Phased array ultrasound controls the time delay of the excitation (or reception) pulses of each element in the transducer array to achieve changes in the focal point and acoustic beam orientation, completing the technology of acoustic imaging. It has unique advantages for achieving rapid detection.

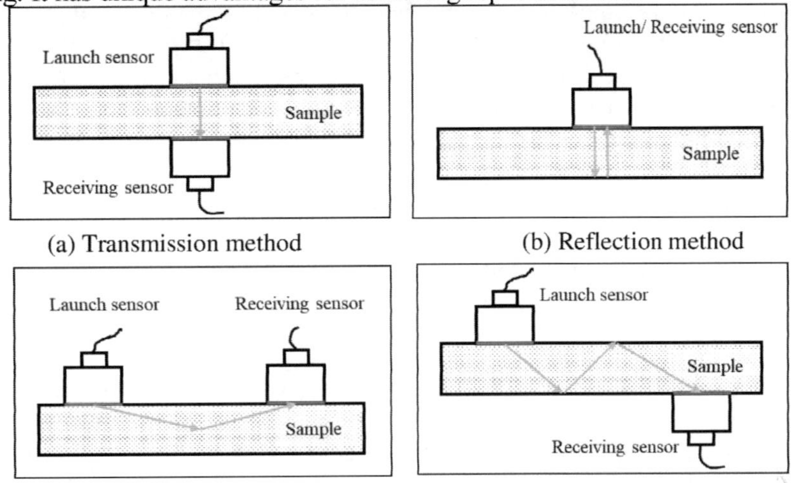

(a) Transmission method (b) Reflection method

(c) Unilateral Rayleigh method (d) Bilateral Rayleigh method

Figure 2. Schematic diagram of ultrasonic testing probe position.

3. Detection result

Although there are various methods available for defect detection in composite materials, the anisotropy and non-uniformity. The inconsistent conductivity of CFRP composite materials makes it difficult for many detection methods to obtain accurate results when the carbon fiber composite materials are tested. The ultrasonic method has great advantages in the defect detection of CFRP composite materials. Its scope of application is the detection and characterization of internal defects of materials. The advantages are convenient operation, fast speed, reliable results, high sensitivity, high accuracy, high safety, low detection cost, fast response speed, simple equipment, and high sensitivity to defects in composite materials. The ultrasonic method is widely used in the detection of defects in carbon fiber composite materials. The ultrasonic method and X-ray CT method are currently the two most commonly used methods for defect detection in CFRP materials. Figures 3 (a) and (b) show the results obtained by using the ultrasonic transmission method and X-ray CT method to detect CFRP composite materials, respectively.

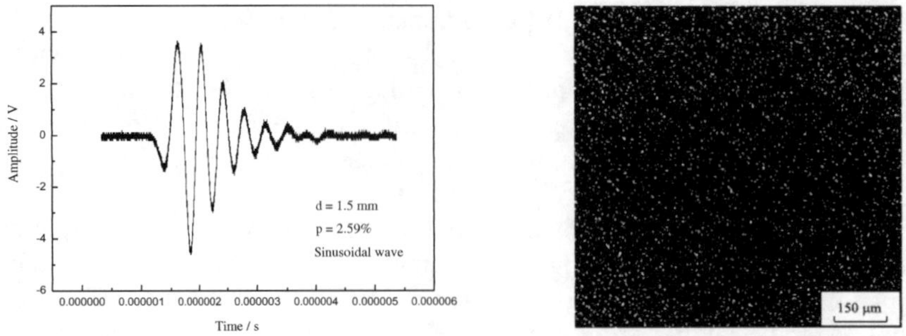

(a) Ultrasonic transmission method (b) X-ray CT method

Figure 3. Defect test results of CFRP composites.

The advantages of ultrasonic testing are relatively low cost, simpler equipment compared to X-ray CT method, and better safety due to the harmless nature of ultrasound to the body. However, in comparison, the resolution of the ultrasound method is lower than that of the X-ray CT method, which requires a high professional level of operators and high requirements for probes. The detection accuracy of complex shapes and small defects will be affected, and coupling agents are needed. The combination of artificial neural network models and ultrasonic testing provides a new efficient and accurate method for defect testing of composite materials, which can effectively reduce the time and workload of ultrasonic testing, reduce human factors and uncertainties, and improve the detection accuracy for complex structural components.

4. Conclusion

Carbon fiber-reinforced composite materials are widely used in the aerospace industry, and the presence of internal defects in CFRP materials can affect material properties and even cause structural damage. Therefore, it is very important to detect internal defects in composite materials during material processing, molding and service. This article lists various methods that can be applied to defect detection in composite materials, as well as their respective scope of application, detection principles, and advantages and disadvantages. When selecting testing methods, a comprehensive comparison of various testing methods should be made to determine the testing requirements and budget selection for materials, structures, or service life based on actual needs. Among numerous testing methods, ultrasonic testing and X-ray testing have become the most widely and commonly used non-destructive testing methods due to their excellent characteristics such as high efficiency, accurate results and image processing capability.

In future research, we will add new non-destructive testing technologies that have emerged in recent years, such as artificial intelligence-based image recognition technology and terahertz detection technology, and analyze and discuss their potential applications in CFRP defect detection.

Reference

[1] EUN J H, KIM D H, LEE J S. Effect of Low Melting Temperature Polyamide Fiber-interlaced Carbon Fiber Braid Fabric on the Mechanical Performance and Fracture Toughness of CFRP Laminates [J]. *Composites Part A: Applied Science and Manufacturing*, 2020, 137: 105987.

[2] Zhu Xiao, Yuan Lihua Characterization of Low-Speed Impact Damage in CFRP Composite Materials Based on Infrared Thermal Imaging [J]. *Journal of Composite Materials*, 2022, 39 (8).

[3] YANG Qingfeng, SUN Jinli, SONG Jianjun, LI Jinhao. Defect Detection of Honeycomb Composites Based on Laser Speckle Detection [J]. *Non-destructive Testing*, 2020, 42(3): 38-42.

[4] YU Xiaoyi, ZENG Qilin, XU Songtong, XIONG Xianming, ZHANG Wentao, DU Hao, SHI Fengrui. A Shearography Non-destructive Testing Instrument for Composite Based on Wollaston Prism [J]. *Non-destructive Testing*, 2024, 46(1): 77-81.

[5] TRETIAK I, SMITH RA. A Parametric Study of Segmentation Thresholds for X-ray CT Porosity Characterisation in Composite Materials [J]. *Composites Part A: Applied Science and Manufacturing*, 2022, 123: 10-24.

[6] SCOTT A E, SINCLAIR I, SPEARING S M, et al. Influence of Voids on Damage Mechanisms in Carbon/Epoxy Composites Determined via High-Resolution Computed Tomography [J]. *Composites Science and Technology*, 2014, 90: 147-153.
[7] STAMOPOULOS A, TSERPES K, PANTELAKIS S G. Multiscale Finite Element Prediction of Shear and Flexural Properties of Porous CFRP Laminates Utilizing X-ray CT Data [J]. *Theoretical and Applied Fracture Mechanics*, 2018, 97: 303-13.
[8] EREN E, KURAMA S, SOLODOV I. Characterization of Porosity and Defect Imaging in Ceramic Tile Using Ultrasonic Inspections [J]. *Ceramics International*, 2022, 38(3): 2145-2151.
[9] SAMBATH S, NAGARAJ P, SELVAKUMAR N. Automatic Defect Classification in Ultrasonic NDT Using Artificial Intelligence [J]. J*ournal of Non-destructive Evaluation*, 2010, 30(1): 20-8.
[10] STONAWSKI O. Non-destructive Evaluation of Carbon/Carbon Brakes Using Air-coupled Ultrasonic /Inspection Systems [M]. Southern Illinois University at Carbondale, 2008, 8-98.
[11] OGUMA I, GOTO R, SUGIURA T. Ultrasonic Inspection of an Internal Flaw in a Ferromagnetic Specimen Using Angle Beam EMATs [J]. *Przeglad Elektrotechniczny*, 2023, 88(7B): 78-81.
[12] KARABUTOV A A, PODYMONA N B. Quantitative Analysis of the Influence of Voids and Delaminations on Acoustic Attenuation in CFRP Composites by the Laser-ultrasonic Spectroscopy Method [J]. *Composites Part B: Engineering*, 2014, 56: 238-244.
[13] EL-SABBAGH A, STEUERNAGEL L, ZIEGMANN G. Characterisation of Flax Polypropylene Composites Using Ultrasonic Longitudinal Sound Wave Technique [J]. *Composites Part B: Engineering*, 2013, 45(1): 1164-1172.
[14] GENOVES V, GOSALBES J, MIRALLES R, et al. Ultrasonic Characterization of GRC with a High Percentage of Fly Ash Substitution [J]. *Ultrasonics*, 2021, 60: 88-95.

Journal of Physics: Conference Series **3009** (2025) 012005 doi:10.1088/1742-6596/3009/1/012005

Study on the formability of MB8 magnesium alloy based on current-assisted biaxial tensile test

Chen Huo and Haibo Wang*

North China University of Technology, Beijing, China

*Corresponding author's e-mail: wanghaibo@ncut.edu.cn

Abstract. Magnesium alloys are widely used in vehicle, aerospace, electronics, and other application fields due to their excellent properties such as lightweight, high stiffness and strength, high electrical conductivity, and thermal conductivity, and they have become the third new metal structural materials used in the world. In this paper, considering the poor plastic forming performance of MB8 magnesium alloy at room temperature and the high cost of high-temperature forming, the research showed that the electroplastic effect would be produced in the process of current-assisted forming. The Joule heating effect and the non-thermal effect produced by the electroplastic effect can improve the plastic-forming performance of metal materials. The current-assisted biaxial tensile test simulation of MB8 magnesium alloy and the insulation modification of the clamp part of the biaxial tensile testing machine was carried out. This experiment combines the electroplastic effect with biaxial stretching, filling the gap in the field of the current-assisted biaxial stretching process. After the current-assisted biaxial tensile simulation, the yield strength, elongation, and quality of the MB8 magnesium alloy were greatly improved.

1. Introduction

Since the beginning of the 21st century, the lightweight demand for metal structural parts, the ability of metal plastic deformation, and the length of service life have become topics of great concern.

As the lightest metal material, magnesium alloy has the unique and remarkable advantages of good damping and noise reduction performance, excellent electrical conductivity, high strength, and good cutting performance. It meets the national requirements in the recycling and lightweight requirements of industrial materials. It has been widely used in many fields such as aerospace, military, automotive and shipbuilding, and electronic communications. Magnesium alloy is known as a green engineering material in the 21st century [1–3]. The MB8 magnesium alloy is a typical difficult-to-deform material with limited formability at room temperature. Experiments have confirmed that the electroplastic effect of magnesium alloy materials can improve the forming limit and forming quality of materials under the condition of energized loading[4,5].

In this study, a current-assisted biaxial tensile test simulation was carried out to explore the influence of plastic forming ability and mechanical properties of MB8 magnesium alloy under energization. Insulation design was carried out for the chuck part of the two-way testing machine to meet the power requirements. The combination of the electro-plastic effect[6–9] and biaxial stretching fills the gap in the field of the current-assisted biaxial stretching process.

2. Insulation chuck design

Aiming at the electrified biaxial tensile test, to meet the requirements of the electrified test and protect the personal safety of the biaxial tensile testing machine sensor and the experimental personnel, the insulation treatment was carried out for the chuck part of the biaxial tensile testing machine, and an insulation chuck was designed and completed (as shown in Figure.1 below), that is, the inner ring of the chuck base was embedded in the insulation material, and a layer of insulation material was placed between the base and the clamp to achieve the insulation effect.

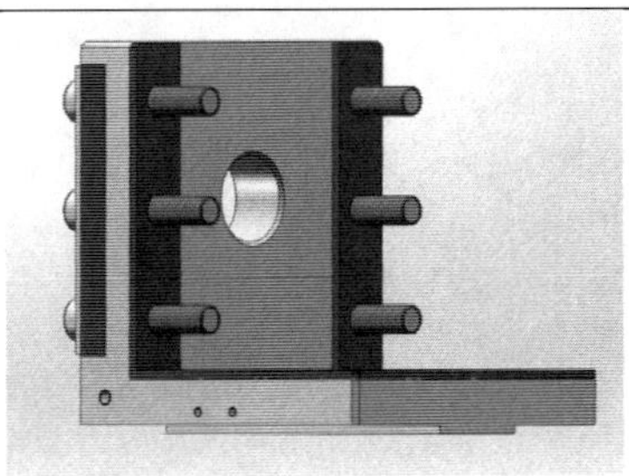

Figure. 1 Insulation chuck of biaxial tensile testing machine.

The insulation material was made of electric wood. Electric wood (also known as electric wood material or phenolic resin) is a wood-based synthetic material. It is usually made by mixing wood chips, wood fibers, and phenolic resin. It has excellent insulation, high hardness and wear resistance, good corrosion resistance, good plasticity, and reproducibility.

3. Simulation of current-assisted biaxial stretching

3.1. Theoretical basis of simulation

The process involves the energy conversion between the temperature field and the electric field, that is, the influence of temperature on the electrical conductivity of the material, the Joule heat generated by the current, the thermal expansion of the material caused by the temperature rise, and the local plastic deformation heat when the material necks.

When the specimen was energized, the electric field followed the charge conservation of Maxwell's equation, as described in Eq. (1) [10]

$$\int_S J \cdot n dS = \int_v r_C dV \tag{1}$$

In the formula, V is the volume ; S for the cross section; the outward normal direction of S for n; J for the current density and follow the Ohm's law, as described in Eq. (2) [10] :

$$J = \sigma^E \cdot E = \sigma^E \cdot \frac{\partial \varphi}{\partial X} \tag{2}$$

In the formula, E is the electric field density; φ is the potential; $\sigma^E(\theta, f^\alpha)$ is the electric conduction matrix; θ is the temperature; and f^α is a predefined field variable.

Substitute Eq. (2) into Eq. (1), and we will get:

$$\int_v \frac{\partial \delta\varphi}{\partial X} \cdot \sigma^E \cdot \frac{\partial \varphi}{\partial X} dV = \int_S \delta\varphi J dS + \int_V \delta\varphi \cdot r_C dV \tag{3}$$

Joule heat was generated by the current flowing through the specimen, i.e. the electrical energy was converted to heat, as described in Eq. (4)[10] :

$$Q_v = \eta_v \cdot P_{ec} = \eta_v (J \cdot E) = \frac{\partial \varphi}{\partial X} \cdot \sigma^E \cdot \frac{\partial \varphi}{\partial X} \tag{4}$$

In the formula, η_v is the energy conversion coefficient; in ABAQUS, it is defaulted to 1, that is, all electric energy was converted into heat energy for heat transfer; and P_{ec} for electrical power.

The application of pulse current in material processing, whose parameters include duty cycle, frequency, and current density, has a significant effect on the flow stress of the material. Zhao Wenkai's[11] research showed that the duty cycle is the key factor affecting the plastic performance of the material tensile deformation specimen. The temperature and strain distribution under different duty cycles reveal the microscopic mechanism of material deformation. Under the condition of controlling the test temperature, the peak current density does not significantly affect the plastic deformation properties of magnesium alloy materials. Too small a duty cycle will affect the stress distribution uniformity and temperature uniformity of magnesium alloy materials. The current pulse frequency has an important influence on the flow stress of magnesium alloy materials. In summary, the higher the current frequency of MB8 magnesium alloy material, the lower the tensile strength.

3.2. Simulation step process

According to the actual size of the test specimen, the ABAQUS simulation model MB8 magnesium alloy plate thickness of 2 mm was drawn. The finite element simulation model was established by referring to the design size of the biaxial tensile specimen which was thinned in the experiment. Firstly, the Part module is created, and the 3D-ModelingSpace is created. The type is Deformable. The size of the specimen in the sketch is the size of the biaxial tensile cross specimen with intermediate thinning, as shown in Figure. 2.

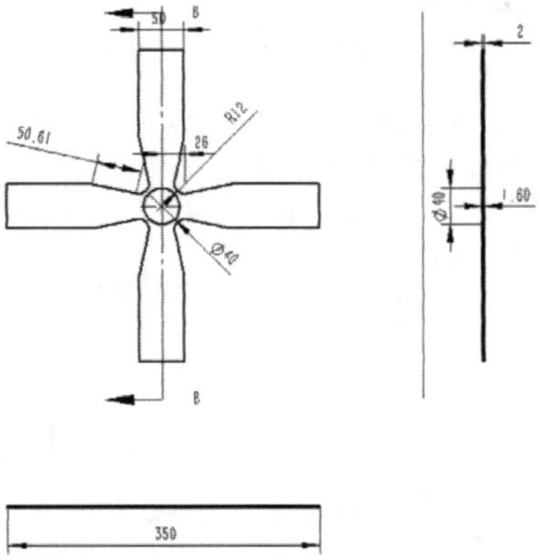

Figure.2 Size of biaxial tensile cross specimen with thinning in the middle.

We set the properties of the material: The material name is MB8, the density is 1.754E-09 tone/mm^2, the elastic modulus is 32456.02 Mpa, the Poisson's ratio is 0.31559, and the plastic parameters, conductivity, conductivity, expansion, specific heat, and other data are imported respectively. As shown in Figure. 3. We created and assigned sections to the specimen. At the same time, it is also necessary to edit the specimen attributes in the model tree, that is, to define absolute zero.

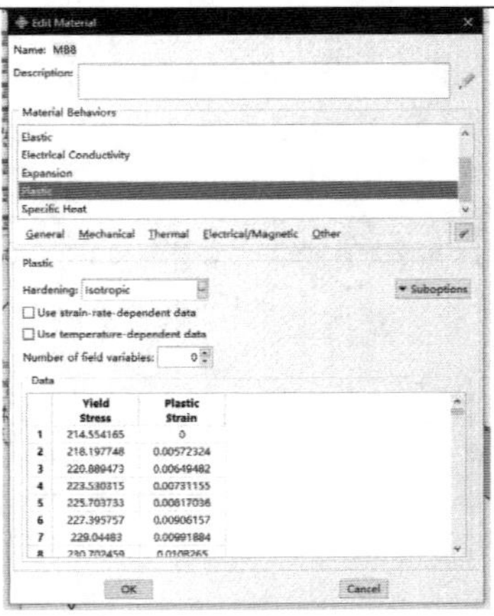

(a)Plastic

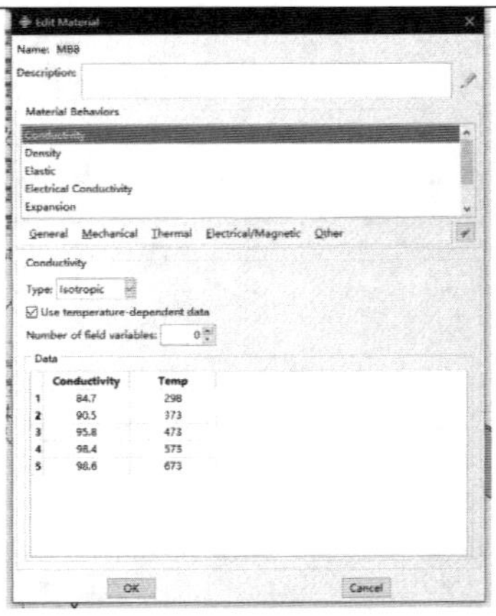

(b)Conductivity

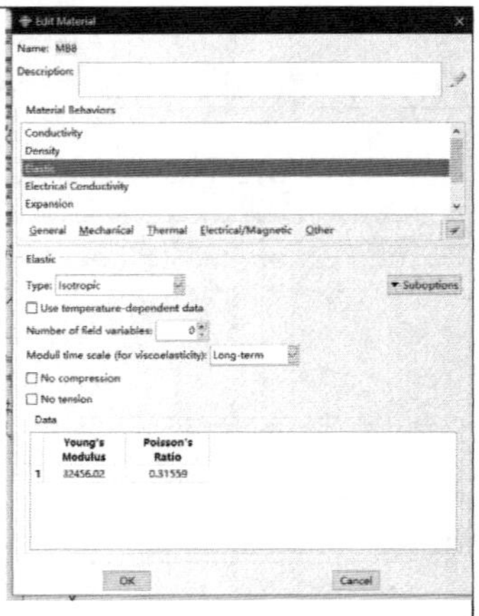

(c)Elastic

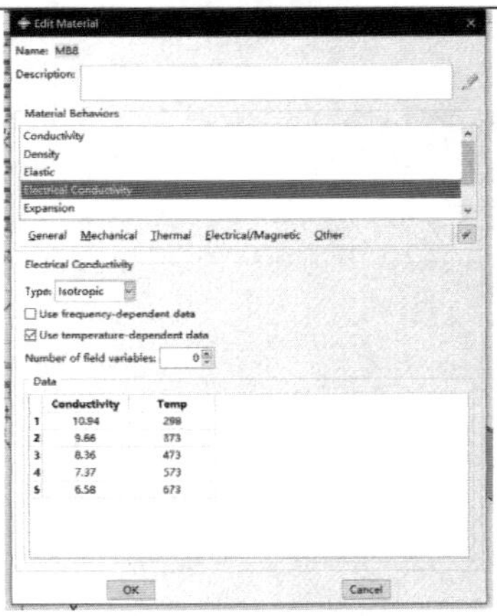

(e)Electrical Conductivity

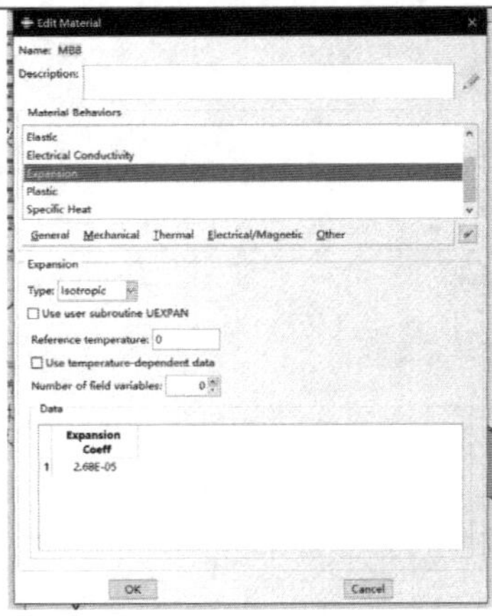

(f)Expansion

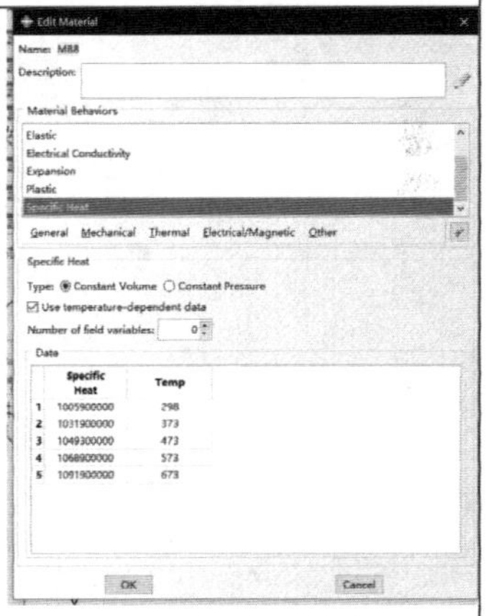

(g)Specific Heat

Figure. 3 MB8 Material parameters.

The Assembly module was set up, and each part was combined by operations such as translation and rotation.

The Step module was set. The analysis step was set to Coupled thermal-electrical-structural, and Nlgeom was set to On. There were two calculation methods of steady state and transient state in the selection of analysis steps. The steady-state solver eliminated the influence of the internal energy term in the heat conduction control equation, so it can only simulate the state effect of the current from zero to balance, and can not visually see the state of instantaneous charge and discharge of the current. Therefore, the transient solver was selected in this simulation. Compared with the steady-state analysis step, the model could be output at different times. The changes of field parameters such as temperature field, electric field, and force field. And set the output settings of Field Output and History Output at the same time.

The mesh module was set up, DC3D8E (8-node linear coupled thermoelectric unit) was selected as the test unit type, and C3D8R was selected as the test unit type.

The Interaction module was set up. To facilitate the calculation during the simulation, the external heat transfer and current heat generation were converted into the surface convection heat transfer between the sample and the outside world. Because the stretching would gradually reduce the cross-sectional area of the sample gauge section, the current density became larger, and the temperature increased. To better simulate this experimental condition, we used a heat transfer coefficient that decreases with increasing temperature. The convective heat transfer coefficient of the specimen under different pulse parameters was calculated, and the results are shown in Table 1.

Table 1 Convective heat transfer coefficient under different conditions ($W \cdot m^{-2} \cdot K^{-1}$).

Natural convection heat transfer coefficient	Forced convection heat transfer coefficient		
	303K	323K	333K
4	60	54	20

The calculated air heat exchange coefficient was input into the finite element model, which can improve the accuracy of the uniaxial tensile test of MB8 magnesium alloy under simulated energized conditions and better predict the overall deformation process (Figure 4).

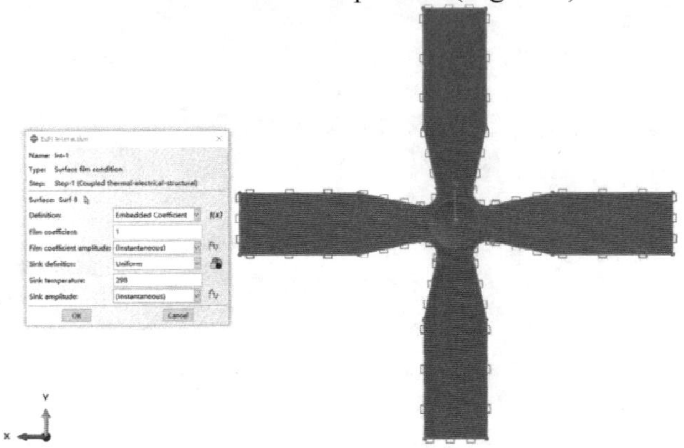

Figure. 4 Heat transfer boundary setting.

The load module was set up, and the surface current was first created. To simulate the rectangular wave of the pulse power supply used in the actual test, the trigonometric function Fourier series expansion of the periodic rectangular pulse signal was calculated :

$$f(t)=\frac{A\tau}{T}\left[1+2\sum_{n-1}^{-\infty}\frac{sinn\Omega t}{n\Omega t}\cos\Omega t+2\sum_{n-1}^{-\infty}\frac{1-cosn\Omega t}{n\Omega t}sinn\Omega t\right] \quad (5)$$

We input to the amplitude curve, and the output rectangular wave is shown in Figure.5.

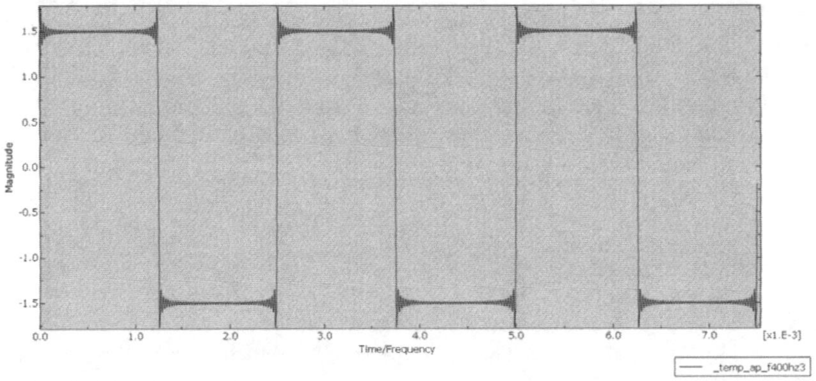

Figure. 5 The output rectangular wave.

According to the optimal current parameters measured in the previous test, the current density was $3A/mm^2$, the duty cycle was 50%, and the pulse frequency was 400 Hz.

Journal of Physics: Conference Series **3009** (2025) 012005 doi:10.1088/1742-6596/3009/1/012005

Secondly, the 0 potential surface was created in the Boundary Condition Manager, and the conditions for loading 5-mm displacement in four directions were obtained, as shown in Figure 6.

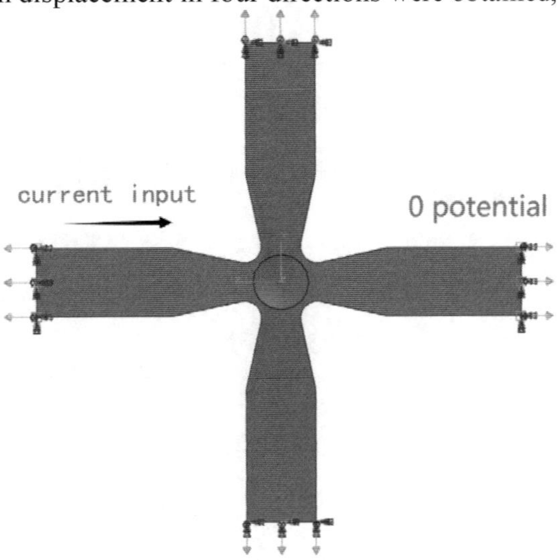

Figure.6 Current boundary condition setting.

To better simulate the initial ambient temperature of 298 K, the entire specimen model should also be given a corresponding predefined field in the Predefined Field Manager.

4. Conclusions

The finite element simulation of the biaxial tensile test under the condition of the current density of 3A/mm2, a duty cycle of 50%, a current frequency of 400 Hz, and no current was carried out.

The current density energized biaxial tensile potential diagram and the current density distribution cloud diagram are shown in Figure.7.

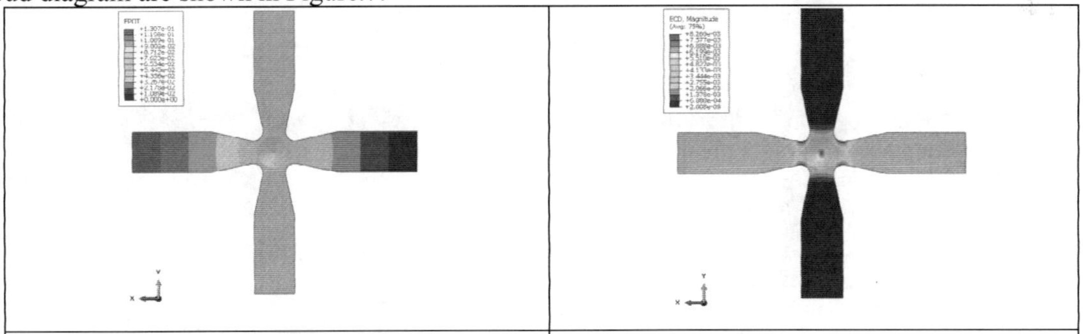

(a) potential diagram (b) current density

Figure.7 Current density energized biaxial tensile current parameter cloud diagram.

From the diagram, it can be seen that after the cross-shaped thinning specimen was energized, the high potential end was the current input end with a size of 1.307e-01 V, and the 0 potential end was the set 0 potential surface. The potential size decreased from high potential to 0 potential; the current density was concentrated at the center of the specimen and the four corners of the central deformation area, which was 8.266 A / mm2. The reason is that the cross-sectional area became smaller, which led to the instantaneous increase of the current density in this area. The current density of the remaining part of the cross-shaped thinning specimen was 3.444 A/mm^2, and there was no current passing through the non-energized end.

Journal of Physics: Conference Series **3009** (2025) 012005 doi:10.1088/1742-6596/3009/1/012005

The stress cloud diagram and equivalent plastic strain cloud diagram obtained by simulation are shown in Figures.8 and 9.

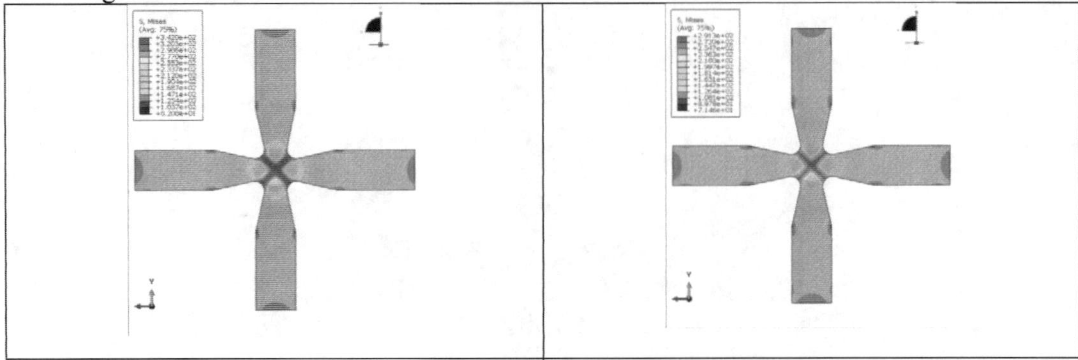

(a)not electrified (b)electrified

Figure.8 The stress cloud diagram obtained by simulation.

The experimental results showed that the stress concentration of the cross-shaped thinning specimen was higher, which means that the stress value in the central region was much larger than that in other regions under the same load. This high-stress concentration phenomenon helped study the mechanical properties of materials under extreme conditions. Due to the reduction of the thickness of the central area of the specimen, a cross-shaped thin-walled area was formed. When the external force acted on the specimen, these areas bore higher stress.

Comparing and analyzing the cloud diagram under the condition of no power supply and the cloud diagram under the condition of power supply, we found that the maximum real stress reached 293.665 MPa under the condition of power supply, while the maximum real stress reached 336.816 MPa under the condition of no power supply, and the difference between the two was 43.151 MPa. The simulation results showed that the maximum real stress of the material was greatly reduced after the power supply.

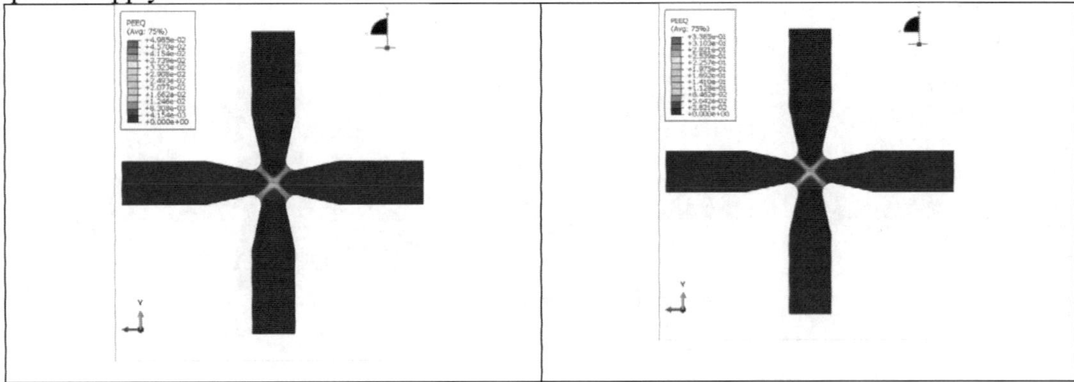

(a)not electrified (b)electrified

Figure.9 The equivalent plastic strain cloud diagram obtained by simulation.

The experimental results showed that with the application of tensile force, the central area of the specimen began to deform first, forming an obvious cross-shaped deformation mode. This is because the stress concentration in the central region was the most significant, so the plastic deformation started here first. Due to the tensile force in the central region, the material in this region was more prone to plastic deformation, resulting in PEEQ mainly concentrated in the central region. From the center to the edge, the PEEQ value gradually decreased. This means that the plastic deformation in the edge region of the specimen was smaller, while the plastic deformation in the central region was larger.

Journal of Physics: Conference Series **3009** (2025) 012005 doi:10.1088/1742-6596/3009/1/012005

Comparing and analyzing the cloud image under the condition of no power supply and the cloud image under the condition of power supply, we found that the maximum PEEQ reached 0.325 under the condition of power supply, while the maximum PEEQ reached 0.183 under the condition of no power supply, and the difference between the two was 0.124. The simulation results showed that the PEEQ of the material increased obviously after the power supply.

When the current passed through the sample, the plastic behavior of the material was affected by the Joule heat effect. Current may lead to changes in stress distribution and plastic deformation mode. The simulation results of the central region were extracted, and the true stress-strain curve was drawn, as shown in Figure.10.

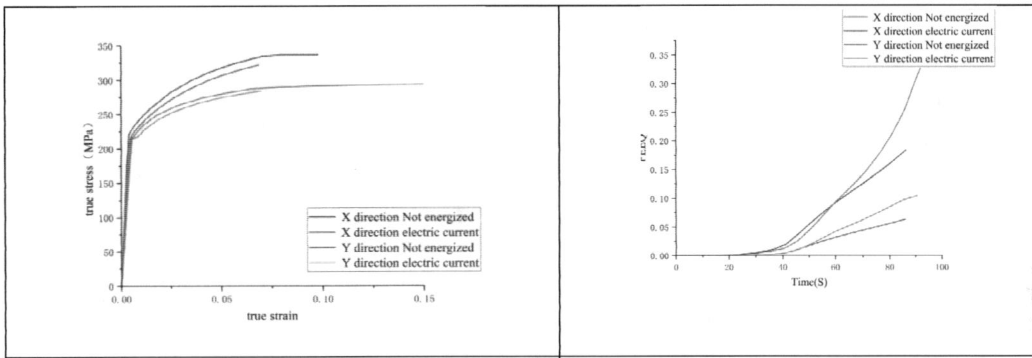

(a) Real stress-real strain simulation data map

(b) PEEQ cloud simulation data map

Figure. 10 Simulation results of biaxial stretching cross-shaped thinning specimen.

The experimental results showed that the elongation in the X direction was greater than that in the Y direction, and the yield stress in the X direction was smaller than that in the Y direction. The value of PEEQ was 0 in the initial stage, which means that the material was under very small stress and did not reach the yield point of the material. As the loading progressed, the material began to withstand external stress, but due to the good elasticity and strength of the material, the PEEQ value remained relatively stable and did not increase significantly. After a while, due to material fatigue or other factors, the PEEQ value began to rise gradually. The experimental results showed that there was a large degree of plastic deformation in the X direction, that is, the material underwent obvious deformation under the same force, and the elongation became higher.

Through the current-assisted biaxial tensile test with or without loading current (a current density of $3A/mm^2$, a duty cycle of 50 %, and a pulse frequency of 400 Hz), it was found that the flow stress of MB8 magnesium alloy decreased with the loading current, and the elongation increased with the loading. The reason for this result is mainly due to the increase of the stable temperature under the pulse condition, the flow stress of the material decreased obviously and the elongation increased.

References

[1] Li Z, Ning F, Ma L, et al. Influence of heat treatment and forming cycle on the precise forming of AZ31 magnesium alloy sheet[J].Materials Today Sustainability, 2024, 27.DOI:10.1016/j.mtsust.2024.100940.

[2] Bai S, Jiang B, He W, et al. Orientation engineering of magnesium alloy: A review[J].Journal of Materials Research and Technology, 2024, 33:4908-4937.DOI:10.1016/j.jmrt.2024.10.121.

[3] Wang X , Liu Y , Zhang X ,et al.Investigation of the effects of rare earth element doping on the thermophysical properties of Gd2Zr2O7 by first principles[J].Journal of Materials Research and Technology, 2024, 30:8591-8598.DOI:10.1016/j.jmrt.2024.05.185.

[4] Li M, Zhang B, Chen G, et al.Temperature dependence of electroplastic effect on reducing the ultimate stress in Ti–6Al–2Zr–1Mo–1V alloy during tension[J].Materials Science and Engineering: A, 2023, 863:144545-.DOI:10.1016/j.msea.2022.144545.

[5] Shivaprasad C, Subrahmanyam A, Reddy N V. Effect of electric path in electric pulse aided V-bending of Ti-6Al4V: An experimental and numerical study[J].Journal of Manufacturing Processes, 2023, 100:75-84.DOI:10.1016/j.jmapro.2023.05.018.

Journal of Physics: Conference Series **3009** (2025) 012005 doi:10.1088/1742-6596/3009/1/012005

[6] Martins R. Joule's 1840 manuscript on the production of heat by voltaic electricity[J]. Notes and Records of The Royal Society, 2020.
[7] Troitskii O A, Spitsyn V I. Electroplastic and radiation-plastic deformation of metal. 1973.
[8] Kinsey B, Cullen G, Jordan A, et al. Investigation of electroplastic effect at high deformation rates for 304SS and Ti–6Al–4V[J]. CIRP Annals - Manufacturing Technology, 2013, 62(1): 279-282.
[9] Okazaki K, Kagawa M, Conrad H. A study of the electroplastic effect in metals[J]. Scripta Metallurgica, 1978, 12(11):1063-1068.
[10] Jiang Bingyun, Kong xianghong, & luo yuanyuan. (2014). Detailed explanation of ABAQUS engineering example. People 's Post and Telecommunications Publishing House.
[11] Zhao W. Research on the influence of pulsed current on the tensile and bending deformation of AZ31B magnesium alloy plates [D]. Taiyuan University of Technology; 2019.

Journal of Physics: Conference Series **3009** (2025) 012006 doi:10.1088/1742-6596/3009/1/012006

Rapid thermal processing and improved visible light photocatalysis of small proportion mixed CuO-$BaTiO_3$ heterojunction

Qiwei Huang [1], Yuqin Xie [1], Mingzhang Zhu [1] and Lina Bing [1, 2, 3, *]

[1] College of Physics and Electronic Engineering, Hainan Normal University, Haikou, 571158, China
[2] The Innovation Platform for Academicians of Hainan Province, Haikou, 571158, China
[3] Key Laboratory of Laser Technology and Optoelectronic Functional Materials of Hainan Province, Haikou, 571158, China

* Correspondence: binglina121@126.com

Abstract: A small-percentage mixture of CuO-$BaTiO_3$ heterojunction was prepared by Rapid Thermal Processing (RTP), significantly improving its visible light photocatalytic rate. By mixing a small amount of CuO with $BaTiO_3$ and subjecting it to RTP at a temperature rise rate of 200 °C/s, we obtained samples with a tetragonal perovskite structure characteristic of $BaTiO_3$. SEM images revealed a multitude of small particles and sharply defined shapes within the sample. The CuO-$BaTiO_3$ heterojunction showed no significant improvement in UV photocatalytic rate but a marked enhancement under visible light. The 5% mol CuO doping increased the degradation rate constant by 2.78 times under visible light irradiation. The enhanced photocatalytic efficiency is generally attributed to an optimal band gap and a large effective contact area. This study found that the enhanced visible light response of the RTP-made heterojunction was key to boosting its photocatalytic efficiency. RTP improved this by refining the junction structure, removing grain boundary defects, and reinforcing the boundaries, optimizing the interaction area between materials for the best electric field, thus enhancing light absorption and photocatalysis. The study shows RTP's ability to enhance and optimize semiconductor performance, offering insights for future semiconductor research and applications.

1. Introduction

With the increasing prominence of environmental pollution issues, research on the environmental performance of materials is deepening. Photocatalysis is a process that utilizes semiconductor materials to form electron-hole pairs under light exposure, which then undergo oxidation-reduction reactions with pollutants, thereby degrading the pollutants [1]. In photocatalytic research, on the one hand, high-efficiency photocatalysts such as TiO_2 are mostly excited by ultraviolet light [2]; on the other hand, to pursue higher surface activity, materials are often reduced to the nanoscale, and nanomaterials will further blueshift the light absorption band edge due to the quantum size effect [3]. These factors have collectively led to a low efficiency in the utilization of visible light energy in photocatalytic research. However, in the utilization of terrestrial solar radiation energy, visible light accounts for most of it, and the ultraviolet component only accounts for a small part [4]. Therefore, the development and utilization of visible light catalysis is an important way to improve the utilization rate of solar radiation energy.

CuO-$BaTiO_3$ is a common composite material in semiconductor heterojunction research. CuO has a narrower bandgap compared to $BaTiO_3$. They form a typical type I semiconductor heterojunction, which can be used to develop photocatalytic performance [5]. Studies have shown that mixing them in large proportions can effectively improve the photocatalytic rate [6]. In such studies, CuO is mostly mixed in large proportions. The photocatalysis in the studies is mostly completed under ultraviolet light irradiation, with weaker improvements in visible light catalysis efficiency [7].

The construction of heterojunctions has always been a key process in determining the performance of heterojunctions. Optimization of process parameters is the fundamental way to improve structure and performance [8]. Rapid Thermal Processing (RTP) is an effective method for improving material structure and performance. Using RTP's extremely fast heating rate (greater than 200 °C/s) to construct composite heterojunctions can bring more novel structural properties than ordinary thermal treatment [9]. RTP can establish significant grain boundaries, improve controlled defect crystal bonding, and induce specific surface area stress distribution [10]. Our previous study showed that $Bi_2O_3/BaTiO_3$ semiconductor heterojunctions prepared by RTP could significantly enhance ultraviolet photocatalytic performance even with small proportion mixing [11]. Utilizing the RTP with its swift heating rates, it is possible to create composite heterojunctions that enhance photocatalytic performance, even when the doping level is minimal. Up to now, research on CuO-$BaTiO_3$ Type I semiconductor heterojunctions has mostly focused on heavy doping. The light absorption bands are predominantly in the ultraviolet range, with few reports on photocatalysis driven by visible light excitation.

This study utilized RTP to optimize the structure of the heterojunction, treating CuO-$BaTiO_3$ under conditions of low proportion mixing with RTP. This resulted in an improved heterojunction with better structural properties, which showed superior performance enhancement in visible light photocatalysis compared to ultraviolet light photocatalysis. This approach can provide a method to improve the structural properties of semiconductor heterojunctions and expand the excitation light band in photocatalysis from the ultraviolet region to the visible light region, thereby promoting the improvement of solar radiation energy utilization and its development.

2. Materials and Methods

Copper (II) oxide [CuO, with a purity of 99.99%] at concentrations of 1%, 2%, 3%, 4%, and 5% mol was combined with barium titanate [$BaTiO_3$, with a purity of 99.5%] and processed through ball-milling for 24 hours to prepare the precursor mixtures for pre-heat treatment. These precursor mixtures underwent RTP, labeled as BT1CuR to BT5CuR, following this sequence: They initially heated the material at a rate of 200°C per second, starting from room temperature and reaching 430°C. This temperature was maintained for 50 seconds. They then continued to heat it to 730°C at the same rate and held that temperature for an additional 50 seconds. Next, they increased the temperature to 900°C and maintained it for 600 seconds before allowing it to cool naturally. A reference sample of pure $BaTiO_3$ (BT) was also included. The photocatalytic activity of these samples was evaluated by their impact on the degradation of methylene blue solution (MB) under stirring and exposure to both a 250W ultraviolet high-pressure mercury lamp and a 500W long-arc xenon lamp, with the former providing UV radiation and the latter visible light. The experiments utilized 10 mg/L MB solutions. For a standard experiment, 0.10 g of the sample was dispersed in a petri dish with 100 mL of a 10 mg/L MB solution. All chemicals and reagents used in this study were sourced from Aladdin Industrial Corporation in Shanghai, China.

RTP was performed using an RTP-300 system, which is manufactured by Beijing Ruis Science and Technology Co., Ltd. The structural and morphological properties of the samples were analyzed with the aid of an X-ray diffractometer from Philips and a field emission scanning electron microscope (FESEM), model JEOL JSM-6700F. To assess the photocatalytic solution's absorption spectra at various stages of degradation under ultraviolet (UV) and visible (V) light, a UV-visible spectrophotometer, Shimadzu UV-3600 plus, from Japan, was utilized.

3. Results

3.1. Crystal structure

Figure 1 displays the XRD patterns for all the samples. It is evident that the predominant crystalline phase in each sample matches the tetragonal perovskite structure characteristic of $BaTiO_3$ (JCPDS, 05-0626). No second phases were detected in the samples that underwent RTP. This means that CuO did not dissolve into the $BaTiO_3$ lattice during this process. No secondary phase was formed by reacting with $BaTiO_3$. Due to the small mixing ratio of CuO and the weak diffraction intensity compared to $BaTiO_3$, the diffraction signal of CuO was not detected in XRD. In the process of material compounding, different materials need time to contact, penetrate, dissolve, and react [12]. In this experiment, we adopted rapid thermal treatment, which means that by significantly shortening the contact time between CuO and $BaTiO_3$, we can inhibit the dissolution and reaction between the two, thus promoting the formation of a heterojunction structure. In composite material photocatalysis research, this heterojunction formed by different semiconductors is considered the key to improving efficiency [13].

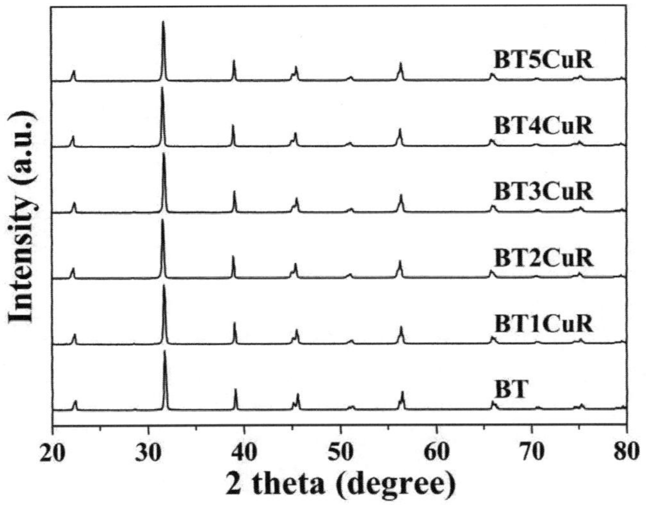

Figure 1. XRD patterns of BT and BT1CuR~BT5CuR.

3.2. Surface morphology

The SEM images of all samples are shown in Figure 2. First, compared with pure phase BT, there are many small particles on the surface of large particles in the BT1CuR~BT5CuR samples. If an oxide is doped into $BaTiO_3$ and subjected to a sufficiently long thermal treatment time, the oxide can gradually dissolve into $BaTiO_3$. However, in RTP, the significantly shortened thermal treatment process can reduce the contact time between the two. This can effectively inhibit the formation of a solid solution and promote the formation of a heterojunction. Therefore, it can be seen that many small CuO particles are attached to the large $BaTiO_3$ particles. Additionally, the BT1CuR to BT5CuR samples exhibit a significant presence of sharp features. This observation is attributed to the RTP, which is capable of enhancing grain bonding at the grain boundaries, reducing defects at these boundaries, and facilitating the diffusion of localized stresses. This leads to the appearance of sharp shapes [14]. These indicate that CuO-$BaTiO_3$ small proportion mixed heterojunctions have been formed in the BT1CuR~BT5CuR samples.

Figure 2. (a)-(f) SEM images of BT and BT1CuR~BT5CuR.

3.3. Photocatalysis

Figure 3(a) presents a comparative analysis of the degradation efficiency of the MB solution by all samples under ultraviolet light irradiation over time, while Figure 3(b) displays the pseudo-first-order kinetic fitting for the degradation efficiency of the MB solution. It can be seen that the small proportion mixing of CuO has little effect on improving the ultraviolet photocatalytic rate of $BaTiO_3$. As the proportion of CuO doping increases, there is no substantial enhancement observed in the samples' ability to degrade MB. This means that the CuO-$BaTiO_3$ small proportion mixed heterojunctions prepared by RTP cannot significantly enhance the ultraviolet photocatalytic rate. This is because, in this experiment, CuO is mixed with $BaTiO_3$ in a small proportion. Although CuO has a narrower bandgap and should have a better light response to ultraviolet light, due to the small amount of CuO and the strong light excitation effect of ultraviolet light in photocatalysis itself, the enhancement of the ultraviolet photocatalytic rate is not significant.

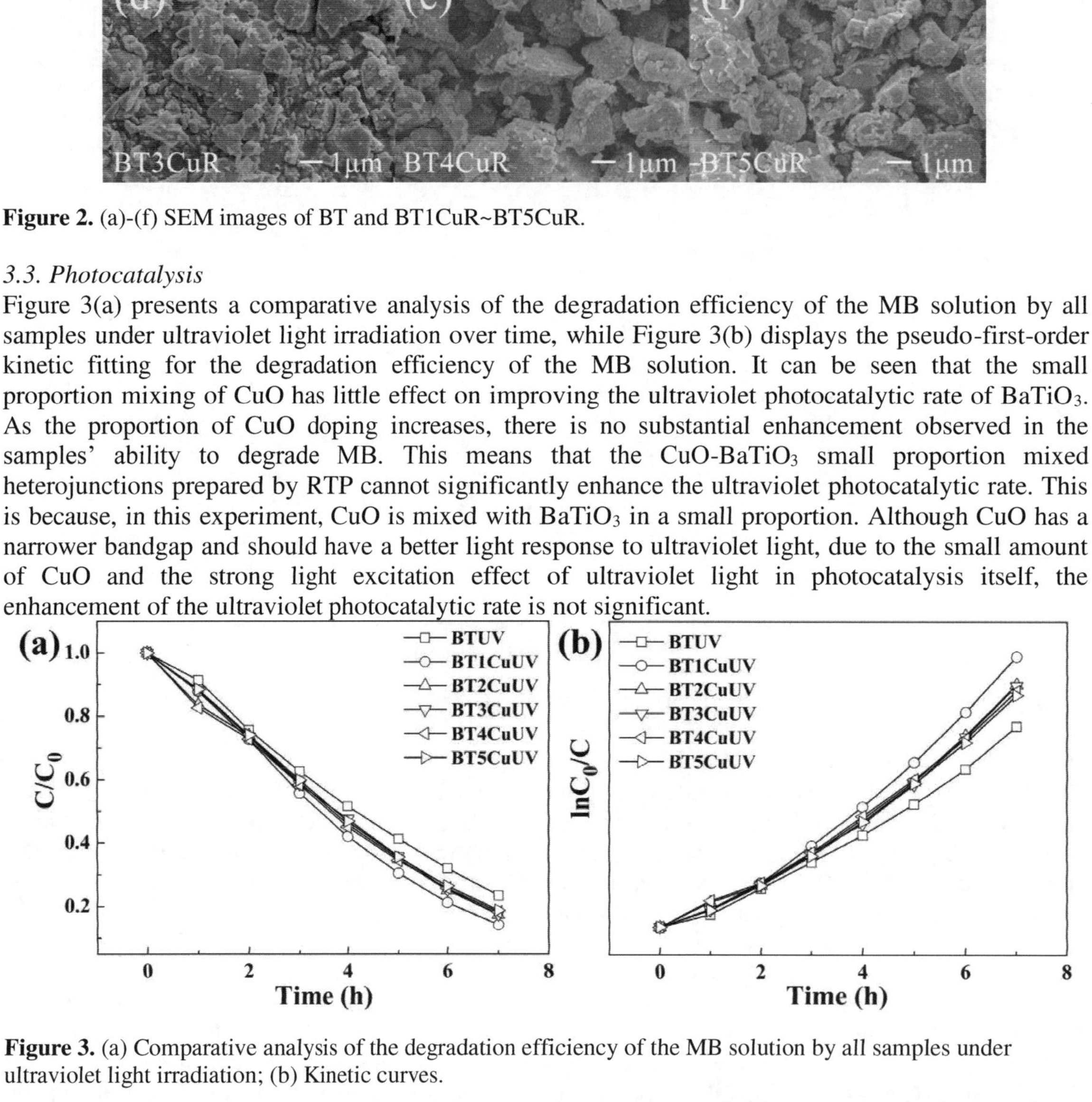

Figure 3. (a) Comparative analysis of the degradation efficiency of the MB solution by all samples under ultraviolet light irradiation; (b) Kinetic curves.

Figure 4(a) presents a comparative analysis of the degradation efficiency of the MB solution by all samples under visible light irradiation over time, while Figure 4(b) displays the pseudo-first-order kinetic fitting for the degradation efficiency of the MB solution. It can be seen from the figure that the small proportion mixing of CuO significantly improves the visible light catalytic rate of $BaTiO_3$. Concomitantly, an escalation in the CuO doping ratio correlates with a rise in the degradation rate of the samples. The degradation rate constants for BT and BT5Cu were 0.134 h^{-1} and 0.506 h^{-1}, respectively, indicating that the 5% mol CuO doping increased the degradation rate constant by 2.78 times. This indicates that the CuO-$BaTiO_3$ small proportion mixed heterojunctions in this experiment can effectively improve the visible light catalytic rate of $BaTiO_3$. Considering the small amount of CuO and the weak enhancement of the ultraviolet photocatalytic rate of the samples, the significant enhancement of the visible light catalytic rate is related to the RTP process constructing the heterojunction.

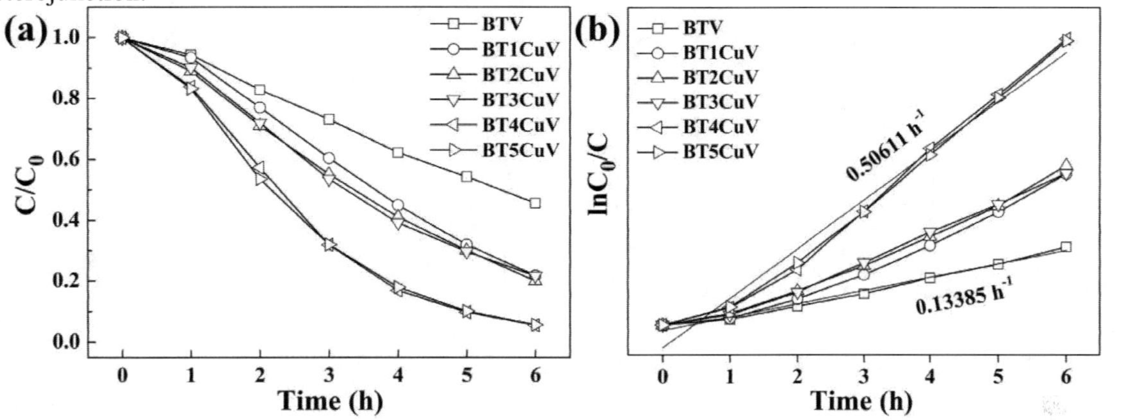

Figure 4. (a) Comparative analysis of the degradation efficiency of the MB solution by all samples under visible light irradiation; (b) Kinetic curves.

4. Discussion

During the photocatalytic reaction, upon exposure to light, a semiconductor can generate unstable electron-hole pairs if the energy of the incoming photons exceeds its band gap. This results in the creation of electrons in the conduction band and holes in the valence band. Typically, these pairs would recombine, but in the presence of pollutants in the solution, they engage in redox reactions with the pollutant ions on the surface, inhibiting the recombination process and leading to the degradation of pollutants [15].

In the field of photocatalysis, several critical factors determine the efficiency of the process: 1. The generation rate of electron-hole pairs [16]. This factor is linked to the bandgap of the semiconductor. A smaller bandgap allows for easier excitation of electron-hole pairs by incident light, thus providing more charge carriers for photocatalytic reactions. This is especially significant in the realm of visible light photocatalysis, where the energy of visible light is lower than that of ultraviolet light, making it challenging for semiconductors to generate electron-hole pairs. Consequently, strategies to enhance the responsiveness of semiconductors to the lower energy of visible light have become a central focus in the development of visible light photocatalysts. 2. The inhibition of electron-hole pair recombination [17]. Photo-excited electrons and holes will recombine if they do not interact with pollutant ions. Excessive recombination reduces the number of electron-hole pairs available for photocatalysis. Therefore, strategies to enhance the separation of electrons and holes have emerged as key to boosting the rate of photocatalytic reactions. 3. The effective contact area [18]. Since photocatalytic reactions occur on the surface, the effective contact area is crucial as it determines the extent to which electrons and holes can interact with pollutant ions to facilitate degradation reactions.

Creating semiconductor heterojunctions is a proven strategy to enhance photocatalytic efficiency. When examining heterojunctions made from two different semiconductors, two primary configurations are identified [19]: In a Type I heterojunction, the bandgap of one semiconductor falls within the bandgap of the other, leading to the accumulation of photo-excited electrons and holes on a single semiconductor during their migration. Conversely, Type II heterojunctions feature an alignment where the conduction band minimum of one material is higher in energy than the valence band maximum of the other, causing photo-excited electrons and holes to be separated across the two materials. As depicted in Figure 5, the heterojunction formed between CuO and $BaTiO_3$, with bandgaps of 1.63eV and 3.2eV respectively, exemplifies a classic Type I heterojunction. Such a semiconductor heterojunction configuration can reduce the effective bandgap, which in turn increases the rate at which electron-hole pairs are excited.

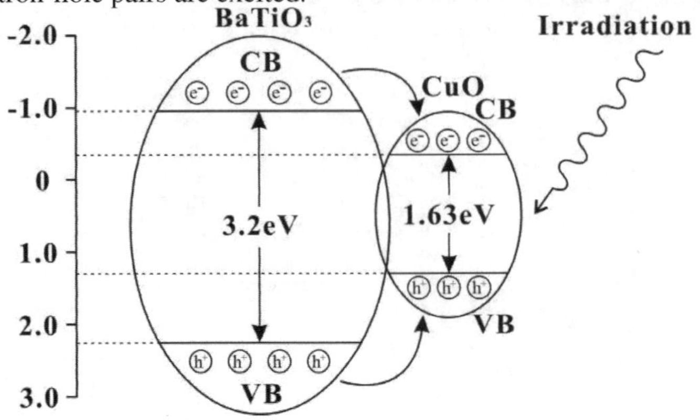

Figure 5. Band potentials and photocatalysis mechanism of CuO-$BaTiO_3$ heterojunction.

Upon exposure to light, the lower position of the conduction band in CuO compared to $BaTiO_3$ results in the formation of an electron trap. Consequently, the electrons excited by photons tend to accumulate on the side of the CuO. Similarly, since the top of the valence band of CuO is higher than that of $BaTiO_3$, a hole trap is formed, and photoexcited holes will also gather on the CuO side. This allows the photoexcited electron-hole pairs of the CuO-$BaTiO_3$ heterojunction to concentrate on CuO [7]. From the SEM photos, it can be clearly observed that a large number of CuO tiny particles are attached to $BaTiO_3$, which makes the photoexcited electron-hole pairs of the CuO-$BaTiO_3$ heterojunction more enriched in the surface area. At the same time, it can be seen from the SEM photos that a large number of small particles are dispersed on the surface of large particles. This structure also greatly increases the effective contact area between the CuO-$BaTiO_3$ heterojunction and pollutant ions.

At the same time, we must also pay attention to the fact that in this study, the ultraviolet photocatalytic rate of the CuO-$BaTiO_3$ heterojunction was slightly improved, while the visible light catalytic rate was greatly improved. The data show that the ultraviolet photocatalytic rate of pure $BaTiO_3$ is high (the degradation rate after 7 hours is 1.40), and the visible catalytic rate is low (the degradation rate after 7 hours is only 0.75). We believe that the CuO-BaTiO3 heterojunction will effectively promote the separation of photoexcited electrons and holes, significantly increasing the effective contact area with pollutant ions. Due to the small total amount of CuO, this method does not significantly improve the ultraviolet photocatalytic rate. However, the visible light catalytic rate of pure $BaTiO_3$ is inherently low, and the visible light catalytic rate of the CuO-$BaTiO_3$ heterojunction is significantly improved. This indicates that the RTP construction of the heterojunction has a larger response enhancement in the visible light excitation band [20]. We consider several key aspects to be crucial. 1. RTP can effectively optimize grain bonding by reinforcing the bonds at heterojunction grain boundaries, thereby eliminating grain boundary defects [9, 21]. In the fabrication of CuO-$BaTiO_3$ heterojunctions, the strength of the heterojunction bonding is pivotal to the final performance. The robust bonding of the heterojunction grains facilitated by RTP can enhance the absorption and utilization efficiency of light energy. 2. Addressing the issue of grain bonding enhancement, the conventional approach involves reducing the heating rate during thermal treatment to minimize surface defects and cracks [22]. However, this approach has a drawback: Prolonged thermal treatment can lead to the diffusion of CuO into $BaTiO_3$ grains, creating a built-in electric field within the heterojunction structure. If the built-in electric field is too broad, it can reduce the field strength. Since the built-in electric field's strength provides the driving force for the separation of electrons and holes, a weakened field can diminish this driving force, impeding the subsequent directional migration of electrons and holes. Thus, the optimal strategy is to confine the interaction area of the two materials to a smaller range to achieve the most favorable built-in electric field width. RTP, with its short-term thermal treatment, can readily accomplish this, further improving the absorption and utilization rate of light energy, thereby broadening the excitation light spectrum and enhancing the visible light photocatalytic efficiency.

To encapsulate, the fabrication of semiconductor heterojunctions through RTP enhances the heterojunction's light energy absorption and utilization and broadens the range of light that can be used for excitation in photocatalysis. It not only boosts the rate of visible light catalysis but also plays a significant role in the enhancement and modulation of semiconductor performance. Further in-depth investigation is warranted in this area.

5. Conclusions

Based on RTP, we constructed CuO-$BaTiO_3$ small proportion mixed heterojunctions and enhanced their visible light catalytic performance. A small proportion of CuO and $BaTiO_3$ were mixed and subjected to RTP treatment. The obtained samples were all $BaTiO_3$ tetragonal perovskite structures. There were many surface small particle distributions and sharp shapes in the samples. The ultraviolet photocatalytic performance of the CuO-$BaTiO_3$ small proportion mixed heterojunction was not significantly improved, but the visible light catalytic performance was significantly improved. The high photocatalytic efficiency of the heterojunction can be attributed to its suitable bandgap and an expanded effective contact area. In this research, the enhanced response to visible light excitation of the RTP-constructed heterojunction was the primary factor contributing to the substantial increase in the visible light catalytic rate. RTP contributes to achieving higher visible light catalytic rates by refining the heterojunction's structure. Firstly, RTP strengthens grain bonding, thereby eliminating grain boundary defects. Secondly, it confines the interaction area of the two materials to an optimal range, resulting in the best width for the built-in electric field. These improvements enhance the heterojunction's light energy absorption and utilization rate, which in turn broadens the excitation light band and enhances the visible light catalytic rate. Additional in-depth research could further enhance the integration and modulation of semiconductor performance.

Acknowledgments

This work was financially supported by the Natural Science Foundation of Hainan Province (Nos. 220RC595 and 521QN240) and the National Natural Science Foundation of China (No. 11304069). Support from the specific research fund of the Innovation Platform for Academicians of Hainan Province (YSPTZX202207) and Key Laboratory of Laser Technology and Optoelectronic Functional Materials of Hainan Province are also acknowledged.

References

[1] Zaid H J, Bassim H G, Saad H A, Doaa A S, Aya A N, Ahmed Y R, and Athraa G T. The latest progress in the design and application of semiconductor photocatalysis systems for degradation of environmental pollutants in wastewater: mechanism insight and theoretical calculations 2024 Mat Sci Semicon Proc 173 108153
[2] Zhang J, Xu Q, Feng Z, Li M, and Li C. Surface phase junctions importance of the relationship between surface phases and photocatalytic activity of TiO_2 2008 Angew. Chem. Int. Ed. 47 1766-69
[3] Tanya G, Samriti, Junghyun C and Jai P Hydrothermal synthesis of TiO_2 nanorods: formation chemistry, growth mechanism, and tailoring of surface properties for photocatalytic activities 2021 Mater. Today Chem. 20 100428
[4] Ke T, Shen S, Yang K and Lin D Construction and visible light photocatalysis of a novel ternary heterostructure BiOI/(001)TiO_2/Ti_3C_2 2020 Nanotechnology 31 1-27
[5] Hua Yang A short review on heterojunction photocatalysts: carrier transfer behavior and photocatalytic mechanisms 2021 Mater. Res. Bull. 142 111406
[6] Yu C, Tan M, Tao C, Hou Y, Liu C, Meng H, Su Y, Qiao L, and Bai Y Remarkably enhanced piezo photocatalytic performance in $BaTiO_3$/CuO heterostructures for organic pollutant degradation 2022 Adv. Ceram. 11 414-26
[7] Shravanti J, Kumar C, Ippolito S, Sabri Y, Esmaielzadeh K, and Bhargava S Straddled band aligned CuO-$BaTiO_3$ heterostructures: role of energetics at nanointerface in improving photocatalytic and CO_2 sensing performance 2018 ACS Appl. Nano Mater 7 3375-88

[8] Cruz-Gonzalez N, Calzadilla O, Roque J, Chale-Lara F, Olarte J, Meléndez-Lira M and Zapta-Torres M Study of the effect of TiO_2 layer on the adsorption and photocatalytic activity of TiO_2-MoS_2 heterostructures under visible-infrared light 2020 Int. J. Photoenergy 8740825
[9] Choi J, Jeong H, Snyder M, Stoeger J, Masel R and Tsapatsis M Grain boundary defect elimination in a zeolite membrane by rapid thermal processing 2009 Sci. 325 590-93
[10] Hori T, Iwasaki H and Tsuji K Electrical and physical properties of ultrathin reoxidized nitrided oxides prepared by rapid thermal processing 1989 IEEE Trans. Electron Devices 36 340-50
[11] Bing LN, Wu W and Shen ZJ Rapid Thermal Processing and Improved Photocatalysis of Bi_2O_3-$BaTiO_3$ Heterojunction 2024 Int. J. Photoenergy 3777201
[12] Lei W, Wen X, Cao G, Yang L, Zhang P, and Zhuge F High performance $MoTe_2$/Si heterojunction photodiodes 2021 Appl. Phys. Lett. 119 131902
[13] Huang L, Liu H, Wang Y, Zhang TC and Yuan S Construction of ternary Bi_2O_3/biochar/g-C_3N_4 heterojunction to accelerate photoinduced carrier separation for enhanced tetracycline photodegradation 2023 Appl. Surf. Sci. 616 156509
[14] Horng RH, Wuu DS and Kung CY Structural and electroluminescent characteristics of sputtered SrS: Ce thin films by rapid thermal process 1997 Thin Solid Films 307 228-32
[15] Goti G, Manal K, Sivaguru J, and Dell'Amico L The impact of UV light on synthetic photochemistry and photocatalysis 2024 Nat. Chem. 16 684-92
[16] Feng JL, Shi JW, Zhi WC and Zhen YL Piezoelectric effect enhanced photocatalytic activity of Pt/$Bi_{3.4}Gd_{0.6}Ti_3O_{12}$ plasmonic photocatalysis 2022 NANOMATERIALS-BASEL 12 1170
[17] Peng C and Yuan X Sulfication induced non radiative electron hole recombination dynamics in graphene quantum dots for tuning photocatalytic performance 2023 Spectrochim Acta A 287 122117
[18] Zhang J, Tian PH, Tang T, Huang GZ, Zeng JH, Cui B, Shen ZJ, Wang HB, Kong Z, Xi JH, and Ji ZG Excellent photoelectrochemical hydrogen evolution performance of $FeSe_2$ nanorod/ZnSe 0D/1D heterostructure as efficiency carriers migrate channel 2020 Int. J. Hydrogen Energy 45 8526-39
[19] Huang GZ, Zhang J, Jiang F, Zhang Z, Zeng JH, Qi X, Shen ZJ, Wang HB, Kong Z, Xi JH, and Ji ZG Excellent photoelectrochemical activity of Bi_2S_3 nanorod/TiO_2 nanoplate composites with dominant {001} facets 2020 J. Solid State Chem. 281 121041
[20] Justicia IP, Ordejon CG, Mozos JL and Figueras A Designed self-doped titanium oxide thin films for efficient visible light photocatalysis 2010 Adv. Mater. 14 1399-1402
[21] Hong YS, Long ZG, Kui XW, Wen HM, and Shao YL Effect of rapid heat treatment on the crystal Defect evolution and electrical properties of highly efficient polycrystalline silicon 2019 SILICON-NETH 11 1083-87
[22] Wei YN, Ji Y, Sun Q, Xia CX and Jia Y Barrier width and built-in electric field effects on hydrogenic impurity in wurtzite GaN/AlGaN quantum well 2011 Physica E 44 511-14

Journal of Physics: Conference Series **3009** (2025) 012007 doi:10.1088/1742-6596/3009/1/012007

Simulation analysis of insulation materials in high-temperature environment based on comsol thermal design

Qiushi Liu, Zhanpeng Xu *, Xinkun Chen and Shihua Li

Power China Huadong Engineering Corporation Limited, Hangzhou 311122, China

*Corresponding author's e-mail: xu_zp@hdec.com

Abstract: In this study, the cooling efficiency of several cooling schemes for photovoltaic modules was studied from the aspects of electrical and structural properties. The water-cooled plate is a metal part of the internal design of the flow path, and the material is usually aluminum alloy or copper alloy. Water-cooled plate flow heat transfer includes both flow and heat transfer and Comsol software can be used for multi-physics coupling of material simulation. An important part of the thermal design of this study is to output the results of thermal simulation and summarize the results of thermal simulation of photovoltaic modules.

1. Introduction

In this study, the cooling efficiency of several cooling schemes for photovoltaic modules was studied from the aspects of electrical and structural properties. At present, the engineering cooling means are mainly air cooling, liquid cooling, and natural heat dissipation [1]. The water-cooled plate is a metal part of the internal design of the flow path, and the material is usually aluminum alloy or copper alloy. The water-cooled plate is a metal part designed for the internal flow path of the liquid cooling method, and the material is usually aluminum alloy or copper alloy [2]. Water-cooled plate flow heat transfer includes both flow and heat transfer and Comsol software can be used for multi-physics coupling of material simulation. An important part of the thermal design of this study is to output the results of thermal simulation and summarize the results of thermal simulation of photovoltaic modules [3].

1.1 Fluid and thermal coupling analysis

The cold plate transfers heat to the coolant through convective heat transfer, and the coolant absorbs heat, causing its temperature to rise (Figure 1). Changes in temperature lead to changes in fluid properties such as density, dynamic viscosity, specific heat capacity, and thermal conductivity [4] of the fluid, thus affecting the flow state of the coolant and resulting in changes in the convective heat transfer state (Figure 2).

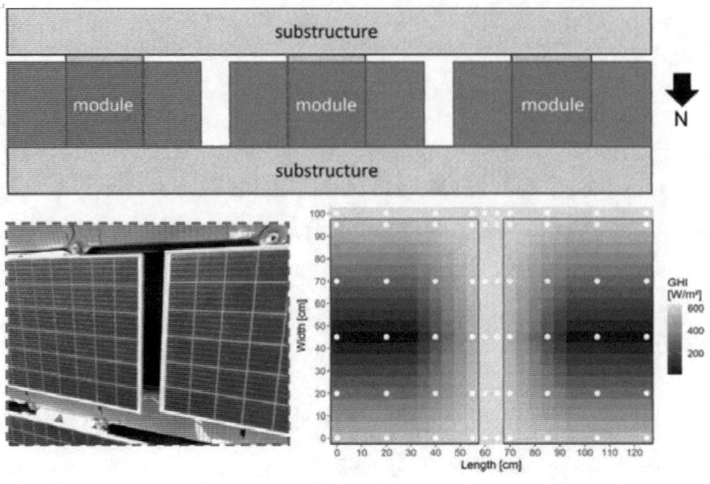

Figure 1. Illustration of the different heat between photovoltaic modules.

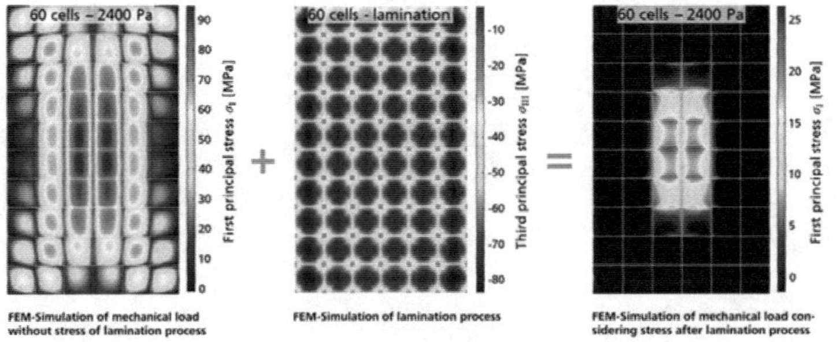

Figure 2. Uneven heating in photovoltaic panels.

1.2 Cold plate heat transfer

In the study of conjugate heat transfer of cold plates, the fluid flow speed [5] is generally considered to be low-speed flow, that is, the Mach number is less than 0.3. Considering that the properties of the fluid change with temperature, especially when the density changes with temperature, the fluid will have the characteristics of compressibility. Therefore, the governing equation of conjugate heat transfer of cold plate can be described as:

$$\nabla(\rho u) = 0 \tag{1}$$

$$\rho(u \cdot \nabla)u = -\nabla\rho + \nabla\{u[\nabla u + \nabla u^T]\} - \frac{2}{3}\nabla(\mu\nabla \cdot u)I \tag{2}$$

$$\rho[u \cdot \nabla(C_p \cdot T)] = \nabla \cdot (\tilde{k}\nabla T) + Q \tag{3}$$

$$u = u_D \qquad on\Gamma_D \tag{4}$$

$$n \cdot (-pI + \mu(\nabla u + \nabla u^T)) = 0 \qquad on\Gamma_N \tag{5}$$

$$T = T_D \qquad on\Gamma_D \tag{6}$$

Journal of Physics: Conference Series **3009** (2025) 012007 doi:10.1088/1742-6596/3009/1/012007

$$-n \cdot k_f \nabla T = 0 \qquad on\Gamma_N \tag{7}$$

where ∇ is the Hamiltonian operator, ρ is the fluid density, μ is the hydrodynamic viscosity, C_p is the specific heat capacity of the fluid at constant pressure, $\tilde{k}$ is the thermal conductivity, u is the velocity field, P is the pressure field, n is the normal unit vector outside the boundary, I is the unit tensor, Γ_D is the first class boundary, and Γ_N the second class boundary.

2. Study on the Structural Properties of Photovoltaic -Comsol

2.1 Combined with 40×40 mm² thermoelectric sheet

We add thermoelectric devices, a photovoltaic cell, a thermoelectric sheet (40×40 mm^2), and a water-cooled plate composite module between the photovoltaic cell and the water-cooled plate [6]. We want to know that with the increase of temperature, the output efficiency of the composite module will occur to what degree of attenuation [7]. Therefore, under the environment of room temperature of 22°C and light intensity of 1, 000 W/m^2, the output characteristics of the solar cell in the whole process were tested during the working time. The resulting image of current and voltage characteristics changing with temperature is shown in Figure 3 (c) as the operating temperature increases. When I_{sc} increases, V_{oc} and filling factor decrease. The changing trend of I_{sc} and V_{oc} over time is shown in Figures 3 (a) and (b).

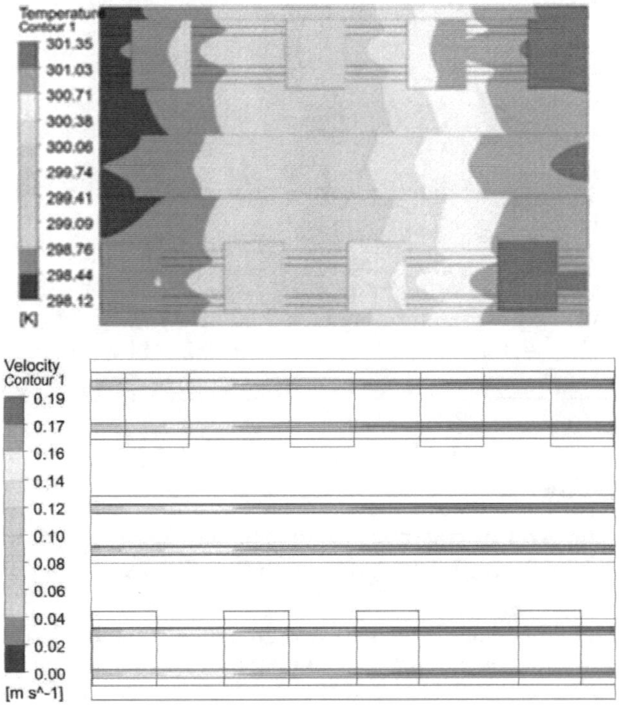

Figure 3. The curve of photovoltaic module characteristics over time.

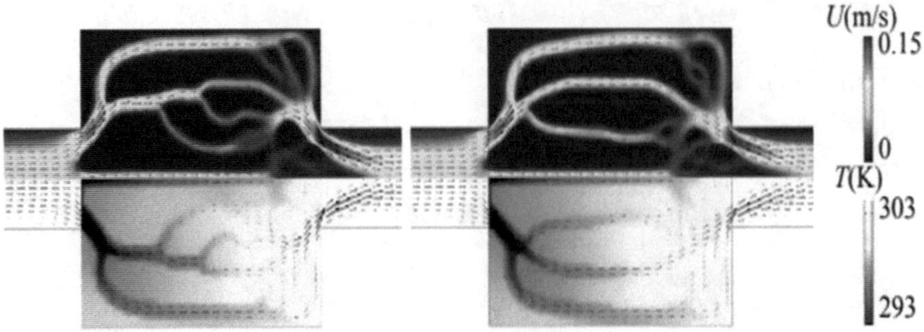

Figure 4. The simulation experiment diagram (distribution of velocity and temperature) at the Reynolds coefficient of 100.

Observing the surface temperature change of photovoltaic cells, it can be found that the surface temperature of photovoltaic cells + thermoelectric sheets (40×40 mm^2) + water-cooled plate composite modules rose approximately linearly for the first 3 minutes, and after reaching 47°C, the temperature rise gradually decreased, and finally stabilized at about 54.1°C. As shown in Figure 4, in addition, the temperature difference between the front and back of the photovoltaic module (T_1-T_2) is always maintained in the range of 18 to 22°C. According to the efficiency calculation formula mentioned in the previous article, the efficiency of each time point is calculated [8]. With the increase in temperature, the efficiency of photovoltaic modules shows a trend of first increasing and then stabilizing, and the highest efficiency is 21.1% when it is about 50°C.

Therefore, through the experiment, it can be obtained that the temperature rise trend is obvious, the heating rate is fast, the steady state temperature is maintained at about 54.1°C, and the efficiency is 20.6%, without significant decline, so that the solar cell can maintain the efficient power generation state [9].

A high temperature and high humidity coupling numerical model was established to evaluate the corrosion rate and bearing capacity of photovoltaic scaffolders, taking into account factors such as material, temperature, humidity, and bearing capacity of photovoltaic scaffolders. Based on the established numerical model, long-term corrosion research and corrosion mechanism analysis of photovoltaic scaffolds under different temperature and humidity working modes [10] were carried out to evaluate the anticorrosive performance of photovoltaic scaffolds, and the material and structure of photovoltaic scaffolds were continuously optimized to improve the anticorrosive performance and load-bearing capacity of photovoltaic scaffolds [11].

3. Experimental Verification

This paper summarizes the above experimental conclusions. Based on Comsol thermal design, the topology optimization results considering fluid-heat coupling at Reynolds numbers of 30, 50, 70, and 100 are experimentally verified.

The flow path topology obtained by topology optimization is milled out on the base plate by the CNC machine tool. The physical picture of the cooling plate is as follows:

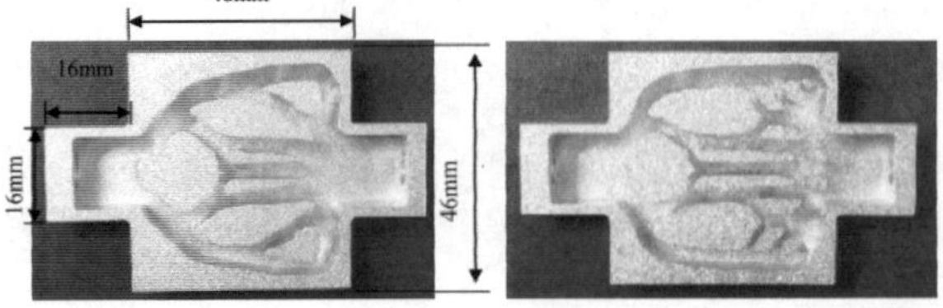

(a) Re=30 (b) Re=50

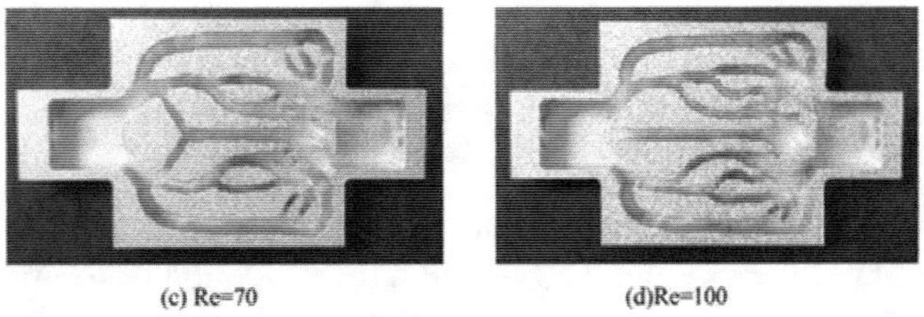

Figure 5. Cooling plate experiment diagram.

As you can see from Figure 5, as the Re value increases, the trench network becomes dense and dendritic.

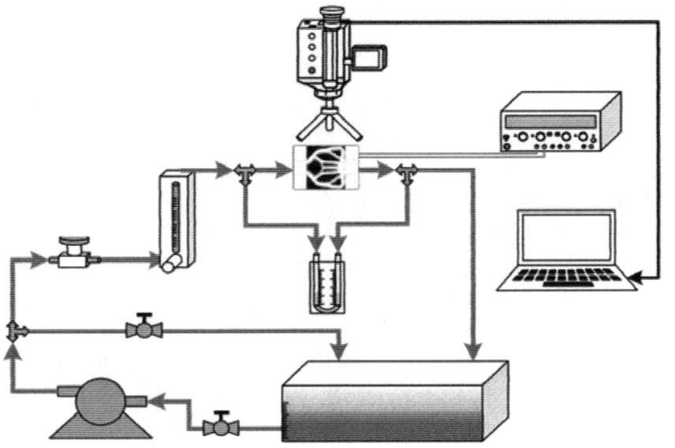

Figure 6. Experimental apparatus.

Figure 6 is our experimental schematic diagram: the blue below is the liquid storage tank. After the liquid is extracted by the instrument, the liquid is pressurized and passed through the experimental mold. Finally, the temperature measurement instrument detects the temperature.

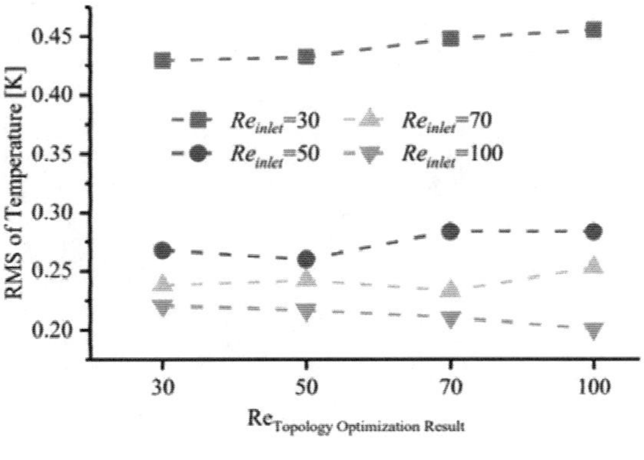

Figure 7. Experimental result.

Through the contrast test (Figure 7), the cooling effect of different cooling lotus root slices is not the same. The table above shows the heat dissipation performance of different heat sinks.

4. Conclusion

It is found that liquid cooling is the best cooling method in high-temperature environments. The water-cooled plate is a metal part designed for the internal flow path of the liquid cooling method, and the material is usually aluminum alloy or copper alloy. Heat conduction equipment is installed on the water-cooling plate. Through the water-cooling plate's internal circulation of the cooling water to take away the heat, it can be controlled by the water-cooling plate inlet cooling water temperature and flow rate to keep the water-cooling plate constant temperature, to provide constant temperature thermal boundary for the heating equipment. Water-cooled plate flow heat transfer includes both flow and heat transfer and Comsol software can be used for multi-physics coupling of material simulation. An important part of the thermal design of this study is to output the results of thermal simulation and summarize the results of thermal simulation of photovoltaic modules.

References

[1] A. J. N. Oliveira et al. "Development of an Optical Library for Coevaporated CdSexTe1−x", in *IEEE Journal of Photovoltaics*, doi: 10.1109/JPHOTOV.2024.3507079.
[2] H. S. Shan et al. "Performance Enhancement of InGaN Laser Photovoltaic Cell with AlGaN Strain Compensation Layer Irradiated by 450 nm Laser", in *IEEE Journal of Photovoltaics*, doi: 10.1109/JPHOTOV.2024.3495024.
[3] K. He, H. Li, J. Chen, K. Yang, Z. Zhang, and L. Hu. "Parameters Investigation and Heat Transfer Path Optimization of IGBT Baseplate in Photovoltaic Inverters", 2024 Energy Conversion Congress & Expo Europe (ECCE Europe), Darmstadt, Germany, 2024, pp. 1-6, doi: 10.1109/ECCEEurope62508.2024.10751948.
[4] S. Marwaha. "Performance Optimization of MA0.9Cs0.1Pb(I0.6Br0.4)3 Based Perovskite Solar Cells Through Numerical Simulations", 2024 IEEE Region 10 Symposium (TENSYMP), New Delhi, India, 2024, pp. 1-5, doi: 10.1109/TENSYMP61132.2024.10752182.
[5] R. Darbali-Zamora and E. I. Ortiz-Rivera. "Implementation of An Analytical Active Power Curtailment Method for Photovoltaic System Applications", 2024 IEEE 52nd Photovoltaic Specialist Conference (PVSC), Seattle, WA, USA, 2024, pp. 0742-0749, doi: 10.1109/PVSC57443.2024.10749215.
[6] S. Gharabeiki, T. Schaaf, F. Lodola, T. Wang, M. Melchiorre, and S. Siebentritt. "Effect of Band Gap Gradient on the Open Circuit Voltage Limit of the Cu(In, Ga)Se2 Absorbers", 2024 IEEE 52nd Photovoltaic Specialist Conference (PVSC), Seattle, WA, USA, 2024, pp. 0615-0615, doi: 10.1109/PVSC57443.2024.10749239.
[7] A. Chougle, M. Almansoori, A. Rezk, and A. Nayfeh. "Bilayer 2D MoS2 Bandgap Tuning as a Buffer Layer HTM for Perovskite Solar Cells Using TCAD Simulations", 2024 IEEE 52nd Photovoltaic Specialist Conference (PVSC), Seattle, WA, USA, 2024, pp. 0135-0137, doi: 10.1109/PVSC57443.2024.10749276.
[8] B. Bonnet-Eymard et al. "Lightweight Solar Modules Implementing Advanced Polymer Materials and Solar Cells", 2024 IEEE 52nd Photovoltaic Specialist Conference (PVSC), Seattle, WA, USA, 2024, pp. 1, 414-1, 414, doi: 10.1109/PVSC57443.2024.10749682.
[9] M. Abrar, I. J. Biswas, S. Datta, and D. Hodges. "A Numerical Investigation of Pb-free Highly Efficient CS2AgInBr6 Double Perovskite Solar Cells using SCAPS", 2024 IEEE 52nd Photovoltaic Specialist Conference (PVSC), Seattle, WA, USA, 2024, pp. 1, 710-1, 712, doi: 10.1109/PVSC57443.2024.10748920.
[10] N. Katakumbura, A. B. Phillips, A. Abasi, E. Bastola, M. J. Heben, and R. J. Ellingson. "Investigating the Role of Temperature on Recombination Current Densities", 2024 IEEE 52nd Photovoltaic Specialist Conference (PVSC), Seattle, WA, USA, 2024, pp. 1, 225-1, 227, doi: 10.1109/PVSC57443.2024.10749077.
[11] T. Colenbrander, J. Li, E. C. Abad, A. Barthel, B. Ramsay, and L. C. Hirst. "Ultra-Thin GaAs Solar Cells Processed on Glass via Low-Temperature Anodic Bonding", 2024 IEEE 52nd Photovoltaic Specialist Conference (PVSC), Seattle, WA, USA, 2024, pp. 0359-0359, doi: 10.1109/PVSC57443.2024.10749642.

Journal of Physics: Conference Series **3009** (2025) 012008 doi:10.1088/1742-6596/3009/1/012008

Performance optimization of forward osmosis membranes modified with TA/FE^{3+} complexes

Haiyang Zhu[1], JinLei Wang*, Tianhao Li, Lianli Wang, Bin Zheng, Li Zhang, Eryong Liu

School of Materials Science and Engineering, Xi' an University of Science and Technology, Xi' an, 710054, China

*Corresponding author's E-mail: jlwang@xust.edu.cn

Abstract. The active surface layer plays a crucial role in the performance of forward osmosis membranes. This study introduced tannic acid-ferric ion (TA/Fe^{3+}) complexes to modify the surface of the polyamide selective layer in forward osmosis (FO) membranes, aiming to enhance their separation and antifouling performance. Modified membranes were fabricated by depositing TA/Fe^{3+} coordination layers in varying numbers (1.5, 2.5, and 3.5 layers) and systematically characterized. The results demonstrated that the PES-PA-$(TA/Fe^{3+})_{2.5}$ (modified membrane) exhibited optimal performance, with its contact angle decreasing from 73° for the PES-PA (unmodified membrane) to 43°. The water flux of the PES-PA-$(TA/Fe^{3+})_{2.5}$ increased significantly by 25.8% compared to the unmodified membrane, while the reverse salt flux decreased by 68.4%, indicating superior separation performance. Antifouling tests revealed that the PES-PA-$(TA/Fe^{3+})_{2.5}$ achieved the lowest flux decline and the highest flux recovery, significantly outperforming the unmodified membrane. The findings indicate that the incorporation of TA/Fe^{3+} complexes markedly enhances the overall performance of FO membranes, providing an effective strategy for optimizing FO membrane properties.

1. Introduction

With the continuous growth of the global population, the issue of water pollution has become increasingly severe, and water scarcity has emerged as one of the major challenges of this century [1]. In recent years, the intensifying problems of water shortages and water quality degradation have driven the rapid development of membrane separation technology in the field of water treatment [2]. Forward osmosis (FO) membrane technology, characterized by low energy consumption, operational simplicity, and high-water recovery rates, has shown significant potential in applications [3]. Furthermore, unlike conventional pressure-driven membranes (e.g., reverse osmosis), the FO process functions at reduced pressure levels, thereby reducing energy consumption to some extent [4]. Notwithstanding the accelerated advancement in FO membrane separation technology [5], numerous challenges remain, including the necessity to enhance separation performance and mitigate membrane fouling [5]. To tackle these challenges, surface modification of membranes has become a research hotspot. Current studies mainly aim to improve membrane performance by introducing hydrophilic groups onto the membrane surface [6], optimizing the interfacial polymerization process [7], or incorporating functional nanomaterials [5]. Cellulose nanocrystals (CNC) extracted from palm trees have been used to modify the selective layer during interfacial polymerization, resulting in a substantial enhancement of membrane hydrophilicity and an improvement in antifouling capabilities [8]. Similarly, the incorporation of hydrophilic groups into the polyamide layer has been demonstrated to enhance water flux and antifouling performance [9]. Despite the success of these surface modification strategies in enhancing the performance of FO membranes, there remains a need to explore more efficient and operable modification strategies for reducing reverse salt flux and enhancing antifouling properties.

Polydopamine, which is formed through the oxidative self-polymerization of dopamine, has been widely applied in material surface modification and functionalization due to its ease of deposition on various substrates. However, challenges such as low dopamine yield and high production costs hinder its large-scale application [10]. In light of the necessity for cost-effectiveness and ease of synthesis, tannic acid (TA), which shares structural similarities with dopamine, has emerged as a promising alternative for material surface modification [11]. TA is a naturally occurring polyphenolic compound that is found in abundance in nature. Its molecular structure, which is rich in phenolic hydroxyl groups, enables it to undergo several processes, including grafting, copolymerization, blending with other compounds, or complexation with metal ions. This broadens the potential applications of TA significantly.

In this study, polyethersulfone (PES) was used as the support layer, and polyamide (PA) served as the selective layer. The layer-by-layer (LbL) self-assembly method was utilized to introduce TA and ferric ions (Fe^{3+}) to the surface of the selective layer, thereby fabricating modified FO membranes. The study systematically investigated the influence of varying numbers of TA/Fe^{3+} coordination layers on the separation performance and fouling mitigation characteristics of the FO membranes. The results obtained demonstrated that surface modification with TA/Fe^{3+} coordination layers led to a significant enhancement in membrane performance, thereby offering significant technical guidance for the practical application of FO membranes.

2. Experimental Section

2.1 Preparation of Forward Osmosis Membranes

(1) Preparation of Unmodified Membrane (PES-PA)

Using PES as the base membrane, the membrane was fixed in a specialized mold with rubber gaskets and clips. An MPD aqueous solution (3.4 wt%) was poured onto the base membrane, allowed to soak for 4 minutes, and then removed. Excess water was removed using filter paper. Subsequently, a hexane solution of TMC (0.15 wt%) was added onto the membrane surface and soaked for 2 minutes, and then removed. The membrane was air-dried for 1 minute before being placed in an oven at 60 °C for 10 minutes. This resulted in the unmodified membrane, designated as PES-PA.

(2) Preparation of TA/Fe^{3+}-Modified Membranes (PES-PA-(TA/Fe^{3+})$_n$)

As illustrated in Figure 1, The modified membranes were fabricated through a sequential LbL deposition process of TA and Fe^{3+}. The unmodified PES-PA membrane was fixed in a specialized mold with rubber gaskets and clips. A 0.17 wt% TA solution was applied to the surface of the membrane, and its surface was in contact with an aqueous solution for 5 min, after which the solution was removed. The membrane surface was then rinsed with deionized water for 30 seconds. Following this, a 0.054 wt% $FeCl_3$ solution was poured onto the surface of the membrane, its surface was in contact with an aqueous solution for 5 min., and removed, followed by a 30-second rinse with deionized water. To achieve a 1.5-layer modification, the sequence was repeated with an additional layer of TA. Similarly, membranes with 2.5 and 3.5 layers were prepared using additional cycles of TA and Fe^{3+}. The resulting membranes were designated as PES-PA-(TA/Fe^{3+})$_{1.5}$, PES-PA-(TA/Fe^{3+})$_{2.5}$, and PES-PA-(TA/Fe^{3+})$_{3.5}$, respectively.

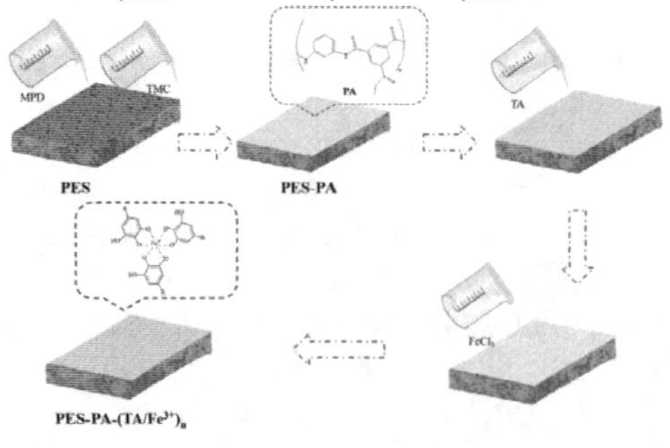

Figure 1. Schematic diagram of the preparation process for the forward osmosis membrane.

2.2 Performance Evaluation

(1) Permeation Performance Evaluation

During the permeation performance experiments, A 1 M NaCl solution served as the draw solution, while deionized water was utilized as the feed solution during the permeation experiments. The membrane's pure water flux (Jw) and reverse salt flux (Js) were evaluated using the following equations:

$$J_w = \frac{\Delta v}{\Delta t}\frac{1}{A} \quad (1)$$

In Equation (1): J_w denotes the water flux of the membrane [L/(m^2·h)], Δv is the change in the volume of the feed solution, A is the effective membrane area, and Δt is the operating time.

$$J_s = \frac{\Delta cv}{\Delta t}\frac{1}{A} \quad (2)$$

In Equation (2): refers to the reverse salt flux of the membrane [g/(m^2·h)], Δcv represents the change in concentration and volume of the feed solution before and after testing.

(2) Antifouling Performance Evaluation

Bovine serum albumin (BSA) and sodium alginate (SA) were selected as typical foulants to investigate water flux decline, flux recovery, and the effects of fouling factors.

Baseline Experiment 1: Deionized water was selected as the feed solution, and 2M NaCl was used as the draw solution. The water flux was tested for 30 minutes and recorded as J_0.

Fouling Experiment: Three different fouling scenarios were tested. The feed solutions were 0.5 g/L BSA,0.5 g/L SA, and 0.25 g/L SA + 1 mM CaI_2. The draw solution remained unchanged from the baseline test. The system was operated under identical conditions for 5 h to observe water flux decline.

Cleaning Experiment: Both the feed and draw solutions were replaced with deionized water. The system was operated for 0.5 h at a higher crossflow velocity.

Baseline Experiment 2: Repeat baseline step 1 and record the water flux as J_e.

The flux recovery ratio (F_{arr}) was defined as the ratio of the water flux from baseline experiment 2 to baseline experiment 1.

3. Results and Discussion

3.1 Chemical Structure

In the spectrum of the unmodified membrane (PES-PA) (Figure 2e), in addition to retaining the characteristic peaks of the support layer (PES), a new peak at 1630 cm^{-1} appeared, attributing to the C=O stretching vibrations of the polyamide selective layer. A broad peak in the 3100–3500 cm^{-1} range was also observed, corresponding to the N-H vibrational modes of the polyamide, confirming the successful formation of the polyamide selective layer. In the spectra of the modified membranes (Figure 2a-c), a new characteristic peak emerged at 1540 cm^{-1}. This peak is primarily attributed to the bending vibrations of N-H and the stretching vibrations of C-N. The TA/Fe^{3+} coordination complex, which is rich in phenolic hydroxyl groups (O-H) and ferric ions, interacts with the carbonyl group or amine group of the PA layer, forming hydrogen bonds and altering the chemical environment of the N-H and C-N groups. This result validates the successful introduction of the TA/Fe^{3+} layer.

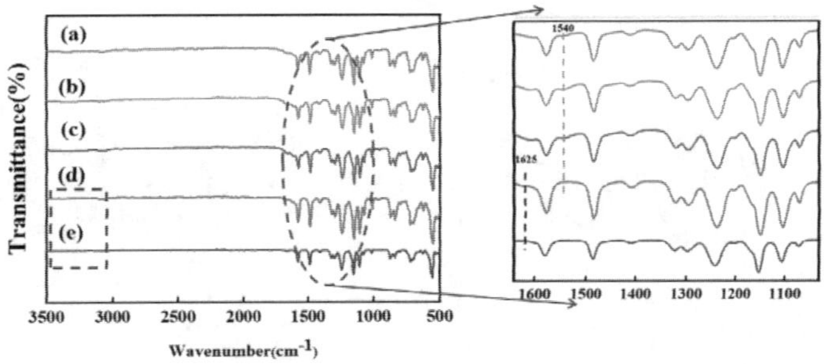

Figure 2. FTIR spectra of the forward osmosis membranes
(a) PES, (b) PES-PA, (c) PES-PA-(TA/Fe^{3+})$_{1.5}$, (d) PES-PA-(TA/Fe^{3+})$_{2.5}$, (e) PES-PA-(TA/Fe^{3+})$_{3.5}$.

3.2 Surface Morphology

The surface morphology of the membranes before and after TA/Fe^{3+} modification is shown in Figure 3. For the unmodified membrane PES-PA (Figure 3a), the porous structure of the substrate is visible, as the ultrathin polyamide (PA) selective layer is tightly adhered to the porous support layer. Figures 3(b-d) show the surface morphology of modified membranes with varying numbers of TA/Fe^{3+} complex layers. The surface of PES-PA-$(TA/Fe^{3+})_{1.5}$ exhibits significant changes, with most of the membrane's porous structure being covered, indicating improved compactness of the membrane (Figure 3b). With an increased number of complex layers, the surface pores of PES-PA-$(TA/Fe^{3+})_{2.5}$ are completely covered, further enhancing membrane compactness (Figure 3c). However, for PES-PA-$(TA/Fe^{3+})_{3.5}$, the surface becomes excessively dense due to the over-deposition of TA/Fe^{3+} complexes Figure 3(d), which likely increases filtration resistance and negatively affects the membrane's permeation performance.

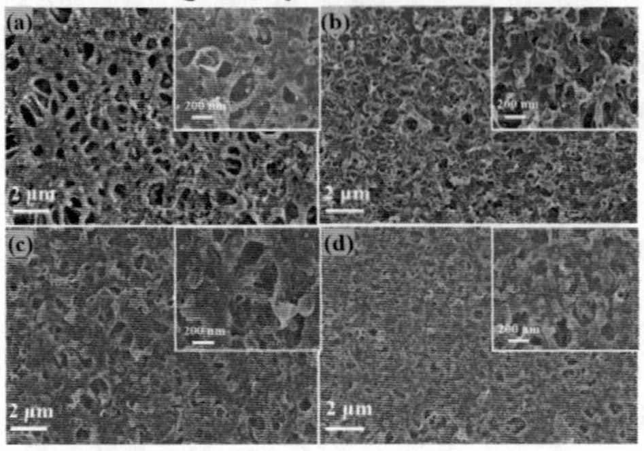

Figure 3. Surface SEM images of membranes
(a) PES-PA; (b) PES-PA-$(TA/Fe^{3+})_{1.5}$; (c) PES-PA-$(TA/Fe^{3+})_{2.5}$; (d) PES-PA-$(TA/Fe^{3+})_{3.5}$.

The surface roughness (Ra) of different membranes is shown in Figure 4. The results indicated that the unmodified membrane (PES-PA) exhibited the highest surface roughness (Ra = 94.31 nm). With the introduction of varying numbers of TA/Fe^{3+} complex layers, Ra of the modified membranes gradually decreased (Ra = 90.86 nm, Ra = 82.13 nm, and Ra = 75.62 nm.), demonstrating that the TA/Fe^{3+} modification layers significantly altered the surface morphology of the membranes. The reduction in roughness, attributed to the deposition of TA/Fe^{3+} complexes, introduces a substantial number of hydrophilic functional groups, which help to reduce interactions between the membrane surface and foulants, thereby improving antifouling performance.

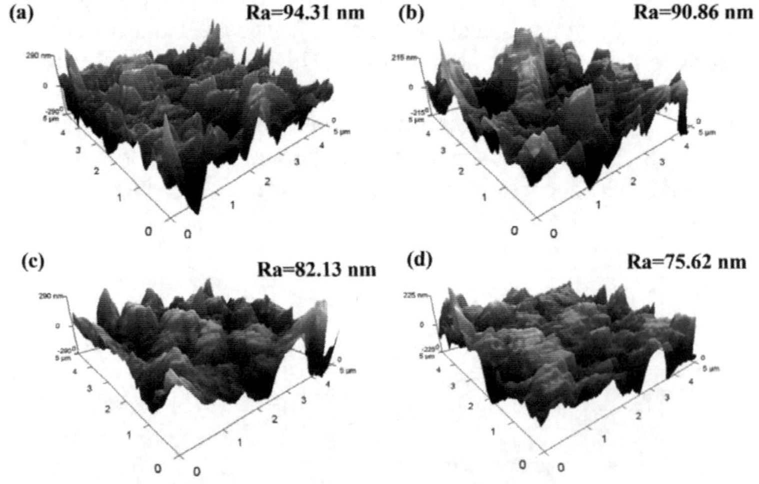

Figure 4. Surface AFM images of membranes (scan area of 5×5 μm)
(a) PES-PA (b) PES-PA-$(TA/Fe^{3+})_{1.5}$ (c) PES-PA-$(TA/Fe^{3+})_{2.5}$ (d) PES-PA-$(TA/Fe^{3+})_{3.5}$.

3.3 Permeation Performance

The permeation performance (Figure 5) is crucial for evaluating the efficiency of forward osmosis membranes. The contact angles of different membranes are shown in Figure 5a. The unmodified membrane exhibited the highest contact angle of 73°. As the number of TA/Fe^{3+} layers increases, the contact angle gradually decreases to 50° (1.5 layers), 43° (2.5 layers), and 31° (3.5 layers). This indicates that the addition of TA/Fe^{3+} complex layers can significantly enhance the hydrophilicity of the membrane.

Water flux testing evaluates the water transport efficiency of the membrane. The water flux of the unmodified membrane was 12.8 LMH (Figure 5b). The water flux increased to 19.3 LMH for PES-PA-$(TA/Fe^{3+})_{1.5}$ and 16.1 LMH for PES-PA-$(TA/Fe^{3+})_{2.5}$, indicating that the enhanced hydrophilicity after modification promoted water diffusion and transport across the membrane. However, PES-PA-$(TA/Fe^{3+})_{3.5}$ dropped to 10.5 LMH, which is lower than the unmodified membrane. This reduction is likely due to the excessive TA/Fe^{3+} layers creating additional transport resistance, thereby hindering water flux.

Reverse salt flux (RSF) testing (Figure 5c) reflects the separation capability of the membrane. The unmodified membrane exhibited the highest RSF of 23.1 g/m^2·h. With the increasing number of TA/Fe^{3+} layers, the RSF decreased significantly to 20.4 g/m^2·h, 7.3 g/m^2·h, and 6.9 g/m^2·h. SEM results support that the modified membranes exhibit a denser surface structure, effectively reducing the pathways for salt ions to permeate in the reverse direction. This confirms that TA/Fe^{3+} complex modification can effectively improve separation performance.

The salt-to-water flux ratio serves as an indicator of the membrane's overall efficiency, where smaller values correspond to superior performance. The unmodified membrane demonstrated the worst performance with a salt-to-water flux ratio of 1.82 (Figure 5d). Upon introducing the TA/Fe^{3+} complexes, PES-PA-$(TA/Fe^{3+})_{1.5}$ exhibited a reduced ratio of 1.06, while PES-PA-$(TA/Fe^{3+})_{2.5}$ achieved the lowest value of 0.45, indicating the ability to maximize water flux while minimizing reverse salt flux. For PES-PA-$(TA/Fe^{3+})_{3.5}$, although the number of modification layers increased, the additional diffusion resistance caused by excessive layers led to a slight increase in the ratio to 0.60.

The results demonstrate that TA/Fe^{3+} modification significantly enhances the permeation performance of forward osmosis membranes. Among them, the 2.5-layer membrane exhibited the best performance and can be used as the optimal modification condition for antifouling performance studies of forward osmosis membranes.

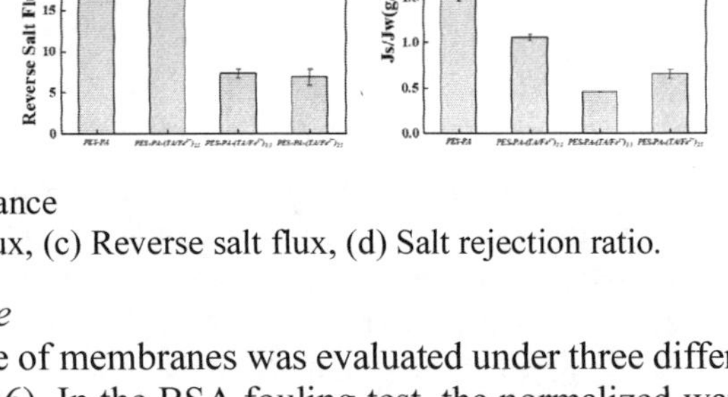

Figure 5. Permeation performance
(a) Contact angle, (b) Water flux, (c) Reverse salt flux, (d) Salt rejection ratio.

3.4 Antifouling Performance

The antifouling performance of membranes was evaluated under three different fouling scenarios: BSA, SA, and SA+$CaCl_2$ (Figure 6). In the BSA fouling test, the normalized water flux of the pristine PES-PA membrane declined slowly at first but then dropped more rapidly over time, reaching 0.89 (Figure 6a) by the end of the test. In contrast, PES-PA-$(TA/Fe^{3+})_{2.5}$ exhibited superior antifouling performance, maintaining a normalized water flux of 0.92 under the same conditions. BSA, as a typical protein foulant, tends to interact with hydrophobic surfaces [12]. The TA/Fe^{3+} modification introduced hydrophilic functional groups that significantly improved membrane surface hydrophilicity, forming a hydration layer that effectively inhibited the adsorption and accumulation of protein molecules.

In the SA fouling test (Figure 6b), PES-PA membrane dropped to 0.62 by the end of the test, while PES-PA-(TA/Fe^{3+})$_{2.5}$ maintained a normalized flux of 0.80. SA, a negatively charged foulant, readily adheres to the relatively hydrophobic surface of PES-PA, leading to faster flux decline. However, the improved hydrophilicity of the TA/Fe^{3+}-modified membrane effectively suppressed the SA.

Under the SA+$CaCl_2$ fouling conditions (Figure 6c), PES-PA experienced the largest flux decline, with its normalized water flux dropping to 0.54 by the end of the test. Meanwhile, PES-PA-(TA/Fe^{3+})$_{2.5}$ showed a much gentler decline, maintaining a normalized water flux of 0.66. Calcium ions (Ca^{2+}) can form stable attachments with the carboxyl groups in SA through the calcium bridging effect, exacerbating fouling on the membrane surface. However, the enhancement in hydrophilicity and increase in the negative charge of the TA/Fe^{3+}-modified membrane significantly mitigated the impact of the calcium bridging effect.

The flux recovery rate after cleaning (Figure 6d) further demonstrated the antifouling performance improvement. For BSA fouling, the flux recovery rate of PES-PA was only 75.1%, while PES-PA-(TA/Fe^{3+})$_{2.5}$ achieved a recovery rate of 92.0%. In the SA fouling scenario, PES-PA showed a flux recovery rate of 72.9%, compared to 98.0% for PES-PA-(TA/Fe^{3+})$_{2.5}$. Similarly, under SA+$CaCl_2$ fouling conditions, PES-PA exhibited a flux recovery rate of 69.4%, while PES-PA-(TA/Fe^{3+})$_{2.5}$ demonstrated a recovery rate of approximately 84.4%. Across all three types of fouling tests (BSA, SA, and SA+$CaCl_2$), PES-PA-(TA/Fe^{3+})$_{2.5}$ consistently showed superior antifouling performance. These results confirm that the TA/Fe^{3+} complex modification significantly enhances the antifouling capacity of forward osmosis membranes.

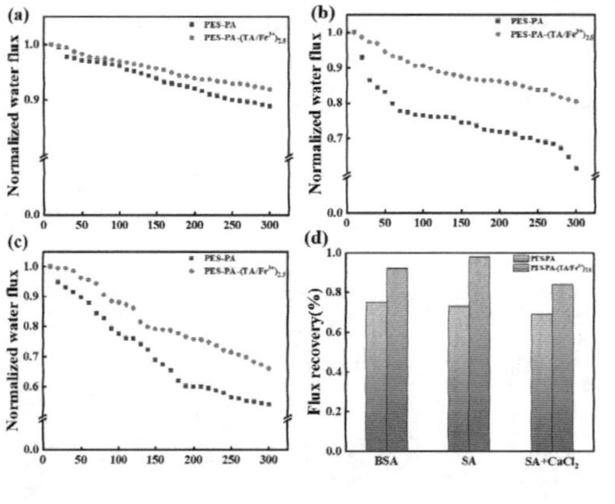

Figure 6. Antifouling performance
(a) Flux decline rate under BSA fouling(b) Flux decline rate under SA fouling(c) Flux decline rate under SA+$CaCl_2$ fouling(d) Flux recovery rate.

4. Conclusion

The PA selective layer of FO membranes was modified with TA/Fe^{3+} coating through the Lb. deposition method in the present study and its impact on permeation performance and antifouling characteristics have been systematically investigated. Infrared spectroscopy confirmed the successful incorporation of TA/Fe^{3+} complexes. AFM and SEM characterizations revealed that the surface roughness of the modified membranes decreased progressively with the number of TA/Fe^{3+} layers, and the surface morphology became denser. The permeation performance results demonstrated that the surface hydrophilicity improved with the addition of TA/Fe^{3+} layers. The PES-PA-(TA/Fe^{3+})$_{2.5}$ membrane exhibited the best overall performance, with water flux increased by approximately 25.8%, reverse salt flux reduced by 68.4%, and the salt-to-water flux ratio reduced by 75.3%, compared to the unmodified

PES-PA membrane. These findings confirm that the TA/Fe^{3+} layers effectively enhance membrane permeation performance. In the antifouling performance tests, PES-PA-(TA/Fe^{3+})$_{2.5}$ demonstrated significantly lower flux decline rates and higher flux recovery rates across all three fouling types, indicating superior antifouling capabilities compared to PES-PA. This study demonstrates that TA/Fe^{3+} complex modification is an efficient and feasible surface modification strategy that significantly improves the permeation and antifouling characteristics of FO membranes, Providing meaningful guidance for the advancement of high-efficiency separation membranes.

References

[1] Kummu M, Guillaume J H A, de Moel H, et al. The world's road to water scarcity: shortage and stress in the 20th century and pathways towards sustainability. *Scientific Reports*, 2016, 6(1): 1-16. https://www.nature.com/articles/srep38495.
[2] Jiang Z, Wang L, Liu Y, et al. Research and application of membrane technology for safe drinking water in rural areas. *Journal of Ecology and Rural Environment*, 2024, 40(11): 1405-1415. 10.19741/j.issn.1673-4831.2023.0948.
[3] Awad A M, Jalab R, Minier-Matar J, et al. The status of forward osmosis technology implementation. *Desalination*, 2019, 461: 10-21. 10.1016/j.desal.2019.03.013.
[4] Zhao, Y., Wang, L., Ma, J., et al. Zwitterionic forward osmosis membrane modified by fast second interfacial polymerization with enhanced antifouling and antimicrobial properties for produced water pretreatment. *Desalination*, 2019, *457*, 114090. 10.1016/j.desal.2019.114090.
[5] Ibraheem B M, Aani S A, Alareer A A, et al. Forward osmosis membrane: review of fabrication, modification, challenges and potential. *Membranes*, 2023, 13(4): 379. https://doi.org/10.3390/membranes13040379.
[6] Sun J, Hu X, Huang Y, et al. 1, 3-Diamino-2-propanol or 2-aminoethanethiol modified active layer of thin-film composite forward osmosis membrane. *Journal of Applied Polymer Science*, 2019, 136(36): 47923. 10.1002/app.47923.
[7] Jiang Y, Zhao P, Xu S, et al. Structurally tuned polyamide membranes via the polyvinylpyrrolidone-modulated interfacial polymerization reaction for enhanced forward osmosis performance. *Desalination*, 2024, 585: 117753. 10.1016/j.desal.2024.117753
[8] Saud A, Saleem H, Khan A W, et al. Date palm tree leaf-derived cellulose nanocrystal incorporated thin-film composite forward osmosis membranes for produced water treatment. *Membranes*, 2023, 13(5): 513. https://doi.org/10.3390/membranes13050513.
[9] Wang F, Zheng T, Wang P, et al. Enhanced water permeability and antifouling property of coffee-ring-textured polyamide membranes by in situ incorporation of a zwitterionic metal–organic framework. *Environmental Science & Technology*, 2021, 55(8): 5324-5334. https://doi.org/10.1021/acs.est.0c07122.
[10] Tian Xinmin, Wang Xuan, Peng Wei, et al. Research Progress in Membrane Surface Modification Technology Based on Dopamine Oxidative Polymerization. *Membrane Science and Technology*, 2021, 41(1): 152-159. 10.16159/j.cnki.issn1007-8924.2021.01.020
[11] Fang, T., Zhang, B. W., Zhang, Z., et al. Tannic acid compound as a corrosion inhibitor: film-forming characteristics and corrosion resistance. *Chinese Journal of Engineering*, 2019, *41*(12), 1527-1535. 10.13374/j. issn2095-9389.2019.06.03.002
[12] Jeyachandran Y L, Mielczarski E, Rai B, et al. Quantitative and qualitative evaluation of adsorption/desorption of bovine serum albumin on hydrophilic and hydrophobic surfaces. *Langmuir*, 2009, 25(19): 11614-11620. https://doi.org/10.1021/la901453a.

Journal of Physics: Conference Series **3009** (2025) 012009 doi:10.1088/1742-6596/3009/1/012009

Effect of Se doping on the electrical properties of $Bi_{0.5}Sb_{1.5}Te_3$ alloy

Yuanqiao Li and Zhiling Bai*

Liupanshui Normal University, Liupanshui, Guizhou, 553004, China

*Corresponding author's e-mail: zhilingbai@lpssy.edu.cn

Abstract. $Bi_{0.5}Sb_{1.5}Te_3$-based alloy is prepared by high-temperature solid-state sintering method, and the electrical transport properties of $Bi_{0.5}Sb_{1.5}Te_3$-based alloy are optimized by changing the sintering temperature and Se element doping amount. The microstructure and electrical properties of the alloys are characterized and analyzed. The results indicate that doping with trace amounts of Se element does not cause significant changes in the crystal structure of the sample, and the basic crystal structure stability of the $Bi_{0.5}Sb_{1.5}Te_{3-x}Se_x$ (x=0.01, 0.04, 0.07, 0.1, 0.13) alloy remains unchanged. The SEM fracture morphology of $Bi_{0.5}Sb_{1.5}Te_{2.96}Se_{0.04}$ alloy shows a layered structure with some pores. The Seebeck coefficients of the five alloys are all positive, indicating that they are all P-type semiconductor materials. The Seebeck coefficients and resistivity of $Bi_{0.5}Sb_{1.5}Te_{3-x}Se_x$ (x=0.01, 0.04, 0.07, 0.1, 0.13) alloys vary with increasing sintering temperature.When the doping amount of Se is 0.01 wt% and the sintering temperature is 833 K, the power factor of the alloy reaches its maximum value of 2459.42 $\mu W/m{\cdot}K^2$.

1. Introduction

Thermoelectric materials can convert thermal energy into electrical energy and have a long history [1]. But in the early stages of research, the performance and conversion efficiency of thermoelectric materials have not been effectively improved, resulting in no outstanding achievements in their research, and only applications in temperature sensing thermocouples. With the rapid development of industry and the rapid improvement of living standards, people's material lives have become unprecedentedly prosperous. Therefore, the electricity demand is increasing day by day, which leads to the massive consumption of resources such as coal, oil, and natural gas. This not only causes serious resource shortages but also results in uncontrolled emissions of carbon dioxide, causing serious pollution to the ecological environment. Energy shortage and environmental pollution have become the hottest topics today, and they are major challenges that cannot be ignored in the long run [2, 3]. Thermoelectric materials have attracted widespread attention from scholars due to their environmental friendliness, recyclability, noiselessness, and ability to work in harsh environments [4, 5]. The research on thermoelectric materials plays a crucial role in the world. Various thermoelectric material systems have been discovered and successfully prepared, thereby improving the relevant theories of thermoelectric materials. The preparation process has been developed and improved, and the thermoelectric properties of various material systems have been significantly enhanced.

Bismuth telluride-based alloys are thermoelectric materials with the best performance in the range of room temperature and have been applied in the commercial field [6-8]. However, its low thermoelectric conversion efficiency limits its application in the thermoelectric field. In this study, $Bi_{0.5}Sb_{1.5}Te_3$-based alloys are taken as the research objects, and the alloys are synthesized by the high-temperature solid-phase sintering method. The effects of different sintering temperatures and Se doping levels on the electrical properties of the alloy were investigated to optimize its electrical properties.

2. Experimental section

Using Se powder, Bi powder, Sb powder, and Te powder with a purity of 99.99% as raw materials, weighing, grinding, forming, and loading into molds, raw materials are sintered in a high-temperature furnace at a temperature range of 713-863 K and a holding time of 50 minutes. TD-2500 X-ray diffractometer is used for phase analysis of alloys. Gemini 300 scanning electron microscope is used for microstructure characterization of alloys. The Seebeck coefficients of the alloys are measured by a self-made calibrated Seebeck tester, and the resistivity values of the alloys are measured by an RTS-9 dual electric measurement four-probe tester.

3. Results and discussion

Figure 1 shows the X-ray diffraction patterns of $Bi_{0.5}Sb_{1.5}Te_{3-x}Se_x$ (x=0.01, 0.04, 0.07, 0.1, 0.13) alloys. It can be seen from Figure 1 that the strong characteristic diffraction peaks of all tested samples align closely with the standard pattern of $Bi_{0.5}Sb_{1.5}Te_3$, with no appearance of diffraction peaks from new Se-containing phases. This outcome signifies the high purity of the prepared samples, unimpaired by impurity phases. As the Se element content increases from 0.01 wt% to 0.13 wt%, despite variations in the alloy's chemical composition, the primary characteristic peak positions and relative intensities in the diffraction patterns remain largely unchanged. This suggests that the incorporation of a trace amount of Se does not induce substantial changes in the sample's crystal structure, preserving the inherent stability of the basic crystal structure of $Bi_{0.5}Sb_{1.5}Te_{3-x}Se_x$ (x=0.01, 0.04, 0.07, 0.1, 0.13) alloys. The absence of additional peaks or significant shifts in the patterns further confirms that no secondary phases have formed or significant crystal phase transitions have occurred in the alloy.

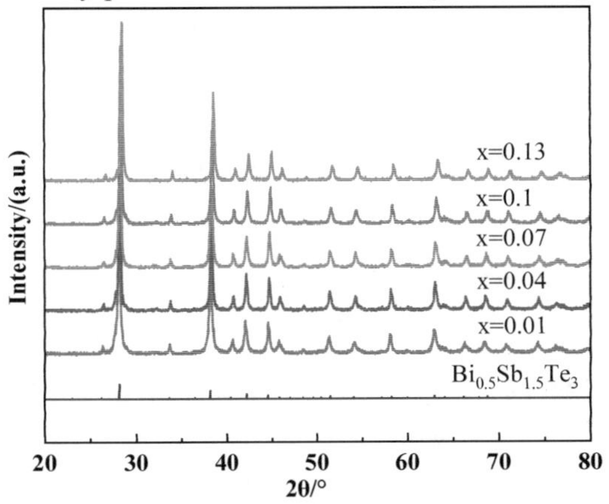

Figure 1. XRD patterns of $Bi_{0.5}Sb_{1.5}Te_{3-x}Se_x$ alloys.

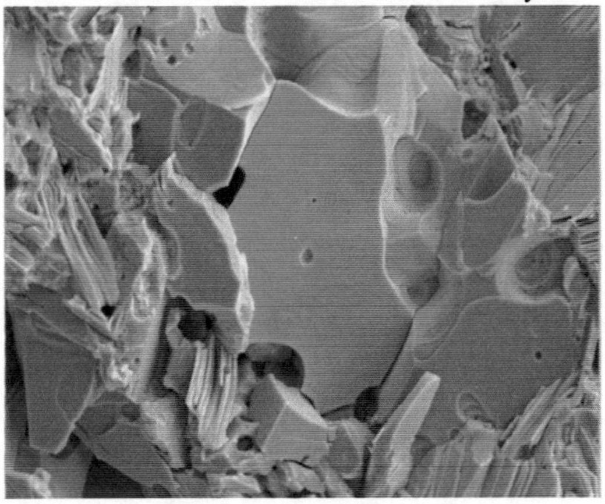

Figure 2. SEM image of $Bi_{0.5}Sb_{1.5}Te_{2.96}Se_{0.04}$ alloy.

Figure 2 is a scanning electron microscope (SEM) image of the fracture surface of the sample $Bi_{0.5}Sb_{1.5}Te_{2.96}Se_{0.04}$ alloy. The $Bi_{0.5}Sb_{1.5}Te_{2.96}Se_{0.04}$ alloy sample is prepared under specific process conditions, that is the sintering temperature is precisely controlled at 833 K and the holding time at this temperature is maintained for 50 minutes to ensure that the alloy can fully react and achieve the expected composition and structure. From Figure 2, it is found that there are some obvious holes in the sample. The formation mechanism of these holes may be relatively complex, but a major reason may be related to the difference in melting and boiling points of the constituent elements of the alloy during high-temperature firing. Since different elements have different physical states under the same temperature conditions, some elements may volatilize when reaching or exceeding their boiling points, thus leaving holes inside the alloy. This phenomenon not only affects the compactness and uniformity of the alloy but also has an adverse effect on its physical and mechanical properties. More compact and uniform block materials are obtained by optimizing parameters during the sample preparation process. In order to better control the porosity and structure distribution in the alloy, the pressure conditions are adjusted during the preparation process. The precise control of sintering temperature and holding time is also crucial, as they directly affect the phase composition and microstructure of the alloy. Judging from the fracture morphology of the $Bi_{0.5}Sb_{1.5}Te_{2.96}Se_{0.04}$ alloy, it presents a lamellar structure, and the lamellae are relatively long and thick. This structural characteristic is usually associated with a brittle fracture mode, indicating that the alloy is prone to sudden fracture without plastic deformation when subjected to external forces. At the same time, the arrangement of the fracture surfaces shows randomness and non-orientation, which further confirms the complexity and non-uniformity of the internal structure of the alloy.

Figure 3 shows the Seebeck coefficients of $Bi_{0.5}Sb_{1.5}Te_{3-x}Se_x$ (x=0.01, 0.04, 0.07, 0.1, 0.13) alloys at different sintering temperatures. From Figure 3, it can be seen that the Seebeck coefficients of all five alloys are positive, indicating that they are P-type semiconductor materials. The Seebeck coefficient of $Bi_{0.5}Sb_{1.5}Te_{3-x}Se_x$ (x=0.01, 0.07) two alloys undergoes a complex process of change with the increase of sintering temperature, that is, it first increases and then decreases, then increases again, and then decreases again, showing a fluctuating change. The Seebeck coefficient of $Bi_{0.5}Sb_{1.5}Te_{3-x}Se_x$ (x=0.04, 0.1) alloys gradually increases with the increase of sintering temperature, reaches a peak, and then begins to decrease. In addition, the Seebeck coefficient of $Bi_{0.5}Sb_{1.5}Te_{2.87}Se_{0.13}$ alloy shows a trend of first decreasing and then increasing, which is different from the Seebeck coefficient changes of the other four alloys.

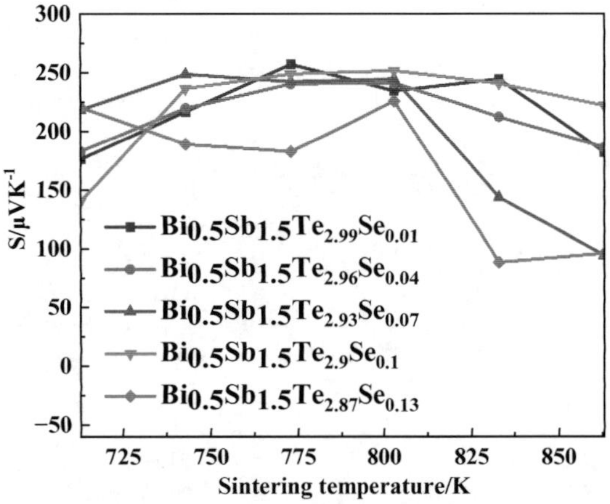

Figure 3. The Seebeck coefficients of $Bi_{0.5}Sb_{1.5}Te_{3-x}Se_x$ alloys.

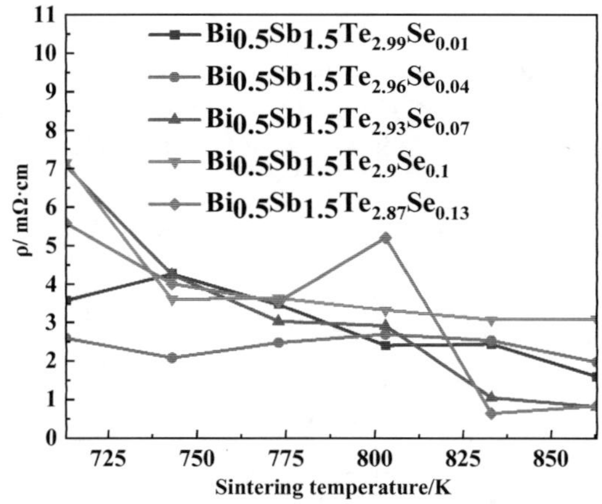

Figure 4. The electrical resistivity of $Bi_{0.5}Sb_{1.5}Te_{3-x}Se_x$ alloys.

When the sintering temperature is 713 K, the Seebeck coefficients of $Bi_{0.5}Sb_{1.5}Te_{3-x}Se_x$ (x=0.01, 0.04, 0.07, 0.1) alloys are generally low, indicating that the thermoelectric conversion efficiency of these alloys prepared at this sintering temperature may be relatively low. However, when the sintering temperature is raised to the range of 743-803 K, the Seebeck coefficients of all four alloys significantly increase and exceed 216 $\mu V \cdot K^{-1}$, demonstrating high thermoelectric conversion efficiency. The maximum Seebeck coefficients of the five alloys $Bi_{0.5}Sb_{1.5}Te_{3-x}Se_x$ (x=0.01, 0.04, 0.07, 0.1, 0.13) are 257.15 $\mu V \cdot K^{-1}$, 240.5 $\mu V \cdot K^{-1}$, 248.72 $\mu V \cdot K^{-1}$, 251.85 $\mu V \cdot K^{-1}$, and 225.98 $\mu V \cdot K^{-1}$, respectively.

Figure 4 shows the electrical resistivity of $Bi_{0.5}Sb_{1.5}Te_{3-x}Se_x$ (x=0.01, 0.04, 0.07, 0.1, 0.13) alloys at different sintering temperatures. By analyzing Figure 4, it can be concluded that while keeping the doping amount constant and only adjusting the sintering temperature, the electrical resistivity of $Bi_{0.5}Sb_{1.5}Te_{2.99}Se_{0.01}$ alloy exhibits a fluctuating change with the increase of sintering temperature. This indicates that its electrical resistivity is highly sensitive to sintering temperature and may be affected by various microstructural changes. For $Bi_{0.5}Sb_{1.5}Te_{3-x}Se_x$ (x=0.04, 0.1, 0.13) alloys, their resistivity exhibits a non-monotonic variation characterized by a decrease followed by an increase and then a decrease, which may be related to the combined effects of grain growth, defects, and carrier mobility inside the material during sintering. The resistivity of $Bi_{0.5}Sb_{1.5}Te_{2.93}Se_{0.07}$ alloy shows a gradually decreasing trend, indicating that at this doping level, increasing the sintering temperature is beneficial for optimizing the resistivity of the alloy.

It is worth noting that the resistivity of $Bi_{0.5}Sb_{1.5}Te_{3-x}Se_x$ (x=0.04, 0.07, 0.1, 0.13) alloy decreased sharply at a sintering temperature of 743 K, which may be due to the optimization of sintering conditions at this time, promoting the optimization of the internal structure of the material and the improvement of carrier transport efficiency. However, within the temperature range of 743-803 K, the resistivity changes of the five alloys were relatively small, indicating that the effect of sintering temperature on resistivity became less significant within this temperature range. As the sintering temperature continues to increase, the electrical resistivity of the four alloys $Bi_{0.5}Sb_{1.5}Te_{3-x}Se_x$ (x=0.04, 0.07, 0.1, 0.13) still shows a decreasing trend, indicating that the microstructure inside the material is still being further optimized at higher sintering temperatures. Within the sintering temperature range of 743~803 K, the changes in electrical resistivity of alloys with different Se doping levels also exhibit a certain regularity. When the Se doping amount is 0.01 wt% and 0.07 wt%, the electrical resistivity of the alloy decreases with the increase of sintering temperature; When the Se doping amount is 0.04 wt%, the alloy resistivity shows an upward trend; For an alloy with a Se doping content of 0.1 wt%, its resistance first increases and then decreases; When the Se doping amount reaches 0.13 wt%, the alloy resistivity first decreases and then increases.

The five alloys $Bi_{0.5}Sb_{1.5}Te_{3-x}Se_x$ (x=0.01, 0.04, 0.07, 0.1, 0.13) achieved the minimum resistivity values under their respective optimal sintering conditions, which were 1.6 mΩ·cm, 1.965 mΩ·cm, 0.82 mΩ·cm, 3.07 mΩ·cm, and 0.64 mΩ·cm, respectively. These results indicate that adjusting the sintering temperature and Se doping amount can significantly affect and optimize the resistivity performance of $Bi_{0.5}Sb_{1.5}Te_3$ based alloys.

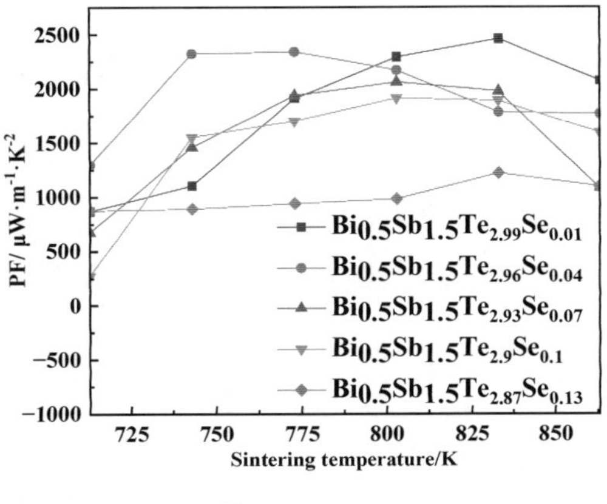

Figure 5 The power factors of $Bi_{0.5}Sb_{1.5}Te_{3-x}Se_x$ alloys.

Figure 5 shows the power factor of $Bi_{0.5}Sb_{1.5}Te_{3-x}Se_x$ (x=0.01, 0.04, 0.07, 0.1, 0.13) alloys at different sintering temperatures. As the sintering temperature gradually increases, the power factor of the alloy first shows an upward trend, and then gradually decreases. When the sintering temperature is 803 K, the power factor of $Bi_{0.5}Sb_{1.5}Te_{2.96}Se_{0.04}$ alloy begins to show a decreasing trend. At a sintering temperature of 833 K, power factors of $Bi_{0.5}Sb_{1.5}Te_{2.93}Se_{0.07}$ and $Bi_{0.5}Sb_{1.5}Te_{2.9}Se_{0.1}$ alloys also show a decrease. When the sintering temperature further increases to 863 K, the power factors of the alloys with doping amounts of 0.01 wt% and 0.13 wt% also decrease successively. In addition, when the doping amount of Se is increased to 0.13 wt%, the change in power factor of the alloy prepared at any sintering temperature is relatively small, and overall, its power factor is generally lower than that of alloys with Se doping amounts of 0.01 wt%, 0.04 wt%, 0.07 wt%, and 0.1 wt%. This phenomenon indicates that excessive doping of the Se element will reduce the power factor of the alloy.

Within the sintering temperature range of 713-863 K, when the doping amount of Se is 0.01 wt% and the sintering temperature is 833 K, the power factor of the alloy reaches its maximum value of

2459.42 μW/m·K^2. This maximum value not only significantly exceeds the power factor of other Se-doped alloys at the same sintering temperature, but also highlights the significant improvement in the electrical properties of the alloy under specific doping and sintering temperature conditions. Therefore, under the condition of Se doping at 0.01 wt%, the electrical properties of the alloy can be optimized through reasonable sintering temperature control.

4. Conclusion

The addition of trace amounts of Se does not cause significant changes in the crystal structure of the sample, and the basic crystal structure stability of the $Bi_{0.5}Sb_{1.5}Te_{3-x}Se_x$ (x=0.01, 0.04, 0.07, 0.1, 0.13) alloy remains unchanged. From the SEM fracture surface morphology of the $Bi_{0.5}Sb_{1.5}Te_{2.96}Se_{0.04}$ alloy, it can be seen that it has a layered structure and some holes. The Seebeck coefficients of the five alloys are all positive, which indicates that they are all P-type semiconductor materials. The Seebeck coefficients and resistivity of the $Bi_{0.5}Sb_{1.5}Te_{3-x}Se_x$ (x=0.01, 0.04, 0.07, 0.1, 0.13) alloys change with the increase of sintering temperature. The maximum Seebeck coefficients of the five $Bi_{0.5}Sb_{1.5}Te_{3-x}Se_x$ (x=0.01, 0.04, 0.07, 0.1, 0.13) alloys are 257.15 μV·K^{-1}, 240.5 μV·K^{-1}, 248.72 μV·K^{-1}, 251.85 μV·K^{-1} and 225.98 μV·K^{-1}, respectively. The five $Bi_{0.5}Sb_{1.5}Te_{3-x}Se_x$ (x=0.01, 0.04, 0.07, 0.1, 0.13) alloys reach the minimum resistivity values at their optimal sintering conditions, which are 1.6 mΩ·cm, 1.965 mΩ·cm, 0.82 mΩ·cm, 3.07 mΩ·cm, and 0.64 mΩ·cm, respectively. Excess Se element doping will reduce the power factor of the alloy. When the Se doping amount is 0.01 wt% and the sintering temperature is 833 K, the power factor of the alloy reaches its maximum value of 2459.42 μW/m·K^2. This maximum value not only significantly exceeds the power factor of other alloys with different Se doping amounts at the same sintering temperature, but also highlights the significant enhancement of the electrical properties of the alloy at a specific doping amount and sintering temperature.

Acknowledgments

This work was financially supported by the National Undergraduate Innovation Training Program Project (202210977025), the Carbon Neutral Engineering Research Center of Guizhou colleges and universities in Coal Industry (Qian Jiao Ji [2023] No. 044), the Materials and Chemical Engineering Team Project (LPSSY2023XKTD06) and Guizhou Province first-class professional construction point (GZSylzy202103).

References

[1] Qiu, W., He, H., Wang, Z., Hu, Q., Cui, X., Wang, Z., Zhang, Y., Gu, L., Yang, L., Sun, Y., Zhao, L., Chen, L., Deng, H., Tang, J., (2021) Enhancing the figure of merit of n-type PbTe materials through multi-scale graphene induced interfacial engineering. Nano Today, 39: 101176. https://doi.org /10.1016/j.nantod.2021.101176.
[2] Soleimani Z., Zoras S., Ceranic B., Shahzad S., Cui Y., (2020) A review on recent developments of thermoelectric materials for room-temperature applications. Sustain. Energy techn. 37:100604. https://doi.org/10.1016/j.seta.2019.100604.
[3] QinY., Qin B., Wang D., Chang C., Zhao L., Qin Y., (2022) Solid-state cooling: thermoelectrics. Energy Environ. Sci., 15: 4527-4541. https://doi.org/10.1039/d2ee02408j.
[4] Le W., Yang W., Sheng W., Shuai J., (2023) Research progress of interfacial design between thermoelectric materials and electrode materials. ACS Appl. Mater. Interfaces,15(10): 12611–12621.https://doi.org/10.1021/acsami.2c21035.
[5] Tang X., Li Z., Liu W., Zhang Q., Uher C., (2022) A comprehensive review on Bi_2Te_3-based thin films: Thermoelectrics and beyond. Interdisciplinary Materials, 1: 88-115. https://doi.org/ 10.1002/ idm 2.12009.
[6] Zhu Y., Wu P., Guo J., Zhou Y., Chong X.,Ge Z., Feng J., (2020) Achieving a fine balance in mechanical properties and thermoelectric performance in commercial Bi_2Te_3 materials.Ceram. Int., 46(10):14994-15002. https://doi.org/10.1016/j.ceramint.2020.03.029.
[7] Liu M., Li W., Pei Y., (2024) Screening metal diffusion barriers for thermoelectric $Bi_{0.5}Sb_{1.5}Te_3$, Sci. China Mater., 67(1): 289-294.https://doi.org/10.1007/s40843-023-2696-1.
[8] Lu Y., Zhou Y., Wang W., et al. (2023) Staggered-layer-boosted flexible Bi2Te3 films with high thermoelectric performance. Nat. Nanotechnol., 18:1281-1288. https://doi.org/10.1038/s41565-023-01457-5.

Effect of 1%sm on corrosion and microstructures behavior of as-cast AT61 alloys

Lewen Wang[1], Yuan Yao[1], Siqi Li[1], Mei Li[1], Zheng Jia[1], Li Fu[1,2,*], Qichi Le[2]

[1] College of Mechanical Engineering, Shenyang University, Shenyang, 110044, China
[2] Key Lab of Electromagnetic Processing of Materials, Ministry of Education, Northeastern University, Shenyang, 110919, China

* Corresponding author: fuli@syu.edu.cn

Abstract. The microstructure and corrosion behavior of AT61 alloy were investigated with the addition of 1% Sm. The results indicated that a new short bar-like $Al_{11}Sm_3$ phase and massive Al_2Sm phase appeared in the ATE611 alloy besides the $Mg_{17}Al_{12}$ and Mg_2Sn phases. At the same time, $Mg_{17}Al_{12}$ phase reduced and changed from continuous distribution along the grain boundaries to discontinuous distribution of bars or lumps. The hydrogen evolution rate of ATE611 alloy is 0.292 $mL \cdot cm^{-2}$, the weight loss rate is 0.483 $mg \cdot cm^{-2} \cdot day^{-1}$, the self-corrosion current density is 7.227 $\mu A \cdot cm^{-2}$, and the polarization resistance is 4405.221 $\Omega \cdot cm^{-2}$. Compared to AT61 alloy, ATE611 alloy has significantly improved corrosion resistance, can be attribute to the cooperative interaction of uniform distribution of the $Mg_{17}Al_{12}$ phase, and the distribution of $Al_{11}Sm_3$ along the grain boundaries to hinder corrosion expansion.

1. Introduction

Magnesium alloys offer numerous benefits, including their large reserves, good strength, good rigidity, and superior damping capabilities, rendering them highly suitable for industrial production advancements. However, their poor corrosion resistance has been an obstacle to their wide application. For example, large numbers of $Mg_{17}Al_{12}$ phases exist in the Mg-Al alloys, which have complex mechanism morphology and higher electrode potential, causing to severe micro galvanic corrosion, and reduced lifespan of magnesium alloy components.

In Mg-Al alloys, the rare earth (RE) tends to preferentially react with Al in the Mg matrix to produce Al-RE second phases, such as Al_2RE, $Al_{11}RE_3$, etc [1]. Wang et al. [2] found that the newly formed Al-RE phases play two roles. First it promotes the grain refinement of the alloy. Secondly, it can make the distribution of $Mg_{17}Al_{12}$ phase more uniform, thereby enhancement the corrosion resistance of Mg-Al alloys. Arrabal et al. [3] found that the addition of heavy rare earth element Gd in AM50 alloy will form Mg-Re, Al-Mn-RE and Al-RE metal intermediate phases in the alloy, reducing the proportion of the $Mg_{17}Al_{12}$ phase and increasing the corrosion resistance by 43%. Liu et al. [4] observed that as the Y content in Mg-Y binary alloys increased from 3% to 7%, there was a corresponding improvement in the alloys' corrosion resistance. The current research mainly focuses on the addition of heavy RE elements in AZ alloys [5].

The role of the light RE Sm in magnesium alloys is similar to that of heavy RE [6]. However, there is a scarcity of reports detailing the changes in corrosion behavior and microstructure of Mg alloys after adding light rare earth Sm. Such as Chen et al. [7] found that the granular Al_2Sm phase in AZ61-Sm alloy promotes the reduction of the $Mg_{17}Al_{12}$ area and improving the corrosion resistance and reducing the grain of the alloy. Chen et al. [8] added Sm to AZ80 alloys, and with the increase of Sm, the course $Mg_{17}Al_{12}$ phase was replaced by Al_2Sm particles, which reduced the number of micro cells formed between $Mg_{17}Al_{12}$ phase and matrix.

Sm exerts a significant impact on the corrosion behavior and microstructure of Mg-Al alloys, yet its influence on Mg-Al-Sn alloys in terms of microstructure and corrosion resistance remains unclear. Consequently, this study aims to explore the effect of incorporating 1% Sm into AT61 alloys. It is hoped that the insights gained from this investigation will play a positive role in continuously improving the properties of Mg-Al-Sn alloys

2. Preparation of experiments and materials

2.1 Materials and testing methods

Mg, Al ingots and Mg-35Sm alloy were melted in the crucible and at 730°C under a mixed protective gas (CO_2: SF_6=99:1 by volume). Anhydrous $MnCl_2$ was added, then followed by No. 5 flux, and it was stirred and refined for 1 min. The preheated mold, at a temperature of 230°C, was filled with the melt to form the ingot. The composition details of the alloys investigated are presented in Table 1. The samples were cut out from the ingots by wire cutting, sanded to a bright finish with SiC sandpaper, polished to a mirror effect and etched with picric acid. Optical microscope (OM, OLYMPUS 8G44018) is used to observe microstructure. Performed phase composition analysis using X-ray diffraction (PANalytical PW3040/60, XRD). The morphology and distribution of phases were detected by electron scanning microscopy (Hitachi-S4800; SEM) and dispersive energy spectroscopy (Bruker Quanax; EDS).

Table 1. Alloys composition

Alloy	Sm	Al	Sn	Mg
AT61	0	6.011	1.042	Bal.
ATE611	1.081	5.751	1.079	Bal.

2.2 Corrosion testing

The corrosion test is divided into two parts: weight-loss test and hydrogen evolution. At ambient temperature, 10 mm × 10 mm × 10 mm specimens were submerged in a solution containing 3.5% NaCl. hydrogen was collected using the test tube above the specimen[9]. After 24 h of immersion, the specimens were using a 200 g/L solution of CrO_3 to eliminate the corrosion products, then weight and calculate the weight loss rate. Experiments were repeated at least three times for accuracy.

2.3 Electrochemical measurements

At ambient temperature, electrochemical testing was conducted utilizing an electrochemical workstation (Chenhua CHI660E), this test comprising two segments: electrochemical impedance spectroscopy (EIS) and polarization curve measurement. Throughout the experiment, the specimens itself as the working electrode, the saturated calomel electrode (SCE) as the reference electrode, and the platinum plate served as the auxiliary electrode. In order to obtain a stable open circuit potential (OCP), the specimens with a test area of 1 cm^2 needs to be submerged in a 3.5% NaCl solution for 1 h in advance. The EIS recording range was from 100 kHz to 0.01 Hz, applying an open circuit voltage disturbance of 5 mV. The polarization curve was generated based on the equilibrium potential, the scanning was conducted within a range of -0.5 V to 0.5 V, with a scanning rate of 1 mV/s. To ensure the reliability of the results, there should have been at least 3 parallel experiments per test. The equivalent circuit was fitted using ZView software.

3. Experimental results and analysis

3.1 Influence of 1% Sm Addition on the Microstructural of AT61 Alloy

Figure 1 is OM photographs of the investigated alloys. In the AT61 alloy, numerous coarse secondary phases are evenly dispersed along the grain boundaries. When rare earth Sm is added, the quantity of second phases within the alloy diminishes, while their size decreases and their distribution becomes discontinuous.

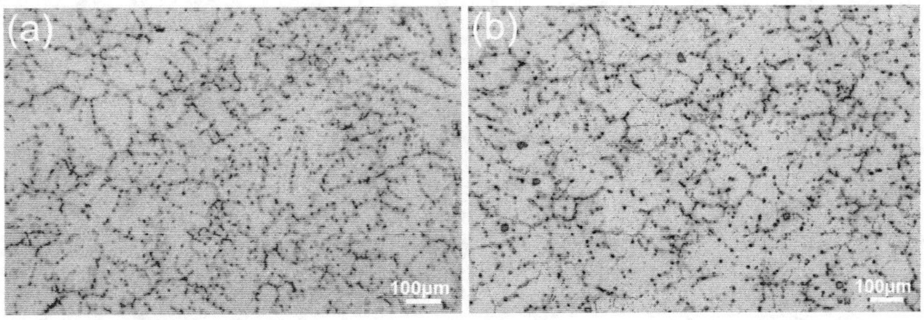

Figure 1. Metallographic images: (a) AT61, (b)ATE611.

Figure 2 displays the XRD pattern of the alloys being investigated. Peaks of Mg and $Mg_{17}Al_{12}$ phases appear in both alloys. After adding 1% Sm, peaks of $Al_{11}Sm_3$ appeared. Peak intensity of $Mg_{17}Al_{12}$ and Mg_2Sn phases significantly decreases in the magnified region. Indicating that the addition of rare earth Sm forms new phases with Al, thereby inhibiting the formation of $Mg_{17}Al_{12}$ and Mg_2Sn phases.

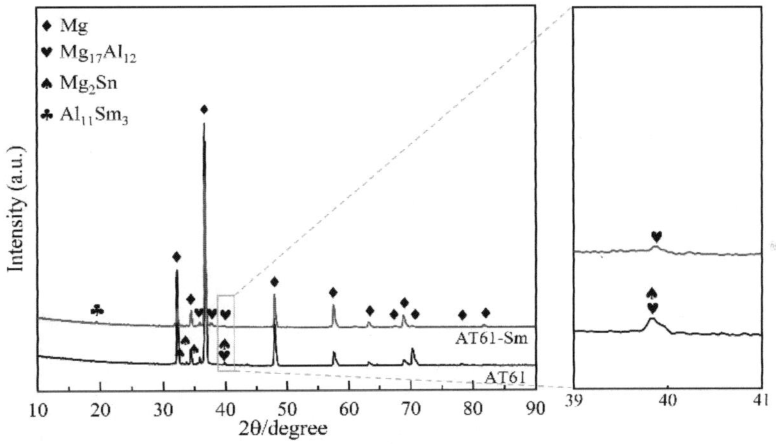

Figure 2. XRD pattern of AT61and ATE611 alloys.

Figure 3 displays SEM images of the alloys, with EDS results of the phases listed in Table 2. As shown in Figure 3 (a), Coarse gray phases and white particles scatter along AT61 alloy grain boundaries. The EDS results indicate that the atomic ratio of Mg to Al in the gray phase at point A currently approximates 17:12, and together with the XRD analysis results, this phase can be characterized as the $Mg_{17}Al_{12}$. The white granular phase at point B contains a large amount of Mg, and a small amount of Al and Sn, which is identified as the Mg_2Sn phase. After the addition of 1% Sm, the $Mg_{17}Al_{12}$ phase exhibits elongation or a blocky appearance, accompanied by a notable decrease in its presence. Concurrently, the ATE611 alloy exhibits rod-shaped phases that are arranged either continuously or semi-continuously along the grain boundaries, as well as polygonal blocky phases and granular phases within the grains. The EDS analysis reveals that the rod-shaped phase at point B has an Al:Sm atomic ratio approximating 11:3, confirming its identity as the $Al_{11}Sm_3$ phase. At point C, The Al:Sm atomic ratio is nearly 2:1, confirming it as the Al_2Sm phase. The granular phase at point D has an Mg:Sm atomic ratio close to 41:5, identifying it as the $Mg_{41}Sm_5$ phase. Notably, the peaks of Al_2Sm and $Mg_{41}Sm_5$ are absent in the XRD pattern, which is attributed to their low content. Figures 3(e-g) present the results of face scanning of the ATE611 alloy's microstructure. It is clear that the precipitation sites of the blocky phase and the rod phase within the ATE611 alloys are primarily linked to the distribution of Al and Sm elements.

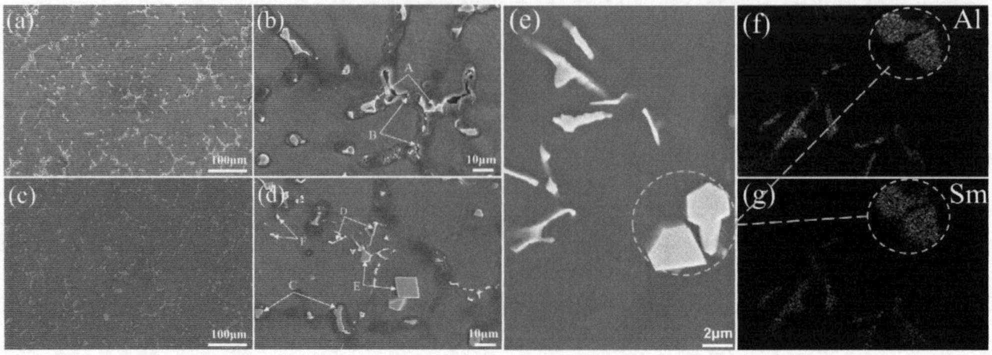

Figure 3. SEM photos: (a, b) AT61, (c, d) ATE611, (e ~ g) EDS mapping of ATE611.

Table 2. EDS results in AT61 and ATE611

Positions	Elements (at. %)				
	Sm	Al	Sn	Mg	Total
A	-	28.7	1.60	69.70	100
B	-	4.92	3.43	71.66	100
C	-	26.26	0.53	73.20	100
D	9.33	33.00	-	57.67	100
E	28.20	56.05	-	15.75	100
F	8.84	22.61	-	68.55	100

3.2 Corrosion behavior analysis

Figure 4 (a) illustrates that the hydrogen evolution curve of the AT61 alloy rises exponentially, while the ATE611 alloy grows approximately linearly. Figure 4(b) shows the macroscopic images of the mass loss corrosion rate and removed corrosion products of the alloy. It is evident that the AT61 alloy surface maintains a relatively smooth appearance, whereas the ATE611 alloy displays a significant area that remains uncorroded. The mean hydrogen evolution rates for AT61 and ATE611 are 7.013 $mL\cdot cm^{-2}$ and 0.292 $mL\cdot cm^{-2}$, respectively, and the mass loss rates are 7.882 and 0.483 $mg\cdot cm^{-2}\cdot day^{-1}$, respectively. This is consistent with the hydrogen evolution. Figures 4(c, d) shows a photo of the investigated alloys for removing corrosion products, this is in agreement with the results obtained from the products. The reason for the improved corrosion resistance of ATE611 is that $Al_{11}Sm_3$ and Al_2Sm phases serve as barriers to hinder corrosion during the process, and the Al_2Sm phases possess a greater surface area. In the initial stages of corrosion, an effective Al_2O_3 protective layer can form, subsequently decreasing the alloy's corrosion rate [10].

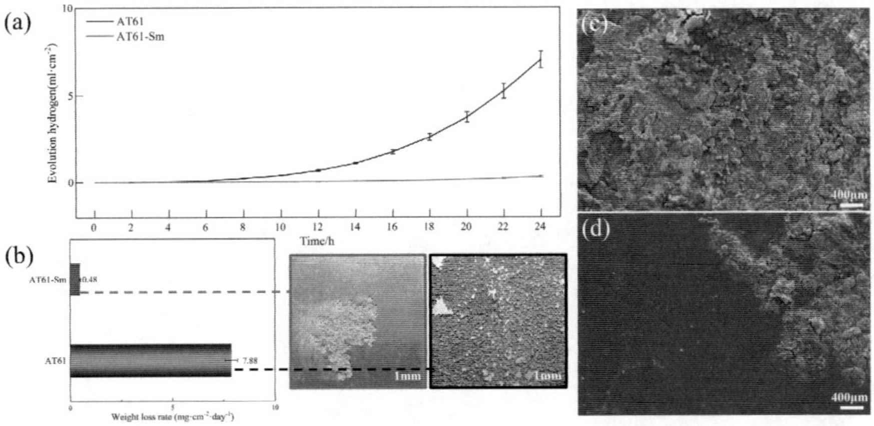

Figure 4. (a) Hydrogen evolution, (b) Weight loss, remove the surface of the corrosion product (c ~ d): Photos of corrosion products on AT61 and AT61-Sm.

3.3 Electrochemical tests

Figure 5 displays the outcomes of the polarization curve tests conducted on the studied alloys, The relevant fitting data are presented in Table 3.. Figure 3 shows the addition of Sm results in a negative shift of the corrosion potential, indicating an increased tendency for self-corrosion. However, as shown in Table 3, the self-corrosion current density of ATE611 has notably decreased to merely 7.227 $\mu A \cdot cm^{-2}$. This reduction is due to its polarization resistance, which reaches 4405.221 $\Omega \cdot cm^{-2}$. This is consistent with the hydrogen evolution test.

The reason why AT61 alloy exhibits high polarization resistance is that the short rod-shaped $Al_{11}Sm_3$ phase and block-shaped Al_2Sm phase are arranged in a semi-continuous to continuous manner along the grain boundaries, leading to a more even corrosion pattern. The micro-electric couple corrosion driving force constructed with the Magnesium matrix is reduced, serving as a hindrance to the spread of corrosion throughout the process. Decreasing the grain size facilitates a more uniform distribution of the $Mg_{17}Al_{12}$ phase, which in turn diminishes the pronounced galvanic corrosion effect of the $Mg_{17}Al_{12}$ phase, ultimately enhancing the corrosion resistance of the alloys.

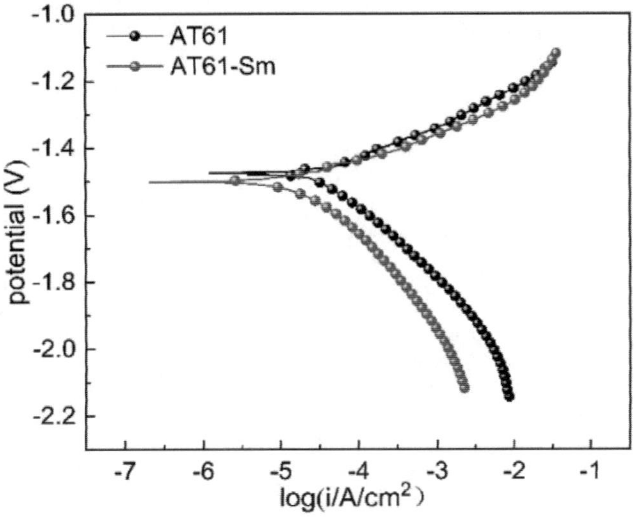

Figure 5. Polarization curves of the AT61 and ATE611.

Table 3. Fitting data.

Alloys	R_p ($\Omega \cdot cm^{-2}$)	β_a ($mV \cdot dec^{-1}$)	β_c ($mV \cdot dec^{-1}$)	E_{corr} (V_{SCE})	I_{corr} ($\mu A \cdot cm^{-2}$)
AT61	968.105	65.09	177.151	-1.48	21.380
ATE611	4405.221	136.259	158.34	-1.515	7.227

4. Conclusion

1. The AT61 alloy consists of a $Mg_{17}Al_{12}$ phase and a trace amount of Mg_2Sn phase. The ATE611 alloy displays the formation of short, rod-shaped $Al_{11}Sm_3$ particles that are distributed in a semi-continuous to continuous manner along the grain boundaries, as well as blocky Al_2Sm phase and granular $Mg_{41}Sm_5$ phase within the grain boundaries.
2. The $Mg_{17}Al_{12}$ phase in ATE611 alloy was changed from a reticulate phase that was distributed continuously along the grain boundaries to a discontinuous bar or massive phase, and the quantity has significantly decreased.
3. The ATE611 alloy exhibits a mass loss rate of 0.483 $mg \cdot cm^{-2} \cdot day^{-1}$, hydrogen evolution of 0.29 $mL \cdot cm^{-2}$, a self-corrosion current density of 7.227 $\mu A \cdot cm^{-2}$, and a polarization resistance of 4405.220 $\Omega \cdot cm^{-2}$ after 24 hours of exposure.
4. The excellent corrosion resistance of ATE611 alloys can be attribute to the cooperative interaction of uniform distribution of the $Mg_{17}Al_{12}$ phase, and the distribution of $Al_{11}Sm_3$ along the grain boundaries to hinder corrosion expansion.

Acknowledgements

The author expresses great gratitude for the support of the China Graduate Education Development Centre, the Liaoning Provincial Department of education, and the Key Laboratory of materials electromagnetic processing of the Ministry of education of Northeastern University [No. ZT-231103515, No. LNYJG2024326, No. NEU-EPM-018].

References

[1] Zhang X P, Fang C F, Wang R, et al.. International Journal of Plasticity, 2025: 104289.
[2] Wang X Y, Liu W F, Dong X, et al.. Physica Scripta, 2025.
[3] Arrabal R, Matykina E, Pardo A, et al. Corrosion Science, 2012, 55: 351-362.
[4] Liu M, Schmutz P, Uggowitzer P J, et al. Corrosion Science, 2010, 52(11): 3687-3701.
[5] Kubásek J, Vojtěch D. Transactions of Nonferrous Metals Society of China, 2013, 23(5): 1215-1225.
[6] Chen X, Jia Y, Le Q, et al. Journal of Materials Research and Technology, 2020, 9(4): 9262-9270.
[7] Chen X, Liao Q, Le Q, et al. Electrochimica Acta, 2020, 348: 136315.
[8] Chen X, Liao Q, Le Q, et al. Electrochimica Acta, 2020, 348:136315.
[9] Wang L, Li S, Li M, et al. International Journal of Metalcasting, 2024: 1-17.
[10] Chen Z, Lai X, Liang Y, et al.. Extreme Materials, 2025.

Journal of Physics: Conference Series **3009** (2025) 012011 doi:10.1088/1742-6596/3009/1/012011

Effect of oxygen flow rate on structure and properties of Cr-based coatings deposited by arc ion plating

Haibo Li [1], Xiaoming Chen [1], Mingli Shen [1], Cean Guo [1,*]

[1] School of Equipment Engineering, Shenyang Ligong University, Shenyang 110159, China

[*] Corresponding author's e-mail: 373055507@qq.com

Abstract: Arc ion plating was employed to conduct high - vacuum and high - speed deposition on the surface of the H13 steel substrate for the purpose of fabricating Cr - based coatings. The effect of different O_2 flow rates (0, 100, 130 sccm) on the microstructure and properties of Cr-based coatings was studied. Scanning electron microscopy, X - ray diffraction, nanoindentation, and a reciprocating friction and wear machine were utilized to conduct an analysis of the surface morphology, phase composition, mechanical characteristics, and tribological behaviors of the coatings. The experimental findings indicated that both the Cr coating and the Cr - O coating with the oxygen flux maintained at 100 sccm exhibited a uniformly distributed microstructure. In terms of thickness, they were comparable. The principal phase detected within the coatings was Cr. Additionally, these coatings exhibited highly consistent hardness and elastic modulus values. Adjust the oxygen flow rate parameter to 130 sccm, the thickness of the coating increases significantly, reaching 23.47 μm. The phase structure of the coating is Cr and Cr_2O_3. Its hardness (4.29 GPa) and elastic modulus (285 GPa) increase significantly respectively. The Cr-based coating with an O_2 flow rate of 130 sccm exhibits the best tribological properties among the three coatings. This is because the Cr_2O_3 phase in the coating itself has certain lubricity and chemical stability, which can effectively reduce the adhesion and ploughing during the friction process. Therefore, this coating has broad application prospects in many fields involving friction and wear conditions, which can effectively improve its working performance and reduce maintenance costs.

1. Introduction

Metallic coatings occupy an important position in the industrial field, so the preparation of high-performance and low-pollution coatings has become an urgent problem to be solved [1]. The electrolyte used in the traditional electroplating method contains harmful substances such as heavy metal ions, which have great limitations in the preparation of wear-resistant coatings. Therefore, the electroplating technology gradually cannot meet the requirements of modern industry [2]. High vacuum arc ion plating technology ranks among the most promising technologies in the fabrication of thick films [3]. In the process of preparing metal coatings by arc ion plating, adding a second component to the coating can change the Material constitution of the surface layer, thereby improving the performance of the metal coating.

Chromium nitrides occupy the basic position in the binary system. Chromium nitride coatings have advantages such as high hardness, good tribological properties, and strong coating adhesion, which are great wear-resistant protective coatings [4]. Zhang et al. [5] studied the phase change law of chromium nitride coating by regulating the flow rate of N_2. The results showed that when the amount of N_2 was less than 4sccm, the phase composition of the coating was Cr and Cr(N) solid solution state. When the amount of N_2 is 4~12sccm, the phase of the coating was mainly in the transition phase of Cr to Cr_2N and Cr_2N to CrN. When the amount of N_2 is greater than 12 sccm, the CrN phase is mainly formed. Shan et al. [6] found that the CrNx coating prepared with 125 sccm N_2 has the lowest wear loss in the simulated marine environment. The key reason is that the dual-phase structure of Cr_2N/CrN provides the coating with good corrosion resistance and high hardness.

Chromium oxide coating is another commonly used binary coating, but the research on chromium oxide coating is still in the initial stage. Consequently, comprehending the mechanism by which O_2 influences the microstructure and properties of chromium coatings holds substantial theoretical value and practical application significance. This understanding is crucial for the popularization and advancement of arc ion - plated chromium coatings and their utilization in high - end industrial sectors. In this experiment, three different Cr-based coatings were prepared on H13 steel (4CrSMoSiV1) by changing the O_2 flow rate using vacuum arc ion plating technology. An investigation was carried out on the phase structure and mechanical attributes of the coatings.

2. Experiment

2.1. Coating preparation

The substrate used in the experiment is H13 steel (4Cr5MoSiV1). Before the preparation of the coating, the substrate was first cut into 20 mm×10 mm×2 mm samples. During the polishing process, 240 # sandpaper was used to remove the rust, and then 600 #, 800 #, 1000 #, 1500 #, and 2000 # water sandpaper were used to polish. Ultimately, the polished specimen is submerged underwent ultrasonic cleaning in ethanol for 10 minutes. Subsequently, it is dried by blowing and then placed in a bag for future utilization.

Coating samples were prepared by DH-7 arc ion plating equipment. Pure Cr and two Cr-based coatings with different O_2 flow rates were finally prepared by introducing 1.5 h of gas (O_2 or Ar).

2.2. Experimental method

An Apreo2C field-emission scanning electron microscope was employed to investigate the surface topography and cross-sectional features of the samples (SEM) manufactured by Thermo Fisher Scientific Inc, an American company. Micro-area composition analysis was analyzed using an UltimMax spectrometer produced by Oxford, UK. Phase analysis of the three coatings was carried out with an UltimaIIV X - ray diffractometer produced by Rigaku, a Japanese company. The device operated at a working voltage of 40 kV and a current of 40 mA, with a Cu target used for the experiment. The experimental setup included a scanning range of 10° to 90° (2θ), a scan velocity of 10°/min, and a step width of 0.02°.

The hardness and elastic modulus of the specimens were determined using a Nano Indenter Agilent G200 nanoindentation device. A load of 20 mN was applied, with the loading duration set to 15 seconds and the holding time set at 10 seconds. In the coating area, a measurement location was designated every 50 μm, and each location was measured six times.

The HSR-2M friction and wear tester was used to test the friction and wear resistance of the samples by using stainless steel balls as the counter body. The experimental conditions were as follows: room temperature 25 °C, load 5 N, friction stroke 5 mm, reciprocal time 300 times/min, and friction time 10 min. The wear rate is calculated by the following formula.

$$W = \frac{d^2 \times h\left\{2\times \arcsin\left(\frac{L}{d}\right) - \sin\left[2\times \arcsin\left(\frac{L}{d}\right)\right]\right\}}{8\times N \times v \times t}$$

where W is the wear rate, $mm^3/(N\cdot m)$; d is the diameter of the grinding ball, mm; h is the length of wear marks, mm; L is the width of wear marks, mm; N is the applied load, N; v is the friction speed, m/s; t is the wear time, s.

3. Experimental results and analysis

3.1. Surface and cross-section morphology of Cr-based coatings

Figure 1 is the surface morphology of the as-deposited Cr-based coatings at distinct O_2 flow parameters. According to the graphical representation in Figure 1 that the surface of the deposited Cr coating prepared without O_2 is relatively smooth and the grain boundary is clear. When the O_2 flow rate increases to 100 sccm, the grain boundary of the coating surface begins to blur, and the grain boundary size begins to increase. Upon reaching an O_2 flow rate of 130 sccm, the coating's morphology undergoes a distinct alteration. The surface becomes flat and sleek, to the extent that the grain boundaries on the coating surface are indistinguishable. As the oxygen flow rate steadily rises, there is a decrease in the quantity of droplets formed during the target evaporation process. This reduction leads to the coating surface becoming more compact and smooth.

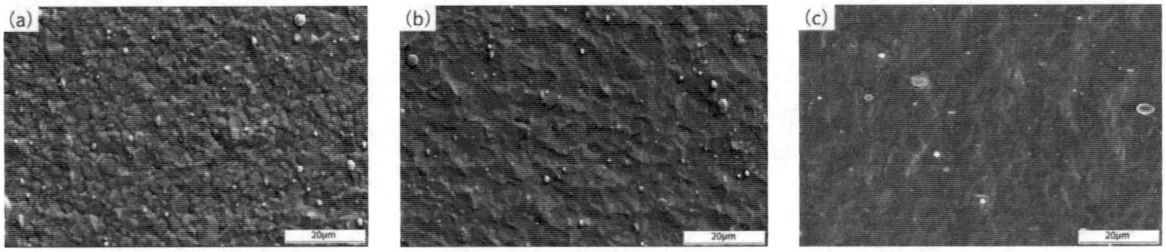

(a) (b) (c)

Figure 1. The surface morphology of Cr-based coatings deposited with different O_2 flow rates: (a) 0 sccm, (b) 100 sccm, (c) 130 sccm.

Figure 2 presents the cross - sectional morphologies of the as - deposited Cr - based coatings under varying O_2 flow rates. As can be observed from Figure 2, the Cr - based coating deposited when the O_2 flow rate is set at 100 sccm exhibits a uniform and dense structure. There are no holes, cracks, or droplet defects in the coating. The thickness (23.47 μm) of the coating with a 130 sccm O_2 flow rate increased significantly. For the Cr-O coating with a 100 sccm O_2 flow rate, the thickness is close to that of the pure Cr coating. Due to the limitation of O_2 supply and slow oxidation reaction rate, more O exists in the form of solid solution in the coating, and the amount of Cr_2O_3 generated is very small, so it is difficult to significantly increase the thickness of the coating. When the O_2 flow rate increased to 130 sccm, sufficient O_2 supply greatly promoted the oxidation reaction rate, so that more Cr reacted with O to form Cr_2O_3, and the formation rate of Cr_2O_3 is significantly increased. Therefore, a thicker Cr_2O_3 layer can be deposited onto the surface of the coating during the same preparation time, resulting in a significantly thicker coating than a pure Cr coating and a 100 sccm Cr-O coating.

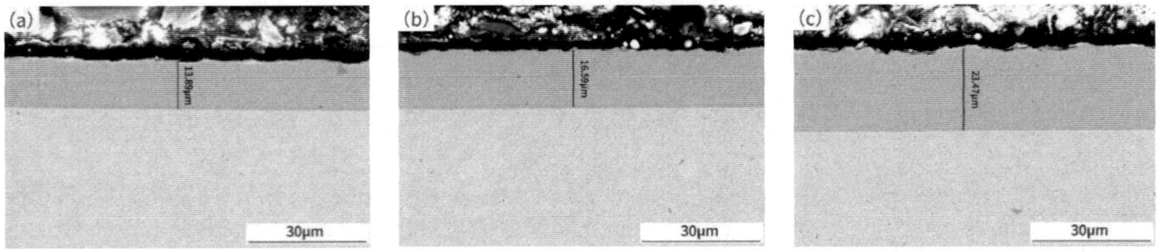

(a) (b) (c)

Figure 2. The cross-sectional morphology of Cr-based coatings deposited with different O_2 flow rates: (a) 0 sccm, (b) 100 sccm, (c) 130 sccm.

3.2. Phase analysis of Cr-based coatings

Shown in Figure 3 are the XRD spectra of Cr - based coatings corresponding to different O_2 flow rates. For the coating with an O_2 flow rate of 100 sccm, the detected phase is Cr. This implies that within the coating, O is prone to existing in the form of a solid solution, or alternatively, the content of the oxide phase is extremely low. Under the condition of 130 sccm O2 supply, Cr and Cr_2O_3 are detected in the coating, but the diffraction peak of Cr_2O_3 is relatively weak. For coatings deposited with a low oxygen flow rate (100 sccm), only the Cr phase was detected, indicating that O may be mainly incorporated into the chromium in the coating in the form of a solid solution, which makes it difficult to detect. When the O_2 flow rate is 130 sccm, Cr and Cr_2O_3 can be detected in the coating, demonstrating the onset of oxide formation at elevated oxygen flow rates. However, the diffraction peak of Cr_2O_3 is relatively weak, which may mean that although Cr_2O_3 is formed, its content is still not very high in the coating, or its crystallinity is relatively poor, resulting in a lower diffraction peak intensity.

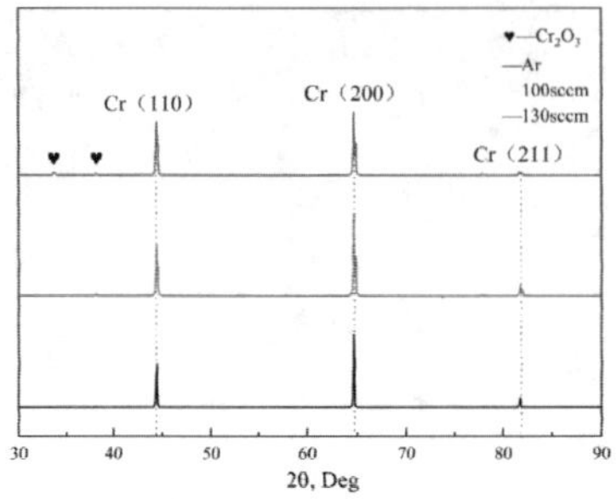

Figure 3. XRD patterns of the as-deposited Cr-based coatings with different O_2 flow rates.

3.3. Mechanical properties of Cr-based coatings

Table 1 displays the mean values of the elastic modulus and hardness of the as - deposited Cr - based coatings under different O_2 flow rates. From an examination of Table 1, it becomes evident that the Young's modulus and indentation hardness characteristics of the unalloyed Cr coating and the Cr - O coating fabricated at an O_2 flow rate of 100 sccm are comparable. For the Cr-O coating prepared with a 100 sccm O_2 flow rate, attributable to the relatively low O_2 flow rate, there may be more incompletely oxidized Cr phase and O element solid solution in the coating, demonstrating overall performance characteristics equivalent to pure Cr coatings. As the O2 flow parameter reaches 130 sccm, the hardness of the coating is obviously improved, which may be related to the oxide dispersion strengthening (Cr_2O_3) and fine-grain strengthening [7].

Table 1. The average values of elastic modulus and hardness of the as-deposited Cr-based coatings with different O_2 flow rates.

	Ar	100 sccm	130 sccm
E/(GPa)	223	242	285
H/(GPa)	2.21	2.35	4.29

3.4. Tribological properties of Cr-based coatings

Figure 4 presents the tribological performance curves of the three different coatings. According to the plotted results, it can be discerned that the friction coefficient of the Cr - based coating with an O_2 flow rate of 100 sccm exceeds that of the pure Cr coating. This phenomenon can be attributed to the relativ ely low O_2 flow rate. Under such conditions, the crystallinity of Cr_2O_3 might be insufficient, which in t urn causes an elevation in the roughness of the coating surface. Once the O_2 flow rate is raised to 130 s ccm, the resulting Cr - based coating features a more compact structure and enhanced crystallinity. Thi s advantageous structural characteristic enables the coating to effectively decrease the friction coefficie nt generated between itself and the counter - body during the rubbing action, thereby leading to an imp rovement in tribological performance. The friction coefficient of the Cr-based coating with a 130 sccm O_2 flow rate is only about 0.2, which suggests that this kind of Cr-based coating has a self-lubricating effect.

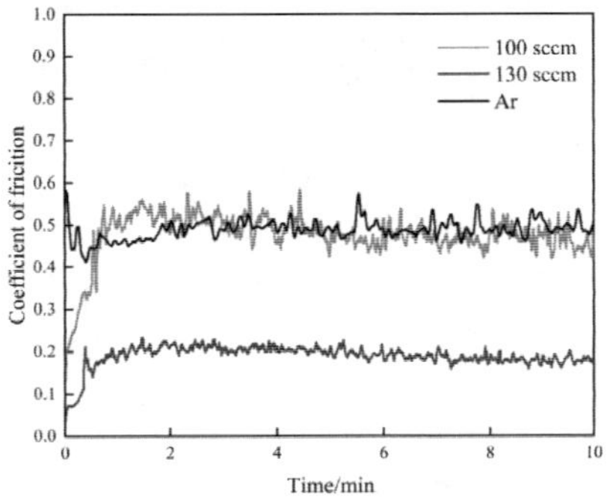

Figure 4. Friction coefficients of three coatings.

Wear rate measurements of the three coating systems are displayed in Figure 5. After rubbing at room temperature, the wear rate of the pure Cr is about 3.7 $10^{-3}mm^3/(N\cdot m)$, that of the Cr-based coating with a 100 sccm O_2 flow rate is about 4.3 $10^{-3}mm^3/(N\cdot m)$, and that of the Cr-based coating with a 130 sccm O_2 flow rate is about 3.2 $10^{-3}mm^3/(N\cdot m)$. This indicates that a higher O_2 flow rate can notably enhance the wear - resistance of the Cr - based coating. During the sliding wear process of the chromium-based coating deposited at 130 sccm oxygen flow rate, a protective Cr_2O_3 tribo-film develops on the surface. This film offers effective lubrication and protection as it serves to prevent the direct contact between the coating and the counterpart material. As a result, it reduces adhesion and wear during the rubbing process, leading to a smaller wear scar width and a lower wear rate.

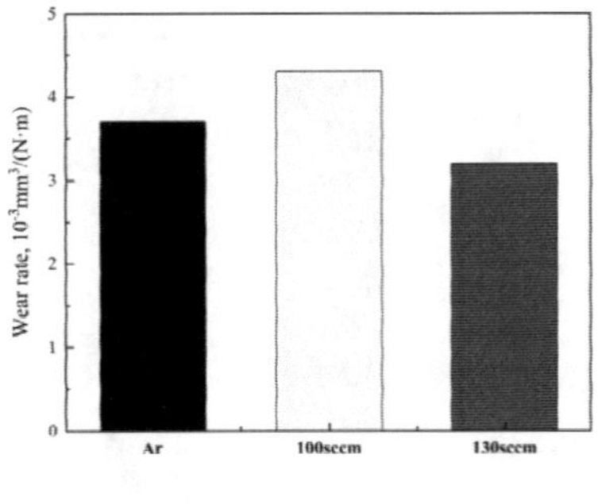

Figure 5. The wear rate of three coatings.

The tribological surface features of the three coatings are displayed in Figure 6. It can be seen from Figure 6 that the pure Cr coating shows an adhesive wear feature. During the rubbing process, the Cr atoms on the coating surface are easy to attract and adhere to the atoms on the surface of the counter body. As the rubbing continues, the adhesion points continue to form and are cut off, forming adhesive wear. For the Cr-based coating prepared with a 100 sccm O_2 flow rate, a substantial amount of wear particles are distributed on the wear track surface. During the rubbing process, the weak area on the coating is prone to crack and peel off under the action of frictional stress, forming wear debris. The Cr-based coating with a 130 sccm O_2 flow rate shows oxidative wear features from EDS results, which is a relatively stable and mild wear mechanism. During the rubbing process, the oxidation reaction mainly occurs between the Cr-based coating and the counter body. The generated oxidation product is relatively fine and has an obvious lubricating effect, which reduces the direct contact and friction between the coating and the counter body, thereby reducing the wear rate and exhibiting better tribological performance.

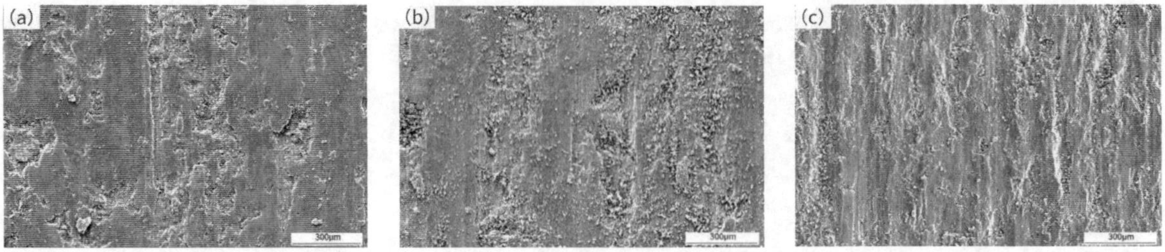

(a) (b) (c)

Figure 6. Worn morphology of three coatings: (a) 0 sccm, (b) 100 sccm, (c) 130 sccm.

4. Conclusion

In light of the results and comprehensive analysis detailed above, the ensuing conclusions can be drawn:

(1) Upon the O_2 flow rate attaining 130 sccm, a significant augmentation in the thickness of the Cr - based coating is observed, with it reaching up to 23.47 μm. As the O_2 flow rate climbs, the phase constitution of the Cr - based coating undergoes a transformation, shifting from a pure Cr phase to a mixed phase incorporating both Cr and Cr_2O_3.

(2) For the Cr - based coating, when the O_2 flow rate is set at 100 sccm, its hardness and elastic modulus exhibit a remarkable similarity to those of the pure Cr coating. However, as the experimental conditions progress and further investigations are carried out, it becomes evident that the hardness and elastic modulus of the same Cr - based coating with an O_2 flow rate of 100 sccm display a significant upward trend under certain specific circumstances, which might be related to factors such as the microstructure evolution and the interaction between different elements during the coating formation process.

(3) Among the three types of coatings under study, the Cr - based coating fabricated with an oxyg en flow rate of 130 sccm demonstrates the most outstanding tribological performance. Specifically, it e xhibits the lowest friction coefficient, approximately 0.2, and the narrowest wear track width, measurin g 678.9 μm. These characteristics clearly set it apart as the top performer in terms of tribology when co mpared to the other two coatings.

Acknowledgments

This work was financially supported by the Light-Selection Team Plan of Shenyang Ligong University (SYLUGXTD5).

References

[1] Qing L, Peng S and Ruiqian Z (2022) Oxidation behavior and Cr-Zr diffusion of Cr coatings prepared by atmospheric plasma spraying on zircaloy-4 cladding in steam at 1300 °C. J. Corrosion Science, 203.

[2] Shichang Z, Shunyan N, and Hefang L (2021) Preparation of ion-exchange resin via in-situ polymerization for highly selective separation and continuous removal of palladium from electroplating wastewater. J. Separation and Purification Technology, 258(P1): 117670-117670.

[3] Chen Y, Yao S and Wang X (2024) Influence of NiCrAlY coating on anti-corrosion of M951 alloy. J. Materials Today Communications, 41: 110766-110766.

[4] Li Z R, Cheng Q C and Pu B J (2024) Effect of CrN coating on the hot salt corrosion fatigue behavior of titanium alloy. J. Corrosion Science, 240: 112448-112448.

[5] Zhang Z, Rapaud O and Bonasso N (2007) Control of microstructures and properties of dc magnetron sputtering deposited chromium nitride films. J. Vacuum, 82(5): 501-509.

[6] Shan L, Wang Y, and Li J (2014) Effect of N_2 flow rate on microstructure and mechanical properties of PVD CrN x coatings for tribological application in seawater. J. Surface & Coatings Technology, 242: 74-82.

[7] Mengde L, Wenju X and Xiaohong L (2022) Tribological behavior and self-repairing performance of chromium oxide films over multiple thermal cycles from 25°C to 1000°C. J. Tribology International, 165.

A High-temperature and corrosion-resistant Co-Al_2O_3 composite superhydrophobic coating for marine exhaust emission systems

Aiguo Chen[1], Zishi Fu[1, 2, a*], Yi Zhang[1], Zan Huang[1], Agustin Valera-Medina[3], Hua Xiao[1, b*], Xiaoyang Wu[1]

[1] School of Ocean Engineering, Guangzhou Maritime University, Guangzhou, China
[2] Guangzhou Lvzhineng Technology Co., Ltd. Guangzhou, China
[3] Net Zero Innovation Institute, Cardiff University, Wales, United Kingdom

[a*]Corresponding author: fujeremy13@gmail.com
[b*]Corresponding author: xiaohua@gzmtu.edu.cn

Abstract: To address the challenges of high-temperature corrosion in marine exhaust systems this study utilizes a one-step electrodeposition technique to fabricate a superhydrophobic Co-Al_2O_3 composite coating on low-carbon steel. A comprehensive analysis was conducted to examine the coating's surface morphology, hydrophobic characteristics, self-cleaning capability, corrosion resistance, and thermal stability. The results demonstrated that the composite coating exhibited remarkable hydrophobicity, with a contact angle (CA) of approximately 163.7°. Furthermore, after heating to 500 °C, the coating maintained superior corrosion resistance, achieving a corrosion protection efficiency (η) of 89.8 % in a 3.5 wt % sodium chloride solution. Electrochemical impedance spectroscopy (EIS) revealed that the coating effectively suppressed electrochemical reactions, thus prolonging the protection of the metal surface. These findings suggest that the Co-Al_2O_3 composite coating provides potential protective capabilities for metals in marine environments, making it suitable for high-temperature corrosion resistance in gas turbines. This approach has the potential to extend equipment lifespan and reduce maintenance costs.

1. Introduction

The world faces critical challenges, including climate change and rising energy costs, making greenhouse gas reduction a shared goal among nations. Ammonia, a carbon-free energy carrier, has gained widespread attention as a fuel [1]. Compared to H_2, NH_3 offers advantages in storage, as it can be maintained at pressures of 0.8–1.0 MPa at ambient temperature or at -33 °C under atmospheric pressure and benefits from a well-established transportation network [2]. Therefore, green ammonia holds great promise as a next-generation fuel for power and energy applications. However, its application as a fuel is hindered by challenges such as a high ignition temperature, a low laminar burning velocity, and the potential release of nitrogen oxides (NO_X) [3]. A key drawback of utilizing ammonia in gas turbines and internal combustion engines is NO_X emissions. Among the current denitration technologies, selective catalytic reduction (SCR) is widely adopted in industry due to its cost-effectiveness and high efficiency, especially in dry flue gas desulfurization [4]. When using SCR for NO_x reduction, a high temperature (400-500 °C) is necessary to activate the catalyst and convert NO_x to N_2 and H_2O [5]. However, at such temperatures, metals undergo thermal corrosion, leading to rapid aging and damage at exhaust outlets of gas turbines and ammonia-blended diesel engines, which consequently reduces equipment life [6]. This study aims to develop a coating to protect the exhaust outlets of marine power systems, extending the lifespan of marine equipment and reducing maintenance costs.

Current metal corrosion protection methods mainly include organic coatings and metal plating. However, organic coatings tend to degrade faster when exposed to UV light, humidity, and salt, often developing cracks and chalking with prolonged sun exposure [7]. Although metal coatings offer excellent mechanical properties such as durability and abrasion resistance, their porous structures can still enable electrochemical reactions between the water film and substrate. Inspired by the lotus leaf effect [CA > 150°, sliding angle (SA) < 10°], there is growing interest in applying superhydrophobic surfaces to metal coatings to mitigate their porosity. Superhydrophobic coatings can form dense "air pockets" between the microstructures and air, creating a barrier to prevent direct contact between metal and corrosive substances [8]. Common ways for fabricating superhydrophobic coatings include spraying, sol-gel processing, and chemical etching. However, these techniques face challenges such as poor mechanical performance, complex procedures, and high production costs [9]. Compared to these approaches, one-step electrodeposition offers advantages in simplicity, durability, and low fabrication cost. Yin *et al.* [10] synthesized and applied Cu-TiO_2 composite coatings to 304 stainless steel by electroplating and fluoridation, which showed excellent corrosion resistance and self-cleaning characteristics.

This study utilizes high-hardness, corrosion-resistant cobalt metal combined with heat-resistant nano-alumina to prepare a Co-Al_2O_3 composite coating. The coating's surface structure, wettability, self-cleaning ability, corrosion resistance, and high-temperature stability are characterized.

2. Methodologies

2.1 Materials

The analytical-grade (AR) chemical reagents used in this study include hexahydrate cobalt chloride ($CoCl_2 \cdot 6H_2O$), boric acid (H_3BO_3), stearic acid (STA), nano-alumina (Al_2O_3), and deionized water with a resistivity of 18.2 MΩ, all of which were purchased from RHAWN.

2.2 Electrodeposition of the Co-Al_2O_3 composite coatings

The substrate selected for this study is low-carbon steel (20 mm × 20 mm × 1 mm) obtained from Shenzhen Guangdiao Industrial Hardware Processing Factory. To ensure uniform adhesion of the coating to the substrate, the steel sheets were initially polished by 320, 800, and 1200 grit silicon carbide sandpaper to remove surface oxides and grease. Following this, the steel sheets were immersed in anhydrous ethanol and acetone solution for ultrasonic cleaning for 15 minutes to further eliminate surface contaminants. Finally, the sheets were rinsed with anhydrous ethanol and oven-dried for 30 minutes.

In the preparation process, the Ti-IrO_2 and CS were placed parallel to each other with a gap of 20 mm and immersed in the electrodeposition solution, which consisted of 100 g/L $CoCl_2 \cdot 6H_2O$, 30 g/L H_3BO_3, 6.7 g/L STA, 2 g/L nano-alumina particles, anhydrous ethanol (C_2H_6O), and deionized water. Under stable temperature conditions of 60 °C and a consistent stirring rate of 300 rpm, a series of samples was collected at a current density of 0.1 A/cm^2 for 10 minutes. The reaction expressions at the cathode and anode are as follows:

Cathode:

$$Co^{2+} + 2e^- \rightarrow Co \tag{1}$$

$$2H^+ + 2e^- \rightarrow H_2 \uparrow \tag{2}$$

$$Co^{2+} + 2CH_3(CH_2)_{16}COOH \rightarrow Co[CH_3(CH_2)_{16}COO]_2 + 2H^+ \tag{3}$$

Anode:

$$4OH^- - 4e^- \rightarrow O_2 \uparrow + 2H_2O \tag{4}$$

2.3 Characterization

Scanning electron microscopy (SEM) was adopted to analyze the surface morphology of the coating. A contact angle measuring instrument (SINDIN SDC-350) was used to measure CA and SA to assess the wettability of the coating. Chalk powder was used to simulate dust particles for evaluating the coating's self-cleaning properties. The samples are heated in a muffle furnace at 500 °C for 30 minutes, then removed and allowed to cool gradually at natural temperature. Following this thermal treatment, CA and SA measurements were repeated using the contact angle measuring instrument. To assess corrosion resistance in a saline environment, the coating was immersed in a sodium chloride solution (3.5 wt %) and evaluated using an electrochemical workstation (DH7000D). The corrosion resistance and CA values obtained post-heating were compared to those of the unheated sample to determine the effects of thermal treatment on the coating's properties.

3. Results and discussion

3.1 Surface morphology
Microstructure and morphology are significant factors influencing surface wettability [11]. Figure 1 shows SEM images of Co-Al_2O_3 and Co coating. The Co coating surface features denser peak-like protrusions compared to bare steel. However, the Co-Al_2O_3 coating exhibits a coral-like structure. This unique microstructure creates mesopores on the surface, trapping more air and shifting the interface state from liquid-solid to liquid-gas-solid. The roughness of the Co-Al_2O_3 coating is greater than Co coatings, indicating significantly enhanced hydrophobicity for this coating.

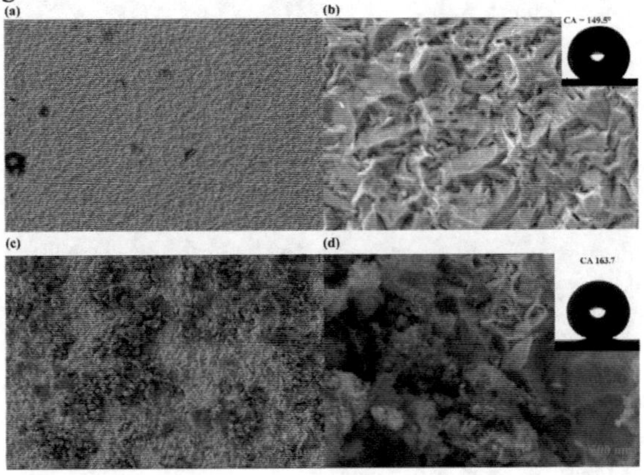

Figure 1. SEM images of coating surfaces.

3.2 Water-repellence and self-cleaning
Figure 2 illustrates the influence of varying current densities on the droplet CA, determined using a contact angle goniometer. In the experiment, 3 μL droplets were placed at three distinct locations on each sample surface, and the average CA was calculated. The results show that the maximum CA of CoAlS reaches 163.7°, demonstrating exceptional water-repellent properties that align with the Cassie-Baxter model [12]:

$$\cos\theta_R = f_1(1 + \cos\theta_0) - 1 \tag{5}$$

where θ_R represents the CA on a rough surface, f_1 signifies the proportion of the solid-liquid contact area, and θ_0 represents the intrinsic CA of the solid-liquid relationship. According to this model, the air retention rate under the droplet on CoAlS is calculated to be 0.04, indicating that approximately 96 % of the area beneath the droplet is occupied by air, trapped within the micro- and nano-scale voids of the Co-Al_2O_3 structure. In contrast, the Co coating exhibits a sticky superhydrophobic surface, where droplets penetrate the rough surface structure, consistent with the Wenzel model [12].

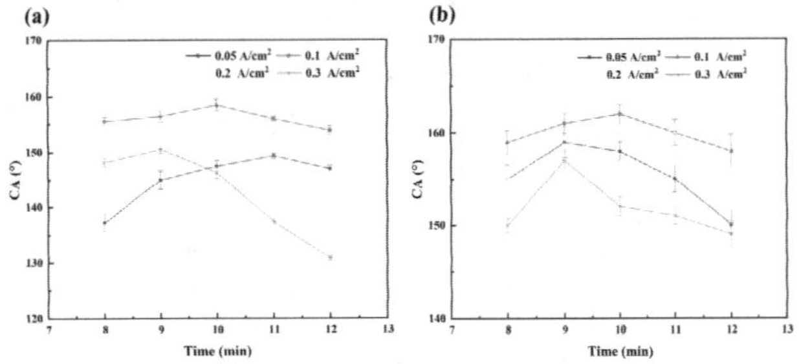

Figure 2. (a) Influence of various current densities on the CA of CoS. (b) Influence of various current densities on the CA of CoAlS.

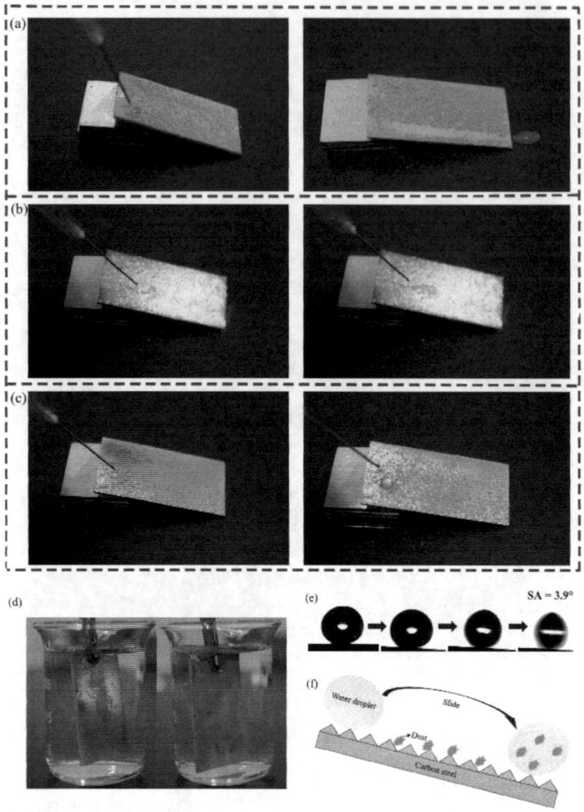

Figure 3. (a-c) Self-cleaning performance of the coatings. (d) Silver mirror effect. (e) SA. (f) Depiction of the self-cleaning mechanism.

To evaluate the self-cleaning performance of the coatings, chalk powder was used as a contaminant. As shown in Figure 3, chalk powder was evenly spread on the inclined coating surfaces. CS is carbon steel, CoS is Co coating surface, and CoAlS is Co-Al_2O_3 composite coating surfaces. On the Co-coated surface, the water droplets removed only a small amount of the dust, and a mixture of dust and water remained on the surface due to inadequate hydrophobicity. Conversely, on the Co-Al_2O_3 composite coating, a clear path was formed as the water droplet rolled down, encapsulating and removing dust along its route. This self-cleaning effect can be attributed to the micro-nano rough structure of CoAlS, which traps a significant amount of air, preventing direct contact between dust and the coating. Underneath these conditions, the adhesion force between the dust and water droplet exceeds that between the dust. This is attributed to the micro-nano rough structure of CoAlS, which traps a significant amount of air, preventing direct contact between dust and the coating. Under these conditions, the adhesion force between the dust and water droplet exceeds that between the dust and the coating.

3.3 High-temperature test and corrosion resistance

The CoAlS samples were heated in a muffle furnace at 500 °C for 30 minutes, then removed and allowed to cool to room temperature. The corrosion resistance of CS, CoAlS, and heat-treated CoAlS samples was assessed using electrochemical corrosion testing by immersing the samples in a sodium chloride solution (3.5 %), recording their Tafel curves and EIS under open circuit potential conditions, as shown in Figure 4. (a–b).

Table 1 presents the corrosion potential (E_{corr}), corrosion current density (I_{corr}), and protective efficiency (η), where η is calculated using the formula:

$$\eta = \frac{I_{corr}(\mathrm{CS}) - I_{corr}(\mathrm{Coating})}{I_{corr}(\mathrm{CS})} \times 100\ \% \qquad (6)$$

Here, $I_{corr}(\mathrm{CS})$ and $I_{corr}(\mathrm{Coating})$ represent the corrosion current density of bare steel and the coated sample, respectively. The CoAlS coating achieved a protective efficiency of 89.8 %, indicating excellent corrosion resistance, while the heat-treated CoAlS retained an efficiency of 84.8 % as well.

Figure 4. (b) shows the EIS spectra and fitted results of the coatings in a sodium chloride solution (3.5 %). In the Nyquist plot, a larger diameter of the capacitive arc corresponds to slower electron transfer rates and better corrosion resistance [10]. The capacitive arc diameter of CoAlS is notably larger than that of CS, and the diameter for heat-treated CoAlS shows only a minor reduction. Figure 4. (c–d) models the coatings equivalent circuits for analyzing the EIS data. In the equivalent circuit, Rs, Rct, and Rf represent solution resistance, charge transfer resistance, and coating resistance, respectively. The constant phase elements CPEdl and CPEf represent the double-layer capacitance and the capacitance of the superhydrophobic layer. Since coatings may have structural inhomogeneities, capacitance is described using constant phase elements (CPE) in the equivalent circuit.

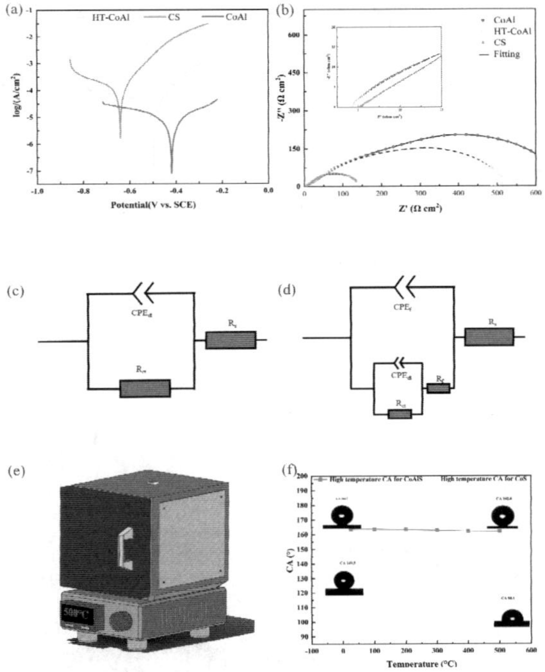

Figure 4. (a) Tafel curves for samples in 3.5wt % sodium chloride solution. (b) EIS Nyquist curves of samples. (c-d) Equivalent circuits. (e) Samples in a muffle furnace at 500 °C for 30 minutes. (f) Changes of CA after high temperature for CoS and CoAlS.

Table 1. Electrochemical parameters for the samples.

Sample	CS	CoAlS	HT-CoAlS
E_{corr} (mV)	-632.5	-426.0	-492.1
I_{corr} (μA/cm^2)	129.5	13.2	19.7
η (%)	/	89.8	84.8

The uniform, dense coating serves as a physical barrier against corrosion, while the nano-alumina attached to the surface effectively prevents the decomposition of the coating in high-temperature environments. When the temperature increased from 25 °C to 500 °C, the CA of CoS decreased from 149.5° to 90.1°, whereas the CA of CoAlS decreased by only approximately 1°. Moreover, the coral-like surface microstructure of the coating captures a significant amount of air, thereby increasing air retention, which is crucial for maintaining its superhydrophobicity under harsh conditions. Additionally, the composition of the coating, including cobalt and nano-alumina, synergistically contributes to corrosion resistance by inhibiting electrochemical reactions and enhancing the stability of the protective layer. This explains the mechanism behind the high-temperature resistance, air retention, and corrosion protection of the Co-Al_2O_3 composite coating.

4. Conclusion

In summary, the Co-Al_2O_3 superhydrophobic composite coating prepared via a one-step electrodeposition method provides an effective solution for corrosion resistance and self-cleaning protection of metals in harsh marine environments. This composite coating exhibits a dense microstructure with pronounced hydrophobic properties, effectively reducing direct contact between corrosive agents and the metal surface. Furthermore, it demonstrates excellent thermal stability and corrosion resistance at temperatures as high as 500 °C, achieving a protective efficiency of 89.8 %, which remains at 84.8 % even after high-temperature treatment. This study indicates significant potential for such coatings in applications like gas turbine exhaust outlets, extending equipment lifespan under marine conditions, reducing maintenance costs, and protecting against corrosion and high-temperature exhausts.

Further research is required to address certain potential challenges and broaden the coating's applicability. The long-term durability and corrosion resistance of the Co-Al_2O_3 composite in highly acidic or alkaline environments need further investigation. Additionally, its high-temperature corrosion resistance under more extreme conditions, such as temperatures ranging from 500 °C to 1000 °C, also requires validation. These challenges provide valuable directions for future research. By optimizing the coating composition and deposition parameters, the mechanical performance and thermal stability of the coating can be further improved.

References

[1] A. Valera-Medina, H. Xiao, M. Owen-Jones, W. I. F. David, and P. J. Bowen, "Ammonia for power," *Prog. Energy Combust. Sci.*, vol. 69, pp. 63–102, Nov. 2018, doi: 10.1016/j.pecs.2018.07.001.

[2] A. Valera-Medina *et al.*, "Review on ammonia as a potential fuel: From synthesis to economics," *Energy Fuels*, vol. 35, no. 9, pp. 6964–7029, May 2021, doi: 10.1021/acs.energyfuels.0c03685.

[3] L. Kang, W. Pan, J. Zhang, W. Wang, and C. Tang, "A review on ammonia blends combustion for industrial applications," *Fuel*, vol. 332, p. 126150, Jan. 2023, doi: 10.1016/j.fuel.2022.126150.

[4] Y. Wei, D. Li, J. Qiao, and X. Guo, "Recovery of spent SCR denitration catalyst: A review and recent advances," *J. Environ. Chem. Eng.*, vol. 11, no. 3, p. 110104, Jun. 2023, doi: 10.1016/j.jece.2023.110104.

[5] Y. Zhu *et al.*, "Combustion and emission characteristics for a marine low-speed diesel engine with high-pressure SCR system," *Environ. Sci. Pollut. R.*, vol. 27, no. 12, pp. 12851–12865, Apr. 2020, doi: 10.1007/s11356-019-04194-2.

[6] X. Zhou *et al.*, "Ammonia marine engine design for enhanced efficiency and reduced greenhouse gas emissions," *Nat. Commun.*, vol. 15, no. 1, p. 2110, Mar. 2024, doi: 10.1038/s41467-024-46452-z.

[7] T. Zhang, T. Zhang, Y. He, Y. Wang, and Y. Bi, "Corrosion and aging of organic aviation coatings: A review," *Chin. J. Aeronaut.*, vol. 36, no. 4, pp. 1–35, Apr. 2023, doi: 10.1016/j.cja.2022.12.003.

[8] W. Zhang *et al.*, "A fluorine-free bioinspired multifunctional slippery coating for ultra-long-term anticorrosion of mg alloy, static/dynamic anti-icing, antibacterial and antifouling," *Chem. Eng. J.*, vol. 500, p. 157516, Nov. 2024, doi: 10.1016/j.cej.2024.157516.

[9] X. Huang *et al.*, "Chemical vapor deposition of transparent superhydrophobic anti-icing coatings with tailored polymer nanoarray architecture," *Chem. Eng. J.*, vol. 454, p. 139981, Feb. 2023, doi: 10.1016/j.cej.2022.139981.

[10] H. Yin *et al.*, "A facile approach to fabricate robust superhydrophobic cu/TiO2 composite coating on stainless steel," *Vacuum*, vol. 195, p. 110699, Jan. 2022, doi: 10.1016/j.vacuum.2021.110699.

[11] S. Wang *et al.*, "Simple and scalable synthesis of super-repellent multilayer nanocomposite coating on mg alloy with mechanochemical robustness, high-temperature endurance and electric protection," *J. Magnes. Alloy.*, vol. 10, no. 9, pp. 2446–2459, Sep. 2022, doi: 10.1016/j.jma.2020.11.024.

[12] H. Y. Erbil and C. E. Cansoy, "Range of applicability of the wenzel and cassie−baxter equations for superhydrophobic surfaces," *Langmuir*, vol. 25, no. 24, pp. 14135–14145, Dec. 2009, doi: 10.1021/la902098a.

Journal of Physics: Conference Series **3009** (2025) 012013 doi:10.1088/1742-6596/3009/1/012013

Synthesis of hierarchical porous Na^+ incorporated MoS_2@NiO/C from Ni-MOF for advanced supercapacitor applications

Sheng Qiang Zheng [1,*], Poi Sim Khiew [2], Siew Shee Lim [2], and Chuan Yi Foo [3,4]

[1,*] School of Automobile and Traffic Engineering, Guangzhou City University of Technology, No. 1, Xuefu Road, Huadu, Guangzhou, China
[2] Center of Nanotechnology and Advanced Materials, Faculty of Science and Engineering, the University of Nottingham Malaysia, Jalan Broga, 43500, Semenyih, Selangor, Malaysia
[3] School of Energy and Chemical Engineering, Xiamen University Malaysia, Sepang 43900, Malaysia
[4] College of Chemistry and Chemical Engineering, Xiamen University, Xiamen 361005, China

*Corresponding author's e-mail: shengqiang.zheng@outlook.com

Abstract. Metal-organic frameworks (MOFs) derived highly porous transition metal oxide and carbonaceous nanosized composites have fascinated huge research interest in advanced supercapacitor applications due to their porous structures, outstanding theoretical capacitance and excellent electronic conductivity. Herein, we have synthesized nickel-based MOF (Ni-MOF)-derived hierarchical porous NiO/C nanostructures through a facile solvothermal-calcination strategy. Meanwhile, using a straightforward hydrothermal approach, the incorporation of Na^+ and layered MoS_2 nanostructures greatly boosted the structural porosity and electrochemical characteristics. Besides, the resulting hierarchical porous nanocomposites exhibited extraordinary electrochemical performance, notably achieving a marvelous capacitance of 2, 540.63 F g^{-1} in an aqueous electrolyte (2M KOH). Furthermore, the recorded outstanding energy density was 36.93 Wh kg^{-1} after the assembly of a symmetric supercapacitor cell. Impressively, the acquired coulombic efficiency and capacitance retention were 97.6% and 111.92 %, respectively over 4, 000 cycles. Hence, these findings explicitly demonstrated the superior electrochemical properties of nanostructured Na^+ incorporated MoS_2@NiO/C nanocomposites, and their potential for utilization as electrode active materials in realizing advanced supercapacitors.

1. Introduction

Energy shortage and environmental pollution have adversely affected human life due to the excessive consumption of nonrenewable energy sources. Meanwhile, the overconsumption of conventional energy, such as fossil fuels, results in serious emission of harmful gases like NO_x and SO_2, which seriously threaten both the environment and human well-being. Hence, it becomes imperative to develop clean and sustainable energy technologies. Electrical energy, a common energy carrier that has been widely used around the world, can be generated from sustainable energy sources like geothermal, wind, and solar energy. Nevertheless, due to the intermittent power output, more efficient and reliable electrochemical energy storage devices are required for practical applications. Moreover, to satisfy the high-power demands of fast charging gadgets, regenerative braking systems, and electrical vehicles, supercapacitors (SCs) are gaining prominence as one of the promising electrochemical energy storage technologies because of the unique features of ultrahigh-power density, long lifespan, and outstanding safe operation [1]. Nonetheless, the relatively low energy density has significantly limited their applications. As a result, considerable endeavors have been dedicated to the design and synthesis of advanced electrode materials, as they are pivotal in determining the overall electrochemical properties of SCs.

In the pursuit of developing advanced supercapacitor electrode active materials, numerous strategies have been implemented recently. Porous materials that possess large specific surface areas and outstanding electronic conductivity are focused. Metal-organic frameworks (MOFs), which possess a highly porous crystalline nature, have manifested significant promise for utilization as high-performing supercapacitors (SCs) electrode materials due to their attractive characteristics like ultra-large specific surface areas, tuneable structures, and high porosity. For instance, nanowire-shaped mesoporous Co-MOF with excellent electrochemical properties was prepared via a simple one-step solvothermal route [2]. Owing to enhanced conductivity and mechanical stability, its electrochemical capacitance could reach 1, 020 F g^{-1} at 5 mV s^{-1}, with a capacitance retention of 81.4% over 3, 000 cycles. Moreover, when a stretchable non-symmetric SC device was assembled by using activated carbon, the achieved highest energy density was 34.4 Wh kg^{-1} while the power density peaked at 7, 522.4 W kg^{-1}, respectively. Electrochemical performance could be enhanced through element doping. Li et al. [3] prepared nanosized molybdenum-doped Ni-MOFs with excellent conductivity through a hydrothermal approach. An appealing capacity of 802 C g^{-1} was obtained at 1 A g^{-1} and 96% of its initial capacity could be retained after 20, 000 GCD cycles in a 3M KOH electrolyte. Besides the direct utilization of pristine MOF as electrode materials for SCs, MOFs can also serve as templates for the synthesis of diverse transition metal oxides and carbonaceous materials. For instance, NiO/Ni/rGO with nanoclusters was prepared via a hydrothermal-calcination strategy [4], which possessed an elephantine surface area of 165.67 m^2 g^{-1}. By taking advantage of the synergistic effect of transition metal oxide and rGO nanosheets, at 3 A g^{-1}, the capacity value of 649.22 C g^{-1} was shown by the synthesized electrode material and 45.7% of the capacity could be held at 30 A g^{-1}. Additionally, the cyclic stability was assessed when the current density was as high as 20 A g^{-1}, and 81% of its original capacity was still there after 5, 000 cycles, showing excellent electrochemical stability. Despite the development and research of numerous MOF-based materials, they are afflicted by low electrical conductivity. To overcome this issue, metal oxides/carbon nanocomposites can be derived from MOFs. Besides, compositing with pseudocapacitance-rich matrixes, including transition metal oxides and dichalcogenides, is an alternative approach to enhance the energy density. Molybdenum disulfide (MoS_2), a typical 2D transition metal dichalcogenide, has been validated to be a prospective contender for advanced energy storage technologies based on the excellent electronic conductivity (10 ~ 100 S cm^{-1}) and theoretical capacitance (1, 504 F g^{-1}) [5, 6]. Sarkar et al. prepared 3D MoS_2/rGO nanoflakes on Mo foil through a hydrothermal route [7], which possessed a volumetric capacity of 343 F cm^{-3} at 10 mV s^{-1}, and 82% (281 F cm^{-3}) of the capacitance could be maintained as the sweep rate was elevated to 1, 000 mV s^{-1}. In addition to the remarkable rate performance, the hybrid nanocomposite could retain 98% of the capacitance over 20, 000 cycles, suggesting remarkable cyclic stability.

In the current study, porous Ni-MOFs were used as templates to derive hierarchical porous NiO and carbon (NiO/C) nanocomposites by calcinating at a mild temperature of 400°C. Subsequently, sodium cations along with layered MoS_2 were incorporated into the synthesized nanocomposites to form Na+-doped porous structures. When the nanocomposite was employed as an electrode active material, the acquired electrochemical capacitance was 2, 540.63 F g^{-1} at 0.5 A g^{-1} in an aqueous electrolyte (2 mol L^{-1} KOH). Moreover, when assembled as a symmetric supercapacitor device, an outstanding energy density of 36.93 Wh kg^{-1} was delivered. Additionally, it was noteworthy that the obtained coulombic efficiency and capacitance retention were 97.6% and 111.92% over 4, 000 cycles. These results confirmed that the as-synthesized hierarchical porous structures can be utilized as potential candidates for advanced electrochemical energy storage devices.

2. Materials and Method

All chemicals of analytical grade were procured from Sigma-Aldrich, Germany, and utilized as obtained without undergoing any purification processes. Deionized (D.I.) water with a resistivity of approximately 18.2 MΩ was employed throughout the entire study. Ni-MOFs were prepared through a solvothermal route as reported [8, 9]. In a typical synthesis, 0.181 mmol of H_2BDC and 1.08 mmol of $Ni(NO_3)_2 \cdot 6H_2O$ were dispersed in a mixture of ethylene glycol (EG) and N, N-Dimethylformamide (DMF) (v/v = 5:8). After being stirred for one hour, the produced homogeneous solution was transferred into a 50 mL autoclave. The reactor was then maintained at a temperature of 160°C for a duration of 12 h. The precipitates were then subjected to centrifugation and rinsed multiple times with DMF and ethanol. A light-green precipitate was acquired after drying it overnight at 60°C in a vacuum oven. Finally, the obtained powder was calcined in air at 400°C for 4 h.

The Na^+-incorporated MoS_2@NiO/C nanostructures were prepared via a hydrothermal approach. Typically, 0.150 g of the fabricated NiO/C was dissolved in 30 mL of D.I. water under magnetic stirring for half an hour. Then, thiourea (NH_2CSNH_2) and sodium molybdate (Na_2MoO_4, n/n = 2:1) were dissolved in the mixed solution and further stirred for an additional hour. The mixture was then transferred into an autoclave and heated at 180°C for 24 h. Thereafter, the mixture was rinsed with D.I. water several times and heated at 80°C for 24 h, and the final nanocomposite was obtained.

3. Results and Discussion

The XRD pattern of the synthesized nanostructure is displayed in Figure 1 (a). The typical reflections detected at 2θ = 37.45, 43.39, 63.03, 75.59, and 79.51° were reflectively associated with the (111), (200), (220), (311), and (222) lattice planes of NiO [10]. Moreover, the diffraction peaks obtained at 2θ values of 14.75, 32.19, 39.65, 49.25, and 72.27° belonged to (002), (010), (013), (015), and (023) plane of MoS_2. These diffractions matched well with the standard JCPDS cards of NiO (47-1049) and MoS_2 (37-1492), which annotated the successful integration of MoS_2 onto NiO. Besides, three diffraction peaks were noticed at 2θ values of 44.61, 52.05, and 76.57°, corresponding to the (111), (200), and (220) planes of nickel (75-1097) [11]. The presence of Nickle could be caused by the partial oxidation of Ni under low calcination temperatures.

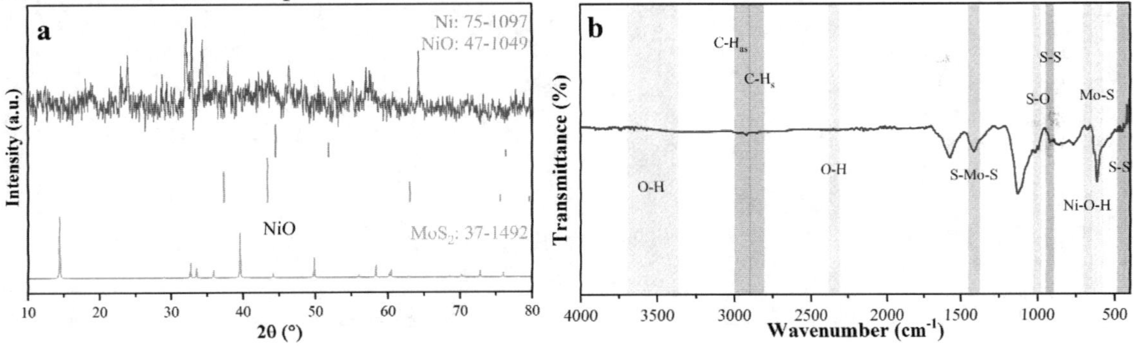

Figure 1. (a) XRD pattern; (b) FT-IR spectra of the prepared Na^+-doped nanostructures.

Meanwhile, **Figure 1 (b)** demonstrates the FT-IR spectrum of Ni-MOF-derived hierarchical porous NiO/C nanostructures. The adsorption peaks were manifested at 1, 662, 2, 356, and 3, 373-3, 660 cm^{-1}, originating from the physically adsorbed water moistures [10]. Two peaks that were observed within the range spanning from 412 to 416 cm^{-1} can be ascribed to the stretching motion of the Ni-O bond. Moreover, the symmetric and asymmetric vibrational bond stretching of C-H bonds was noticed at 2, 827-2, 881 cm^{-1} and 2, 920-2, 994 cm^{-1}, which could be owing to the presence of Ni-MOF residues resulting from incomplete transformation processes. Additionally, the stretching vibrations of S-O and S-Mo-S were witnessed at approximately 1, 113 and 1, 407 cm^{-1}, respectively [12]. The peaks that appeared at 472 and 921 cm^{-1} could be correlated with the vibrational stretching of the S-S bond [12]. Besides, the appearance of multiple peaks between 695 and 1, 200 cm^{-1} was on account of the vibrational bond stretching of sulfur groups. These results were in line with the results of XRD.

FESEM and TEM were used to evaluate the morphologies and structures of the prepared nanocomposites, respectively, as demonstrated in **Figure 2**. It could be observed that the as-synthesized nanostructure exhibited hollow porous structures composed of nanoparticles and nanobars. Such architecture could be favorable to offer electrochemical reactive sites for electrochemical reactions during the charge storage process. Meanwhile, the existence of a porous structure could provide more channels for ion diffusion. The elemental composition was displayed by EDX and elemental mapping (**Figure 2d** and **g**). The existence of Ni, O, and C annotated the fabrication of Ni-MOF-derived hierarchical porous NiO/C nanostructures. Additionally, the display of Na, Mo, and S indicated the successful incorporation of Na cations and MoS_2 onto NiO/C nanostructures. **Figures 2e** and **f** display the low-resolution TEM images of the as-prepared sample, and the existence of nanoparticles and nanobars could be observed, which was in good agreement with the result from SEM images. FFT analysis was studied in the picked areas. Amorphous structures were observed near the edges, and the uneven distributions of the lattice fringes implied the polycrystalline structures. On the other hand, the measurement revealed lattice spacings of 1.58 and 1.16 Å, which aligned with the (-321) plane of MoS_2 and the (-120) plane of NiO, respectively.

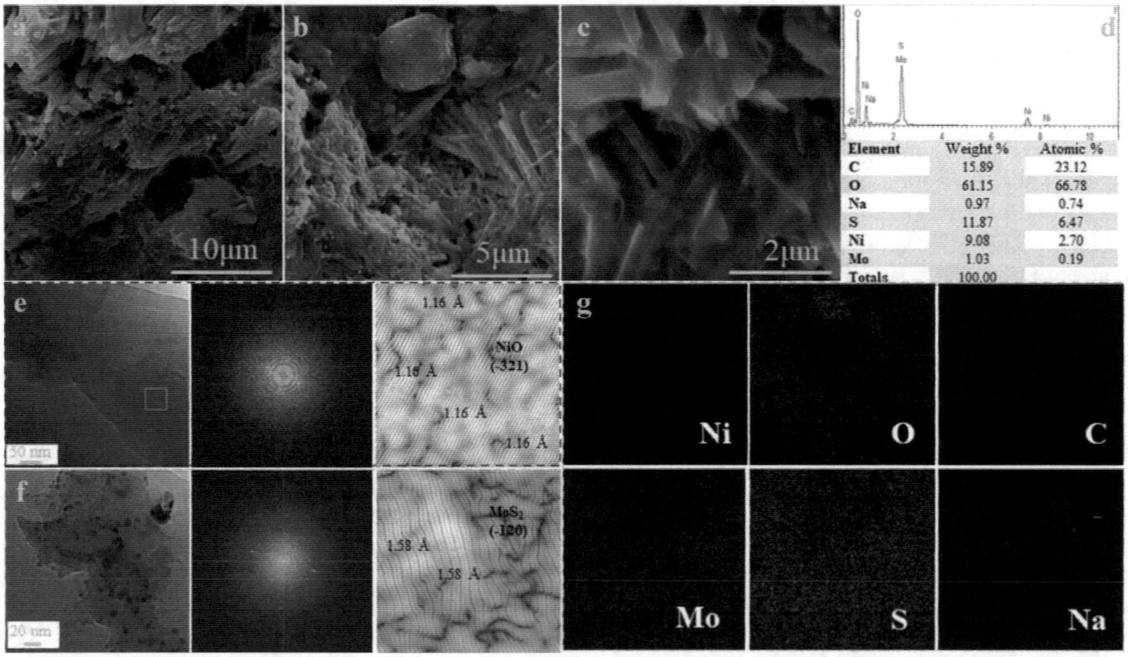

Figure 2. (a-c) SEM; (d) EDX images; (e, f) TEM images with fast Fourier transform and lattice fringes; (g) Corresponding elemental mapping.

For the evaluation of the electrochemical properties, a three-electrode configuration was adopted, utilizing KOH aqueous electrolyte (2 mol L^{-1}) as the electrolytic medium. **Figure 3** demonstrates the CV and GCD profiles. For all CV profiles evaluated at the potential window from 0 to 0.5 V (vs. Ag/AgCl), a pair of redox peaks were discernible, originating from the interactions between NiO and OH anions (Equation (1)). In particular, the anodic peak was caused by the oxidation of NiO to NiOOH, while the cathodic peak could result from the reduction of NiOOH to NiO [13]. Moreover, it should be emphasized that the interactions between MoS_2 and potassium cations (K^+) exerted a significant influence on enhancing the overall capacitance (Equation (2)). Upon increasing the rates of sweeping from 2 to 50 mV s^{-1}, a shift in the cathodic and anodic peak, along with an elevation in current densities, was observed. The transference of redox peaks was mainly owing to the existence of internal resistance (IR), and the augmented current densities were a result of the high mobility of cations (K^+) [14]. Nevertheless, the peak intensity tended to decrease. This can be ascribed to the insufficient duration provided where the electrolyte ions could not penetrate deep pores when a high sweeping rate was applied.

$$NiO + OH^- \leftrightarrow NiOOH + e^- \quad (1)$$

$$MoS_2 + K^+ + e^- \leftrightarrow MoS - SK \quad (2)$$

To comprehensively examine the electrochemical energy storage properties of the synthesized MOF-derived NiO/C nanocomposites, current-controlled charge/discharge analysis was meticulously carried out at a diverse range of current densities ranging from 0.5 to 3.0 A g^{-1}, as illustrated in **Figure 3b**. The asymmetric nature of the GCD curves indicated the capacitance was predominantly sourced from Faradaic redox reactions. Moreover, a slight voltage drop was viewed at each of the GCD profiles, showing the existence of internal resistance. The electrochemical capacitance of Na^+-doped MoS_2@NiO/C was 2, 540.63 F g^{-1} when measured at 0.5 A g^{-1}. Such exceptional electrochemical capacitance was comparable to the performance of most NiO-based nanostructures.

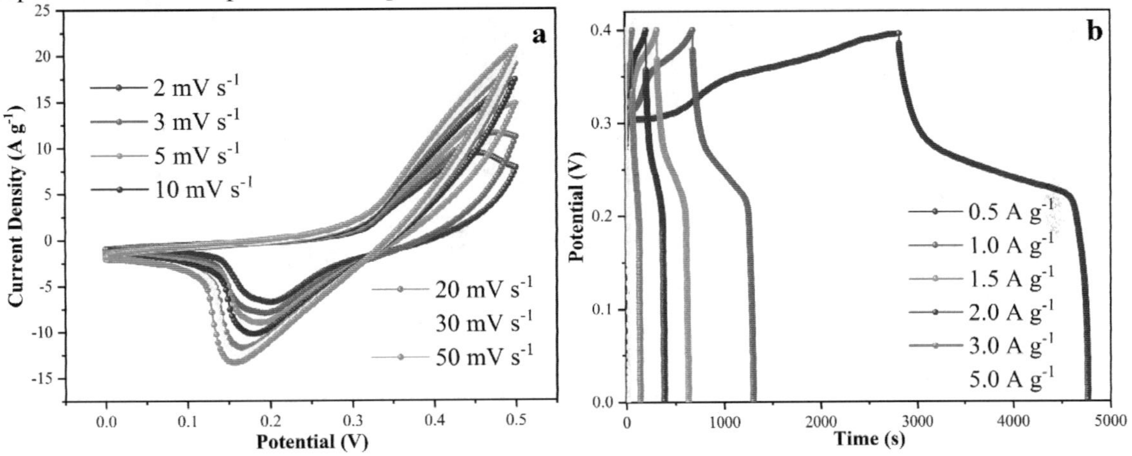

Figure 3. (a) CV; (b) GCD profiles of MoS_2 incorporated MOF-derived NiO/C nanostructures.

Besides, a supercapacitor cell with a symmetrical design was assembled to investigate its viability for real applications. CV and GCD were conducted in 2M Na_2SO_4 neutral aqueous electrolyte. **Figure 4a** depicts the CV curves at various scanning rates (50-500 mV s^{-1}). Interestingly, a pair of broad and weak peaks were observed for all CV profiles, which could be owing to the intercalation and de-intercalation of electrolyte ions. Meanwhile, all enclosed CV curves showed similar symmetrical shapes, and the loop areas increased with increasing sweeping rates, suggesting the excellent reversibility and rate performance of the Ni-MOF-derived hierarchical porous NiO/C electrodes.

To explore the potassium storage mechanism, CV at sweeping rates ranging between 5 and 500 mV s^{-1} was utilized to investigate the kinetic mechanism. **Figures 5a** and **b** present the diagrams of current density versus sweep rate (i vs. v) and the square root of sweep rate (i vs. $v^{0.5}$). The corresponding current densities vary directly with scan rates for the electric double-layer process, while for a pseudocapacitive process, current densities are proportional to the square root of sweeping rates [9]. The regression values (R^2) for the relationship between i and $v^{0.5}$ were significantly higher compared to those for i and v, suggesting the preliminary dominance of the diffusion-controlled process. To further prove it, the power-law method was evaluated according to Equation (3). It could be seen that the logarithm of current densities had a good linear relationship with the logarithm of the sweeping rates (**Figure 5c**). The value of b was found to be between 0.631 and 0.672, manifesting that the charge storage process was mainly limited by a diffusion-controlled process. To quantitatively examine the capacitance contributions from non-diffusion- and diffusion-limited processes, Equation (4) was applied, which could be rearranged into Equation (5).

$$i = av^b \quad (3)$$

$$i = k_1 v + k_2 v^{0.5} \quad (4)$$

$$\frac{i}{v^{0.5}} = k_1 v^{0.5} + k_2 \quad (5)$$

where $k_1 v$ and $k_2 v^{0.5}$ denote the current contributions from the electric double layer and diffusion-limited intercalation process. The fitted CV profiles suggested that the capacitance from the diffusion-controlled process was 88.18% at a low scanning rate of 50 mV s^{-1}, and 70.37% at a high sweeping rate of 500 mV s^{-1}. As the sweeping rates increased from 100, 200, to 300 mV s^{-1}, the energy capacity originating in the electric double layer process rose from 13.76 to 18.28 to 22.28%, respectively. This could be ascribed to the inadequate duration provided, preventing the electrolyte from fully penetrating all pores.

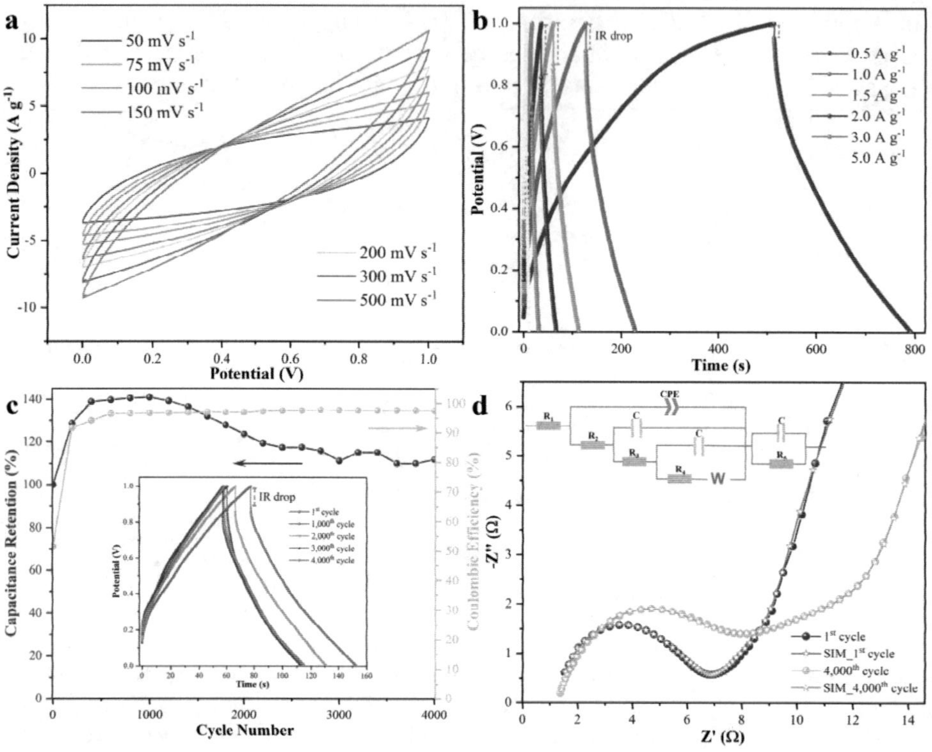

Figure 4. (a) Cyclic voltammograms; (b) GCD; (c) Cyclic stability; (d) Nyquist plot of the synthesized nanocomposites.

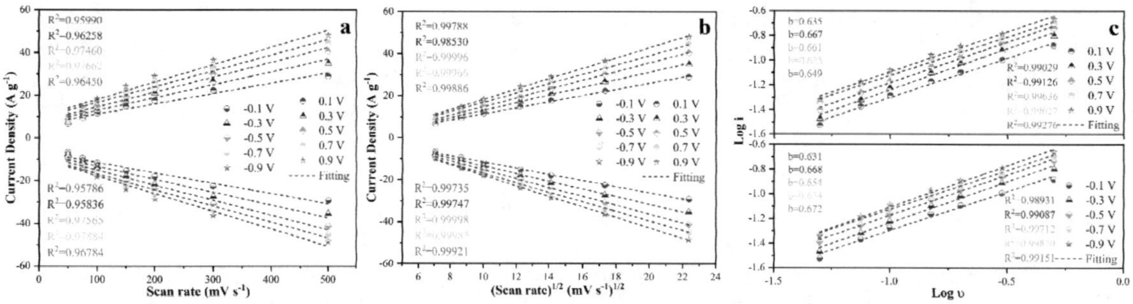

Figure 5. Plots of (a) i against v; (b) i against $v^{0.5}$; (c) Log (i) against log (v).

Figure 4b shows the GCD profiles at a variety of current densities. It could be seen the electrode active material exhibited almost symmetrical GCD profiles, indicating the fabricated SC cell possessed appealing electrochemical reversibility. Meanwhile, the yielded respective electrochemical capacitance was 292.14, 225.27, and 116.76 F g^{-1} at 0.5, 1.0, and 3.0 A g^{-1}. The gradual moderation in capacitive performance with an increasing current density was mainly caused by the sluggish movement of electrolytes. Notably, the recorded energy density was remarkable at 36.93 Wh kg^{-1} at a power density of 477 W kg^{-1}. When a high-power density of 2, 200 W kg^{-1} was applied, the retained energy density was 8.88 Wh kg^{-1}. Besides the excellent energy density, the assembled symmetric SC device also demonstrated superior recyclability. The delivered capacitance retention, after 4, 000 consecutive charge-discharge cycles, was 111.92% at a current density of 1.5 A g^{-1} (**Figure 4c**). In addition, over 4, 000 cycles, the acquired coulombic efficiency was 97.6%, highlighting the remarkable cycling stability of the as-fabricated MOF-derived hierarchical porous NiO/C electrode active materials. In addition, the Nyquist diagrams with fitting and equivalent circuits before and after performing cyclic tests were displayed in **Figure 4d**. The intercept (intrinsic resistance (R_s)) did not change significantly (from 1.11 to 1.19 Ω) after 4, 000 cycles. Moreover, the diameters of the semicircles (R_{ct}), which represent the charge transfer resistance, exhibited a modest increment from 5.08 to 6.97 Ω resulting from a minor loss of active materials from the electrodes. This observation corresponded to a decrease in coulombic efficiency.

Overall, the elevated electrochemical characteristics of Na-doped MoS_2@NiO/C hybrid structures could be primarily attributed to three reasons: (i) Porous NiO derived from Ni-MOFs and the incorporation of 2D MoS_2 exhibited high theoretical capacitance, which is capable of substantially augmenting the electrochemical capacitance and energy density; (ii) Carbonaceous materials and the addition of Na^+ could meliorate the overall electronic conductivity; (iii) The existence of a hollow porous structure could offer more channels for electrolyte ion diffusion and tolerate the volume variation of transition metal oxides, thereby enhancing cyclability.

4. Conclusion

In summary, hierarchical porous Na^+ and MoS_2 incorporated MOF-derived nickel oxides and carbonaceous nanostructures were fabricated via a facile hydrothermal-calcination strategy, which showed appealing electrochemical properties. The recorded electrochemical capacitance reached a high value of 2, 540.63 F g^{-1} at 0.5 A g^{-1}. Upon assembling a symmetric supercapacitor cell, an impressive specific energy density of 36.93 Wh kg^{-1} was achieved by using an aqueous electrolyte. Additionally, after 4, 000 cycles, the fabricated cell could retain 111.92% of its initial capacitance, and the acquired outstanding coulombic efficiency was 97.6% as well. All of these outcomes implied that the as-synthesized nanocomposites have the potential to serve as promising candidate materials for electrodes in the development of high-performing SCs. Meanwhile, this study also provides an originative strategy for architecture design for advanced electrochemical capacitor applications.

Acknowledgments

The research was financially supported by grants (FRGS/1/2019/STG02/UNIM/01/1 and FRGS/1/2023/STG07/UNIM/01/1) from the Ministry of Higher Education Malaysia.

References

[1] Zheng S. Q., et al. *Recent Progress on the Applications of Carbonaceous and Metal-Organic Frameworks Nanomaterials for Supercapacitors.* Frontiers in Materials, 2021: p. 500.
[2] Zhang H., et al. *Wire spherical-shaped Co-MOF electrode materials for high-performance all-solid-state flexible asymmetric supercapacitor device.* Journal of Alloys and Compounds, 2021. **879**: p. 160, 423.
[3] Li Q., et al. *Self-assembled Mo-doped Ni-MOF nanosheets-based electrode material for high-performance battery-supercapacitor hybrid device.* International Journal of Hydrogen Energy, 2020. **45** (41): p. 20, 820-20, 831.
[4] Gowdhaman A., et al. *Ni-MOF derived NiO/Ni/r-GO nanocomposite as a novel electrode material for high-performance asymmetric supercapacitors.* Journal of Energy Storage, 2023. **61**: p. 106, 769.
[5] Liang Z., et al. *Molecular Understanding of Charge Storage in MoS_2 Supercapacitors with Ionic Liquids.* ENERGY & ENVIRONMENTAL MATERIALS, 2021. **4** (4): p. 631-637.
[6] Bello I. T., et al. *Non-modulated synthesis of cobalt-doped MoS_2 for improved supercapacitor performance.* International Journal of Energy Research, 2022. **46** (7): p. 8, 908-8, 918.

[7] Sarkar D., et al. *Expanding Interlayer Spacing in MoS_2 for Realizing an Advanced Supercapacitor.* ACS Energy Letters, 2019. **4** (7): p. 1, 602-1, 609.
[8] Zheng S. Q., et al. *Fabrication of sodium and MoS_2 incorporated NiO and carbon nanostructures for advanced supercapacitor application.* Journal of Energy Storage, 2023. **63**: p. 106, 980.
[9] Qiang Zheng S. et al. *Probing the Effect of Solvents on the Electrochemical Performance of Graphene Incorporated Nickel-Based Metal-Organic Frameworks.* Journal of Electroanalytical Chemistry, 2022 (925): p. 116, 860.
[10] Dhas S. D., et al. *Synthesis of NiO nanoparticles for supercapacitor application as an efficient electrode material.* Vacuum, 2020. **181**: p. 109, 646.
[11] Mao H., et al. *Structure-activity relationship toward electrocatalytic nitrogen reduction of MoS_2 growing on polypyrrole/graphene oxide affected by pyridinium-type ionic liquids.* Chemical Engineering Journal, 2021. **425**: p. 131, 769.
[12] Lalithambika K. C., K. Shanmugapriya, and S. Sriram. *Photocatalytic activity of MoS_2 nanoparticles: an experimental and DFT analysis.* Applied Physics A, 2019. **125** (12): p. 817.
[13] Ahmed R. and G. Nabi. *Enhanced Electrochemical Performance of Cr-doped NiO Nanorods for Supercapacitor Application.* Journal of Energy Storage, 2021. **33**: p. 102, 115.
[14] Barzegar F., et al. *Investigation of different aqueous electrolytes on the electrochemical performance of activated carbon-based supercapacitors.* RSC Advances, 2015. **5** (130): p. 107, 482-107, 487.

doi:10.1088/1742-6596/3009/1/012014

Impact of water addition on the properties of calcium hexaaluminate-based insulating castables for steel rolling reheating furnaces

Yufeng Xia[1], Ru Ding[2], Cuijiao Ding[3], Wei Luo[3], Wenjie Yuan[1,4*]

[1] State Key Laboratory of Advanced Refractories, Wuhan University of Science and Technology, Wuhan 430081, China
[2] Hot Rolling Plant, Wuhan Iron and Steel Corporation Ltd., Wuhan 430083, China
[3] Energy and Environment Research Institute, Baosteel Central Research Institute, Wuhan 430080, China
[4] National-Provincial Joint Engineering Research Center of High Temperature Materials and Lining Technology, Wuhan University of Science and Technology, Wuhan 430081, China

*Corresponding author: yuanwenjie@wust.edu.cn

Abstract The quality of the casting construction directly depends on the amount of water added to the castables. To ensure the construction and service performance of calcium hexaaluminate (CA_6) based insulating castables in the water-cooled beams and posts of reheating furnaces, the effects of various water additions on the properties of CA_6-based insulating castables were evaluated. The results indicate that the flowability of CA_6 castables gradually increases with the increase in water addition, while the flowability is reduced by the incorporation of a thickening agent. The cold modulus of rupture (CMOR) of CA_6 castables decreases as the amount of water increases, whereas the cold crushing strength (CCS) exhibits a non-monotonic variation. The thickening agent can enhance the strength of samples with the same water addition. Overall, the optimal water addition for castables is 10.3%, and the flowability of T0, T30, and T60 is measured as 175%, 160%, and 155%, respectively. CCS of precast samples after heat treatment at 1300°C reaches 100.23 MPa, and thermal conductivity at 1000°C is as low as 0.559 W/m·K.

1. Introduction

"Energy conservation" represents the primary battleground for future corporate competition in the steel industry, with steel rolling reheating furnaces being among the most energy-intensive processes in steel plants. Consequently, requirements for castables used in steel rolling reheating furnaces are becoming increasingly stringent [1]. Castables with high strength and low thermal conductivity are expected to be applied in water-cooled beams and posts [2]. Calcium hexaaluminate ($CaA_{12}O_{19}$, CA_6), the compound with the highest Al_2O_3 content in the CaO-Al_2O_3 system, exhibits low wettability to molten metals, excellent volume stability at high temperatures, and strong alkali corrosion resistance. Additionally, it can significantly enhance the toughness and thermal shock resistance of castables [3-7]. Therefore, potential applications of CA_6 are highly concerned in high-temperature industries. CA_6-based castables composed of aggregates, fines, and fine alumina powders as primary were formulated for the permanent lining of ladles [8]. It is demonstrated that CA_6-based castables can offer superior thermal insulation compared to high-alumina castables.

Due to the short construction duration and the significant workload attributes of steel rolling heating furnaces, quality control in construction is even more imperative. The excellent flowability and setting properties of castables are closely related to the appropriate amount of water addition [9]. Excessive water can cause the separation of aggregates and binders in castables, compromising their performance and reducing their service life. Conversely, insufficient water can degrade the workability, resulting in practical construction difficulties and issues such as inadequate later-stage strength. A significant number of investigations have documented the impact of water addition on various types of low-cement castables [9-10], but the correlation between water addition and the properties of CA_6-based insulating castables has yet to be elucidated.

The water content is a multifaceted parameter that is predominantly affected by various factors such as the type of raw materials, the particle size distribution, as well as the type and amount of cement and water-reducing agents [11-13]. Due to the lack of a calculation formula or standard for determining water addition, multiple testing and optimization during the experiments are required. According to relevant works of literature on Al_2O_3-CaO castables and construction practice experience, the water content of castables generally ranges from 4% to 10% when CA_6 is selected as the primary raw material [8,14-15].

In order to achieve the performance of designed castables in practical application, the impact of water addition on the properties of CA_6-based insulating castables should be comprehensively evaluated. In this work, the optimal water amount for CA_6-based insulating castables has been determined by comparing the properties including flowability, permanent linear change, apparent porosity, strength, and thermal conductivity.

2. Experiments

2.1 Raw materials

The primary raw materials for the experiments include CA_6 coarse aggregates (< 5 mm) and powers (≤0.074 mm, Shengchuan, China), alumina powders (AMA-40 and AMA-10, Smile, China), calcium aluminate cement (CAC, Secar71, Imerys, France), dispersant (WSM-M, Smile, China), and hydroxyethyl methyl cellulose (WALOCELT MT400PFV, Dow, America) as thickening agent.

2.2 Calculation of water addition

Considering no clear regulations about water addition in the standard for self-flowing castables (YB/T 4197-2009), the specification for mix proportion design of ordinary concrete (JGJ 55-2011) was selected as a reference for calculations in this study. The theoretical range of water addition was determined to be between 8.05% and 10.35%. It was identified that casting could only be carried out when the water addition reached 9.3% or above by preliminary experiments. Due to the interference of human and environmental factors that interfere during the construction process, the most suitable addition of water shall be determined by general comparison.

2.3 Sample preparation

Samples were prepared according to the standard (YB/T 5202.1-2003). All raw materials were weighed according to the formulation shown in Table 1 and then underwent pre-mixing. They were subsequently placed in a cement mortar mixer and dry-mixed for 1 minute. Deionized water was gradually added within 1 minute, followed by mixing for another 1 minute. The mixing was paused briefly, and after cleaning the mixing blades and the bucket wall, the mixture was continuously stirred for an additional 2 minutes. The stirred mixture was then poured into molds of 25×25×150 mm and ϕ180×20 mm for self-flowing molding. Samples were maintained in an environment with a temperature of 25°C and humidity of 80% for 24 hours, followed by drying at 110°C for 24 hours. The dried samples were heat-treated at 1000°C and 1300°C for 3 hours, respectively.

Table 1 Formulation of CA_6-based castables (wt.%).

Raw materials	Samples				
	A	B	C	D	D1
CA_6 (<5mm)	78	78	78	78	78
Calcined alumina (AMA40,D_{50}=5μm)	8	8	8	8	8

Reactive alumina (AMA-10,D_{50}=1.5μm)	7	7	7	7	7
CAC (Secar71)	7	7	7	7	7
Dispersant (WSM-M)	0.2	0.2	0.2	0.2	0.2
Thickening agent (MT400PFV)	0	0	0	0	0.01
Water addition	9.3	9.8	10.3	10.8	10.8

2.4 Test
The flowability of the castables was measured according to GB/T 4513.4-2017, with flow values at 0 min, 30 min, and 60 min of mixing recorded as T0, T30, and T60, respectively. Permanent linear change (PLC) was calculated by the variation of the length according to GB/T 5988-2007. The CCS of samples after treatment at different temperatures was tested according to GB/T 5072-2008. The CMOR was measured using the three-point bending method specified in GB/T 3001-2017. The apparent porosity and bulk density of the post-fired samples were measured according to GB/T 2997-2015. The thermal conductivity of the samples was measured using a plate thermal instrument (PBDR-03P, Precondar, Luoyang, China) according to YB/T 4130-2005.

3. Results

3.1 Flowability
Figure 1 shows the flowability of CA_6-based castables with varying amounts of water addition. T0, T30, and T60 represent the test results of the samples immediately after mixing, and after being sealed and allowed to stand for 30 minutes and 60 minutes, respectively. It can be observed that the flowability of the samples increases with the increase in the amount of water added. The flowability (T60) of sample B decreases the most, by 40%. For samples C and D, the flowability substantially reaches maximum value, with minimal changes in the flow value over time. When the water addition is 10.8%, the incorporation of the thickening agent reduces the flowability (T0) and ensures that the flowability (T30) remains almost unchanged compared to T0. Concurrently, it substantially reduced the stratification phenomenon, which was characterized by a distinct separation between the slurry and aggregate that occurred when the mixed sample was permitted to remain undisturbed. This phenomenon was exacerbated by the relatively high volume of water added.

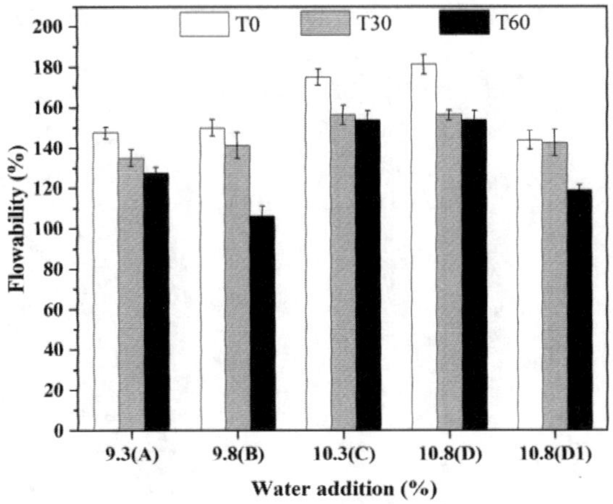

Figure 1 Flowability of CA_6-based castables with different water additions.

3.2 General performance
Figure 2 shows the PLC of CA_6-based castables with different amounts of water addition after heat treatment. Post-fired castables with varying water additions exhibit a trend of shrinkage at 1000 °C and expansion at 1300 °C. The sintering among the components at high temperatures leads to volumetric shrinkage. The reaction between the corundum and calcium aluminate cement in castables

results in the formation of CA_6 at 1300°C accompanied by a volumetric expansion of 3.01% [16]. Notably, sample B also exhibits a very slight shrinkage at 1300°C. Sample D1 with the thickening agent addition of 0.01% presents greater expansion after being fired at 1300°C. In all samples, sample C with a water addition of 10.3% exhibits the lowest volumetric shrinkage of 0.15% after treatment at 1000°C and the highest volumetric expansion of 0.36% after treatment at 1300°C.

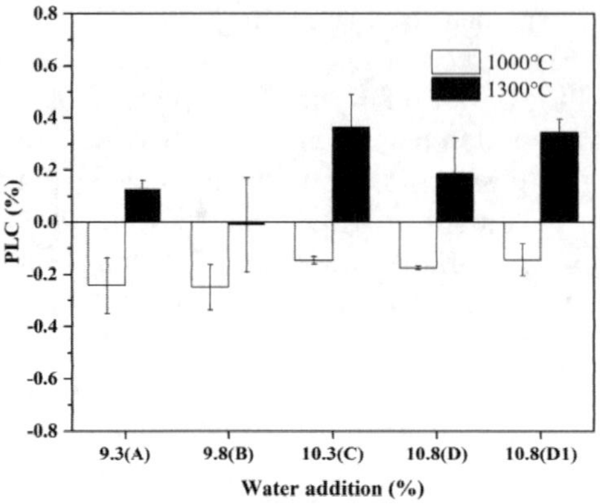

Figure 2 Permanent linear change of CA_6-based castables with different water additions.

The apparent porosity and bulk density of CA_6-based castables with different water additions are shown in Figure 3. As shown in Figure 3(a), the apparent porosity of all samples dried at 110°C is lower than that after treatment at 1000°C and 1300°C, which is consistent with the changes in their bulk density (seen in Figure 3(b)). High water addition leads to an increased amount of water filling the voids between the aggregates, and upon evaporation of this water, a significant number of pores are left behind. After drying at 110°C, sample D with a water addition of 10.8% exhibits the highest apparent porosity (24.6%). However, the apparent porosity of sample D1 with the addition of 0.01% of a thickening agent is lower than that of sample D. The incorporation of the thickening agent induces the formation of closed pores, which consequently results in a reduction of the apparent porosity of castables. After treatment at 1000°C and 1300°C for 3 hours, sample D1 has the highest apparent porosity (28.84% and 28.88%, respectively), indicating that the addition of the thickening agent may enhance the apparent porosity of castables after firing at high temperatures.

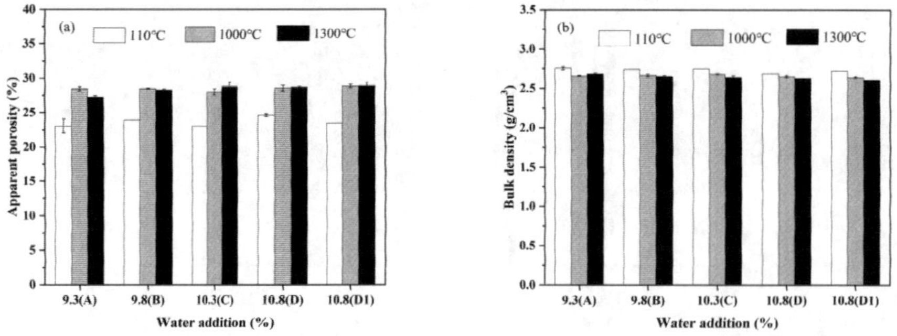

Figure 3 Apparent porosity (a) and bulk density (b) of CA_6-based castables with different water additions.

3.3 Strength

Figure 4 shows the strength of CA_6-based castables with different water additions. As shown in Figure 4(a), the CMOR of samples after drying at 110 °C first increases and then decreases with increasing water addition, and the strength increases after the addition of the thickening agent. The CMOR of CA_6-based castables after treatment at 1300°C decreases with increasing water addition, while the

CMOR of sample D1 with the addition of 0.01% thickening agent is higher than that of samples without the thickening agent. Although sample B exhibits the highest CMOR and CCS (9.77 MPa and 41.3 MPa, respectively), the strength presents a relatively lower value after treatment at 1000°C (8.45 MPa and 56.3 MPa, respectively). Sample D has the lowest CMOR and CCS compared to other samples regardless of the treatment temperature. From Figure 4(b), it can be seen that the CCS of CA_6-based castables after treatment at 1000°C and 1300°C exhibits a non-monotonic trend. The CCS of sample D1, after being subjected to various temperature treatments, ranks second among its peers. Furthermore, the addition of the thickening agent can also contribute to enhancing the strength of castables with the same amount of water added. A thickening agent may be incorporated in instances where there is an excess of water present during the construction process on-site. It is noteworthy that generally, a higher apparent porosity implies lower strength [17]. Differences in pore size and structure within the samples may result in samples with high apparent porosity and high strength simultaneously.

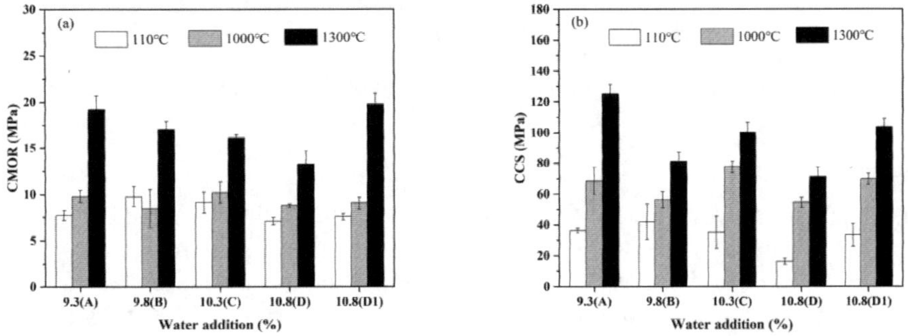

Figure 4 Cold modulus of rupture (a) and cold crush strength (b) of CA_6-based castables with different water additions.

3.4 Thermal conductivity

The variations in thermal conductivity of CA_6-based castables with different water additions are shown in Figure 5. Both changes in water addition and the thickening agent have significant effects on the thermal conductivity of castables. The impact of the thickening agent is more prominent. Except for sample A, the thermal conductivity of castables decreases as the temperature increases. The thermal conductivity at 1000°C of each sample is ranked in ascending order of C, D, A, D1, and B. At the same temperature, the apparent porosity of the castables in each group is ranked in descending order of C, A, B, D, and D1. In contrast, sample C exhibits the most superior thermal insulation performance in high-temperature service environments (with a thermal conductivity of 0.559 W/m·K at 1000°C) and has the smallest variation in thermal conductivity at different temperatures. Conversely, sample B has the greatest thermal conductivity tested above 800°C.

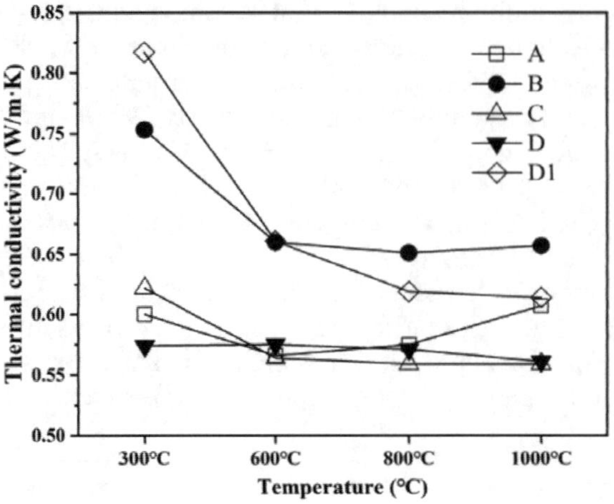

Figure 5 Thermal conductivity of CA_6-based castables with different water additions.

4. Conclusions

The variations of properties for CA_6-based castables with water additions were evaluated. The results can be summarized as follows: The flowability of CA_6-based castables increases continuously with the increase in water addition. The CMOR of CA_6-based castables decreases as the water addition increases, while CCS exhibits a non-monotonic trend. The addition of a thickening agent can enhance the strength of castables with the same water addition. The thermal conductivity of CA_6-based castables varies non-monotonically with water addition. In general, more water addition correlates with a greater formation of pores, resulting in a reduction of thermal conductivity. Considering all factors, CA_6-based castables with a water addition of 10.3% are most suitable for practical construction. Their flowability at T0, T30, and T60 are 175%, 160%, and 155% respectively. CCS after drying at 110 °C, and firing at 1000 °C and 1300 °C is 35.27 MPa, 77.63 MPa, and 100.23 MPa, respectively. The lowest thermal conductivity of 0.559 W/m·K at 1000°C can be achieved.

References

[1] Liu S, Zhao J X and Sun Z Y. (2023) Analysis of existing energy-saving technologies for heating furnaces in steel plants from the perspective of refractory materials. *Metallurgical management*, 4: 101-103.
[2] Xu P K. (2023) Development history and technological progress of refractory materials for rolling steel heating furnaces. Industrial Furnace, 45(1): 17-21. DOI:10.3969/j.issn.1001-6988.2023.01.005
[3] Hallstedt B. (1990) Assessment of the CaO-Al_2O_3 system. *Journal of the American Ceramic Society*, **73(1)**: 15-23. DOI: 10.1111/j.1151-2916.1990.tb05083.x.
[4] Luz A P, Consoni L B, Pagliosa C, et al. (2018) Sintering effect of calcium carbonate in high-alumina refractory castables. *Ceramics International*, **44(9)**: 10486-10497. DOI: 10.1016/j.ceramint.2018.03.066.
[5] Qu J, Ding D, Xiao G, et al. (2021) Preparation of Al_2O_3@$CaCO_3$ aggregates and its effects on the thermal shock resistance of Al_2O_3-MgO castables. *International Journal of Applied Ceramic Technology*, **18(4)**: 1379- 1391. DOI: 10.1111/ijac.13745.
[6] Schnabel M, Buhr A, Büchel G, et al. (2011) Advantages of calcium hexaluminate in a corrosive environment. *Refractories Worldforum*, **3**:87-94.
[7] Vázquez B A, Pena P, de Aza A H, et al. (2008) Corrosion mechanism of polycrystalline corundun and calcium hexaluminate by calcium silicate slags. *Journal of the European Ceramic Society*, **29(8)**: 1347-1360. DOI:10.1016/j.jeurceramsoc.2008.08.031.
[8] Xu G T, Wang Z Q, Qin S M, Liu L, Zhang Y W and Zhou W Z. (2024) Development and application of new generation calcium hexaaluminate castables. *Naihuo Cailiao*, **58(1)**:53-56. DOI: 41.1136.TF.20231102.1519.002.
[9] Cai Z E, Jing X J and Li Z B. (2019) Effect of water addition on construction performance of castable for rolling heating furance. *Shangdong Metallurgy*, **41(1)**:29-31+34. DOI:10.16727/j.cnki.issn1004-4620.2019.01.013.
[10] Guo H X, Li X Y and Li L Z. (2011) Influence of water quantity on low cement castable performance. *Industrial Furnace*, **33(5)**:52-54.

[11] Wang Y L, Zhang L J and Chen Y Q. (2006) Development and application of self-flowing castable refractory materials for the bottom of large steel ladles. *In: The 12th Academic Exchange Conference on Refractory Materials Among Hebei, Shandong, Henan, Shanxi, and Beijing Provinces (Cities) in 2006*. Chengde, 2:62-63.
[12] Ma X J. (2006) Development and research of corundum-based self-flowing castable refractory materials. *In: The 12th Academic Exchange Conference on Refractory Materials Among Hebei, Shandong, Henan, Shanxi, and Beijing Provinces (Cities) in 2006*. Chengde, 3: 50-52.
[13] Wang J T, Li Z B, Yu B B, Ren J X and Sun X Y. (2023) Development of explosion-proff quick-drying self-flowing castables for water-cooled beam wrapping. *Shangdong Metallurgy*, **45(6)**:54-56. DOI:10.16727/j.cnki.issn1004-4620.2023.06.012.
[14] Chen Y C, Pan L P, Li Y W, Liao N and Dai Y J. (2022) Effect of calcium hexaluminate aggregate content on mechanical properties and fracture behavior of corundum-calcium hexaluminate castables. *Journal of the Chinese Ceramic Society*,**50(12)**:3293-3304. DOI:10.14062/j.issn.0454-5648.20220480.
[15] Xu Y C, Zhang S H, Qin H B, Xu Y S and Huan Z L. (2022) K_2CO_3 corrosion resistance of CA_6, MA or β-Al_2O_3 castables. *Naihuo Cailiao*, **56(5)**:412-415. DOI:10.3969/j.issn.1001-1935.2022.05.009.
[16] Braulio M A L, Milanez D H, Sako E Y, et al. (2007) Expansion behavior of cement-bonded alumina-magnesia refractory castables. *American Ceramic Society Bulletin*, **86(12)**: 9201-9206.
[17] Xia Y F, Jia Z G, Jiang Y C, Yuan W J. (2023) Effects of rare earth oxides on in-situ reaction and properties of alumina-magnesia castables. Modern Chemical Research, **22**: 32-34. DOI:10.20087/j.cnki.1672-8114.2023.22.010.

Journal of Physics: Conference Series **3009** (2025) 012015 doi:10.1088/1742-6596/3009/1/012015

Preparation and anti-icing performance of fluorosilicate superhydrophobic anti-icing coating

Yueju Zhao[1*], Huibo Ma[1], Jinbiao Shi[1], Xiaoyan Yang[1], Dahang Zha[1], Xianzhong Tan[1], Qiaowei Hu[1], Xiwei Xie[1], Xinhua Wang[1], Yuanzhu Cai[1], Xiaodong Gong[1], Rui Liu[1], Weimin Peng[1]

[1] Beijing Guodian Futong Science and Technology Development Co., LTD, PR China.

*Corresponding author's e-mail: 18201669737@163.com

Abstract. In response to the issues of complex and cumbersome preparation and coating processes for various superhydrophobic materials, as well as insufficient superhydrophobic and anti-icing performance, we have developed a superhydrophobic anti-icing coating based on fluorosilicone rubber. By using hydroxyl silicone oil and silicone oil as base materials, and incorporating nano powders, fillers, and modified additives, we have developed a fluorosilicone superhydrophobic anti-icing coating that cures at room temperature. The performance of this coating has been characterized, and the results show that it has good storage stability, with a contact angle of up to 150° and a rolling angle of less than 10°. In terms of anti-icing performance, the ice adhesion strength on the coated surface is only 40-80 kPa. Regarding durability, the coating still possesses good superhydrophobic capabilities and anti-icing performance after being subjected to a period of wet and cold environment spraying. In simulated icing tests, the delay in ice formation on the coated wire surface can reach 30-40 minutes. This coating is an excellent choice for anti-icing coatings.

1. Introduction

Electric power safety is the foundation of energy security. The threats posed to electric power safety by disasters caused by the variable climate environment, particularly ice accretion disasters, are extremely severe [1]. China's transmission lines are the longest in the world in terms of length and scale, and the highest in voltage level. A significant portion of these transmission lines pass through environments with low temperatures and high humidity, mainly concentrated in the central and southern regions of the country, where ice accretion disasters are highly likely to occur. This can lead to problems such as transmission line breakage and tower collapse, resulting in large-scale power outages or even system collapse, threatening the safe and stable operation of the power grid [2]. Furthermore, with the rise of the new energy industry, the development of wind power is also expanding. These regions rich in wind energy resources often come with cold and humid climates that are prone to ice accretion, which increases the demand for anti-icing technologies on wind turbine blades [3].

At present, the mainstream anti-icing technologies are divided into active de-icing and passive anti-icing. Active de-icing includes mechanical means for ice breaking and electric power melting ice, which suffer from poor timeliness and harsh construction environments. Passive anti-icing mainly relies on anti-icing materials such as superhydrophobic materials, photothermal materials, and low surface energy materials. These materials are applied to the surfaces of power equipment to reduce ice accretion and delay freezing. The advantages lie in the ability to pre-treat and prevent ice accretion before it occurs, and the simplicity of construction methods, making them the focus of attention in various research institutes in recent years [4-6]. Despite this, research in this area is still relatively fundamental. There are many challenges to overcome before it can fully address the icing problems of outdoor equipment, such as the durability of material surfaces, maintaining the anti-icing performance of coatings in low-temperature environments, and the need for multiple components and multiple spray applications for the coating.

Fluorosilicone rubber is an organo-inorganic hybrid polymer based on Si-F and Si-C fundamental units. Due to its direct applicability and use in combination with additives, fluorosilicone rubber coatings possess excellent heat resistance, cold resistance, high voltage endurance, resistance to acids, alkalis, and oils, as well as resistance to weathering [7]. At the same time, the introduction of fluorine groups is more conducive to providing the low surface energy properties required for superhydrophobic coating surfaces. In recent years, the research on preparing anti-icing coatings using matrix materials with fluorine groups has become a hot topic and has been applied in the fields of power transmission and transformation, wind power, and aerospace [8-11].

This study presents the preparation of a fluorosilicone superhydrophobic anti-icing coating using hydroxyl fluorosilicone oil and silicone oil as the base materials through a relatively simple preparation process. The coating exhibits excellent superhydrophobic properties and outstanding durability, while also providing effective anti-icing performance. Compared to traditional superhydrophobic anti-icing materials, this coating is simple to prepare and apply, and it offers good anti-icing effects.

2. Experimental Methods

2.1. Raw Materials and Equipment

Raw materials and equipment are listed in Table 1.

Table 1. Raw materials and Equipment.

Raw materials	Equipment
18000 viscosity hydroxyl fluorosilicone oil	Triple-roll mill
Nano aerosilnanoscale fumed silica	Disperser
Nanoscale aluminum hydroxide	Contact angle meter
Ethyl acetate	Viscometer
Methyltri-n-butylketoxime silane	Temperature and humidity chamber
Tetramethyl orthosilicate	Ice adhesion strength testing platform
Dibutyl tin dilaurate	Simulated icing testing platform
Coupling agent	Scanning electron microscope
Tetrahydrofuran	-
Deionized water	-
Anhydrous ethanol	-
Hexamethyldisilazane	-

2.2. Material Preparation

According to the ratios provided in Table 2, prepare Components A and B as follows:

Table 2. Coating Raw Material Ratios.

Component A		Component B	
Material	Proportional Mass/g	Material	Proportional Mass/g
Nano Aerosil	10	Nano Aerosil	10
Nanoscale Aluminum Hydroxide	8	KH-570	1
Hydroxyl Fluorosilicone Oil	40	Hexamethyldisilazane	1
Hydroxyl Silicone Oil	10	Tetrahydrofuran	40

Hexamethyldisilazane	1	Ethanol	40
Tetramethyl Silane	1		
KH-550/KH560	1/1		
Ethyl Acetate	40		

Here are the specific steps for preparing the coating based on the provided material ratios: First, nano powder is preliminarily mixed with hydroxyl fluorosilicone oil and hydroxyl silicone oil, then subjected to vacuum kneading, dispersion, and grinding to obtain the base adhesive; the base adhesive is mixed and dispersed with solvent ethyl acetate and additives, and stirred to prepare Component A; nano aerosol is dispersed in a mixture of tetrahydrofuran and ethanol, and hydrophobic modifying agents are added for hydrophobic modification, after which Component B is obtained after uniform dispersion; finally, Component A is mixed and dispersed with Component B, and a curing agent and catalyst are added to obtain the final superhydrophobic anti-icing coating.

2.3. Sample Preparation

The aluminum plate of 150 mm × 70 mm × 0.8 mm and the steel-core bare stranded wire are cleaned with cotton dipped in alcohol, ensuring the surface is free of contaminants. After drying the aluminum plate and wire, the superhydrophobic anti-icing coating is applied on the surface of the steel plate with a brush, resulting in several superhydrophobic anti-icing coating samples and anti-icing wires. Then, they are allowed to dry for 3 days before conducting related tests. At the same time, the cleaned aluminum plate samples and steel-core aluminum stranded wire are used as the control group.

2.4. Testing and Characterization

2.4.1 Superhydrophobic Coating Performance Test. A contact angle meter is used to record the contact angle on the coating surface. With the injection needle on the contact angle meter, 5 μL of deionized water is injected onto the sample surface at each instance, and the contact angle values at five different positions on the same sample surface are recorded. The average of these five position values is taken as the test result for the water contact angle on the superhydrophobic sample surface.

2.4.2 Ice Adhesion Strength Test. The ice adhesion strength test is conducted using a self-built ice adhesion strength testing platform. The test results of the ice adhesion strength at five different positions on the same sample surface are recorded, and the average value of these five positions is taken as the test result for the ice adhesion strength of the superhydrophobic sample. The ice adhesion strength testing platform consists of a low-temperature chamber, icing mold, fixed stage, force measuring device, sliding rail, and control system, as shown in the structural schematic in Figure 1.

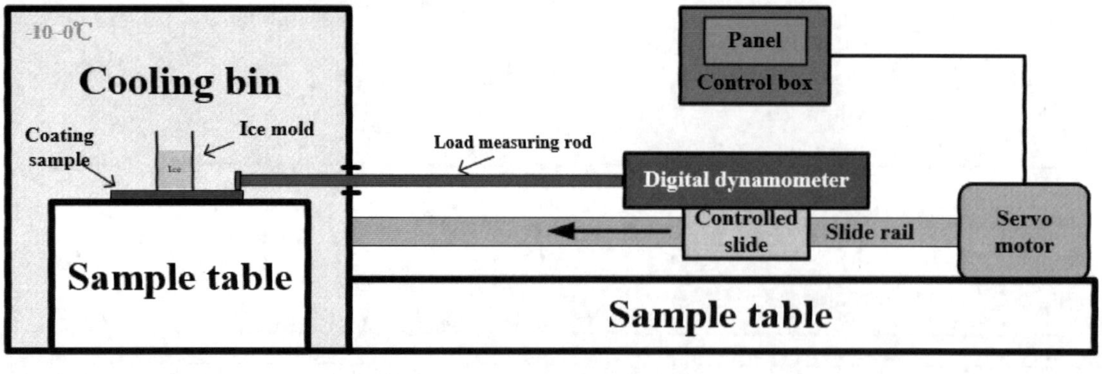

Figure 1. Schematic Diagram of Ice Adhesion Strength Test Platform.

2.4.3 Coating Sample Surface Morphology Analysis. The surface of the cured coating is observed using SEM (Scanning Electron Microscope), and through the observation of the microstructure, a comprehensive analysis is conducted in combination with other properties.

2.4.4 Coating Viscosity and Storage Property Test. The viscometer is used to test the viscosity of the prepared superhydrophobic anti-icing coating. The viscosity test results of the same batch of samples are recorded at different times: immediately after preparation, 7 days after sealed storage, 14 days after sealed storage, 30 days after sealed storage, and 60 days after sealed storage. For each test, the viscosity of 5 different samples is measured.

2.4.5 Effect of Wet and Cold Environment on Coating Performance. To compare the differences in coating performance before and after the influence of low-temperature rainwater, the following tests were conducted. The prepared coating samples were placed in the simulated icing platform to be continuously sprayed in a wet and cold environment, to compare the performance differences between the initial coating samples and the coating samples after different durations of spraying. The spraying conditions are as follows

- Ambient temperature: 0°C;
- Ambient humidity: 98%RH;
- Peristaltic pump water flow rate: 120 ml/min;
- Temperature-controlled cooling bath water temperature: 3.5°C;
- Wind speed: 2 m/s;
- Spraying time: 30 min; 60 min; 90 min; 120 min.

2.4.6 Simulated Icing Test. The test is conducted using a self-built simulated icing test platform. The test results of the simulated icing for superhydrophobic anti-icing wires and ordinary wires are recorded simultaneously. The simulated icing test platform consists of a refrigerated temperature-controlled tank, a peristaltic pump, a humidifier, a blower, an environmental chamber, and a spray system, as shown in Figure 2. The refrigerated temperature-controlled tank is responsible for providing cooled spray water; the peristaltic pump is responsible for transporting the cooled water to the environmental chamber for spraying through a spray nozzle; the humidifier is responsible for providing humidity; the blower is responsible for providing cold air in a low-temperature environment; and the environmental chamber is responsible for providing the low-temperature environment for icing.

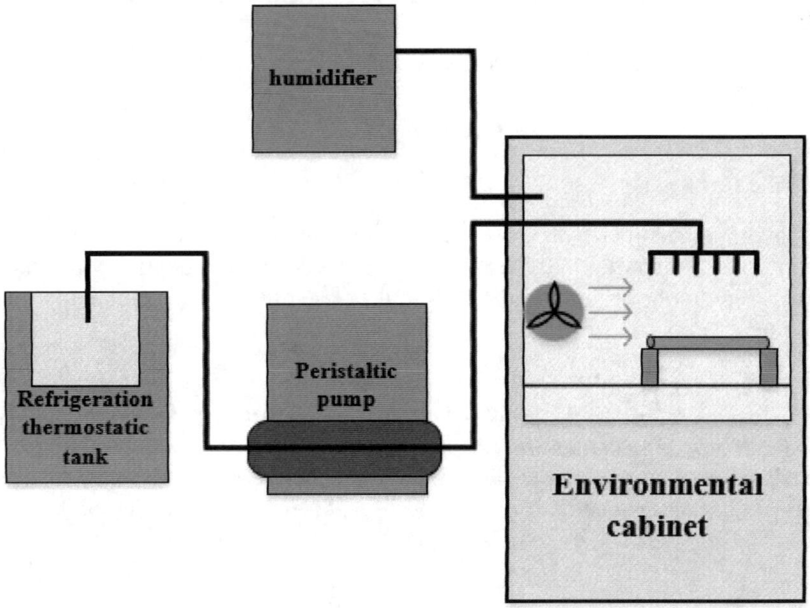

Figure 2. Schematic Diagram of Ice Accretion Simulation Test Platform.

The parameters set for the icing process are as follows
- Ambient temperature: -7°C;
- Ambient humidity: 90%RH;
- Peristaltic pump water flow rate: 120 ml/min;
- Temperature-controlled cooling bath water temperature: 3.5°C;
- Wind speed: 2 m/s.

3. Results and Discussion

3.1. Analysis of Superhydrophobic Results of Coatings

The anti-icing coating was tested for contact angle and roll angle, and the test results are shown in Figure 3.

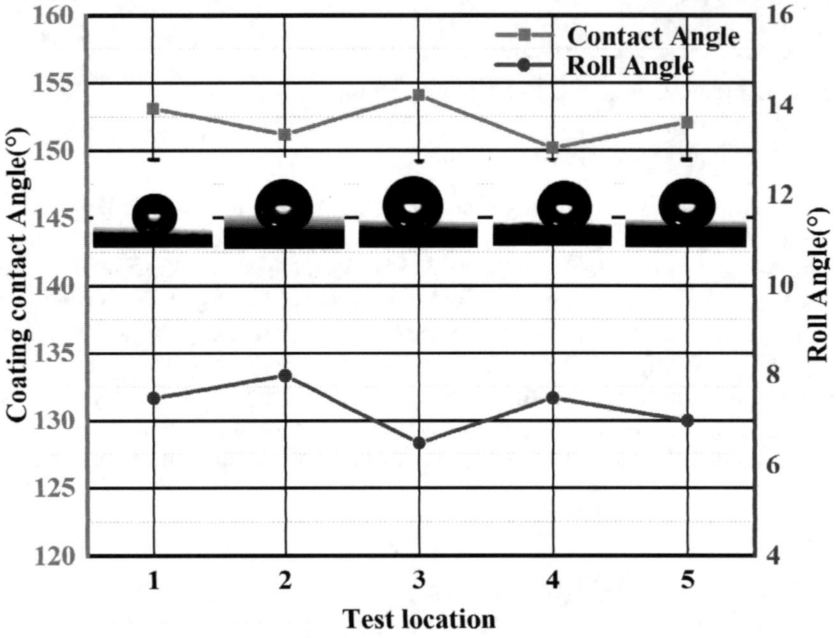

Figure 3. Rolling Angle Test Results.

The results show that the superhydrophobic anti-icing coating has a contact angle of over 150°and a rolling angle below 10°, indicating that the coating possesses excellent superhydrophobic properties. The nanoparticles and hydrophobic agents in the coating play a role, being uniformly dispersed on the coating surface to form micro-nano structures.

3.2. Analysis of Ice Adhesion Strength Results of Coatings

Tangential ice adhesion strength tests were conducted on anti-icing samples, and based on this, multiple ice adhesion strength tests were performed at the same position to evaluate the effect of icing frequency on ice adhesion strength. At the same time, ice adhesion strength tests were also conducted on aluminum sheets of the same specification. The test results are shown in Figure 4.

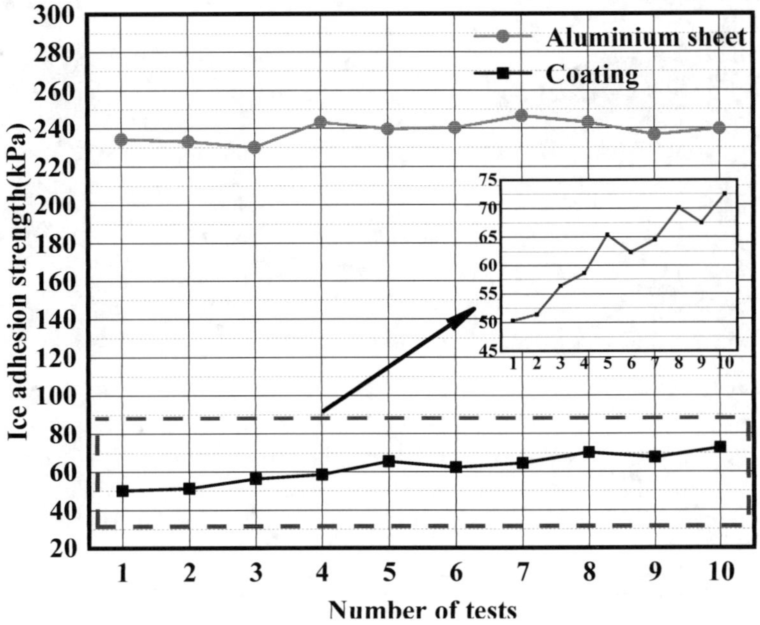

Figure 4. Ice Adhesion Strength Test Results.

The test results indicate that compared to the aluminum control group, the coating significantly reduces the ice adhesion strength of the ice layer. Additionally, by locally magnifying the data line of the coating, it can be observed that as the icing frequency at the same position on the coating increases, there is a trend of increasing ice adhesion strength for the superhydrophobic anti-icing coating. This suggests that the surface structure of the coating plays a role in reducing ice adhesion strength. However, during the process of multiple icing and de-icing cycles, the effective structure on the coating surface is gradually lost due to the shearing action of the ice layer, which diminishes its effect on reducing ice adhesion strength. Nevertheless, the overall ice adhesion strength remains lower than that on the surface of the aluminum sheet.

3.3. Microscopic Morphology Analysis of Coatings

The test results are shown in Figure 5.From the SEM micrographs, it can be observed that after the surface is coated with a superhydrophobic anti-icing coating, the coating surface exhibits a porous micro-nano structure, as indicated by the parts enclosed by the black frame in the figure. Upon further magnification, there are even finer micro-nano morphologies on its surface, as indicated by the parts enclosed by the white frame in the figure. These structures together form the micro-nano structure on the surface of the anti-icing coating, thereby creating a superhydrophobic effect. It is known from existing research that the size of water vapor can often reach several micrometers or even below 1 micrometer. Compared with the scale bar in the figure, the size of the porous structure on the coating surface is approximately 1 micrometer. This means that if the coating is exposed to a humid environment for a long time, small-sized water vapor may infiltrate into these nanostructures. With the icing temperature at low temperatures, this water vapor will gradually transform into fine ice structures, thereby becoming fixed within these micro-nano structures, causing damage to the micro-nano structures, reducing the hydrophobic ability, and increasing the ice adhesion strength.

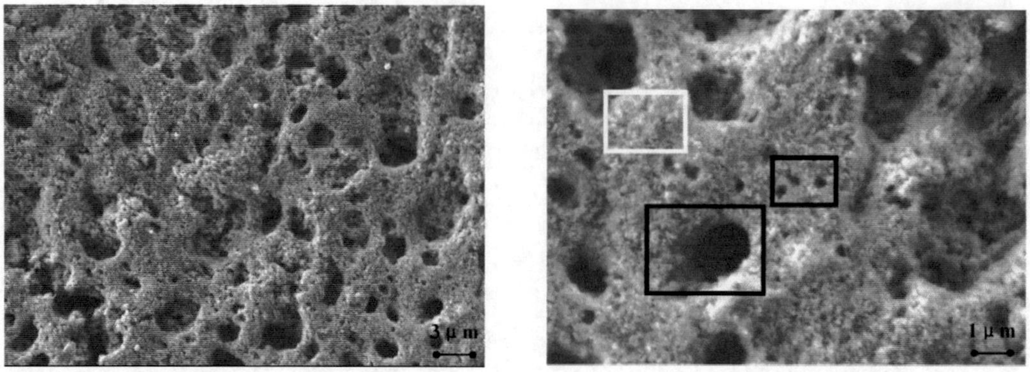

Figure 5. SEM picture.

3.4. Analysis of Coatings Analysis of Viscosity and Storage Performance Results of Coatings

The viscosity test results are shown in Figure 6. The viscosity increases with the progression of storage time and gradually stabilizes after 14 days, indicating that the overall storage performance of the coating is good.

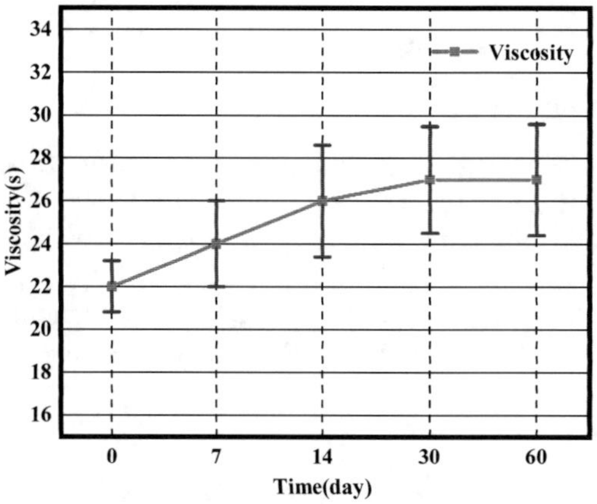

Figure 6. The Process of Paint Viscosity Changing Over Time.

To further explore whether the properties of the coating change significantly during storage, samples from various storage times were taken for spread coating evaluation, including contact angle, rolling angle, and ice adhesion strength. The specific test results are shown in Figure 7.

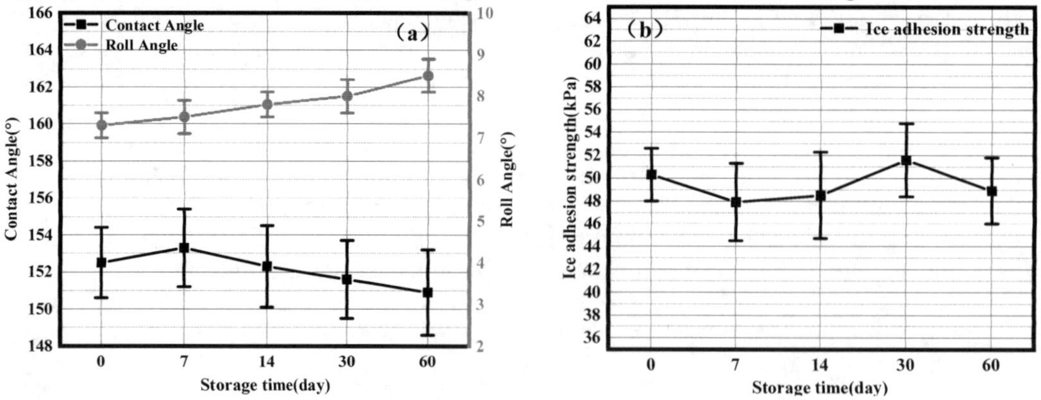

Figure 7. Performance Comparison of Paint Samples Stored for Different Durations.

The test results show that as the storage time increases, the contact angle of the coating decreases and the rolling angle increases, indicating that storage time has a certain tendency to degrade the final micro-nano structure of the coating. However, the overall decrease is relatively low, with the contact angle still reaching above 150° and the rolling angle below 150°, meeting the performance requirements for superhydrophobicity. In terms of ice adhesion strength, there is no significant increase or decrease trend for coatings of different storage times, indicating that storage time does not significantly affect ice adhesion strength, and the active components in the coating do not undergo additional reactions before the coating is opened.

3.5. Analysis of the Impact of Wet and Cold Environment on Coating Performance

The coating samples were tested for the contact angle, rolling angle, and ice adhesion strength under different treatment conditions. The test results are shown in Figure 8.

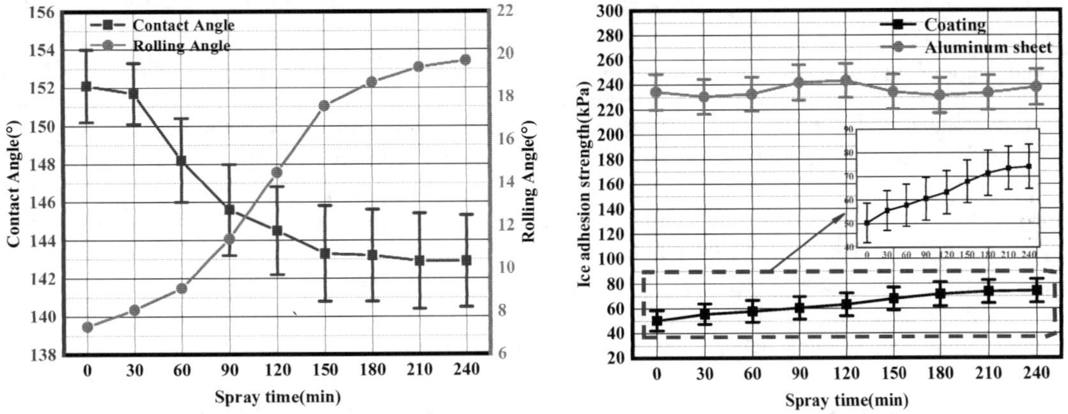

Figure 8. Coating Performance After Spraying.

From the test results, it can be observed that as the coating is sprayed in the wet and cold environment for an extended period, the contact angle of the coating decreases significantly, and the rolling angle increases, indicating a clear reduction in its hydrophobicity, which tends to level off after 150 minutes. The test results for ice adhesion strength show that spraying the coating in a wet and cold environment increases its ice adhesion strength, but the increase is limited, with the overall ice adhesion strength ranging from 40 kPa to 80 kPa. Prolonged exposure to wet and cold conditions and icing environments will directly affect the surface structure of the coating, thereby reducing its anti-icing performance.

3.6. Analysis of Simulated Icing Results

Two 30 cm-long wires were placed parallel on the experimental rack in the environmental chamber to ensure consistent spray effects. One wire was treated with the coating, and the other was a regular wire; water was pre-cooled in the temperature-controlled water bath until it reached the set water temperature of 3.5°C; the humidifier, blower, and environmental chamber were turned on, with the temperature box temperature set to 0°C; after reaching the set temperature, the time was recorded as T0, then the peristaltic pump was turned on, and the environmental chamber temperature was set to-7°C; subsequently, continuous cooling and spraying were carried out until the environmental chamber temperature reached-7°C, after which all equipment continued to operate. The icing conditions of both wires were observed simultaneously. The time when ice crystals formed on the surface of the regular wire was recorded as T1, and the time when ice crystals formed on the surface of the coated wire was recorded as T2. The icing time is T1-T0, and the delayed icing time is T2-T1. The test results are listed in Table 3.

Table 3. The simulated icing test results.

Number of Tests	T1-T0 (min)	T2-T1 (min)
1	36	44
2	37	41
3	35	35
4	36	33
5	36	34

The test results show that the wire coated with the coating can delay the icing time by twice compared to the regular wire. However, for the same coating, the delayed icing time T2-T1 gradually decreases after continuous icing processes. This indicates that the superhydrophobic structure on the coating surface was damaged to some extent after multiple icing events, thereby reducing the coatings' hydrophobic ability against freezing droplets in low-temperature environments.

4. Conclusions and Outlook

This study developed a fluorosilicone superhydrophobic anti-icing coating and conducted tests on hydrophobicity, ice adhesion strength, storage stability, durability in wet and cold environments, and simulated icing performance. The test results show that the coating has a contact angle of up to 150°and a rolling angle of up to 10°, meeting the standards for superhydrophobic coatings; the ice adhesion strength is between 40 kPa and 80 kPa; it has good storage stability, maintaining good coating performance even after 60 days of storage; the coating exhibits good durability in wet and cold environments; and in simulated icing tests, it can delay icing by more than 30 minutes. This coating is a type of anti-icing coating with excellent superhydrophobicity and anti-icing durability, and its preparation process is simple.

In future research, it is necessary to further explore the mechanism of coating degradation over time, and combine actual icing environment experiments to further optimize the coatings' anti-icing durability and increase the delay in icing, making the coating more suitable for actual icing disasters.

Acknowledgments

This study was supported by the 2024 Beijing Science and Technology Association Golden Bridge Project and the State Grid Corporation headquarters technology project. We would like to thank the editors and the anonymous reviewers for their work.

References

[1] Wang B, Zhang X, Li B, et al. Boron Nitride Modified Superhydrophobic Anti-Icing Coating for Electrical Equipment Protection [J]. ChemNanoMat, 2024, 10(8): e202400001. DOI: 10.1002/cnma.202400001.

[2] Hao S, Zhan Y, Li W, et al. Preparation of Epoxy Resin/Carbon Black Superhydrophobic Photothermal Materials with Shape Memory Capacity for Anti/De-Icing [J]. Advanced Engineering Materials, 2024, 26(7): 2301948. DOI: 10.1002/adem.202301948.

[3] Fang X, Liu Y, Lei S, et al. Novel SLIPS based on the photo-thermal MOFs with enhanced anti-icing/de-icing properties [J]. RSC Advances, 2022, 12(22): 13792-13796. DOI: 10.1039/D2RA02046G

[4] Wang J, Long M, He J, et al. Experimental study on ice-covered samples of composite material tower [J]. IEEE Transactions on Dielectrics and Electrical Insulation, 2017, 24(5): 2937-2944. DOI: 10.1109/TDEI.2017.005983.

[5] Miao B, Yuan L, Zhu C L. Electromechanical ice protection system: de-icing capability prediction considering impedance matching effect [J]. The Aeronautical Journal, 2024: 1-21. DOI: 10.1017/aer.2024.15.

[6] Zhang Z, Zhang H, Yue S, et al. A Review of Icing and Anti-Icing Technology for Transmission Lines [J]. Journal, 2023, 16(2). DOI: 10.3390/en16020601.

[7] Cao Z, Gong Z, Ma X, et al. A superhydrophobic/photothermal synergistic anti-icing mesh with active/passive anti-icing function [J]. Materials Research Express, 2024, 11(7): 075009. DOI: 10.1088/2053-1591/ad59ef.

[8] Lo T N H, Lee J, Hwang H S, et al. Nanoscale Coatings Derived from Fluoroalkyl and PDMS Alkoxysilanes on Rough Aluminum Surfaces for Improved Durability and Anti-Icing Properties [J]. ACS Applied Nano Materials, 2021, 4(7): 7493-7501. DOI: 10.1021/acsanm.1c01526.

[9] B. Wu, X. Cui, H. Jiang, et al. A superhydrophobic coating harvesting mechanical robustness, passive anti-icing and active de-icing performances. Journal of Colloid and Interface Science. 2021, 590: 301-310. DOI: 10.1016/j.jcis.2021.01.054.
[10] Lv C, Hao P, Yao Z, et al. Departure of condensation droplets on superhydrophobic surfaces. Langmuir, 2015, 31(8): 2414-2420. DOI: 10.1021/la504638y.
[11] Li K, Xu S, Chen J, et al. Viscosity of interfacial water regulates ice nucleation. Applied Physics Letters, 2014, 104(10): 101605. DOI: 10.1063/1.4868255.

Journal of Physics: Conference Series **3009** (2025) 012016 doi:10.1088/1742-6596/3009/1/012016

Effects of Br-Doped and Ligand-Symmetry on the Properties of $APb(I_{1-x}Br_x)_4$ (A=PeDA/DPA) Perovskites

Zhonghai Lin[1*], Mengdi Liu[1], Pingjian Wang[1], Huitian Du[1], Jing Sun[1*], Guangfen Wei[1], Ruiyang Zeng[1] and Xiaoyan Yang[1]

[1]Shandong Technology and Business University, Yantai, 264005, PR China

[*]Corresponding author's e-mail: zhlin@sdtbu.edu.cn(Z. Lin), sunjing@sdu.edu.cn (J. Sun)

Abstract. In this paper, the properties of perovskites based on symmetric (1,5-Pentanediamine (PeDA)) and asymmetric (1,4-pentanediamine (DPA)) ligands are systematically tuned by substituting I with Br. This paper employs the method of density functional theory to explore the characteristics of $APb(I_{1-x}Br_x)_4$ (A = PeDA/DPA; x = 0.00, 0.25, 0.50, 0.75, 1.00). The analysis reveals that bromine doping results in a noticeable reduction in the bond lengths within the metal-halide framework. The electronic structure analysis shows that with an increasing Br doping ratio, the bandgap of $PeDAPb(I_{1-x}Br_x)_4$ increases linearly from 2.88 eV to 3.64 eV, and $DPAPb(I_{1-x}Br_x)_4$ exhibits an overall upward trend from 2.31 eV to 3.49 eV. Optical property analysis demonstrates that when the ligand is DPA with a Br doping ratio of 0.50, the material exhibits excellent light absorption in the visible region.

1. Introduction

As global demand for energy steadily rises, the shift from conventional fossil-based fuels to renewable energy sources has turned into an unavoidable trajectory. Among various renewable energy sources, solar power stands out as a highly promising alternative owing to its superior efficiency, environmental friendliness, and abundant availability [1]. Perovskite-based solar cells, characterized by outstanding power conversion capabilities (PCE) and low production cost, have garnered considerable interest in the photovoltaic sector over recent years. In the most recent studies, the efficiency of perovskite solar cells has recently achieved 26.41%, nearing the levels seen in some silicon-based counterparts [2].

Three-dimensional (3D) ABX_3 perovskite materials are characterized by favorable properties, including suitable optical bandgaps, long carrier diffusion lengths, and low production costs [3]. However, these materials encounter notable drawbacks, such as poor thermal stability and rapid degradation in ambient conditions. Researchers have shifted their focus to two-dimensional (2D) perovskites to overcome these limitations, which exhibit superior moisture resistance, enhanced chemical stability, and improved phase stability. To further enhance the performance and stability of perovskite materials, doping has been identified as a promising strategy for regulating their properties. For instance, Wu et al. demonstrated that Br-doped effectively enhanced the stability of $Cs_2AgInCl_6$, emphasizing the role of compositional engineering in improving material robustness [4]. Lei et al. successfully modulated the bandgap of $(CMA)_2MAPb_9I_{28}$ through Br-doped, demonstrating the potential of halogen substitution in tuning electronic structure [5]. Furthermore, Zhang et al. revealed that br-doped in $PDAPbI_4$ enhanced optoelectronic performance by affecting the crystal structure and bond lengths [6]. These studies collectively highlight the versatility of Br-doped strategies in optimizing the properties of perovskite materials.

We find that previous research on Br doping in 2D DJ-phase perovskites based on symmetric/asymmetric ligands remains limited. To address this gap, this paper systematically investigates the effects of Br doping on the properties of $APb(I_{1-x}Br_x)_4$ (A = PeDA, DPA; x = 0.00, 0.25, 0.50, 0.75, 1.00), where PeDA and DPA represent symmetric (1,5-Pentanediamine (PeDA)) and asymmetric (1,4-pentanediamine (DPA)) ligands, respectively. The results reveal the mechanisms of Br doping on octahedral structural stability and bandgap tuning in perovskites, providing critical theoretical support for experimental design and the optimization of optoelectronic materials.

2. Computational method

All computations presented in this paper are carried out through the CASTEP module embedded in the Materials Studio (MS) platform [7]. The methodology utilizes density functional theory (DFT), applying the Perdew-Burke-Ernzerhof (PBE) exchange-correlation function under the generalized gradient approximation (GGA) scheme [8]. To consider the effects of intermolecular interactions on the computed outcomes, the Tkachenko-Scheffler (TS) dispersion correction was incorporated.

3. Results and analysis

3.1 Geometric Structure and Stability

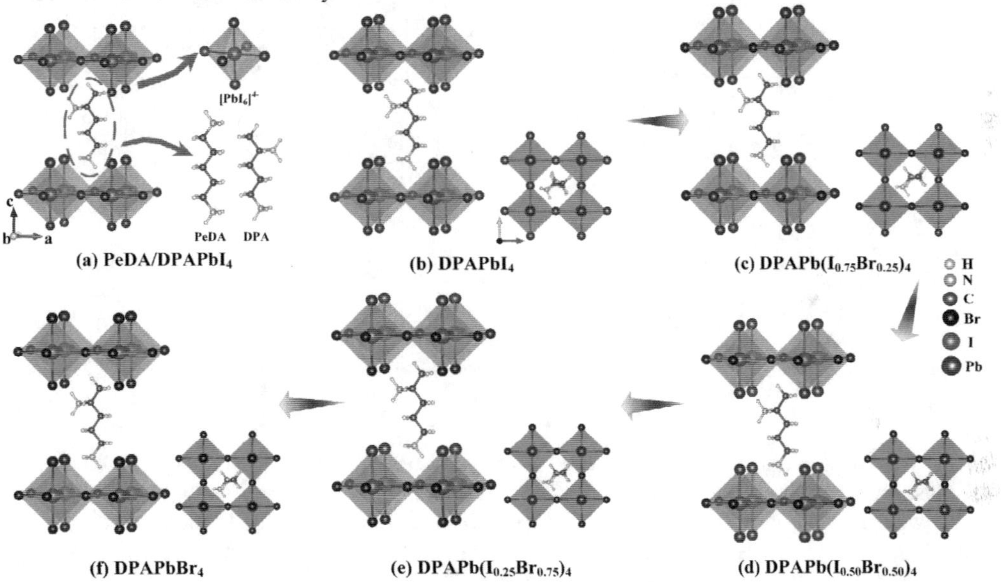

Figure 1. (a) illustrates the structure of PeDA/DPAPbI4 models. (b)-(f) illustrates both the frontal and top perspectives of the PDAPb($I_{1-x}Br_x$)$_4$ (x=0.00, 0.25, 0.50, 0.75, 1.00) models.

Figure 1 presents the crystalline configurations of APb($I_{1-x}Br_x$)$_4$ perovskites, where A represents PeDA or DPA, and the Br doping ratio (x) varies from 0.00 to 1.00. The changes in the mean Pb–X (X=I, Br) bond lengths for the APb($I_{1-x}Br_x$)$_4$ structures are depicted in Figures 2(a) and 2(b). The Pb–I bonds exhibit average lengths ranging from 3.16 to 3.30 Å (PeDA: 3.16–3.27 Å, DPA: 3.18–3.30 Å), which are notably longer compared to the Pb–Br bonds (PeDA: 2.91–2.96 Å, DPA: 2.80–3.01 Å). This disparity arises from the Shannon ionic radii, with I^- measuring 2.06 Å and Br^- being smaller at 1.82 Å [3]. Generally, smaller ionic radii result in electron clouds closer to the nucleus, leading to higher electronegativity, stronger electron-attracting capacity, and shorter bond lengths.

To further investigate the bonding strength and structural stability of APb($I_{1-x}Br_x$)$_4$ perovskites, the formation energy (E_f) is determined based on Equation (1):

$$E_f = \{E_t - N_A E_A - N_{Pb}E_{Pb} - (4-m)E_I - mE_{Br}\}/28 \\ m = (0,1,2,3,4) \tag{1}$$

Here, E_t denotes the overall energy of the APb($I_{1-x}Br_x$)$_4$ system, while E_i (where i refers to PeDA/DPA, Pb, I, or Br) represents the energy of each atom, and N_i specifies the count of corresponding atoms. The parameter m defines the doping ratio of Br, ranging from 0 to 4, corresponding to Br contents of 0.00, 0.25, 0.50, 0.75, and 1.00. The E_f of these materials are negative, indicating that the formation is an exothermic reaction (figure 2(c)(d)). The results show that the absolute values of E_f for these two types of DJ-phase perovskites are relatively close. However, within each type of DJ-phase perovskite, the absolute value of E_f increases with the Br doping ratio. And the E_f of $PeDAPb(I_{1-x}Br_x)_4$ follows a linear fit function of y(x) = -0.019x - 4.822. According to the principle of minimum energy, a greater absolute value of E_f implies enhanced structural stability. This suggests that compared to $APbI_4$, $APb(I_{1-x}Br_x)_4$ ($0 < x \leq 1$) exhibits lower E_f values, meaning that Br doping enhances the structural stability of perovskites, and the higher the Br content, the more stable the material becomes.

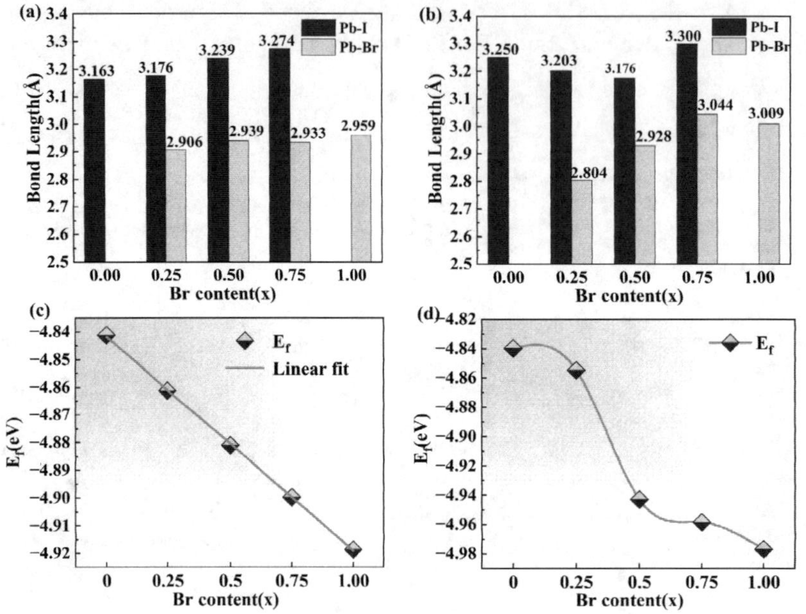

Figure 2. (a) shows the average bond lengths of Pb–X (X=I, Br) in $PeDAPb(I_{1-x}Br_x)_4$ (x=0.00, 0.25, 0.50, 0.75, 1.00). (b) illustrates the average Pb–X bond lengths (X=I, Br) in DPAPb $(I_{1-x}Br_x)_4$ (x=0.00, 0.25, 0.50, 0.75, 1.00). (c) shows the variation trend of E_f of $PeDAPb(I_{1-x}Br_x)_4$. (d) illustrates the variation trend of E_f for $DPAPb(I_{1-x}Br_x)_4$.

3.2 Electronic properties

We calculated the bandgap value of $PeDAPbI_4$ (2.878 eV), which is consistent with previous studies [9]. Furthermore, in $PeDAPb(I_{1-x}Br_x)_4$ (2.878-3.642 eV) and $DPAPb(I_{1-x}Br_x)_4$ (2.308-3.490 eV), the bandgap exhibits an overall increasing trend as x increases from 0 to 1. This effect can be attributed to the disparity in ionic radii between I^- and Br^-, where the larger radius of I^- compared to Br^- results in stronger Pb–Br bonds. These stronger bonds lower the position of the valence band maximum and increase the energy required for electronic transitions, thereby leading to a larger bandgap. Notably, for $PeDAPb(I_{1-x}Br_x)_4$ models, the bandgap value increases linearly with the Br doping ratio, and the bandgap change for $PeDAPb(I_{1-x}Br_x)_4$ follows a linear relationship, represented by the equation y(x)=2.707+0.197x (as shown in Figure 3(a)). In contrast, for $DPAPb(I_{1-x}Br_x)_4$ models, the bandgap reaches a minimum when the Br doping ratio is 0.50(figure 3(b)). We speculate that this may result from the asymmetry of the ligand, which alters the interactions between the halogen atoms, ultimately causing modifications to the band structure, as depicted in Figure 4. In summary, Br doping has a significant regulatory effect on the bandgap of DJ-phase perovskites, with the regulation being more pronounced in perovskites based on asymmetric ligands (PeDA: ΔE_g=0.746 eV, DPA: ΔE_g=1.182 eV). This approach facilitates the improvement of material properties and offers valuable theoretical insights. It serves as a foundational guide for designing and advancing next-generation optoelectronic systems.

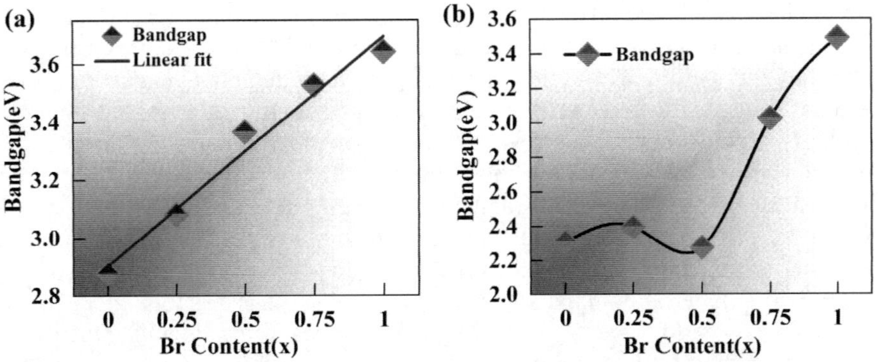

Figure 3. (a) The bandgap variation of PeDAPb($I_{1-x}Br_x$). (b) The bandgap variation DPAPb($I_{1-x}Br_x$).

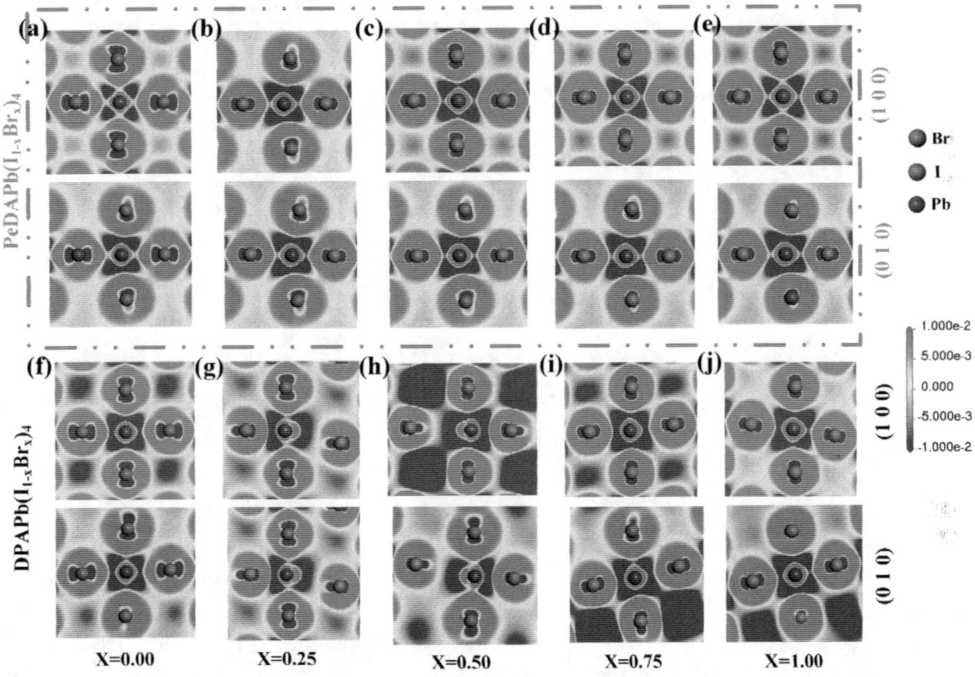

Figure 4. (a)-(e) The calculated charge density differences on the (1 0 0) and (0 1 0) plane of PeDAPb($I_{1-x}Br_x$)$_4$. (d)-(f) The calculated charge density differences on the (1 0 0) and (010) plane of DPAPb($I_{1-x}Br_x$)$_4$.

3.3 Optoelectronic properties

To explore the optical characteristics of the materials, this paper focuses on determining the optical absorption coefficient, denoted as $\alpha(\omega)$, for APb($I_{1-x}Br_x$)$_4$. The calculation of $\alpha(\omega)$ provides valuable insights into the optical properties of these compounds. The $\alpha(\omega)$ can be calculated from $\varepsilon_1(\omega)$, with its specific form provided in Equations (2) and (3), where ω, e, and $\hbar$ represent the angular frequency of the photons, the elementary charge, and the reduced Planck constant, respectively [10]. The $\alpha(\omega)$ characterizes both the propagation distance of light within the material and its absorption efficiency.

$$\varepsilon(\omega) = \varepsilon_1(\omega) + i\varepsilon_2(\omega) \tag{2}$$

$$\alpha(\omega) = \sqrt{2}e\left[\sqrt{\varepsilon_1(\omega)^2 + \varepsilon_2(\omega)^2} - \varepsilon_1(\omega)\right]^{\frac{1}{2}} / \hbar c \tag{3}$$

Figure 5(a)(b) illustrates the α(ω) of $APb(I_{1-x}Br_x)_4$. In the $PeDAPb(I_{1-x}Br_x)_4$ model, as the Br doping ratio increases, the absorption spectra exhibit a gradual blue shift, which corresponds to the bandgap variation (2.818-3.642 eV). Similarly, in the $DPAPb(I_{1-x}Br_x)_4$ model, the overall absorption spectra also shift toward higher energy with the increase in Br doping ratio from 0 to 1, and the bandgap values increase accordingly (2.275-3.490 eV). However, it is noteworthy that at a Br doping ratio of 0.5, the absorption spectrum shows a significant red shift compared to other doping ratios, and the α(ω) value in the visible light range is markedly superior to that observed in other materials. In summary, the optical absorption properties of perovskite materials can be effectively tuned by substituting atoms with Br atoms and adjusting the ligand symmetry. Specifically, when the ligand symmetry is low, and the Br doping ratio is 0.5, the material exhibits superior optical absorption performance, making it a more suitable candidate for absorption layers in optoelectronic devices. Consequently, during the synthesis of DJ-phase perovskites of the $APb(I_{1-x}Br_x)_4$ type, both Br doping and ligand symmetry should be carefully considered to optimize the optical properties.

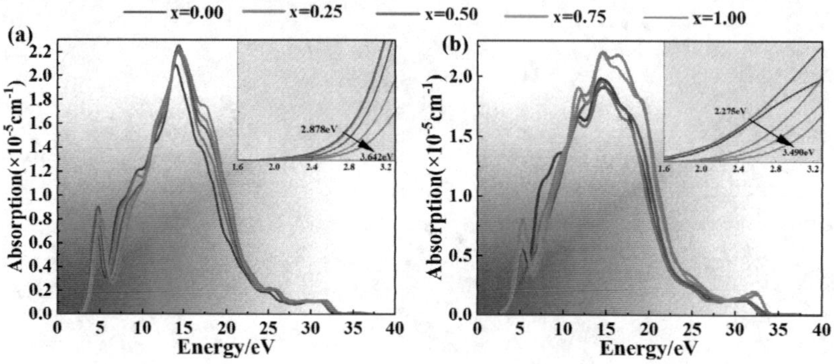

Figure 5. (a) The calculated α(ω) of $PeDAPb(I_{1-x}Br_x)_4$ model (the yellow inset indicates the absorption edge variation under different Br ratios). (b) The calculated α(ω) of $DPAPb(I_{1-x}Br_x)_4$ model (the yellow inset indicates the absorption edge variation under different Br ratios).

4. Conclusions

First-principles calculations are employed in this paper to systematically investigate properties of $APb(I_{1-x}Br_x)_4$ (A = PeDA/DPA; x = 0.00, 0.25, 0.50, 0.75, 1.00) perovskites. The findings reveal that with an increase in the Br doping ratio, the average bond length within the metal-halide framework of $APb(I_{1-x}Br_x)_4$ decreases, while the structural stability is enhanced. Furthermore, Br doping effectively adjusts the bandgap value of $APb(I_{1-x}Br_x)_4$, with a stronger adjustment observed in perovskites based on asymmetric ligands. Specifically, the bandgap of $PeDAPb(I_{1-x}Br_x)_4$ increases linearly from 2.88 eV to 3.64 eV (ΔE_g = 0.58 eV), whereas that of $DPAPb(I_{1-x}Br_x)_4$ varies from 2.31 eV to 3.49 eV (ΔE_g=1.18 eV). Optical analysis further demonstrated that the absorption edge of $APb(I_{1-x}Br_x)_4$ is consistent with the bandgap variation. Notably, when A = DPA and x = 0.5, the material exhibits superior light absorption performance. These findings underscore the tunability of the optoelectronic properties of perovskites through precise adjustments of halide composition and ligand symmetry, providing valuable insights for the design and development of high-performance photovoltaic devices based on $APb(I_{1-x}Br_x)_4$ perovskites.

Acknowledgments

The authors gratefully acknowledge financial support from The National Natural Science Foundation of China (62173213, 61804089), The Natural Science Foundation of Shandong Province (ZR2019MF069, ZR2020QA055), The Shandong Provincial Science and Technology Support Program of Youth Innovation Team in College (2019KJN041 and 2020KJN005), The Yantai Key R&D Program

(2020LJRC119,2022JCYJ033, 2022JCYJ034, 2022JCYJ1044), The Science Fund of Shandong Laboratory of Advanced Materials and Green Manufacturing at Yantai (AMGM2023A16).

References

[1] Pourasl H. H, Brenji R. Vand Khojastehnezhad V. M. Solar Energy Status in the World: A Comprehensive Review. *Energy Reports* 2023, **10**, 3474–3493. https://doi.org/10.1016/j.egyr.2023.10.022.
[2] Zhou J, Tan L, Liu Y, Li H, Liu X, Li M, Wang S, Zhang Y Jiang C and Hua R. Highly Efficient and Stable Perovskite Solar Cells via a Multifunctional Hole Transporting Material. *Joule* 2024. https://doi.org/10.1016/j.joule.2024.02.019.
[3] Lin Z, Lei J, Wang P, Zhang X, Xu L, Chen M, Kang Y and Wei G. Density Functional Study of Structural, Electronic and Optical Properties of Bromine-Doped CsPbI3 with the Tetragonal Symmetry. *Journal of Alloys and Compounds* 2022, **892**, 162165. https://doi.org/10.1016/j.jallcom.2021.162165.
[4] Lin Z, Wu Q, Sun C, Zeng R, Liu M, Zhang X, Wang P, Wei G and Qiao Z. Study on Cs2AgIn (Cl1- xBrx) 6 Perovskite Based on the Balance of Stability and Optical Absorption Properties. *Journal of Alloys and Compounds* 2023, **968**, 172138. https://doi.org/10.1016/j.jallcom.2023.172138.
[5] Wei Y, Sun Y, Liu Y and Song J. Impact of Bromide Incorporation on Strain Modulation in 2D Ruddlesden-Popper Perovskite Solar Cells. *Cell Reports Physical Science* 2023, **4** (12), 101739. https://doi.org/10.1016/j.xcrp.2023.101739.
[6] Zhang X, Wang P, Wu Q, Xu L, Chen M, Kang Y, Sun C, Wei G, Qiao Z and Lin Z. DFT Study of the Structural, Electronic and Optoelectronic Properties of PDAPb(I1−xBrx)4 Perovskites. *Semicond. Sci. Technol.* 2022, **37** (9), 095026. https://doi.org/10.1088/1361-6641/ac86ee.
[7] Clark S. J, Segall M. D, Pickard C. J, Hasnip P. J, Probert M. I. J, Refson K and Payne, M. C. First Principles Methods Using CASTEP. *Zeitschrift für Kristallographie - Crystalline Materials* 2005, **220** (5–6), 567–570. https://doi.org/10.1524/zkri.220.5.567.65075.
[8] Hohenberg P and Kohn W. Inhomogeneous Electron Gas. *Phys. Rev.* 1964, **136** (3B), B864–B871. https://doi.org/10.1103/PhysRev.136.B864.
[9] Shtarev D. S, Balanov M. I, Major A. J, Gerasimenko A. V, Kevorkyants R, Zharovov D. A, Bulanin K. M, Pankin D. V, Rudakova A. V and Chaplygina D. A. The Effect of the Halogen on the Structural, Optoelectronic, and Luminescent Properties of Hybrid (1, 5-Pentanediamine) PbX 4 (X= Cl, Br, I) Perovskites. *Journal of Materials Chemistry C* 2024, **12** (1), 262–275. https://doi.org/10.1039/D3TC02767H.
[10] Ming W, Shi H and Du M.-H. Large Dielectric Constant, High Acceptor Density, and Deep Electron Traps in Perovskite Solar Cell Material CsGeI 3. *Journal of Materials Chemistry A* 2016, **4** (36), 13852–13858. https://doi.org/10.1039/C6TA04685A.

Journal of Physics: Conference Series **3009** (2025) 012017 doi:10.1088/1742-6596/3009/1/012017

Numerical study on vibration mode of a single blade of carbon fiber composite propeller based on laminated structure

Ruipin Li [1], Zeyu Shi [2], and Jinlong Wang [2] *

[1] People's Liberation Army of China, No. 32604, Wuwei 733202, China
[2] Marine Engineering College, Dalian Maritime University, Dalian 116026, China

*Corresponding author's e-mail: wjl19890806@dlmu.edu.cn

Abstract: To clear the vibration characteristics of carbon fiber composite propellers, the numerical simulation of vibration modes for composite propeller blades is conducted. This article takes self-designed composite material propeller blades as the research object, establishing a finite element model of a single composite material propeller blade, and a modal simulation of composite material propeller blade is conducted to analyze the natural frequency and deformation distribution characteristics of the blade under different order conditions. The numerical simulation showed that the maximum vibration deformation occurred at the tip of the leaf tip in Orders 1, 3, 5, 7, 9, and 10. In contrast, in Orders 2, 4, and 8, the maximum vibration deformation appeared on both sides of the leaf tip, and the vibration deformation increased gradually from the root to the tip. Under the excitation of sine and cosine loads, the maximum stress of self-designed laminated composite propeller blades caused by random vibration appears in the middle of the blade, and the stress caused by cosine excitation is greater than that caused by sine excitation. The results of this study can provide theoretical support for the performance evaluation and engineering application of composite material propellers.

1. Introduction

The propeller is a key component in ship propulsion. Traditionally, it is made of metals or alloys such as copper alloys, stainless steel, titanium alloys, and aluminum alloys, which offer high strength and corrosion resistance. However, due to prolonged exposure to water corrosion and complex loading conditions, metal propellers are prone to issues such as cracking, damage, poor acoustic damping, and excessive noise, all of which affect navigation safety and efficiency [1-2]. Carbon fiber composites, known for their high strength, corrosion resistance, and low noise properties, have been successfully applied in aerospace and energy industries. As a result, the design, manufacturing, and performance analysis of carbon fiber composite propellers have also garnered significant attention in the maritime industry [3-6].

Random vibrations during the service of propellers affect their lifespan and structural strength, and this is also true for composite propellers [7-8]. Vibrations induced by high-speed rotation can lead to structural failure. Gong et al. [9], Huang et al. [10], and Jing et al. [11] have investigated the effects of damping, layer orientation, number of layers, and fiber angles on the vibration characteristics of composite propellers through modal analysis, frequency response functions, vibration noise assessment, and finite element analysis. Their studies revealed that the natural frequency of carbon fiber propellers is lower than that of copper propellers, but the structural damping is higher. Kishore and Behera [12] examined how variations in layer orientation, number of layers, and fiber angles affect the vibration properties of composite propellers. Additionally, Hong et al. [13] studied the vibration and damping characteristics of composite blades in different environments, finding that the natural frequency of composite blades in water is lower than in air, although their modal characteristics remain similar.

To address the insufficient research on the modal and vibration characteristics of laminated composite propellers, this paper focuses on a composite propeller designed in-house. A single-blade model is established for numerical simulation to analyze the higher-order modes and the stresses and deformations caused by random vibrations. The results aim to provide a theoretical basis for the design, manufacturing, and maintenance of composite propellers.

2. Carbon Fiber Composite Blade Model Based on ACP

2.1 Structure of the Carbon Fiber Composite Blade

This study focuses on a four-blade propeller designed in-house, the blades are constructed from T300 carbon fiber/epoxy resin prepreg, with a fiber volume fraction of 67% and a resin content of 33%. The nominal thickness of each layer is 0.145 mm. The single blade and its varying cross-sectional characteristics were fabricated by using a lamination process, as shown in Figure 1 (a). The propeller blade exhibits different cross-sectional shapes at varying radii, with an overall characteristic of both variable width and thickness. Figure 1 (b) shows the variation of the chord length c and maximum thickness t of the cross-section with radius r. The blade thickness decreases as the radius increases, while the blade width initially increases with the radius, reaching a maximum of around 0.7R, before gradually decreasing.

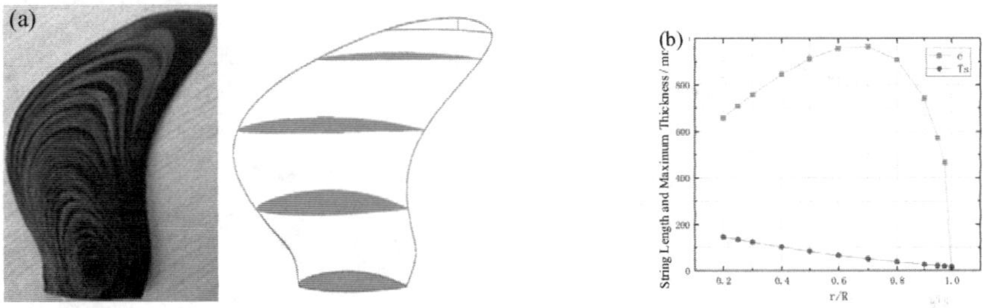

(a) Composite propeller and its cross-section (b) Variable cross-section characteristic curve of composite propeller blade

Figure 1. Composite propeller and its variable cross-section characteristics.

Modeling of composite materials using the ACP module typically starts with the shell model of the object. For composite propeller blades, the pressure (water-facing) and suction (rear-facing) sides of the blade are chosen as the enveloping shell layers. An asymmetric lamination approach is adopted, where the stacking proceeds symmetrically from the central layer towards both the pressure and suction sides. Each layer has a thickness of 0.145 mm, with fiber orientations of [0°/45°/-45°/90°], as shown in Figure 2.

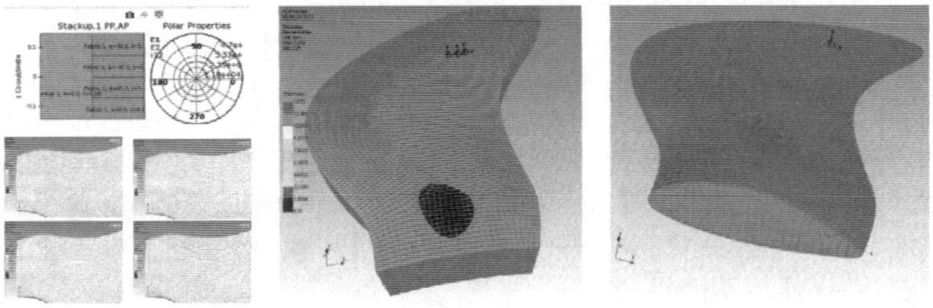

(a) Ply angle of each layer (b) Composite propeller foundation model (c) Composite material propeller solid model

Figure 2. Composite propeller model based on the ACP.

3. Modal Simulation of Composite Blade

3.1 Modal Analysis Equation

Before conducting modal analysis on the propeller, it is essential to first understand the modal analysis theory in Workbench. The equation of motion for a linear structure with damping is given by Equation (1):

$$[m]\{\ddot{q}\}+[k]\{q\}=0 \tag{1}$$

The structural displacement of the object can be decomposed into a superposition of simple harmonic motions. It is assumed that the motion of the structure is simple harmonic, and the displacement and velocity matrices are given by Equations (2) and (3), respectively:

$$\{q\}=\{\phi\}_i\sin(\omega_i+\theta_i) \tag{2}$$

$$\{q\}=-\omega_i^2\{\phi\}_i\sin(\omega_i+\theta_i) \tag{3}$$

where $\{\varphi\}_i$ represents the non-zero amplitude vector, ω_i is the angular frequency, and θ_i is the initial phase angle. By substituting the displacement and velocity of the structure into the equation of motion, we obtain Equation (4):

$$\omega_i^2[m]\{\phi\}_i=[k]\{\phi\}_i \tag{4}$$

3.2 Modal Analysis Results and Discussion of the Composite Propeller Blade

Figure 3 shows the natural frequencies and variation trends of the 1st to 10th mode shapes obtained from the modal simulation of the composite propeller blade. As seen in Figure 3, the natural frequency increases with the mode order and can be divided into two stages: 1-3 modes, and 6-10 modes, where the slope of the fitted curve for the natural frequency versus mode order is relatively large. In contrast, the slope of the fitted curve for Modes 4-5 is smaller.

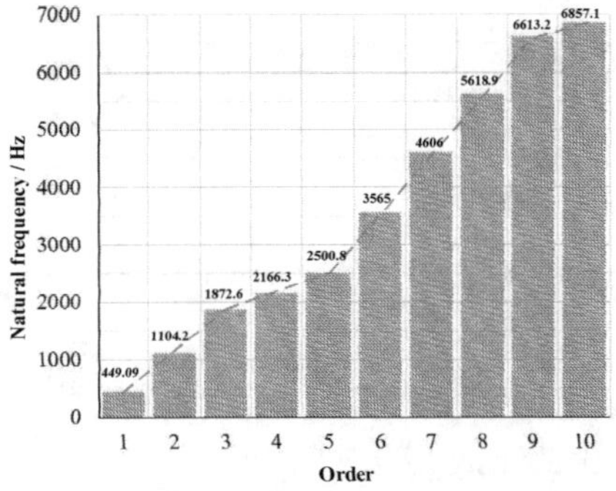

Figure 3. Variation law of intrinsic frequency of composite blades of Orders 1-10.

Figure 4 shows that in the 1st mode, the natural frequency is 449.09 Hz, and the blade experiences oscillations along the z-axis, with the deformation increasing from the root to the tip of the blade. In the 2nd mode, the deformation becomes more concentrated at the tip, and the deformation magnitude increases significantly. The deformation patterns for the 4th, 6th, and 8th modes are similar, with the maximum deformation occurring on both sides of the blade tip. The natural frequency increases from 2, 166.3 Hz to 5, 618 Hz. For the 5th, 7th, 9th, and 10th modes, the deformation is most pronounced at the tip. As the mode order increases, noticeable vibrations appear in the middle of the blade, with the natural frequency rising from 2, 500.8 Hz to 6, 857 Hz. Due to the laminated structure of the composite material, the number of layers at the blade tip is minimal, resulting in a lack of interlayer support and connections. As a result, the deformation at the tip is the greatest, mostly taking the form of torsional deformation, which makes the blade tip highly susceptible to deformation and failure.

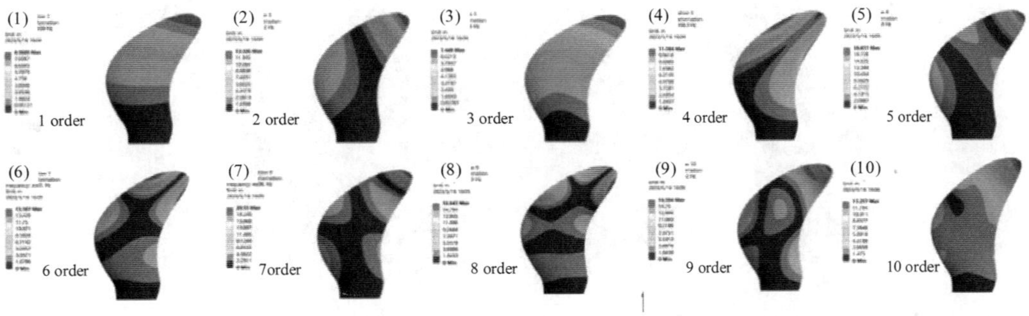

Figure 4. Modal simulation of composite blade cloud map of deformation distribution of each order vibration mode.

Modal testing is conducted on the composite blade specimen shown in Figure 1 to obtain the natural frequency of the 1st model to the 5th model. Comparing the experimental results with the simulation results, the maximum error is 5.3% and the minimum error is 2.5%. The errors are in the acceptable range, and the simulation method is proved to be suitable for analyzing the vibration characteristics of composite blades.

4. Vibration of the Propeller Blade and Analysis

4.1 Power Spectral Density of Random Vibration

The Power Spectral Density (PSD) is a statistical measure that describes the characteristics of random vibration excitation. In random vibration, energy is proportional to the square of the amplitude, and the PSD can characterize how energy is distributed across frequencies during the vibration process. Two different power spectral densities are used to represent the corresponding external random vibration loads, as shown in Equations (5) and (6).

$$PSD_1 = 1000\sin\left(\frac{H}{100} - 2\pi\right) + 1500 \tag{5}$$

$$PSD_2 = 1000\cos\left(\frac{H}{100} - 2\pi\right) + 1500 \tag{6}$$

where PSD_1 represents the sine-characteristic excitation, while PSD_2 represents the cosine-characteristic excitation. Based on the actual operating conditions, nine frequency values are selected for the composite propeller blade during its service life. This results in two different power spectral densities, as shown in Table 1.

Table 1. Power spectral density.

H / Hz	PSD_1 (m/s^2)^2/Hz	PSD_2 (m/s^2)^2/Hz	H / Hz	PSD_1 (m/s^2)^2/Hz	PSD_2 (m/s^2)^2/Hz	H / Hz	PSD_1 (m/s^2)^2/Hz	PSD_2 (m/s^2)^2/Hz
628	1, 500	2, 500	865	2, 207	739	1, 099	500	1, 500
706	2, 207	2, 207	942	1, 500	500	1, 177	739	2, 207
785	2, 500	1, 500	1, 020	739	739	1, 256	1, 500	2, 500

4.2 Random Vibration Simulation Results and Discussion

In this study, the direction of the propeller blade is selected as the excitation direction for the random vibration load, with a damping ratio set to 0.01. The random vibration simulation results of the composite propeller under two different excitation conditions are presented in Figures 5 and 6. The maximum stress and the maximum deformation in different directions, along with their simulation results and variation patterns, are shown in Figure 7.

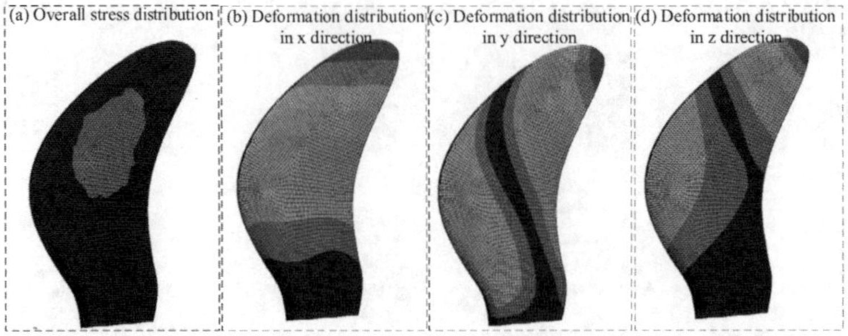

Figure 5. Simulation results of random vibration of composite propeller blades under sinusoidal excitation PSD_1.

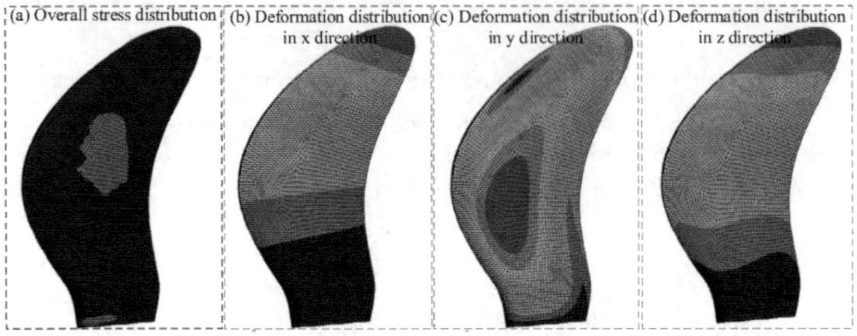

Figure 6. Simulation results of random vibration of composite propeller blades under sinusoidal excitation PSD_2.

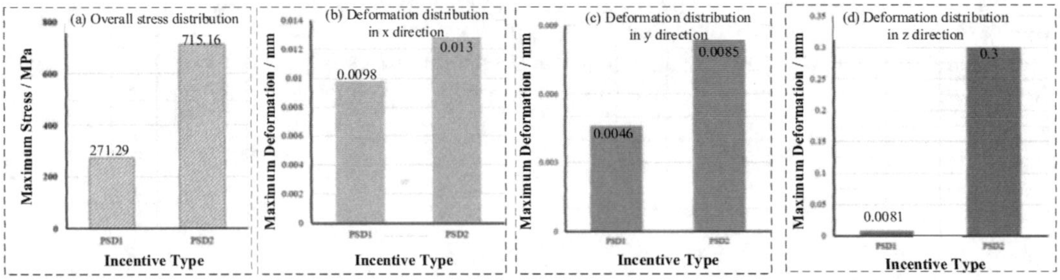

Figure 7. Variation of maximum stress and deformation in three directions of composite propeller blades under two kinds of excitation.

Under the condition where the Power Spectral Density (PSD) is PSD_1, the maximum stress induced by external excitation in the propeller blade's random vibration is approximately 271.29 MPa, located in the central region and near the root of the blade. The maximum deformation in all three directions occurs at the blade tip, with the largest deformation in the x-direction and the smallest in the y-direction, which is consistent with the results from the modal analysis. However, there are noticeable differences in the deformation distribution patterns across the three directions. In the x-direction, the deformation gradually increases from the root to the tip, while in the y-and z-directions, the deformation spreads outward from the root, with two concentrated deformation regions forming at the middle of the blade.

When the Power Spectral Density is PSD_2, the maximum stress caused by external excitation in the blade's random vibration increases to approximately 715.16 MPa, again located in the central region and near the root of the blade, which is similar to the situation with PSD_1. The maximum deformations in the x-direction and z-direction are about 0.013 mm and 0.3 mm, respectively, showing a significant increase compared to the PSD_1 case. In the y-direction, the maximum deformation occurs in the middle of the blade, at approximately 0.0085 mm, and decreases toward the root, tip, and edges of the blade.

5. Conclusion

Based on the modeling and simulation analysis of the composite propeller, the following conclusions can be drawn:

1. The first-order natural frequency of the propeller blade, designed independently, is 449.09 Hz, with its vibration deformation gradually increasing from the root to the tip of the blade.
2. In the 1st, 3rd, 5th, 7th, 9th, and 10th mode shapes, the maximum deformation occurs at the tip of the blade. In contrast, for the 2nd, 4th, and 8th mode shapes, the maximum deformation occurs on both sides of the blade tip. As the mode order increases, the natural frequency of the composite propeller blade gradually increases.
3. Under both sine and cosine excitation loads, the maximum stress induced by random vibration in the independently designed layered composite propeller blade occurs in the central region of the blade. Moreover, the stress induced by cosine excitation is greater than that induced by sine excitation.
4. Under sine excitation, the maximum deformation of the composite propeller blade in different directions occurs at the blade tip. Under cosine excitation, the maximum deformation in the x- and z-directions occurs at the blade tip, while the maximum deformation in the y-direction occurs in the central part of the blade.

Acknowledgments

This work was financially supported by the National Natural Science Foundation of China (52205515).

References

[1] ZHOU X., TONG Z. Y., YANG W. K., Wei J. Q., and Yi X. D. Design and Molding Process of Large Thickness Marine Composite Propeller Blade. Ship engineering, 2022, 44 (9): 24-28.
[2] CONG Q., LIN F. S., CENG Q. Y., Sui X. H., and Li G. Y. Research on the manufacturing technology of large-size composite propeller. New Mechanical Materials, 2022, 50 (S1): 465-469.
[3] DU X. W. and MAN X. H. Study on the Application of Composites in Marine Ships. Fiber Composites, 2023, 40 (02): 107-110.
[4] QIN T. T., DING Y. L., GUO J., Cai W. J., Han X. B., and Tan K. Influence of composite layer angle on the mechanical characteristics of the propeller. Ship science and technology, 2023, 45 (08): 20-25.
[5] CHEN Q., JIA Z. K., and ZHANG G. J. Analysis of Vibration Response Characteristics of Ship Composite Material Shafting. Journal of Wuhan University of Technology (Transportation Science & Engineering), 2023, 47 (03): 472-478.
[6] FENG Y. R., WANG J. L., ZHANG B. J., Liu Z. Y., Li J. Y., and Zhang B. S. Modeling of the composite blade and analysis of the effect of ply parameters on strength. Composites Science and Engineering, 2022 (07): 25-31.
[7] CHEN W. Q., WU R., and LIU H. Test of propeller blade natural frequency. Journal of Shanghai Ship and Shipping Research Institute, 2021, 44 (03): 6-10.
[8] WANG K. Research on the damping characteristics of composite propellers. Harbin: Harbin Institute of Technology, 2022.
[9] GONG H. S., MEI Z. Y., HUANG Z., and Zhang Y. B. Calculation and Analysis of Influence of Damping on the Vibration Characteristics of Composite Blade. Ship engineering, 2022, 44 (09): 1-6+13.
[10] HUANG Z., XIONG Y., and YANG G. Composite propeller's strain modal and structural vibration performance. Chinese journal of ship research, 2016, 11 (02): 98-105.
[11] JING J., HU Y. F., and DING G. P. Simulation Research on Modal Analysis of CFRP Propeller. Digital Manufacture Science, 2022, 20 (01): 29-33.
[12] Kishore ML P. and Behera R. K. Base Line Study for Determination of Effect of Stacking Sequence on Vibration Characteristics of Composite Propeller Blade. Aquatic Procedia, 2015. 4: 458-465.
[13] Hong Y., He X. D., and Wang R. G. Vibration and damping analysis of a composite blade. Materials & Design, 2012. 34: 98-105.

Journal of Physics: Conference Series **3009** (2025) 012018 doi:10.1088/1742-6596/3009/1/012018

A high-performance electrochromic Mo-doped WO3 film towards smart window

Tianzi Xiang [1], Quanyuan Xu [*, 1, 2], Haorui Liu [1], Ying Lu [1]

[1]College of Big Data and Intelligent Engineering, Southwest Forestry University, 300 Bailong Temple, Kun Ming, Yunnan Province, China 650224;
[2]Key Laboratory of Forestry and Ecological Big Data State Forestry Administration on Southwest Forestry, Kun Ming, Yunnan Province, China 650224.

* Corresponding author: xuqy@swfu.edu.cn

Abstract. Mo-doped WO_3 (Mo@WO_3) thin film electrode was prepared through the hydrothermal method with convenient and inexpensive steps. The microstructure and characterization of Mo@WO_3 thin film were studied through SEM, XRD, and EDS. The Cyclic voltammetry, Chronocoulometry, Transmission spectrum, and Optical transmittance response were studied by electrochemical workstation and optical fiber spectrometer. The results show that compared with single WO_3 film, Mo@WO_3 thin film improves the electrochromic properties, which provides a novel idea for electrochromic smart windows.

1. Introduction

With the progress of social civilization, the deficiency of energy scarcity and ecological degradation are turning into a grave crisis [1, 2]. Smart windows have been designed, which can effectively modulate light and heat through smart windows with low energy consumption [3]. Tungsten trioxide (WO_3) is the most typical metal oxide for smart windows with significant colour transformation [4-8]. But tungsten trioxide has some disadvantages, like low electronic conductivity, long response time, and slow rate of ion insertion and extraction [9], which impact the electrochromic properties of the smart windows. To solve these disadvantages, researchers enhance the microstructure and conductivity of WO_3 by doping it with metals, which improves the electrochromic of the film [10]. Among these metals, Mo possesses high electron transmission efficiency. Meanwhile, Mo atoms have similar characteristics to W atoms. Madhavi et al. prepared Mo@WO_3 film through the magnetron sputtering technique method. The electrochromic of Mo-doped WO_3 thin films is obviously improved compared with single WO_3 thin films [11]. Jittiarporn et al. use the sol-gel method to prepare MoO_3-WO_3 film. The MoO_3-WO_3 film creates porosity in the film, which enhances electrochromic properties [13]. O-Rueda et al. use the spray pyrolysis method to prepare Mo-doped WO_3 films in FTO [12]. The microstructure was modified by doping Mo, which can improve the H^+ transmission rate in electrolytes.

However, the preparation methods of the above research have some defects, such as a complex manufacturing process and the need for professional machines, which lead to long preparation times and high costs. However, the hydrothermal is easy to prepare, and the prepared films usually have a special structure. The film with a three-dimensional structure can be produced through the hydrothermal method, which can improve the transport capacity of the electrolyte, thus improving the electrochromic properties. In this paper, Mo@WO_3 films were prepared by one-step hydrothermal. Meanwhile, their electrochromic properties were studied. The preparation of Mo@WO_3 film by hydrothermal method provides a cleaner preparation process and reduces the preparation cost of smart windows in the future.

2. Experimental

2.1. Materials

Nitrogen (N_2) was purchased from Anqiu Hengan Gas LTD. Sodium tungstate ($Na_2WO_4 \cdot 2H_2O$) was purchased from Wujiang Tengxiang Chemical LTD. Oxalic acid ($C_2H_2O_4$) was purchased from Qingzhou Xinsheng Chemical LTD. Sodium molybdate (Na_2MoO_4) was purchased from Nanjing Chemical Reagent LTD. Conductive glass (FTO) was purchased from Beijing Tebo Wande Technology LTD. Acetone (CH_3COCH_3) was purchased from Yangzhou Xinyuan Chemical LTD. Lithium perchlorate ($LiClO_4$) was purchased from Anhui Jinyue Guanxin Material Technology LTD. Isopropanol ($(CH_3)_2CHOH$) was purchased from Shanghai Xinlong Yuteng LTD. Hydrochloric acid (36~38%) and methanol (CH_3OH) were all purchased from McLean Biotech LTD. Deionized water was purchased from Tangshan Jingsu Water Treatment LTD.

2.2. Preparation of WO_3 and Mo@WO_3 films

Initially, 8.2465 g of $Na_2WO_4 \cdot 2H_2O$ was added in 150 mL of deionized water. The pH was modulated to 1.4 using Hydrochloric acid. The solution was stirred continuously for five minutes. Subsequently, 0.5673 g of $C_2H_2O_4$ was dissolved in 50 mL of deionized water. Thereafter, $Na_2WO_4 \cdot 2H_2O$ was added to this solution in quantities corresponding to Mo to W atomic ratios of 2% with stirring maintained for five minutes. Both solutions were then combined and stirred for an additional ten minutes to formulate a solution. Then, 4.1 mL of the prepared solution was added to a container. The reaction was carried out in 150°C for 1.5 h. Upon rapid cooling reached 27°C, the samples were removed and extensively rinsed through deionized water. The films were then dried at 80°C for ten minutes and subsequently put at 420°C for 1.2 h. For ease of identification, Mo-doped WO_3 film was called Mo@WO_3.

3. Electrochromic performance of WO_3 and Mo@WO_3 samples

The XRD patterns are shown in Figure 1. The XRD patterns reveal distinct peaks at 2θ, which correspond to the hexagonal WO_3 (JCPDS No. 75-2187), respectively. Notably, no diffraction peak attributable to Mo was detected in the XRD patterns. This absence may be explained by the structural similarity between Mo^{6+} and W^{6+}, which makes the Mo peaks undetectable. Additionally, the concentration of Mo might be below the detection accuracy of the technique.

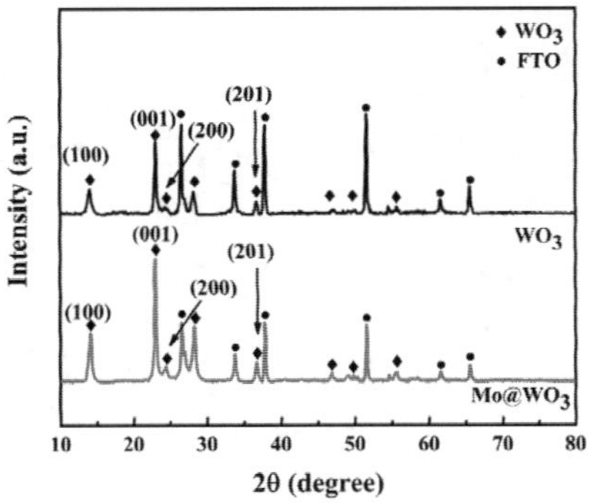

Figure 1. XRD pattern.

Figure 2 shows the SEM image. The WO_3 film surfaces of samples are characterized by a clustered morphology consisting of aggregated particles. The Mo@WO_3 film surfaces of samples are characterized by a regular nanowire.

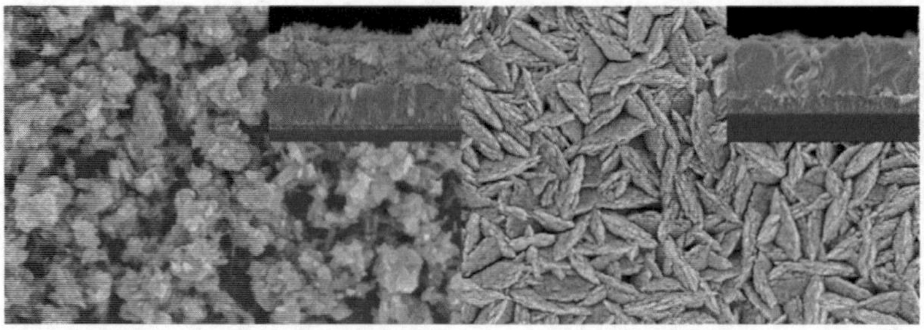

Figure 2. SEM of the WO_3 and Mo@WO_3 samples.

(a) SEM of WO_3 film; (b) SEM of Mo@WO_3 film.

Figure 3 presents the EDS spectrum. The spectrum identifies the presence of Mo elements. The results show that Mo is doped into thin films WO_3. The combined analyses from XRD, EDS, and SEM confirm that the Mo@WO_3 film has been prepared.

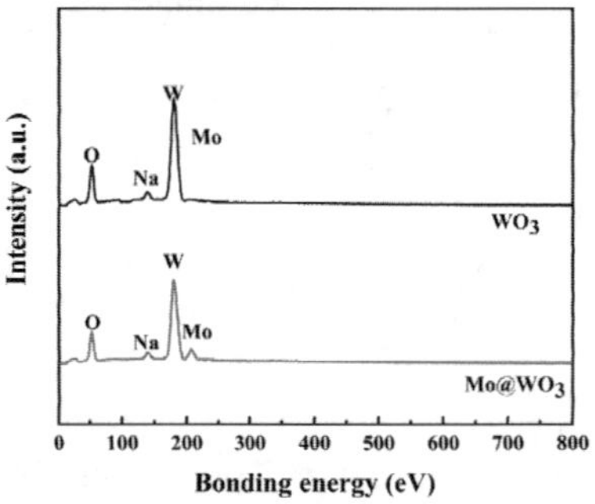

Figure 3. EDS spectrum.

Figure 4 displays the CV curves. The CV profiles exhibit similar shapes. Notably, the Mo@WO_3 film demonstrates a larger peak current and a larger area. This indicates that the Mo@WO_3 film possesses better electrochromic performance.

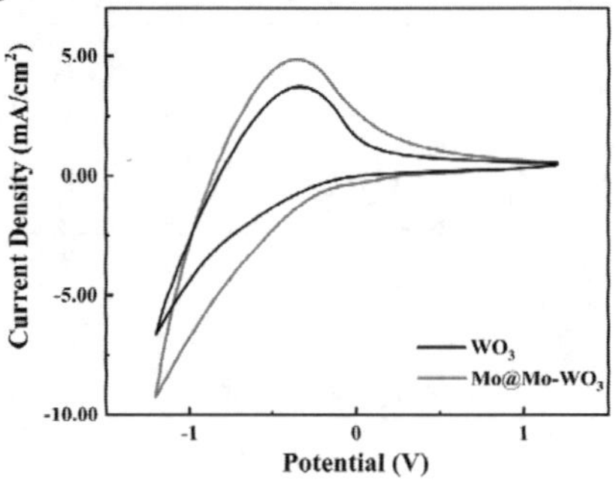

Figure 4. Cyclic Voltammetry (CV) curves.

Figure 5 presents the chronocoulometry (CC) curves, measured at 50-second intervals. The reversibility of the films (Q_{di}/ Q_i) can be derived from these curves. Compared with the WO_3 film (77.10%), the reversibility of Mo@WO_3 film has been enhanced (95.64%).

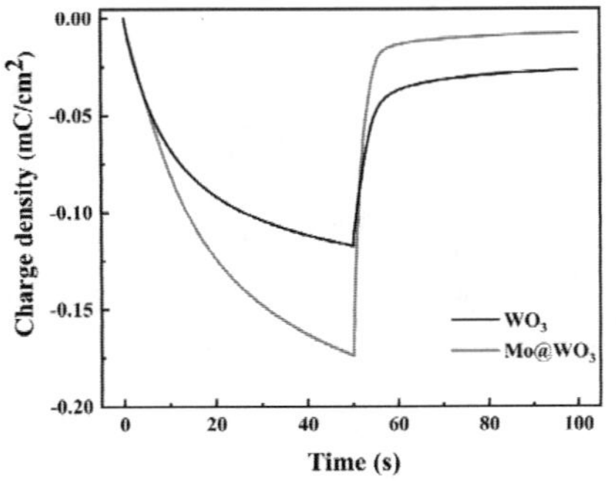

Figure 5. CC curves with 50 s intervals.

Figure 6 illustrates the transmission spectra. The optical modulation range of Mo@WO_3 film is 64.95%, nearly 1.49 times that of WO_3 film (43.49%) in 633 nm. Meanwhile, the photo of Mo@WO3 film is shown in Figure 5, with the bleached state above and the colored state below.

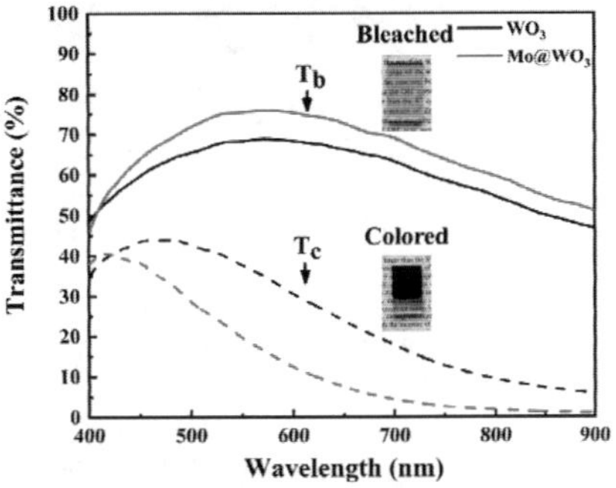

Figure 6. The transmission spectrum between 400 nm to 900 nm.

Figure 7 depicts the optical transmittance response, recorded at 50-second intervals. The Mo@WO_3 films exhibit a faster color conversion rate (t_c/t_b is 9.25 s/30.54 s) than WO_3 film (t_c /t_b is 10.31 s/46.23 s).

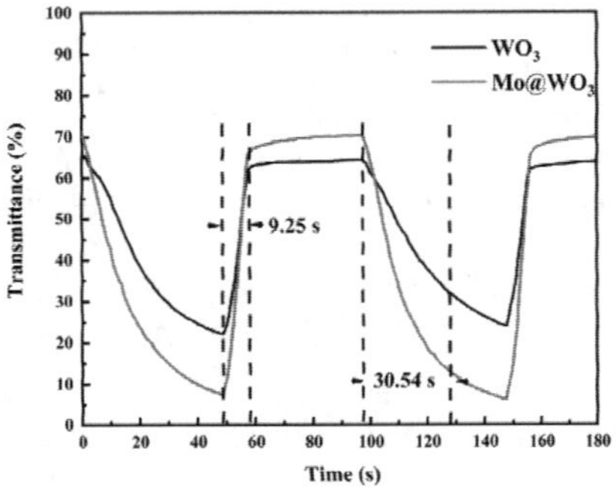

Figure 7. Optical transmittance response.

The electrochromic properties of Mo@WO_3 films exhibit significant enhancement. This improvement is primarily concluded that doping with Mo markedly improves the microstructure and crystallinity of WO_3 films. These enhancements facilitate the formation of efficient transport channels that support the movement of Li ions and electrons.

4. Conclusion

Mo@WO_3 film was fabricated by hydrothermal. The Mo@WO_3 film exhibited marked improvements in properties. The Mo@WO_3 film demonstrated superior reversibility (95.64%), the fastest color switching time (t_c/t_b is 9.25 s/30.54 s), and the widest optical modulation range (64.95% in 633 nm). This not only introduces a simple preparation method but also broadens the potential application of electrochromic films in smart windows.

Acknowledgments

This work was supported by the Education Department Project of Yunnan Province (Grant No. 2022J0493)

References

[1] Amalia D. (2024) The Integration of Eco-Literacy in Teaching-Learning Activities as An Effort to Reduce Environmental Crisis. Madinah: Jurnal Studi Islam, 11: 257-274. DOI: 10.58518/madinah.v11i2.2962
[2] Teguh, A., Syam, F. (2024) Evaluation of Legal Policy Implementation in Addressing Environmental Crisis: A Case Study on Plastic Waste Management. Indonesian Journal of Interdisciplinary Research in Science and Technology, 2: 405-414. DOI: 10.55927/marcopolo.v2i4.8757
[3] Chen, H., Chang, G., Lee T H., Min, S., Nam, S., Cho, D., Ko, K., Bae, G., Lee, Y., Feng, J., Zhang, H., Kim, J., Shin, J., Hong, J., Jeon, S. (2024) Compression-sensitive smart windows: inclined pores for dynamic transparency changes. Nature Communications, 15: 8074. DOI: 1 0.1038/s41467-024-52305-6
[4] Bai, J., Zhang, Y., Pan Z., Yin, X., Yang, G., Zhang, D., Zhou, H., Liang, X. (2024) Optimization of electrochromic Mo doping WO3 films: A study on dual-phase stacked structures for energy-efficient smart windows. Electrochimica Acta, 508: 145295. DOI: 10.1016/j.electacta.2024.145295
[5] Maho A. (2024) Plasmonic ITO and WO3-x nanostructures as innovative electrochromic materials for dual-band VIS-NIR smart windows. In: 5th International Conference on Emerging Advanced Nanomaterials-ICEAN 2024. Sydney. DOI: 10.2139/ssrn.5050857
[6] Netto, M. S., Lemos, R. M. J., Rodrigues, M. P., Porto, J. S., Brum, R. S., Bulhôes, L. O. S., Avellaneda, C. O. (2024) Characterization and Energy Performance of WO3 Doped and Undoped Photochromic Films. Physica status solidi (a), 2400425. DOI: 10.1002/pssa.202400425
[7] Santosa, A. S. S., Wardani, N. K., Albab, M. F., Jahandar, M., Heo, J., Chang, D. W., Kim, S., Lim, D. C. (2024) Scalable flexible electrochromic devices with WO3-Based ion storage layer for enhanced optical modulation and stability. Applied Surface Science, 2024: 162101. DOI: 10.1016/j.apsusc.2024.162101
[8] Koay, H. H., Thien, G. S. H., Au, B. W. C., Abdelhamed, A. H. E., Balachandran, R., Ng, Z. N., Chan, K. Y. (2024) An IoT-enabled monitoring and control system for electrochromic smart windows. Discover the Internet of Things, 4: 1-18. DOI: 10.1007/s43926-024-00086-1
[9] Zhang, Y., Wang, Q. (2024) Acceleration photochromic performance in tungsten oxide. Optical Materials, 157: 116365. DOI: 10.1016/j.optmat.2024.116365
[10] Li, R., Lan, F., Tang, L., Huang, Ya., Zhao, S., Wang, B., Han, Y., Gao, D., Jiang, Q., Zhao, Y., Zhao, Z., Wang, F., Zhang, R. (2024) A Local-Dissociation Solid-State Polymer Electrolyte with Enhanced Li+ Transport for High-Performance Dual-Band Electrochromic Smart Windows. Advanced Functional Materials, 2024: 2419357. DOI: 10.1002/adfm.202419357
[11] Madhavi V., Kondaiah P., Hussain O.M., Uthanna S. (2014) Structural, optical and electrochromic properties of RF magnetron sputtered WO_3 thin films. Physica B: Condensed Matter, 454: 141-147. DOI: 10.1016/j.physb.2014.07.029
[12] O-Rueda de León J.M., Acosta D.R., Pal U., Castañeda L. (2011) Improving electrochromic behavior of spray pyrolysed WO_3 thin solid films by Mo doping. Electrochimica Acta, 56: 2599-2605. DOI: 10.1016/j.electacta.2010.11.038
[13] Jittiarporn P., Sikong L., Kooptarnond K., Wirach T., Stefan S., Simona B., Vo-Van T. (2017) Electrochromic properties of MoO3-WO3 thin films prepared by a sol-gel method, in the presence of a triblock copolymer template. Surface and Coatings Technology, 327: 66-74. DOI: 10.1016/j.surfcoat.2017.08.012

Study on the wear resistance and process parameters of FeCoNiCrAl high entropy alloy coating by laser cladding

Maolong Chen [1]; Kaiguo Zhang [2]; Yifeng Li [2]; Antao Zhao [2]; Qilin Zhang [2]; Lubin Sang [2]; Wangliang Zhao [2]; Xuefeng Yang [3*]

[1] School of Mechanical Engineering, University of Jinan, Jinan City, CHINA, email: chenmaolong2022@163.com;
[2] Shandong Xingneng Thermal Energy Technology Co., Ltd., Tai'an City, CHINA;
[3] School of Mechanical Engineering, University of Jinan, Jinan City, CHINA

[*]Corresponding author's e-mail: wshytzh@163.com

Abstract: FeCoNiCrAl high-entropy alloy coatings formed via laser cladding were examined to improve wear resistance. An orthogonal experimental design adjusted laser power, scanning speed, and powder flow rate. Higher laser power and moderate scanning speeds tripled the coating's hardness compared to the substrate. Refined microstructures and uniform phase distribution enhanced resistance to adhesive, abrasive, and fatigue wear. Optimal parameters (1400 W, 600 mm/min, and 1.0 L/min) yielded superior performance, providing a basis for future advancements in high-performance coatings.

1 Introduction

The ability to withstand wear is a vital attribute for materials employed in a multitude of industrial sectors, such as aerospace, automotive, and manufacturing, where components are consistently exposed to mechanical stress and friction [1, 2]. Conventional materials often fail to meet the stringent requirements of high-wear environments, which has driven the investigation into more advanced materials like high-entropy alloys.

High-entropy alloys (HEAs) is a special multicomponent alloy that has at least five or more miscible principal elements, where each element's atomic fraction falls between 5 and 35% [3, 4]. Compared with traditional alloys, HEAs are highly homogeneous in composition and have no obvious primary and secondary elements. The inclusion of multi-principal components in HEAs helps to attain higher mechanical characteristics because of the combination effect of sluggish diffusion, high entropy, cocktail effect, and lattice distortion. FeCoNiCrAl is one of the most widely studied high-entropy alloy systems. The large atomic radius of the Al element is beneficial to the lattice distortion effect, while Co and Ni are beneficial to the formation of FCC solid solution. At the same time, the Co element can improve the plasticity and wear resistance of the alloy, and Ni can improve the corrosion resistance and high-temperature oxidation resistance of the alloy. The random arrangement of these elements in the solid solution of FeCoNiCrAl HEA improves the strength and hardness of the material and makes it exhibit stronger wear resistance.

Due to its complex alloy composition and high melting point, HEAs may encounter certain difficulties in the processing and forming process. For example, phase separation or crack problems are prone to occur during casting and welding. Laser cladding is a sophisticated surface modification technique that provides precise control over the deposition process, resulting in coatings with exceptional adhesion, uniformity, and minimal dilution with the underlying substrate [5, 6]. This technology is particularly suitable for HEA coatings because it can achieve high-density energy beams and rapid cooling rates to ensure the microstructure and properties of HEA [7].

Although previous studies have shown that laser cladding HEA coatings can significantly improve the wear resistance of various substrates, few scholars have systematically studied the process parameters of laser cladding HEA coatings [8]. The laser cladding process parameters have a direct impact on the bonding quality and microstructure of the HEA coating, which affects the performance of the HEA wear-resistant coating. In this study, the FeCoNiCrAl HEA with excellent performance was studied. The orthogonal experiment of laser cladding process parameters was designed to study the influence of process parameters on the wear resistance of HEA coating, and the optimal process parameters were obtained.

2 Orthogonal experiment

Table 1 gives the three levels of the three most critical parameters affecting coating quality during laser cladding: laser power, scanning speed, and powder flow rate. Table 2 is an L9 (33) orthogonal array constructed based on Table 1.

Table 1 Factor levels.

Factor	1	2	3
Laser power A (W)	1200	1400	1600
Scan speed B (mm/min)	500	600	700
Powder flow C (L/min)	0.8	1.0	1.2

Table 2 Orthogonal array.

Sample	A	B	C
#1	1	1	1
#2	1	2	2
#3	1	3	3
#4	2	1	2
#5	2	2	1
#6	2	3	3
#7	3	1	3
#8	3	2	1
#9	3	3	2

3 Wear resistance analysis

3.1 Harness

Figure 1 describes the surface hardness of the HT250 substrate and the laser cladding FeCoNiCrAl HEA coating on the HT250 surface. The hardness of the FeCoNiCrAl HEA coating is the lowest in the #2 coating, but it is still twice that of the substrate, and the highest hardness is the #8 coating, which is three times that of the substrate. From Figure 1, it can be seen that #1, #2, and #3 are coatings with a laser power level of 1200 W. The hardness is 114~247 $HV_{0.2}$ lower than that of the medium laser power level (1400 W) and the high laser power level (1600 W). This may be because when the laser power is low, the energy density is small, resulting in insufficient heat input on the substrate surface, small molten pool depth, and width, poor metallurgical bonding quality between the cladding layer and the substrate, and insufficient diffusion of alloy elements. Simultaneously, the grain growth of the cladding layer is insufficient and the microstructure uniformity is poor, ultimately leading to lower hardness. Low-level laser power may a cause low dilution rate of the cladding material and insufficient diffusion of alloy elements, resulting in a small amount of intermetallic compound generation and uneven distribution. Moreover, the temperature of the molten pool rises slowly at low power, and the depth of the molten pool is shallow, which cannot produce sufficient convection and stirring effects and will also cause uneven mixing of components and reduce hardness. In addition, areas with insufficient local melting may form coarse columnar crystals or inter-grain segregation, further weakening the hardness. At higher power, the melting depth and width of the cladding layer are greater, and the alloy elements are fully melted and diffused, forming a better metallurgical bond with the substrate.

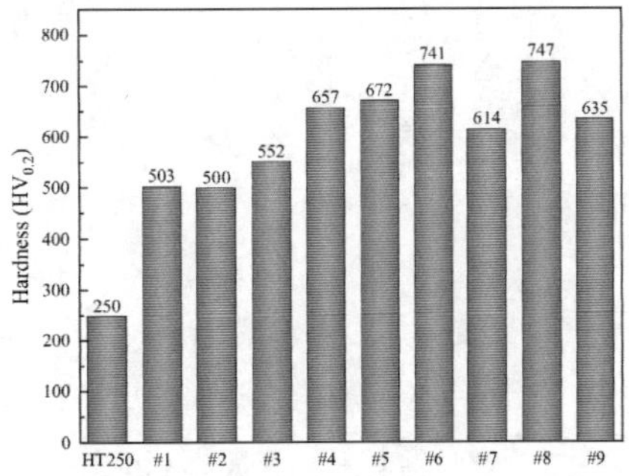

Figure 1 Surface hardness of substrate and coating.

3.2 Abrasion morphology

In Figure 2, spalling pits and large patches of adhesive layer were discovered on the surface. Under the action of alternating stress, the gray cast iron with low hardness and good plasticity experienced severe deformation along the sliding direction, resulting in the observation of deep and narrow dimples under high-magnification SEM. The gray cast iron's surface fractures and undergoes some elastic shrinkage when the deformation is beyond the tensile limit. The diameter of these dimples represents the shrinkage distance. These dimples are continuously expanded along the sliding direction to create elongated dents during the ensuing wear process. These features show that adhesive wear, fatigue wear, and mild abrasive wear are the wear mechanisms and that the substrate has a low wear resistance.

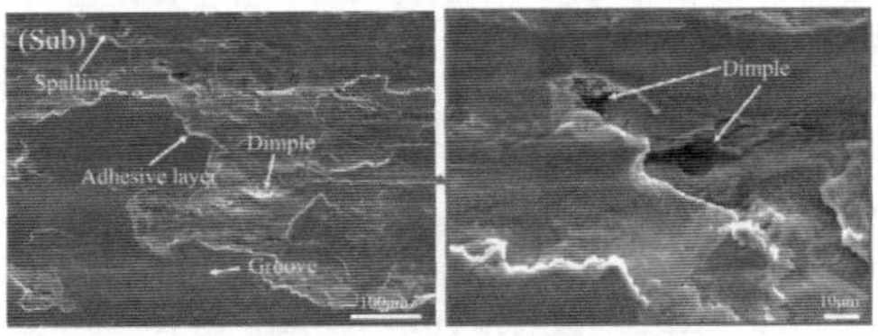

Figure 2 High-power SEM friction and wear morphology of the matrix.

In Figure 3, there are different degrees of groove-like wear marks and adhesive layers in coatings #1, #2, and #3. The grooves of coating #2 are deeper and the spalling pits are shallower. There is a large number of irregular wear debris in coating #1, and coating #3 has similar but shallow dimples with the substrate. This indicates that the overall wear resistance of coating #1 and coating #3 is lower than that of coating #2, and the wear resistance of coating #3 is similar to that of the substrate. The wear mechanisms of coatings #1 and #2 are adhesive wear and abrasive wear, and the wear mechanisms of coating #3 are adhesive wear and fatigue wear.

When the laser power is maintained at 1200 W, a low scanning speed (such as 500 mm/min) will lead to excessive heat input, which will cause coarse grains, enlarged heat-affected zone, and residual stress concentration inside the coating, eventually weakening the coating performance. With the increase in scanning speed (600-700 mm/min), the cooling rate of the molten pool is accelerated, the grains are refined, the coating structure is more compact, and the wear resistance is significantly improved. A low powder flow rate (such as 0.8 L/min) may lead to insufficient coating thickness or high porosity in the coating, which reduces the wear resistance of the coating. Properly increasing the powder flow rate (such as 1.0-1.2 L/min) can improve the thickness and density of the coating, reduce the peeling and crack formation of the coating during wear, and improve the wear resistance. The laser power remains unchanged, and the heat input is jointly adjusted by the scanning speed and powder flow rate. A higher scanning speed (such as 600 mm/min) and a higher powder flow rate (such as 1.0 L/min) can reduce the defects of the coating, enhance its density and uniformity, and thus improve the wear resistance.

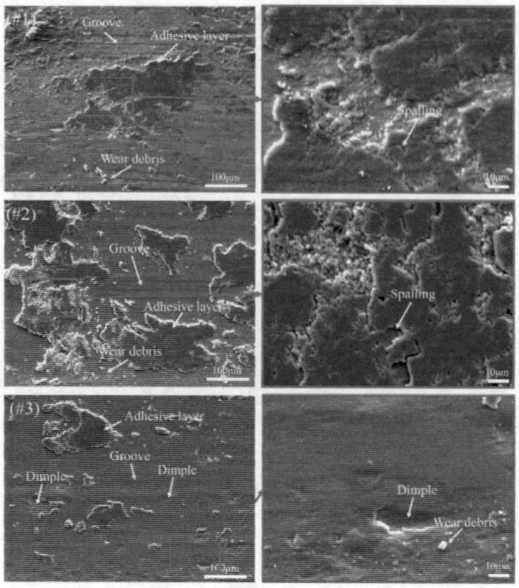

Figure 3 High-power SEM friction and wear morphology of HEA coating #1, #2, and #3.

In Figure 4, the adhesive layer and spalling microcracks and pitting pits were observed in coating #4, and the wear mechanism was adhesive wear and fatigue wear. Adhesive layer and grooves were observed in coating #5, and the wear mechanism was mainly adhesive wear, supplemented by abrasive wear. The wear surface of coating #6 is relatively flat, and the adhesive layer is less. It is mainly characterized by furrows parallel to the sliding direction, which indicates that its wear resistance is relatively good. However, the grooves on the surface of coating #6 are relatively deep, and there are still large spalling pits and microcracks at the edge of the spalling pits. The wear mechanism is abrasive wear and a small amount of adhesive wear and fatigue wear.

At medium laser power, the coatings have less wear debris and a thinner adhesion layer. The low scanning speed (500 mm/min) of coating #4 has a small number of microcracks because the molten pool has a long residence time and a large heat input, which may cause grain coarsening in the coating microstructure and poor density in local areas. The internal residual stress of the coating increases, increasing the risk of cracks and grooves. The medium scanning speed (600 mm/min) of coating #5 provides moderate heat input, allowing the molten pool to achieve a balance between cooling and melting. Rapid cooling helps to refine the grains, improve the hardness and wear resistance of the coating, reduce the expansion of the heat-affected zone, and reduce residual stress. The obvious microcracks on the surface of coating #6 are because of the molten pool residence time at high scanning speed. The grains are small but the bonding force is insufficient, which is prone to cracks and brittle peeling.

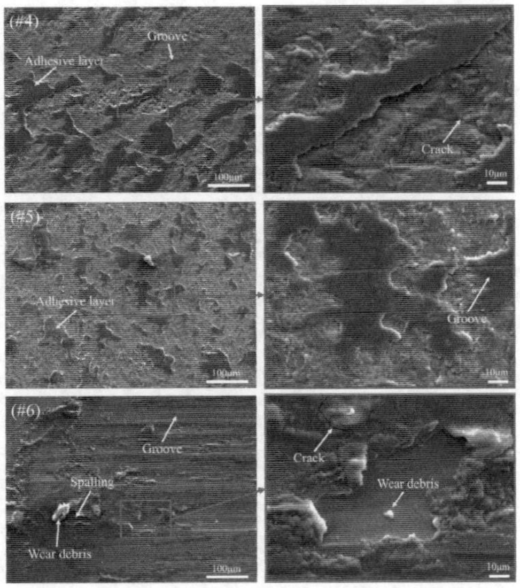

Figure 4 High-power SEM friction and wear morphology of HEA coating #4, #5, and #6.

In Figure 5, large spalling pits and adhesive layers were observed in coating #7, and long grooves appeared on the spalling surface, accompanied by a large amount of wear debris, which indicated that coating #7 was adhesively worn at the initial stage of wear. With the increase of friction and wear time, abrasive wear began to appear on the peeling surface, and the wear mechanism was a combination of adhesive wear and abrasive wear. A relatively thin adhesive layer and shallow furrows were observed in coating #8, and some small spalling pits and microcracks that had not yet expanded were observed. The wear mechanism of coating #8 was abrasive wear and slight adhesive wear. Thick and large spalling pits appear on coating #9, and microcracks appear in the depth of the spalling pits, which indicates that the spalling continues to expand in the depth, which indicates that the bonding of coating #9 is not good, and the wear mechanism is mainly fatigue wear and a small amount of adhesive wear.

The low scanning speed (500 mm/min) and high powder flow rate (1.2 L/min) of coating #7, the high heat input of the laser, and the high powder flow rate together lead to a decrease in the bonding strength of the coating and an increase in surface defects, which become the starting point for crack propagation and spalling. The medium scanning speed (600 mm/min) of coating #8 provides moderate heat input and cooling rate to form a dense coating. At the same time, the low powder flow rate (0.8 L/min) makes the molten pool fluid, reduces unmelted particles, and forms a dense coating. Coating #9 has a fast cooling rate at high laser power (1600 W) and high scanning speed (700 mm/min), which may produce high residual stress inside the coating, triggering crack propagation. In addition, the insufficient bonding strength of the coating leads to the shedding of more particles and the formation of an adhesive layer.

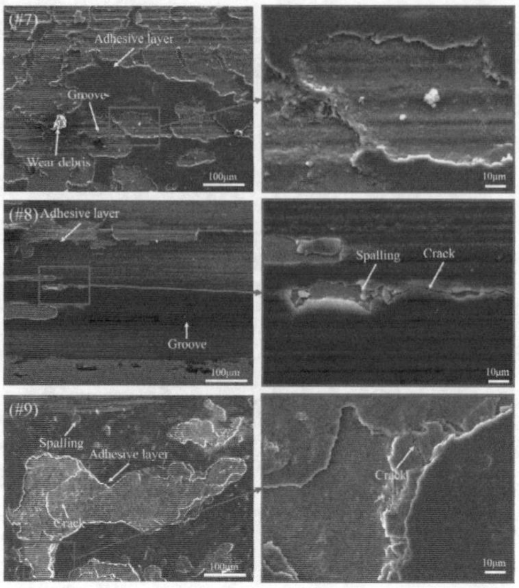

Figure 5 High-power friction and wear morphology of HEA coating #7, #8, and #9.

3.3 Three-dimensional morphology

The friction and wear depth morphology were observed by confocal microscopy, and the volume wear data of the substrate and the coating were obtained. Figure 6 shows the wear depth of the substrate and coating. On the left is the wear depth morphology, and on the right is the depth profile curve measured at the deepest wear depth. It can be seen that the wear depth of the substrate is deeper, reaching about 32 μm. Among the coatings, the wear depths of coatings #1 and #3 are deeper than the substrate, reaching about 34 μm and 57 μm respectively. The rest of the coatings show a smaller wear depth than the substrate. Among them, the wear depth of coating #8 is the shallowest, only about 5 μm.

Table 3 shows the measured wear volume and the wear rate of each coating calculated by the volume wear rate formula. By observing the overall trend, it can be found that as the scanning speed increases, the wear rate generally shows a downward trend. At the same or similar scanning speed, with the increase of laser power, the wear rate shows a downward trend. The effect of powder flow rate on wear rate is not as obvious as that of scanning speed and laser power. Increasing the scanning speed helps to reduce the wear rate, which is due to the higher scanning speed reducing the residence time of the material under the action of the laser, thereby reducing the heat-affected zone and wear. At higher laser power, especially at higher scanning speeds, the wear rate is significantly reduced, which is because the higher laser power improves the melting and recrystallization quality of the material, reducing defects and wear.

Table 3 Wear volume and volume wear rate of substrate and coating.

Sample	Laser power	Scanning speed	Powder flow	Rate of wear
#1	1200	500	0.8	3.08×10^{-4}
#2	1200	600	1.0	7.12×10^{-5}
#3	1200	700	1.2	5.32×10^{-4}
#4	1400	500	1.0	2.50×10^{-5}
#5	1400	600	0.8	1.90×10^{-5}
#6	1400	700	1.2	5.20×10^{-5}
#7	1600	500	1.2	4.60×10^{-5}
#8	1600	600	0.8	1.23×10^{-5}
#9	1600	700	1.0	7.36×10^{-5}

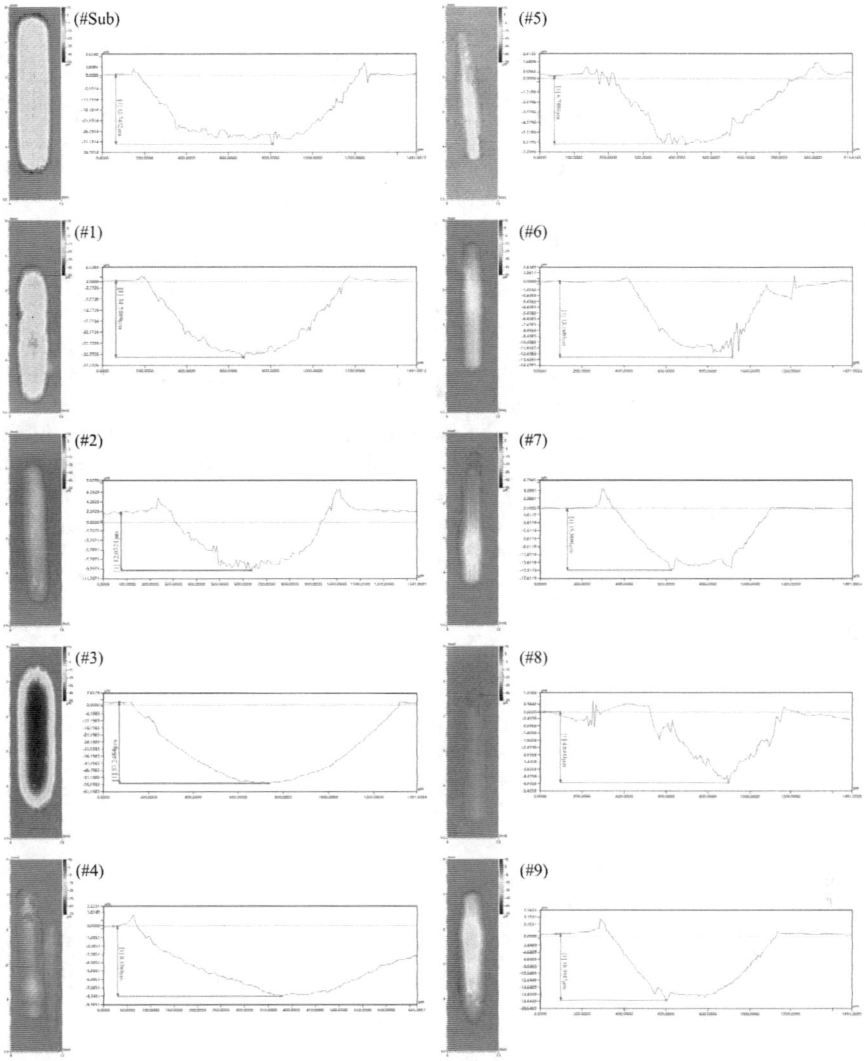

Figure 6 Depth of substrate and coating.

3.4 Mathematical analysis

3.4.1 Extremum difference analysis. Table 4 is the result of the range analysis in response to wear rate. The values of K1, K2, and K3 in the table represent the sum of the wear rates of laser power (A), scanning speed (B), and powder flow rate (C) at different levels. The values of k1, k2, and k3 represent the average wear rates of each factor at different levels. R is the range, which is the difference between the maximum and minimum wear rates of laser power (A), scanning speed (B), and powder flow rate (C) at different levels. It can be seen from the table that the influence of laser power (A) and powder flow rate (C) on the wear rate of the coating is similar, and the influence of powder flow rate (C) is slightly larger. With the increase in laser power, the wear rate began to decrease significantly and then increased slightly. With the increase of powder flow rate, the wear rate also decreases first and then increases, but the range is larger. The smaller the wear rate is, the better the wear resistance of the coating is. Through the range analysis of the wear rate of each factor, it is concluded that (A2, B2, C2), that is, the laser power is 1400 W, the scanning speed is 600 mm/min, and the powder flow rate is 1.0 L/min is the best process parameter for the wear resistance of the coating.

Table 4 Results of range analysis of the orthogonal experiment of wear rate.

Output factor		Laser power A	Scanning speed B	Powder flow C	Optimum parameter
Wear rates	K1	9.11×10^{-4}	3.79×10^{-4}	3.39×10^{-4}	A2, B2, C2
	K2	9.60×10^{-5}	1.03×10^{-4}	1.70×10^{-4}	
	K3	1.32×10^{-4}	6.58×10^{-4}	6.30×10^{-4}	
	k1	3.03×10^{-4}	1.26×10^{-4}	1.13×10^{-4}	
	k2	3.20×10^{-5}	3.43×10^{-5}	5.67×10^{-5}	
	k3	4.40×10^{-5}	2.19×10^{-4}	2.10×10^{-4}	
	R	2.71×10^{-4}	1.85×10^{-4}	1.53×10^{-4}	

3.4.2 Analysis of variance. In the orthogonal experiment, the influence intensity of variance analysis factors on the results was further used. The results of the orthogonal experimental analysis of variance were obtained by consulting the F critical value comparison table, calculating the sum of squared deviations, df, and F ratio, as shown in Table 5. It can be seen from the table that when the dilution rate is the output variable, the order of the F ratio of each factor is: F (A) > F (B) > F (C). That is, the influence intensity of laser power is the largest, about 60%. The scanning speed is about 24%. The powder flow rate has the least influence on the wear rate of the coating, and the influence strength is about 17%.

Table 5 Wear rate orthogonal experiment variance analysis results.

Output factor	d f	Quadratic sum (SS)	Mean square (MS)	F	F marginal value
Laser power A	2	1.30×10^{-7}	6.50×10^{-8}	5.28	19.00
Scanning speed B	2	5.13×10^{-8}	2.57×10^{-8}	2.09	19.00
Powder flow C	2	3.63×10^{-8}	1.82×10^{-8}	1.48	19.00
Error	2	2.46×10^{-8}	1.23×10^{-8}		
Grand total	8	2.12×10^{-7}			

4 Discussion

4.1 Effects of laser power

As can be seen from Figures 1, 3, 4, 5, and 6, from 1200 W to 1600 W, as the laser power increases, the hardness shows an overall upward trend. This is because high power increases the temperature of the molten pool and promotes more complete melting and rapid cooling, resulting in rapid solidification of the coating, grain refinement, and thus increased hardness. Under 1200 W power (#1, #2, #3), a larger bonding layer and cracks appear, and #3 has obvious dimples, indicating that the molten pool in the laser cladding process does not produce sufficient bonding force during bonding. Under 1400 W (#4, #5, #6), the coating uniformity improves and the cracks are significantly reduced. Under 1600 W (#7, #8, #9), obvious grooves appear on the surface of #7 and #8 coatings, showing good wear resistance, but inappropriate combinations (such as #9) show worse wear resistance. This is because low powder flow at high power and high scanning speed will lead to excessive dilution of the substrate, which will also lead to increased stress and crack propagation, thereby reducing the wear resistance of the coating.

Taking all factors into consideration, 1400 W laser power can provide higher heat input to fully melt the powder and substrate, improve the density and bonding of the coating, and avoid defects caused by excessive dilution of the substrate and excessive increase in coating stress; 1400 W laser power is a relatively suitable laser power parameter.

4.2 Effects of scanning speed

It can be seen from Figures 1, 3, 4, 5, and 6 that at each power level, as the scanning speed increases, the hardness also shows an increasing trend. This is because a faster scanning speed reduces heat input, reduces the size of the heat-affected zone, and increases the rapid cooling and solidification rate of the coating, thereby refining the grains and improving the structure. At a higher scanning speed, the wear rates of coatings #3, #6, and #9 are all high, and the peeling phenomenon is obvious (#3, #9), indicating that the molten pool is not fully melted and combined during the laser cladding process, and the coating is prone to defects such as pores and cracks. These defects expand with the action of alternating stress during the wear process and eventually lead to peeling. When the scanning speed is moderate (600 mm/min), the surface density is optimal, cracks and wear debris are very few, and the wear rate is low.

4.3 Effects of powder feeding rate

From the variance analysis in Table 5, it can be seen that changes in powder flow rate have no significant impact on the wear rate. But in Table 1, increasing powder flow also seems to help increase hardness. In Figures 3, 4, and 5, it can be found that under medium (1400 W) and high (1600 W) laser power, as the powder flow rate increases, the hardness first increases and then decreases. It can be found in Tables 3, 4, and 5 that a high powder flow rate usually introduces more wear debris and spalling (#3, #7), which indicates that incompletely melted powder easily falls off during friction. Under medium powder flow, the wear morphology of the coating shows a large area of adhesive layer. This is because, under medium powder flow, the surface bonding strength and plasticity of the coating are relatively moderate, resulting in adhesion during the wear process. Low powder flow (e.g. #8) exhibits better surface friction wear morphology, reducing stickiness and debris in wear.

5 Conclusion

This study focuses on the effect of laser cladding parameters on the wear resistance of FeCoNiCrAl HEA coatings. Through SEM image analysis and confocal microscopy image analysis of friction and wear experiments, it was observed that the process parameters have a significant impact on the wear morphology and wear depth as well as the wear rate. The results of variance analysis show that laser power has the most significant impact on the wear resistance of FeCoNiCrAl HEA coating, followed by scanning speed, and the powder flow rate has the smallest impact. Through range analysis, the optimal combination of process parameters for wear resistance was found (laser power: 1400 W, scanning speed: 600 mm/min, and powder flow rate: 1.0 L/min). This research has practical significance and application prospects for optimizing the wear resistance of laser cladding high-entropy alloy coatings.

Acknowledgements

This work is supported by Key Technologies and Industrial Applications for Design and Manufacturing of Plate-type Liquid Cooling Devices for New Energy Battery Packs (2024TSGC0646) (2024TATSGC008).

References

[1] Smith, J. A., & Lee, K. H. (2020). Wear Mechanisms in Industrial Components: A Comprehensive Review. Journal of Materials Engineering, 45(3), 123-145.
[2] Cai, Y., Manladan, S. M., Luo Z. Tribological behavior of the double FeCoNiCrCux middle-entropy alloy coatings, Surface Engineering, 35 (2019) 14-21. https://doi.org/10.1080/02670844.2018.1444548.
[3] Yeh, J. W., Liaw, P. K., Chen, S. K., & Chang, Y. K. (2004). Nanostructured High-Entropy Alloys with Multiple Principal Elements: Novel Alloy Design Concepts and Outcomes. Advanced Engineering Materials, 6(5), 299-303.
[4] Gao, Y., Zhang, Y., & Wang, S. (2021). Microstructural Evolution and Mechanical Properties of FeCoNiCrAl High Entropy Alloys. Metallurgical and Materials Transactions A, 52(8), 4567-4578.
[5] Kumar, P., Singh, R., & Sharma, M. (2022). Effect of Aluminum Addition on the Wear Resistance of FeCoNiCr High Entropy Alloys. Surface and Coatings Technology, 398, 127345.
[6] Li, X., & Huang, Q. (2018). Laser Cladding Techniques for Surface Modification: A Review. Journal of Laser Applications, 30(6), 061008.
[7] Nguyen, T. H., & Pham, D. T. (2020). Influence of Laser Cladding Parameters on the Microstructure and Properties of High Entropy Alloy Coatings. Applied Surface Science, 518, 146487.
[8] Chen, R., Liu, F., & Zhao, Y. (2023). Wear Behavior of Laser-Cladded CoCrFeMnNi High Entropy Alloy Coatings. Tribology International, 178, 108253.

A magnetic responsive chitosan gel for highly efficient CU (II) adsorption

Ya Wang [1], Junjie He [1], Nan Xiao [2], Yilong Ma [1] and Xianfu Luo [1*]

[1] School of Metallurgy and Power Engineering, Chongqing University of Science and Technology, Chongqing 401331, China
[2] School of Chemistry and Chemical Engineering, Chongqing University of Science and Technology, Chongqing 401331, China

*Corresponding author's e-mail: 2021031@cqust.edu.cn

Abstract: A magnetic chitosan aerogel adsorbent modified by polyethyleneimine (CS-PEI@Fe_3O_4) was successfully synthesized through a simple method. Chitosan (CS) was used as the carrier, glutaraldehyde (GA) as the crosslinking agent, and polyethyleneimine (PEI) as the grafting material, while magnetic Fe_3O_4 particles were incorporated. The CS-PEI@Fe_3O_4 aerogel was fabricated via freeze-drying. Comprehensive characterization was conducted using SEM, VSM, XRD, FTIR, and TG. The effects of PEI molecular weight, Fe_3O_4 content, and pH on adsorption performance were investigated. The statistically averaged particle size of the synthesized Fe_3O_4 was 0.88 μm. The VSM characterization revealed a magnetization intensity of 91.82 emu/g. At the optimal pH of 5, the equilibrium adsorption capacity of PEI-CS for copper (II) ions reached 34.02 mg/g. Kinetic fitting revealed that Cu (II) adsorption by CS-PEI@Fe_3O_4 adhered to second-order kinetics, and the isothermal data were consistent with the Langmuir model. The adsorption process was predominantly chemical adsorption.

1. Introduction

Water resource depletion and environmental pollution have emerged as critical global challenges. Industrial activities and energy production have released metal ions, including Cu, Ni, Pb, and Co, into aquatic ecosystems [1]. These ions accumulate in the human body through biological cycles, posing significant health risks. Various techniques, including chemical precipitation, photocatalysis, electrodialysis, and adsorption, are utilized for heavy metal removal. Adsorption stands out for its simplicity, affordability, and eco-friendliness, attracting significant attention. Recently, nanomaterials have demonstrated high adsorption capacity and rapid adsorption rates, but their limited recoverability and high costs restrict their application in water treatment [2].

Aerogels, characterized by their low density, large surface area, and high adsorption capacity, are considered ideal adsorbents. However, conventional aerogels often lack functional groups, limiting their adsorption effectiveness. Functionalizing natural polysaccharide-based aerogels, such as chitosan, significantly enhances their ability to remove heavy metals. Chitosan, which contains amino ($-NH_2$) and hydroxyl (-OH) functional groups, readily interacts with heavy metal ions to form stable complexes [3]. Nevertheless, its flexible molecular structure and low density of functional groups hinder its adsorption efficiency. PEI, a polyamine with a high charge density, improves chitosan's adsorption capacity through electrostatic interactions and coordination complexation [4]. Traditional gel adsorbents are challenging to recover. However, the integration of magnetic particles like Fe_3O_4 enables efficient retrieval using external magnetic fields. Fe_3O_4 is extensively utilized because of its simple synthesis, non-toxic nature, and biocompatibility. Combining Fe_3O_4 with chitosan enhances the adsorbent's specific surface area, mechanical properties, and adsorption capacity [5].

In this study, a magnetic chitosan aerogel adsorbent (PEI-CS@Fe_3O_4) was synthesized through the cross-linking of PEI with glutaraldehyde and the incorporation of magnetic Fe_3O_4 particles. The effects of PEI molecular weight, Fe_3O_4 content, and pH on adsorption performance were systematically evaluated. Additionally, adsorption kinetics and isothermal adsorption models were applied to investigate the adsorption behavior, performance, and mechanisms of Cu (II) removal from water using PEI-CS@Fe_3O_4.

2. Experiment

2.1. Reagents and materials
Chitosan (CS), polyethyleneimine (PEI), glutaraldehyde (GA), $Cu(NO_3)_2$, ferrous sulfate ($FeSO_4 \cdot 7H_2O$), KOH, HCl, NaOH, and acetic acid were all analytical grade. All experiments were carried out using deionized water prepared in the laboratory.

2.2. Preparation of Fe3O4
Fe_3O_4 was prepared through a hydrothermal co-precipitation method under a nitrogen atmosphere. The Fe^{2+} solution, prepared using $FeSO_4 \cdot 7H_2O$, had a concentration of 33.5 mg/mL, while the Cu^{2+} solution, prepared using $CuSO_4$, had a concentration of 6 mg/mL. The OH^- solution, prepared using KOH, had a concentration of 60 mg/mL. Subsequently, 20 mL of the Fe^{2+} solution, 2 mL of the Cu^{2+} solution, and 10 mL of the OH^- solution were quickly measured, mixed, and transferred to a high-temperature reaction vessel for 2 hours. After cooling, the product was washed with deionized water to obtain Fe_3O_4.

2.3. Preparation of PEI-CS and PEI-CS@Fe_3O_4
1 g of chitosan was completely dissolved in an acetic acid solution, and then the mixture was transferred to a 250 mL round-bottom flask. We dissolve 3 g of polyethyleneimine (PEI) in 10 mL of the 2% acetic acid solution. We add it to the chitosan solution and stir the mixture for 1 hour. We prepare the PEI-CS solution by gradually adding 2 mL of a 25% glutaraldehyde solution at 40°C, followed by stirring for 4 hours.

Subsequently, we incorporate 0.5 g of Fe_3O_4 dispersion (ultrasonicated for 0.5 hours) into the PEI-CS solution and stir for 1 hour to prepare PEI-CS@Fe_3O_4. We add 1 mol/L NaOH solution dropwise to induce gel formation, then filter and wash the gel with deionized water and ethanol until neutral. We freeze the gel for 8 hours and vacuum freeze-dry it at -70°C and 1 Pa for 24 hours to obtain PEI-CS and PEI-CS@Fe_3O_4.

2.4. Adsorption performance evaluation
To assess the Cu (II) adsorption efficiency of PEI-CS@Fe_3O_4, static adsorption experiments were conducted. PEI-CS@Fe3O4 was introduced into a series of solutions containing Cu (II), and the adsorption process was systematically analyzed. The experimental setup included 50 mg of adsorbent, 20 mL of solution, and 100 mg/L Cu (II) solution, a temperature maintained at 30 °C, and an adsorption period of 12 hours.

The saturated adsorption capacity (q_e) of Cu (II) was measured using an atomic absorption spectrometer and calculated using the following formula:

$$q_e = \frac{(C_0 - C_1)}{m} \times V \tag{1}$$

Here, C_0 represents the initial concentration of Cu (II) in the solution (mg/L), C_1 denotes the residual concentration of Cu (II) at adsorption equilibrium (mg/L), V indicates the total volume of the solution (mL), and m stands for the mass of the adsorbent (mg).

3. Results and discussion

3.1. Analysis of Materials

3.1.1. Scanning electron microscopy (SEM). SEM was employed to examine the surface morphology of PEI-CS and PEI-CS@Fe_3O_4. Figures 1(a) and (b) illustrate the surfaces of PEI-CS without Fe_3O_4 and PEI-CS@Fe_3O_4, respectively. The surface of the PEI-CS aerogel is relatively flat, while PEI-CS@Fe_3O_4 shows a similar appearance. SEM analysis further reveals the dispersion of Fe_3O_4 particles on the surface (Figure 1(c)), accompanied by numerous folds and pores of varying sizes. This dispersion results from the uniform embedding of Fe_3O_4 particles within the aerogel matrix, creating a rough and porous surface that facilitates Cu (II) adsorption.

3.1.2 Vibrating sample magnetometry (VSM). VSM was employed to assess the magnetic properties of Fe_3O_4. The saturation magnetization was found to be 91.86 emu/g (Figure 1(d)). Particle size distribution analysis was performed on 40 randomly selected samples (Figure 1(e)), revealing an average particle size of 0.88 μm. PEI-CS@Fe_3O_4 aerogels were doped with Fe_3O_4 at concentrations of 5%, 10%, and 15%, respectively. The measured saturation magnetization values were 3.80, 7.08, and 10.06 emu/g (Figure 1(f)), indicating the successful integration of magnetic Fe_3O_4 into the aerogel. The hysteresis curves showed no typical characteristics of remanence and coercivity, indicating that the nanoparticles exhibit superparamagnetism [6].

3.1.3. X-ray diffraction (XRD). Figure 1(g) displays XRD patterns of Fe_3O_4, PEI-CS, and PEI-CS@Fe_3O_4. PEI-CS without magnetic Fe_3O_4 exhibits broad peaks and lacks distinct crystalline structures. In contrast, PEI-CS@Fe_3O_4, containing magnetic Fe_3O_4, exhibits distinct diffraction peaks at 18.285°, 30.076°, 35.426°, 37.057°, 43.053°, 53.411°, 56.935°, and 62.520°, which correspond to the (111), (220), (311), (222), (400), (422), and (511) planes.

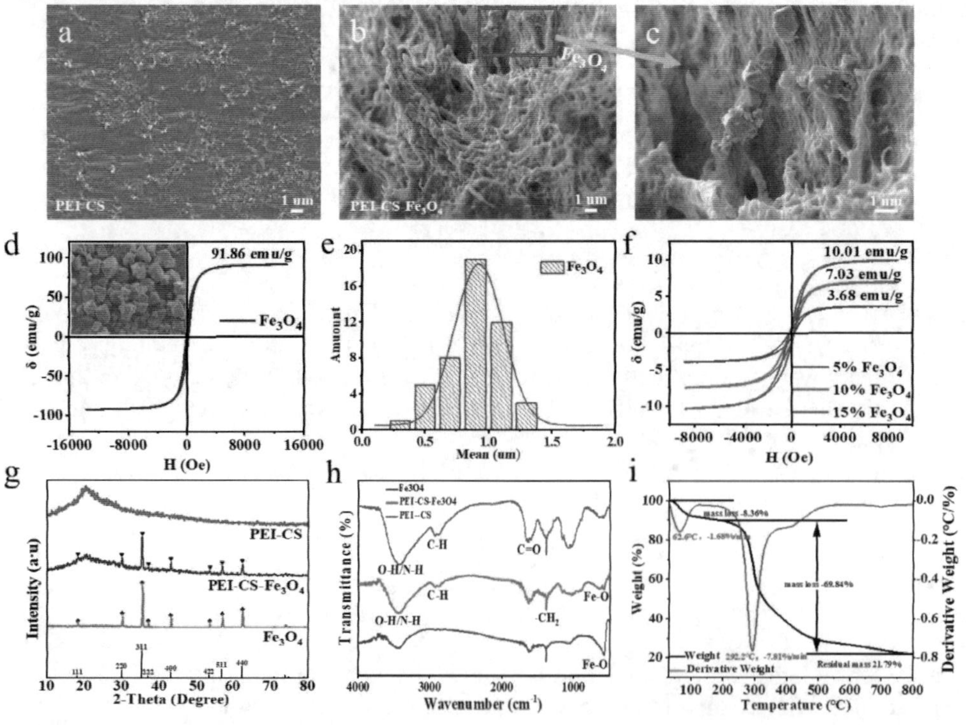

Figure 1 SEM images: (a) PEI-CS, (b) PEI-CS@Fe_3O_4, (c) Fe_3O_4 particles attached to PEI-CS@Fe_3O_4; (d) VSM analysis of Fe_3O_4; (e) The average particle size distribution of Fe_3O_4; (f) VSM analysis of PEI-CS with varying Fe_3O_4 content; (g) XRD patterns of Fe_3O_4, PEI-CS, and PEI-CS@Fe_3O_4; (h) FTIR spectrum of PEI-CS@Fe_3O_4; (i) TG and DTG analysis of PEI-CS@Fe_3O_4.

3.1.4. Fourier-transform infrared (FTIR). Functional group characterization of materials was performed using FTIR, as illustrated in Figure 1(h). Fe_3O_4 exhibited a Fe-O stretching peak at 586 cm^{-1}. PEI-CS and PEI-CS@Fe_3O_4, peaks corresponding to O-H/-NH_2 stretching vibrations appeared at 3432 cm^{-1}; C-H stretching vibrations appeared at 2927 cm^{-1} and 2865 cm^{-1}; and a shear vibration of -NH_2 was noted at 1633 cm^{-1}. Bending vibrations of O=C/-NH_2 and C-H appeared at 1455 cm^{-1} and 1383 cm^{-1}, respectively. Peaks at 1071 cm^{-1} and 1030 cm^{-1} were attributed to C-O stretching vibrations. The peak at 584 cm^{-1} was assigned to Fe-O vibrations. These results indicate that chitosan, PEI, and Fe_3O_4 are the main components of the aerogels [7].

3.1.5. TG analysis. Thermogravimetric analysis (TGA) of PEI-CS@Fe_3O_4 was conducted using a thermal stability analyzer. The TG and derivative thermogravimetric (DTG) curves are shown in Figure 1(i). The mass loss profile consists of three main stages. In the first stage, between 0 and 150°C, the aerogel experiences a mass loss of 8.36%, primarily due to the evaporation of bound water and partial decomposition of PEI. A peak mass loss rate of 1.68%/min is observed at 62.6°C [8]. In the second stage, within the 150-800°C range, a significant mass loss of 69.84% occurs, resulting from the decomposition of PEI and chitosan at higher temperatures. The peak mass loss rate of 7.81%/min is observed at 292.2°C. Finally, approximately 21% of the mass remains in PEI-CS@Fe_3O_4, consisting mainly of carbon and Fe_3O_4 particles that resist decomposition under heat.

3.2. Cu (II) adsorption performance

3.2.1 Analysis of the Effect of PEI Molecular Weight on Adsorption. The adsorption performance of PEI-CS@Fe_3O_4 for Cu (II) is influenced by the molecular weight of PEI. As the molecular weight of PEI increases, the aerogel's adsorption capacity for Cu (II) improves. This improvement is attributed to the increased number of -NH_2 groups on the aerogel's molecular chain, which can coordinate with more Cu (II) ions as the molecular weight of PEI increases (Figure 2(a)).

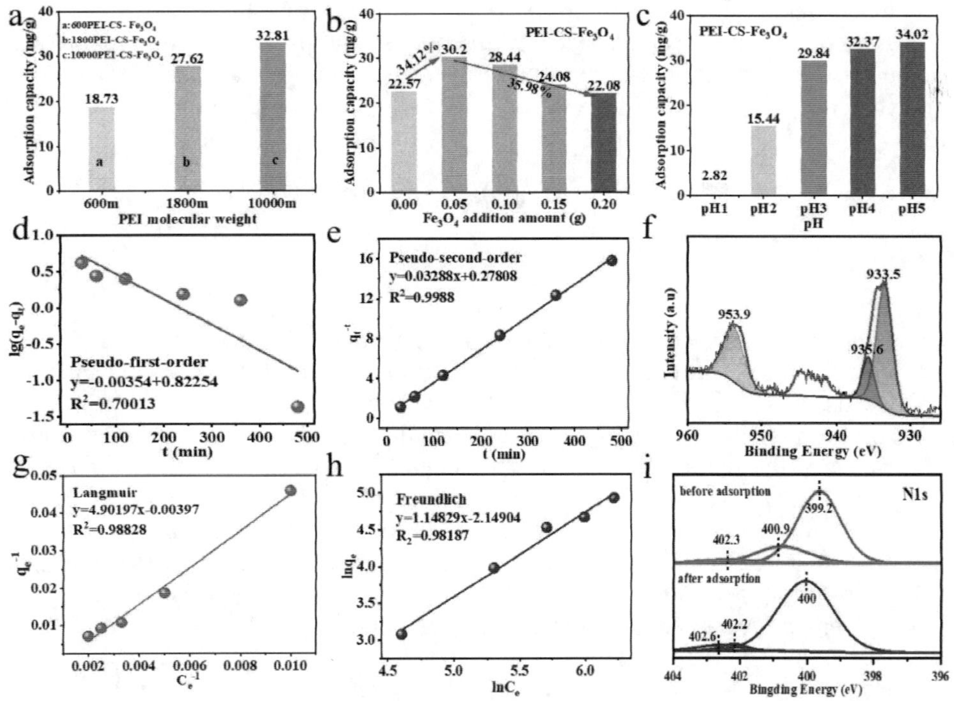

Figure 2 illustrates the influence of various conditions on PEI-CS@Fe_3O_4: (a) PEI molecular weight, (b) Fe_3O_4 content, (c) pH, pseudo-kinetics model: (d) and (e) for Cu (II) adsorption by PEI-CS@Fe_3O_4, (f) Cu 2p core-level spectra, isotherm fitting model (g) and (h) and (i) XPS spectrum of N 1s.

3.2.2. Analysis of the influence of Fe_3O_4 addition on adsorption. The addition of an appropriate amount of Fe_3O_4 enhances the Cu (II) adsorption performance of PEI-CS@Fe_3O_4. As shown in Figure 2(b), Cu (II) adsorption reaches its maximum when the Fe_3O_4 content is 5%, resulting in a 34.13% improvement in adsorption performance compared to the absence of Fe_3O_4. This enhancement results from the higher density of active sites, facilitated by the presence of Fe_3O_4, as well as the enhanced electrostatic attraction. However, as the Fe_3O_4 content increases further, the adsorption performance of PEI-CS@Fe_3O_4 decreases. Specifically, when the Fe_3O_4 content reaches 20%, a 35.98% reduction in adsorption performance is observed compared to the absence of Fe_3O_4. This reduction is caused by an excessive amount of Fe_3O_4, which interferes with the crosslinking of PEI and chitosan (CS), which reduces the density of -NH_2 groups.

3.2.3. Effect of Ph. The adsorption performance of PEI-CS@Fe_3O_4 for copper ions is significantly influenced by the pH level of the solution, as H^+ ions and heavy metal ions compete for the same sites. When the pH exceeds 5, Cu (II) begins to precipitate, with substantial precipitation observed at pH 6 [9]. Therefore, the experimental study focused on evaluating the adsorption behavior of PEI-CS@Fe_3O_4 for Cu (II) within a pH range of 1 to 5. The results demonstrate that the adsorption capacity of PEI-CS@Fe_3O_4 for Cu (II) is strongly influenced by the solution's pH. As the pH increases from 1 to 5, the adsorption capacity rises significantly, reaching a peak value of 34.02 mg/g at pH 5. This trend can be attributed to the activation of -NH_2 and -OH groups at higher pH levels, which significantly enhances the adsorption efficiency of PEI-CS@Fe_3O_4 for copper ions.

3.3. Adsorption fitting of Cu (II)

3.3.1. Pseudo dynamics. The adsorption kinetics model is essential for understanding and optimizing adsorption processes. Its primary purpose is to assess the adsorption rate, elucidate the underlying mechanisms, and identify optimal conditions for achieving efficient adsorption [2]. The pseudo-first-order (PFO) model describes adsorption processes primarily governed by physical interactions, such as van der Waals forces or electrostatic attraction. In contrast, the pseudo-second-order (PSO) model is more suitable for characterizing adsorption processes dominated by chemical interactions, including electron transfer or covalent bonding between adsorbents and adsorbates, with their kinetic equations as follows:

$$\lg(q_e - q_t) = \lg q_e - \frac{K_1}{2.303}t \tag{2}$$

$$\frac{t}{q_t} = \frac{1}{K_2 q_e^2} + \frac{t}{q_e} \tag{3}$$

In these equations, t represents time (min), K_1 and K_2 are the rate constants, and q_t indicates the adsorption capacity at time t (mg/g). q_e corresponds to the equilibrium adsorption capacity when the system reaches adsorption equilibrium (mg/g).

The experimental curves for the pseudo-kinetic models are shown in Figures 2(d) and 2(e), while Table 1 summarizes the corresponding kinetic parameters. The calculated equilibrium adsorption capacity (q_e, cal) of PEI-CS@Fe3O4 for Cu (II), derived from the PFO model, was 2.28 mg/g (R^2 = 0.700). In contrast, the q_e, cal for Cu (II) based on the PSO model was 30.41 mg/g (R^2 = 0.998), which closely matches the experimental equilibrium adsorption capacity (q_e,exp). These results demonstrate that Cu (II) adsorption onto PEI-CS@Fe3O4 follows pseudo-second-order kinetics, indicating a predominantly chemical process driven by chemical interactions.

Table 1 Pseudo-kinetic models for Cu (II) adsorption by PEI-CS@Fe_3O_4.

Metal ions		PFO model			PSO model		
	$q_{e,exp}$/(mg.g^1)	K_1 (min^{-1})	$q_{e,cal}$ (mg.g^1)	R^2	K_2 (min^{-1})	$q_{e,cal}$ (mg.g^1)	R^2
Cu (II)	30.37	0.00354	2.28	0.700	0.00389	30.41	0.9988

3.3.2. Adsorption isotherm. Adsorption isotherms are essential tools for characterizing the adsorption performance of adsorbents in both gaseous and liquid phases [4]. In this work, adsorption isotherm models were employed to investigate how the initial mass concentration of the adsorbate influences the adsorption capacity of the adsorbent. The Langmuir model assumes that the interactions between adsorbate molecules are negligible. The equation is expressed as:

$$\frac{C_e}{Q_e}=\frac{C_e}{Q_m}+\frac{1}{Q_m K_L} \tag{4}$$

where Q_m represents the saturated adsorption capacity at equilibrium, K_L denotes the Langmuir constant (L/mg), and C_e refers to the equilibrium concentration of Cu (II) in the liquid phase under stable conditions.

The Freundlich adsorption model, on the other hand, describes adsorption on a heterogeneous surface with multi-layer adsorption. Its equation is presented as Equation:

$$\ln Q_e=\ln K_F+\frac{1}{n}\ln C_e \tag{5}$$

K_F represents the Freundlich constant [(mg/g) (l/mg)^(1/n)], while n denotes the non-uniformity factor. A value of $n<1$ suggests predominance of chemical adsorption, whereas $n>1$ indicates physical adsorption predominance.

The adsorption data for Cu (II) isotherms were simulated using Langmuir and Freundlich adsorption isotherms (Figure 2(g) and (h)), with the calculated data presented in Table 2. The R^2 value of the Langmuir model is 0.989, which is higher than the Freundlich model's R^2 value of 0.982, suggesting that the adsorption of Cu (II) by PEI-CS@Fe_3O_4 is more accurately described by the Langmuir model, suggesting a predominantly chemical adsorption mechanism.

Table 2 The isotherm of PEI-CS@Fe_3O_4 adsorption on copper (II) ions.

	Langmuir			Freundlich		
Metal ions	q_m(mg/g)	K_L(Lmg^{-1})	R^2	K_F(mg/g)	1/n	R^2
Cu (II)	200.4	0.001	0.989	0.1167	1.148	0.982

3.4. Adsorption mechanism

Figure 2(f) displays the XPS spectra of PEI-CS@Fe_3O_4 after Cu (II) adsorption. The Cu 2p peak is clearly visible, confirming the successful adsorption of Cu (II) by PEI-CS@Fe_3O_4 [5]. The peak observed at 933.5 eV corresponds to Cu $2p_{3/2}$, which may indicate NH_2Cu^{2+} or $(NH_2)Cu^{2+}$. The peak at 935.6 eV is significantly higher than the binding energy of Cu $2p_{3/2}$ and is likely attributed to the satellite peaks of Cu (II). These satellite peaks result from electronic correlation effects and unfilled d-orbitals, generally observed on the higher-energy side of the main peak [7]. The peak at 953.9 eV is associated with the Cu $2p_{1/2}$, indicating pure physical adsorption. These findings confirm the successful adsorption of Cu (II) by PEI-CS@Fe_3O_4.

Figure 2(i) illustrates the N 1s peak position in the XPS spectrum of PEI-CS@Fe_3O_4. The peak position of $-NH_2$ corresponds to 399.2 eV. The peak at 400.9 eV is typical for $-NH_2$. The peak positions at 402.2 eV, 402.3 eV, and 402.6 eV are attributed to the protonation of $-NH_3^+$ [8]. After Cu (II) adsorption, these key positions shift significantly: The coordination of N-Cu leads to the emergence of a peak at 400 eV. The significant shift in the N 1s peak position, reflecting the strong coordination of $-NH_2$ with Cu (II) [9], demonstrates that PEI-CS@Fe_3O_4 exhibits excellent adsorption performance for Cu (II).

4. Conclusion

An efficient magnetic chitosan aerogel adsorbent (PEI-CS@Fe_3O_4) was successfully synthesized by grafting polyethylenimine (PEI) onto the chitosan surface and incorporating Fe_3O_4 magnetic particles. The preparation process was straightforward, utilizing freeze-drying technology to create a highly porous structure. The microstructure and functional properties of the adsorbent were analyzed through multiple techniques, including SEM, VSM, XRD, FTIR, and TGA. Analysis of the experimental data demonstrated that the adsorbent exhibited a high Cu (II) adsorption capacity, reaching a peak of 34.02 mg/g at pH 5. The kinetic analysis and isotherm fitting results indicate that the adsorption process aligns with a monolayer chemical adsorption mechanism. These results highlight the adsorbent's excellent performance and recyclability, offering a promising approach for remediating heavy metal-contaminated water.

Acknowledgment

This study received funds from the Natural Science Foundation of China (No. 52102089), the Science and Technology Research Program of Chongqing Municipal Education Commission (No. KJQN202301508), the Research Foundation of Chongqing University of Science and Technology (No. 20240304) and the graduate innovation project of Chongqing University of Science and Technology (No. YKJCX2320221).

References

[1] Guo Z, Yang R, Yang F, et al. (2021) Fabrication of polyethylenimine functionalized magnetic cellulose nanofibers for the sorption of Ni (II), Cu (II) and Cd (II) in single-component and multi-component systems [J]. International Journal of Biological Macromolecules, 184: 68-78. https://doi.org/10.1016/j.ijbiomac.2021.06.041

[2] Araucz K, Aurich A, Kołodyńska D. (2020) Novel multifunctional ion exchangers for metal ions removal in the presence of citric acid [J]. Chemosphere, 251: 126331. https://doi.org/10.1016/j.chemosphere.2020.126331

[3] Rong N, Chen C, Ouyang K, et al. (2021) Adsorption characteristics of directional cellulose nanofiber/chitosan/montmorillonite aerogel as adsorbent for wastewater treatment [J]. Separation and Purification Technology, 274: 119120. https://doi.org/10.1016/j.seppur.2021.119300

[4] Xiao C, Liu X, Mao S, et al. (2017) Sub-micron-sized polyethylenimine-modified polystyrene/Fe_3O_4/chitosan magnetic composites for the efficient and recyclable adsorption of Cu (II) ions [J]. Applied Surface Science, 394: 378-385. https://doi.org/10.1016/j.apsusc.2016.10.116

[5] Zhang C, Ni C, Liu Z, et al. (2023) Polyethyleneimine (PEI) and Chitosan (CS) grafted with L-cysteine were used as effective materials for Hg (II) adsorption [J]. https://doi.org/10.1016/j.matlet.2023.134614

[6] Sun Y, Wang G, Guo Y, et al. (2024) An eco-friendly wheat straw adsorbent functionalized with mercaptopropionyl for the removal of Cu (II) from aqueous solution [J]. Chemical Engineering Science, 2024, 299: 120454. https://doi.org/10.1016/j.ces.2024.120454

[7] Ali A, Shah T, Ullah R, et al. (2021) Review on recent progress in magnetic nanoparticles: Synthesis, characterization, and diverse applications [J]. Frontiers in chemistry, 9: 629054. https://doi.org/10.3389/fchem.2021.629054

[8] Yang J, He W. (2012) Synthesis of lauryl grafted sodium alginate and optimization of the reaction conditions [J]. International journal of biological macromolecules, 50(2): 428-431. https://doi.org/10.1016/j.ijbiomac.2011.12.027

[9] Wu Z, Deng W, Zhou W, et al. (2019,) Novel magnetic polysaccharide/graphene oxide@Fe_3O_4 gel beads for adsorbing heavy metal ions [J]. Carbohydrate polymers,216: 119-128. https://doi.org/10.1016/j.carbpol.2019.04.020

Journal of Physics: Conference Series **3009** (2025) 012021 doi:10.1088/1742-6596/3009/1/012021

Effects of sandblasting angle on coating removal

Yuqin Li [1*], Hangchao Wang [1], Weizhuo Hua [1*], Jinyu Fan [1], Hang Wang [2], Yuchang Qing [2]

[1] Notional Key Lab of Aerospace Power System and Plasma Technology, Air Force Engineering University, Xi'an 710038, Shaanxi, China;
[2] State Key Laboratory of Solidification Processing, School of Material Science and Engineering, Northeastern Polytechnical University, Xi'an 710072, Shaanxi, China

*Corresponding author's e-mail: *yuqin511@163.com; *hwz1991@sina.com

Abstract. Surface coatings may experience wear, peeling, and other forms of damage during service, which severely impacts their absorbing performance. To restore the performance of damaged coatings, it is crucial to remove coatings. This paper investigated the effect of sandblasting angle on the removal of damaged coatings from TC4 titanium alloy surfaces and determined the optimal sandblasting angle. When SiC is selected as the abrasive, the removal effect is best when the blasting angle is 60°, taking into account the comprehensive sandblasting effect and the degree of damage to the substrate. The sandblasting removal mechanism of surface coatings was explored and this study provides a procedural pathway for the effective removal of surface coatings.

1. Introduction

During the use of surface coatings, cracks, peeling, and other damages may occur, leading to a gradual decline or even failure of their absorbing performance[1]. To repair damaged coatings, it is necessary to promptly remove absorbing coatings with severe performance degradation. TC4 titanium alloy has the characteristics of high strength, good toughness, and excellent welding performance. It is usually used to manufacture key components such as aircraft landing gear. However, the biggest disadvantage of titanium alloy is poor wear resistance [2]. During the coating removal process, it is inevitable that the surface roughness of the TC4 titanium alloy substrate will increase. Appropriate surface roughness can help improve the bonding strength of the coating, but excessive roughness can cause damage to the TC4 titanium alloy substrate.

At present, sandblasting is widely recognized as the most effective surface removal method to improve the bonding strength of coatings [3]. During sandblasting, the impact and abrasion of the sand particles are less likely to create beneficial "hook-like," "crater," and "scratch" features on the surface for mechanical interlocking of the coating and substrate. However, it can remove weaker graphite present on the titanium alloy substrate surface, thus significantly improving the bonding strength between the coating and the substrate[4]. These factors, along with the complexity of the sandblasting process itself, make the sandblasting treatment of titanium alloy substrates complex and necessitate systematic research.

The surface treated by sandblasting is anisotropic, and the microscopic state of the sandblasted surface is influenced by factors such as sand particle characteristics (including shape, hardness, mass, composition, type, etc.), workpiece characteristics, and sandblasting process specifications[5]. During sandblasting, the energy of the sand particles impacting the workpiece surface is primarily consumed in cutting the surface material and causing plastic deformation of the surface. Different combinations of sandblasting distance, angle, and pressure inevitably lead to variations in the energy of the sand particles, the amount used for cutting, and the amount used for plastic deformation, ultimately resulting in differences in the obtained surface condition[6]. During arc spraying, molten spray material particles are ejected at high speeds onto the substrate surface, impacting, deforming, and spreading on the substrate surface, thereby forming a coating through mechanical interlocking with the substrate surface. Therefore, differences in substrate surface roughness, activation level, and surface morphology will result in variations in the bonding strength between the coating and substrate, which is primarily based on mechanical interlocking[7]. Additionally, there is limited research on the effect of sandblasting combined with radar-absorbing coating treatment on the coating removal effectiveness for titanium alloy surfaces.

To address this research gap, many researchers have started focusing on the impact of sandblasting angle on coating removal effectiveness. Some studies indicate that different sandblasting angles significantly affect coating removal efficiency and surface roughness. For example, Li et al.[8] found that a sandblasting angle of 45° results in the best coating removal effect and the lowest surface roughness. Another study by Zhang et al.[9] indicated that sandblasting angles between 60° and 75° are more effective for removing coatings from titanium alloy surfaces.

Sandblasting pressure and the type of blasting media also have significant effects on coating removal efficiency. Wang et al.[10] noted that increasing sandblasting pressure significantly improves the coating removal rate, but excessively high pressure may cause over-damage to the substrate surface. Conversely, lower pressure, while protecting the substrate, may be less effective in removing coatings. Regarding blasting media, Chen et al.[11] found that using harder media such as silicon carbide significantly increases coating removal efficiency but also raises the risk of substrate damage. Softer media such as glass beads, while better protecting the substrate, offer lower coating removal efficiency. A combination of different media might find a balance between substrate protection and removal efficiency.

Recent studies have also explored optimizing sandblasting process parameters. Wu et al.[12] showed that adjusting sandblasting time and flow rate can significantly affect coating removal effectiveness. Li et al.[13] proposed a new method for optimizing sandblasting parameters by considering the interaction between various parameters, which improved coating removal efficiency. Zhao et al.[14] investigated the residual stress distribution on the substrate surface after sandblasting and found that appropriate sandblasting parameters can reduce residual stress and avoid stress concentration.

Other research has focused on changes in surface morphology and physical-chemical properties after sandblasting. Xu et al.[15] used scanning electron microscopy (SEM) and energy-dispersive spectroscopy (EDS) to analyze the surface after sandblasting, finding that sandblasting significantly alters the micro-morphology and improves surface activation, which is beneficial for coating re-adhesion. Zhang et al.[16] studied the effect of different sandblasting media on surface oxide films, finding that harder media more effectively remove oxide films but may introduce new surface defects.

Moreover, there is research on the combined application of sandblasting with other surface treatment methods.

Among previous research, it can be seen that there have been no public reports on the removal of absorbing coatings by sandblasting. Finding a good balance between the removal efficiency of absorbing coatings and the protection of the substrate is of great significance for the application of sandblasting in the removal of absorbing coatings. Therefore, this article studied the influence of sandblasting angle on the removal effect of TC4 titanium alloy surface coating, determined the optimal sandblasting angle, explored the sandblasting removal mechanism of absorbing coating, and provided a process path for the effective removal of absorbing coating.

2. Materials and Methods

2.1. Materials

The experimental substrates are TC4 titanium alloy plates measuring 30 mm × 30 mm × 5 mm, with a self-made radar-absorbing coating approximately 1 mm thick. The sandblasting abrasive used is SiC, provided by Shaanxi Xinghua Chemical Co., Ltd.

2.2. Methods

The sandblasting method used is dry sandblasting. The equipment, as shown in Figure.1, consists mainly of six systems: the structural system, the abrasive power system, the piping system, the control system, and the dust removal system.

Figure 1. Sandblasting equipment.

To achieve better removal efficiency and a shorter removal time, the experiment uses SiC with a higher Vickers hardness (20.4–33.3) as the abrasive.

In this experiment, to study the effect of different sandblasting angles on the removal of radar-absorbing coating from TC4 titanium alloy surfaces, the sandblasting distance is fixed at 100 mm, the sandblasting pressure is standardized at 0.3 MPa, and the sandblasting angles are set at 60°, 75°, and 90°. The schematic diagram of the sandblasting experiment is shown in Figure.2.

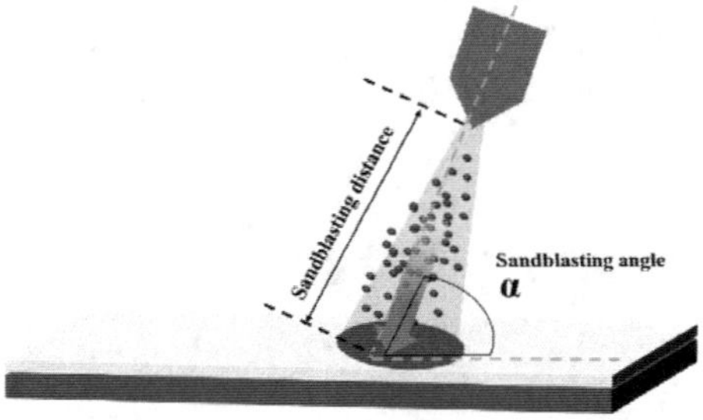

Figure 2. Schematic diagram of sandblasting experiment principle.

2.3. Performance Testing and Characterization

The structure and surface morphology of the coating are analyzed using a JSM-6390A scanning electron microscope (SEM). The average surface roughness (Ra) is measured using an SJ-301 roughness tester. The range of experimental factors is: sandblasting angles from 60° to 90°.

3. Results and Analysis

3.1. Effect of Sandblasting Angle on Removal Efficiency

To study the effect of sandblasting angle on the removal efficiency, sandblasting is performed at a pressure of 0.3 MPa with angles of 60°, 75°, and 90°, and the surface morphology and 3D profile of the substrate are observed. The results are shown in Figure.3. It can be observed that with constant sandblasting pressure, as the sandblasting angle increases, the substrate surface transitions from a large number of sloped morphologies to a morphology dominated by pits and grooves. The number and size of pits and grooves increase, leading to a rise in the average surface roughness of the substrate (as shown in Figure.3 and Figure.4). As the sandblasting angle increases, the vertical component of the abrasive velocity becomes larger, resulting in a stronger impact on the TC4 substrate, while the horizontal cutting component is reduced. This ultimately leads to more and deeper pits on the substrate surface, increasing the roughness.

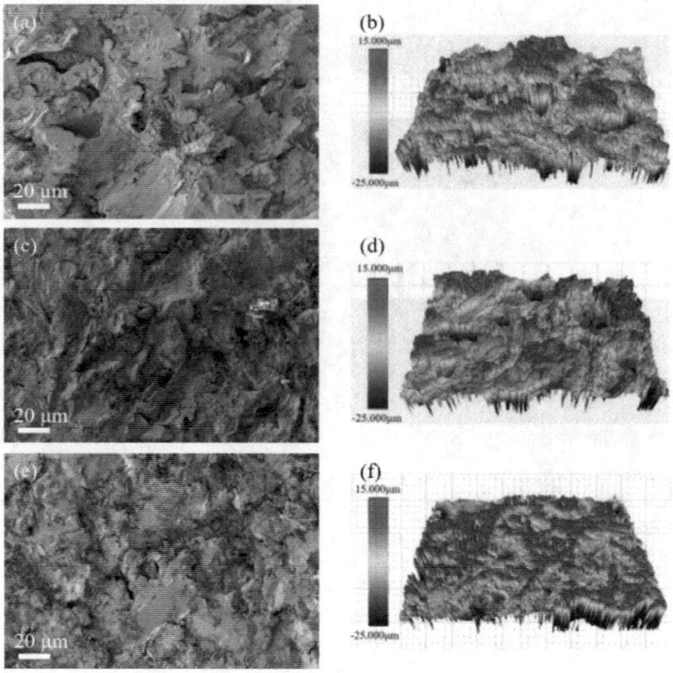

Figure 3. The micro and three-dimension morphologies of the substrate surface after coating removal at different sandblasting angles with a sandblasting pressure at 0.3 Mpa.

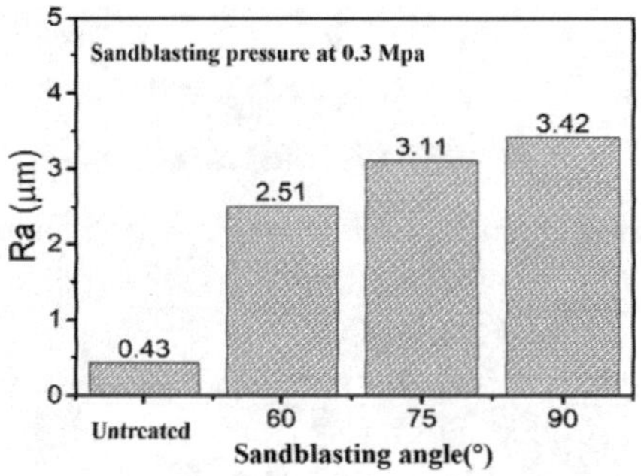

Figure 4. The micro and three-dimension morphologies of the substrate surface after coating removal at different sandblasting angles with a sandblasting pressure at 0.3 Mpa.

In summary, sandblasting can effectively remove surface coatings. When other process parameters are kept constant, increasing the sandblasting angle causes the substrate surface to transition from a morphology dominated by sloped features to one characterized by pits and grooves. The number and size of these pits and grooves increase, leading to higher average surface roughness of the substrate. Therefore, considering both removal efficiency and substrate damage, the optimal sandblasting angle is 60° .

3.2. Mechanism of Radar Absorbing Coating Removal by Sandblasting

It is important to note that during sandblasting, even at a 90° angle, because the abrasives are ejected in a jet-like manner, some abrasives still have a horizontal component when striking the material surface.

Thus, the velocity of the abrasive emerging from the nozzle should be divided into vertical and horizontal components under any condition.

The sandblasting removal process is illustrated in Figure.5 and can be divided into three stages:

Initial Removal Stage: The mechanism of interaction between the abrasives and the coating involves two aspects. On one hand, when the abrasive impacts the coating at a high angle, the impact effect is dominant, and the tangential force causes the abrasive to plow and wear the coating surface, generating debris that is quickly removed under the impact. On the other hand, due to the poor impact resistance of the radar-absorbing coating, cracks are likely to form during impact, leading to peeling under continuous impact. Thus, the removal of the coating is a coupled effect of impact and abrasion.

Second Stage: When the sandblasting abrasive reaches the interface between the coating and the substrate, the TC4 titanium alloy substrate, being ductile, undergoes minor plastic deformation under sustained impact. The radar-absorbing coating, with poor ductility, experiences stress concentration at the interface between the substrate and the coating, leading to delamination and peeling. The abrasive's impact rapidly removes the radar-absorbing coating.

Third Stage: Before the coating is completely removed, some sandblasting abrasives will strike the substrate and interact with it. At this point, the metal substrate undergoes some plastic deformation, which increases its surface hardness and creates small cracks under abrasive impact. Sandblasting removes the coating through physical action without causing phase changes in the metal surface.

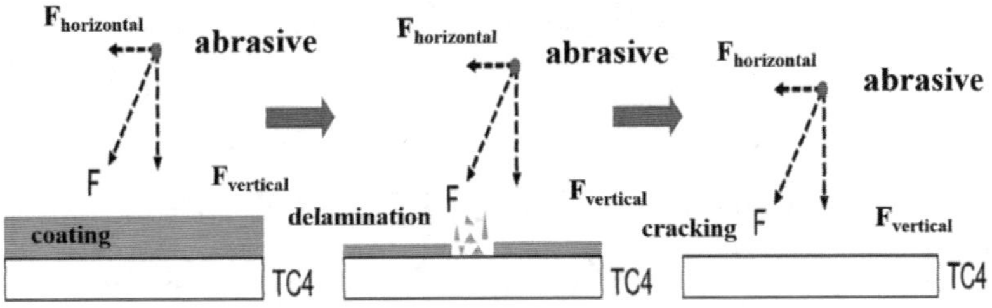

Figure 5. The three stages of sandblasting for coating removal.

4. Conclusion

The study focused on sandblasting to remove surface coatings from TC4 titanium alloy surfaces. The influence of sandblasting angle on coating removal was investigated, and the changes in surface morphology, 3D profile, and surface roughness of the substrate after coating removal were analyzed. The process and mechanism of coating removal were explored, leading to the following main conclusions.

1) The sandblasting angle significantly affects the micro-morphology and average roughness of the TC4 titanium alloy surface. As the sandblasting angle increases, the substrate surface morphology transitions from a combination of plowing and pits to a dominant presence of pits and grooves. The number and size of these pits and grooves increase, leading to higher average surface roughness.

2) Considering both the effectiveness of sandblasting and the extent of substrate damage, the use of SiC as the abrasive at a sandblasting angle of 60° yields the best removal results.

3) During the sandblasting removal process, the removal of the coating involves a coupling effect of impact and abrasion. When the abrasive reaches the interface between the coating and the substrate, the TC4 titanium alloy substrate, due to its ductility, undergoes minor plastic deformation under sustained impact. The radar-absorbing coating, which has poor ductility, experiences stress concentration at the substrate-coating interface, leading to delamination and peeling. The impact of the abrasive results in the rapid removal of the radar-absorbing coating.

Acknowledgments

This research was funded by the National Science and Technology Major Project of China, grant number 2019-VI-0015-0130.

References

[1] Zhang, W., Li, Y., Wang, X. Research on the Damage and Degradation of Radar Absorbing Coatings[J]. Journal of Materials Science, 2018, 33(4): 567-573.
[2] Liu, J., Chen, Q. Characteristics and Applications of TC4 Titanium Alloy[J]. Journal of Aerospace Engineering, 2019, 45(2): 102-108.
[3] Du B., Zhou X., Li Q., Liu J., et al. Surface Treat Method to Improve the Adhesion between Stainless Steel and Resin: A Review[J]. ACS Omega 2023, 8(43): 39984-40004.
[4] Li, F., Sun, Y. The Role of Sandblasting in Improving Coating-Substrate Bond Strength. Journal of Materials Protection, 2017, 34(3): 115-120.
[5] Rudawska A., Danczak I., Müller M., Valasek P. The effect of sandblasting on surface properties for adhesion. International Journal of Adhesion and Adhesives, 2016, 70: 176-190.
[6] Chen, L., Zhang, M. Effects of Sandblasting Parameters on Surface Characteristics and Coating Performance. Journal of Mechanical Engineering, 2018, 55(5): 234-242.
[7] Rakhadilov B., Magazov N., Kakimzhanov D. Influence of Spraying Process Parameters on the Characteristics of Steel Coatings Produced by Arc Spraying Method. Coatings, 2024, 14(9): 1145.
[8] Li, X., Wang, Y., Zhang, Y. Effect of Sandblasting Angle on Coating Removal Efficiency. Surface and Coatings Technology, 2020, 385: 125364.
[9] Zhang, L., Chen, J., Liu, H. Optimization of Sandblasting Parameters for Effective Coating Removal on Titanium Alloys. Journal of Materials Processing Technology, 2019, 271: 293-301.
[10] Wang, J., Li, M., Xu, P. Influence of Sandblasting Pressure on Coating Removal Efficiency and Substrate Damage. Journal of Surface Engineering, 2021, 37(4): 456-464.
[11] Chen, Q., Sun, H., Zhao, L. Effect of Abrasive Media on Sandblasting Efficiency and Substrate Integrity. International Journal of Advanced Manufacturing Technology, 2022, 119(3): 1123-1133.
[12] Wu, H., Liu, B., Zhang, C. Influence of Sandblasting Time and Flow Rate on Coating Removal and Surface Roughness. Surface Engineering, 2021, 37(5): 478-485.
[13] Li, J., Ma, L., Xu, K. Optimization of Sandblasting Process Parameters for Coating Removal Efficiency. Journal of Manufacturing Processes, 2020, 56: 214-223.
[14] Zhao, Q., Huang, Y., Guo, Z. Residual Stress Analysis of Titanium Alloy Substrate after Sandblasting Treatment. Journal of Materials Science and Technology, 2021, 39(2): 456-462.
[15] Xu, L., Tang, S., Zhang, J. Surface Characterization of Titanium Alloy after Sandblasting Using SEM and EDS. Materials Characterization, 2019, 154: 109-115.
[16] Zhang, Y., Liu, H., Wang, J. Effect of Different Abrasive Media on Surface Oxide Layer Removal. Journal of Surface Science and Technology, 2022, 48(1): 97-104.

Study on the stabilization and antioxidant activity of nanosuspension with a tiny amount of stabilizer

Wenhui Zhou [1], Tingting Yang [2] and Tao Liu [*1]

[1] College of Chemical Engineering, Qingdao University of Science and Technology, Qingdao, 266042, China
[2] School of Life Sciences, Yunnan University, Yunnan, China

* Corresponding author's E-mail: taoyanghe@hotmail.com

Abstract: Resveratrol nanosuspension was prepared using sodium alginate as a biologically active agent instead of a traditional chemical stabilizer. The nanosuspensions were prepared using the wet milling technique and optimized by the response surface with respect to the content of the drug and surfactant as well as the milling time. Solubility, stability, and antioxidant properties were explored. Molecular simulation indicated that sodium alginate contributed to overcoming the effect of external milling forces and exerted the migration resistance of drug molecules to prevent the occurrence of agglomeration and sedimentation. Stability results showed that solubility and particle size did not undergo any significant change after 180 days of storage. The ability of the nanosuspension to scavenge DPPH oxygen radicals was also not disrupted.

1. Introduction

Sodium alginate was a natural polysaccharide and a natural surfactant that could exhibit polyanionic properties in aqueous solution [1]. Sodium alginate showed a certain viscosity dissolved in water, which could improve the stability of the suspension [2]. Due to good biocompatibility, sodium alginate has been widely used in the pharmaceutical field.

Nanosuspensions were characterized by high drug loading, high bioavailability, and targeted drug delivery [3]. Media milling and high-pressure homogenization were the common techniques used in the preparation of nanosuspensions [4]. Surfactants were generally synthetic surfactants such as polyoxyethylene amide, tween, etc. Such substances were only suspending aids and might have potential side effects and other effects on the human body [5].

The aim of this paper was to prepare a nanosuspension with a tiny amount of sodium alginate as a stabilizer, which avoided the effects of chemically synthesized surfactants on the human body and also greatly improved the proportion of the drug due to the extremely low content of sodium alginate. The effect of the ratio of resveratrol and sodium alginate and grinding time on drug dissolution was studied. The intermolecular interactions between the drug and stabilizer were analyzed using molecular dynamics simulations. The stability and antioxidant capacity of the nanosuspensions were examined.

2. Methods

2.1. Preparation of nanosuspensions

Resveratrol and sodium alginate of a certain amount were added to the grinding chamber by wet milling technique using 10 mL of pure water and 5 mL of grinding balls. The speed was set to 2000 rpm/min, and the nanosuspension was obtained after milling for a certain time.

Journal of Physics: Conference Series **3009** (2025) 012022 doi:10.1088/1742-6596/3009/1/012022

3. Results and discussion

3.1. Process optimization

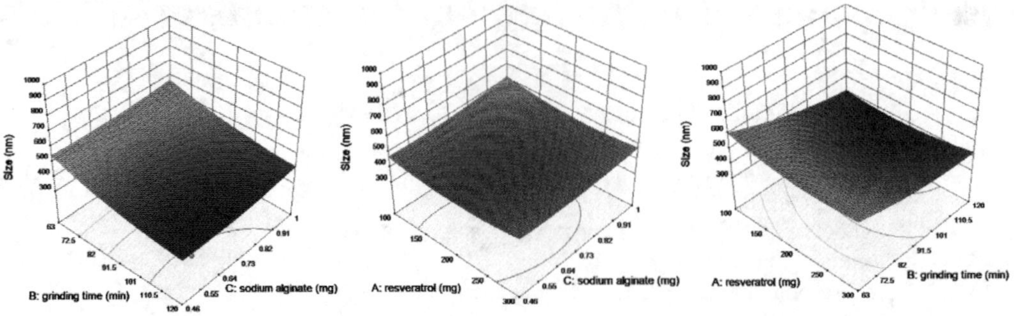

Figure 1. Three-dimensional response surface for drug content, stabilizer content, and grinding time.

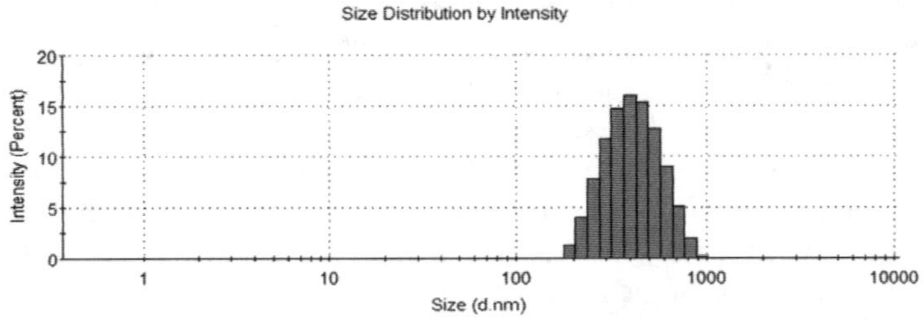

Figure 2. The particle size of the optimal formulation.

The 3D plot of the response surface was plotted through the regression equation. Figure 1 shows the 3D response surface with the presence of the lowest point, indicating the presence of a minimum value of particle size. The optimal processing conditions designed by Design-Expert software were: 178.04 mg of resveratrol, 118.3 min of grinding, and 0.46 mg of sodium alginate, at which the particle size reached a minimum of 321.74 nm. Considering the comprehensive factors, the final processing conditions were adjusted to 200 mg of resveratrol, 0.5 mg of sodium alginate, and a grinding time of 120 min. The adjusted formulation was verified using a particle size determination. Figure 2 shows that the average particle size of the formulation was 356.9 nm, and the PDI was 0.141. It was very close to the predicted value, which proved that the model was well-fitted.

3.2. In vitro dissolution

The dissolution curve of the nanosuspension is shown in Figure 3. The sample with sodium alginate as a stabilizer reached the highest value of 88.4% at 3 min. The particle size of the sample was 356.9 nm, and PDI was 0.141 after milling. The sample without stabilizer of particle size was 633.8 nm, PDI was 0.18, and the sample reached 77.52% dissolution at 5 min. Raw resveratrol rose more slowly and reached the maximum value around 30 min. According to the Noyes-Whitney and Freundlich-Ostwald equations, when the particle size of the drug was reduced to the nanometer level, the surface area was greatly increased, and the dissolution rate and saturation solubility increased with the increase of surface area [6]. This concluded that the reduction of particle size could greatly improve the dissolution rate of the drug. Sodium alginate as a stabilizer has also been shown to be able to be used in the preparation of nanosuspensions and contribute to the reduction of particle size.

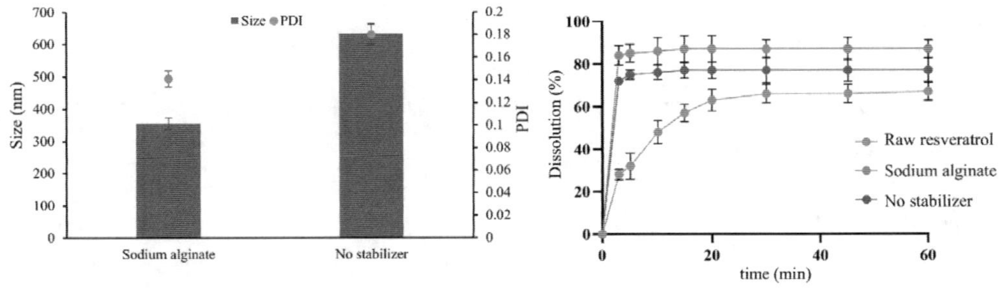

Figure 3. Particle size and dissolution of raw resveratrol and nanosuspensions.

3.3. Molecular dynamics simulation

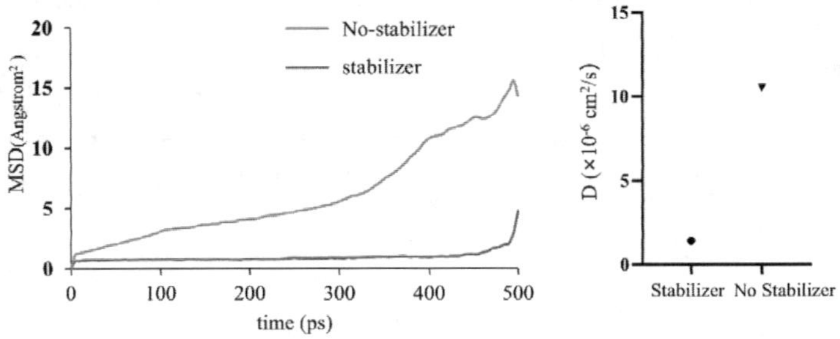

Figure 4. Molecular dynamics simulations. Mean square displacement (MSD) values and diffusion coefficients (D).

The mean square displacement (MSD) value was analyzed to probe the migration behavior of drug molecules [7]. In addition to MSD, the diffusion (D) coefficient could also be used to quantitatively analyze and evaluate the migratory behavior of particles [8]. The MSD values of the nanosuspensions with stabilizers were smaller than those without stabilizers. This indicated that resveratrol was more resistant to migration in the sodium alginate matrix (Figure 4). In order to further quantify the diffusion behavior of resveratrol, D values were calculated. Figure 4 showed that the migration and diffusion behavior of resveratrol in the space containing sodium alginate was significantly inhibited [9]. In addition, water molecules tended to replace the drug and carrier materials to form hydrogen bonds [10]. This further attenuated the interaction between the stabilizer and the drug.

3.4. Stability of nanosuspension

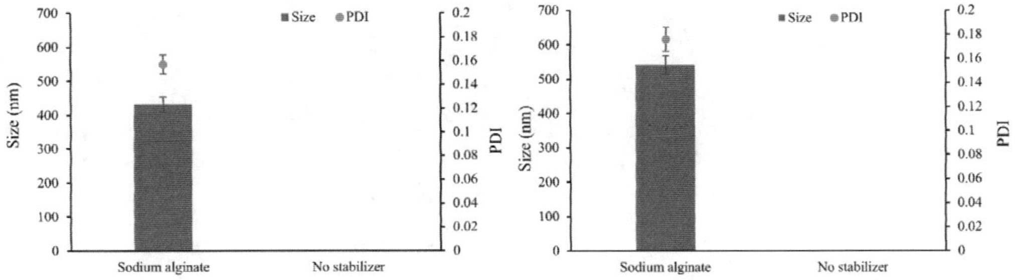

Figure 5. Particle size of nanosuspensions on days 30 (left) and 180 (right).

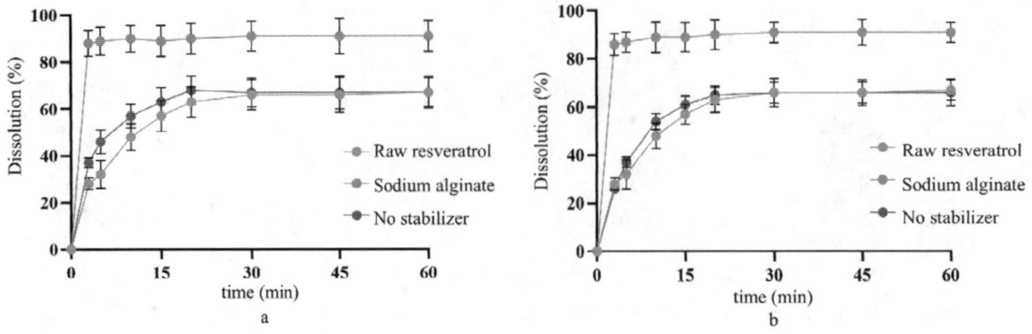

Figure 6. Dissolution curve of raw resveratrol and nanosuspensions on days 30 (a) and 180 (b).

Figure 5(a) shows that the average particle size was 432 nm. There was no significant increase from the initial particle size. Figure 6(a) shows that the solubility reached a maximum value of 88.3% in 3 min, which proved that the sample could be stabilized within 30 days. The sample without stabilizer showed sedimentation and could not recover the homogeneous system, whose particle size was not detectable. Figure 6(a) shows that the maximum solubility of 69% was reached at 30 minutes.

Figure 5(b) reveals that the particle size of the nanosuspension containing sodium alginate increased by 193 nm from the initial particle size. The PDI was 0.186. The sedimentation occurred in the sample containing the stabilizer after 180 days. The sedimentation disappeared after simple shaking. The dissolution value remained consistent with the initial value. Related studies have shown that the increase in particle size affected nanosuspension delamination and sedimentation [11]. The sample without a stabilizer showed significant sedimentation. Vitro dissolution experiments showed that a value of 67% was reached in 30 min (Figure 6(b)). After 180 days of storage, the samples containing sodium alginate showed an increase in particle size and PDI, but it did not affect the dissolution. This further proved that sodium alginate could be used as a stabilizer for resveratrol nanosuspensions and could keep the nanosuspensions stable for 180 days.

Molecular dynamics simulations also demonstrated that sodium alginate inhibited further agglomeration and diffusion, thus preventing aggregation and settling of the nanosuspension. This might be the reason why nanocrystals containing resveratrol could be stably stored for 180 days.

3.5. Antioxidant of nanosuspensions

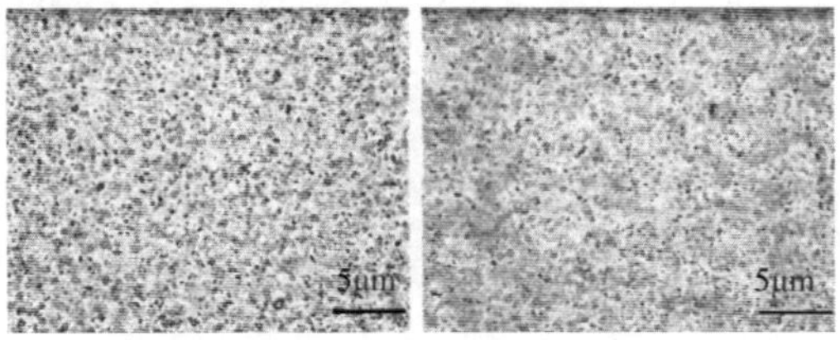

Figure 7. Microscope pictures of nanosuspensions (day 0 and day 180).

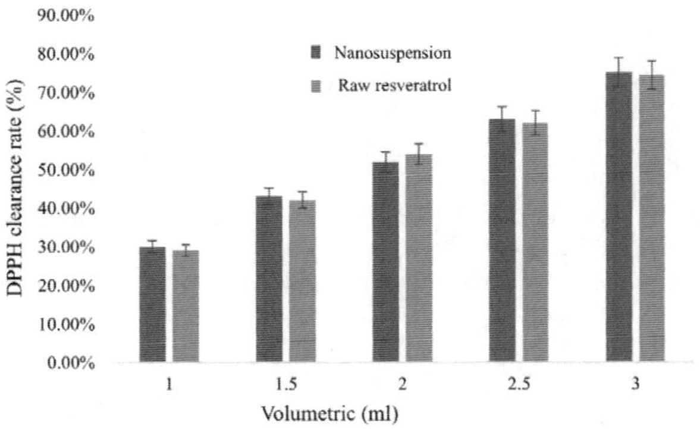

Figure 8. DPPH radical-scavenging activity.

As shown in Figure 7, there was no obvious difference between the nanosuspensions containing sodium alginate after 180 days of storage (right) and the photo of the nanosuspensions on day 0 (left). The nanosuspensions exhibited a uniform particle size distribution without significant aggregation.

DPPH was violet-black in organic solvents and the color of the solution became gradually lighter as the single electron of DPPH was captured [12]. The absorbance at the maximum absorption wavelength of 517 nm decreases linearly. Figure 8 shows the gradual increase in DPPH scavenging rate with increasing resveratrol concentration. The nanosuspension containing sodium alginate maintained the same DPPH oxygen radical scavenging rate as raw resveratrol. This indicated that the antioxidant properties of the nanosuspension were not disrupted.

4. Conclusion

The nanosuspensions prepared with a tiny amount of sodium alginate as a stabilizer could achieve smaller particle sizes and faster dissolution rates. Stable storage for 180 days did not show any significant change in dissolution rate. Meanwhile, molecular simulations showed that sodium alginate might inhibit the migration and diffusion of resveratrol and prevent drug agglomeration and sedimentation. Meanwhile, the antioxidant capacity of the nanosuspension was not different from that of the raw resveratrol.

References

[1] Li H, Chen X, Sun Y, et al. (2024). Construction and characterization of sodium alginate/polyvinyl alcohol double-network hydrogel beads with surfactant-tailored adsorption capabilities for efficient tetracycline hydrochloride removal. International journal of biological macromolecules, 280: 135879.
[2] Fatemeh Heidari, et al. (2024). Evaluation of sodium alginate sponge infused bromelain spray and Helichrysum italicum nanoemulsion to accelerate wound healing. International Journal of Biological Macromolecules, 283: 137799.
[3] Y. Liu, et al. (2024). Advances in Nanotechnology for Enhancing the Solubility and Bioavailability of Poorly Soluble Drugs, Drug design, development and therapy, 18: 1469-1495.
[4] O.A. Aksoy, et al. (2023). Preparation of Piroxicam nanosuspensions by high pressure homogenization and evaluation of improved bioavailability, Drug Development and Industrial Pharmacy, 49: 715-722.
[5] AlObaidi Hisham, et al. (2022). Antifungal nanosuspensions with surfactants and silver for the treatment of onychomycosis, Official journal of Arbeitsgemeinschaft für Pharmazeutische Verfahrenstechnik e.V., 179: 194-205.
[6] Gungor D, Aytekin E, Akdag Y, et al. (2024). Optimization of Glyburide-Loaded Nanosuspensions via Ball Milling and Homogenization Techniques: A Central Composite Design Approach for Enhanced Solubility, Current pharmaceutical design, 31: 377-390.
[7] Sidian, Z., et al. (2022). Study on the Effect of Polymer Excipients on the Dispersibility, Interaction, Solubility, and Scavenging Reactive Oxygen Species of Myricetin Solid Dispersion: Experiment and Molecular Simulation. ACS omega, 7(1): 1514-1526.
[8] Zhang, S., et al. (2024). Screening of Organic Small Molecule Excipients on Ternary Solid Dispersions Based on Miscibility and Hydrogen Bonding Analysis: Experiments and Molecular Simulation. AAPS PharmSciTech, 25(1): 21.
[9] Aulifa, D.L., et al. (2024). Elucidation of Molecular Interactions Between Drug-Polymer in Amorphous Solid Dispersion

by a Computational Approach Using Molecular Dynamics Simulations. Advances and applications in bioinformatics and chemistry: AABC, 17: 11-19.
[10] Toziopoulou, F., et al. (2017). Production of aprepitant nanocrystals by wet media milling and subsequent solidification. International Journal of Pharmaceutics, 533(2): 324-334.
[11] Wang Y, Zheng Y, Zhang L, et al. (2013). Stability of nanosuspensions in drug delivery, Journal of Controlled Release,172(3): 1126-1141.
[12] Md. Waziur Rahman Chy, et al. (2024). paya peel using response surface methodology and artificial neural network. Applied Food Research, 2: 100591.

Effect of temperature on volatilization behavior of re in alkaline earth metal perrhenates during calcination stage of simulated high-level liquid waste

Xiaofan Mao [1], Zhiheng Wang [1], Hua Zhang * [1], Xiaorui Li [1], Huilin Sun [1], Zheng Li [1], and Cheng He [1]

[1] China Institute of Atomic Energy, Beijing, China

* Corresponding author's e-mail: zhanghua_ciae@163.com

Abstract. During the process of cold crucible two-step treatment of high-level liquid waste, volatilization of the radionuclides Tc, Cs, Sr, etc. will occur. Because of the similarity between the physicochemical properties of Re and Tc, we choose Re instead of Tc to avoid radioactive exposure in the calcination stage volatility experiment. Re exists as ReO_4^- in the liquid environment, while alkali earth metal cation is added, and then calcination is finished at high temperature. The perrhenic acid combines with the cation to form perrhenate. There are two volatile forms of Re in alkali earth metal perrhenates, in which one is volatilized in the form of perrhenate and the other is volatilized by Re_2O_7 after the decomposition of perrhenate. Re in $Mg(ReO_4)_2$ is almost all decomposed and volatilized after 900°C. The volatilization rate of Re in $Ca(ReO_4)_2$, $Sr(ReO_4)_2$, and $Ba(ReO_4)_2$ is about 60% at 1, 200°C, and these two volatilization modes are parallel.

1. Introduction

Since the 1950s, with the gradual development of nuclear technology around the world, the extensive use of nuclear energy has resulted in a large amount of spent fuel accumulation, which has brought challenges to the nuclear fuel cycle [1]. For relatively depleted uranium countries, the "closed fuel cycle" scheme using nuclear fuel is more in line with the medium and long-term planning of nuclear power than the "once-through fuel cycle" [2]. In the "closed fuel cycle", water treatment of spent fuel will produce high-level liquid waste [3], which has high radioactivity levels, complex compound composition, and large toxicity difference. The water and organic matter will be released to produce a large number of flammable and explosive gases during storage [4, 5]. At present, there are four main methods for the safe treatment of HLLW (glass vitrification, ceramic vitrification, glass ceramic vitrification, and separation-transmutation-glass vitrification), among which only glass vitrification has been achieved in industrial applications [6, 7].

The first step of the two-step cold crucible glass vitrification process is the rotary calcining of the high-level waste liquid, which is carried out by the rotary calcining furnace with stable operation and easy control. At this stage, low-temperature evaporation denitrification and high-temperature calcination are carried out in accordance with a certain proportion of the high-level waste liquid and additives, and the metal salt components are calcined at high temperature with the water and nitric acid in the waste liquid are removed [8, 9]. The second step is the mixed melting of calcined matter and glass. The cold crucible induction melting process is adopted, which has a higher melting temperature, less corrosivity, and longer service life of the equipment because the molten glass is not in direct contact with the metal tube wall [10]. In the calcination stage, high temperature causes significant volatilization of a large number of radionuclides such as ^{99}Tc, ^{137}Cs, and ^{106}Ru, which will pollute the exhaust pipeline and even cause radioactive leakage after volatilization. Due to its special chemical properties and volatile properties, ^{99}Tc has attracted much attention.

The content of ^{99}Tc in spent nuclear fuel is 1, 090 g· $T^{-1}U$ (burnup 40 GWd/t) and 1, 350 g ·$T^{-1}U$ (burnup 70 GWd/t) [11]. In the oxidizing environment, Tc exists as the form of TcO_4^-, which is most stable in aqueous solution [12]. Because of the similarity of atomic radius, possible valence state, ionic radius, and the structure of the compound between Tc and Re, Re is the most preferred alternative element for Tc [13, 14]. Therefore, Re is used to replace Tc in this study. Research shows that ReO_4^- can combine with alkali metals to form alkali perrhenate $MReO_4$ (M is Li, Na, K, Rb, and Cs) and volatilize in the form of its monomer or dimer at a certain temperature [15, 16], which may be due to the different vapor pressures of different $MReO_4$. $CsReO_4$ has the strongest volatile capacity [17]. In this research, the effect of temperature change on the volatilization behavior of Re and alkali earth metal cation in alkali earth metal perrhenates was studied by simulating the liquid environment of HLLW, and the specific volatilization process was concluded.

2. Experiments and Results

2.1 Experiments

The main reagents used in the experiments are: HNO_3, Re_2O_7, NaOH, $NaNO_3$, $CsNO_3$, MgO, $Ca(NO_3)_2·4H_2O$, $Sr(NO_3)_2$, and $BaNO_3$. The experimental device includes three parts: the high-temperature furnace heating device, condensate recovery device, and tail gas absorption device.

The experimental process mainly includes three parts: (1) Preparation of simulated liquid waste. As the concentration of Tc in the HLLW is about 1.99 g/L (burnup is 55 GW), to make the experimental phenomenon more obvious, 1 g Re_2O_7 and alkali earth metal nitrate or oxide, and nitric acid are added to form simulated liquid waste; (2) In the calcination experiment, the box-type high-temperature furnace is set to the specified temperature, and the configured waste liquid is heated up with the furnace and calcined for 3 h. We collect the calcined materials when it is cooled to room temperature; (3) We analyze the composition of the obtained calcined material, mainly including the specific composition, the content of Re, alkaline earth metal cation, and the image structure.

The experimental scheme is mainly divided into two parts. The first part is the pre-experiment in the combination of thermal-heavy element content analysis (TG-ICP) to detect the volatilization of Re. The sample is alkaline metal perrhenate, and the temperature range is 25°C to 1, 200°C; the second part is the experiment in the high-temperature furnace. The alkaline metal cation M (M: Mg, Ca, Sr, Ba) is added with ReO_4^- to form $M(ReO_4)_2$, which was identified as a reactant in the subsequent experiments. The specific conditions are shown in Table 1, and the calcination time is 3 h.

Table 1. The effect of the volatilization capacity.

Cation (M^{n+})	$C(M^{n+})$ (mol/l)	$C(ReO_4^-)$ (mol/l)	Temperature (°C)	$C(HNO_3)$ (mol/l)	V (ml)
Mg^{2+} Ca^{2+} Sr^{2+} Ba^{2+}	0.0205	0.041	700/800 /900/1, 000 /1, 100/1, 200	2.5 M	50

At the end of the experiment, the solid calcination was collected for ICP testing to calculate the concentration of Re and M^{n+}, and then indirectly calculate the volatile rate of Re and M^{n+}. The calculation formula of the volatilization rate **α** of Re/M^{n+} is shown as follows:

$$N_x = C_x * V_x \tag{1}$$

$$\alpha = N_x / N_0 \tag{2}$$

where the content of Re/ M^{n+} after dissolution of the calcined matter is N_x, the volatilization rate is recorded as **α**, and N_0 is the total content of Re/ M^{n+} in the original solution involved in the reaction. The calcined products were analyzed by XRD and SEM-EDS, and the specific materials and micromorphology of the calcined products were obtained. The influence of temperature change on the reaction was analyzed, and the volatilization mechanism of Re and M^{n+} under such conditions was obtained.

2.2 Results

2.2.1 TG-ICP

Alkaline earth metal perrhenate was configured according to the concentration shown in Table 1, and TG-ICP analysis was performed on the obtained perrhenate solids. The TG and ICP graphs obtained are shown in Figure 1.

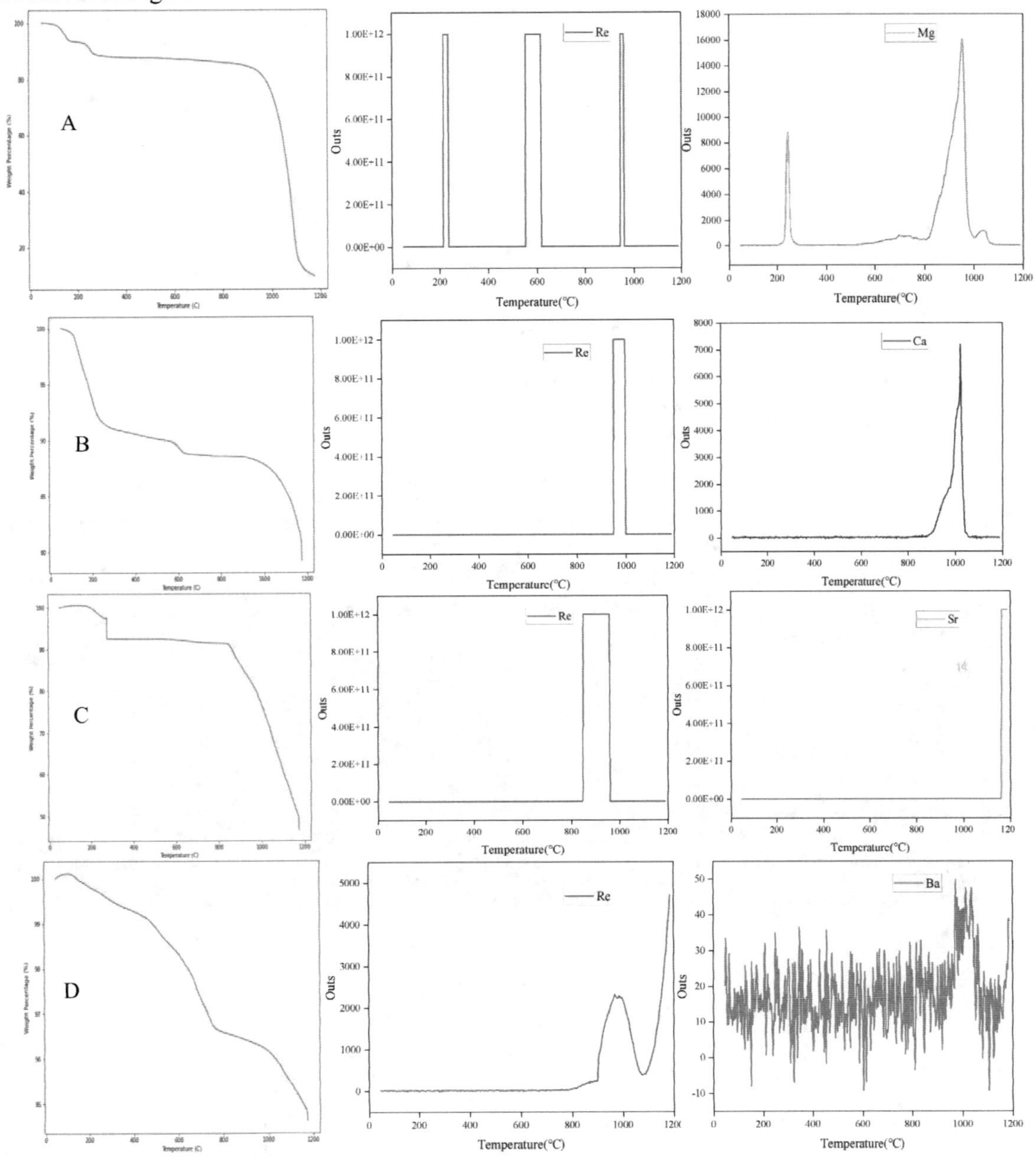

Figure 1. TG (left)-ICP (right) of $M(ReO_4)_2$.

(A is $Mg(ReO_4)_2$, B is $Ca(ReO_4)_2$, C is $Sr(ReO_4)_2$, and D is $Ba(ReO_4)_2$)

2.2.2 High-temperature furnace experiment

(1) Effect of temperature on the volatility of $M(ReO_4)_2$ calcined product

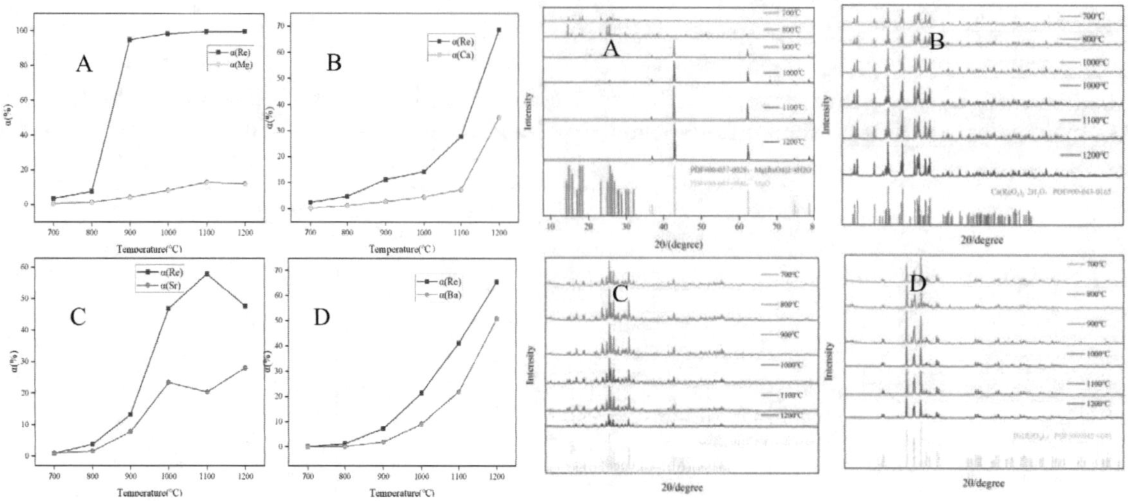

Figure 2. Effect of temperature on volatilization rate (left) and XRD (right) of $M(ReO_4)_2$.

(A is $Mg(ReO_4)_2$, B is $Ca(ReO_4)_2$, C is $Sr(ReO_4)_2$, and D is $Ba(ReO_4)_2$)

(2) SEM-EDS of $M(ReO_4)_2$ calcined product

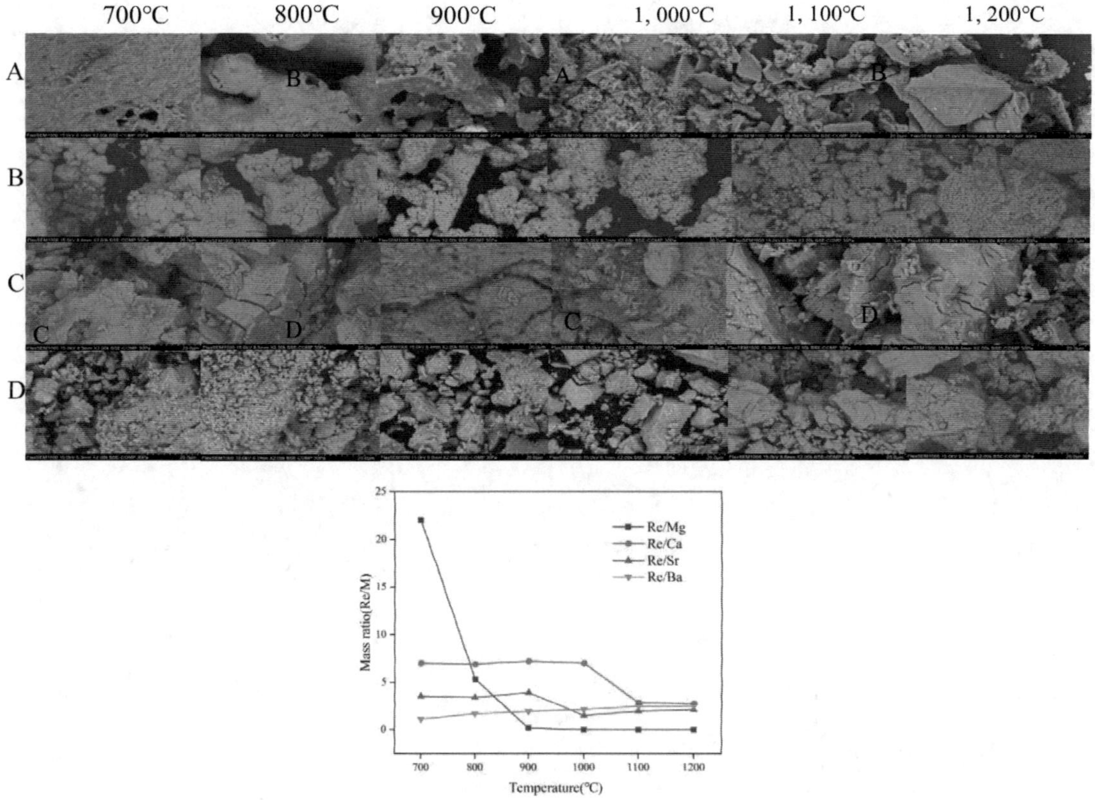

Figure 3. SEM of $M(ReO_4)_2$ (up) and mass ratio of Re/M (down).

(A is $Mg(ReO_4)_2$, B is $Ca(ReO_4)_2$, C is $Sr(ReO_4)_2$, and D is $Ba(ReO_4)_2$)

According to the high-temperature furnace experiment as shown in Table 1, the diagram of the temperature influence on the volatilization rate(α) of $M(ReO_4)_2$ was obtained. The calcined products were analyzed by XRD and SEM-EDS, and the diagram was obtained as shown in Figures 2 and 3.

3. Discussion

The influence of coexisting alkali earth metals on the volatilization ability of Re is studied at 700°C-1, 200°C. The results show that there are two volatilization modes of Re in $M(ReO_4)_2$. Mode (1) is that Re is decomposed into MO and Re_2O_7, and then volatilized at high temperature. Mode (2) is $M(ReO_4)_2$ volatilized in its form.

$$M(ReO_4)_2\,(s) \rightarrow MO(s) + Re_2O_7(s/g) \quad (3)$$

$$M(ReO_4)_2\,(s) \rightarrow M(ReO_4)_2\,(s/g) \quad (4)$$

At 900°C, Re in the $Mg(ReO_4)_2$ sample is almost completely volatilized, and the calcined product is only MgO, which proves that Mode 1 plays a dominant role and Mode 2 almost does not occur at this time. At 1, 000°C -1, 200 °C, there is a small amount of Mg volatilization, and Re volatilization completely, which may be due to that at this high-temperature state Modes 1 and 2 simultaneously occur and Mode 1 plays a leading role. The XRD and SEM-EDS prove that the calcined product changes from $Mg(ReO_4)_2$ to MgO.

The effect of coexisting alkali earth metal Ca on the volatilization capacity of Re was studied at 700°C-1, 200°C. It is concluded that there are two volatilization modes of Re in $Ca(ReO_4)_2$. In the temperature range of 700°C-900°C, the volatilization rate is lower than that in the temperature range of 1, 000°C-1, 200°C. There are no obvious CaO structures found in the XRD pattern, but combined with SEM-EDS, a different crystal structure appears at 1, 100°C to 1, 200°C, and the ratio of Ca to Re is different from that of the other temperature ranges. These all prove that Re in $Ca(ReO_4)_2$ is volatilized according to Modes 1 and 2.

While the process in the experiment of $Sr(ReO_4)_2$ and $Ba(ReO_4)_2$ is similar to that of $Ca(ReO_4)_2$, the difference is that there are no obvious changes in the ratio of Ba/Re found in the $Ba(ReO_4)_2$ SEM-EDS diagram, which may be due to the dominance of Process (1) in $Ca(ReO_4)$ and $Sr(ReO_4)_2$, while Process (2) in $Ca(ReO_4)_2$.

After the addition of alkaline earth metals Mg, Ca, Sr, and Ba, the addition of Mg has a stronger volatilization promotion effect on Re, while the addition of Ca, Sr, and Ba has a similar effect. It may be because the electrostatic force is inversely proportional to the distance between atoms (the electrostatic force between ions $F = k*q^{+}*q^{-}/L^2$, q^+ and q^- is the effective electric carried by positive and negative ions respectively, L is the ionic distance, and k is the constant). The oxide ion spacing of alkaline earth metals is: MgO (210 pm), CaO (240 pm), SrO (257 pm), and BaO (277 pm). The electrostatic force gradually decreases with the increase of ionic spacing, that is, the electrostatic force of MgO is the largest. Compared with other alkali perrhenates, the decomposition tendency of $Mg(ReO_4)_2$ to form MgO is greater, so the decomposition of $Mg(ReO_4)_2$ makes Re the most volatile.

4. Conclusion

From the above research, it can be seen that the Re volatilization law of alkali earth metal perrhenate in the first stage of calcination of the two-step cold crucibles process is that: ①Volatilization in $M(ReO_4)_2$ form; ② After $M(ReO_4)_2$ decomposition, Re_2O_7 is volatilized. Among them, $Mg(ReO_4)_2$ is mainly volatilized in Mode 2, and the rest is volatilized in both ways. The volatilization rate of Re also increases with the increase in temperature. $Mg(ReO_4)_2$ has the highest volatilization rate, which may be due to its greatest tendency to decompose into MgO.

Journal of Physics: Conference Series **3009** (2025) 012023 doi:10.1088/1742-6596/3009/1/012023

Reference

[1] Li Qiang et al. (2023). The world's present situation of uranium resources and the countermeasures and suggestions for resource security of nuclear power development in China. J. China Mining Magazine, 32 (03): 1-9.
[2] Zhao mi, et al. (2023). Economic study of spent fuel reprocessing in China. J. Science & Technology Information, 21 (12): 252-256.
[3] Li Li (2019). Analysis of the classified management of radioactive waste. J. Science and Technology Innovation Herald, 16 (07): 112-114.
[4] Li Yusong, et al. (2020). Research and development of CIAE high-level liquid waste calcination technology. J. Atomic Energy Science and Technology, 54 (S1): 126-136.
[5] Jiang Yunlu (2023). High-temperature performance and irradiation evolution of stainless steel welded joints in high-level radioactive waste liquid cured container. D. University of Science and Technology of China.
[6] Carol M. Jantzen (2012). Radioactive Waste Conditioning, Immobilization, and Encapsulation Processes and Technologies: Overview and Advances. Radioactive Waste Management and Contaminated Site Clean-up: Processes, Technologies, and International Experience (SRNL-STI-00407): 2, 8.
[7] Ruan Jiajian, et al. (2023). Research progress of glass calcination and artificial rock calcination technology of high-level waste. J. Nuclear science and engineering 43 (01): 225-232.
[8] Li Zheng, et al. (2022). Effect of sucrose denitration on the properties and structure of calcined products from simulated dynamic stacking liquid waste. J. Atomic Energy Science and Technology, 56 (03): 434-442.
[9] Larson D. E., et al. (1976). High-level waste vitrification by spray calcination/in-can melting: BNWL-2092. R. Washington: Pacific Northwest Laboratories, 21-25.
[10] Liang Xuan, et al. (2023). Research status and development prospect of two-step cold crucible glass curing technology. J. Contemporary Chemical Industry, 08: 14-16.
[11] O. A. Ustinov, et al. (2021). Technetium in Gas Emissions of Radioactive Waste Vitrification Technology (Scientific and Technical Information Overview). J. Radiochemistry, 63 (4): 295-400.
[12] Wang Jing (2021). Study on the thermal stability of Tc/Re doped mineral phase during glass calcination. D. Hu Bei: Wuhan University of Technology.
[13] Darab J. G. and Smith P. A. (1996). Chemistry of Technetium and Rhenium Species during Low-Level Radioactive Waste Vitrification [J]. Chemistry of materials, 8 (5), 1, 004- 1, 021.
[14] Shannon R. D. (1976). Revised effective ionic radii and systematic studies of interatomic distances in halides and chalcogenides. J. Acta Crystallographica A-Foundation and Advances, 32 (5): 751-767.
[15] G. Rouschias (1974). Recent advances in the chemistry of rhenium. J. Chemical Reviews journal, 74 (5): 531-566.
[16] S. M. Shugurov, et al. (2008). Thermochemical study of gaseous salts of oxygen-containing acids: XXIV.1 polymers of alkali metals perrhenates. J. General Chemistry, 78 (10): 1, 882-1, 888.
[17] F. Baumgärtner, et al. (1987). On the emission of fission products and transuranium nuclides from highly radioactive waste between 200 and 1, 100°C. 51 (3): 151-156.

Journal of Physics: Conference Series **3009** (2025) 012024 doi:10.1088/1742-6596/3009/1/012024

Study on the effects of different types and concentrations of copper salts on the texturing of p-type single crystal silicon

Yuxin Zou[1], Huan Liu[1,2], Ting Yang[1], Xingwei Yang[3*]

[1]Department of Intelligent Manufacturing, Xiangtan Institute of Technology, Xiangtan, China
[2]Hunan geely Automobile College, Xiangtan Institute of Technology, Xiangtan, China
[3]Faculty of Metallurgical and Energy Engineering, Kunming University of Science and Technology, Kunming, China

* Corresponding author: yangxingwei18@mails.ucas.ac.cn

Abstract—Using diamond wire sawing P-type single crystal silicon as raw material, the study examines the effects of different types and concentrations of copper salts on texturing the surface of silicon wafers. The research results indicate that under three different copper salt etching systems, the controlled preparation of various structures such as grid-like, porous, and inverted pyramid on the silicon wafer surface is achievable. The etched morphologies on the silicon wafer surface are primarily due to the combined effects of the oxidizing nature and complexing ability of the anions in the copper salts, as well as the interaction forces between the cations and anions affecting the deposition of copper nanoparticles on the silicon surface. This theoretical research can lay the foundation for the texturing of high-efficiency silicon solar cells.

1. Introduction

In recent years, the growing energy demand, scarcity of traditional energy sources, and the large emissions of CO_2 gas have made energy and environmental issues one of the hot topics of concern for all humanity[1].Vigorously developing renewable energy has gradually become the consensus for the future development of energy[2-3]. Solar energy, due to its infinite reserves, no geographical restrictions, cleanliness, and non-pollution, has become one of the most promising renewable energy sources. Under the strong promotion of countries around the world, the development of solar photovoltaics has been particularly rapid in the past 20 years. By the conclusion of 2023, over 3870GW of photovoltaics had been installed worldwide, making up 37% of the total renewable energy capacity. Projections suggest that by the mid-21st century, this figure will surpass 8500 GW, positioning solar energy as a significant global energy source[4-7].

Solar cells are the core component in photovoltaic power generation for photovoltaic conversion. Low-cost, high-efficiency solar cells have always been the goal pursued by photovoltaic companies. Silicon, due to its abundant reserves and suitable band structure, has become the main material for solar cells. Nowadays, silicon-based solar cells occupy more than 95% of the market share of solar cells[8]. In recent years, with the introduction of diamond wire cutting technology and the rapid development of PERC technology, the market share of single crystal silicon has gradually surpassed that of polycrystalline silicon[9-10].In single crystal cells, texturing the surface of the silicon wafer is a key step in the preparation of efficient solar cells. A high-quality matte surface can effectively improve the light capture ability of the silicon wafer surface, increase the photocurrent density, and thereby enhance the conversion efficiency of the solar cell[11-12].

Currently, in the industrialization of textured diamond wire sawing single crystal silicon, metal-catalyzed chemical etching (copper-catalyzed chemical etching) technology occupies a large intermediate market. This technology is simple, low-cost, and compatible with existing industrial production lines[13-14].Using a metal copper-catalyzed chemical etching system, different nanostructures have been successfully prepared on the surface of silicon wafers. Hong Shihao[15] and others, through studying the mechanism of copper-catalyzed chemical etching, have prepared uniformly structured inverted pyramid structures on single crystal silicon substrates, with reflectivities below 10%.Nevertheless, the mechanism of metal copper-catalyzed chemical etching and the principles behind the formation of different nanostructures require further investigation.

In response to the above issues, copper-catalyzed chemical etching is applied to diamond wire sawing single crystal silicon. The study investigates the effects of different types and concentrations of copper salts on texturing the surface of silicon wafers. Uniform grid-like, porous, and inverted pyramid structures have been successfully prepared on the surface of P-type single crystal silicon wafers. Further research has been conducted on the deposition distribution of copper nanoparticles on single crystal silicon surfaces in three different etching systems.

2. Experimental sections

Select sample size of 15.6×15.6 cm², thickness of 0.19~0.21 cm, and resistivity of 1~10 Ω·cm for the double-sided polished <100> single crystal silicon wafers.Sequentially immerse the single crystal silicon wafers in acetone, ethanol, and deionized water for ultrasonic cleaning for 10 minutes, then dry with nitrogen gas for further use.Three etching systems are used in the experiment:HF-H_2O_2-$CuSO_4$ etching system, HF-H_2O_2-$CuCl_2$ etching system, HF-H_2O_2-$Cu(NO_3)_2$ etching system (specific experimental parameters are shown in Tables 1 to 5).First, place the cleaned silicon wafers in 5% HF (volume fraction) to react for 5 minutes to remove the oxide layer.After room temperature etching in the three etching systems, quickly transfer the reacted silicon wafers into HNO_3 solution to remove residual Cu particles, then rinse three times with DI water, and finally dry with N_2.

Use a Scanning Electron Microscope (SEM, FEI, USA) to observe the changes in microstructure and copper nanoparticle deposition on the silicon wafer surface before and after etching.

3. Result and discussion

3.1 HF-H_2O_2-$CuSO_4$ etching system

By maintaining the concentration ([HF]/[H_2O_2] = 4.6M/1.0M) constant, and by adjusting the concentration of copper sulfate for etching for 15 minutes, the silicon wafer surface was characterized by SEM post-texturing. Specific experimental conditions are shown in Table 1.From Figure 1, it can be seen that as the concentration of copper sulfate increases, the surface of the single crystal silicon wafer gradually shows a uniform grid-like indentation. Etching begins on the surface of the silicon wafer at a copper sulfate concentration of 0.03M. When the copper sulfate concentration reaches 0.07 M, the grid pattern obtained is most uniform, with the diameters of the structures ranging between 3-7 μm.

Table 1. The experimental conditions of the <100>-oriented c-Si wafers etching under different CuSO4 concentration.

	[HF]/M	[H_2O_2]/M	[$CuSO_4$]/M	Etching time/min	Morphology
Etching	4.6	1.0	0.01	15	1 (a)/(e)
	4.6	1.0	0.03	15	1 (b)/(f)
	4.6	1.0	0.05	15	1 (c)/(g)
	4.6	1.0	0.07	15	1 (d)/(h)

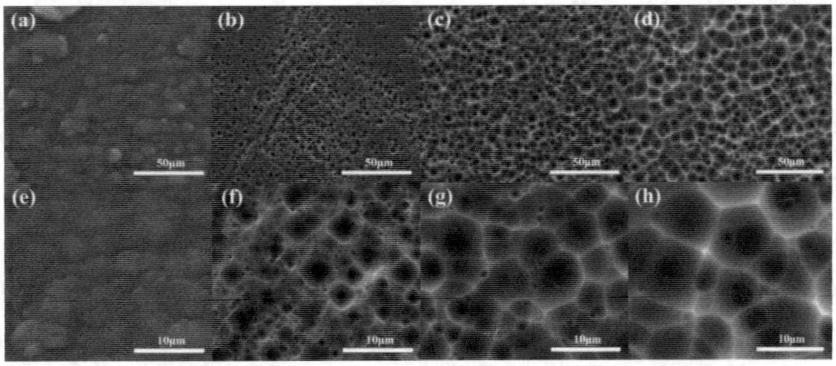

Figure 1. SEM images of <100>-oriented c-Si wafers etching under different $CuSO_4$ concentration: (a,e) 0.01M; (b,f) 0.03M; (c,g) 0.05M; (d,h) 0.07M.

3.2 HF-H_2O_2-$CuCl_2$ etching system

Figure 2 shows SEM images of silicon wafers textured under different concentrations of $CuCl_2$ in the HF-H_2O_2-$CuCl_2$ etching system. The images reveal that as the concentration of copper chloride increases, a sparse porous structure appears on the silicon surface with uneven flatness. At a low concentration of copper chloride (0.01M), there is no significant etching on the single crystal silicon surface. At a higher concentration (0.07M), the surface of the silicon wafer exhibits alternating sparse and dense porous structures.

Table 2. The experimental conditions of the <100>-oriented c-Si wafers etching under different CuCl2 concentration.

	[HF]/M	[H_2O_2]/M	[$CuCl_2$]/M	Etching time/min	Morphology
Etching	4.6	1.0	0.01	15	2 (a)/(e)
	4.6	1.0	0.03	15	2 (b)/(f)
	4.6	1.0	0.05	15	2 (c)/(g)
	4.6	1.0	0.07	15	2 (d)/(h)

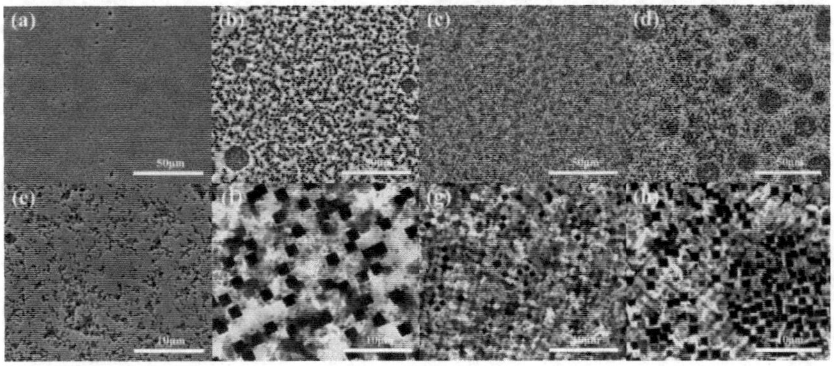

Figure 2. SEM images of <100>-oriented c-Si wafers etching under different $CuCl_2$ concentration: (a,e) 0.01M; (b,f) 0.03M; (c,g) 0.05M; (d,h) 0.07M.

3.3 HF-H_2O_2-$Cu(NO_3)_2$ etching system

Figure 3 provides images showing changes in the surface morphology of silicon wafers at different concentrations of copper nitrate in the HF-H_2O_2-$Cu(NO_3)_2$ etching system.The textured single crystal silicon surfaces primarily show inverted pyramidal structures.With increased etching time, the uniformity of the inverted pyramids on the silicon surface shows a trend of improvement.At a copper nitrate concentration of 0.07M, the corresponding silicon surface displays the most regular and uniform inverted pyramidal structures.As shown in Figure 3(d), the size of the inverted pyramid structures ranges from 0.8 to 1.2 μm.

Table 3. The experimental conditions of the <100>-oriented c-Si wafers etching under different Cu(NO3)2 concentration.

	[HF]/M	$[H_2O_2]$/M	$[Cu(NO_3)_2]$/M	Etching time/min	Morphology
Etching	4.6	1.0	0.01	15	3 (a)/(e)
	4.6	1.0	0.03	15	3 (b)/(f)
	4.6	1.0	0.05	15	3 (c)/(g)
	4.6	1.0	0.07	15	3 (d)/(h)

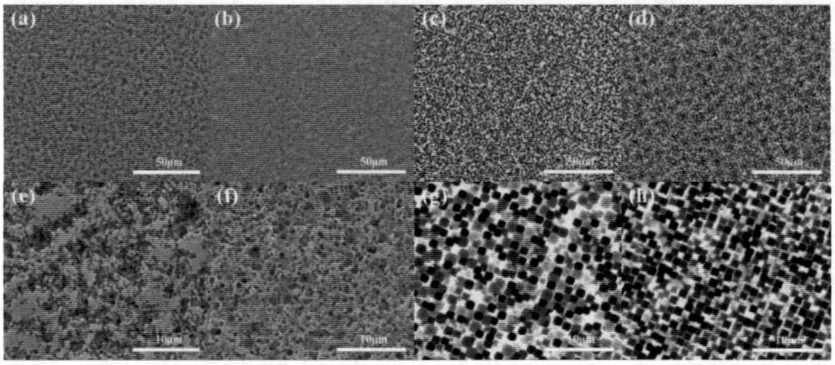

Figure 3. SEM images of <100>-oriented c-Si wafers etching under different $Cu(NO_3)_2$ concentration: (a,e) 0.01M; (b,f) 0.03M; (c,g) 0.05M; (d,h) 0.07M.

3.4 Depositional systems

Under the HF-H_2O_2-$CuSO_4$ etching system, the reflectivity of the silicon wafer after fuzzing is shown in Figure 4(a). The reflectivity of the etched silicon wafer is significantly reduced compared to the original wafer. The best anti-reflective effect on the silicon surface occurs at a copper sulfate solution concentration of 0.03 mol/L.Figure 4(b) shows the reflection spectrum of multicrystalline silicon cut by diamond wire under the HF-H_2O_2-$CuCl_2$ etching system. In the 300~1100nm wavelength range, the best reflectivity is observed in a 0.07 mol/L copper chloride solution. Additionally, under this condition, the silicon wafer SEM image shows the most uniform worm-like etch pits.In the HF-H_2O_2-$Cu(NO_3)_2$ etching system, the average reflectivity of the silicon wafer initially decreases and then increases as the etching time progresses, as shown in Figure 4(c). The overall best anti-reflective effect is achieved when the copper nitrate concentration is at 0.05 mol/L. The corresponding silicon surface displays the most regular and uniform inverted pyramid nanostructures. The anti-reflective capability of the wafer continuously declines over time.

Table 4. The experimental conditions of the <100>-oriented c-Si wafers deposition under different depositional systems (with H2O2).

	[HF]/M	$[H_2O_2]$/M	$[H_2SO_4]$/M	[Cu]/M	Etching time/min	Morphology
Deposition 1	5.6	0.5	0.0	0.02	15	4 (a)
	5.6	0.5	0.1	0.02	15	4 (b)
	5.6	0.5	0.3	0.02	15	4 (c)
	5.6	0.5	0.5	0.02	15	4 (d)
Deposition 2	[HF]/M	$[H_2O_2]$/M	[HCl]/M	[Cu]/M	Etching time/min	Morphology
	5.6	0.5	0.0	0.02	15	4 (e)

	5.6	0.5	0.1	0.02	15	4 (f)
	5.6	0.5	0.3	0.02	15	4 (g)
	5.6	0.5	0.5	0.02	15	4 (h)
	[HF]/M	[H_2O_2]/M	[HNO_3]/M	[Cu]/M	Etching time/min	Morphology
Deposition 3	5.6	0.5	0.0	0.02	15	4 (i)
	5.6	0.5	0.1	0.02	15	4 (j)
	5.6	0.5	0.3	0.02	15	4 (k)
	5.6	0.5	0.5	0.02	15	4 (l)

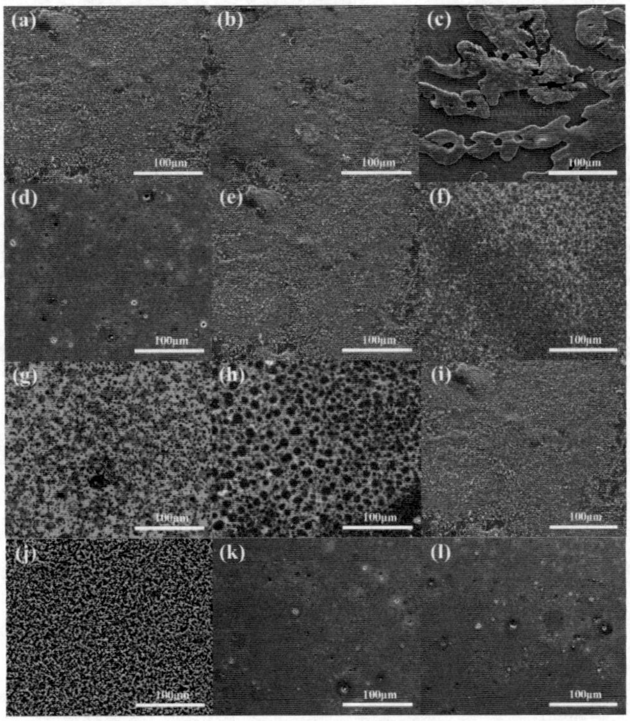

Figure 4. SEM images of <100>-oriented c-Si wafers deposition under three depositional systems (with H_2O_2): (a,b,c,d) HF-H_2O_2-H_2SO_4-Cu; (e,f,g,h) HF-H_2O_2-HCl-Cu; (i,j,k,l) HF-H_2O_2-HNO_3-Cu.

Figure 5 shows SEM images of copper deposition on <100> single crystal silicon in three different deposition systems (without hydrogen peroxide added), with experimental conditions as shown in Table 5. From figures 5a to 5d, it can be seen that under different sulfuric acid concentrations, the silicon wafer surface is almost not covered by copper particles, with only very few copper particles present. Similarly, under the hydrochloric and nitric acid systems, the silicon wafer surface is almost not covered by copper particles, with only trace amounts of copper particles present.

Table 5. The experimental conditions of the <100>-oriented c-Si wafers deposition under different depositional systems (without H2O2).

	[HF]/M	[H_2SO_4]/M	[Cu]/M	Etching time/min	Morphology
Deposition 1	5.6	0.0	0.02	15	5 (a)
	5.6	0.1	0.02	15	5 (b)
	5.6	0.3	0.02	15	5 (c)
	5.6	0.5	0.02	15	5 (d)
Deposition 2	[HF]/M	[HCl]/M	[Cu]/M	Etching time/min	Morphology
	5.6	0.0	0.02	15	5 (e)

	5.6	0.1	0.02	15	5 (f)
	5.6	0.3	0.02	15	5 (g)
	5.6	0.5	0.02	15	5 (h)
Deposition 3	[HF]/M	[HNO_3]/M	[Cu]/M	Etching time/min	Morphology
	5.6	0.0	0.02	15	5 (i)
	5.6	0.1	0.02	15	5 (j)
	5.6	0.3	0.02	15	5 (k)
	5.6	0.5	0.02	15	5 (l)

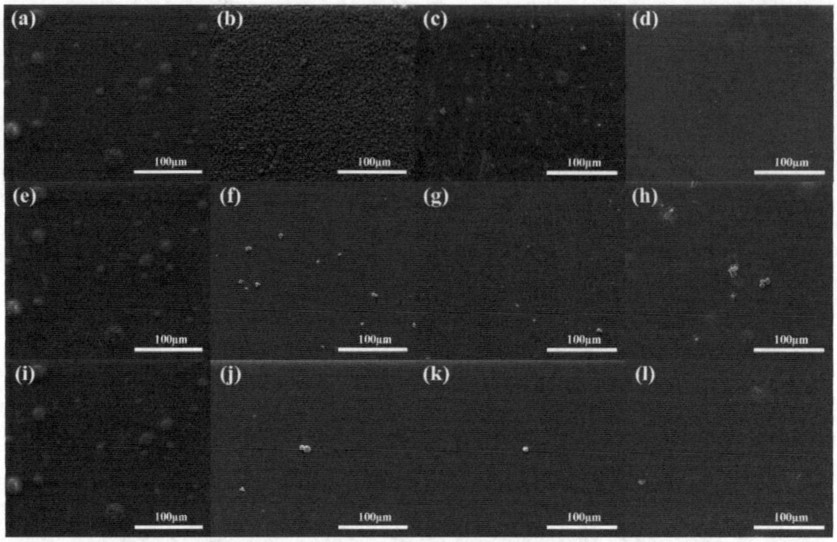

Figure 5. SEM images of <100>-oriented c-Si wafers deposition under three depositional systems (without H_2O_2): (a,b,c,d) HF-H_2SO_4-Cu; (e,f,g,h) HF-HCl-Cu; (i,j,k,l) HF-HNO_3-Cu.

3.5 Working principle of Copper-catalyzed chemical etching

In the etching solution, the interface between silicon and copper nanoparticles forms a spontaneous primary cell structure. The potential of copper and copper ions is greater than that of silicon, making it easier for copper nanoparticles to capture electrons from the silicon surface and become negatively charged. To maintain charge balance in the solution, the silicon surface loses electrons.As etching time increases, copper nanoparticles sink deeper into the silicon, continuously oxidizing the underlying silicon atoms as they lose electrons. The oxidized silicon dioxide is dissolved and removed by hydrofluoric acid, forming nanostructures on the silicon wafer surface[16].

The morphology of nanoscale copper particles is influenced by the oxidizing nature and complexing ability of the anions in the copper salts, as well as the bonding strength between the cations and anions, ultimately leading to different etching behaviors. In Figure 6, Unlike the weaker interactions in the SO_4^{2-} system, the stronger electrostatic attraction in the NO^{3-} system helps slow down the rapid deposition of copper particles.Compared to the aggregation behavior caused by strong complexation in the Cl^- system, the oxidizing nature of NO^{3-} is more favorable for improving the dispersity and size uniformity of deposited copper particles.

In the HF/$CuSO_4$ solution, the attraction of SO_4^{2-} to Cu^{2+} is the weakest, which facilitates the reduction deposition of Cu^{2+}, minimizing the impact on the copper deposition on the silicon wafer surface. In the HF/$Cu(NO_3)_2$ system, due to the oxidizing nature of NO^{3-}, some of the deposited copper particles are oxidized back to copper ions, reducing the amount of copper deposition, while the oxidizing action of NO^{3-} alleviates the aggregation of copper particles. In the HF/$CuCl_2$ system, chlorine and copper predominantly exist in the form of covalent bonds, making it harder for Cu to be freed for deposition, and in a Cl^- rich solution, Cu^{2+} and deposited copper particles tend to form $[CuCl_2]^-$ complexes, leading to the precipitation of $CuCl_2$. The electrostatic attraction between cations and anions, as well as the oxidizing nature and complexing ability of anions in different copper salt solutions, all affect copper deposition, ultimately impacting the etching behavior and etched morphology[17].

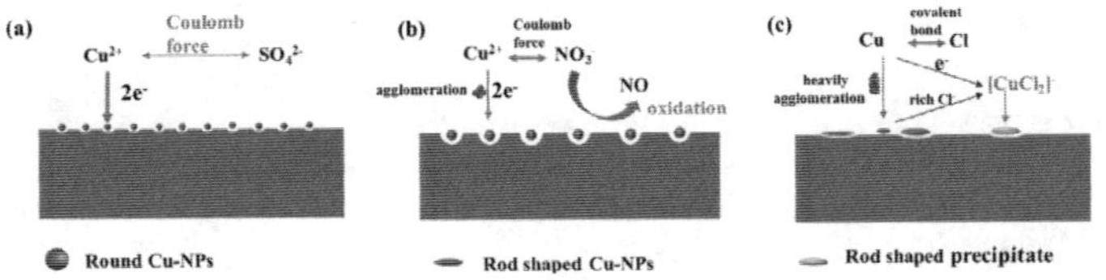

Figure 6 Schematic diagram of copper particle deposition in different copper salt systems.

$$\text{Cathode reaction:} \quad Cu^{2+}+2e^{-} \rightarrow Cu \qquad (1)$$

$$H_2O_2+4H^{+}+2e^{-} \rightarrow 2H_2O+2h^{+}+H_2 \qquad (2)$$

$$\text{Anode reaction:} \quad Cu+H_2O_2+2H^{+} \rightarrow H_2O+Cu^{+} \qquad (3)$$

4. Conclusions

By controlling the types and concentrations of metal copper salts, it is possible to achieve controlled preparation of different nanostructures (grid-like, porous, inverted pyramid) on the silicon wafer surface. In three deposition systems with added oxidizing agent (H_2O_2), as the concentrations of H_2SO_4, HCl, and HNO_3 increase, the content of Cu nanoparticles deposited on the single crystal silicon surface shows a decreasing trend. In the three deposition systems without added oxidizing agent (H_2O_2), the changes in Cu particles deposited on the silicon wafer surface are not significant. This is due to the combined effects of the oxidizing nature and complexing ability of the anions in the copper salts, as well as the binding force between cations and anions, on the deposition of Cu nanoparticles on the silicon wafer surface. This ultimately leads to different etched morphologies of the silicon wafer during the velvet-making process.

Acknowledgment

Financial support of this work from Hunan Province Education Science "14th Five-Year" Plan Project (XJK22CZY104) and Education Department of Hunan Province of China（24C1223）.

References

[1] Owusu P A, Asumadu-Sarkodie S. (2016) A review of renewable energy sources, sustainability issues and climate change mitigation.J. Cogent Engineering. 3(1), 1167990.

[2] Pourasl H H, Barenji R V, Khojastehnezhad V M.(2023) Solar energy status in the world: A comprehensive review.J. Energy Reports, 10: 3474-3493.

[3] Naveen S, et al.(2023) A review on solar energy intensified biomass valorization and value-added products production: practicability, challenges, techno economic and lifecycle assessment.J. Journal of Cleaner Production, 405 : 137028.

[4] Hasan M M, et al.(2023) Harnessing solar power: a review of photovoltaic innovations, solar thermal systems, and the dawn of energy storage solutions.J. Energies 16(18): 6456.

[5] Mohammad A, Mahjabeen F.(2023) Revolutionizing solar energy: The impact of artificial intelligence on photovoltaic systems.J. International Journal of Multidisciplinary Sciences and Arts, 2(1).

[6] Manisha, et al.(2023) Recent developments on photovoltaic thermal drying systems: a clean energy production.J. Clean Technologies and Environmental Policy, 25(7):2099-2122.

[7] Bouraima, Mouhamed B, et al.(2024) An integrated intelligent decision support framework for the development of photovoltaic solar power.J. Engineering Applications of Artificial Intelligence,127:107253.

[8] Addonizio M L, A Antonaia.(2023) Textured p-type crystalline silicon surfaces obtained by multi-step plasma process for SHJ solar cells.J. Vacuum, 215:112284.

[9] Huang Z, Zhao L, Li T, Kang J, Zhou X, Li S, Ren Y. (2024) Study on the effect of melt flow behavior on the uniformity of phosphorus do** during the growth of large-size n-type monocrystalline silicon by Czochralski method.J. Materials Science in Semiconductor Processing, 175, 108261.

[10] Korichi H, Boucheham A, Bensdira A, Kazerane M.(2024) Advancements in Passivation and Metallization Techniques for n-Type Monocrystalline Silicon Solar Cells.J. Solar Energy and Sustainable Development Journal, 13(2),151-173.
[11] Zhang H, et al.(2022) Electrochemical oxidation pre-treatment for wet texturing of monocrystalline silicon solar cells.J. Applied Surface Science 599:154059.
[12] Huo C L.(2024) Controllable Fabrication of Silicon Solar Cells with Inverted or Upright Pyramid Structures by a One-Step Copper Catalysis Method.J. ECS Journal of Solid State Science and Technology 13(6): 065002.
[13] Srivastava, Avritti, et al.(2024) Rapid and low temperature micro inverted pyramid texturing of silicon via copper catalyzed chemical etching for high efficiency PEDOT: PSS/silicon hybrid solar cells.J. Surfaces and Interfaces, 46: 104066.
[14] Dridi R B, Khelifi W, Aouida S, Bouaïcha, M.(2024) Tuning surface reflectance and carrier lifetime of cone-shaped macropore structures obtained by platinum-assisted chemical etching of silicon.J. Applied Physics A, 130(8), 565.
[15] Zou Y, Liu X, Wang M, Song Y, Liu H, Hong S, et al. (2025) Study on the stable preparation and optimization treatment of DWS N-type single-crystal silicon pyramid arrays.J. Materials for Renewable and Sustainable Energy, 14(1), 1-9.
[16] Hong S, Zheng D, Ma L, et al.(2022) Study on the Influence of Hydrogen Peroxide on Nano-Cu Catalyzed Etching of n-Type Single Crystal Silicon.J.Surface Technology, 51(1): 257-264.
[17] Hong S, Zou Y, Ma L,et al.(2022) Surface Texturing Behavior of Nano-Copper Particles under Copper-Assisted Chemical Etching with Various Copper Salts System.J.Silicon, 14(12):7169-7177.

A new Tm^{3+}-doped $ZnLaB_5O_{10}$ borate blue-emitting phosphor for w-LEDs

Juanjuan Wen, Lingling Zheng*, and Jiayi Wu

Guangzhou Vocational College of Technology & Business, Guangzhou 511442, Guangzhou, China

*Corresponding author's e-mail: zsuzhengll@163.com

Abstract. A blue-luminescent Tm^{3+}-activated $ZnLaB_5O_{10}$ material was initially synthesized via thermal solid-state reactions. Structural characterization and optical evaluations confirmed phase homogeneity and radiative characteristics of the crystalline products. Experimental data identified 3 mol% Tm^{3+} doping (x=0.03) as the critical concentration threshold before emission efficiency decline. Mechanistic studies revealed that multipolar electrical interactions dominate energy dissipation processes in over-doped systems. The optimized compound exhibited intense blue emission with CIE chromaticity values (0.1803, 0.1007), demonstrating suitability as a blue-conversion material in w-LED architectures.

1. Introduction

The lighting sector has undergone transformative advancements through white light-emitting diodes (w-LEDs), which now dominate markets by superseding conventional fluorescent technologies due to superior energy efficiency, ecological compatibility, and enhanced optical quality [1, 2]. Current w-LED manufacturing predominantly employs blue GaN semiconductor chips coated with yellow-emitting cerium-doped yttrium aluminum garnet (YAG:Ce^{3+}). Nevertheless, such configurations exhibit critical deficiencies, particularly inadequate color reproduction accuracy stemming from insufficient red spectral emissions [3]. While alternative strategies utilizing trichromatic phosphor blends address these issues, emergent challenges include parasitic absorption of blue wavelengths by complementary red/green emitters, resulting in diminished blue output and compromised photonic efficacy [4, 5]. These constraints highlight the critical demand for innovative ultraviolet-excitable blue-emitting materials capable of boosting spectral conversion efficiency while minimizing energy losses in next-generation w-LED systems.

Borate is known for its outstanding physical properties and chemical stability, which has led numerous research teams to explore its structure and luminescence properties [6, 7]. $K_6Li_{0.9}Na_{0.1}CaSc_2(B_5O_{10})_3$:$Cr^{3+}$ exhibits near-red light emission and possesses excellent thermal stability, making it suitable for the production of near-infrared light-emitting diodes [8]. Polyborate $ZnLa_{1-x}Eu_xB_5O_{10}$ exhibits a delayed concentration quenching, enhanced thermal stability, and an optimal red-to-orange emission ratio ranging from 2.7 to 3.1. These attributes render it an auspicious contender for applications in warm w-LEDs [9]. The introduction of samarium ions (Sm^{3+}) into the $ZnLaB_5O_{10}$ host lattice yields exceptional spectral conversion properties, making it an ideal candidate for high-fidelity lighting systems demanding superior color fidelity, particularly in white light-emitting diodes requiring enhanced chromatic reproduction accuracy [10]. This performance highlights the material's viability as a core component in novel phosphor architectures for next-generation illumination devices. In this research, $ZnLaB_5O_{10}$:Tm^{3+} phosphors were prepared using a solid-state reaction method. The phase purity of the synthesized materials was verified through X-ray diffraction (XRD) analysis. Additionally, detailed measurements of the photoluminescence excitation and emission spectra, as well as the CIE color coordinates, were conducted to thoroughly assess their optical performance.

Journal of Physics: Conference Series **3009** (2025) 012025 doi:10.1088/1742-6596/3009/1/012025

2. Experiment

2.1. Synthesis
A high - temperature solid - state reaction synthesized Tm^{3+}-doped $ZnLaB_5O_{10}$ phosphors. ZnO, La_2O_3, H_3BO_3, and Tm_2O_3 (A.R.) were raw materials, used as - is. For 0.03 mol doping, the materials were weighed, ground in an agate mortar, then transferred to a quartz one for 20 - min grinding. Each sample, weighing about 1.0 g, was heated to 830°C for 5 h in a muffle furnace, then reground at room temperature for 20 min. The chemical reaction involved in the synthesis can be represented by the following equation:

$$(1\text{-}x)/2La_2O_3 + ZnO + 5\ H_3BO_3 + x/2Tm_2O_3 \xrightarrow{830^\circ C\ ,\ 5\ h} ZnLa_{(1\text{-}x)}Tm_xB_5O_{10}$$

2.2. Characterization method
The crystal structure and composition of phosphor materials are critically dependent on phase purity, making it a key factor in their performance. To evaluate the phase purity of $ZnLaB_5O_{10}:Tm^{3+}$, X-ray diffraction (XRD) analysis was conducted using a radiation wavelength of λ = 0.154184 nm. The measurements were performed with a D2 PHASER instrument, a high-precision device produced by Bruker, Germany. Additionally, the photoluminescence excitation and photoluminescence spectra were measured using an FLS 980 spectrometer equipped with a 450 W xenon lamp and a spectral resolution of 0.05 nm.

3. Results and discussion

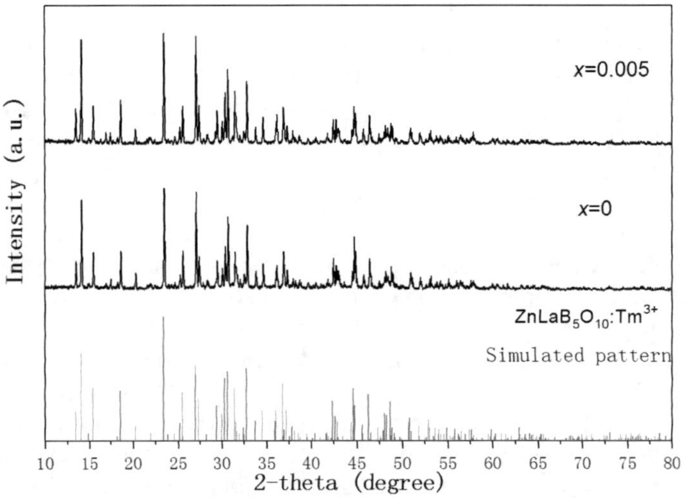

Figure 1. The XRD patterns of the $ZnLaB_5O_{10}:xTm^{3+}$ (x = 0 and 0.005 mol) phosphors.

As shown in Figure 1, the XRD pattern of $ZnLaB_5O_{10}:Tm^{3+}$ aligns precisely with the peak positions of the simulated reference pattern for the host material $ZnLaB_5O_{10}$, with no observable peak shifts. This confirms that the dopant ions have successfully integrated into the host lattice without disrupting the crystal structure. Specifically, Tm^{3+} ions have substituted La^{3+} ions within the $ZnLaB_5O_{10}$ matrix. The preservation of the host crystal structure can be attributed to the comparable ionic radii of Tm^{3+} (r = 0.88 Å) and La^{3+} (r = 0.1032 Å), which facilitates the formation of a single-phase product.

Formula (1) is utilized to calculate the radius percentage difference (D_r) between Tm^{3+} and the probable substituent ions in $ZnLaB_5O_{10}$:

$$D_r = \frac{R_m(CN)\text{-}R_d(CN)}{R_m(CN)} \times 100\% \quad (1)$$

The ionic substitution radius (R_m) is governed by substituting cations, whereas the dopant ionic radius corresponds to Tm^{3+}, and the structural coordination index (*CN*) reflects atomic bonding geometry. Quantitative analysis reveals cation mismatch percentages (D_r) of 15.91% for Zn^{2+} (*CN*=6)–Tm^{3+} (*CN*=6) pairs and 14.53% for La^{3+} (*CN*=6)-Tm^{3+} (*CN*=6) configurations, values calculated using ionic radius differential formulas. These critical parameters determine lattice stability in doped crystalline systems. Given that La^{3+} and Tm^{3+} share the same charge, the likelihood of La^{3+} being replaced is significantly higher.

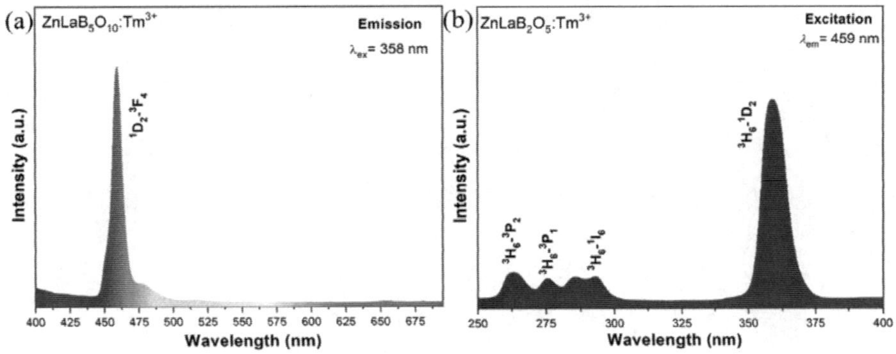

Figure 2 (a-b). The excitation spectra and emission spectra of sample $ZnLaB_5O_{10}:Tm^{3+}$.

Under ambient conditions, the radiative output of $ZnLaB_5O_{10}:Tm^{3+}$ under 358 nm photostimulation (Figure 2a) exhibits monochromatic emission characteristics. Spectral analysis reveals a singular radiative feature at 455 nm, arising from intra-4*f* electronic reorganization within Tm^{3+} centers, specifically corresponding to the $^1D_2 \rightarrow {}^3F_4$ electronic state transition. As can be seen from Figure 2 (b), when $ZnLaB_5O_{10}:Tm^{3+}$ is monitored with λ_{em}= 459 nm, the excitation peaks of the matrix $ZnLaB_5O_{10}$ manifest within the 250-400 nm range of the excitation spectra. At 358 nm, there is a distinct single excitation peak, which is attributed to the $^3H_6 \rightarrow {}^1D_2$ transition.

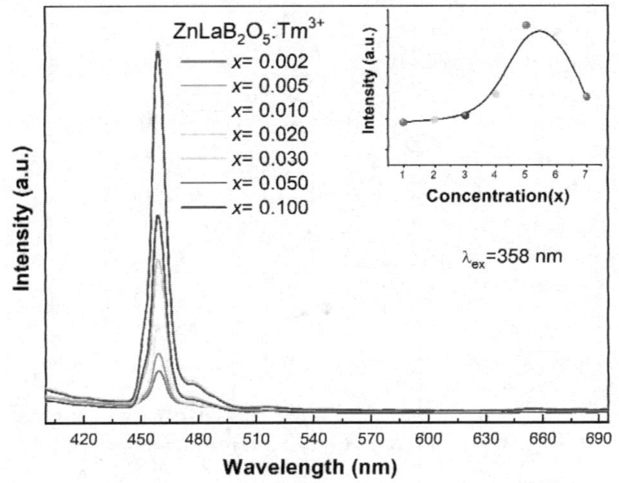

Figure 3. Photoluminescence spectra of $ZnLaB_5O_{10}:xTm^{3+}$ phosphors ($0.002 \leqslant x \leqslant 0.1$) are presented. Inset: The trend of intensity variation with Tm^{3+} doping concentration.

As shown in Figure 3, the luminescence intensity of $ZnLaB_5O_{10}$:$x$$Tm^{3+}$ (0.002 ⩽ x ⩽ 0.1) phosphors initially rises and then declines as the concentration of Tm^{3+} ions changes. The initial enhancement in emission intensity is attributed to the growing number of luminescent centers resulting from the increased Tm^{3+} concentration. However, at higher concentrations, the reduced distance between Tm^{3+} ions promotes energy transfer among these luminescent centers, increasing the probability of non-radiative energy transfer and causing concentration quenching. Based on the data in Figure 3, the optimal doping concentration for $ZnLaB_5O_{10}$:Tm^{3+} phosphors is approximately 0.03 mol.

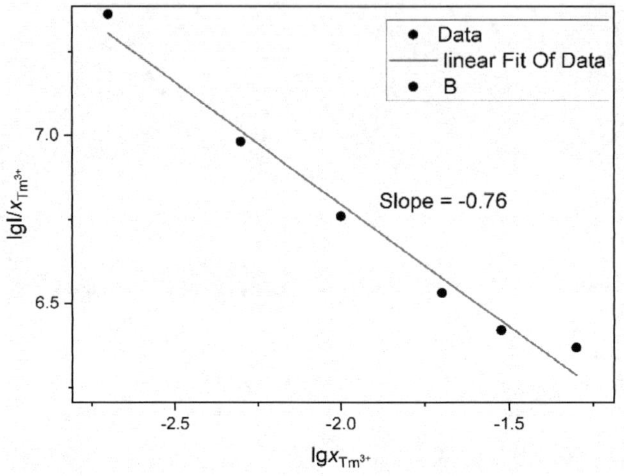

Figure 4. The linear correlation between lg$x$ and lg(I/x) for $ZnLaB_5O_{10}$:$x$$Tm^{3+}$ (0.002 ≤ x ≤ 0.1) phosphors.

In the study of rare earth-doped luminescent materials, concentration quenching is typically attributed to two primary mechanisms: multipolar interactions and exchange interactions. These mechanisms are governed by a critical distance, denoted as R_c. Multipolar interactions play a dominant role when R_c is less than 5 Å; beyond this distance, their influence diminishes significantly. The value of R_c can be calculated using Equation (2) [11]:

$$R_c \approx 2\left(\frac{3V}{4\pi\ x_c N}\right)^{1/3} \tag{2}$$

The substitution capacity N quantifies replaceable cation sites (N=4), while V defines the crystallographic unit volume (641.3Å^3), and x_c marks the ideal dopant molar ratio. Through systematic computation employing these variables, the critical separation distance (R_c) yields 21.72 Å. In solid-state physics, R_c>20 Å explicitly confirms electric multipole dominance. The Q parameter quantifies multipolar coupling types, adopting discrete values (3, 6, 8, 10) corresponding to distinct energy transfer mechanisms. A derived mathematical framework models these multilevel energy-exchange processes involving interdependent quantum states within crystalline lattices. This equation offers a mathematical framework to analyze the dynamic transitions and interactions between these energy levels, providing valuable insights into the system's behavior and properties. Through this analysis, a deeper understanding of the underlying mechanisms governing the system can be achieved [12]:

$$\frac{I}{x} = K\left[1+\beta\left(x\right)^{Q/3}\right]^{-1} \tag{3}$$

The mathematical model defines x as the Tm^{3+} dopant molar ratio, I as radiative output intensity, with K and β serving as empirical scaling parameters. Graphical analysis in Figure 4 demonstrates logarithmic correlations between lg(x) and lg(I/x) under 358 nm irradiation. Linear regression yields a gradient of -0.76, translating to an interaction parameter Q=2.28, statistically proximate to the integer 3. This quantitative alignment implies dominant short-range energy transfer mechanisms driving emission quenching, the behavior congruent with observations in structurally analogous Tm^{3+}-activated systems such as Y_2MgTiO_6 and K_3YF_6 hosts [13, 14].

To evaluate the luminescent properties of materials, CIE chromaticity coordinates serve as essential indicators. As shown in Figure 5, the color coordinates of $ZnLaB_5O_{10}$:$x$$Tm^{3+}$ ($0.002 \leqslant x \leqslant 0.10$) consistently fall within the blue spectral region. Specifically, $ZnLaB_5O_{10}$:0.03Tm^{3+} demonstrates color coordinates of (0.1803, 0.1007). The color purity of these materials can be determined using Equation (4), which provides a quantitative measure of the color's purity.

$$Color\ purity = \sqrt{\frac{(x-x_i)^2+(y-y_i)^2}{(x_d-x_i)^2+(y_d-y_i)^2}} \times 100\% \tag{4}$$

Within the CIE 1931 color space, (x_i, y_i) corresponds to the standard white point (0.333, 0.333), while (x_s, y_s) characterizes chromaticity variations across dopant levels. The extrapolated trajectory linking these coordinates intersects the spectral locus periphery at (x_d, y_d), defining gamut boundary coordinates. Spectral mapping (Figure 5) confirms all synthesized specimens exhibit chromatic parameters localized within the deep-blue spectral domain, consistent with high-purity emission profiles. For $ZnLaB_5O_{10}$:$x$$Tm^{3+}$ ($0.002 \leqslant x \leqslant 0.10$), the color coordinates are (0.2396, 0.1982), (0.2273, 0.1982), (0.2075, 0.1478), (0.1944, 0.1258), (0.1803, 0.1007), (0.1791, 0.0979), and (0.1814, 0.1021), respectively. In the case of $ZnLaB_5O_{10}$:0.03Tm^{3+}, the calculated color purity reaches 78.4%.

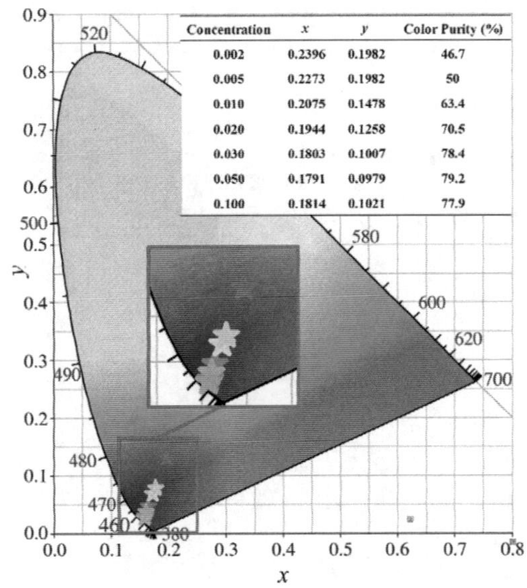

Concentration	x	y	Color Purity (%)
0.002	0.2396	0.1982	46.7
0.005	0.2273	0.1982	50
0.010	0.2075	0.1478	63.4
0.020	0.1944	0.1258	70.5
0.030	0.1803	0.1007	78.4
0.050	0.1791	0.0979	79.2
0.100	0.1814	0.1021	77.9

Figure 5. Chromaticity coordinates chart of the $ZnLaB_5O_{10}$:$x$$Tm^{3+}$ ($0.002 \leq x \leq 0.1$) phosphors.

4. Conclusions

For the first time, $ZnLaB_5O_{10}$:$x$$Tm^{3+}$ ($0.002 \leqslant x \leqslant 0.1$) phosphors were successfully prepared using a conventional high-temperature solid-state reaction method. When excited at 358 nm, these phosphors displayed strong emission peaks at 459 nm. The optimal doping concentration of Tm^{3+} in $ZnLaB_5O_{10}$:$x$$Tm^{3+}$ was determined to be 0.03 mol. The observed concentration quenching is attributed to interactions between nearest-neighbor ions, with a critical distance calculated to be approximately

21.72 Å. The CIE chromaticity coordinates of the phosphors are located close to the standard blue region. Notably, $ZnLaB_5O_{10}:Tm^{3+}$ exhibits intense blue emission with a high color purity of 78.4%. These properties highlight the significant potential of $ZnLaB_5O_{10}:Tm^{3+}$ phosphors for future applications in solid-state lighting, particularly in scenarios requiring high efficiency and excellent color rendering.

Acknowledgments

This work was supported by the Tertiary Education Scientific Research Project of the Guangzhou Municipal Education Bureau (2024312294).

References

[1] B. Han, L. Shi, Y. Lv, W. Li, J. Hu, Y. Liu, X. Han, Y. Wang, J. Lumin. **281**, 121142 (2025) [CrossRef]
[2] X. Wei, S. Huang, Z. Xu, Y. Wang, X. Xing, M. Shang, Opt. Mater. **160**, 116792 (2025) [CrossRef]
[3] X. Yang, Q. Sun, Y. Liu, X. Wang, J. Jing, H. Zou, Y. Song, Y. Sheng, Ceram. Int. **51**, 1508 (2025) [CrossRef]
[4] Z. Zhang, R. Cui, J. Zhang, C. Deng, Ceram. Int. **51**, 6906 (2025) [CrossRef]
[5] W. Fang, Y. Yang, Y. Liu, D. Ma, J. Huang, B. Song, L. Xia, Inorg. Chem. **64**, 4121 (2025) [CrossRef]
[6] S. Cao, F. Su, Z. Liu, J. Yuan, T. Liang, L. Hou, Y. Wang, Y. Dai, J. Chen, Mater. Today Chem. **43**, 102500 (2025) [CrossRef]
[7] K.J. Albert, S.M.M. Kennedy, J. Lumin. **280**, 121122 (2025) [CrossRef]
[8] M. Hua, S. Liu, L. Zhou, J.-C. Bünzli, M. Wu, Chem. Sci. **16**, 2089 (2025) [CrossRef]
[9] M.N. Huang, Y.Y. Ma, F. Xiao, Q.Y. Zhang, Spectrochim. Acta, Part A **120**, 55 (2014) [CrossRef]
[10] Y. Du, H. Su, Z. Zou, D. He, Y. Shang, L. Zhou, J. Chen, S. Cheng, Y. Ran, S. Xie, R. Yu, J. Mater. Sci: Materials Electronics **34**, 1781 (2023) [CrossRef]
[11]G. Blasse, B. Grabmaier, Radiative return to the ground state: Emission, in: Luminescent Materials, Springer, 1994, pp. 33. [CrossRef]
[12]S. Ding, C. Li, M. Zhang, C. Zhang, H. Hu, Y. Zou, X. Tang, W. Liu, J. Alloys Compd. **1014**, 178571 (2025) [CrossRef]
[13]L. Wang, J. Huang, L. Chen, Z. Xiong, J. Guo, J. Lumin. **280**, 121114 (2025) [CrossRef]
[14]L. Ke, K. Ren, X. Cai, Y. Zhang, Ceram. Int. **51**, 8012 (2025) [CrossRef]

Journal of Physics: Conference Series **3009** (2025) 012026 doi:10.1088/1742-6596/3009/1/012026

Comparative study of exciton properties in $CsSnX_3$(X=Cl, Br, I) halide perovskite thin films

Zhonghai Lin[1*], Ruiyang Zeng[1], Pingjian Wang[1], Guangfen Wei[1], Huitian Du[1] and Xiaoyan Yang[1]

[1] Shandong Technology and Business University, Yantai 264005, PR China

*Corresponding author's e-mail: zhlin@sdtbu.edu.cn (Z. Lin)

Abstract: $CsSnX_3$ (X = Cl, Br, I) inorganic perovskites exhibit excellent stability and optoelectronic properties. This study establishes a quantitative correlation the regulation of their fundamental physical properties by layers. As X-site atoms increase, the lattice becomes constant, and optimized volume grows. The band gap decreases and absorption spectra shift red as layer thickness increases, without altering the direct bandgap nature. Exciton properties are influenced by X atoms and layer number, with the maximum exciton binding energy decreasing and the Bohr radius increasing as the X atom size grows.

1. Introduction

All-inorganic perovskite thin films have demonstrated breakthrough potential for applications in next-generation optoelectronic devices, with advantages stemming from the unique synergy between quantum confinement effects and defect engineering mechanisms. This is validated by experimental data from transient absorption spectra, showing Stokes shifts.[1]. These materials offer the advantage of solution-processable fabrication, which simplifies the manufacturing process and allows for scalable production. This characteristic is particularly valuable for large-area devices, as it eliminates the need for costly vacuum deposition techniques typically used in other materials[2]. Furthermore, these perovskite thin films exhibit superior charge transport characteristics, which are essential for high-performance electronic devices such as solar cells, light-emitting diodes (LEDs), and photodetectors[3]. Their high electrical conductivity further supports efficient charge extraction and transport, enhancing the overall efficiency of optoelectronic devices.

The crystallographic architecture of these materials is highly tunable, allowing for precise control over their structural properties. This tunability enables engineers and researchers to manipulate the material's electronic and optical behaviors, thus optimizing them for specific applications. By adjusting the composition of the perovskite, such as varying the halide ions or altering the cation structure, the materials can be tailored to achieve desired characteristics, such as wide or narrow optical bandgaps. This level of control significantly improves the performance of devices across a broad spectral range, enabling them to be used for various applications from visible to infrared light absorption and emission.

Compared to organic-inorganic perovskite films ($CH_3NH_3PbI_3$), all-inorganic perovskite films ($CsPbI_3$, $CsSnI_3$) offer significant advantages in terms of stability. Studies have demonstrated that $CsPbI_3$ films exhibit superior thermal and optical stability under high humidity and elevated temperature conditions, thereby expanding their potential applications in optoelectronic devices[4]. Furthermore, the electron mobility of all-inorganic perovskite films is generally higher than that of their organic-inorganic counterparts[5]. These outstanding properties endow all-inorganic perovskites with a distinct advantage in high-efficiency optoelectronic devices. By observing charge distribution, we calculated the excitonic interactions, which facilitate the prediction of the material's photovoltaic properties.

2. Calculation method

The electronic structure is calculated by DFT with the PWscf code in Quantum Espresso, employing the PBE functional for the exchange-correlation potential and optimized norm-conserving Vanderbilt pseudopotentials for the core electrons[6]. The GW method is applied for quasiparticle corrections, and exciton effects are considered via the Bethe-Salpeter equation[7]. The pseudopotential method with norm-conserving approximations is used to handle the valence electron wavefunctions of Cs, Sn, Cl, Brand I, with energy convergence tested at 65 Ry. Structural optimization was performed based on dual convergence criteria. The optical properties are computed by Yambo software.

3. Results and conclusion

3.1 Structural properties

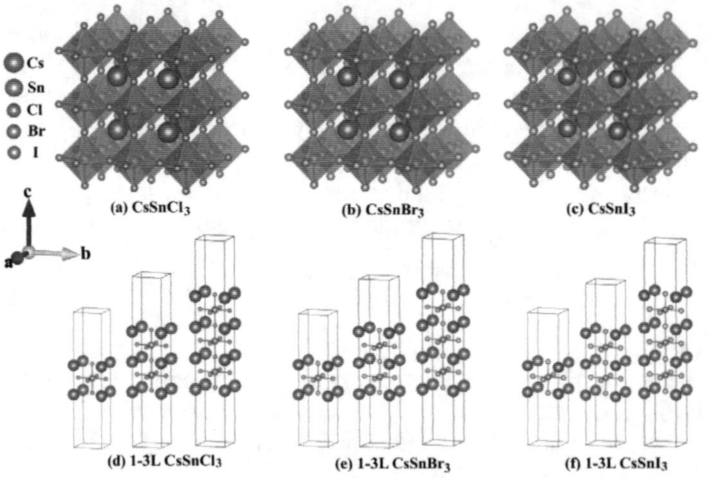

Figure 1. Crystal structures of (a) cubic $CsSnCl_3$, (b) cubic $CsSnBr_3$, (c) cubic $CsSnI_3$, (d) 1~3L $CsSnCl_3$, (e) 1~3L $CsSnBr_3$, (f) 1~3L $CsSnI_3$.

The 1-3 layer thin film models of the α-cubic phase (Pm-3m) $CsSnX_3$ (X = Cl, Br, I) system were constructed along the (1 0 0) crystal plane and relaxed to reveal surface excitonic behavior. A 20 Å vacuum layer induced anisotropic lattice expansion, with volume increasing as layer number (1 - 3) and halogen atomic number (Cl - I) rose, linked to halide ion radius changes (Cl^-: 1.81 Å - I^-: 2.20 Å). Structural evolution is shown in Figure 1 and Table 1.

Table 1 Calculated the lattice parameter, crystal volume V, E_{total}, E_f, E_b.

Materials	Lattice parameter			V(Å^3)	E_{total}(eV)	E_f (eV)	E_b(eV)
	a(Å)	b(Å)	c(Å)				
1L-$CsSnCl_3$	5.504	5.504	25.504	772.618	-5212.121	-18.110	-2.587
2L-$CsSnCl_3$	5.504	5.504	31.007	939.356	-9338.020	-31.276	-2.606
3L-$CsSnCl_3$	5.504	5.504	36.512	1106.094	-13463.872	-32.641	-2.620
1L-$CsSnBr_3$	5.910	5.910	25.190	879.838	-5138.519	-21.531	-3.075
2L-$CsSnBr_3$	5.910	5.910	31.820	1111.502	-9209.249	-37.306	-3.108
3L-$CsSnBr_3$	5.910	5.910	37.385	1255.463	-13279.884	-52.987	-3.116
1L-$CsSnI_3$	6.205	6.205	26.205	1009.197	-4832.738	-22.431	-3.204
2L-$CsSnI_3$	6.205	6.205	32.411	1248.186	-8674.089	-38.850	-3.237
3L-$CsSnI_3$	6.205	6.205	38.617	1487.168	-12503.638	-55.221	-3.248

Furthermore, the energy evolution characteristics (E_f)and the interface bonding strength parameter (E_b)., of 1~3layer $CsSnX_3$(X = Cl, Br, I) were calculated in **Table 1**. Both E_b and E_f were calculated using the following equations to investigate the bond strength and structural stability.

$$E_b = \frac{E_{total} - N(\mathrm{Cs})E(Cs) - N(Sn)E(Sn) - N(\mathrm{X})E(X)}{N_{total}} \quad (1)$$

$$\mathrm{E_f} = E_{total} - N(Cs)E(CsX) - N(SnX_2)E(SnX_2) \quad (2)$$

Here, E_{total} refers to the total energy, E(i) represents the DFT energy of i, and N(i) indicates the count of atoms in i. Lower formation energy (E_f) corresponds to greater stability. When the number of layers is constant, halogen substitution induces a gradual decay in the binding energy gradient, exhibiting a stability sequence of Cl > Br > I. When the halogen composition remains constant, the system's formation energy increases linearly with the number of layers, indicating how interlayer coupling contributes to improved thermodynamic stability.

3.2 Electronic properties

As shown in **Table 2**, the E_g of $CsSnX_3$ with varying layers were obtained using GW approximations. The E_g calculated through GW calculations is consistent with previous studies[8].

Table 2 Bandgap (E_g) (eV) values calculated using the GW approach.

Materials	Bandgap(eV)		
	1L	2L	3L
$CsSnCl_3$	2.786	2.247	1.994
$CsSnBr_3$	2.292	1.826	1.453
$CsSnI_3$	1.573	1.262	1.056

The band gap reduces because of the shift in the conduction band[8]. The energy bands become more densely packed, and the bandgap decreases (in the order from 1 to 3 layers). The VBM is predominantly formed by the X-p and Sn-s states, whereas the CBM is derived from the p states of Sn. The electron and hole distributions on the surface of 2L-$CsSnX_3$ (1 0 0) are shown in **Figure 3 (a-f)**. Electrons primarily surround Sn atoms, with fewer around X atoms, reflecting Sn's lower electronegativity. No electrons are found around X atoms in the film's center, likely due to longer Sn-X bond lengths. As halogen atomic radius increases, electronegativity decreases, reducing larger halogens' ability to attract electrons. Holes are mainly around X atoms, indicating an X-p state in the VBM[10].

The interfacial charge redistribution between tin and halogen atoms within the perovskite inorganic framework was visualized through differential charge density analysis (shown in **Figure 3 (g-i)**), with the computational methodology expressed as:

$$\Delta\rho = \rho_{\mathrm{Sn_n}X_{3n+1}} - \rho_{\mathrm{Sn}_n} - \rho_{X_{3n+1}} \quad (3)$$

where n represents the number of layers and $\rho_{\mathrm{Sn_nX_{3n+1}}}$ denotes the density of the inorganic Sn_nX_{3n+1}. Similarly, $\rho_{X_{3n+1}}$ and ρ_{Sn} represent the densities of the 3n+1 X atoms and the n Sn atoms, respectively, after the removal of Sn. Differential charge density analysis shows that Sn-X bonding is predominantly covalent with ionic character. The yellow region (electron accumulation) between Sn and X indicates electron sharing, while the blue region around Sn suggests electron depletion, highlighting the polar nature of the bond. For more electronegative X atoms, the ionic character becomes more prominent.

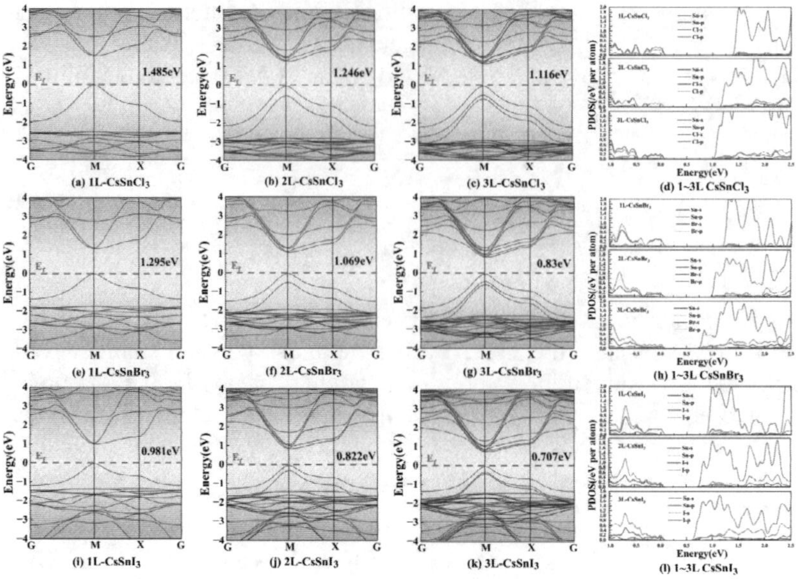

Figure 2. Band structures and PDOS calculated using DFT for 1~3LCsSnX$_3$.

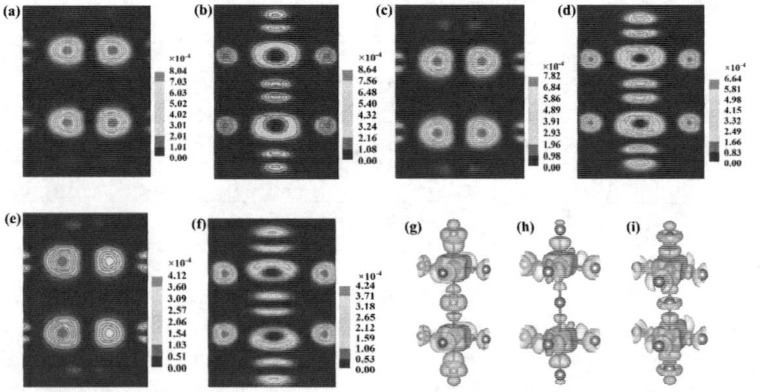

Figure 3. (a-b) Electron and hole distributions of (a-f) 2L-CsSnX$_3$, (g-h) Differential charge density maps of 2L-CsSnX$_3$.

3.3 Optoelectronic properties

The optoelectronic properties of CsSnX$_3$ perovskites with varying layer thicknesses were calculated using the GW + BSE method, as shown in **Figure 4**. In perovskites, the exciton energy, determined by the transition from the CBM to the VBM, typically dictates the main features of the absorption spectrum. Upon excitation, an absorption peak forms as the electron moves to the conduction band. As the film thickness decreases, excitonic absorption becomes more pronounced. The calculated relative dielectric constants (ε_r) for 1-3 layer CsSnX$_3$ are: CsSnCl$_3$: 1.91, 2.35, 2.63; CsSnBr$_3$: 1.90, 1.99, 2.81; CsSnI$_3$: 2.27, 2.69, 3.10. The maximum exciton binding energies for 1-3L CsSnCl$_3$ are 0.82, 0.15, 0.01 eV; for CsSnBr$_3$: 0.66, 0.55, 0.42 eV; for CsSnI$_3$: 0.61, 0.43, 0.34 eV.

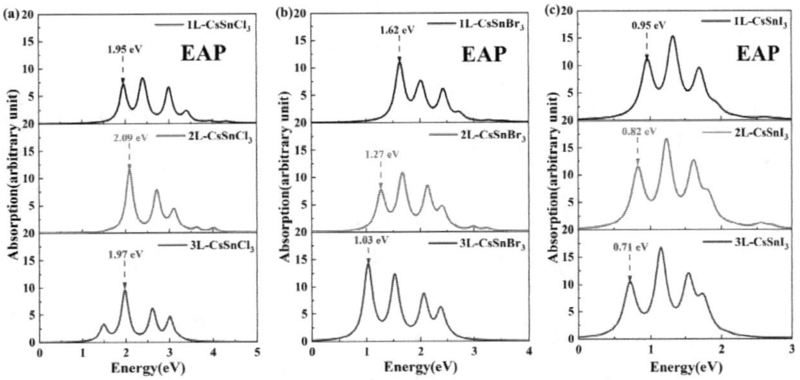

Figure 4. Calculated the optical absorption of $CsSnX_3$, arrows indicate the exciton absorption peak(EAP).

The Bohr radius is calculated using the following formula:

$$a = \frac{e^2}{8\pi\varepsilon_0\varepsilon_r E} \tag{4}$$

Here, e denotes the elementary charge, and ε_0 corresponds to the vacuum permittivity. This set of parameters forms the core equations describing the exciton binding state, where ε is the material's relative permittivity and r is the electron-hole separation distance, revealing the mechanism by which the dielectric screening effect regulates exciton stability. The Bohr radius of 1-3L $CsSnX_3$ increases with layer number: $CsSnCl_3$ (4.54,19.56, 143.57 Å), $CsSnBr_3$ (5.67, 6.52, 6.04 Å), and $CsSnI_3$ (5.12, 6.09, 6.78 Å). A larger Bohr radius enables broader light absorption, and the reduced exciton binding energy with more layers enhances carrier mobility.

4. Conclusions

In Summary, the GW approximation gives bandgap values in agreement with experimental data. As film thickness increases, the bandgap decreases, and electron/hole effective masses reduce. A larger dielectric constant leads to smaller exciton binding energy. Optical properties show a redshift in absorption, and thicker films enhance photoelectric efficiency. The results emphasize the role of exciton interactions in enhancing stability and spectral modulation, demonstrating strong potential for optoelectronic applications.

Acknowledgments

The authors gratefully acknowledge financial support from The National Natural Science Foundation of China (62173213, 61804089), The Natural Science Foundation of Shandong Province (ZR2019MF069, ZR2020QA055), The Shandong Provincial Science and Technology Support Program of Youth Innovation Team in College (2019KJN041 and 2020KJN005), The Yantai Key R&D Program (2020LJRC119, 2022JCYJ033, 2022JCYJ034, 2022JCYJ044), The Science Fund of Shandong Laboratory of Advanced Materials and Green Manufacturing at Yantai (AMGM2023A16).

References

[1] Ullah S, Wang J, Yang P,et al.Lead-ree Cs_2SnI_6 Perovskites for Optoelectronic Applications: Recent Developments and Perspectives[J].Solar RRL, 2021. DOI:10.1002/solr.202000830.

[2] Bimli S, Manjunath V, Mulani S R, et al. Theoretical investigations of all inorganic Cs_2SnI_6 double perovskite solar cells for efficiency~ 30%[J]. Solar Energy, 2023, 256: 76-87.DOI:10.1016/j.solener.2023.03.059.

[3] Mallick Z, Saini D, Sarkar R, et al. Piezo-phototronic effect in highly stable lead-free double perovskite Cs_2SnI_6-PVDF nanocomposite: Possibility for strain modulated optical sensor[J]. Nano Energy, 2022, 100: 107451.DOI:10.1016/j.nanoen.2022.107451.
[4] Tang Y, Lesage A, Schall P. $CsPbI_3$ nanocrystal films: towards higher stability and efficiency[J]. Journal of Materials Chemistry C, 2020, 8(48): 17139-17156.DOI:10.1039/D0TC04475J.
[5] Liu B, Long M, Cai M Q, et al. Influence of the number of layers on ultrathin $CsSnI_3$ perovskite: from electronic structure to carrier mobility[J]. Journal of Physics D: Applied Physics, 2018, 51(10): 105101.DOI:10.1088/1361-6463/aaa7ca.
[6] Kresse G, Hafner J, Needs R J. Optimized norm-conserving pseudopotentials[J]. Journal of Physics Condensed Matter, 1992, 4(36):7451. DOI:10.1088/0953-8984/4/36/018.
[7] Marini A. The Yambo Code: A Comprehensive Tool to Perform Ab Initio Simulations of Equilibrium and Out-of-equilibrium Properties [J]. APS Meeting Abstracts, 2016. https://api.semanticscholar.org/CorpusID:124928423.
[8] Xu L, Wang P, Kang Y, et al. Structural, Electronic, and Optical Properties of Cs_2SnX_4 (X=Cl, Br, and I) Multilayers: A Density Functional Theory Study[J]. Physica status solidi, B. Basic solid state physics, 2023, 260(1):n/a-n/a. DOI:10.1002/pssb.202200220.
[9] Xiang G, Wu Y, Li Y, et al. Structural and Optoelectronic Properties of Two-Dimensional Ruddlesden-Popper Hybrid Perovskite $CsSnBr_3$[J]. Nanomaterials, 2021(8). DOI:10.3390/nano11082119.
[10] Xiao Z, Meng W, Wang J,et al. Searching for promising new perovskite-based photovoltaic absorbers: the importance of electronic dimensionality[J].Materials Horizons, 2017, 4. DOI:10.1039/C6MH00519E.

Effect of additive amount of *n*-WS_2 particles on tribological properties of Ni-Co-Al_2O_3-WS_2 composite coating prepared by physical-field-assisted electroplating

Songru Yang [1], Quan Liu [1*], Biao Lv [2], Shaohu Tao [1], Cean Guo [1], Jian Zhang [1], and Yong Chen [3]

[1] Shenyang Ligong University, Shenyang, Liaoning, 110159, China
[2] Taiyuan Satellite Launch Center, Xinzhou, Shanxi, 036304, China
[3] Chengdu Lingchuan Characteristic Industrial, Chengdu, Sichuan, 610100, China

[*] Corresponding author's e-mail: qliu@sylu.edu.cn

Abstract: In this thesis, a self-developed flexible friction-assisted device was used for the pulsed electrodeposition of Ni-Co-Al_2O_3-WS_2 nanocomposite coatings on the surface of gun steel (PCrNi3MoVA). Scanning electron microscopy (SEM), electron spectroscopy (EDS), X-ray diffraction (XRD), nanoindentation, and friction and wear tests were used to investigate the effects of different additions of WS_2 nanoparticles on the microscopic morphology and chemical composition of the composite layer, the physical phase composition, the nanohardness, and the friction coefficient and wear rate. The results showed that with the increase of the *n*-WS_2 concentration in the plating solution from 1g/L to 10-15g/L, the surface of the composite plating layer became denser and denser, and the surface defects of the plating layer increased due to the agglomeration of *n*-WS_2 when it was 20g/L. Therefore, the quality of the plating layer was improved. The quality of the plating layer was optimal when the addition amount of *n*-WS_2 was 10-15g/L. At 10g/L, the nano-hardness value (8.34GPa), Young's modulus (241.3GPa), and elastic-plasticity index (0.0315) were higher, the average coefficient of friction (0.46) and wear rate ($0.91\times10^{-4}mm^3\cdot(N\cdot m)^{-1}$) was the lowest, and the plating layer had the best wear resistance.

1. Introduction

Electrodeposition is one of the common methods to prepare nickel-based composite coatings [1]. The main problem faced in the preparation of advanced nickel-based composite coatings is how to make the composite particles uniformly distributed and the amount of composite more, by combining with the pulse to improve the electrodeposition process is a common method. In this paper, a self-developed flexible friction-assisted device is used in combination with the electrodeposition process to maintain the concentration of the plating solution particles through the reciprocating motion of the brush plate and the plated parts and to improve the effect of the co-deposition of the insoluble second-phase particles with different properties. Nano-Al_2O_3 and WS_2 particles (*n*-Al_2O_3, *n*-WS_2) are introduced into Ni-Co alloys to take advantage of the effect of the diffusely distributed *n*-Al_2O_3 particles on the dislocation strengthening and strain strengthening effects [2], to increase the movement resistance of dislocations in the slip surface [3], and to enhance the strengthening effect of grain boundaries, thus improving the hardness of the plating layer [4]. It is to utilize the self-lubricating properties of *n*-WS_2 particles [5-6], which can be easily dislodged and encapsulated on the surface of friction sub-surface during friction wear to produce a large number of slip surfaces, thus reducing the adhesive wear [7], and increasing the wear-resistant properties of the plating layer [8]. The friction medium speed is 13m/min, the average current is 0.8A, the plating solution pH is 4, the duty cycle is 0.1, the frequency is 400Hz, the temperature is 50°C, the time is 60min, the *n*-Al_2O_3 concentration is 5g/L and other parameters of the electroplating process were controlled constant to explore the effect of different plating solution *n*-WS_2 particle concentration on the coating.

2. Test

2.1. Plating preparation

The test cathode substrate is gun steel (PCrNi3MoVA), processed by wire cutting method with the size of 20mm×50mm×3mm. Its chemical composition is shown in Table 1; the double anodes are titanium mesh coated with iridium-tantalum plating (IrO_2+Ta_2O_5), with a size of 90mm×100mm×0.5mm.

The experiments were carried out using a self-developed flexible friction-assisted pulse electrodeposition device, which consists of a speed control component, a driving mechanism, a brush plate, an anode, a cathode, its fixing bracket, a plating tank, and other components. After contacting the cathode, the plated parts were placed on the fixing frame and clamped in the plating tank. The brush plate was controlled by the drive mechanism to do vertical reciprocating motion in the plating tank. The specimen is sanded with 240#, 600#, 1000#, and 1500# sandpaper in order to remove the oxides formed when the specimen is wire-cut, and rinsed with water after sanding is completed. The surface of the specimen is blown dry with a high-pressure air gun; the sanded specimen is put into ethanol and acetone for ultrasonic cleaning for 10min in order, and then placed in a drying dish to be used; the cleaned specimen is put into an alkaline degreasing solution at 80°C for 10min. We immerse the specimen in an 80°C alkaline degreasing solution for 10 minutes. After degreasing, we rinsed the specimen with hot deionized water followed by cold deionized water to remove any residual alkaline solution from its surface. Then we degreased the completed specimen into the acid activation solution for 30s, after the completion of the activation into the cold deionized water to remove the residual acid wash on the surface of the specimen.

Al_2O_3 particles (average particle size 30-50nm) and WS_2 particles (average particle size 100nm) with purity higher than 99.9% were selected from Shanghai McLean Biochemical Technology Co. The composition of the plating solution was as follows: nickel sulfate hexahydrate ($NiSO_4$-$6H_2O$) 180-200g/L, cobalt sulfate heptahydrate ($CoSO_4$-$7H_2O$) 30-35g/L, nickel chloride hexahydrate ($NiCl_2$-$6H_2O$) 40-45g/L, boric acid (H_3BO_3) 30-35g/L, saccharin ($C_7H_5NO_3S$) 2-2.5g/L, sodium dodecyl sulfate ($C_{12}H_5NO_3S$) 2.5g/L, sodium dodecyl sulfate ($C_{12}H_5NO_3S$) 1.0mg/L, sodium sulfate ($C_{12}H_{25}NaSO_4$) 0.08-0.11g/L, cetyltrimethylammonium bromide (CTAB) 0.08-0.11g/L, *n*-Al_2O_3 particles 5g/L, and *n*-WS_2 particles 5g/L. The formulation of the plating solution for the preparation of Ni-Co alloy plating is shown in Table 2, and that of the composite plating for the preparation of Ni-Co-Al_2O_3-WS_2 is shown in Table 3. It is necessary to add *n*-WS_2 particles and *n*-Al_2O_3 particles on the basis of the plating solution for Ni-Co alloy plating, and use a dispersant to make it better dispersed.

Table 1. Chemical composition of gun steel PCrNi3MoVA (wt.%).

Elemental	C	Mn	Ni	Si	Cr	Mo	V	P	S	Fe
Quantity contained	0.40	0.41	3.14	0.25	1.28	0.37	0.20	0.012	0.001	Headroom

2.2. Testing and characterization

X-ray diffractometer was used to analyze the physical composition of the coating; the surface morphology and cross-section morphology of the chemical coating were observed by Hitachi S-4800 scanning electron microscope. The micro-area composition was analyzed by Oxford X-max50 spectrometer; the microhardness test was carried out on HV-1000 digital Vickers hardness tester with a load of 100g and a holding time of 15s; the average value of the 5points was taken as the test result. The abrasion test was carried out on UT-3000 friction and wear tester. The friction and wear load of the coating was 500N and the rotary speed of the turntable was 500r/min.

3. Test results and analysis

3.1. Microscopic morphology

Figure 1 shows the surface morphology of Ni-Co-Al_2O_3-WS_2 composite coatings deposited under different concentrations of *n*-WS_2 particles in the plating solution. The addition of *n*-WS_2 particles in the plating solution up to 5g/L resulted in irregular "cauliflower-like" nodules on the layer. As the concentration increases, the nodules are agglomerated and the surface of the plated layer becomes denser. Since the *n*-Al_2O_3 particles were generally covered by the Ni-Co alloy grains and encapsulated inside the plating layer, no obvious *n*-Al_2O_3 particles were seen, while the effect of *n*-WS_2 particles on the surface morphology of the composite plating layer was more obvious [9]. At *n*-WS_2 particle concentration in the interval of 10-15g/L, the plating layer was homogeneous and dense. At 20g/L, the surface reappeared to be porous. The analyzed surface was basically saturated with *n*-WS_2 particle content. The deposition of elemental W on the plated surface at 10g/L reached the highest in the same group at 6.6Wt%.

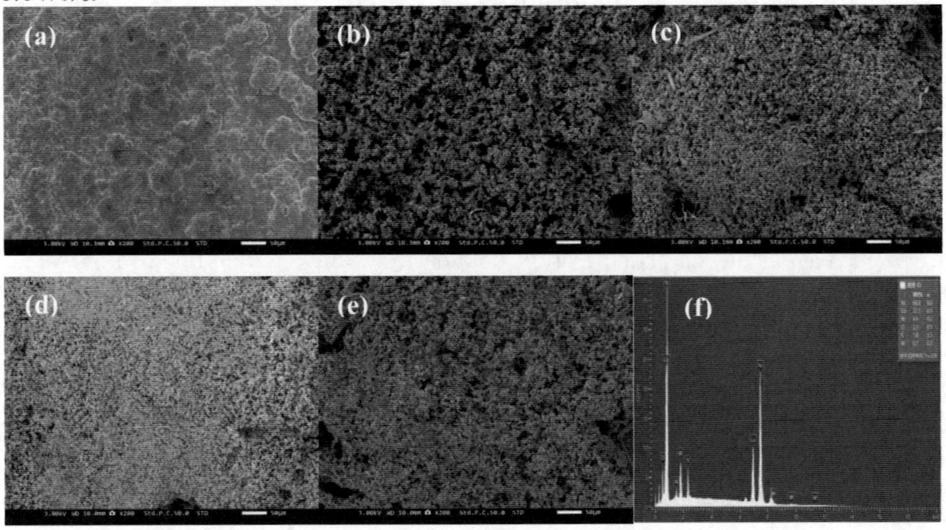

Figure. 1. SEM and EDS images of composite coatings at different *n*-WS_2 concentrations.
(a) 1g/L, (b) 5g/L, (c) 10g/L, (d) 15g/L, (e) 20g/L, and (f) SED diagram of 10g/L coating.

3.2. Phase structure

Figure 2 shows the X-ray diffraction analysis results of the composite plating layer. Comparative analysis shows that the surfaces of composite coatings with different concentrations of *n*-WS_2 particles in the plating solution are mainly composed of the matrix Ni-Co phase. The addition of *n*-WS_2 particles will not change the phase composition of the chemical plating layer, but the intensity of the diffraction peaks of the densely arranged hexagonal structure of the crystalline surfaces of the (002) crystal surface is gradually enhanced with the increase of the concentration of *n*-WS_2. It reaches a higher value at a concentration of 10g/L and decreases after the concentration of 15g/L increases. The intensity of the diffraction peaks of *n*-WS_2 gradually increased with an increase of the concentration of *n*-WS_2, reaching a high value at 10g/L, and then decreased after the concentration increased to 15g/L. When the concentration of *n*-WS_2 particles is increasing, the concentration of metal ions near the cathode surface that is effectively supplemented is higher, the nucleation rate of the grains increases, the number of nuclei increases, the grains are finer, and the strength and hardness of the plating layer are increased. Although the diffraction peak was reached when the concentration was increased to 20g/L, the excess *n*-WS_2 particles accumulated on the surface of the coating and did not reach the effect of co-deposition with other particles. The bonding force of the coating was not good, so it was judged that it was more suitable to control the concentration of *n*-WS_2 at 10g/L under the conditions of this process. The results were also verified in the subsequent tests.

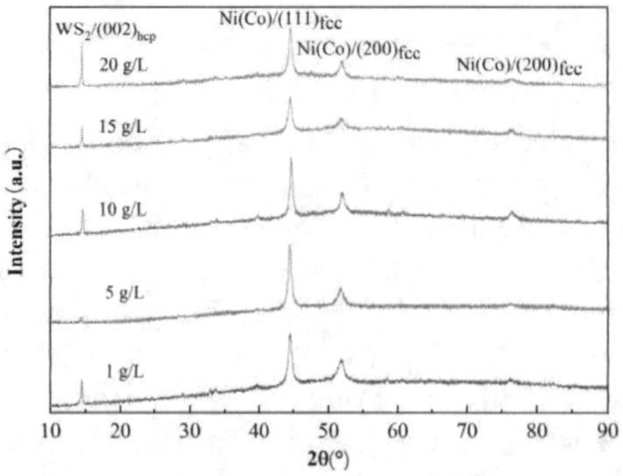

Figure 2. XRD patterns of composite coatings under different n-WS_2 concentration conditions.

3.3. Nanomechanical properties and elastic-plastic properties

Table 2 shows the nano-hardness, elastic modulus, and elastic-plastic properties of the composite plating layers under different n-WS_2 concentration conditions. It can be analyzed that the hardness of the composite plating layer was gradually enhanced with the increase of n-WS_2 concentration in the plating solution. In particular, the hardness of the composite plating layer was significantly increased to 9.45GPa when the n-WS_2 concentration reached 20g/L, which showed that the n-WS_2 concentration had a significant positive effect on the hardness of the composite plating layer. Cobalt atoms enter into the lattice of nickel coating to form a replacement solid solution through replacement, causing lattice distortion and increasing the number of grain boundaries, which is helpful to improve the surface quality of Ni-Co composite coatings and realize grain refinement. In addition, n-WS_2 particles are diffusely distributed in the formed composite coatings with their unique morphology [10], showing two key functional properties: first, in the electrodeposition process, n-WS_2 acts as the nucleation core of nickel (Ni) and cobalt (Co) matrix metals and participates in the growth mechanism of the alloy crystals. By providing a stable nucleation environment, it effectively promotes the increase in the number of grains, thus improving the uniformity and strength of the coating. Secondly, n-WS_2 can also play a role in the growth process of nickel (Ni) and cobalt (Co) base metals themselves, preventing excessive growth of grains, and thus achieving the effect of fine grain strengthening.

Table 2. Hardness, modulus of elasticity, elastic index, and resistance to plastic deformation of composite coatings at different concentrations of n-WS_2.

WS_2 concentration (g/L)	Hardness (GPa)	Modulus of elasticity (GPa)	Elasticity index (H/E*)	Resistance to plastic deformation (H^3/E^{*2})
0	4.81	191.3	0.0229	0.0025
1	6.07	248.8	0.0222	0.0030
5	6.52	241.8	0.0245	0.0039
10	8.34	241.3	0.0315	0.0083
15	8.69	196.3	0.0403	0.0141
20	9.45	189.3	0.0454	0.0195

3.4. Tribological properties

Figure 3 shows the friction coefficient curves of the composite plated layers under different n-WS_2 concentrations. Table 3 shows the volumetric wear rate of the same group of plated layers. It can be analyzed that, with the increasing concentration of n-WS_2 particles in the plating solution, the amount of n-WS_2 particles attached to the plating layer also increased gradually. The density of the plating layer increased gradually. At 20g/L, due to the high n-WS_2 particle adhesion, the wear-resistant effect of n-Al_2O_3 particles was not reflected, and the anti-friction performance was not ideal. While the concentration of n-WS_2 particles at 10g/L had stronger anti-friction performance, the average friction coefficient and wear rate were the lowest in the same group, which were 0.46 and $0.91\times10^{-4}mm^3\cdot(N\cdot m)^{-1}$, respectively.

Figure 3 shows the wear morphology of composite coatings with different n-WS_2 contents. When the concentration of n-WS_2 particles is 1g/L, the n-WS_2 particles attached to the surface of the coating are less, which fails to play the role of anti-wear. The resistance to wear is mainly borne by the Ni-Co coating. With the increasing concentration of n-WS_2 particles, when the concentration reaches 15g/L, the abrasion marks are smooth. The distribution of n-WS_2 particles and n-Al_2O_3 agglomerated particles is dense and uniform. The friction coefficient curve tends to be stabilized in the late stage, which proves that the lubrication and wear reduction of n-WS_2 particles and the wear-resistant property of n-Al_2O_3 particles improve the tribological performance of the coating. The anti-wear performance of the coating is better under this concentration. In friction wear, the n-WS_2 particles on the coating surface will soften under the load force and enter into the depression area, forming a discontinuous lubrication film with low shear strength. With the increase of friction time, the discontinuous lubrication film becomes continuous and complete, which in turn reduces the friction coefficient and wear rate of the material.

Table 3. The volume wear rate of composite coatings at different n-WS_2 concentrations.

WS_2 concentration (g/L)	Attrition rate ($mm^3 \cdot (N \cdot m)^{-1}$)
1	3.68×10^{-4}
5	2.50×10^{-4}
10	0.91×10^{-4}
15	1.37×10^{-4}
20	1.84×10^{-4}

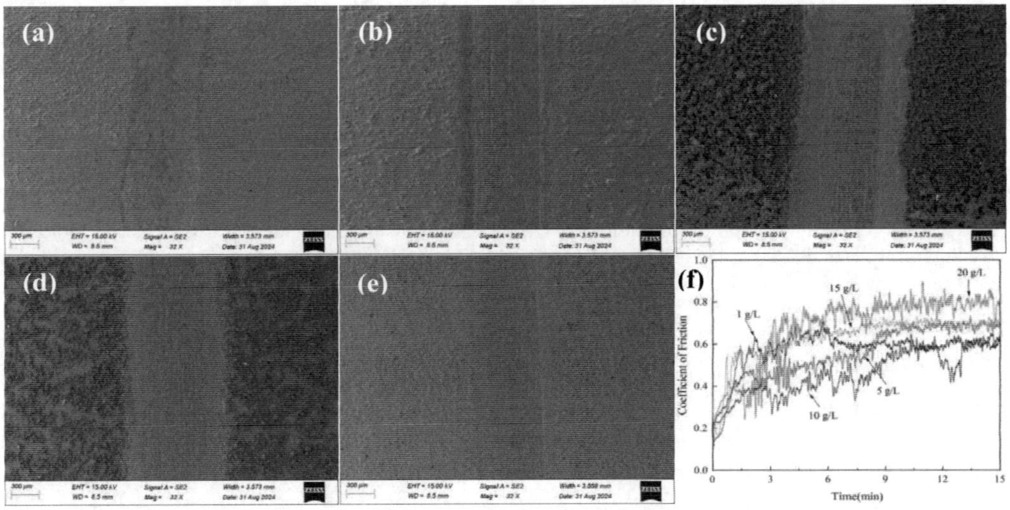

Figure 3. Wear morphology of composite coatings and Friction coefficient curves of composite coatings at different n-WS_2 concentrations.
(a) 1g/L, (b) 5g/L, (c) 10g/L, (d) 15g/L, (e) 20g/L, (f) Friction coefficient curve

4. Conclusion

In this paper, Ni-Co-Al_2O_3-WS_2 composite coatings were prepared on the surface of gun steel (PCrNi3MoVA) by electrodeposition technique supplemented with flexible friction action. The influence of a single factor, n-WS_2 concentration in the plating solution, on the composite coating preparation process was investigated. The following conclusions were formed:

(1) In the composite plating layer, the morphology of the plating layer was changed by continuously increasing the concentration of n-WS_2 particles in the plating solution. The surface of the plating layer showed nodular growth. When the concentrations of n-WS_2 and n-Al_2O_3 particles were

15g/L and 5g/L, respectively, the deposited plating layer was more likely to form a wear-reducing lubrication film, with flat and smooth abrasion marks, and the lowest average coefficient of friction and wear rate of 0.46 and $0.91\times10^{-4}mm^3\cdot(N\cdot m)^{-1}$, respectively, and the nano-hardness and modulus of elasticity of 8.34 and 241.3GPa, respectively. The elasticity index of higher The high elasticity index also showed excellent elasticity and plasticity properties. The best wear resistance of the plated layer was obtained under this process condition.

(2) Composite coating wear reduction mechanism is that *n*-WS_2 particles in the coating will fall off under the action of shear and fill in the friction surface of the depression and gap parts, forming a lubricating film with low shear strength, improving the tribological performance of the coating; *n*-Al_2O_3 particles support the friction surface to play the role of the "micro-roller ball" to reduce the actual friction between the friction partner. *n*-Al_2O_3 particles support the friction surface to play the role of "micro roller ball", reducing the actual contact area between the friction pair. At the same time, together with *n*-WS_2 particles to fill in the friction surface depressions and gaps in the formation of lubricating film, we play a synergistic role in reducing wear and improving the wear resistance of plating.

Acknowledgments

It is supported by the Basic Research Project of the Education Department of Liaoning Province of China (No. JYTMS20230191), the Applied Basic Research Program Projects of the Department of Science and Technology of Liaoning Province of China (No. 2023020243-JH2/1013), and Optical-Selection Growth Plan (SYLUGXRC14) and Optical-Selection Team Plan (SYLUGXTD5) from Shenyang Ligong University.

References

[1] Wang, W.X. Preparation of Ni-MoS_2-Y_2O_3 composite coating and its properties [D]. Shenyang University, 2024.
[2] Raghavendra C R, Basavarajappa S, Sogaland I and Naik K. Study on Wear Mechanism and Contact Temperature against Dry Sliding Wear of Ni-Al_2O_3 Nanocomposite Coating [J]. Transactions of the Indian Ceramic Society, 2020, 79(3): 1-5.
[3] Karthik R, Mani R, Manikandan P. Tribological studies of Ni-SiC and Ni-Al_2O_3 composite coatings by Pulsed Electrodeposition [J]. Materials Today: Proceedings, 2021, 37(2): 701-706.
[4] Wang T. Toughening of nickel-based deformed high-temperature alloys strengthened by cold rolling deformation-induced "twin + γ' phase" [D]. Lanzhou University of Science and Technology, 2023.
[5] Yu C T, Chen M H, Wang F H. Study on mechanical and tribological properties of FHG97/WS_2 self-lubricating composites [J]. Surface Technology, 2023, 52(03): 143-150+227.
[6] HE, Y., et al. An electrodeposited Ni-P-WS2 coating with combined super-hydrophobicity and self-lubricating properties. Electrochimica Acta, 2017, 245: 872-882.
[7] CAI H C, XUE Y J, LI H, et al. Friction and wear behavior of non-equilibrium magnetron sputtered La-Ti/WS_2 self-lubricating thin films (in English) [J]. Rare Metal Materials and Engineering, 2021, 50(08): 2708-2714.
[8] Fan Y L, Lu Y F. Analysis of wear debris during friction [J]. Lubrication Engineering, 2010, 35(12): 17-21.
[9] Baskaran I, Narayanan T, Stephen A. Effect of accelerators and stabilizers on the formation and characteristics of electroless Ni-P deposits [J]. Materials Chemistry and Physics, 2006, 99(1): 117-126.
[10] LIU X B, LIU H Q, MENG X J, et al. High-temperature tribological behavior of laser-melted NiCr/Cr_3C_2-WS_2 self-lubricating wear-resistant coatings [J]. Materials Engineering, 2013, (11): 26-31+37.

Wear behaviour of magnetron sputtered Cr and Cr(Ta) coatings on gun steel

Jifei Yu [1], Minghuang Bi [1], Cean Guo [1,*] and Jian Zhang [1]

[1] School of Equipment Engineering, Shenyang Ligong University, Shenyang Liaoning 110159, China

[*] Corresponding author's e-mail: 373055507@qq.com

Abstract: Cr and Cr(Ta) coatings were deposited on the surface of gun steel by using magnetron sputtering technology. The two coatings were analyzed for their morphological features, phase composition, hardness, and elasticity modulus using scanning electron microscopy (SEM), X-ray diffraction (XRD), and nanoindentation techniques. The wear resistance differences between the two coatings were evaluated using a friction and wear testing machine. The results showed that the Cr coating has a typical columnar crystal structure, while the Cr(Ta) coating has an equiaxed crystal structure. Both the Cr and Cr(Ta) coatings show preferred orientation on the Cr(110) crystal plane. After analyzing the coefficient of friction (COF) and the morphology of the wear tracks, it was found that the Cr(Ta) coating exhibits superior wear resistance compared to the Cr coating. This demonstrates that the incorporation of the Ta element enhances the wear resistance of the Cr coating.

1. Introduction

Extending the life of a gun barrel is crucial for improving the comprehensive performance of artillery. This demand is particularly urgent, especially in the future battlefield. At present, the life-extension technology of gun barrels by using chromium (Cr) coating still has many persistent problems [1][3]. Therefore, it is critical to develop new coating preparation technologies. Magnetron sputtering technology is one of the three major technologies of physical vapor deposition that is rapidly developing and widely used to deposit hard coatings with high hardness and excellent wear resistance. Unlike electroplating, magnetron sputtering deposits coatings with lower substrate temperature, less damage to the coating, better coating density, a wide range of platable targets, and easy control of the coating composition [4][5].

Introducing a second component to Cr coatings has been proven to be an effective way to improve their properties. The refractory metal tantalum (Ta), which has excellent with excellent ductility and wear resistance, is an ideal additive element to combine the performance advantages of Cr and Ta, Therefore preparing Cr(Ta) binary coatings will further enhance the coating wear-resistant properties to meet the increasingly stringent service requirements of gun barrels [6][8].

In this work, the Cr and Cr(Ta) coatings were prepared on the surface of gun steel by adjusting the target composition by using magnetron sputtering technology. The wear resistance of the two coatings was investigated by analyzing the microscopic morphology and phase composition of the two coatings, as well as hardness and elastic modulus tests and friction wear experiments.

2. Specimen preparation and experimental methods

2.1. Sample preparation

The substrate material is PCrNi3MoVA steel used for gun barrels with the nominal chemical composition shown in Table 1. Firstly, the PCrNi3MoVA steel was cut into a size of 20 mm×10 mm×5 mm prior to the preparation of the coating. Then, the samples were mechanically polished using a diamond polishing paste after grinding to 1500 # with aqueous sandpaper. Finally, the polished samples were ultrasonically cleaned in a 3:1 alcohol-acetone mixture for 15 min and then dried by a hair dryer.

Table 1. Nominal chemical compositions of PCrNi3MoVA (Wt.%).

Fe	Cr	Ni	Mo	V	C	Mn	Si	S	P
Balanced	1.28	3.14	0.37	0.2	0.4	0.41	0.25	0.001	0.012

In view of the fact that dual-target magnetron sputtering Cr-Ta coatings can obtain variable Cr and Ta elemental ratios by controlling the sputtering power, the elemental content is not fixed [9]. In this paper, we used a pure metal Cr target (99.95 at.%) and a Cr(Ta) target (Cr 96 at.%, Ta 4 at.%), and used the PVD V201-8 hard-coating equipment for the preparation of the two coatings. During the deposition process, the target current was set at 8 A, the bias voltage at 150 V, the chamber pressure at 4.5×10^{-1} Pa, the argon flow rate at 220 sccm, and the sputtering time at 720 minutes.

2.2. *Characterisation and experimental methods*

The surface and cross-sectional microstructures, as well as the elemental composition of the two coatings, were characterized using a Gemini 360 scanning electron microscope equipped with an X-Max energy spectrometer. The phase compositions of the coatings were analyzed using a D8 DISCOVER X-ray diffractometer with a Cu-Kα radiation source (λ=0.1542 nm). The measurements were performed over a diffraction angle range of 20°-90° at a scanning speed of 20 °/min.

The hardness (*H*) and elastic modulus (*E*) of the two coatings were measured by using a G200 nanoindentation tester with a test load of 20 mN and an indentation depth of 1000 nm. The coefficient of friction and wear resistance of the coatings were evaluated using an HSR-2M high-speed reciprocating friction and wear tester. A 304 stainless steel ball with a diameter of 6 mm served as the friction counterpart. The test parameters included a reciprocating frequency of 300 cycles per minute, an applied load of 4 N, a friction stroke length of 5 mm, and a total wear duration of 10 minutes. Both tests were conducted at room temperature.

3. Experimental results and analysis

3.1. *Coating morphology*

Figure 1 illustrates the surface and cross-sectional morphologies of the Cr and Cr(Ta) coatings. As shown in Figure 1(a)(a1) and (b)(b1), the Cr and Cr(Ta) coatings have a dense surface with good quality and no obvious defects. The surfaces of the two coatings show a 'cauliflower' shape, which is a structural characteristic of several tiny particles forming one large particle.

Figure 1(a2) and (b2) show the cross-sectional morphology of the two coatings. It is evident that both coatings are fully developed and exhibit strong adhesion to the substrate. They are slightly curved internally with a few gaps and uneven bonds. Additionally, the Cr(Ta) coating exhibits an equiaxed crystal structure, whereas the Cr coating displays a characteristic columnar crystal structure.

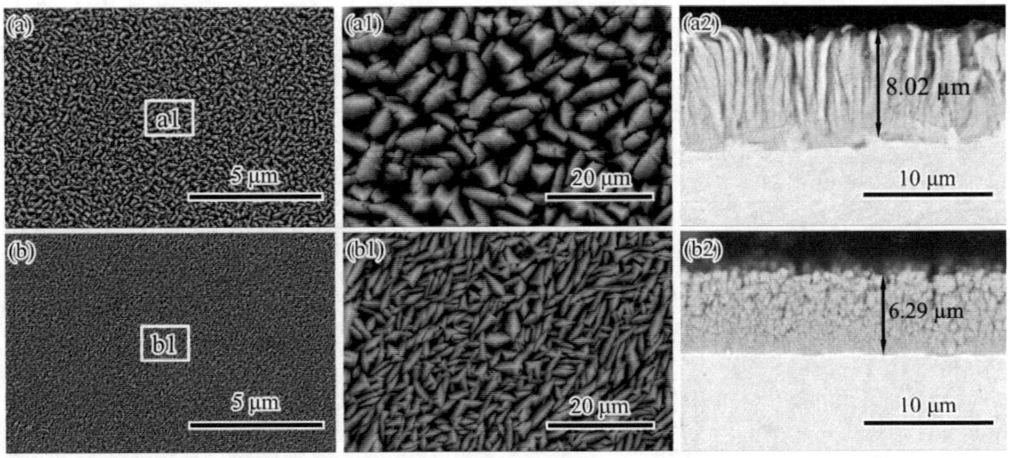

Figure 1. Surface and cross-sectional morphology: Cr coating (a), (a1), (a2); Cr(Ta) coating (b), (b1), (b2). Here, (a1) and (b1) are magnified images corresponding to the regions marked in (a) and (b), respectively.

3.2. XRD physical analysis of coatings

Figure 2 shows the XRD patterns of the Cr and Cr(Ta) coatings. It can be seen that Cr(110), Cr(200), and Cr(211) are detected for the two coatings. Both Cr and Cr(Ta) coatings are α-Cr phases and exhibit preferred orientation on the Cr(110) crystal plane. It is also evident that the diffraction peak intensity at Cr(110) is significantly higher for the Cr(Ta) coating compared to the Cr coating. This is attributed to the increased number of diffracting crystal planes resulting from the incorporation of the Ta element, which enhances the preferred orientation of the Cr(Ta) coating along the Cr(110) crystal plane. Furthermore, the addition of Ta induces lattice distortion, causing a shift in all three diffraction peaks of the Cr(Ta) coating.

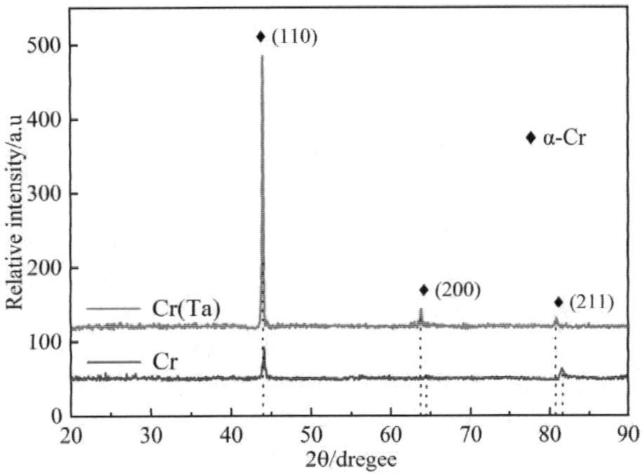

Figure 2. XRD patterns of the Cr and Cr(Ta) coatings.

3.3. Analysis of nanomechanical properties of coatings

Table 2 presents the nanomechanical properties of the two coatings, including both the H/E and H^3/E^2 ratios, which are calculated from the hardness (H) and elastic modulus (E) values. As shown in Table 2, the hardness and elastic modulus of the Cr coating are higher than that of the Cr(Ta) coating, resulting from the addition of the Ta element. H/E is a parameter that can reflect the abrasion resistance of the material. The higher the value of H/E is, the better the wear resistance is. In addition, H^3/E^2 is positively correlated with yield pressure. The higher its value is, the better the coating's resistance to plastic deformation is. It can also be seen from Table 2 that H/E and H^3/E^2 of Cr(Ta) coatings are higher than those of Cr coatings. Consequently, the Cr(Ta) coatings exhibit superior wear resistance and greater resistance to plastic deformation compared to the Cr coatings.

Table 2. Nanoindentation test results of the Cr and Cr(Ta) coatings.

Coating	*E*/GPa	*H*/GPa	*H*/*E*	H^3/E^2
Cr	262.4	9.03	0.034	0.0107
Cr(Ta)	253.5	10.81	0.043	0.0197

3.4. Wear resistance analysis

The coefficient of friction (COF) of the two coatings under a 4 N load is shown in Figure 3. In the course of friction wear, the COF exhibits a fluctuating process due to the sample surface state. When the sample and counter body fit to reach a better level, the COF will be maintained in a relatively stable state. Therefore, the value of the COF at the stabilization stage is often selected for comparison. Generally, the smaller the COF is, the better the wear resistance is. The COF of the Cr coating is about 0.68, while that of the Cr(Ta) coating is about 0.58, which is much less than the COF of the Cr coating, showing better wear resistance.

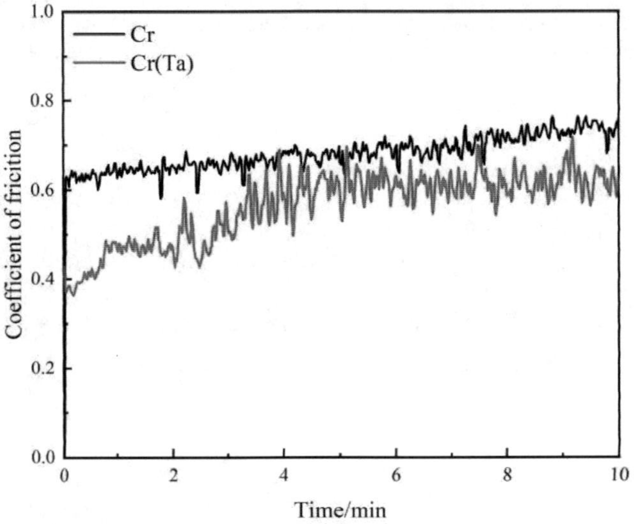

Figure 3. COF of the Cr and Cr(Ta) coatings under a 4 N load.

Figure 4 illustrates the worn morphologies of the Cr and Cr(Ta) coatings, while Table 3 presents the EDS results of the worn surfaces corresponding to Figure 4. As shown in Figures 4(a) and 4(c), the width of the wear track on the Cr(Ta) coating is significantly narrower than that on the Cr coating, indicating that the Cr(Ta) coating demonstrates superior wear resistance compared to the Cr coating.

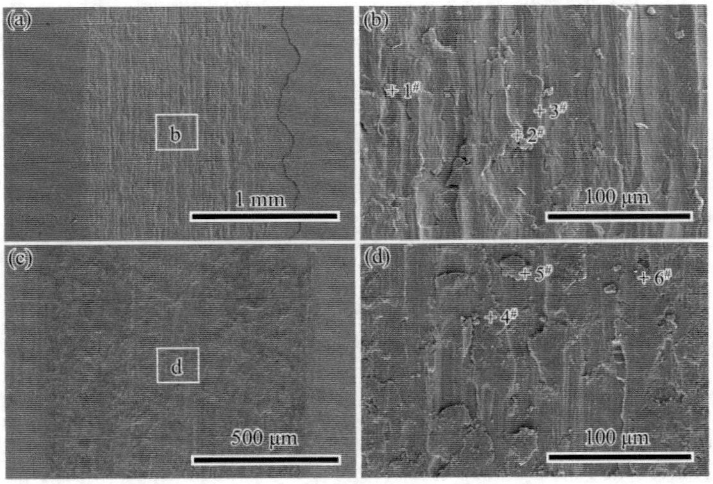

Figure 4. Worn morphologies of the Cr and Cr(Ta) coatings, where (b) and (d) are magnified images corresponding to the regions marked in (a) and (c), respectively.

As illustrated in Figures 4(a) and 4(b), the surface of the Cr coating exhibits severe wear, characterized by flaking and the presence of abrasive debris. The wear tracks appear as furrow-like grooves with small pits. Based on the EDS results in Table 3, which show the concentrations of Fe, Cr, and O, it is concluded that the wear mechanism of the Cr coating involves abrasive wear combined with oxidative wear.

As illustrated in Figures 4(c) and 4(d), the surface of the Cr(Ta) coating is relatively smooth. The wear tracks exhibit distinct furrow-like patterns with numerous abrasive chips and white oxide particles adhering to the surface. Based on the EDS results provided in Table 3, the wear mechanism of the Cr(Ta) coating is also determined to be abrasive wear accompanied by oxidative wear.

Table 3. EDS results of worn surfaces of the Cr and Cr(Ta) coatings in Figure 4 (At.%).

Content	Location					
	1#	2#	3#	4#	5#	6#
Fe	91.2	85.27	16.51	97.78	31.00	33.06
Cr	3.8	3.06	48.09	2.22	49.53	21.33
O	-	11.66	35.4	-	9.59	44.74
Ta	-	-	-	-	2.09	0.87

4. Conclusion

Based on the results and analysis presented above, the conclusions are obtained as below:

(1) The Cr coating shows a typical columnar crystal structure, while the Cr(Ta) coating shows an equiaxial crystal structure.

(2) Both the Cr and Cr(Ta) coatings exhibit preferred orientation on the Cr(110) crystal plane.

(3) The COF (about 0.58) of the Cr(Ta) coating is smaller than that (about 0.68) of the Cr coating, as well as the width of the abrasion mark, showing better wear resistance.

(4) The addition of Ta element could enhance the wear resistance of the magnetron sputtering Cr coatings. Cr(Ta) coating can be used in place of Cr coatings to meet the increasingly stringent requirements of gun barrels.

Acknowledgments

This work was financially supported by the Light-Selection Team Plan of Shenyang Ligong University (SYLUGXTD5).

Journal of Physics: Conference Series **3009** (2025) 012028 doi:10.1088/1742-6596/3009/1/012028

References

[1] Huang J F, Zhang J and Chen J Y (2023) Failure Mechanisms of Gun Barrels and the Development of Gun Steel. J. Journal of Gun Launch & Control. 44(01): 10-18+29.
[2] Mao B Q, Zhao Q J and Bai X H (2023) Review and Prospect of Life Extension Technology for Gun Barrels. J. ACTA ARMAMENTARII, 44(03): 638-655.
[3] Mishra S and Bharagava R N (2016) Toxic and genotoxic effects of hexavalent chromium in environment and its bioremediation strategies. J. Journal of Environmental Science and Health, Part C, 34(1): 1-32.
[4] Gui B H, Zhou H and Liu X G (2021) Influence of N_2 flow rate on microstructure and properties of CrNx ceramic films prepared by MPP technique at low temperature. J. Ceramics International, 47(15): 20875-20884.
[5] Wang D, Lin S S and Shi Q (2021) Microstructure effects on fracture failure mechanism of CrAl/CrAlN coating. J. Ceramics International, 47(3): 3657-3664.
[6] Li C, Sun G and Ma G J (2024). Effects of Tantalum Doping on Friction and Corrosion Behavior of Multilayer Ta-DLC Films. J. Surface Technology, 53(08): 63A-73.
[7] Kainz C, Pohler M and Gruber G C (2021) Influence of bias voltage on microstructure, mechanical properties and thermal stability of arc evaporated Cr0.74Ta0.26N coatings. J. Surface and Coatings Technology, 417, 127212.
[8] Saha R, Inturi R B and Barnard J A (1996) Structural and mechanical characterization of Cr-Ta-N hard coatings prepared by reactive magnetron sputtering. J. Surface and Coatings Technology, 82(1-2): 42-47.
[9] Li C Y, Guo C A and Liu Q (2022) Study on Properties of Tantalum Chromium Amorphous Alloy Coatings preperd by Magnetron Sputtering Method. J. Journal of Shenyang Ligong University, 41(02), 31.

Journal of Physics: Conference Series **3009** (2025) 012029 doi:10.1088/1742-6596/3009/1/012029

Research and analysis of the dynamic corrosion behaviour of superhydrophobic magnesium alloys based on electroplated/de-alloyed composite coatings

Dingbang Chen and Yongjun Zhang*

School of Mechanical & Automotive Engineering, South China University of Technology, Guangzhou, 510640, China

* Corresponding author: koujie@ldy.edu.rs

Abstract: Based on the development of China, the transformation and upgrading of the automotive industry and 3C industry, and the significant improvement of China's economic status, the magnesium alloy industry makes the market bullish. Among them, the automotive industry's lightweight, environmental protection needs, especially the development of new energy vehicles, as well as the continuous progress of magnesium alloy R & D technology and recycling technology will be good news for promoting the widespread application of magnesium alloys. In addition, as a sub-industry of the non-ferrous metal alloy industry, the magnesium alloy industry is upgrading China's manufacturing industry to benefit from this. As the capital and material-intensive industry, the stability of raw material prices and lower levels, casting parts industry integration and concentration, technological research and development progress will be more conducive to the development of the magnesium alloy industry, so the market is more favourable. Research magnesium alloy superhydrophobic surface modification technology to improve the corrosion resistance of magnesium and its alloys still has important practical significance. Taking the AZ91 magnesium alloy as the research object, this paper successfully prepared the superhydrophobic surface of magnesium alloys through Zn-Ni electrodeposition/de-alloying composite technology and chemical modification technology, which greatly improved the corrosion resistance of magnesium alloys. In this paper, the preparation process and its influence law were systematically studied and optimised, and typical samples were analysed and characterised on the basis of the optimised process.

1. Introduction

Generally speaking, the process of physical and chemical interaction between metal and external environmental media, which makes the metal surface or chemical properties change, is called metal corrosion [1]. In addition, metal corrosion is pervasive, hidden and sudden. Metal corrosion not only consumes resources and pollutes the environment, but also reduces the service life of mechanical equipment, causes serious accidents, and endangers people's life, health, and property safety [2]. Therefore, the research on metal corrosion protection is of great significance. Since the 21st century, with the improvement of people's living standards and the development of industrialisation, the transport industry has become more and more important. For vehicles such as cars, planes, and trains, lightweight and energy consumption is undoubtedly an important indicator of the pursuit of mankind, so magnesium alloys are considered one of the best structural metals. While meeting the processing needs and performance requirements, the weight of components such as the body, engine, driver's stick elements, and gearboxes can be significantly reduced by displacing ferrous alloys and aluminium alloys, resulting in reduced transportation costs and energy savings [3]. It is estimated that when the weight of a car is reduced by 10 percent, the efficiency of its engine can be increased by 5.5 percent; when magnesium alloys are used in the aerospace field, their weight reduction effect is even more obvious. In the same weight reduction situation, the average commercial aircraft is far more energy-saving than the average car. It is because of this that the application of magnesium alloys in transport is increasing day by day. Alone in China, the application of magnesium alloys in automotive parts and components is growing at a high rate of 15% per year. In aerospace, Boeing uses magnesium alloys to manufacture the components of the main transmission system of certain models [4].

2. Relevant theory

2.1 Superhydrophobicity theory

(1) Young's model

As early as 1805, Young established Young's model based on the study of the wettability of solid surfaces, as shown in Figure. 1, and proposed the far-reaching Young's equation:

$$\cos\theta = \frac{\gamma_{SV} - \gamma_{SL}}{\gamma_{LV}} \tag{1}$$

where γsv is the solid-gas surface tension; γsl is the solid-liquid surface tension; γlv is the liquid-gas interfacial tension; θ is the contact angle of the solid surface.

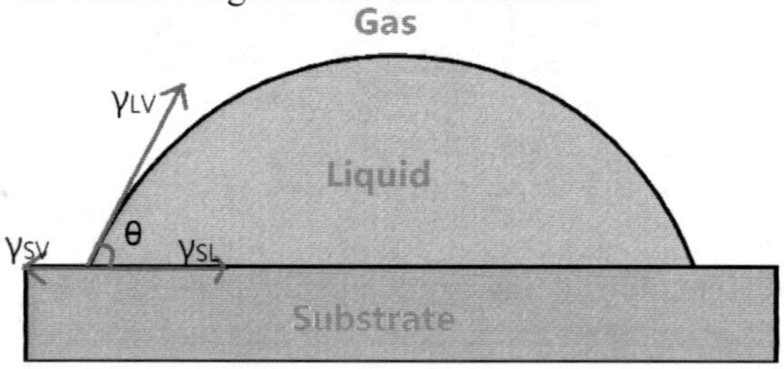

Figure 1. Young's theory.

Young's equation provides a theoretical basis for the calculation of contact angle. However, the conditions for the application of Young's equation are extremely stringent, requiring that the solid surface must be ideal, i.e., smooth, flat, and isotropic. However, ideal surfaces are difficult to apply in real production and life, and general solid surfaces usually have rough and uneven microstructures [5].

(2) Wenzel model

Wenzel's theory suggests that when a droplet falls on a rough surface, due to the unevenness of the rough surface, the droplet enters the concave portion of the rough surface, which results in the de facto solid-liquid contact area being significantly larger than the apparent solid-liquid contact area [6], as shown in Figure. 2.

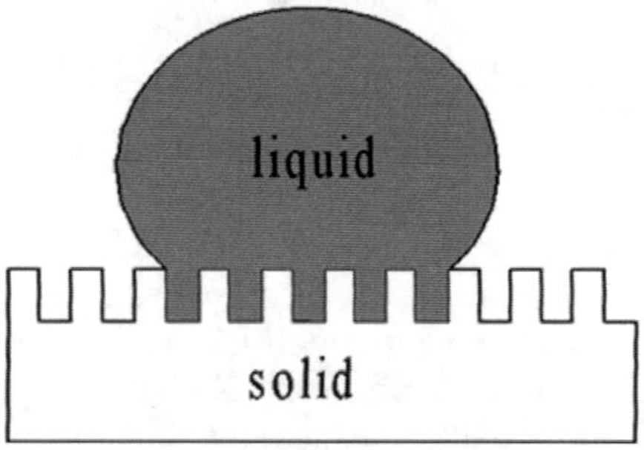

Figure 2. Wenzel theory.

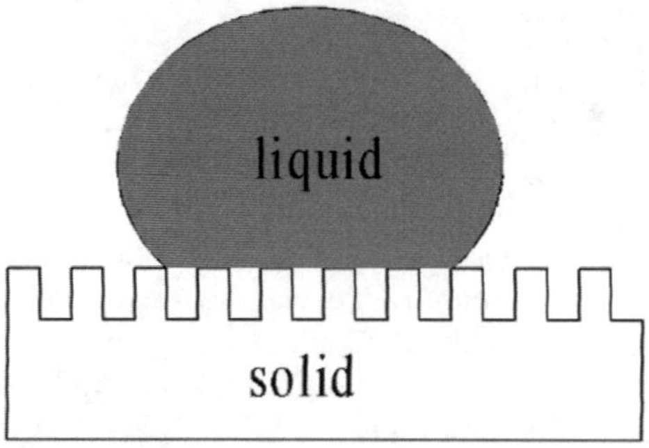

Figure 3. Cassie's theory.

Based on Wenzel's model theory, Wenzel took Young's equation one step further and proposed the Wenzel equation:

IOP Publishing
doi:10.1088/1742-6596/3009/1/012029

$$\cos\theta_w = \frac{r(\gamma_{SV}-\gamma_{SL})}{\gamma_{LV}} = r\cos\theta \tag{2}$$

where r is the roughness factor; θ is the intrinsic contact angle; θ_w is the apparent contact angle. From Wenzel's equation, for hydrophilic materials ($90>\theta_0$), θ_w decreases as r increases, whereas for hydrophobic materials ($90<\theta_0<180$), θ_w increases as r increases. In short, the rough structure of a solid surface can enhance its hydrophilicity or hydrophobicity. However, the researchers found that Wenzel's equation could not be applied to all scenarios.

(3) Cassie model

In order to break through the limitations in the application of Wenzel's equation, Cassie and Baxter further optimised Wenzel's equation and based on it, Cassie's model was developed [7]. Cassie's theory assumes that when a droplet falls onto a rough solid surface, there is trapped air in the depression of some surfaces, which prevents the droplet from completely filling the groove, so the solid-liquid junction actually consists of 3 parts: solid, gas and liquid. Cassie's model is shown in Figure. 3:

Based on the Cassie model theory, Cassie et al. proposed Cassie equation:

$$\cos\theta_w - f_1\cos\theta_1 = f_2\cos\theta_2 \tag{3}$$

where f_1 and f_2 represent the contact area fraction at the solid-liquid and liquid-gas interfaces, respectively; θ_w is the contact angle; θ_1 and θ_2 are the actual contact angles of solid-liquid and liquid-gas. In general, it can be assumed that $\theta_2 = 180°$ and $f_1 + f_2 = 1$, and then Eq. 3 can be transformed into Eq. 4.

$$\cos\theta_w = f_1(\cos\theta_1 + 1) - 1 \tag{4}$$

From this equation, it can be seen that as f_1 decreases, θ_w increases, and thus the better the hydrophobicity will be for a constant θ. Therefore, it becomes possible to construct micro-nano-structures on hydrophilic surfaces to obtain hydrophobic surfaces.

The three theoretical models of wettability are inherited and developed. Both the Wenzel theory and the Cassie theory are developed on the basis of Young's theory. While Wenzel theory is a special state of Cassie theory, so Cassie equation has a wider range of applications [8].

According to Cassie and Wenzel's theory, the roughness of a solid surface is one of the most important reasons that affect the wettability properties of a solid.

3. De-alloying formulations and process optimisation

3.1 De-alloying temperature

The temperature of de-alloying has a significant impact on the de-alloying process, and the effect of de-alloying temperature on the properties of the de-alloyed layer was investigated using the de-alloyed layer wettability and corrosion resistance as evaluation indexes. Taking the de-alloying temperature as a variable, the test was conducted in steps of 10.0 °C, gradually increasing from the minimum value of 30.0 °C to 90.0 °C. Other parameters were fixed as follows: ① de-alloying solution formula: 168.0 g/L aqueous KOH solution; ② de-alloying time: 24 h. The static contact angle (referred to as SCA) and rolling angle (referred to as SA) and the change rule of the corrosion resistance are shown in Figures. 4 and 5. The static contact angle (referred to as SCA) and rolling angle (referred to as SA) and corrosion resistance change rule are shown in Figures. 4 and 5.

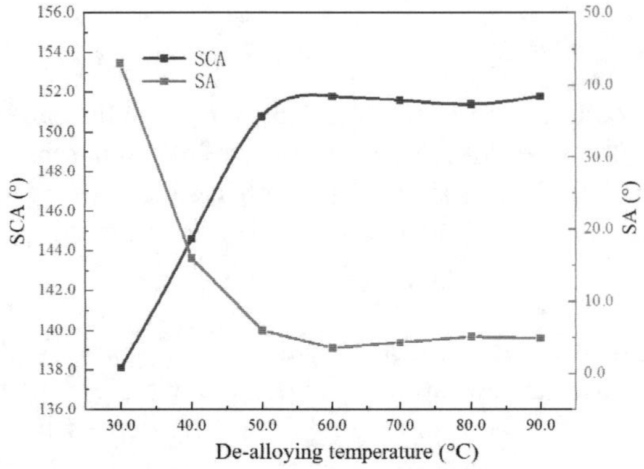

Figure 4. Effect of de-alloying temperature on the wettability of de-alloyed layers.

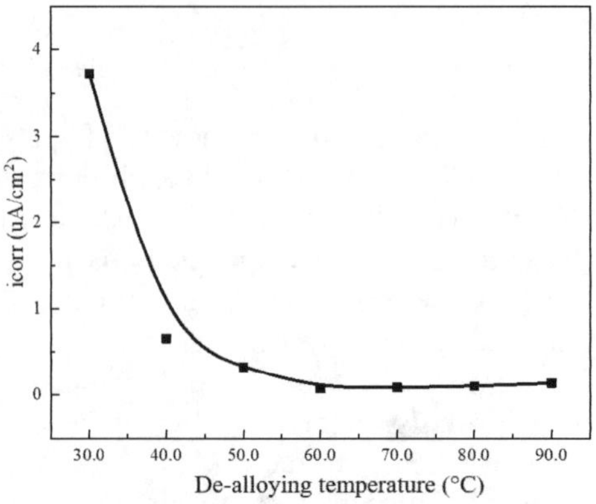

Figure 5. Effect of de-alloying temperature on the corrosion resistance of de-alloyed layers.

As seen in Figure. 4, the change of SCA on the surface of the de-alloyed layer exhibits an obvious phase: it is mainly divided into two phases, in the temperature interval of 30.0 °C-50.0 °C. The SCA increases nearly linearly with the temperature, from 138.1° to 150.8°, which follows a functional relationship SCA=119.2+0.635T, where R=0.99; in the temperature interval of 60.0- 90.0 °C temperature interval, the SCA, on the other hand, is not affected by temperature and stabilises around 151.8±1°.

As can be seen from Figure. 5, the de-alloying temperature has a large influence on the corrosion resistance of the de-alloyed layer, and the corrosion current density (i_{corr}) decreases with increasing temperature from 3.76 uA/cm^2 to 0.07 uA/cm^2 in the temperature interval of 30.0 °C-50.0 °C. This is because when the temperature is low, the de-alloyed layer does not acquire superhydrophobicity and therefore has poor corrosion resistance, while the hydrophobicity of the plated layer gets better and better as the temperature increases.

3.2 De-alloying time

Taking de-alloying time as a variable, the de-alloying time was controlled to be 5 min, 10 min, 30 min, 1 h, 4 h, 8 h, 12 h, 24 h, 48 h, and other parameters were fixed as follows: ① de-alloying temperature: 60.0 °C; ② de-alloying solution: 200.0 g/L KOH aqueous solution. The change rule of surface hydrophobicity and corrosion resistance of the de-alloyed layer was investigated, and the results were shown in Figures. 6 and 7.

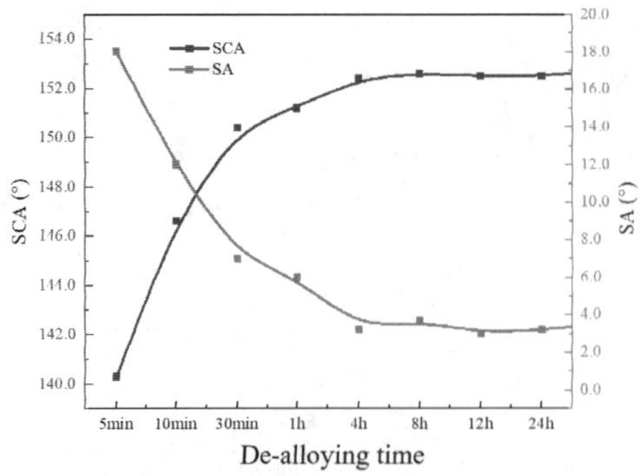

Figure 6. Effect of de-alloying time on de-alloyed hydrophobic properties.

As can be seen in Figure. 6, for a shorter period of time (5 min-30 min), the de-alloying time was increased and the SCA of the de-alloyed layer followed a rapid increase from 140.3±1° to 150.4±1°. Continuing to increase the de-alloying time (30 min-4 h), the SCA showed a decelerated increase and its value further increased to 152.4±1°. Further increasing the de-alloying time, the SCA stabilised at a level of about 152.5±1°, while the rolling angle, an indicator of adhesion, changed in the opposite direction to the hydrophobicity angle, decreasing rapidly in the range of 5 min-12 h from 18.0° to 3.0±1°. When the de-alloying time exceeded 4 h and it was extended, the de-alloyed layer specimens' contact angle and rolling angle tended to be stable and did not change much.

4. Hydrophobicity analysis

4.1 Hydrophobicity of the coating

After stearic acid treatment, Blank, Zn-Ni electrodeposited specimens and de-alloyed specimens were subjected to hydrophobicity measurements, and the side photographs of the measured surface droplets are shown in Figure. 7.

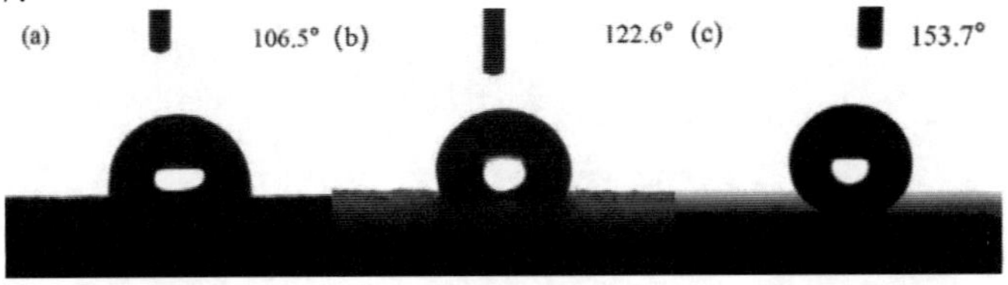

(a) Blank:106.5° (b) Pr: 122.6° (c) DC: 153.7°

Figure 7. Optical photographs of water droplets on different specimen surfaces.

As seen in Figure. 7(a), the Blank specimen surface was treated with stearic acid and its SCA was only 106.5°; the distilled water droplets showed a hemispherical shape on the surface of the AZ91D magnesium alloy; Figure. 7(b) shows the side-view optical image of the droplets of a typical Zn-Ni electrodeposited layer [9]. For the blank specimen after Zn-Ni alloy electrodeposition, the surface of magnesium alloys has been coated with a Zn-Ni electrodeposited layer. The hydrophobicity was enhanced after stearic acid modification, and its SCA measurement reached 122.6°, but it is still far from superhydrophobic, as shown in Figure. 7(c). The Zn-Ni electrodeposited layer was de-alloyed, and the surface of the plating layer was constructed with a micro-nano-layered structure. The obtained de-alloyed layer was chemically modified with distilled water droplets that were approximately spherical on its surface. Its SCA was up to 153.7°.

4.2 Corrosion resistance analysis

Total immersion corrosion tests were carried out on Blank, Zn-Ni electrodeposited specimens and superhydrophobic de-alloyed specimens, and the experimentally relevant results are shown in Table 1.

Table 1. 48 h average corrosion rate of typical specimens.

Specimen type	Blank	Pr	DC
Average corrosion rate ($mg \cdot m^{-2} \cdot h^{-1}$)	13.3	7.6	2.1

As can be seen from Table 1, the corrosion rate of the superhydrophobic de-alloyed layer is much lower than the other two, while the Zn-Ni electrodeposited specimen, i.e. Pr, again has a lower corrosion rate than Blank. After the magnesium alloy specimen has been chemically plated with Ni-P and electrodeposited with Zn-Ni, there are two protective layers on its surface. However, the Zn-Ni electrodeposited layer belongs to the sacrificial anode plating layer, so the Zn-Ni plating layer on the surface of the Zn-Ni electrodeposited specimen will use its plating layer as an anode, as well as the chemically plated Ni-P as a cathode, which can realise good protection for the Ni-P plating layer [10].

4.3 Polarisation curve

Figure. 8 shows the kinetic potential polarisation curves for typical specimens (including Blank, Zn-Ni electrodeposited specimens and superhydrophobic de-alloyed specimens, all three of which were optimised) in 3.5 wt% NaCl aqueous solution, and the corresponding electrochemical parameters are listed in Table 2.

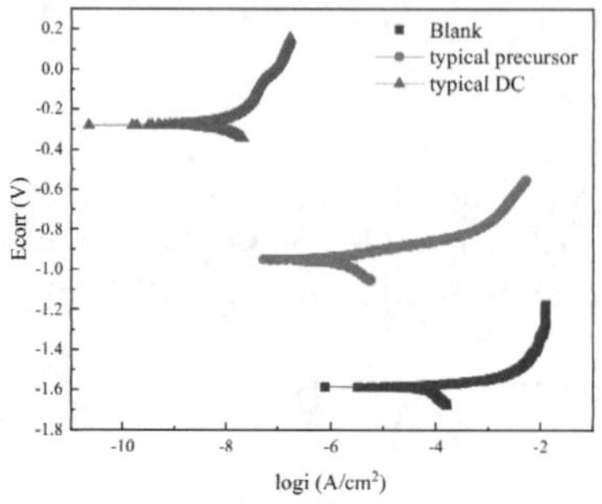

Figure 8. Tafel curves for typical specimens.

Table 2. Electrochemical corrosion parameters for tafel curves of typical specimens.

Sample	E_{corr} (V)	i_{corr} ($\mu A/cm^2$)	η_p (%)
Blank specimen	-1.602	75.06	/
Zn-Ni electrodeposition specimen	-0.951	2.08	97.23
De-alloyed specimens	-0.339	1.33×10^{-2}	99.98

As can be seen from Table 2, the E_{corr} of the superhydrophobic de-alloyed layer is 1263 mv higher than that of Blank and 612 mv higher than that of the Zn-Ni electrodeposited layer, and the i_{corr} of the DC specimen is about 0.02% of that of Blank and 0.64% of that of the Zn-Ni electrodeposited layer, which suggests that the corrosion resistance of the superhydrophobic de-alloyed specimen is much higher than that of the magnesium alloys and the Zn-Ni electrodeposited specimen. The corrosion inhibition efficiency P can be obtained according to the following equation.

$$P=\left(1-i_{corr}/i_{corr}^{0}\right)\times100\% \tag{5}$$

where i_{corr}^{0} is the corrosion current density of the substrate, i.e. Blank, and i_{corr} is the corrosion current density of the plated specimens (DC specimen and Pr specimen). The retardation efficiency of the superhydrophobic de-alloyed samples is up to 99.98% and that of the Zn-Ni electrodeposited layer is up to 97.23%. This indicates that both of them have good protection functions for magnesium alloy substrates.

5. Conclusion

In this paper, the AZ91D magnesium alloy was selected as the research object. Due to the poor corrosion resistance of magnesium alloy in the process of magnesium alloy application, Zn-Ni electrodeposition was innovatively used to coat the surface of magnesium alloy with a layer of Zn-Ni electrodeposited layer, and then chemical de-alloying was carried out by KOH solution to construct a micro-nano roughened structure on the surface of magnesium alloys, and finally combined with the chemical modification of stearic acid to successfully prepare a superhydrophobic magnesium alloy surface.

References

[1] Zhao K, Yang B P, Zhang J Y. Study on Preparation and Anti-icing Performance of Super-Hydrophobic Surface on Aluminum Alloy [J]. Gongneng Cailiao/Journal of Functional Materials, 2023, 41:80-83+88.

[2] Saffar M A, Eshaghi A, Dehnavi M R. Fabrication of Superhydrophobic, Self-cleaning and Anti-icing ZnO/PTFE-SiO2 Nano-composite Thin Film [J]. Materials Chemistry and Physics, 2021, 259 (1–2):124085.
[3] Long J, He Z, Zhou P, et al. Low-Cost Fabrication of Large-Area Broccoli-Like Multiscale Micro-and Nanostructures for Metallic Superhydrophobic Surfaces with Ultralow Water Adhesion and Superior Anti-Frost Ability [J]. Advanced Materials Interfaces, 2022: 1800353.
[4] Lee K, Lyu S, Lee S, et al. Characteristics and Self-cleaning Effect of the Transparent Superhydrophobic Film Having Nanofibers Array Structures [J]. Applied Surface Science, 2019, 256 (22): 6729-6735.
[5] Ding S, Xiang T, Li C, et al. Fabrication of Self-cleaning Superhydrophobic Nickel/graphene Hybrid Film with Improved Corrosion Resistance on Mild Steel [J]. Materials & Design, 2021, 117 (Mar): 280-288.
[6] Anjum A S, Sun K C, Ali M, et al. Fabrication of Coral-reef Structured Nano-silica for Self-cleaning and Superhydrophobic Textile Applications [J]. Chemical Engineering Journal, 2020, 401: 125859.
[7] Feng L, Zhang H, Wang Z, et al. Superhydrophobic Aluminum Alloy Surface: Fabrication, Structure and Corrosion Resistance [J]. Colloids and Surfaces A: Physicochemical and Engineering Aspects, 2023, 441 (3): 319-325.
[8] Chang Ling, Fang Wenbin, Wang Jianming. Preparation and Electrochemical Capacitance Performance of Porous Nickel-based Composite Membrane Electrodes [J]. Chinese Science and Technology Paper, 2016, 11 (6): 4.
[9] Gnanamuthu R, Mohan S, Saravanan G, et al. Comparative Study on Structure, Corrosion and Hardness of Zn–Ni Alloy Deposition on AISI 347 Steel Aircraft Material [J]. Journal of Alloys & Compounds, 2022, 513 (none): 449-454.
[10] Bahadormanesh B, Ghorbani M, Kordkolaei N L. Electrodeposition of Nano-crystalline Zn/Ni Multilayer Coatings from Single Bath: Influences of Deposition Current Densities and Number of Layers on Characteristics of Deposits [J]. Applied Surface Science, 2021, 404 (MAY 15): 101-109.

Study on the remaining life of the high-temperature steam pipeline in long-term service of power plants

Lukuan Yang *, Shangyu Lu, Minjin Tang, Yingcong Zhu, Xianxi Xia, Haitao Dong

Suzhou Nuclear Power Research Institute, Suzhou Xihuan Road 1688, Suzhou 215004, China

* Corresponding author's e-mail: yanglukuan@126.com

Abstract: The purpose of this article is to explore the remaining service life of the main steam pipeline in power plants that have been in operation for an extended period. The main steam pipeline, which operates in a high-temperature and high-pressure environment for a long time, experiences a series of microstructural transformations in its materials (such as 10CrMo910 steel), including pearlite spheroidization and carbide precipitation, leading to a significant decline in the mechanical properties of the pipeline, thereby affecting the safe and stable operation of the power plant. This study conducts an in-depth investigation into the microstructure, mechanical properties, and their relationship with the remaining service life of the main steam pipeline through methods such as metallographic examination, mechanical performance analysis, and high-temperature creep life assessment. Based on the data from high-temperature sustained strength performance tests, a life assessment model was established by using the L-M parameter equation to estimate the remaining service life of the high-temperature steam pipeline. The results indicate that after long-term service, there is a noticeable occurrence of pearlite spheroidization and carbide precipitation in the pipeline's microstructure, and the remaining service life of the pipeline meets the requirements for subsequent operation. This research provides a theoretical basis and technical support for the safe operation of high-temperature steam pipelines in power plants, aiding in the formulation of scientific and reasonable maintenance strategies and replacement plans to ensure the safe and efficient operation of the power plant. In the future, further exploration of more precise life prediction methods and the application of new materials will continue to enhance the durability and safety of high-temperature steam pipelines in power plants.

1. Introduction

The high-temperature steam pipeline, as one of the important pressure-bearing components of power plants, has extremely harsh service conditions and operates in high-temperature and high-pressure environments for a long time. The reliability and safety of pipelines are directly related to the overall operational efficiency and safety performance of power plants. Therefore, conducting in-depth research on the remaining life of the high-temperature steam pipeline in long-term service of power plants has extremely important practical significance and engineering application value [1]-[3].

The high-temperature steam pipeline is usually made of low-alloy heat-resistant steel, such as 10CrMo910 steel, which has good welding performance, oxidation resistance, and structural thermal stability, and is the preferred material for manufacturing high-temperature and high-pressure pipelines [4]-[5]. However, during long-term high-temperature service, pipeline materials undergo a series of structural transformations, such as pearlite spheroidization, carbide aggregation, and growth, which significantly affect the mechanical properties and creep behavior of the material, leading to gradual deterioration or even failure of pipeline performance.

At present, research on the remaining life of steam pipelines both domestically and internationally is mostly focused on analyzing the causes of cracking in furnace tubes and studying creep performance. However, there is relatively little research on the organizational transformation of materials after long-term high-temperature service and their impact on the mechanical properties and remaining life of pipeline materials. Therefore, analyzing the creep performance of the high-temperature steam pipeline in power plants after long-term high-temperature service and studying the relationship between pipeline characteristics and remaining life after service is of great significance for preventing pipeline degradation and ensuring the safe operation of power plants [6]-[8].

The purpose of this study is to conduct sampling analysis on the high-temperature steam pipeline and reheated steam pipeline of a power plant, using methods such as wall thickness verification, internal pressure stress, and remaining life calculation to systematically investigate the remaining life evaluation method of the pipeline after long-term high-temperature service. Through this research, theoretical basis and technical support can be provided for the safe operation of the high-temperature steam pipeline and reheated steam pipeline in power plants, which helps to develop scientific and reasonable maintenance strategies and replacement plans, ensuring the safe and efficient operation of power plants.

2. Life assessment method

2.1. Parameters of the research object

The research object is a domestically produced 300 MW subcritical coal-fired unit, and the boiler is a coal-fired drum boiler with a subcritical pressure, natural circulation, fully suspended outdoor layout, four-corner tangential combustion, and solid-state slag discharge. The evaporation section of the boiler water circulation system is equipped with a forced circulation pump and a steam-water separator. The furnace shaft is suspended in sequence from front to back, including a full-screen superheater, a screen superheater, a medium-temperature reheater, a high-temperature reheater, a high-temperature superheater, a horizontal flue wall tube, a tail shaft wall tube, a low-temperature superheater, and an economizer. The equipment is suspended by internal suspension pipes, and the heating surface header is suspended by external suspension pipes. Further research will be conducted on the remaining life of the cylinder bodies of the main steam pipeline and reheat steam pipeline of the above-mentioned research objects. The sampled steam pipelines are shown in Figure 1.

Figure 1. Sampling diagram of the main steam pipeline and reheated steam pipeline.

2.2. Wall thickness verification and metallographic structure inspection

The minimum required wall thickness for a straight cylindrical body δ_m can be calculated by using the following formula for wall thickness verification.

$$\delta_m = \frac{pD_o}{2\phi_{\min}[\sigma]^t + p} + C_1 \tag{1}$$

where D_o is the outer diameter of the pipeline; p is the pipeline calculation pressure; $\phi_{\min}$ is the allowable stress correction factor; C_1 is corrosion allowance, generally taken as 0.5 mm; $[\sigma]^t$ is the allowable stress for steel at service temperature.

After checking the wall thickness, we conduct a metallographic examination on the submitted tube and observe it under a Zeiss Axiover 200MAT metallographic microscope.

2.3. Mechanical properties

The mechanical performance tests for the sampling tube include hardness tests, room-temperature tensile performance tests, high-temperature tensile performance tests, impact performance tests, bending performance tests, and ductile-brittle transition temperature measurement tests.

The surface of the hardness sample is polished with metallographic sandpaper and subjected to polishing treatment, using a fully automatic Brinell hardness tester with a test load of 187.5 kg and a load holding time of 10 seconds. The room temperature tensile performance test is conducted on an AG-IC Shimadzu electronic tensile testing machine, with a pre-yield loading strain rate not exceeding 0.00025/s and a post-yield loading strain rate not exceeding 0.0067/s. The conditional yield strength is measured by using a strain gauge, with the main steam and reheating steam pipelines using φ10 round bar proportional tensile samples, and the heated surface using arc-shaped samples, with the sampling direction being longitudinal. The high-temperature tensile performance test is also conducted on the AG-IC Shimadzu electronic tensile testing machine, with a pre-yield loading strain rate not exceeding 0.000083/s and a post-yield loading strain rate not exceeding 0.0033/s, using φ10 mm standard samples, with the conditional yield strength measured by using a strain gauge. The impact performance test is conducted on an AMSLER RKP 450 pendulum impact testing machine, with a hammer edge radius of 2 mm and a pendulum angle of 150°. The parameters involved in the bending performance test are as follows: bending head diameter of 40 mm, support roller spacing of 71.21 mm, and bending angle of 180°. The ductile-brittle transition temperature measurement is conducted on the AMSLER RKP 450 pendulum impact testing machine, with a hammer edge radius of 2 mm, a pendulum angle of 150°, and an impact energy of 450 J.

2.4. Internal pressure converted stress

The maximum circumferential stress caused by internal pressure in the pipeline can be calculated by using the following formula.

$$\sigma_\theta = \frac{P(D_o - S)}{2S} \tag{2}$$

where P is the working pressure of the pipeline, D_o is the outer diameter of the pipeline, and *S* is the wall thickness of the pipeline.

2.5. Larson Miller's method

For components with creep as the main failure mode, the remaining life can be calculated by using the Larson-Miller parameter method. The Larson-Miller parameter is a combination of time and temperature, represented by LMP, and has the following relationship:

$$LMP(\sigma) = (T+273.15)(C + \lg t_r) \tag{3}$$

where T is the Kelvin temperature, K; t_r is creep fracture life, h; σ is creep stress, MPa; C is the Larson Miller constant.

The L-M parameter can be used to perform multiple linear regression on the experimental data according to the following formula to obtain the *C* value.

$$\lg t_r = C_{LM} + \left(C_1 \lg\sigma + C_2 \lg^2\sigma + C_3 \lg^3\sigma + C_4 \lg^4\sigma + ... + C_0\right)/(T+273.15) \tag{4}$$

where C_0, C_1, C_2, C_3, and C_4 are fitting coefficients.

3. Calculation results

3.1. Wall thickness verification results and microstructure

The main steam outlet temperature/pressure is 540°C/17.45 MPa, and the reheated steam outlet temperature/pressure is 540°C/3.53 MPa. The measured wall thickness of the main steam sampling pipe of Unit 1 is 87.52 mm, the measured wall thickness of the reheated steam sampling pipe of Unit 1 is 37.86 mm, the measured wall thickness of the main steam sampling pipe of Unit 2 is 87.54 mm, and the measured wall thickness of the reheated steam sampling pipe of Unit 2 is 38.91 mm. The design wall thickness of the main steam pipeline is 85.34 mm, and the design wall thickness of the reheated steam pipeline is 37.5 mm. The minimum wall thickness measured by the sampling tube meets the requirements of GB/T 16507.4-2022.

The metallographic results are shown in Figure 2. The metallographic results indicate that the microstructure of the base material is ferrite+bainite, with a spheroidization level of 4 and a grain size of 4~7. The metallographic structure of the weld seam is bainite, with good fusion and no welding defects found.

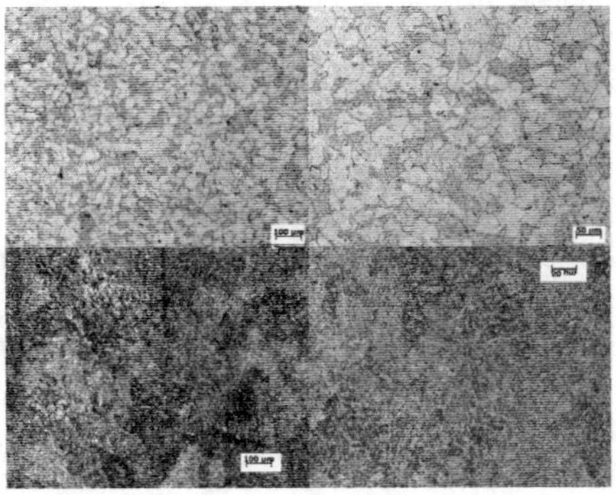

Figure 2. Metallographic structure.

3.2. Calculation results of internal pressure converted stress

The calculation of converted stress generated by internal pressure in the main steam pipeline and reheated steam pipeline is shown in Table 1 and Table 2.

Table 1. Calculation of converted stress in 10CrMo910 main steam pipeline.

Parameter	Name	Unit	Calculation or value basis	Value
/	Texture of material	/	10CrMo910	/
P	Design pressure	MPa	/	17.45
D_o	Pipe OD	mm	Φ401.6×64.3	401.6
			Φ457.2×65	457.2
			Φ540.8×85.34	540.8
S	Wall thickness	mm	Φ401.6×64.3	64.3
			Φ457.2×65	65
			Φ540.8×85.34	85.34
σ_θ	Internal pressure converted stress	MPa	Φ401.6×64.3	45.769
			Φ457.2×65	52.645
			Φ540.8×85.34	46.565

Table 2. Calculation of converted stress in 10CrMo910 reheated steam pipeline.

Parameter	Name	Unit	Calculation or value basis	Value
/	Texture of material	/	10CrMo910	/
P	Design pressure	MPa	/	3.74
D_o	Pipe OD	mm	Φ561.6×26	561.6
			Φ609.6×25	609.6
			Φ756.5×37.5	756.5
S	Wall thickness	mm	Φ561.6×26	26
			Φ609.6×25	25
			Φ756.5×37.5	37.5
σ_θ	Internal pressure converted stress	MPa	Φ561.6×26	38.522
			Φ609.6×25	43.728
			Φ756.5×37.5	35.854

3.3. Mechanical properties

A hardness test, room temperature tensile performance test, high-temperature tensile performance test, impact performance test, bending performance test, and ductile-brittle transition temperature measurement test were conducted on the sampling tube. The test results are as follows.

The average hardness of the main steam pipe base material is 141.7 HBW, the average hardness of the main steam pipe weld is 163.4 HBW, the average hardness of the reheated steam pipe base material is 141.8 HBW, and the average hardness of the reheated steam pipe weld is 163.7 HBW. The qualified range for the hardness of the 10CrMo910 base material is between 130 and 197 HBW and the qualified range for the hardness of the weld seam is between the hardness of the base material and 241 HBW. Therefore, the hardness value of the inspection tube meets the requirements of the DL/T 438-2023 standard. The $R_{p0.2}$ for room temperature stretching of the main steam pipe is 215 Mpa (longitudinal sampling), and the $R_{p0.2}$ for room temperature stretching of the reheated steam pipe is 222 Mpa (longitudinal sampling). The qualified range of $R_{p0.2}$ for room temperature tensile testing of 10CrMo910 base material is not less than 20 Mpa, so the room temperature tensile performance of the tested tube meets the requirements of GB/T 5310-2017. The $R_{p0.2}$ for high-temperature stretching of the main steam pipe is 147 Mpa (longitudinal sampling), and the $R_{p0.2}$ for high-temperature stretching of the reheated steam pipe is 155 Mpa (longitudinal sampling). The qualified range of $R_{p0.2}$ for high-temperature tensile testing of 10CrMo910 base material is not less than 161 Mpa, so the high-temperature tensile performance of the tested tube does not meet the requirements of GB/T 5310-2017. The minimum measured impact energy of the main steam pipe and reheated steam pipe is 81 J, and the qualified range is not less than 40 J (longitudinal sampling) or not less than 27 J (transverse sampling), so the impact performance meets the requirements of GB/T 5310-2017 standard.

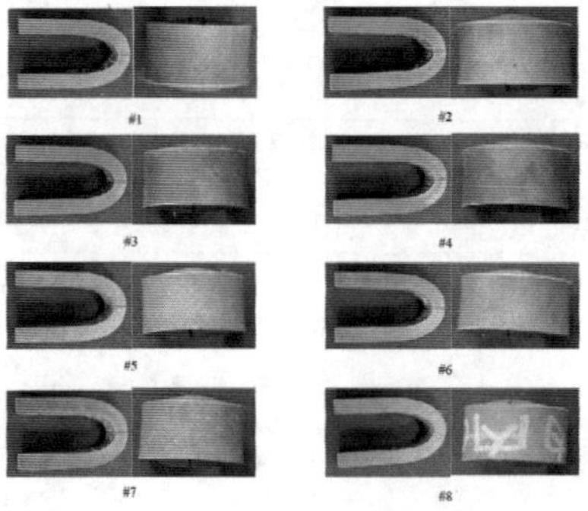

Figure 3. Bent specimen.

The macroscopic image of the bent sample is shown in Figure 3. There are no visible cracks on the outer surface, which meets the requirements of the GB/T 232-2010 standard.

When measuring the ductile-brittle transition temperature, the ductile-brittle transition temperature of transverse sampling of the main steam pipeline is 50.0°C, and the ductile-brittle transition temperature of transverse sampling of the reheated steam pipeline is 36.0°C. The corresponding transition temperature curve is shown in Figure 4.

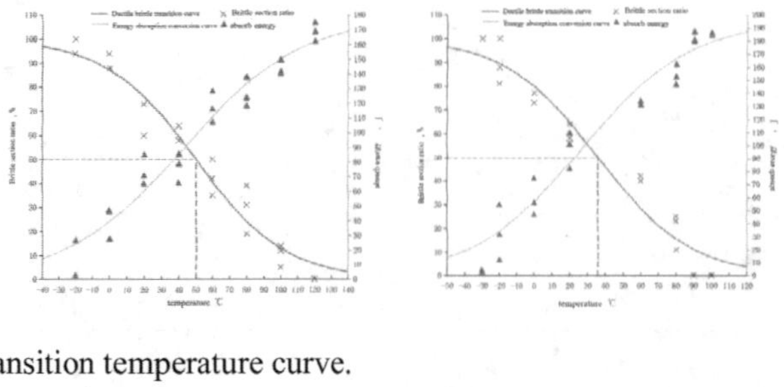

Figure 4. Ductile brittle transition temperature curve.

3.4. Remaining creep life calculation results

Based on ECCC and EN 10216 statistical data on the durability of 12CrMoV, X20CrMoV121, and ASME T/P91 steels, the $LMP(\sigma)-\sigma$ curve of 10CrMo910 is shown by least squares fitting. Using the LM method and taking 1.5 times the stress coefficient, the remaining life calculation results of the main steam pipeline and reheated steam straight pipe section are shown in Table 3.

Table 3. The remaining creep life of the main steam straight pipe section and reheated steam straight pipe section.

Name	Position	Internal pressure converted stress	Stress coefficient	Temperature	Residual life
Steam pipe	Φ401.6×64.3	45.769	1.5	542°C	$>10^5$ h
	Φ457.2×65	52.645	1.5	542°C	$>10^5$ h
	Φ540.8×85.3	46.565	1.5	542°C	$>10^5$ h
Hot reheat pipe	Φ561.6×26	38.522	1.5	540°C	$>10^5$ h
	Φ609.6×25	43.728	1.5	540°C	$>10^5$ h
	Φ756.5×37.5	35.854	1.5	540°C	$>10^5$ h

4. Conclusion

Based on ECCC and EN 10216 statistical data on the durability performance of 10CrMo910 steel, the LM method is used to calculate the wall thickness by using the minimum allowable deviation between the measured wall thickness and the nominal wall thickness, with a stress coefficient of 1.5 times and a calculated temperature of 540°C. The remaining creep life of the 10CrMo910 main steam pipe and the 10CrMo910 reheated hot section straight pipe is greater than 100, 000 hours.

References

[1] Halyna Krechkovska, Oleksandra Student, Olha Zvirko, Myroslava Hredil, Lesya Svirska, Ivan Tsybailo, and Petro Solovei. Substantiation of the critical structural and mechanical state of low-alloy heat-resistant steel from steam pipelines of the thermal power plant, Engineering Failure Analysis, Volume 150, 2023.
[2] Xue Wang, Conghao Du, Kai Zhang, Yalin Zhang, and Qiaosheng Huang. Steam oxidation behavior of T92 steel tube for high-temperature reheater after long-term service in ultra-supercritical boilers, Corrosion Science, Volume 221, 2023.
[3] M. A. Mohtadi-Bonab, Vahid Javaheri, and F. Totonchi-Maleki. Rupture analysis of 12Cr1MoV super-heater tubes in power plants, Engineering Failure Analysis, Volume 153, 2023.
[4] R. Bonetti, A. Morris, P. H. Shipway, and W. Sun. Requirements for developing high-temperature creep life models for aging pipework systems using power plant condition monitoring and inspection data, International Journal of Pressure Vessels and Piping, Volume 188, 2020.
[5] Sam J. Root, Porter Throckmorton, Jonathan Tacke, Jacob Benjamin, Michael Haney, and R. A. Borrelli. Cyber hardening of Nuclear Power Plants with real-time nuclear reactor operation, 1. Preliminary operational testing, Progress in Nuclear Energy, Volume 162, 2023.
[6] Jorma Salonen, Pertti Auerkari, Olli Lehtinen, and Mikko Pihkakoski. Experience on in-service damage in power plant components, Engineering Failure Analysis, 14: 6 (970-977), 2007.
[7] Xiaoxiang Sun, Chang Che, Gong Qian, and Xue Wang. Microstructure, hardness, and creep properties for P91 steel after long-term service in an ultra-supercritical power plant, International Journal of Pressure Vessels and Piping, Volume 212, Part A, 2024.
[8] Robert Krivanek. Long-term operation of nuclear power plants – IAEA SALTO peer review service and its results, Nuclear Engineering and Design, 280 (99-104), 2014.

Journal of Physics: Conference Series **3009** (2025) 012031 doi:10.1088/1742-6596/3009/1/012031

Research on hyperspectral stealth design based on forest background

Xiaodong Ma [1], Xiaolong Weng [1] *, Le Yuan [2], and Yaqin Wang [2]

[1] School of Electronic Science and Engineering, University of Electronic Science and Technology of China, Chengdu 611731
[2] Key Laboratory of Fluid and Power Machinery of Ministry of Education, Xihua University, No. 999 Jinzhou Road, Chengdu 610039, China

[*] Corresponding author's e-mail: wengxl59@163.com

Abstract. This study analyzed the characteristic spectra of the canopy and leaves of the same plant species. Although the spectral curves of the vegetation were similar in shape, the reflection intensity differed between the leaf and canopy spectra. This difference is attributed to the scattering and absorption of solar radiation by the vegetation canopy and the atmosphere, affecting the sensor's energy. Additionally, detection distance and angle also play a role in these variations. When designing hyperspectral stealth materials, it is crucial to consider these factors that influence the spectral characteristics of forests. Moreover, the spectral reflection intensity of camouflage coatings can be adjusted by modifying the content and particle size of layered double hydroxides (LDHs) within the coating.

1. Introduction

In recent years, the application of hyperspectral imaging technology in military detection has become increasingly extensive. This technology can detect target spectral details from both aircraft and satellite platforms. Hyperspectral imaging can simultaneously capture two-dimensional spatial information, which describes the distribution of ground objects, and one-dimensional spectral information, which characterizes their spectral properties. This dual capability allows for integrating both image and spectral data, enhancing the precision and effectiveness of military surveillance and reconnaissance [1]. Many countries have initiated projects to explore the military applications of hyperspectral imaging technology. For instance, Canada launched the Hyperspectral Image Exploitation (HYMEX) technology demonstration project, and Europe initiated the CEPAJP 8.10 project [2, 3]. These initiatives reflect a growing recognition of the strategic advantages provided by hyperspectral imaging in military operations.

The Airborne Visible/Infrared Imaging Spectrometer (AVIRIS) is one of the most advanced hyperspectral imagers in the world, with a wavelength range of 0.41 to 2.45 μm and a spectral resolution of 10 nm [4]. The Advanced Responsive Tactically-Effective Military Imaging Spectrometer (ARTEMIS) was the primary payload of Tactical Satellite-3. It was utilized in tactical reconnaissance, offering capabilities for target detection and recognition [5]. Many researchers have been focused on developing recognition algorithms. A target detection and recognition system has been proposed utilizing various fusion algorithms, enabling the identification of military targets against complex backgrounds [6]. The HSI anomaly detection and target recognition were also studied based on the neural network method [7].

Hyperspectral imaging has been applied in the military, posing a significant threat to traditional optical stealth technology [8]. Consequently, it is essential to study the changes in forest background spectra and develop camouflage coating to counteract this threat.

2. Experiments and methods

2.1 Device and methods

The Analytica Spectra Devices (ASD) spectrometer offers a spectral resolution of 3 nm at 700 nm and 8 nm at 1, 400 nm and 2, 100 nm, covering a wavelength range from 350 to 2, 500 nm. There are two primary methods for using the ASD spectrometer:

a) Contact Type: In this method, a halogen lamp serves as the light source to collect the spectrum of vegetation leaves. This approach involves direct contact with the sample, ensuring precise spectral measurements of individual leaves.

b) Handle Type: This method utilizes sunlight as the light source and employs a 25° lens to collect the canopy reflection spectrum. The spectrometer is positioned approximately 1 meter above the vegetation, allowing for non-contact measurements of the canopy's spectral properties.

The microstructure of the particles was examined by using scanning electron microscopy (SEM; FEI Inspect F50).

The tensile strength was evaluated according to the standard GB/T 528-2009 "Determination of tensile stress-strain properties of vulcanized rubber or thermoplastic rubber" using an HD-B607-S universal mechanical properties testing machine.

2.2 Experiment

The process of preparing a hyperspectral camouflage coating in this study is as follows:

1) Mixing: Raw rubber (MP5000), reinforcing agents (MH180, R812s), hyperspectral pigments (Mg-Al LDHs) [9], and other additives (thickening agent MA9100, coupling agent KH560) are sequentially added into an internal mixer according to the designed proportions.

2) Curing Control: Catalyst (ACS-pt-10) and inhibitor (MA8300) are used to control the curing rate of the coating.

3) Compounding: After thorough mixing, the compound is cooled with a thin pass on an open mill for further compounding to obtain mixed rubber.

4) Preforming: The mixed rubber is then rolled into preforms.

5) Vulcanization Molding: The preforms are placed in a mold and subjected to vulcanization molding by using a vulcanizing machine with controlled time and temperature parameters.

The specific mixing formula can be found in Table 1.

Table 1. Coating process formulation.

Ingredients	MP5000	MH180	R812s	ACS-pt-10	MA8300	MA9100	KH560	LDHs
Mass ratio	38.462	2.883	7.693	0.385	0.192	0.192	0.192	25/50/75

3. Results and discussion

3.1 Characteristic spectrum analysis of vegetation

Figure 1 shows the spectral curve of the canopy and leaf of the same planting of Conyza canadensis, rice, reed, and paulownia flower (1-xx: leaf spectrum, 2-xx: canopy spectrum, the same below). The spectral characteristics of the same vegetation were consistent. However, the spectral intensities varied among different types of vegetation.

Conyza canadensis, rice, and reed had small leaves that were densely distributed, resulting in a more complex canopy structure, as shown in Figures 2 (a), 2 (b), and 2 (c). The reflected energy of the ground objects received by the sensor was minimal due to the significant scattering of solar radiation by the complex canopy structure. This led to the intensity of the canopy reflection spectrum being lower than that of the leaf spectrum. In contrast, the spectral difference between the leaves and canopy of Paulownia flowers was small, because the leaves of Paulownia flowers were larger and the canopy structure was simpler, as shown in Figure 2 (d).

Journal of Physics: Conference Series **3009** (2025) 012031 doi:10.1088/1742-6596/3009/1/012031

The reflectance spectrum is primarily influenced by the canopy structure of vegetation. Therefore, it is crucial to consider the variation in reflection intensity caused by different canopy structures when designing camouflage coatings.

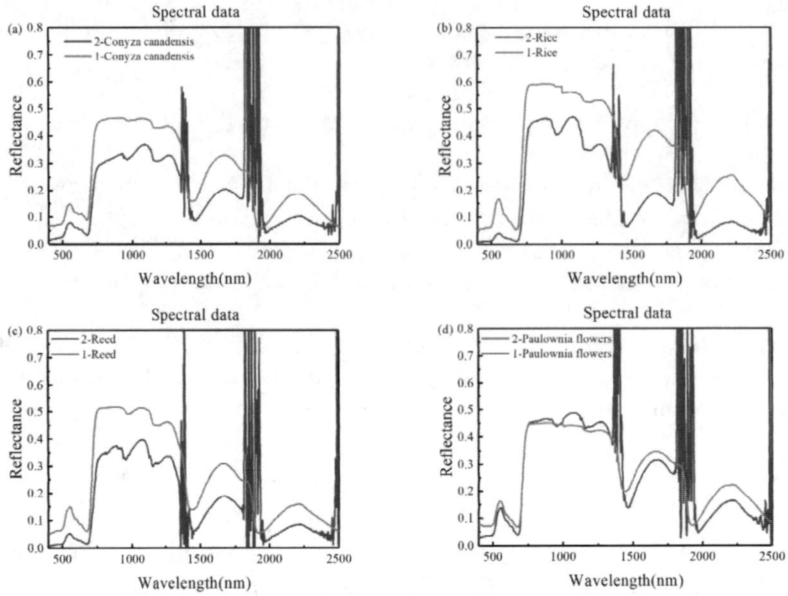

Figure 1. Spectrum curve of vegetation canopy and leaf spectral data: (a) Conyza canadensis; (b) Rice; (c) Reed; (d) Paulownia flower.

Figure 2. Vegetation canopy image: (a) Conyza canadensis; (b) Rice; (c) Reed; (d) Paulownia flower.

3.2 Study on coating preparation

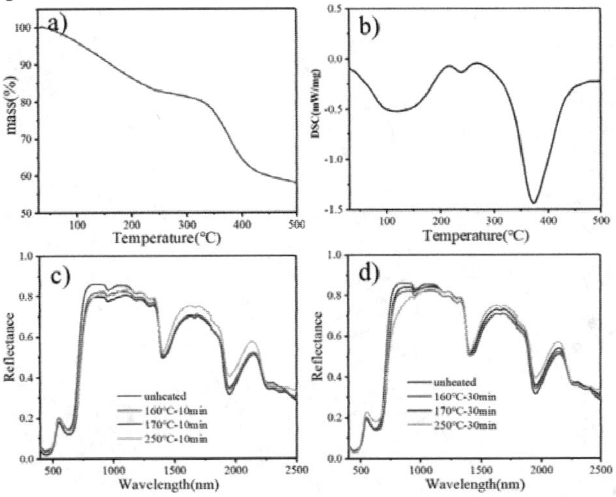

Figure 3. Thermal stability curves of LDHs: (a) TG; (b) DSC; (c)~(d) Reflectance after different heat treatments.

The vulcanization of silicone rubber requires a certain temperature, but excessively high temperatures can adversely affect the spectral performance of coatings in the 1.2-2.5 μm wavelength range. We have conducted tests on the thermal stability of powders. According to the TG/DSC curves of LDHs (Figures 3a-b), the first weight loss stage from 50°C to 220°C can be attributed to the removal of free water molecules and residual organics adsorbed on the surface of LDH powders. Similarly, the DSC curve displays a broad endothermic peak in this temperature range. The second weight-loss stage between 220°C and 270°C can be ascribed to the decomposition of organic pigments. Meanwhile, as seen in Figure 3d, the "red slope" of the LDHs reflectance curve exhibited a decrease after being heated at 250°C for 30 minutes. The third weight loss stage between 270°C and 650°C can be attributed to the dehydroxylation of the LDH layers, the removal of interlayer structural water molecules, and the decomposition of nitrate ions, corresponding to an endothermic peak centered at 380°C in the DSC curve [10].

Figures 3c-d show the reflectance curves of LDHs after different heat treatments. The temperature ranges from 160°C to 170°C, with a duration of 10 to 30 minutes. However, no significant changes are observed in the reflection spectrum of LDHs under these conditions.

Table 2. Tensile strength of coating under different vulcanization conditions.

Temperature \ Time	5 min	10 min	20 min	30 min
160°C	1.5 MPa	1.9 MPa	2.2 MPa	7.8 MPa
170°C	7.7 MPa	7.6 MPa	7.8 MPa	7.8 MPa

The test results of tensile strength for various vulcanization coatings are presented in Table 2. It can be observed from the table that a longer vulcanization time is required for the coating to exhibit satisfactory mechanical properties at a vulcanization temperature of 160°C. Ultimately, we employed a vulcanization temperature of 170°C and a vulcanization time of 5 minutes to successfully prepare the hyperspectral camouflage coating.

3.3 Characteristics of coating

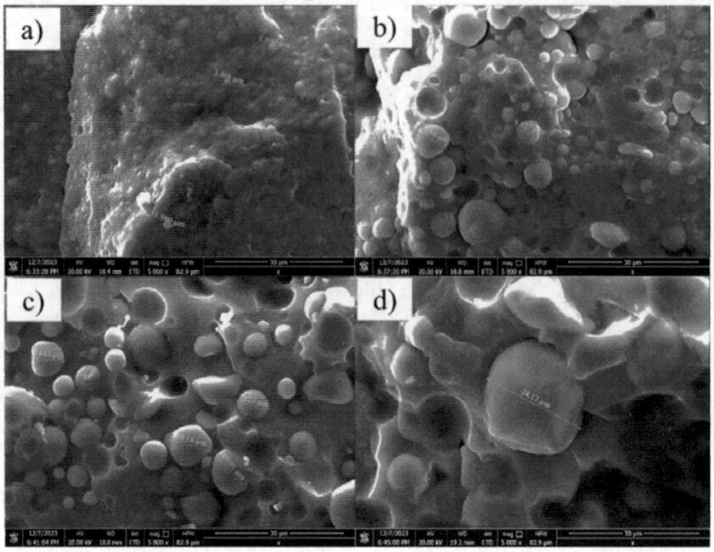

Figure 4. SEM cross-section images of the coating under different LDH size.

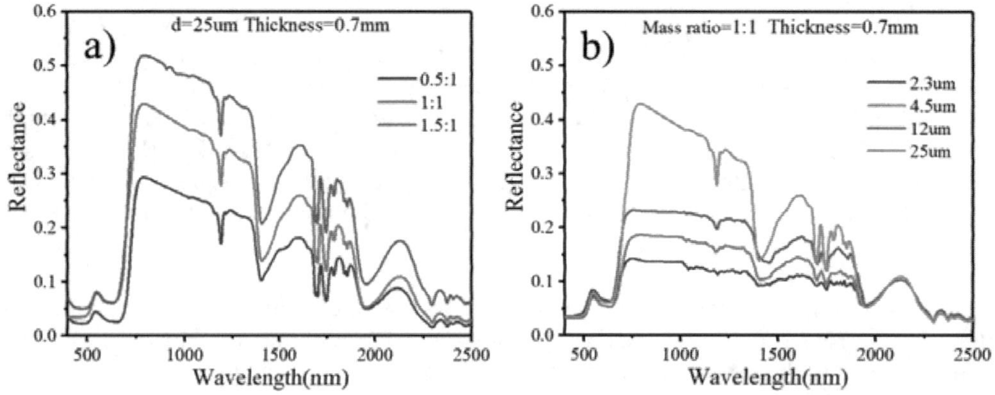

Figure 5. Reflection curves of coating: (a) Different mass ratio; (b) Different LDH size.

The SEM cross-section of the coating under various LDH particle sizes is illustrated in Figure 4. The LDHs prepared by spray granulation exhibit a particle size ranging from 2.3 to 25 μm, with a spherical shape and uniformity. Additionally, they possess a smooth surface and are embedded in silicone rubber. The reflection curves of coatings with varying mass ratios and particle sizes are depicted in Figures 5a and 5b, respectively. The reflection intensity increases with the increase in mass ratio and LDH size, attributed to the enhanced scattering ability of the coating. Consequently, by adjusting the mass ratio and particle size of the coating, one can effectively control its reflection intensity without compromising its absorption characteristics.

4. Conclusion

In this study, the characteristic spectra of canopy and leaves under the same planting conditions were analyzed. For the same vegetation, the spectral curve shape is similar, but there are noticeable differences in reflection intensity between the spectral data of leaves and canopy. Additionally, the spectral reflection intensity of the camouflage coating is controlled by adjusting the content and particle size of LDHs in the coating.

In conclusion, this research is important for the study of forest hyperspectral stealth design within the 0.4~2.5 μm range.

Acknowledgments

This project is supported by the National Natural Science Foundation of China (Grant number: 51972276).

References

[1] Xue T. ,Wang C. , Xie H. , & Wang Y. (2023). A novel method based on GPU for real-time anomaly detection in airborne push-broom hyperspectral sensors. *Remote Sensing*, 15 (18).
[2] Ardouin J. P. , Levesque J. , & Rea T. A. (2007). Demonstration of hyperspectral image exploitation for military applications. *IEEE.*
[3] Briottet X. , Boucher Y., Dimmeler A. , Malaplate A., & L. Klasén (2006). Military applications of hyperspectral imagery - art. No. 62390b. *Proceedings of SPIE – The International Society for Optical Engineering.*
[4] Groeneveld D. P. , Baugh W. M. , Silverman J. B. , Barz D. D. , & Prescott M. B. (2022). Non-imaging airborne spectroscopy for calibrating satellite imagery.
[5] Straight S. D., Doolittle C., Cooley T., Gardner J., Armstrong P., & Nadile R., et al. (2010). Tactical satellite-3 mission overview and initial lessons learned (postprint).
[6] Zhuang X., Li D., Wang Y., & Li K. (2024). Military target detection method based on efficient and generative adversarial network. *Engineering Applications of Artificial Intelligence: The International Journal of Intelligent Real-Time Automation* (Jun.), 132.
[7] A Ruhan, Shen Danyao, Liu Lijing, Yin Juanjuan, & Lin Renpu (2024). Hyperspectral anomaly detection based on a beta wavelet graph neural network. *IEEE MultiMedia.*
[8] Lin Y., Ren L., Yang X., & Yuan H. (2024). Design and application of bionic camouflage materials simulating spectral reflection characteristics of plants: a review. *Applied Sciences (2076-3417), 14 (11).*
[9] Xie Z., et al. (2024). Hygroscopic Mg-Al LDHs composite microspheres for highly efficient hyperspectral camouflage in the VIS and NIR wavebands. Sci. Rep. 14 (1): 15, 649.
[10] Yuan L., et al. (2020). Synthesis and fine spectroscopy tuning of the hyperspectral simulation material based on organic anions intercalated Mg-Al layered double hydroxide. Infrared Physics & Technology 107: 103, 328.

Design of a gear-rack biomass honeycomb coal forming machine

Tingzhou Ning[1], Zhanying Feng[1], Xuewu Qian[1*], Lei Zhang[2]

[1]Jinan Vocational College, Jinan, Shandong, China
[2]Geshi Town People's Government, Taian, Shandong, China

[*]Corresponding author: sdqxwu@163.com

Abstract: To compress and compact the crushed biomass material into a honeycomb shape and improve the combustion efficiency of biomass solidified fuel, a gear-rack biomass honeycomb coal forming machine was designed. This forming machine adopts a gear-rack meshing system, which can efficiently compress biomass materials such as sawdust, straw, and rice husks into standard honeycomb coal shapes, suitable for the production and processing of various biomass materials. The biomass solidification forming method proposed in this article has certain theoretical value and practical significance for promoting the promotion and application of biomass solidification forming equipment and fuels.

1. Introduction

Biomass cold-pressing forming technology, referred to as CZSN technology, compresses agricultural and forestry waste such as straw, weeds, shrub branches, and fruit shells into high-density fuel rods or pellets with a heat value ranging from 2850 to 4500 kcal at room temperature. This technology offers combustion efficiency four times higher than traditional methods, providing a clean, high-quality energy source that closely resembles coal in combustion characteristics and heat value while having almost no pollutant emissions. It presents an environmentally friendly, economical, and safe green consumption option for urban, rural, and industrial energy use [1-5].

Traditional biomass-forming machines mainly produce solid biomass fuels [6–10] (as shown in Figure 1). These solid fuels do not adequately come into contact with air during combustion, leading to incomplete burning and lower combustion efficiency. Therefore, a gear-rack biomass honeycomb coal-forming machine was proposed.

Figure 1. Solid Biomass Fuels.

Compared with traditional biomass solidification equipment, the gear-rack biomass honeycomb forming machine adopts a gear-rack compression structure [11–15]. With the cooperation of the forming device (forming barrel, push rod, die, etc.), the crushed agricultural and forestry waste is compressed and compacted into a honeycomb coal shape, thereby increasing the effective contact area between biomass solidification forming fuel and oxygen during combustion and improving combustion efficiency.

2. Forming machine structure

The forming machine consists of a drive device, crushing device, forming device, and other components. The specific components are shown in Figure 2.

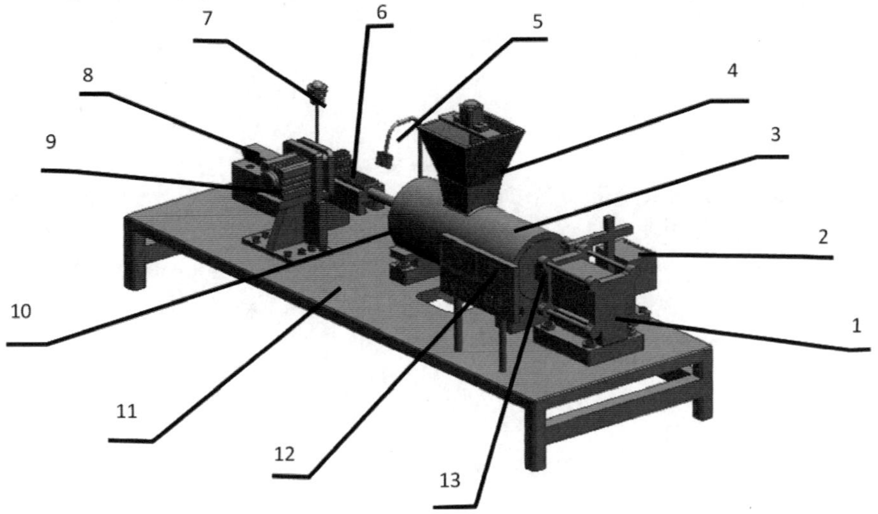

1-Individual Push Rod, 2-Sensor, 3-Forming Barrel, 4-Spiral Wall-Breaking Machine, 5-Infrared Monitoring Sensor, 6-Gear and Rack Device, 7-Multi-Layer Indicator Light, 8-Guide Rail, 9-Servo Motor, 10-Push Rod, 11-Workbench, 12-Control Panel, 13-Die

Figure 2. Structure of the Forming Machine.

3. Working principle of the forming machine

When the forming machine operates, biomass materials enter the spiral wall-breaking machine for crushing and refinement. The processed biomass material is then transported to the forming barrel. At this point, the push rod drives the pressing plate and the die in the forming barrel to perform a reciprocal motion. During this process, the individual push rod pushes the die from the rear end, causing the biomass material inside the forming barrel to gradually increase. The material is then compressed. When the pressure inside the barrel reaches a predetermined value, the control system intervenes, and the baffle inside the barrel automatically opens, allowing the honeycomb coal to slide out. Once the biomass honeycomb coal slides out, the infrared sensor automatically closes the baffle, and the forming machine continues processing until the biomass material is used up and the machine stops working.

4. Key component design

4.1 Drive device design

4.1.1 Drive device scheme selection and analysis. The following three mechanisms were analyzed:

- Mechanism 1 (as shown in Figure 3) is simple, with low manufacturing and maintenance costs, and offers ease of operation.

Journal of Physics: Conference Series **3009** (2025) 012032 doi:10.1088/1742-6596/3009/1/012032

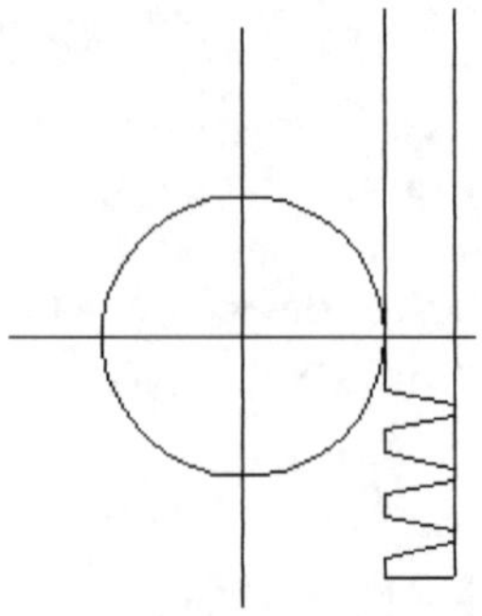

Figure 3. Gear-Rack Mechanism.

- Mechanism 2 (as shown in Figure 4) is more complex and involves more components, which increases the production cost and the risk of mechanical jams.

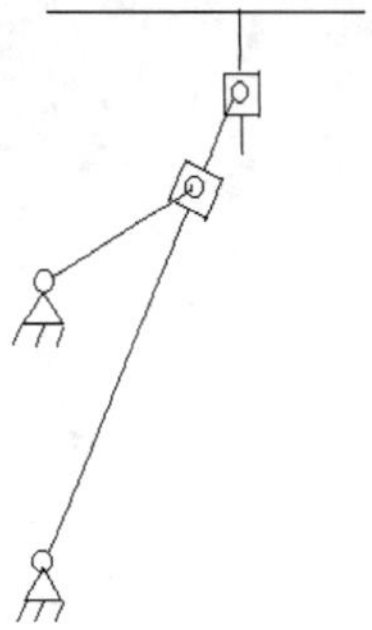

Figure 4. Push Rod and Slider Mechanism.

- Mechanism 3 (as shown in Figure 5) has a simple structure and low processing costs but requires vertical movement, which is not suitable for the intended use.

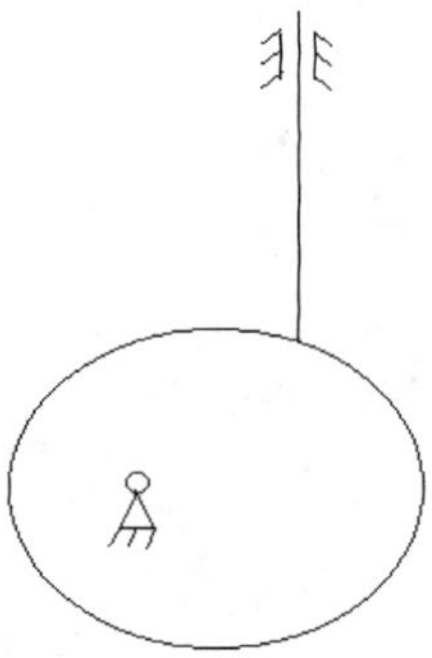

Figure 5. Cam Mechanism.

Based on the above analysis, mechanism 1 was selected due to its simplicity, cost-effectiveness, and ease of maintenance.

4.1.2 Drive device. The drive device is shown in Figure 6. Upon powering on, the forming machine receives a working signal, and the servo motor rotates, driving the gear rack to perform a linear compression motion. The rack and guide rail cooperate to produce reciprocating motion, allowing the forming machine to compress the biomass into honeycomb coal. The speed of the servo motor controls the forming speed, and the forward or reverse rotation of the motor controls the direction of movement.

Figure 6. Drive Device.

4.2 Crushing device design

Before processing the biomass into honeycomb coal, the biomass material must be crushed and refined. Therefore, a crushing device is added, consisting of a servo motor, auger, and hopper. The motor drives the auger to crush and refine the material, which then automatically enters the forming barrel, as shown in Figure 7.

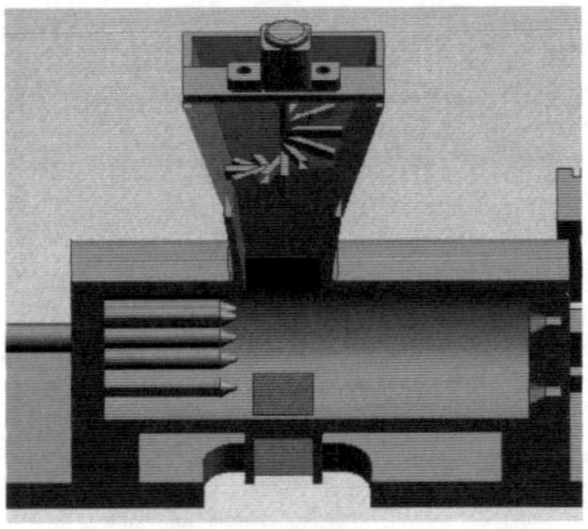

Figure 7. Crushing Device.

4.3 Forming device design

4.3.1 Forming device working principle. Once the gear-rack biomass honeycomb coal-forming machine is in operation, the biomass material is crushed in the crushing device. When the biomass particles enter the forming barrel, the push rod and die perform a dual-axis compression on the biomass particles. The push rod penetrates the biomass particles, shaping them into the honeycomb coal form (as shown in Figure 8). To improve combustion efficiency, ten punches are designed, each with a draft angle to facilitate the retraction of the push rod, thereby enhancing processing efficiency.

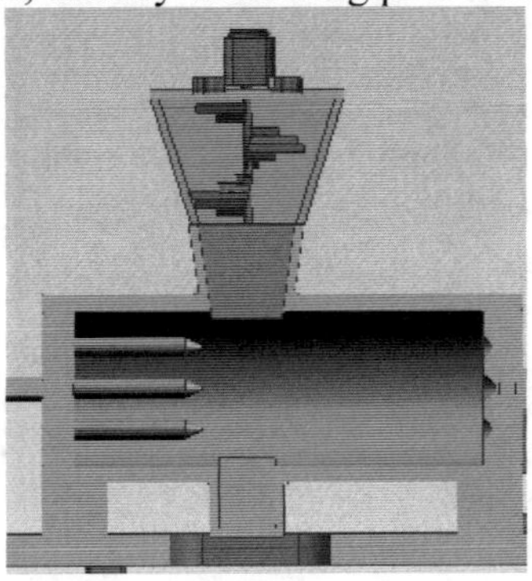

Figure 8. Forming Device.

4.3.2 Push rod. The push rod (as shown in Figure 9) is designed with good plasticity and bending performance. After heat treatment, it achieves high hardness and excellent mechanical properties. It is installed on the rack, and through gear-rack transmission, it performs compression movements. The needle portion of the push rod is designed with a 30° draft angle, facilitating the compression of the biomass and improving retraction speed, increasing processing efficiency and fuel yield.

Figure 9. Push Rod.

4.3.3 Die. The die, made of 305 stainless steel, is a key part of the forming machine. It has a sloped chamfer at the mouth for easier coordination with the push rod, improving forming efficiency. After forming, the honeycomb coal is expelled from the forming barrel by the force exerted by the die, facilitating subsequent sorting (as shown in Figure 10).

Figure 10. Die.

4.3.4 Pressure sensor. The pressure sensor (as shown in Figure 11) is installed inside the forming barrel. As biomass particles gradually accumulate, the pressure inside the forming barrel increases due to the pressing force. When the pressure reaches a preset value, the pressure sensor activates the controller, causing the baffle inside the forming barrel to automatically open. This allows the biomass honeycomb coal to slide out. Once the process is complete, the pressure sensor automatically resets.

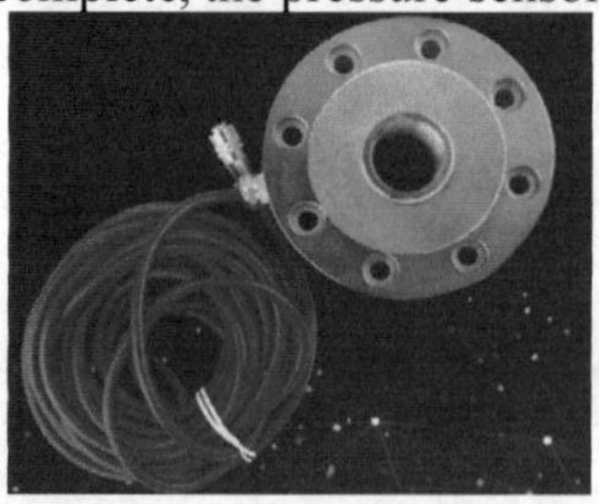

Figure 11. Pressure Sensor.

4.4 Other components

4.4.1 Individual push rod. The individual push rod is a new type of linear actuator, consisting of a motor-driven push rod and control mechanism (as shown in Figure 12). It is used in complex working conditions or mechanical equipment, providing advantages such as long-distance control, digital control, and intelligent control.

Figure 12. Individual Push Rod.

4.4.2 Multi-layer indicator light. The automatic operation function is a unique feature of the gear-rack biomass honeycomb coal-forming machine. To distinguish the working status of the forming machine, a three-color indicator light, also known as a three-color tower light, is designed. The three colored signal lights represent the following statuses:

- Red: Indicates a malfunction in the forming machine, such as the activation of the emergency stop switch or a limit fault in one of the working components.
- Yellow: Indicates a waiting state, meaning the machine is not processing and the spiral wall-breaking machine is without biomass material.
- Green: Indicates that the machine is running and processing biomass material into honeycomb coal.

The signal indicator can provide alerts to the operators, ensuring smooth communication regarding the machine's operating status. The multi-layer indicator light is shown in Figure 13.

Figure 13. Multi-Layer Indicator Light.

4.4.3 Guide rail. The guide rail, as shown in Figure 14, is installed on the workbench to facilitate parallel sliding motion. The rack's dovetail slot and the guide rail's dovetail slot work together to allow linear reciprocating motion, improving the efficiency of gear-rack transmission.

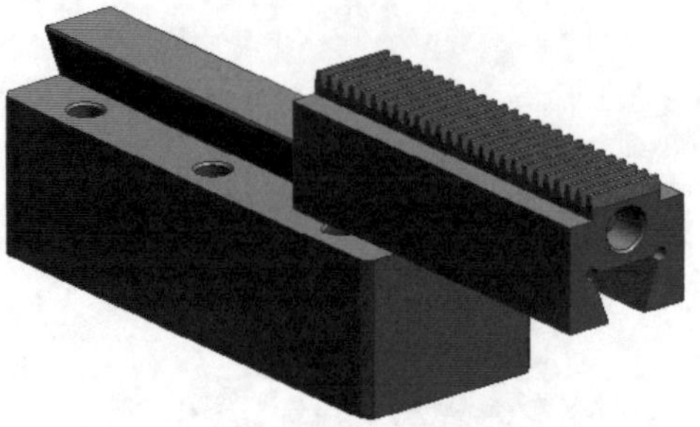

Figure 14. Guide Rail.

5. Conclusions

The gear-rack biomass honeycomb coal forming machine is an efficient, environmentally friendly, and energy-saving biomass solidification forming device. Compared to traditional biomass forming machines, it compresses biomass materials into honeycomb shapes, increasing the effective contact area with air during combustion and improving combustion efficiency. This has certain theoretical value and practical significance for promoting the promotion and application of biomass solidification forming equipment and fuels.

References

[1] Mingli Di, Yunkui Huang, Yuefeng Chen, et al. "Biomass molding technology and design of LH-600 biomass granulation molding machine,"Agricultural Engineering,2024,14(1):89-94.
[2] Jiaping Hou, Wen Wang, Leixin Zhang, et al. "Modern biomass energy technology system and its industrial application trend,"Modern Chemical Industry,2022,42(5):7-13.
[3] Tongqiong Xiang, Liming Zhou, Banghua Zheng, et al. "Analysis of Energy Efficiency and Energy-Saving and Emission Reduction Technology of Biomass Briquette Boiler,"China Special Equipment Safety,2022,38(4):82-86.
[4] Yishui Tian, Ming Dan, Geng Kong, et al. "Development Strategy of Biomass Economy in China," Strategic Study of CAE,2021,23(1), 133-140.
[5] Yunusa, S. U., Mensah, E., Preko, K., et al. "A comprehensive review on the technical aspects of biomass briquetting," Biomass Conversion and Biorefinery,2024,14(18),21619-21644.
[6] Wei Yang, Shilong Feng, Shanzhi Xin, et al. "Analysis of Particulate Matter Emission Characteristics of Commercial Biomass Pellet Combustion,"Journal of Zhengzhou University(Engineering Science), 2024,45(6):137-142.
[7] Yuanqing Yang, Qiankun Sun, Hongliang Jian, et al. "Optimization analysis of hot pressing forming process solid fuel from sophora japonica forest waste,"Acta Energiae Solaris Sinica,2024,45(3):178-185.
[8] Rong Liu, Haifeng Luo, Dengyu Xiong, et al. "Research status of biomass combustion technology,"Agricultural Engineering and Equipment,2022,49(1):8-14.
[9] Daixiao Wei, Yatai Men, Yaojie Li, et al. "Environmental health benefits and cost analysis of using biomass pellets for space heating,"Ecology and Environmental Sciences,2024,33(6):927-934.
[10] Sarker, T. R., Nanda, S., Dalai, A. K., et al. "A review of torrefaction technology for upgrading lignocellulosic biomass to solid biofuels," BioEnergy Research,2021,14,645-669.
[11] Jing Liu, Dong Liang, Liu Yang, et al."Research on damping mechanism of cable with pinion rack type inertial viscous damper," Engineering Mechanics,2023,40(1):155-167.
[12] Bin Meng, Hailiang Xu, Chunhe Wang."Design simulation and experimental research on the front wheel turning mechanism with rack and pinion,"Machine Design and Manufacturing Engineering,2023, 52(10):34-40.
[13] Feng Li, Shaohua Wang, Xie Xie. "Guiding device optimization for truss framed-gear rack type spud leg," Ship Science and Technology,2024, 46(3):121-125.
[14] Shouying Li, Chaoyu Huo, Weiyang Mao, et al. " FEM Numerical Study of Eddy Current Damping-Rack and Gear Wall,"Journal of Hunan University(Natural Sciences), 2022, 49(9):100-107.
[15] Nicolalde, J. F., Martínez-Gómez, J., Vallejo, J. "Multicriteria Decision Making of a Life Cycle Engineered Rack and Pinion System," Processes,2022,10(5), 957.

Journal of Physics: Conference Series **3009** (2025) 012033 doi:10.1088/1742-6596/3009/1/012033

Design and development of multi-dimensional adaptive soft cigarette packaging forming system

Ling Zhang, Jiang Ma, Chengbo Yang, Xiang Xu, Sen Yang, Jinjiang Han, Jun Wang, Qingping Wang and Mao Zheng*

Hongta Tobacco (Group) Co., Ltd., Yuxi, 653100, Yunnan, China

*Corresponding author's e-mail: 973340521@qq.com

Abstract: In recent years, with the prominence of personalized consumer demand, tobacco industry enterprises in order to meet the market demand, urgently need a flexible intelligent production line for cigarette special-shaped packaging, in order to better achieve the purpose of rapid response to market demand. At present, the packaging of cigarette products of different sizes and specifications in cigarette factories is mainly realized by manual class, which has the problems of low production efficiency and poor product quality consistency. In order to overcome the above problems, this paper aimed at the molding process of small box packaging of cigarette products, using the concept of flexibility, through modular design, developed and produced a set of production systems that can meet the requirements of different sizes and specifications of cigarette product packaging. The packaging forming system includes cigarette group arrangement, inner lining paper forming, inner frame paper forming, trademark paper forming, etc. At the same time, it involves the supply process of auxiliary materials required by various processes. The main innovation of this paper is that it can meet the requirements of various size specifications of cigarette products, and can realize the rapid switching between different size specifications of cigarette products. Through the modular design concept, a set of flexible intelligent manufacturing systems for cigarette-shaped packaging is designed, which realizes the production speed > 10 packets/min, and realizes the rapid switching of different specifications and packaging methods, so as to meet the needs of the group brand in the corresponding market quickly. At the same time, compared with other competing enterprises, it maintains the advantages of shaped and flexible packaging, highlighting the technical strength of the enterprise and reflecting the brand value.

1. Introduction

Cigarette specifications include differences in the diameter of the cigarette and the number of cigarettes in a single pack. Cigarette diameter classification: coarse branch $\varphi > 7.8$mm, medium branch φ6-7.3mm, fine branch φ5.4-6mm, ultra-fine branch $\varphi < 5.4$mm, the number of cigarettes in a single box: 5, 10, 16, 20, X+l, etc. Packaging methods include fully open, side open, drawer type, double inner package, multiple inner packages, and so on.

At present, there are three ways to achieve special-shaped packaging: ① To buy a traditional cigarette packaging machine to transform and produce special-shaped packaging cigarettes. ② To directly purchase the existing special-shaped packaging machine. ③ To design and develop a flexible intelligent manufacturing system for cigarette special-shaped packaging [1-3].

To transform the traditional cigarette packaging machine, it needs to invest 40-50 million yuan to achieve good results, and it can not adapt to the production of special-shaped packaging with multiple specifications and various packaging methods. One of the ways is the direct purchase of the existing special-shaped packaging machines such as the double aluminum packaging machine ZB418 jointly developed by Shanghai Tobacco Machinery Co., LTD and Shanghai Tobacco Group, and the special-shaped packaging machine of GIMA TT in Italy. A set requires more than 60 million coins, and ZB418 has a three-year technical protection period to block cigarette companies. Italy GIMA TT special packaging machine is more expensive, only the delivery time is more than a year. If you want to convert the packaging specifications and packaging methods, each set will cost nearly 10 million more.

In terms of mold design and manual completion of special-shaped packaging, the technology of Yuxi Cigarette Factory of Hongta Group is relatively mature. Through the design, manufacturing, and assembly of the mold of each link of cigarette packaging, Yuxi Cigarette Factory has formed a certain scale of manual cigarette packaging line and obtained related patents. In terms of the design and manufacture of special-shaped cigarette packaging equipment of specific specifications, Shanghai Tobacco Machinery Co., Ltd. and Shanghai Tobacco Group jointly developed a set of double-aluminum packaging machine ZB418 equipment. The cigarette size of ZB418 is φ6.36 ×89mm, the cigarette arrangement is (5-5) ×2, the single aluminum paper size is 208 ×50mm, and the paper size of the inner frame of the single is 35.9×l mm. The shape of the hard box is 94×66.2×14.7mm, and the shape of the hard strip is 31×333×95mm [4]. In summary, the research on the flexible production of special-shaped packaging outside the industry is mainly limited to the packaging process of cigarette packaging that does not involve cigarette packaging. Domestic research on flexible production of irregular packaging is mainly limited to the transformation of existing equipment and the development of manual molds. There is a research gap in the development of fully autonomous flexible production platforms for irregular packaging with complete process flow.

2. Design objective

This paper urgently needs to solve the problem of designing a set of flexible production systems to meet different sizes and specifications of cigarette products, so the design objective of the device is shown in Table 1.

Table 1. The design objective of a flexible production system for cigarette-shaped packaging.

(1) To meet the cigarette factory full-size specifications of cigarette products small box packaging molding;
(2) The production speed of the product is not less than 10 packets/minute.

3. Overall process route design

3.1 Technical route of the flexible production system

The formulation of the special-shaped small box flexible packaging process, referring to the traditional machine tool small box packaging process, combined with the requirements of multi-specification flexibility, respectively designs the smoke warehouse part, lining part, inner frame forming part, and trademark forming part of the process, using synchronous transportation or asynchronous double transportation of semi-finished products to connect each part of the process. A complete set of special-shaped small box packaging processes is formed. Under the guidance of the overall process route, the main research and development of the small box flexible packaging system can be divided into four main functional modules [5]: cigarette storage part, lining forming part, inner frame forming part, trademark forming part, plus intermediate product conveyor belt auxiliary module part, with a total of five design links.

According to the R&D consensus of "Total score→total score" and "Key breakthrough→radiation periphery→correlation summary", the total R&D technical route is set as shown in Figure 1.

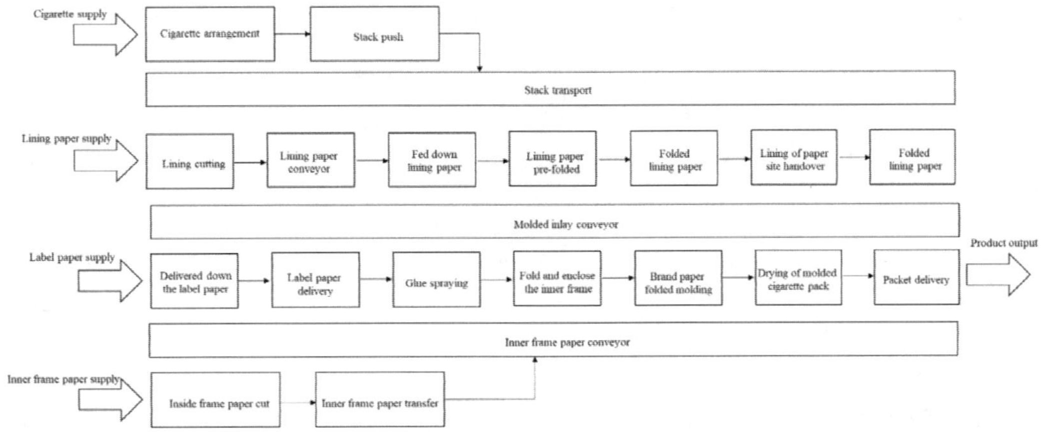

Figure 1. Technical route of the flexible production system.

3.2 Component of flexible production system

According to the flexible packaging process of special-shaped small boxes and the requirements of multi-specification flexibility, the functional modules of the flexible production system for cigarette special-shaped packaging are divided into a tobacco storage part, an inner lining forming part, an inner frame forming part, and a trademark forming part. The semi-finished product conveying method of synchronous or asynchronous double conveying is used to connect the various parts of the process, forming a complete special-shaped small box packaging production system, as shown in Figure 2.

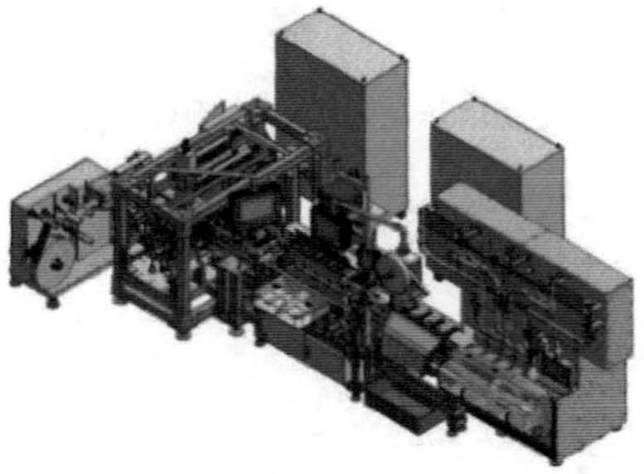

Figure 2. Composition of flexible production system for irregular packaging.

4. Design of each functional module

4.1 Design of smoke storage section

On the basis of inheriting the traditional vertical storage capacity reduction screening method, the tobacco warehouse will make several targeted innovations according to the requirements of flexible multi-specification adaptability. The backbone architecture adopts dual material libraries in parallel and four independent channels for output. In terms of compatibility, the tobacco storage unit needs to have sufficient adaptability to changes in cigarette specifications, and should also have multiple specifications for arranging and outputting cigarette packs [6,7]. After summarizing and analyzing, the project team has determined the flexibility range of this part as follows: ① Acceptable cigarette specifications range cigarette length L45~L100 mm, cigarette diameter φ 5.4~φ 7.8 mm. ② Output cigarette group arrangement modes: 1 × n single row, 2 × n double row, 3 × n triple row, n-m-n triple row, etc; The range of arrangement of cigarettes and cigarette packs above ($2 \leq n \leq 10$, m=n-1) basically covers the characteristic categories of current mainstream cigarette products. The schematic diagram of the tobacco warehouse is shown in Figure 3.

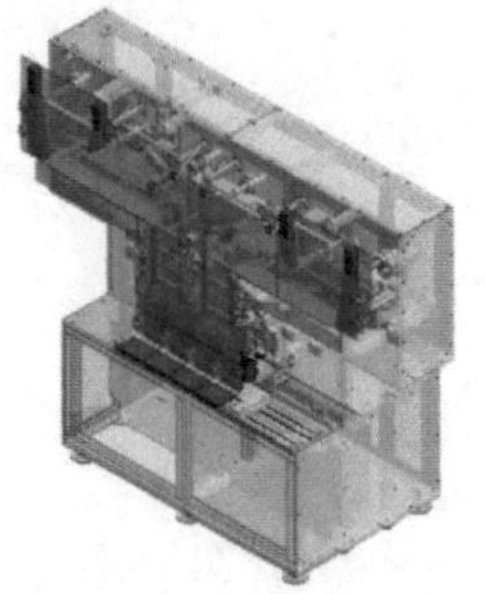

Figure 3. Diagram of the tobacco warehouse

4.2 Design of inner lining packaging forming part

According to the requirements of flexible multi-specification adaptability, referring to the traditional mechanized forming process of small box inner packaging, and determining the benchmark selection of the inner packaging forming station in the length, width, and height directions [8,9], the inner lining small packaging forming process flow is set as shown in Figure 4. The lining materials are generally in the form of rolls. This part of the process flow includes unwinding the inner lining roll material, cutting to a fixed length, cutting with a zipper, and transporting it in place according to a 4-channel design, as shown in Figure 5.

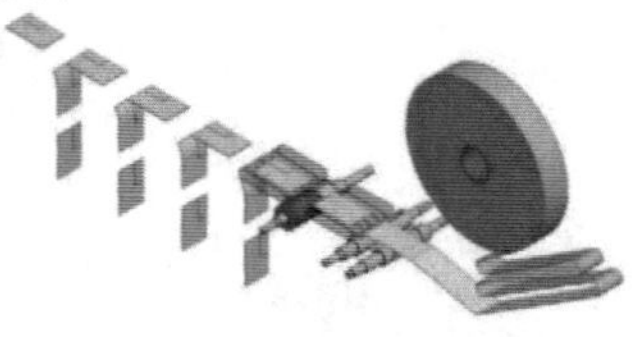

Figure 4. Standard lined four-channel paper feeding process flow.

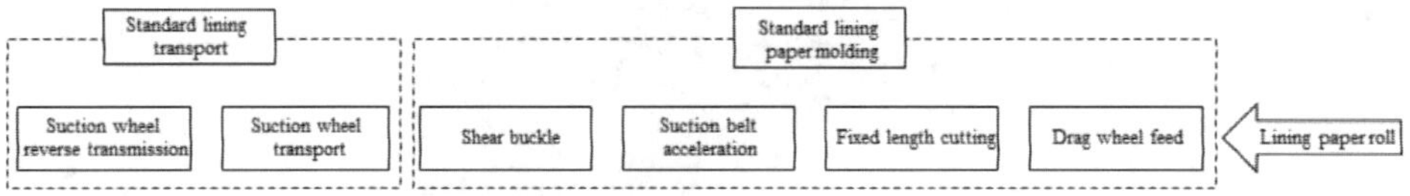

Figure 5. Standard lining paper four-channel supply process.

The transition unit of the smoke group is designed as a transition station, as shown in Figure 6. The function is to synchronously push the (four-channel) boxed cigarette group transported from the cigarette storage unit, and prepare for the subsequent integration of the lining and the cigarette group into packets. One station folding unit of the inner lining package accepts the smoke pack and standard inner lining accessories pushed in front, as shown in Figure 7. Through the combination of a series of folding actions, the fusion of the smoke group and the lining is completed. Because there is action nesting with the subsequent cross station, the action nested into this station will be inserted according to the action sequence for convenience of description.

Figure 6. Operation principle and mechanism of transition unit of the cigarette group.

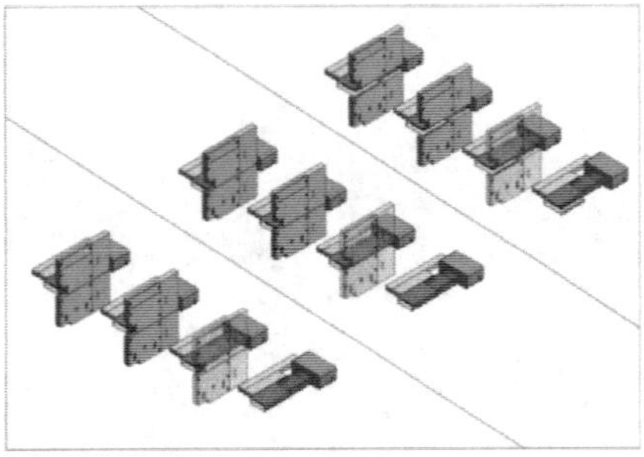

Figure 7. Lining lining-folding action principle.

4.3 Design of inner frame forming part

In order to ensure the continuity of the supply process of the inner frame paper, the inner frame paper of the hard-wrapped trademark is generally supplied by the reel type, and the corresponding counterweight overwheel is set between the reel and the feed port to maintain a certain tension during the supply process of the inner frame paper, as shown in Figure 8.

Figure 8. A diagram of inner frame paper supplies the cutting mechanism.

In the process of inner frame coil conveying, it is necessary to roll two indentation lines at the right position in the middle and pre-fold at a small Angle on both sides of the edge to ensure the profile quality for the subsequent folding of the inner frame of the sheet [10]. The indentation line and the fold edge are realized by matching the roller with the corresponding hobbing cutter and the folding Angle, and the roller cutting (roller pressing) is formed, as shown in Figure 9.

Figure 9. Structure diagram of paper indentation mechanism in the inner frame.

The folding process of the inner frame paper is shown in Figure 10. When the main manipulator grabs the bag, the inner frame paper is picked up by the negative pressure floppy disk and absorbed by the inner frame paper. When the folding mechanism of the two-way linear module completes the folding, the inner frame paper is transferred to the upper part of the soft inner envelope mold box through the transmission of the connecting rod mechanism.

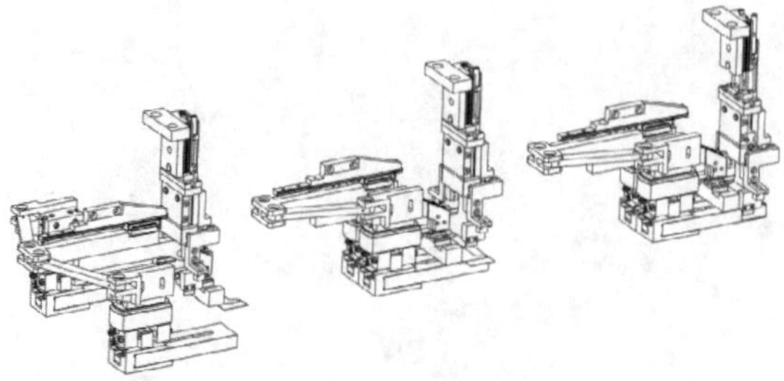

Figure10. Inner frame paper folding process.

4.4 Trademark forming part design

Firstly, based on the flexibility requirements of the design and referring to the traditional small box trademark packaging method, the benchmarks for the length, width, and thickness directions of the trademark are set. The folding process of the trademark is designed as shown in Figure 11.

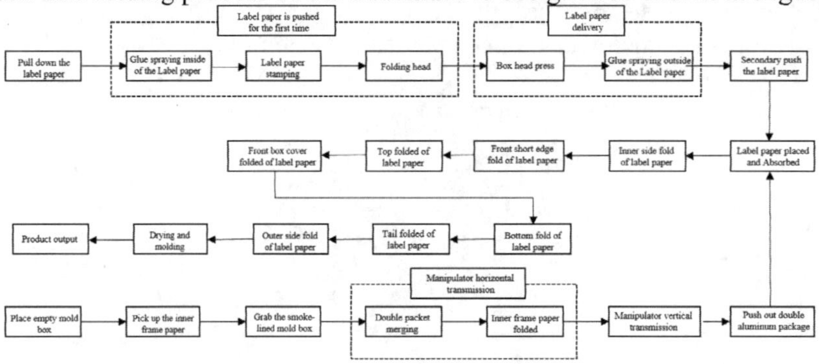

Figure 11. Label paper folding action flow.

According to the trademark forming process, the trademark forming part is divided into a trademark conveying unit, a soft inner package picking transfer unit, and a trademark folding forming unit. Specifically, the trademark conveying unit includes the trademark paper unit, the trademark pushing unit, the embossing unit, the folding box head unit and the glue-spraying unit, the master manipulator unit includes the master hand horizontal lifting unit, the soft inner wrapping unit and the inner frame picking folding unit, the trademark folding forming unit includes the trademark folding unit, the auxiliary manipulator wrapping unit and the drying conveying unit [11,12]. Figure 12 shows the functional unit division of the trademark forming part.

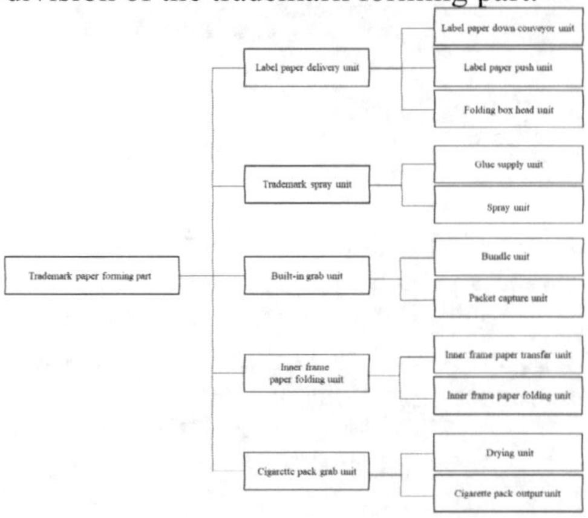

Figure 12. Trademarks form part of the functional unit division.

The main functions of the trademark folding part are the folding process and the drying molding process. As shown in Figure 13, the folding process is mainly divided into trademark adsorption in place, trademark inner side folding, trademark front short edge folding, trademark top folding, trademark front cover folding, trademark back short edge folding, trademark bottom folding, trademark back cover folding, trademark outer side folding, and trademark drying molding.

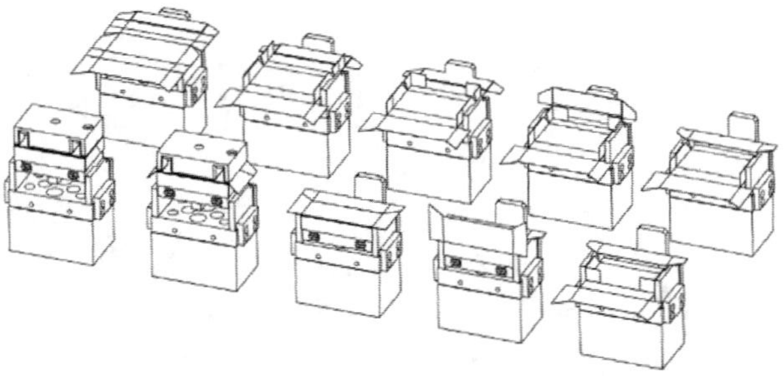

Figure 13. Folding process of Trademark.

The material transfer mechanism of the whole special-shaped packaging cigarette product packaging system is shown in Figure 14, which mainly includes the transfer of the mold box of the cigarette group, the transfer of the soft inner bag, and the transfer of the inner frame paper. The whole material transfer is converged in the forming part of the trademark paper, and the molding of the cigarette pack is finally completed [13]. In order to improve the operating efficiency of the packaging system, the transmission device is designed as a double conveyor belt circulation.

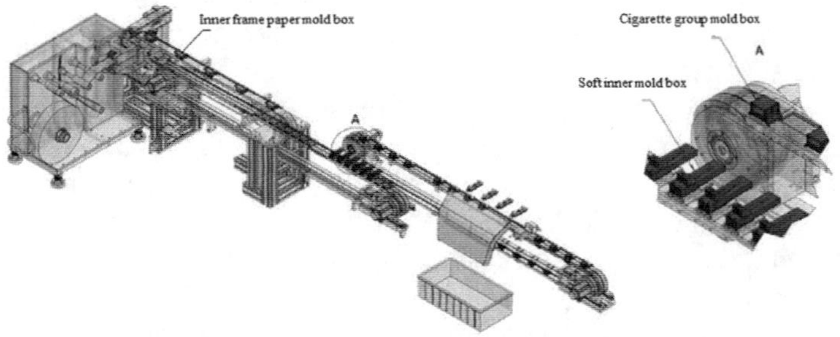

Figure 14. Schematic diagram of material transfer mechanism of special-shaped packaging system.

5. Control system design

5.1 Electrical control system design

According to the division of the functional modules of the special-shaped packaging flexible production system, the electrical control system is designed for each module, and the overall control of the flexible production system of heterogeneous packaging is realized through the communication of the control signals of the connecting actuator of each functional module [14]. The electrical control diagram of the flexible production system is shown in Figure 15.

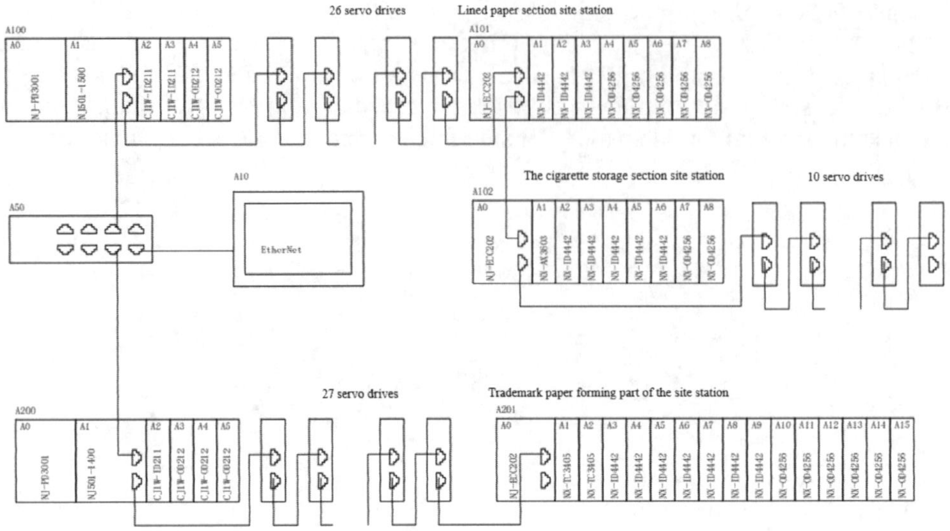

Figure 15. The electrical control diagram of the flexible production system.

There are two types of networks in the whole system: the yellow line in the figure represents the industrial fieldbus EtherCAT network, which realizes high-speed data exchange between the controller the servo drive, and the field IO. The blue line represents industrial Ethernet, which enables communication between the touch screen and the controller, and a small amount of interlocking data exchange between the two controllers.

The system is composed of three sections: cigarette storage, lining packaging, and trademark packaging. The control system uses two controllers to complete the control function, and one controller is used to control the cigarette storage section and lining packaging section.

The control procedures of the tobacco storage section and the lining packaging section are logically separated, using their own variable space. If these two sections need to be taken out separately to connect with other devices, as long as a controller is added to the smoke depot section, the control program can be split and used with little setting adjustment.

For the convenience of debugging, the system provides two touch screens as man-machine interfaces. One of them uses the same system as the controller, which can be quickly configured or modified to meet the needs of repeatedly modifying the process and picture of new product development; The other one uses an industrial computer, which can be expanded in the future to do database, formula management, ERP, MES system interface.

The wiring methods of each servo driver and motor are basically the same, as shown in Figure 16.

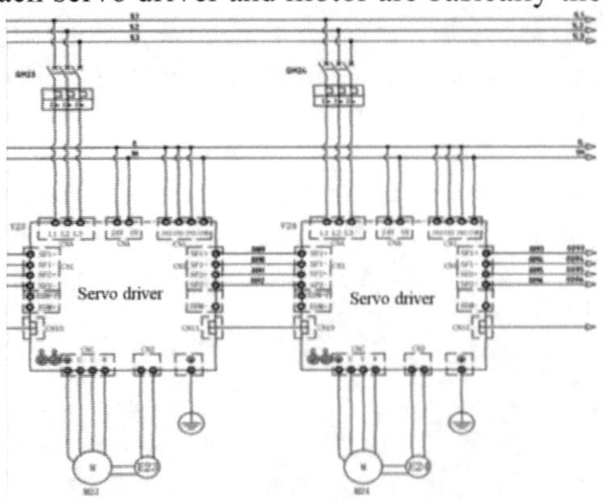

Figure 16. Servo motor wiring diagram.

5.2 Pneumatic control system design

The hardware components of the gas control system mainly include positive pressure solenoid valve, negative pressure solenoid valve, positive pressure air source, negative pressure air source, terminal, pressure control dial, gas exhaust, pressure gauge, pressure regulator, oil mist, muffler, double-acting cylinder, etc. The gas path control principle is shown in Figure 17.

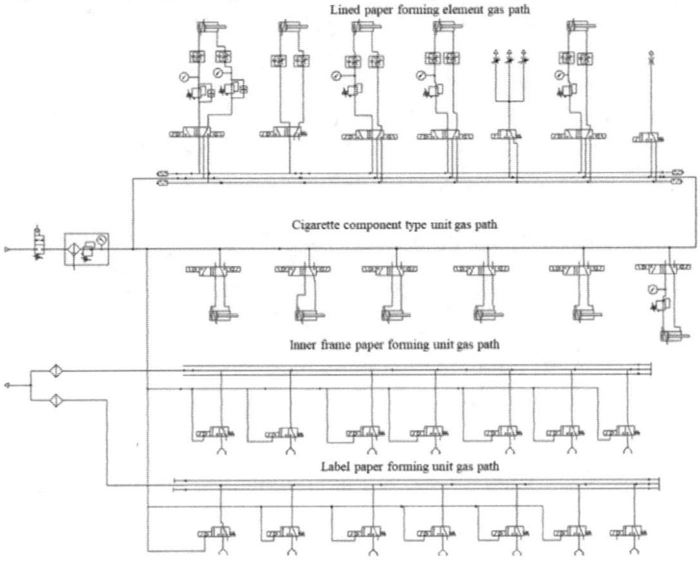

Figure 17. Gas path design of packaging molding system.

6. Experimental verification

After completing the installation and debugging of the equipment, material testing is carried out, which is divided into two stages: testing each part separately and testing the whole machine's linkage with the material. Figure 18 shows a picture of the material testing site for each part.

Figure 18. Combined debugging and testing with materials.

Materials: "MC" heating and non-burning new tobacco products (specification 7.3 mm × 45 mm) cigarettes, lining paper, inner frame paper, and label paper [provided by Yuxi Cigarette Factory of Hongta Tobacco (Group) Co., Ltd.];

Device: Self-developed heating and non-burning label paper folding molding device with double aluminum package.

Methods: After debugging, the self-developed brand folding molding device for heating and non-burning double aluminum packages is operated at a production speed of 10 packages/min. According to GB5606.3-2005 "Cigarettes Part 3: Packaging, Rolling Technical Requirements, and Storage and Transportation" [15], the appearance quality of cigarettes is tested, and the qualified rate of cigarette packaging is counted. The testing period is 3 months, and the average value is taken.

It can be seen from Table 2 that the qualified rate of appearance quality of label paper packaging of heated non-burning cigarette double aluminum package produced by self-developed heating and non-burning double aluminum package label paper folding molding device is 96.8% on average, which can meet the production demand.

Table 2. Quality qualification rate of special-shaped cigarette products.

No.	Output/Piece	Qualification Rate/%
1	746	95.4
2	725	96.6
3	719	98.4
Average	730	96.8

7. Conclusion

This paper uses the flexible concept and modular design to complete the design and development of the flexible production system of heterogeneous cigarette packaging, mainly designing the cigarette group arrangement, lining forming unit, inner frame paper forming unit, and trademark forming unit. The production speed reaches 10 packets/min. The specific innovations are as follows:

(1) We completed the independent research and design of a flexible production system of cigarette special-shaped packaging in the domestic cigarette production field. Through the independent design of hard box small box-shaped packaging flexible production process route and technical demonstration, a set of flexible production systems of shaped packaging that can meet the requirements of various packaging methods and sizes, and can realize the rapid switching of different sizes and specifications is designed and manufactured.

(2) The design and manufacture of a complete set of flexible production systems for hard and small boxes of cigarettes was realized. We complete the design, manufacturing, and integration of five modules (cigarette storage part, lining soft bag forming part, inner frame forming part, trademark forming part, conveying part), and realize several flexible features of hard box small box production. The length and thickness of the cigarette can be adjusted continuously, the arrangement of cigarette group can be selected in multiple specifications, the length, width, and height of the lining soft bag can be adjusted continuously, and the trademark packaging has a variety of packaging methods and the size specifications can be adjusted continuously, and the molding size of the inner frame can be adjusted continuously.

(3) Design and application of cigarette storage structure with double flavor cigarette output capacity. The structure of the system is a parallel double library, which can store one flavor of cigarettes. When producing the double-packed small box of cigarette products, with the logic control of the push of the cigarette channel, the double-packed double flavor can be realized

(4) An asynchronous double conveyor belt wheel press box device is developed and used, which supports the operation of two coaxial conveyor belts in an asynchronous manner (that is, not synchronous). By means of alternating docking, the problem of wake-matching between the smoke storage unit and the lining unit and the execution actions related to the conveyor belt is eliminated, and the flexibility of their respective actions is realized.

Journal of Physics: Conference Series **3009** (2025) 012033 doi:10.1088/1742-6596/3009/1/012033

References

[1] Lu Yuzhen, Ma Li, Lv Xiaobo, et al. Design of cigarette packet turning device for YB65A cartoner [J]. Tobacco Science and Technology,2019,52(8):95-100.
[2] Zhang Hao, Li Shoucheng, Lv Xiaobo, et al. Design and application of automatic cigarette packet collating and conveying device for YB55 overwrapper [J]. Tobacco Science and Technology, 2022,55(6):106-111.
[3] Yang Mingquan, Lv Xiaobo, Pan Yonghua, et al. Design of folding and wrapping device for the inner lining paper of FOCKE350 Packaging Unit [J]. Tobacco Science and Technology, 2019,53(1):107-111.
[4] Su Qi, Wu Jianjun. Design of inner liner cutting device for YB418 double aluminum foil cigarette packer[J]. Tobacco Science & Technology,2017,50(6):81-87.
[5] Li Kun. Flexible Improvement of Cigarette Packaging Production Line and Packaging Equipment [J]. Packaging and Food Machinery,2019,35(6):45-48.
[6] Lu Zhi-Min, Zhang Xiao-Hang, Su Ling, et al. Design of Flexible Connection System for YF17A Cigarette Storage and Conveying Device [J]. Tobacco Science and Technology,2023,56(6):108-112. (in Chinese)
[7] Sun Baotong, Du Weidan, Zhu Yixi, et al. Design of automatic positioning and splicing System for Lining paper in YB48 Hard Box Packaging Machine [J]. Tobacco Science and Technology,2024,57(11):82-87.
[8] Tan Shiwei, Fan Zhiyuan, Wu Jianjun, et al. Design of automatic Filling device for Lining Paper cigarette bag of YB417C Side Open Hard Box Packaging Machine [J]. Tobacco Science and Technology, 2019,57(7):97-103.
[9] Huang Yi-Qing, Cao Wen-Zhi, Huang Fu, et al. Application of single-head transfer lining paper combination folding Technology in GDX2 packaging unit [J]. Tobacco Science and Technology,2011(6):20-22,38.
[10] Zhu Tiantian. Optimization Design and Simulation Analysis of the Main Mechanism of Carton Conveying Molding Machine [D]. Hubei: Hubei University of Technology,2021.
[11] Liu Yongbin, Zhang Jiangtao, Gao Yan, et al. Design of Microbending device for the inner tongue of YB45 Packaging Machine [J]. Packaging Engineering,2024,45(17):135-141.
[12] Hu Jingchen, Meng Kun, Yin Guohui. Design and Application of Preforming Device for FOCKE350 Middle Equipment [J]. China Machinery,2023(11):7-11.
[13] Li Rui, Tian Chengxin, Wang Hang, et al. Research on Mechanical Design and Key Technology of Automatic Production Line of Circuit Board Special-shaped Automatic plug-in System [J]. Automation & Instrumentation,2019(1):50-52.
[14] Jiang Zi-Lan, XU Shi-Xu, ZHANG Hao-Lin. Control System Design of Pillow Packing Machine based on electronic CAM flying shear [J]. Packaging Engineering,2023,44(13):197-207.
[15] Standardization Administration of China. GB5606.3-2005 Cigarette-Part 3: Technical Requirements for packing, making, storage, and transportation [S].

Multi-objective optimization of CNC milling parameters based on MOEAD

Xuemin Wang[1], Sen Jiang[2], Yang Shen[1], Tianlong Qian[1], Shulong Mei[3], Yiqun Dai[3], Yang Xie[*3] and Jinfeng Liu[3]

[1]COSCO SHIPPING Heavy Industry (Yangzhou) Co., Ltd., Yangzhou, China
[2]Ningbo Yi Ji Value Manufacturing Co., Ltd., Ningbo, China
[3]School of Mechanical Engineering, Jiangsu University of Science and Technology, Zhenjiang, China

[*]Corresponding author: xiey_just@just.edu.cn

Abstract: This research addresses the inherent conflict between energy efficiency and productivity in CNC milling by developing a dual-objective optimization framework. The model simultaneously minimizes machining power and specific cutting energy through strategic adjustment of spindle speed, feed rate, cutting depth, and milling width parameters. Employing the MOEAD algorithm, the study successfully identifies Pareto-optimal parameter configurations that balance energy conservation with operational effectiveness. The testing results show that using the optimal combination of milling settings reduces the machine tool's milling power and cutting ratio by 10.5% and 26.27%, respectively, as compared to the original values.

1. Introduction

As a fundamental industrial process in critical sectors including aerospace engineering, automotive manufacturing, and precision tooling, CNC milling necessitates rigorous optimization of operational variables. Critical operational variables including spindle rotation rate, feed velocity, cutting depth, and tool engagement width substantially affect both power utilization and specific cutting energy during material removal processes [1]. Conventional parameter determination methodologies based on empirical operator experience exhibit limitations in addressing multi-dimensional optimization requirements, particularly when reconciling conflicting technical and economic objectives.

Scholars have extensively explored multi-objective optimization strategies for machining parameters through diverse methodological approaches. In addressing surface quality, productivity, and energy efficiency. Li's team [2] developed an optimization framework integrating surface roughness, material removal rate, and energy consumption metrics, employing an enhanced teaching-learning optimization mechanism. Zhao's research [3] established a power consumption prediction model incorporating cutting variables and tool characteristics, subsequently applying NSGA-II evolutionary computation to resolve temporal-energy tradeoffs. Feng et al. [4] formulated an energy-surface integrity bi-objective optimization architecture utilizing Pareto dominance principles for parameter selection. Expanding optimization dimensions, another Li-group study [5] implemented an elitist genetic algorithm to reconcile production efficiency, environmental impact, and surface finish requirements within a tri-objective optimization paradigm. These investigations collectively demonstrate the academic community's progression from conventional single-objective optimization towards sophisticated multi-criteria decision-making frameworks in advanced manufacturing contexts.

Focusing on milling operations as the research subject, this study establishes a dual-objective optimization framework that integrates machining power and specific cutting energy metrics, while incorporating practical manufacturing constraints. The investigation employs a decomposition-based multi-objective algorithm to systematically explore the solution space and identify Pareto-optimal parameter configurations under operational constraints. This methodological approach enables comprehensive analysis of energy-efficiency tradeoffs while maintaining compliance with practical machining requirements.

2. Multi-objective optimization model

2.1 Milling power of machine tools

The whole process of machine tool milling includes standby stage, air cutting stage and milling stage, and the total power P_c during milling processing can be expressed as:

$$P_c = P_{stand} + P_{air} + P_a + P_{cutting} \tag{1}$$

where P_{stand}, P_{air}, P_a, and $P_{cutting}$, respectively, represent standby power, air cutting power, loss of power, milling power, W.

Standby power P_{stand} is related to the performance of the machine tool and is a constant. P_{air} is the power of the machine when milling has not started and can be expressed as follows:

$$P_{air} = f(n) + g(v_f) \tag{2}$$

where both $f(\cdot)$ and $g(\cdot)$ represent quadratic relationships.

The loss power P_a is related to the milling power $P_{cutting}$ [6], which can be expressed according to the empirical formula:

$$P_a = b_m P_{cutting} \tag{3}$$

$$P_{cutting} = Kn^{-w} v_f^{\,y} a_p^{\,x} a_e^{\,u} \tag{4}$$

where b_m is the power loss coefficient; K is the correction coefficient; x, y, u and w are the index of the influence degree of process parameters.

The total energy consumption P_c in the milling stage is expressed as:

$$P_c = P_{stand} + f(n) + g(v_f) + (b_m + 1)Kn^{-w} v_f^{\,y} a_p^{\,x} a_e^{\,u} \tag{5}$$

2.2 Modelling of specific cutting energy

Within CNC milling operations, specific cutting energy (*SCE*) serves as a critical metric for assessing energy utilization efficiency during material removal processes. This parameter quantitatively evaluates the energy expenditure per unit volume of machined material, serving as a fundamental parameter for process optimization strategies aimed at enhancing energy-efficient material removal operations. The systematic analysis of *SCE* enables manufacturers to identify energy-saving potential while maintaining required production specifications[7]:

$$SCE = \frac{P_c \times T}{V_m} \tag{6}$$

In the above equation, T is the milling time, s, $T=L/v_f$, L is the machining length of the workpiece, mm; V_m is the volume of the resected material, mm^3, $V_m = L \times a_p \times a_e$.

2.3 Milling constraints

In actual processing, the selection of milling parameters is often subject to the constraints of milling conditions, mainly including milling power constraints and milling parameter constraints. Milling power constraints refer to the milling power in the milling process that cannot exceed the rated power of the machine tool spindle, that is:

$$Kn^{-w} v_f^{\,y} a_p^{\,x} a_e^{\,u} \le \eta P_{\max} \tag{7}$$

where η is the milling efficiency; $P_{\max}$ is the spindle rated power, W.

The milling parameter constraints are mainly the constraints of the optimization variable milling parameters, i.e:

$$\begin{cases} n_{\min} \le n \le n_{\max} \\ v_{f\min} \le v_f \le v_{f\max} \\ a_{p\min} \le a_p \le a_{p\max} \\ a_{e\min} \le a_e \le a_{etool} \end{cases} \tag{8}$$

In the above equation, $n_{\min}$, $n_{\max}$, $v_{f\min}$, $v_{f\max}$, $a_{p\min}$, $a_{p\max}$, $a_{e\min}$ and a_{etool} represent the maximum and minimum spindle speed, feed rate, milling width and milling depth respectively..

Combining the above energy consumption and specific cutting energy models as well as the constraint functions, the multi-objective optimization model for this study can be obtained as shown in Equation (9):

$$F\left(n, v_f, a_p, a_e\right) = \left(\min P_c, \min SEC\right) \tag{9}$$

3. Moead optimization algorithm

A decomposition-based method based on ideal points and scalarized subproblems with Chebyshev weight functions is the MOEAD optimization algorithm. The core idea is to treat the objective function of a multi-objective optimization problem as multiple objectives in a multi-dimensional space. By defining a weight vector, the algorithm combines each objective function with the weights to form a series of subproblems. The goal of MOEAD is to approach the Pareto optimal solution set by progressively optimizing these subproblems. It has been shown that MOEAD converges faster than traditional GA and PSO in multi-objective optimization . Comparative analysis reveals distinct algorithmic competencies across optimization scenarios: MOEAD's decomposition mechanism demonstrates superior performance in high-dimensional optimization, while GA exhibits enhanced global exploration in low-dimensional contexts. PSO maintains computational efficiency advantages in solving low-complexity optimization problems through rapid convergence characteristics.

Finding the maximum of all weighted objective deviations is the fundamental idea behind the Chebyshev weight function. This procedure attempts to balance the optimization of many objectives, ensuring that the subproblem solution process is not unduly skewed towards one. The formula for the Chebyshev weight function is:

$$g^{te}(x \mid w, z^*) = \max_{1 \le i \le m} \{\lambda_i \cdot | f_i(x) - z_i^* |\} \tag{10}$$

where x represent the current solution; m denotes the number of objective functions; $w = (\lambda_1, \lambda_2, \ldots, \lambda_m)$ is the weight vector, which represents the trade-offs between the different objectives; $z^* = (z_1^*, z_2^*, \ldots, z_m^*)$ is the ideal point, i.e., the theoretical optimum of each objective function; and $g^{te}(x \mid w, z^*)$ is the scalarised objective value of the sub-problem, which measures the quality of the solution by maximising all the weighted objective deviations.

The optimization steps of the MOEAD optimization algorithm :

(1) Initialization: A set of initial solutions P are randomly generated, which are evenly distributed in the feasible domain. Weight vectors w are defined for each objective function that is evenly distributed over the unit hyperplane and are usually generated using the method of uniform distribution, ensuring that each weight vector reflects the tradeoff between different objectives.

(2) Neighborhood definition: A neighborhood N(i) is defined for each solution that contains other solutions similar to the weight vector of the solution i. The definition of a neighborhood can use Euclidean distance or other similarity measures to select similar solutions.

(3) The following steps are repeated within the specified number of iterations: First, each solution xi in the current population P is traversed. Then one or more solutions are randomly selected from the neighborhood N(i). A new solution x' is generated using crossover and mutation operations. This can

be achieved through a variety of genetic algorithm operations, such as simulated binary crossover (SBX) and polynomial variation. In addition, the objective function value fj(x') for the new solution is computed (for all objective functions j). The quantization target is calculated using the Chebyshev weight function. Finally, the new solution is compared with the solution in the neighborhood. According to the domination relation (i.e. whether one solution is better than another), the solution in the neighborhood is updated, and if the new solution x' is better than some solution xj in the neighborhood, xj is replaced with x'. The corresponding target value and the quantified target value are updated.

(4) The optimization procedure terminates upon satisfying either of two stopping criteria: attainment of predefined iteration limits or detection of solution convergence below specified thresholds, typically characterized by minimal metric variation in population diversity.

(5) After the end of the algorithm, all non-dominated solutions are collected to form an approximate Pareto optimal solution set. These solutions represent optimal solutions that strike a balance between different objectives. This systematic framework, diagrammatically represented in Figure 1.

Table 1. CY-VMC850Le Machine parameter.

CY-VMC850Le	
Table parameter (mm×mm)	1000×500
Maximum allowable load of table (Kg)	600
Spindle speed range (r/min)	50~8000
Maximum output torque of spindle (N ·m)	35.8
Spindle motor power (KW)	7.5~11
Machine weight (Kg)	5800
Machine profile dimensions (L × W × H) (mm×mm×mm)	3300×2400×3000

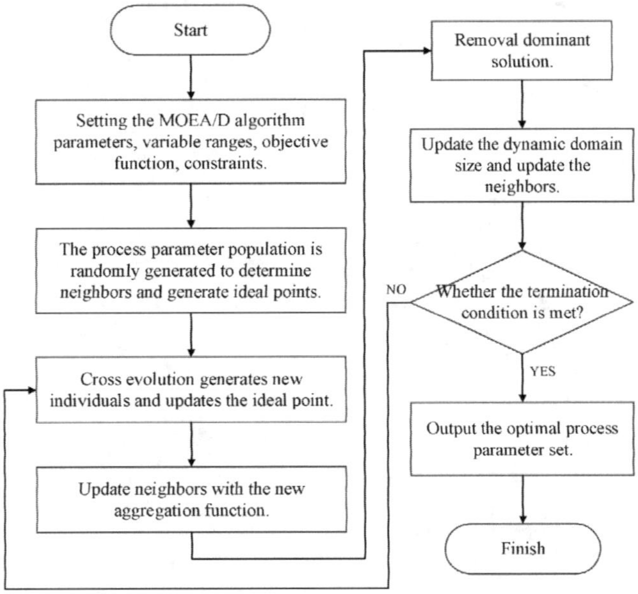

Figure 1. Flowchart of MOEAD optimization algorithm.

4. Optimization example

The CY-VMC850Le CNC milling machine has been chosen as the experimental platform, and its processing power is measured using the WB9128 power sensor. The tool used in the experiment is a high-carbon steel end mill with four teeth and a tool diameter of 10 mm. The experimental material is 304 stainless steel, measuring 30 mm by 60 mm by 90 mm. Figure 2 displays the experimental apparatus. The machine parameters are shown in Table 1.

Figure 2. Diagram of experimental equipment.

Extending the experimental program through multifactor experimental design can effectively cover a wide range of scenarios in the parameter space. This design method can cover a wide range of machining conditions with a limited number of experiments, providing stronger robustness support for the model. By extending the cutting parameter combinations and machining conditions, the experimental data will reflect the actual machining scenarios more comprehensively, thus providing stronger support for the reliability and robustness of the optimization results.

The spindle speeds of 700 r/min, 800 r/min and 900 r/min, the feed speeds of 100 mm/min, 150 mm/min and 200 mm/min, the milling depths of 0.4 mm, 0.6 mm and 0.8 mm, and the milling widths of 10 mm, 20 mm and 30 mm were chosen to conduct the four-factor three-level experiment. The machining mode was selected as the milling plane, and the cooling mode was wet cutting. Multifactor experimental designs have been widely used for multi-objective optimization to ensure the validity of experimental results in a wide range of scenarios .

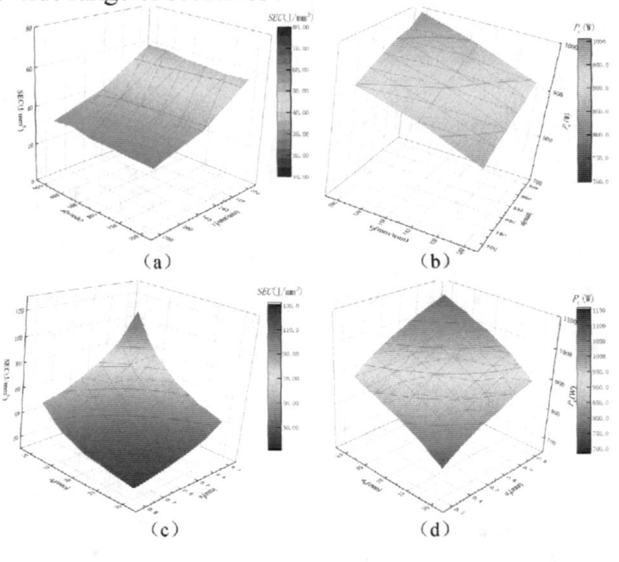

Figure 3. Analysis of the influence of process parameters on the optimization objective.

According to the experimental results, analyzing the influence of each factor on the optimization objective, it is found that the milling energy consumption P_c increases with the increase of spindle speed n, feed rate v_f, milling depth a_p and milling width a_e. The spindle speed n has less influence on the specific cutting energy *SCE*, which will slowly increase with the increase of spindle speed *n*. *SCE* decreases with the increase of feed speed v_f, milling depth a_p and milling width a_e. Provide guidance for practical operation.

Utilize MOEAD to optimize the experimental data in order to solve the optimization process. The optimization of the rear portion of the decomposition set is displayed in Table 2 and occurs when the optimization objective and optimization parameters are iterated to 200 generations. The optimization results are generally stable.

Table 2. Selected MOEAD-Pareto Solution Sets.

Serial No.	n	v_f	a_p	a_e	P_c	*SCE*
1	481.01	159.82	0.32	15.01	586.57	141.03
2	674.73	148.55	0.63	12.52	807.78	23.95
3	578. 12	183.51	0.74	14.14	727.84	39.53
4	669.36	211.37	0.79	19.78	966.92	10.17

The parametric analysis derived from Table 2's optimization results indicates substantial variations in machine tool power demand and specific cutting energy across different input configurations. Experimental data reveals an inverse correlation between these two performance metrics when modifying spindle velocity, feed rate, cutting depth, and milling width. Particularly, elevated machining power levels typically correspond to reduced specific energy values, a phenomenon attributable to enhanced material removal rates under higher power conditions. This operational characteristic effectively diminishes energy expenditure per unit volume of removed material through intensified cutting action.

As the spindle speed and feed rate increase, the machining power and total energy consumption show a gradual increase. This is due to the fact that higher spindle speeds and feed rates increase the material removal rate, which leads to a rise in energy consumption per unit time and a rise in the specific energy of cutting. Experiments show that proper adjustment of the depth of cut and milling width can control energy consumption while increasing the material removal rate, thus realizing higher machining efficiency. Reasonable adjustment of feed rate helps to control the level of energy consumption while reducing machining time, reflecting the value of process parameter optimization in efficiency. The analysis of the effects of process parameters on energy consumption and machining efficiency further reveals the complex relationship between process parameters and optimization objectives, providing theoretical support for the rationality of the optimization strategy. This understanding helps to apply the optimization strategy to different machining scenarios, thus improving the practicality and applicability of the optimization scheme.

The Pareto solution set plot obtained from the MOEAD iterative solution is shown in Figure 4. From Figure 4, it can be seen that the two, P_c and *SCE*, show a typical mutual constraint relationship, which means that in practice, it is not possible to minimize both of them at the same time. In the actual production process, it is necessary to make a reasonable trade-off between the two according to the specific machining requirements in order to select the optimal combination of milling parameters.

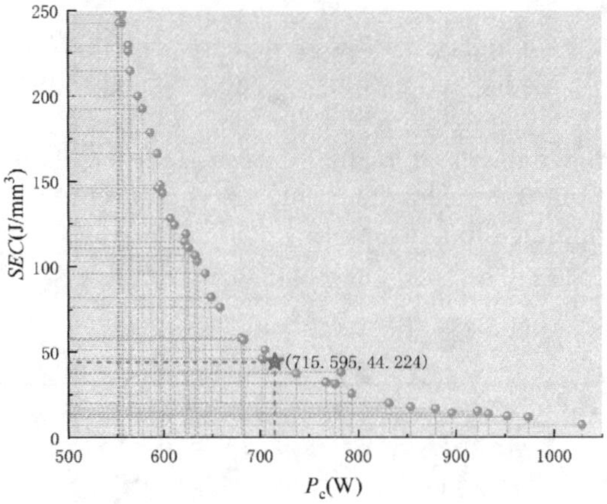

Figure 4. Pareto optimal solution set diagram.

This research does a comparative examination of the machine's milling parameters before and after optimization, together with their related milling power and specific cutting energy, in order to further verify the optimization effect. The results are displayed in Table 3.

Table 3. Comparison Of Parameters Before And After Optimization.

Parameters	*Pre-optimization*	*Post-optimization*
n (r/min)	700	451.51
v_f (mm/min)	200	346.07
a_p (mm)	0.4	0.51
a_e (mm)	10	5.01
P_c (W)	799.58	715.595
SCE (J/mm^3)	59.98	44.224

Table 3 illustrates how the MOEAD algorithm optimization results in a 10.5% reduction in milling power P_c and a 26.27% reduction in specific cutting energy *SCE*. This demonstrates that the ideal combination of milling parameters increases energy efficiency greatly while simultaneously improving power.

5. Conclusions

This study systematically investigates multi-objective process optimization in CNC milling operations employing the MOEAD framework, with dual optimization objectives of minimizing power consumption and specific cutting energy. The research explores the influence mechanism of four key process variables - spindle velocity, feed rate, cutting depth, and milling width - on operational performance metrics. Through the development and implementation of a multi-objective optimization architecture, the methodology generates comprehensive Pareto frontiers representing optimal parameter configurations. Empirical validation confirms that the refined parameter configurations maintain product integrity while establishing an effective equilibrium between energy demands and production throughput.

In particular, following optimization, the machine tool's milling power and specific cutting energy are decreased by 10.5% and 26.27%, respectively. This demonstrates how the MOEAD algorithm handles the multi-objective problem with great adaptability and robustness, which not only successfully lowers the energy consumption mark optimization during CNC milling machining but also greatly increases machining efficiency. This study's findings offer a fresh concept for the clever optimization of CNC machining parameters, as well as a conceptual framework and useful example for implementing green manufacturing.

Acknowledgment

This study is supported by the Demonstration of a digital re-engineering application for ship final assembly construction (CBG01N23-05-01).

References

[1] Xiao Y, Jiang Z, Gu Q, Yan W&Wang R. (2021). A novel approach to CNC machining center processing parameters optimization considering energy-saving and low-cost. Journal of Manufacturing Systems, 59, 535-548. https://doi.org/10.1016/j.jmsy.2021.03.023
[2] Li B, Tian X&Zhang M. (2022). Modeling and Multi-objective Optimization Method of Machine Tool Energy Consumption Considering Tool Wear. International Journal of Precision Engineering and Manufacturing-Green Technology, 9 (1), 127-141.10.1007/s40684-021-00320-z
[3] Zhao J, Li L, Nie H, Chen X, Liu J&Shu X. (2021). Multi-objective integrated optimization of tool geometry angles and cutting parameters for machining time and energy consumption in NC milling. The International Journal of Advanced Manufacturing Technology, 117 (5), 1427-1444.10.1007/s00170-021-07772-2
[4] Feng C, Qin E, Li W, Li M, Su J&Xu X. (2024). Energy consumption and surface roughness modelling for multi-objective optimization of machining processes. The International Journal of Advanced Manufacturing Technology, 133 (9), 5065-5087.10.1007/s00170-024-14043-3
[5] Li C, Zhao G, Meng F, Yu S, Yao B&Liu H. (2024). Multi-objective optimization of machining parameters in complete peripheral milling process with variable curvature workpieces. Journal of Manufacturing Processes, 117, 95-110. https://doi.org/10.1016/j.jmapro.2024.03.004
[6] Li C, Cui L, Liu F&Li L. (2013). Multi-objective NC Machining Parameters Optimization Model for High Efficiency and Low Carbon. Journal of Mechanical Engineering, 49 (09), 87-96.
[7] Li J, Wang Y, Liu K, Zhao D, Jiang S, Yang Y&Yu Q. (2023). Tough-brittle transition mechanism and specific cutting energy analysis during cryogenic machining of Ti–6Al–4V alloy. Journal of Cleaner Production, 383, 135533. https://doi.org/10.1016/j.jclepro.2022.135533

Manufacturing process generation based on multiple kernel Markov chain

Qingming Song [1*], Siyuan Shen [1], and Peiyan Wang [2]

[1] Manufacturing Engineering Technology Center, COMAC Shanghai Aircraft Manufacturing, Shanghai 201324, China
[2] School of Computer Science, Shenyang Aerospace University, Shenyang 110136, China

[*] Corresponding author: Qingming Song (E-mail: songqingming@comac.cc)

Abstract. Research on the generation of manufacturing process problems in process design has been conducted, and a method based on multiple kernel Markov chains has been proposed. This method utilizes the sequential logical properties of processing methods in the process to establish a processing method Markov chain and generates subsequent processing methods based on the previous processing methods. The engineering information is introduced into the Markov chain through kernel functions from the aspects of part name, part number, material, and thermal surface treatment requirements. Compared with existing methods, this method fully utilizes historical process outline documents, does not require manual determination of feature expression symbol sets, nor does it require labeled data, effectively avoiding the "knowledge bottleneck" problem. On 2, 360 aircraft sheet metal process manufacturing outline data (including 251 processing methods), the accuracy of the proposed method can reach 0.7963, verifying the effectiveness of the method.

1. Introduction

The process planning design arranges the processing steps from raw materials to finished products, representing one of the core aspects of process design. Process planning design must consider factors such as part types, materials, heat and surface treatment requirements, and equipment. Among various possible qualified processes, the most optimal or suitable process needs to be selected, making process planning design the task that most reflects the capability of process engineers. Therefore, automatic process planning design is the primary objective of Computer-Aided Process Planning (CAPP) systems [1].

Process generation is one of the key technologies in CAPP, which aims to automatically generate the corresponding process according to the given manufacturing factors. With the development of machine learning and knowledge graph, new solutions have emerged for the automatic generation of process sequences. Natarajan and Gokulachandran [2] published an article introducing the deep learning methods for process generation in recent years and pointed out that the understanding of manufacturing factors and generation of the processing method sequence is a crucial problem to be solved. Xu et al. [3] proposed a deep learning method for process generation, applied convolution networks as an encoder to understand and represent the manufacturing factors, and applied LSTM to generate the sequence of processing method sequence. However, deep learning does not use the knowledge of manufacturing. To address knowledge accumulation, inflexible reuse, and fragmented reasoning in machining process design, Hua et al. integrated the structured representation characteristics of the knowledge graph and the perceptual reasoning capabilities of deep reinforcement learning [4]. Shi et al. [5] proposed a knowledge graph-based structured representation of assembly process planning that combines expert experience and a large amount of existing process information. Christian et al. [6] presented a support system, which transfers the knowledge into ontologies and uses these to support the planning. The research [7-8] utilized an enhanced version of Blockchain so that functionalities at the process level can be certified as per a particular scheme. Zhou et al. [9] proposed a knowledge graph-driven assembly process generation and evaluation method for complex components.

Both knowledge graph-based feature-based methods and ontology-based methods fundamentally belong to the knowledge engineering approach. In these methods, a formal symbolic set and representation for process features are manually defined, and historical process data or 3D CAD models are annotated based on this symbolic set. During recommendation, engineering information is parsed according to this symbolic set, assessing the applicability of process plans and methods to enable recommendations. As a knowledge engineering method, the definition of formal symbolic sets, representation methods, and the association between symbolic sets and process plans or methods presents a bottleneck. This issue represents the “knowledge bottleneck” problem in traditional knowledge engineering methods as applied to processing method recommendations. Wang et al. [10] proposed a process reuse method based on multiple kernel learning. This approach does not require manually defining a symbolic set of features or data annotation; instead, it utilizes historical process outline data to learn the relationship between engineering information features and process plans. Additionally, it confirms the effectiveness of easily parsed features, such as process outline names and codes, in process reuse. Ke et al. [11] proposed a vector space model and example-by-example reasoning. However, these methods only reuse historical processes and cannot generate new processes based on current task requirements.

To address the generation of process plans in process design, a multiple kernel Markov chain-based approach is proposed. This method leverages the sequential logic of processing methods within process plans by constructing a Markov chain for processing methods, recommending subsequent processing methods based on preceding ones. Kernel functions introduce engineering information, such as part name, part code, material, and heat or surface treatment requirements, into the Markov chain. Compared with existing methods, this approach fully utilizes historical process outline data, avoiding the need for manual feature symbolic sets and data annotation, thus effectively circumventing the “knowledge bottleneck” problem. The proposed method’s effectiveness was validated by using 2, 360 aircraft sheet metal manufacturing process outline data (encompassing 251 processing methods). The contributions of this paper include: demonstrating the sequential logical relationship of processing methods, showing that the current method depends on a finite sequence of preceding methods and can be modeled with a Markov chain, and applying kernel methods to the process recommendation problem. Kernel methods, as effective similarity measurement techniques, reflect the analogy-based methodology in process design from another perspective.

Section 1 introduces the proposed process generation method, including definitions of the multiple kernel Markov chain, kernel functions, and parameter learning methods. Section 2 describes the experimental dataset, evaluation metrics, and analysis of the experimental results to validate the proposed method. Section 3 summarizes the paper and discusses promising future research directions.

2. Processing method recommendation

2.1. Problem description

It is assumed that the process design task space X and the set of processing methods $O=\{o_1,o_2,...,o_n\}$. A process plan $s=o_1o_2,...o_l$ is a sequence of processing methods $o_i \in O$. The formal definition of the process generation problem is: Given a task $x \in X$, a sequence of processing methods $s=o_1o_2,...o_l$ is generated; given the task x, the optimal sequence $p\left(s|x;\boldsymbol{\theta}\right)$ of processing methods s^* can be found, where $\boldsymbol{\theta}$ represents the parameters.

$$s^* = \arg\max_{s \in S} p\left(s|x;\boldsymbol{\theta}\right) \tag{1}$$

The generation of processing methods starts from the first operation and proceeds step by step, meaning that, given the already known x and generated portions of the process plan $s_{0,i} = o_0 o_1 o_2 ... o_i$, the next operation o_{i+1}^* is generated. Therefore, $p(s|x;\boldsymbol{\theta})$ is defined in Equation (2):

$$o_i^* = \arg\max_{s \in S} p(o_i | x, o_1 o_2 ... o_{i-1}; \boldsymbol{\theta}) \tag{2}$$

2.2. *Multiple kernel Markov chain*

The Markov model was proposed by the Russian mathematician Andrey Markov in 1907 [12]. A process flow is an ordered sequence composed of operations. In this paper, it is assumed that the sequence of operations follows the Markov assumption, meaning that the subsequent operation methods are determined solely by the preceding finite steps. For example, if the preceding operations are "pre-drilling, reaming, honing, final drilling", then the subsequent operations could be "burring" or "deburring". For an operation s_i, it is influenced by the previous n steps operation $s_{i-1,...,}s_{i-n}$, satisfying the condition of a conditional probability $p(s_i|s_{i-1...i-n})$. The $p(s_i|s_{i-1...i-n})$ calculation method is as follows:

$$p(s_i | s_{i-1,...,}s_{i-n}) = \frac{\sum_{O \in D} 1(O, s^n)}{Z} \tag{3}$$

where $1(O,s^n)$ represents the frequency of the sequence s^n occurring in the process outline sequence O; $\sum_{O \in D} 1(O, s^n)$ is the total number of occurrences of the sequence s^n in all process outline documents; Z is the normalization factor; n is the order of the Markov chain, meaning the length of the processing method sequence.

In addition to the preceding operation sequence, engineering information also plays a critical role in determining the order of operations within a process sequence, such as part name, part number, material, and heat treatment requirements. Therefore, when modeling the operation Markov chain, these factors must be considered. Some researches indicate that in new product development, approximately 40% involves reusing past component designs, about 40% involves slight modifications to existing component designs, and only around 20% represents entirely new designs [12]. Therefore, the more similar the engineering information is, the more similar the process sequences are. Based on this principle, a multiple kernel learning-based Markov chain model is proposed, which is formalized as follows:

$$\underbrace{p(s_i | s_{i-1...i-n})}_{s^n} = \frac{\sum_{O \in D} \sum_{p=1}^{l} \eta_p k_p((O, s^n), (O', s^n))}{Z}, \eta \in \mathbb{R}^+, \ \sum_{p=1}^{P} \eta_p = 1 \tag{4}$$

where $k_p((O,s^n),(O',s^n))$ represents different kernel functions, which can be used to measure the similarity of engineering information between two operation sequences based on aspects such as part name, part number, material, and heat treatment requirements; η_p represents the weight of the similarity for different aspects; $\sum_{O \in D} \sum_{p=1}^{l} \eta_p k_p((O,s^n),(O',s^n))$ represents the weighted summation of kernel function values that compute the similarity between the current design task's O' engineering information and the operation outline's O engineering information containing the sequence S^n; Z is the normalization factor. Compared to traditional Markov chains, the proposed method replaces the counting function $1(O,S^n)$ with the kernel function between the process outline's engineering information, allowing the inclusion of engineering information as guidance.

Higher-order Markov chains consider longer preceding operation sequences when generating the current operation, leading to stricter constraints and potentially higher recommendation accuracy. However, this inevitably causes data sparsity, meaning that the likelihood of not finding a matching preceding sequence during the recommendation process increases. Lower-order Markov chains, on the other hand, consider shorter preceding sequences with looser constraints, which can compensate for the sparsity issue in higher-order models. Therefore, a linear combination of different-order probabilities is used, as formulated below:

$$p\left(s_i \middle| s_{i-1\ldots i-n}\right) = \sum_{j=1}^{n} \lambda_j p\left(s_i \middle| s_{i-1\ldots i-j}\right), \lambda \in \mathbb{R}^+, \quad \sum_{j=1}^{n} \lambda_j = 1 \tag{5}$$

2.3. *Kernel function design*

Based on the reference method [10], kernel functions are defined by using easily accessible and analyzable process information, avoiding complex features such as geometric structures. The kernel functions are defined with respect to four aspects of information: part name, part number, material, and processing area.

2.3.1. Part name kernel. The part name is a specialized term that briefly expresses the type of part, such as "window frame decorative cover bracket" or "water service board corner plate". From a semantic perspective, although technical terms are highly scientific vocabulary, their meaning is derived from the abstraction and generalization of human experience. [13] The part name reflects the understanding of the part's position, shape, and function. The methods of part naming can be categorized into the following types, as shown in Table 1:

Table 1. Naming methods for parts.

Naming method	Example
Shape/form	Rod, tube, plate, sheet, frame, box
Function	Connector, skin, support
Function + shape	Reinforcement plate, compensation sheet, closure cover plate
Position + shape	Web plate, busbar
Position + function + shape	Left runway turn lamp bracket, switch bracket

From this perspective, the part name kernel is proposed, where parts with more similar names (e.g., "left runway turn light bracket" and "right runway turn light bracket") are considered to have more similar process plans. The part name kernel is defined as follows:

$$k_n\left(C, C'\right) = \frac{\left|C \cap C'\right|}{\sqrt{|C|}\sqrt{|C'|}} \tag{6}$$

where C represents the set of characters that constitute the part name, and $|C|$ represents the number of elements in the set. The higher the value $k_n\left(C, C'\right)$ is, the more shared characters there are between the two-part names, and the higher the degree of process plan reusability is. According to the calculation method of $k_n\left(C, C'\right)$, the computed value for the "left runway turn light bracket" and "right runway turn light bracket" is 0.875.

2.3.2. Part number kernel. Every official process document must have a unique identifier, and the same identifier can only be assigned to one process document. The numbering of process documents follows a defined standard and has an encoding system. The national standard "GB/T 24735-2009 Method for Numbering Mechanical Manufacturing Process Documents" specifies the numbering methods for management process documents, process equipment documents, and production process documents. Figure 1 illustrates the encoding method for production process documents as defined in the GB/T 24735-2009 standard.

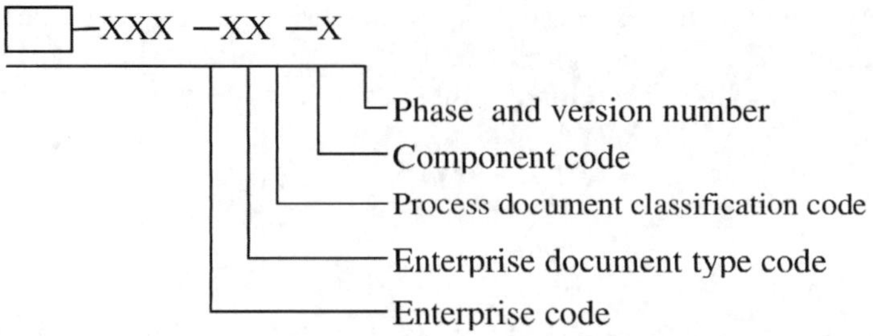

Figure 1. Coding method for production process documents as specified in GB/T 24735-2009.

The part number contains the company's management methods and systems for products and components, implicitly reflecting the relationships between product parts. Experienced process engineers are often able to identify the part type and its location on the aircraft based on the part number. Therefore, the part number kernel is defined to measure the similarity between part numbers. The part number kernel is defined as follows:

$$k_d(\mathbf{x},\mathbf{x}') = \frac{1}{l}\sum_{i=1}^{l}\exp\left(-\left|\mathbf{x}_i - \mathbf{x}_i'\right|\right) \tag{7}$$

where $\mathbf{x} = [\mathbf{x}_1, \mathbf{x}_2, ..., \mathbf{x}_l]$ and $\mathbf{x}_i$ denotes the i segments of the part number, and $l = 5$ according to the national standard GB/T 24735-2009. The higher the value $k_d(\mathbf{x},\mathbf{x}')$ is, the closer the two segments of the part numbers are to each other.

2.3.3. Material kernel. Material information is one of the key factors in determining the process plan. Zhang et al. [14] proposed that material information is categorized as a feature of the process design task, while Sun et al. [15] classified it as an item of process design intent. Material information includes: material name, material grade, material specifications, and material standards. Table 2 lists examples of material information.

Table 2. Material information example.

Material name	Material grade	Material specification	Material standard
Clad aluminum plate	2024-O	0.1"	AMS-QQ-A-200/11
Clad aluminum plate	7075-O	0.063"	
Non-clad aluminum plate	2024-T351	0.063"	

The material kernel compares the similarity of material information based on the four aspects mentioned above. The material kernel is defined as follows:

$$k_m(\mathbf{m},\mathbf{m}') = \sigma\left(m_n, m_n'\right) + \sigma\left(m_c, m_c'\right) + \sigma\left(m_s, m_s'\right) + \sigma\left(m_d, m_d'\right)$$
$$\sigma(x,x') = \begin{cases} 0.25 & x = x' \\ 0 & x \neq x' \end{cases} \tag{8}$$

where $\mathbf{m}$ represents the material information, which includes material name m_n, material grade m_c, material specifications m_s, and material standards m_d.

2.3.4. Heat treatment and surface treatment requirement kernel. The heat and surface treatment requirements describe processes such as painting, coating, heat treatment, and chromic acid anodizing for parts. These requirements are described in natural languages, such as: "Apply a colorless conversion coating to the outer surface, chromic acid anodize the inner surface, and apply fluid-resistant primer to the inner surface", or "Heat treat 2024-O to 2024-T42, fluorescent penetrant inspection". The Bag-of-Words (BOW) vector is used to represent these requirements, first by tokenizing the text and then calculating the weight of each word using TF-IDF. The heat and surface treatment requirement kernel is defined as follows:

$$k_a(\mathbf{a},\mathbf{a}') = \frac{\langle \mathbf{a},\mathbf{a}' \rangle}{\sqrt{\langle \mathbf{a},\mathbf{a} \rangle}\sqrt{\langle \mathbf{a}',\mathbf{a}' \rangle}} \tag{9}$$

where $\mathbf{a}$ represents the Bag-of-Words vector for the heat and surface treatment requirements.

2.4. Parameter optimization method

In summary, the multiple kernel Markov chain model is defined as follows, where λ and η are unknown parameters. An optimization method is used to determine λ and η. This method is an Expectation-Maximization (EM) algorithm [16].

$$p(s_i|s_{i-1\ldots i-n}) = \sum_{j=1}^{n} \lambda_j p(s_i|s_{i-1\ldots i-j}), \lambda \in \mathbb{R}^+, \ \sum_{j=1}^{n} \lambda_j = 1 \tag{10}$$

$$\underbrace{p(s_i|s_{i-1\ldots i-j})}_{s^j} = \frac{\sum_{O\in D}\sum_{p=1}^{l} \eta_p k_p\left((O,s^j),(O',s^j)\right)}{Z}, \eta \in \mathbb{R}^+, \ \sum_{p=1}^{P} \eta_p = 1 \tag{11}$$

We set process outline documents as D. To maximize the likelihood of all processing method sequences, the following optimization problem is defined:

$$\begin{aligned} &\max \sum_{s^n\in D} \log p(s_i|s_{i-1\ldots i-n}) \\ &s.t. \quad \eta \in \mathbb{R}^+, \ \sum_{p=1}^{P} \eta_p = 1, \ \lambda \in \mathbb{R}^+, \ \sum_{j=1}^{n} \lambda_j = 1 \end{aligned} \tag{12}$$

We calculate the gradients with respect to λ and η separately.

$$g_\eta = \frac{\partial \sum_{s^n\in D} \log p(s_i|s_{i-1\ldots i-n})}{\partial \eta_p} = \sum_{s^n\in D} \frac{\sum_{j=1}^{n}\sum_{S^j\in D} \lambda_j k_p\left((O,s^j),(O',s^j)\right)}{\sum_{j=1}^{n}\sum_{S^j\in D}\sum_{p=1}^{l} \lambda_j \eta_p k_p\left((O,s^j),(O',s^j)\right)} \tag{13}$$

$$g_\lambda = \frac{\partial \sum_{s^n\in D} \log p(s_i|s_{i-1\ldots i-n})}{\partial \lambda_j} = \sum_{s^n\in D} \frac{\sum_{S^j\in D}\sum_{p=1}^{l} \eta_p k_p\left((O,s^j),(O',s^j)\right)}{\sum_{j=1}^{n}\sum_{S^j\in D}\sum_{p=1}^{l} \lambda_j \eta_p k_p\left((O,s^j),(O',s^j)\right)} \tag{14}$$

Based on Equations (13) and (14), the parameter λ and η optimization algorithm can be derived as follows, which is illustrated in Figure 2:

Input: Process outline document set, number of iterations T

Output: λ, η

Initialization: $\lambda = \frac{1}{n}$, $\eta = \frac{1}{P}$

REPEAT

Compute Equations (13) and (14);

Obtain $\lambda^{(t)} = \lambda^{(t-1)} + g_\lambda^{(t)}$;

Obtain $\eta^{(t)} = \eta^{(t-1)} + g_\eta^{(t)}$;

Figure 2. Multiple kernel Markov chain parameter optimization algorithm.

3. Experimental results and analysis

3.1. Experimental setup

The dataset comprises 2, 360 documents related to the production of sheet metal components for a specific aircraft. Given that the process method descriptions contain duplicates, this paper first removes redundant entries from the process method content. Following this deduplication process, the dataset encompasses 251 distinct processing methods. Each document contains pertinent information about the manufacturing of the components, including details such as process outline numbers, names, required materials, and the process of manufacturing the components. A 10-fold cross-validation approach is used, where 10% (236 documents) are selected as the test sample set, representing the "new" processes to be designed. The remaining 2, 124 process outline documents are used as the training set for model parameter training. For the test documents, the current process is generated sequentially based on the preceding processes, with 10 results being pushed at each step. The evaluation method follows the approach [10], calculating the accuracy of the generated process relative to the reference process in the test data, as shown in Equation (15), where $LD(Y,\bar{Y})$ represents the edit distance between the generated process Y and the reference process in the test set $\bar{Y}$, and $len(Y)$ and $len(\bar{Y})$ represent the number of processing methods Y and $\bar{Y}$ in the generated and reference processes, respectively. The value of Acc ranges from $[0.0,1.0]$, where 1.0 indicates complete agreement between Y and $\bar{Y}$, and 0.0 indicates complete dissimilarity or reversed order. $Acc(Y,\bar{Y})$ indicates that fewer insertions, deletions, or replacements are needed to convert the generated process to the reference process, thus implying a higher applicability of the generated process.

$$Acc(Y,\bar{Y}) = 1 - \frac{LD(Y,\bar{Y})}{\max(len(Y), len(\bar{Y}))} \tag{15}$$

3.2. Single kernel and multiple kernel equal weight combination experiments

Figure 3 presents the experimental results of the multiple order equal weight combination Markov chain ($\gamma_1 = \gamma_2 = ... = \gamma_n = \frac{1}{n}$) for all single kernels, two kernels, three kernels, and four kernels equal weight combinations ($\eta_1 = \eta_2 = ... = \eta_l = \frac{1}{l}$), where "NU" represents the part number kernel, "MA" represents the material kernel, "NA" represents the part name kernel, "FA" represents the heat treatment information kernel, and "NK" represents the results based solely on the Markov chain of the process sequence without any kernel function. Figure 3 shows the paired t-test results based on 10-fold cross-validation with a confidence level of 0.05. The value "1" in Figure 3 indicates that the two methods have a statistically significant difference at a confidence level of 0.05. For example, if the value inside the cell for "NU" and "NK" is "1", it means that there is a significant difference between "NU" and "NK", with the "NU" method being superior to "NK".

From Figures 3 and 4, it can be observed that: 1) The minimum accuracy (Acc) value is 0.6705 (FA), and the maximum Acc value is 0.7922 (NU + MA + FA). According to the definition of Acc, this indicates that 79% of the processing methods do not require modification, demonstrating the effectiveness of the proposed method; 2) Except for "MA", "FA", and "NK", which show no significant differences, all other methods outperform "NK". This highlights the necessity of incorporating the part name, part number, material, and heat treatment requirements information in the process Markov chain; 3) Multiple kernel combinations outperform single kernel methods, with three kernels and four kernel combinations performing better than two kernel combinations. Furthermore, the combined kernel functions generally perform better than the individual results of the separate kernels, indicating the importance of integrating multi-faceted information. This also suggests that multiple kernel combinations are not merely a simple aggregation of individual kernel results but can achieve effective complementarity and integration of diverse information.

 IOP Publishing
doi:10.1088/1742-6596/3009/1/012035

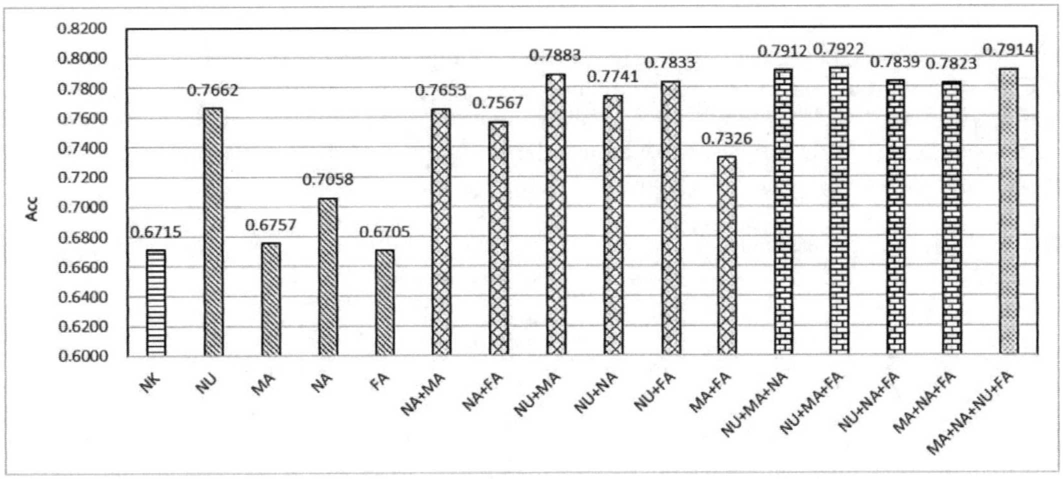

Figure 3. Experimental results of single kernel and multiple kernel equal-weight combination in the case of multiple order equal weight combination Markov chains.

	NK	NU	MA	NA	FA	NA+MA	NA+FA	NU+MA	NU+NA	NU+FA	MA+FA	NU+MA+NA	NU+MA+FA	NU+NA+FA	MA+NA+FA
NU	1														
MA	0	1													
NA	1	1	0												
FA	0	1	0	0											
NA+MA	1	0	1	1	1										
NA+FA	1	0	1	1	1	1									
NU+MA	1	1	1	1	1	1	1								
NU+NA	1	1	1	1	1	0	1	1							
NU+FA	1	1	1	1	1	1	1	0	1						
MA+FA	1	0	1	0	1	0	0	1	1	1					
NU+MA+NA	1	1	1	1	1	1	1	1	1	1	1				
NU+MA+FA	1	1	1	1	1	1	1	1	1	1	1	0			
NU+NA+FA	1	1	1	1	1	1	1	0	1	0	1	1	1		
MA+NA+FA	1	1	1	1	1	1	1	0	0	0	1	1	1	0	
MA+NA+NU+FA	1	1	1	1	1	1	1	0	1	1	1	0	0	1	1

Figure 4. Paired T-Test results of experimental data.

3.3. Comparison experiment of different orders and multiple order combined Markov chains

Table 3 presents the experimental results of different-order Markov chains and multiple order equally weighted combined Markov chains under the four kernels equal-weight combination ($\eta_1 = \eta_2 = ... = \eta_l = \frac{1}{l}$).

The multiple-order equally weighted combined Markov chain significantly outperforms other methods, confirming the complementarity of the models at different orders discussed earlier. In the experiment, the second-order model performs better than other-order models. The reason is that for sheet metal processes, the operation step is usually followed by an inspection step. The second-order model considers two preceding steps, which adequately capture the operational steps prior to the inspection, where the sequential logical relationship between operation steps is stronger, thus resulting in better push performance.

Table 3. Experimental results of Markov chains with different orders and multiple order equal-weight combination Markov chains.

	1st order	2st order	3st order	4st order	EP (MA+NA+NU+FA)
Acc	0.6964	0.7226	0.6535	0.5755	**0.7914**

3.4. Comparison experiment with other methods
Table 4 shows a comparison between parameter learning (OURS) and the process reuse method (VSMCBR) [10] and the process generation method (BGLA) [17]. From Table 4, it can be found that our method outperforms VSMCBR and BGLA. This result reflects the effectiveness of our method, which can effectively generate the process by learning the parameters of multiple kernel combinations and multiple-order Markov chain combinations.

Table 4. Comparison experimental results.

	OURS	**VSMCBR**	**BGLA**
Acc	**0.7963**	0.7011	0.7668

4. Conclusion
A method based on multiple kernel Markov chains is proposed to address the process generation problem in process design. This method leverages the sequential nature of process steps, using a Markov chain to model the sequential logical relationships between process steps. Engineering information such as part name, part number, material, and surface treatment requirements is introduced into the process Markov chain through kernel functions, enabling the generation of process flows guided by engineering information. The effectiveness of the proposed method was validated through experiments involving 2, 360 aircraft sheet metal process outline documents and 251 processing methods. The results demonstrate that: there exists a sequential logical relationship between processing methods in the process flow, with the current processing method depending on the preceding finite steps, which can be modeled by using a Markov chain; similar engineering information leads to similar process flows, and kernel methods are effective for measuring similarity and integrating engineering information into the Markov chain; for process generation problems, the multiple kernel learning method is an effective information fusion approach, significantly outperforming methods based on single information types. Furthermore, the fusion strategy is not merely the combination of results but effectively achieves complementarity and integrative use of diverse information.

Although the proposed method performs well in current applications, it has limitations in handling complex process scenarios. Future research will explore introducing more contextual information (such as environmental conditions and equipment status) and using more sophisticated model structures (such as graph neural networks) to enhance the model's performance and applicability, thereby advancing the development of smart manufacturing technologies.

Acknowledgment
This work is supported by the Applied Basic Research Program of Liaoning Province (2022JH2/101300248).

References
[1] Adapa S. K. and Jagadish (2023). An exhaustive review of intelligent computer-aided process planning in context with various optimization techniques. J. International Journal of Materials & Product Technology, 66 (2): 209-231.
[2] Natarajan K. K. and Gokulachandran J. (2021). Application of artificial neural network techniques in computer-aided process planning-a review. J. International Journal of Process Management and Benchmarking, 11 (1): 80-100.
[3] Xu W. W., Wang P. Y., Zhang G. P., et al. (2022). Automatic Generation Method of Manufacturing Process on Deep Learning. J. Computer Applications and Software, 12: 96-101 (in Chinese).
[4] Hua Y. W., Wang R., Wang Z. X., et al. (2024). Knowledge graph with deep reinforcement learning for intelligent generation of machining process design. Journal of Engineering Design, 4-17.
[5] Shi X. L., Tian X. T., Ma L. P., et al. (2024). A knowledge graph-based structured representation of assembly process planning combined with deep learning. J. International Journal of Advanced Manufacturing Technology, 133 (3-4): 1, 807-1, 821.

[6] Christian F., Simon S., Werner H., et al. (2021). An Approach for Knowledge-Driven, Flexible Process Generation. Procedia CIR. 104: 1, 239-1, 244.
[7] Alexios P., Harry B., Christos M., et al. (2021). On the generation of validated manufacturing process optimization and control schemes. Procedia CIR. 96: 57-62.
[8] Qian J. H., Zhang Z. J., Chao S., et al. (2021). Assembly sequence planning method based on knowledge and ontostep. Procedia CIR. 97: 502-507.
[9] Zhou B., Bao J. S., Chen Z. Y., et al. (2022). KGAssembly: Knowledge graph-driven assembly process generation and evaluation for complex components. International Journal of Computer Integrated Manufacturing, 35 (11): 1, 151-1, 171.
[10] Wang P., Zhang G., Zhai S., et al. (2018). Multiple kernel learning-based assembly process reuse. J. Computer Integrated Manufacturing Systems, 24 (7): 1, 850-1, 857.
[11] Ke C., Jiang Z., Zhang H., et al. (2020). An intelligent design for remanufacturing method based on vector space model and case-based reasoning [J]. Journal of Cleaner Production, 277.
[12] Serfozo R. (2009). Basics of applied stochastic processes. Springer Science & Business Media.
[13] Wang D. and Zhang Z. (2009). The Three Theoretical Foundations of Terminology Semantics. J. Linguistic Research, (4): 4-11.
[14] Zhang B., Jia X., Zhou D., et al. (2016). Research on process knowledge push based on the semantic feature list. J. Modern Manufacturing Engineering, (9): 63-74.
[15] Sun P., Hou J., Shi Q., et al. (2016). Research on knowledge push method for 3D process design. J. Manufacturing Automation, 38 (9): 96-105.
[16] Li H. (2019). Statistical Learning Methods. Tsinghua University Press, Beijing.
[17] Tang W. J., Wang P. Y., Cai D. D., et al. (2023). Encoder-Decoder Based Method for Process Generation. Journal of Computer Integrated Manufacturing Systems, 29 (11): 3, 656-3, 668.

Journal of Physics: Conference Series **3009** (2025) 012036 doi:10.1088/1742-6596/3009/1/012036

Design and analysis of an external energy launching mechanism intermittent mechanism

Zifeng Cao *, Lixin Wei, Hongliang Yang

No.1 Research Office, No.713 Research Institute of CSSC, Zhengzhou, China

*Corresponding author's e-mail: 1412674020@qq. com

Abstract. This paper investigates the design of intermittent mechanisms for external energy launching mechanisms, focusing on analysing parallel indexing cam mechanisms. Common laws of motion, including the improved sine and cosine laws of motion, are introduced, and the cosine law of motion and the improved sine law of motion are compared. Simulation analysis was also carried out, and the angular acceleration of the improved sinusoidal motion law is 10%of the pre-optimisation one under high-speed and heavy load environment, which improves the stability and service life of the intermittent mechanism. The paper is of reference value for the design of parallel cams and the optimisation of the intermittent mechanism of external energy chain launching mechanisms.

1. Introduction

Externally powered chain launching mechanisms rely on an electric motor to rotate the chain, which in turn moves the latch body back and forth to achieve an automated cyclic operation. In this mechanism, the core component that ensures precise intermittent movement of the latch is the intermittent mechanism. The four commonly used mechanisms to achieve intermittent motion are: ratchet mechanism, grooved wheel mechanism, Cam-type intermittent motion mechanism and incomplete gear mechanism [1] . Among them, the ratchet mechanism and the groove wheel mechanism are relatively common and simple mechanism, suitable for light load, low-speed situation, in the low-speed situation only has better mechanical properties, the higher speed of the case of stopping and starting there is a large impact force, and its movement in the high-speed state has a lower degree of accuracy, this mechanism cannot be applied to the chain launching mechanism high speed and heavy load of the mechanism. Compared with the traditional intermittent mechanism, the indexing cam mechanism has better rigidity, higher reliability and other advantages, parallel indexing cam can also be based on different transmission requirements, different application scenarios, using different laws of motion to design the cam's working curve, parallel indexing cam profile has excellent kinematics and dynamics characteristics [2].

The key to the design of parallel indexing cam mechanism is to accurately design the cam profile curve and determine the indexing motion law [3] . In the external energy transmission mechanism, the cam mechanism needs to maintain stability under high-speed and heavy-load conditions while reducing wear on key components. Therefore, when designing the cam profile curve, it is necessary to focus on the load changes and transient responses in the actual working environment to ensure that the mechanism can meet performance requirements in various complex scenarios. At present, domestic and foreign research is mostly focused on the theoretical modelling of plane indexing cams, and the motion simulation research on parallel indexing cam mechanism is relatively scarce.More researches focus on the motion simulation of cylindrical roller arc plane indexing cam mechanism, in addition, the related literature of intermittent mechanism and cam mechanism is also more abundant. Literature [4] proposed a method of parametric design of cam profile curves based on UG NX software, and the designed curves were analysed for continuity to ensure their smoothness and continuity. Literature [5] concluded that parallel indexing cam has better mechanical properties by comparing ratchet mechanism with parallel indexing cam mechanism. Literature [6] describes various types of indexing cam mechanisms and commonly used cam motion laws, as well as general methods for designing cam mechanisms. Literature [7] establishes a mathematical model of a parallel indexing cam and proposes a method for calculating the cam contour lines.In the analysis and design process, qualitative analysis and study of rigid or flexible impacts and their boundary conditions during high and low speed motions of the mechanism are needed to increase the accuracy and reliability of the simulation results.

Compared with traditional intermittent mechanisms, parallel indexing cam mechanisms have unique advantages in external energy transmission systems. These advantages include higher rigidity, improved reliability in high-speed and heavy-load environments, and the ability to adapt to a variety of motion requirements. These features significantly improve the stability of the mechanism and extend its service life, making it an ideal choice for meeting the needs of demanding industrial applications.

For an external energy chain launching mechanism system, the selection of the cam mechanism's law of motion is very important, and the following section will analyse and optimise the common law of motion of the cam mechanism.

2. Common Laws of Motion Analysis

When selecting the motion law of the parallel indexing cam, it should be ensured that the speed and acceleration of the cam change smoothly throughout the motion without sudden changes. At the same time, the speed and acceleration of the slave rotor of the parallel indexing cam should be zero at the beginning and end of the movement. In addition, the law of motion should be chosen so that the maximum speed, the maximum acceleration and the jump value are as small as possible.

In high-speed and heavy-duty working conditions, the design of the motion law of the parallel indexing cam is crucial because it directly affects the motion characteristics of the follower (e. g. , the indexing disc), including displacement, velocity and acceleration. From the point of view of practical application, the cam motion law should be simple and effective. Considering the above factors, it can be seen that the improved sinusoidal motion law, and the cosine motion law [8] both meet the requirements.

Improved sinusoidal motion pattern Displacement changes in a similar way to sinusoidal motion, but adjusted to make the motion pattern smoother. It is suitable for applications where smoothness of motion is important, such as systems that require less vibration and higher accuracy.

Law of Cosine Motion Displacement varies with angle in the form of cosine squared. It is suitable for applications with high accuracy and stability. The cosine motion law provides a more uniform motion characteristic where sudden changes in motion need to be avoided.

The improvement of the sinusoidal acceleration curve when analysing the motion of an indexing cam mechanism is usually expressed in terms of dimensionless parameters is represented as follows:

$$\begin{cases} T = \dfrac{t}{t_i} = \dfrac{\theta}{\theta_i} \\ S = \dfrac{\phi_i}{\phi_\mathrm{f}} \\ V = \dfrac{\mathrm{d}S}{\mathrm{d}T} \\ A = \dfrac{\mathrm{d}V}{\mathrm{d}T} \\ J = \dfrac{\mathrm{d}A}{\mathrm{d}T} \end{cases} \tag{1}$$

In Eq. (1), T, S, V, A, and J represent dimensionless time, displacement, velocity, accel eration, and leaps, respectively; t denotes the time of shaft operation; tf is the length of the i ndexing period of the cam; θ represents the angle of rotation of the cam; θf is the angle of rotation of the cam during the indexing period; and ϕi and ϕf represent the angle of rotation of the carousel at any moment and during the indexing period, respectively. The following are the formulas for the cosine acceleration law of motion and the optimised sinusoidal law:

Laws of motion for cosine acceleration:

$$A = \frac{\pi^2}{2}\cos\pi T \tag{2}$$

Improvement of the law of motion for sinusoidal acceleration:

$$\begin{cases} A1 = \frac{4\pi^2}{4+\pi}\sin(4\pi T) 0 < \mathrm{T} < \frac{1}{8} \\ A2 = \frac{4\pi^2}{4+\pi}\sin\frac{\pi+4\pi T}{3} \frac{1}{8} < \mathrm{T} < \frac{7}{8} \\ A3 = \frac{4\pi^2}{4+\pi}\sin(4\pi T) \frac{7}{8} < \mathrm{T} < 1 \end{cases} \tag{3}$$

Equations (2)and (3)were programmed using the numerical computing software Matlab to obtain the dimensionless acceleration versus dimensionless time curves as shown in Figure 1.

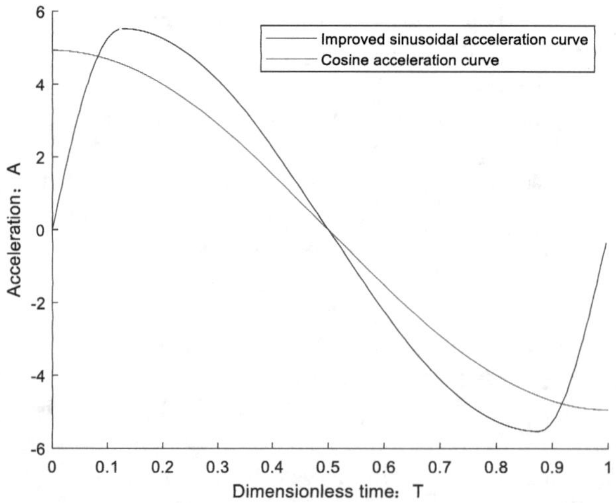

Figure 1. Acceleration comparison curves for cosine and sine.

Figure 1 demonstrates the acceleration comparison curves for the cosine law of motion and the modified sinusoidal law of motion, from which both images have good continuity and the cosine acceleration law of motion Am is smaller than the modified sinusoidal acceleration law of motion Am, which is adopted as the equation of the cam's contour in the chain launching mechanism.

3. Cam motion simulation analysis

3. 1. Cam parameter establishment

When designing the parallel indexing cam, it is necessary to determine the relevant important parameters, and the relevant important parameters of this parallel indexing cam are set as shown in Table 1 below, in which the indexing period angle needs to be set with reference to the timing diagram of the automatic machine to set the specific value. Next, the numerical calculation software is used for programming [9] calculations, which can obtain the motion profile of the cam as shown in Figure 2:

doi:10.1088/1742-6596/3009/1/012036

Table 1. Design Parameters for Parallel Indexing Cams.

name (of a thing)	nicknames
indexes of a turntable	I
headcount	H
Number of turntable rollers	Z
centre distance	C/mm
Pitch radius of cam	Rp/mm
Turntable indexing period angle	Kf/ (°)
Installation phase angle of two-piece cam	θp/ (°)

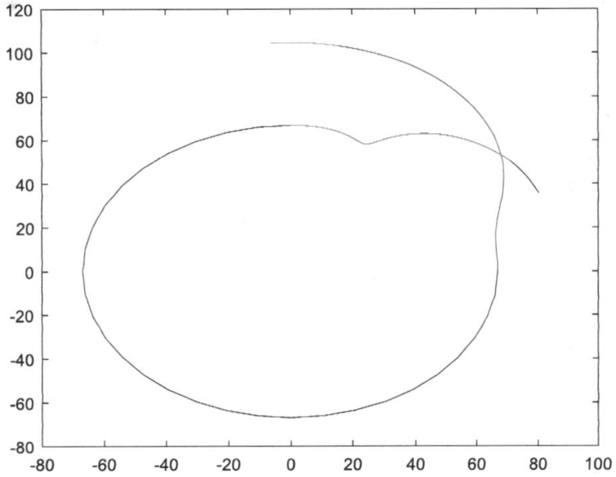

Figure 2 Cosine law of motion contours.

3. 2. 3D Modelling and Simulation. The 3D model of the parallel indexing cam can be obtained by modelling, and then assembled with the feed shaft, camshaft and other components to obtain the model of the intermittent mechanism as shown in the Figure 3:

Figure 3 Cosine intermittent mechanism model.

Using kinematics simulation software to simulate the movement of the mechanism to analyse [10-11] its movement, set the rotational speed of the drive shaft, i. e. camshaft, in the software to 8πrad/s,

Journal of Physics: Conference Series **3009** (2025) 012036
doi:10.1088/1742-6596/3009/1/012036

simulate two cycles, set the simulation time length to 0. 6s, and the simulation step length to 0. 001s, and obtain the angular velocity change and angular acceleration change curves of the feed shaft as shown in Figure 4~Figure 5:

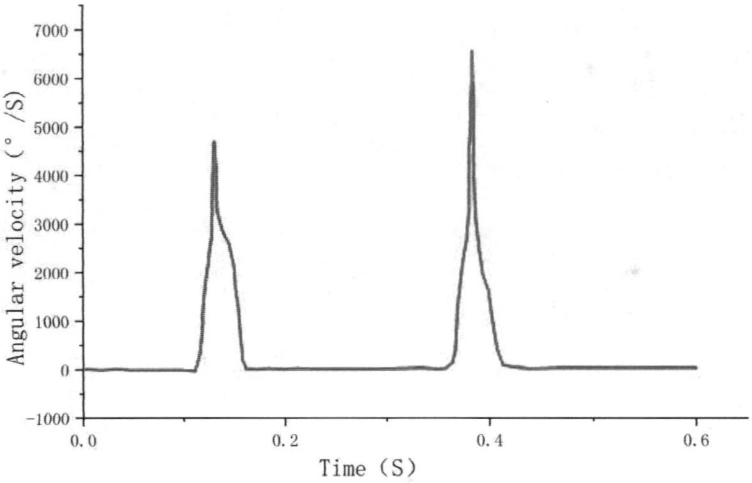

Figure 4 Angular velocity curve of the feed axis (cosine).

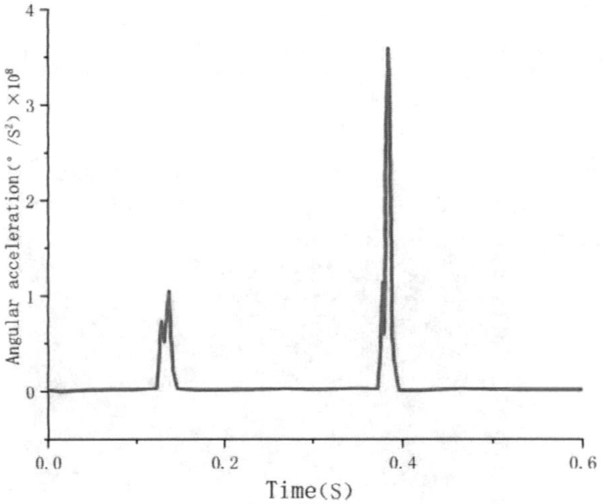

Figure 5 Angular acceleration curve of the feed axis (cosine).

Figure 4 shows that in the two cycles of cam rotation, (let the data speak) the feed shaft contacts the cam twice, the angular velocity and angular acceleration fluctuations are too large, and there are also some positions with large sudden changes, this is due to the cosine curve contour changes faster, resulting in more changes between the cam and the feed shaft, the whole process is not smooth. In practice, the use of the cosine rule of motion manufactured from the parallel indexing cam also brings heavy wear to the roller on the indexing disc, each roller on the different degrees of wear, affecting the life of the mechanism. Therefore, it is necessary to optimise the design of this parallel indexing cam.

4. Optimized analysis and design

The cam designed by using the cosine law of motion in the simulation calculation and actual use of the contact force is large, that is, the angular acceleration of the indexing disc is too large, so the optimisation goal is to reduce the contact force between the cam and the indexing disc, that is, to reduce the angular velocity and angular acceleration of the rotation of the indexing disc.

Considering that the acceleration characteristics of both the cosine and the improved sine laws of motion meet the requirements for the selection of the laws of motion, the differences between the two in terms of displacements, velocities and leaps are analysed as shown in the Figure 6:

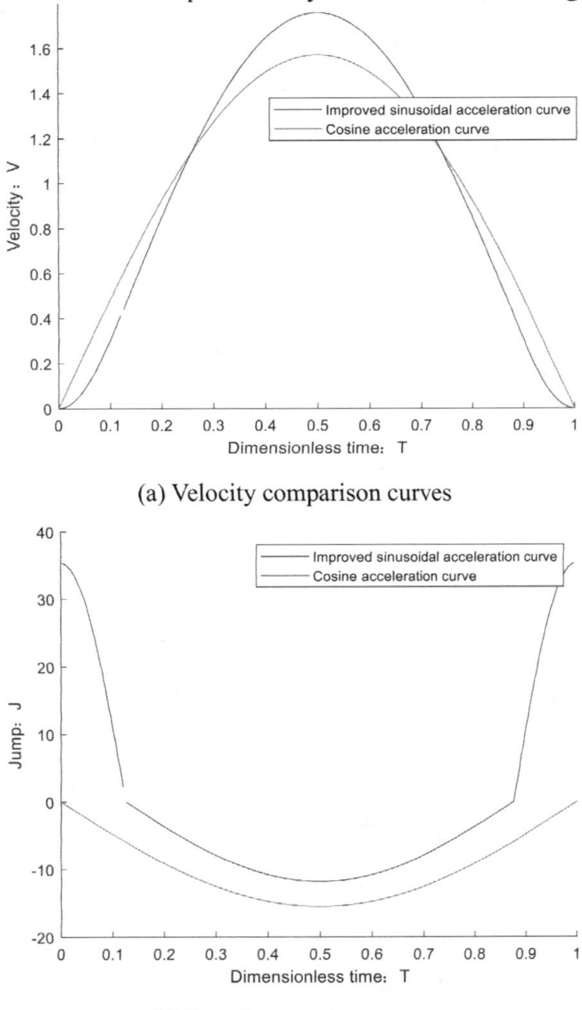

(a) Velocity comparison curves

(b) Leapfrog contrast curves

Figure 6 Velocity vs.

Analyzing the two curves, the AV (dynamic load torque) [12] of the improved sinusoidal motion law is larger compared to that of the cosine motion law, but in terms of the leaps, the leaps of the cosine motion law tend to infinity when T tends to 0, and the flexural shock will occur at $J = \infty$. Therefore, under high speed and heavy load conditions, it is safer to choose the improved sinusoidal motion law.

The convex contour lines are drawn using the software as shown in Figure.7:

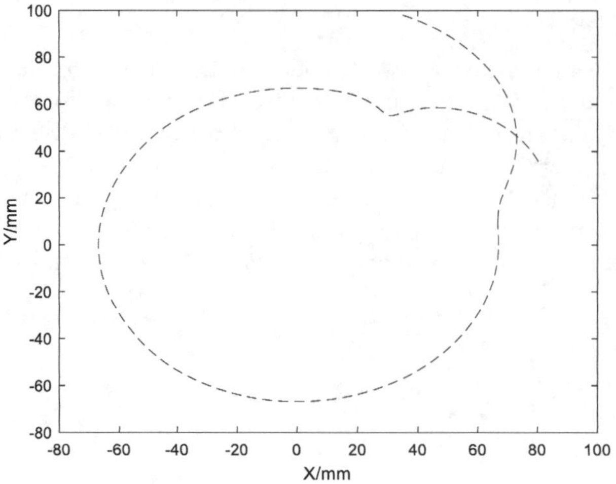

Figure 7 Improved sinusoidal law of motion contours.

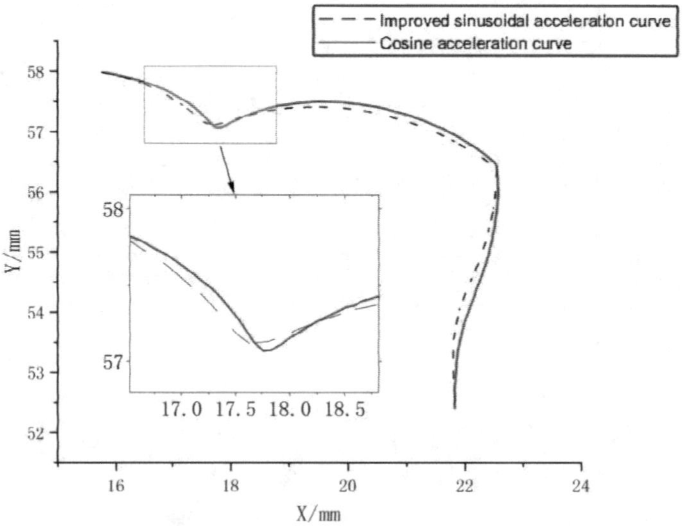

Figure 8 Comparison curves of the contours of the two laws of motion.

Figure 8 shows the comparison of the contour curves of the improved sine motion law and the cosine motion law. It can be seen from the curve that the contour changes of the improved sine motion law at the beginning and end stages are smoother, which significantly reduces the violent jump phenomenon during the movement. This optimized design reduces the contact impact between the cam and the roller, and further extends the service life of the mechanism.The kinematics software is used to simulate the motion of the parallel indexing cams with the modified sinusoidal law of motion to analyse the differences more significantly.

Dynamics simulation software is used to carry out dynamic's simulation analysis of the parallel indexing cam drive mode, remove other irrelevant parts, and import the simplified drive mode components into the software for constraint addition, as shown in Figure 9.

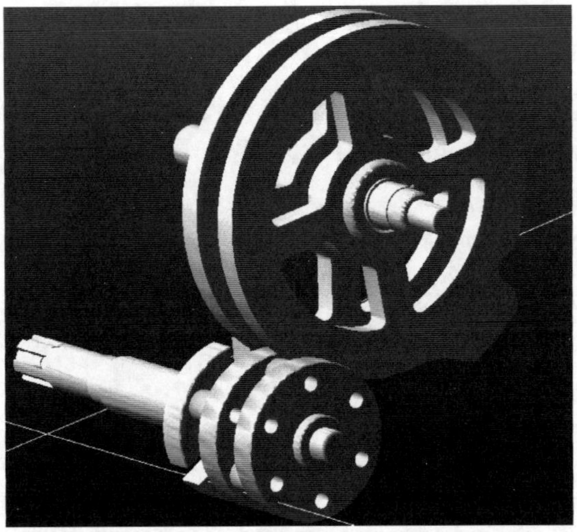

Figure 9 Improved sinusoidal intermittent mechanism model.

The angular velocity change and angular acceleration change curves of the feed axis are shown in Figure 10~Figure 11.

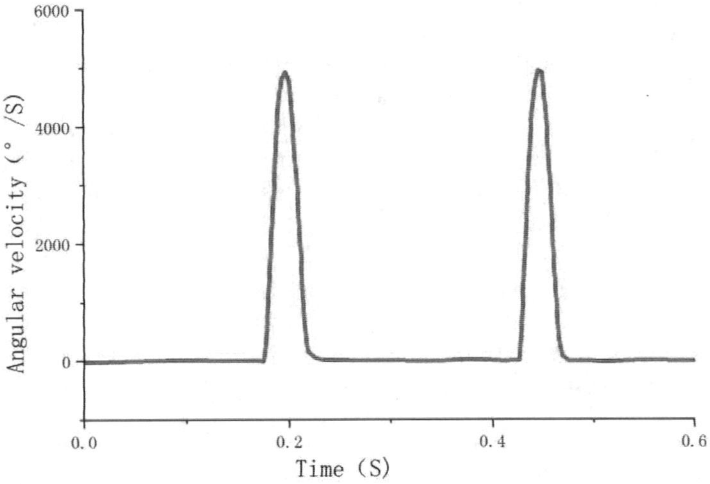

Figure 10 Angular velocity profile of the feed axis (improved sine).

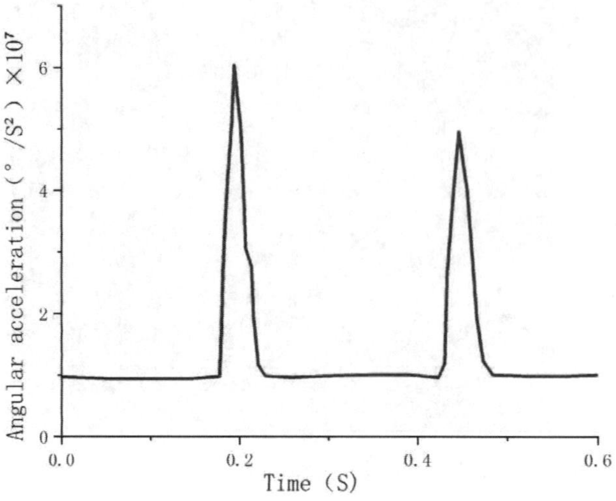

Figure 11 Angular acceleration curve of the feed axis (improved sine).

As can be seen from Figure. 10 and Figure. 11, the angular velocity change when the cam with improved sinusoidal motion is driving the feed shaft is smoother than that of the cosine motion, and there are no spikes and mutations, and the angular acceleration is much smaller than that of the cam with cosine motion, and the maximum value of the angular acceleration is 10%of the original one, and the small fluctuation appearing in the middle of the image is a normal result of the contact of the rigid body, and it also shows that the parallel indexing came with improved sinusoidal motion has better mechanical properties compared with that of the cam with improved sinusoidal motion. The small fluctuation in the middle of the image is the normal result of the contact of the rigid body, which also shows that the parallel indexing cam with improved sinusoidal motion law has better mechanical properties compared with the other cams, and it is more suitable for high-speed and heavy-duty working environment.

The dynamic performance of the cam mechanism is simulated and analyzed at different speeds for the improved sinusoidal and cosine motion laws. The study found that under given parameter conditions, the improved sinusoidal law has lower acceleration fluctuations and a smoother acceleration change trend, as shown in Figure 10 and Figure 11. At the same time, the angular velocity and angular acceleration change curves under different motion laws further illustrate the significant impact of the acceleration jump value on mechanical wear and stability.

5. Conclusion

The intermittent mechanism of an external energy chain launching mechanism was studied to investigate the motion of parallel cams under different motion laws, and the cosine motion law in the original scheme was changed to the improved sine motion law, and the dynamics simulation was carried out for comparison, and the simulation results showed that:

(1) The angular velocity of the parallel indexing cam with the improved sinusoidal motion law changes smoothly, and there is no sudden change in the angular velocity value, which indicates that the collision generated by the improved sinusoidal parallel cam is less than that of the cosine parallel cam.

(2) The contact force generated by the improved sinusoidal parallel indexing cam when driving the feed shaft is much smaller than that of the cosine parallel indexing cam, indicating that the driving process of the improved sinusoidal parallel indexing cam is smoother. Compared with the pre-optimisation, the angular acceleration is 10%of the original.

(3) It is of some reference significance for the drawing of parallel indexing cam and the selection of external energy chain launching mechanism intermittent mechanism.

The simulation results verify the significant advantages of the improved sinusoidal motion law in dynamic performance, but the results may be biased because the material elasticity and lubrication conditions are not considered in the simulation. In subsequent research, the simulation model can be modified in combination with experimental data, and the influence of material properties and lubrication conditions on the performance of the mechanism can be explored to further optimize the design.

References

[1] Huang Li-Jing. CAD/CAM system and motion simulation of parallel indexing cam drive mechanism [J]. Journal of Luoyang Normal College, 2021, 40 (02):12-15. DOI:10. 16594/j. cnki. 41-1302/g4. 2021. 02. 004.
[2] ZHAO Shitian, FU Yingying, ZHENG Lei, et al. Research on modelling and motion simulation analysis of parallel indexing cam mechanism [J] . Mechanical Transmission, 2020, 44 (03):52-56. DOI:10. 16578/j. issn. 1004. 2539. 2020. 03. 009.
[3] FAN Yan.CAD of Parallel indexing Cam Mechanism Based on Matlab[C]//Proceedings of 2010 International Conference on Circuit and Signal Processing ＆ 2010 Second IITA International Joint Conference on Artificial Intelligence(Volume 2). [S.l.]:[s.n.],2010.
[4] Chen Jiting, Yi Zhenqin. Application of UG in the design of parallel indexing cam [J] . Hebei Agricultural Machinery, 2019, (03):65-66. DOI:10. 15989/j. cnki. hbnjzzs. 2019. 03. 049.
[5] WANG Liangwen, LI Ansheng, TANG Weigang, et al. Parametric design of ratchet mechanism [J] . Machine Mechanical Transmission, 2010, 34 (12):27-29.
[6] XIA T,HUANG Y,JIANG X Y, et al.Parametric design of parallel indexing cam based on VB [J].Advanced Materials Research,2011,411： 50-53.
[7] WANG Gang, REN Ziwen, ZHOU Kui. Design method and simulation analysis of polynomial law of motion for cam mechanism [J] . Mechanical Design and Manufacturing Engineering, 2020, 19 (08):23-26.
[8] Zou Huijun, Yin Hongliang. Intermittent motion mechanism design and application innovation [M] . Beijing:Machinery Industry Press, 2008:100-160.
[9] Zhou Yinhe. Research on the motion law of naval gun shell drum based on arc surface indexing cam mechanism [J] . Ship Science and Technology, 2024, 46 (08):169-173.
[10] ZHANG Mengfei, YANG Zhen, LIU Wanchuan, et al. Improved design and simulation of an intermittent mechanism of a rifled launching mechanism [J] . Mechanical Design and Manufacturing Engineering, 2022, 51 (12):6-10.
[11] LIU Mingtao, YUAN Shengnan, CHEN Gang, et al. Dynamic simulation of a new type of two-stage cam-type indexing mechanism [J] . Mechanical Design, 2024, 41 (02):96-101. DOI:10. 13841/j. cnki. jxsj. 2024. 02. 016.
[12] LONG Yongyun. Motion simulation analysis of curved surface cam mechanism based on UG software [J] . Precision Manufacturing and Automation, 2023, (01):37-39+43. DOI:10. 16371/j. cnki. issn1009-962x. 2023. 01. 007.

Journal of Physics: Conference Series **3009** (2025) 012037 doi:10.1088/1742-6596/3009/1/012037

Process optimization of ultra-precision inclined-axis grinding by tungsten carbide aspheric mould

Jing Huang, Rong Gao [*], Han Zheng, Changyan Mou

School Of Intelligent Manufacturing, Chengdu Technological University, Chengdu, China;

* Corresponding author: grong1@cdtu.edu.cn

Abstract: In order to improve the quality of the surface in ultra-precision grinding process by the tungsten carbide mould with complex optical curved surface, this research explore the impact of technological parameter on the accuracy of the inclined-axis ground surface roughness by this particular mould. This paper also analyzes the effect law of, workpiece speed, feed rate and grinding depth on the accuracy of surface roughness by aspheric Tungsten carbide inclined-axis grinding mould.It further discusses the reasonable range of the technological parameter of this grinding process.

1. Introduction

Glass molding uses high-precision molding dies under specific conditions of temperature and pressure to replicate the surface morphology of the mold onto the glass preform. This enables the direct and efficient production of optical glass lenses with functional properties. The method is characterized by high efficiency, high precision, and low cost[1-3]. Compared to traditional processing techniques such as ultra-precision machining, etching, and laser processing, it has the characteristics of high precision, high efficiency, low pollution, and low cost.[4-7] However, to achieve ultra-precision molding of aspheric glass lenses, it is essential first to manufacture high-precision optical molds, as good quality and mechanical properties of tungsten carbide are crucial for obtaining optimal process parameters. Tungsten carbide is a typical ceramic material, and current research is mold accuracy is a prerequisite for obtaining high-precision aspheric lenses[8].

Zhang et al. [9] developed a finite element model for the grinding of SiC f /SiC composites with a single diamond grain and conducted a simulation study on the ultrasonic vibration-assisted grinding process from a microscopic perspective. The simulation results demonstrated that the primary removal mechanism for the composites is brittle fracture.Yin Bo et al.[10] performed simulation analyses on the influence of grinding process parameters on the surface roughness of tungsten carbide alloys. Using the optimized process parameters, they achieved an average surface roughness of 3.379nm for tungsten carbide molds. Han Chen et al.[11] designed a piezoelectric ceramic-driven flexure hinge micro-feed mechanism to precisely control the wheel's cutting depth movement. By employing a single-point grinding method with a cup-shaped diamond wheel, they processed tungsten carbide complex surfaces with a surface roughness Ra of 15nm and a form accuracy RV of 0.25μm.

In this paper, through the analysis of the grinding mechanism of brittle hard materials, the influence of process parameters on the Ra value of surface roughness is studied for the ultra-precision inclined-axis grinding of tungsten carbide molds, and on this basis, the reasonable range of process parameters for tungsten carbide alloy inclined-axis grinding is determined, and the Taguchi method is used to optimize the experimental research to determine the optimal combination of process parameters for tungsten carbide alloy oblique axis grinding.

2. Grinding Mechanism of Hard and Brittle Materials

Understanding the basic grinding mechanism and analyzing the impact of the grinding process on the surface quality and mechanical properties of tungsten carbide are crucial for obtaining optimal process parameters. Tungsten carbide is a typical ceramic material, and current research on the grinding mechanism of ceramic materials is mostly focused on the indentation fracture mechanics model and the cutting model approximation.

2.1 Indentation Fracture Mechanics Model

The indentation fracture mechanics model approximates the interaction between abrasive grains and the workpiece in ceramic grinding as a small-scale indentation phenomenon. By analyzing the material deformation and microcrack formation during the penetration and exit of abrasive grains, an indentation model is established. This model features an expanded plastic zone surrounded by an elastic matrix under a sharp tetrahedral indenter. The relationship between the stable crack size C after loading and unloading under the sharp tetrahedral indenter and the applied load P is given by:

$$\frac{P}{C^{3/2}} = K_c\left\{\zeta(cot\psi)^{\frac{2}{3}}(\frac{E}{H})^{\frac{1}{2}}\right\} \quad (1)$$

where H is the material hardness; E is the material elastic modulus; K_c is the material fracture toughness; ψ is the angle between the indenter face and edge; ζ is a constant.

Equation (1) applies only when the applied load exceeds the critical condition for central crack formation in the material. The critical load P' is given by:

$$P' = 54.5(\frac{\alpha}{\eta^2\gamma^2})(\frac{K_c^4}{H^3}) \quad (2)$$

where α, η, and γ are constants; for a Vickers indenter, $\alpha = \frac{2}{\pi}$, $\eta \approx 1$, $\gamma \approx 0.2$.

This model indicates that during grinding, when the applied load is less than the critical load, material removal occurs primarily through plastic deformation, resulting in an ideal surface roughness.

2.2 Approximation of Cutting Model

The approximation of the cutting model is a common research approach in ceramic grinding mechanics, encompassing measurements of cutting forces, studies of chip morphology, and investigations into processed surface morphology. The cutting model approximation confirms that material removal in ceramics typically occurs through brittle fracture formation, yet a significant portion of energy consumption is closely associated with plastic deformation, specifically the plowing process in ceramic grinding deformation. The model analysis indicates that when the grinding depth is less than the critical depth of cut for the material, plastic flow becomes the primary mode of material removal, leading to improved grinding surface quality. According to the model analysis, the critical depth of cut ac corresponding to the critical fracture load for a single abrasive grain grinding ceramic materials can be expressed as:

$$a_c = \beta(\frac{E}{H})(\frac{K_c}{H})^2 \quad (3)$$

where β is a constant. For critical feed ranges from 3 to 200, assuming the grinding depth equals the machine feed rate, β=0.15.[11]

From the indentation fracture mechanics model and the cutting model approximation, it is understood that the surface quality of tungsten carbide grinding is closely related to the critical load and critical depth of cut. In practical grinding, the influence of critical load and critical depth of cut on the surface quality of tungsten carbide is manifested through process parameters, directly affecting the grinding surface quality. Therefore, studying the influence of process parameters on the surface quality of tungsten carbide grinding and determining a reasonable range of process parameters has practical engineering significance for guiding industrial applications.

2.3 Mold Geometric Parameters

The equation for the non-spherical profile is:

$$Z = \frac{x^2/R}{1+\sqrt{1-(1+K)x^2/R^2}} + A_4x^4 + A_6x^6 + A_8x^8 \quad (4)$$

where: Z is the non-spherical axial coordinate, x is the non-spherical horizontal coordinate, R is the radius of the lens surface, K is the quadratic curve constant, A_4、A_6、A_8 are coefficients for surface shape.

The specific parameters for the non-spherical mold profile equation are shown in Table 1.

Table 1. Workpiece Non-Spherical Geometric Parameters.

Parameters	Concave Non-Spherical
R	32.4048
K	-6.8992
$A_4/10^{-5}$	7.5003
$A_6/10^{-6}$	-1.5188
$A_8/10^{-7}$	1.5294

3. Inclined-Axis Grinding Experimental System and Scheme

3.1 Properties of Tungsten Carbide Material

In this experiment, R07 tungsten carbide material was used. The material properties are shown in Table2.

Table 2. R07 Tungsten Carbide Material Properties.

Material Properties	Property Values
Density $\rho(\mathrm{g/cm^3})$	14.7
Elastic Modulus (GPa)	630
Hardness HRA	94
Fracture Toughness $K_c/(\mathrm{MPa/m^{0.5}})$	6.5
Poisson's Ratio	0.215

3.2 Inclined-Axis Grinding Experimental System

The grinding machine used in the experiment is a four-axis ultra-precision grinding machine.The workpiece spindle has a maximum rotational speed of 1500r/min, and the grinding wheel spindle can reach up to 80,000r/min. The rotational accuracy is less than 0.05μm, with both axial and radial runouts controlled within 15nm.

Figure 1. Aspheric Grinding Experimental System.

3.3 Grinding Wheel Parameters

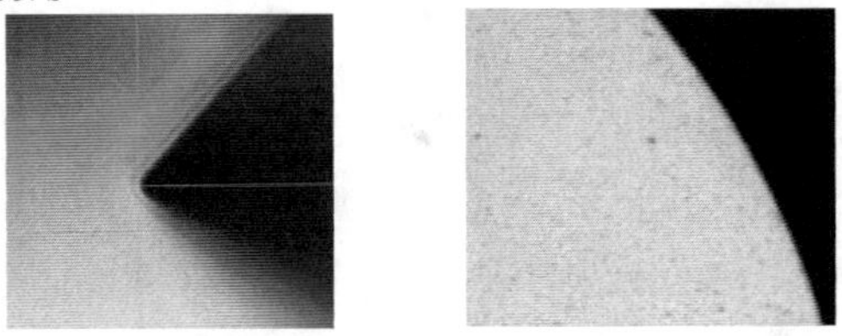

(a)Profile of the Grinding Wheel (b) Surface Topography of the Grinding after Dressing Wheel after Dressing

Figure 2. Profile and Surface Morphology of the Dressed Grinding Wheel.

The grinding wheel used in the experiment is a cylindrical cubic boron nitride (CBN) wheel, with a diameter of 14mm and a grit size of 1.5μm. Before grinding, a PCD (polycrystalline diamond) wheel with a grit size of 20μm is used to dress the CBN wheel. After dressing, the wheel features sharp and complete right-angle edges, with no clogging or other issues, as shown in Figure 2.

4. Analysis of the Influence of Process Parameters on Surface Roughness

In ultra-precision grinding, the surface roughness of tungsten carbide molds is closely related to multiple process conditions, including the grinding wheel, process parameters, andmachining environment. Among these, the key factor influencing the surface roughness of tungsten carbide molds is the process parameters. This paper utilizes the inclined-axis grinding experiment for aspheric tungsten carbide molds, with the parameter ranges shown in Table 3, to study the effects of feed rate, workpiece speed, and grinding depth on the surface roughness (Ra value) of tungsten carbide molds. This allows for the preliminary determination of the reasonable selection ranges for each process parameter.

Table 3. Analysis of the Influence of Process Parameters on Surface Roughness.

Control Factors	**Values 1**
Spindle Speed n(r/min)	20000、25000、30000、35000、40000
Workpiece Speed n_2(r/min)	100、200、300、400、500、600
Feed Rate v(mm/min)	0.1、0.5、1.0、1.5、2.0、2.5
Grinding Depth a_p(μm)	0.3、0.5、0.9、1.3、1.7、2.1

4.1 Effect of Spindle Speed on Surface Roughness

Figure 3 illustrates the variation trend of surface roughness Ra with spindle speed under the conditions of feed rate v=1.0 mm/min, workpiece rotational speed n2=200r/min, and grinding depth ap=0.3μm. According to the curve in the figure, it is evident that as the spindle speed n varies between 10000 r/min and 40000 r/min, the Ra value decreases. When the grinding depth is constant, increasing the spindle speed reduces the material removed per revolution of the grinding wheel, resulting in a thinner cutting thickness. Hence, within this speed range, the Ra value shows a decreasing trend. Therefore, in oblique-axis grinding of non-spherical tungsten carbide molds, it is advisable to use higher spindle speeds within the allowable range.

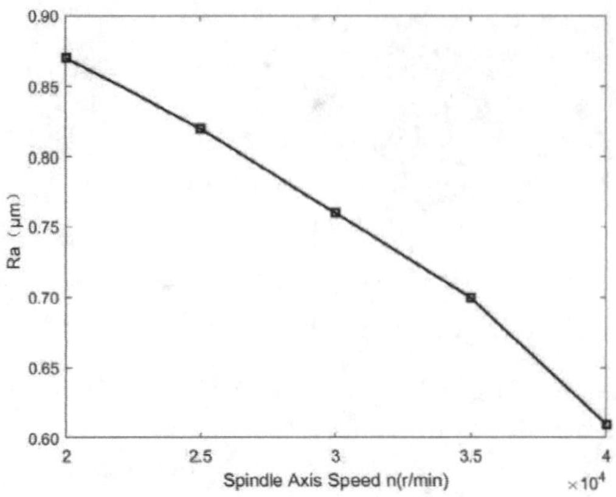

Figure 3. Trend of Roughness with Spindle Speed.

4.2 Influence of Workpiece Speed on Surface Roughness

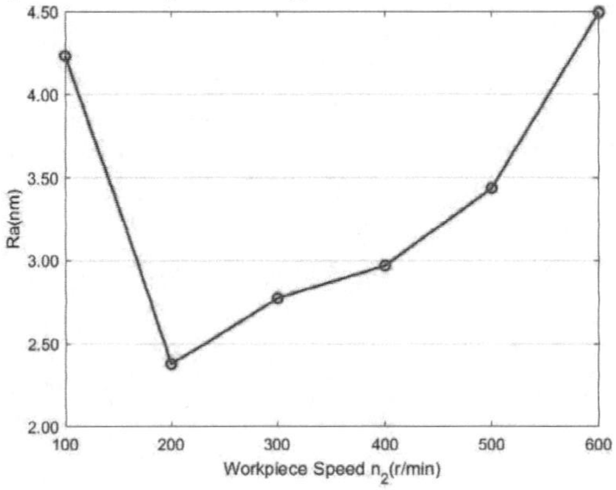

Figure 4. Trend of Roughness with Workpiece Speed.

Figure 4 depicts the variation trend of surface roughness Ra with spindle speed is set at n=40000r/min, feed rate v=1.0mm/min, and grinding depth ap=0.5μm. According to the curve, as the workpiece speed increases from 100r/min to 600r/min, the surface roughness Ra first decreases and then increases. The minimum Ra value occurs when the workpiece speed is 200r/min. At lower feed rates, the grinding particles participate in the grinding process more frequently, allowing material to be easily removed due to plastic deformation, leading to better grinding quality. However, when the feed rate is too low, chips are not easily removed, and excessive heat leads to an increase in surface roughness.[10] Based on these observations, the optimal workpiece speed for inclined-axis grinding of tungsten carbide is around 200r/min.

4.3 Influence of Feed Rate on Surface Roughness

Based on the previous two sets of experiments, with spindle speed n=40000r/min, workpiece speed n_2=200r/min, and grinding depth a_p=0.5μm. Figure 5 shows the trend of surface roughness (Ra value) as the feed rate from 0.5mm/min to 2.5mm/min changes. According to the curve, as the feed rate increases from 0.1mm/min to 2.5mm/min, the surface roughness Ra increases. At lower feed rates, the grinding particles participate more frequently in the grinding process, and material is easily removed due to plastic deformation, resulting in better grinding quality. However, when the feed rate is too low, it becomes difficult to remove chips, and excessive heat can increase the surface roughness.[10] In actual processing, taking efficiency into account, the typical optimal feed rate range for precision inclined-axis grinding is between 0.5mm/min and 1.0mm/min.

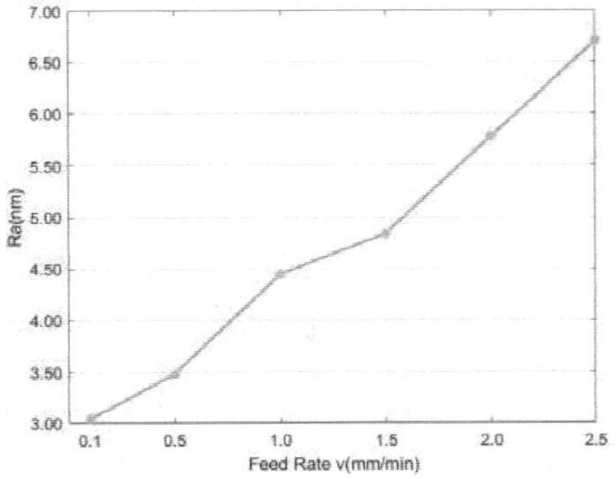

Figure 5. Trend of Roughness with Feed Rate.

4.4 Influence of Grinding Depth on Surface Roughness

In summary, with spindle speed n=40000r/min, workpiece speed n_2=200r/min, and feed rate v=1mm/min. Figure 6 shows the trend of surface roughness (Ra value) as grinding depth a_p increases from 0.3μm to 2.1μm changes. According to the curve, the surface roughness Ra increases. When the cutting depth increases, the cutting force also increases, which leads to higher stress on the spindle, causing greater deformation and an increase in surface roughness. In practical processing, the grinding depth should be selected based on the required surface roughness. For precision inclined-axis grinding, a typical grinding depth of 0.3μm is often chosen for fine finishing.

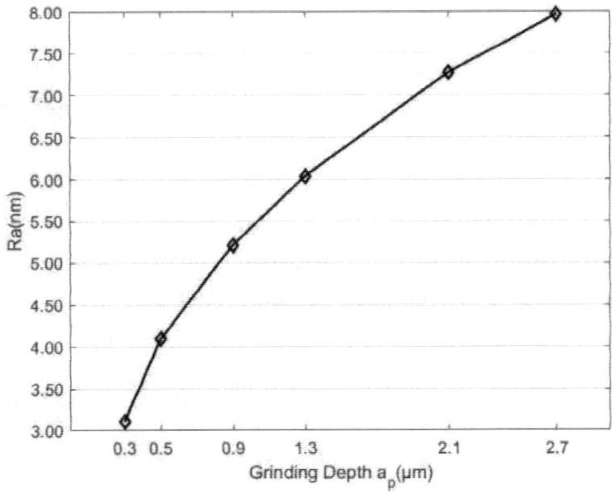

Figure 6. Trend of Roughness with Grinding Depth.

4.5 Orthogonal Experimental of the technological parameter

Based on the process parameter range determined from previous, orthogonal experiments were conducted using a non-spherical Ø14 mm tungsten carbide mold at a spindle speed of 40000 r/min to optimize the process for inclined -axis grinding. The obtained surface roughness Ra values at the center is presented in orthogonal tables, as shown in Table 4.

Table 4. Orthogonal Experimental Data for Non-Spherical Tungsten Carbide Mold Components.

Serial Number	*Grinding Depth* a_p(μm)	*Feed Rate* v(mm/min)	*Workpiece Speed* n_2(rpm /min)	*Center of Workpiece Ra Value* (nm)
1	0.3	0.1	100	3.6116
2	0.3	0.5	200	2.7869
3	0.3	1.5	300	3.7623
4	0.8	0.1	200	5.9117
5	0.8	0.5	300	3.5175

6	0.8	1.5	100	4.8703
7	1.3	0.1	300	3.8974
8	1.3	0.5	100	4.9958
9	1.3	1.5	200	6.5730

From the data in the table, it is evident that the optimal process parameter combination for achieving the best surface roughness includes: spindle speed (n) of 40000 r/min, workpiece speed (n_2) of 200 r/min, feed rate (v) of 0.5 mm/min, and grinding depth (a_p) of 0.3 μm. These parameters correspond to the second set of parameters listed in Table 4. Comparison analysis with other experimental groups demonstrates that this parameter set yields superior surface roughness results, affirming the effectiveness of orthogonal experimentation in optimizing process parameters.

5. Conclusion

This study investigated the influence of process parameters on the surface roughness of aspheric tungsten carbide molds. The effects of workpiece speed, feed rate, and grinding depth on surface roughness were analyzed. Comparative analysis between simulation and experimental results revealed that for ultra-precision inclined axis grinding of tungsten carbide, higher spindle speeds, lower feed rates, and lower grinding depths are optimal. Workpiece speeds around 200r/min were found to be favorable for achieving better surface roughness.

Acknowledgement

This research was funded in part by the Provincial College Student Innovation and Entrepreneurship Training Program of Chengdu Technological University, grant number 202411116047

References

[1] Zhang, Z.Y.; Zhou, CC. Grinding mechanism and mechanical chemical grinding experiment of binderless tungsten carbide at different linear speeds. Aeronautical Manufacturing Technology, 2022, 9, 36–42.
[2] Zhou, S.Q.; Simulation and Experimental Study of Glass Molding for Small Aspherical Lens. Master's Degree, Hunan University, Hunan, 2016.
[3] Yu, T.B.; An, J.H.; Yang, X.Z., etal. The study of ultrasonic vibration assisted polishing optical glass lens with ultrasonic atomizing liquid[J]. Journal of Manufacturing Processes, 2018, 34: 389–400.
[4] Wang, Q.L.; Yao P. Simulation of precision glass molding for aspherical cylindrical microlens array[J/OL]. Chinese Journal of Lasers, 2024: 20240415.
[5] Luo, H.; Zhang, Y.Y., etal. Additive subtractive and formative manufacturing of glass based functional micro/nanostructures: a comprehensive review[J].Materials & Design, 2023:112285.
[6] Zhou, T. F.; He, Y.P.; Wang, T. X., et al. A review of the techniques for the mold manufacturing of micro/nanostructures for precision glass molding[J]. International Journal of Extreme Manufacturing, 2021, 3(4): 042002.
[7] Wang, F.L.; Hu, Y.W. Rapid manufacturing technology for aspheric optical elements[J]. Opto-Electron Eng, 2024, 51(1): 230171
[8] Xu, Z.Q.; Study on Key Technology of Inclined Magnetorheological Polishing for Micro Aspherical Surface[D].Hunan University, Hunan,2015.
[9] Zhang, Z.K.; Yuan, S.M., et al. FEM simulation investigation of ultrasonic vibration-assisted grinding of SiCf/SiC composites[J].Journal of Physics: Conference Series, 2022, 2348(1): 012011.
[10] Yin, B.; Xue, C.X.; Li, C. Grinding simulation and process optimization method of tungsten carbide alloy. Infrared and Laser Engineering, 2023. 07, 284-293.
[11] Han, C.; Dai, Y.F. Guan C.L. A Study on Complex Surfaces of WC Hermetic Parts in Precision Grinding Process. Hydraulics Pneumatics & Seals, 2011, 09, 14-18.

Implementation of a magnetophoretic microfluidic chip for microparticle detection

Jia Lou [1], Shuyue Jiang [1], Wensi Zhang [1], Kai Zhao [*2], Tiger H. Tao[1], Kaihuan Zhang[*1]

[1]2020 X-Lab & the State Key Laboratory of Transducer Technology, Shanghai Institute of Microsystem and Information Technology, Chinese Academy of Sciences, Shanghai, China, 200050
[2]Department of Information Science and Technology, Dalian Maritime University, 116026 Dalian, China

[*]Corresponding author's
*kai.zhao@dlmu.edu.cn; phone +86 150 4063 9607
*kzhang@mail.sim.ac.cn; phone +86 136 8358 8303

Abstract: A novel microfluidic chip based on Fe-polydimethylsiloxane magnetophoresis was designed for continuous separation and detection of metal abrasive particles in oil. Firstly, COMSOL Multiphysics software is used to simulate the designed Fe-PDMS magnetophoresis chip, analyze the influence of flow field, structure and other factors on the trajectory of metal abrasive particles, and optimize the chip structure and parameters to improve the separation efficiency and accuracy of metal abrasive particles. Secondly, the Fe-PDMS magnetophoresis chip was processed by lithography, including the preparation of PDMS microchannels and the preparation of Fe-PDMS magnetic colloids. Finally, the experimental system platform of oil metal abrasive particle separation was constructed. Ferromagnetic particles (iron particles) and non-ferromagnetic particles (aluminum particles) were used for experiments, and the experimental results were analyzed. Through the online detection of wear particles in oil, the reliability of the machine can be improved, and the maintenance cost and safety risk can be reduced. This method provides a powerful tool for wear particle detection and provides an early warning for aging and wear fault diagnosis of mechanical transmission systems. Compared with the traditional sensor detection method, this method has higher detection accuracy and smaller size, which can be more easily applied to practical industrial applications. Therefore, the online detection method based on magnetophoresis separation has a high application prospect, and it will be possible to provide a more reliable machine health monitoring method for the industrial field.

1. Introduction

Microfluidic chips exhibit significant advantages in manipulating micro- and nano-scale objects, typically achieved by employing various external fields such as electric fields [1-2], magnetic fields [3], optical fields [4], and acoustic fields [5] to control the motion of microscopic or nanoscale objects. Among these, magnetic manipulation technology stands out as an intriguing and widely applied technique with several notable advantages. Magnetic fields can penetrate non-magnetic or weakly magnetic materials (e.g., glass and polydimethylsiloxane) as well as biological substances, enabling the effective non-contact manipulation of micro- and nano-scale objects in closed environments. Magnetic fields can also be utilized to manipulate and differentiate between magnetic and non-magnetic objects with distinct physical properties [6]. For example, non-magnetic objects can be manipulated either via magnetic labeling or through a label-free approach using ferrofluids.

Additionally, particles of different shapes may exhibit varying magnetic forces and torques in a magnetic field, resulting in differences in their motion behavior within microfluidic channels [7-9]. On small scales, magnetic fields of various types can be generated and precisely controlled, enabling magnetic manipulation to be realized in diverse ways [10].

Magnetophoretic mixing is one of the most widely applied uses of magnetic fields in microfluidic systems, with applications in chemical reactors, bioanalysis, chemical synthesis, and polymerization [11]. Ferrofluids prepared via magnetic mixing techniques can be used to manipulate non-magnetic particles. By leveraging the magnetic force gradients generated by externally applied non-uniform magnetic fields, fluid flow can be disrupted, thereby enhancing mixing performance. Experimental studies commonly employ permanent magnets and ferrofluids containing magnetic nanoparticles. During the mixing process, the ferrofluid experiences motion induced by magnetic body forces or additional flows generated by the motion of magnetic particles, thereby improving mixing performance. This technique, which controls fluid flow by manipulating magnetic forces, holds significant application potential. In this work, a microfluidic chip based on the Fe-PDMS magnetophoresis for the continuous separation and detection of metallic wear particles in oil. The developed microfluidic chip provides a powerful tool for detecting wear particles, featuring high detection precision, miniaturization, and integration, offering early warnings for aging and wear fault diagnosis in mechanical transmission systems. The novel Fe-PDMS composite material used in this study combines the ferromagnetic properties of iron with the excellent adhesion and plasticity of PDMS. It can be easily positioned on one side of the microchannel, generating a gradient magnetic field under the magnetization of a permanent magnet. As wear particles flow near the Fe-PDMS region, they are subjected to magnetophoretic forces, causing trajectory deviation and enabling the separation of different metallic particles. Experimental results demonstrate that this method effectively separates ferromagnetic and non-ferromagnetic particles and achieves the separation of ferromagnetic particles of different sizes.

2. Design and fabrication

Magnetophoretic separation is a method that utilizes a gradient magnetic field to separate particles based on their varying magnetic properties and forces. This paper aims to design a microfluidic chip based on the Fe-PDMS magnetophoresis principle to separate ferromagnetic particles of different sizes and to distinguish between ferromagnetic and non-ferromagnetic particles in oil. Specifically, the chip operates by introducing a gradient magnetic field perpendicular to the fluid flow direction. This gradient field generates differential forces on particles depending on their size or magnetic properties, causing varying degrees of trajectory deflection during their flow, thus achieving particle separation. To achieve efficient particle separation, this study uses Fe-PDMS material to fabricate the microfluidic chip. Fe-PDMS possesses excellent ferromagnetic properties and plasticity, enabling the generation of a significant magnetic field gradient within the microfluidic channel for enhanced particle separation.

IOP Publishing
doi:10.1088/1742-6596/3009/1/012038

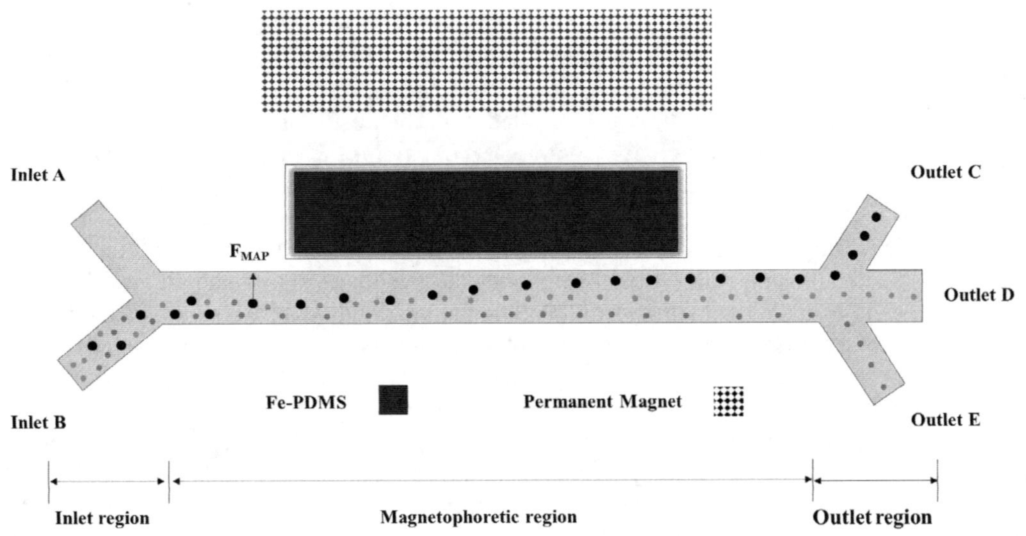

Figure 1. Schematic illustration of the Fe-PDMS based microfluidic channel.

The microfluidic chip consists of an inlet region, a magnetophoretic region, and an outlet region. The inlet region includes Inlet A and Inlet B. Inlet A connects to the sheath fluid flow, and Inlet B connects to the sample flows. The sheath fluid flow directs the sample flow to the side of the microchannel away from the Fe-PDMS material. On one side of the magnetophoretic region, Fe-PDMS and a magnet are placed. Under the influence of the magnet, Fe-PDMS generates a secondary magnetic field, which deforms the distribution of magnetic field lines when superimposed with the original magnetic field. Due to the close placement of Fe-PDMS within the microchannel, a gradient magnetic field is formed at the Fe-PDMS edge inside the channel. The outlet region comprises three exits: C, D, and E. Figure 1 shows a top view of microparticles separated through magnetophoresis in the microchannel. The figure clearly illustrates the trajectories and separation of microparticles in the microchannel. In the mixture of metal wear particles, ferromagnetic iron particles experience a positive magnetophoretic force, causing trajectory deviations. Since the magnetophoretic force on particles is proportional to their size, larger iron particles are deflected closer to the magnet compared to smaller particles. As a result, larger and smaller particles are directed to separation outlets C and D, respectively, while aluminum particles, unaffected by magnetophoretic forces, exit through outlet E. The chip's structure adopts a symmetric design. Apart from the rectangular Fe-PDMS and the magnet, the channel is symmetrical about the central axis of the microfluidic channel. The distance between the rectangular Fe-PDMS and the microchannel is fixed at 30 μm, with the Fe-PDMS center aligned with the center of the microchannel. The entire microchannel has a depth of 60 μm. The horizontal main channel has a length and width of 2000 μm and 200 μm, respectively, while the inlet channels (A and B) and outlet branches (C-E) have a length and width of 800 μm and 100 μm, respectively. The permanent magnet used in this study has dimensions of 3 mm × 2 mm × 1 mm (length × width × height) and a remanent flux density of 1.18 T. Simulations and experimental analyses demonstrate that the gradient magnetic field generated by the novel Fe-PDMS composite material within the microchannel effectively separates wear particles of different sizes and magnetic properties in the oil. The microfluidic oil wear particle separation device is simple to fabricate and delivers noticeable results, laying a favorable foundation for subsequent particle counting and other detection processes in oil analysis.

The microfluidic channel and Fe-PDMS are both fabricated using standard photolithography. The photolithography process involves masks for shaping, and the final microchannel and Fe-PDMS are bonded using a self-designed alignment system combined with a microscope. After plasma bonding treatment, the Fe-PDMS-based magnetophoretic microfluidic chip is obtained. The rectangular Fe-

Journal of Physics: Conference Series **3009** (2025) 012038 doi:10.1088/1742-6596/3009/1/012038

PDMS fabrication process is shown in Figure 2, and the schematic diagram of the magnetophoretic microfluidic chip structure is shown in Figure 3. Fe-PDMS is fabricated on glass substrates coated with DuPont dry film, a negative photoresist. After exposure, the unexposed parts are removed during development, leaving grooves. The grooves are filled with the Fe-PDMS mixture, which is cured, and the remaining DuPont film is washed away with NaOH, leaving the rectangular Fe-PDMS structures on the glass substrate. A glass substrate coated with DuPont dry film (DuPont, SD238, USA) is prepared. The protective plastic films on both sides are removed to expose the green DuPont dry film surface for subsequent operations. The DuPont dry film-coated side of the glass is aligned with the Fe-PDMS rectangular mask (non-transparent black film). Exposure is performed at 10 mW/cm^2 for 10 seconds. After exposure, the dry film turns blue in the exposed regions, while unexposed regions remain green. The exposed glass substrate is immersed in a 1% Na_2CO_3 solution, dissolving the unexposed film to form grooves. The development quality is checked under a microscope, and adjustments are made as necessary. The developed glass substrate is heated at 80°C for 10 minutes to strengthen the remaining dry film. Fe-PDMS is prepared by mixing PDMS (10:1 base-to-curing agent ratio) with 1 μm iron particles at a 75% weight fraction. The mixture is stirred for 15 minutes and left to settle for 2 hours. The Fe-PDMS mixture is placed into the grooves using a glass rod, compressed with a coverslip, and cured at 120°C for 35 minutes. Excess Fe-PDMS is cleaned off, leaving a rectangular Fe-PDMS structure. The remaining dry film is removed using a 3% NaOH solution. The cleaned rectangular Fe-PDMS structures are transferred to a new glass slide and bonded after heating at 80°C for 10 minutes.

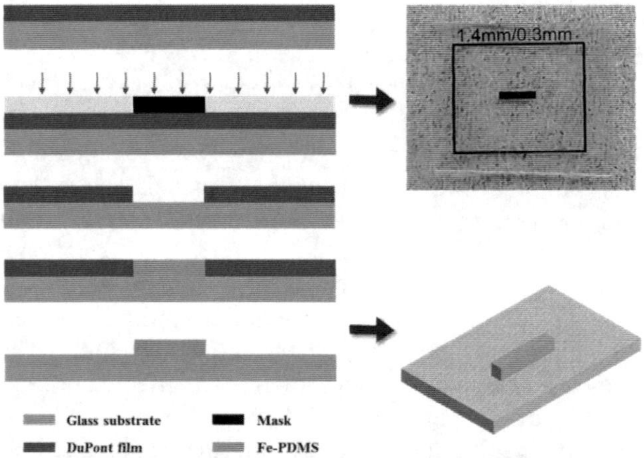

Figure 2. Schematic illustration of the Fe-PDMS microfluidic chip fabricated by standard photolithography.

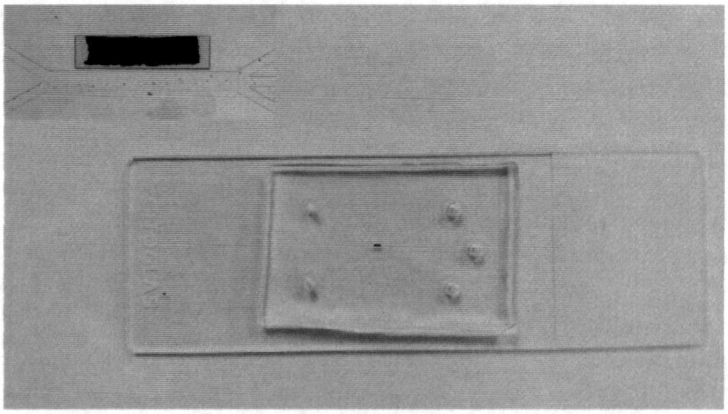

Figure 3. Schematic diagram of magnetophoretic microfluidic chip structure.

3. Results and discussion

The effect of the chip structure on particle trajectories was investigated. Given the numerous variables in the chip structure, including channel dimensions, a symmetrical design was adopted for simplicity. This symmetry, apart from the rectangular Fe-PDMS and the magnet, ensured that the channel remained symmetric along the central axis of the microfluidic centerline. On this basis, the distance between the rectangular Fe-PDMS and the microchannel was fixed at 30 μm, and the center position of the Fe-PDMS was aligned with the center of the microchannel. Following this, the effects of Fe-PDMS length and the magnet distance from the microchannel were studied. It should be noted that regardless of changes in the rectangular Fe-PDMS length, its center position remained aligned with the center of the microchannel, while the magnet distance referred to the distance between the lower edge of the magnet and the wall of the microchannel.

To investigate the effect of Fe-PDMS length on particle trajectories, all other conditions were held constant. By varying the Fe-PDMS length, changes in the size of the non-uniform magnetic field region and particle trajectories in the microchannel were observed, as shown in Figure 4. The results indicate that the smaller the length of the rectangular Fe-PDMS, the larger the non-uniform magnetic field region produced in the microchannel. This is because the magnet's length is finite, and asymmetry becomes more pronounced when the rectangular Fe-PDMS length is small relative to the magnet. Within a certain range, a longer rectangular Fe-PDMS results in smaller particle deflection angles, as shown in Figure 4(a), where particles exit from outlet E. Conversely, a shorter Fe-PDMS results in larger deflection angles, as shown in Figure 4(b), where particles exit from outlet D. Therefore, the deflection angle of particles can be effectively controlled by adjusting the Fe-PDMS length, enabling the desired separation effect.

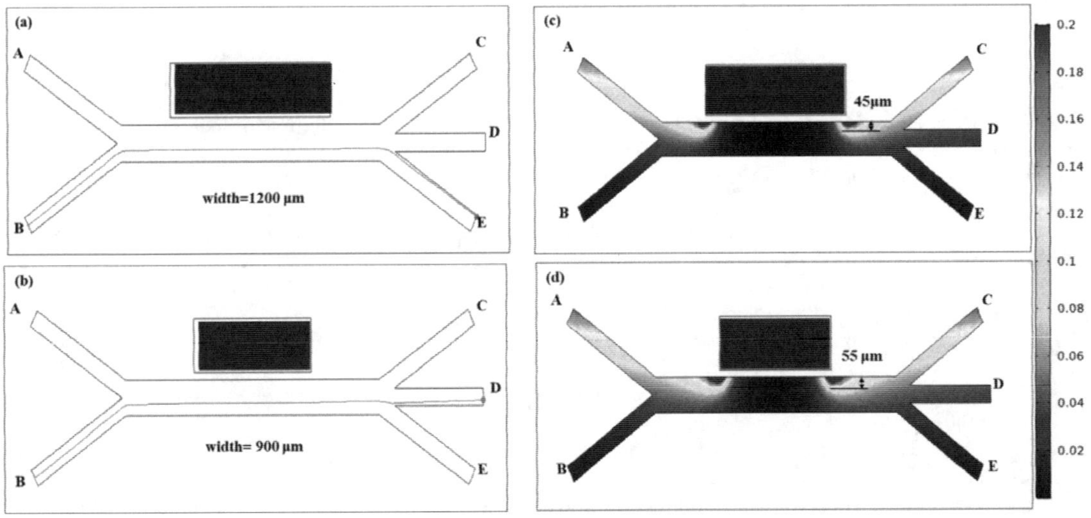

Figure 4. Schematic illustration of the particle trajectories with the Fe-PDMS width of (a) 1200 μm and (b) 900 μm, and the magnetophoretic gradient with the Fe-PDMS width of (c) 1200 μm and (d) 900 μm.

The performance of the proposed chip was compared to existing microfluidic systems. The novel Fe-PDMS composite material combines the ferromagnetism of iron with the excellent adhesion and plasticity of PDMS, enabling easy integration into microchannels. Unlike conventional chips, the developed system achieves enhanced precision in separating ferromagnetic and non-ferromagnetic particles and enables size-based separation of ferromagnetic particles, which is critical for applications like wear particle detection in lubricating oils. Additionally, its high integration and miniaturization offer significant advantages for early fault diagnosis in mechanical systems. These features underscore its superior performance in terms of precision, versatility, and practical utility.

4. Conclusion

This study designs a microfluidic chip based on the Fe-PDMS magnetophoresis principle for the continuous separation and detection of metallic wear particles in lubricating oil. The developed microfluidic chip provides a powerful tool for detecting wear particles with high precision, miniaturization, and integration, offering early warning for aging and wear fault diagnosis in mechanical transmission systems. The novel Fe-PDMS composite material used in this study exhibits both the ferromagnetism of iron and the excellent adhesion and plasticity of PDMS. This allows it to be easily positioned on one side of the microchannel and generate a gradient magnetic field under the magnetization of a permanent magnet. As wear particles flow through the vicinity of the Fe-PDMS with the fluid, they experience magnetophoretic forces that cause trajectory deflection, thereby enabling the separation of different metallic particles. Experimental results demonstrate that this method effectively separates ferromagnetic particles from non-ferromagnetic particles and achieves size-based separation of ferromagnetic particles.

Acknowledgement

This work was supported by grants from the National Key Research and Development Program of China (Grant No. 2018AAA0103100), the National Science Fund for Excellent Young Scholars, and the CAS Young Talents Training Program. This work was supported by grants from the National Natural Science Foundation of China (62401555). This work was supported by the Postdoctoral Fellowship Program of CPSF under Grant Number GZC20232838.

References

[1] Xuan X. Recent advances in direct current electrokinetic manipulation of particles for microfluidic applications [J]. Electrophoresis, 2019, 40(18-19): 2484-2513.

[2] Jianhe, Guo, Jeremie, et al. Electric-Field-Guided Precision Manipulation of Catalytic Nanomotors for Cargo Delivery and Powering Nanoelectromechanical Devices [J]. ACS nano, 2018, 12(2): 1179-1187.

[3] Leong S S, Ahmad Z, Low S C, et al. Unified view of magnetic nanoparticle separation under magnetophoresis [J]. Langmuir, 2020, 36(28): 8033-8055.

[4] Gao D, Ding W, Nieto-Vesperinas M, et al. Optical manipulation from the microscale to the nanoscale: fundamentals, advances and prospects [J]. Light Science & Applications, 2017, 6(9): e17039.

[5] Baron V O, Chen M, Hammarstrom B, et al. Real-time monitoring of live mycobacteria with a microfluidic acoustic-Raman platform [J]. Communications Biology, 2020, 3(1): 236.

[6] Su P, Ren C, Fu Y, et al. Magnetophoresis in microfluidic lab: Recent advance [J]. Sensors and Actuators A: Physical, 2021, 332: 113180.

[7] Zhou Y, Xuan X. Diamagnetic particle separation by shape in ferrofluids [J]. Applied Physics Letters, 2016, 109(10).

[8] Zhou R, Bai F, Wang C. Magnetic separation of microparticles by shape [J]. Lab on a Chip, 2017, 17(3): 401-406.

[9] Matsunaga D, Meng F, Zöttl A, et al. Focusing and sorting of ellipsoidal magnetic particles in microchannels [J]. Physical review letters, 2017, 119(19): 198002.

[10] Cao Q, Fan Q, Chen Q, et al. Recent advances in manipulation of micro-and nano-objects with magnetic fields at small scales [J]. Materials Horizons, 2020, 7(3): 638-666.

[11] Dalili A, Samiei E, Hoorfar M. A review of sorting, separation and isolation of cells and microbeads for biomedical applications: microfluidic approaches [J]. Analyst, 2019, 144(1): 87-113.

Vacuum ultrasonic welding of the CF/PEEK rods by small-sized on-orbit welding end

Chen Ai[1], Zhanjiao Gao[2], Jinqi Pan[2*], Licheng Liu[2], Jiayong Yan[2] and Yun Tian[2]

[1]China Academy of Space Technology, Beijing, China
[2]Beijing Satellite Manufacturing Factory Co., Ltd , Beijing, China

* Corresponding author, e-mail address: 17862701823@163.com

Abstract: High-performance thermoplastic composites are one of the main materials for large truss structures in space. In this paper, for the on-orbit rapid connection of thermoplastic composite truss structures, a lightweight and small-sized ultrasonic welding end with a weight reduction of 64% which is suitable for vacuum high and low-temperature environments, was designed, and a welding control strategy was proposed. Moreover, the test platform for the vacuum welding process was established to complete the experimental validation. The experimental results showed that the ultrasonic welding method presented good feasibility, high welding efficiency, and ease of miniaturization of the equipment, and could be applied to the space welding of 2 mm CF/PEEK rods. The load capacity of the CF/PEEK joint could reach 83 N. Furthermore, the microstructure of the joints showed that the carbon fibers could be tightly bonded to the resin, and almost no voids could be discovered at the welding interface. The application of this technology can promote technical development for the rapid connection of large space truss structures on-orbit manufacturing.

1. Introduction

Thermoplastic composites have very broad application prospects in aerospace, electronics, and other fields due to their lightweight, high specific strength, adjustable thermal conductivity, recyclability, excellent electromagnetic shielding, and moldable complex structures. At present, thermoplastic composites are mainly divided into short-fiber composites (SFT), long-fiber composites (LFT), glass fiber mat reinforced thermoplastic sheet (GMT), fabric prepregs and unidirectional continuous fiber reinforced thermoplastic prepregs (CFRTP). Their connection technology has experienced the development of mechanical connections, adhesive connections, and new connection technologies. In the 1960s, NASA started research on the mechanical connection of thermoplastic materials. However, the mechanical connection destroyed the fiber structure of the composite material and greatly reduced its performance. By the 1990s, the technology of adhesive joining of composite materials was developed, but the peel strength of the adhesive tip was low, and the performance was easily affected by the spatial environment. Now, many researchers are committed to new joining technologies for thermoplastic composites, and welding technology gradually stands out due to its lower cost and high efficiency [1-3].

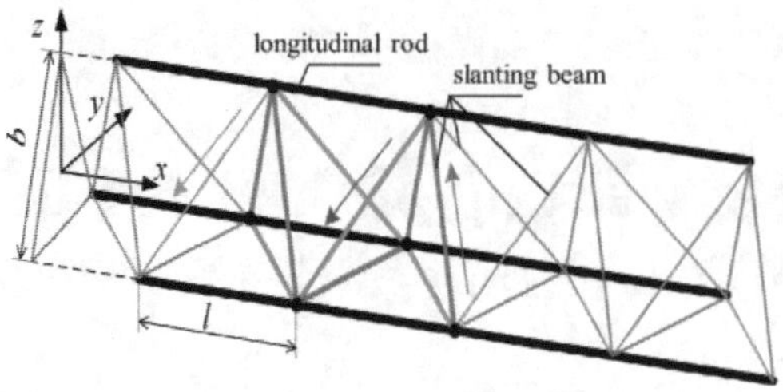

Figure 1. Efficient and high-strength connection of space truss structure.

There are many welding methods for thermoplastic composites, including hot plate welding, infrared welding, induction welding, resistance welding, two-body resin welding, dissolution welding, vibration welding, friction stir welding, and ultrasonic welding [4-7]. However, for the on-orbit manufacturing of thermoplastic composite trusses, large-scale welding equipment cannot meet the actual demand, and the lightweight and small-scale ultrasonic welding, which is suitable for vacuum and high-low temperature environments, is considered to be an excellent welding method for on-orbit manufacturing due to lightweight, small welding surface, fast heating speed, high joint strength, low power consumption, ease of automation, and no high-temperature damages. It has been widely used in the manufacturing of thermoplastic composite trusses and the aerospace field. In this paper, a lightweight and small-scale ultrasonic welding end is designed and developed for the on-orbit connection of carbon fiber-reinforced thermoplastic composite truss structure, as shown in Figure 1. Besides, the CF/PEEK rods are tested to fabricate the space truss structure by the ultrasonic welding end, and the welding quality of the joints is evaluated, which lays the foundation for the application of ultrasonic welding technology on-orbit manufacturing.

2. Design of lightweight and compact ultrasonic welding ends

The ultrasonic welding machine adopts a split welding machine. The power supply is separated from the welding handle, which makes it easy to fix the handle on the loading mechanism, and the power supply is placed in the control cabinet. The ultrasonic vibration welding machine consists of four parts: vibration cover 1, vibration sleeve, transducer and welding head, and vibration cover 2, as shown in Figure 2.

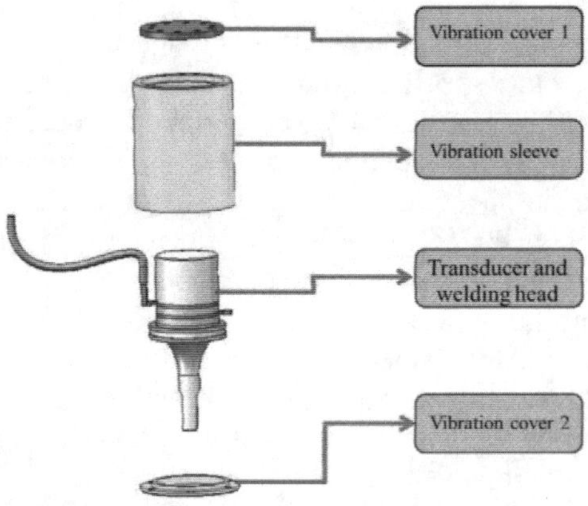

Figure 2. The diagram of the welding machine.

Figure 3 shows the welding process. The tip of the welding head is equipped with a guiding surface at a certain angle, which guides the longitudinal rod deviating from the welding position to the welding position, accommodates the positional deviation of the slanting beam on the longitudinal rod, and ensures the consistency of the spacing and the welding points. When the two rods come under the welding head, the welding head moves downward through the welding loading assembly to weld the slanting beam. After welding, the welding machine moves upward through the loading assembly to complete a welding action.

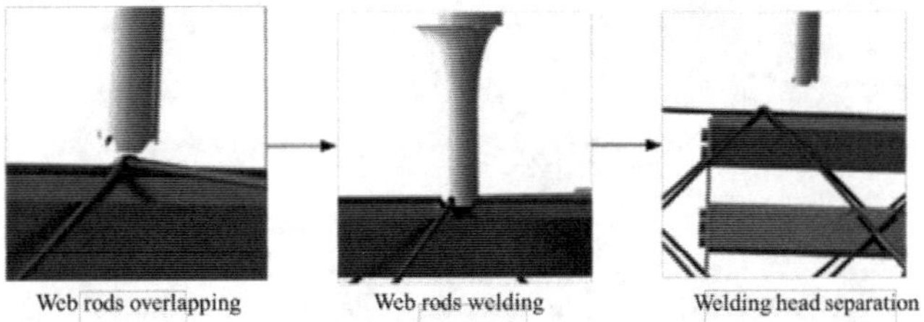

Figure 3. Welding working diagram.

2.1 Ultrasonic generator design

An ultrasonic generator transforms the industrial frequency (50 Hz) current into ultrasonic frequency (15~60 kHz) oscillating current through the output transformer and transducer matching. There are electron tube amplification, transistor amplification, thyristor inverter, transistor inverter, and other road forms. The electron tube type efficiency is low (30-45%) and has been replaced by the transistor amplification type. Currently, the most widely used is the transistor-amplified generator. In the ultrasonic generator as a welding application, automatic tracking frequency is necessary. Some generators are also equipped with a constant amplitude controller to ensure that the mechanical amplitude of the acoustic system remains constant. At this point, the selection of a suitable amplitude sensor will become a technical key. This ultrasonic generator adopts a self-developed control system, which can automatically adjust the frequency as well as control the amplitude, and is an important part of the ultrasonic system.

2.2 Acoustic system design

The acoustic system of the welding machine consists of three components: the transducer, the amplifier, and the welding head, which are mainly used for energy conversion and transmission. The transducer is used to convert high-frequency and high-voltage electrical energy into mechanical energy. The amplitude modulator increases or decreases the vibration according to the transmission ratio. The welding head strengthens and transmits the vibration (amplitude) to the welded parts. To reduce the overall length of the welding machine, the acoustic system adopts the design of abandoning the amplitude modulator, the transducer, and the welding head to form a duo group, and the welding head is made of titanium alloy material. The 2D dimensions of the transducer and welding head are shown in Figure 4.

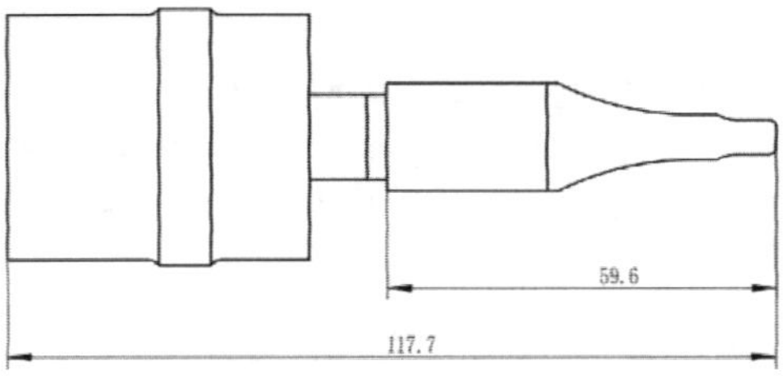

Figure 4. The 2D dimensions of the transducer and welding head.

According to the formula λ = c/f, where c is the propagation speed of the sound wave in a particular tissue and f is the frequency. The length of the welding head or the transducer can only be an integer multiple of the half wavelength, and according to the requirement of the minimum length, both the transducer and the welding head take up one-half wavelength. The actual design requires the use of finite element analysis to assess the structural reliability, as shown in Figure 5, and the final solution yields a weld head length of 59.6 mm and a total acoustic system length of 117.7 mm.

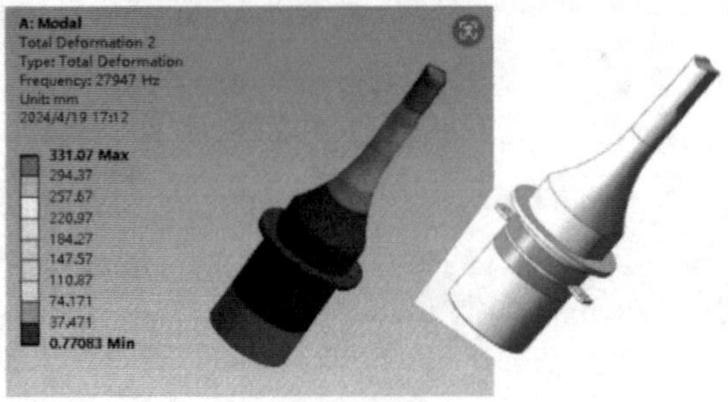

Figure 5. Simulation and analysis diagram of miniaturized acoustic system.

2.3 Piezoelectric transducer design

The ultrasonic transducer will ultrasonic generator output high-frequency AC signal excitation piezoelectric ceramics using the inverse piezoelectric effect, which will be converted to the same frequency mechanical vibration signal, and ultimately through the welding head will be the energy effect on the welded parts. The transducer is used to generate electromagnetic oscillation for the ultrasonic generator at the same frequency as mechanical vibration. The transducer consists of a shell, front driver, piezoelectric ceramic, electrodes, rear driver, cooling holes, and other parts, as shown in Figure 6. The piezoelectric ceramic disc transducer is made of PZT-5 piezoelectric material polarized in the thickness direction.

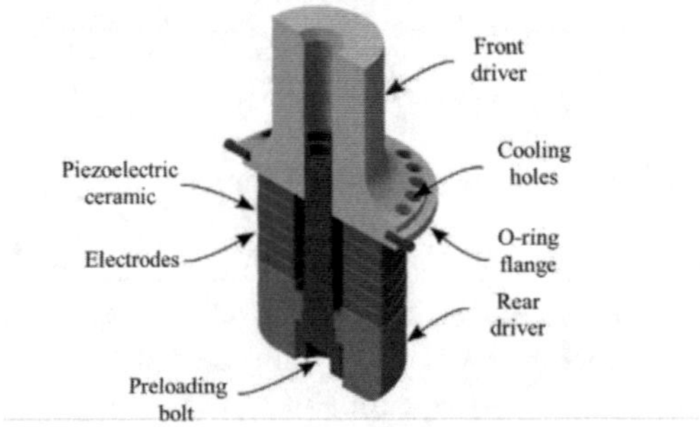

Figure 6. The schematic diagram of the transducer.

2.4 Welding head design

After analyzing material winding conditions and considering the material thickness, position error, and other influencing factors, a variety of welding head end optimization designs were carried out to meet the requirements of the variable width design and ensure the quality of the weld surface. Finally, the dot matrix welding head configuration was selected.

2.5 Prototype development

According to the requirements, the on-orbit ultrasonic welding equipment was designed, which adopted the abolition of the amplitude lever, as shown in Figure 7, and the use of a "generator + welding head" directly assembled mode to shorten the length of the welding gun. Compared with the ground welding equipment, the weight of on-orbit welding decreases by 64%, which can significantly improve space welding efficiency.

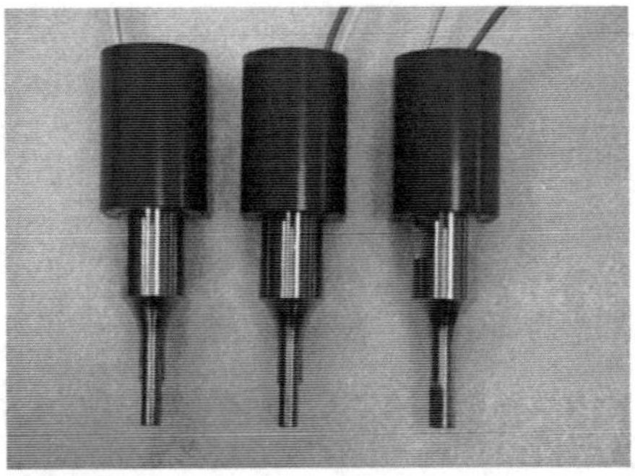

Figure 7. The ultrasonic welding equipment.

3. Ultrasonic welding test of the cf/peek rods under a vacuum environment

The ultrasonic welding test equipment is constructed to carry out the ultrasonic welding test of thermoplastic composite materials under a vacuum environment. The system mainly includes a vacuum loading system, pressure loading system, ultrasonic welding machine, welding support base, temperature measurement system, and pressure sensor. The limit of the vacuum loading system is 0.1 Pa, which can be carried out in a 0-1.0 Pa vacuum environment using thermoplastic composite materials for an ultrasonic welding test. The pressure sensors were set at the bottom of the installation to detect and control the welding pressure. The welding power is 500 W, and the welding frequency is up to 35 kHz, in which the transducer in the ultrasonic handle adopts a sandwich-type piezoelectric transducer, and the amplitude change rod is a stepped amplitude change rod.

A 35 kHz prototype was used to weld 2 mm+2 mm diameter carbon fiber reinforced thermoplastic composites (CF/PEEK rods), and the space truss structure is shown in Figure 8. The real-time frequency, amplitude, power, voltage, displacement, and pressure change curves were collected during the ultrasonic welding process to collect the welding parameter changes. The welding process uses energy mode, triggered by the displacement depth, which sets the welding energy at 60 J and the amplitude at 50%. Three joints (marked 1#, 2#, and 3#) were selected to analyze the welding quality using a scanning electron microscope (SEM). After the tensile test, the load capacity is 83 N, which basically meets the truss strength requirements (>50 N).

Figure 8 Macroscopic morphology of the CF/PEEK rods space truss structure by ultrasonic welding.

Figure 9 shows the SEM figures of three welded joints to analyze the fusion on the interface of the vertical cross-section. At the 1# and 3# welded joints of the rods, the carbon fibers are tightly bonded to the resin, and no voids can be discovered on the interface, indicating that the resin is melted and well connected to the carbon fiber during the ultrasonic welding process. However, a small number of voids can be observed at the welding interface of the 2# joint. The reason for this phenomenon can be attributed to the high strength of CF/PEEK rods at room temperature, which shows large stress concentrations at the joint. Although these stress concentrations are largely released during ultrasonic welding, they are difficult to eliminate completely. And the complete elimination of residual stresses and voids can be achieved by secondary welding. Besides, no impurities can be found at the welding interface, indicating that airborne impurities do not enter the weld interface and affect the weld properties of the material when ultrasonic welding in a space vacuum.

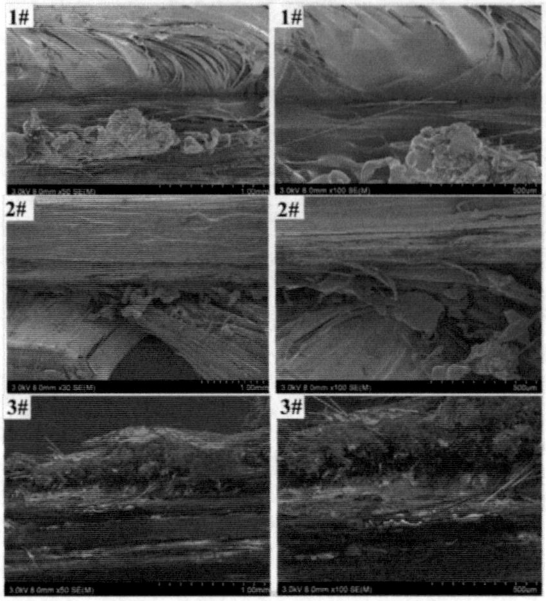

Figure 9. The SEM figures of different ultrasonic joints on CF/PEEK rods.

The welding surface of 1#~3# joints is presented in Figure 10 to evaluate the melting and solidification of the CF/PEEK rods by tearing off the joints. These results show that the rods are well integrated with each other without any gap at the interface. The welding head transmits the ultrasonic energy to the weld zone, which generates localized high temperatures due to the high acoustic resistance at the interface between the two rods. The resin of the contact surface melted due to the energy transmitted by high-frequency vibrations and wrapped the carbon fibers under pressure. When the ultrasound stops working, the resin gradually solidifies during the pressure holding time, resulting in an effective bonding of the rods.

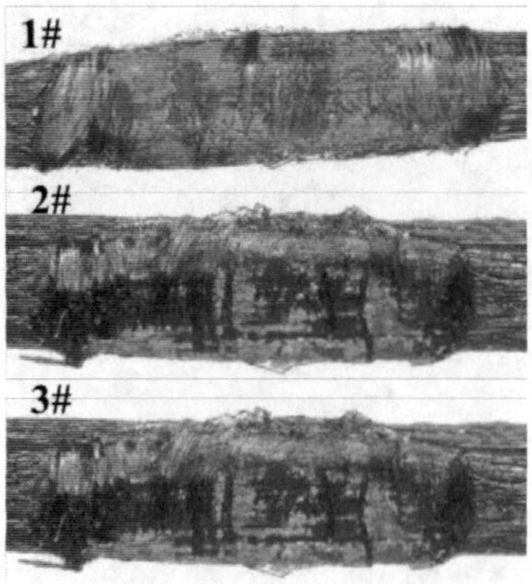

Figure 10. The welding surface of 1#~3# joints.

4. Conclusions

This paper researched vacuum ultrasonic welding technology and equipment for longitudinal rods and slanting beams of the space truss using CF/PEEK rods. Moreover, a lightweight and small-sized ultrasonic welding end, which was suitable for vacuum high and high-low temperature spatial environments, was designed with a weight reduction of 64%, and the entangled vacuum welding of CF/PEEK longitudinal rods and slanting beams of the truss structure was tested. The results showed that the load capacity of the CF/PEEK joint was 83 N. The SEM figures also showed that the carbon fibers could be tightly bonded to the resin, and almost no voids or impurities could be discovered on the welding interface. These works can promote the development of the on-orbit manufacturing technology of thermoplastic composite material in the future.

References

[1] F. V. Irene, V. M. Regis, Ultrasonic welding of carbon/epoxy and carbon/PEEK composites through a PEI thermoplastic coupling layer, Composites Part A, vol.109, pp. 75-83, 2018.
[2] I. F. Villegas, B. V. Grande, H.E.N. Bersee, R. Benedictus, A comparative evaluation between flat and traditional energy directors for ultrasonic welding of CF/PPS thermoplastic composites, Composite Interfaces, vol. 22, pp. 717–729, 2015.
[3] G. Palardy, I. F. Villegas, On the effect of flat energy directors thickness on heat generation during ultrasonic welding of thermoplastic composites, Composite Interfaces,vol. 24, pp. 203–214, 2017.
[4] J. Goncalves, J. F. D. Santos, L. B. Canto, Friction spot welding of carbon fiber-reinforced polyamide 66 laminate, Materials Letters, vol. 159, pp. 506-509, 2015.
[5] Y. Huang, X. Meng, Y. Xie, Friction spot welding of carbon fiber-reinforced polyetherimide laminate, Composite Structures, vol.189, pp. 627, 2018.
[6] W. F. Qin, J. W. Fu, X. Y. Wang, Study on eddy current heating property of carbon fiber composite based on vacuum Ni plating, New Chemical Materials, vol. 48, pp. 69-72, 2020.
[7] L. Zhou, Z. W. Qin, S. Liu, Progress on joining technology of thermoplastic resin matrix composites, Materials Review, vol. 33, pp. 3177-3183, 2019.

Research on the design of composite fixtures for thin-walled shell parts

Jun Han, Zhenyao Zhao*, Zexi An, Junwei Zhu and Yang Zhang

Inner Mongolia University of Science and Technology, School of Mechanical Engineering, Baotou, Inner Mongolia 014010, china

*Corresponding author's e-mail: zhaozhenyao99@163.com

Abstract. Aiming to address the problem of traditional fixtures in thin-walled parts machining not effectively inhibiting the deformation of parts machining, a special fixture combining negative pressure adsorption and multi-bit support is designed. The fixture consists of a bottom-wall adsorption section and a side-wall support section. A vacuum adsorption disk positions the workpiece and limits the degrees of freedom, and a cylinder and linkage system provides support to counteract the milling forces during machining. Through finite element simulation, the research results show that the fixture can effectively suppress the deformation of thin-walled parts, reducing the deformation to 0.04208mm. The design enhances the fixture's stability and reliability in part machining and improves work efficiency and product quality.

1. Introduction

Thin-walled parts are widely used in aerospace, the automotive industry, electronic products, and other high-tech fields because of their advantages, such as their lightweight and high strength. In the thin-walled parts machining process, because of its weak structure and ease of deformation, the traditional fixture design methods often can not meet modern manufacturing requirements for high precision and high efficiency. Therefore, the research and development of new fixture design methods, especially fixtures that can be adapted to the special needs of thin-walled parts, is an important topic in the current manufacturing field.

As a key component of the manufacturing process, the design and research of specialized fixtures have received extensive attention from researchers at home and abroad due to their strong relevance and wide range of applications. MM Natarajan et al.[1] uses a fixture with a cylindrical head for end milling, which improves surface roughness and suppresses distortion. Mori K et al.[2] proposed a method for estimating the frequency response of the workpiece and optimizing fixture placement to suppress vibration in thin-wall milling. Liu Y et al.[3] proposed an evolutionary algorithm to optimize the fixture layout to minimize the deflection and torque of thin-walled workpieces. S Rathaur et al.[4] designs fixtures for clamping tapered workpieces to ensure that the fixtures perform well in machining. ML Sonjaya et al.[5] designs fixturing aids that significantly improve product productivity. The clamping adaptive digital twin (CADT) model proposed by Y Guo[6] effectively solves the problem of clamping deformation of formed blades. A Naeem designed a milling fixture pallet system that increases productivity and reduces costs. V Karpus et al.[7] proposes to optimize the fixture design of parts such as levers using structural codes to improve the flexibility and efficiency of CNC multi-axis machining.

In order to solve the problem of excessive local machining error of thin-walled shell parts, a special fixture was designed, and the fixture was simulated and analyzed by using finite element software, and the designed fixture was verified to inhibit the deformation of parts machining through test milling.

2. Analysis of the structure and clamping method of thin-walled shell parts

As shown in Figure 1, the part is made of 7075-T651 aluminum alloy material, the blank size is 150 mm × 80 mm × 16 mm. 7075 aluminum alloy has excellent mechanical properties and good corrosion resistance, but its rigidity is poor.7075 aluminum alloy has excellent mechanical properties and good corrosion resistance, but its rigidity is poor. When machining thin-walled parts, due to the lack of rigidity of the aluminum alloy, the parts are prone to deformation, and the traditional clamping method makes the parts deformed in the machining process. This deformation is concentrated at the bottom of the sidewalls of the part cavities, making it difficult to keep the machined dimensions and geometry within tolerance.

The traditional machining clamping method of the part is shown in Figure 2. The blank is clamped with a flat vise for cavity milling and then bolted to the outer wall for milling due to the thin wall thickness in the outer wall milling process, and the cavity is not supported, resulting in deformation of the part by the radial extrusion of the milling cutter, the deformation amount of up to 0.264mm, the size of the accuracy is difficult to ensure.

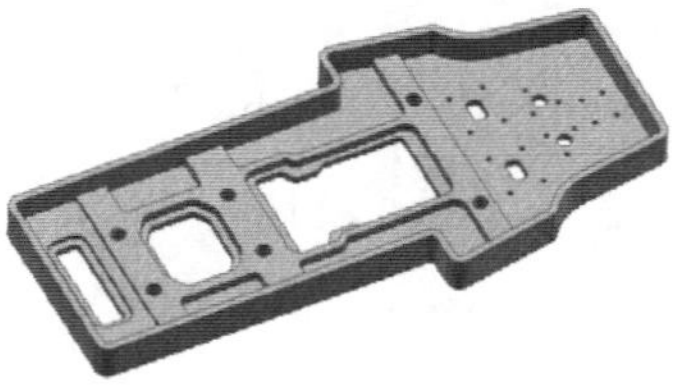

Figure 1. Parts drawing.

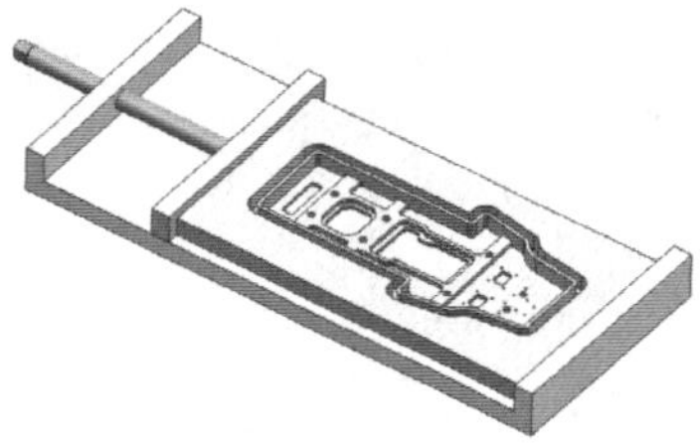

Figure 2. Conventional clamping.

The thinnest point of the part after machining is only 1.5mm, the wall thickness is small, and the aluminum alloy material has low stiffness, which are other factors that make it difficult to ensure the machining accuracy of the workpiece during clamping and machining. Since the parts are involved in mass production, the pneumatic method of clamping is used in the machining to save the time of manual clamping, which helps to speed up the machining progress. Therefore, the design of a composite pneumatic fixture using a combination of adsorption and support layout is reasonable, and the clamping force is appropriate.

3 Composite fixture design

3.1 Fixture structure

The special fixture design adopts the bottom wall adsorption and side wall support to solve the machining deformation of thin-walled shell parts. The fixture consists of two parts, respectively, the bottom wall adsorption part and the inner wall support part, the vacuum adsorption disk for the positioning part of the workpiece and provides the adsorption force part, the cylinder connecting rod support plate to form the support part. The composite fixture design is shown in Figure 3. Cylinder through the connecting rod for the support plate to provide support force, used to offset the milling force of the parts machining process, to reduce the deformation of the parts, and to ensure machining accuracy. The bottom fixture base is for realizing a stable and smooth process, improving clamping speed, reducing the clamping process, and improving work efficiency. The exploded view of the specialized composite fixture is shown in Figure. 4.

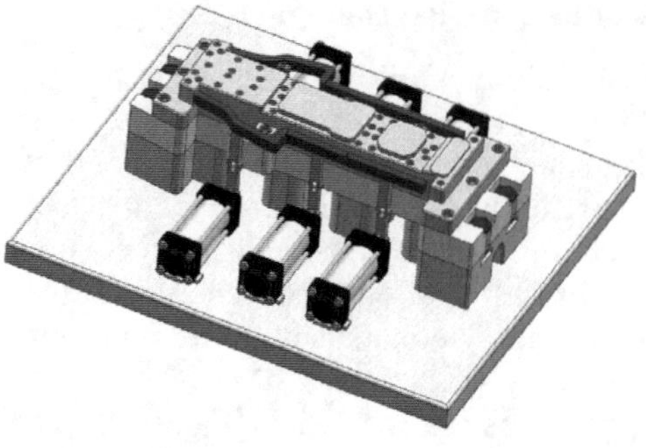

Figure 3. Fixture Design.

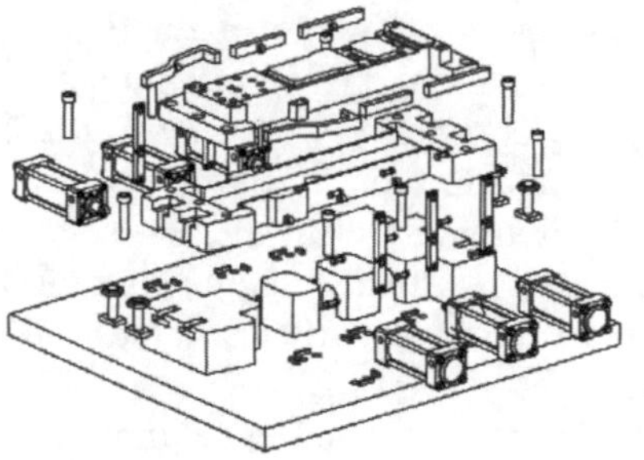

Figure. 4. Exploded view of the fixture.

3.2 Calculation of milling force

Clamping force and milling force need to meet the equilibrium constraints. The parts will be deformed by the milling force during machining, so the finite element analysis of the special fixture clamping in the finishing and semi-finishing stages will obtain the deformation of thin-walled shell parts. To ensure that the designed special fixture program can meet the requirements of the dimensional accuracy of the part, the generalized formula for the milling force is as equation (1)[8]:

$$F = Cv^x a_p^y f^m a_w^n d \tag{1}$$

This can be obtained by taking logarithms on both sides of the equal sign of equation (1) above and converting to a linear equation:

$$lgF = lgC + xlgv + ylga_p + mlgf + nlga_e + lgd \tag{2}$$

Transformation according to the quadratic linear model leads to a linear equation (2) of:

$$y=b_0+b_1x_1+b_2x_2+b_3x_3+b_4x_4 \tag{3}$$

The exponential formula for the milling force in the *x*, *y*, and *z* directions is obtained from the regression analysis of the data[9].

$$\begin{aligned} F_x &= 9.105a_p^{0.897} f^{0.0271} v^{-0.4727} a_w^{0.7519} dk \\ F_y &= 15.8a_p^{0.9190} f^{0.0255} v^{-0.4895} a_w^{0.7983} dk \\ F_z &= 5.487a_p^{0.843} f^{0.0022} v^{-0.5413} a_w^{0.7277} dk \end{aligned} \tag{4}$$

where $a_p = 5.5\text{mm}, f = 600\text{mm/min}, v = 68\text{m/min}, a_w = 1.5\text{mm}, d = 10\text{mm}$.Bringing these da ta into equation (4) and calculated to get the result.$F_X = 93.39N, F_y = 156.12N, F_z = 32.05N$.

Theoretically, the cutting force and the theoretical restraining force have to reach a static equilibrium, but in actual machining, the milling force calculation formula needs to be multiplied by the safety coefficient K. The actual restraining force formula is equation (5)(6)[10]:

$$F_C = KC_F k_F a_w^{0.86} a_f^{0.72} d^{-0.86} z a_p \tag{5}$$

$$K = K_0 K_1 K_2 K_3 \tag{6}$$

where $C_F = 167$, $k_F = 1.0$, $a_w = 0.6mm$, $a_f = 0.1mm$, $d = 10mm$, $z = 4$, $a_p = 4mm$, $K_0 = 1.5$, $K_1 = 1.2$, $K_2 = 1.3$, $K_3 = 1.0$.Bringing these data into equation (5)(6)and calculated to get the result $F_C = 105.99N$.

3.3 Design of bottom wall adsorption part

Vacuum adsorption is a work method of fixation by creating a negative pressure environment to apply clamping force. This part has several places where the dimensional accuracy requirements are stringent, the smallest tolerance of ± 0.05mm, the wall thickness is thin, and easy to deform, taking into account the location of the adsorption point and the size of the adsorption force on the deformation of the part also has an impact. The distribution of adsorption pores is shown in Figure 5. Therefore, under the premise of satisfying reasonable positioning and support, it is also necessary to calculate the suitable adsorption force, and the force of milling force is shown in Figure 6.

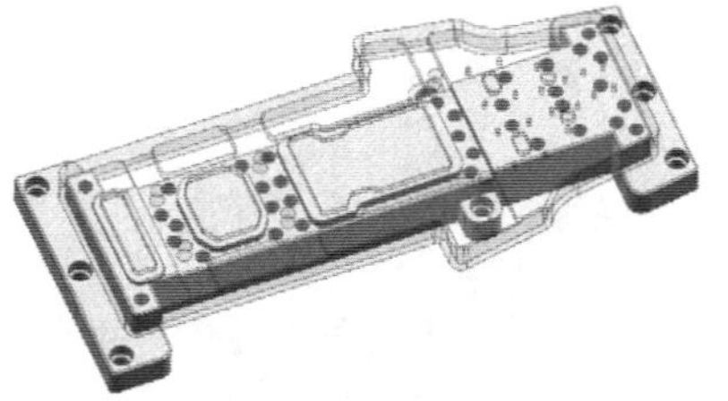

Figure. 5. Adsorption pore distribution.

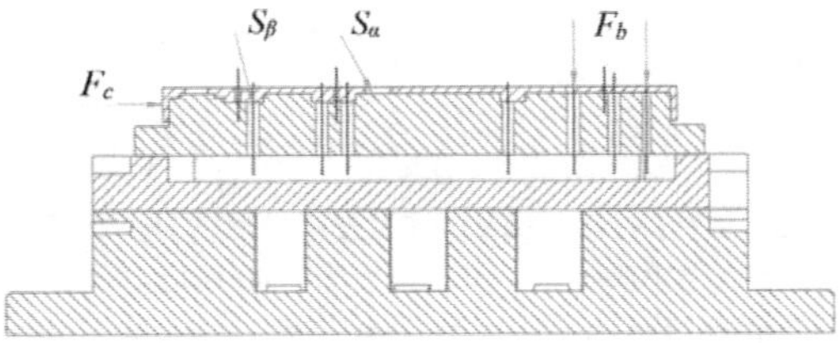

Figure. 6. Force diagram.

According to the adsorption force calculation formula, combined with the structure of the thin-walled shell parts analyzed. The formula is as equation (7) (8):

$$P = P_a - F_C/(0.15S_\beta) \tag{7}$$

$$S_\beta = \pi D^2 n/4 \tag{8}$$

Where $F_C = 105.99N$,P_a=1.01MPa,D=3mm,n=34, Bringing these data into equation (7)(8)and calculated to get the result P=1.93MPa.

Considering the processing environment adsorption plate material selection has high strength, high corrosion resistance, and high temperature-resistant 316 stainless steel.

3.4 Design of inner wall support section

The actuator of the special fixture selects the cylinder, which can realize automatic clamping and release, shorten the clamping time, and improve the production efficiency; it can precisely control the clamping force and ensure the machining accuracy; the structure is simple, convenient for maintenance and replacement and has good adjustability.

When the parts are clamped, the cylinder carries out pneumatic pressurization and depressurization behavior, and it is necessary to ensure the stability of the movement when clamping, so a single rod double-acting cylinder is chosen as the execution element.

So that the support plate has a good support effect, the profile of the support plate and the inner wall profile of the part closely fit, and the segmented design ensures multi-position support. The distribution of the support plate is shown in Figure 7.

The cylinder transmits the support force to the support plate through the connecting rod, which is used to counteract the milling force by supporting the side wall of the part from the inside, in which the principle of leverage plays an important role. The design of the support section is shown schematically in Figure 8.

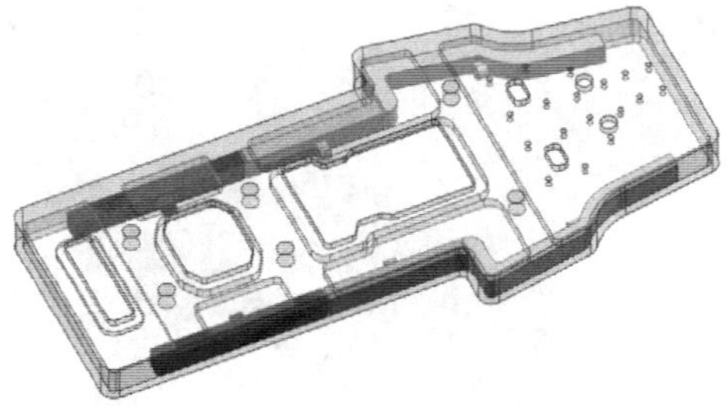

Figure. 7. Support plate distribution position.

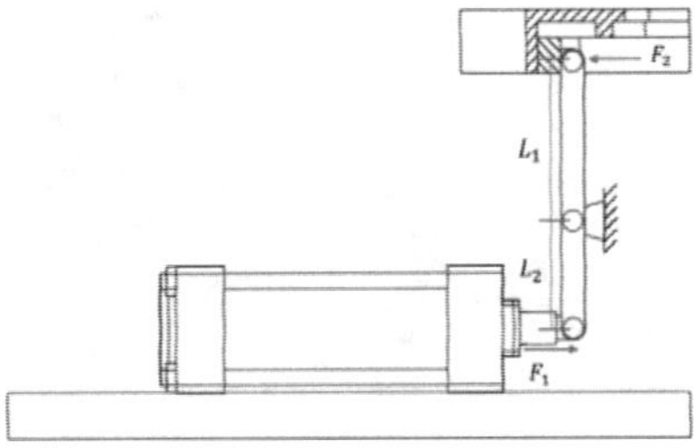

Figure. 8. Support section design.

Use the principle of leverage to calculate the selection of a suitable cylinder by equation (9):

$$F_1 \times L_1 = F_2 \times L_2 \tag{9}$$

Where $F_2 = 105.99\text{N}, L_1 = 50\text{mm}, L_2 = 35\text{mm}$ .Bringing these data into equation (9)and calculated to get the result $F_1 = 74.19\text{N}$.

The formula for calculating the cylinder bore is as equation (10):

$$\dot{D} = \sqrt{\frac{4}{\pi} \times \frac{F}{P\mu}} \tag{10}$$

Where $\mu = 0.6$; $P = 0.3$MPa.The cylinder has to push the connecting rod. $F \geq F_1$.Bringing these data into equation (10)and calculated to get the result $\dot{D}$=22.91mm.

After consulting the manual, the standard values of inner diameter and piston rod diameter were obtained as $D = 40$ mm,$d = 14$mm.

The formula for calculating cylinder air consumption is as equation (11):

$$Q = \frac{\pi \times (\frac{D-d}{2})^2}{1000} \times L \times f \tag{11}$$

Where f =60, L =2mm, $D = 40$mm , $d = 14$mm .Bringing these data into equation (11)and calculated to get the result $Q = 13.2$ L/min.

Based on the data obtained from the above calculations, SMC CQ2B40-10D was selected as the fixture cylinder by considering the bore, stroke, structure, and air consumption per minute of the cylinder.

4. Specialized fixture finite element analysis

4.1 Physical Model Creation

The thin-walled shell parts with six support plates are geometrically modeled using UG modeling software and imported into finite element analysis software for force analysis.

The material of the thin-walled housing parts is 7075-T651 aluminum alloy and the material of the support plate elements is Q235.

The meshing is performed for the parts as well as the support plates, and the cell type selected for the parts is an eight-node linear hexahedral cell (C3D8R) with a total of 11,571 cells. The cell type selected for the six support plates is a four-node linear tetrahedral cell (C3D4). The meshing results of the parts and support plates are shown in Figure 9.

4.2Binding settings

Apply constraints on the part and the support plate, the part is adsorbed on the vacuum suction cup, so the vacuum hole position of the part cavity is subjected to adsorption of fixed constraints, and the support plate applies a support force of 105.99N transmitted to the side wall of the part cavity, and the radial milling force of the milling cutter is 156.12N. The constraints applied to the part and the support plate are shown in Figure. 10.

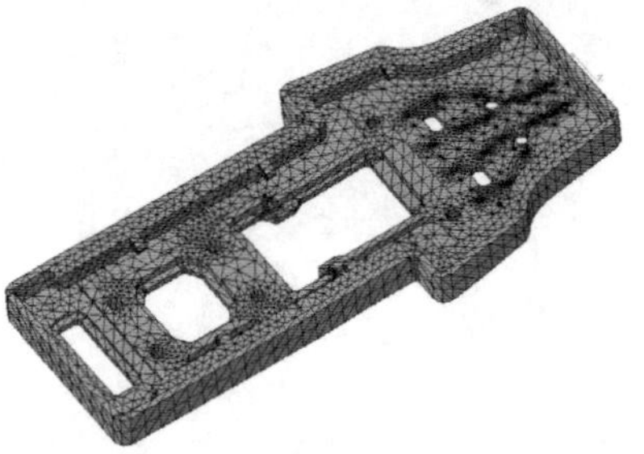

Figure. 9. Grid division diagram.

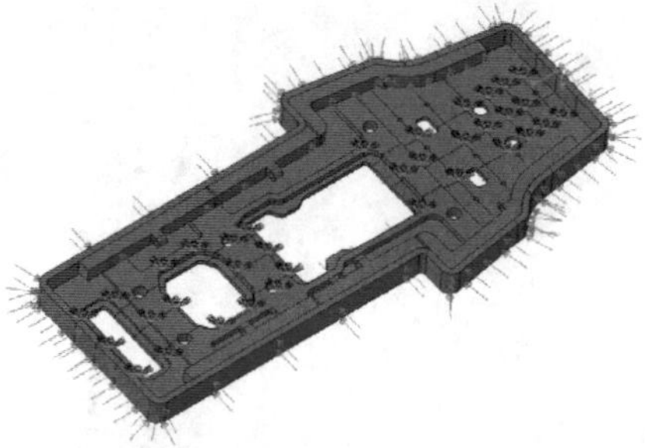

Figure. 10. Apply constraints diagram.

4.3Simulation machining results analysis

The results of finite element simulation analysis after machining with special fixtures are shown in Figure 11. The maximum deformation of 0.04208mm using special composite fixtures for machining, which is a significant reduction in deformation compared to the deformation of 0.265mm with the traditional fixtures for machining.

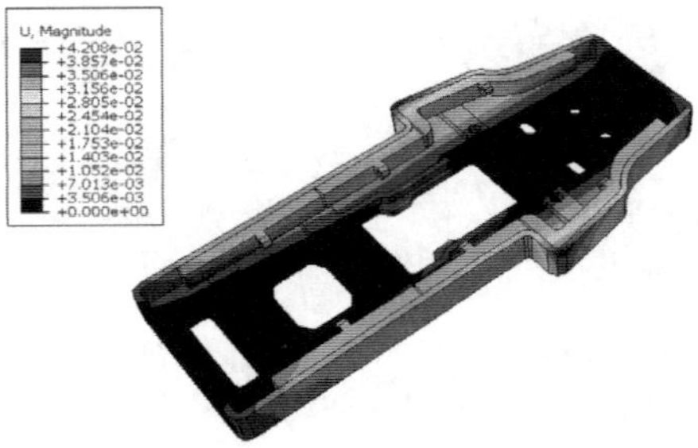

Figure 11. Deformation analysis chart.

5. Conclusion

For thin-walled shell parts processing deformation, control of its sidewall dimensional accuracy is particularly important in this paper through the adsorption and support of the two existing holding methods to the combination of the design of a composite fixture for the adsorption to limit the axial displacement of parts, combined with the support plate to inhibit the side wall of the radial force extrusion deformation, the design is feasible through finite element simulation software to verify the measured error of the parts after machining in the 0.05 mm to meet the accuracy requirements. Precision requirements to solve the problem of parts processing error are too large.

References

[1] Natarajan, M. M., Chinnasamy, B., Alphonse, B. H. B.(2022) Investigation of Machining Parameters in Thin-Walled Plate Milling Using a Fixture with Cylindrical Support Heads. Strojniški vestnik - Journal of Mechanical Engineering., 68: 746-756. https://doi.org/10.5545/sv-jme.2022.273.
[2] Mori, K., Matsubara, A.(2022) Estimation of supporting fixture receptance for thin-walled milling. CIRP Annals., 71: 333-336. https://doi.org/10.1016/j.cirp.2022.04.038.
[3] Liu, Y., Guan, S., Zhao, H., et al.(2022) Evolutionary Design of Machining Fixture Layout for Thin-Walled Structure. Mathematical Problems in Engineering., 2022:5216966. https://doi.org/10.1155/2022/5216966.
[4] Rathaur, S., Sanoria, R., Abidi, S. H. A., et al. (2021) Design and Fabrication of CNC Lathe Fixture for Cone. In: International Conference on Mechanical and Energy Technologies. Singapore. pp.177-186. https://link.springer.com/chapter/10.1007/978-981-19-1618-2_17.
[5] Sonjaya, M. L., Massaguni, M., Amaluddin, M. N. H.(2023) Reducing Cycle Time in the Manufacturing Process of Circular Pattern Interpolar Products (CPIP) Using Special Fixtures on a 3-Axis CNC Machine. INTEK: Jurnal Penelitian., 10: 131-136. https://doi.org/10.31963/intek.v10i2.4439.
[6] Guo, Y., Ren, J., Liang, Y., et al. (2023) Construction of digital twin for clamped near-net-shape blade in adaptive manufacturing. Journal of Manufacturing Processes., 108: 12-25. https://doi.org/10.1016/j.jmapro.2023.10.055.
[7] Karpus, V., Ivanov, V., Dehtiarov, I., et al. (2019) Technological assurance of complex parts manufacturing. In: Advances in Design, Simulation and Manufacturing. Sumy, Ukraine. pp:51-61. https://link.springer.com/chapter/10.1007/978-3-319-93587-4_6.
[8] Han, J., Shen, W. D., Zhao Z. Y., et al. (2023) Research on flexible fixture of annular thin-walled parts based on clamping layout. Manufacturing Technology & Machine Tool., 73: 101-106. https://link.cnki.net/doi/10.19287/j.mtmt.1005-2402.2023.11.015.
[9] YI, C.H.,(2023) Machining Process Handbook. Chemical Industry Press, Beijing. https://thinker-cnki-net-s.webvpn.imust.edu.cn:8118/bookstore/book/bookdetail?bookcode=9787122405869000&type=book.
[10] Yang, Y., Zhang, L., Liu, C. J., et al. (2019) Research on the whole CNC machining technology of complex frame parts in a certain type of cabin. Manufacturing Technology & Machine Tool., 69: 13-16. https://link.cnki.net/doi/10.19287/j.cnki.1005-2402.2019.10.001.

Study of robotic underwater anticorrosion coating process

Heng Zhong[1 #], Yibiao Liu[2 #], Shilun Du[2], Jun Hu[1], Yukai Jin[2], Zhenghu Chen[1], Guangyao Chen[1], Jingxuan Meng[1], Yong Lei[2*]

[1]Three Gorges Hydropower Plant, Yichang, Hubei, China, 443000
[2]State Key Lab of Fluid Power & Mechatronic Systems, Zhejiang University, Hangzhou, Zhejiang, China, 315000
[#] These authors contributed equally to this work and should be considered co-first authors

[*] Corresponding author: ylei@zju.edu.cn

Abstract: To achieve automated maintenance of the anticorrosion coating layer on underwater steel structure surfaces, it is essential to design and study the effects of the parameters of the automated coating process. Based on the mechanical design of the coating unit, an experimental platform was developed to investigate the effects of various coating parameters on the coating quality. Several case studies were conducted to study the effects of the design parameters on coating quality. The painting tool experimental results demonstrated that for automated painting mechanisms, the roller brush is more suitable than the brush typically used in manual operation. Extrusion speed experiment found that high or lower extrusion speed could cause various coating defects. Further investigation into the effects of tool movement speed revealed that higher tool movement speed leads to higher pull-off strength. Finally, an optimized coating tool and parameters were found for the automated maintenance mechanism. Future studies could be conducted on the effects of different underwater environments or on different coating surfaces.

1. Introduction

In large underwater structures, during the operation, factors such as water flow corrosion and bio-erosion can progressively degrade and dislodge the protective coatings on the steel surface, leading to rust formation. Hence, to prolong the lifespan of the steel structures and ensure designed service life, regular inspections, and maintenance of the underwater facility's surface coatings are essential, including repairs to damaged areas. Currently, maintenance tasks for underwater facilities predominantly rely on manual diving operations or the creation of dry work zones. However, manual operation is costly, inefficient, prone to quality variability, and carries significant safety risks. Consequently, many underwater facilities, such as underwater steel piles, often miss scheduled maintenance, posing considerable safety hazards. Therefore, the development of a robot with an anticorrosion coating unit that is suitable for coating repair in underwater facilities, achieving automated inspection and maintenance of these coatings is crucial.

In recent years, research on anticorrosion coatings for underwater steel structures, both domestically and internationally, has primarily focused on anticorrosion materials, with limited studies on automated maintenance solutions. Zhao et al. designed an auto-mated maintenance device for the coating repair of dock steel piles, based on the under-water dry area construction method. This device creates a dry environment by removing water from the robot's clamping area, allowing for the spraying of anti-corrosion coatings [1]. Similarly, for the anti-corrosion repair of steel pipe piles, Zhang and Li developed an automated wrapping solution that does not require a dry environment, enabling the application of anticorrosion materials directly onto the steel pipe surface [2]. However, the automated maintenance equipment in [1] is constrained by its reliance on dry construction methods, resulting in a complex system that necessitates a series of water extraction and drying devices. This complexity significantly limits the flexibility and efficiency of the repair device. In contrast, while the anti-corrosion wrapping solution in [2] does not demand a dry environment, it is constrained by the materials used, which are more expensive and involve complex application methods, making it unsuitable for anti-corrosion repairs in intricate structures. Thanks to the will-developed underwater anti-corrosion coating material technology, directly applying coatings in underwater environments has become feasible [4] [5]. Kim et al. carried out a study on the bonding performance of epoxy coatings using different coating devices and came to the conclusion that the difference in bonding performance can be attributed to the type of coating equipment rather than the coating itself [3].

As can be seen from the literature, there remains a substantial gap in research on automated maintenance robots for anti-corrosion coatings in underwater non-dry environments. Therefore, the aim of this work is to conduct a systematic study on anti-corrosion coatings unit design for underwater robots. By analyzing the extrusion and application processes of underwater coatings, we define several factors that influence coating layer quality will be studied, and corresponding experiments are designed and conducted. This study will provide essential information for the design of an optimized coating strategy that is suitable for underwater automated robotic anti-corrosion repair with high quality and minimal material waste.

The rest of this paper is organized as follows: In Section II, the extrusion process of the coating is analyzed, and the factors affecting coating quality are listed. In Section III, experiments setup are stated for investigating the effects of different influencing factors on coating application quality. Section IV presents experimental results and discussions. Conclusions are stated in Section V.

2. Materials and methods

In this work, we designed a wall-climbing robot with an anti-corrosion unit for automatic coating on steel surfaces, as shown in Figure 1. The anti-corrosion unit is attached to the robot with several brushers and the coating material is extruded to the brushers and with the movement of the robot, the coating material can be applied to the surface. However, suitable coating unit parameters and factors, such as brusher type, extrusion speed, and unit coating tool movement speed, need to be determined for robot design.

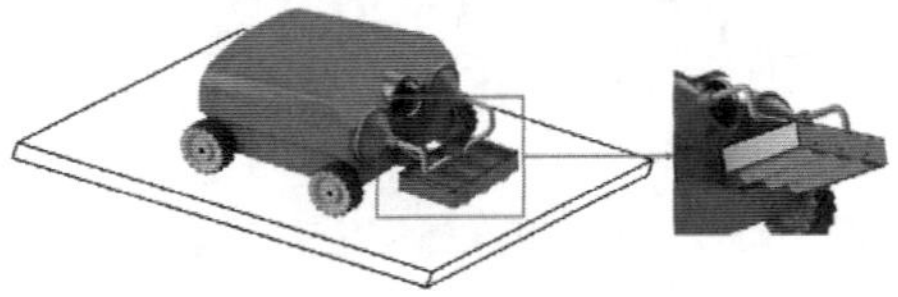

Figure 1. Automated maintenance robot and coating unit.

Referring to established manual coating methods [6] [7] [8], the underwater coating process can be simplified into two parts: material extrusion and application. Material extrusion involves expelling the pre-mixed anti-corrosion coating through a nozzle onto the painting tool under specific pressure and flow conditions. Factors influencing the coating extrusion process include pressure, flow rate, and nozzle shape. During underwater operations, the extrusion pressure must be greater than the water pressure at the outlet for successful material expulsion. The extrusion flow rate must be appropriately coordinated with the tool movement speed to achieve a standard layer thickness while minimizing material waste. The application process involves evenly spreading the extruded coating onto the target surface using the application tool. Factors influencing the painting process include the type of tool and the tool movement speed. The choice of painting tool directly affects efficiency, cost-effectiveness, and the smoothness of the coating layer [9] [10]. Additionally, the tool movement speed must be considered in conjunction with the extrusion speed to ensure both effective application and operational efficiency.

Experiments were conducted to investigate the effect of factors such as brush type, material extrusion speed, and tool movement speed, as shown in Figure 2. In this unit setup, different painting tools could be fixed on a sliding table driven by an electric motor, and the material is extruded through the nozzle by another speed-adjustable electric motor.

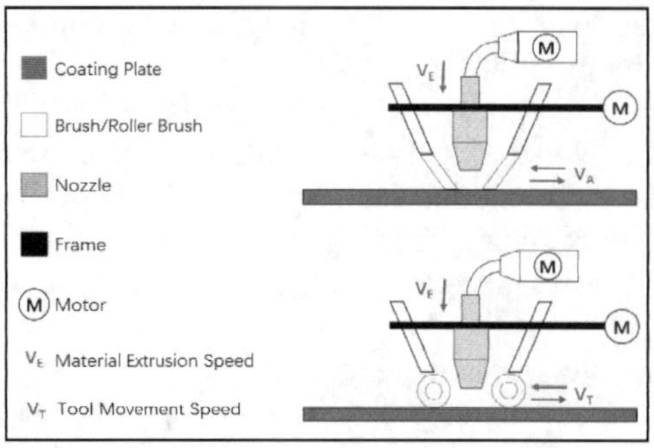

Figure 2. Experiment unit design.

To investigate the effects of those variables, an experimental platform simulating the underwater application process was built as shown in Figure 3. The platform consisted of an aluminum profile frame, a sliding table, a drive motor, an electric extrusion device, and a test steel plate. The sliding table was equipped with a fixture for the painting tool and a coating extrusion nozzle.

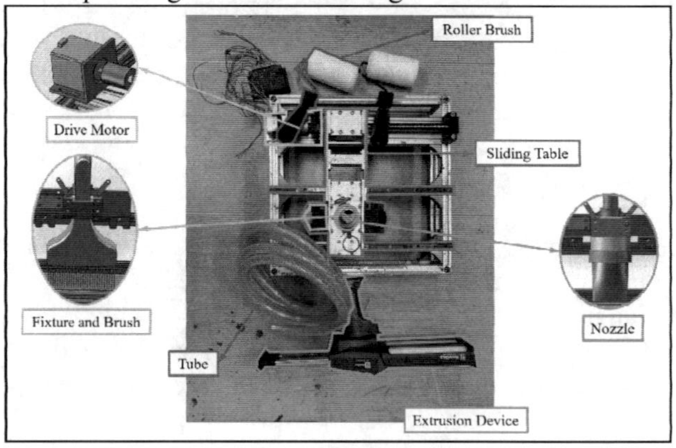

Figure 3. Experimental platform.

The experiment utilized UM-ES210 two-component solvent-free epoxy coating. Before application, the base and curing agent were thoroughly mixed at a 4:1 ratio.

3. Experiment DESIGN

Three sets of experiments were designed considering three factors on coating quality measures, as listed in Table 1. The range of extrusion speed and tool movement speed were selected according to the design constraints and cost limits.

Table1. experiment design table.

Experiments	Factor compared	Dependent Variable and Measure methods
Painting tools comparison	Type of tool	The coating surface measured by a microscope
Extrusion speed comparison	Extrusion speed	Surface observation and film thickness measured by micrometer
Tool movement speed comparison	Tool movement speed	Film thickness measured by micrometer and pull-off strength test, in accordance with the GB/T 5210-2006 standard

3.1 Painting tool study

In manual coating methods, brushes are commonly used to ensure uniform application and reduce material loss. Compared to manual operation, robotic coating has limited flexibility. But once the robot adheres to the surface, its movement during the application process becomes more stable, making roller brushes potentially more suitable for automated coating solutions.

Figure 4 shows the painting tool experiment procedure. The experiment used the type of painting tool as the independent variable. Prior to testing, the selected painting tool was securely fastened to the fixture for preparation. During the experiment, the coating was uniformly extruded through the nozzle to the surface of the painting tool by the extrusion mechanism. The painting tool was fixed to the sliding table and, driven by the motor, moved at a constant speed across the surface of the steel plate for the application process.

After the application, the sample was placed in the experimental water tank for 72 hours, as per the coating specifications, to allow the coating to cure. An industrial micro-scope was employed to capture images of the sample surface before and after the experiment. The effectiveness of different painting tools was evaluated based on the smoothness of the coated surface after application.

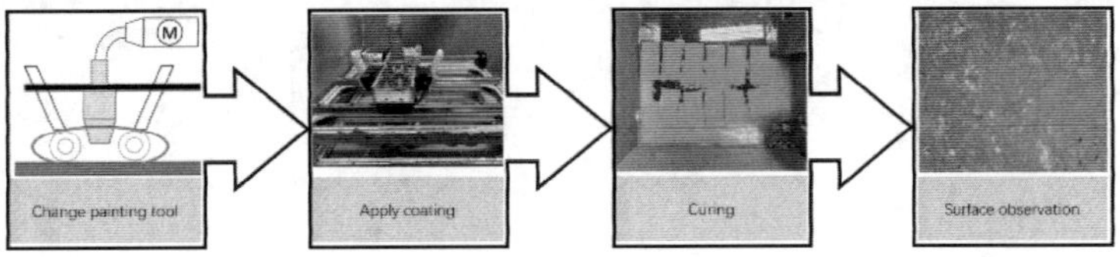

Figure 4. Painting tool experiment procedure.

3.2 Extrusion speed study

To investigate the effects of extrusion speed on coating quality, material extrusion speed experiments were conducted using the same test platform. The extrusion speed could be adjusted via a knob on the extrusion device, which had speed levels 1-6, and it was divided into three groups: low extrusion speed (level 1: 5 mm/s), medium extrusion speed (level 3: 15 mm/s), and high extrusion speed (level 5: 25 mm/s). The quality of the coating application was assessed by measuring the coating film thickness and observation of the surface. The test surface consisted of an acrylic frame holding three q235 steel plates.

Figure 5 shows the extrusion speed experiment procedure. In the experiment, the extrusion speed of the coating from the extrusion mechanism was treated as an independent variable. Before the experiment, the independent variable was set by adjusting the extrusion speed of the electric extrusion device to prepare for the experiment. During the test, the coating material was uniformly extruded through the nozzle onto the surface of the roller brush, which moves at a constant speed across the sample surface, driven by the sliding table and motor.

After the application, the sample was placed in the experimental water tank for 72 hours, as per the coating specifications, to allow the coating to cure. Subsequently, the sample was subjected to coating thickness measurement.

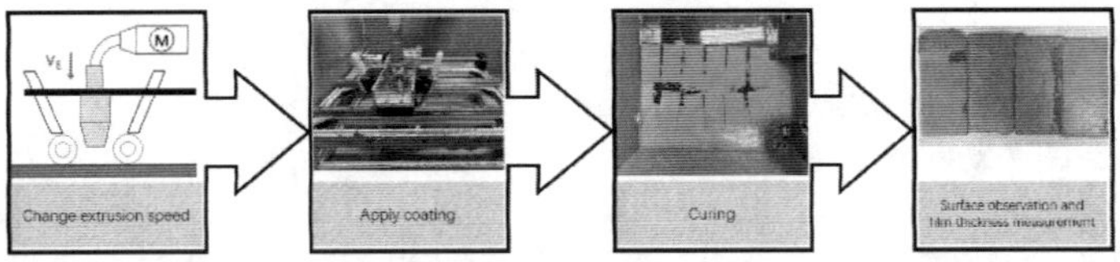

Figure 5. Extrusion speed experiment procedure.

3.3 Tool movement speed study

To investigate the effects of tool movement speed on coating quality, tool movement speed experiments were conducted using the same test platform. The sliding table's drive motor is a stepper motor, with its speed adjusted by inputting different pulse counts through coding. The tool movement speed, as an independent variable, represented by the sliding table's speed, was set to low speed (75 mm/s), medium speed (225 mm/s), and high speed (375 mm/s) separately. The quality of the coating layer was assessed using film thickness and surface pull-off strength test, in accordance with the GB/T 5210-2006 standard. The test surface consisted of an acrylic frame holding three q235 steel plates, each measuring 150 mm × 70 mm × 3 mm, according to the test standard.

Figure 6 shows the extrusion speed experiment procedure. During the test, the coating material was uniformly extruded through the nozzle onto the surface of the roller brush, which moves at different speeds across the sample surface, driven by the sliding table and motor.

After the application, the sample was placed in the experimental water tank for 72 hours, as per the coating specifications, to allow the coating to cure. Subsequently, the sample was subjected to coating thickness measurement and pull-off strength testing.

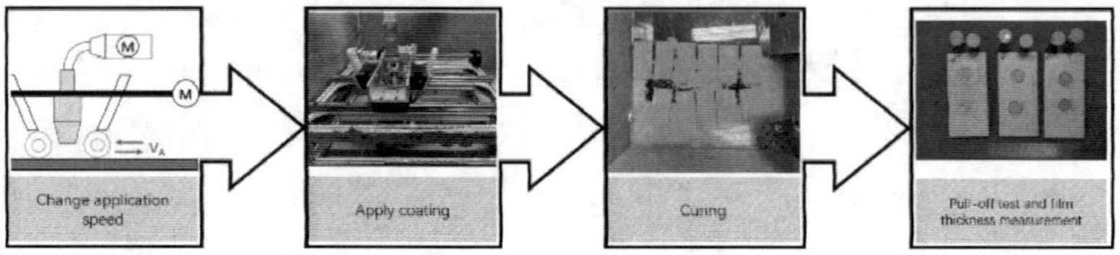

Figure 6. Tool movement speed experiment procedure.

4. Results and Discussion

4.1 Effect of painting tools on coating uniformity

The coating quality of different painting tools is evaluated by examining the uniformity and smoothness of the coating layer. The results of the painting tool selection experiment are shown in Figure 7.

Figure 7. Results of painting tool experiment.

The experimental results reveal that the coating layer applied by the brush tool shows substantial deficiencies in both surface smoothness and uniformity. Due to the thinner coating layer, the brush-coated surface contains numerous voids, resulting in insufficient coverage and failing to meet anti-corrosion requirements. In contrast, the roller brush achieves a significantly more uniform and smoother coating layer. Consequently, the roller brush is identified as a more suitable tool for automated underwater coating applications.

4.2 Effect of extrusion speed on coating quality

Experiment shows that excessively high or low extrusion speeds significantly impact coating quality. Both the low and high extrusion speed experimental groups exhibited notable surface defects. Due to the significant defects in the coated surfaces under both conditions, which fail to meet the anti-corrosion requirements and the specifications for the pull-off strength test samples, further pull-off strength testing was not conducted on these two samples.

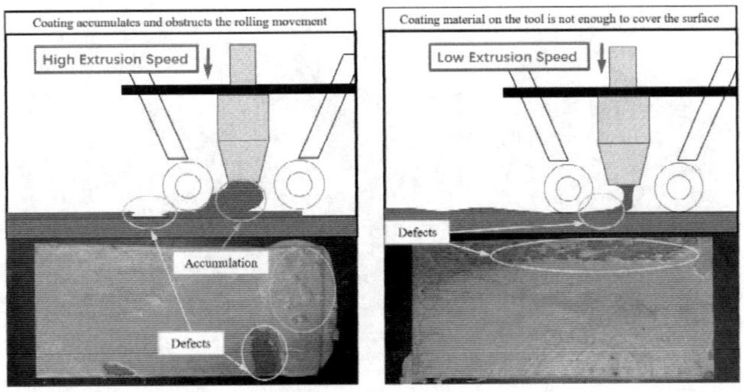

Figure 8. Results of extrusion speed experiment.

As shown in Figure 8, when the extrusion speed is relatively low, instances may occur where the coating does not fully cover the sample surface. This is attributed to the roller brush not picking up an adequate amount of coating material as it passes over the area. When the extrusion speed is relatively high, there may also be instances where the coating does not fully cover the surface. This occurs because the coating material has a relatively high viscosity. At higher extrusion speeds, excess coating accumulates in front of the roller brush, obstructing its rolling movement and thereby worsening the coating quality.

Experimental results are shown in Figure 9. The coating thickness tends to increase with increasing extrusion speed. Additionally, despite the film defects, the low extrusion speed group and high extrusion group show obvious inequality in film thickness. Also, the high extrusion group shows an obvious thicker film, as a result of the coating accumulation. Therefore, in this case, a medium extrusion speed is more suitable for automated coating operation.

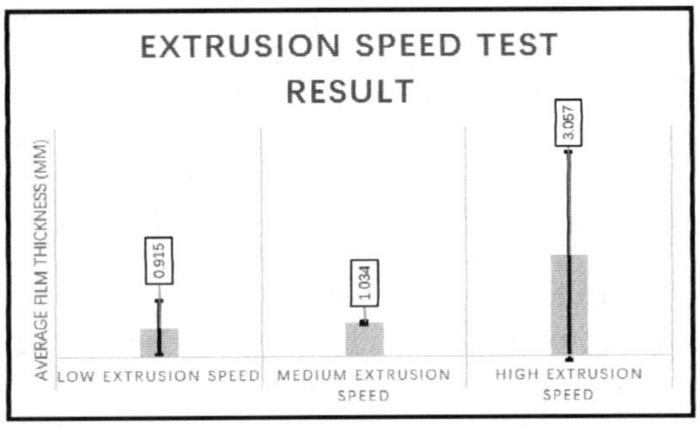

Figure 9. Extrusion speed test result.

4.3 Effect of tool movement speed on coating quality

At the medium extrusion speed, surfaces were coated at different tool movement speeds, resulting in three sets of samples that were uniformly covered with no significant large-scale defects as shown in Figure 10. However, in the scenario with high tool movement speeds, small voids may appear in the coating layer, potentially adversely affecting the service life of the anti-corrosion layer. This issue arises from the excessive tool movement speed, which prevents the viscous coating from fully adhering to the steel surface, leading to its removal by the roller brush. As shown in Figure 11, the three batches of samples underwent further testing for surface pull-off strength and coating layer thickness to evaluate the coating effectiveness more comprehensively.

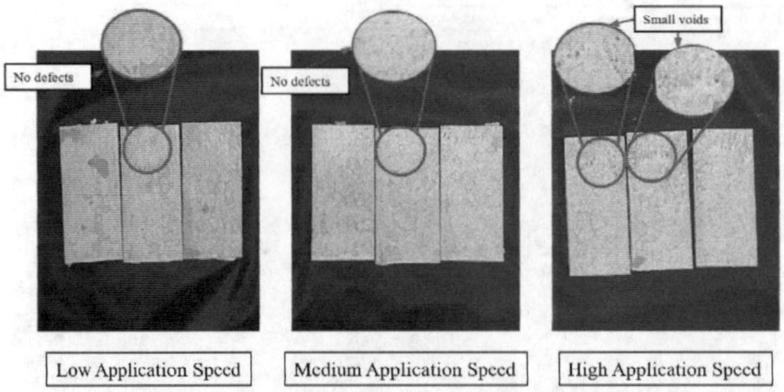

Figure 10. Samples of tool movement speed experiment.

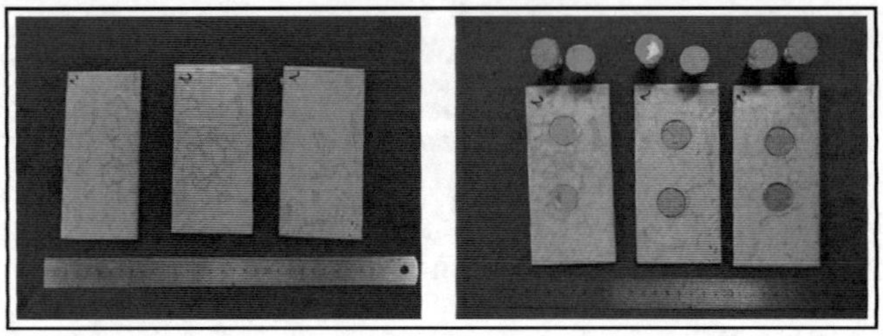

Figure 11. Samples of Pull-off strength test.

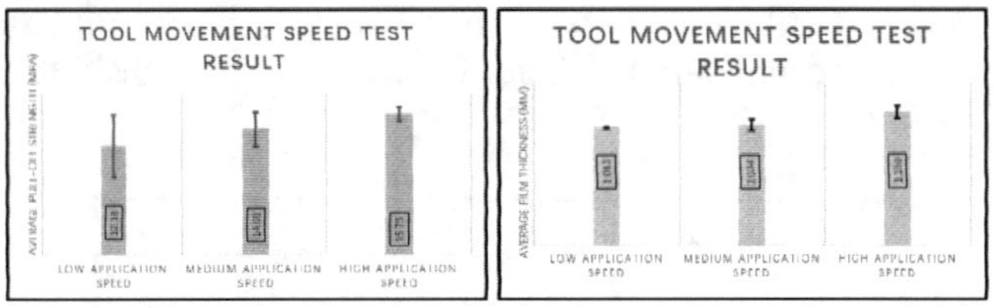

Figure 12. Tool Movement speed test results.

Figure 12 shows the experimental results of tool movement speeds. The outcomes of the three groups reveal a clear trend. With a fixed extrusion speed, both the thickness and tensile strength of the coating layer increase as the coating speed increases. This indicates that within the motor's operational speed range, and given an appropriate extrusion rate, a faster tool movement speed is generally preferable. However, considering that small surface holes appear at a higher speed coated surface, a medium coating speed is recommended for practical operations to ensure optimal coating quality.

The overall experimental results are summarized in Figure 13, where an optimized coating strategy can be derived: for the coating material and mechanism used in this study, roller brush painting tool, medium extrusion speed (15 mm/s), and medium tool movement speed (225 mm/s) would be used for automated underwater coating repair robot.

	Low application speed(75mm/s)	Medium application speed(225mm/s)	High application speed(375mm/s)
Low Extrusion speed(5mm/s)	Obvious film defects	Obvious film defects	Obvious film defects
Medium Extrusion speed(15mm/s)	Smooth surface Low pull-off strength	Smooth surface Medium pull-off strength	Small void defects High pull-off strength
Medium Extrusion speed(25mm/s)	Obvious film defects	Obvious film defects	Obvious film defects

Figure 13. Overall results.

5. Conclusions

In this study, experiments were conducted to systematically study an optimized coating strategy for robotic underwater anti-corrosion coating procedures. The surface quality, film thickness, and pull-off strength were selected as coating quality performance indices. Experiments are designed and conducted to systematically analyze the effects of the brusher selection, extrusion speed, and tool movement speed. The experimental results show that using a roller brush with medium extrusion and tool movement speed is suitable in an underwater environment. Future work needs to be conducted on the coating quality in different underwater flow speeds and on different water content to improve coating applicating performance. Also, an integrated underwater coating repair robot will be designed and built based on these studies.

Acknowledgment

This work is sponsored by China Yangtze Power Co., Ltd (2123020040).

References

[1] Zhao, Z.; Liang, Q.; Wang, S.; et al. Research on automated anti-corrosion treatment process and equipment for jetty steel piles in tidal zone and wave splash zone (in Chinese). Mechanical Science and Technology for Aerospace Engineering, 2008, (08):1077-1080. Zhang, J.; Li, J. Automated construction technology and equipment program for underwater anticorrosion repair of steel piles by cladding method (in Chinese). China Harbour Engineering, 2021, 41(08):52-55.
[2] Zhang, J.; Li, J. Automated construction technology and equipment program for underwater anticorrosion repair of steel piles by cladding method (in Chinese). China Harbour Engineering, 2021, 41(08):52-55.
[3] Kim, S.; Hong, H.; Han, T.H.; Kim, M.O. Early-Age Tensile Bond Characteristics of Epoxy Coatings for Underwater Applications. Coatings 2019, 9, 757. https://doi.org/10.3390/coatings9110757
[4] Peng, Z. Research on Underwater Construction Coatings and Their Coating Process (in Chinese). Development and Application of Materials, 2004, (02):27-30. DOI: 10.19515/j.cnki.1003-1545.2004.02.009.
[5] Sachin Sharma Ashok Kumar, Shahid Bashir, K. Ramesh, S. Ramesh. A comprehensive review: Super hydrophobic graphene nanocomposite coatings for underwater and wet applications to enhance corrosion resistance. FlatChem, Volume 31, 2022, 100326, ISSN 2452-2627, https://doi.org/10.1016/j.flatc.2021.100326.
[6] Wang, J. New program for corrosion protection and repair of in-service steel pipe piles (in Chinese). Plant Maintenance Engineering, 2008, (08):14-15.
[7] Orlikowski, J.; Żakowski, K.; Szociński, M.; Igliński, P.; Jażdżewska, A.; Gaweł, Ł. Underwater Electrochemical Offshore Tests of a Paint Coating Applied in Water on the Legs of an Oil Production Platform. Materials 2024, 17, 3580. https://doi.org/10.3390/ma17143580
[8] He, P.; Qu, J.; Li, S.; Sun, R.; Huang, K.; Zhao, J.; Mo, J. Study on the construction of durable superhydrophobic coatings using simple spray coating method and their underwater drag reduction performance. Colloids and Surfaces A: Physicochemical and Engineering Aspects, Volume 701, 2024, 134858, ISSN 0927-7757, https://doi.org/10.1016/j.colsurfa.2024.134858.

Journal of Physics: Conference Series **3009** (2025) 012041 doi:10.1088/1742-6596/3009/1/012041

[9] Tripp, Julia L. Underwater Coatings Testing for INEEL Fuel Basin Applications. United States. https://doi.org/10.2172/910618
[10] Nichols, N. W., & Khan, R. Structural integrity management system (SIMS) implementation within PETRONAS' operations. Journal of Marine Engineering & Technology, 2015, 14(2), 61–69. https://doi.org/10.1080/20464177.2015.1096612.

Mixing effect analysis of a novel agitator based on MPS solid-liquid interaction method

Wei Li [1*], Zhen Li [1], Kun Wang [1], Guangshuai Sun [2], Xianchao Sun [2]

[1] Qingdao Construction Management Co., Ltd. of Shandong Hi-speed, Qingdao, 266590, China
[2] Shandong Hi-speed Engineering Construction Group Co., Ltd, Ji'nan, 250000, China

*Corresponding author's e-mail: liwei234023@163.com

Abstract. A novel bidirectional agitator has been designed to enhance mixing performance. Utilizing the Moving Particle Simulation (MPS) method, a visual computational model of the new agitator based on solid-liquid interaction has been established. The impact of altering the mixing parameters of the new agitator on the flow field distribution was analyzed by varying the mixing speed, mixing height, and mixing mode. The results indicate that when both the main shaft and the mixing wheel of the agitator rotate simultaneously, the fluid flow range is significantly broader. As the speed of the agitator increases, the mixing range of solidified particles in the mixing tank gradually expands. Within the selected range, changes in mixing height have minimal impact on the mixing range and the mixing effect of the flow field. As the speed and height of the mixer increase, the loss of stirring power gradually increases.

1. Introduction

The agitator is a crucial mechanical stirring device utilized across various industries, including chemical, petroleum, pharmaceuticals, food processing, and paint manufacturing. By facilitating the two-phase mixing of solid particles and liquids, the agitator enhances the dispersion and suspension of solid particles within the liquid medium, thereby creating a homogeneous solid-liquid mixture [1,2]. Traditionally, the design process and performance evaluation of agitators have heavily relied on experimental validation. However, with advancements in computational fluid dynamics (CFD) and computing technology, numerical simulation methods have emerged as cost-effective solutions that offer high precision while significantly reducing time requirements. Consequently, these methods have become an invaluable tool for investigating the stirring performance of agitators [3–5].

Zhou et al. [6] employed the Laminar model to investigate the influence of various structural parameters on fluid flow characteristics, diffusion properties, unit volume stirring power, mixing time, and effective stirring volume fraction of the agitator during the stirring process using Computational Fluid Dynamics (CFD) methods. Tan et al. [7] examined the velocity field distribution in a radial flow double-layer agitator and analyzed the distribution of solid particle volume at different rotational speeds utilizing Fluent software. Their findings revealed that a significant number of solid particles accumulated at the bottom of the agitator when operating at low rotational speeds. Based on the finite volume method, Zhou Zhen et al. [8] investigated the mixing effects of various combinations of double-layer agitators in a general liquid fermentation tank. By analyzing the velocity field cloud maps and stirring power characteristics within the stirred tank under different configurations, they identified the optimal impeller combination. Wang Yinghui et al. [9] investigated the effects of various rotational speeds, blade geometries, rotational directions, and flow patterns on the mixing characteristics of disc turbine agitators. Yao Chenming et al. [10] conducted transient simulation analyses of gas-liquid two-phase flows across different stirring models, highlighting that the velocity fields and power losses associated with distinct stirring combinations varied significantly. They noted that the flow field distribution became more uniform when a radial flow impeller was employed as the upper impeller.

Journal of Physics: Conference Series **3009** (2025) 012042 doi:10.1088/1742-6596/3009/1/012042

Zhang Huimin et al. [11] compared and analyzed the flow field characteristics between single-layer and double-layer turbine agitators within stirred tanks. Their findings indicated that the fluid flow range for single-layer turbine agitators was narrower than that for double-layer agitators; furthermore, achieving an equivalent stirring effect required higher power consumption in single-layer systems. Min et al. [12] developed a novel agitator model that integrates both stirring and spraying functions. Through an investigation of the effects of various operating conditions on the desulfurization efficiency within the molten iron tank, it was observed that this new agitator significantly enhances the flow dynamics of the fluid field, resulting in a broader and more uniform mixing range.

In this paper, a novel stirring device is introduced, featuring four mixing wheels mounted on the apparatus. These wheels perform rotational stirring, while the stirrer body itself achieves multifunctional circumferential rotation. Compared to existing single-structure stirrers, this design significantly enhances both mixing efficiency and quality. The newly proposed multi-degree-of-freedom agitator serves as the research subject. The Moving Particle Simulation (MPS) method is employed to conduct a simulation analysis of the solid-liquid mixing flow field characteristics within the agitator. This study investigates the velocity flow field and solid particle distribution under various rotational speeds, fluid immersion depths, and rotational modes, as well as the impact of these operating conditions on the power loss incurred during the mixing process.

2. MPS numerical calculation method

The MPS method constitutes a meshless numerical analysis technique for dealing with incompressible fluids. Its fundamental principle entails replacing the fluid within the computational domain with a particle swarm. Then, based on the Lagrange equation, the discrete basic flow equation is solved in line with each interaction model. Unlike the conventional CFD method, the MPS method eliminates the necessity for elaborate meshing, thus making it easier to simulate complex moving boundary problems.

2.1. Control equation

The MPS method postulates that the fluid is both incompressible and viscous. Subsequently, the continuity equation (mass conservation law) and the Navier-Stokes equation (momentum conservation law) are formulated in Lagrangian form, serving as the fundamental governing equations, which are presented as follows :

$$\frac{D\rho}{Dt} = 0 \tag{1}$$

$$\frac{D\vec{u}}{Dt} = -\frac{\nabla P}{\rho} + \nu\nabla^2\vec{u} + \vec{g} \tag{2}$$

In the formula, u is the fluid velocity, t is the time, P is the pressure, ρ is the density, v is the kinematic viscosity coefficient, and g is the acceleration of gravity.

2.2. Kernel function

In the MPS method, the interaction among particles is accomplished through a kernel function. Moreover, each mathematical model is actualized by employing the kernel function as a weight function. The underlying principle dictates that as the distance between particles increases, the value of the kernel function diminishes, leading to a weaker effect. Specifically, when the particle spacing exceeds the effective radius of the particles, no interaction occurs between them. In this paper, the kernel function model is presented as follows :

$$w(r_p) = \begin{cases} \frac{r_e}{r_p} - 1, 0 < r_p < r_e \\ 0, r_p \geq r_e \end{cases} \tag{3}$$

In the formula, r_e is the effective radius of the particles, and r_p is the distance between the particles.

2.3. Particle number density and particle interaction model

The particle number density is defined as the superposition of the function values of a particle and its neighboring particles within the scope of the kernel function. In a particle arrangement that fulfills the incompressible condition, the particle number density assumes a constant value. The gradient model and the Laplace model, which are grounded on the discrete control equations of the kernel function, have their expressions presented as follows :

$$n = \sum_{j \neq i} w(|r_j - r_i|) \tag{4}$$

$$< \nabla \phi >_i = \frac{d}{n_0} \sum_{j \neq i} \frac{\phi_j - \phi_i}{|r_j - r_i|^2} (r_j - r_i) w(|r_j - r_i|) \tag{5}$$

$$< \nabla^2 \phi >_i = \frac{2d}{\lambda n_0} \sum_{j \neq i} \frac{\phi_j - \phi_i}{|r_j - r_i|^2} w(|r_j - r_i|) \tag{6}$$

In the formula, ϕ is the calculated particle physical parameters ; d is the space dimension, 2 in 2D and 3 in 3D; n_0 is the initial state particle number density constant; i, j are particle codes; r_j, r_i are the coordinate vectors of the particles; and λ is the correction factor.

2.4. Boundary condition

To avert the boundary distortion of particles at the boundary, it is essential to enclose the boundary with more than three layers of particles within the model. The boundary pressure particles are in direct contact with the fluid, while the boundary non-pressure particles are situated two or more layers beyond the boundary pressure particles. The particle arrangement is shown in Figure 1.

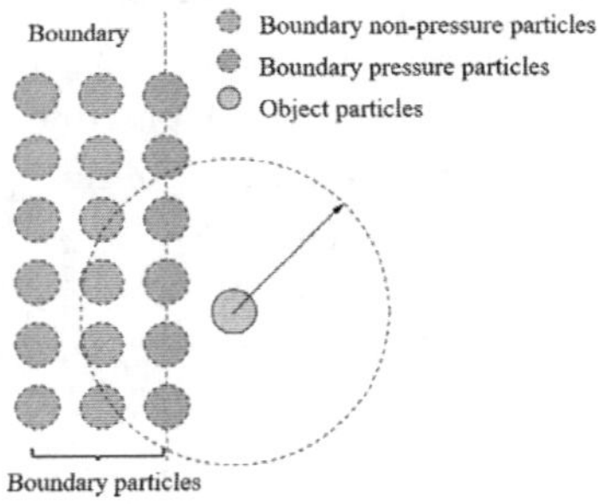

Figure 1. Boundary particle arrangement form of MPS method [13].

3. Simulation settings

3.1. Analysis of stirring condition

To conduct a more comprehensive assessment of the agitator's stirring performance, this paper undertakes an analysis of the flow field distribution of the agitator and the power loss of the stirring wheel under diverse working conditions, including different rotational speeds, varying liquid heights, and distinct rotation modes. The rotation mode is categorized into normal rotation and overall rotation. The specific parameters are presented in Table 1.

Table 1. work condition.

Working condition	Revolution speed /(rad·s^{-1})	Liquid height /mm	Rotating way
1	41.5	283.4	Normal rotation
2	49.5	283.4	Normal rotation
3	56	283.4	Normal rotation
4	66	283.4	Normal rotation
5	49.5	263.4	Normal rotation

6	49.5	243.4	Normal rotation
7	49.5	223.4	Normal rotation
8	11.5	283.4	Total rotation
9	13.8	283.4	Total rotation
10	15.6	283.4	Total rotation
11	18.3	283.4	Total rotation

3.2. Parameter setting
By leveraging RecurDyn, a stirred tank model and an agitator kinematics model are constructed, as illustrated in Figure 2. In the model, the motor is stationary in relation to the earth, the stirring tank is also fixed with respect to the earth. The spindle and the shell rotate in reference to the earth, while the planetary gear and the gears within the lower shell rotate relative to the shell. Among the meshing gears at all levels, the spindle serves as the driving component. The period 0 ~ 0.5s is the acceleration stage, and 0.5s ~ 2s is the uniform speed stage.

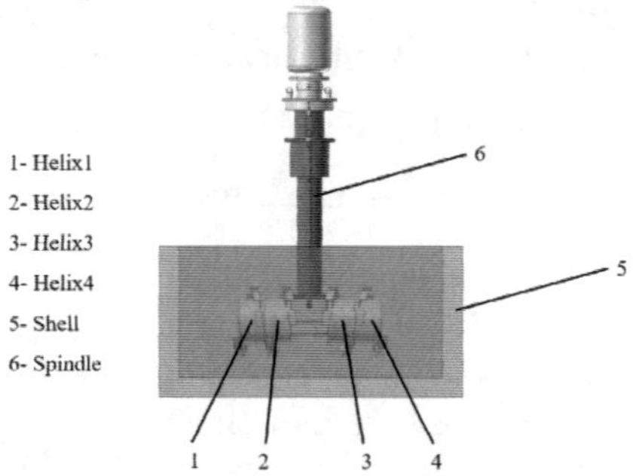

Figure 2. Three-dimensional model of agitator.

The dynamic model of the agitator is imported into the MPS calculation software, thereby establishing the fluid-solid coupling co-simulation model of the agitator. The fluid particle diameter is configured as 6 mm, and the solid particle diameter is set to 10 mm. The gravitational acceleration is defined in the y-axis direction as - 9.8 m/s^2. The implicit iterative model is applied for pressure calculation, while the explicit model is utilized for viscosity. The simulation time is designated as 2 s, and both the file output interval and the torque output interval are set to 0.01s.

4. Analysis of flow field characteristics of stirred tank

4.1. Analysis of liquid velocity field
From Figure 3, when the spindle and the stirring wheel rotate concurrently, the fluid velocity in the vicinity of the stirring wheel is relatively high, and there is scarcely any dead zone. With the position of the stirring wheel as the center, the closer the fluid in the stirring tank is to the wall of the tank, the lower the velocity becomes. Consequently, the liquid near the stirring wheel can be promptly stirred and made to flow around. Simultaneously, the stirring wheel rotates about its connecting shaft, agitating the liquid in the tangential direction of the wheel, and effectively dispersing the solidified particles. The velocity flow fields of the agitator at different liquid heights are analogous (Figure 4), which further indicates that the liquid of the mixed solidified particles flows rapidly around the stirring wheel.

As depicted in Figure 5, when the agitator rotates in its entirety, the liquid within the stirred tank merely flows towards the tank wall, lacking any vertical stirring motion. At a low rotational speed, the stirring wheel is only capable of agitating a minor portion of the liquid, resulting in a large stirring dead zone and rendering it difficult to effectively disperse the solidified particles. As the rotational speed escalates, a greater volume of liquid is set in motion under the influence of the stirring wheel. However, in comparison to the normal rotation condition, the velocity flow field remains less disordered, thereby leading to a suboptimal dispersion effect of the solidified particles in the vertical direction.

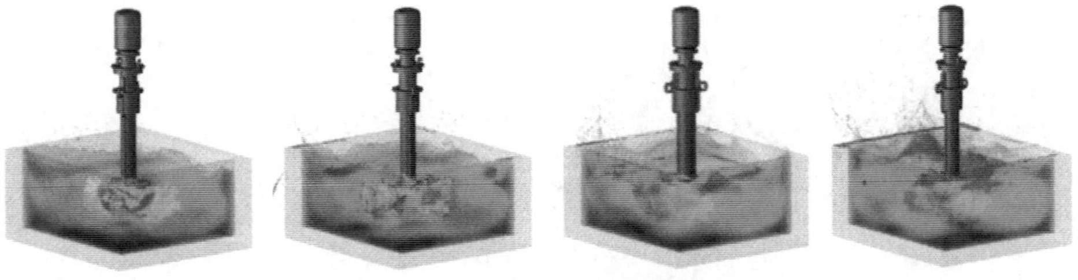

(a) Working condition 1 (b) Working condition 2 (c) Working condition 3 (d) Working condition 4

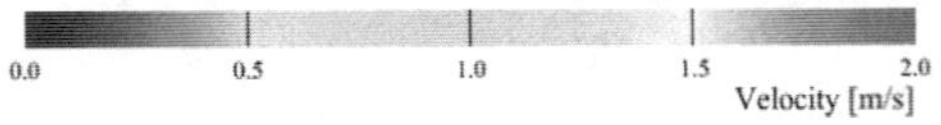

Figure 3. Fluid velocity field at different speeds under normal rotation.

(a) Working condition 5 (b) Working condition 6 (c) Working condition 7 (d) Working condition 8

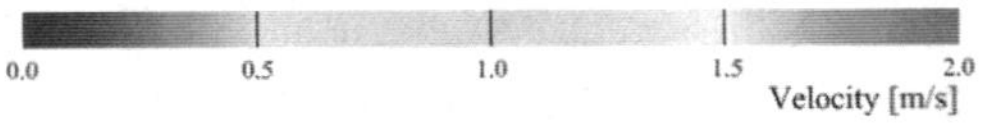

Figure 4. Fluid velocity field at different fluid heights under normal rotation.

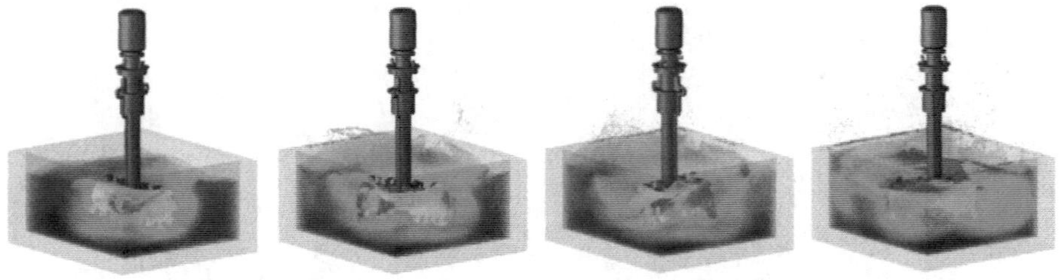

(a) Working condition 9 (b) Working condition 10 (c) Working condition 11 (d) Working condition 12

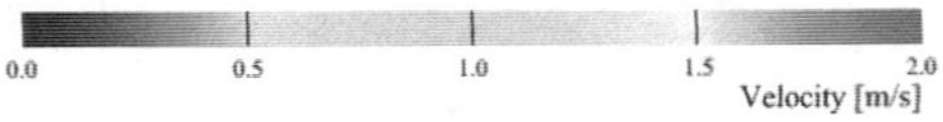

Figure 5. Fluid velocity field at different rotational speeds under overall rotation.

5. Power loss analysis

As depicted in Figure 6, in the initial stage, the torque loss is at a minimum level and then shows an oscillatory upward tendency as time elapses. When the agitator starts to rotate, a significant sudden change appears in the torque loss. This can be attributed to the introduction of solidified particles at the very beginning. As these particles sink into the liquid, the impact causes an instantaneous disturbance to the liquid, thus creating a substantial resistance against the stirring wheel. Since the stirring wheel rotates around the spindle while in operation, it makes periodic contact with a larger amount of liquid. Consequently, after the increase in torque loss during the acceleration phase, a wave-like pattern characterized by repeated rises and falls emerges in the uniform speed stage, and finally reaches a dynamically stable state. Moreover, it is clear that as the input speed of the spindle increases, the torque loss of the stirring wheel also rises. And the temporal trend of the resistance torque of the stirring wheel at different speeds remains consistent.

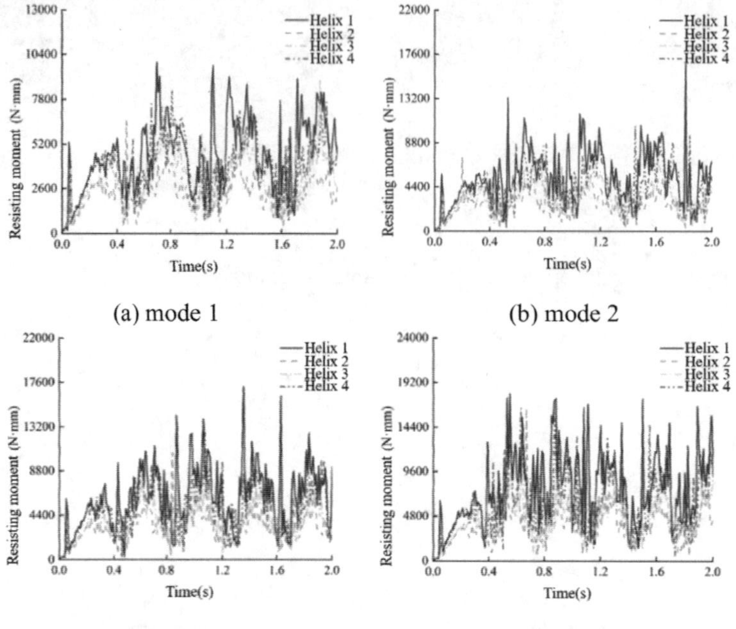

(a) mode 1 (b) mode 2

(c) mode 3 (d) mode 4

Figure 6. Stirring resistance torque at different speeds under normal rotation.

As can be observed from Figure 7, the variation trend of the resistance torque loss of the stirring wheel at different liquid heights is analogous to that shown in the figure. It also varies periodically with time. Specifically, the higher the initial liquid height, the larger the torque loss. Moreover, the lower the initial liquid height, the smaller the abrupt change in torque loss when the agitator initiates rotation.

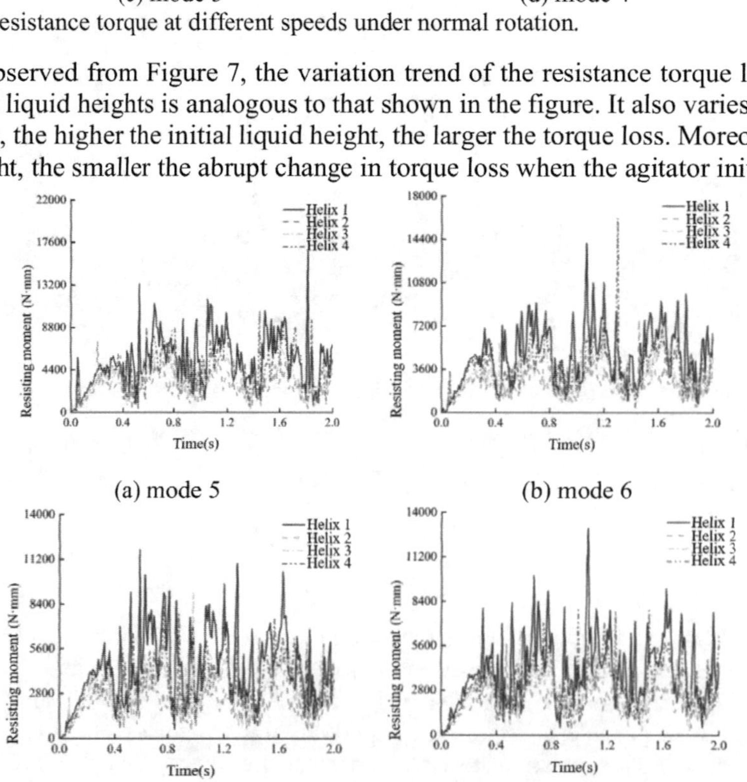

(a) mode 5 (b) mode 6

(c) mode 7 (d) mode 8

Figure 7. Stirring resistance torque at different liquid heights under normal rotation.

As is evident from Figure. 8, when the components of the agitator rotate in unison, the torque loss curve exhibits a periodic variation, characterized by a sudden increase during the steady acceleration stage, followed by a sudden decrease to a steady state and then another sudden increase. This indicates that during the overall rotation of the agitator, the stirring wheel propels the liquid of the mixed solidified particles toward the inner wall of the stirring tank, leading to a reduction in the amount of liquid in the vicinity of the stirring wheel and a consequent decrease and stabilization of the torque loss. Subsequently, under the influence of gravity, the liquid near the inner wall of the tank flows back towards the center of the tank. The stirring wheel then comes into contact with a significant volume of liquid, causing the torque loss to spike abruptly. Thereafter, the stirring wheel displaces the liquid and solidified particles, and the torque loss plummets suddenly back to a stable state. This cycle repeats, thereby manifesting the trend depicted in the figure. Additionally, when the agitator rotates as a whole, the torque loss also escalates with an increase in rotational speed.

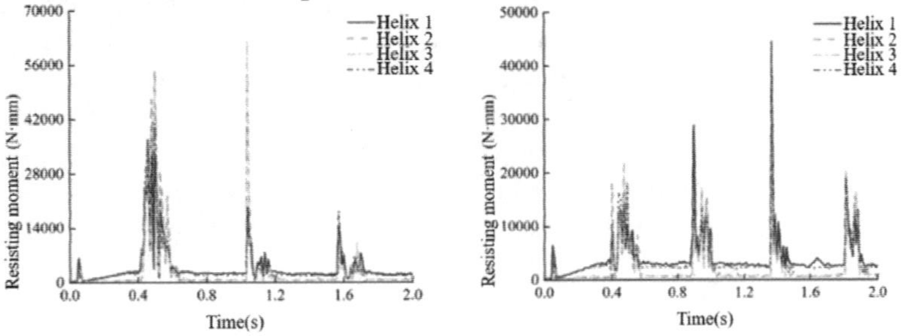

(a) mode 9 (b) mode 10

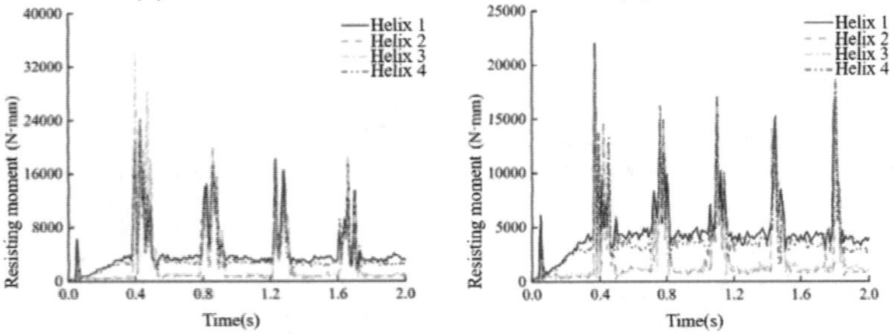

(a) mode 11 (b) mode 12

Figure 8. Stirring resistance torque at different rotational speeds under overall rotation.

The stirring power loss formula of the agitator is the product of the resistance moment of the stirring wheel and the rotational angular velocity :

$$P = M \times \omega \tag{7}$$

where P is the power loss, w; M is the stirring resistance moment, N·m; ω is the angular velocity of the stirring wheel, rad/s.

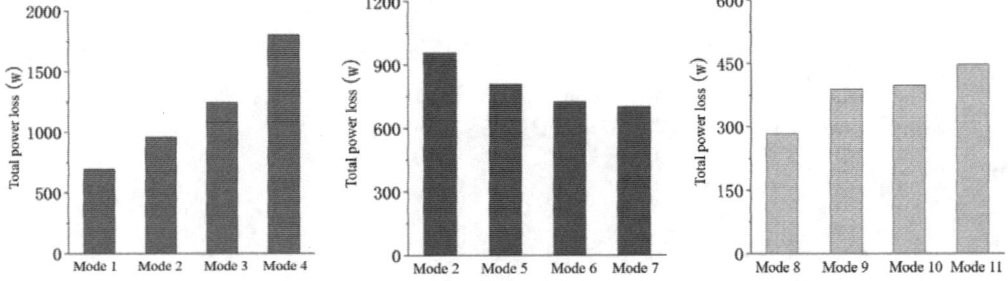

(a) Different speed (b) Different liquid heights (c) Different ways of rotation

Figure 9. Comparison of power loss under different working conditions.

Based on the time interval subsequent to the agitator attaining a uniform speed, the resistance moment data of stirring wheels 1 through 4 are extracted. Subsequently, the total stirring power loss of the agitator under diverse working conditions is calculated in accordance with the aforementioned formula, as illustrated in Figure 9.

6. Conclusion

(1) The rotation of the main shaft-stirring wheel, which functions as a discrete body, is capable of enhancing the dispersion of solidified particles within the stirred tank. This leads to a more uniform spatial distribution and a more favorable mixing outcome.

(2) When the spindle-stirring wheel rotates in its entirety, the stirring range of the stirring claw is rather limited. As a result, the solidified particles tend to accumulate more, which is unfavorable for the mixing of the solidified particles and the liquid.

(3) The stirring power loss of the agitator escalates in tandem with the augmentation of the rotational speed. The impact of varying rotational speeds on the stirring effect is obvious. Specifically, a higher rotational speed corresponds to a broader stirring range.

(4) The stirring power loss of the agitator rises as the liquid height increases. However, the variation in liquid height exerts a relatively minor influence on the stirring effect. Selecting an appropriate stirring height can, to a certain degree, enhance the stirring effect.

References

[1] Chen D. (2008) Development of agitators and pressure vessels with agitators. Pressure Vessel Technology, (02): 33-41+46.
[2] Yang M., Lai Y (2011) Experimental investigation on solid-liquid suspension characteristics in stirred tanks, Journal of Mechanical Engineering, 47(06): 186-192.
[3] Chen J., Li H (2011) Study on the research hotspots of mixer in chemical industry, Modern Chemical Industry, 31(10): 12-16.
[4] Jia H., Tang K., et al. (2013) Numerical simulation and experimental study of solid-liquid internal flow field in stirred tank, Machinery Design & Manufacture, (05): 170-172.
[5] Jia H., Qi Y., et al. (2014) CFD research for influence of disk turbine blade on mixing characteristics in stirred tank, Chemical Industry and Engineering Progress, 33(05): 1118-1122+1169.
[6] Zhou C., Qian S., et al. (2024) Research on the influence of agitator structural parameters on stirring characteristics based on fluid simulation, 41(03):124-134.
[7] Tan H., Zheng W., et al. (2022) Numerical simulation of solid-liquid two-phase flow in agitator based on fluent, 51(09): 191-193.
[8] Zhou Z., Wang M., et al. (2022) Optimization of double-layer hybrid turbine agitator based on CFD simulation, Light Industry Machinery, 40(04): 79-84+94.
[9] Wang Y., Hao L., et al. (2023) Comparison and analysis of stirring performance stirred by disc turbine agitators, Chemical Industry and Engineering, 40(02): 123-132.
[10] Yao C., Shi X., et al. (2021) Research of multiphase flow field with different stirring combinations based on CFD, Light Industry Machinery, 39(04): 20-25.
[11] Zhang H., Liao H., et al. (2016) Flow field analysis and comparison of single and double turbine agitator, Journal of Chinese Agricultural Mechanization, 37(09): 270-273+284.
[12] Min C., Dan B., et al. (2020) Numerical analysis of flow field based on the new desulfurization agitator of KR method, Steelmaking, 36(05): 14-20.
[13] Liu H., Xie C., et al. (2021) Flow field distribution of splash lubrication of gearbox and churning gear torque loss, Journal of Zhejiang University (Engineering Science), 55(05): 875-886.

Journal of Physics: Conference Series **3009** (2025) 012043 doi:10.1088/1742-6596/3009/1/012043

Generative adversarial network with AdaIN feature for 3D reconstruction of battery electrode microstructure under different processing parameters

Dongfang Fan[1], Bo Zhao[1], Yelin Deng[2,3*], and Xiang Chen[2,3]

[1] China Datang Corporation Science and Technology General Research Institute Co., Ltd. Central South Electric Power Test Research Institute, Zhengzhou, China
[2] School of Rail Transit, Soochow University, Suzhou, China
[3] Intelligent Urban Rail Engineering Research Center of Jiangsu Province, Suzhou, China

[*]Corresponding author's e-mail: yelin.deng@suda.edu.cn

Abstract. The manufacturing process of the battery electrode, consisting of homogenization, drying and calendaring, is the decisive element in the final performance of the lithium battery. Different design and process parameters lead to different physical and texture including tortuosity, porosity, percolation in the electrode, which greatly affect the electrochemical performance of the porous electrode. Therefore, it is of great importance to understand the microstructure of the battery electrode and thus identify possible influences of the manufacturing parameters on the microstructure. In this study, we have developed a generative adversarial network (GAN) with adaptive instant normalization (AdaIN) to reconstruct the 3D microstructure with orthogonal 2D images under different calendaring parameters: 0bar, 300bar and 2000bar. Then, we validated this computed microstructure with the real one obtained from the X-ray tomography for these calendaring parameters. Our results show that the GAN-AdaIN derived structure have showed very close tortuosity, percolation, and porosity distribution with real data. Finally, we have analyzed the impact of different parameter of active material content on the battery microstructure evolution.

1. Introduction

The global push to reduce greenhouse gas emissions has driven significant advancements in lithium-ion batteries, essential for electric vehicles and stationary energy storage systems. By 2022, global production capacity reached 550 GWh [1]. This rapid growth emphasizes the need to optimize battery performance metrics such as energy density, cycle life, power density, and thermal stability. Among these factors, the microstructure of battery electrodes has emerged as a critical area of research due to its profound impact on electrochemical performance.

Battery electrodes, including cathodes and anodes, are porous composite materials. Cathodes typically consist of metal oxide particles, conductive carbon additives, and binders, while anodes often use graphite [2]. These three-dimensional porous structures, infused with electrolyte, serve as channels for lithium-ion transport during charging and discharging. Key microstructural parameters such as porosity, tortuosity, particle size distribution, and connectivity determine electrolyte wettability, active material utilization, and ionic transport. Poorly optimized microstructures can lead to significant capacity losses, uneven degradation, and reduced cycle life. For instance, insufficient contact between the active materials and electrolyte reduces reaction efficiency, while high tortuosity impedes ion diffusion, lowering power density.

Given the strong correlation between microstructure and performance, precise characterization is essential. Traditional imaging methods such as Scanning Electron Microscopy (SEM), Focused Ion Beam SEM (FIB-SEM), and X-ray computed tomography (CT) allow for detailed microstructural analysis [3]. However, these techniques face challenges. They are expensive, time-intensive, and require sophisticated image processing. Moreover, while 2D imaging provides insights into structural attributes, only 3D reconstructions can accurately capture complex features such as anisotropy, connectivity, and diffusion pathways, which are critical for realistic performance simulations and optimization [3].

Recent advancements in artificial intelligence (AI) offer innovative solutions to these challenges. Generative models such as SliceGAN, 3D-GAN, and convolutional neural networks (CNNs) have been applied to reconstruct 3D microstructures from 2D images [5]. However, these approaches often face limitations in achieving high fidelity and precision, particularly for complex battery electrode architectures. Moreover, existing studies primarily focus on reconstructing 3D structures without considering the influence of processing conditions [8]. However, linking manufacturing parameters to microstructural attributes is critical for designing optimized electrodes and improving battery performance. A precise process-informed 3D model can enable predictive insights into diffusivity, tortuosity, and percolation, guiding the development of advanced energy storage technologies.

To bridge the gap between microstructure characterization and manufacturing optimization. the GAN AdaIN (Adaptive Instance Normalization) method presents a breakthrough, offering unique advantages in generating high-resolution 3D microstructures. By incorporating style transfer capabilities, GAN AdaIN excels in maintaining geometric and physical accuracy while enabling the integration of manufacturing parameters as controllable inputs. This adaptability makes it particularly suited for modeling the intricate microstructures of battery electrodes [6][7].

This study aims to develop a GAN AdaIN-based framework for synthesizing 3D battery electrode microstructures from 2D X-ray images while incorporating manufacturing parameters. By validating the fidelity of generated models against real data and analyzing their impact on battery performance, this research offers a transformative approach to advancing battery technology and achieving sustainable energy solutions [9].

2. Materials and Method

The overall flowchart for this research can be seen in the following Figure1. First, the different levels of the calendaring pressures are applied to the electrodes and result in different electrode compression levels. Then, the CT scan is performed on these samples to obtain the real 3D texture data for battery electrodes. The reconstruction algorithm is then applied with the inputs of three slices of orthogonal positions. The obtained synthetic 3D structure is then compared with the real data from the CT scan to evaluate its effectiveness and accuracy. Finally, the calendaring compression pressure, which is the control parameter for the reconstruction process, is changed to examine how the structure will be produced and the accuracy of the generated structure.

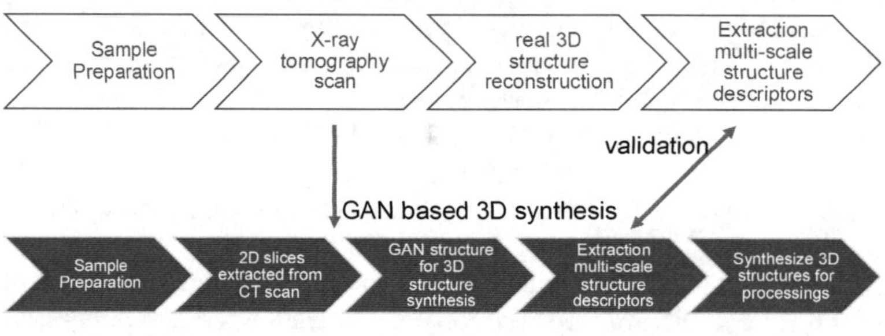

Figure 1 the overall structure and steps.

2.1. Battery electrode preparation and characterization.

The battery electrode data are sourced from the literature [10]. The datasets include battery electrode processing parameters, electrochemical characterisation and the CT scan information. The electrodes were based on the NCM333 (BC-618K, 3M), C65 carbon black (TIMCAL), and PVDF binder (Kynar

765, Arkema). Other materials include the N-methyl pyrrolidone solvent (NMP, from Sigma-Aldrich) and the LP30 electrolyte, which consists of 1 M LiPF6 in a 1:1 mass-ratio mixture of ethylene carbonate (EC) and dimethyl carbonate(DMC). The electrode slurry with weight ratios for active materials, carbon black and binder at 94:3:3 in NMP solvent were prepared by 5 minutes mixing using a high-shear disperser, and then casted onto an 12μm aluminum foil at 200 micro meters. The films were dried in vacuum for 8 hours. The final calendaring were prepared with three levels of calendaring pressures: 0, 300, 2000 bars to evaluate its potential influence on the battery microstructure (see Figure 2). The structure of the dataset also include synchrotron radiation X-ray tomographic microscopy (SRXTM) to obtain statistically significant volume($\sim 700 \times 700 \times 70$ micro meter) at TOMCAT beamline of the Swiss Light Source at the PaulScherrer Institute (Villigen, Switzerland).

(a) 300 bar calendaring compression (b) 2000 bar calendaring compression

Figure 2 the CT reconstructed 3D graphs for the NMC electrode.

2.2. The generative adversarial network architecture.
This study mostly follows the works from the literature [9, 11] with same modifications to embed the processing information into GAN structure (see Table 1 for detailed information).

Table 1 The detailed architecture of the SliceGAN.

Generator				Critic			
Layer	Function	Kernel, Stride, Padding	Output Shape	Layer	Function	Kernel, Stride, Padding	Output Shape
Input	-	-	64 × 4 × 4 × 4	Input	-	-	1×64×64
1	ConvTranspose3d	4, 2, 2	512 × 6 × 6 × 6	1	Conv2d	4, 2, 1	64×32×32
	BatchNorm3d	-			ReLU	-	
	ReLU	-		2	Conv2d	4, 2, 1	128×16×16
2	ConvTranspose3d	4, 2, 2	256 × 10 × 10 × 10		ReLU	-	
	BatchNorm3d	-		3	Conv2d	4, 2, 1	256×8×8
	ReLU	-			ReLU	-	
3	ConvTranspose3d	4, 2, 2	128 × 18 × 18 × 18	4	Conv2d	4, 2, 1	512×4×4
	BatchNorm3d	-			ReLU	-	
	ReLU	-		5	Conv2d	4, 2, 0	1×1×1
4	ConvTranspose3d	4, 2, 2	64 × 34 × 34 × 34				
	BatchNorm3d	-					
	ReLU	-					
5	ConvTranspose3d	4, 2, 3	1 × 64 × 64 × 64				
	tanh	-					

The brief introduction to the Kench&Cooper's work, dubbed as the "SliceGAN". The 64×64 pixel images firstly cropped from a input 2D image and compared with 64×64 pixel images sliced from a generated fake 3D image by the critic. The SliceGAN uses five layers of transpose convolution to

Journal of Physics: Conference Series **3009** (2025) 012043 doi:10.1088/1742-6596/3009/1/012043

generate a 1 channel × 64 × 64 × 64 voxel 3D fake image from 64 sets of latent variables in the format of 4 × 4 × 4. With this "SliceGAN" technique, it is possible to reconstruct the isotropic 3D structure with an input of single 2D image and anisotropic structure with three 2D orthogonal images.

2.3. The adaptive instance normalization.

As introduced before, the main drawback for the original SliceGAN architecture is that it does not allow a parameter input (e.g. electrode manufacturing parameters) to create the processing – structure linkage and further allow a controllable synthesis of the 3D structure for a preset manufacturing parameters. On the other hand, the development of GAN structure in these years have provided the tool to inject certain attributes to the images and thus enables controlling structure generation compatible to the inputted parameter. One of the most famous and widely used examples is styleGAN [12], where adaptive instance normalization (AdaIN) is adopted to inject style at certain layers of the generator. Chung et al. proposes incorporating the AdaIN in 3D structure generation to disentangle the feature space [13]. Following the idea in Chung et al., the AdaIN feature vector is injected into the architecture of SliceGAN and Adaptive instance normalization substitutes the batch normalization as in the original SliceGAN(see Figure 3) [13]. With the introduction of the AdaIN feature vector, the processing conditions can be used as inputs and trained with the corresponding settings. The obtained generator can be used to synthesis microstructure under the specified processing conditions.

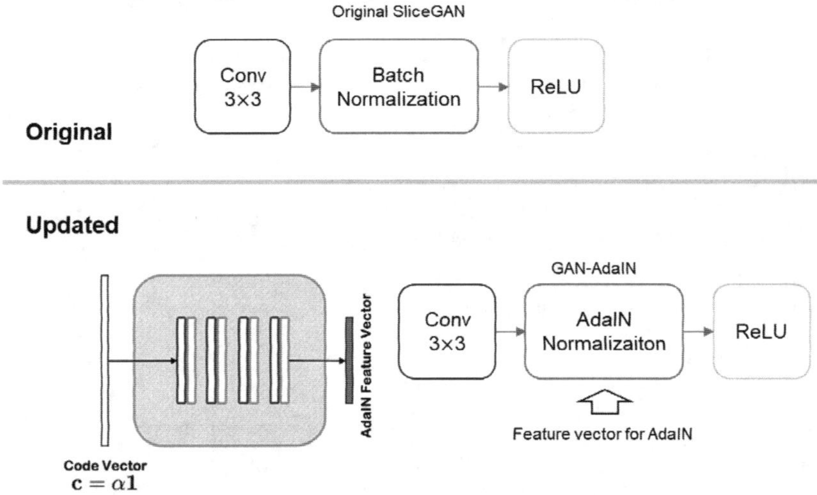

Figure 3 Architecture of the GAN-AdaIN.

3. Result

Figure 4 presents the details of training scheme, loss graph, inputted slices and their respective derivative of 3D structure. The input datasets for 3D reconstruction are three orthogonal slices. The slices are taken from the CT result. However, in actual practices, they can be well obtained from sectionized SEM graphs. The training for the GAN-AdaIN maximumly mimics the protocol documented in the SliceGAN. In one epoch, the discriminators, allocated for XY, YZ, and XZ images, respectively, are trained fives before the 3D generator updated one time. After repeated training, 64 sets of latent variables in the format of 4 × 4 × 4 are input to the 3D generators to obtain a larger 3D synthetic structure with a 64 × 64 × 64 voxel multiplied by 1 channel. The method uses 32 subdivided datasets (batch size) for generator training, performed only once. Meanwhile, 8 subdivided datasets are used for critic training, which is repeated five times in one iteration. The optimizer used is Adam with learning rate to be 0.0001 for both the generator and discriminator.

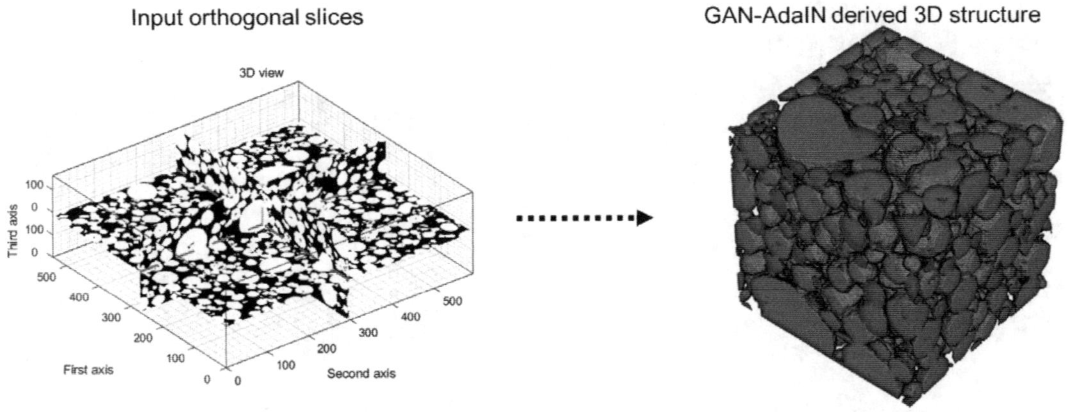

Figure 4 Reconstruction 3D structures for NMC electrode through 2000 bar calendaring.

3.1. Validation between the Reconstructed 3D structures and the real CT data

Since the 3D structure of the NMC electrode are available, it is thus highly interesting to examine the diffusive properties including tortuosity, percolation, porosity, and active surface area between the 3D reconstructed. The tau-factor software is employed to calculated this results [14]. The two-points correlation is calculated and presented in the Figure5.

From Table 2, it is clear that the GAN-AdaIN derived structure can fully capture the intrinsic statistical distribution for the porous structure, and therefore recreate the 3D results highly resembled the ones obtained from the CT. Meanwhile, it can be seen that GAN-AdaIN has reproduced the influences from the different levels of calendaring pressures. The porosity levels decrease from the around 55% to 40% with the increase in calendaring pressures. The corresponding tortuosity level has been elevated from 1.41~1.8. The active surface area (measured as per voxel) also rises from 140 voxels to 183 voxels. Surprisingly, these variations are also resembled and presented through reconstructed structures.

Table 2 Comparison between the synthetic reconstructed structure and CT derived real structure.

	0bar		300bar		2000bar	
	Reconstructed structure	CT obtained structure	Reconstructed structure	CT obtained structure	Reconstructed structure	CT obtained structure
Porosity	53.7%	55.5%	51.6%	52.8%	38.3%	41.0%
Tortuosity	1.46	1.41	1.48	1.48	1.74	1.8
Percolation	99.8%	100%	99.8%	99.8%	99.7%	99.8%
Active surface area (/voxel)	144	140	144	155	177	183

4. Discussion

The GAN structure is well-known to be hard to achieve a converge. To improve the robustness of the GAN-AdaIN model, we generally follows the original "SliceGAN" structure [9] using the Wasserstein distance as the loss function (Eq.1).

$$L = D_w(f_s) - D_w(r) + \lambda(\|\nabla_k D_w(k)\|_2 - 1)^2 \quad (1)$$

Where D_w is the discriminator of the GAN-AdaIN model, f_s and r stands for the fake images and true images, respectively. The Wasserstein distance, which is represented by the last term in Eq.6, calculates the penalty for any the L2 norm of the gradients of discriminator (D_w) deviating from the unit level. Thus, the third term in the loss function helps to maintain the gradients to a relatively stable level. Such a loss function has shown remarkable effects for creating a stable structural generation for the

GAN-AdaIN. The snapshots of the sectionized images for the 3D structure have also been provided in the courses of the training process (Figure 5).

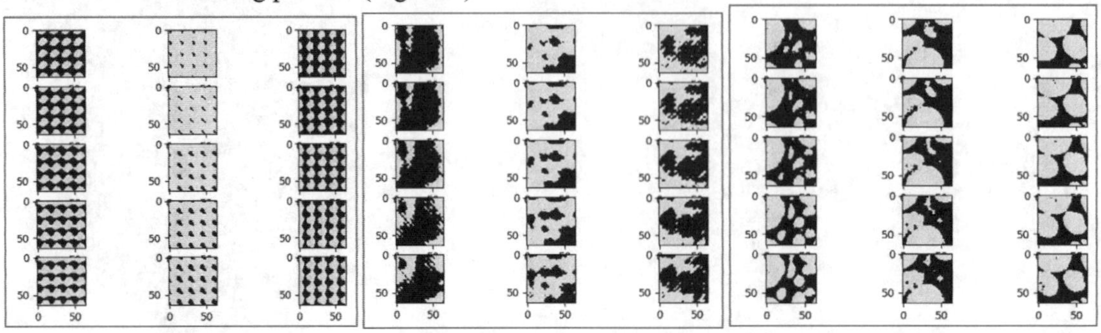

(a) Epoch=20 (b) Epoch=200 (c) Epoch=2000

Epoch	Observations	Understanding
20	Randomly generated patterns, no clear structure visible.	Images are formed from latent vectors with numbers randomly generated. No discernible structure is observed at this stage
200	Porous electrode pattern begins to emerge, partially recognizable.	Training progresses to a point where a recognizable porous electrode pattern begins to emerge. The structure hints at alignment with the intended features.
2000	Fully developed porous structure, closely matching NMC electrode features.	After extensive training, the generated structure closely resembles the porous structure of an NMC electrode, demonstrating high fidelity to the target pattern

Figure 5 the prognostic structural generation during the training process.

Figure 5 illustrates the intermediate results generated by GAN-AdaIN at various training epochs, highlighting the progression in structural detail and fidelity. At the initial stage (Epoch 20), the images appear to be random patterns generated from latent vectors, with no discernible structure. As training advances to Epoch 200, a recognizable porous electrode pattern begins to emerge, indicating the model's increasing ability to capture structural features. By Epoch 2000, the generated structure demonstrates a high level of refinement, closely resembling the porous structure of an NMC electrode. This progression showcases the effectiveness of GAN-AdaIN in gradually learning and reproducing complex structural features with extended training.

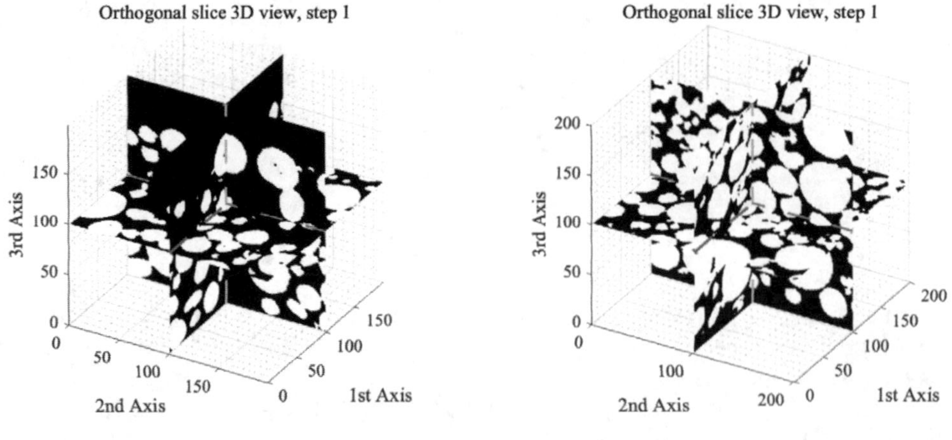

(a) Weight percentage 90% NMC at 0 bar (b) Weight percentage 94% at 2000 bar

Figure 6 The generated structure from GAN-AdaIN with different processing conditions.

4.1. The influence of different material content

The GAN-AdaIN model have also shown a great compatibility toward different styles of manufacturing processing conditions. With the inputs of NMC weight percent and compression forces, a combination of weight percentage and compression forces are inputted to derive the 3D structures (Figure 6). From the Orthogonal slice 3D view of the generated structures, though no real experimentation results are performed, the pore space presented for the Figure 6(a) is much smaller than that in Figure 6(b). Thus, the results fit the reality in a general term.

4.2. The influence of the controllable calendaring pressure

Since GAN-AdaIN parameterize the calendaring pressure as in the input for model development. Thus, we can assign an arbitrary calendaring pressure value, for instance, 600 bar, to see response of the trained model. From Figure 7, it can be seen that the deep learning model do present a reaction with the 600 bar calendaring parameter encoded as the input. The model export a 48.4% porosity, 1.55 tortuosity and 99.8% percolation and 175 active surface area, which are in between 300 bar and 2000 bar. This characteristic is particularly significant as it demonstrates the GAN-AdaIN method's ability to extrapolate and predict microstructures for unseen manufacturing conditions. Such a capability is crucial for practical applications, where a wide range of processing conditions may be encountered, and comprehensive experimental data for all scenarios may not be feasible. By enabling reliable predictions for conditions outside the training range, the GAN-AdaIN model can accelerate the design and optimization of battery electrodes, reduce reliance on costly experiments, and provide valuable insights into the relationship between manufacturing parameters and microstructural properties.

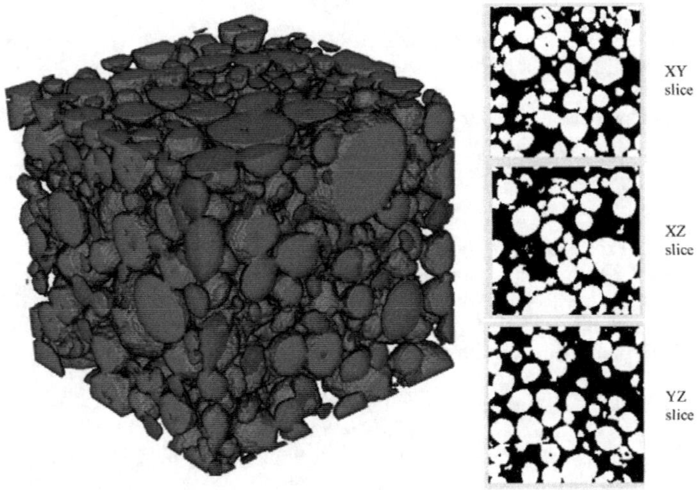

Figure 7 The GAN-AdaIN derived 3D structure with 600 bar pressure as the input.

5. Summary

In this study, we have adapted the SliceGAN model in literature with AdaIN normalization to encoder the battery manufacturing parameter, i.e., the calendaring pressure. The deep learning model can effectively reconstruct the three-dimensional battery electrode structure with high fidelity. For battery NMC electrode with 0bar, 300bar and 2000bar calendaring pressures, the real porosity according to the CT scan, is 55.5% to 41%. With the orthogonal images as the input, the reconstructed 3D synthetic structures possess a porosity in range of 53.7% to 38.3%, leading to a high accuracy. For other 3D-correlated performance metrics such as tortuosity, percolation, the calculated structure and real data also present a high degree of matching and overlapping, which all prove the effectiveness of this model in the course of the 3D structure generation. In the final experiment, by assigning different processing

Journal of Physics: Conference Series **3009** (2025) 012043 doi:10.1088/1742-6596/3009/1/012043

conditions of NMC content, the GAN-AdaIN model produces a physically meaningful results in all critical performance metrics.

Acknowledgments

This work is supported by the National Science Foundation of China (52477218, 51905361) and the Intelligent Urban Rail Engineering Research Center of Jiangsu Province(SDGC2413). Support from research contract of Energy Storage Industry Development Research Technical Service (HZS||FW[2024]0605) is appreciated acknowledged.

References

[1] IEA Global EV Outlook 2023, IEA, Paris https://www.iea.org/reports/global-ev-outlook-2023
[2] Scharf, J., Chouchane, M., Finegan, D.P. et al. Bridging nano- and microscale X-ray tomography for battery research by leveraging artificial intelligence. Nat. Nanotechnol. 17, (2022) 446–459.
[3] Phang, J.T.S., Lim, K.H. & Chiong, R.C.W. A review of three dimensional reconstruction techniques. Multimed Tools Appl 80, (2021), 17879–17891.
[4] Maken, P., Gupta, A. 2D-to-3D: A Review for Computational 3D Image Reconstruction from X-ray Images. Arch Computat Methods Eng 30, (2023), 85–114.
[5] Kench, S., Squires, I., Dahari, A. et al. MicroLib: A library of 3D microstructures generated from 2D micrographs using SliceGAN. Sci Data 9, (2022), 645.
[6] Kasten Y, Doktofsky D, Kovler I End-to-end convolutional neural network for 3d reconstruction of knee bones from bi-planar X-ray images. Machine learning for medical image reconstruction. Springer, Cham, (2020), 123–133.
[7] Dixit S, Pai VG, Rodrigues VC, Agnani K, Priyan SRV (2019) 3d reconstruction of 2d X-ray images. In: 2019 4th International Conference on Computational Systems and Information Technology for Sustainable Solution (CSITSS), 1–5.
[8] Cang, H. Li, H. Yao, Y. Jiao, Y. Ren, Improving direct physical properties prediction of heterogeneous materials from imaging data via convolutional neural network and morphology-aware generative model, Computational Materials Science 150 (2018) 212-221.
[9] Kench, S., Cooper, S.J. Generating three-dimensional structures from a two-dimensional slice with generative adversarial network-based dimensionality expansion. Nat Mach Intell 3, (2021) , 299–305.
[10] Ebner, M., Geldmacher, F., Marone, F., Stampanoni, M. and Wood, V., X-Ray Tomography of Porous, Transition Metal Oxide Based Lithium Ion Battery Electrodes. Adv. Energy Mater., 3, (2013), 845-850.
[11] Sugiura, K.; Ogawa, T.; Adachi, Y.; Sun, F.; Suzuki, A.; Yamanaka, A.; Nakada, N.; Ishimoto, T.; Nakano, T.; Koizumi, Y. Big-Volume SliceGAN for Improving a Synthetic 3D Microstructure Image of Additive-Manufactured TYPE 316L Steel. J. Imaging 2023, 9, 90.
[12] Karras, T., Laine, S. & Aila, T. A style-based generator architecture for generative adversarial networks. In Proceedings of the IEEE/CVF Conference on Computer Vision and Pattern Recognition, (2019), 4401–4410
[13] Chung, H., Ye, J.C. Reusability report: Feature disentanglement in generating a three-dimensional structure from a two-dimensional slice with sliceGAN. Nat Mach Intell 3, (2021), 861–863
[14] S.J. Cooper, A. Bertei, P.R. Shearing, J.A. Kilner, N.P. Brandon, TauFactor: An open-source application for calculating tortuosity factors from tomographic data, SoftwareX, 5, (2016), 203-210

Journal of Physics: Conference Series **3009** (2025) 012044 doi:10.1088/1742-6596/3009/1/012044

Design and research of telescopic arm of platform truss robot for building machine

Jiangnan Huang[1], Xiaoliang Shi[1], Haibin Yin[1,*], Han Pan[2]

[1]School of Mechanical and Electrical Engineering, Wuhan University of Technology, Wuhan, 430070, Hubei, China
[2]China Construction Third Engineering Bureau Cloud Structure Robot Co., Ltd, Shenzhen, 518132, Guangdong, China

*Corresponding author's email: chinaliuyin@whut.edu.cn

Abstract. To improve the performance of the truss robot and guide the subsequent design of the telescopic arm of the truss robot under different working conditions, the working principle of the telescopic arm of the truss robot is introduced, and a mathematical model for the size optimization of the telescopic arm is established under a given working condition with the goal of minimizing the mass of the telescopic arm and ensuring that the telescopic arm has a certain stiffness. According to the optimized dimension parameters, the prototype of the telescopic arm of the platform truss robot is manufactured and run.

1. Introduction

Building machine is a large-scale intelligent combination of construction machinery construction equipment platform, mainly composed of the top canopy system, concrete distributing machine system, steel platform system, formwork system, hanger system and support power system, and equipped with hydropower facilities, control room, portable toilet and other ancillary systems, and can be carried out according to the overall construction progress[1-2]. Compared with the traditional construction method, its most significant advantage is that all the processes required for building are concentrated, and the construction is carried out layer by layer in the air to realize the secondary structure, decoration and other processes at the same time[3-4]. Wakisaka et al. developed the BIG CANOPY building machine for reinforced concrete construction and applied it to actual construction sites[5]. Unlike BIG CANOPY, Ikeda et al. developed a building machine for steel structures called ABCS[6]. Under the premise of improving the level of automation, the Japanese Shimizu company has designed and built a highly automated building machine, SMART, for high-rise buildings[7].

As a typical industrial robot, the truss robot has been widely used in many industries[8]. The truss robot is an automatic industrial equipment based on the rectangular coordinate system, also known as the rectangular coordinate robot, which is mainly through the linear motion on the X, Y and Z axes to complete the work[9]. The truss robot studied in this paper combines the truss robot with the platform of the building machine for the first time. Its main function is to serve as the base of a six-axis robot arm, which is used to assist in the process required for building construction and improve work efficiency.

2. Building machine truss robot

The structure of the building machine truss robot is shown in Figure 1, which is mainly composed of a cart device, a beam device, a trolley device and a telescopic arm device. Its workflow is as follows:

Step 1: After the truss robot is powered on for the first time, the reset movement is carried out first. After the reset movement is completed and the information required for the work is received, the cart devices at both ends move directly above the target position;

Journal of Physics: Conference Series **3009** (2025) 012044 doi:10.1088/1742-6596/3009/1/012044

Step 2: After the cart device moves directly above the target position, the cart device and the vertical telescopic arm device move to the target position;

Step 3: After the trolley device and the vertical telescopic arm device move to the target position, the end of the vertical telescopic arm device of the truss robot is in the target position. At this time, the control system sends the position signal to the six-axis robot arm, and the six-axis robot arm starts to work;

Step 4: After the six-axis robot arm completes its work here, it sends a feedback signal to the control system to complete the work, after which the vertical telescopic arm device moves to the highest point and proceeds to the next cycle.

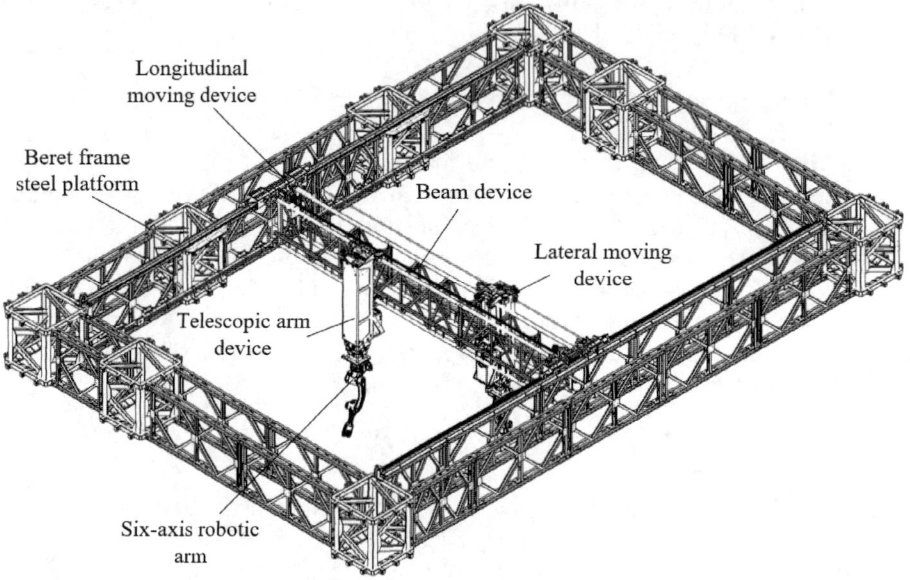

Figure 1. Building machine truss robot structure.

3. Telescopic arm principle

In order to facilitate the demonstration of the working principle of the telescopic arm, a three-section arm is used to introduce the telescopic arm mechanism, as shown in Figure 2. The working principle is as follows: n+1 arm is fixed, and the rest of the arm is movable, in which n arm is the active arm and the rest is the driven arm. Ball screw drives the expansion and contraction of the n arm and drives the expansion and contraction of the n-1 arm by stretching and repositioning the synchronization belt. In theory, the movable section arm can have multiple sections and can be cascaded multiple times on the basis of the three-section telescopic arm.

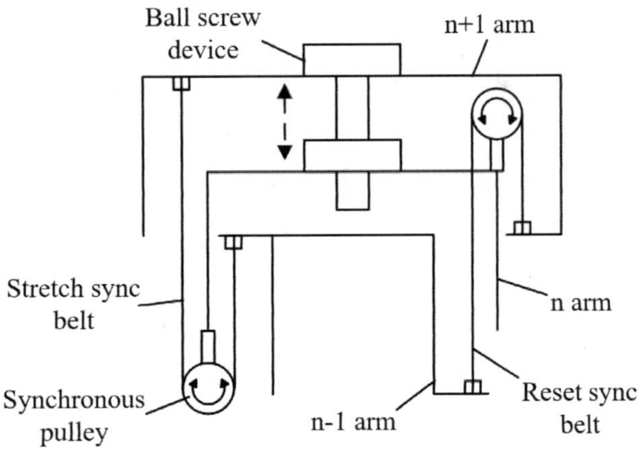

Figure 2. Schematic diagram of telescopic arm mechanism.

4. Motion analysis and size optimization of the telescopic arm

To determine the optimal number and height of the telescopic arm on the truss robot under the working condition of the building machine platform, the motion of the telescopic arm is analyzed. The structural dimension diagram of the telescopic arm is shown in Figure 3. Assuming that the number of the movable arm of the telescopic arm is n, there are n+1 arms, the outermost arm is the n+1 arm, the innermost arm is the first arm, then the n+1 arm is the fixed arm, the n arm is the active arm, and the rest are the driven arm. The lifting belt pulley and its supporting synchronization belt can be regarded as a movable pulley mechanism at work, and the free end movement of the rope is twice the movement of the movable pulley. Therefore, when the expansion quantity of the n section arm is z, the expansion quantity of the n-1 section arm is 2z, and the expansion quantity of the subsequent section arm is similar, and the expansion quantity of the x section arm is expressed as follows:

$$L_x = (n - x + 1)z \tag{1}$$

where L_x is the extension of the x arm, mm; n is the number of movable segments; x is the arm of any segment; z is the expansion of the active pitch arm, mm.

When the telescopic arm is working, in order to ensure that it has a certain stiffness in the horizontal direction, it is necessary to ensure that there is a certain overlap between the knuckle-arm and the knuckle-arm. The joint degree of the knuckle-arm is defined as the ratio of the overlapping part of the two adjacent knuckle-arms to the height of the previous knuckle-arm. The ξ expression of the joint degree is as follows:

$$\xi = \frac{h - z}{h} \tag{2}$$

where ξ is the joint degree of the knuckle arm; h is the height of the pitch arm, mm. To reduce the weight of the telescopic arm, the following aims to minimize the total mass M_s of the telescopic arm and ensure that the telescopic arm has a certain stiffness. The number of movable arm n and the height of the movable arm h are designed.

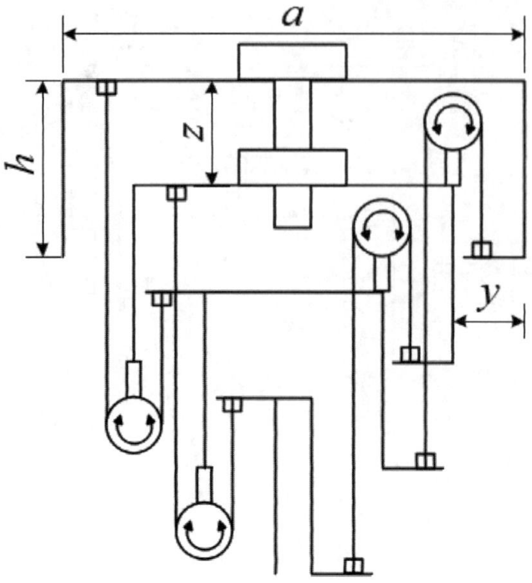

Figure 3. Dimensions of the telescopic arm.

The expression for the mass of the x segment arm (M_x) is:

$$M_x = \{4b[a-2y(n-x+1)]+4b^2\}\rho h \tag{3}$$

where a is the inner wall size of the n+1 section arm, mm; y is the distance between adjacent arm segments of the telescopic arm, mm; b is the wall thickness of the movable segment arm, mm; M_x is the mass of the x segment arm, kg; ρ is the material density of the pitch arm material, kg/mm3.

The total mass M_s of the movable segment arm of the telescopic arm is expressed as follows:

$$M_s = \sum_{x=1}^{n}\{4b[a-2y(n-x+1)]+4b^2\}\rho h \tag{4}$$

Under the working condition of the building machine platform, the travel of the telescopic arm is 6500mm. Considering the stiffness of the telescopic arm, it is required that the joint degree of the arm is not less than 0.25. In order to enable the six-axis mechanical arm to be installed on the end flange of the first arm, it is specified that n is not greater than 7. In order to have enough space on the trolley to place the control cabinet of the six-axis robot arm, the specified h is not less than 2000. The size of the inner arm a is 462mm, the distance between the inner walls of the adjacent section arms y is 26mm, and the wall thickness of the movable section arms is 3mm. The mathematical model of telescopic arm size optimization is established, which can be expressed as:

$$\min M_s = \sum_{x=1}^{n}\{4b[a-2y(n-x+1)]+4b^2\}\rho h$$

$$s.t.\begin{cases} nz \geq 6500 \\ 0.25 \leq \xi \leq 1 \\ h \geq 2000 \\ n \in [2,3\cdots 7] \end{cases} \tag{5}$$

To solve this Optimization problem, the fmincon function in the Optimization Toolbox in MATLAB software can be used[10-11]. We take an integer for n respectively and use the Optimization Toolbox in MATLAB software to find the optimal solution for n taking different integers. The solution results are shown in Table 1.

Table 1. Optimization results of the telescopic arm.

n	2	3	4	5	6	7
M_S(kg)	315.95	294.72	273.49	291.08	319.90	338.93

According to the optimization results, when the number of movable joint arms n is 4, the total mass M_s of movable joint arms is the smallest, which is 273.49 kg. In this case, the corresponding height h of the movable joint arms is 2166 mm, and the expansion capacity z of the active joint arms is 1625mm. Considering the processing cost, safety and other factors, it is determined that the height of the pitch arm h is 2200 mm, the expansion capacity of the active pitch arm z is 1625 mm, and the expansion arm has five arms.

5. Building machine platform truss robot testing machine

The testing machine of building the machine platform truss robot is shown in Figure 4. The specific structural form and processing dimensions of the telescopic arm of the truss robot are processed according to the structural form and size designed above. After a period of trial operation, the results show that the truss robot runs smoothly and meets the operation requirements under the given working conditions.

Figure 4. Testing machine of platform truss robot for building machine.

6. Conclusion

The main content of this paper is the design of the telescopic arm of a truss robot under the condition of the building machine platform. The workflow of the truss robot and the working principle of the telescopic arm of the building machine are introduced. The motion characteristics of the telescopic arm are analyzed, and the size optimization design of the telescopic arm is carried out to minimize the mass of the telescopic arm on the premise of ensuring the rigidity and adaptability of the working condition. Under the working condition of the building machine platform, the optimal number of pitch arms is 5, and the optimal height of pitch arms is 2166 mm. The testing machine of the telescopic arm of the platform truss robot for the building machine is manufactured and run. The results show that it runs smoothly and meets the requirements.

References

[1] Li J Q, Yang X, Wang S, et al. Design of lightweight integrated platform of Wuhan Yangtze River Center [J]. Construction Technology (English and Chinese), 2022, 51(10): 25-28+116.
[2] Luo F J, Su S W, Yu Z Q, et al. Key construction technology of residential building machine; 2022 National Civil Engineering Construction Technology Exchange Conference, Beijing, China, F, 2022 [C].
[3] Wang J Q, Zhang L M, Chen K. Research on intelligent construction application framework and key technologies for aerial building machine; 2023 Smart City and Intelligent Construction High-end Forum and Annual Meeting of Intelligent Construction Academic Committee of Architectural Society of China, Wuhan, Hubei, China, F, 2023 [C].
[4] Xiang Y H, Lu A G, Xie J W, et al. Analysis of key technologies and advantages of intelligent construction integrated platform; 2023 National Engineering Construction Industry Construction Technology Exchange Conference, Beijing, China, F, 2023 [C].
[5] Wakisaka T, Takemoto Y, Matsunami T, et al. Automatization of Reinforcement Work in High-Rise Reinforced Concrete Buildings [J]. Proc of 7th ISARC, Bristol, 1990: 79-86.
[6] Ikeda Y, Harada T. Application of the automated building construction system using the conventional construction method together; proceedings of the Proceedings of the 23rd International Symposium on Automation and Robotics in Construction, Tokyo, Japan, F, 2006 [C].
[7] Wakisaka T, Furuya N, Inoue Y, et al. Automated construction system for high-rise reinforced concrete buildings [J]. Automation in Construction, 2000, 9(3): 229-250.
[8] Jiao H C, Wu W K, Zhao H T, et al. Application and development status of truss robot technology [J]. Heilongjiang Science, 2024, 15(22): 94-97.
[9] Liu C L. Mechanical analysis and structural optimization of high-speed truss robot [D], 2017.
[10] Abu-Reesh I M, Alnaizy R. Optimal design of multi-stage bioreactors performing wastewater treatment using the MATLAB optimization function (fmincon) [J]. International Journal of Environmental Engineering, 2014, 6(4): 403-414.
[11] Le Luong H T, Messine F, Henaux C, et al. Comparison between fmincon and NOMAD optimization codes to design wound rotor synchronous machines [J]. International Journal of Applied Electromagnetics and Mechanics, 2019, 60: S87-S100.

Journal of Physics: Conference Series **3009** (2025) 012045 doi:10.1088/1742-6596/3009/1/012045

Optimal design and simulation of beam rib structure

Yulong Chen [1,2], Gang Rao [1,2,*] and Jiahao Yuan [1,2]

[1] Hubei Key Laboratory of Mechanical Transmission and Manufacturing Engineering, Wuhan University of Science and Technology, Wuhan 430081, China
[2] College of Machinery and Automation, Wuhan University of Science and Technology, Wuhan 430081, China

*Corresponding author's e-mail: 17356423234@163.com

Abstract: Regarding the optimization issue of the rib plate structure in machine tool beams, this paper addresses the irregularities and difficulties in processing associated with the optimization results of the rib plate structure in machine tool beam topology optimization. A method is proposed that can effectively optimize the design of the rib plate structure in machine tool beams. Firstly, the structural characteristics of the topological model were analyzed. The key characteristic parameters required for the structural formation of the stiffeners were found, such as the thickness of the stiffeners, the lightweight, the number of stiffeners, and the positional parameters. Secondly, the characteristic parameters were combined with the finite element method. The corresponding characteristic parameters were iterated step by step through the thickness parameter discrete method, the mesh sequence reconstruction method, and the enumeration method. The model was reconstructed with the characteristic parameters, so as to realize the multi-level optimization of the rib structure and obtain the ideal rib structure model. Finally, taking the beam as an example for verification, it is found that the method can obtain a stiff plate structure that meets the design requirements. The topological structure model and the stiff plate structure model are simulated and compared. It is found that the natural frequency of the optimized model of the stiff structure is increased by 20%.

1. Introduction

Processing efficiency and precision are important technical indicators for measuring the performance level of machining centers. In order to meet the demands of modern high-speed, high-precision, five-axis, and composite processing, a bridge structure with a moving spindle beam has emerged. This design aims to reduce the mass of moving components, increase structural symmetry, minimize thermal deformation, and improve the efficiency and precision of the machine tool [1]. As the beam of a five-axis machine, it supports the sliding table above and connects to the guide rails below, making it a crucial moving part. Therefore, the design of the beam layout, the arrangement of ribs, and the lightweight structure directly impact the machining performance and manufacturing costs of the five-axis machine.

Many scholars have conducted research on beam-column structures. Zhou et al. [2] performed static and dynamic characteristic analyses of four types of rib plates using finite element methods, concluding that the "O"-shaped rib plate beam exhibited the best performance. They varied the number of "O" shapes in the "O"-shaped beam and conducted comparative static and dynamic analyses on seven beams with different numbers of "O" shapes, ultimately determining the optimal rib plate structure for beam performance. Zhang et al. [3], while keeping the overall dimensions of the beam unchanged, increased the number of "rice"-shaped rib plates from 5 to 10, resulting in a new beam structure with significant improvements in overall performance. Ju et al. [4] conducted a multi-objective optimization design using the beam rib plate structure, rib plate thickness, and the incline angle of the upper guide rail supporting the rib plate as design variables. They employed an orthogonal experimental method to design a 3-factor, 4-level orthogonal test; the optimal levels were determined using a comprehensive balancing method, leading to the best experimental scheme for rib plate structure. Shen et al. [5] used an adaptive growth method based on the natural growth mechanism of support branches to design the layout of internal stiffeners in structures, achieving a new dynamic design optimization method for overall machine tool structures. Zhuang et al. [6] identified the key design parameters for transverse rib plates through finite element parameter analysis.

Currently, the mainstream optimization methods for machine tool beam ribs mainly rely on finite element analysis to assist in manual design. Through the design of rib plates, it is possible to effectively enhance the static and dynamic performance of the overall beam. However, in the design process of rib plates, the structure of the rib plate is typically separated from the main body of the beam. First, several rib plate structures are designed. Then they are assembled onto the beam for simulation analysis. Although the rib plates are redesigned, it is not possible to determine the optimal solution for the rib plate structure; improvements can only be made within a specified range.

In order to address the structural issues of the machine tool beam rib, an analysis of the ribbed plate structure was conducted, leading to the proposal of a structural design method specifically for the ribbed plate. This method enables the determination of the ribbed plate structure within the beam structure, resulting in a rational ribbed plate structural model under constrained conditions.

2. Machine Tool Beam Structure Analysis and Modeling

2.1. Basic Components of a Machining Center

With the continuous development of modern machining centers, their application range is becoming increasingly broad and diverse. Based on the classification of the machine tool spindle placement, machining centers can be divided into vertical, horizontal, or universal machining centers, which correspond to the spindle being positioned vertically, horizontally, or rotatable [7]. The universal machining center, also known as the five-sided machining center, combines the functionalities of both vertical and horizontal machining centers. After the workpiece is clamped once, it can complete machining on all sides and the top surface except for the mounting surface. Universal machining centers offer higher machining accuracy and positioning precision, good adaptability, and high machining efficiency, enabling them to process various complex surfaces. They are currently a primary research focus in the production and development of high-precision machine tools.

The main structural components of a machining center include the bed, crossbeam, and spindle box. The sliding saddle and spindle box are generally suspended on one side of the crossbeam. During the machining process, significant torque is generated, which can easily cause the crossbeam to bend and deform, thereby reducing machining accuracy. As the main structural connection part of the machining center, the primary supporting components on the crossbeam are the sliding saddle and spindle box. A simplified representation of the crossbeam assembly is shown in Figure 1.

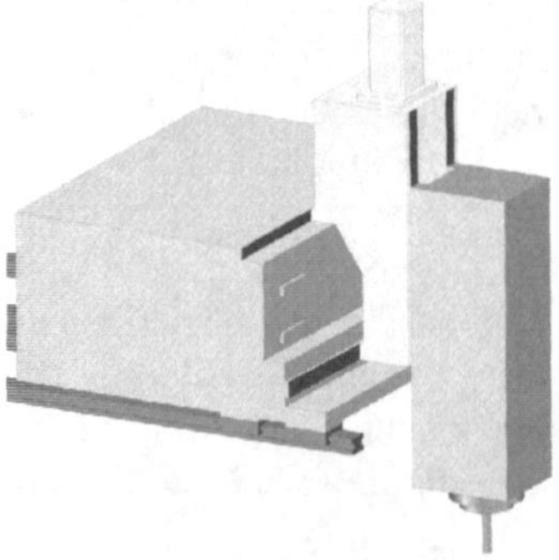

Figure 1. Simplified geometry of the machining center beam assembly.

Journal of Physics: Conference Series **3009** (2025) 012045 doi:10.1088/1742-6596/3009/1/012045

2.2. Beam Topology Model

The initial model of the beam is shown in Figure 2, retaining the contact surfaces with components such as the sliding saddle and the bed. The overall dimensions are $L \times D \times H = 1505 \times 1340 \times 730$ mm. Assuming the machine tool is in working condition, the mechanical properties of the material are in the linear elastic range, using En-GJL-300 material, with Young's modulus $E = 120$ Mpa, Poisson's ratio $\mu = 0.26$, and density $\rho = 7250\ kg/m^3$.

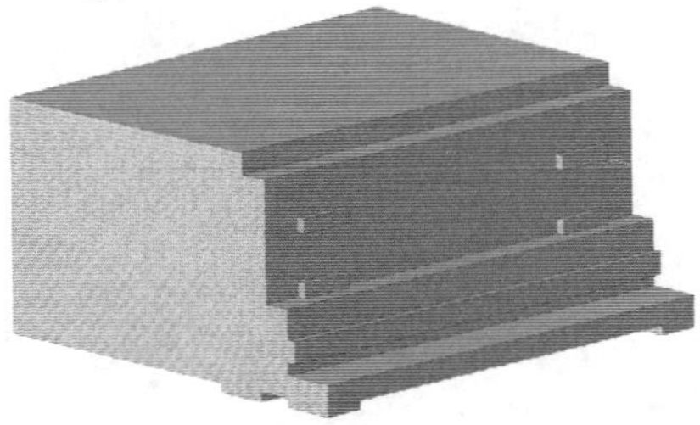

Figure 2. The initial model of the beam.

Based on the topology optimization method, the continuous space is divided into a finite number of discrete grid regions. The presence or absence of material in a specified region $X(i)$ can be represented using discrete values of 0 and 1. In the process of structural topology optimization, the element density of regions from which material needs to be removed is set to 0, while the element density of regions where material is retained is set to 1. Finally, by deleting the grid elements with a density of 0, a topological design grid model is obtained. Taking the bidirectional evolutionary structural optimization method as an example [8], the optimization model is represented as Equation (1), where $x(i)$ is the element density, $C(X)$ is the objective function representing strain energy, F is the load vector, U is the displacement vector of the structure, K is the stiffness matrix of the structure, p is the penalty factor (p = 3), n is the total number of elements, V^* is the upper limit of material volume, and V_i is the volume of the element.

$$\begin{cases} find: & x(i)\ (i = 1, 2, \dots, n) \\ Min: & C(x) = \frac{1}{2} U^T K U \\ s.t.: & \sum_{i=1}^{n} V_i x(i) \geq V^* \\ & x(i) = x_{min}\ or\ 1 \end{cases} \tag{1}$$

The topology optimization test for the machine tool beam model is shown in Figure 2. The topology optimization results are shown in Figure 3.

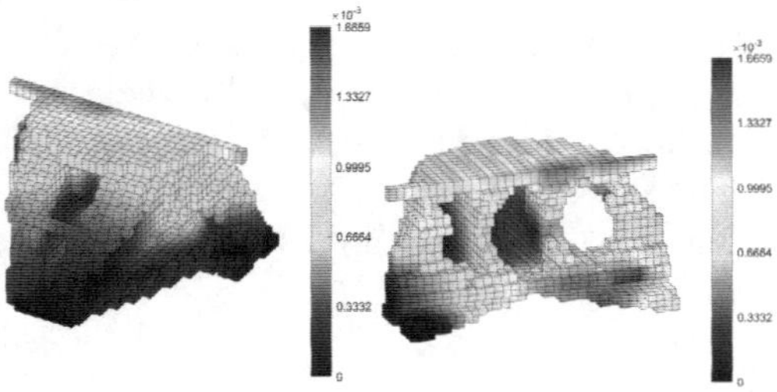

Figure 3. Machine tool Beam topology optimization model.

3. Design and Theory of Machine Tool Beam Rib Structure

3.1. Overall Variable

For example, based on the topology optimization results shown in Figure 3, topology optimization design is commonly used in the initial stages of structural design, especially in cases where the constraints are complex and empirical designs fail. It can help find the maximum stiffness structural shape within a limited spatial region while ensuring a predetermined material volume.

In topology optimization methods, although a preliminary topological model can be obtained, the discretization of elements leads to each element being treated as an independent optimization variable during the optimization process. The result tends to be irregular, making it difficult to process and often requiring secondary design [9]. Therefore, based on research into topology optimization and in conjunction with the ribbed plate structure type, this paper proposes an overall variable structure optimization design method that treats groups as design variables for optimizing the machine tool beam rib plate structure.

The overall variable connects discrete independent unit variables through feature grouping by analyzing the geometric characteristics of the model. It relates the unit variable x in topology optimization to the overall variable X in the optimization of the beam and plate structure, forming an internal continuous and external discrete overall variable, as shown in Figure 4.

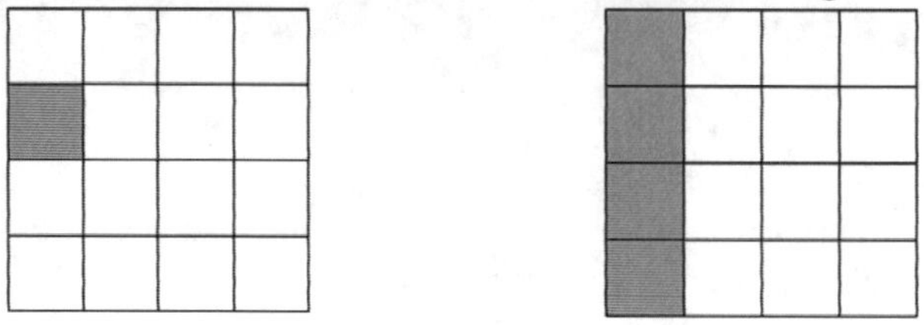

Figure 4. Independent unit variable x and overall variable X.

3.2. Multi-Resolution Refinement

The multi-resolution method decouples the finite element mesh and the density mesh, using a coarser mesh (coarse elements) for finite element analysis calculations, while the finer mesh (fine elements) serves as design variables for optimization design [10]. This approach allows for a significant reduction in finite element computation time while achieving refinement of the optimization model.

The main methods of mesh subdivision can be divided into interpolation methods and approximation methods, among which the most popular subdivision schemes are the Catmull-Clark method, Octree method, and Loop method. However, algorithms like Catmull-Clark have complex structures, low computational efficiency, and are difficult to apply [11]. Therefore, this paper proposes a point-insertion-based mesh subdivision scheme using MATLAB, leveraging its unique matrix processing capabilities. This subdivision scheme reads the node information of the element mesh, performs point insertion between the element nodes to achieve mesh subdivision, and, based on MATLAB's characteristics, realizes the correspondence between coarse and fine grids as well as the sharing of element and node information.

The simulation analysis and the preliminary design of the rib plate use larger-sized higher-order elements to reduce the number of elements, while the grid model participating in the final structural formation employs smaller-sized lower-order elements to enhance the resolution of the structural design results [12]. The smaller elements share common nodes with the higher-order elements, ensuring that the information at the common nodes is identical, thus establishing consistency between the coarse and fine meshes, as shown in Figure 5.

Figure 5. Multi-resolution mesh refinement.

3.3. Optimization Expression for Machine Tool Beam Rib Plate Structure

Based on the aforementioned theoretical research and combined with the topology optimization method of finite element theory, the continuous space is divided into finite mesh regions of the overall rib plate. The identification of material presence or absence in the rib plate region $X(i)$ is determined using characteristic parameters, represented by discrete numerical values of 0 and 1 [13]. During the optimization process of the structure, multiple iterations of calculations are required to achieve a satisfactory solution. Each solution yields a set of characteristic parameters for the rib plate structure, referred to as an iteration step. Within each iteration step, the two main processes involve finite element solving of the mesh model and updating the rib plate structure. The optimization model for the rib plate structure based on characteristic parameters is formulated as follows: the element density for areas where material needs to be removed is set to 0, while the element density for areas where material is retained is set to 1. Finally, the rib plate structure design mesh model is obtained by deleting the mesh elements with a density of 0. The mathematical model for the optimization of the beam structure is shown in Equation (2).

$$\begin{cases} find: \quad X(i) \ (i = 1, 2, \dots, n) \\ \quad max: \quad fre \\ \quad s.t.: \quad distm < L \\ \qquad dist(c_i, c_j) < k \\ \qquad X(i) = 0 or 1 \end{cases} \tag{2}$$

In the formula, $X(i)$ represents the overall variable, which is the stiffened plate; fre is the natural frequency; $distm$ is used to calculate the thickness of the stiffened plate; $dist(c_i, c_j)$ is used to calculate the radius of the lightweight hole.

4. Feature Parameter Stepwise Optimization Method

4.1. Enumeration Method for Optimizing Design of Ribbed Plates

Based on the topology optimization model and retaining only the ribbed box structure, the ribbed box model is reconstructed according to the shape of the machine tool's beam ribbed box. Combining the characteristics of the mesh model, vertical ribs with a fan-shaped distribution are used as the style for the ribs. The machine tool's beam box is grouped to determine the ribbed model with overall variables.

The overall variable grouping and numbering are performed based on the grid model of the rib plate box shown in Figure 6, denoted as $X(1), X(2) \dots X(n)$. Each overall variable is treated as a predefined shape of the rib plates, meaning that all grid points $x(i)$ in each group $X(i)$ are considered as a single overall variable. This way, the overall variables simultaneously contain both the element grid information and the rib plate position information. As illustrated in Figure 6, $X(1)$ represents the leftmost rib plate, while $X(n)$ represents the rightmost rib plate.

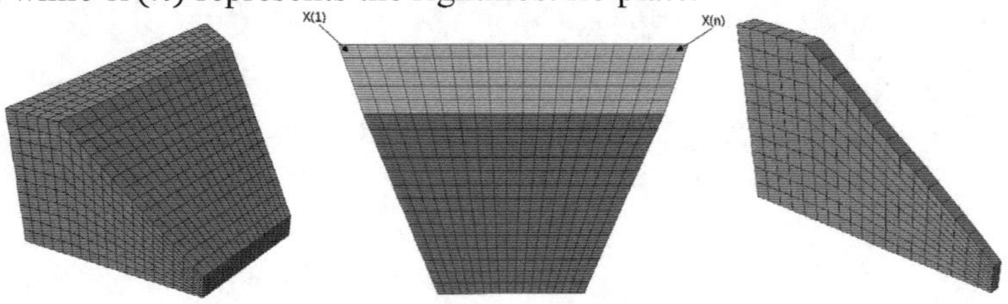

Figure 6. Rib plate grouping diagram.

By using characteristic grouping, the total number of rib plates and their position parameters are determined, transforming the problem of forming the rib plate structure into a selection problem for the rib plates. The enumeration method is employed to select combinations of the overall variable X, obtaining the number and position information of the rib plates. The model's rib plate structure combinations are traversed, with a maximum rib plate count of 5 as an example. Using MATLAB, tests

Journal of Physics: Conference Series **3009** (2025) 012045 doi:10.1088/1742-6596/3009/1/012045

are conducted on the selection of rib plates, as shown in Figure 7.

As shown in Figure 8, there are 5 rib plates, with one set of position parameters being 2, 6, 11, 15, and 19, which correspond to the overall variable $X(2), X(6), X(11), X(15), X(19)$.

	1
1	[7,3,9,20,10]
2	[7,13,3,19]
3	[10,15,13,16]
4	[4,8,16,5,19]
5	[11,1,20,9,19]
6	[1,2,10,19]
7	[5,18,9,16,19]
8	[4,8,15,14]
9	[11,8,12,20]
10	[1,19,4,2]

Figure 7. Matlab stiff plate selection test.

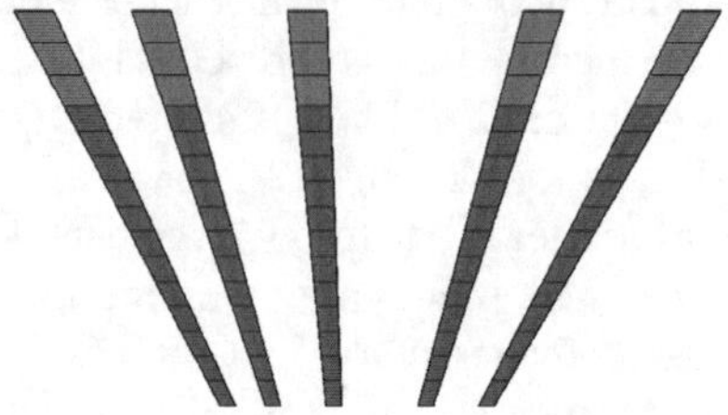

Figure 8. Five ribs.

4.2. Grid Sequence Reconstruction

In order to achieve a smooth and clear outline of the ribbed structure, a smaller mesh size is often used to discretize the optimization area. Therefore, a multi-resolution method is employed to refine the mesh. Based on the selected ribbed plate $X(m)$, the center coordinates of the included fine mesh are processed separately. The center points are coupled with the elements, using points to represent the elements. The center nodes, or fine mesh elements, are sequenced and selected based on the values of x, y or z, completing the reconstruction of the mesh sequence. Based on the fine mesh numbering and the center coordinates $c_i(x, y, z)$, the retention and removal of elements are realized, as shown in Figure 9.

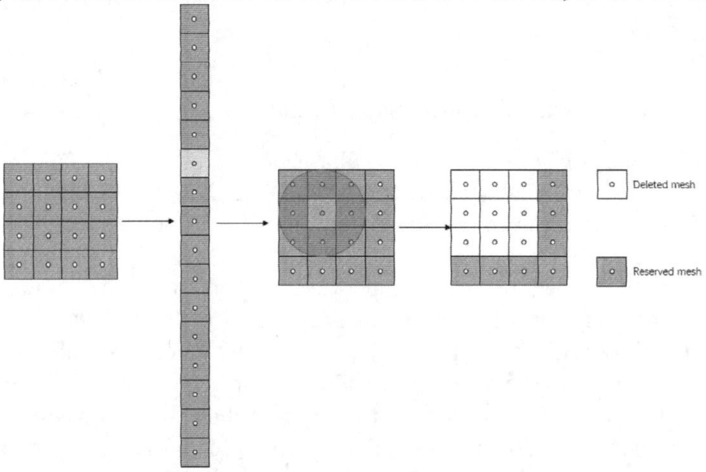

Figure 9. Schematic diagram of a lightweight hole.

The lightweight expression, as shown in Equation (3), is based on the sequence reconstruction of units. It allows for the customization of lightweight hole centers to specify the referred units, using the coordinates of the center point as the center of the hole, with k as the radius of the hole.

$$\begin{cases} find: \ \ x(i)\ (x(i) \in X(m)) \\ subject\ to: dist(c_i, c_j) < k \\ \qquad\qquad x(i) \rightarrow c_i \\ \qquad\qquad x(j) \rightarrow c_j \end{cases} \tag{3}$$

4.3. Thickness Parameter Discretization Design

The thickness parameter of the stiffener plate is a continuous variable and cannot be applied to discrete grid variables. Therefore, based on a multi-resolution method for refining the grid, the continuous thickness parameter is transformed into discrete grid parameters. By using fine grid units, the thickness issue of the stiffener plate is converted into a problem of the number of fine grid layers. The minimum unit size of the fine grid is considered as one unit of thickness. The optimization of the number of grid layers for the stiffener plate grid is carried out. As shown in Figure 10, one layer of the grid is considered to have a thickness of 1, and two layers of the grid are considered to have a thickness of 2. The optimization model for the thickness of the stiffener plate is shown in Equation (4):

$$\begin{cases} find: & x(i)\ (i = 1, 2, \ldots, n) \\ \text{minimize:} & vol \\ subject\ to: & distm < L \\ & vol < volfrac * VOL \end{cases} \tag{4}$$

Among them, $x(i)$represents the fine mesh element, vol is the target volume, L is the thickness of the ribs, $volfrac$ is the target volume fraction, and VOL is the initial volume of the model.

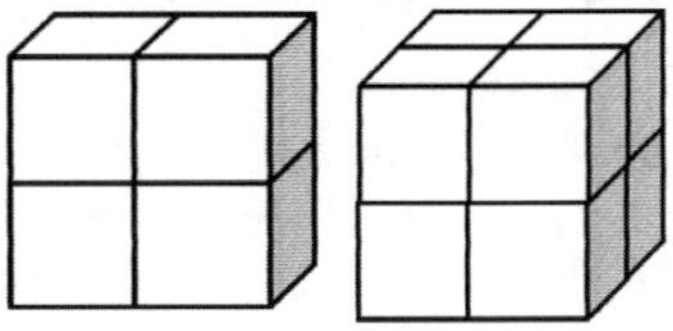

Figure 10. Correlation between thickness and number of layers.

Taking the multi-resolution refined rib plate mesh as a reference, one layer of the mesh is used as the base thickness mesh. Based on the coordinates of the central node of the base thickness mesh, it is denoted as $x(i)(x, y, z)$. Three nodes are selected with coordinates $x(j)(x1, y1, z1)$, $x(k)(x2, y2, z2)$, $x(l)(x3, y3, z3)$. Using the three-point formula, the initial mesh plane for thickness calculation based on the base thickness mesh is obtained.

$$p_1 = xj - xk \tag{5}$$

$$p_2 = xj - xl \tag{6}$$

$$\vec{n} = p_1 \times p_2 = \begin{vmatrix} j & k & l \\ x2 - x1 & y2 - y1 & z2 - z1 \\ x3 - x1 & y3 - y1 & z3 - z1 \end{vmatrix} = aj + bk + cl = (a, b, c) \tag{7}$$

Among them, p_1 and $p_2$are plane vectors, $\vec{n}$ is the normal vector, and (a, b, c) are the parameters of the plane equation.

We calculate the parameters of the reference plane equation for the rib plate thickness based on the above plane equation.

$$a = (y2 - y1) * (z3 - z1) - (y3 - y1) * (z2 - z1) \tag{8}$$

$$b = (z2 - z1) * (x3 - x1) - (z3 - z1) * (x2 - x1) \tag{9}$$

$$c = (x2 - x1) * (y3 - y1) - (x3 - x1) * (y2 - y1) \tag{10}$$

Treating a single rib plate as a whole variable, we traverse all the center nodes $c_0(x0, y0, z0)$ of the fine mesh sub-variables. Using the following formula, we calculate the point-to-surface distance to replace the thickness dimension. By calculating the distance and applying volume constraints, we determine the required thickness for the elements, thus achieving control over the rib plate thickness, as shown in Figure 11.

$$distm = \frac{|ax_0 + by_0 + cz_0 + d|}{\sqrt{a^2 + b^2 + c^2}} \tag{11}$$

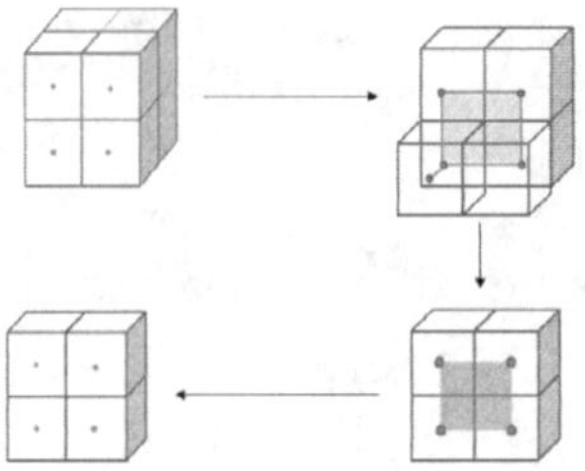

Figure 11. Schematic diagram of thickness optimization.

5. Optimization and Verification Analysis of Ribbed Structure

5.1. Optimization Method Validation

To verify the feasibility of the optimized design method for ribbed plate structures, the aforementioned topology beam model structure is used as the test object. The model is constructed, as shown in Figure 12, to validate the feasibility of the ribbed plate structure optimized by the method proposed in this paper. (To ensure the intuitiveness of the structural optimization process, the top cover of the beam is hidden during the optimization.)

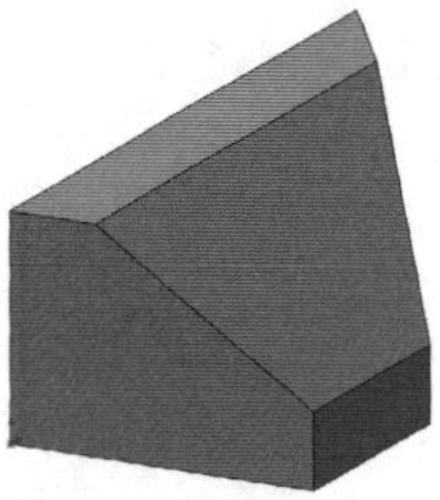

Figure 12. Initial model of the ribbed box.

The machine tool beam in this structural design uses pearlitic gray cast iron (HT300). Its material parameters are shown in Table 1.

Table 1. HT300 material properties.

Parameter	Value
Material	HT300
Density Kg·m-3	7.25x10^-9
Elastic modulus E(Pa)	1.2x10^5
Poisson's ratio μ	0.26

We take the design of 100 iterations for this model as an example, as shown in Figure 13:

(a) (b) (c)

Figure 13. The 10th, 50th, and 100th iterations of the model were refactored.

With the continuous iteration of the feature parameters, the structure of the rib plate also changes accordingly. By continuously retaining the structural models with higher inherent frequencies, a structural model of the rib plate with the maximum inherent frequency and reasonable structure is ultimately obtained. The natural frequencies of the different structures are shown in Figure 14.

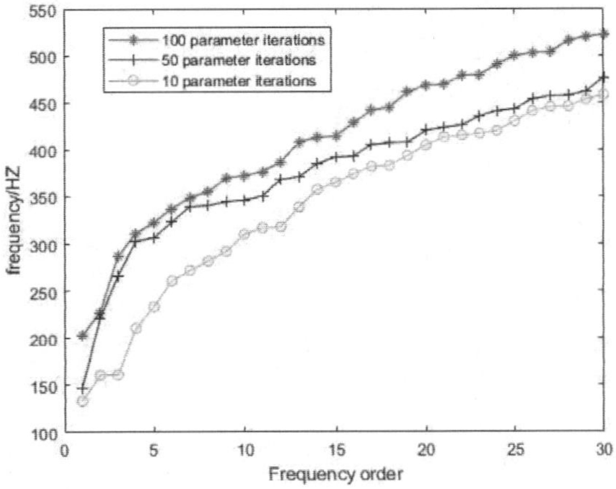

Figure 14. Comparison of frequencies of different structural designs.

5.2. Structural Simulation Verification

To further verify the effectiveness of the feature parameter-based stiffener plate structure optimization design method proposed in this paper for the design of beam stiffener structures, a comparative modal analysis was conducted between the beam model optimized using this method and the model that underwent only topological optimization. The structural models are shown in Figure 15 and Figure 16.

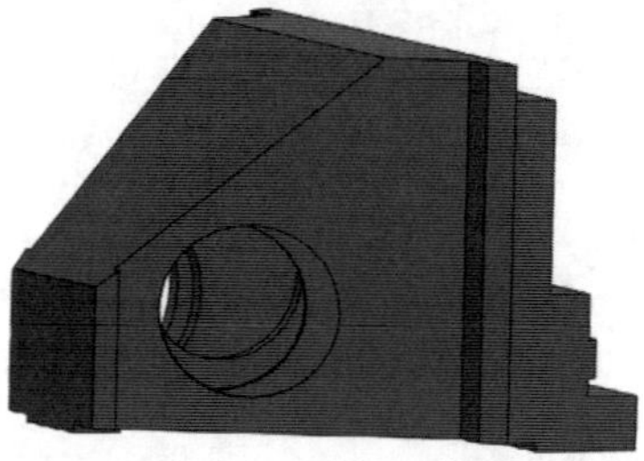

Figure 15. Optimization model of rib structure.

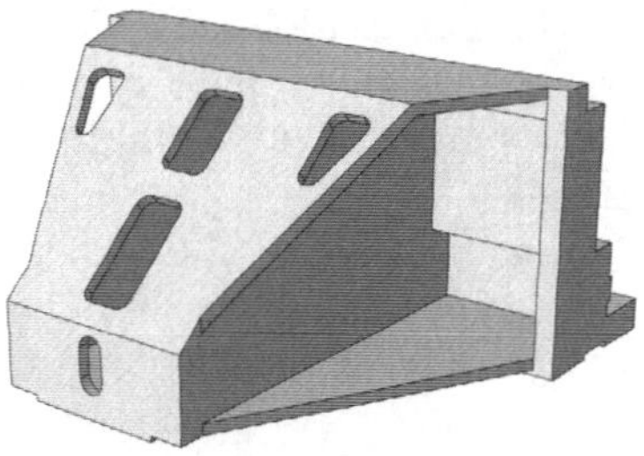

Figure 16. Topology optimization model.

The first three modes' natural frequencies and mode shapes are shown in the cloud diagram below.

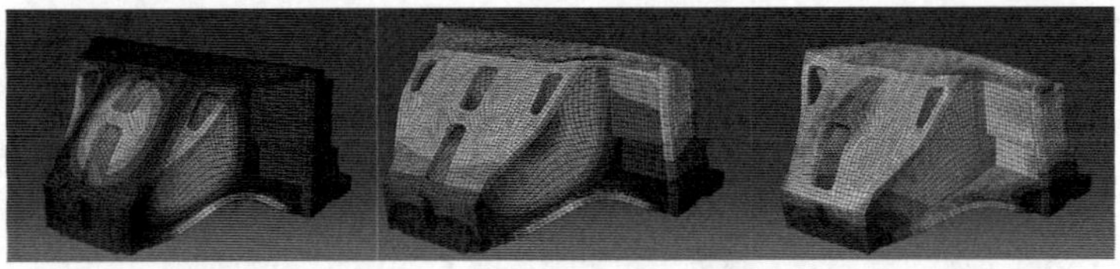

Figure 17. Modal analysis of the first three orders of the topological model.

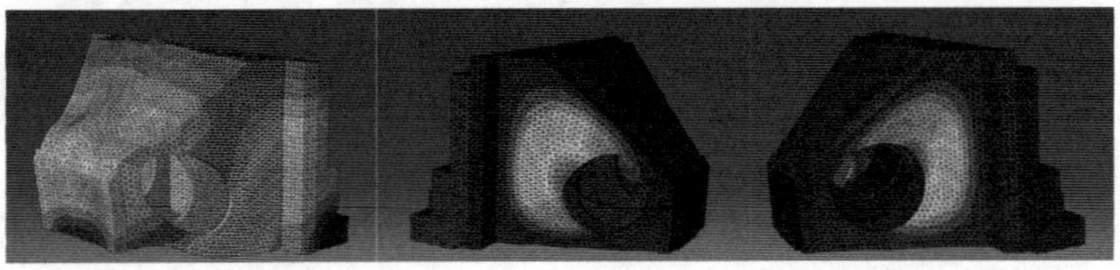

Figure 18. Modal analysis of the first three stages of the stiff plate structure model.

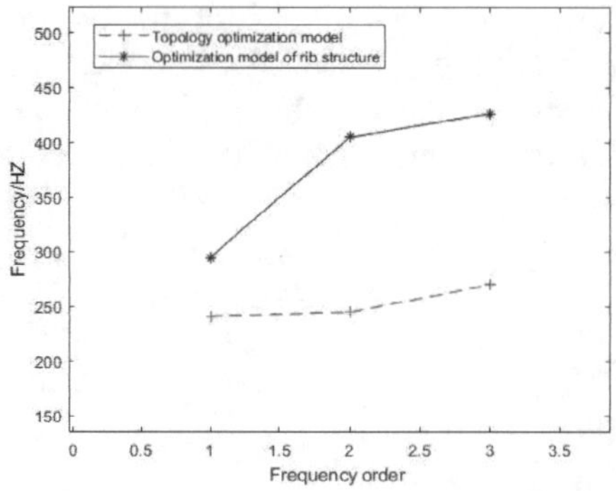

Figure 19. Comparison of natural frequencies.

The comparison of simulation results, as shown in Figure 19, along with the modal analysis diagrams of the model shown in Figure 17 and Figure 18, reveals that after the optimization of the ribbed plate structure, the model's natural frequencies have significantly increased. The minimum frequency after the optimization of the ribbed plate structure is 294.14 Hz. The maximum frequency is 426.34 Hz, whereas the minimum frequency after only topological optimization is 241.04 Hz, and the maximum frequency is 269.72 Hz.

6. Conclusion

This article is based on the rib plate structural model of the machine tool beam, proposing a stepwise optimization method for the rib plate structure with feature parameters as the iterative objects, and establishing a joint simulation platform using ABAQUS and MATLAB. This approach not only achieves the separate optimization design of individual features of the rib plate but also facilitates the overall optimization design of the rib plates in the beam structure, effectively addressing the issues of irregularities and difficulties in processing caused by topology optimization. The conclusions are as follows:

(1) In terms of feasibility, the proposed method optimizes the rib plate structure model based on feature parameters, allowing for multi-level design of each feature in the beam rib plate structure. This results in a rib plate structure that meets design requirements. Furthermore, by employing a multi-resolution method, it generates models with higher resolution and clearer boundaries.

(2) Regarding effectiveness, simulations of the optimized model were conducted using ABAQUS software. A comparative analysis of the rib plate structure's performance before and after optimization was performed, validating the rationality of the proposed method. Compared to the initial model, the overall frequency of the model after rib plate optimization increased by at least 20%, effectively enhancing the overall natural frequency of the machine tool and ensuring its processing accuracy and stability.

References

[1]. Frank Schneider, Jayanti Das, Benjamin Kirsch, Barbara Linke, and Jan C Aurich 2019 Sustainability in ultra precision and micro machining: A review. International Journal of Precision Engineering and Manufacturing-Green Technology, 6: 601-610.

[2]. Zhou L, Yuan J T, and Wang Z H 2014 Analysis and Optimization of the Structural of Beam in the Planer Type Machines. Machinery Design & Manufacture, 1: 15-17

[3]. Zhang X Q, Lin J, Zheng J W and Zhang D S 2016. Lightweight Design to the Beam Structure of the Milling Machine. Mechanical Research & Application, 29(4): 106-109

[4]. Jun J Q, Qiu Z X, Cui D Y, Ren D and Shang A K 2018 MULTI-TARGETS OPTIMIZATION DESIGN FOR MOVING CROSSBEAM OF MACHINE TOOL BASED ON ORTHOGONAL EXPERIMENTAL METHOD. Journal of Mechanical Strength, 40(2): 356-362.

[5]. Lei S, Ding X H, Li T J, Kong X Z and Dong X H
2019. Structural dynamic design optimization and experimental verification of a machine tool. Int J Adv Manuf Technol 104: 3773–3786.
[6]. Zhuang, H, Fang, S and Zhou, M and Mostafa Fahmi Hassanein
2021. Numerical Study on the Mechanical Properties of the Novel Wide Box Girders Stiffened by Transverse Ribs. KSCE J Civ Eng 25: 3448-3457.
[7]. Feng H B, Song D Z, and Wu F H and Jin Z L 2021. High rigidity and lightweight design for cross beam of large span gantry woodworking machine tool. Manufacturing Technology & Machine Tool, 8: 127-131.
[8]. Hanush S S and Manjaiah M 2022. Topology optimization of aerospace part to enhance the performance by additive manufacturing process. Materials Today: Proceedings, 62: 7373-7378.
[9]. Jeroen P. Groen, Florian C. Stutz, Niels Aage, Jakob A. Bærentzen and Ole Sigmund 2020. De-homogenization of optimal multi-scale 3D topologies. Computer Methods in Applied Mechanics and Engineering, 112979.
[10]. Herrero-Pérez D and Castejón P J M 2021 Multi-GPU acceleration of large-scale density-based topology optimization. Advances in Engineering Software, 157: 103006.
[11]. Hamza Yusuf Fatihu and Lin Hong Wei 2022 Conjugate-Gradient Progressive-Iterative Approximation for Loop and Catmull-Clark Subdivision Surface Interpolation. Journal of Computer Science and Technology (2), 487-504.
[12]. Ortigosa R and Martínez-Frutos J 2021. Multi-resolution methods for the topology optimization of nonlinear electro-active polymers at large strains. Computational Mechanics, 68(2): 1-23.
[13]. Chen L, Zhang H, Wang W and Zhang Q L 2023 Topology Optimization Based on SA-BESO. Applied Sciences, 13(7): 456.

Journal of Physics: Conference Series **3009** (2025) 012046 doi:10.1088/1742-6596/3009/1/012046

Self-floating porous PVDF/CNT microspheres for solar desalination

Xiaoqiao Li[1], Luyao Zhang[1] and Pingping Liang*[1]

[1]Jilin Agricultural Science and Technology University Jilin, Jilin, China.

*Corresponding author's e-mail: liangpp0408@163.com (Pingping Liang)

Abstract. The application of photothermal materials for achieving continuous and efficient solar interfacial evaporation possesses substantial research significance in solar desalination. To realize continuous and efficient evaporation, interfacial evaporation materials must integrate high light absorption capacity, superior photothermal conversion performance, sufficient water supply, and buoyancy. In this study, polyvinylidene fluoride/carbon nanotubes (PVDF/CNT) self-floating porous microspheres, suitable for solar interfacial water evaporation, were synthesized via a convenient phase transformation method. The microspheres' distinct composition and porous structure enable effective sunlight absorption, sufficient water flow, and quick steam discharge. Additionally, the performance of the microspheres is enhanced as the content of CNTs increases. These microspheres can absorb up to 93.8% of incident sunlight and elevate the water surface temperature to 41.5 °C within 2 minutes. Under one-sun illumination, the water evaporation rate (WER) reaches 1.485 kg/m²·h, with a solar energy conversion efficiency (SECE) of 93.1%.

1. Introduction

With the escalating global population and intensifying environmental pollution, energy and freshwater resource scarcity has emerged as a significant impediment to human societal development [1]. Solar energy, being an inexhaustible renewable resource, offers a promising solution. Among various water evaporation technologies, solar-driven interfacial water evaporation (SIWE) stands out for its efficiency in producing fresh water using solar energy [2]. This technology employs photothermal materials that float at the gas-liquid interface, capturing solar energy and converting it into localized heat, thereby minimizing heat loss associated with conventional water heating methods and enhancing evaporation efficiency [3]. Sunlight comprises approximately 3% ultraviolet light, 45% visible light, and 52% near-infrared light. An ideal photothermal material should exhibit strong absorption across the full solar spectrum (300nm-2500nm). In recent years, diverse photothermal materials such as carbon-based materials [4], metal-based nanoparticles [5], metal oxides [6], polymers [7], and biomaterials [8] have been developed for SIWE applications. Among these materials, carbon nanotubes (CNTs) have garnered significant attention owing to their remarkable properties., including ultra-wideband absorption, high light absorption coefficients, superior thermal conductivity, excellent thermal stability, and relatively low cost. CNTs are renowned for being one of the darkest materials, absorbing nearly all visible light with minimal reflection [9]. Currently, CNTs are predominantly utilized in two-dimensional SIWE systems. However, three-dimensional materials, particularly porous structures, offer enhanced spatial utilization and increased effective evaporation area. Moreover, three-dimensional materials facilitate more efficient sunlight utilization.

In this study, CNTs were employed as efficient photothermal conversion materials [10], while polyvinylidene fluoride (PVDF) served as a self-floating support and insulating material. PVDF/CNT self-floating porous microspheres were synthesized via a phase transformation method. This technique is characterized by its simplicity, rapidity, minimal equipment requirements, and suitability for industrialization with short production times and high yields. The impact of CNT content on microspheres' light absorption characteristics, photothermal performance, and solar water evaporation efficiency was systematically examined.

2. Materials and Methods

2.1. Preparation of PVDF/CNT microspheres

Various concentrations of CNTs were ultrasonically dispersed in 20 mL of DMF. Thereafter, 2.4 g of PVDF was dissolved in the aforementioned dispersion to prepare the precursor solution. Ultimately, the mass fractions of CNTs in the precursor solution were set at 0wt%, 1.0wt%, and 2.0wt%, respectively. PVDF/CNT microspheres were prepared by vertically dispensing the precursor solution into high-purity water using a syringe. The microspheres were then subjected to three alternating washes with high-purity water and ethanol, followed by vacuum drying for storage. The resulting microsphere samples were designated as PVDF, PC-1, and PC-2.

2.2. Water evaporation performance

The PVDF/CNT microspheres were placed in a Teflon container filled with high-purity water. A solar simulator, outfitted with an AM 1.5 filter and delivering a power density of 1 kW/m² (equivalent to one sun), was utilized as the light source. The illumination experiments were conducted under controlled ambient temperature conditions at 25 ± 0.5 °C and relative humidity at 40% ± 10%. An electronic analytical balance with a precision of 0.0001 g was utilized to monitor the mass changes during the water evaporation process, whereas an infrared camera documented the temperature variations. Each experiment was repeated ten times under identical conditions. Following each cycle, the PVDF/CNT microspheres were washed and dried for subsequent use.

2.3. Solar desalination

A custom-built solar desalination device was utilized to collect fresh water following solar desalination to evaluate the performance of solar interface desalination. Real Bohai Seawater samples were collected in Dalian, China, for desalination testing. The concentrations of Na+ in the four simulated seawater samples and various ions in the real seawater before and after solar desalination were analyzed using ICP-OES.

3. Results and discussion

PVDF/CNT microspheres were prepared using a simple phase transformation method to obtain high-efficiency solar interfacial evaporation materials. Figure 1 is a SEM image of PVDF/CNT microspheres at different magnifications. As illustrated in the figures, millimetre-scale teardrop-shaped microbeads were synthesized via phase transformation, exhibiting a surface characterized by approximately ordered nanogullies and disordered submicron pores. The internal structure contains a significant number of pores. In the higher magnification SEM images [Figures 1(B) and 1(D)], linear protrusions with dimensions of approximately 18 nm were observed on both the surface and interior of the microbeads, consistent with the size of carbon nanotubes.

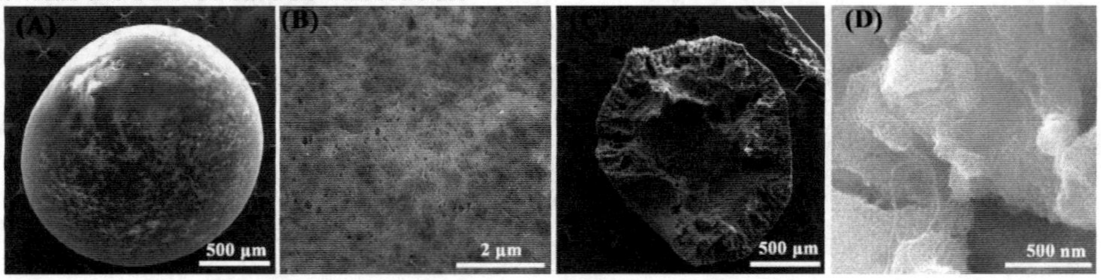

Figure 1. Low-magnification SEM image of PC-2 (A), enlarged image of the surface (B), cross-section image (C), and enlarged cross-section image (D).

Figure 2(A) illustrates XRD patterns of PVDF, CNTs, and PVDF/CNT composite microspheres. PVDF exhibits four prominent diffraction peaks at 18.4°, 20.6°, 36.5°, and 41.1°, corresponding to the (020) and (110) crystal planes of the α phase, as well as the (020) and (111) crystal planes of the β phase, respectively. The graphite-structured CNTs display characteristic diffraction peaks at 25.6° and 42.9°, attributed to the (002) and (100) crystal planes, respectively. The XRD pattern of the PVDF/CNT composite microspheres not only encompasses the aforementioned characteristic diffraction peaks of both PVDF and CNTs but also confirms the successful incorporation of CNTs into the PVDF matrix, resulting in stable composite microspheres.

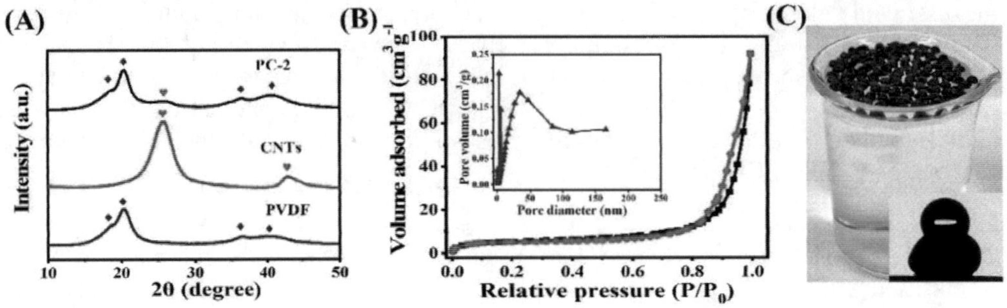

Figure 2. XRD patterns of different samples (A). N2 adsorption-desorption isotherms of PVDF/CNT microspheres, with corresponding pore size distribution shown in the figure (B). Photo of PVDF/CNT microspheres floating on water, illustration showing a 1 μl water drop on the surface of PVDF/CNT microsphere (C).

The pore structures of PVDF/CNT microspheres were further characterized using nitrogen adsorption-desorption isotherms. As illustrated in Figure 2(B), the isotherm exhibits a Type IV curve, indicating the presence of mesoporous characteristics within the PVDF/CNT microspheres. The BET-specific surface area of PVDF/CNT microspheres was determined to be 18.2 m^2/g. The pore size distribution analysis reveals that, in addition to a significant number of mesopores, there are also some micropores and macropores present, with a total pore volume of 0.148 cm^3/g and an average pore diameter of 29.3 nm, as determined by the characterization method. This unique three-dimensional porous structure facilitates efficient sunlight capture and provides water transport and steam escape pathways. Figure 2(C) displays an image of the PVDF/CNT microspheres floating on the water's surface. The microspheres clearly exhibit hydrophobic properties, as evidenced by their ability to remain buoyant on the water surface, as shown in Figure 2(C). This self-floating behavior facilitates heat confinement at the gas-liquid interface, thereby promoting more favorable conditions for interfacial water evaporation.

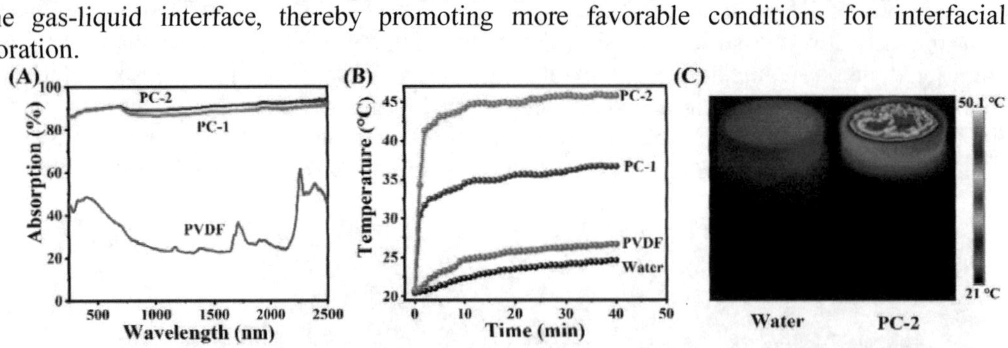

Figure 3. The absorption spectra of different samples (A). Surface temperatures of different samples for different times (B). IR thermal images of pure water and PVDF/CNT microspheres after light irradiation for 30 min (C).

It is widely recognized that water evaporation efficiency is significantly influenced by the photothermal materials' capability to capture sunlight. To assess this, we conducted an analysis of the reflectance and transmittance spectra for PVDF microspheres and PVDF/CNT microspheres across a wavelength range of 300-2500 nm. As illustrated in Figure 3(A), the average light absorption capacity of PVDF microspheres was considerably lower than that of PVDF/CNT microspheres, reaching approximately 32.2%. The light absorption capacity of PVDF/CNT microspheres increased progressively with the addition of CNTs, achieving optimal performance at the CNTs content of 2%, with an absorption rate of approximately 93.8% of sunlight. In the near-infrared band, the light absorption rate of the material reached approximately 91.2%.

Efficient light absorption facilitates the acquisition of superior photothermal properties. We monitored the effect of the time-varying carbon nanotube content of PVDF/ carbon nanotube microspheres on the surface temperature distribution of the microspheres in pure water under a solar illumination condition. Based on the data presented in Figure 3(B), the surface temperatures of pure water and PVDF microbeads exhibited a steady upward trend over the 40-minute period, ultimately reaching 24.6°C and 26.7°C, respectively. In contrast, the surface temperatures of PVDF/CNT microspheres containing 1% and 2% CNTs rose rapidly to 31.7°C and 41.5°C, respectively, after just 2 minutes of irradiation. With prolonged irradiation, the rate of temperature increase slowed, and the maximum temperatures were achieved after 40 minutes. At this point, the peak surface temperature of PC-2 reaches 45.9°C, whereas the surface temperature of pure water is merely 24.6°C [Figure 3(C)]. These findings demonstrate that PVDF/CNT microspheres can efficiently harness sunlight and transform it into thermal energy, thereby enhancing solar-driven water evaporation.

Evaluating the performance of a photothermal material in converting absorbed solar energy into heat for driving water evaporation typically relies on two critical parameters: the rate of water evaporation and the efficiency of solar energy conversion. These evaluations offer valuable insights into the material's efficacy in harnessing and utilizing solar energy. WER (v_e) and SECE (η) are determined using the following equations:

$$v_e = \frac{dm}{S \times dt} \quad (1)$$

$$\eta = \frac{v_e H_{LV}}{Q_i} \quad (2)$$

In this context, m represents the mass lost due to water evaporation, S denotes the opening area of the container, t signifies the duration of evaporation, H_{LV} is the latent heat of vaporization for water transitioning from liquid to gas phase, and Qi indicates the optical power density of the solar simulator. As illustrated in Figure 4(A), PVDF/CNT microspheres with varying CNT contents exhibited a faster rate of mass loss compared to PVDF microspheres and pure water under one sunlight condition, which aligns with the observed trend in surface temperature changes. The microbeads' water evaporation rate and solar energy conversion efficiency progressively increased with higher CNT content in the microspheres [Figure 4(B)]. When the CNT content reached 2%, WER peaked at 1.485 kg/m²·h, and SECE of the microbeads was 93.1%, representing a 3.1-fold increase over pure water and a 1.68-fold increase over CNTs alone. To evaluate the long-term stability of the microspheres, PC-2 was subjected to 10 irradiation cycles under simulated solar illumination, with each cycle lasting 1 hour. The microspheres maintained a high WER of approximately 1.472 kg/m²·h, demonstrating excellent cyclic stability and reusability.

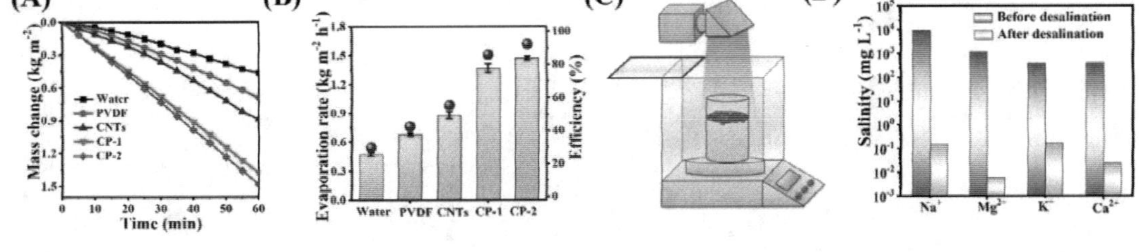

Figure 4. Cumulative water weight loss as a function of irradiation duration (A). WER and SECE of different samples (B). Schematic illustration of a solar interface desalination system utilizing PVDF/CNT microspheres (C). The concentration of major ions in actual seawater prior to and following solar desalination (D).

Considering the substantial potential of PVDF/CNT microspheres for achieving sustained and efficient solar desalination, we developed a laboratory-scale solar interface desalination system to assess their desalination performance [Figure 4(C)]. The system is composed of three main components: a sunlight simulator, an evaporator, and a condensing chamber. Specifically, the evaporator comprises a polytetrafluoroethylene container with a diameter of 24 cm and a capacity of 25 ml filled with seawater. The surface of the seawater is covered by PVDF/CNT microspheres.

To evaluate the efficacy of solar desalination using PVDF/CNT microspheres, a series of seawater models represented by NaCl solutions with varying salinities were employed. Desalination tests were conducted on seawater samples from the real Bohai Sea (Dalian, China). The water quality obtained by evaporation was evaluated by inductively coupled plasma spectrometry (ICP). After testing the concentration of four main ions (Na+, Mg2+, K+, and Ca2+) in the condensed water obtained after evaporation of actual seawater, it was found that the ion concentration was reduced by about three orders of magnitude, which met the standards for safe drinking water [Figure 4(D)]. After desalination of both simulated and real seawater, the ion concentrations in the condensed water are lower than those achieved through traditional reverse osmosis and conventional distilled seawater desalination methods.

4. Conclusions

Self-floating PVDF/CNT porous microspheres were synthesized via a phase conversion method. The self-floating capability of these microspheres satisfies the requirements for interfacial evaporation applications. The synergistic effect of the macroscopic three-dimensional spatial structure and the porous architecture of the PVDF/CNT microspheres significantly enhances light absorption capacity, expands the effective evaporation area, and achieves full-spectrum sunlight absorption up to 93.8%, thereby increasing the WER and SECE to 1.485 $kg/m^2 \cdot h$ and 93.1%, respectively. After solar desalination of real seawater, the concentration of ions in the collected condensate was significantly reduced, meeting drinking water standards set by the World Health Organization and the US Environmental Protection Agency. This demonstrates the superior performance of PVDF/CNT porous microspheres in seawater desalination. Furthermore, this approach offers a green and efficient solution to address the freshwater crisis. The simple, scalable, and cost-effective manufacturing process ensures that the microspheres can meet industrialization requirements for rapid production and high yield, with broad prospects for large-scale application.

Acknowledgment

This work was supported by "The National Undergraduate Innovation and Entrepreneurship Training Program" (Project number: S202411439037).

References

[1] Jing X X, Wang W D, et al. Research progress on photothermal materials and their application in solar desalination [J]. Chinese Journal of Inorganic Chemistry, 2024, 40(6):1033-1064.
[2] Liang P P, Liu S, Li H Y, et al. S A self-floating electrospun nanofiber mat for continuously high-efficiency solar desalination [J]. Chemosphere, 2021, 280:130719.
[3] Xiao X, Pan L, Chen T, et al. Scalable Core–Sheath Yarn for Boosting Solar Interfacial Desalination Through Engineering Controllable Water Supply [J]. Engineering, 2023,30(11):153-160.
[4] Ma X, Fang W, Guo Y, et al. Hierarchical porous SWCNT stringed carbon polyhedrons and PSS threaded MOF bilayer membrane for efficient solar vapor generation[J]. Small, 2019, 15(15):1900354.
[5] Zhao L P, Du C, Zhou C, et al. Structurally ordered AgNPs@C3N4/GO membranes toward solar-driven freshwater generation[J]. ACS Sustainable Chem. Eng., 2020, 8(11):4362-4370
[6]Shi L, He Y, Huang Y, et al. Recyclable Fe3O4@CNT nanoparticles for high-efficiency solar vapor generation[J]. Energy Convers. Manage., 2017, 149:401-408.
[7]Zhang Y, Yin X, Yu B, et al. Recyclable polydopamine-functionalized sponge for high-efficiency clean water generation with dual-purpose solar evaporation and contaminant adsorption[J]. ACS Appl Mater Interfaces, 2019, 11(35):32559-32568.
[8] Guo X K, Xu Y C, Sha J G, et al. Research progress of biomass-based carbon materials for seawater desalination [J].

China Paper & Paper, 2019,43(09):47-55.
[9] Song Y M, Qiu S X, Feng S X, et al. A review of carbon nanotubes in modern electrochemical energy storage [J]. New Carbon Materials,2024,39(06):1037-1074.
[10] Kou H, Liu Z X, Zhu B, et al. Recyclable CNT-coupled cotton fabrics for low-cost and efficient desalination of seawater under sunlight[J]. Desalination, 2019, 462:29-38.

Journal of Physics: Conference Series **3009** (2025) 012047 doi:10.1088/1742-6596/3009/1/012047

A multi-fluid heat transfer spiral coil microchannel heat exchanger based on CFD

Xingyu Liu [1*] and Guangji Li [1]

[1] Shanghai Institute of Technology, Shanghai, China

[*] Corresponding author: yuyuyu_0106@163.com

Abstract: Currently, most microchannel heat exchangers are limited to providing heat transfer between one or two working fluids. However, in fields such as waste heat recovery, and solar thermal power generation, there is a need for energy conversion between multiple media. Consequently, this study introduces a microchannel heat exchanger with a spiral coil structure designed to facilitate simultaneous heat transfer with two or more working fluids. The heat exchange units of this microchannel heat exchanger feature two semicircular cross-section spiral channels that are alternately wound around a circular central channel; the flow channels are tightly attached to the central channel and arranged in an alternating spiral pattern, which increases the heat exchange area. The direction of the working fluid flow within the spiral channels continuously changes, reducing areas of uneven heat transfer. This research establishes a physical model of the spiral coil microchannel heat exchanger and simulates the cooling effect of hot fluids in two arrangements within a SCO_2 Brayton cycle. These findings provide a reference for the study of novel structures in microchannel heat exchangers.

1. Introduction

Heat exchangers, as crucial mechanical equipment in the energy and chemical industries as well as in petroleum engineering, significantly impact energy utilization and waste heat recovery. Microchannel heat exchangers, with their high heat transfer efficiency, compact design, material diversity, and adaptability to multiple working fluids, hold broad application prospects in fields such as solar thermal power generation, chemical pharmaceuticals, air conditioning refrigeration, and aerospace. Microchannel heat exchangers have diameters that vary between 200 μm and 3 mm [1]. Common varieties include Printed Circuit Heat Exchanger (PCHE), Plate Fin Heat Exchanger (PFHE), and Micro Shell and Tube Heat Exchanger (MSTE). The microchannel heat exchanger is noted for its extensive heat transfer surface area, compact structure, and capability to disrupt fluid boundary layers.

For flow and heat transfer issues, there are typically three research methodologies: theoretical analysis, experimental research, and numerical simulation. With the evolution of modern computer technology, numerical solutions can be computed using Computational Fluid Dynamics (CFD) software, which is not constrained by experimental conditions and can effectively guide experimental studies, simulating operating conditions that are challenging to achieve experimentally [2]. The thermal performance of microchannels is contingent upon their geometric parameters and flow conditions, and the combination of CFD analysis with multi-objective genetic algorithms can determine the optimal geometric parameters for microchannel heat exchangers [3]. Mogaji et al. [4] numerically investigated heat transfer augmentation in internally grooved spiral pipes, and contrasted smooth tubes with those featuring spiral grooves of varying pitches. Finding the intricate geometry of the grooved spiral pipes was particularly efficient in evaporative application. Pan et al. [5] numerically investigated the heat transfer and pressure drop of a spiral rectangular channel heat exchanger under oscillating conditions, finding that spiral rectangular channel heat exchangers perform better at high frequencies and speeds.

CO_2 has a relatively low critical point (Pc=7.38 MPa, Tc=304.13K), making its supercritical state readily attainable. Supercritical CO_2 (SCO_2) is favored in areas such as nuclear reactors [6], solar thermal power generation, coal power generation [7] and novel refrigeration due to its unique physical properties. The SCO_2 Brayton cycle is characterized by its environmental friendliness, compact machinery, and good economic performance [8]. Compared to other cycles, the advantages of the SCO_2 Brayton cycle stem from the unique thermophysical properties of SCO_2, which can achieve high thermal efficiency under moderate pressure and temperature conditions. Microchannel heat exchangers, due to their efficient heat transfer and compact structure, have become vital heat transfer equipment in SCO_2 systems. The heat transfer performance of SCO_2 flowing within channels is notably affected by factors like heat flux, mass flow rate, and inclination angle. An increase in heat flux or a reduction in mass flow rate can augment the buoyancy effect, thereby giving rise to more pronounced heat transfer non-uniformity [9]. In microchannel, single-phase convective heat transfer involves single-phase media including gas and liquid. Keniar and Garimella [10] conducted experimental research on the flow phenomena of carbon dioxide in a single microchannel. The experiments revealed that heat transfer coefficients increase with increasing pressure drop and heat flux density. Liu et al. [11] established a distributed parameter model for counter-flow and cross-flow plate-fin heat exchangers (PFHE), investigating the thermodynamic characteristics of parallel-flow, counter-flow, and cross-flow with different thermal boundary conditions set at both ends of the separator plates. The SCO_2 Brayton cycle enjoys high energy conversion efficiency and reduced construction time and costs; however, its most significant drawback is the potential emergence of pinch points in the heat exchangers within the loop [12]. Pinch points refer to locations within a heat exchanger where the minimum temperature difference occurs between cold and hot fluids. Son et al. [13] utilized a novel approach to address and mitigate the issue of internal constriction within SCO_2 regenerators. This is achieved by establishing a revised criterion for assessing the performance of heat exchangers, centered on the concept of zero internal pinch point.

2. Physical model

This study introduces a microchannel heat exchanger with a spiral coil structure. Figure 1 illustrates the heat exchange unit of the spiral coil structure, where the central tube is a straight channel, and the channel section is circular. Two spiral channels, each initiated at a 180° phase difference, are alternately coiled around the outer surface of a central channel, with their flow path cross-sections shaped like semicircles. A single unit for heat exchange is made up of one central channel and two spiral channels. The spiral coil structure facilitates the simultaneous flow and heat exchange of two or more fluids, suitable for the concurrent cooling or heating of various working fluids, and collecting and transferring heat from multiple different sources, thereby enhancing energy utilization efficiency. The study examines the flow characteristics and heat transfer performance of this heat exchange unit under two configurations: (A) the spiral channels are fed with cold fluid, while the straight central channel is fed with hot fluid; (B) the spiral channels are fed with hot fluid, while the straight central channel is fed with cold fluid.

The spiral coil structure of the flow path allows the working fluid to fully develop within the channel, enhancing turbulence and creating strong vortices. The spiral motion of the fluid creates secondary circulation, enhancing heat transfer and markedly diminishing zones of flow stagnation. This design also extends the fluid's travel within the channel, boosting the heat transfer surface area and enhancing the compactness of the heat exchanger. As the fluid flows through the narrow, curved channels, its velocity and direction change continuously, leading to intense turbulence even at relatively low speeds. This promotes the disturbance of the fluid boundary layer, improves the heat transfer coefficient, and intensifies the heat exchange process. Additionally, it minimizes the chances of pinch points or uneven thermal distribution within the channels, providing an innovative design strategy for microchannel heat exchangers. Counter-flow heat exchange between cold and hot fluids maintains a larger temperature difference throughout the process, minimizing the probability of pinch points and reducing pressure drop.

Adding a straight section at the inlet and outlet ends ensures a smooth transition, guarantees fluid flow stability, prevents stress concentration and deformation, and reduces resistance losses caused by corners.

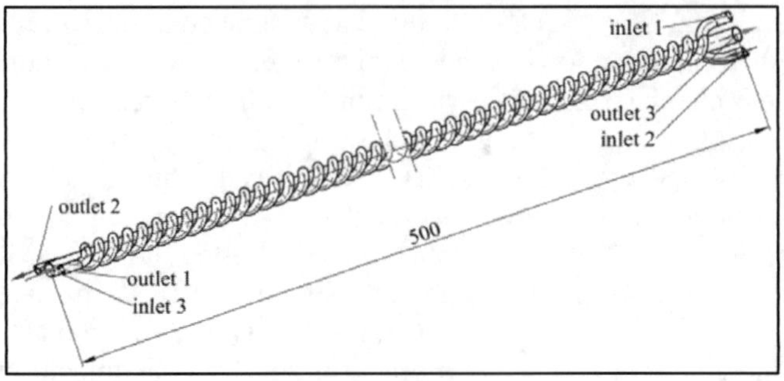

Figure 1. Heat transfer unit of spiral coil structure.

2.1 Simplified model

The core arrangement of the whole heat exchanger is orderly and periodic, so a single heat transfer beam is selected as the main research object. The spiral coil structure has asymmetry in space, so the 3D model of the spiral coil structure is established, and the central channel radius (r_c), the spiral channel radius (rs), wall thickness (δ), screw pitch (P), number of loops (N), spiral diameter (d_s), diversion section (l_d) are shown in Table 1

Table 1. Heat transfer unit size.

Name	Heat transfer unit						
	r_c	r_s	δ	P	N	d_s	l_d
Data	1.3 mm	1 mm	0.1 mm	7 mm	69	5 mm	3.9 mm

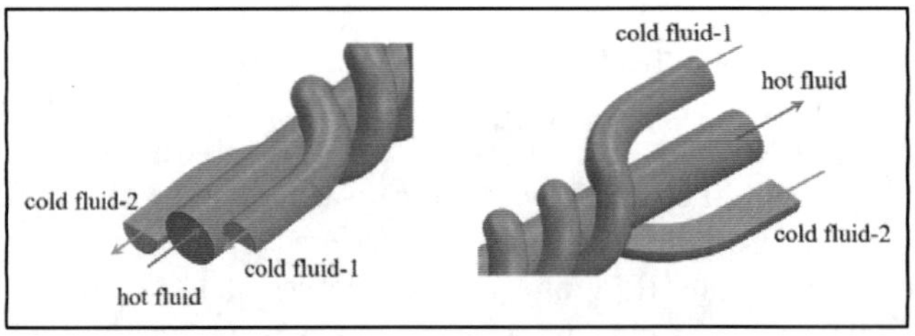

Figure 2. Arrangement (A).

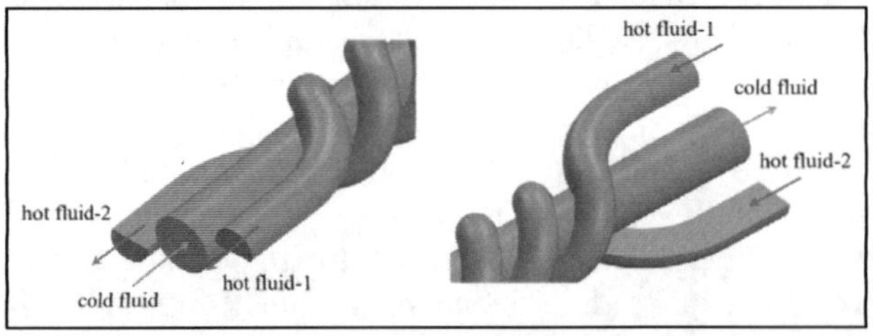

Figure 3. Arrangement (B).

The cold and hot fluids flow in opposite directions, and the fluid conducts convection heat transfer through the wall shell. Figures 2 and 3 show the working fluid flow of the arrangements (A) and (B). Common materials of microchannel heat exchangers include copper, aluminum, stainless steel, silicon and ceramics. Stainless steel has become commonly used due to its great corrosion resistance, high strength, and long service life. Among them, stainless steel 316L has excellent corrosion resistance, good high temperature strength and low temperature toughness, suitable for a variety of working environments.

2.2 Equations

The fundamental concepts governing fluid dynamics include the conservation of mass, momentum, and energy. These concepts are represented by the continuity equation, the momentum equation, and the energy equation, respectively.

Continuity Equation:

$$\frac{\partial\rho}{\partial t}+\nabla\cdot(\rho\boldsymbol{u})=0 \tag{1}$$

where ρ is the fluid density, t is time, and $\boldsymbol{u}$ is the fluid velocity vector.

Momentum Conservation Equation:

$$\rho\left(\frac{\partial\boldsymbol{u}}{\partial t}+\boldsymbol{u}\cdot\nabla\boldsymbol{u}\right)=-\nabla p+\mu\nabla^2\boldsymbol{u}+\boldsymbol{f} \tag{2}$$

where p is the static pressure, μ is the dynamic viscosity, and $\boldsymbol{f}$ is the body force per unit mass (such as gravity).

Energy Conservation Equation:

$$\rho\left(\frac{\partial h}{\partial t}+\boldsymbol{u}\cdot\nabla h\right)=\frac{\partial p}{\partial t}+\nabla\cdot(k\nabla T)+\Phi \tag{3}$$

where h is enthalpy, k is the thermal conductivity, ∇T is the temperature gradient, and Φ is the viscous dissipation term.

Heat Conduction Equation:

$$\frac{\partial T}{\partial t}=\frac{k}{\rho c_p}\left(\frac{\partial^2 T}{\partial x^2}+\frac{\partial^2 T}{\partial y^2}+\frac{\partial^2 T}{\partial z^2}\right)+Q \tag{4}$$

where c_p is the specific heat capacity at constant pressure, (x, y, z) is position, and Q is the heat source term.

The Reynolds number is a dimensionless quantity used to predict flow patterns in different fluid flow situations:

$$Re=\frac{\rho vL}{\mu} \tag{5}$$

where v is the characteristic velocity, and L is the characteristic length.

3. Numerical simulation

3.1 Turbulence model

The SST k-ω model, proposed by Menter, combines the robustness of the k-ω model in the near wall region with the free flow independence of the k-ε model in the far field.

The SST k-ω model encompasses the transport equations for both turbulent kinetic energy (k) and the specific dissipation rate (ω):

$$\frac{\partial k}{\partial t}+u\cdot\nabla k=\nabla\cdot\left(\left(v+\frac{v_t}{\sigma_k}\right)\nabla k\right)+P_k-\beta^*\rho\omega k \tag{6}$$

$$\frac{\partial\omega}{\partial t}+u\cdot\nabla\omega=\nabla\cdot\left(\left(v+\frac{v_t}{\sigma_\omega}\right)\nabla\omega\right)+P_\omega-\beta\rho\omega^2 \tag{7}$$

where P_k and P_ω are the production terms for k and ω respectively, v_t is the turbulent viscosity, σ_k, σ_ω, β^* and β are model constants.

3.2 Mesh generation

Figure 4 illustrates the model with mesh generation performed using Ansys Mesh, where the growth rate is set to 1.2, resulting in a total of 2,311,267 grid cells. The maximum characteristic length is 0.50807 mm, and the maximum cell quality is 0.99999, indicating a well-meshed model.

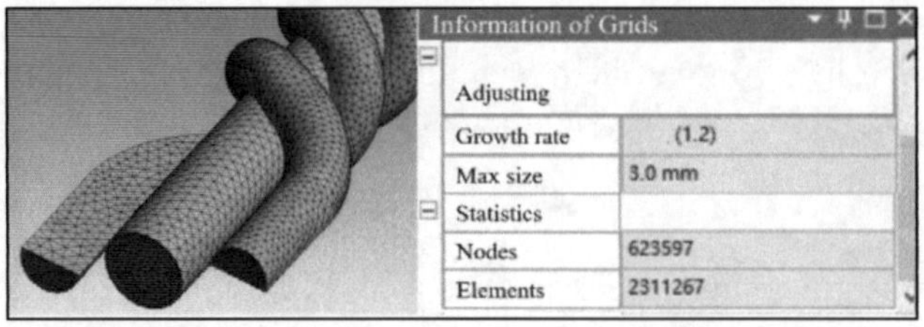

Figure 4. Grid division local graph.

Due to the complex and spatially asymmetric nature of the spiral coil structure, it is challenging to mesh using structured grids. Therefore, a tetrahedral meshing scheme is employed. Tetrahedral grids offer high flexibility and can more easily adapt to complex geometries, enabling the generation of high-quality grids within domains featuring irregular shapes and complex boundary conditions. Additionally, local mesh refinement has been applied to the inlets and outlets of the three channels to achieve more detailed and accurate results. The model includes boundary layer meshing for all wall surfaces.

3.3 Boundary condition

FLUENT is used as the numerical solver for the simulation, with double precision calculations and gravity effects. The heat transfer working fluid is considered to be SCO_2 in an ideal gas state. The inlet is set with mass flow boundary conditions, and the outlet is set as a pressure outlet condition. The heat exchanger tube walls are made of 316L stainless steel, configured as isothermal, no-slip walls, with the cold and hot channel walls set as coupled interfaces. The SST k-ω turbulence model employs an enhanced wall treatment as its standard near-wall handling, targeting a boundary layer y+ value around 1.1. The solution equations use a coupled pressure-velocity coupling method (Coupled), which allows for the simultaneous solution of all variables (including velocity, pressure, temperature, etc.). The momentum and energy equations are discretized using Second Order Upwind, aiming for a solution convergence where residuals are less than 10^{-5}. The solution is initialized using a hybrid initialization approach. The boundary conditions are shown in Table 2

Table 2. Boundary Conditions.

Name	Data
Hot fluid inlet temperature ($T_{i\text{-hot}}$)	600 K
Cold fluid inlet temperature ($T_{i\text{-cold}}$)	310 K
Hot fluid inlet mass ($m_{i\text{-hot}}$)	0.5 kg/s
Cold fluid inlet mass ($m_{i\text{-cold}}$)	0.5 kg/s
Outlet pressure (P_{out})	12 MPa
Turbulence intensity (I)	5%
Reflux turbulent viscosity ratio ($\frac{\mu_t}{\mu}$)	10
Operating temperature (T_o)	310 K
Wall thickness (δ)	0.1 mm
Solid material	Stainless steel 316L
Material convective thermal conductivity (k)	16.5 W/(m^2·K)

Table 3. Thermophysical Properties Of Sco_2.

Temperature (K)	Pressure (MPa)	Density (kg/m^3)	Enthalpy (kJ/kg)	Cv (kJ/(kg*k))	Cp (kJ/(kg*k))	Viscosity (μPa*s)
600	10	89.941	779.27	0.90169	1.1561	28.943
600	11	99.044	777.3	0.90313	1.1642	29.104
600	12	108.15	775.35	0.90454	1.1723	29.277
600	13	117.25	773.43	0.90594	1.1804	29.462

600	14	126.35	771.53	0.90732	1.1884	29.658
560	12	117.8	728.46	0.88649	1.1735	27.919
580	12	112.73	751.91	0.89551	1.172	28.598
620	12	103.99	798.81	0.91353	1.1739	29.955
640	12	100.18	822.31	0.92242	1.1764	30.631

In this study, SCO_2 is chosen as the working fluid, and its partial thermophysical properties within the temperature range of 300 to 700 K and simulation pressures of 10 to 14 MPa are presented in Table 3. This study is based on the following assumptions: (a) only convective and conductive heat transfer are considered; (b) the fluid flow and heat transfer are in a steady state; (c) the supercritical fluid is incompressible during heat transfer; (d) the fluids are assumed to be in either an ideal gaseous or liquid state.

3.4 Results and discussion

3.4.1 Results of arrangement (A). Figure 5 displays the residual curves generated after 700 iterations by the solver, with each curve tending towards convergence. Additionally, the temperature and pressure contour plots at the hot fluid's outlet are shown in Figure 6, $T_{o\text{-}hot}$\{max, min\}=\{518.894, 533.56\} K, $P_{o\text{-}hot}$\{max, min\}=\{13.2333, 12.7227\} MPa. Figure 7 presents the velocity vector of the hot fluid at the inlet and outlet, obtained through post-processing with CFD. The fluid velocity is high at the center of the straight central channel and decreases near the walls. Within the spiral channel, the direction of fluid velocity is in continuous flux.

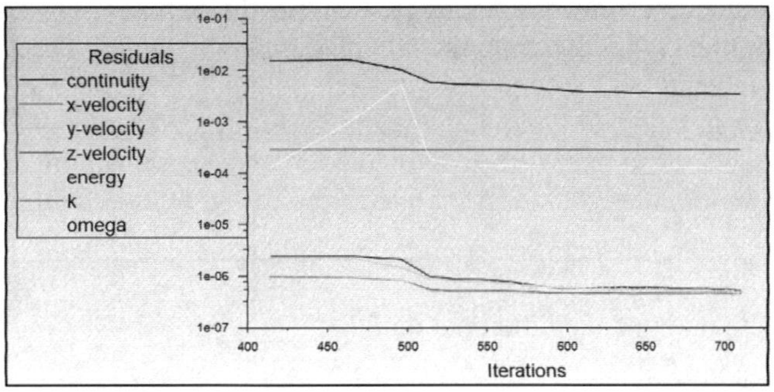

Figure 5. Residuals of arrangement (A).

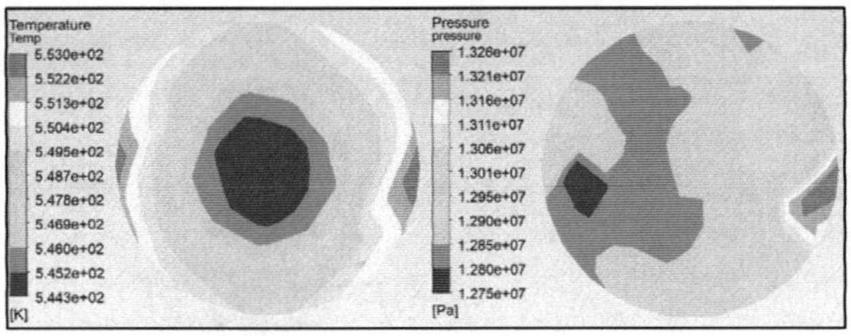

(a) Temperature (b) Pressure

Figure 6. (a) Temperature and (b) pressure at the outlet of hot fluid.

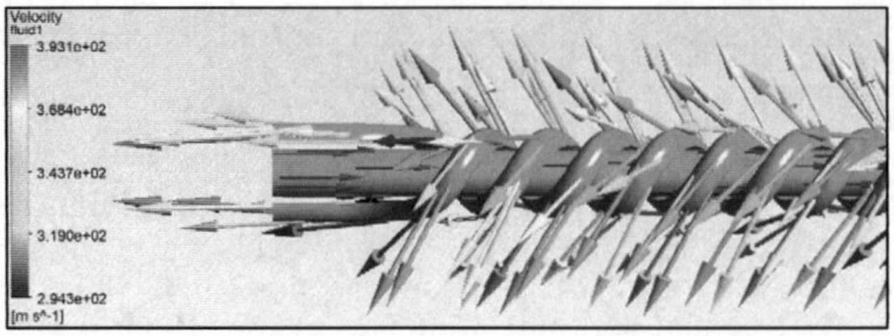

(a) Cold fluid outlet 1, 2 and hot fluid inlet velocity vector

(b) Cold fluid inlet 1, 2 and hot fluid outlet velocity vector

Figure 7. Inlet and outlet velocity vector of arrangement (A).

3.4.2 Results of arrangement (B). Figure 8 presents the residual curves after 700 iterations by the solver, indicating convergence for each curve. Figure 9 depicts the temperature distributions at the two semicircular hot fluid outlets, $T_{o1\text{-}hot}$\{max, min\}=\{525.631, 508.382\}K, $T_{o2\text{-}hot}$\{max, min\}=\{523.356, 507.505,\}K. Figure 10 details the pressure at outlet 1 and outlet 2, $P_{o1\text{-}hot}$\{max, min\}=\{30.446, 19.027\}MPa, $P_{o2\text{-}hot}$\{max, min\}=\{30.436, 18.351\}MPa. Figure 11, using CFD Post processing to obtain thermal fluid inlet and outlet velocity vector. In the curved sections, the fluid, influenced by the centrifugal force within the channel, the fluid deviated from the mainstream direction of the flow passage to the outer side.

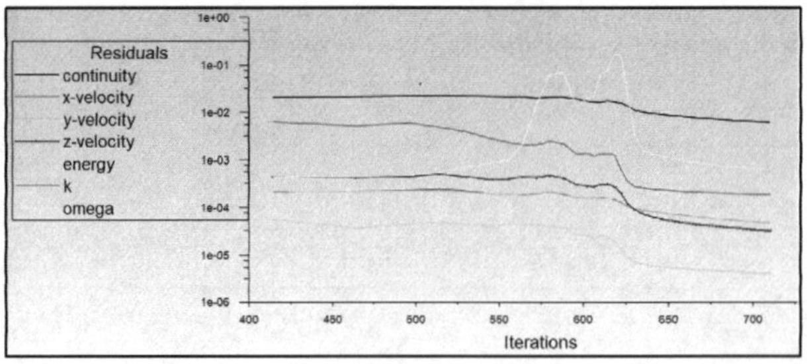

Figure 8. Residuals of arrangement (B).

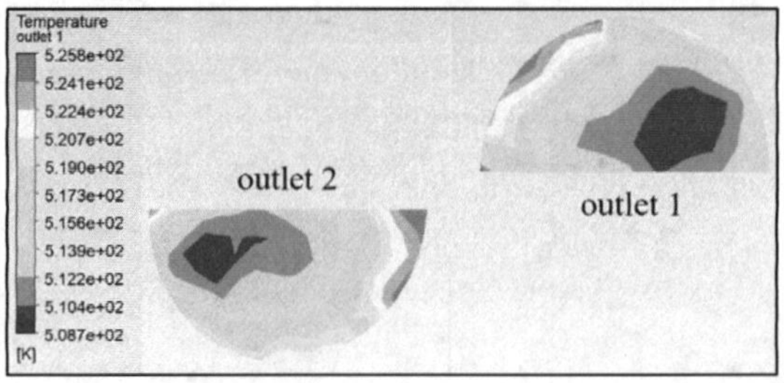

(a) Outlet 1 section temperature (b) Outlet 2 section temperature

Figure 9. Outlet 1 and Outlet 2 temperature of arrangement (B).

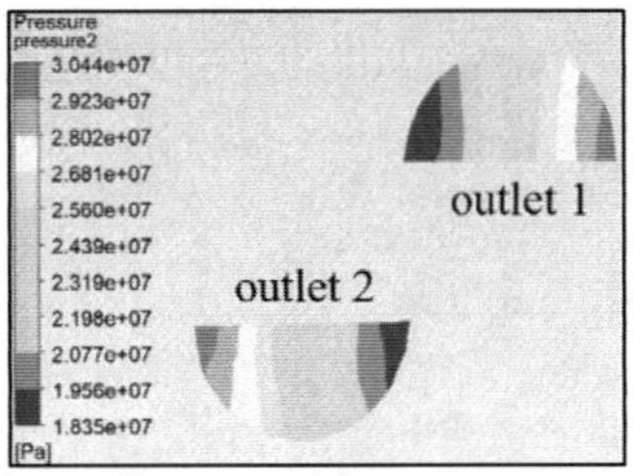

Figure 10. Outlet 1 and outlet 2 pressure of arrangement (B).

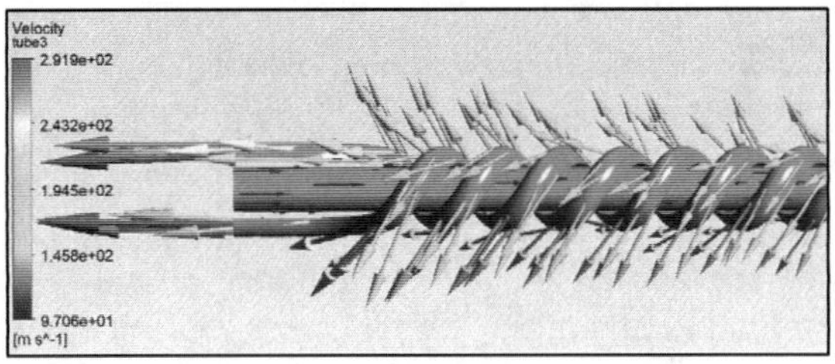

(a) Hot fluid outlet 1, 2 and cold fluid inlet velocity vector

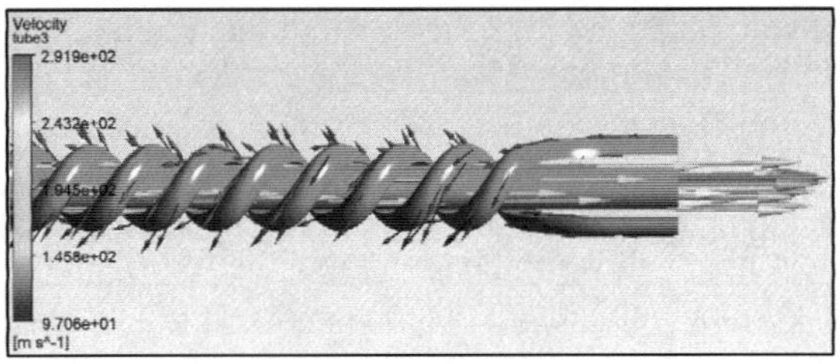

(b) Hot fluid inlet 1, 2 and cold fluid outlet velocity vector

Figure 11. Inlet and outlet velocity vector of arrangement (B).

3.4.3 Comparison of arrangements A and B. Based on the simulation outcomes of the two layout designs, Figure 12 illustrates how the Reynolds number varies along the centerline when cold and hot fluids enter the central channel. The flow characteristics in the straight central channel are comparable for both arrangements. Figure 13 depicts the temperature variations as the hot fluid cools in each arrangement. Since the inlets of arrangements (A) and (B) are arranged in opposition, they exhibit inverse flow patterns in the plane. Arrangement (B) records a bigger temperature differential of ΔT_B=84.14 K compared to ΔT_A=65.39 K in arrangement (A). Nonetheless, the pressure distribution of arrangement (B) is less consistent than that of arrangement (A).

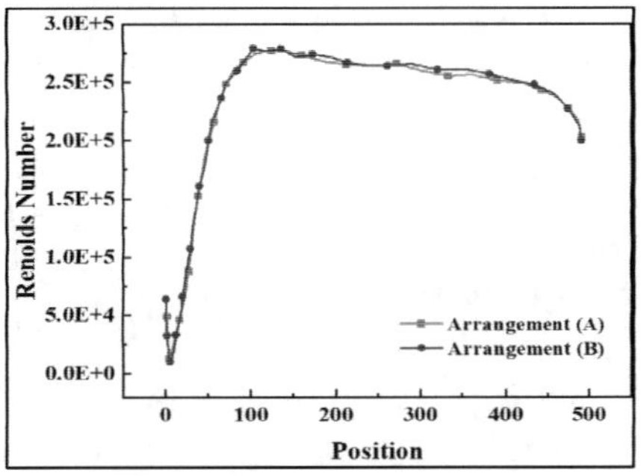

Figure 12. Reynolds number of arrangement (A) and (B).

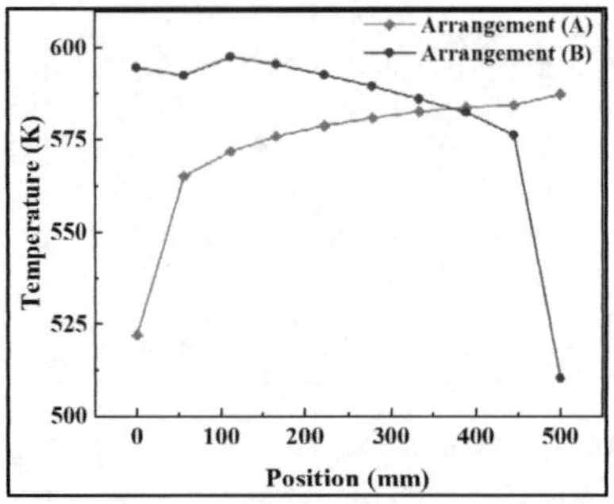

Figure 13. Hot fluid temperature variation in arrangement (A) and (B).

4. Conclusions

This research concentrates on the thermal exchange component of a microchannel heat exchanger that incorporates a spiral coil configuration, employing CFD to perform numerical simulations of the fluid dynamics and thermal transfer within the spiral and the central channel. The subsequent findings were ascertained:

- Enhanced heat transfer: The spiral coil structure of the heat exchanger allows the working fluid to fully develop within the channels, enhancing disturbances and creating strong turbulence. The spiral flow of the fluid generates secondary circulation, which significantly strengthens heat transfer and greatly reduces flow stagnation areas.
- Simultaneous heat exchange of multiple fluids: A single heat transfer unit contains two spiral channels and one straight central channel, enabling the simultaneous heat exchange of three fluid streams. This design collects and transfers heat from multiple sources, thereby improving energy utilization.
- Variability in arrangement: The three channels can be arranged in different arrangements for cold and hot fluids to meet various heat transfer requirements. This paper compares two counter-flow arrangements, providing temperature, velocity, and pressure distributions, which serve as a reference for numerical simulations of microchannel heat exchangers.

References

[1] Dwivedi, Akash, Khan Mohammad Mohsin. Pali Harveer Singh. A comprehensive review of thermal enhancement techniques in microchannel heat exchangers and heat sinks. Journal of Thermal Analysis Calorimetry, 2023, 148: 13189-13231.
[2] Aibo Wei, Xuan Ren, Shifeng Lin, Xiaobin Zhang. CFD analysis on flow and heat transfer mechanism of a microchannel Ω-shape heat pipe under zero gravity condition. International Journal of Heat and Mass Transfer, 2020, 163: 120448.
[3] Tabatabaei Malazi Mahdi, Kaya Kenan, Çolak Andaç Batur, et al. CFD and ANN analyses for the evaluation of the heat transfer characteristics of a rectangular microchannel heat sink with various cylindrical pin-fins. Heat Mass Transfer, 2024 60: 1393-1411.
[4] Taye Stephen Mogaji, Abdullahi O. Olapojoye, et.al. CFD study of heat transfer augmentation and fluid flow characteristics of turbulent flow inside helically grooved tubes. Journal of the Brazilian Society of Mechanical Sciences and Engineering, 2022(44): 90.
[5] Changzhao Pan, Tong Zhang, Junjie Wang, Yuan Zhou. CFD study of heat transfer and pressure drop for oscillating flow in helical rectangular channel heat exchanger. International Journal of Thermal Sciences, 2018(129): 106-114.
[6] Qiu, Yu, Erqi E, Qing Li. Triple-Objective Optimization of SCO_2 Brayton Cycles for Next-Generation Solar Power Tower. Energies, 2023, 16(14): 5316.
[7] Daniel Kriz, Petr Vlcek, Otakar Frybort. Numerical studies of sCO_2 Brayton cycle. Energy, 2024, 296: 131129.
[8] Patel, Raj C., Diego C. Bass, Ganza Prince Dukuze, Angelina Andrade, Christopher S. Analysis and Development of a Small-Scale Supercritical Carbon Dioxide (sCO_2) Brayton Cycle. Energies. 2022, 15(10): 3580.
[9] M. Monjurul Ehsan, Zhiqiang Guan, A.Y Klimenko. A comprehensive review on heat transfer and pressure drop characteristics and correlations with supercritical CO_2 under heating and cooling applications. Renewable and Sustainable Energy Reviews, 2018(92): 658-675.

[10] Khoudor Keniar, Srinivas Garimella. Enhanced condensation heat transfer for dielectric fluid within Analytical modeling of laminar annular flow condensation in an arbitrarily shaped microchannel heat sink. International Journal of Heat and Mass Transfer, 2020(162):120334.
[11] Yuce Liu, Ke Li, Jian Wen, Simin Wang. Thermodynamic characteristics of counter flow and cross flow plate fin heat exchanger based on distributed parameter model. Applied Thermal Engineering, 2023, 219(B): 119542.
[12] V. Pandey, P. Kumar. Modelling and analysis of pre-cooler for a sCO_2 Brayton cycle with different banking configurations using a stack-based model. Applied Thermal Engineering,2024, 242: 122466.
[13] Seongmin Son, Jin Young Heo, Jeong Ik Lee. Prediction of inner pinch for supercritical CO_2 heat exchanger using Artificial Neural Network and evaluation of its impact on cycle design. Energy Conversion and Management, 2018,163: 66-73.

Topology design of lightweight vehicle-mounted stretch equipment

Zhen Chen [1, 2, *], Jianmin Wu [1, 2], Hao Chang [1, 2], Honglei Qu [1, 2], Naiwen Xu [1, 2], Bo Yu [1, 2], Hailong Li [3]

[1] Emergency Rescue Center of State Grid Liaoning Electric Power Supply Co., Ltd, Liaoning Shenyang 110021, China
[2] Liaoning Power Transmission and Transformation Engineering Co., Ltd, Liaoning Shenyang 110021
[3] State Grid Liaoning Electric Power Supply Co., Ltd, Liaoning Shenyang 110021

[*]Corresponding author's e-mail: dsbcz@126.com;

Abstract. This paper investigates the topological design of lightweight vehicle-mounted tensioning and stringing equipment. Vehicle-mounted tensioning and stringing devices serve as crucial tools for the construction and maintenance of transmission lines, necessitating efficient mobility, rapid site transition capabilities, and user-friendly operational characteristics. To meet the stringent requirements of high strength, high reliability, and lightweight design for the working platform, this study adopts the design concept of topological optimization. Finite element analysis (FEA) is conducted to thoroughly assess the stress distribution and deformation of the working platform, ensuring that the design conforms to established standards. The study implements a topological optimization strategy and compares and analyzes the performance of the working platform in terms of stress and deformation before and after optimization. The results indicate that the working platform achieves significant weight reduction while fulfilling operational demands. Furthermore, engineering trials validate the practical application effectiveness of the first domestically produced self-propelled, license-plate-equipped, road-legal, centralized-control, visualized tensioning and stringing electric power engineering vehicle, demonstrating a substantial enhancement in construction capabilities. This research provides empirical support for the design optimization of the working platform of vehicle-mounted tensioning and stringing equipment and establishes a theoretical foundation for subsequent improvements.

1 Introduction

The vehicle-mounted tensioning device is a tool specifically designed for power transmission line construction, featuring efficient mobility, rapid site conversion capabilities, and user-friendly operational characteristics. Consequently, it has been widely adopted in the construction and maintenance of power transmission lines. The working platform of this device undergoes complex mechanical forces during transportation. To meet the stringent requirements for high strength, high reliability, and lightweight design of the working platform, an advanced design concept of topology optimization was employed for structural design. The application of this method significantly reduced the dependence on human experience in the design process and effectively avoided potential over-design issues that may arise from traditional empirical-based design approaches [1].

Among the numerous methods of topological optimization, the homogenization method, evolutionary structural optimization (ESO) technology, and variable-density approach are widely adopted strategies [2]. Notably, the material interpolation model within the variable-density approach has gained universal application in engineering practice, ensuring the accuracy of computational results through the adoption of various interpolation schemes, such as the Solid Isotropic Material with Penalization (SIMP) model [3], the Rational Approximation of Material Properties (RAMP) model [4], and the Voigt upper bound model [5] [6]. Extensive and in-depth research has been conducted on material interpolation theory and topological optimization design both domestically and internationally, yielding fruitful outcomes. Ma et al. [7] conducted global topological optimization of a vehicle frame using the variable-density approach, obtaining the material distribution of the frame, thereby achieving efficient material utilization and enhancing the torsional stiffness of the frame. Liu et al. [8] performed finite element analysis (FEA) on the topological optimization design domain of a high-speed train set, reconstructing and modeling each component of the bodywork in a truss-like manner, resulting in a weight reduction of approximately 18% for the body structure, with the reliability and safety of the design verified through FEA. Nguyen et al. [9] proposed utilizing topological optimization methods for designing products intended for additive manufacturing technology, aiming to create optimal product structures with minimal material usage while ensuring the required mechanical properties. Zhu and Chen [10] introduced a heuristic approach utilizing engineering constraints in the process of topological optimization sensitivity, utilizing FEA software to achieve cost savings and improved mechanical performance while meeting structural design requirements. Liao et al. [11] conducted topological and sizing optimization on an existing subframe, altering the original design structure to achieve a lightweight and reliable structure that meets technical and assembly requirements, thereby achieving the optimization objectives.

This paper focuses on the overall structural design of vehicle-mounted tensioning equipment and delves into the structural strength optimization of its working platform. During the research, finite element software was employed to conduct a thorough analysis of the working platform, with the aim of ensuring that the design conforms to established standards while optimizing stress distribution to enhance overall performance. Specifically, a topology optimization strategy was implemented, and a comparative analysis of the working platform before and after optimization was conducted in terms of stress and deformation. This comparison not only validates the accuracy of the optimization direction but also vividly illustrates the optimization results: the working platform achieves significant weight reduction while fulfilling operational requirements. This study not only provides empirical support for the design optimization of the working platform of vehicle-mounted tensioning equipment but also establishes a solid theoretical foundation for subsequent improvement efforts.

2 Overall structure design of vehicle-mounted traction equipment

Vehicle-mounted tensioning equipment is a highly efficient and multifunctional device specifically designed for the construction and maintenance of transmission lines. Its overall structural design aims to meet the requirements of rapid movement, convenient operation, high-intensity work, and lightweight needs, ensuring stable and safe execution of tasks in complex and varied construction environments. The following is a detailed elaboration of the overall structural design of the vehicle-mounted tensioning equipment, with the overall structure illustrated in Figure 1.

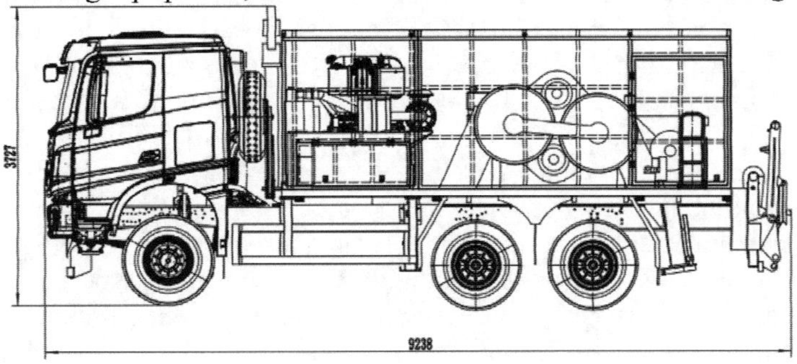

Figure 1. Structure design of vehicle-mounted traction equipment.

The vehicle-mounted tensioning equipment integrates an independent chassis module onto the second-class chassis of a cargo truck and has undergone professional vehicle modification and acceptance to obtain roadworthiness certification. This design offers significant advantages over traditional trailer-mounted equipment. Firstly, the power systems, control systems, and execution systems are all rationally distributed on the vehicle chassis, facilitating easy and safe operation while eliminating the need for cranes or lifting equipment for loading and unloading, thereby greatly enhancing work efficiency.

In terms of mobility, the vehicle-mounted tensioning equipment is equipped with a dedicated second-class off-road chassis, enabling the equipment to be moved anytime and anywhere without the need for a separate transport vehicle. Its robust off-road capability ensures normal operation on adverse terrains, bumpy and muddy roads, or unpaved surfaces, and demonstrates strong roadworthiness. Additionally, some vehicles are equipped with nighttime lighting systems to meet the needs of nighttime construction or emergency repairs, further enhancing the practicality and flexibility of the equipment.

In terms of structural strength, the working platform of the vehicle-mounted tensioning equipment experiences complex forces and requires high strength and lightweight characteristics. Therefore, modern design methods such as topological optimization were adopted during the development process to optimize the structural design of the working platform. Detailed assessments of the stress distribution and deformation of the working platform were conducted through finite element analysis and other means to ensure its structural safety and stability under harsh conditions such as heavy-load off-road travel. Meanwhile, the optimized working platform achieved a lightweight design, reducing the overall weight of the equipment and enhancing the convenience of transportation and construction.

Specifically, the working platform of the vehicle-mounted tensioning equipment is primarily composed of components such as the engine, hydraulic oil tank, drum assembly, and intelligent instrument box. These components are rationally arranged on the chassis to ensure the compactness and stability of the overall structure. The engine and hydraulic oil tank are longitudinally positioned at the front of the chassis, facilitating the arrangement of electrical wiring and hydraulic pipelines. The drum assembly is located in the middle of the chassis, directly above the rear wheels of the vehicle, to ensure uniform force distribution during operation. The intelligent instrument box is situated in a position easily observable and operable by the operator, allowing for real-time monitoring of equipment status and construction progress.

Additionally, the vehicle-mounted tensioning equipment is equipped with a remote control box that enables remote centralized control of multiple tensioners or tractors by a single operator. This design not only reduces the number of operators and mitigates the impact of noise on them, but also improves the operating environment and better ensures their safety. Simultaneously, the use of the remote control box enhances construction efficiency, enabling the equipment to perform tasks more flexibly and efficiently in complex and varied construction environments.

In terms of parameters, the specific parameters of the vehicle-mounted tensioning equipment vary depending on the model and configuration. Taking a certain model of a three-axle vehicle-mounted tractor as an example, it is equipped with a 6×6 off-road chassis produced by Beiben Heavy-Duty Truck Group Co., Ltd., with a maximum permissible gross mass of 24 tons, a curb weight of 23.87 tons, and body dimensions of 924025003740 mm. The equipment also features high maximum approach and departure angles, as well as a high maximum climbing gradient, ensuring its maneuverability and stability on complex terrains. Meanwhile, its maximum speed does not exceed 80 km/h, and the cab accommodates two people, meeting the basic requirements during construction.

The overall structural design of the vehicle-mounted tensioning equipment fully considers mobility, operational convenience, structural strength, and lightweight aspects. By adopting advanced chassis technology, optimized design of the working platform, and intelligent means such as remote control, the equipment ensures efficient, stable, and safe operation during transmission line construction and maintenance.

Journal of Physics: Conference Series **3009** (2025) 012048 doi:10.1088/1742-6596/3009/1/012048

3 Topology optimization of the working platform structure

Vehicle-mounted tensioning equipment plays a pivotal role in the field of transmission line construction and maintenance. This type of equipment integrates numerous advanced technologies aimed at enhancing construction efficiency and ensuring operational safety. In particular, the design of the working platform has become a focus of research due to its need to withstand complex and varied loads, while also requiring high strength and lightweight characteristics.

Regarding the specific design of the working platform, as a crucial component supporting the traction and tension modules, it experiences extremely complex force conditions. In the working state, the working platform must not only bear the gravitational force from the traction and tension modules but also cope with dynamic loads generated during vehicle movement, such as impact forces, vibrations, and overturning moments. These factors impose stringent requirements on the structural strength and stability of the working platform.

3.1 Finite-element modeling

In order to ensure that the design of the working platform meets the actual needs, this paper uses the finite element analysis method to carry out in-depth simulation research. Finite element analysis is a numerical analysis method, which can simulate the stress of the structure in the real environment, and calculate the stress, strain, and displacement of the structure. By using Solidworks to establish the three-dimensional model of the working platform and importing it into finite element analysis software such as ANSYS Workbench, the fine division of the working platform structure and the accurate setting of the loading conditions can be realized.

The material of the working platform of the vehicle-mounted traction equipment is ordinary carbon steel, and its physical properties are shown in Table 1.

Table 1. Physical properties of materials.

Materials	Density/(kgm^{-3})	Elastic Modulus/GPa	Poisson Ratio	Yield Strength/MPa
Plain carbon steel	7850	206	0.3	235

A three-dimensional model of the working platform was established using Solidworks and imported into ANSYS Workbench for meshing. Local mesh refinement was conducted on the webs and edges of both the upper and lower surfaces. Based on the structural design of the working platform, fixed support constraints were applied to the contact surface between the platform and the vehicle chassis. Corresponding forces were applied to the mounting surface in contact with the working equipment, along with the standard gravitational force. On this basis, researchers conducted a static simulation analysis of the working platform, obtaining its stress-strain contour plots, as shown in Figure 2. The analysis results indicate that both the maximum deformation and the maximum equivalent stress of the working platform in its static state fall within safe limits, meeting the requirements for practical use.

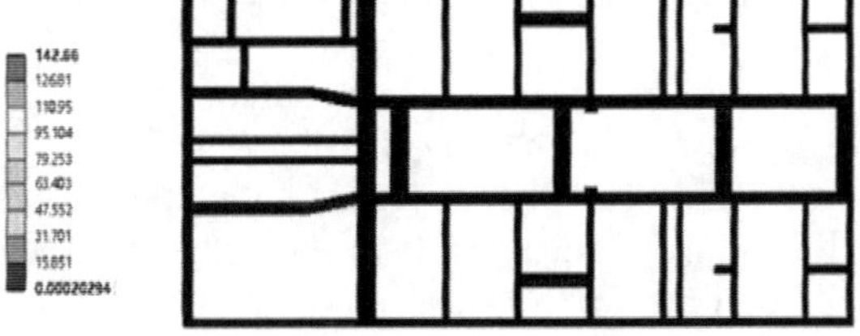

(a) Equivalent stress cloud diagram of working platform

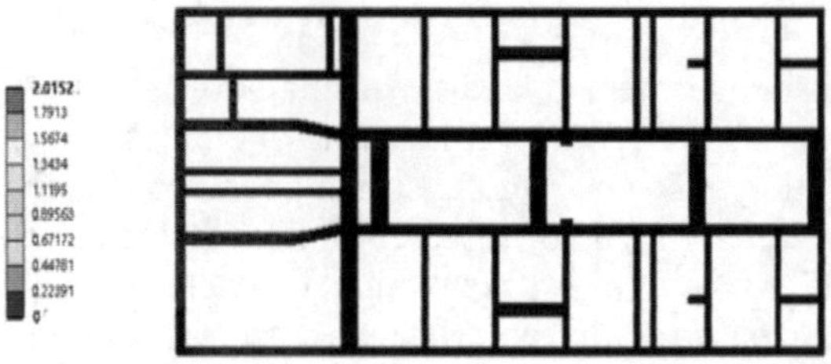

(b) Deformation cloud of working platform

Figure 2. Stress-strain diagram of the working platform.

3.2 Structural topology optimization

To meet the requirements of lightweight design, this paper further introduces the method of topology optimization. Topology optimization is an advanced structural optimization technique that can reduce the mass of a structure by altering the material distribution while ensuring structural performance. During the topology optimization process of the working platform, this paper adopts a topology optimization strategy based on the variable-density method. This paper analyzes the topology optimization of the working platform using the variable-density method. The variable-density method assumes that materials are variable and employs a density penalty approach to enable the relative density of structural elements to continuously vary between 0 and 1. This establishes an interpolation relationship between the relative density of elements and material properties, transforming the structural topology optimization into a material optimization problem. With the objective of minimizing structural compliance (maximizing stiffness), the mathematical model for variable-density topology optimization is constrained by the percentage of mass retention. The mathematical model is formulated as follows:

$$
\begin{aligned}
&\textit{find} && \rho_i (i = 1,2,\cdots n) \\
&\min && \max_{i \le j \le p} C_j \\
&s.t. && \frac{\sum_{i=1}^{n} \rho_i \rho_0 V_{oi}}{m_0} \le \alpha \\
& && 0 \le \rho_i \le 1
\end{aligned}
\tag{1}
$$

In the equation: ρ_i denotes the relative density of the i-th unit; C_j denotes the total structural compliance; p denotes the total number of loading conditions; ρ_0 denotes the original density of the unit; V_{oi} denotes the volume of the i-th unit; m_0 denotes the total mass of the structure when all units have their original densities; α denotes the percentage of mass retention. Topology optimization is conducted based on the static analysis in ANSYS Workbench, with the optimization target being the entire working platform. The response constraint is set as mass, and the retained mass is set to 80%.

Following the topology optimization analysis, researchers obtained the topology density contour plot of the working platform, as shown in Figure 3. The plot reveals significant material redundancy in the connecting beams and webs between the upper and lower surfaces of the working platform, indicating potential for optimization. Consequently, they redesigned the structure of the working platform based on the topology optimization results, removing some connecting beams and slotting the webs. This design not only reduces the amount of material used but also decreases the weight of the working platform, achieving the goal of a lightweight design.

(a) Topological density cloud diagram of the working platform

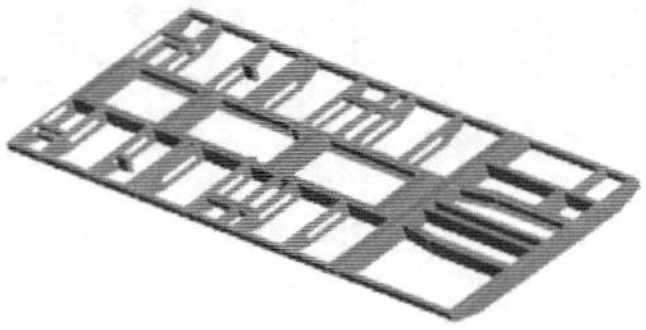

(b) The structure of the working platform after topology optimization

Figure 3. Topology optimization of the working platform.

The optimized working platform structure undergoes mesh generation and static analysis. When in a static state, the maximum deformation of the working platform structure is 2.07 mm, and the maximum equivalent stress is 164.91 MPa. A comparison of the static characteristics of the working platform before and after optimization is presented in Table 2. After optimization, the maximum equivalent stress remains below the yield strength of the material. As a support structure, the increased deformation is acceptable and meets the requirements.

Table 2. Comparison of static characteristics of working platform before and after optimization.

contrast characteristics	before optimization	optimized	variation/%
maximum deflection/mm	2.02	2.07	2.4
maximum stress/MPa	142.66	164.91	15.6

After redesigning through topology optimization, the mass of the working platform is 1484 kg, representing a 20% reduction compared to the pre-optimization mass of 1867 kg. The maximum equivalent stress after optimization is lower than the yield strength of the material, and the maximum deformation increases by approximately 2%. The increased deformation is still within acceptable limits for use. In summary, the structural stiffness of the working platform after topology optimization meets the requirements. Meanwhile, the significant reduction in mass after lightweight design achieves the objectives of safety and economy in design.

4 Engineering test

On November 15, 2024, China's first self-propelled, centralized-controlled, and visualized tensioning and stringing electric power engineering vehicle, equipped with a license plate and authorized for road use, successfully underwent application testing at the stringing construction site of the 500 kV line project between Tieling and Shiling in Liaoning Province, as shown in Figure 4. This test vehicle was predominantly developed by the Liaoning Electric Power Employee Innovation Studio, marking a significant breakthrough in the field of lightweight vehicle-mounted tensioning and stringing equipment in China.

Figure 4. Stretch line electric power engineering vehicle.

The novel electric power engineering vehicle comprises several core modules: an independent centralized control system, a visual monitoring system, a tensioning and stringing system, a lighting system, and an off-road vehicle chassis module. This integrated design enables the electric power engineering vehicle to achieve autonomous mobility and the ability to move and stop as needed without relying on additional transport vehicles. Particularly during nighttime construction, the vehicle's lighting system ensures the safety and visibility of the working environment. Furthermore, the visual monitoring system enables remote monitoring of the entire stringing line, while the centralized control system supports the collective control of multiple devices, significantly enhancing construction efficiency and management convenience.

The electric power engineering vehicle is suitable for various complex construction scenarios, including but not limited to three-span operations, construction in complex terrain conditions, and emergency rescue missions. Its flexible mobility not only reduces the need for cranes and transport vehicles but also effectively lowers safety risks and labor intensity during the construction process. Through validation in this engineering trial, the electric power engineering vehicle has demonstrated significant improvements in construction capabilities, providing a more efficient and secure solution for the construction of electric power projects in China.

5 Conclusion

This paper focuses on the issues of over-design and material redundancy in the structures of large-scale equipment, with the working platform of vehicle-mounted tensioning and stringing equipment selected as the specific research object. In the research process, an accurate three-dimensional model of the working platform was first created using Solidworks software. Subsequently, topological optimization techniques were employed to identify areas with minimal impact on the overall structural performance of the working platform, and targeted optimization design strategies were implemented accordingly.

After applying topological optimization, the weight of the working platform was successfully reduced by 20%. To evaluate the optimization effects, static analysis was conducted on both the optimized and pre-optimized working platforms. By comparing changes in static performance indicators, it was found that the optimized structure still met the established stiffness requirements, thereby achieving cost economization while ensuring performance.

This optimization design practice not only verifies the effectiveness of topological optimization in guiding the structural design of working platforms for large-scale equipment but also reveals the design redundancy issues that may arise from traditional empirical design methods. By optimizing the structure, design quality was improved, and unnecessary material consumption was significantly reduced. Notably, the optimized working platform still has a certain safety margin in terms of stress values, indicating that its structure still has potential for further optimization. In the future, we can continue to deepen the optimization design by adjusting material types, panel configurations, and thickness parameters, aiming to achieve more ideal structural performance and economic benefits.

References

[1] Ribeiro, Tiago P., Bernardo, Luís F. A., Andrade, Jorge M. A. (2021) Topology Optimisation in Structural Steel Design for Additive Manufacturing. Applied Sciences, Vol.11 (2112): 2112. DOI: 10.3390/app11052112.
[2] Bendsoe, M. P., Diaz, A. R., Lipton, R. P., et al. (1994) Prediction of extremal material properties for the optimal design of topology, shape, and material. Proceedings of SPIE - The International Society for Optical Engineering, 2192: 414-425.
[3] Zhu, Lei, Zuo, Tongxing, Wang, Chong, Wang, Qianglong, Yu, Zhengdong, Liu, Zhenyu. (2024) Hole Appearance Constraint Method in 2D Structural Topology Optimization. Mathematics, Vol.12(17): 2645. DOI: 10.3390/math12172645.
[4] Wang, Jun, Liu, Jikai, Li, Lei. (2023) Topology Optimization for Digital Light Projector Additive Manufacturing Addressing the In-Situ Structural Strength Issue. Polymers, Vol.15(17): 3573. DOI: 10.3390/polym15173573.
[5] Swan, Colby C., Kosaka, Iku. (1997) Voigt-Reuss topology optimization for structures with nonlinear material behaviors. International Journal for Numerical Methods in Engineering, Vol. 40(20): 3785-3814.
[6] Fan, Weichun, Zhang, Zhifei, Xu, Zhongming, He, Yansong. (2024) An adaptive material-field series-expansion method for structural topology optimization. Structures, Vol.70: 107693. DOI: 10.1016/j.istruc.2024.107693.
[7] Ma, Weiyang, Lu, Yanhui, Wang, Pengyu, Wang, Yongjia, Wang, Jiahao. (2023) Double Optimization Design of the Formula Racing Car Frame Based on the Variable Density Method and the Joint Variable Method. Applied Sciences (Switzerland), Vol.13(18): 10155. DOI: 10.3390/app131810155.
[8] Liu, Chunyan, Ma, Kai, Zhu, Tao, Ding, Haoxu, Sun, Mou, Wu, Pingbo. (2024) A New Car-Body Structure Design for High-Speed EMUs Based on the Topology Optimization Method. Applied Sciences-Basel, Vol.14(3): 1074. DOI: 10.3390/app14031074.
[9] Nguyen, Dinh Son, Vignat, Frédéric. (2017) Topology optimization as an innovative design method for additive manufacturing. In: 2017 IEEE International Conference on Industrial Engineering and Engineering Management, IEEM 2017. 304-308. DOI: 10.1109/IEEM.2017.8287375.
[10] Zhu, J., Chen, X. (2017) An engineering constraint method for continuum structural topology optimization. Advances in Mechanical Engineering, 9: 1-6. DOI: 10.1177/1687814017692233.
[11] Liao, Jun, Lan, Shan, Liu, Jing-Zhong, Hu, Wang-Yue. (2013) Topology-Optimization of Fuel-Cell-Car's Vice-Frame Structure. Applied Mechanics and Materials, Vol.278-280: 256-260. DOI: 10.4028/www.scientific.net/AMM.278-280.256.

Journal of Physics: Conference Series **3009** (2025) 012049 doi:10.1088/1742-6596/3009/1/012049

Research on zero position measurement technology of infrared thermal imager based on CCD imaging technology

Miao Cao [1], Yanyu Su [2] *, and Luyao Tian [3]

[1, 2, 3] Department of Life Science and Technology, Changchun University of Science and Technology, Changchun 130022, China. E-mail: caomiao@cust.edu.cn; suyanyu1230228@163.com; 1945152439@qq.com

* Correspondence: suyanyu1230228@163.com

Abstract: The infrared thermal imager zero momentum of the detection method is currently in the stage of subjective measurement, and the detection efficiency and detection accuracy are not high. This paper starts from the theory of zero position measurement of an infrared thermal imager with a new five-hole target and infrared long wave collimator. It is simulated to the infinite distant beat and combined with CCD imaging and image processing technology, and the zero position of the infrared thermal imager and beat momentum test system is designed. The experimental results show that the zero-detection accuracy of the infrared thermal imaging system is better than 0.05 mil.

1. Introduction

With the development of infrared thermal imaging technology, it is more and more widely used in the field of night view, ranging, detection, and other fields [1-3]. It has become a kind of high technology in the future high-tech war. In the current high-intensity electronic warfare conditions, infrared cameras, as infrared imaging technology-derived products, can detect camouflaged targets to achieve long-range weather and observation, which are widely used in many fields. However, the performance parameters of infrared thermal imaging systems are very backward, and some parameters are still in the stage of subjective measurement, which cannot meet the requirements of the high-performance infrared thermal imaging system [4-6].

A scheme for the objective measurement of the amount of infrared camera beating zero is proposed, and it is tested for its capacity to move around zero, aiming to ensure that the thermal imager performance plays into full play.

2. Zero position measurement theory

2.1 Analysis of the generation of zero position

The zero change of the infrared thermal imaging system is the baseline of the infrared thermal imaging instrument aiming at the baseline due to the impact of the use of shock and so on, and the amount of changes generated before and after the baseline. Zero change is one of the important parameters in the evaluation of infrared thermal imaging systems [7-8].

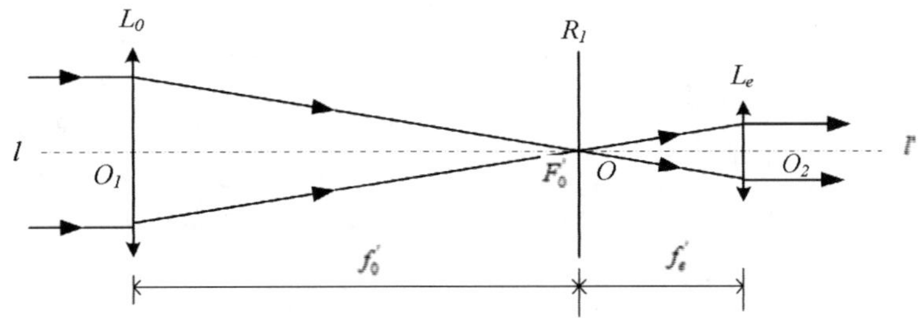

Figure 1 Ideal working state of infrared sight.

According to Figure 1, the infrared ray aims at the center, the center of the line OO_2 is aimed at the baseline, and the zero change of the quantity is the corresponding change of the baseline position. When the infrared sight after the experiment is completed, the results in the relative position change.

As shown in Figure 2, the partition plate in the experiment after the partition plate center shifts, deviating from the angle of β aimed at baseline OO_1 and theory of the optical axis ll'. Now infrared aiming with zero variation relative to baseline skew angle value is: $\theta = \beta$.

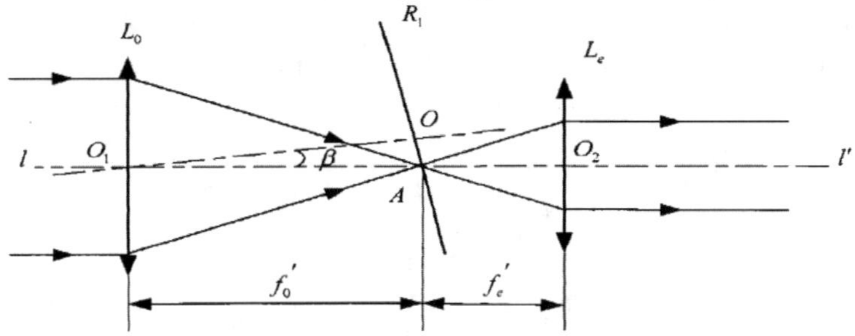

Figure 2 The sketch map of the target baseline.

As shown in Figure 3, infrared sight with partition plate and lens shifting in the test process leads to infrared sight with the optical axis O_1O_2 and ideal optical axis ll' relative angle. It aimed at baseline OO_1 and theory of optical axis deviation angle of β, showing infrared sights of zero variation with respect to the baseline deflection from every angle relationship. The angle value is: $\theta = \alpha + \beta$.

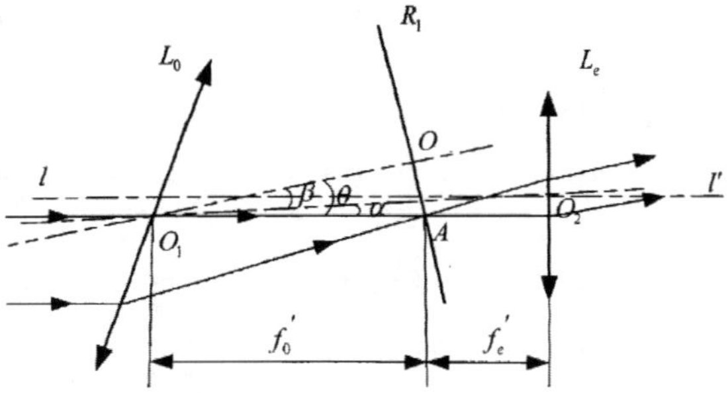

Figure 3 Schematic diagram of the line of sight of the dividing plate and the objective lens.

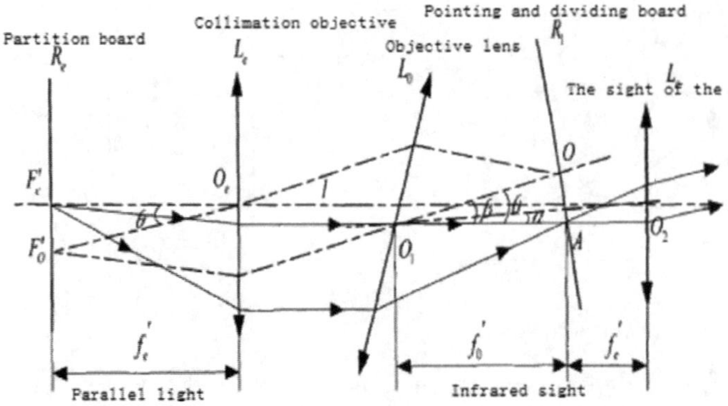

Figure 4 Principle of zero position change detection.

2.2 Zero position change detection principle

According to Section 2.1, it is known that the parallel light tube is needed to simulate an infinite target in the measurement of the zero-position change of the infrared sight [9]. As shown in Figure 4, L_e is the parallel light tube lens, R_e is the partition board, F'_e is the partition plate center through a parallel light tube into Position A, and O is the thermal imager partition plate midpoint corresponding to the parallel light pipe partition board on the point F'_O. The light tube optical axis $F'_e O_e$ is used for the detection of the reference axis ll'. According to the geometric relationship of the icon:

$$\theta = \angle F_e{}' O_e F_o{}'$$

$$\tan\theta = \frac{F'_E F'_O}{F'_e O_e} = \frac{F'_E F'_O}{f'_e}$$

$$\theta = \arctan\left(\frac{F'_e F'_O}{f'_e}\right)$$

The θ type is very small and can be approximated as:

$$\theta = \frac{F'_e F'_O}{f'_e}$$

where $F'_e F'_O$ is the partition plate center in infrared sight with the partition board on the image relative to the infrared sight graticule plate center offset and f'_e is the light pipe.

3. Infrared thermal imager zero position measurement system components

The system's off-axis parabolic mirror is the main body to construct a Newtonian reflecting collimator system, the word "+" pinhole integrates the target, and the infrared thermal imager provides infinite far-infrared observations of the target [10-12]. The target is located in an off-axis parabolic mirror at the focus of the infrared sub-reticle and partition plate behind the higher control accuracy surface blackbody form. Infrared thermal imager will be an infinitely far infrared target image displayed on the display by the computer through the video capture card to capture the video signal analysis and processing, to achieve zero beating parameters. The structure of the system composition is shown in Figure 5.

This system names zero-beat quantity interpretation software, according to 5 pin holes in the measured infrared sight within the field of view with respect to its partition location. Its own points

Journal of Physics: Conference Series **3009** (2025) 012049 doi:10.1088/1742-6596/3009/1/012049

row is given in the angle measuring instrument coordinate position, namely the test materials were aimed at baseline zero. After the shooting stress and zero amount of beating are defined, the stress is clamped in the instrument before the fixed sights on the mounting seat. It is sentenced to read sight aimed at the baseline value zero, according to the test aimed at baseline zero coordinate value variation of measurement, and it can be obtained by the sight of the zero momentum.

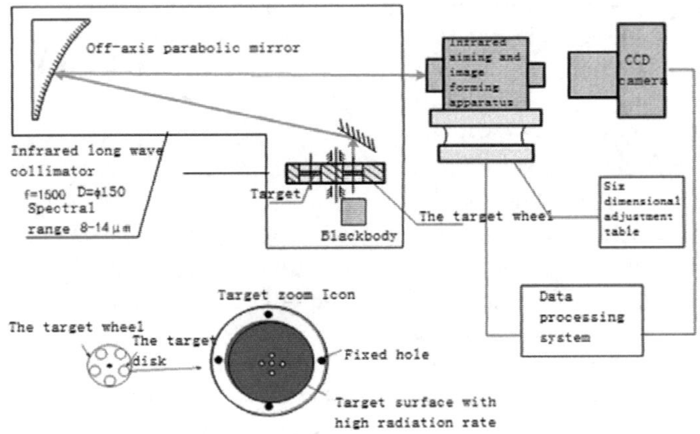

Figure 5 Zero position measurement system of infrared sight.

3.1 Surface blackbody

We use precision temperature measurement technology and temperature control technology to design and manufacture, to achieve the following performance:

Differential temperature range: -10°C-50°C; Effective radiation area: ≥Φ30 mm; Temperature resolution: 0.01°C; Temperature control precision: ±0.3°C; Temperature uniformity: 0.1°C (ΔT≤±5°C, 2/3 central region); Temperature stability: ±0.01 °C/h; Effective emissivity: ≥0.95.

3.2 CCD imaging system

According to the test accuracy requirement of zero position, the measurement camera uses over 1, 000×1, 000 resolution CCD [13-14]. Therefore, the system uses a Lu175m camera made by Canadian Lumenera Corporation, whose technical indicators are shown in Table1.

Table 1. Lu175m camera technology index.

Image sensor	Micron MT9M001 COMS $1/2''$ format, mono, 6.6×5.3 array
Effective pixels	1, 280×1, 024, 5.2 μm square pixels
Frame rate	30 fps at 1, 280×1, 024, 100 fps at 640×480
Sensitivity	High
Dynamic range	60 dB
Auto exposure	Automatic/Manual
White balance	Automatic/Manual
Dimensions (W×H×D)	2.25×3.85×1.56 inches (enclosed) 2.00×2.50×1.63inches (board level)
Mass	~150 g/300 g
Power requirement	USB bus power, or external 6VDC, 500 mA
Power consumption	~2.5 W
Operating temperature	0 to +50°C

4. Test results

For the zero-momentum test, the first measured infrared thermal imager is installed in a three-dimensional turntable, and a three-dimensional table should be regulated. It enables the collimation of optical systems with measured infrared thermal imager axis to be consistent. In the system test target disk, a five-hole target is provided.

The CCD camera lens keeps a thermal infrared imager of the exit pupil position in the measured and alignment of the infrared parallel light pipe of the main axis. Zero-beat momentum test software is opened, and the CCD camera aperture and focal length are adjusted. The five-hole target and thermal imager get the sighting reticle clear image on the display.

After the test of the infrared thermal imaging equipment, the measurement is to the center of the aimed coordinates. The upper, lower, middle, left, and right five circle spot center coordinates are measured, the measurement coordinate system is established, and then we measured at the coordinates of the measurement coordinates. The measured test of the thermal imager is clamped once again. The above operations are repeated, and they can be used to measure thermal infrared imager infrared aiming with zero momentum. 6 inferior precision measurement results are shown in Table 2 shows.

Table 2. Infrared thermal imager zero position measurement datasheet

Measuring times	Measured value (mil)	
	△X	△Y
1	2.537	-1.130
2	2.523	-1.122
3	2.510	-1.117
4	2.538	-1.129
5	2.512	-1.134
6	2.519	-1.113
s	0.0117	0.0088

The standard uncertainty of the measurement results is:

$$u_c = \delta / \sqrt{n} = 0.0117 / \sqrt{6} = 0.0048$$

Extended uncertainty: $U_c = 3 \times u_c = 3 \times 0.0048 = 0.015$

5. Conclusion

Infrared sights zero beat momentum measurement precision mainly depends on the resolution of the CCD camera, more than 100-million-pixel camera resolutions are chosen, and the general zero momentum indicators qualified for not more than 0.7 mil, which is 15.12 corners/second. So, the zero momentum of the measuring range is to be set to ±1°. In this way, since the resolution of the selected CCD is 1, 280×1, 024, the corresponding angle of each pixel is 4°/1, 024=0.033 mil. The weighted algorithm can achieve sub-pixel interpretation. The infrared aiming at the baseline of zero-point interpretation accuracy can reach 0.05 mil. From the experimental results, this design meets the system requirements.

Acknowledgement

This research was funded by the Jilin Province Science and Technology Development Plan Project, China (Grant number: 20230201056GX).

Journal of Physics: Conference Series **3009** (2025) 012049 doi:10.1088/1742-6596/3009/1/012049

References

[1] JIANG Qi, HE Dingfan, and JIANG Yupeng. Application of Infrared Thermal Imaging Detection Technology in Intelligent Operation of Substation. Techniques of Automation & Applications, 2024.
[2] YOU Hongbo, LI Xing, and FAN Xiang. Research and analysis on center deviation and assembly and adjustment method for the infrared optical system. LASER & INFRARED, 2022.
[3] ZHANG Jing. Research on infrared image segmentation and target extraction technology. Intelligent Computer and Applications, 2024.
[4] Development of Calibration Specification for Laser Marker and Technical Discussion of Existing Calibration Standard [J]. China Standardization, 2021.
[5] LIANG Juanfeng, CHEN Shuo, and FAN Xiang. Construction of dimension chain and tolerance design based on infrared continuous zoom mechanism. LASER and INFRARED, 2024 (8).
[6] SHI Zhang, HE Yuan, and PAN Dong. Research on the structural design and assembly process of continuous zoom thermal imager. LASER & INFRARED, 2024.
[7] GENG Yufei, ZHOU Kuan, and LIU jizhou. Moving target recognition and tracking algorithm based on infrared image processing. LASER & INFRARED, 2024.
[8] Fu Y., Wang Z., Jiang G., et al. A novel three-dimensional shape measurement method based on a look-up table [J]. Optik-International Journal for Light and Electron Optics, 2014.
[9] Dongsheng W., Zongjun T., Lida S., et al. Research Status and Development of Laser Cladding Technology [J]. Applied Laser, 2012.
[10] Li W., Li F., and Jiang Z. A machine vision-based radial circular run out measurement method. Int. J. Adv. Manuf. Technol., 2023.
[11] Hao M., Yu H., and Li D. The measurement of fish size by machine vision – A review. In IFIP Advances in Information and Communication Technology; Springer: Berlin/Heidelberg, Germany, 2016.
[12] Ji T., Zhao Z., and Zhao N. A Machine Vision Measurement Method for Large Plates Based on Reference Point Assistance. In Proceedings of the 5th International Conference on Electronics and Electrical Engineering Technology, 2022.
[13] General Specification for China National Standard Level Television Cameras, China National Standard, GB/T15465-1995.3.3.
[14] Prakash N., Manconi A., and Loew S. Mapping Landslides on EO Data: Performance of Deep Learning Models vs. Traditional Machine Learning Models. Remote Sens., 2020.

Journal of Physics: Conference Series **3009** (2025) 012050 doi:10.1088/1742-6596/3009/1/012050

Laser scanning path planning based on LPBF molding AlSi10Mg stress analysis

Bo Zhang[1,2] **and Yan Shi**[1,2,*]

[1]School of Mechanical and Electric Engineering, Changchun University of Science and Technology, Changchun 130022, PR China
[2]National Base of International Science and Technology Cooperation for Optics, Changchun 130022, PR China

[*]Corresponding author: shiyan@cust.edu.cn

Abstract: To reduce residual stresses and warpage in large parts molded by LPBF (Laser Powder Bed Fusion), this paper adopts numerical simulation analysis to partition the molding area for planning, and analyses the stress-strain under different scanning strategies when 3d printing molding single-layer AlSi10Mg powder. In addition, a comparative analysis of the area of the delineated area is carried out, and finally a reasonable scanning strategy is optimized, which reduces the residual stress and deformation inside the molded part. The results show that the residual thermal stresses and distortions inside the molded part can be reduced by using the "spaced zone scanning" strategy compared with the large-area, same-direction scanning strategy. It also makes the internal thermal stress range more balanced. The smaller the divided area is, the smaller the internal residual stress will be, but at the same time, the production cycle time will increase. The results of the study can provide new ideas for deformation control in the forming process of large parts and so on.

1. Introduction

LPBF laser additive manufacturing technology is an important part of "Made in China 2025" in the field of high-end intelligent manufacturing [1], which is a technology that constructs an object based on a digital model file by printing layer by layer using powdered metal or bondable materials. Compared with traditional manufacturing technology, LPBF additive molding of metal structures has high molding precision, good surface quality, and high material strength. This processing method utilizes a high rate of material, is not subject to the limitations of the complexity of the model, and greatly expands product design ideas. However, this manufacturing method involves melting the powder particles and finally moulding the structural parts in a very short period of time with a high-power laser beam. In this kind of rapid cooling and heating manufacturing environment [2-3], the molded parts have a large residual thermal stress and thermal deformation cannot be released in a timely manner, resulting in the molding of thin-walled structures or large structural parts produced warp deformation and other defects. How to reduce or even avoid these defects has become one of the key issues that needs to be solved.

With LPBF additive manufacturing technology, during the molding of large-scale parts, the temperature gradient at the ends of the sweeping line increases dramatically with the increase in the single-layer planar filler powder volume. When this layer-by-layer accumulation of thermal stress exceeds the yield strength of the molding material, it will cause large local warping deformation, resulting in internal defects in the molded part or even print molding failure, for which scholars at home and abroad have done a lot of research in the 3D printing partitioning strategy and other aspects. Zrnic et al. [4] investigated the effect of different scanning angles on part deformation under three scanning modes through thermo-indirect structural coupling simulations. The 15° rotational scanning strategy was found to produce the least amount of stress, while the various scanning strategies had little effect on the melt pool size. Deng et al. [5] compared the accumulation of residual stresses in the layer under s-shape scanning and zonal scanning modes. The effectiveness of the partitioning strategy to reduce the tensile stress at the part boundary is validated, but the validation model scale is small and not universal. Lee et al [6-7] used finite element techniques to simulate the temperature and stress fields under Z-scan and S-scan modes, and investigated the effects of scan line length and scan area size on the temperature and stress fields. It was found that the S-shape scanning method was effective in reducing the residual stresses and that the oscillation frequency of the temperature field depended on the scanning line length. When the scanning area is large, the average temperature of the temperature field is lower, but larger residual stresses accumulate, and it is recommended that scanning lines that are too long should be avoided as much as possible when LPBF molding. The above study mainly focuses on the simulation and analysis process within a small area, which is mainly limited to the simulation of the angular change of the fill line, and it is difficult to carry out large-scale calculations in terms of macroscopic scale or block scanning sequence simulation, so it is also necessary to plan and analyze the scanning strategy by planning and analyzing the scanning strategy on a larger area.

In this paper, based on previous research, for the problem of large internal residual thermal stresses in the molding of large parts, numerical simulation is used to propose the partitioning of the molding area and the comparison of the stresses on the size of the area of the partitioned area. Finally, a reasonable scanning strategy is optimized to regulate the residual stress inside the molded part within a certain range, to avoid deformation and warping of the molded part or printing failure due to large residual thermal stress.

2. LPBF molding AlSi10Mg powder bed calculation model

2.1. Experimental method

In this paper, finite element simulation analysis is carried out using Ansys APDL command flow.

First, the powder bed is modeled, and the model is simplified to save the arithmetic example, as shown in Figure 1. The purple area is the powder bed substrate model, and the green area is the powder bed model, in which the size of the powder bed is 20 mm*20 mm, and the size of the substrate is 50 mm*50 mm. In order to improve the solution accuracy, the powder bed model was meshed using the "vsweep" command flow (swept-abstract meshing method), which produces high-quality hexahedral mesh cells with a mesh size control of about 0.2 mm. The "mshkey" command flow is used to specify the substrate mesh model delineation method as mapped mesh delineation. The "mshape" command flow is used to specify the shape of the substrate mesh model as a tetrahedral mesh cell, which can greatly reduce the cost of computing, and its size is controlled at 0.2 mm-2 mm with fixed constraints on the bottom of the substrate.

Figure 1. Mesh model of substrate and powder bed.

2.2. Material preparation

Materials are assigned to the model created. The project team uses JMatPro software to calculate the thermophysical parameters of the required materials, where the powder material is AlSi10Mg, and the substrate material is 6061 aluminum alloy. The values of the parameters of the different properties of the material at different temperatures are derived and brought into the calculation. The specific parameters are shown in Table 1.

Table 1. Corresponding thermophysical parameters for powder bed modeling.

Temp (°C)	25	200	400	550	600	650	700	900	1300
Density (g/cm³)	2.69	2.66	2.62	2.58	2.46	2.45	2.43	2.36	2.21
Heat conductivity (W/m*K)	168	167	158.8	147	79.9	81.1	82.7	89.1	101.9
Specific heat (J/g*k)	0.9	0.98	1.05	10.26	1.55	1.14	1.145	1.148	1.15
Poison's ratio	0.32	0.33	0.34	0.356	0.498	0.5	0.5	0.5	0.5
Bulk modulus (GPa)	71	67.6	62.4	57.2	51.9	51.2	50.6	48	43
Expansion coefficient (10e-6 /K)	20.78	22.2	24.1	26.2	52.28	52.3	52.2	52.7	56.1

2.3. Scan path division

After the material model is given, as shown in Figure 2, four scanning strategies, namely, "unidirectional scanning", "reciprocating scanning", "zonal scanning" and "interval-zonal scanning", are simulated and analyzed. The four scanning strategies of "one-way scanning", "reciprocating scanning", "partition scanning" and "interval partition scanning" are simulated and analyzed. The initial parameter values for the simulation of the four scanning strategies were used with a laser power of 340 w, a scanning speed of 1870 mm/s, and a scanning pitch of 120 μm. The Gaussian surface heat source model is used for the heat source model in the simulation process with the following expression:

$$q(r) = \frac{Q}{\pi r^2} e^{-\frac{r^2}{a^2}} \tag{1}$$

where Q is the maximum heat flow density of the heat source, r is the length from the center of the heat source, and a is the standard deviation of the Gaussian distribution.

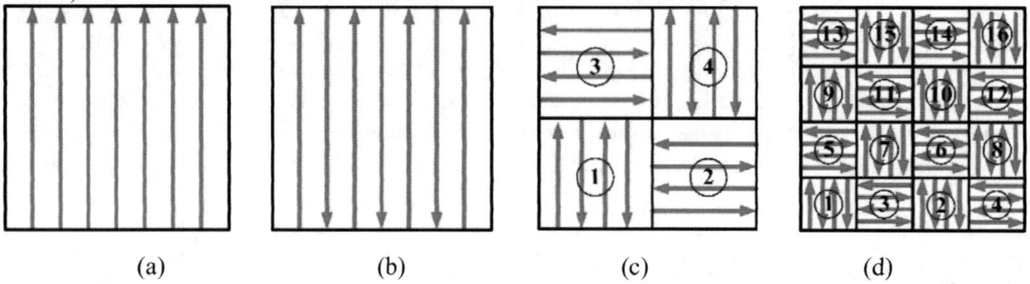

Figure 2. Schematic diagram of scanning strategy: (a) Unidirectional; (b) Reciprocal; (c) Partition; (d) Interval partition.

3. LPBF molding AlSi10Mg stress-strain calculation results

3.1. Residual thermal stress analysis with four scanning strategies

Figure 3 shows the thermal stress simulation analysis results of the four scanning strategies. Firstly, from the thermal stress simulation cloud diagram, the thermal stresses of these four scanning strategies are decreasing in order. The maximum thermal stress is 239 MPa for unidirectional scanning and 237 MPa for reciprocating scanning, whereas the maximum thermal stress is mitigated by using zonal scanning and interval zonal scanning strategies to 234 MPa and 231 MPa, respectively.

Although the reduction of thermal stresses using the scanning strategy of interval partition scanning is not numerically significant, it is worth noting that it commences from the distribution area of the highest thermal stresses in Figure. On the one hand, the use of interval partition scanning can significantly reduce the range of the highest thermal stress distribution, and on the other hand, compared with other scanning strategies, interval partition scanning will lead to a relatively uniform distribution of thermal stresses in the overall model. The thermal stresses around the powder-ceramic bond are low from the stress cloud diagram, which can effectively mitigate the risk of warpage cracking.

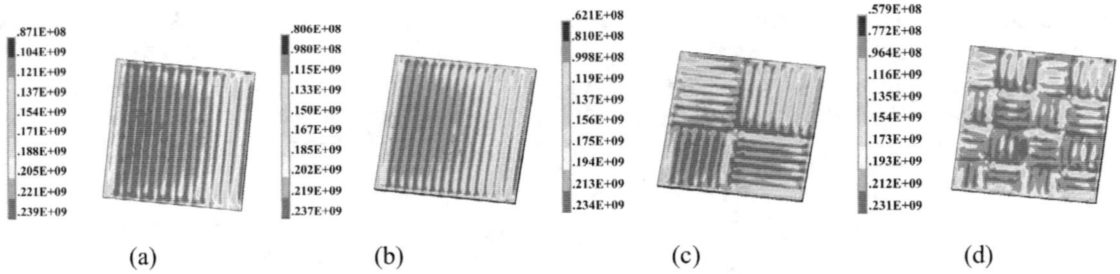

Figure 3. Thermal stress cloud of powder bed under different scanning strategies: (a) Unidirectional scanning thermal stress; (b) Reciprocal scanning thermal stress; (c) Partitioned scanning thermal stress; (d) Interval zonal scanning for thermal stress.

3.2. Stress effects of different areas in spaced partitioned scans

Thermal stress analysis of small area molded surfaces of different sizes under interval partition scanning strategy is shown in Figure 4.

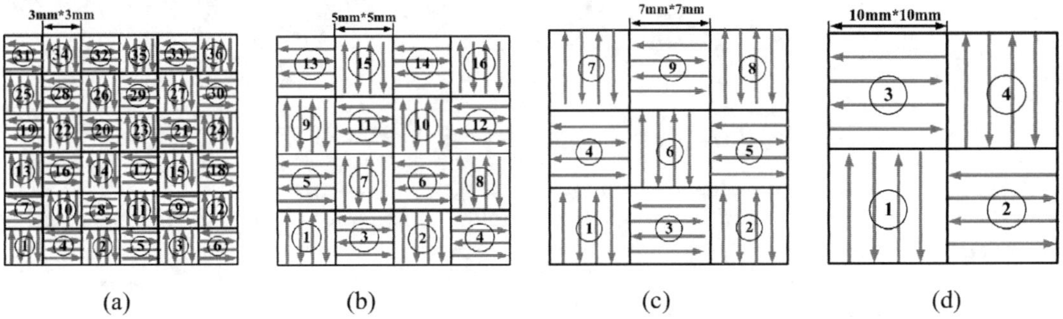

Figure 4. Scanning strategy for different "square" sizes: (a) 3 mm*3 mm; (b) 5 mm*5 mm; (c) 7 mm*7 mm; (d) 10 mm*10 mm.

The time-dependent values of stresses at the same node location were extracted from the resultant files of the four different "grid" size scanning strategies, as shown in Figure 5.

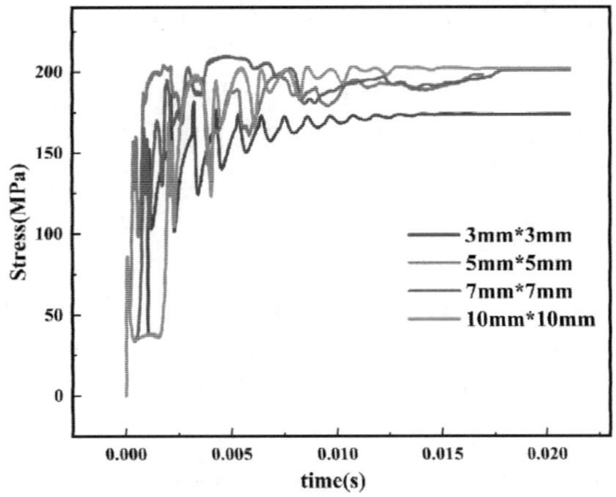

Figure 5. Nodal stress variation at the same location with different "square" dimensions.

As can be seen from Figure 5, since the thermal conductivity of aluminum alloy is relatively high compared with that of general metals, when the heat source is around the node, it will melt some of the powder at the node, and then when the heat source is far away from the node the powder solidifies to produce a certain amount of stress, the powder will solidify to produce a certain amount of stress. When the heat source moves right above the node, the powder above the node is completely melted, and at this time, the stress is suddenly reduced because it is not solidified, but there is some stress on the solidified powder near the node. So, the stress at this node at 0-0.005 s is not stable and changes repeatedly. After 0.01 s, the stress at the node starts to stabilize at this point because the heat source is far away from the node location and the powder at the node basically changes from liquid to solid.

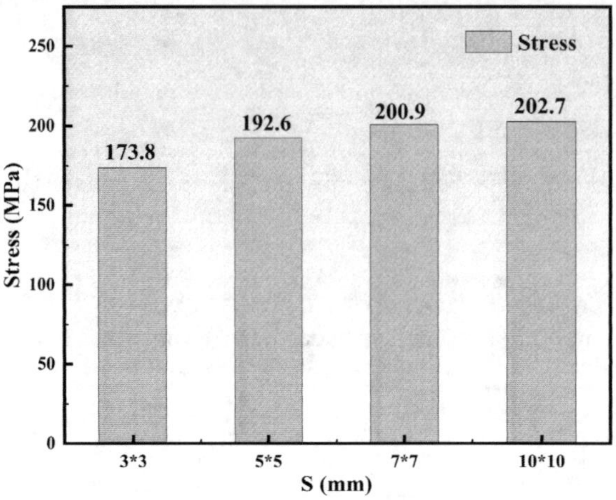

Figure 6. Maximum post-cooling stresses at nodes of the same position for different "grid" sizes.

The results in Figure 6 show that the area of the partitioned "squares" has a significant effect on the stress when using the interval partitioning scanning strategy. The larger the area is, the greater the stress is generated. Therefore, after selecting the interval partitioning scanning strategy, as many areas as possible should be partitioned for printing and molding.

4. Conclusion

(1) The use of interval zonal scanning strategy for scanning and printing molding of large area surface of AlSi10Mg can reduce its internal residual thermal stress after molding.

(2) Adopting the interval zonal scanning strategy to scan and print molding of large surface areas of AlSi10Mg can lead to a more uniform distribution of thermal stresses, and risks such as local warping and deformation can be avoided.

(3) For the area of the "square grid" in the interval partition scanning, its area is positively correlated with the thermal stress, but too low an area will reduce productivity, so it is necessary to select the appropriate area according to the actual situation.

Acknowledgments

This work was supported by the Equipment Pre-Research Common Technology Project of China (grant no. 50923030512). The authors wish to express their thanks for this funding.

Reference

[1] Qi X, Liang X, Wang J, et al. (2024) Microstructure tailoring in laser powder bed fusion (L-PBF): Strategies, challenges, and future outlooks. Journal of Alloys and Compounds, 970:172564-172584. DOI: 10.1016/j.jallcom.2023.172564.

[2] Zhao M R, Liu J, Yan S. (2024) Influence of laser mode on size effect in manufacturing AlSi10Mg mini-structures by laser powder bed fusion technology [J]. Optics & Laser Technology, https://doi.org/10.1016/j.optlastec.2024.111748.

[3] Zheng Z, Sun B, Mao L. (2024) Effect of Scanning Strategy on the Manufacturing Quality and Performance of Printed 316L Stainless Steel Using SLM Process [J]. Materials, 17. DOI:10.3390/ma17051189.

[4] Zrnic D S, Melnikov V M, Doviak R J, et al. (2017) Scanning Strategy for the Multifunction Phased-Array Radar to Satisfy Aviation and Meteorological Needs [J]. IEEE Geoscience & Remote Sensing Letters, 12(6): 1204-1208.DOI:10.1109/LGRS.2014.2388202.
[5] Deng S S, Yang Y Q, Li Yang, et al. (2016) Partition scanning path planning and its influence on residual stress distribution of SLM molded parts [J]. China Laser, 43 (12): 9. DOI: 10.3788/CJL201643.1202003
[6] Tsai C Y, Cheng C W, Lee A C, et al. (2019) Synchronized Multi-Spot Scanning Strategies for the Laser Bed Fusion Process [J]. Additive Manufacturing, DOI: 10.1016/j.addma.2019.02.009.
[7] Huang Q Y, Zhong Y, Wang W Y. (2024)3D printing slicing algorithm and partition scanning strategy for big data [J]. Intelligent Decision Technologies, 18(2):731-741.

Journal of Physics: Conference Series **3009** (2025) 012051 doi:10.1088/1742-6596/3009/1/012051

Mechanical properties of gradient foam aluminum under high strain rate loading

Hang Gao[1], Fengjun Zhou[1*], Feng Li[1], Qingfeng Liang[1] and Pengsheng Ma[1]

[1] Defense Engineering Institue, AMS, PLA, Beijing, 100036, China

*Corresponding author's e-mail: m18613873088@163.com

Abstract. The mechanical characteristics of closed-cell gradient foam aluminum under high strain rates were investigated. An enhanced direct impact Hopkinson pressure bar (DIHB) was employed to examine the impact response of cylindrical closed-cell foam aluminum specimens across a broad spectrum of strain rates, with the peak strain rate reaching $1300s^{-1}$. The dynamic deformation process was documented using a high-speed camera and analyzed via the digital image correlation (DIC) technique. The full-field strain evolution and distribution of the gradient foam aluminum during compression were acquired, and its deformation process, energy absorption mechanism, and damage mode were explored. The internal microstructure of the samples was inspected by X-ray micro-computed tomography (X-CT). Based on the experimental findings, the failure mechanism of the foam aluminum and the specific locations of internal structural damage were analyzed. The results indicate that the deformation behavior of foam aluminum is highly dependent on the loading rate. At elevated strain rates, the plateau stress rises, and alterations occur in the formation and propagation of the deformation front.

1. Introduction

Foam metal materials are effective impact protection materials widely used in aerospace, mechanical, and civil engineering[1-5]. The higher porosity provides the materials with excellent lightweight and high strength and energy absorption capacity to protect objects from impact and explosive loads, and in particular, the mechanical response behaviour of foam metal materials has been extensively studied by researchers[6-8]. The mechanical behaviour of the materials is highly dependent on the strain rate and hence it is important to carry out a comprehensive characterisation of the materials under various loading states. There are three strain rate loading modes for foam metal materials: quasi-static, transitional dynamic, and impact. The dynamic response of material deformation in the quasi-static mode is mainly determined by the equilibrium of forces, and the effect of inertia can be neglected[9]. The results of quasi-static tests are usually used as a basis for evaluating material deformation modes at different strain rates. As the strain rate increases, the effect of inertia begins to dominate and the material shows localised deformation and stress enhancement, which is more pronounced in the shock state. This usually affects the conclusions of the split Hopkinson bar test (SPHB) and leads to improper design of the foam material[10].

In order to overcome these limitations, researchers have improved the direct impact of Hopkinson bar (DIHB) from the conventional SPHB equipment, while using digital image correlation (DIC) methods to refine the measurement of strain, displacement, and velocity changes in the specimen[11-13].Lopatnikov[14] used the Taylor cylinder-Hopkinson method , bonding an aluminium foam specimen and back-end material as a Taylor launching cylinder, impinging on a stationary metal rod, and investigating the unbalanced deformation of the foam material under the action of a rapid impact. Govender et al[15] used a direct impingement Hopkinson pressure rod, fixing the specimen at the end of the output rod, and using the incident rod impingement, which is expected to obtain the stress time course on the two surfaces of the specimen. Li Lang et al[16], Yi-Nan Lv[17], and TAnja Mauko[18] investigated the mechanical properties of density gradient closed-cell and open-cell aluminium foams at high strain rates by using the direct impact Hopkinson rod impact method, where the specimen was directly launched as a projectile to impact a rigid rod, respectively.

It has been found that the mechanical properties of gradient foam aluminium materials under impact loading depend on the relative density and cell morphology[19]. Therefore it is crucial to characterise the material under quasi-static and dynamic impacts. X-ray computed tomography allows in-situ observation of the specimen, scanning reconstruction, and non-destructive acquisition of realistic three-dimensional visualisation images of the internal damage of the material under different operating conditions[20-22].

In the present paper, continuous density gradient foam aluminium materials were fabricated via the melt foaming technique. Subsequently, their pore structure was characterized using X-ray tomography (X-CT), enabling the acquisition of morphological parameters, namely cell wall thickness and porosity. The mechanical attributes of the density gradient foam aluminium materials were examined through quasi-static compression and dynamic impact tests. Concurrently, a full-field strain analysis of the material's deformation process was executed by integrating the digital image correlation (DIC) methodology. In combination with the experimental outcomes, the failure mechanism of the aluminium foam and the precise localization of damage within the structure were analyzed.

2. Experimental component

2.1. Aluminium foam preparation

The functional gradient closed-cell aluminium foam specimens used in this test were provided by the Shanghai Yirong Material Centre. Aluminium foam blocks were used as raw materials and processed into density gradient cylindrical aluminium foam projectiles with a length of 100 mm and a diameter of 80 mm using EDM cutting. The density gradient foam aluminium blocks were prepared by the melt foaming method, which melts industrially pure aluminium in a steel mould, adds 1.5 wt% calcium particles to the melt, stirs, and then adds an appropriate amount of TiH_2 powder to make it dispersed into the aluminium melt, and cools the aluminium melt before it is completely foamed. Relative density is the density of the aluminium foam material divided by the density of the substrate and is usually used to measure the mechanical properties of the material. The overall relative density of the gradient aluminium foam specimens prepared in this study was 0.3.

2.2. X-ray tomography

The mechanical properties of density gradient closed-cell aluminium foam specimens of aluminium foam are mainly affected by a combination of porosity, pore size, and specimen size. In order to accurately reconstruct the pore characteristics, the samples were scanned by X-CT. It was tested with a FF35 X-CT system with a detector pixel resolution of 139 nm, a detection area of 249 mm × 302 mm, an image displaying 16-bit grey scale data, a resolution of 2,146 × 2,146 per X-CT image, a pixel resolution of 3 nm, and a number of slices of 1,080 slices. The voltage of the X-ray tube was 200 V, and the current was 230 A. Figure 1 demonstrates the closed pore X-ray mCT image processing, segmentation, and reconstruction process of the foam specimen. The tested slices were analysed by ImageJ software and vg studio software to determine the distribution pattern of different pore sizes and pore sizes along the thickness direction of the aluminium foam.

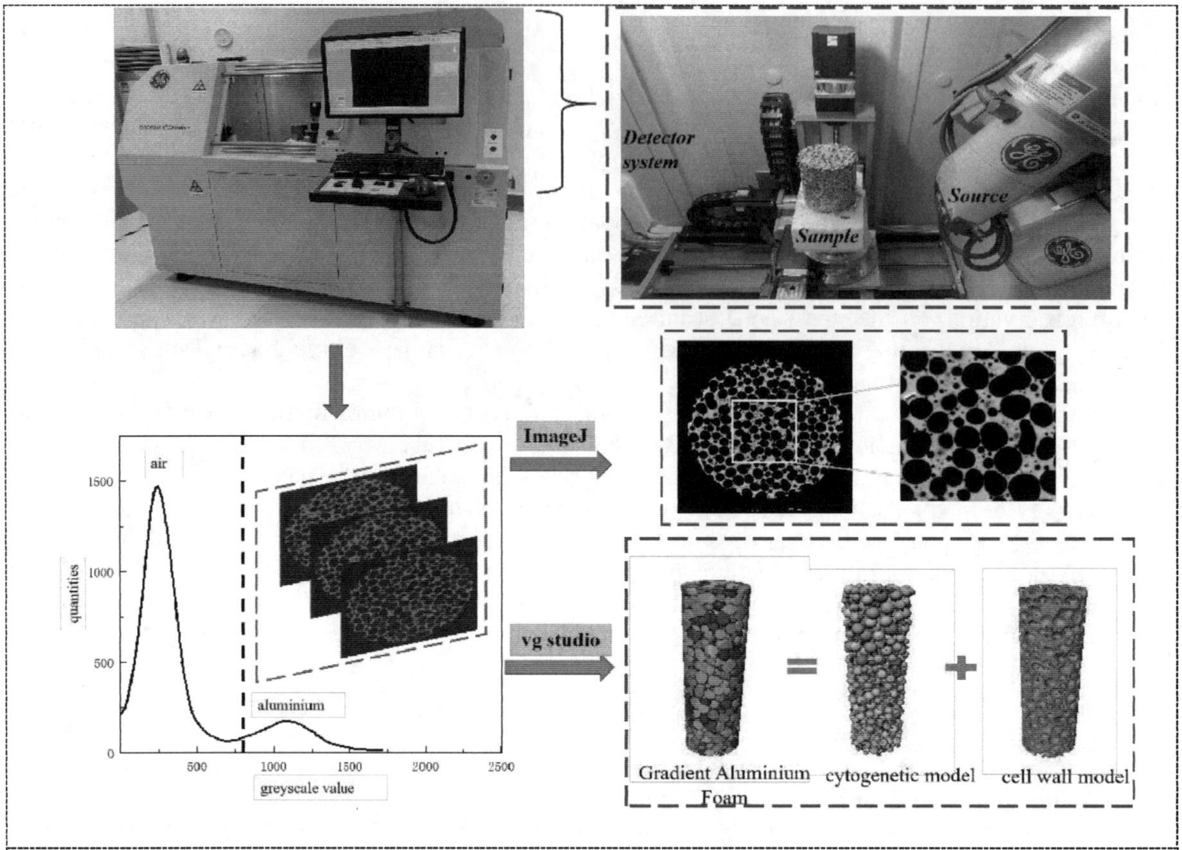

Figure 1. Application of X-CT for geometric reconstruction and damage detection in aluminium foam materials.

2.3. Quasi-static compression

The basic material properties of aluminium foam specimens were investigated by performing quasi-static compression tests on an INSTRON-8803 electronic universal testing machine. The machine is capable of delivering a maximum load of 500 kN with a set loading speed of 5 mm/min during the test.

2.4. Dynamic impact

The high-speed experiments were carried out using the DIHB device, which consists of an incident rod with a length of 5,000 mm and a diameter of 80 mm. The performance parameters of the incident aluminium rod ARCHIMEDES-ALT1000 are listed in Table 1.

Table 1. Parameters of Incident Rod.

E(GPa)	v	ρ(kg/m^3)	c(m/s)	f(MHz)
71	0.33	2850	5000	0.1

A foam specimen is placed as a projectile in a gun barrel, and the pressure of the releasing gas tank accelerates the firing specimen into the incident rod. The experiment was modeled on the principle of Taylor's experiment, where the impact section of the incident rod is treated as a rigid wall. The deformation of the specimen during a collision with the incident rod was recorded using a high-speed video camera, Phontron Fastcam SA-Z, with a resolution of 384 px × 160 px and a frame rate of 210,000 fps. The main objective was to accurately record the change in the impact velocity of the specimen and the propagation of deformation fronts.

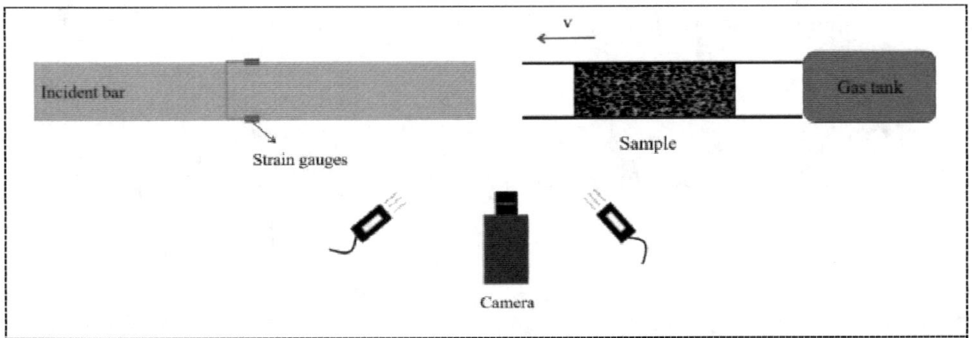

Figure 2. DIHB Test Set Schematic.

Two strain gauges were installed on the opposite side of the center of the incident rod as illustrated in the figure 2, with the test point positioned 2500 mm away from the impact end. The measurements taken by the strain gauges were primarily utilized to acquire the propagation of the deformation wave within the incident rod. The strain gauges were connected to a Wheatstone circuit and the output voltage of the Wheatstone circuit was documented. Then the output voltage values were amplified to make up for the relatively low amplitude points recorded during the experiment. The value of strain in the incident rod can be calculated based on equation (1).

$$\varepsilon_{bar} = \frac{2V_0}{G_f A_g (1+v)(V_E - V_0)} \tag{1}$$

Where, excitation voltage V_E = 10 V, strain gauge factor G_f = 2.09, amplification factor A_g = 200, Poisson's ratio $v = 0.33$.

It is assumed that the strain pulses recorded by the strain gauges and the strain pulses at the interface of the specimen in contact with the rod are equal[23], while the impact stresses on the specimen can be calculated from equation (2):

$$\sigma_s = \left(\frac{A_{bar}}{A_s}\right) E_{bar}\varepsilon_{bar} = \left(\frac{R_{bar}^{\ 2}}{R_s^{\ 2}}\right) E_{bar}\varepsilon_{bar} \tag{2}$$

Where A_{bar} and A_s are the cross-sectional areas of the incident rod and specimen, respectively.

The mechanical properties of the aluminium foam specimens were analysed by converting the results into engineering stresses using the measured strain values. The plateau region is characterised by two significant parameters, the plateau stress (i.e. the average stress value between 0.2 and 0.4 strain) and the densification stress (the magnitude of the value is 1.3 times that of the plateau stress). The densification strain corresponds to the densification stress. The specific energy absorption SEA of the specimen before the densification strain can be calculated by the following formula.

$$\varepsilon_d = 1 - 1.4\left(\frac{\rho'}{\rho_{ce}}\right) \tag{3}$$

$$SEA = \frac{\int_0^x F dx}{m} \tag{4}$$

2.5. Digital image technology

Digital image correlation (DIC) is a type of kinematic measurement technique in which two images of an object's surface, taken before and after deformation, are correlated to achieve a measurement of the object's deformation field. The DIC technique was used to obtain full-field strains for quasi-static and impact experiments at room temperature according to the ASTM C1275-15 test standard. The aluminium foam specimen itself is metal and reflects the light source more strongly, so no further scattering spray was required. Afterwards, the specimens were calibrated and the coordinate system was established by the technique in reference[24]. A subset of about 30 to 35 pixels and a step size of 4 to 8 were used to obtain the full-field deformation, which was analysed by the incremental correlation method in the Vic-2D 6 software.

3. Pore structure analysis

For the CT images of gradient foam aluminium, the porosity is calculated as shown in equation (5).

$$\mathrm{Pr} = \frac{V_{cell}}{V_{total}} \tag{5}$$

Specifically, it is obtained by dividing the number of black pixel dots in the binarised picture (represented by the cellular elements) by the total number of pixels. As shown in Figure 3, the porosity of the aluminium foam specimens prepared according to the gradient design increased from 62% to 78% from the bottom to the top, with an overall linear increase.

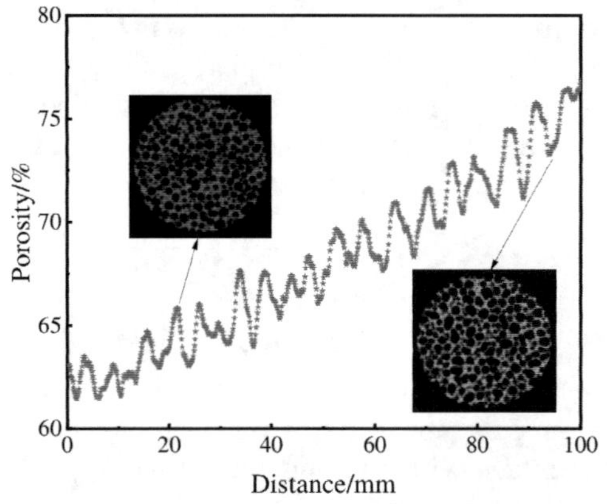

Figure 3. Porosity distribution.

Figure 4 illustrates the distribution of the wall thickness of the cell elements in the aluminium foam sample. The frequency of its distribution exhibits the characteristic of normal distribution, implying that most of the solid thickness in the sample is concentrated in a relatively thin region. Moreover, this feature becomes more significant as the porosity increases.

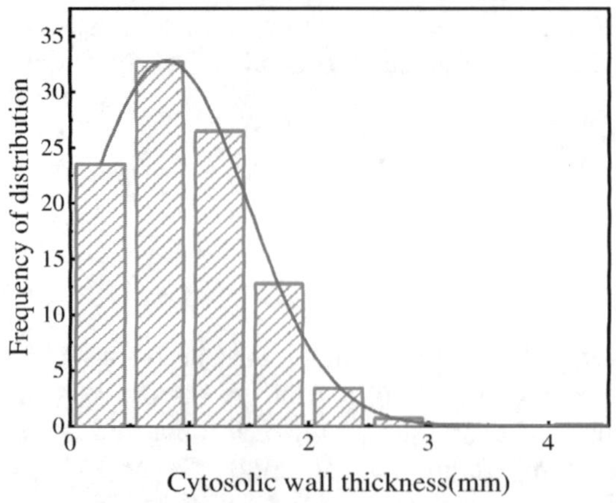

Figure 4. Distribution of cytosolic wall thickness.

4. Experimental results analysis

4.1. Quasi-static compression

The results regarding the stress-strain relationship from the quasi-static tests are depicted in Figure 5, presenting the typical three stress phase patterns: an initial elastic phase where there is a substantial increase in strain yet minimal alteration in stress; a relatively long plateau phase; and a densification phase characterized by a sharp rise in stress. The elastic phase is rather brief, and Young's modulus of the aluminium foam specimen at this stage amounts to 40.97 MPa. The deformation within the elastic phase persists until reaching a peak stress of 13.6 MPa, after which it enters a plateau phase with an average stress of 11.29 MPa. Minor and insignificant stress fluctuations might occur in the plateau region due to the random geometry of the specimen, the unstable deformation of the underlying material, and specimen defects. When the densification strain reaches 0.57, the densification stress attains 8.33 MPa and the specific energy absorption (SEA) value is 7.67 J/g.

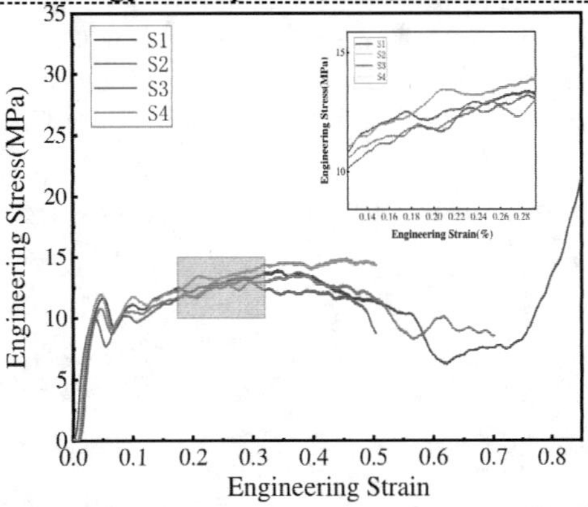

Figure 5. Quasi-static stress-strain curve.

The internal structure of gradient foam aluminium is characterized by anisotropy and this property leads to the manifestation of extremely intricate damage patterns upon damage. It is hypothesized that the disparity between its upper structural yield strength and lower yield strength gives rise to the collapse of pores. In this paper, when processing the collected data, a series of strain-specific points in Figure 6, like peaks and troughs, were selected to conduct a comprehensive analysis of the digital image correlation (DIC) test results. The DIC analysis determines the displacements and strains by contrasting the reference image with the deformation image acquired during the compression test.

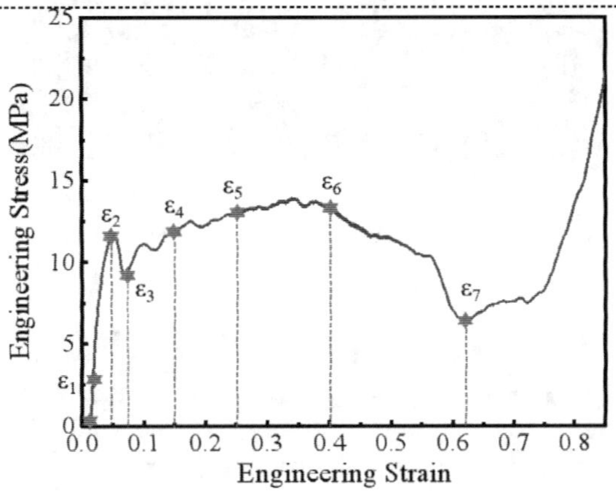

Figure 6. Quasi-static compressive strain nodes.

Journal of Physics: Conference Series **3009** (2025) 012051 doi:10.1088/1742-6596/3009/1/012051

As shown in Figure 7, the cloud diagram of strain field distribution in the vertical direction during the compression of gradient foam aluminium. The vertical strain ε_{yy}, i.e., the strain generated along the axial direction at each point on the surface of the specimen, with positive values indicating tensile strain and negative values indicating compressive strain. The first peak after the elastic phase occurs when the strain is between 0.04 and 0.06, indicating that the aluminium foam starts to yield. The purple and blue parts are the preferentially strained low-density foam aluminium part, which reaches the highest point of the first peak when the strain reaches 0.05, and the foam aluminium has less strain and is in the initial stage of yielding. At this time, the gradient foam aluminium is the overall deformation state, the compression strain is distributed in bands, the high-density part of the deformation is small, and the low-density part is the main deformation region. When the strain variable is 0.07, the stress-strain curve appears as the first trough. The density change before and after compression of the gradient foam aluminium makes the yield value change at each stage. Subsequently, the whole foam aluminium enters densification, and the densification compaction phenomenon is more significant in the upper foam aluminium with lower density and thinner cell walls. After densification, the strain field no longer changes significantly. As compression continues to increase, the broken aluminium foam material stretches outwards, which is the main source of energy absorbed during compaction.

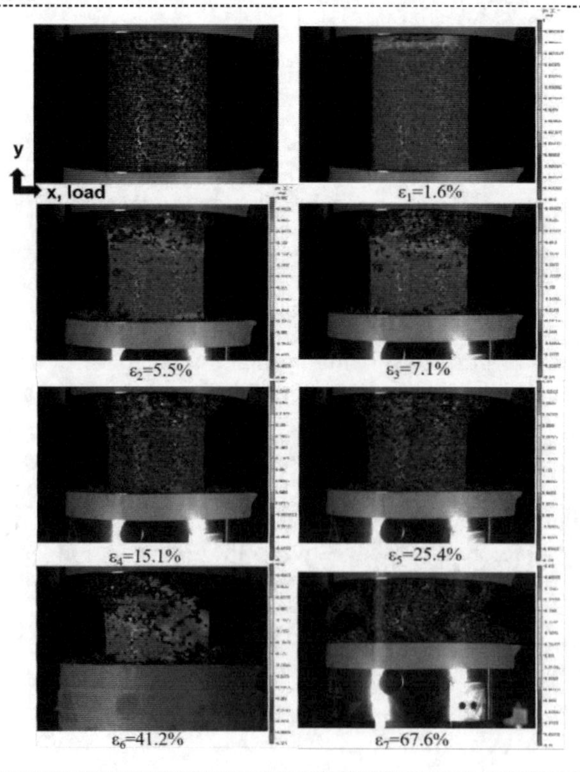

Figure 7. Quasi-static compressive strain cloud measured by DIC.

4.2. Dynamic impact

The real-time deformation of gradient foam aluminium at a high-velocity impact of 88 m/s is shown in Figure 8. The aluminium foam specimen was not fully compacted at the specified impact velocity, indicating that it had a sufficient length to absorb the energy. Plastic deformation occurs at the impact end and propagates forward in the form of a plane wave; the plane propagation is such that the deformation band occurs strictly in the form of a plane, the pore structure in front of the deformation band is not deformed at all, and the pore structure behind the deformation band is completely densely compressed.

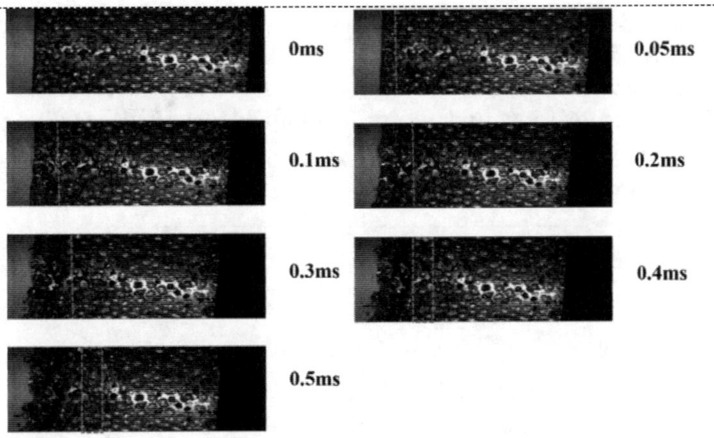

Figure 8. Real-time deformation maps of gradient foam aluminium for high-speed impacts.

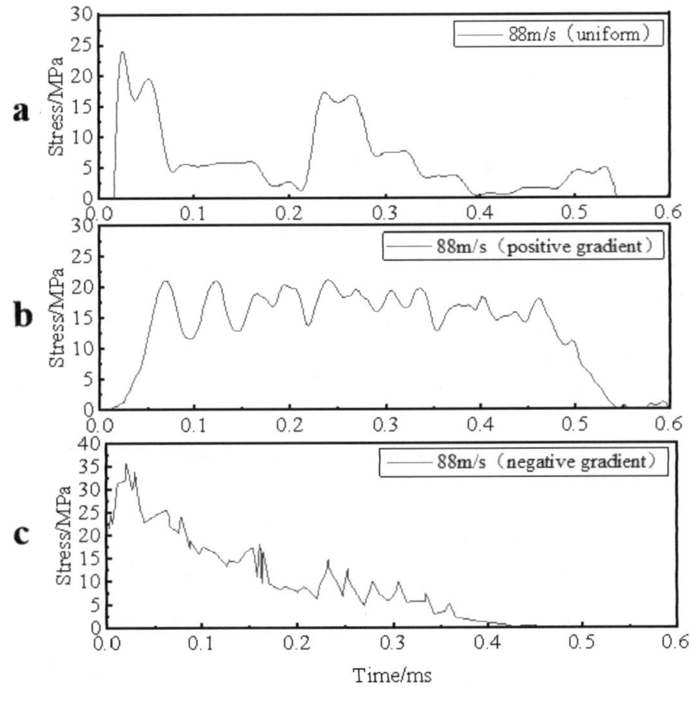

Figure 9. Impact end stress versus time curve for aluminium foam.

A comparison of the pressure time-course curves of homogeneous aluminium foam with approximately the same relative density and positive gradient (high porosity end as the impact end) and negative gradient (low porosity end as the impact end) aluminium foams at the same impact velocity is shown in Figure 9. The pulse shape of the negative gradient aluminium foam exhibits a triangular shape, whereas the pressure time-course curves of the positive gradient and homogeneous aluminium foam materials exhibit a similar rectangular shape. The impact stress of the homogeneous aluminium foam material quickly reaches the highest point, i.e., the peak stress, in a short period of time and then gradually decreases with time. The shape of the stress-strain curve of the positive gradient aluminium foam is changed, the peak stress reaches a constant level and fluctuates up and down, and the energy-absorbing effect is more stable. It can be seen that at the same impact velocity, the utilisation of positive gradient aluminium foam is higher, the impact stress is more constant and the peak stress level is lower than that of uniform aluminium foam and negative gradient aluminium foam materials.

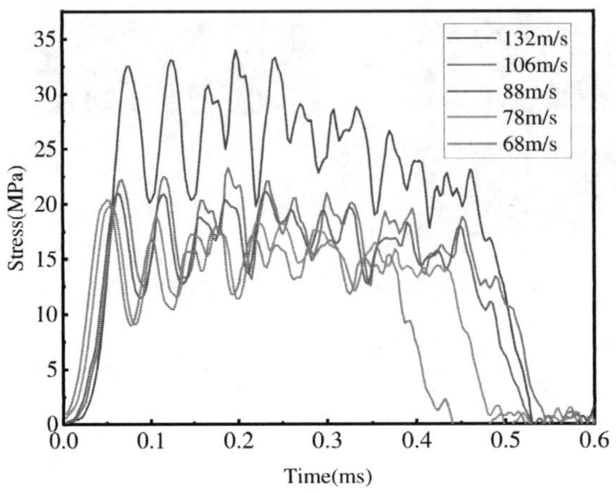

Figure 10. Dynamic impact stress time course curve.

As shown in Figure 10, the impact test result curves of the positive gradient foam aluminium are shown, and the difference in strain rate is controlled by setting different impact speeds and thus the strain rate. When the impact velocity is set to 67 m/s, 77 m/s, and 87 m/s, the peak stress of the impact is more stable, with less fluctuation, which is approximate to a constant, and the stable stress platform at the impact end of the specimen under the respective strain rate can be obtained by taking the average value of the data within 0.1~0.48 ms, which is approximately 15.1 MPa, 16.8 MPa and 17.5 MPa. With the increase of the impact velocity, the decrease of the stress level of the specimen gradually slows down, which is related to the pore structure of the deformation zone.

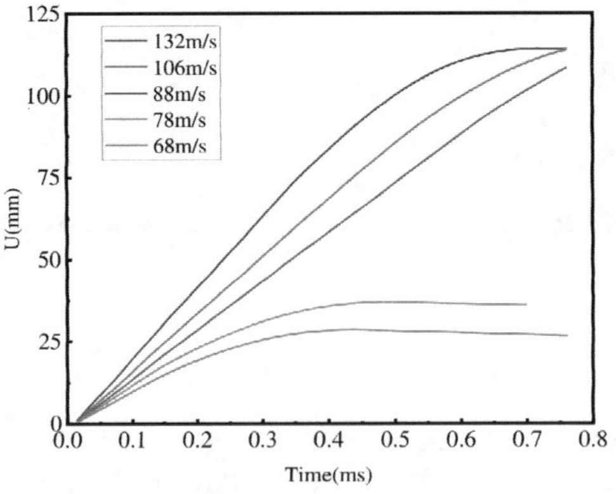

Figure 11. Compression time curve.

With the increase of the impact velocity, the decrease of the stress level of the specimen gradually slows down, which is related to the change of the pore structure of the deformation zone, and the increase of the density due to the compression of the cytosolic elements at the late stage of the impact, which makes the initial density and the density gradient distribution of the specimen change. When the impact velocity distribution is increased to 106 m/s, 132 m/s, 144 m/s, the gas in the cell may also increase the strength under high strain rate loading, and the peak stress of the impact increases significantly, and the stress decreases at a faster rate, which may be caused by the densification strain of the cell elements at the impact end.

The time course curves of the compression volume of the positive gradient foam aluminium at different impact velocities are shown in Figure 11. With the increase of the impact speed, the compression volume of the material shows an overall rising trend and the corresponding compression platform appears; when the speed of impact is 88 m/s, the compression volume change platform appears to be delayed. Impact speed of 132 m/s and 106 m/s specimens, in the time of 0.65 ms, the compression amount of change tends to be consistent, at this time, the density gradient foam aluminium in the high-speed impact environment, close to the impact end of the side of the low-density foam and away from the impact end of the free end of the high-density foam at the same time deformation, the overall synergistic deformation of the better, so that the compression change tends to stabilise.

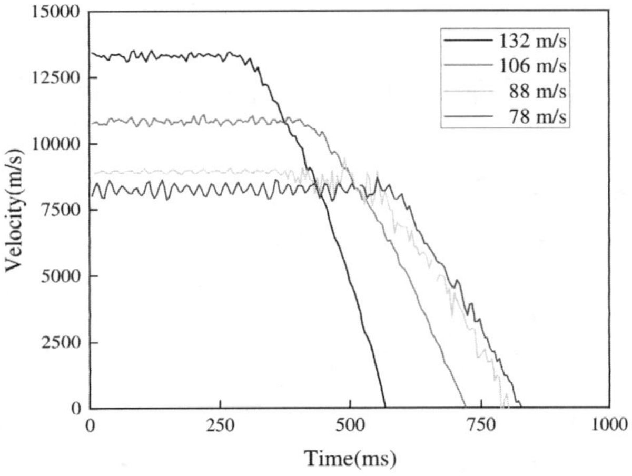

Figure 12. Velocity distribution of the specimen after initial impact.

The DIC method was used to determine the velocity distribution of the specimen after the initial impact. As shown in Figure 12, the continuous deformation pulse indicates that the specimen has a constant impact response. A constant velocity plateau can be observed throughout the experiment up to the pre-densification stage, which is attributed to the complete compression of the specimen and the inertial influence that the incident rod possesses causing the specimen velocity to drop sharply. A more pronounced isokinetic plateau is observed at 78, 88, and 106 m/s. At 132 m/s the plateau phase is of short duration due to the high impact velocity and high strain rate during the initial impact, making it less visible in the graph. The gradient foam aluminium specimens were not fully compacted at the specified impact velocity, indicating that there was excess porous structure to collapse and absorb the energy.

The horizontal and tangential deformation evolution of the gradient foam aluminium material at an impact velocity of 132 m/s is shown in Figure 13 and Figure 14. A large compressive strain is generated at the impact end at 0.1 ms and the specimen starts to compress. At 0.2 ms, the free end of the rear end produces a small portion of tensile strain due to the reflected tensile wave effect, and the strain at the free end decreases as the compression of the specimen gradually increases. Under dynamic compression conditions, the deformation of the gradient foam aluminium exhibits significant directional characteristics. The low-density layer is prone to collapse damage in the process of resisting the shock wave due to its more fragile structure. The reason for this phenomenon is that during the dynamic compression process, the weakest region in the structure will be the first to experience stress concentration, which will then reach plastic yield and deform, thus acting as a buffer and energy absorption.

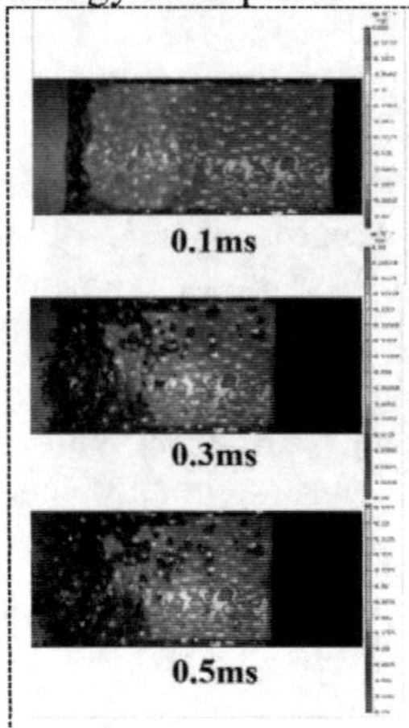

Figure 13. Horizontal strain field distribution for 132 m/s impact of gradient aluminium foam.

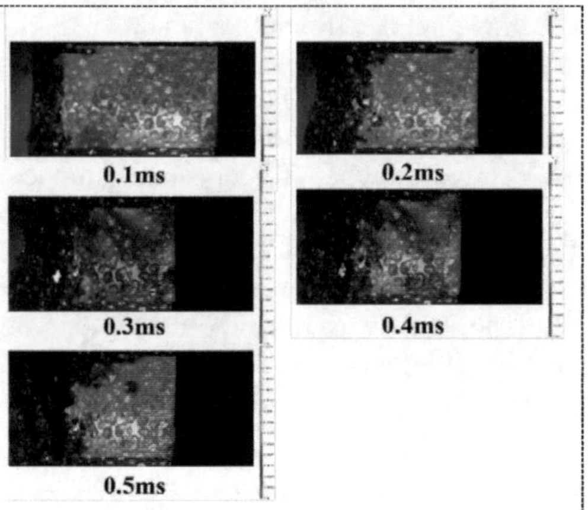

Figure 14. Tangential strain field distribution for 132 m/s impact of gradient aluminium foam.

It can be deduced from this that density gradient foam aluminium can effectively protect the internal load when subjected to external impact forces. As the loading proceeds, each layer of material at the impact end is gradually compacted and the impact stress is transferred to the next layer with the wavefront, and the subsequent material collapses layer by layer. As the impact progresses, the tangential strain gradually produces a pulling effect on the free end, and there is no obvious wavefront generated.

5. Conclusions

The author selected density gradient foam aluminium as the research subject and acquired its solid material thickness as well as porosity distribution through X-ray tomography. Subsequently, quasi-static compression tests and dynamic impact tests under high strain rates were performed on it. With the assistance of digital image technology, the full-field strain of the material deformation process was obtained in a non-destructive manner. The conclusions are as follows:

(1) X-ray tomography facilitated the quantitative characterization of the material's porosity arrangement within the range of 62% - 78%. Morphological parameters such as solid material thickness and porosity were obtained, and the frequency distribution of the solid material thickness of the specimens exhibited a normal distribution trend.

(2) Under quasi-static loading conditions, the compressive deformation of the gradient foam aluminium was initially initiated in the low-density region and then propagated towards the high-density region. When the pore structure in the low-density region began to deform, no significant deformation of the pore structure in the high-density region was detected.

(3) The behavior of gradient aluminium foam under high strain rate loading differed from that of homogeneous aluminium foam, and the damage patterns also varied. As the impact velocity increased, the compression of the aluminium foam demonstrated an overall upward trend along with a corresponding compression plateau. For the specimens with impact velocities of 132 m/s and 106 m/s, the compression changes seemed to converge at 0.65 ms. At this moment, the low-density foam near the impact end and the high-density foam at the free end away from the impact end deformed

simultaneously, and the overall synergistic deformation was more pronounced, resulting in a stabilization phenomenon in the compression changes.

In the future, a 3D model will be reconstructed based on the X-ray tomography method, and the impact of ultra-high strain rates on the deformation patterns of gradient foam aluminium will be further investigated through numerical simulations.

References

[1] Ji C, Huang H, Wang T and Huang Q (2023) Recent advances and future trends in processing methods and characterization technologies of aluminum foam composite structures: A review. *Journal of Manufacturing Processes*, **93**: 116-152.
[2] Wang E, Yao R, Li Q, Hu X, Sun G (2023) Lightweight metallic cellular materials: A systematic review on mechanical characteristics and engineering applications. *International Journal of Mechanical Sciences*, **270**: 108795.
[3] Duan Y, Zhao X, Liu Z, Hou N, Liu H, Du Bing, Hou B and Li Y (2020) Dynamic response of additively manufactured graded foams. *Composites Part B: Engineering*, **183**: 107630.
[4] Chen J, Zhang P, Cheng Y and Liu J (2020) On the crushing response of the functionally graded metallic foams based on 3D Voronoi model. *Thin-Walled Structures*, **157**: 107085.
[5] Gao C, Iwamoto T and Kusaka T (2024) A DIC-based Taylor impact test by measuring inertia force from acceleration distribution to obtain uniaxial stress–strain behavior of pure aluminum. *International Journal of Impact Engineering*, **184**: 104834.
[6] Shi D, Wang Z, Marrow J, Liu C, Wan and Yang X (2024) Piecewise damage model for SiC/SiC composites with multilevel experimental validation. *Composites Part A: Applied Science and Manufacturing*, **178**: 107952.
[7] Gupta N (2007) A functionally graded syntactic foam material for high energy absorption under compression. *Materials Letters*, **61(4-5)**: 979-982.
[8] MAUKO A, SARIKAYA M, GÜDEN M, DUARTE I, Borovinšek M, Vesenjak M and Ren Z (2023) High strain-rate deformation analysis of open-cell aluminium foam. *Journal of Materials Research and Technology*, **25**: 1208-1221.
[9] Wang Y, Chen F, Wang X, Liu Y, Sun J, Gu Y and Yu Y (2023) Micro-CT in the mechanical properties and energy absorption of closed-cell aluminium foam.*Materials Today Communications*, **37**: 106962.
[10] Hu S, Wang L, Song L and Zhang L (2014) A review of Hopkinson press bar technology in China. *Explosions and impacts*, **34(6)**: 641-657.
[11] Krstulović-Opara L,Surjak M, Vesenjak M, Tonković Z, Kodvanj J and Domazet Ž (2015) Comparison of infrared and 3D digital image correlation techniques applied for mechanical testing of materials. *Infrared Physics and Technology*, **73**: 166-174.
[12] Deng F, Liu Y, Fan J, Wei S and Nie J (2021) Bending behaviour of aluminium foam composite structures based on digital image correlation techniques. *Rare metal*, **45(03)**: 297-305.
[13] Deng Y, Hu X, Yang X, Yu H and Zheng Y (2024) Quasi-static compressive response and energy absorption mechanism of all-CFRP honeycomb core using digital image correlation (DIC). *Thin-Walled Structures*, **198**: 111641.
[14] Lopatnikov S L, Gama B A and Jahirul Haque Md (2003) Dynamics of metal foam deformation during Taylor cylinder–Hopkinson bar impact experiment. *Composite Structures*, **61(1-2)**: 61-71.
[15] Govender R A, Curry R J. (2016) The "Open" Hopkinson Pressure Bar: Towards Addressing Force Equilibrium in Specimens with Non-uniform Deformation. J*ournal of Dynamic Behavior of Materials*, **2(1)**: 43-49.
[16] Li L, Han B and He S (2019) Shock loading simulation using density-graded metallic foam projectiles. *Materials & Design*, **164**: 107546.
[17] Liu H, Ng B F. (2022) Dynamic response of density-graded foam subjected to soft impact. *Composite Structures*, **284**: 115145.
[18] Mauko A, Sarikaya M and Güden M (2023) High strain-rate deformation analysis of open-cell aluminium foam. *Journal of Materials Research and Technology*, **25**: 1208-1221.
[19] Qiao P, Wei J and Pan J (2021) Aluminium foam porosity calculation based on multi-energy spectral CT with NC-POCS reconstruction algorithm. *CT Theory and Application Research*, **30(01)**:71-80.
[20] Dai K, Wu D, Zhang Z and Chen P (2021) Experimental and three-dimensional finite element simulations of the compressive mechanical properties of closed-cell aluminium foam. *Journal of Beijing Institute of Technology*, **41(06)**:579-587.
[21] Lang Y, Liang Z and Dong Z (2024) Three-dimensional finite element simulation and reconstruction of jointed rock models using CT scanning and photogrammetry. *Journal of Rock Mechanics and Geotechnical Engineering*, **16(4)**: 1348-1361.
[22] Wang A, Yu X, Wang H, Li Y, Zhang J and Fan X (2022) Dynamic Response of Sandwich Tubes with Continuously Density-Graded Aluminum Foam Cores under Internal Explosion Load. *Materials*, **15(19)**: 6966.
[23] Liu K, Kang S and Gao S (2023) Experimental and analytical study on impact response of stainless steel-aluminium foam-alloy steel sandwich panels. *International Journal of Impact Engineering,* **179**: 104661.
[24] Linul E, Galatanu S, Marsavina L and Kováčik J (2024) Crushing behavior of closed-cell metallic foams: Anisotropy and temperature effects. *Journal of Materials Research and Technology*, **30**: 1436-1449.

Journal of Physics: Conference Series **3009** (2025) 012052 doi:10.1088/1742-6596/3009/1/012052

Molecular dynamics simulation of gold sintering with different nano-sizes

Xia Cai*

Shanghai Technical Institute of Electronics & Information, Shanghai, China

*Corresponding author's e-mail: 458603447@qq.com

Abstract: In this paper, the sintering process of Gold nanoparticles was simulated by using molecular dynamics simulation to observe the changes in microstructure during the sintering process. The fitted relationship between the sintering neck growth width and angle and the sintering time was obtained. The changes in the crystalline structure of the sintered system model of nanoparticles with different sizes at different times as well as different shapes at the time of sintering completion were observed. The results show that as the size of the nanoparticles becomes smaller and smaller, the more it tends to be a sphere after the sintering is completed, the more the sintering angle tends to be 180° after the sintering is completed. The Fcc structure of the Gold particles is transformed to an amorphous structure during the sintering process. The structural transformation of nanoparticles during sintering is a complex process, which is affected by a variety of factors such as particle size, sintering temperature, sintering time and inter-particle interactions. The nanosized effect gives Gold an effect of low-temperature sintering.

1 Introduction

Electronic pastes are an indispensable building block in many high-tech fields, from high-end aerospace technology to everyday consumer electronics[1-2], and Gold pastes are one of the most common electronic pastes due to their outstanding properties. Solid-phase sintering is a crucial part of the process of converting electronic pastes into circuits or solder joints[3-4]. It refers to a holding process at a specific temperature below the melting point of the metal, which induces the formation of a sintered neck[5-6] between the metal particles and gradual densification as the process advances. Known for its excellent mechanical properties, biocompatibility and ease of processing, Gold is a high-performance material that exhibits high electrical and thermal conductivity. However, its extremely high melting point makes it difficult to use Gold as a joining material in practical applications[7].

Gold nano paste is a connecting material prepared by mixing Gold nanoparticles with a particle size between 10 and 100 nm with a variety of organic substances[8]. Due to its high specific surface area, Gold nanopaste exhibits a series of unique properties such as surface effect, volume effect, quantum size effect, macroscopic quantum tunneling effect, and dielectric confined domain effect. This material achieves the special process requirement of low-temperature sintering[9] but high-temperature service. Nanopaste interconnect materials are receiving increasing attention in the areas of low-temperature interconnect technology, multilevel interconnect structures and high-power chip connectivity[10].

2 Simulation

2.1 Simulation Methods

Due to the tiny size and complex changes in the microstructure during the sintering process, it is difficult to capture these changes by conventional experimental means. In order to deeply investigate the motion of nanoscale atoms and their influence on material properties, we adopt a molecular dynamics simulation method. Given that the sintering process of nano paste is essentially a high-temperature-driven atomic diffusion behaviour, the sintering process can be effectively simulated by accurately simulating the motion trajectories of atoms. The properties of nano-Au pastes at different sintering temperatures, particle sizes, and sintering durations were compared to simulate the changes in crystalline structure during the sintering process of the nanoparticles, the shapes of the sintered particles after the completion of sintering, and the change in sintering necks of the microscopically complex structures after sintering and the change in sintering temperatures were observed using Ovito software, with a view to reveal whether the nano-size significantly affects the macroscopic properties of the materials.

2.2 Selection of Potential Function

Molecular dynamics simulation of micro- and mesoscale atomic motions is often difficult to verify the correctness of the simulation by experimental means, and the difference in the simulation results mainly comes from the selection of the potential function. Before the simulation starts, the correctness of the potential function is firstly verified, according to the sintering law of the main nanoparticles and the thermal conductivity problem of the microstructure, and in the verification process, the main focus is on the ability of molecular dynamics to better simulate the size effect of the material in terms of the thermal conductivity, and verify it. In order to describe the interactions between atoms in the Gold nanospheres, the EAM potential function was chosen. This potential function can reflect the complex interactions between metal atoms more accurately and is suitable for the simulation needs of this study.

2.3 Simulation Process

To gain a deeper understanding of the interconnection mechanism of nanoparticles during sintering, simulations of the melting point and melting process were carried out on Gold nanospheres with different radii. These Gold spheres all exhibit a complete Fcc structure with a lattice constant of 4.08 Å and a relative atomic mass of Gold of 196.96656. In the simulation setup, a very small-time step, i.e., 0.001 ps, has been used to ensure the accuracy and stability of the simulation. Periodic boundary conditions were applied to simulate an infinite crystal system to reduce the influence of boundary effects on the simulation results. Also, the simulations were carried out under the NPT system, which allowed us to observe the behaviour of the system at constant pressure and temperature.

In order to investigate the melting point of Gold nanorods, the sintering temperature range in this study was set between 300 K and 1300 K, which is a wide interval covering the range of possible melting temperatures of Gold nanorods. During the simulation process, a sintering step number of 100,000 was set to ensure that the system had enough time to reach equilibrium and exhibit its melting behaviour.

To investigate the melting point of Gold nanospheres, the sintering temperature range in this study was set between 300 K and 1300 K, which is a wide interval covering the range of possible melting temperatures of Gold nanospheres.

We choose four sets of simulation data:

The first set models two Gold nanoparticles with a radius of 7.0 nm. In order to minimise the potential influence of the surrounding virtual particles in the periodic boundary conditions on the actual simulation results, a distance buffer of 2.9 nm is deliberately set between these two Gold nanoparticles and the boundary of the simulation environment, and in the crystallographic coordinate system, the left sphere: the x-axis is oriented in the (71)-crystalline direction; the y-axis is oriented in the (0) crystal direction, while the z-axis is oriented to the (0) crystal direction; right ball: x-axis is oriented to the (-71) crystal direction; the y-axis is oriented to the (0) crystal direction, while the z-axis is oriented to the (0) crystal direction. Such a design aims to ensure the accuracy and reliability of the simulation results and avoid the interference of boundary effects on the simulation of the melting process of Gold nanoparticles. Three other sets of Gold nanoparticles with radii of 3.5 nm, 2.4 nm, and 1.4 nm were constructed sequentially, and these systems all contained two spherical powder particles of the same size.

At an initial temperature of 300 K, we initiated the simulation process of Gold nanoparticles and allowed them to reach equilibrium after 100,000 time steps of evolution, a process that ensured that nanoparticles of different sizes were in a stable configuration before the start of the simulation. Subsequently, the temperature of the system is increased at a constant rate of 1 K/ps, and the energy state and atomic configuration of the system are recorded at 1 ps intervals until the temperature reaches 1300 K. Through this series of recordings and analyses, we succeeded in obtaining the sintering neck data and the crystalline shape change characteristics of the system. In the simulation flow, we use colour coding to visualise these structures: the Fcc structure is given in blue; the Hcp structure is marked in red; the Bcc structure is given in green; and the amorphous structure is shown in yellow.

In summary, we have systematically investigated the melting point and melting process of Gold nanospheres with different radii in the temperature range from 300 K to 1300 K by simulating them with EAM potential functions, periodic boundary conditions and NPT coefficients.

3 Results and Discussion

3.1 *Sintered Finished Shape*

The two modelled Gold nanoparticles are sintered for 100 ps at 300~1300 K. As shown in Figure 1(a), the initial distance between the two 7 nm Gold nanoparticles is 2.9 nm, and the atoms of the two particles diffuse with each other as sintering proceeds, and the morphology of the nanoparticles gradually transforms from bispherical to ellipsoidal rods, and the width of sintered necks is 9 nm, which is slightly larger than the original radius of 7 nm. As shown in Figure 1(b), the initial distance between two 3.5 nm Gold nanoparticles is 6.4 nm; the morphology of the particles gradually changes from bispherical to ellipsoidal rods; the width of the sintered neck is 6 nm, which is close to twice of the original radius of 3.5 nm. As shown in Figure 1(c), the initial distance between two 2.4 nm Gold nanoparticles is 7.5 nm; the particle morphology is gradually transformed from bispherical to spherical; the width of the sintered neck is 5.5 nm, which is greater than twice of the original radius of 2.4 nm. As shown in Figure 1(d), the initial distance between two 1.4 nm Gold nanoparticles is 8.5 nm; the particle morphology gradually changes from bispherical to spherical, with a sintered neck width of 3.6 nm, which is two times larger than the original radius of 2.4 nm. There is a strong attraction between two nanoparticles so that nanoparticles of various particle sizes form sintering necks of a certain width when they are in contact. As can be seen from Figure 1, as the nanoparticle size becomes smaller, the more it converges to a sphere after sintering is completed, where the sintered system of 2.4 nm and 1.4 nm is already a sphere. This is due to the fact that small-sized particles have higher surface energy, which makes them more inclined to form a spherical structure with the lowest energy during the sintering process.

3.2 *Changes in Sintered Necks*

Figure 2 shows the variation of the sintering neck width with increasing temperature for the four sintered systems. When two nanoparticles close to each other are sintered at the beginning of sintering, a sintering neck is formed rapidly. For the Gold particles with a line of 7 nm, the sintering neck rapidly rises to about 2.5 nm at the beginning of sintering. The sintering neck gradually rises to about 5 nm when the temperature exceeds 400 K and then decreases. When the temperature is further increased to about 900 K, the sintering neck increases dramatically, and the atoms inside begin to sinter, which is due to the fact that the temperature has reached the melting point of the nanoparticles. This is due to the fact that the melting point of the nanoparticles has been reached at this temperature. For the Gold particles with a line of b of 3.5 nm, the sintering neck increases rapidly to about 2 nm at the beginning of sintering and then increases slowly. When the temperature exceeds 500 K, the sintering neck increases gradually, which indicates that this temperature is the starting sintering temperature of the nanoparticles. When the temperature is further increased to about 950 K, the sintering neck increases drastically, which indicates that the melting point of the nanoparticles has been reached at this temperature. The Gold particles with line c of 2.4 nm, which have a sintering neck of 2.4 nm, have a

Journal of Physics: Conference Series **3009** (2025) 012052 doi:10.1088/1742-6596/3009/1/012052

sintering neck of 2.4 nm and then decrease. For Gold particles with 2.4 nm in line c, the sintering neck rises rapidly to about 1.4 nm at the beginning of sintering. The two spheres slowly come together without sintering the atoms inside them, and then the sintering neck stays basically unchanged. When the temperature is raised to near 950 K, the width of the sintering neck reaches the size of the diameter of the sphere particles. For Gold particles with 1.4 nm in line d, the sintering neck rises rapidly to about 1 nm at the beginning of sintering (more than 1/2 the radius of the sphere) and then rises steadily. After that, it rises steadily. When the temperature is further increased to around 850 K, the sintered width reaches the size of the diameter of the spherical particles. This is due to the nano-size effect that gives Gold its low-temperature sintering effect.

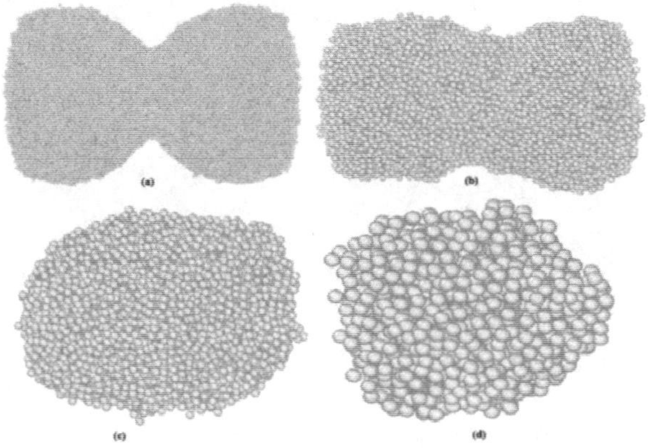

Figure 1. Sintering completion diagrams for four sintering regimes: (a)7 nm, (b)3.5 nm, (c)2.4 nm, (d)1.4 nm.

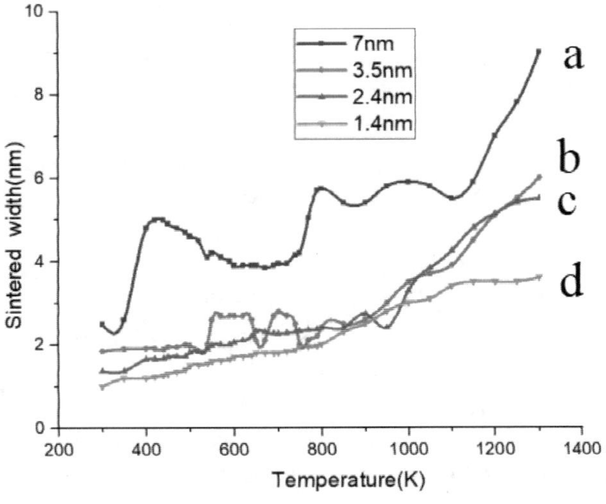

Figure 2. Plots of sintering neck variation with increasing temperature for four sintering Regimes: (a) 7 nm (b) 3.5 nm (c)2.4 nm(d)1.4 nm.

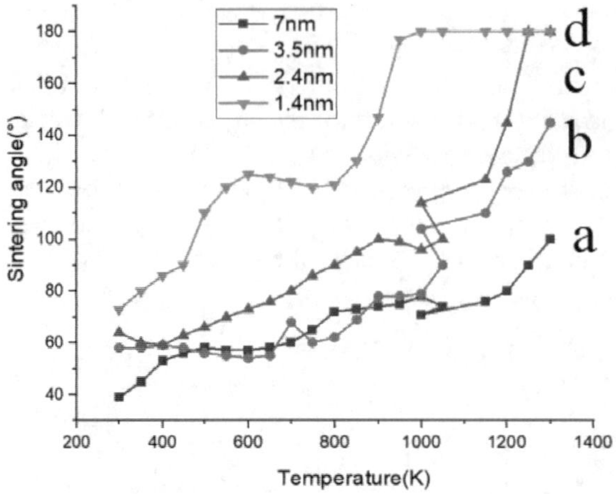

Figure 3. Plot of sintering angle variation with increasing temperature for four sintering regimes :(a)7 nm, (b)3.5 nm (c)2.4 nm(d)1.4 nm.

Figure 3 shows the variation of sintering neck angle with increasing temperature for the four sintered system sizes. Line a shows 7 nm Au particles with an initial sintering angle of 39° .Line b shows 3.5 nm Au particles with an initial sintering angle of 58° . As the sintering temperature increases, the angle increases steadily up to 100° .Line b shows 3.5 nm Au particles with a 58° angle at the beginning of sintering. As the sintering temperature increases, the angle rises steadily, and then rapidly at 550 K (which corresponds to the width of the sintering neck) to a maximum of 145° . Line c is a 2.4 nm Au particle with an initial angle of 64° . Line d is a 2.4 nm Au particle with an initial angle of 64° . As the sintering temperature increases, the angle rises steadily up to 180° . Line d is a line of 1.4 nm Au particles with an initial sintering angle of 73° . As the sintering temperature increases, the angle increases steadily, and the sintering angle reaches 180° around 900 K and remains at 180° thereafter.The sintering angle tends to be closer to 180° as the particle size decreases, which further demonstrates the high degree of fusion of small-sized particles during the sintering process.

3.3 Changes in Lattice Structure

The lattice structure of the nanoparticles was further analysed and observed during the sintering process, and the post-processed files were used for lattice identification in Ovito software, as shown in Figure 4 for the sintering process of 3.5 nm Gold particles at different times. According to the interatomic coordination number in each frame, the lattice structure is determined. According to the comparison analysis of the simulation results, the lamellar red particles identified by the simulation software are not the Hcp structure but are the grain boundaries formed on both sides after the formation of the sintering neck, which is mistaken as Hcp structure by the software due to the unique coordination structure. The yellow colour is the uncalibrated lattice structure, which exists in two cases during the simulation, one is the atoms that are detached from the lattice sites due to diffusion during the warming process, while the other is the atoms that are on the surface of the particles at the beginning of the simulation because they do not satisfy the coordination number of face-centered cubic, and are judged as an uncalibrated lattice structure. The atoms at the surface of the particles undergo sufficient diffusion and all move to positions farther away from the initial sites, while the atoms in the interior are displaced a small distance, and the atoms filling the gap between the two particles mainly come from the surface atoms of the particles on both sides. The main mode of atomic diffusion during the formation of the sintered neck structure is surface diffusion.

The sintering angle tends to be closer to 180° as the particle size decreases, which further demonstrates the high degree of fusion of small-sized particles during the sintering process.

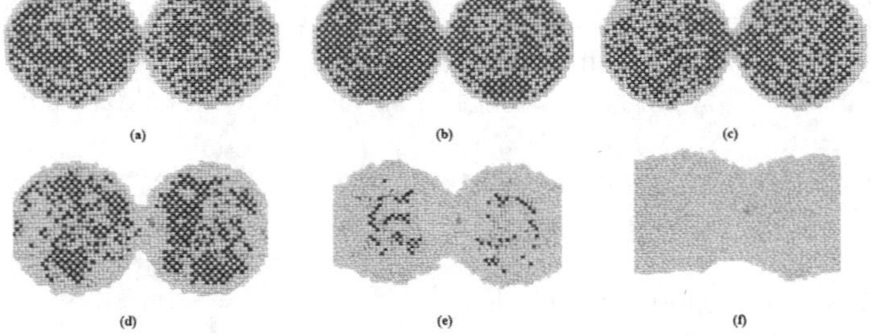

Figure 4. Sintering process of 3.5 nm Gold particles at different times: (a) 0 ps (b) 20 ps (c) 40 ps (d) 60 ps (e) 80 ps (f) 100 ps.

3.4 Laws of Crystallographic Changes During Sintering of Metal Nanoparticles

The migration of atoms or the melting of metal during the sintering process will inevitably lead to a change in the atomic arrangement in the sintered system, which in turn will lead to a change in the crystalline structure. In this paper, the transformation of the crystal structure of Gold nanoparticles of different sizes during the sintering process is investigated. Figure 5 shows the crystal structure changes during the sintering process of 7 nm, 3.5 nm, 2.4 nm and 1.4 nm Gold particles, where blue, red and yellow colour represent the Fcc, Hcp and undefined structures, respectively. As can be seen from Figure 5, with the increase of temperature, the atomic thermal motion becomes more and more intense. The Fcc structure in Gold nanoparticles decreases, and the yellow undefined structure increases, while the red Hcp structure appears in all sintered systems.

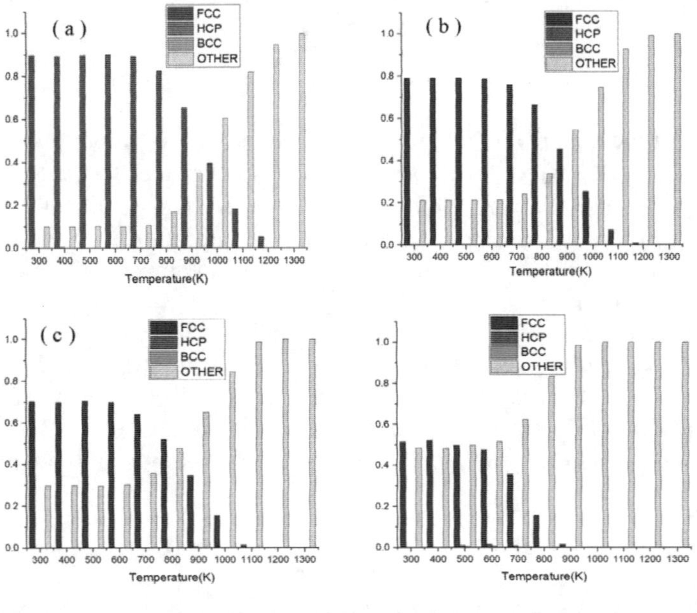

Figure 5. Changes in the crystalline structure of Gold nanoparticles during the sintering process:(a)7 nm, (b)3.5 nm (c)2.4 nm(d)1.4 nm.

From Figure 5(a), it can be seen that 7 nm Gold particles form Hcp structures at 300 K~500 K and 800 K, respectively; 3.5 nm Gold particles form Hcp structures only at 900 K; 2.4 nm Gold particles form Hcp structures at 400 K and 800 K~1000 K, respectively; 1.4 nm Gold particles form

Hcp structures at 500 K~700 K, respectively. Hcp structure is formed at 500 K~700 K for 1.4 nm Gold particles. This may be due to two reasons: for 1.4 nm and 2.4 nm nanoparticles to form Hcp structures, due to the very small number of atoms, the role of dislocations cannot be ignored at this time, and one possible mechanism is that the Hcp structure is formed by the slip and rearrangement of the grain surface during the sintering process; the other more likely mechanism is that the Hcp structure is formed by the dislocations in the grains or by a large number of small slips of the atomic surface at the grain boundaries. When the size of nanoparticles is 3.5 nm, the process of atom migration is mainly controlled by the combination of grain boundary (or dislocation) slip, surface diffusion, and viscous flow. At this stage, the effects of grain boundaries and surfaces are particularly significant due to the small size of the particles, and they provide important channels for the migration of atoms. Grain boundary slip and dislocation movement allow for rapid expansion of the contact area between particles, while surface diffusion facilitates the rearrangement of atoms on the particle surface, and viscous flow further promotes inter-particle fusion.

As the particle size increases to about 7 nm, the main mechanisms of atomic migration shift to grain boundary diffusion, surface diffusion and viscous flow. Compared to 3.5 nm particles, the contribution of grain boundary diffusion becomes more prominent, although surface diffusion and viscous flow still play an important role. This is due to the fact that as the particle size increases, the proportion of grain boundary regions increases relatively, providing more paths for the migration of atoms.

When the particle size increases further, then the mechanism of atom migration relies mainly on grain boundary diffusion and surface diffusion. At this stage, due to the larger particle size, the influence of viscous flow is gradually weakened, while grain boundary diffusion and surface diffusion become the main forces dominating atomic migration. These two mechanisms work together to allow the sintering neck to grow steadily until the entire sintering process is completed.

Nanoparticles of different sizes have different atom migration and sintering neck growth mechanisms during the sintering process. Together, these mechanisms determine the rate of the sintering process and the morphology of the final product.

4 Conclusions

The smaller the size of the nanoparticles, the more the shape of the sintered finish tends to be close to a sphere. The sintering angle tends to be close to 180° as the size of the particles decreases, which is due to the nano-size effect that allows Gold to have the effect of low-temperature sintering.

Nanoparticles of different sizes have different atomic migration mechanisms and sintering neck growth mechanisms during the sintering process. These mechanisms together determine the rate of the sintering process and the morphology of the final product.

Overall, the structural transformation of nanoparticles during sintering is a complex process, which is affected by a variety of factors such as particle size, sintering temperature, sintering time and inter-particle interactions. For nanoparticles of different sizes, the mechanism of structural transformation may be different but usually involves processes of atomic migration and rearrangement. These processes ultimately determine the structure and properties of the sintered products.

Author Resume:

Xia Cai, Lecturer, M.S., Integrated Circuits and Electronic Products, 458603447@qq.com.

References

[1] Ketkar S A, Umarji G G, et al. Lead-free photo imageable silver conductor paste formulation for high-density electronic packaging. Materials Science and Engineering: B, 2006, 132(1/2): 215.

[2] Lu G G, Xuan T P, et al. Development tendency and research progress of the electronic paste. Met Funct Mater, 2008, 15(1): 48.
[3] Liao Q W, Hou W, et al. Solid-phase sintering and vapor-liquid-solid growth of BP@ MgO quantum dot crystals with a high piezoelectric response. Journal of Advanced Ceramic, 2022, 11:1725–1734.
[4] Rosenblad L, Staf H, et al. Parametric dependency of a constitutive model describing solid state sintering of cemented carbides. Powder Technology.2022, 403:117407
[5] Liu ZJ, Cheng Q, et al. Sintering neck growth mechanism of Fe nanoparticles: A molecular dynamics simulation. Chemical Engineering Science, 2020, 218: 115583.
[6] D. Demirskyi, D. Agrawal, et al. Neck growth kinetics during microwave sintering of copper. Scripta Materialia, 2010, 62(8): 552-555.
[7] Antioxidant high-conductivity copper paste for low-cost flexible printed electronics. Hong S Q, Liu C M, et al. Flexible Electronics, 2022, 6 (17).
[8] Liu H, Ma R, et al. Application of silver nanoparticles in electrically conductive adhesives with silver micro flakes [J]. China Welding, 2022, 31(1): 23 − 28.
[9] YANG J L, DONG C C, et al. Development of Low-temperature Sintered Nanoscale Silver for New PowerDevice Packaging. Materials Reports, 2019.
[10] Yang D S, Zhang H, et al. Research progress on micro/nano joining technologies and failure behaviors in electronic packaging [J]. Transactions of the China Welding Institution, 2022, 43(11): 126 − 136.

Mechanical analysis of hydro-bulging process of bimetal clad pipe

Ge Zhang[1*], Lanyun Li[1, 2], Chenzhengzhe Yan[1], Zhibo Zhao[3] and Tingzhen Yao[1]

[1] School of New Energy, Xi'an Shiyou University, Xi'an, 710065, China
[2] Key Laboratory of Materials Processing Engineering, School of Material Science and Engineering, Xi'an Shiyou University, Xi'an, 710065, China
[3] Xi'an Survey and Mapping Institute, Xi'an, 710054, China

[*]Corresponding author's e-mail: 190609@xsyu.edu.cn

Abstract: Bimetal clad pipes manufactured by the hydro-bulging technique have the advantages such as good mechanical properties and excellent corrosion resistance. The residual contact pressure on the interface between the inner pipe (i.e., liner pipe) and the outer pipe (i.e., base pipe) is an important technical index to measure the forming quality of the clad pipe and is affected significantly by the bulging pressure. Based on the displacement compatibility condition and the plane strain hypothesis, the relationship between residual contact pressure and bulging pressure is established. In addition, the ranges of the bulging pressure when the base pipe only produces elastic deformation and when the base pipe produces tiny plastic deformation are determined. The comparisons between the analytical value and the experimental results verified that the theoretical analysis in this paper is valid.

1. Introduction

Higher demands for strength, corrosion resistance, and high-temperature resistance are put forward to the pipelines because of the diversification of the transport mediums and the complications of the transportation environments. For the bimetal clad pipe, the base pipe is used for strength support and the liner pipe is used for corrosion resistance or abrasion resistance, thus the bimetal clad pipe takes full advantage of the materials of the two pipes and has been widely used in the oil and gas industry [1, 2]. The bimetal-clad pipe manufactured by the hydro-bulging process has no scratches or damages on the inner surface and its wall thickness is uniform and has excellent corrosion resistance [3], so it has great development potential and gains extensive attention from the researchers. However, the investigation on the hydraulic bulging technology of the clad pipe is still insufficient compared to the traditional manufacturing technology.

When hydraulic forming bimetal clad pipes, the surface-treated liner and base pipe are first assembled. Then, pressure is applied to the inner wall of the liner pipe through a hydraulic press, causing the liner pipe to swell and come into contact with the base pipe. As the bulging pressure increases, the base pipe also undergoes elastic deformation, and even a small amount of plastic deformation occurs. Then, as the rebound of the base pipe is greater than that of the liner pipe, the base pipe eventually tightly holds the liner pipe, resulting in residual contact pressure (RCP for short) at the interface, forming a bimetal clad pipe.

The RCP is an important measurement to evaluate the forming quality of the bimetal clad pipe, and improving it is one of the key technologies of the hydraulic bulging process of bimetal clad pipes. Research on the RCP of the bimetal-clad pipe has been carried on in recent years. In [4], equivalent yield strength is proposed to consider the strain hardening of stainless steel. Fan et al. [5] derived the theory of the expansion of composite pipes and obtained expansion methods and related parameter calculation formulas. Akisanya et al. [6] showed that the strain of the liner pipe accumulated before the two tubes were in contact is ignored in SCC and proposed that the practical unloading process will be described more really if the elastic recovery amounts of the radial displacement of the two pipes are equal (i.e., displacement compatibility condition, DCC for short). According to the DCC, Tang et al. [7] analyzed the force loading status, but they just considered the situation that the base tube only produced elastic deformation. However, in the process of hydraulic bulging, a more elastic recovery amount of the base pipe will be gained and its performance will also not be affected if the base pipe produces tiny plastic deformation, which is really the key point to obtain the maximum residual contact pressure [8].

Based on the DCC, the range of the hydro-bulging pressure is determined and the relationship between the bulging pressure and the RCP is established, considering the two cases in which one the base pipe only produces elastic deformation and the other the base pipe produces tiny plastic deformation.

2. Graphical method of the equivalent yield strength

The graphical method of the equivalent yield strength based on the true stress-strain curve is shown in Figure 1. The curve OAH is the true stress-strain curve of the liner pipe, and point A corresponds to the initial yield strength ($\boldsymbol{\sigma}_{s0}$). We set the strain of point B equal to $\frac{\delta}{R_{o,L}}$ (where $\boldsymbol{\delta}$ is the initial gap between the liner pipe and the base pipe, and $\mathbf{R}_{o,L}$ is the outer radius of the liner pipe). We make a line BH parallel to OA and intersect with OAH at the point H. The stress at the point H is taken as the equivalent yield strength $\boldsymbol{\sigma}_{seql}$ of the liner pipe.

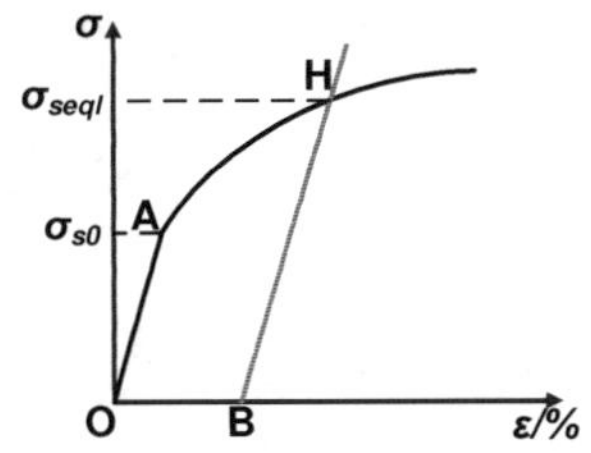

Figure1 Graphical explanation of the equivalent yield strength.

3. Mechanical analysis process of hydro-bulging process of bimetal clad pipe

3.1. Basic assumptions

(1) The material of the base pipe is elastic-perfectly plastic and that of the liner pipe is elastic-perfectly plastic with equivalent yield strength, the Von Mises yield criterion is followed in plastic deformation.

(2) Before contacting with the base pipe, the liner pipe has yielded completely.

(3) The deformation of the base pipe and liner pipe belongs to plane strain deformation.

(4) The reverse yield of the liner pipe cannot be allowed during unloading.

Based on the above assumptions, the cylindrical coordinate system is adopted in the mechanical analysis. In this case, the stress, strain, and displacement components of the pipes are only the functions of the radial coordinate and are independent of the tangent coordinate. The radial stress, hoop stress, and axial stress are all the principal stresses and the shearing stress components are zero.

3.2. Mechanical analysis

According to characteristics in the hydro-bulging process of the bimetal clad pipe, the forming process can be divided into four stages: Stage 1 is the elastic deformation stage of the liner pipe; Stage 2 is the elastoplastic deformation stage of the liner pipe; Stage 3 is the deformation stage after the two pipes contact; Stage 4 is the elastic recovery stage during unloading.

The stress components and radial displacement of the pipe can be obtained by the Lamé formula:

$$\begin{cases} \sigma_r = \dfrac{P_i}{K^2-1}\left(1-\dfrac{R_o^2}{R}\right) & \text{(1a)} \\ \sigma_\theta = \dfrac{P_i}{K^2-1}\left(1+\dfrac{R_o^2}{R}\right) & \text{(1b)} \\ U_r = \dfrac{P_i}{E(K^2-1)}\left[(1-\mu)R+(1+\mu)\dfrac{R_o^2}{R}\right] & \text{(1c)} \end{cases}$$

where $\sigma_r$and σ_θ are the radial and hoop stresses of the pipe, respectively; P_I is the bulging pressure; K is the ratio of R_o to R_I, i.e., $K = \frac{R_o}{R_I}$ (where R_o and R_I are the inner and outer radius of the pipe, respectively); R is the radius of the pipe; μ is the Poisson's ratio of the pipe; E is the elastic modulus of the pipe, and U_r is the radial displacement of the pipe.

Stage I: At the beginning of the hydro-bulging process, the liner pipe produces elastic deformation and has not contacted with the base pipe, so the base pipe has not deformed in this stage. According to Equation (1c), the hoop strain $\varepsilon_{\theta,L}$ at the inner surface of the liner pipe can be obtained as:

$$\varepsilon_{\theta,L} = \frac{U_{r,L}|_{R_L=R_{i,L}}}{R_{i,L}} = \frac{P_{i,1}[(K_L^2+1)+\mu_L(K_L^2-1)]}{E_L(K_L^2-1)} \quad (2)$$

where the subscript L denotes the liner pipe (the same as below); $P_{i,1}$ is the bulging pressure in Stage 1.

Stage 2: With the increase of the bulging pressure, the inner surface of the liner pipe will produce elastoplastic deformation when the stress reaches its equivalent yield strength (σ_{seql}). If the correction factor is used to consider the effect of intermediate principal stress (σ_z), the Von Mises yield criterion [9] is expressed as:

$$\sigma_{\theta,L} - \sigma_{r,L} = \frac{2}{\sqrt{3}}\sigma_{seql} \quad (3)$$

As we known, the differential equation of stress equilibrium in axisymmetric problems is:

$$\frac{d\sigma_r}{dR} + \frac{\sigma_r - \sigma_\theta}{R} = 0 \quad (4)$$

According to Equation (3) and Equation (4), the following equation can be deduced:

$$\sigma_{r,L} = \frac{2\sigma_{seql}}{\sqrt{3}}\ln R_L + Q \quad (5)$$

where Q is the integral constant. It is known that the radial stress on the inner surface of the liner pipe is equal to the bulging pressure, i.e., $\sigma_{r,L}|_{R_L=R_{i,L}} = P_{i,2}$. Substituting this condition into Equation (5), the value of the integral constant Q can be obtained:

$$Q = P_{i,2} - \frac{2\sigma_{seql}}{\sqrt{3}}\ln R_{i,L} \quad (6)$$

Combining Equation (5) and (6), the radial stress of the liner pipe in Stage 2 ($\sigma_{r,L,2}$) can be get:

$$\sigma_{r,L,2} = \frac{2\sigma_{seql}}{\sqrt{3}}\ln\frac{R_L}{R_{i,L}} + P_{i,2} \quad (7)$$

Substituting Equation (7) into Equation (3), the hoop stress of the liner pipe in Stage 2 $\sigma_{\theta,L,2}$ can be obtained:

$$\sigma_{\theta,L,2} = \frac{2\sigma_{seql}}{\sqrt{3}}\left(1+\ln\frac{R_L}{R_{i,L}}\right) + P_{i,2} \quad (8)$$

Because the two pipes are not in contact with each other in this stage, the radial stress on the outer surface of the liner pipe is equal to zero, $\sigma_{r,L,2}|_{R_L=R_{o,L}} = 0$. Substituting it into Equation (7), the bulging pressure of the liner pipe in Stage 2 can be obtained:

$$P_{i,2} = \frac{2}{\sqrt{3}}\sigma_{seql}\ln K_L \quad (9)$$

Combining Equations (9) and (1c), the radial displacement ($U_{r.L,2}$) at the outer surface of the liner pipe in **Stage 2** can be get:

$$U_{r,L,2} = \frac{2}{\sqrt{3}}\frac{\sigma_{seql}R_{o,L}}{E_L} \quad (10)$$

In the practical condition, for the liner pipe, the diameter expansion is very small and the wall thickness can be considered to be unchanged during the whole hydro-bulging process [10]. The bulging pressure remains to be Equation (9) until the gap between the liner pipe and the base pipe disappears.

Once the liner pipe comes into contact with the base pipe, the interface pressure between the two pipes P_c will occur, and the deformation of the liner pipe will be restrained by the base pipe. Thus, the bulging pressure has to be increased further if the liner pipe produces further plastic deformation.

The bulging pressure ($P_{i,3}$) in Stage 3 can be divided into two parts: one part makes the liner pipe produce plastic deformation (marked as $P_{i,3,p}$) and the other part conquers contact pressure on the interface between the liner pipe and the base pipe. The expression of $P_{i,3,p}$ is the same as $P_{i,2}$, i.e.,

$$P_{i,3,p} = \frac{2}{\sqrt{3}}\sigma_{seql} \ln K_L \tag{11}$$

So $P_{i,3}$ can be expressed as: $P_{i,3} = P_{i,3,p} + P_c = \frac{2}{\sqrt{3}}\sigma_{seql} \ln K_L + P_c$ (12)

In **Stage 3**, the deformation process is analyzed in two cases:

Case 1: The base pipe only produces elastic deformation.

Case 2: The base pipe produces tiny plastic deformation.

The deformation amount at which the inner surface of the base pipe is about to yield is both the upper limit in Case 1 and the lower limit in Case 2. We set the bulging pressure at this critical state as the critical bulging pressure $P_{i,cri}$.

Case 1: The base pipe is only produced elastic deformation

According to the generalized Hooke's law $\varepsilon_z = \frac{1}{E}[\sigma_z - \mu(\sigma_r + \sigma_\theta)]$ and Equations (1a) and (1b), the axial strain at the outer surface of the base pipe is:

$$\varepsilon_{z,B} = -\frac{2\mu_B P_c}{E_B(K_B^2-1)} \tag{13}$$

where the subscript B denotes the base pipe (the same as below).

The hoop strain $\varepsilon_{\theta,B}$ at the outer surface ($R_B = R_{o,B}$) of the base pipe according to Equation (1c)

$$\varepsilon_{\theta,B} = \frac{U_{r,B}}{R_{o,B}} = \frac{2P_c}{E_B(K_B^2-1)} \tag{14}$$

The interface pressure reaches its maximum value in Case 1 when the inner surface of the base pipe is about to yield. According to the modified Von Mises yield criterion, Equation (1a) and (1b), and the maximum value of contact pressure in Case 1, the critical contact pressure ($P_{c,cri}$) can be obtained.

$$P_{c,cri} = \frac{\sigma_{s,B}(K_B^2-1)}{\sqrt{3}K_B^2} \tag{15}$$

where $\sigma_{s,B}$ is the yield strength of the base pipe.

Combining Equations (9), (11) and (15), the maximum value of the bulging pressure in Case 1, i.e., the critical bulging pressure ($P_{i,cri}$) can be derived:

$$P_{i,cri} = \frac{\sigma_{s,B}(K_B^2-1)}{\sqrt{3}K_B^2} + \frac{2}{\sqrt{3}}\sigma_{seql} \ln K_L \tag{16}$$

Case 2: The base pipe produces tiny plastic deformation.

When the bulging pressure exceeds its critical value $P_{i,cri}$, the contact pressure between the liner pipe and the base pipe is greater than the yield strength of the base pipe, resulting in partial or even complete plastic deformation of the base pipe [8]. In order to prevent excessive plastic deformation in the base pipe, the deformation when the outer surface of the base pipe is about to yield is set as the upper limit in Case 2. Similar to the derivation of Equation (7), the radial stress of the base pipe when the outer surface of the base pipe is about to yield ($\sigma_{r,B}$) can be expressed as:

$$\sigma_{r,B} = \frac{2\sigma_{s,B}}{\sqrt{3}} \ln \frac{R_B}{R_{i,B}} - P_c \tag{17}$$

Similar to the derivation of Equation (8), the hoop stress of the base pipe when the outer surface of the base pipe is about to yield ($\sigma_{\theta,B}$) can be expressed as:

$$\sigma_{\theta,B} = \frac{2\sigma_{s,B}}{\sqrt{3}}\left(1 + \ln \frac{R_B}{R_{i,B}}\right) - P_c \tag{18}$$

The outer surface of the base pipe is free surface, so the stress at this surface is $\sigma_{r,B}|_{R_B=R_{o,B}} = 0$. Substituting this known condition into Equation (17), the maximum contact pressure ($P_{c,max}$) of the base pipe can be obatained:

$$P_{c,max} = \frac{2}{\sqrt{3}_{s,B}} \ln K_L \tag{19}$$

Combining Equations (11) and (19), the maximum bulging pressure of the base pipe ($P_{i_{max}}$) is:

$$P_{imax} = P_{c,max} + P_{i,3,p} = \frac{2}{\sqrt{3}}\sigma_{s,B} \ln K_L + \frac{2}{\sqrt{3}}\sigma_{seql} \ln K_L \tag{20}$$

Stage 4:

Case 1: The base pipe only produces elastic deformation

For the bimetal clad pipe, the contact pressure after completely unloading (i.e., RCP, marked as P_c^*) has to be greater enough to tightly bond the liner pipe with the base pipe [6]. During unloading, the instantaneous values of the bulging pressure and the contact pressure are denoted as $P_{i,u}$ and $P_{c,u}$, respectively.

During the bulging pressure drops from $P_{i,_{max}}$ to $P_{i,u}$, the elastic radial displacements of the liner pipe and the base pipe has to be equal to ensure the contact between two pipes, i.e., $\Delta U_{r,L} = \Delta U_{r,B}$ (21)

where $\Delta U_{r,L}$ and $\Delta U_{r,B}$ are the radial displacement recovery amounts of the liner and base pipes, respectively. According to Equations (1c) and (21), the instantaneous contact pressure between the two pipes can be obtained:

$$P_{c,u} = P_{i,max} - P_{i,p} \frac{2R_{o,L}(P_{imax}-P_{I,u})}{E_L(K_L^2-1)(MR_{o,L}+NR_{I,B})} \tag{22}$$

where $M = \frac{(K_L^2+1)-\mu_L(K_L^2-1)}{E_L(K_L^2-1)}$ and $N = \frac{(K_B^2+1)+\mu_B(K_B^2-1)}{E_B(K_B^2-1)}$.

When $P_{i,u} = 0$, the unloading is finished. Substituting $P_{i,u} = 0$ into Equation (22), the RCP in Case 1 is:

$$P_{c,1}^* = \left[1 - \frac{2R_{o,L}}{E_L(K_L^2-1)(MR_{o,L}+NR_{i,B})}\right] P_i - \frac{2}{\sqrt{3}}\sigma_{seql} \ln K_L \tag{23}$$

The minimum bulging pressure has to ensure the RCP exits between the liner and base pipes, i.e., $P_{c,1}^* \geq 0$.Combining $P_{c,1}^* \geq 0$ with Equation (23), the minimum bulging pressure can be obtained:

$$P_{i,min} = \frac{2}{\sqrt{3}}\sigma_{seql} \frac{2E_L(K_L^2-1)(MR_{o,L}+NR_{i,B})}{E_L(K_L^2-1)(MR_{o,L}+NR_{i,B})-2R_{o,L}} \tag{24}$$

According to Equations (16) and (24), the range of bulging pressure if the base pipe only produces elastic deformation is:

$$\frac{2}{\sqrt{3}}\sigma_{seql} \frac{2E_L(K_L^2-1)(MR_{o,L}+NR_{i,B})}{E_L(K_L^2-1)(MR_{o,L}+NR_{i,B})-2R_{o,L}} \leq P_{i,1} \leq \frac{\sigma_{s,B}(K_B^2-1)}{\sqrt{3}K_B^2} + \frac{2}{\sqrt{3}}\sigma_{seql} \ln K_L \tag{25}$$

Case 2: The base pipe produces tiny plastic deformation

The DCC has also to be satisfied in Case 2. When unloading is finished, $Pi,u = 0$. Substituting $P_{i,u} = 0$ into Equation (22), the RCP in Case 2 can be obtained:

$$P_{c,2}^* = \left[1 - \frac{2R_{o,L}}{E_L(K_L^2-1)(MR_{o,L}+NR_{i,B})}\right] P_{i,max} - \frac{2}{\sqrt{3}}\sigma_{seql} \ln K_L \tag{26}$$

According to Equations (16) and (20), the range of the bulging pressure in Case 2 is:

$$\frac{\sigma_{s,B}(K_B^2-1)}{\sqrt{3}K_B^2} + \frac{2}{\sqrt{3}}\sigma_{seql} \ln K_L \leq P_{i,2} \leq \frac{2}{\sqrt{3}}\sigma_{s,B} \ln K_L + \frac{2}{\sqrt{3}}\sigma_{seql} \ln K_L \tag{27}$$

4. Example

In the examples, the 316L stainless steel pipe and the X70 pipeline steel pipe are selected as the liner pipe and base pipe, respectively; the length of the pipes are 1000mm; the outer radius of the base pipe is $R_{o,B} = 100$mm and its inner radius is $R_{i,B} = 76$mm; the outer radius of liner pipe is $R_{o,L} = 74$mmand and its inner radius is $R_{i,L} = 71$mm; and the initial gap between the two pipes is $\delta = 2$mm. The mechanical properties of the liner pipe and the base pipe materials are shown in Table 1. The constitutive curve of the liner pipe is presented in Figure 2. Prior to the start of the hydro-bulging process, strain gauges were attached to various locations on the outer surface of the base pipe. These gauges were oriented in both the axial and hoop directions to measure axial strain and hoop strain, respectively. It is important that the areas where the strain gauges are applied are clean and smooth.

In the experimental bulging process, the bulging pressure ascended gradually to its maximum value, i.e., 120MPa, and then it remained unchanged for a while. At last it descended gradually from 120MPa to 0, at which the process was finished.

The axial strain and hoop strain at the outer surface of the base pipe were measured and recorded during the loading procedure. The mean of the axial strains at different places is taken as the experimental result, as well as the hoop strain. Figure 3 shows the real products of bimetallic clap pipe.

Table 1 Mechanical properties of the bimetallic clad pipe [7].

Material	Density/(kg·m^{-3})	Yield strength /MPa	Tensile strength /MPa	Elastic modulus /MPa	Poisson's ratio
X70	7980	532.2	626.8	207×10^3	0.3
316L	7800	297.9184	639.2764	195×10^3	0.26

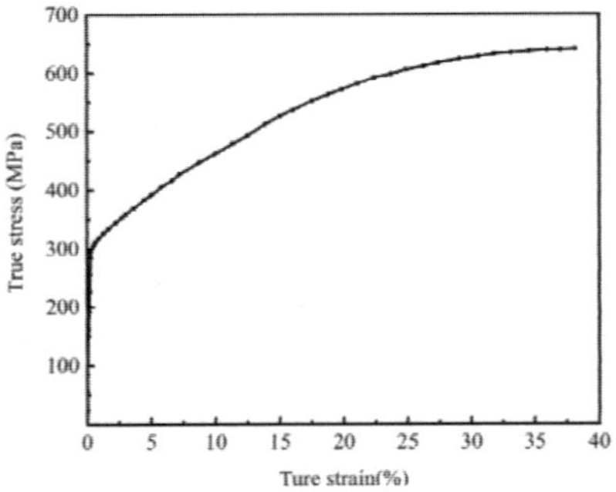

Figure2 The true stress and strain curve of 316L [7].

Figure3 The real products of bimetallic clad pipe.

According to the graphical method of the equivalent yield strength discussed above, the equivalent yield strength of the liner can be obtained as σ_{seql}=386.16MPa.

Figure 4 and Figure 5 show the variations of the axial strain and the hoop strain at the outer surface of the base pipe with the bulging pressure. The experimental results are from [7]. By comparing the theoretical predicted values of axial strain and circumferential strain with experimental values, it can be clearly seen that the axial strain and hoop strain considering the strain hardening of the liner pipe are closer to the experimental values than those without considering the strain hardening of the liner pipe. The maximum errors are 6.07% and 1.14%, respectively. Therefore, it proves that the elastic-plastic mechanics formula proposed in this study is more in line with experiments.

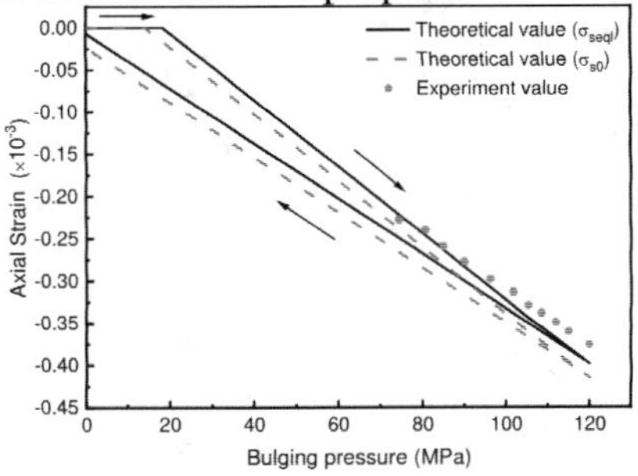

Figure 4 Axial strain variation at the outer surface of the base pipe during the bulging process.

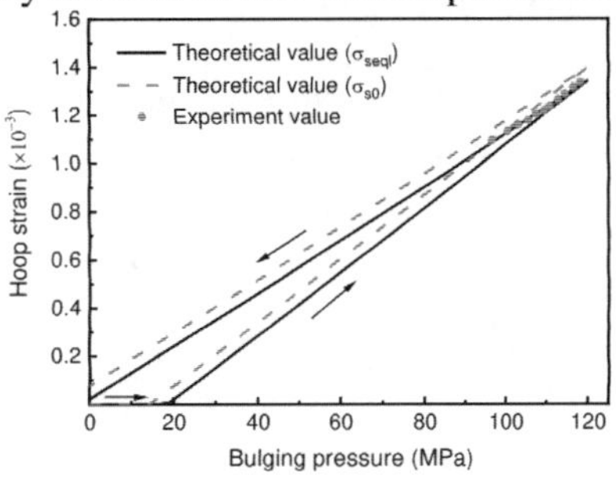

Figure 5 Hoop strain changes on the outer surface of the base pipe during the bulging process.

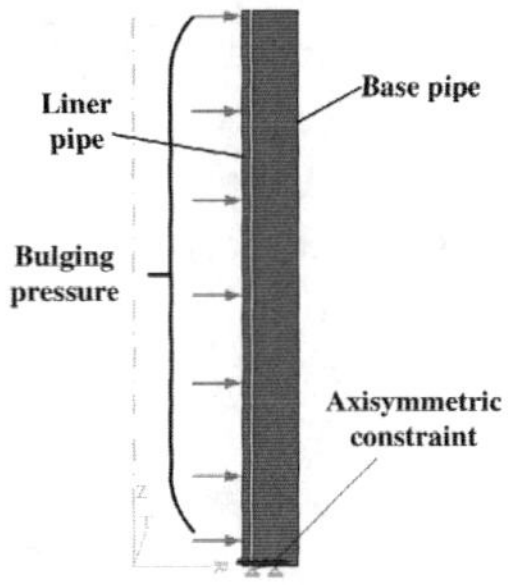

Figure 6 Schematic diagram of finite element model.

Since it is very troublesome and costly to test the bulging pressure range, the theoretical values are compared with the finite element simulation results to verify whether the bulging pressure range given by Equations (25) and (27) is reasonable or not.

During the hydro-bulging process, the geometries of the liner and base pipes as well as the bulging pressure are axisymmetric, thus, it can be simplified as a two-dimensional axisymmetric problem in the finite element simulation. Furthermore, only half of the pipes in the axial direction are considered and a symmetry constraint is set on the axial symmetry plane because of the symmetry of the pipe blank and bulging pressure along this direction. The static implicit algorithm is selected in order to calculate precisely the elastic recovery of the pipes during unloading. The established model is shown in Figure 6. The contact stress distribution on the interface is shown in Figure 7.

The hoop strain on the outer surface of the base pipe obtained by FEM is compared with that obtained by the experimental method shown in Figure 8. Obviously, the value obtained by FEM is very closer to the experimental value. The maximum relative error between them is less than 1.77%, which illustrates that the FE model built is precise enough to predict the actual hydro-bulging process. Thus, the bulging pressure range obtained through the FE simulation can be adopted to verify whether the theoretical range is reliable.

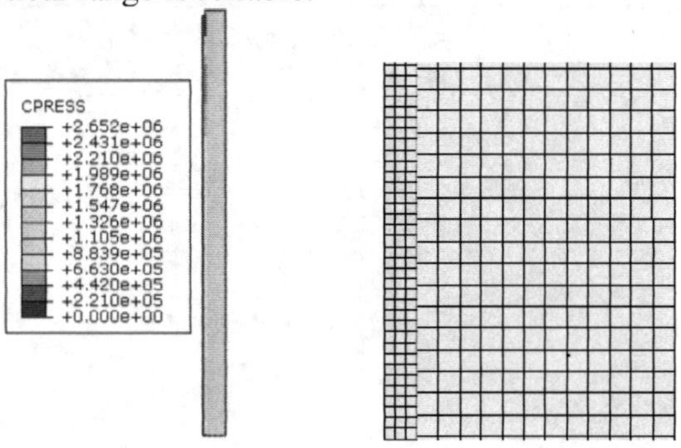

(a) The whole clad pipe **(b)** A part of the clad pipe

Figure 7 The contact pressure distribution on the interface.

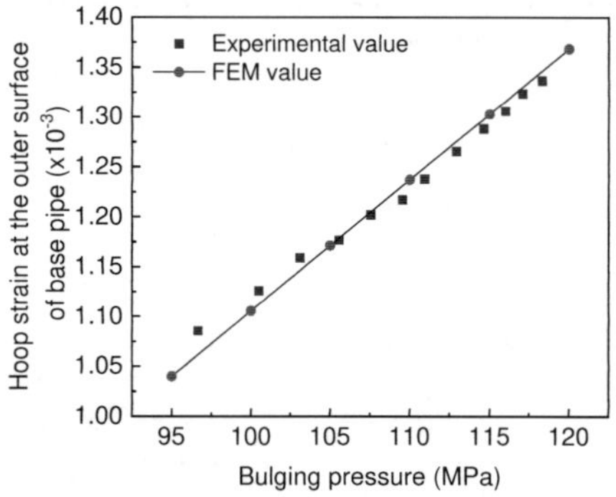

Figure 8 Variation of hoop strain on the outer surface of base pipe during the loading process.

The minimum, critical, and maximum values of the bulging pressure are shown in Figure 9. It can be clearly seen that the three values obtained by the analytical method and finite element method are in good agreement with the errors of 5.4%, 3.9%, and 4.3%, respectively.

The change of the RCP with the bulging pressure is shown in Figure 10. Since the residual contact pressure on the interface is varied a little along the axial direction, as shown in Figure 7, the average value along this direction is adopted as the finite element simulation value in Figure 10. The theoretical values of the RCP on the interface are obtained according to Equations (23) or (26). From Figure 10, a good agreement between the curves can be observed and the maximum error is less than 7%.

In conclusion, since a novel algorithm of σ_{seql}, the modified Von Mises yield criterion and the displacement compatibility are employed in the mechanical analysis carried out in this paper, the strain of the pipes and the range of the bulging pressure can be predicted very well. Moreover, according to the relation between residual contact pressure and bulging pressure given in Equations (23) or (26), the residual contact pressure can be predicted more precisely, which is very useful for engineering applications.

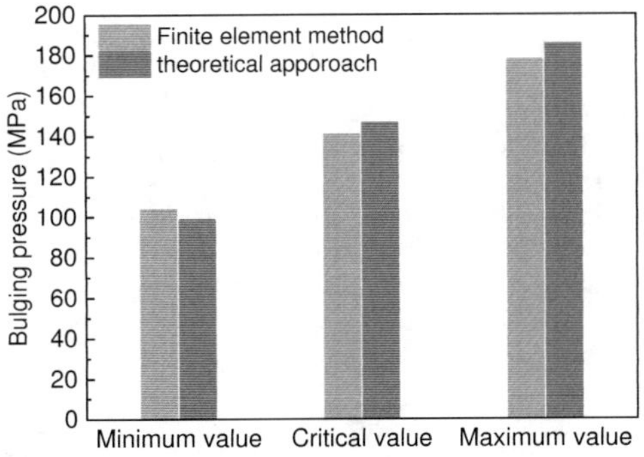

Figure 9 Comparison between the analytical results and finite element results of the bulging pressure.

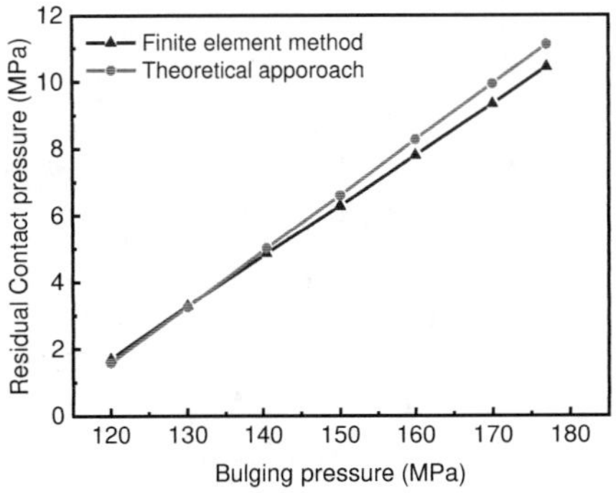

Figure 10 Variation of residual contact pressure with the bulging pressure.

5. Conclusions

(1) The relation between the residual contact pressure and the bulging pressure in the hydro-bugling process of bimetal clad pipe is established and the range of the bulging pressure is determined in two cases, in which one the base pipe only produces elastic deformation and the other one the base pipe produces tiny plastic deformation.

(2) This study has improved the graphical method for equivalent yield strength. Based on the displacement compatibility condition and the plane strain assumption, an elastoplastic mechanical analysis was conducted on the hydroforming of bimetal clad pipes, and the calculation formulas for relevant parameters in each stage were derived.

(3) By comparing the analyzed values with experimental values and finite element simulation results, it was verified that the range of bulging pressure and the analytical expression describing the relationship between residual contact pressure and bulging pressure are reasonable and can fully satisfy engineering requirements.

This study can assist engineers in determining more appropriate processing parameters and provide a scientific basis for the optimization of the manufacturing process. It has important theoretical significance and application value for improving the quality of bimetal-clad pipes manufactured by hydroforming, as well as enhancing the success rate and production efficiency of composite pipe manufacturing.

References

[1] Wang, L., Wu, D., Xu, M. et al (2023) Microstructure and mechanical properties of L415/N08825 bimetallic composite pipe welded joint using GTAW + SMAW. *Archiv. Civ. Mech. Eng* 23, 18. https://doi.org/10.1007/s43452-022-00554-x.

[2] Amirkhanlou, S., Ji, S. (2019). A review on high stiffness aluminum-based composites and bimet- allics. *CRIT REV SOLID STATE*, 45(1): 1–21. https://doi.org/10.1080/10408 436.2018.1485550

[3] Jie Zhang, Te Hu (2022) Buckling failure mechanism of liner pipe in bimetal mechanical clad pipe under complex loading, *Ocean Engineering*, 262, https://doi.org/10.1016/j.oceaneng.2022.112091.

[4] Ida Mascolo, Federico Guarracino, Sina Sarfarazi, Gaetano Della Corte (2024) A proposal for a simple characterization of stainless steel connections through an equivalent yield strength, *Structures*, 68, https://doi.org/10.1016/j.istruc.2024.107043.

[5] Fan X, Wang X, Zhai S, et al. Study on Forming Theory of Inner Strong & Outer Weak Strong Bi-metal Pipe [C]//IOP Publishing Ltd. *IOP Publishing Ltd*, 2021: 012009 (8pp). DOI: 10.1088/ 1755-1315/693/1/012009.

[6] A. R. Akisanya, F. U. Khan, W. F. Deans, P. et al (2011) Cold hydraulic expansion of oil well tubulars.*INT J PRES VES PIP*, 88(s11–12): 465-72. https://doi.org/10.1016/j.ijpvp.2011.08.003

[7] Tang Y, Ni XJ, Wang YQ (2018) Mechanism of hydraulic expansion forming for bimetal composite tube 316L/X70. *FST* 1:90-96. DOI: 10.13330/j.issn.1000-3940.2018.01.016.

[8] Chen HY (2006) Application and Development of Plastic Forming Technique for Double Metal Combined Pipe. J *Process Equip. Piping* 43: 16-18.

[9] Mehrdad Negahban. (2012) Finite deformation mechanical theory of plasticity, CRC Press, Boca Raton. https://doi.org/10.1201/b12050-15.
[10] Zeng DZ, Deng KH, Shi TH, et al (2014) Theoretical and Experimental Study of Bimetal-Pipe Hydroforming. *J PRESS VESS-T ASME.* 136:1-10. https://doi.org/10.1115/1.4026976.

Research on error correction algorithms for Coriolis flowmeters

Yajun Liu [1*], Junfeng Zeng [1], and Qinglin Liu [1]

[1] School of Mechanical and Automotive Engineering, South China University of Technology, Guangzhou 510000, China

*Corresponding author's e-mail: yajun@scut.edu.cn

Abstract. The Coriolis mass flowmeter is a high-precision flowmeter that can directly measure the mass flow value. Its working principle ensures that the measurement results of the Coriolis mass flowmeter have good reproducibility and high measurement accuracy. However, when dealing with a flow medium composed of gas and liquid in two phases, the Coriolis flowmeter will no longer possess stability and high precision under single-phase flow measurement. The research collected the vibration signal data of the Coriolis flowmeter under gas-liquid two-phase working conditions, used the power spectrum algorithm to extract its internal frequency domain features such as phase values and frequency values, and then combined the pressure data in the flow pipeline to conduct data modeling on the Coriolis flowmeter. Experiments have demonstrated that the GRU (Gated Recurrent Unit) model has the best training effect on the aforementioned feature vector data. In the dynamic tracking research of the flowmeter, the instantaneous measurement accuracy has reached 94.41%, and the cumulative mass flow measurement error within 60 seconds is only 2.63%.

1. Introduction

In the past half-century, flow measurement technology has experienced rapid development. As an applied measurement discipline, it has been widely utilized in multiple aspects, such as process control, trade settlement, energy metering, and waste discharge monitoring. Meanwhile, flow measurement is also a complex problem in the field of industrial measurement. Its complexity is principally exhibited by the multiplicity of measurement media, the mutability of the flow process, and the perturbation from external environmental factors. Therefore, it is necessary to continuously improve and design flow measurement instruments, especially in the current era when the automation and precision levels of the manufacturing industry are constantly increasing and the energy system is transforming.

In the late 1970s, Micro Motion in the United States developed the Coriolis flowmeter. The Coriolis meter is not sensitive to the pressure, viscosity, and velocity distribution of the fluid medium, and it can directly measure the mass of the fluid with high precision. However, when dealing with the flow medium composed of gas and liquid in two phases, there is a complex interaction at the phase interface between the gas and the liquid. This interaction further leads to a complex interaction between the pipe wall and the fluid, causing the Coriolis flowmeter to no longer possess the stability and high precision it has under single-phase flow measurement. To solve the problem of inaccurate measurement of Coriolis mass flowmeters under two-phase flow working conditions, research usually focuses on digital signal processing methods or establishing error correction models to reduce measurement errors under gas-liquid two-phase flow. Stadler [1] and Denis [2] proposed a Coriolis mass flowmeter (CMF) signal processing method based on quadrature demodulation. Through quadrature demodulation, the orthogonal components of the difference signal between two CMF signals are obtained and used to calculate the phase difference. The accuracy of the phase difference calculation depends on the performance of the low-pass filter used in this method. Due to the delay in the low-pass filtering stage, the response speed of the quadrature demodulation algorithm is relatively slow. Tu and Zhang [3] took into account the influence of negative frequencies of signals and proposed a negative frequency recursive DTFT (Discrete Time Fourier Transform) algorithm, which shortened the convergence time of the discrete Fourier transform and ensured the real-time performance of the signals. Zhang et al. [4] put forward an improved correlation analysis method. Firstly, the relationship between the phase difference and the correlation function was deduced with the help of the Hilbert transform, and then the correlation analysis method was used to solve the phase difference. Li et al. [5] compared the performance of phase extraction and frequency estimation of the Hilbert method, the quadrature demodulation method, and the sliding Fourier transform method under single-phase and two-phase flows.

In terms of algorithm model research, the research group led by Professor Xu Kejun from Hefei University of Technology has conducted calibration studies by using the algorithms of the BP neural network, Elman neural network [6], and support vector machine [7] respectively. When the density drops by less than 15%, the online calibration error does not exceed 1%. Olga et al. [8] used the Support Vector Regression (SVR) algorithm on the two-phase flow data set and discussed how to construct an error correction model with a correction effect through the SVR algorithm based on the prediction results. However, in all previous studies, the experimental scenarios were under relatively stable conditions, that is, measurements and corrections were carried out under fixed gas flow rates and liquid flow rates. This does not conform to the situation in real industrial production where the gas and liquid contents fluctuate and more attention needs to be paid to the flowmeter's flow dynamic tracking performance (this research focuses on the measurement accuracy of the Coriolis flowmeter when the real-time flow variation range is relatively large, which is referred to as the dynamic tracking performance). Traditional digital signal processing algorithms rely heavily on the calculation accuracy of frequency when calculating the phase difference, and have requirements on whether the signal conforms to an integer-period signal. To improve the calculation accuracy, it is often necessary to collect a large amount of data information and perform a large number of calculations before the accurate phase difference can be extracted. It is difficult to track the measurement parameters promptly in practical applications. The algorithm model driven by big data can establish the mapping relationship between information such as vibration signals and mass flow during the training process. The trained model deployed in applications can respond to the collected data more quickly. Although traditional digital signal processing makes it difficult to meet high requirements in terms of calculation accuracy, it can still effectively extract the frequency domain information in the data. When the input data of the algorithm model contains physical characteristics, there is a greater chance to ensure the measurement accuracy and repeatability of the algorithm model. Therefore, the research innovatively proposes to extract the frequency domain information of the vibration signal of the flowmeter by using digital signal processing methods first, and then build the algorithm model. Experiments have proved that the model can ensure the stability and accuracy of flow measurement during continuous measurements within one minute.

2. Experimental Platform and Digital Signal Processing

2.1. Experimental Platform

The Coriolis flowmeter contains an electromagnetic exciter inside. Once it is activated, the flow tube vibrates at its natural frequency. While the fluid in the U-shaped tube flows along the pipe, it also vibrates along with the tube. As a result, the mass of the fluid is subjected to a Coriolis force. The Coriolis force causes the inlet and outlet of the U-shaped tube to deform in opposite directions, and the magnitude of the deformation is detected by the detector. When there is no fluid passing through, the signals output by the two detectors overlap each other. When there is fluid passing through, due to the opposite-direction deformations at the inlet and outlet, the signals output by the two detectors will generate a phase difference. Its appearance and structure are shown in Figure 1.

(a) (b)

Figure 1. (a) Coriolis mass flowmeter; (b) Schematic diagram of the measuring principle.

According to the principle of the Coriolis force, the corresponding mass flow can be calculated from the phase difference and the Coriolis flowmeter can measure mass flow with high precision [9]. The exciter drives the flow tube to perform simple harmonic motion, and the form of the signal output by the detector is a sinusoidal signal. When the flow medium is single-phase, the two-channel vibration signals are standard signals with equal amplitudes and equal frequencies. However, when the medium is changed to a two-phase flow, the vibration signals are more in line with the characteristic shape of the random walk model as shown in Figure 2 (b), which leads to an increase in the difficulty of extracting the phase difference from them and a decrease in the measurement accuracy of the flowmeter under two-phase flow.

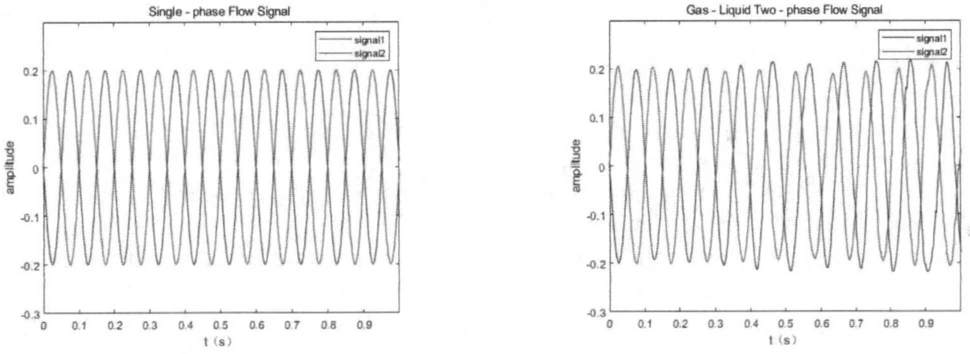

(a) (b)

Figure 2. (a) Vibration signal of single-phase flow; (b)Vibration signal of two-phase flow.

The gas-liquid two-phase flow experimental platform consists of a water tank, a throttle valve, a variable frequency pump, a compressed air cylinder, a gas-liquid mixer, a gas standard flowmeter, a liquid standard flowmeter, and a Coriolis mass flowmeter. The air cylinder provides compressed air at 0.6 MPa, and the mass flow of the gas phase is collected in real time through the gas standard flowmeter (Alicat-MB Series Alicat gas mass flowmeter). Similarly, after the mass flow of the liquid phase is measured by the liquid standard flowmeter, it forms a gas-liquid two-phase mixed flow with the gas phase through the gas-liquid mixer and is then transported to the Coriolis mass flowmeter. Each phase of gas and liquid is measured separately before mixing to ensure accuracy, and the sum of their mass flows is taken as the reference value of the Coriolis flowmeter for analysis and research.

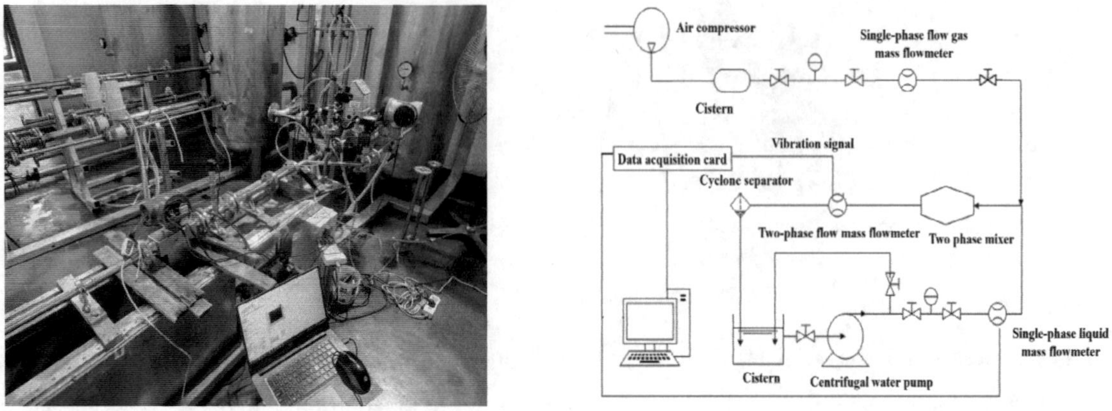

(a) (b)

Figure 3. (a) Experimental platform; (b) Experimental flow chart

The NI-9234 acquisition card with the synchronous sampling function is selected and the sampling mode of direct current coupling is adopted. In this way, the voltage signals output by the sensors at both ends of the measuring tube of the Coriolis flowmeter in Figure 3 can be collected, that is, the vibration signals of the measuring tube. The working frequency of the Coriolis flowmeter is 300 Hz, and the sampling frequency of the acquisition card is 3, 000 Hz. Meanwhile, the experimental setup can synchronously collect the measured values of the pressure gauges and standard meters on the platform, which ensures the accuracy and validity of data collection.

2.2. Extraction of Signal Frequency Domain Features

As can be seen from the previous chapters, under the two-phase flow, the vibration signals of the Coriolis flowmeter have a lot of noise due to the interaction of the medium. It is difficult for the Coriolis flowmeter to select an appropriate digital signal processing method from the vibration signals with noise to accurately extract the phase difference, thus resulting in a significant decrease in measurement accuracy. However, after using the original vibration signals to train the algorithm model, it was found that the effect was still not good. The model had great difficulty in mining the phase difference features in the massive data and was hard to learn the mapping relationship from the vibration signals to the mass flow. To make the subsequent algorithm model itself more physically interpretable, the research first used the method of cross-power spectrum in digital signal analysis to collect the phase difference of the vibration signals as a main data set for the subsequent model.

2.2.1. Cross Spectrum. The cross-spectrum is a frequency-domain function between the signals of two channels. It is supposed that there are two sets of signals $x(t)$ and $y(t)$, of a certain system, and their Fourier transforms are $X(t)$ and $Y(t)$ respectively. $X(f)$ and $Y(f)$ are generally complex numbers, so the expressible forms of their corresponding power spectra are as follows:

$$G_{xx} = |X(f)|^2 = (\mathrm{Re}[X(f)])^2 + (\mathrm{Im}[X(f)])^2 \tag{1}$$

Since $X(f)$ is a complex number, when the conjugate function of $X(f)$ is denoted as $X(f)^*$, the power spectrum can be expressed as:

$$G_{xx}(f) = X(f)^* \times X(f) \tag{2}$$

Similarly, the power spectrum of the signal $y(t)$ can be expressed as:

$$G_{yy}(f) = Y(f)^* \times Y(f) \tag{3}$$

The power spectrum represents the power component at each frequency and is a real function. Therefore, it can be represented by a rectangular coordinate system (Cartesian coordinate system), with the abscissa representing the frequency and the ordinate representing the power (the square of the amplitude). Next, the function that describes the relationship between the two sets of signals, the Cross Spectrum $G_{xy}(f)$, will be explained, and its definition is as follows:

$$G_{xy}(f) = X(f)^* \times Y(f) \tag{4}$$

We divide the result obtained from Formula (4) into the real part $CR(f)$ and the imaginary part $CI(f)$. Then the cross-spectrum can be expressed by the corresponding phase spectrum as follows:

$$\theta(f) = \arctan(\frac{CI(f)}{CR(f)}) \tag{5}$$

2.2.2. Extraction of the Phase Difference Using the Cross-Spectrum. We use simulation experiments to verify the phase extraction performance of the cross-power spectrum method under a two-phase flow. The vibration signals under two-phase flow conform to the random walk model. We use MATLAB to generate a group of sinusoidal signals with a signal-to-noise ratio of 50 dB and a phase difference variation range of approximately from 29.8° to 30.2°. We conduct relevant calculations by using the power spectrum, with a calculation period of 100 points, and its mean square error value is 0.0261°. As can be seen from Figure 4, the power spectrum has good dynamic performance and excellent phase difference extraction performance.

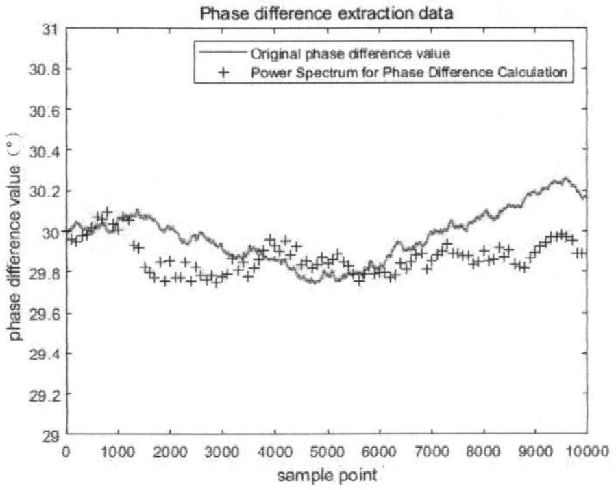

Figure 4. Power spectrum phase difference extraction diagram

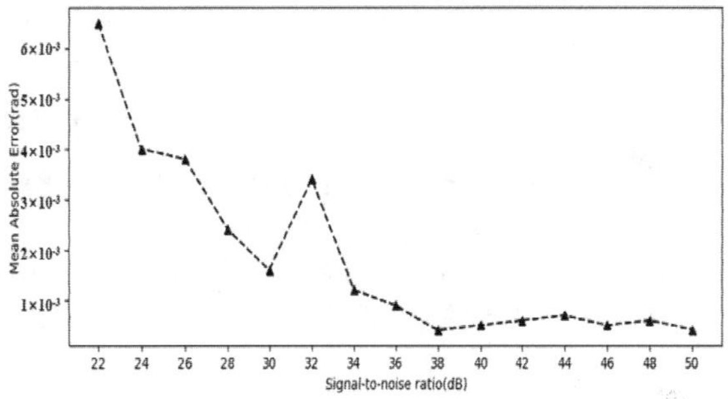

Figure 5. Phase accuracy of power spectrum extraction under different signal-to-noise ratios

Figure 5 shows the mean error of phase difference calculation for the power spectrum algorithm under different signal-to-noise ratios. The higher the signal-to-noise ratio of a signal is, the smaller the amplitude of the influence of noise on the signal will be. Therefore, as the signal-to-noise ratio increases in the graph, the mean error shows a decreasing trend. However, due to the internal disorder of the random walk signal, in the experiment, the phase of one signal may fluctuate in a decreasing trend while the phase of another group of signals fluctuates in an increasing trend. To some extent, this amplifies the noise interference in the signal. Therefore, fluctuations of abnormal values will appear in Figure 5. This is also the shortcoming of traditional digital signal processing in this research. The algorithm is sensitive to the fluctuation trend of the signal itself during calculation and it is rather difficult to adapt to scenarios where the phase fluctuates greatly.

3. Model Algorithm

3.1. Gate Recurrent Unit Neural Network

The vibration signals of the Coriolis flowmeter are time series signals. The Gate Recurrent Unit network (GRU) is a neural network architecture specifically designed for processing sequence data [10], and it can effectively capture the dynamic change characteristics of signals in the time series. The fluctuations of the vibration signals of the Coriolis flowmeter caused by the complexity of the two-phase flow at different time points can be well learned and analyzed by the GRU network. The measured value of the Coriolis flowmeter at the current time point fluctuates around the measured value at the previous time point. Therefore, not only are the signal characteristics at the current moment important but also the signal state at previous moments will have an impact on the current measurement accuracy. The internal structure of the GRU network, through the mechanisms of the update gate and the reset gate, can handle long-term and short-term dependence relationships well.

The parameters contained within the GRU model are as follows: the input data $x(t)$, the reset gate $r(t)$, the update gate $z(t)$, the hidden state $h(t-1)$, the hidden state $h(t)$ at the current moment, and the candidate hidden state $\tilde{h}(t)$. The reset gate determines the degree of combination between the content at the previous moment and the current input data. Its calculation formula is as follows:

$$r_t = \sigma(W_r \cdot [h_{t-1}, x_t]) \tag{6}$$

$$\tilde{h}_t = \tanh(W \cdot [r_t * h_{t-1}, x_t]) \tag{7}$$

The smaller the value of r_t is, the more previous the information is forgotten. Conversely, the influence of the previous moment on the current one is more significant.
The update gate directly calculates the information transmission of the current layer and determines the information content to be transmitted in the future. Its calculation formula is as follows:

$$z_t = \sigma(W_z \cdot [h_{t-1}, x_t]) \tag{8}$$

$$h_t = (1 - z_t) * h_{t-1} + z_t * \tilde{h}_t \tag{9}$$

Figure 6 shows the system framework and operation process of GRU model. It can be seen from the calculation formula and figure 6 that the GRU (Gated Recurrent Unit) model can remember the important historical information in the signal sequence and can effectively cope with the irregular changes of the flowmeter vibration signal over time.

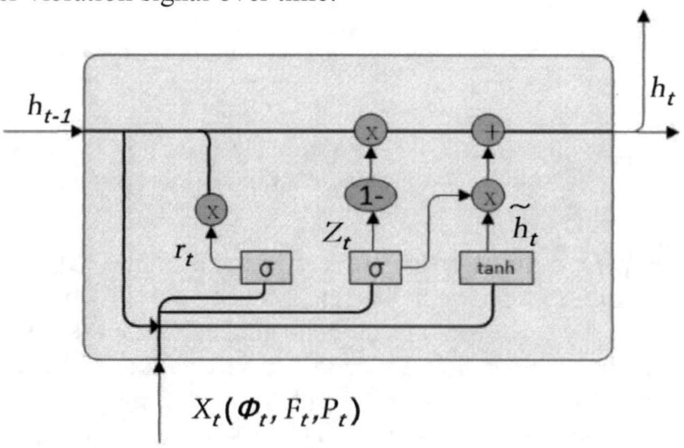

Figure 6. Framework of the GRU model structure

3.2. Model Structure

There are three types of input data for the model, namely the phase difference of the vibration signal of the Coriolis flowmeter, the vibration frequency, and the pressure data of the flow pipeline. The phase difference is collected from the vibration signal through the aforementioned power spectrum. The frequency information is extracted by the Hilbert algorithm. Since the impact of the frequency information on the model is relatively small, the details of the extraction will not be elaborated. The pressure is obtained by measuring the pipeline with a pressure gauge. The pressure is selected as a feature vector because the mainstream correction model in the two-phase flow model takes pressure as an important consideration factor. As described earlier, the vibration frequency of the flowmeter is 297

Hz, and the sampling frequency of the equipment is 3, 000 Hz. The vibration signal can be accurately collected, but a distortion of the working condition under a two-phase flow may also occur. The research mainly makes a judgment by extracting and analyzing the frequency values of the signals within 10 cycles (in this research, 10 cycles are taken as the extraction unit for the phase difference) to check whether the range is between 295 Hz to 305 Hz. When the frequency is not within this range, this group of signals will be discarded.

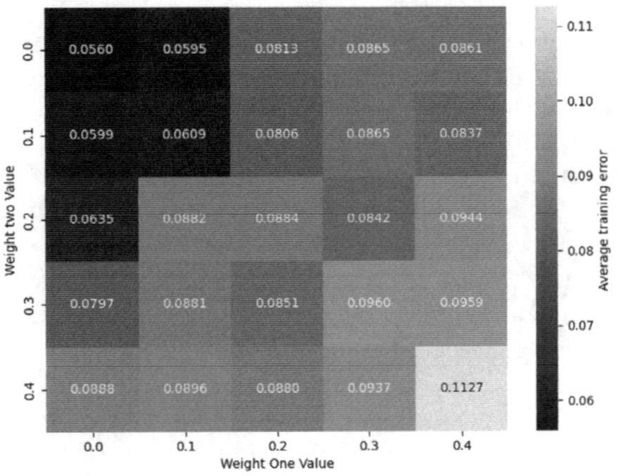

Figure 7. Mean error of training with different weights of feature data

The ranges of the above feature data vary greatly, and their influence factors on the model are also different. Firstly, it is necessary to carry out the normalization processing for them respectively as shown in Formula (10). To determine the appropriate weight proportions of different feature data in the model, the grid search method is used to systematically traverse the combination space of parameters and obtain the convergence situations of the model under different weight combinations. Figure 7 shows the mean error of the final calculation of the model under different weight values of pressure and frequency data. When the values of these two are smaller, that is, when the weight of the phase difference data is larger, the model converges better, indicating that the phase difference data can best determine the training results of the model. However, even when the phase difference data is used as the sole data source, the convergence effect of the model increases very little. Therefore, to increase the dimension of the data and improve the generalization ability of the model, the weight proportions finally assigned to the phase difference, vibration frequency, and pressure are 0.8, 0.1, and 0.1 respectively.

$$x' = \frac{x_i - \overline{x}}{x_{\max} - x_{\min}} \tag{10}$$

After the data has been preprocessed, different feature data are trained in different branches of the GRU layers. The data was further divided into a training set, a test set, and a validation set, which were set to a ratio of 0.8:0.1:0.1 according to the general model proportion. Then they are respectively passed to the fully connected layers for linear transformation to obtain scalar outputs. Next, the outputs of the three branches are weighted and averaged according to the weights to obtain the predicted result of the model, and this is one round of the training process.

The form of the model's loss function combines the Mean Squared Error (MSE) with L2 regularization, as specifically shown in Formula (11). Adding L2 regularization to the loss function is equivalent to adding a penalty term, which restricts the complexity of the model to some extent and enhances the generalization ability of the model at the same time.

$$Loss = \frac{1}{n}\sum_{i=1}^{n}(\hat{y}_i - y_i)^2 + \frac{\lambda}{n}\sum_{j} w_j^2 \quad (11)$$

The model simultaneously uses the Adam optimizer to control the learning rate, making the training process more stable. For the linear layers, the ReLU activation function is selected to increase the nonlinear characteristics during the training process, enabling the model to learn more complex patterns.

3.3. Analysis of Algorithm Results and Feasibility Testing

The experiment collected multiple sets of vibration data from the Coriolis flowmeter and pipeline pressure data under two-phase flow. Among them, the mass flow rate range of the liquid (water was selected as the research object in this study) was from 15 kg/min to 25 kg/min, the volumetric flow rate of the gas (air was selected in this study) was stabilized at 1 SLPM (Standard Liters per Minute), 1.25 SLPM, 1.5 SLPM, and 2 SLPM, and the gas void fraction range of the medium during the flow process was approximately from 3.84% to 11.76%.

The experiment constructed ANN (Artificial Neural Network) and CNN (Convolutional Neural Network) models, which are also neural networks, for comparison with the GRU model. Under the condition that the settings such as the activation function and the learning rate control function of the models were consistent, Figure 8 shows the changes in the loss function values of the three algorithms during the training process. Figure 9 shows the mean square error calculated by the model on the validation set. As can be seen from the box plot, the median, upper, and lower whisker values of the GRU model are significantly smaller than those of the other two models. Although it has more outliers, the fluctuation range is small and the maximum value of the outliers does not exceed 0.065. It performs best on the validation set.

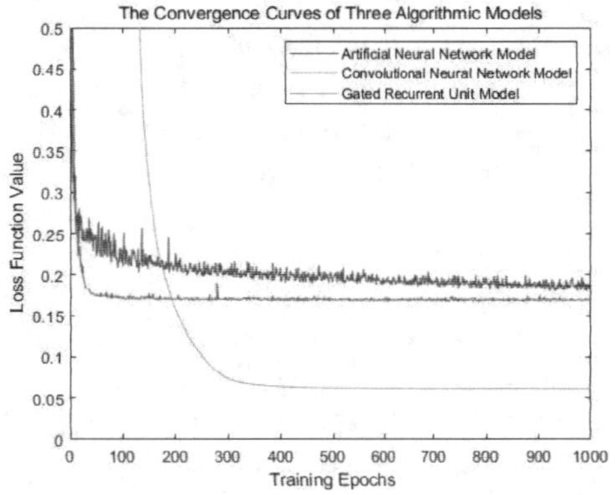

Figure 8. Comparison of the training effects of multiple sets of models

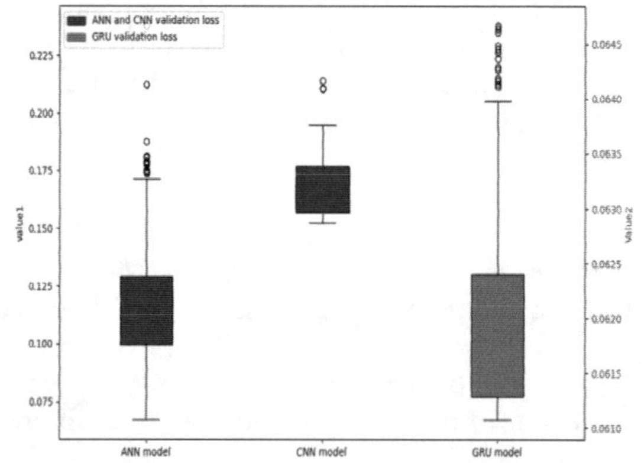

Figure 9. The mean error of the validation sets for each algorithm

Figure 10 shows the prediction error of the GRU model under different gas-liquid ratio working conditions. Among them, the experimental group with a gas volumetric flow rate of 1.25 SLPM was never used during the training process, but its relative prediction error does not exceed 2%, which reflects the generalization ability of the model. In addition, the model has undergone a large amount of training for the working conditions of the other three groups of experimental data. When facing newly collected experimental data under the same working conditions, the model performs more stably and has higher prediction accuracy. To conduct a further test on the generalization ability of the GRU model, the model is required to make continuous predictions on flow measurements within a certain period. The true value curve of the flow rate and the predicted curve are shown in Figure 11.

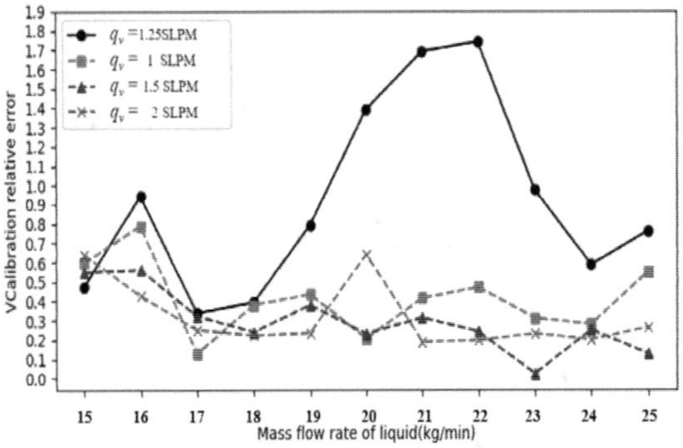

Figure 10. Relative error of GRU model in multiple sets of experimental predictions

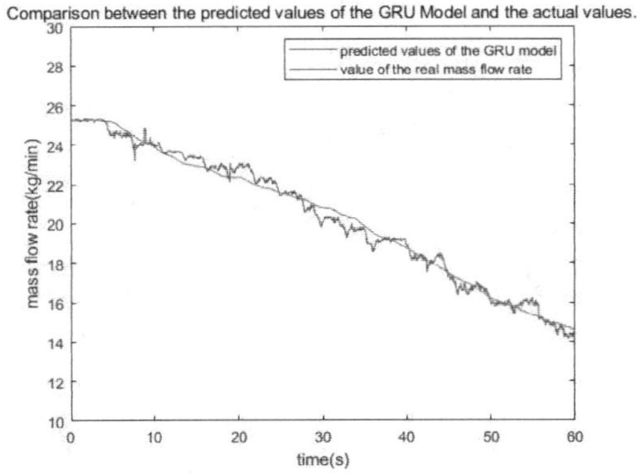

Figure 11. Flow measurement by GRU model and standard meter

It can be seen from Figure 11 that the GRU model can effectively learn the mapping related to mass flow from the data. When applied to an untrained scenario, the initial experimental mass flow is relatively stable, and the model can accurately predict a stable value. When the mass flow begins to change more significantly, the prediction error of the model also increases, which is related to the more fluctuating characteristics of the vibration signal data. However, the predicted value of the model only fluctuates around the true value. After conversion of the results, it is proven that: under this state, the instantaneous measurement error of the mass flow does not exceed 0.92 kg/min, and the instantaneous measurement accuracy has reached 94.41%. The cumulative mass flow measurement error within 60 seconds is only 2.63%.

4. Conclusion

This paper puts forward a scheme of combining digital signal processing to extract feature vectors and then constructing an algorithm model to improve the measurement accuracy of the Coriolis flowmeter under two-phase flow conditions.

Firstly, a measurement platform for the Coriolis flowmeter under two-phase flow was built through research, and the accurate acquisition of multiple types of data was realized by means of hardware. The power spectrum algorithm in digital signal processing was chosen to calculate the phase difference in the small cycle of the vibration signal, and the results showed the excellent phase difference tracking performance of the algorithm. Then, the applicability of the characteristics of the GRU model itself in flow measurement was expounded, and a network structure capable of integrating three groups of feature vectors was established. Through horizontal comparison with other deep learning models, the performance advantages of the GRU model on this issue were further demonstrated, which proved that the model could learn the mapping relationship between data types such as vibration signals and mass flow. Finally, the algorithm model was used to calculate the experimental data that had not been trained to test the generalization ability of the model. The experiments showed that the predicted values of the model could track the real values relatively well and had excellent performance in situations where the two-phase flow fluid medium changed dynamically. All of the above indicate the applicability of this research to the measurement of the Coriolis flowmeter under two-phase flow conditions.

References

[1] D. Stadler. Measuring and operating circuit of a Coriolis-type mass flowmeter [P]. U.S. Patent: 6, 073, 495, 2000-6-13.
[2] H. Denis. Multi-rate digital signal processor for vibrating conduit sensor signals [P]. WIPO Patent: 0 101 083, 2001-1-4.
[3] Tu Yaqing and Zhang Haitao. Method for CMF signal processing based on the recursive DTFT algorithm with negative frequency contribution [J]. IEEE Transactions on Instrumentation and Measurement, 2008, 57 (11): 2, 647-2, 654.

[4] Zhang Y. J., Liu Y. J., Liu Z. D., et al. Developing a Long Short-Term Memory-based signal processing method for Coriolis mass flowmeter [J]. Measurement, 2019, 148 (2019): 106, 896.
[5] Li Chunhui, et al. Improvement of signal processing in Coriolis mass flowmeters for gas-liquid two-phase flow [J]. Frontiers of Information Technology & Electronic Engineering, 2021, 22 (2): 272-286.
[6] Hou Q. L., Xu K. J., Fang M., et al. Gas-Liquid Two-Phase Flow Correction Method for Digital CMF [J]. IEEE Transactions on Instrumentation and Measurement, 2014, 63 (10): 2, 396-2, 404.
[7] Yue J., Xu K. J., Liu W., et al. SVM-based measurement method and implementation of gas-liquid two-phase flow for CMF [J]. Measurement, 2019, 145: 160-171.
[8] Ibryaeva Olga L., Denis K. Lebedev, and Manus P. Henry support vector machine modeling applied to benchmark data set for two-phase Coriolis mass flow metering [J]. Flow Measurement and Instrumentation, 2021, 81: 102, 014.
[9] V. A. Romanov. Effects of dissipative forces on Coriolis flowmeter readings [J]. Flow Measurement and Instrumentation. Doi: 10.1016/j.flowmeasinst.2022.102202.
[10] Walaa Saber Ismail. Optimized feature-based image sentiment analysis using the refined gated mechanism in GRU neural network model [J]. International Journal of Information Technology. Doi: 10.1007/S41870-024-02346-3.

Numerical simulation study of interlayer properties on the stress state and performance of solder joints in ni3al nickel-based alloy

Maksym Matviienko[1,2,#], Boyi Wu[1,#], Zhihong Cai[1], Shiyi Gao[1], Qi Li[1] and Haitao Gao[1,*]

[1] China-Ukraine Institute of Welding, Guangdong Academy of Sciences, Guangdong Provincial Key Laboratory of Advanced Welding Technology, Guangzhou, 510650, China.
[2] Kherson Educational–Scientific Institute, Admiral Makarov National University of Shipbuilding, Kherson, 73003, Ukraine.

Co-first author
*Corresponding author's e-mail: gaoht@gwi.gd.cn

Abstract. The influence of interlayer characteristics on the stress state and performance of nickel-based alloy diffusion welded joints is still unclear, which limits its application. In this paper, the effect of the interlayer properties of transient liquid phase diffusion welding on the joint stress was studied using numerical simulation. The results show that when the relative thickness of the solder layer is between 0.0005 and 0.01, complex volumetric stress states will occur in the base material near the joint outer surface and inside the solder layer. This phenomenon leads to plastic deformation in the micro-region of the transient liquid phase diffusion welded joint. When mechanical loads are applied to the welding point, the degree of softening or hardening of the metal in the welding area depends on the comparison between Young's modulus of the interlayer and that of the base material. This study provides theoretical support for stress regulation in transient liquid phase diffusion joints.

1. Introduction

Aero-engines are advancing toward higher thrust-to-weight ratios, requiring aero blades to endure demanding service temperatures [1]. This evolution presents significant challenges for the design, manufacturing processes, and performance of aero-engine blade materials [2]. Ni_3Al Nickel-based alloys exhibit high mechanical strength, excellent oxidation resistance, and robust resistance to gas corrosion at elevated temperatures, positioning them as the most promising candidates for aero-engine blades. To reduce the operating temperature of blade materials, modern aero blades incorporate hollow designs with high-speed gas cooling structures and thermal barrier coatings [3]. These complex hollow configurations necessitate machining the blades in halves and subsequently joining them, introducing new challenges to the welding process. Conventional fusion welding connection techniques encounter difficulties when joining single-crystal nickel-based alloys [4-6]. The uncontrolled crystallization during fusion welding inevitably leads to the formation of equiaxed grain structures, brittle phases, coarse microstructures, and significant residual stresses within the joint, thereby diminishing the high-temperature mechanical properties of the weld [7-8]. Conversely, brazing techniques suffer from drawbacks, including weld seams having service temperatures over 100 °C lower than the parent material and reduced mechanical strength.

Diffusion welding establishes joints through the mechanism of solid-state diffusion, characterized by low heating temperatures and the absence of melting in the base materials. This process is marked by a straightforward workflow and a reduction in defects within the welded joints. Diffusion welding offers irreplaceable advantages over fusion welding techniques, such as laser welding, TIG (Tungsten

Inert Gas) welding, and electron beam welding, particularly for materials that present challenges during conventional welding processes. It is especially well-suited for joining novel materials, including dissimilar metals, heat-resistant alloys, ceramics, intermetallic compounds, and composite materials. So far, the poor weldability and inferior strength of nickel-based Nickel-based alloy weld zones remain critical challenges for the joining technology of hollow aero blades [9]. Diffusion welding, with its advantages of high joint strength and preservation of the parent material's crystal structure, emerges as the most promising candidate process for hollow aero blade joining [10-11]. This study employs numerical simulation and computational methods to examine the variations in stress and strain of Ni_3Al alloys during the diffusion welding process. The findings of this work are significant for addressing the challenges associated with joining hollow aero blades made from nickel-based alloys in aero-engines.

2. Simulation methods

The finite element analysis of the cylindrical joint in transient liquid phase diffusion welding was conducted by using simulation software. As illustrated in Figure 1, the height of the solder layer ranges from s = 10 to 200 μm (with a relative thickness ratio s/d ranging from 0.0005 to 0.01). The axisymmetric two-dimensional element was selected for modeling the stress-strain state. Given the pronounced stress gradients present in the narrow region adjacent to the solder layer, a gradient mesh refinement approach was employed, progressively increasing finite element sizes as one moves away from the connection zone where stresses and strains are concentrated (see Figure 1b). In both the solder interlayer material and that being joined within this connection zone, square-shaped finite elements with an aspect ratio of 1:1 were utilized. The dimensions of these finite elements were determined such that there were at least twenty elements across the thickness of the solder layer. By exploiting symmetry about half its thickness concerning the solder interlayer, we constructed a finite element model representing only its upper half (the upper cylinder along with half of the interlayer thickness). Model nodes along this section were constrained along the Y-axis (cylinder axis) to respond to movements along the X-axis (radial direction), thereby mitigating any "small hole" effect; additionally, constraints were applied at the lower edge corresponding to mid-thickness in Y-direction (refer to Figure 1c).

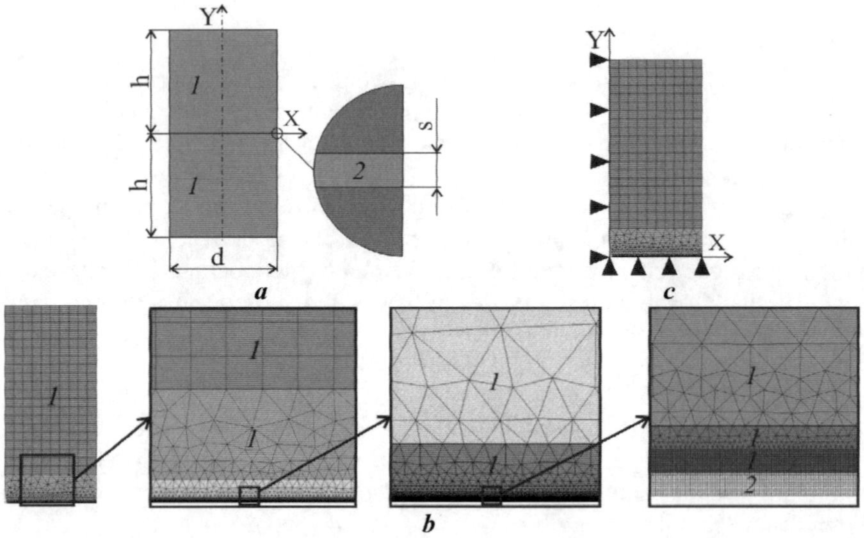

Figure 1. Schematic diagram of finite element meshing (a), finite element model With gradient breakdown (b), and finite element model fixation diagram (c).

3. Results and discussion

The joint stress of the sandwich structure in transient liquid phase diffusion welding is studied. The results pertaining to the elastic modulus are presented in Figure 2, while the findings related to the linear thermal expansion coefficient are illustrated in Figure 3. Due to the differences in Young's modulus and thermal expansion coefficient between the substrate material and the intermediate layer material, a complex stress state is formed in the area near the intermediate layer.

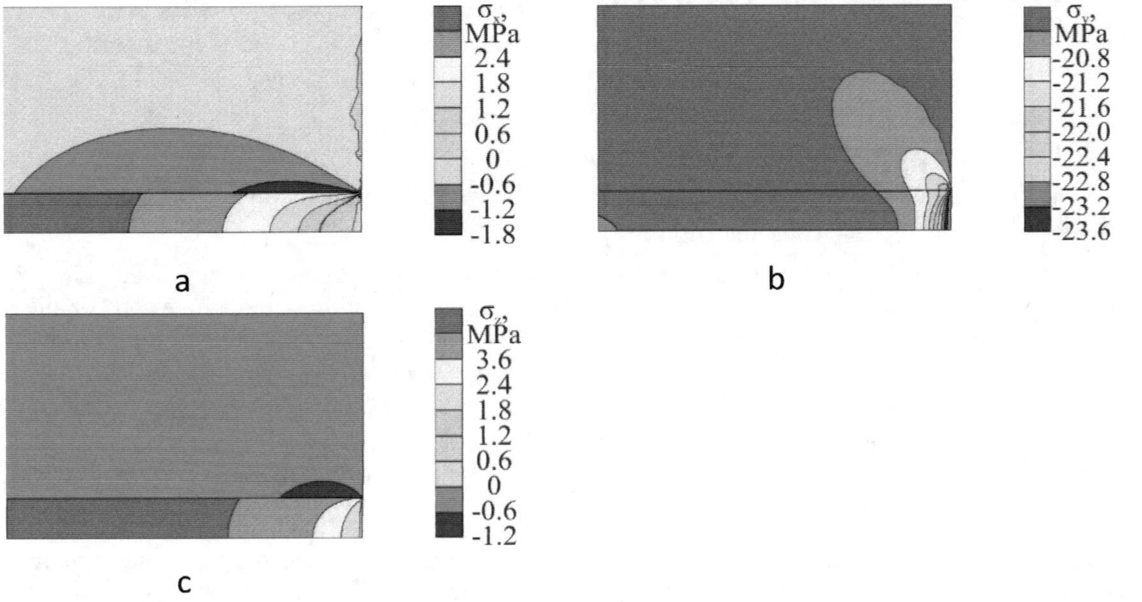

Figure 2. Fields of radial σ_x (a), axial σ_y (b), and circumferential σ_z (c) stresses under force loading of a brazing unit.

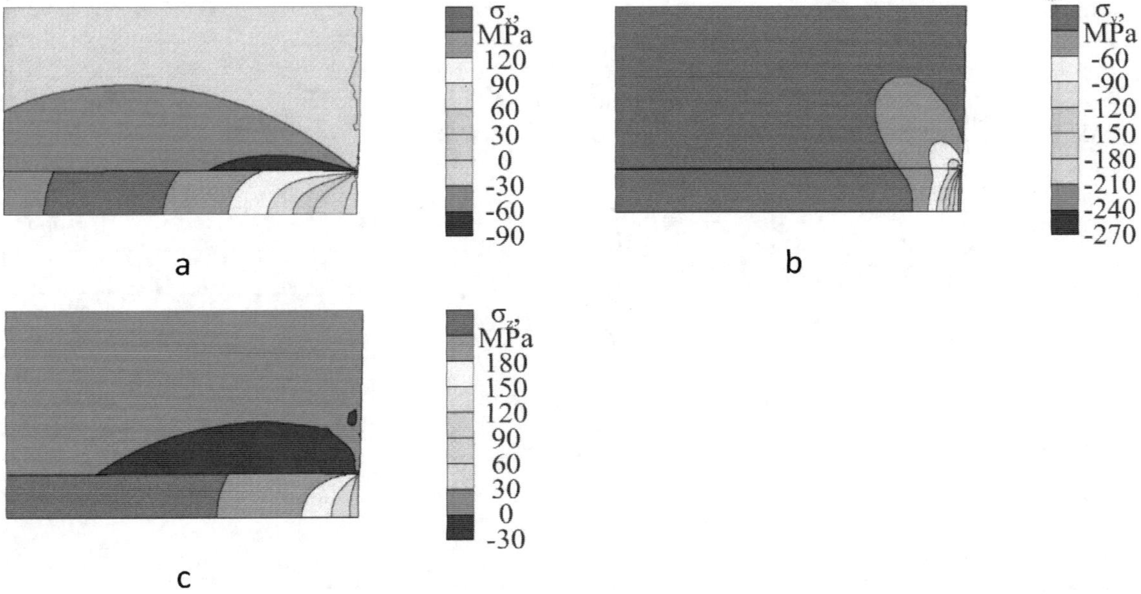

Figure 3. Radial σ_x (a), axial σ_y (b), and circumferential σ_z (c) stress fields during solder assembly cooling.

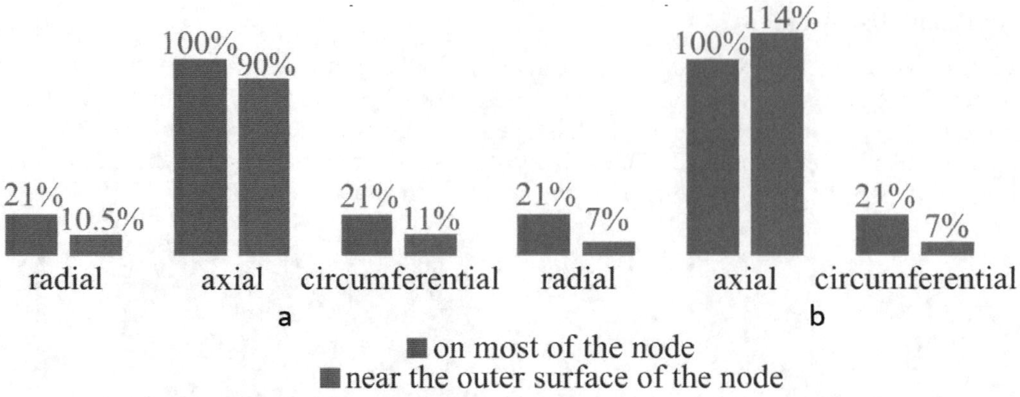

Figure 4. Stress levels (% of the applied axial load) in the interlayer material under the force loading of the soldered unit: (a) soft interlayer; (b) hard interlayer.

As illustrated in Figure 4, the solder layer exhibits an uneven and complex stress state along its entire length when subjected to applied force loading. Additionally, within a small zone comparable to the layer's thickness located at the joint edge, the stress distribution differs from that in the majority of the layer. Specifically, radial and circumferential stresses account for 21% of the total stress throughout most of the solder layer. At the joint edge, axial stresses decrease when a soft layer is used and increase when a hard layer is employed.

As shown in Figure 5, when a welded joint with a "soft" intermediate layer is subjected to compressive force loading, the stress generated at the joint has little effect on the tensile strength and deformation of the joint. The trend reverses when soldered joints with a "hard" interlayer are subjected to force loading. On the outer side of the cylindrical body formed by the welded joint, the strength of the joint is enhanced, and the stiffness coefficient increases to 1.03 (Figure 5a). The metal is softened across most of the interlayer's joint length, reducing the stiffness coefficient to between 0.90 and 0.91 (Figure 5b). This means that the metal in the "hard" layer is softened, diminishing its rigidity. Changes in the thickness of the solder layer, within the range considered in this study, have little effect on the nature of the stress state and the distribution of k_s values. However, the size of the zone exhibiting a complex volumetric stress state changes in direct proportion to the thickness. This leads to an increase in the unevenness of the stress state in the joint zone and a possible decrease in the reliability of the soldered joint with an increase in the thickness of the interlayer.

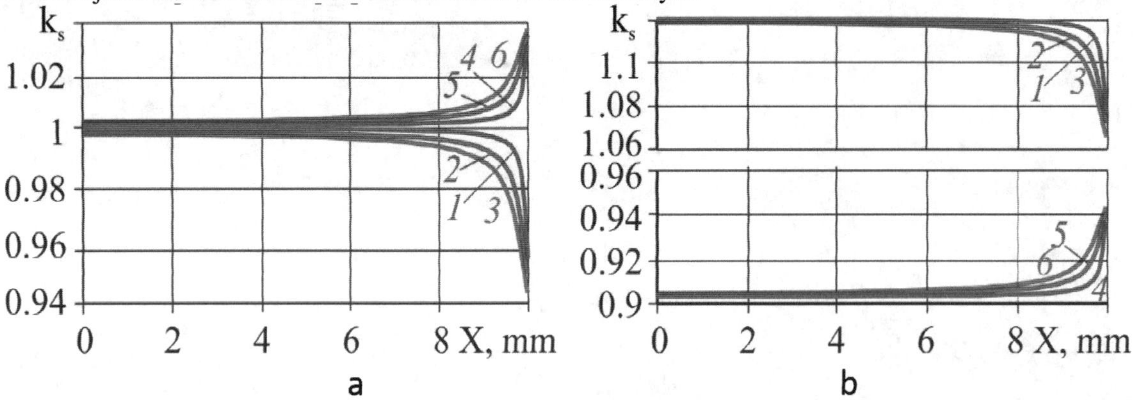

Figure 5. Stiffness coefficient variation with the hardness of the intermediate layer: (a) base material; (b) interlayer.

As shown in Figure 6, the degree of softening or hardening of the metal in the joint area depends on the relative relationship between Young's modulus of the interlayer material and that of the matrix material (E_{in}/E_{bm}). This relative relationship directly affects the stress distribution. On the outer side of the joint cylinder, $k_s = -0.09\cdot(E_{in}/E_{bm})^2+0.38\cdot Ein/Ebm+0.71$, while on the inner side of the joint, $k_s = 0.165\cdot(E_{in}/E_{bm})^2 -0.8\cdot E_{in}/E_{bm}+1.635$.

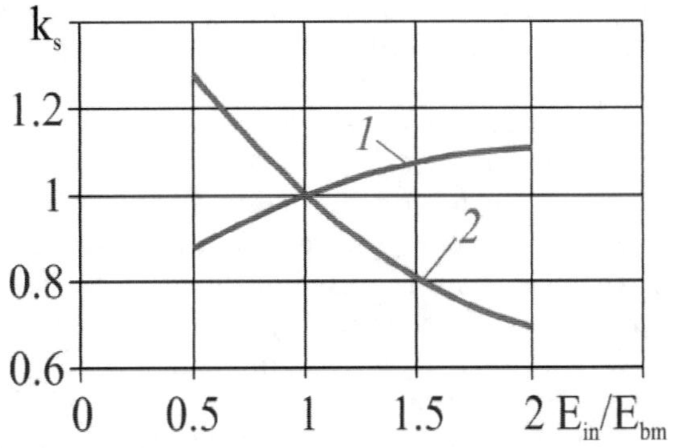

Figure 6. Dependence of the stiffness coefficients of the interlayer 1 and the base material 2 on the relative stiffness ratio E_{in}/E_{bm}.

The application of pressure during cooling significantly affects the stiffness coefficient k_s and the degree of hardening or softening. In joints with a small CLTE, as one moves away from the axis toward the joint edge, the interlayer k_s first decreases from 0.95 to 0.55 and then increases to 1 at the edge itself (Figure 7 (1a)), similar to cooling without pressure but near the edge. Conversely, in joints characterized by a significant coefficient of linear thermal expansion (CLTE) interlayer, the stiffness coefficient exhibits an increase from 1.05 to 2 as it approaches the edge. Subsequently, it decreases back to 1 at the very edge (Figure 7 (2a)). That is, applying compressive force during cooling, from the point of view of the stress state arising in the base metal, is advisable only in the case of using an interlayer with a TCLE lower than that of the base metal. To increase the reliability of the connection in the interlayer zone, applying compressive force is advantageous only when the TCLE of the interlayer is higher than that of the base metal.

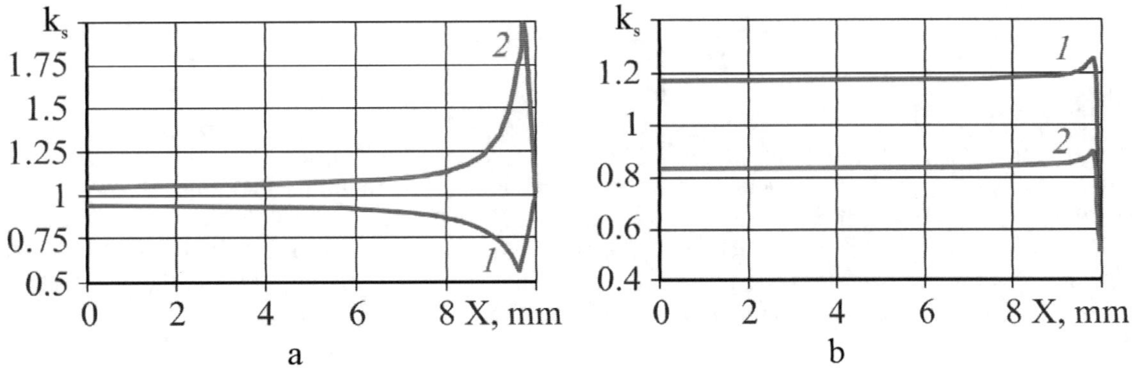

Figure 7. Variation of the stiffness coefficient ks at the junction of the intermediate layer (1) with a smaller linear thermal expansion coefficient and the intermediate layer; (2) with a larger one: (a) matrix metal; (b) interlayer.

These studies' results were considered when soldering a batch of test samples from the IC10 alloy. Tests of soldered joints after heat treatment showed that mechanical properties close to those of the base metal were provided. The tensile strength of the joints with solder was 98% (268 MPa) of the strength of the base metal (275 MPa). The durability requirements were also met at the level of the requirements for the base metal (102 hours) at a test temperature of 1100 °C.

4. Conclusions

1. In welded assemblies where the relative thickness of the solder layer (s/d) ranges from 0.0005 to 0.01, complex volumetric stress states are generated both in the base metal near the outer surface of the joint and within the solder layer itself. This phenomenon results in plastic deformation occurring in the micro-region of the transient liquid phase diffusion welded joint.

2. During the mechanical loading of soldered joints, the extent of softening or hardening of the metal within the joint area is dependent on the relative relationship of Young's modulus of the intermediate layer to that of the matrix material (E_{in}/E_{bm}). When Young's modulus of the interlayer is comparatively low, it has a minor impact on the stress in the joint area, only causing a slight weakening of the micro-region near the outer surface. Conversely, when there is an interlayer with a larger Young's modulus, the area near the outer edge of the joint cylinder is strengthened, while the interlayer is softened due to its relatively smaller Young's modulus.

3. Applying pressure at varying temperatures significantly impacts the degree of hardening or softening in the micro-regions of the joint material. The base material exhibits maximum softening at its edges in joints featuring a small coefficient of linear thermal expansion (CLTE) interlayer. Conversely, in joints with a large CLTE interlayer, strengthening occurs instead. Furthermore, in cases where the interlayer possesses a high CLTE, the base metal is also strengthened.

4. Within the interlayer itself, the stiffness coefficient remains largely stable across approximately 95% of the joint area, maintaining values around 1.18 for regions with low CLTE interlayers and 0.83 for those with high CLTE interlayers. This observation suggests that applying pressure when utilizing an interlayer with a lower CLTE than that of the base metal enhances the strength of this interlayer metal, thereby increasing susceptibility to cracking. In contrast, when employing an interlayer characterized by a higher CLTE, the application of pressure diminishes the stiffness of this material under stress.

5. By modulating applied pressure, it becomes feasible to control how stiffness coefficients are distributed; however, determining optimal pressure levels must be conducted through case-by-case modeling. Applying compressive force during cooling on the side of the base metal is advisable only when using an interlayer with a lower CLTE than that of said base metal. Conversely, applying compressive force on the side containing an interlayer is advantageous solely when its CLTE exceeds that of the base metal.

This research is conducive to enhancing the reliability of transient liquid phase diffusion bonding.

Acknowledgements

This work is supported by the Chinese Ministry of Science and Technology Project "Study on interlayer design, preparation, and thermal deformation control in transient liquid phase diffusion welding of Ni_3Al Nickel-based alloy (No. G2023030019L)", the GDAS' Project of Science and Technology Development (2022GDASZH-2022010109, 2020GDASZH-2020030100l), the Science and Technology Planning Project of Guangdong Province (2022B1212050001), and the National Key Research and Development Program of China (No. 2020YFE0205300).

Reference

[1] Wei, J., et al. (2024) Advances in resin matrix composite fan blades for aircraft engines: A review. Thin-Walled Structures., 202: 112058. https://doi.org/10.1016/j.tws.2024.112058.
[2] Yang, P., et al. (2022) Review of damage mechanism and protection of aero-engine blades based on impact properties. Engineering Failure Analysis., 140: 106570. https://doi.org/10.1016/j.engfailanal.2022.106570.
[3] Yan, S., et al. (2024) Advances in aeroengine cooling hole measurement: A comprehensive review. Sensors., 24(7): 2152.https://doi.org/10.3390/s24072152.
[4] Jappes, J. W., et al. (2022) Welding on Ni based super alloys–A review. Materials Today: Proceedings., 60: 1656-1659. https://doi.org/10.1016/j.matpr.2021.12.208.

[5] Peng, J., et al. (2023) High-temperature mechanical properties and microstructure of welded joint in GH4169/IC10 dissimilar nickel-based superalloys by vacuum electron beam welding. Materials Science and Engineering: A., 884: 145561.https://doi.org/10.1016/j.msea.2023.145561.
[6] Azari, M., et al. (2021) Investigation of temperature distribution and melt pool microstructure in laser fusion welding of Inconel 625 superalloy. Journal of Laser Applications., 33: 022015. https://doi.org/10.2351/7.0000376.
[7] Kataria, R., et al. (2021) Welding of super alloys: a review. Materials Today: Proceedings., 38: 265-268. https://doi.org/10.1016/j.matpr.2020.07.198.
[8] Perka, A. K., et al. (2022) Advanced high-strength steels for automotive applications: arc and laser welding process, properties, and challenges. Metals., 12(6): 1051. https://doi.org/10.3390/met12061051.
[9] Sonar, T., et al. (2024) An overview of microstructure, mechanical properties and processing of high entropy alloys and its future perspectives in aeroengine applications. Materials Science for Energy Technologies., 7: 35-60. https://doi.org/10.1016/j.mset.2023.07.004.
[10] Zhao, Y., et al. (2021) Deformation behavior and creep properties of Co-Al-W-based superalloys: A review. Progress in Natural Science: Materials International., 31(5): 641-648. https://doi.org/10.1016/j.pnsc.2021.09.009.
[11] Motamedi, M., et al. (2024) Molecular dynamics simulation of superalloys: A Review. Archives of Computational Methods in Engineering., 31(4): 2417-2429. https://doi.org/10.1007/s11831-023-10051-w.

Journal of Physics: Conference Series **3009** (2025) 012056 doi:10.1088/1742-6596/3009/1/012056

Effect of Mg doping on mechanical property and shape memory behavior of Ni-Fe-Ga alloy

Mingguang Wang *, Yu Shan

Key Laboratory for Anisotropy and Texture of Materials (Ministry of Education), Northeastern University, Shenyang, PR China

*Corresponding author's e-mail: wangmg@imp.neu.edu.cn

Abstract: We prepared Ni-Fe-Ga alloys and investigated the effect of Mg doping on transformation temperatures, and mechanical and shape memory properties of these alloys by using thermal analysis (DSC), optical microscopy (OM), X-ray diffraction (XRD), and compression testing. The transformation temperatures decrease with the increase of Mg content. Both compressive strength and plasticity of the alloys increase with Mg doping. In addition, the addition of trace Mg is conducive to shape memory property.

1. Introduction

NiFeGa is a new shape memory alloy and has a variety of excellent properties such as superelasticity (pseudoelasticity) [1], shape memory effect [2], and elastocaloric effect [3]. The recoverable strain in NiFeGa alloy is comparable to that in NiTi alloy. Moreover, NiFeGa alloy possesses a larger temperature range of superelasticity and outstanding functional stability. Therefore, NiFeGa alloy is expected to be a substitute for NiTi alloy. So far, these good properties can be obtained in the monocrystalline NiFeGa alloy, considering the high price and time-consuming preparation of the material in the single crystalline state, the polycrystalline alloy deserves much attention. However, the brittleness of the polycrystalline in NiFeGa alloy has always been an obstacle to their application.

Until now, several researchers have found ways to address the shortcomings of brittleness. For example, the introduction of a ductile phase in single-phase alloys can effectively improve the ductility of polycrystalline alloys, as demonstrated in two-phase Ni-Fe-Ga-Co, Co-Ni-Si and Co-V-Ga alloys [4-6]. However, the results showed that the precipitation of an excessive second phase causes a decline in shape memory properties because the second phase can produce plastic deformation during deformation. Recently, doping of the Mg element in master alloys has been found to improve the mechanical properties of the alloys. At the same time, controlling the amount of Mg does not harm the functional properties of the alloys [7]. In the present study, we prepared a new type of quaternary shape memory alloy by adding a trace amount of Mg into the NiFeGa alloy. The influence of trace Mg on mechanical properties and shape memory effect in polycrystalline state was investigated.

2. Experimental procedures

Polycrystalline $(Ni_{55}Fe_{18}Ga_{27})_{100-x}Mg_x$ (at.%) (x = 0, 0.1, 0.3) alloys, named Mg0, Mg0.1, Mg0.3, respectively, were prepared by arc melting under high purity argon atmosphere. During the melting process, each sample was turned over and melted four times to ensure the uniformity of the chemical composition of the ingot and then removed after cooling. Subsequently, these ingots were sprayed into 10 mm rods in diameter. The rods were further annealed at 1150°C for 72 h in a vacuum, followed by water quenching in order to obtain good chemical homogeneity. The homogenization rods were annealed at 800°C for 1 h, water quenching. Some 3 mm × 3 mm × 5 mm samples were cut from the annealed samples for mechanical property testing.

The phase structure was identified at room temperature by using an X-ray diffractometer with Cu-K_{α} radiation. The characteristic temperatures of the alloys during martensitic transformation were measured by differential scanning calorimetry with a cooling/heating rate of 10°C min^{-1}. Room temperature uniaxial compression tests were performed with a strain rate of 0.1 mm/min. The compression strength and SME of the alloys were obtained from the stress-strain curves. The calculation method of SME and the recovery rate can refer to our previous work [8].

3. Results and discussion

XRD patterns of the annealed alloys at ambient temperature are shown in Figure 1. The diffractions can be indexed as a mixture of 10M, 14M, NM martensite, and the precipitated γ phase for all alloys. The appearance containing martensite and γ phase microstructure is independent of the amount of Mg addition. We calculated the diffraction patterns of the observed martensitic structures and γ phase by using the software PowderCell [9]. The calculated lattice constants of the three martensites mentioned above and γ phase are given in Table 1.

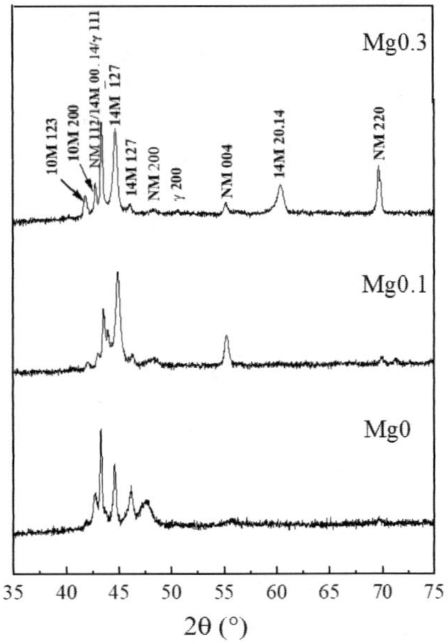

Figure 1. XRD patterns of $(Ni_{55}Fe_{18}Ga_{27})_{100-x}Mg_x$ alloys.

Table 1. Lattice constants of martensite and γ phase.

Phase	Lattice constants
10M (S.G. P2/m)	a = 0.426 nm, b = 0.543 nm, c = 2.10 nm, β = 90.2°
14M (S.G. I2/m)	a = 0.422 nm, b = 0.551 nm, c = 2.94 nm, β = 93.0°
NM (S.G I4/mmm)	a = 0.380 nm, c = 0.654 nm
γ phase (S.G. $Fm\bar{3}m$)	a = 0.362 nm

Due to the high supercooling in the spray casting process, the alloy forms a supersaturated solid solution after solidification and shows a martensitic phase microstructure. After heat treatment at 800°C for 1 h, γ phase precipitates from the alloy matrix. As illustrated in Figure 2, for Mg0 or Mg0.1 alloy, γ phases precipitate in a stripe form along the grain boundaries, while for Mg0.3 alloy, γ precipitates located at grain boundaries and almost completely covers the grain boundaries. The precipitate phase is rarely observed in the grains because the alloys were annealed at high temperatures.

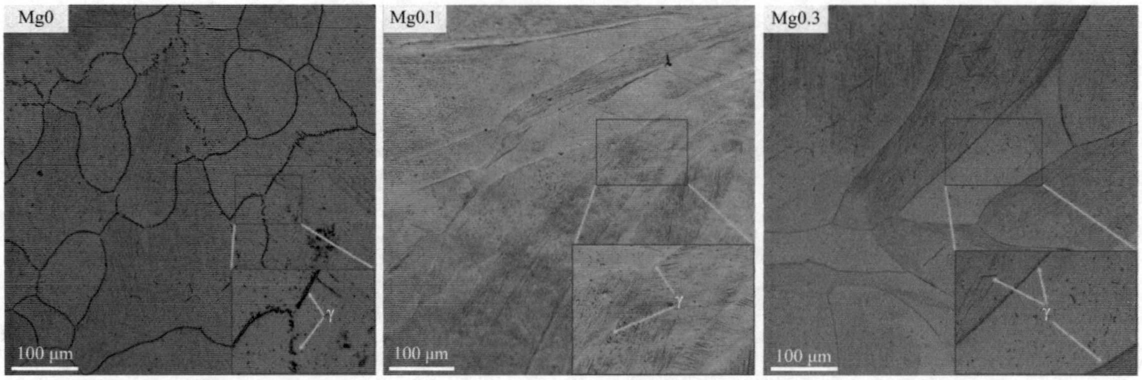

Figure 2. Metallographic microstructure of $(Ni_{55}Fe_{18}Ga_{27})_{100-x}Mg_x$ alloys.

From the DSC curves of the three alloys (Figure 3(a)), one can observe that all the samples possess a reversible martensitic transformation. The transformation peaks show the obvious thermal hysteresis which is inherent to a first-order transition. The doping of a trace amount of Mg into the NiFeGa matrix is very effective on the transformation temperatures. The doping of 0.1%Mg reduces the peak value of phase transformation by more than 15°C. The temperature hysteresis, $\Delta T = A_f - M_s$, is also calculated as 8.89 °C for Mg0, 7.31 °C for Mg0.1, and 9.56 °C for Mg0.3, respectively. Because Mg has two valence electrons, the valence concentration (e/a) of alloys decreases regardless of replacing any of the elements Ni, Fe, and Ga with Mg. The decrease in transformation temperatures is expected, as shown in Figure 3(b), which is consistent with the obtained results in NiFeGa alloys. The influence of the γ phase on martensitic transformation in alloys is complicated. After the precipitation of the γ phase, the composition of the matrix will be changed, thus affecting the martensitic transformation temperatures. In addition, the γ phase will hinder the migration of the parent-martensitic phase interface and increase the fraction between grains, leading to the increase of phase transition hysteresis. However, the difference in the proportion of γ phase in the studied alloys is relatively small, indicating that the influence of γ phase on the transition temperatures is not obvious.

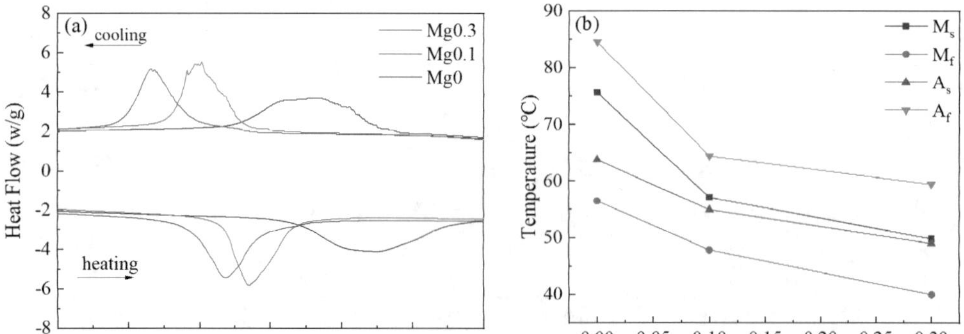

Figure 3. (a) DSC curves (b) Dependence of transformation temperatures on Mg content in $(Ni_{55}Fe_{18}Ga_{27})_{100-x}Mg_x$.

Considering the brittleness of the studied alloys, a compression test was conducted to assess the mechanical properties. All the alloys were compressed to fracture during the compression process. The compressive stress-strain curves of $(Ni_{55}Fe_{18}Ga_{27})_{100-x}Mg_x$ alloys at room temperature are shown in Figure 4(a). Without the addition of the Mg element, the compressive strength and maximum compressive strain are approximately 1050 MPa and 14.8%, respectively. With 0.1% Mg doping, the strength and strain slightly improved compared to the undoped Mg sample, i.e., 1225 MPa and 17.4%, respectively. When the Mg content is increased to 0.3%, the compressive strength and strain are increased obviously, reaching 1891 MPa and 35.4%, respectively. The morphology of compression fracture of the studied alloys was observed by scanning electron microscope. The fractography of Mg-free or 0.1% doped Mg alloy reveals mixed morphology of intergranular fracture and cleavage fracture (Figure 4(b)), which shows the typical brittle fracture along the grain boundaries. In addition, some tearing edges are also observed in some particular regions, which suggests that plastic deformation occurs before fracturing. When the Mg content is increased to 0.3%, as shown in Figure 4(c), the amount of ductile tearing ridge remarkably increases, which manifests that Mg0.3 alloy possesses relatively higher strength and larger compressive strain. The prior studies showed that a soft phase in the grain interior remarkably improves alloy ductility, because this kind of soft material cannot produce martensitic transition, and thus may interfere with the reversible transformation of shape memory alloys, reducing the overall recoverable strain. However, when the precipitated phases are mainly distributed along grain boundaries, an optimal effect on ductility can be obtained due to the cushion action of the precipitated phases on grain boundaries.

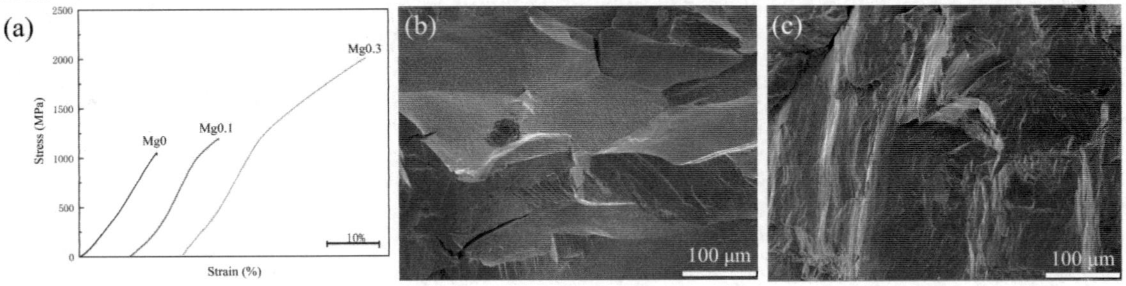

Figure 4. The compressive stress-strain curves (a), and the fractography of Mg0 alloy (b) and Mg0.3 alloy (c).

The compression experiment of $(Ni_{55}Fe_{18}Ga_{27})_{100-x}Mg_x$ was conducted to investigate the relationship between Mg doping and the shape memory effect of alloys. A pre-strain of 4% and a stress rate of 0.1 mm/min were selected for the compression tests. Afterward, the samples were heated to 100°C and resided at the temperature for 15 min to obtain the recovery ratio. Figure 5 shows the stress-strain curves of the alloys investigated at 4% pre-strain. As can be seen from the figures, when the samples are compressed by 4% pre-strain and then unloaded down to zero, they will spontaneously recover a strain greater than 2%. After heating, the martensite will reversely transform and the alloys will return to the state close to that before compression, which is the embodiment of the shape memory behavior of the alloys. As illustrated in Table 2, the Mg0.3 sample shows the highest shape recovery rate among the three samples, which may result from a continuous layer of the precipitates along grain boundaries compared to the Mg0 or Mg0.1 sample. According to the available research results, precipitates along grain boundaries can alleviate grain boundary constraints and enhance local strain compatibility, lowering the critical stress for transformation and enhancing strain recovery of polycrystalline SMAs [10]. The microscopic mechanism of the improvement of strain recovery with Mg doping deserves further investigation.

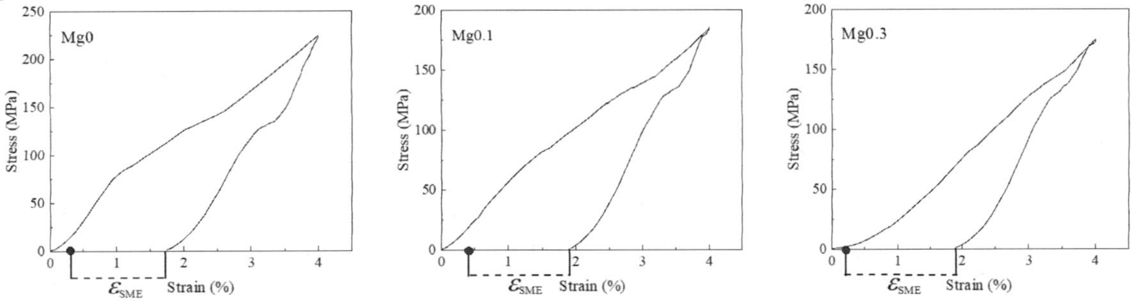

Figure 5. Shape memory property of $(Ni_{55}Fe_{18}Ga_{27})_{100-x}Mg_x$ alloys.

Table 2. SME testing at 4% pre-strain.

Sample	SME strain (%)	Recovery ratio (%)
Mg0	1.42	79.89
Mg0.1	1.44	82.45
Mg0.3	1.62	87.41

4. Conclusions

$(Ni_{55}Fe_{18}Ga_{27})_{100-x}Mg_x$ (x = 0, 0.1, 0.3) alloys were prepared and the effect of trace Mg addition on transformation temperatures, mechanical property, and shape memory behavior of the alloys was investigated.

(1) The as-cast alloys have a martensite-phase microstructure, after a heat treatment at 800°C for 1 h, γ phases precipitate from the martensitic matrix and distribute along with the grain boundaries. Doping of a small amount of magnesium into the alloys tends to reduce the phase transition temperatures.

(2) Proper Mg addition and heat treatment can improve both the mechanical property and SME of the alloys.

References

[1] Daroczi, L., Kamel, S. M., Tahara, M., et al. (2024) Acoustic emission during approaching the critical point on stress-temperature diagram of martensitic transformation in $Ni_{48}Fe_{20}Co_5Ga_{27}$ (at.%) single crystal. Intermetallics, 173: 108432.
[2] Timofeeva, E. E., Dmitrienko, M. S., Panchenko, E. Yu, et al. (2024) Thermal and cyclic stability of two-way shape memory effect in $Ni_{44}Fe_{19}Ga_{27}Co_{10}$ single crystals. Russ. Phys. J., 67: 1273-78.
[3] Cheng, F., Dong, T. J., Wang, Y., et al. (2024) Elastocaloric effect with plateau-shape adiabatic temperature change in Ni-Co-Fe-Ga strain glass alloy. J. Phys. D: Appl. Phys., 57: 13LT02.
[4] Timofeeva, E. E., Panchenko, E. Yu, Zherdeva, M. V., et al. (2025) High-temperature thermoelastic martensitic transformations in $Ni_{44}Fe_{19}Ga_{27}Co_{10}$ single crystals. Mater. Lett., 380: 137744.
[5] Liao, Q., Wang, D., Yang, X., et al. (2024) Larger recovery strains in a CoNiSi alloy due to enhanced reversibility of FCC (sic) HCP martensitic transformation. J. Mater. Sci. Technol., 28: 235-243.
[6] Liu, C., Liu, H. W., Li, Z. B., et al. (2024) Phase transition and internal friction behaviors in Mn- doped Co-V-Ga shape memory alloys. Appl. Mater. Today, 40: 102426.
[7] Liu, Y. J., Hou, J. S., Jiang, C. H., et al. (2024) Theoretical study of synergistic effect of P and Mg on the cohesive properties of Ni_3Al grain boundaries. Comput. Mater. Sci., 237: 112891.
[8] Wang, M. G., Gao, Y. Q., Yang, J. F., et al. (2022) Effect of Zr doping on precipitated phase and mechanical properties of NiFeGa alloy. J. Phys.: Conf. Ser., 2263: 012012.
[9] Kraus, W., Nolze, G. (1996) POWDER CELL—A program for the representation and manipulation of crystal structures and calculation of the resulting X-ray powder patterns. J. Appl. Cryst., 29: 301–303.
[10] Dar, Rebecca D., Chen, Y. (2017) Nanoscale martensitic phase transition at interfaces in shape memory materials. Appl. Phys. Lett., 110: 041906.

Journal of Physics: Conference Series **3009** (2025) 012057 doi:10.1088/1742-6596/3009/1/012057

Numerical simulation of residual stress in dissimilar brazed joints of Nb-1Zr/GH4068

Shuhan Yao [1], Mingyue Wen * [1], Shiyang Wang [2] and Yuan Sun [2]

[1] Shenyang Ligong University, Shenyang, China
[2] Institute of Metal Research, Chinese Academy of Sciences, Shenyang, China

*Corresponding author's e-mail: mywen@sylu.cn

Abstract: This study focuses on the cracking behavior of GH4068 and Nb-1Zr alloy joints brazed with different filler materials. It employs finite element numerical simulation techniques to investigate the distribution of residual stresses in the brazed joints of Nb-1Zr and GH4068 filled with two brazing alloys. The experimental results demonstrate that using BNi-2 filler results in a heightened susceptibility to cracking in the Nb-1Zr joint and the diffusion-affected region of the brazed seam. In contrast, the Au-18Ni filler effectively mitigates crack formation. The findings indicate that Mises stress in the Nb-1Zr/GH4068 dissimilar alloy brazed joint is predominantly concentrated in the brazed seam region and the interfacial reaction layer between the base metal and the brazed joint. Gold (Au) in the Au-18Ni filler imparts exceptional plastic deformation characteristics, enabling it to alleviate the residual stresses present in the Nb-1Zr/GH4068 brazed joint.

1. Introduction

Nb-Zr alloys exhibit exceptional cryogenic plasticity, high-temperature oxidation resistance, radiation resistance, and neutron absorption capabilities, making them extensively used in aerospace, nuclear energy, and biomedical fields [1-2]. The GH4068 superalloy is characterized by low density, a high degree of alloying, excellent microstructural stability, creep resistance, and hot workability [3-4]. Due to the significantly higher melting point and thermal conductivity of Nb-1Zr alloys compared to nickel-based superalloys, combined with the fact that the thermal expansion coefficient of Nb-1Zr is only half that of nickel-based alloys [5], residual stresses inevitably arise in the dissimilar material joints during the cooling process post-brazing. These stresses can trigger crack formation, subsequently reducing the load-bearing capacity and fatigue resistance of the brazed joints and compromising the quality of the connection [6-8].

Currently, investigations into the internal stress field distribution of dissimilar joints after brazing predominantly employ numerical simulation methods, with commonly used finite element software such as ABAQUS, ANSYS, and SYSWELD. Qiao et al. [9] analyzed the distribution of residual stresses in YG8/GH4169 brazed joints using ANSYS finite element software, revealing significant residual stresses in the brazed joints of hard alloys and high-temperature alloys, noting that the incorporation of intermediate layers of Cu or Mo can effectively relieve these residual stresses. Zhang et al. [10] selected eutectic filler materials, Au80Sn20 and Sn63Pb37, for brazing 6061 aluminum alloy to 4J34 Kovar alloy and found that the residual stresses in the joints were predominantly tensile, with higher residual stresses located near the Kovar alloy side of the joint. Wang et al. [11] established a moving Gaussian heat source model of TC4/6061 dissimilar alloys using finite element methods to explore the temperature field distribution under varying laser powers. They observed an uneven temperature field distribution in titanium/aluminum dissimilar alloy joints, but further research on the distribution and effects of residual stresses within the joints was not conducted.

2. Experimental Methods

The base materials used in the experiments are Nb-1Zr alloy and GH4068 high-temperature alloy. The dimensions and chemical compositions of the materials used in this work are shown in Figure 1, Table 1, and Table 2. The brazing fillers used are BNi-2 and Au-18Ni. To ensure that the overall performance of the brazing filler is not adversely affected, the filler foil is prepared using the spinning method, with a thickness of 100μm. The Nb-1Zr alloy, brazing filler, and GH4068 are assembled in a sandwich configuration and brazed using a vacuum brazing furnace. The Nb-1Zr/GH4068 brazed joint is heated to 1050°C and then cooled to room temperature with the furnace, with a holding time set between 5 to 15 minutes.

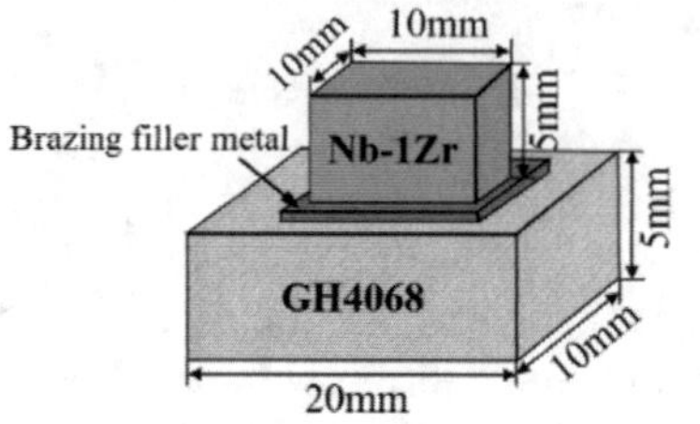

Figure 1. Structure diagram of brazed joint.

Table 1. Main element composition of BNi-2 filler (at%).

Ni	Cr	Si	B	Fe
Bal.	7.0	3.2	4.5	3

Table 2. Chemical compositions of base materials for testing (wt.%).

Base material	Nb	Zr	Ni	Co	Cr	Mo	W	Ti	Al
Nb-1Zr	Bal.	0.89	-	-	<0.0014	0.02	0.02	<0.0014	-
GH4068	-	0.03	Bal.	20-27	13-15	2.4-2.8	1.1-1.3	3.8-4.4	3.8-4

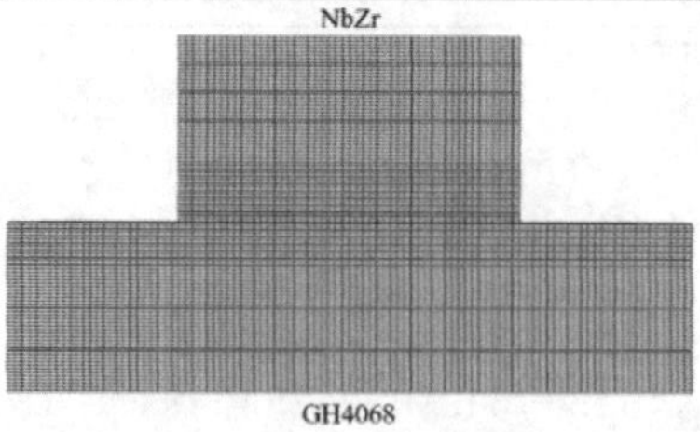

Figure 2. Mesh division diagram of finite element model.

Based on the ABAQUS finite element software, a three-dimensional nonlinear finite element simulation of the brazed joint is established. The stress field calculation is performed using a temperature-stress sequential coupling method. During the model meshing process, the mesh type selected is C3D8RT elements, with a total of 77, 500 elements and 95, 727 nodes. Figure 2 shows the finite element model mesh division.

In the experiments, the effects of variations in material properties—such as elastic modulus, expansion coefficient, yield strength, and Poisson's ratio—with temperature on the residual stresses in the joint were considered. As shown in Figure 3, the material properties of the base materials and brazing fillers change with temperature. Based on the performance parameters illustrated in the figure, a material model was established to calculate the residual stresses in the Nb-1Zr/GH4068 brazed joint.

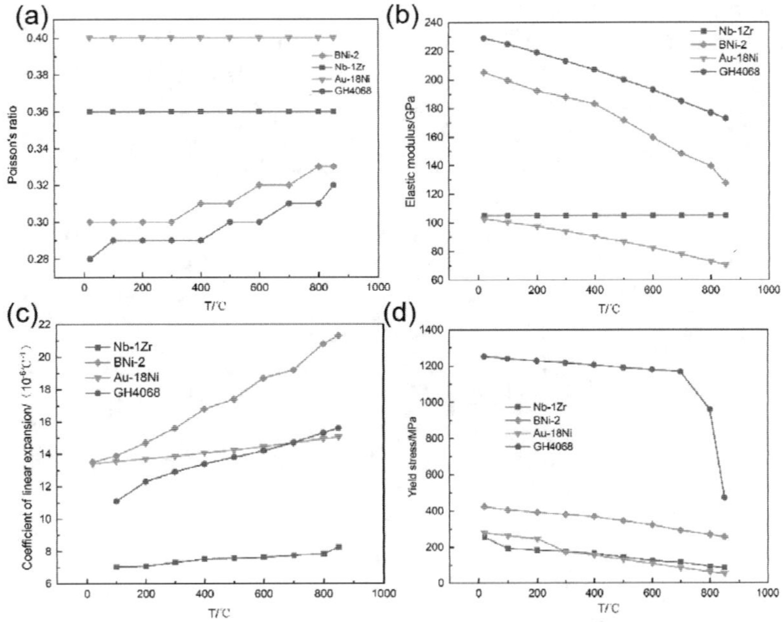

Figure 3. Change of material properties with temperature (a) Poisson's ratio; (b) Elastic modulus; (c) Coefficient of linear expansion; (d) Yield stress.

3. Discussion

3.1. Microstructure of brazed joint

The microstructural characteristics of the brazed joint between Nb-1Zr alloy and GH4068 using BNi-2 and Au-18Ni filler metals are illustrated in Figure 4. As shown in Figure 4(a), the brazed joint utilizing the BNi-2 filler metal exhibits cracks in the interfacial reaction layer. A substantial amount of intermetallic compounds is present within the brazing seam. In the case of the Au-18Ni filler metal, there is an infinite solid solution of $(Au, Ni)_{ss}$ formed due to the unlimited solubility between Au and nickel (Ni), as shown in Figure 4(b). The interface formed under this condition is well-defined, with no micro-cracks or porosity defects. Thus, it can be concluded that by adding Au to nickel-based filler metal, the mutual diffusion of niobium (Nb) and nickel (Ni) within the brazing seam can be suppressed, leading to a decrease in the content of brittle phases, a reduction in the size of precipitated phases, and an improvement in the ductility of the joint.

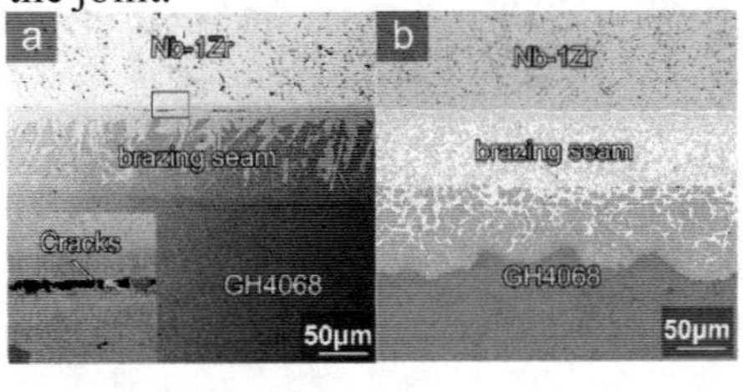

Figure 4. Overall morphology diagram of the brazed joint (a) BNi-2; (b) Au-18Ni

3.2. Von mises stress distribution of brazed joint

Figure 5 presents the Von Mises equivalent stress distribution contours and path stress curves for the brazed joint using BNi-2 as the filler metal. It can be observed that the stresses on the Nb-1Zr side and the GH4068 side exhibit an asymmetrical distribution. To analyze the entire joint, a cross-section is shown in Figure 5(b). The results indicate that the Von Mises stress primarily concentrates in the brazing seam region and the interfacial reaction layer between the base material and the filler. Moreover, the

Journal of Physics: Conference Series **3009** (2025) 012057 doi:10.1088/1742-6596/3009/1/012057

residual stress within the interfacial layer is higher than that in the surrounding base material. From the AB path illustrated in Figure 5(c), it is evident that the maximum residual stress at the interface between GH4068 and the BNi-2 filler occurs at the corner, reaching 340 MPa. This indicates that this region may be the most susceptible to crack initiation within the interfacial reaction layer. The stress variation trends along Paths AB and CD are relatively similar. Ultimately, the maximum equivalent stress for the entire joint is concentrated on the lower side of the GH4068 base material, reaching 628 MPa.

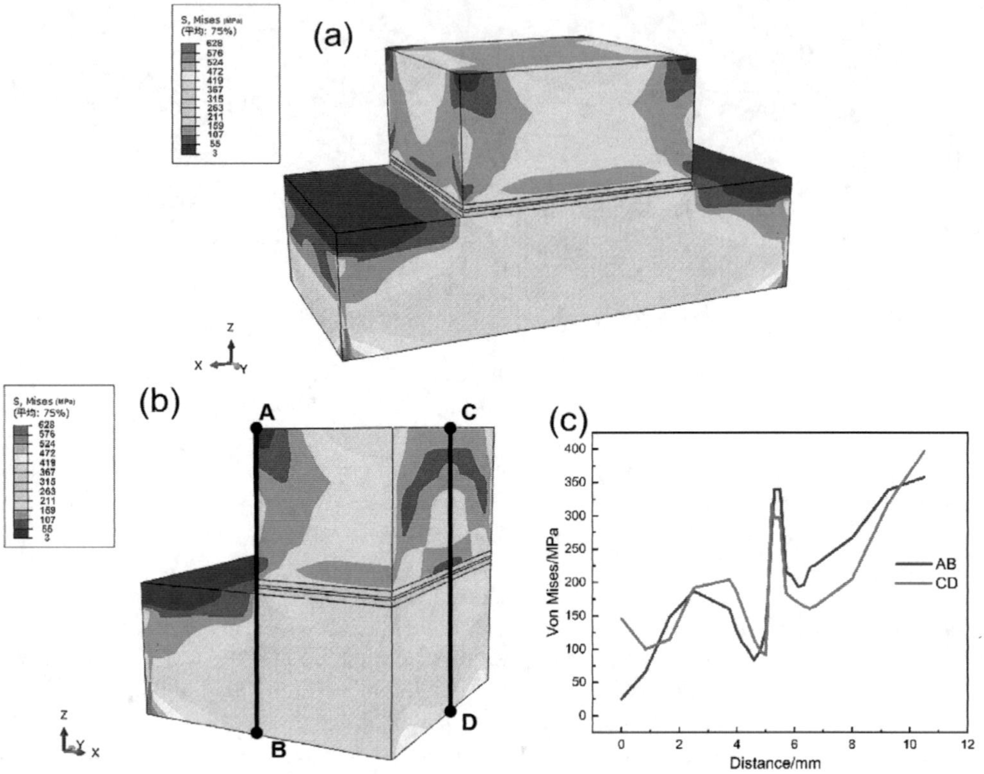

Figure 5. BNi-2 joint of (a) Von Mises stress contour map; (b) Cross-sectional stress contour map; (c) Stress variation curves along paths AB and CD.

Figure 6 presents the Von Mises stress simulation results for the joint when using Au-18Ni as the filler metal. The overall distribution characteristics of Von Mises stress within the joint show similarities to the equivalent stress observed in the joint that used BNi-2 as the filler metal. It is evident that the closer the path is to the brazing surface and the interfacial reaction layer, the higher the equivalent stress is and the steeper the rate of increase is. The maximum equivalent stress value for the Au-18Ni brazed joint is 589 MPa, which is 39 MPa lower than the maximum equivalent stress observed in the BNi-2 brazed joint. This reduction is attributed to the unlimited mutual solubility between gold (Au) and nickel (Ni) within the Au-18Ni filler metal, as well as the excellent plastic deformation capability of Au. Consequently, the residual stress in the material is alleviated through plastic deformation [12-13].

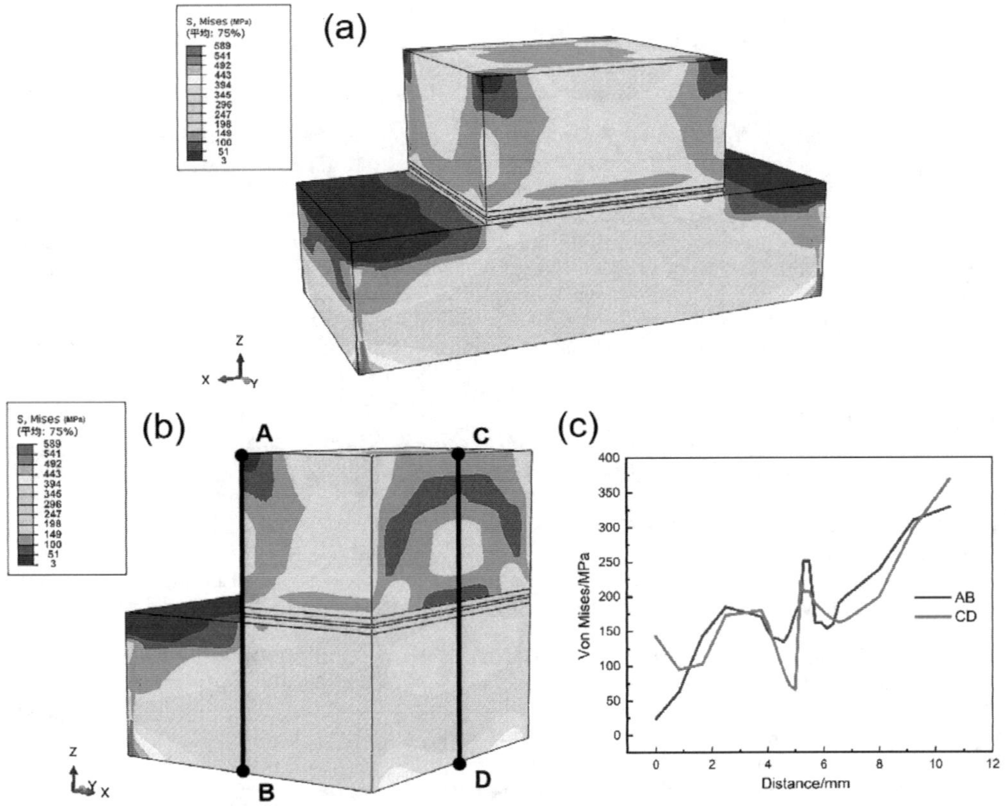

Figure 6. Au-18Ni joint of (a) Von Mises stress contour map; (b) Cross-sectional stress contour map; (c) Stress variation curves along paths AB and CD.

4. Conclusion

Whether using BNi-2 brazing material or Au-18Ni brazing material, the Von Mises stress in the Nb-1Zr/GH4068 dissimilar alloy brazed joint is mainly concentrated in the brazed joint area and the interface reaction layer between the base metal and the brazed joint. The maximum equivalent stress value of the brazed joint using Au-18Ni brazing material is 589MPa, which is 39MPa lower than that of the BNi-2 brazed joint. The Au element in Au-18Ni brazing material has excellent plastic deformation ability and can release residual stress in Nb-1Zr/GH4068 brazed joints.

Acknowledgments

National Natural Science Foundation (52301098, 52201054); National Doctoral Program (2023 M743576); Basic Research Projects of Liaoning Provincial Department of Education (LJKQZ20 222306); Research Support Funding for High Level Talents Introduced by Shenyang Ligong U niversity (1010147001117).

References

[1] BEHERA A N, CHAUDHURI A, KAPOOR R, et al. High temperature deformation behavior of Nb–1wt.%Zr alloy [J]. Materials & Design, 2016, 92: 750-9. https://doi.org/10.1016/j.matdes.2015.12.077

[2] RUBITSCHEK F, NIENDORF T, KARAMAN I, et al. Microstructural stability of ultrafine-grained niobium–zirconium alloy at elevated temperatures [J]. Journal of Alloys and Compounds, 2012, 517: 61-8. https://doi.org/10.1016/j.jallcom.2011.11.150

[3] CUI C, BAI Y, ZHANG R, et al. Study on Forming Process and Mechanical Properties of GH4068 Alloy Pipe [J]. Aeronautical Manufacturing Technology, 2022, 65(10): 36-43. https://doi.org/ 10.16080/j.issn1671-833x.2022.10.036

[4] ZHANG R, LIU P, CUI C, et al. Present Research Situation and Prospect of Hot Working of Cast & Wrought Superalloys for Aero-Engine Turbine Disk in China [J]. Acta Metallurgica Sinica, 2021, 57(10): 1215-28. https://kns.cnki.net/kcms2/article/abstract?v=YMwpULBJqz515LZA5X44HQaLolr_gfFNq4Ly_lFiHhtNnsMMmncU0cwUxvtRaKhRYlBhuyj0qlWsomp9mtvoCbSm0QC9aBNYOounJvqDoLOYKMjnOxGghCowGxrXw3zlrhVOXCc1arOQA3Oc4rlZ5ubavXzRLO9vJ7rmIYL-t457wnUqPzfR6A==&uniplatform=NZKPT&language=CHS

[5] ZHANG G, CHEN G, CAO H, et al. Electron beam offset welding to ameliorate metallurgical compatibility and mechanical performance of refractory metal/Ni-base superalloy dissimilar alloys: Nb/GH3128 [J]. Materials Science and Engineering: A, 2022, 840: 142966. https://doi.org/ 10.1016/J.MSEA.2022.142966

[6] FENG J. Research progress on dissimilar materials joining [J]. Acta Aeronautica et Astronautica Sinica, 2022, 43(02): 6-42+457. https://link.cnki.net/urlid/11.1929.V.20211209.2033.014

[7] SONG X, QIAN J, ZHONG S, et al. Research Progress of Brazing Connection of Nickel-based Superalloy GH4099 [J]. Electric Welding Machine, 2023, 53(08): 122-30. https://kns.cnki.net/kcms2/article/abstract?v=YMwpULBJqz54dqhWvIYPzhgcEaKJFzMa1OgwK7htzsdlcXk9E0e6K5dwDIvUa-h5a7WLSwAkC_1xlf6zblGlo_wFMhlY7C0AD5JnSd5XlUF846OZPTw97JoXEcXhGQk5tUNVsMnuGXJnjf2uU04QBT9eg11rQlrn7BP8p6wy9fxYY4-qdIW2Pw==&uniplatform=NZKPT&language=CHS

[8] ARDALANI A, NAFFAKH-MOOSAVY H. Investigating the microstructure and mechanical properties in furnace brazing Ti–6Al–4V to 17-4 PH stainless steel dissimilar joint with BNi-2 filler metal [J]. Journal of Advanced Joining Processes, 2024, 9: 100215. https://doi.org/10.1016/J.JAJP.2024.100215

[9] QIAO R, LONG W, QIN J, et al. Numerical simulation of residual stress in YG8/GH4169 dissimilar material brazed joints [J]. Transactions of the China Welding Institution, 2024, 45(03): 68-74+132-3. https://link.cnki.net/urlid/23.1178.TG.20231114.1035.002

[10] ZHANG Y, HU Z, WANG M, et al. Analysis of residual stress in 6061 Aluminum alloy and Kovar alloy soldered joints [J]. Electric Welding Machine, 2020, 50(01): 15-20+124. https://kns.cnki.net/kcms2/article/abstract?v=YMwpULBJqz5M7egT9zlDej7OtNECLTYOgiGphbXzMmle68GMq1HchxA1PInBskGdjusF6T5qt_q_AyeottFrDFaOoyQpzGvjVdSfXAVxQhYQGeWcaHBBYuWu7U5R8CO_0pjSr5UV9n4NsW2ye3PLpSVEzWvfDOHGLGuaYRf36pVZTOv2D8zGNg==&uniplatform=NZKPT&language=CHS

[11] WANG Z, LIU Y, ZHANG H, et al. Numerical Simulation of Temperature Fields in Autogenous Laser Welding-brazing of TC4/6061 Dissimilar Alloy [J]. New Technology & New Process, 2024, (01): 69-72. https://doi.org/10.16635/j.cnki.1003-5311.2024.01.013

[12] COLLETTE R, WU Y, RACK P D. Correlating the optical property evolution in the Au-Ni binary thin films: From metastable solid solution to phase separated alloy [J]. Journal of Alloys and Compounds, 2019, 793: 695-704. https://doi.org/ 10.1016/j.jallcom.2019.04.207

[13] HAICHUAN S, ZHAOLONG L, YANCHUN Z, et al. Microstructure, mechanical properties and fracture behavior of GH3535 alloy joints brazed with Au–18Ni filler metal [J]. Journal of Materials Research and Technology, 2023, 26: 9024-38. https://doi.org/ 10.1016/J.JMRT.2023.09.147

Analysis of mechanical structure operating deflection shape based on modal weighted interpolation

Qilong Hu[1*], Heng Zhang[1], Weijun Zhang[1], Songnan Geng[2] and Feihu Zhou[2]

[1]Xi'an Thermal Power Research Institute Co., Ltd. Xi'an, China
[2]Liaoning Hongyanhe Nuclear Power Co., Ltd. Dalian, China

[*] Corresponding author: huqilong@tpri.com.cn

Abstract: Operating deflection shape (ODS) analysis is widely used in mechanical fault diagnosis, vibration source identification, modal analysis, and structural damage identification. However, obtaining an accurate ODS requires a large number of measuring points to be arranged on the structure, which can be challenging in engineering. To improve the calculation accuracy of ODS, an analysis method based on modal weighted interpolation is proposed. The modal matrix and weighted interpolation coefficient matrix can be acquired by the finite element method or modal test. The overall ODS of the structure can then be calculated using vibration data from a limited number of points. To demonstrate the effectiveness of this method, a cantilever beam is used as an example. Results show that the calculated ODS corresponds well with the measured ODS, with a 9.30% reduction in relative error compared to the linear interpolation method. It is verified the feasibility of the proposed method.

1. Introduction

Vibration is an important factor affecting the safety of large and complex machinery or engineering structures [1][2]. Accurately identifying the operation vibration shapes of large and complex machinery or engineering structures is of great significance for their state monitoring.

Operating deflection shape (ODS) analysis is a method used to identify and visualize the vibration shapes of structures in operation [3]. It is particularly useful for identifying the response of structures at specific frequencies or times. ODS analysis can begin with any force motion and track the deformation of the structure at different points in time by analyzing the time-domain waveform of the measured structural response. Additionally, by calculating a series of frequency response signals from the time-domain response signals, the deformation shape of the structure at a specific frequency can be observed. This technology connects the monitored vibration signals with lines or surfaces by arranging numerous measuring points on the structure. The resulting vibration shape provides a global view of the structure's vibration. The ODS analysis can find out the maximum vibration point and weak areas of the structure, which is helpful for structure fault diagnosis and treatment. It offers the advantage of displaying vibration shapes in an intuitive manner and has been widely utilized in mechanical fault diagnosis, vibration source identification, modal analysis, and structural damage identification [4][5][6].

Dossing [7] pointed out that ODS can offer abundant and valuable information into the absolute dynamic behavior of a whole structure, a machine and a component, and proposed using a two-channel measurement of complex transmissibility, with one transducer fixed as a reference and the other sequentially moved to all points and directions of interest on the structure, to achieve ODS measurements. Richardson *et al.* [8] studied the relationship and differences between ODS and mode shape, and presented the test method of ODS on the basis of traditional modal test. Vold *et al.* [9] and Schwarz *et al.* [10] presented two post-processing methods that enable the display of ODSs from data which are measured under nonstationary operating conditions. In response to the time-varying excitation force, a scale factor is introduced to correct the response data, including vibration amplitude and phase, in order to obtain reasonable ODS FRF data. If the excitation force is a function of the rotational speed of rotating machinery, order tracking is used to correct the response data, and order tracked responses can be displayed as ODSs of the structure or machine. Schwarz *et al.* [11] utilized a combination of a finite element model and ODS data to accurately calculate stress and strain in a mechanical structure, a machine or a component, which provides a new method for predicting structural damage, and can be applied to online state monitoring and early warning of structural stress and strain.

ODSs can be represented by a set of two or more transmissibilities. Devriendt *et al.* [12] introduced the application of transmissibility functions in the field of ODS analysis. On the basis of transmissibility measurements, a new post-processing method was proposed for modal parameter estimation and mode shape identification, particularly for unscaled mode shapes. Yuan *et al.* [13] developed a new 3D continuously scanning laser Doppler vibrometer (CSLDV) system to achieve overall ODS measurement of mechanical structures. The study results indicate that the 3D CSLDV system has higher accuracy and a more efficient scanning rate than the traditional CSLDV system. Gorjup *et al.* [14] introduced the spectral optical flow imaging method to measure high-frequency three-dimensional vibrations. Compared to conventional multiview image-based methods, this method utilizes still-frame cameras instead of multi-cameras and high-speed imaging hardware, which greatly reduces data size and post-processing time. Xu *et al.* [15] proposed that 3D laser scanning technology could be used to measure the longitudinal ODSs of beams. Currently, the majority of studies use linear or spline interpolation methods to calculate ODS. However, these methods often result in significant errors due to the uneven vibration changes of the structure within the calculation area. To enhance the accuracy of linear or spline interpolation, a large number of measurement points must be arranged.

Because of the limits of the working environment, available space and monitoring cost, it may not be possible to arrange a large number of measuring points on the structure. According to the mode superposition method, the ODS of a structure is a result of the superposition of its multiple mode shapes. In this paper, a method based on modal weighted interpolation is proposed to determine the ODSs of mechanical structures.

2. Method for calculating ods based on modal weighted interpolation

The structure experiences forced vibration under external excitation, its vibration differential equation is as follows:

$$\boldsymbol{M}\ddot{\boldsymbol{x}} + \boldsymbol{C}\dot{\boldsymbol{x}} + \boldsymbol{K}\boldsymbol{x} = \boldsymbol{F}(t), \tag{1}$$

where $\boldsymbol{M}$, $\boldsymbol{C}$ and $\boldsymbol{K}$ denote mass, damping and stiffness matrices respectively; $\ddot{\boldsymbol{x}}$, $\dot{\boldsymbol{x}}$ and $\boldsymbol{x}$ are vibration acceleration, velocity and displacement vectors, respectively; $\boldsymbol{F}(t)$ is the external excitation vector.

Without considering damping and external excitation, Equation (1) can be written as:

$$\boldsymbol{M}\ddot{\boldsymbol{x}} + \boldsymbol{K}\boldsymbol{x} = \boldsymbol{0}. \tag{2}$$

The particular solutions of Equation (2) are as flows:

$$x_j = A_j \sin(\omega t + \theta), \quad (j = 1,2,...,n), \tag{3}$$

where ω is modal frequency, θ is the initial phase, and A_j (j =1, 2, …, n) is vibration amplitude at each coordinate within the system.

Equation (3) can also be written in matrix form:

$$\boldsymbol{x} = \boldsymbol{A}\sin(\omega t + \theta). \tag{4}$$

where $\boldsymbol{A}$=(A_j) is an n-order array composed of vibration amplitude at each coordinate.

Substitution of Equation (4) into Equation (2) results in the following generalized eigenvalue problem:

$$(\boldsymbol{K}-\omega^2\boldsymbol{M})\boldsymbol{A}=\boldsymbol{0}. \tag{5}$$

The sufficient and necessary condition for the existence of a non-zero solution to Equation (5) is as follows:

$$|\boldsymbol{K}-\omega^2\boldsymbol{M}|=0. \tag{6}$$

If all the modal frequencies obtained by solving Equation (6) are distinct, namely, $0\le\omega_1\le\omega_2\le\ldots\le\omega_n$. The eigenvectors corresponding to each modal frequency ω_i (i =1, 2, …, n) satisfy the orthogonal relations as follows:

$$(\boldsymbol{K}-\omega_i^2\boldsymbol{M})\boldsymbol{A}^{(i)}=\boldsymbol{0},\quad (i=1,2,\ldots,n). \tag{7}$$

The ith mode can be obtained by normalizing $\boldsymbol{A}^{(i)}$. The mode shapes and modal frequencies of the structure can be obtained through the finite element method or vibration theory.

The working frequency of the structure is ω_0. According to vibration theory [16], the mode shapes with modal frequencies less than $1.5\omega_0$ have a significant impact on the vibration under the working frequency, while other mode shapes have minimal effects and can be ignored. Thus, only mode shapes with a frequency of $\omega_i\le 1.5\omega_0$ should be selected for the analysis, and denote m as the number of influential mode shapes.

To accurately represent the ODS of the structure, a minimum of k (where $k\ge m$) vibration measurement points must be strategically arranged on the structure. According to the mode superposition method, the vibration at each measuring point under the working frequency of ω_0 can be expressed as:

$$\begin{Bmatrix} x_1 \\ x_2 \\ \vdots \\ x_k \end{Bmatrix}=\begin{bmatrix} \varphi_{1,1} & \varphi_{1,2} & \cdots & \varphi_{1,m} \\ \varphi_{2,1} & \varphi_{2,2} & \cdots & \varphi_{2,m} \\ \vdots & \vdots & & \vdots \\ \varphi_{k,1} & \varphi_{k,2} & \cdots & \varphi_{k,m} \end{bmatrix}\begin{Bmatrix} q_1 \\ q_2 \\ \vdots \\ q_m \end{Bmatrix}, \tag{8}$$
$$\boldsymbol{X}=\boldsymbol{\Phi}\cdot\boldsymbol{Q},$$

where $\varphi_{i,j}$ represents the amplitude of the jth mode shape at measurement point i, q_j is the coefficient of the jth mode shape component at the working frequency, $\boldsymbol{\Phi}$ is the modal matrix, and $\boldsymbol{Q}$ is the matrix composed of the amplitudes of each mode shapes.

The vibrations at these measurement points can be obtained by sensors arranged on the structure. By solving Equation (8) with a least-square method, the matrix $\boldsymbol{Q}$ can be obtained:

$$\boldsymbol{Q}=[\boldsymbol{\Phi}^T\boldsymbol{\Phi}]^{-1}\boldsymbol{\Phi}^T\boldsymbol{X}. \tag{9}$$

Thus, the vibration at any position on the structure can be expressed as:

$$x_i=\begin{Bmatrix}\varphi_{i,1} & \varphi_{i,2} & \cdots & \varphi_{i,m}\end{Bmatrix}\left[\boldsymbol{\Phi}^T\boldsymbol{\Phi}\right]^{-1}\boldsymbol{\Phi}^T\boldsymbol{X}. \tag{10}$$

Furthermore, Equation (10) can be rewritten as:

$$x_i=\boldsymbol{\Theta X}, \tag{11}$$

where $\boldsymbol{\Theta}$ is the weighted interpolation coefficient matrix corresponding to each vibration measurement point,

Journal of Physics: Conference Series **3009** (2025) 012058 doi:10.1088/1742-6596/3009/1/012058

$$\boldsymbol{\Theta} = \{\varphi_{i,1} \quad \varphi_{i,2} \quad \cdots \quad \varphi_{i,m}\}\left[\boldsymbol{\Phi}^T \boldsymbol{\Phi}\right]^{-1} \boldsymbol{\Phi}^T. \tag{12}$$

For any position on the structure, the weighted interpolation coefficient matrix can be calculated by Equation (12), and then the vibration at this specific position can be determined. By connecting the vibrations of adjacent points in space based on the vibrations at all positions on the structure, the ODS can be obtained.

3. Case study and analysis

A cantilever beam with a rectangular cross-section and made of steel material, as shown in Figure 1, is taken as an example to validate the accuracy of the proposed modal weighted interpolation method. Table 1 lists the parameters of the uniform section beam. The left end of the cantilever beam is fixed, and the right end is free.

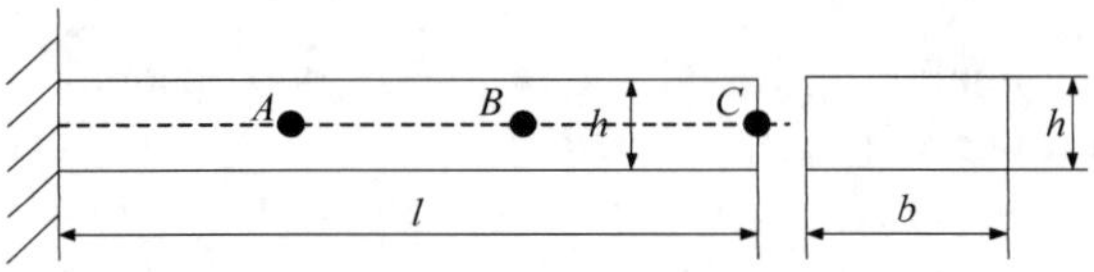

Figure 1. Cantilever beam model.

Table 1. The parameters of the cantilever beam.

Parameters	**Value**
Section width b (mm)	20
Section height h (mm)	1
Length l (mm)	700
Elastic modulus E (N/m^2)	2.06×10^{11}
Density ρ (kg/m^3)	7850

3.1 Dynamics Analysis

According to vibration theory [16], the modal frequencies of a cantilever beam can be calculated using the following equation:

$$\omega_i = \frac{(\beta_i l)^2}{l^2}\sqrt{\frac{EJ}{\rho A}}, \quad (i = 1, 2, 3, ..., n), \tag{13}$$

where J is the area moment of inertia, and A is the cross-sectional area. $\beta_i l$ is the coefficient corresponding to each modal frequency, and the coefficient values corresponding to the first four modal frequencies are 1.875, 4.694, 7.855, and 10.996, respectively.

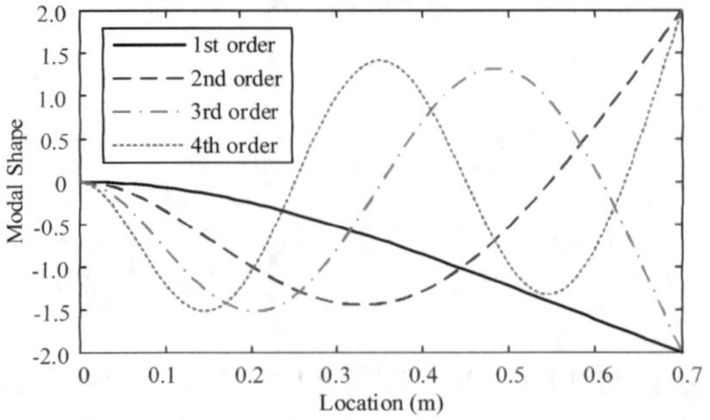

Figure 2. The first four mode shapes of the cantilever beam.

Thus, the first four modal frequencies of the cantilever beam can be calculated to be 10.61 Hz, 66.50 Hz, 186.21 Hz, and 364.91 Hz, respectively. Figure 2 depicts the first four mode shapes of the beam.

3.2 Selection of the number of mode shapes for analysis
The structure operates at a vibration frequency of 50 Hz. According to the principle of $\omega_i \leq 1.5\omega_0$, the first two mode shapes can be selected for analysis, that is, $m = 2$.

3.3 Selection of the number of vibration measurement points
According to the principle that the number of vibration measurement points is greater than or equal to the number of mode shapes, three vibration measurement points are selected, that is, $k = 3$. Without loss of generality, three vibration measurement points are evenly distributed on the cantilever beam (Figure 1).

3.4 Calculation of weighted interpolation coefficient matrix
In order to verify the effectiveness of the proposed modal weighted interpolation method, the actual ODS of the cantilever beam is obtained by testing on a test rig, as shown in Figure 3. The vibration amplitudes at three measurement points are -0.0718 mm, -0.0708 mm and 0.0300 mm, respectively, then

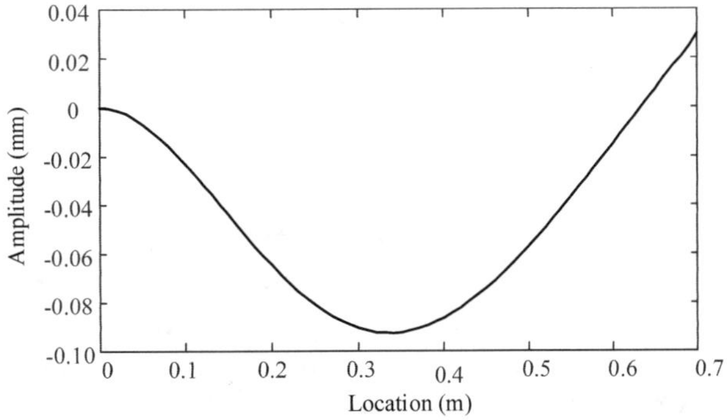

Figure 3. The measured ODS.

$$\boldsymbol{X} = \begin{Bmatrix} -0.0718 \\ -0.0708 \\ 0.0300 \end{Bmatrix} \text{ mm.}$$

Based on the amplitudes of the first two vibration mode shapes at three measurement points shown in Figure 2, the corresponding modal matrix can be constructed:

$$\boldsymbol{\Phi} = \begin{bmatrix} -0.297 & -1.109 \\ -1.069 & -0.901 \\ -2.000 & 2.000 \end{bmatrix}.$$

According to Equation (12), the weighted interpolation coefficient matrix between the vibration at any position on the cantilever beam and the existing vibration measurement points can be calculated. Taking the position 0.5 m away from the left end of the cantilever beam as an example, the resulting weighted interpolation coefficient matrix is as follows:

$$\boldsymbol{\Theta}\big|_{z=0.5} = \{0.3868 \quad 0.6149 \quad 0.2237\}.$$

Then, the vibration at any position on the cantilever beam can be calculated based on Equation (11). The vibration amplitude at the position 0.5 m away from the left end of the beam is:

$$x\big|_{z=0.5} = \boldsymbol{\Theta}\big|_{z=0.5} \boldsymbol{X} = -0.0646 \text{ mm.}$$

As shown in Figure 3, the measured ODS reveals a vibration amplitude of -0.0569 mm at the position 0.5 m away from the left end of the cantilever beam. The result suggests that the calculated vibration amplitude corresponds well with the measured value basically.

The same method can be applied to calculate the vibrations at other positions on the cantilever beam. By connecting the vibrations of all adjacent points in space, the ODS of the cantilever beam can be determined. Figure 4 compares the ODS calculated using the linear interpolation method and the modal weighted interpolation method to the measured ODS. It can be observed from Figure 4 that the ODS obtained through the modal weighted interpolation method closely matches the measured ODS, with a maximum relative error of 13.53%. In comparison, the ODS calculated using the linear interpolation method has a maximum relative error of 22.83%.

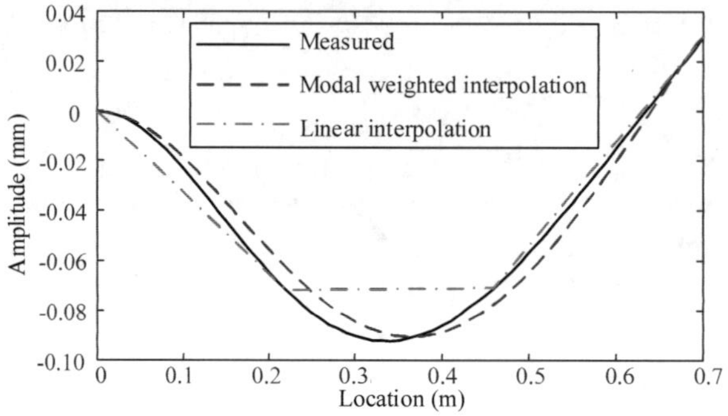

Figure 4. ODS comparison.

The comparison results indicate that the modal weighted interpolation method has a higher accuracy than the linear interpolation method with an equal number of vibration measurement points. It is important to note that increasing the number of vibration measurement points can further improve the accuracy of the modal weighted interpolation method, which will result in a closer match between the calculated ODS and the actual ODS. In practical engineering, due to various factors such as working environment, available space, and monitoring cost, it is necessary to choose the optimal number and position of measuring points based on the characteristics of the structure.

4. Conclusion

In this study, a modal weighted interpolation method is proposed to analyze the operating deflection shapes (ODSs) of mechanical structures based on the mode superposition method. The overall ODS of a mechanical structure or a machine can be calculated utilizing vibration data from a limited number of points. To demonstrate the feasibility of this method, a cantilever beam is used as an example and the results are compared to the ODS obtained from experimental tests. The results indicate that the modal weighted interpolation method for calculating ODS is more accurate than the linear interpolation method. For a complex structure, the modal matrix and weighted interpolation coefficient matrix can be obtained through finite element analysis or modal test. The overall ODS of the structure can then be calculated using vibration data measured by a small number of measurement points. This method is particularly useful for monitoring ODSs of mechanical structures under operating conditions.

Acknowledgment

The authors would like to express their gratitude for the financial support provided by Xi’an Thermal Power Research Institute Co., Ltd., China. Additionally, we would like to acknowledge Jiangang Yang from the National Engineering Research Center of Power Generation Control and Safety at Southeast University, China for the help on the experimental tests.

References

[1] J. Q. Wang, D. M. Shen, Y. M. Chen, J. G. Yang and P. Tan, "Equipment operation deformation analysis based on limited reference points", *Noise and Vibration Control*, vol. 43, no. 1, pp. 265-269, 2023.
[2] R. Corbally and A. Malekjafarian, "Bridge damage detection using operating deflection shape ratios obtained from a passing vehicle", *Journal of Sound and Vibration*, vol. 537, pp. 117225, 2022.
[3] Y. H. Shi, S. C. Ji, F. Zhang, Y. L. Dang and L. Y. Zhu, "Application of operating deflection shapes to the vibration-based mechanical condition monitoring of power transformer windings", *IEEE Transactions on Power Delivery*, vol. 36, no. 4, pp. 2164-2173, 2021.
[4] H. Jiang, Y. Y. Jiang and J. W. Xiang, "Quantitative detection of multiple damages in wind turbine blade based on the operating deflection shape and natural frequencies", *Ocean Engineering*, vol. 278, pp. 114387, 2023.
[5] E. J. Obrien, D. P. McCrum, M. A. Khan and L. J. Prendergast, "Wavelet-based operating deflection shapes for locating scour-related stiffness losses in multi-span bridges", *Structure and Infrastructure Engineering*, vol. 19, no. 2, pp. 238-253, 2023.
[6] Y. F. Xu and J. S. Kim, "Baseline-free structural damage identification for beam-like structures using curvature waveforms of propagating flexural waves", *Sensors*, vol. 21, no. 7, pp. 2453, 2021.
[7] O. Dossing, "Structural stroboscopy - measurement of operational deflection shapes", *Sound and Vibration*, vol. 22, no. 1, pp. 18-26, 1988.
[8] M. H. Richardson, "Is it a mode shape, or an operating deflection shape?", *Sound and Vibration*, vol. 31, no. 1, pp. 54-61, 1997.
[9] H. Vold, B. Schwarz and M. Richardson, "Display operating deflection shapes from nonstationary data", *Sound and Vibration*, vol. 34, no. 6, pp. 14-18, 2000.
[10] B. Schwarz and M. Richardson, "Measurements required for displaying operating deflection shapes", *Proceedings of the 22nd International Modal Analysis Conference (IMAC XXII)*, Dearborn, Michigan, 2004.
[11] B. Schwarz, S. Richardson and M. Richardson, "Calculating stress and strain from experimental ODS data", *Sound and Vibration*, vol. 50, no. 5, pp. 12-15, 2016.
[12] C. Devriendt, G. Steenackers, G. D. Sitter and P. Guillaume, "From operating deflection shapes towards mode shapes using transmissibility measurements", *Mechanical Systems and Signal Processing*, vol. 24, no. 3, pp. 665-677, 2010.
[13] K. Yuan and W. D. Zhu, "In-plane operating deflection shape measurement of an aluminum plate using a three-dimensional continuously scanning laser Doppler vibrometer system", *Experimental Mechanics*, vol. 62, no. 4, pp. 667-676, 2022.
[14] D. Gorjup, J. Slavic, A. Babnik and M. Boltezar, "Still-camera multiview spectral optical flow imaging for 3D operating-deflection-shape identification", *Mechanical Systems and Signal Processing*, vol. 152, pp. 107456, 2021.
[15] W. Xu, R. Q. Li and W. D. Zhu, "Schemes of local fitting for damage detection of bars using longitudinal operating deflection shapes measured by 3D laser scanning", *Measurement*, vol. 233, pp. 114681, 2024.
[16] H. Y. Hu, "Vibration mechanics: a research-oriented tutorial", *Beijing: Science Press*, pp. 402-426, 2022.

Journal of Physics: Conference Series **3009** (2025) 012059 doi:10.1088/1742-6596/3009/1/012059

Research on time-varying meshing stiffness of spiral bevel gears based on mathematical modeling method

Jun Ma[1], Rui Shen[1], Chunlei Wang[1], Mozhi Lei[1], Enming Wang[2], Fajia Li[*1,2]

[1] AECC Hunan Aviation Powerplant Research Institute, Zhuzhou, Hunan, China
[2] School of Mechanical Engineering, University of Jinan, Jinan, Shandong, China

[*]Corresponding author's e-mail: me_lifj@ujn.edu.cn

Abstract. To enhance operational longevity and mitigate vibration/noise in spiral bevel gear transmissions, this investigation focuses on the dynamic characteristics of such systems with emphasis on their time-varying meshing stiffness. The analytical framework integrates multiple stiffness components: Hertzian contact stiffness is derived through elastic contact mechanics, while bending, shear, and axial compressive stiffness are computed via strain energy principles. Crucially, gear body compliance at the tooth root is incorporated into the comprehensive stiffness model. A coupled parametric formulation is developed to characterize the dynamic mesh stiffness variation pattern. Through computational modeling analysis, the time-varying meshing stiffness evolution over a 10-millisecond operational cycle is quantified, revealing a parabolic trajectory that peaks at 1.966×10^8 N/m before gradual attenuation.

1. Introduction

In power transfer systems, toothed gear mechanisms constitute a fundamental methodology for torque conversion and motion control. It has many forms and is widely used. It has the following advantages: The high efficiency of gear transmission is incomparable to other mechanical transmissions; Under the same working conditions, the gear transmission has a compact space structure and a smaller size; For developing, producing, and sustaining high-performance gear transmission systems, the service life can be up to twenty years, and the operation is very stable; Gear transmission has a stable transmission ratio, making it widely used in the market.

Because the spiral bevel gear can not only realize the commutation in the transmission process, but also has the advantages of stable transmission, strong bearing capacity, and compact structure, it meets the needs of a lightweight aviation reducer, so it is widely used in the transmission of aviation field [1-4]. The spiral bevel gear demonstrates exceptional load-bearing performance, enabling efficient transfer of substantial engine torque while maintaining precise synchronization in rotorcraft drivetrain operations. Its operational dependability proves critical for ensuring aerial vehicle safety under demanding flight conditions. Approximately 98% of helicopter main reducers employ spiral bevel gears in high-speed stages to accomplish three core functions: multi-source power distribution, kinematic coordination, and angularly adaptive cross-axis power delivery, as shown in Figure 1. This transmission solution also finds extensive implementation across marine propulsion, automotive differentials, heavy equipment, and precision machining systems.

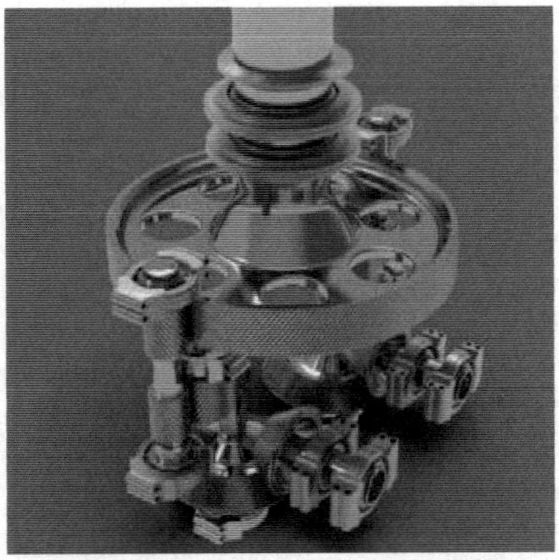

Figure 1. Helicopter transmission gear set.

The spiral bevel gear was proposed by French geometer T·Olivier in the early 19th century. He not only gave the envelope surface method of conjugate tooth surface but also proved the possibility of obtaining the conjugate surface of line or point contact through the auxiliary surface [5]. In 1874, Gleason's founder, William Gleason, invented the world's first bevel gear planning machine. In 1910, Paul Bottcher, a German, obtained a technology patent for spiral bevel gears and published a series of technical patents on spiral bevel gears. In 1913, James Gleason, an engineer from the American Gleason Company, invented the first spiral bevel gear end face machining machine. In 1919, Gleason began to try to improve the previous processing method under the condition of obtaining the patent of Paul Bottcher. The concept of tapered teeth was discovered and put forward in the coincidence of opportunity and applied to the actual production. Then Baxter used the space geometry method to derive the calculation formula of the tooth surface curvature. The tooth profile of spiral bevel gear is studied and analyzed accurately, which lays a certain foundation for future scholars to carry out in-depth analysis and research [6].

According to established academic consensus, bevel gears are fundamentally categorized into two distinct groups based on tooth trace characteristics: straight bevel gears and curved bevel gears. This classification system further delineates straight bevel gears into standard straight and helical variants, while curved bevel gears encompass spiral configurations (Gleason-type, Figure 2) and cycloid configurations (Oerlikon-type, Figure 3) [7]. The theory of spiral bevel gear technology was first proposed by Wildhaber, Baxter, and others of Gleason Company in the United States. Later, the Oerliko Company in Switzerland and the Klingelnbe Company in Germany also owned spiral bevel gear technology and formulated their own standards. Gleason company uses the circular arc gear system, and the latter two companies use the extended epicycloid gear system, which is the most widely used in the domestic and international Gleason bevel gear.

Figure 2. Gleason spiral bevel gear.

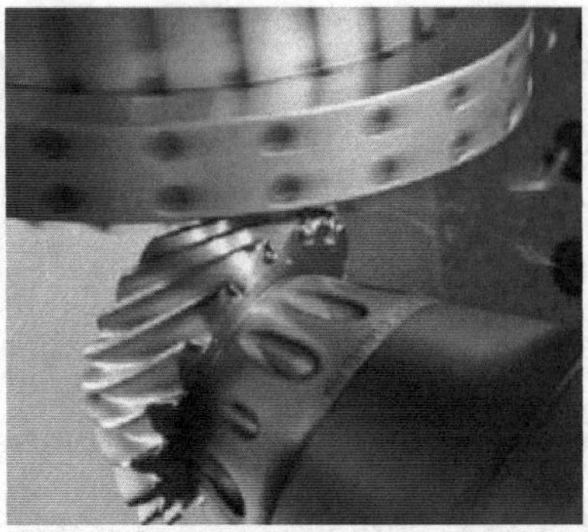

Figure 3. Oerlikon spiral bevel gear.

Contemporary helicopter propulsion systems demonstrate unprecedented performance parameters, with next-generation engine rotational velocities reaching 20000 r/min and power output exceeding 3000 kW. As critical drivetrain components, spiral bevel gearss are subjected to unprecedented dynamic demands - their operational durability directly governs the service life of helicopter main reducers. This performance escalation necessitates enhanced dynamic characterization of spiral bevel gear systems, particularly regarding time-varying meshing stiffness behavior. Consequently, our investigation focuses on Gleason-format spiral bevel gears, developing a parametric formulation to quantify transient meshing stiffness dynamics through rigorous computational modeling.

2. Research Methods and Theoretical Derivation

This investigation employs an analytical mechanics framework grounded in material mechanics principles, with computational implementation through the energy conservation methodology. The analytical protocol rigorously applies Weber-Banascheck deformation theory coupled with potential energy conservation laws to derive time-varying meshing stiffness characteristics, establishing closed-form solutions through variational energy minimization processes.

The analytical framework evolved through three critical developmental phases rooted in Weber-Banascheck's deformation theory. The foundational stage (1985-1987) saw Yang's pioneering establishment of strain energy partitioning in spur gear systems [8], which systematically decomposed total potential energy into three constitutive components: Hertzian contact energy, bending strain energy, and axial strain energy, ultimately deriving stiffness formulations via variational energy minimization. This methodology underwent dual advancements in 2004: Tian's contribution enhanced the model through shear strain energy integration, formalizing the primitive time-varying meshing stiffness computation paradigm, while Sainsot's concurrent work [9] addressed gear foundation compliance by incorporating fillet deformation compensation mechanisms, thereby completing the fundamental stiffness calculation architecture. Recent methodological refinements have systematically incorporated gear foundation compliance into meshing stiffness computations. Building upon this foundation, Zhang et al. [10, 11] advanced the analytical framework through two critical enhancements: resolving root circle-base circle eccentricity effects, and optimizing the variable cross-section cantilever beam model via strain energy recalibration. Parallel developments by Zhang Tao [12] established a friction-inclusive meshing stiffness algorithm, demonstrating through comprehensive parametric analysis the coupled influence of fundamental design parameters and operational torque on stiffness temporal characteristics.

The analytical framework presumes ideal lubrication conditions for spiral bevel gearss, with manufacturing/assembly tolerances and kinematic errors excluded from the current investigation. Stiffness determination follows the displacement-load quotient principle, where total meshing rigidity constitutes a composite of five discrete components: Hertzian contact stiffness, bending stiffness, axial compression stiffness, shear stiffness, and fillet-foundation stiffness by using force divided by deformation ($k = \mathrm{d}F/\mathrm{d}u$).

2.1. Hertzian Contact Stiffness

Assuming that the gear pair is a homogeneous elastomer, according to Hertzian theory: the contact between the elastomers can be approximated by the contact of two adjacent parabolas, and the error is less than 0.5% in this approximation. In the contact process of the spiral bevel gears, the contact stress is distributed according to the law of the ellipsoid in the contact area, and the meshing of the spiral bevel gear can be simplified as a point contact model of two ellipsoids, as shown in Figure 4.

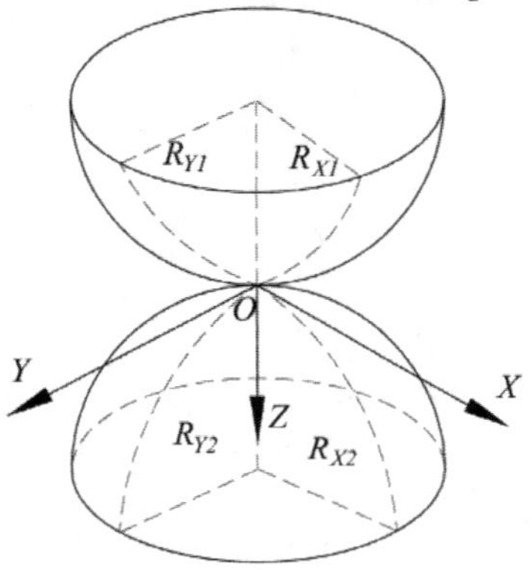

Figure 4. Ellipsoid point contact equivalent model.

Hertzian contact stiffness k_h can be expressed as:

$$k_h = \frac{4E_x\sqrt{R_x}}{\pi} \tag{1}$$

$$E_x = \frac{1}{\frac{1-v_z^2}{E_z} + \frac{1-v_c^2}{E_c}} \tag{2}$$

In the formula, R_x is the equivalent curvature radius of the gear, m; E_x is the equivalent elastic modulus of the two materials, Pa; E_z and E_c are the elastic modulus of elasticity of the driving wheel and the driven wheel respectively, Pa; v_z and v_c are the Poisson's ratios of the driving wheel and the driven wheel, respectively.

2.2. Bending Stiffness, Axial Compression Stiffness, and Shear Stiffness

According to the energy method, the bending energy U_b, shear energy U_s, and axial compression energy U_a in the spiral bevel gear can be expressed as:

$$\begin{cases} U_b = \dfrac{F^2}{2k_b} \\ U_s = \dfrac{F^2}{2k_s} \\ U_a = \dfrac{F^2}{2k_a} \end{cases} \tag{3}$$

In the formula, F is the interaction force at the meshing point, and it can be solved by the finite element method generally.

In the contact process of the spiral bevel gears, the contact stress is distributed according to the law of the ellipsoid in the contact area, and the interaction force F at the meshing point is expressed as:

$$F = \frac{2\pi abP_F}{3} \tag{4}$$

In the formula, P_F is the contact pressure of the contact area, Pa; a and b are the length of the semi-major axis and the semi-minor axis of the contact ellipse, m.

According to the Hertzian contact theory, the calculation formulas of the semi-major axis a and the semi-minor axis b of the contact ellipse can be derived:

$$\begin{cases} a = \left(\dfrac{3F \dfrac{R_{X1}R_{X2}}{R_{X1}+R_{X2}}}{2\pi E_x} \right)^{\frac{1}{3}} \\ b = \left(\dfrac{3F \dfrac{R_{Y1}R_{Y2}}{R_{Y1}+R_{Y2}}}{2\pi E_x} \right)^{\frac{1}{3}} \end{cases} \tag{5}$$

The computational methodology employs finite element discretization of bevel gear teeth, with each discretized segment approximated as an equivalent spur gear sector. This enables stiffness characterization through a beam-theoretic framework originally developed for cylindrical gear systems. For deformation analysis under load, the tooth segment spanning from root to contact point is partitioned into n differential elements. The aggregate deformation across these infinitesimal constituents determines the total cantilever beam displacement. Assuming uniform material properties within each beam segment, displacement solutions derive from continuum mechanics principles governing elastic deformation.

Based on the beam theory and assuming that the cantilever beam does not rotate, the expression of bending strain energy can be obtained and the expressions of shear strain energy and axial compression energy can be derived.

To characterize the angular-dependent stiffness evolution in driver/driven component interactions, angular parametrization replaces linear displacement metrics - this coordinate transformation enhances analytical tractability while directly correlating tooth parametric characteristics with instantaneous rotational positions. Consequently, derived bending, shear, and axial compliance stiffness matrices emerge as angularly parametrized functions establishing a differentiable formulation framework essential for variational calculus implementation in dynamic system modeling.

2.3. Fillet-Foundation Stiffness

The spiral bevel gear body constitutes a deformable continuum exhibiting elastic compliance during meshing engagements. To enhance model fidelity, stiffness characterization must rigorously account for gear body compliance through constitutive equation integration, where material deformation tensors are coupled with time-dependent contact kinematics in the parametric formulation.

When the gear teeth are subjected to meshing force, the elastic matrix will undergo elastic deformation, which can be expressed as σ_f. The coefficient in the formula is approximated by a polynomial function. According to the time-varying meshing stiffness theory derived by Fakher Chaari, the coefficient in the formula is fitted to obtain the elastic deformation of the matrix and the expression of the matrix stiffness is obtained:

$$k_f = \frac{F}{\sigma_f} \tag{6}$$

3. Results & Discussion

By coupling the Hertzian contact stiffness k_h, bending stiffness k_b, shear stiffness k_s, axial compression stiffness k_a, and matrix stiffness k_f obtained above, the spiral bevel gears stiffness k_g can be obtained [13].

$$k_g = \frac{1}{\frac{1}{k_h}+\frac{1}{k_{bz}}+\frac{1}{k_{bc}}+\frac{1}{k_{sz}}+\frac{1}{k_{sc}}+\frac{1}{k_{az}}+\frac{1}{k_{ac}}+\frac{1}{k_{fz}}+\frac{1}{k_{fc}}} \quad (7)$$

In the formula, k_g is the stiffness of the gear pair, N/m; the subscript z and subscript c represent the driver component and the driven component respectively.

When the gear is running, the periodic excitation force generated by the time-varying meshing stiffness of the gear teeth has a significant impact on the life of the helicopter main reducer in many aspects. First of all, this periodic excitation force will cause the continuous vibration of the internal parts of the main reducer, and the long-term vibration will cause the connection between the parts to be loose, such as bolts, nuts, and other connecting parts gradually relax, destroying the stability of the whole system. Secondly, the vibration will aggravate the wear of the parts, especially the wear of the tooth surface will be greatly increased. The wear makes the tooth surface roughness change, further affects the meshing state of the gear, forms a vicious circle, and leads to the overall performance of the main reducer. Furthermore, the periodic excitation force may also cause resonance. Once the natural frequency of the main reducer is close to the frequency of the excitation force, the huge stress generated by the resonance will cause fatigue cracks in key components such as gear shafts and boxes. Over time, the cracks continue to expand, eventually leading to the failure of the parts, which greatly shortens the service life of the helicopter main reducer.

To enhance main reducer durability, this investigation employs computational modeling to characterize spiral bevel gear time-varying meshing stiffness. This analytical approach offers cost efficiency, computational tractability, and parametric repeatability advantages over empirical testing protocols. Leveraging the developed constitutive equations, the study constructs a parametric model quantifying stiffness dynamics in spiral bevel gearss. Implementation is demonstrated through an aircraft intermediate reducer system case study, the parameters of the spiral bevel gears are shown in Table 1.

Table 1. Parameters of Spiral Bevel Gears.

Parameter	Numerical Value
Elastic Modulus, E/Pa	2.07×10^{11}
Poisson Ratio, v	0.25
Tooth Number, Z_z/Z_c	23/115
Pressure Angle, α/°	25
Mean Spiral Angle, β_m/°	35
Shaft Angle, $\sum$/°	90
Modulus, m/mm	6.0
Tooth Width, B/mm	60.0
Quality, (m_z/m_c)/kg	5.148/165.65
Driving Wheel Rotational Speed, n_z/(r/min)	500/1000/1500

The comprehensive curvature radius of the spiral bevel gears at different meshing moments is shown in Figure 5. The comprehensive curvature radius of spiral bevel gear refers to the value calculated by combining the main curvature radius of two tooth surfaces according to certain rules at the meshing interface characteristics of spiral bevel gears critically influence contact fatigue resistance. Enhanced equivalent curvature dimensions correlate with reduced localized tooth deflection and improved contact stress uniformity, effectively extending gear service life while augmenting load capacity. Kinematic analysis reveals progressive curvature amplification along the contact path under operational conditions, particularly pronounced in systems operating above unity speed ratios. Since the transmission ratio of the spiral bevel gears is greater than one, the comprehensive curvature radius gradually increases as the meshing line continues to move. It is worth noting that the larger the transmission ratio is, the more obvious the change trend is. Through the spiral bevel gear parameters given in Table 1, the time-varying meshing stiffness of the spiral bevel gears can be calculated, as shown in Figure 6.

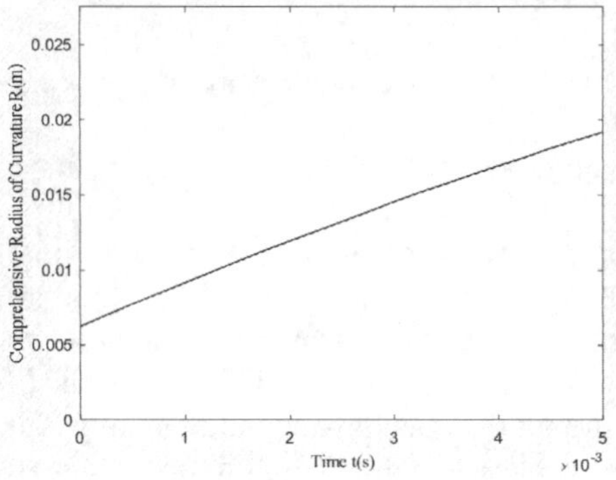

(a) 500 r/min

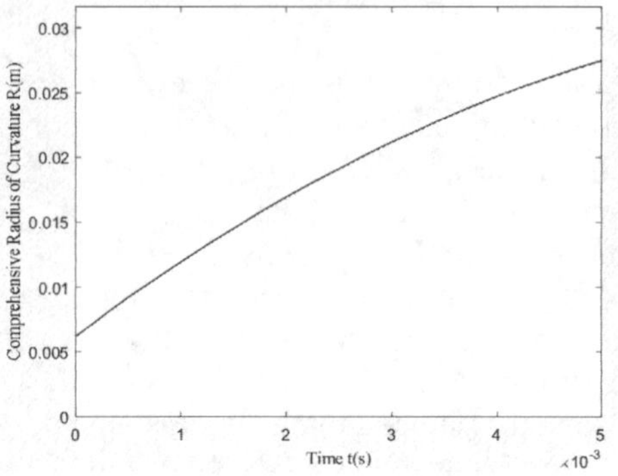

(b) 1000 r/min

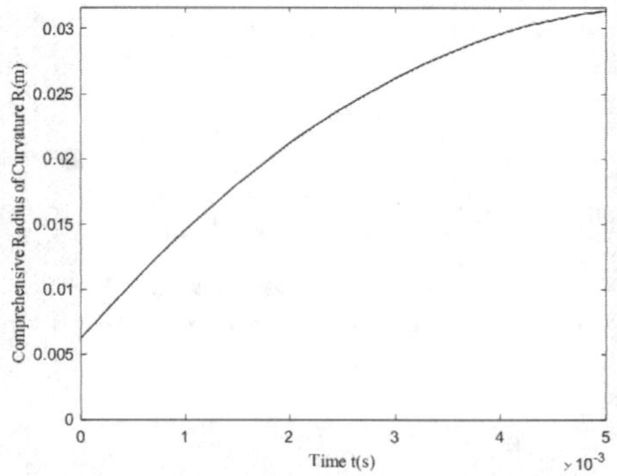

(c) 1500 r/min

Figure 5. Comprehensive radius of curvature.

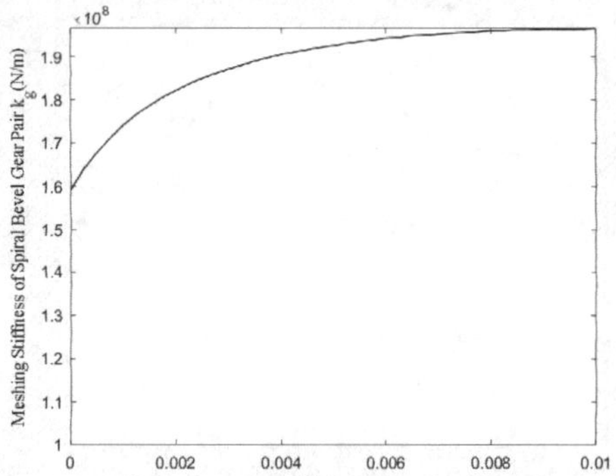

(a) 500 r/min

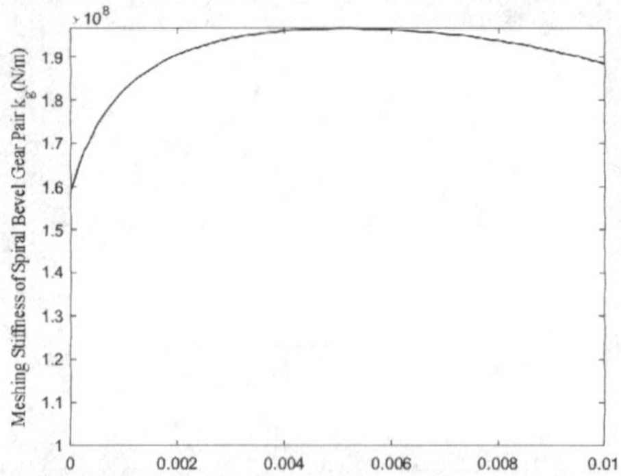

(b) 1000 r/min

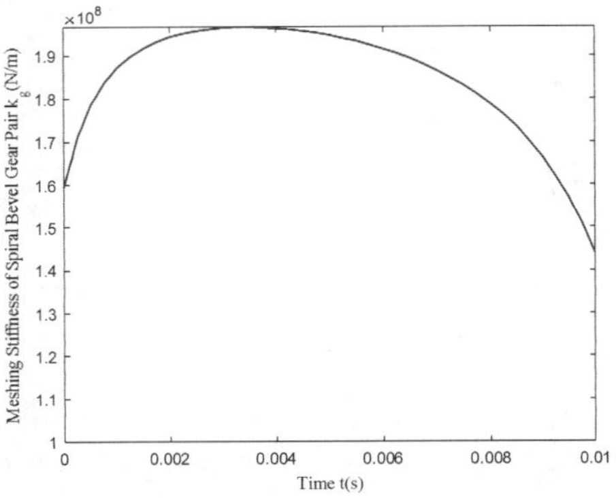

(c) 1500 r/min

Figure 6. Time-varying meshing stiffness of spiral bevel gears.

Kinematic analysis of curvature evolution (Figure 5a-c) demonstrates progressive curvature intensification under speed escalation, where elevated rotational velocities induce amplified curvature dimensions at equivalent temporal intervals. This curvature modulation directly governs transient stiffness dynamics, as evidenced in corresponding engagement profiles (Figure 6a-c). Parametric evaluation reveals a characteristic parabolic stiffness trajectory during acceleration phases, peaking at 1.966×10^8 N/m within the critical 10-millisecond operational window before subsequent attenuation. Maximum stiffness magnitude correlates with optimal curvature geometry during transitional speed regimes.

4. Conclusions

This investigation establishes a computational framework for characterizing time-varying meshing stiffness in spiral bevel gear systems through synthesized integration of five critical deformation mechanisms:

Hertzian contact rigidity governing localized surface interactions, flexural and axial deformation resistance under bending/compressive loads, shear strain accommodation during tangential displacements, and tooth root compliance effects from foundation deformation. The analytical methodology combines strain energy principles for bending/shear/axial stiffness derivation with Hertzian contact theory formulations, explicitly incorporating gear body compliance dynamics. Through systematic coupling of these complementary deformation modes, the developed model captures operational-phase stiffness modulation characteristics, achieving peak transient stiffness magnitudes of 1.966×10^8 N/m during critical engagement intervals.

The investigation of transient meshing stiffness in spiral bevel gear systems fundamentally advances drivetrain dynamic analysis. This critical parameter governs three operational optimizations: enhanced kinematic performance through refined load distribution, extended fatigue resistance via stress cycle management, and vibration/noise attenuation achieved through harmonic suppression. These collective improvements facilitate superior meshing behavior while maximizing transmission system longevity. Furthermore, the methodology provides essential theoretical guidance for precision engineering applications in power transmission design.

Acknowledgments

The experimental investigations and computational analyses were made possible by research grants provided under Shandong Natural Science Foundation funding initiatives: Research on the Method of Digital Twin System for the Health Status of High-End Computerized Numerical Control (CNC) Equipment (ZR2021ME101). Contemporary smart manufacturing ecosystems impose elevated operational imperatives on precision CNC systems, necessitating enhanced functional safety protocols, sustained reliability benchmarks and optimized maintenance feasibility frameworks, while the study of the time-varying meshing stiffness of the curved bevel gears, as a key part of high-end CNC equipment, helps to solve the problems of safety, reliability and maintainability. This project carries out the construction of a digital twin prototype system for high-end CNC equipment through side cloud collaboration, adaptive extraction of health state information embedded in the bevel gear, real-time data-driven intelligent diagnosis modeling, and establishment of intelligent assessment methods based on big data, and pioneers systematic investigations into the implementation of intelligent remote O&M frameworks for precision computer numerical control (CNC) machinery.

References

[1] Gu B. (2024) Research on the vibration calculation method of spiral bevel gear transmission system considering bearing rigidity. Chongqing University of Technology, Chongqing. DOI: 10.27753/d.cnki.gcqgx.2024.000871.
[2] Wu Z, Jiang Z, Lei Y, Wang X, Hu L. (2024) Dynamic transmission error analysis of spiral bevel gears for tooth surface wear model. Machinery and Automation, 53(06): 74-77. DOI:10.19344/j.cnki.issn1671-5276.2024.06.014.
[3] Luan X, Wang S, Yang G, Sha Y, Zhang M. (2024) Traveling wave resonance dynamic response characteristics and test of aviation high-speed thin-web spiral bevel gear. Aeronautical dynamics,1-11. https://doi.org/10.13224/j.cnki.jasp.20240452.
[4] Sun Y. (2023) Contact analysis and vibration characteristics of spiral bevel gears. Shenyang University of Technology, Shenyang. DOI: 10.27322/d.cnki.gsgyu.2023.000084.
[5] Shen T. (2019) Research on Digital Analysis and Manufacturing Technology of Spiral Bevel Gear. Tianjin University, Tianjin. https://kns.cnki.net/kcms2/article/abstract.
[6] Long J. (2019) Calculation of bearing capacity of spiral bevel gear and test of tooth root bending stress. Hunan University, Hunan. https://kns.cnki.net/kcms2/article/abstract.
[7] Wang Y, Zhang Y, Wang S, Ruan D. (2004) Overview of domestic bevel gear technology development. In: Proceedings of the 4th China-Japan International Conference on Mechanical Technology History and Mechanical Design. Beijing. 159-166. https://kns.cnki.net/kcms2/article/abstract.
[8] He Z, Yi F, Yang Z, Wu H, Pei S. (2024) Analytical algorithm of time-varying meshing stiffness of gear pair considering impact and load. Journal of Hunan University (Natural Science Edition), 51(12): 38-46. https://kns.cnki.net/kcms2/article/abstract.
[9] Zhou W. (2022) Study on the effect of elastohydrodynamic lubrication on the dynamics of cylindrical gears. Nanjing University of Aeronautics and Astronautics, Nanjing. https://kns.cnki.net/kcms2/article/abstract.
[10] Zhang L, Xu Y, Xu F, Qing Y, Chen X. (2015) Improved meshing stiffness algorithm and dynamic response analysis of non-standard gear with tooth root crack. Journal of Northwestern Polytechnical University, 33(6): 956-961. https://kns.cnki.net/kcms2/article/abstract.
[11] Wan Z, Zi Y, Cao H, He Z, Wang S. (2013) Time-varying meshing stiffness algorithm correction and dynamic modeling of tooth root crack. Chinese Journal of Mechanical Engineering, 49(11):153-160. https://kns.cnki.net/kcms2/article/abstract.
[12] Zhang T, He Z, Yin S, Sun S. (2019) Meshing stiffness calculation of spur gear pair considering time-varying friction and its influencing factors analysis. Journal of Mechanical Transmission, 43(09):54-59+72. https://kns.cnki.net/kcms2/article/abstract.
[13] Liu W, Zhu R, Zhou W, Shang Y. (2022) Analysis of meshing stiffness and dynamic characteristics of tooth surface spalling fault based on image processing, 44 (03): 525-530. https://kns.cnki.net/kcms2/article/abstract.

First-principle analysis of phase and mechanical properties of Cu-Zn-Sn alloys

Xiaoyu Yang[1,3], Tianhao Zhao[1,3,*], Jinshi Li[2], Ruolan Wang[1,3], Jingjie Kang[1,3], Xihui Xu[2] and Xin Yunxia[4*]

[1]China Ordnance Science Research Institute Ningbo Branch, Ningbo; 315000, China
[2] SPIC Nuclear Power Technology Services Co., LTD, Haiyang; 265100, China
[3]Inner Mongolia Metallic Materials Research Institute, Baotou; 014034, China
[4] Weihai City Investment Real Estate Co., LTD, Weihai 264200, Shandong, China

*Corresponding author: Tianhao Zhao, Xin Yunxia. Email:
zhaotianhao@mail.nwpu.edu.cn; 736770861@qq.com

Abstract. Copper alloy is an important material broadly used in the fields of new-generation information technologies, new energy vehicles, rail transit equipment, aerospace ocean engineering, etc. Tin brass alloy is a kind of material popular in engineering and manufacturing fields. It has good mechanical properties, excellent cold working properties, and can withstand hot rolling and hot extrusion. It is widely used in ships and thermal power plants with high-strength corrosion-resistant condensing pipes. In this paper, the first-principle calculation of the phase diagram of Cu-Zn-Sn alloy and the crystal model of each alloy phase is carried out. The results show that Cu-Zn-Sn alloy mainly contains five compounds: ZnCu, Zn8Cu5, Cu3Sn, Zn3Sn, and Cu6Sn5 phases. Cu6Sn5 phase shows excellent compressive resistance, shear strain resistance, and rigidity in all Cu-Zn-Sn alloy phases, but the anisotropy of the Cu3Sn phase (Orthorhombic) and the Cu6Sn5 phase has a large difference in each index. Therefore, Cu6Sn5 and Cu3Sn phases (Orthorhombic) reduce the mechanical properties of brass alloys.

1. Introduction

Tin brass alloy is a kind of material widely used in engineering and manufacturing fields. It has good mechanical properties, excellent cold working properties, and can withstand hot rolling and hot extrusion. It is widely used in ships and thermal power plants with high-strength corrosion-resistant condensing pipes[1-3]. This alloy is prone to stress corrosion fracture, so cold-worked tubes and rods must be annealed at low temperatures to eliminate stress [4-7]. Two-phase tin brass (a+β), such as HSn60-1, has good mechanical properties, a soft tensile strength of 380 MPa, and an elongation of 40%; it can be subjected to cold and hot pressure processing, features easy cutting, good weldability, and high corrosion resistance in seawater, making it suitable for Marine parts, seawater, gasoline, and other contact parts. For the first time, this article systematically calculates the detailed phase diagram of Cu-Zn-Sn alloy and the mechanical properties of each alloy phase through first principles, especially revealing the advantages and disadvantages of the mechanical properties of Cu_6Sn_5 phase and Cu_3Sn phase (orthogonal) and their influence on the overall properties of the brass alloy.

Journal of Physics: Conference Series **3009** (2025) 012060 doi:10.1088/1742-6596/3009/1/012060

2 Experimental methods

2.1 Phase diagram calculation results and analysis

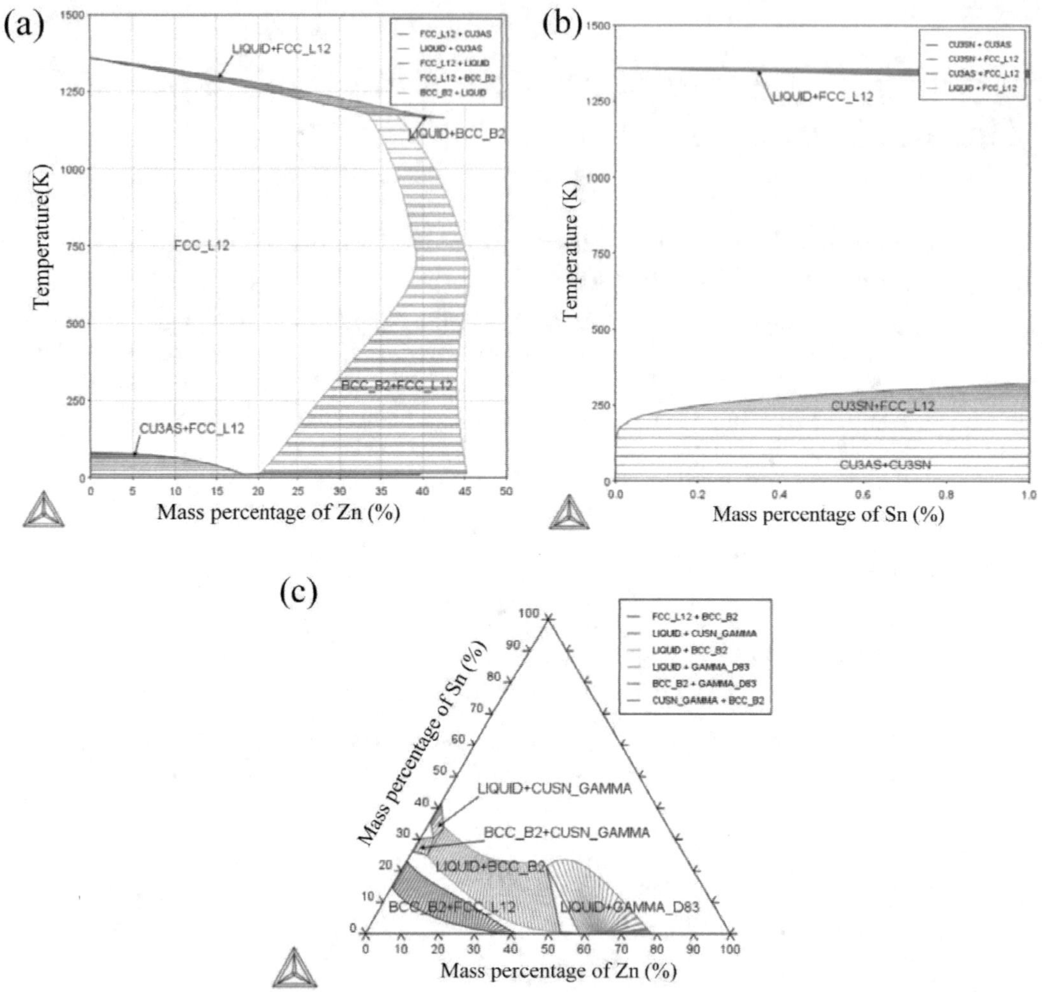

Figure 1. Phase diagram calculated by thermo-calc software: (a) Cu-Zn phase diagram; (b) Cu-Sn phase diagram; (c) Cu-Zn-Sn phase diagram.

As shown in Figure 1, (a) the calculated results of the Cu-Zn phase diagram show the different phase structures of alloys under different temperature and composition conditions. The phase diagram of Cu-Zn alloy mainly includes the following phases: ① α phase: copper-rich phase, with face-centered cubic (FCC) structure; ② FCC-L12(Zn_8Cu_5) phase: stable at room temperature, with a cubic structure; ③ BCC-B2 (ZnCu) phase: stable at room temperature and has a body-centered cubic (BCC) structure. The phase composition of the alloy at different temperatures and composition conditions can be determined by the phase diagram. For example, under low Zn content and high-temperature conditions, Cu-Zn alloys mainly exist in the α phase. At high Zn content and low temperature, β phase and γ phase appeared. When the content of Zn is low, the FCC-L12 phase mainly exists, and when the content of the Zn element increases, the coexistence of the BCC-B2 phase and the FCC-L12 phase occurs. Figure 1 (b) shows the phase diagram of Cu-Sn alloy, showing the phase transition behavior at different temperatures and compositions. The increase in the Sn content will affect the melting point and solid phase structure of the alloy. At a low Sn content, Cu-Sn alloys mainly behave as α phase (copper-rich solid solution). With the increase of Sn content, the Cu_3Sn phase and the Cu_6Sn_5 phase will be formed. Based on the above, the phase diagram of Cu-Zn-Sn ternary alloy was further calculated, as shown in Figure 1(c). Since the content of Zn in the copper alloy is less than 40%, and the content of Sn is less than 1.5%, the main compounds in Cu-Zn-Sn alloy are mainly BCC-B2 phase, FCC-L12 phase, and Cu_6Sn_5 phase, and no new ternary phase is formed. The common stable phase in the Zn-Sn phase diagram is Zn_3Sn, so the Cu-Zn-Sn alloy mainly contains five compounds: ZnCu, Zn_8Cu_5, Cu_3Sn, Zn_3Sn and Cu_6Sn_5 phases.

2.2 First-principle calculation results and analysis

The crystal structure models of five main intermetallic compounds in Cu-Zn-Sn alloy were constructed according to the phase diagram calculation, and the electronic calculation was carried out, as shown in Figure 2, in which Cu_3Sn has three crystal structures.

The lattice constants and properties of compounds in Cu-Zn-Sn alloys at 0 K and 0 GPa were analyzed and calculated. The calculated enthalpy of formation of the Cu_3Sn binary compound in the ground state optimized by first principles and the comparison with other calculated values are listed in Table 1. It is well known that the enthalpy of formation and binding energy of a compound can be used to judge the stability of an alloy compound. The energy absorbed or released when a compound is formed from a stable element is the enthalpy of formation, and it can be seen that the enthalpy of formation is related to the stability of the compound and is the most important thermodynamic parameter. When the process of compound formation is an endothermic reaction, the total energy of the compound generated is greater than the sum of the energy of the stable single substance, and the enthalpy of formation is greater than zero. In this case, the compound is unstable and easy to decompose. On the contrary, when the process of compound formation is an exothermic reaction, the total energy of the compound produced is less than the sum of the stable energy of the single substance, the enthalpy of formation is less than zero, and the compound is stable. Moreover, the smaller the enthalpy of formation, the more stable the compound. The energy released by atoms when they form compounds from the free state is called binding energy. In general, the value of the binding energy of a compound is negative, and the smaller the value, the greater the energy released by the free atoms when bonding. The formation enthalpy and binding energy of the alloy are calculated as follows:

$$\Delta H = \frac{1}{x+y}\left(E_{tot} - xE_{solid}^{A} - yE_{solid}^{B}\right)$$

Where ΔH is the formation enthalpy of Cu_xZn_y or Cu_xSn_y. For Cu_xZn_y or Cu_xSn_y binary alloys, E_{tot} is the total energy, and E_{solid}^{A} and E_{solid}^{B} are the energies of Cu and Zn (or Sn) atoms in the crystal structure, respectively; x and y are the numbers of Cu and Zn (or Sn) atoms in the crystal structure of the intermetallic compound, respectively.

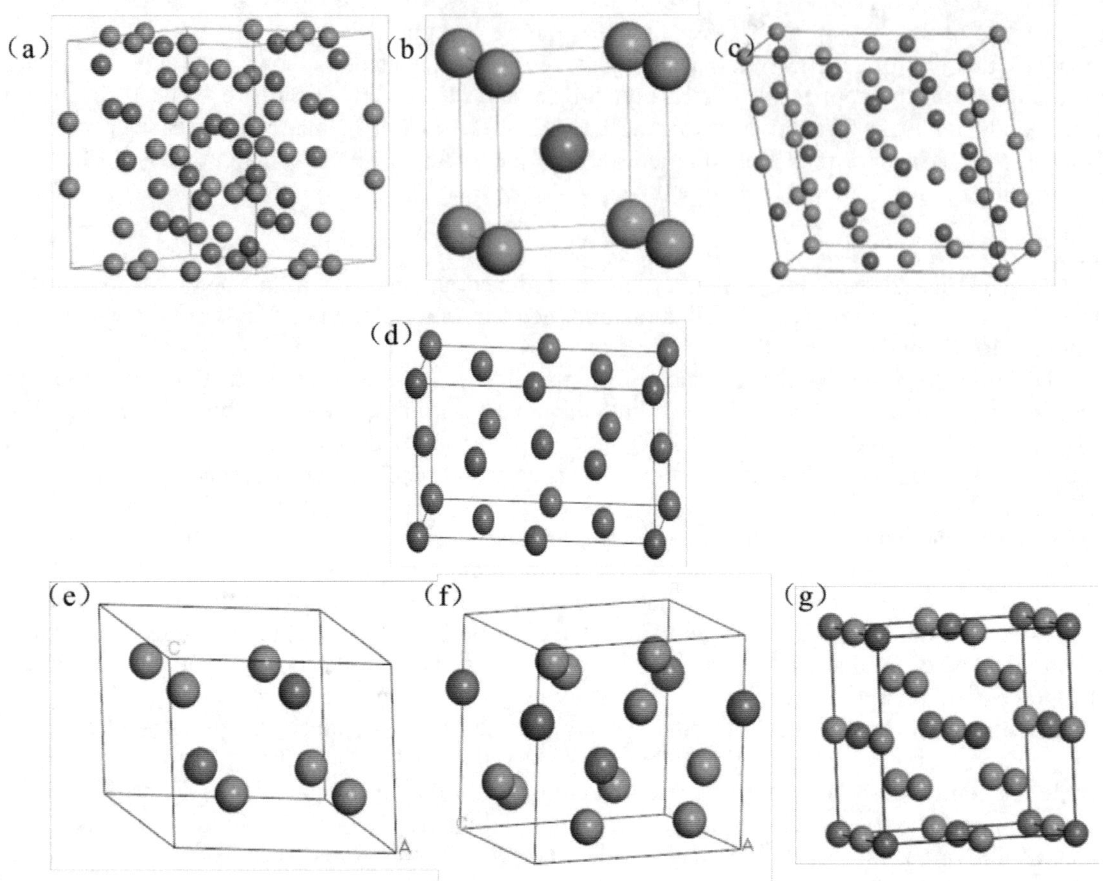

Figure 2. Crystal structure model of each phase: (a)Zn_8Cu_5, (b) ZnCu, (c) Cu_6Sn_5, (d) Zn_3Sn, (e~g) Cu_3Sn.

It can be seen from Table 1 that the calculated formation enthalpies of Cu-Sn compounds are all greater than 0, indicating that all Cu-Sn binary compounds are unstable. Therefore, compared with the BCC-B2 phase and FCC-L12 phase of Cu-Zn alloy, the addition of Sn element does not form a new stable phase in copper alloy, on the contrary, it reduces the number of BCC-B2 phase and FCC-L12 phase and thus reduces the performance of the copper alloy.

In this paper, the elastic constants of these compounds are calculated, as shown in Table 2. Due to the different structural types of the compounds, the obtained elastic constants are also different. It can be found from the table that the calculated results are in line with the mechanical stability conditions of Orthorhombic, Hexagonal, Cubic, and Monoclinic crystal structure systems respectively.

To study the mechanical properties of Cu-Zn-Sn intermetallic compounds in detail, we obtained the bulk modulus (B), shear modulus (G), Young's modulus (E), Poisson's ratio (σ) and hardness (H) of the crystal compounds by Viogt-Reuss-Hill (VRH) approximation method. The calculation results are shown in Table 3.

Table 1. Cu-Zn-Sn compounds: lattice constant(a,b,c), volume(V), enthalpy of formation(ΔH).

	Space group	a (Å)	b (Å)	c (Å)	V ($Å^3$)	ΔH (eV/atom)
Cu_3Sn	Orthorhombic	4.32	4.76	5.43	111.91	0.033

Cu_3Sn	Hexagonal	5.51	5.51	4.27	112.21	0.036
Cu_3Sn	Cubic	6.17	6.17	6.17	234.43	0.070
Cu_6Sn_5	Monoclinic	10.97	7.30	9.84	777.56	0.023
Zn_3Sn	Cubic	4.14	4.14	8.73	149.98	0.124
ZnCu	Cubic	2.92	2.92	2.92	24.91	-0.091
Zn_8Cu_5	Cubic	8.74	8.74	8.74	668.65	-0.108

As Table 3 shows, the Cu_6Sn_5 phase has the largest volume modulus, showing excellent compression resistance, while Zn_3Sn has the smallest volume modulus, indicating low compression resistance. Compared with the BCC-B2 phase and FCC-L12 phase, the Cu_6Sn_5 phase still has higher compression resistance. At the same time, the Cu_6Sn_5 phase also has the largest shear modulus value and Young's modulus value, which are 92.9GPa and 237.4GPa, respectively. Therefore, the Cu_6Sn_5 phase shows excellent compressive resistance, shear strain resistance, and rigidity in all Cu-Zn-Sn alloy phases. However, Poisson's ratio of the Cu_6Sn_5 phase is small, which indicates that the transverse deformation of the Cu_6Sn_5 phase is small.

Three-dimensional models of shear modulus, Young's modulus, Poisson's ratio, and hardness of alloy compounds are shown in Figure 3~6. It can be seen from Figure 3 that the 3D diagram of the shear modulus of metal compounds deviates from sphericity. According to the degree of deviation, the shear modulus anisotropy of the ZnCu phase is the largest, while that of the Zn_8Cu_5 phase is the least. The values of the two compounds in the three crystal directions [100], [010], and [001] are equal, indicating that these compounds have the same shear deformation resistance in these three directions, which is related to the structure of the cubic crystal system. The shear modulus anisotropy directly determines the rigidity, flexibility, and fatigue resistance of the material. Therefore, during the design process, it is necessary to observe the value difference of the compound in the three crystal orientations [100], [010] and [001]. It can be seen from the figure that in addition to the Cu_3Sn (Orthorhombic) phase, The values of the other compounds in the crystal direction [100], [010], and [001] are equal, indicating that the anisotropy of Cu_3Sn (Orthorhombic) phase is large, which will reduce the fatigue resistance of the material, and then reduce the mechanical properties of the brass alloy.

Table 2. Elastic constant C_{ij} (GPa) of Cu-Zn-Sn compounds.

	C_{11}	C_{12}	C_{13}	C_{15}	C_{22}	C_{23}	C_{25}	C_{33}	C_{35}	C_{44}	C_{46}	C_{55}	C_{66}
Cu_3Sn	205.9	107.1	79.4		174.6	46.3		108.4		36.2		45.2	67.6
Cu_3Sn	196.0	82.2	116.2					227.8		63.2			
Cu_3Sn	94.9	63.1								55.6			
Cu_6Sn_5	647.8	121.4	115.5	-54.4	1074	71.9	-22.4	87.3	-2.6	16	-9.3	45.8	8
Zn_3Sn	94.8	30.9								40.2			
ZnCu	125.28	94.45								113.24			
Zn_8Cu_5	193.2	74.2								15.3			

Table 3. Cu-Zn-Sn alloy compounds: bulk modulus, shear modulus, Young's modulus, Poisson's ratio and hardness.

	Space group	B(GPa)	G(GPa)	E(GPa)	σ	H(GPa)
Cu_3Sn	Orthorhombic	90.0	82.8	190.2	0.14	19.1
Cu_3Sn	Hexagonal	114.4	77.4	189.4	0.22	12.8
Cu_3Sn	Cubic	84.3	21.3	69.8	0.64	
Cu_6Sn_5	Monoclinic	177.6	92.9	237.4	0.27	10.9
Zn_3Sn	Cubic	51.2	7.2	20.7	0.43	
ZnCu	Cubic	115.0	10.68	110.55	0.57	
Zn_8Cu_5	Cubic	113.72	10.62	30.90	0.45	

As Figure 4 shows, the 3D diagram of Young's modulus shows that these compounds have varying degrees of anisotropy. The Cu_6Sn_5 phase has the highest Young's modulus anisotropy, while the Zn_3Sn phase has the lowest Young's modulus anisotropy. Since the anisotropy of Young's modulus directly determines the strength and stiffness of the material as well as the elastoplastic deformation ability, it is necessary to observe the value difference of the compound in the three crystal directions [100], [010], and [001] during the design process. It can be seen from the figure that the anisotropy of the three directions of the Cu_6Sn_5 phase is very different, so the introduction of the Cu_6Sn_5 phase will reduce the mechanical properties of the brass alloy.

According to Figure 5, the Poisson's ratio anisotropy values of the Cu_6Sn_5 phase and Cu_3Sn phase (Orthorhombic) in the three crystal orientations [100], [010], and [001] of all compounds are significantly different, indicating that the Cu_6Sn_5 phase and Cu_3Sn phase (Orthorhombic) phase will cause deformation stress in the material. This reduces its strength and stiffness. In addition, as shown in Figure 6, it can be seen that the hardness anisotropy values of the Cu_3Sn phase in the crystal direction [100], [010], and [001] are very different, which further proves that the addition of Sn elements can reduce the mechanical properties of brass alloy by Cu_6Sn_5 and Cu_3Sn phases (Orthorhombic).

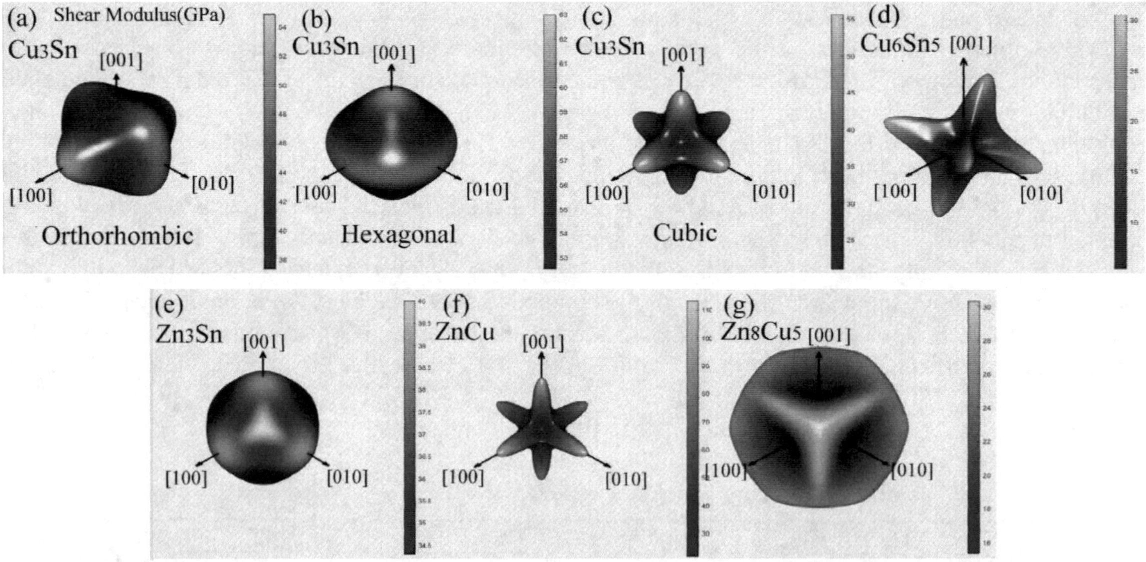

Figure 3. Three-dimensional anisotropic surface diagram of shear modulus: (a) Cu3Sn phase (Orthorhombic), (b) Cu3Sn phase (Hexagonal), (c) Cu3Sn phase (Cubic), (d) Cu6Sn5 phase, (e) Zn3Sn phase, (f) BCC-B2 phase (ZnCu), (g) FCC-L12 phase (Zn8Cu5).

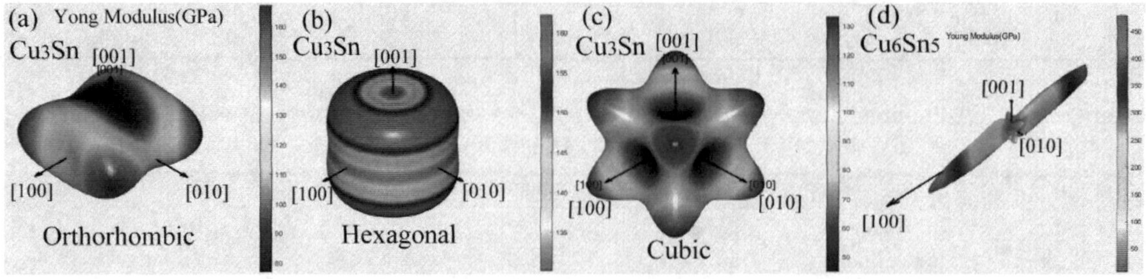

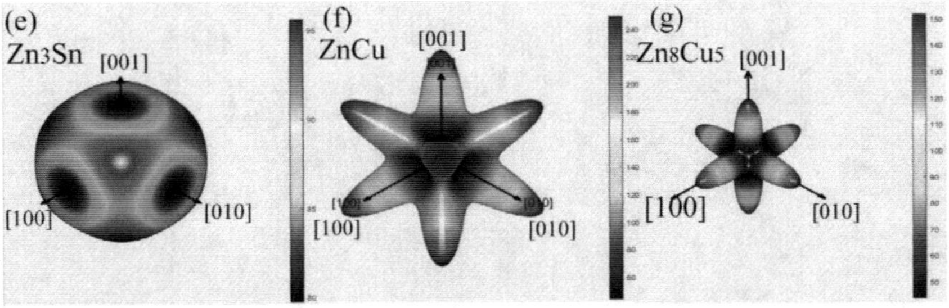

Figure 4. Three-dimensional anisotropic surface diagram of Young's modulus: (a) Cu3Sn phase (Orthorhombic), (b) Cu3Sn phase (Hexagonal), (c) Cu3Sn phase (Cubic), (d) Cu6Sn5 phase, (e) Zn3Sn phase, (f) BCC-B2 phase (ZnCu), (g) FCC-L12 phase (Zn8Cu5).

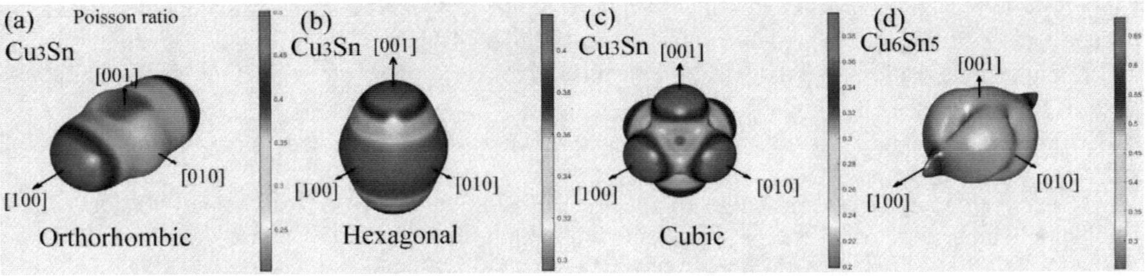

Figure 5. Three-dimensional anisotropic surface diagram of Poisson's ratio: (a) Cu3Sn phase (Orthorhombic), (b) Cu3Sn phase (Hexagonal), (c) Cu3Sn phase (Cubic), (d) Cu6Sn5 phase, (e) Zn3Sn phase, (f) BCC-B2 phase (ZnCu), (g) FCC-L12 phase (Zn8Cu5).

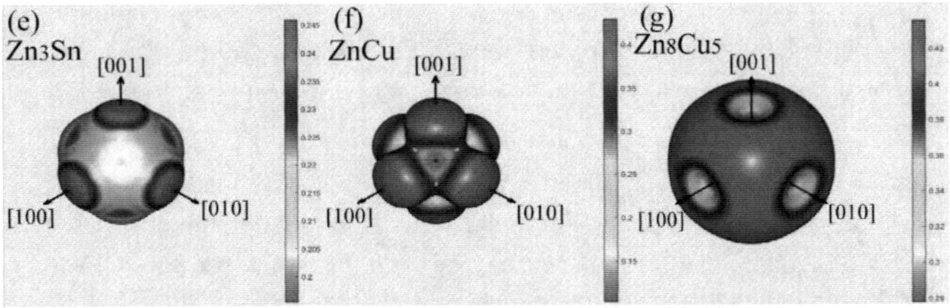

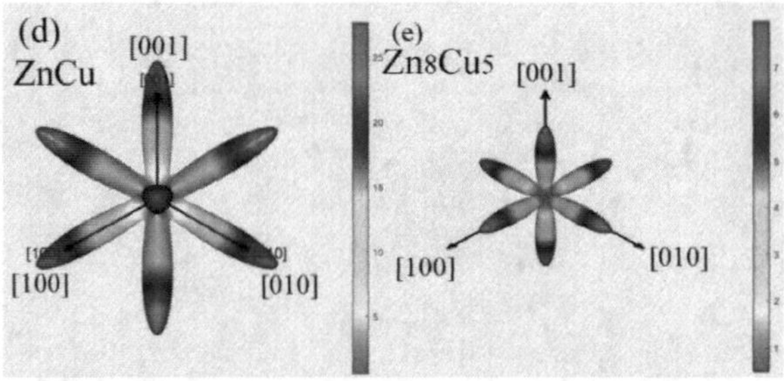

Figure 6. Three-dimensional anisotropic surface diagram of hardness: (a) Cu_3Sn phase (Hexagonal), (b) Cu_3Sn phase (Cubic), (c) Zn_3Sn phase, (d) BCC-B2 phase (ZnCu), (e) FCC-L12 phase (Zn_8Cu_5).

3 Conclusions

According to the characteristics of Cu-Zn-Sn ternary alloy, the phase diagram calculation and first-principle calculation were carried out, and the results and analysis were as follows:

(1) Cu-Zn binary alloys mainly exist in three phases: α rich copper phase, FCC-L12(Zn_8Cu_5) phase, BCC-B2(ZnCu) phase;

(2) The Cu-Zn-Sn alloy mainly contains five compounds: ZnCu, Zn_8Cu_5, Cu_3Sn, Zn_3Sn and Cu_6Sn_5 phases;

(3) There are three crystal structures of Cu_3Sn. The addition of the Sn element will not form a new stable phase in the brass alloy species, but will reduce the number of BCC-B2 phase and FCC-L12 phase, thus reducing the performance of copper alloy;

(4) The Cu_6Sn_5 phase shows excellent compressive resistance, shear strain resistance, and rigidity in all Cu-Zn-Sn alloy phases, but the anisotropy of the Cu_3Sn phase (Orthorhombic) and the Cu_6Sn_5 phase has a large difference in each index. Therefore, Cu_6Sn_5 and Cu_3Sn phases (Orthorhombic) reduce the mechanical properties of brass alloys.

Acknowledgments

This research was supported by Zhejiang Provincial Natural Science Foundation of China under Grant No. LTGC24E050003， and Natural Science Foundation of Inner Mongolia Autonomous Region (2024FX09), and Yuyao City science and technology innovation project (2023JH03010013).

References

[1] Y K Kim, S H Park and K A Lee. Effect of post-heat treatment on the thermophysical and compressive mechanical properties of Cu-Ni-Sn alloy manufactured by selective laser melting. Mater. Charact. 2020, 162, 110194.

[2] Y Kang. China nonferrous metals industry development report. Nonferrrous Metals Processing. 2006, 3:2-13.

[3] L N Huang, X K liu, P liu, X H Chen, et al. Research on Microstructures and Corrosion Resistance of Modified HSn70-1 Brass, Foundry. 2016, 08:718-723.

[4] H W Pickering, C Wanger. Stress corrosion cracking of brass in ammonia solution. Joumal of the Electrichemical Society. 1967, 114(7):698.

[5] F.Z. Qi, G.H. Wu, X. Tong, M. Sun, L. Zhang, J.W. Sun, Y.J. Guo, Insight into the microstructure configuration and strengthening mechanisms of a cast in-situ Al3Zr reinforced Al–Li matrix composite, Compos. Commun. 45 (2024).

[6] X.Y. Chen, M.X. Guo, D.X. Qiao, L.K. Liu, K.C. Yu, L.Z. Zhuang, Simultaneously enhanced strength, ductility and formability of Al-Mg-Si-Cu-Zn-Fe-Mn alloy sheets by coupling distribution of solute, phases and grains, Mater. Char. 209 (2024).

[7] Y Shi, C Guo, J Chen, et al. Recrystallization behavior and mechanical properties of a Cu–15Ni–8Sn(P) alloy during prior deformation and aging treatment. Mater. Sci. Eng. A Struct. Mater. 2021, 826, 142025.

A point-of-load micro-module based on HTCC technology for aerospace applications

Qi Tang*, Guoshuai Zhang, Mingming Ji, Xingzheng Pei, Xiaofeng Sun, Yuhuan Zhang, Qiong Wu, Jianping Wang and Chenlei Cao

Beijing Spacecraft, China Academy of Space Technology, Beijing, China

*Corresponding author's e-mail: power_tng@163.com

Abstract. This paper puts forward a point-of-load (POL) micro-module that combines high power density with enhanced reliability by leveraging high-temperature co-fired ceramic (HTCC) technology. In order to meet the growing power demands of aerospace electronic load systems, the conventional approach of constructing POL converters through the use of discrete devices and printed circuit board (PCB) fails to simultaneously achieve a small converter volume, high output power, and excellent heat dissipation capabilities. In contrast to the traditional methods, the POL micro-module based on HTCC technology proposed in this paper capitalizes on the outstanding heat dissipation and high wiring density features of the HTCC process. It is capable of supplying a power of up to 10A to the load within the tiny area of a single discrete device chip, which measures merely $18\times16\times5(\text{mm}^3)$. Meanwhile, it possesses good heat dissipation capabilities and structural strength, rendering it highly suitable for aerospace and other high-reliability applications. Comprehensive analyses, simulations, and test results are provided in this paper to conduct a thorough assessment of the performance of this miniaturized, high power-density, and high-reliability POL micro-module.

1. Introduction

As the final stage of the aerospace power supply system, the point of load (POL) converter directly interfaces with the electrical load systems, precisely providing the power required by the load [1].

The conventional approach to constructing a POL converter is by utilizing discrete components and devices such as MOSFETs, driver chips, controller chips, and inductors. Nevertheless, in the face of the escalating power requirements of aerospace large-scale microprocessors and other high-performance electrical loads, the traditional POL converter employing discrete components has increased in size and weight [2]. An effective way to reduce the size of the POL converter is to adopt smaller and lighter components and concentrate them within a limited area [3 - 4]. However, this will result in heat concentration, which significantly reduces the reliability of the space power supply system. High-temperature co-fired ceramic (HTCC) boasts advantages like good heat dissipation, high structural strength, excellent chemical stability, and high wiring density, rendering it suitable for miniaturization design in high-reliability systems [5 - 8].

Consequently, a POL micro-module based on HTCC technology is investigated in this paper. The central aim of this research is to propose a novel methodology for designing a POL converter that combines high power density with enhanced reliability.

The proposed POL micro-module uses synchronous buck topology. In section 2, following the analysis and simplification of the circuit, the HTCC integrated structure is designed and the manufacturing process flow of the micro-module is elaborated. In order to verify the reliability of the micro-module structure based on the HTCC technology, thermal simulation, and mechanical simulation are conducted. Section 3 gives the experimental and test results of the proposed micro-module. Section 4 presents the conclusions of this paper.

2. Micro-module design

2.1. Circuit design

Figure 1 demonstrates the system block diagram of the proposed POL micro-module using synchronous buck topology. The module consists of MOSFETs M_1 and M_2, inductor L, input capacitor C_1, output capacitor C_2, load resistor R, PWM controller, and driver.

The two MOSFETs, PWM controller and driver are integrated into one chip so that the circuit of this converter is simplified and the volume of the micro-module can be small. Additionally, the integrated chip has anti-irradiation capability, making the converter suitable for aerospace applications.

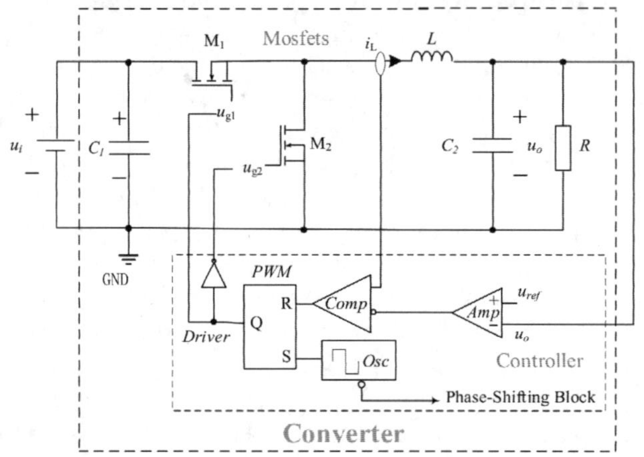

Figure 1. System block diagram of the proposed micro-module.

A periodic square wave signal is provided by the oscillator inside the chip. The main operating logic of the synchronous buck topology is as follows: During a switching cycle, with different combinations of the switching states of MOSFETs M_1 and M_2, energy is stored and released by the inductor L, thus realizing the variation between the input and output voltages [9].

According to the law of conservation of energy, the efficiency of the micro-module cannot reach 100%. The loss of efficiency will lead to difficulties in heat dissipation of the micro-module. The losses of the micro-module are mainly concentrated on the two constantly switching MOSFETs and the inductors used for energy storage. By rationally adjusting the turn-on and turn-off times of MOSFETs and balancing their switching losses and conduction losses, the overall losses of MOSFETs can be minimized. At the same time, the simplified circuit design creates the condition for increasing the area that the inductor can occupy. This not only expands the effective area of the magnetic circuit of the inductor but also significantly reduces the losses of the inductor.

2.2. Structure and process flow design

Figure 2 shows the structure of the POL micro-module proposed in this paper. The micro-module structure mainly consists of a metal cover, Kovar frame, HTCC substrate, and BGA solder balls.

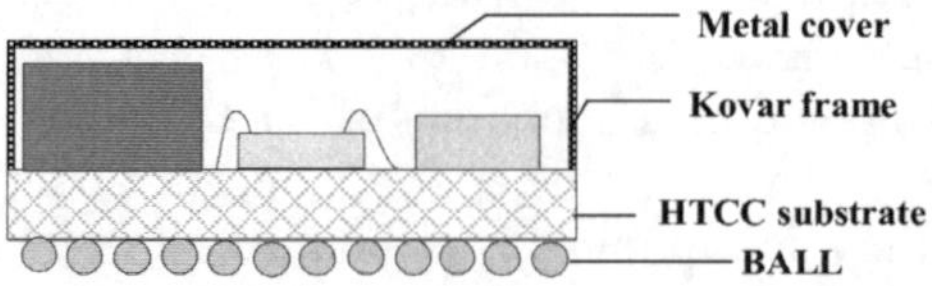

Figure 2. The structure of the proposed micro-module.

The processing flow of the micro-module is as follows:

(1) The HTCC substrate is mainly composed of Al_2O_3 ceramic, and the internal conductor material is tungsten. Moreover, multi-layer conductors and multi-layer ceramic are sintered together at high temperatures to achieve electrical interconnection of the circuit proposed in section 1. Besides, between the surface of the HTCC substrate and the bottoms of the inductors and MOSFETs, a layer of copper conductor with a high thermal conductivity is added by electroplating to reduce the thermal resistance of the heat-generating components inside the micro-module.

(2) The HTCC substrate is brazed with the Kovar frame.

(3) All the components including the integrated chip, inductor, capacitors, and resistors are soldered onto the HTCC substrate. Notice that the melting point of brazing solder is higher than that of component-welding solder.

(4) The metal cover is seam-welded to the Kovar frame.

(5) BGA solder balls are soldered onto the back side of the HTCC substrate.

Additionally, nitrogen gas is filled into the cavity of the module, so that all the components can work in a vacuum environment, avoiding the low-pressure-discharge effect. Therefore, the structure this paper designed in this paper can improve the operational reliability of electronic components in aerospace applications.

By using HTCC technology for circuit integration, the POL micro-module can be integrated into the volume of $18\times16\times5(mm^3)$. The optimized design of circuits and structural processes have been put forward to solve the heat consumption concentration problem. Thermal and mechanical properties of the micro-module must be studied to verify that the micro-module designed in this paper can adapt to the harsh aerospace application environment.

2.3. Thermal simulation analysis

In the vacuum environment, the heat generated by the components in the micro-module is mainly transferred to the outside by heat conduction [10].

Figure 3 gives the comparison diagram of the thermal resistance models between the proposed POL micro-module and the traditional POL converter. It can be clearly seen from Figure 3 that the structure of the proposed POL micro-module has a shorter heat conduction path than the traditional POL converter.

The thermal resistance path is shown in Figure 4. It can be derived from Figure 3 and Figure 4 that the main heat conduction path of the components inside the module is as follows: components-welding materials-substrate -welding materials-BGA balls.

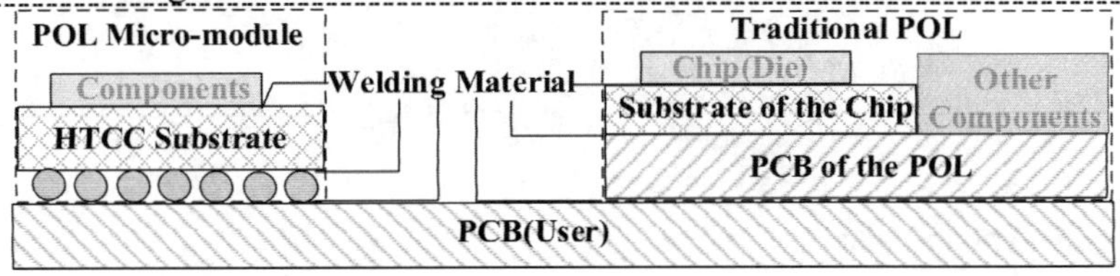

Figure 3. Comparison diagram of the thermal resistance models between the proposed POL micro-module and traditional POL converter.

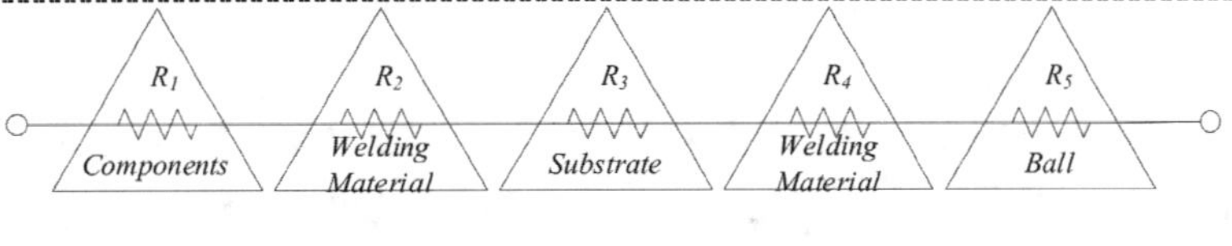

Figure 4. The thermal resistance path of the proposed module structure.

As can be seen from Figure 4, the total thermal resistance of the components inside the module is equal to the sum of the thermal resistance of each material in the heat dissipation path between the components and the package shell.

$$R_{all} = R_1 + R_2 + R_3 + R_4 + R_5 \tag{1}$$

According to the principle of heat divergence at a standard 45° angle, the heat dissipation area is gradually increased from top to bottom. The thermal resistance of components is analyzed.

$$R_x = \frac{L_x}{K_x \times S_x} \tag{2}$$

In (2), x represents the number '1' to '5' in (1), L represents the path length of the components in the heat conduction direction, K represents the thermal conductivity of the heat conductor, and S represents the cross-cutting area perpendicular to the heat flow.

Table 1 shows the thermal conductivity of the materials used in the proposed micro-module. Substitute the parameters in Table 1 into (1) and (2), and the thermal resistance of the components in the micro-module can be calculated.

When the micro-module operates at the full-load current of 10A, the losses of the internal components are the largest. Thermal simulation is taken under the condition of maximum heat consumption. The thermal simulation analysis of the micro-module at the shell temperature of 90℃ is carried out using ANSYS software as is shown in Figure 5.

Table 1. Thermal conductivity of commonly used materials in the module.

Material Name	Thermal conductivity (W / ℃ • cm)
Silicon（Chip）	0.91
Kovar	0.14
Al_2O_3(HTCC substrate)	1.27
Molybdenum-Copper (Electroplated layer)	2
Welding material	0.91

Figure 5. Thermal simulation result.

As can be seen from the analysis results shown in Figure 5, the maximum temperature rise of the components inside the micro-module proposed in this paper does not exceed 6.7℃, which verifies that the thermal design of the internal components is in good condition.

2.4. Mechanical simulation analysis

The HTCC structural substrate of the micro-module is mainly composed of Al_2O_3, which has the characteristics of lightweight, high ultimate strength, and easy processing. Its typical physical properties are shown in Table 2.

Table 2. Physical properties of structural substrates.

Materials	Density (kg/m^3)	Elastic Modulus (MPa)	Poisson's Ratio (μ)	Ultimate Strength (MPa)
Al_2O_3 (type: B210)	3700	280000	0.3	340

By importing the parameters of Table 2 into ANSYS software, this paper established the mechanical model of the micro-module and completed the random vibration simulation according to the test conditions required by GJB548. The mechanical simulation results are shown in Figure 6 and Figure 7.

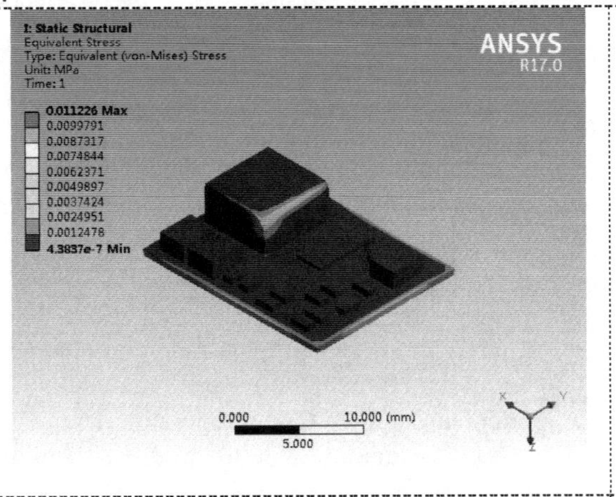

Figure 6. Equivalent stress response (Z-direction).

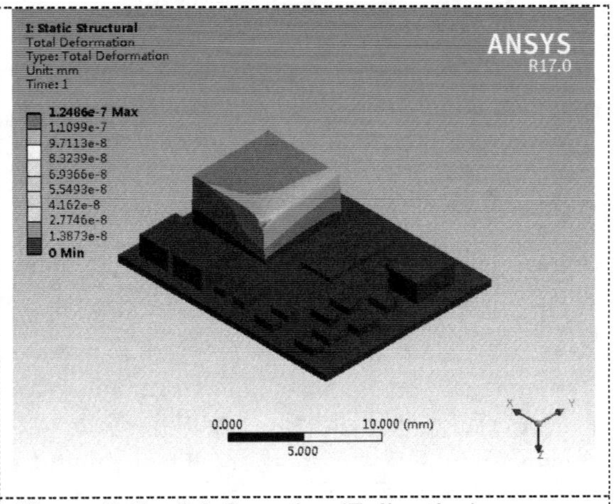

Figure 7. Total deformation (Z-direction).

As can be seen from the mechanical simulation results, the micro-module structure studied in this paper has relatively small deformations of the HTCC substrate and a high safety margin for the maximum equivalent stress, which can meet the requirements of aerospace applications.

3. Experimental Results and Test Verification

After the previous analysis and design, in a volume of 18×16×5(mm^3), a POL micro-module is completed. The micro-module is designed to work and tested under the following conditions: input voltage: 0V~7V; output voltage: 2.5V; output current: 10A. Figures 8-11 are the test results of the module.

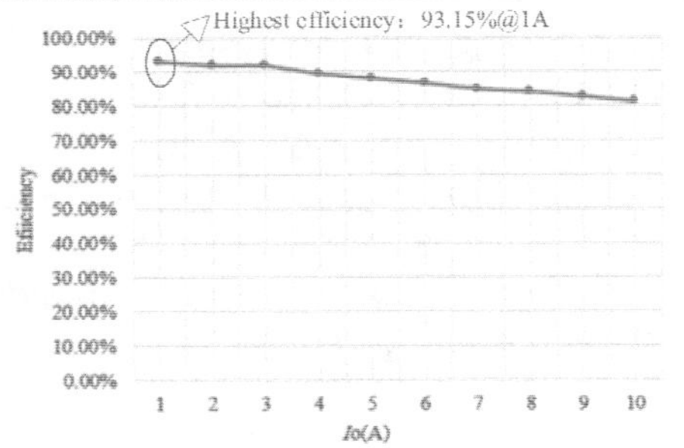

Figure 8. The test result for the proposed micro-module efficiency .

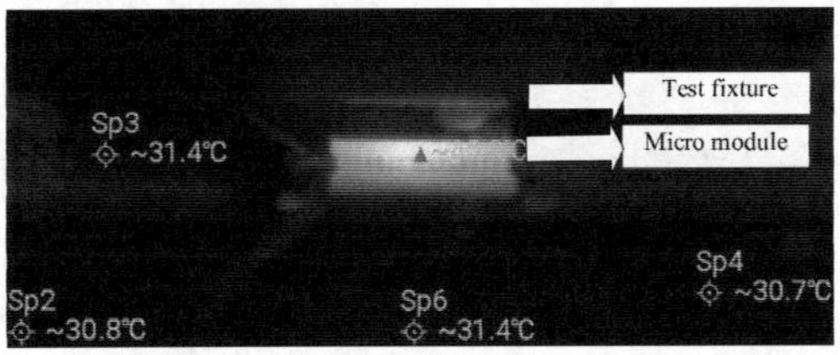

Figure 9. Thermograph of the proposed micro-module .

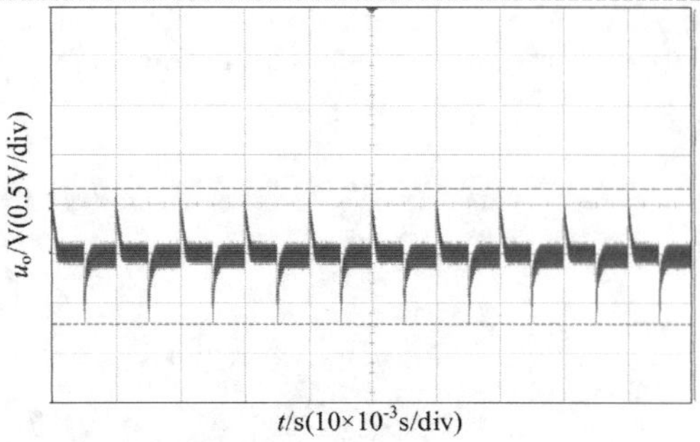

Figure 10. Load step waveform of the proposed micro-module.

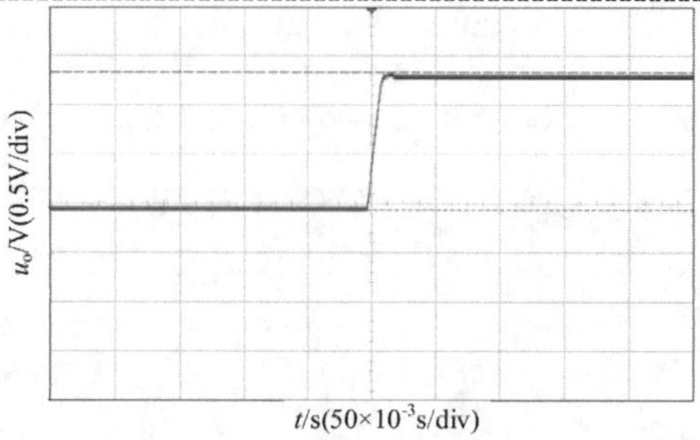

Figure 11. Startup waveform of the proposed micro-module.

It can be observed from Figure 8 that the micro-module power supply proposed herein exhibits a high efficiency, with the highest efficiency reaching up to 93.1% (when the output current is 1A). Consequently, the thermal dissipation of the internal components is low, and the temperature rise of the module is minor. The thermograph in Figure 9 indicates that the maximum temperature rise of the module is merely 13.9°C, which is a little higher than the result of the thermal simulation. This is attributed to the fact that no thermal conduction design has been implemented for the module test fixture, and the conditions are worse than those simulated with a cold plate in the thermal simulation. This outcome validates that the micro-module based on the HTCC technology possesses superior heat dissipation capability, enabling the module to have an extremely high operational reliability.

Figures 10 and 11 represent the load step waveform and startup waveform of the proposed micro-module, respectively. It can be seen from the figures that the module possesses superior dynamic and static characteristics, verifying that the micro-module proposed in this paper can ideally realize the function of the load point power supply.

4. Conclusion

This study has yielded favorable outcomes and made remarkable contributions by comprehensively exploring a POL (point-of-load) micro-module based on the HTCC (high-temperature co-fired ceramic) technology. By taking full advantage of the outstanding features of HTCC, specifically its excellent thermal conductivity, strong structural strength, and high wiring density, a novel POL micro-module has been successfully developed. Simplifying the circuit at the initial stage has proven to be a critical step, allowing the micro-module to achieve a significantly higher power density in contrast to traditional designs. Subsequently, a customized HTCC structure was put forward, along with a detailed processing flow, which laid a solid foundation for the practical implementation of the module. Thermal simulations, mechanical simulations, as well as experimental and test results have all verified that the proposed micro-module possesses excellent thermal conductivity, high mechanical strength, good electrical properties, and is well-suited to the high-reliability usage environment. Overall, this research provides a valuable reference and paves the way for future progress in high-reliability microelectronics, stimulating further investigations and innovations in the design and application of POL micro-modules.

Acknowledgments

This research was supported in part by the National Key R & D Program of China (Grant Nos. 2023YFB3611900) .

References

[1] Wang, S. (2023) Design and Implementation of Low-Power and Miniaturized Point-of-Load Power Supply. University of Electronic Science and Technology of China, Chengdu

Journal of Physics: Conference Series **3009** (2025) 012061 doi:10.1088/1742-6596/3009/1/012061

[2] Zhi, S.B. (2023) Research on magnetic integration technology of half-bridge converter for aerospace application. In: CEETC 2023. Chengdu
[3] You, X.A., Sun, X.F., et al. (2019) A Gallium-Nitride Point-of-load DC-DC converter for Space Applications. ESPC2019. Juan-les-Pins.
[4] Guan, L.S., Yi, C., et al. (2023) A 10 MHz High-Frequency DC-DC Power Converter and Its Synchronous Rectification Technology. Transactions of China Electrotechnical Society, **38**:15-20
[5] Jia, M.G., Xiong, J.J., et al. (2021) Design and Simulation Research of High-Temperature Resistant HTCC Passive Vibration Sensor. Sensor and Microsystem, **40**:119
[6] Mo, L., Hua, T.T, et al. (2016) Design of a Miniaturized Down-Conversion Circuit Based on HTCC. Radar Science and Technology, **14**:671-673
[7] Yang, Z.N, Wang, Y., et al. (2022) A Design of Hybrid Integrated Power Supply Structure Based on Top Heat Sink. Application of Electronic Technique, **48**:40-42
[8] Yang, Y.C., Liu, Y., et al. (2021) SnO_2 thin film NO_2 gas sensor based on HTCC sensitive core. Transducer and Microsystem Technologies, **40**:90-92
[9] Chen, G.P., Chen, L., et al. (2016) Zero-Voltage-Switching Synchronous Buck Converter with Coupled Inductor. Transactions of China Electrotechnical Society, **31**:103-105.
[10] Chen, T., Zhang, J., et al. (2017) Heat Dissipation Design of Surface-Mounted Power Devices Based on Finite Element Simulation. Electronics Process Technology, **38**:14.

Synthesis of hierarchically porous carbon doped with molybdate for capturing the greenhouse gas SF_6

Qiukai Lin, Yingzhe Cheng*, Dexing Chen

EHV Branch Company of State Grid Fujian Electric Power Co.Ltd, Fuzhou, 350108, China.

* Corresponding author: GWchenyingzhe@163.com

Abstract: Biomass carbon materials are economical and effective for preparing gas adsorbents. In this study, tung nut shells were pyrolyzed to create three types of porous structural carbonized biomass doped with molybdate for capturing SF_6. The material prepared at 800°C (BioM8) possessed abundant micropore structures that were close to sulfur hexafluoride's kinetic diameter. BioM8 exhibited the highest SF_6 adsorption capacity (44.36 cm^3/g at 0.1 bar and 298 K in ambient conditions). Multiple analysis techniques revealed that the biomaterial co-doped with molybdate improved porosity formation and enhanced the absorption of SF_6. Furthermore, theoretical calculations confirmed that molybdate could enhance the interaction energy between the biomaterial and SF_6. The method proposed in this study may provide an economical way to produce porous biomaterials for the absorption of efficient SF_6.

1. Introduction

SF_6 is widely used in high-voltage power transmission systems due to its unique properties, including high electrical insulating strength, good chemical stability, and low dielectric loss [1, 2]. The density of SF_6 gas is approximately five times that of air, making it hard to decompose even at high temperatures. Furthermore, SF_6 has a powerful cooling effect on arcs, with arc-extinguishing capabilities over 100 times stronger than air. SF_6 gas also has no corrosive effect on metals and insulation materials, making it an ideal insulation and arc-extinguishing medium in power equipment [3, 4]. A report indicated that in China, SF_6 emissions significantly grew from 2.6 (with a 68% uncertainty range of 2.3-2.7) Gg yr^{-1} in 2011 to 5.1 (ranging from 4.8 to 5.4) Gg yr^{-1}. Without practical control measures, this rising emission trend may hinder the realization of carbon neutrality by 2060 [5].

However, SF_6 gases have an extremely high global warming potential (GWP). It is estimated that on a 100-year time scale, their GWP is 2.35×10^4 times more than that of carbon dioxide [6]. This projected estimation implies that SF_6 more significantly contributes to the earth's greenhouse effect than carbon dioxide with the same mass. Therefore, even a small amount of SF_6 leaking into the atmosphere can significantly impact global climate change. SF_6 has an extremely long lifetime in the atmosphere, lasting for approximately 3,200 years [6], meaning its environmental impact persists for a long time once released into the atmosphere. However, alternatives to SF_6, including fluorocarbons and nitrogen fluoride compounds, are being researched and developed. These compounds have similar insulating properties as SF_6 but have a lower GWP. The excellent dielectric constant and arc-extinguishing properties of SF_6 have made it the most important insulating gas in the power industry, thus making it challenging to be replaced by other compounds.

Recovery of SF_6 for re-disposal is currently the main treatment route in high-voltage transmission systems. The main SF_6 adsorption materials are traditional porous adsorbents (such as activated carbon and zeolite molecular sieves) and new porous materials (such as metal-organic framework organics) [7-10]. To achieve the adsorption and separation of SF_6, materials with high porosity and specific surface area, systematically adjustable pore size, and charge attraction are used as absorbents. Kim et al. (2014) reported a material of X-MOF-74 (where X labels as Zn, Co, and Mg) that had a pore of 11 Å and a lot of adsorption sites to selectively adsorb SF_6 over nitrogen. The maximum adsorption SF_6 capacity of these materials demonstrated was 6.42 mmol/g under the conditions of 1 bar and 298 K) [10]. Wang et al. (2021) offered a calcium-based MOF with channels functionalized by multi-aromatic rings, achieving an impressive selectivity of 727 for an SF_6/N_2 mixture relevant to industrial applications (with a volume concentration of SF_6 at 0.03%) at a temperature of 298 K and pressure of 1 bar [9]. Furthermore, Wang et al. (2022) reported that the Ni (ina)2 could be used to capture SF_6, attaining the highest selectivity for SF_6/N_2 (375.1) at a temperature of 298 K and pressure of 1 bar. Meanwhile, its capacity of SF_6 reached 53.5 cm^3/g in ambient conditions (the temperature was 298 K and the pressure was 1 bar) [8].

However, chemical synthesis approaches have several drawbacks, including complex processes, low production yields, and the use of environmentally harmful strong acids. Biomass-derived carbon materials possess natural three-dimensional (3D) structures with nitrogen groups and carbon rings formed after pyrolysis, making them suitable for gas adsorption [11, 12]. T These materials are biodegradable, and their carbonization process is non-toxic or of low toxicity. For example, Chiang et al. (2010) discovered that multi-walled carbon nanotubes (MWNTs) could adsorb between 278 mg/g and 497 mg/g of SF_6 at 25°C [13]. Theoretically, electron groups combined with the porous structure of biomaterials can enhance gas adsorption. However, few studies have explored the use of various biochars for SF_6 adsorption.

Therefore, this study aimed to develop a porous carbon material as SF_6 adsorbent. First, the structure, elemental content, specific surface area, and shape of the material were examined. Second, its adsorption capacity was determined. Finally, the possible adsorption mechanism was analyzed. The data obtained will help create a porous biomaterial capable of SF_6 adsorption, with applications in high-voltage transmission systems.

2. Materials and Methods

2.1 Preparation of Biomaterials

Shell of tung nut (STN) was corrected from a farm in Wuyishan City, Fujian Province, China (E118.0012, N27.7385). All the reagents were bought from Sigma Aldrich China Co. Ltd. SF_6 was bought from Dehe Chemical Co. LTD. in Fuzhou, China. The biomaterials preparation process is shown in Figure 1. First, the STN was rinsed with deionized water and mixed with Na_2MoO_4 (1 mol/L), followed by pyrolysis in an N_2 atmosphere. Then it was heated from room temperature to different high-temperature conditions: 600°C (low), 800°C (medium), and 1000°C (high). The heating rate was set at 2°C/min, and the samples were held at the target temperature for 1 hour [14]. After pyrolysis, the products were washed with HCl (2 mol/L) or NaOH (2 mol/L) until the final pH of the product was approximately 7. The materials were subsequently dried at 60°C overnight. The prepared samples were labeled according to their pyrolysis temperatures: BioM6 (600°C), BioM8 (800°C), and BioM10 (1000°C). An STN sample which was directly pyrolyzed at 800°C without being mixed with molybdate was utilized as CK.

2.2 Physical characterizations

The microstructure of the prepared biomaterials was observed using a Hitachi S-4800 scanning electron microscope (SEM) (Hitachi, Ltd., Japan) and a Krios G4 Cryo-TEM (Thermo Fisher Scientific Inc., USA). The specific surface areas of the materials were measured using ASAP2460 (Micromeritics, USA). The elemental chemical bonds were analyzed by X-ray photoelectron spectroscopy (XPS) using the Kratos XSAM800 (Kratos Analytical, USA). The bond energy types and distributions were examined through Fourier transform infrared (FT-IR) spectroscopy (Nicolet iS5; Thermo Fisher Scientific Inc., USA). SF_6 adsorption experiments at 298 K were conducted using BSD PM (Beishide

Instrument Technology, China). A constant-temperature bath circulatory system was employed to maintain precise temperature control during adsorption measurements.

2.3 Theoretical calculation

The theoretical calculations based on density functional theory (DFT) were performed using the Vienna Ab-initio Simulation Package (VASP). The Perdew–Burke–Ernzerhof (PBE) functional was selected [15]. The ionic cores were analyzed by projected augmented wave (PAW), and the valence electrons (450 eV) were determined by a plane wave basis set. The geometry (1×10^{-5} eV) was optimized by a force convergency less than 0.02 eV/Å. The method of Monkhorst Pack (2×2×1) was used in the calculations. The adsorption energy (Ea) was computed as follows:

$$Ea = E(BARE+SF_6) - E(BARE) - E(SF_6).$$

Additionally, the Bader charge analysis, charge density mapping, and projected density of states (PDOS) plots were generated using a 4×4×1 Monkhorst–Pack mesh.

2.4 Statistical analysis

Three duplicates were made in each experiment, and the average value was presented. Graphs were plotted using Origin 2024.

3. Results and discussion

3.1 Preparation and characterization of biomaterials

High-temperature carbonization is a cost-effective method for producing porous materials from biomass [14, 16, 17]. The STN was utilized as the raw biomass material, which was first soaked in a molybdate solution and then carbonized at high temperatures (Figure 1). The STN exhibits a natural three-dimensional (3D) surface structure (Figure 2A-B). After carbonization, numerous microporous structures formed on the surface of the biomaterial samples at different pyrolysis temperatures (600°C, 800°C, and 1000°C) (Figure 2C–H), indicating the effectiveness of the activating agent (Na_2MoO_4). The salt crystals tend to vaporize and volatilize during high-temperature pyrolysis, creating a porous structure within the biochar material [17-19]. However, as the carbonization temperature increased, the pore size also grew larger. Compared with BioM6 and BioM8, BioM10 (pyrolyzed at 1000°C) exhibited larger pores on its surface, consistent with previous studies [14, 19].

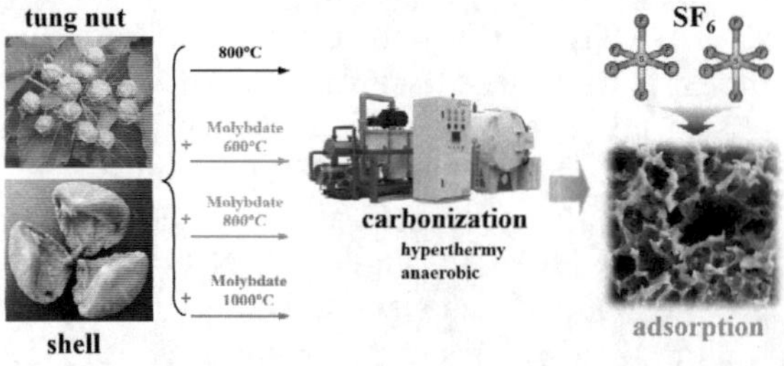

Figure 1 shows the synthesis process of the biomaterials and their application on SF_6 adsorbent.

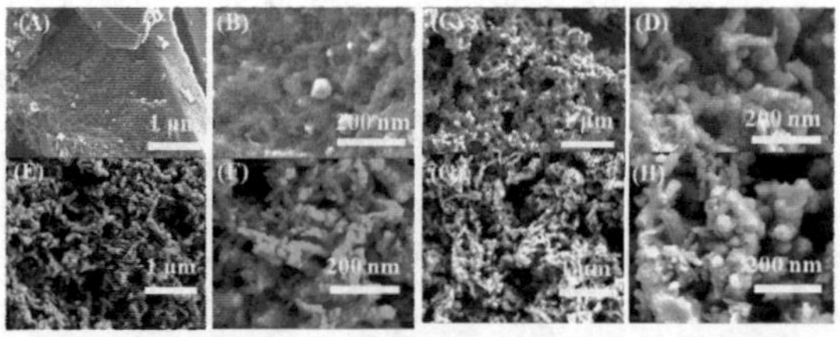

Figure 2 shows SEM images of biomass raw material (A, B) and pyrolysis product (BioM6 (C, D), BioM8 (E, F), and BioM10 (G, H)), and it can be observed that many microporous structures have formed on the surface of the pyrolysis products.

Among all the biomaterials, the adsorption process was consistent with Langmuir adsorption (Figure 3A), indicating that gas molecules were adsorbed inside the micropore structures. The Brunauer–Emmett–Teller (BET) surface areas of CK, BioM6, BioM8, and BioM10 were 219.497, 545.864, 989.627, and 647.86 m^2/g, respectively. The primary micropore distribution ranged from 1 to 30 nm (Figure 3B). The average pore diameters were 19.1422 nm (CK), 5.0689 nm (BioM6), 5.2814 nm (BioM8), and 10.8902 nm (BioM10). These results indicate that pore size is closely related to both the carbonization temperature and the presence of molybdate. Excessively high carbonization temperatures result in larger pore structures, consistent with SEM observations. BioM8 exhibited the largest BET surface area, while CK had the lowest, suggesting that a higher abundance of micropores contributes to an increased surface area. The SEM and TEM analyses further confirmed the presence of molybdate deposits on the surface of the biomaterials, forming a core-shell structure that enhances adsorption efficiency.

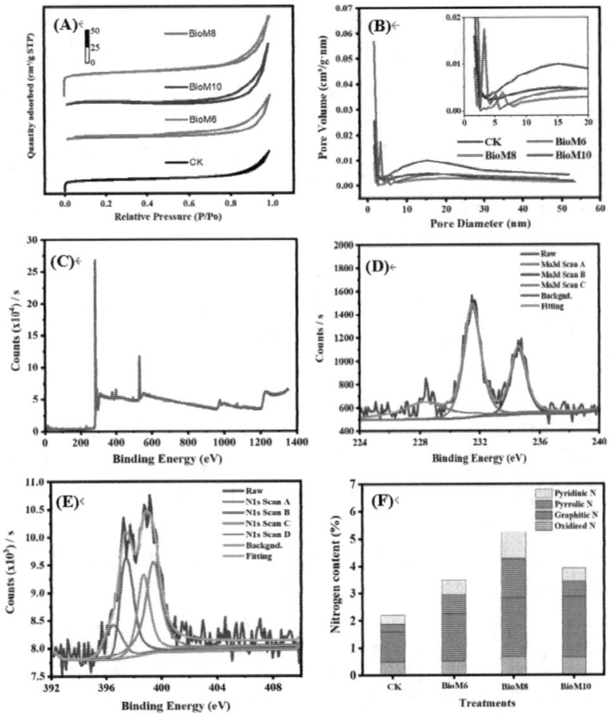

Figure 3. A shows the area different measured by isothermal adsorption curve; B shows the pore distribution of prepared carbonized material at various pyrolysis temperature, indicating that pore size is closely related to pyrolysis temperature; C- E shows the XPS spectrum of full-scan, Mo3d, N1s from BioM8, suggesting the doping of these element; F shows the four types of nitrogen content in the BioM8, suggesting the different kind forms of nitrogen.

The XPS full-scan spectrum revealed that the primary elements in CK were nitrogen, carbon, oxygen, and molybdenum. In comparison, the other three samples primarily contained carbon, nitrogen, and oxygen (Figure 3D–E). The Mo 3d spectrum of BioM8 mainly consisted of three peaks at 234.6 eV, 231.5 eV, and 229.3 eV (Figure 3D). Peaks between 229.1 and 232.3 eV corresponded to Mo^{4+}, while the peak at 234.6 eV corresponded to Mo^{6+}, confirming the presence of molybdenum ions in the sample [20, 21]. The N 1s spectrum comprised four peaks: graphitic-N (398.3 eV), pyridinic-N (400.1 eV), pyrrolic-N (400.9 eV), and another pyridinic-N (403.5 eV) (Figure 3E) [22, 23]. The types of nitrogen groups formed at various pyrolysis temperatures were also analyzed. The graphitic-N content increased from 600°C to 800°C and then decreased from 800°C to 1000°C, with the highest content observed in BioM8 (Figure 3F). Graphitic-N contributes to promoting electron transport due to its delocalized electrons. Similarly, the C 1s spectrum of the prepared biomaterials showed peaks at 284.6 eV (graphitic carbon), 285.6 eV (C-N single bond), 286.5 eV (C-O single bond), and 288.9 eV (C=O double bond). The data suggested that the graphitic carbon content increased with the rise in carbonization temperature. These findings indicate that higher pyrolysis temperatures promote graphitization, implying that elevating the pyrolysis temperature is beneficial for increasing graphitic content.

3.2 Adsorption performance

The SF_6 adsorption performance of the four biomaterials was measured at 298 K under 100 kPa pressure. Among them, BioM8 showed the highest SF_6 uptake (38.91 cm^3/g), approximately 3.5 times higher than that of CK (11.12 cm^3/g) (Figure 4A). The BioM6 and BioM10 samples exhibited similar adsorption capacities of 24.90 cm^3/g and 20.31 cm^3/g, respectively. The adsorption performance of these biomaterials is comparable to that of many chemically synthesized materials [8, 9]. However, the method used in this study is more environmentally friendly and sustainable. This superior adsorption property of BioM8 can be attributed to its larger BET surface area and suitable pore structure. Wang et al. (2022) created a new Ni(ina)2 material, which possessed a pore structure of 6 Å. This pore was close to the equivalent diameter size of SF_6 (5.2 Å) [8]. Most of the microporous distribution was between 5 and 7.5 Å (Figure 3B). As a result, the microporous structure enhanced the BET of the prepared biomaterial and promoted the adsorption of SF_6.

The sorption isotherm curve demonstrated a sharp increase at extremely low pressures (<10 kPa) at 298 K, followed by a plateau, indicating strong interactions between SF_6 molecules and the prepared biomaterials. The SF_6 adsorption capacity of BioM8 reached 44.49 cm^3/g at 298 K, while the total SF_6 adsorption capacity of CK was much lower at 12.36 cm^3/g (Figure 4B). The high adsorption capacity of BioM8 suggests that it may serve as a promising and environmentally friendly adsorbent for capturing SF_6.

Further analysis of the FT-IR spectrum revealed no SF_6 stretching peak in CK. However, in the SF_6-loaded state, an obvious stretching peak at 928 cm^{-1} was observed in BioM8 (Figure 4C-D). This peak did not appear before SF_6 adsorption. This peak was the strongest among those materials, indicating that more SF_6 gases adsorbed onto the biomaterial of BioM8.

The performance of BioM8 was lower than that of many chemically synthesized metal structure materials. However, BioM8 was obtained from an economical biomass material, the synthesis of which is simple and eco-friendly.

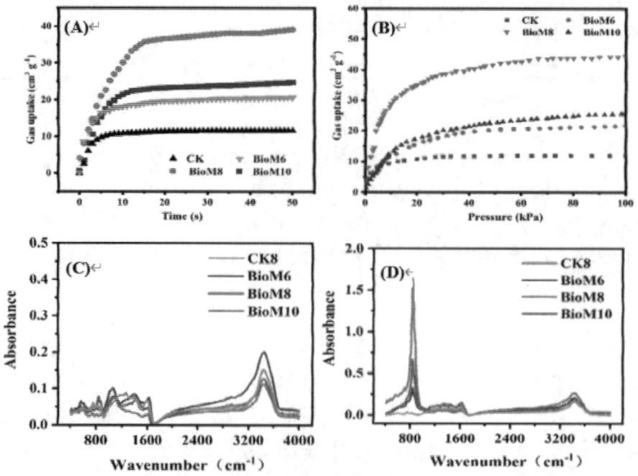

Figure 4. The Kinetic sorption measurements in **Figure** 4A exhibits the sorption characteristics, indicating that BioM8 had the maximum absorption capacity for SF_6 adsorption. B shows sorption characteristics at different pressure at 298 K, suggesting the BioM8 had the maximum absorption capacity for SF_6 adsorption; FT-IR spectra of biomaterials before adsorbing SF_6 (C) and after SF_6 absorption (D), showing that the characteristics peak (928 cm^{-1}).

3.3 Mechanism analysis

The host-guest binding interactions were investigated using DFT calculations [15]. The adsorption energy between biochar and SF_6 was calculated to be -0.205 eV, while the adsorption energy between Mo-modified biochar and SF_6 was -0.37 eV. The results showed that doping biochar with Mo atoms enhanced the adsorption capacity for SF_6. Differential charge non-quantitative analyses showed that biochar doped with Mo reduced the Bader charge to -0.498159 eV, which was lower than without Mo doping (-0.019401 eV) (Figure 5A-B), showing that the doping of Mo promoted the electronic interaction ability between SF_6 and the biochar surface, thus improving the adsorption of SF_6. Furthermore, the calculation of PDOS revealed that the resonance peaks between Mo and F were formed, which led to the enhancement of adsorption (Figure 5C-D).

Therefore, it could be concluded that the excellent adsorption performance of the BioM8 on SF_6 is attributable to the following three main reasons: First, BioM8 has a large amount of surface area, providing more adsorption points for SF_6. Second, a suitable pore structure of BioM8, which is close to the equivalent diameter of SF_6. Third, the doping of Mo further enhanced the adsorption energy between the biochar and SF_6.

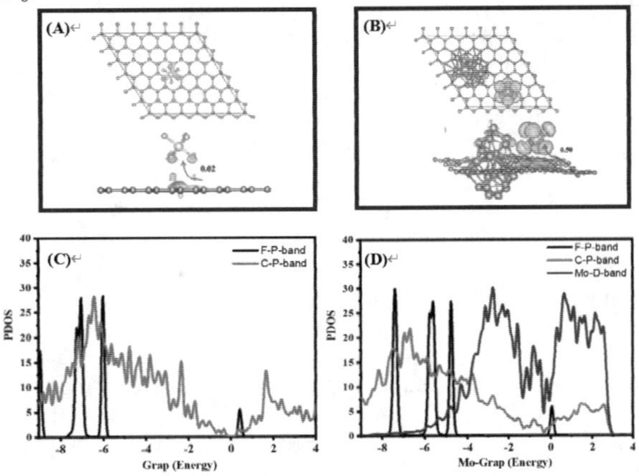

Figure 5 shows the adsorption energy, differential charge density, and other results obtained from DFT calculations, revealing the mechanism by which molybdenum doping affects adsorption performance. The differential charge non-quantitative analyses without Mo doping (A), with Mo doping (B), the calculation of PDOS without Mo doping (C), with Mo doping (D).

4. Conclusions

All above, a novel porous biomaterial was successfully synthesized from the shell of a tung nut through high-temperature pyrolysis. The biomaterial that was pyrolyzed at 800°C and coped with Mo atom (BioM8) exhibited a higher adsorption performance for SF_6. The main reason was possible because of the suitable size of pore structure, large area, and Mo doped groups co-existing. The prepared adsorbent exhibits excellent performance in SF_6 gas adsorption and has broad application prospects. This study provides insight into the development of an excellent carbonized material for adsorption of SF_6.

Acknowledgment

This research was funded by State Grid Fujian Electric Power company (SGFJCG00JXWT2310239)

Reference

[1] Tsai W. The decomposition products of sulfur hexafluoride (SF6): Reviews of environmental and health risk analysis. Journal of Fluorine Chemistry, 2007, 128, 1345-1352.

[2] Xiao S., Zhang X., Tang J., Liu S. A review on SF6 substitute gases and research status of CF3I gases. Energy Reports, 2018, 4, 486-496.
[3] Kozyrev A.V., Kozhevnikov V.Y., Semeniuk N.S. Theoretical simulation of high-voltage discharge with runaway electrons in sulfur hexafluoride at atmospheric pressure. Matter and Radiation at Extremes, 2016, 1, 264-268.
[4] Okubo H., Beroual A. Recent trend and future perspectives in electrical insulation techniques in relation to sulfur hexafluoride (SF6) substitutes for high voltage electric power equipment. IEEE Electrical Insulation Magazine, 2011, 27, 34-42.
[5] An M., Prinn R.G., Western L.M., Zhao X., Yao B., Hu J., Ganesan A.L., Mühle J., Weiss R.F., Krummel P.B. et al Sustained growth of sulfur hexafluoride emissions in China inferred from atmospheric observations. Nature Communications, 2024, 15, 1997.
[6] Jo E., Yang S., Kim D.W., Kang D.W. Greenhouse sulfur hexafluoride gas capture and separation using emerging porous solids. Coordination Chemistry Reviews, 2024, 515, 215958.
[7] Hasell T., Miklitz M., Stephenson A., Little M.A., Chong S.Y., Clowes R., Chen L., Holden D., Tribello G.A., Jelfs K.E. et al Porous Organic Cages for Sulfur Hexafluoride Separation. Journal of the American Chemical Society, 2016, 138, 1653-1659.
[8] Wang S.M., Mu X.T., Liu H.R., Zheng S.T., Yang Q.Y. Pore-Structure Control in Metal-Organic Frameworks (MOFs) for Capture of the Greenhouse Gas SF. Angewandte Chemie (International ed in English), 2022, 61, e202207066.
[9] Wang T., Chang M., Yan T.A., Ying Y.P., Yang Q.Y., Liu D.H. Calcium-Based Metal–Organic Framework for Efficient Capture of Sulfur Hexafluoride at Low Concentrations. Industrial & Engineering Chemistry Research, 2021, 60, 5976-5983.
[10] Kim M., Lee S., Lee C.Y., Bae Y. High SF6 selectivities and capacities in isostructural metal-organic frameworks with proper pore sizes and highly dense unsaturated metal sites. Microporous and Mesoporous Materials, 2014, 190, 356-361.
[11] Bamdad H., Hawboldt K., MacQuarrie S. A review on common adsorbents for acid gases removal: Focus on biochar. Renewable and Sustainable Energy Reviews, 2018, 81, 1705-1720.
[12] Chen Y.Q., Zhang X., Wei C.Z., Yang H.P., Chen H. The structure evolution of biochar from biomass pyrolysis and its correlation with gas pollutant adsorption performance. Bioresource Technology, 2017, 246, 101-109.
[13] Chiang Y., Wu P. Adsorption equilibrium of sulfur hexafluoride on multi-walled carbon nanotubes. Journal of Hazardous Materials, 2010, 178, 729-738.
[14] Tang J.H., Wang Y.J., Zhao W.Q., Zeng R.J., Liu T., Zhou S.G. Biomass-derived hierarchical honeycomb-like porous carbon tube catalyst for the metal-free oxygen reduction reaction. Journal of Electroanalytical Chemistry, 2019, 847, 113230.
[15] Wang Z., Xu R.J., Ye Q.T., Jin X.Y., Lu Z., Yang Z.B., Wang Y., Yan T., Liu Y.P., Pan Z.J. et al Tailoring First Coordination Sphere of Dual-Metal Atom Sites Boosts Oxygen Reduction and Evolution Activities. Advanced Functional Materials, 2024, 34, 2315376.
[16] Yuan H.R., Deng L.F., Tang J.H., Zhou S.G., Chen Y., Yuan Y. Facile Synthesis of MnO2/Polypyrrole/MnO2 Multiwalled Nanotubes as Advanced Electrocatalysts for the Oxygen Reduction Reaction. ChemElectroChem, 2015, 2.
[17] Yuan Y., Bolan N., Prévoteau A., Vithanage M., Biswas J.K., Ok Y.S., Wang H. Applications of biochar in redox-mediated reactions. Bioresour Technol, 2017, 246, 271-281.
[18] Zhou L.H., Fu P., Wen D.H., Yuan Y., Zhou S.G. Self-constructed carbon nanoparticles-coated porous biocarbon from plant moss as advanced oxygen reduction catalysts. Applied Catalysis B: Environmental, 2016, 181, 635-643.
[19] Zhou L.H., Fu P., Cai X.X., Zhou S.G., Yuan Y. Naturally derived carbon nanofibers as sustainable electrocatalysts for microbial energy harvesting: A new application of spider silk. Soil & Tillage Research, 2016, 188, 31-38.
[20] Zhang Q.Q., Li X.S., Ma Q., Zhang Q., Bai H., Yi W.C., Liu J.Y., Han J., Xi G.C. A metallic molybdenum dioxide with high stability for surface enhanced Raman spectroscopy. Nature Communications, 2017, 8, 14903.
[21] Baltrusaitis J., Mendoza-Sanchez B., Fernandez V., Veenstra R., Dukstiene N., Roberts A., Fairley N. Generalized molybdenum oxide surface chemical state XPS determination via informed amorphous sample model. Applied Surface Science, 2015, 326, 151-161.
[22] Ayiania M., Smith M., Hensley A.J.R., Scudiero L., McEwen J.-S., Garcia-Perez M. Deconvoluting the XPS spectra for nitrogen-doped chars: An analysis from first principles. Carbon, 2020, 162, 528-544.
[23] Ohta R., Lee K.H., Saito N., Inoue Y., Sugimura H., Takai O. Origin of N 1s spectrum in amorphous carbon nitride obtained by X-ray photoelectron spectroscopy. Thin Solid Films, 2003, 434, 296-302.

Molecular dynamics simulation of the effect of silane coupling agent-modified SiO_2 on the diffusion of water molecules in polypropylene

Jiang Guo*, Mengyi Xu, Hui Li, Xiang Huang, Hao Lin, Shengbao Jiang, Shang Chen and Cheng Chen

State Key Laboratory of Power Grid Environmental Protection, China Electric Power Research Institute, Wuhan 430074, China

*Corresponding author's e-mail: 15832269286@163.com

Abstract: Polypropylene (PP), as an environmentally friendly thermoplastic material, is widely employed in high-voltage cables. However, the cables are prone to form water trees when they are in a humid environment for a long time, leading to the degradation of insulation properties. The diffusion of water molecules seriously restricts the rapid advancement of PP. In addition, SiO_2 nanoparticles have been proven to enhance the insulation properties of the material, but the mechanism underlying the effect of nanoparticles modified with silane coupling agents on the water molecule diffusion remains inadequately studied. To address this, an interface model of SiO_2 nanoparticles/water-containing PP, both unmodified and modified with different coupling agents (KH570 and KH792), was constructed, respectively, by using molecular dynamics simulation to investigate the intrinsic mechanism at the molecular level in this article. The results indicated that the effect of different coupling agent modifications on the diffusion of water molecules was different, among which KH792 modified SiO_2 nanoparticles/water-containing PP interfacial structure had the largest interaction energy between PP and water molecules with the smallest free volume fraction, which had a better inhibition effect on the movement of water molecules in PP. It provided theoretical support for further research on nanoparticle-modified PP composite materials.

1. Introduction

With the increasing global attention to environmental issues, polypropylene (PP) based insulated cables are gradually being developed and applied. PP has the advantages of low density, high melting point, high mechanical strength, low production cost, and wide utilization as a thermoplastic polymer material [1]. At the same time, there is no cross-linking process in the production of PP, and it can be recycled after use. So, PP is an environmentally friendly insulation material for high voltage direct current (HVDC) cables with promising applications.

However, PP has drawbacks, such as low thermal deformation temperature and poor creep resistance in practical applications [2]. When PP is used for HVDC cables, its performance must be improved to meet the requirements of cable processing, laying, and use. Currently, the performance improvement of PP is mainly realized through modification methods such as copolymerization, blending, and the addition of nanoparticles [3]. Wherein, the addition of nanoparticles is an effective and simple method to improve the overall mechanical, electrical, and insulation properties of PP, and certain research results have been obtained. Cao et al. [4] found doping magnesium oxide (MgO) nanoparticles in PP can suppress the space charge accumulation, and MgO/PP composites with low nanofiller concentration had better electrical insulation properties. Gao et al. [5] found that adding SiO_2 nanoparticles can prolong the electrical aging life of PP insulation.

The statistics on distribution network faults show that moisture is one of the main reasons for the damage of internal insulation in the accessories for the long-term operation of cables in a humid environment [6]. Water not only invades the interior of PP insulation but also accumulates into a water film at the interface between PP and other accessories, which will accelerate the generation of water trees under the electric field, thus leading to the degradation of insulation properties [7, 8].

Although numerous research has been conducted on the modification of PP by doping nanoparticles, the macroscopic experiments mentioned above make it difficult to explain the modification mechanism, especially at the molecular level. With the improvement of computational arithmetic, molecular simulation has developed rapidly. The molecular simulation has been applied to study the elastic modulus and water saturation of semi-crystalline PP, and it is found that the higher the crystallinity of PP, the higher the stiffness and strength. Pei et al. [9] studied the effect of surface coupling agents on the mechanical behavior of SiO_2/PP composites. The presence of coupling agents can improve the mechanical strength of the composites, and APTES exhibited a more pronounced effect. However, there are few reports on the effect of different coupling agent-modified nanoparticles on the diffusion of intruded water in PP and the microscopic mechanism.

Therefore, SiO_2/PP composite models modified with different silane coupling agents were constructed in this article. The microscopic properties such as thermodynamic motion, interaction energy, and hydrogen bond number of water molecules in each model were calculated using molecular dynamics. The microscopic modification mechanism on the effect of different coupling agents on the diffusion behavior of water molecules in PP was interpreted at the molecular level, which provided a certain theoretical basis for evaluating and selecting surface modification coupling agents to improve the performance of composites.

2. Molecular model construction and simulation

In this article, the initial model of the SiO_2/PP interface was constructed in Materials Studio (MS) software using molecular dynamics. A series of structural optimization of the corresponding model, relaxation process, and data collection of the final dynamic simulation process was accomplished under the COMPASS II force field using the Forcite module.

2.1. Molecular model of the PP layer

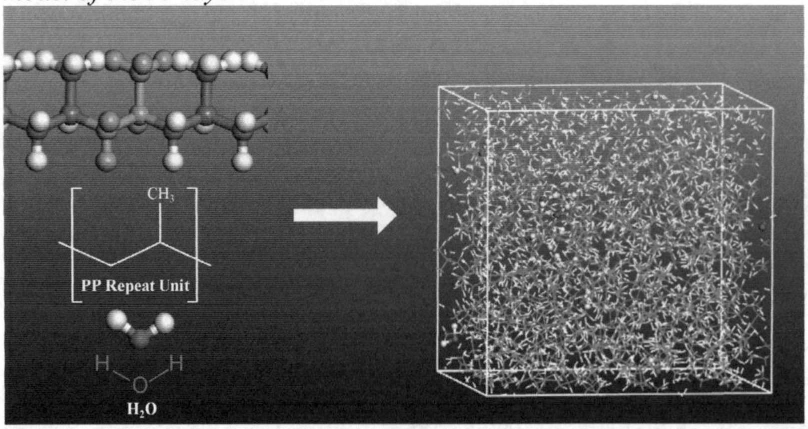

Figure 1. Molecular model of PP layer.

PP has two kinds of structures at the microscopic level: crystalline and amorphous, of which the amorphous refers to the disordered and irregular arrangement of molecular chains. The cable works in a humid environment for a long time. Moisture will invade the free volume cavities inside the material through the amorphous region of PP, forming water trees [10], which ultimately leads to the degradation of the overall insulation properties. So, the amorphous structure of PP was chosen as the research object in this article. Considering both practical engineering applications and computational quantities, and the structural sTableility of amorphous PP with a polymerization degree of 24-72 [7], PP chains (polymerization degree of 50) were constructed using the Build Polymers module in MS software. To reduce the vacuum region between the PP layer molecular model and the SiO_2 layer molecular model, the amorphous structure model was esTablelished using a Confined Layer in an Amorphous Cell module as shown in Figure 1, which contained 10 PP chain and 1 wt.% water molecules. The initial model of PP layer molecules was structurally optimized, with a model density of 0.91 g/cm^3.

2.2. Molecular model of SiO_2 layer

The molecular model of the SiO_2 layer was carried out using the SiO_2 cell provided by MS software. The Cleave Surface module was used for the sectioning (001), and the cell expansion was performed on the basis. Two silane coupling agents (KH 570 and KH 792) with good modification effect in the experiments were grafted on Si atoms on the surface of SiO_2, and the molecular structure of the characteristic functional groups was shown in Figure 2. The remaining unsaturated Si atoms on the surface were subjected to the hydrogenation treatment. The molecular model of the SiO_2 layer was also optimized.

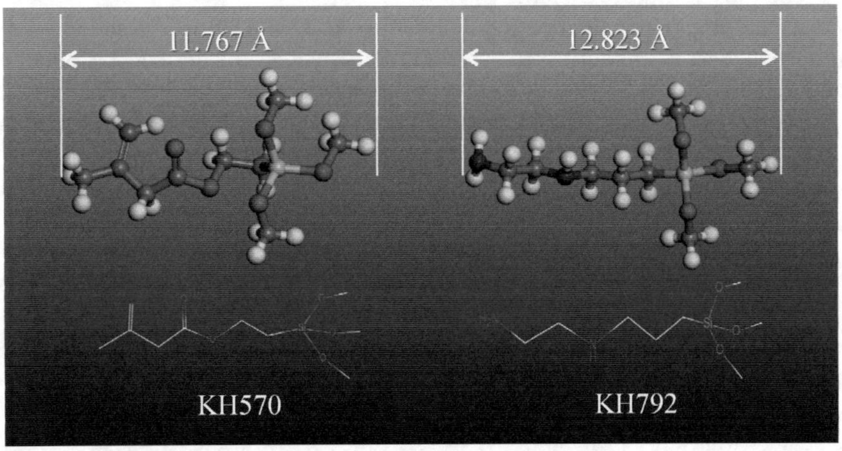

Figure 2. Molecular structure of functional groups of silane coupling agents.

2.3. SiO_2/PP interface model and simulation

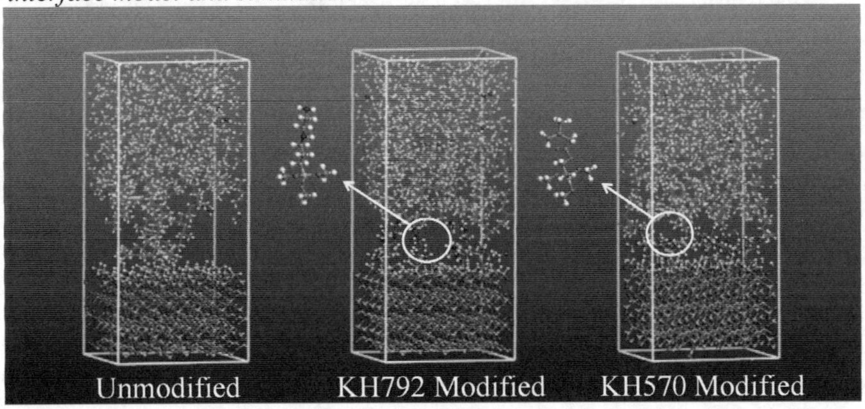

Figure 3. Schematic diagram of SiO_2/PP interface model.

The normal particle size of SiO_2 nanoparticles used for PP in the experiments is usually above 30 nm, while the scale of molecular simulation is 10^{-10} m. Therefore, local regions on the surface of nanoparticle spheres can be modeled as a plane. The water-containing PP model and the molecular model of the SiO_2 layer with different silane coupling agents were merged using the Build Layers module to construct the

SiO_2/PP interface model as shown in Figure 3, where the upper layer was the PP layer and the lower layer was SiO_2 layer.

2.4. Simulation

Firstly, the establelished SiO_2/PP interface model was subjected to geometry optimization, five annealing cycles, and a series of relaxation optimizations. The smart algorithm was used for the geometry optimization, and the relaxation process included the NVT process with 100 ps and the NPT process with 100 ps. Finally, the model obtained after the relaxation optimization was subjected to a kinetic simulation with 300 ps. During the simulation process, Nosé-Hoover was chosen as the temperature controller for NVT and Berendsen as the pressurizer controller for NPT. The temperature and pressure in the model were 298 K and 0.0001 GPa, respectively. The integration step was 1 fs, and the data was collected once every 1000 steps for the subsequent result analysis.

3. Diffusion and distribution of water molecules

3.1. Diffusion coefficient of water molecules

The diffusion coefficient (D) can characterize the diffusion ability of a substance in a system and is a physical quantity that describes the motion intensity of a substance. D can be obtained from Mean Square Displacement (MSD), as shown in Equation (1), which determines whether or not the diffusion, transmission, or hopping of particles occurs by the numerical value of its motion trajectory.

$$MSD = \left\langle \left| \vec{r}_i(t) - \vec{r}_i(0) \right|^2 \right\rangle$$

$$D = \frac{1}{6N} \lim_{t \to \infty} \frac{d}{dt} \sum_{i=1}^{n} \left| \vec{r}_i(t) - \vec{r}_i(0) \right|^2 = \frac{1}{6} a \tag{1}$$

where, $\vec{r_1}(t)$ and $\vec{r_1}(0)$ denote the coordinate vectors of water molecules at t and initial time, respectively. $< >$ denotes the averaging of all water molecule-related values, D (Å^2/ps) denotes the diffusion coefficient, N denotes the number of molecules, and a denotes the fitting slope of the MSD curve.

Figure 4 shows MSD curves of water molecules in the three SiO_2/PP interface models and the fitting slopes. It can be seen that the MSD slope of water molecules was the highest in the model without functional group modification. The fitting slopes of water molecules in the model modified with KH 570 and KH 792 were significantly reduced and the fitting slope of the MSD curve for water molecules in the SiO_2/PP interface model modified with KH 792 was most obviously decreased.

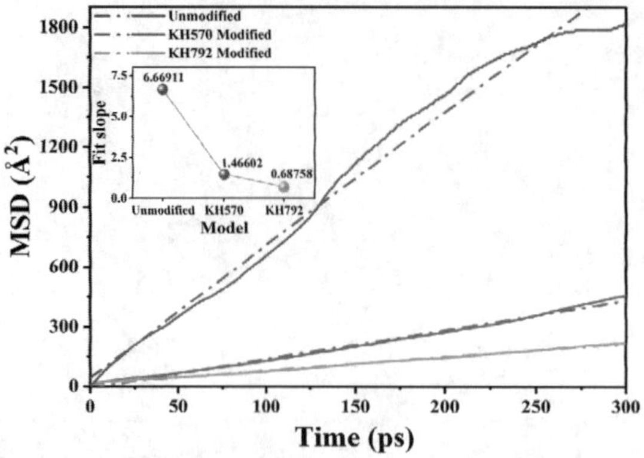

Figure 4. MSD curves of water molecules in the three models.

To further quantitatively analyze the effect of different silane coupling agents modified SiO_2 on the diffusion of water molecules in PP, the diffusion coefficients of water molecules in the three interface models were calculated and compared, and the results are shown in Table 1.

According to Table 1, the diffusion coefficient of water molecules in the unmodified interface model was the highest, with a value of 1.11152 Å2/ps. The diffusion coefficient in the KH 792 modified model was the smallest, with a value of only 0.11460 Å2/ps, which was about 1/10 of the diffusion coefficient in the unmodified model. In summary, SiO_2 modified with silane coupling agents can effectively suppress the diffusion of water molecules in PP, and KH 792 had the most obvious inhibitory effect. KH 792 modified SiO_2 can prevent the aggregation of water molecules to form water molecule clusters, reducing the probability of water tree formation in PP to effectively improve the insulation properties of PP cables.

Table 1. Fitting slope of MSD curve for water molecules and the diffusion coefficient.

	Unmodified	KH 570 modified	KH 792 modified
a	6.66911	1.46602	0.68758
D (Å^2/ps)	1.11152	0.24434	0.11460

3.2. Centroid trajectory of water molecules

The centroid trajectory can directly observe the position changes of the selected molecules over time, and can also characterize the diffusion ability of molecules. The centroid trajectories of water molecules in the three interface models are shown in Figure 5. The red part represented the position of the water molecule centroid each time, and the blue part was the projection of the centroid trajectory on the YZ and XZ planes. To better quantitatively describe the motion range of the water molecule centroid, the maximum motion range distance in the three directions of X, Y, and Z was calculated, and the results are shown in Table 2.

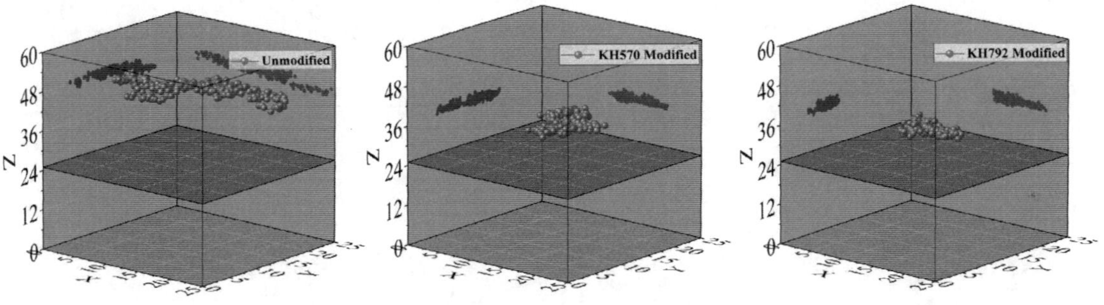

(a) Unmodified interface model.

(b) KH 570 modified interface model.

(c) KH 792 modified interface model.

Figure 5. Centroid trajectory of water molecules.

Table 2. Maximum motion range distance of the water molecule centroid.

	X-axis (Å)	Y-axis (Å)	Z-axis (Å)	Motion range volume (Å^3)
Unmodified	21.28340	15.69332	5.28266	1764.44651
KH 570 modified	8.90995	11.45998	7.76711	793.08289
KH 792 modified	8.18389	5.77366	3.60809	170.48585

From Figure 5, the centroid trajectories of water molecules in KH570 and KH792 modified models were closer to the SiO_2/PP interface in the Z-axis direction compared with the unmodified interface model. According to the maximum motion range distance of the water molecule centroid in the three models, the maximum motion range distances of water molecules on the X and Y axes had been significantly reduced after the modification with KH 570 and KH 792. It was decreased from 21.28340 Å without the modification to 8.90995 Å and 8.18389 Å in the X-axis direction. It decreased from 15.69332 Å to 11.45998 Å and 5.77366 Å in the Y-axis direction. However, the maximum motion range after modification with KH 570 increased in the Z-axis direction. Taking into account the three directions of the X, Y, and Z axes, the motion range volume was calculated, and it was found that the motion range volume of water molecules in the SiO_2/PP interface model modified with KH792 was the smallest, only 170.48585 Å3, which was about 1/10 of the value in the unmodified SiO_2/PP interface model. This corresponded to the variation law of the water molecule diffusion coefficient calculated above.

3.3. Relative molecular concentration of water molecules

The relative molecular concentration of water molecules in the Z-axis direction of the SiO2/PP interface was analyzed through the Concentration profile in the Forcite module to more intuitively demonstrate the effect of silane coupling agent-modified SiO_2 on the diffusion and migration of water molecules in PP, and the results are shown in Figure 6. By comparing the distribution of water molecules in the three interface models before and after simulation, the spatial position variation of water molecules in PP under the effect of SiO_2 interface can be obtained in different interface models.

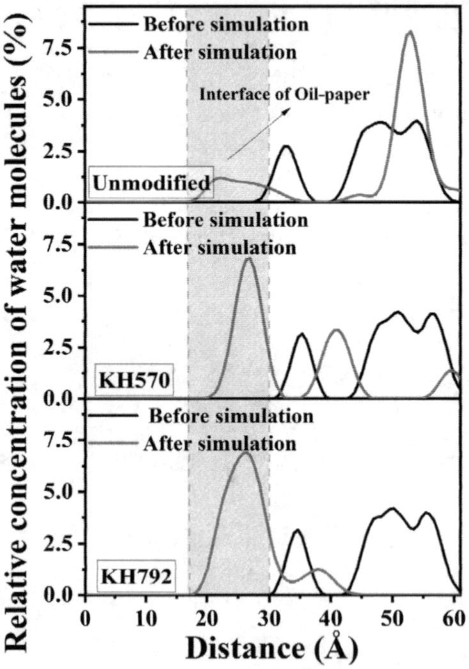

Figure 6. Relative molecular concentration of water molecules in the Z-axis direction.

In Figure 6, the blue area represents the location of the SiO_2 interface, the left dashed line represents the SiO_2 surface, the right dashed line represents the molecular chain terminal of the surface grafted silane coupling agent, the black line represents the relative molecular concentration distribution of water molecules before simulation just when the model is esTablelished, the red line represents the relative molecular concentration distribution after simulation. After the molecular simulation, the concentration of water molecules in the three interface models shifted towards the SiO_2 molecular layer model, reflecting the migration of water molecules towards SiO_2 due to the interaction. In addition, in the unmodified model, except for a part of water molecules diffusing to the interface, a part of water molecules aggregated near 53 Å, where water molecule clusters were formed. And, compared with the KH 570 modified model, more water molecules entered the interface in the KH 792 modified model, with a wider peak width and a lower relative concentration of water molecules outside the interface. This phenomenon was consistent with the above analysis. KH792 had a stronger adsorption effect on the water molecules, thereby inhibiting the diffusion movement of water molecules in PP.

4. Intrinsic mechanism of the effect of different silane coupling agents on the water molecule diffusion

4.1. Interaction energy

The diffusion coefficient variation of water molecules in PP is related to the interaction energy between the two. The interaction energy is a physical quantity that characterizes the interaction strength between molecules, and the energy in a system is mainly composed of the non-bonding interaction energy and bonding interaction energy. The interaction energy was calculated from the Eq. (2):

$$E_{inter} = E_{total} - (E_{Water} + E_{Nano}) \tag{2}$$

where E_{inter} represents the total interaction energy between two chemical substances in the model, E_{total} represents the total potential energy of the model, E_{Water} and E_{Nano} represent the potential energy of water molecules and SiO_2 nanoparticles, respectively. There is no interaction between two substances when $E_{inter} = 0$. It represents mutual repulsion between two substances when $E_{inter} > 0$. On the contrary, it represents the mutual attraction between two substances when $E_{inter} < 0$, and it indicates a stronger binding effect between the two substances when E_{inter} is a negative value, and its absolute value is larger.

Table 3. The interaction energy between water molecules and SiO_2 surface.

E (kcal/mol)	Unmodified	KH 570 modified	KH 792 modified
E_{total}	-41363.72	-41420.14	-40703.44
E_{Water}	-34.79	-35.99	-42.47
E_{Nano}	-41314.43	-41362.88	-40635.32
E_{inter}	-14.50	-21.26	-25.65
E_{vwd}	-13.01	-12.59	-8.56
E_{elect}	-0.75	-7.87	-16.31

The magnitude and main components of the interaction energy between water molecules and SiO_2 surface (with surface modification characteristic functional groups) in the three SiO_2/PP interface models are given in Table 3. Firstly, the interaction energy between water molecules and SiO_2 surface in the KH 792 modified model was the largest, with a value of -25.65 kcal/mol, the interaction energy in the KH 570 modified model was the second, with a value of -21.26 kcal/mol, and the interaction energy in the unmodified model was the smallest, with a value of only -14.50 kcal/mol. The above values indicated that SiO_2 before and after modification can adsorb water molecules in the three interface models. The addition of a silane coupling agent enhanced the diffusion energy towards the water molecules, and KH 792 had the best effect.

Meanwhile, the COMPASS II force field does not have a hydrogen bonding term, which describes the electrostatic interaction by the atomic charges. Therefore, the hydrogen bonding term is delineated in electrostatic interactions. Based on the analysis of the main components of interaction energy, it was found that Evwd and Eelect represented the van der Waals interaction and electrostatic interaction, respectively, and the van der Waals force and electrostatic force played a major role in the interaction energy between water molecules and SiO_2 surfaces. As the interaction energy increased, the proportion of electrostatic force gradually increased. The proportion of electrostatic force in the unmodified model was only 5.17%, while the proportion increased to 63.59% in the KH 792 modified model. This was mainly because the introduction of silane coupling agents enhanced the polarity of the SiO_2 surface, and the water molecules are also polar substances, which easily formed hydrogen bonds between the two, thereby increasing the proportion of electrostatic interaction and strengthening the interaction.

4.2. Average number of hydrogen bonds

The variation of interaction energy is closely related to the number of hydrogen bonds formed in the system. The more hydrogen bonds in a model indicate the stronger binding ability between molecules in the system. In molecular simulation, geometrical determination is usually used to determine the generation of hydrogen bonds. When the distance between a hydrogen atom and another electronegative atom is less than 0.25 Å, and the angle formed between the electronegative atom and the hydrogen atom donor is greater than 100°, hydrogen bonds are formed. The overall average number of hydrogen bonds and the average number of hydrogen bonds between water molecules and SiO_2 surface were calculated in the three SiO_2/PP interface models, as shown in Table 4.

Table 4. Average number of hydrogen bonds between molecules.

	Unmodified	KH 570 modified	KH 792 modified
The overall average number of hydrogen bonds (number)	7.77	9.52	13.42
The average number of hydrogen bonds between water molecules and SiO_2 surface (number)	0.30	2.08	4.41
Proportion (%)	3.86	21.85	32.86

As seen from Table 4, the overall average number of hydrogen bonds and average number of hydrogen bonds between water molecules and SiO_2 surface in the model both increased with the introduction of a silane coupling agent. Moreover, the average number of hydrogen bonds in the KH 792 modified model was the highest, and the proportion of hydrogen bonds between water molecules and SiO_2 surface was also the highest, up to 32.86%, further indicating that KH 792 had the best modification effect.

The variation of various hydrogen bond numbers in the three interface models was closely related to the introduction and type of silane coupling agents. Firstly, both silane coupling agents had polar groups, which made it easy to form hydrogen bonds with polar water molecules, enhancing the adsorption of water molecules on the surface of SiO_2. Secondly, the KH 570 molecule contained only one polar group (-COOR), while the KH 792 molecule contained two polar groups (-NH2, -NH-). This ensured that when the surface was modified with the same number of coupling agents, KH792 could introduce more polar groups and provide more hydrogen bonding sites, thus making it easier to form hydrogen bonds with water molecules in PP. This was also the main reason for the difference in the average number of hydrogen bonds between water molecules and SiO_2 surface in the KH 570 and KH 792 modified models.

4.3. Free volume fraction

The free volume fraction is another important parameter that characterizes the strength of molecular diffusion ability. The total volume of polymers includes free volume (Vo) and occupied volume (Vf). The existence of free volume provides molecular motion space. The more free volume, the stronger the molecular motion ability in the system. This phenomenon can be characterized by the free volume fraction, which can be calculated by the ratio of free volume to total volume, as shown in Eq. (3).

$$FFV = \frac{V_f}{V_o + V_f} \times 100\% \tag{3}$$

Figure 7 illustrates the free volume fraction of three SiO2/PP interface models. The occupied volume of the model modified with silane coupling agents increased while the free volume decreased compared with the unmodified model. The free volume fraction in the KH 792 modified model was the smallest, indicating that the free volume cavities in the system were relatively small, providing a reduced space for the movement of water molecules, thereby suppressing the diffusion movement of water molecules. The reason for the decrease in free volume fraction was that compared to the molecular chain length of KH570 (11.767 Å), the chain length of KH 792 (12.823 Å) increased, and it had more polar groups. The surface-grafted silane coupling agent molecular chain can be better blended with PP, and enhance the interaction between PP and SiO_2, which coupled with the stronger adsorption effect of silane coupling agent on the water molecules can effectively reduce the volume gap between SiO_2 and PP containing micro water, thereby further suppressing the diffusion behavior of water molecules in PP, ultimately inhibiting the formation of water trees in PP, and prolonging the service life of PP cables.

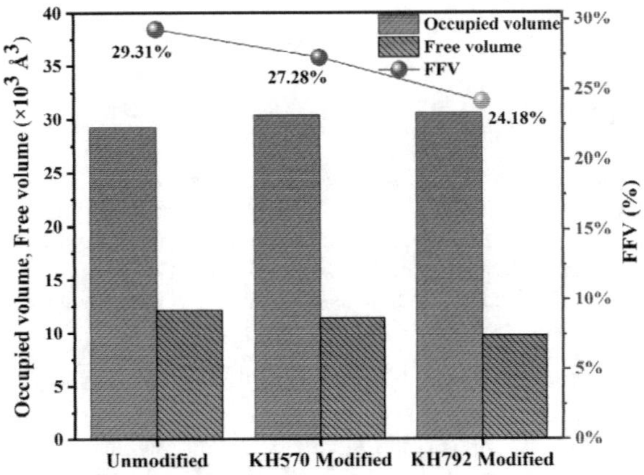

Figure 7. Free volume fraction in the three SiO_2/PP interface models.

5. Conclusion

In this article, the effect of different silane coupling agent-modified SiO_2 on the diffusion of water molecules in PP was investigated using molecular dynamics, and the mechanism was explained at the molecular level. The conclusions were as follows:

(1) The surface of SiO_2 modified with silane coupling agents can enhance the inhibitory effect on the diffusion of water molecules. The diffusion coefficient of water molecules in the KH 792 modified model was significantly reduced, with a value of only 0.11460 Å2/ps, which was about 1/10 of that in the unmodified model.

(2) The centroid trajectory of water molecules in the model modified with KH 792 decreased and was closer to the interface. More water molecules were adsorbed at the SiO_2 interface in the KH 792 modified model, which can reduce the formation of water molecule clusters in the model.

(3) The introduction of silane coupling agents enhanced the polarity of the SiO_2 surface, which can better restrain the diffusion behavior of water molecules in PP. Moreover, KH 792 had more polar groups than KH 570, resulting in the largest interaction energy, the largest average number of hydrogen bonds, and the best modification effect. At the same time, the long chains of silane coupling agent molecules occupied a certain amount of free space in the system, reducing the number of cavities for water molecule diffusion, which was conducive to further inhibiting the diffusion of water molecules.

The intrinsic mechanism of different silane coupling agents modified SiO_2 on the diffusion behavior of water molecules in PP at the molecular level was interpreted, which provided certain theoretical support for inhibiting the formation of water trees in PP, thereby further improving the insulation properties of PP cables and prolonging the service life.

Acknowledgments

This work was supported by State Grid (5500-202155350A-0-0-00).

References

[1] ZHI Y, LI J C, DENG Z X, et al. Progress in the Development and Commercial Application of Polypropylene Based Cable. Insulating Materials, 2023, 56(7): 1-8.
[2] HU S Z, LU J, LIU Z H, et al. Exploration of Localization Application of Environmentally Friendly Polypropylene Insulated Power Cable. Wire & Cable, 2023(3): 30-32.
[3] DU G, YANG J, CAI H X, et al. Application Prospect of Polypropylene-based DC Cable Insulation Materials in Humid and Hot Environment. High Voltage Apparatus, 2022, 58(6): 9-16.
[4] CAO W, LI Z, SHENG G, et al. Insulating Property of Polypropylene Nanocomposites Filled with nano-MgO of Different Concentration. IEEE Transactions on Dielectrics and Electrical Insulation, 2017, 24(3): 1430-1437.

[5] GAO J G, ZHANG G W, LIU Y L, et al. Evaluation of Electrical Aging Life of Nano SiO2/PP Composites. Acta Materiae Compositae Sinica, 2024, 41(3): 1270-1280.
[6] POMPILI M, CALCARA L, D'ORAZIO L, et al. Joints Defectiveness of MV Underground Cable and the Effects on the Distribution System. Electric Power Systems Research, 2021, 192: 107004.
[7] LI Y S, WANG J M, XIA Y, et al. Molecular Dynamics Simulation of Thermodynamic Properties of Polypropylene Modified by Nano-ZnO. Acta Materiae Compositae Sinica, 2024, 41(1): 485-494.
[8] GAO P, ZHAO J, WANG Z Y, et al. Research Progress of Polypropylene Materials in Application of Power Cables. Insulating Materials, 2023, 56(8): 1-10.
[9] PEI Q X, SORKIN V, LIU P, et al. Effect of Surface Coupling Agents on the Mechanical Behaviour of Polypropylene/Silica Composites: A Molecular Dynamics Study. Journal of Polymer Research, 2021, 28(2): 29.
[10] WANG L, SUN Y, WANG H P, et al. Status-Quo of the Study on Water-Tree Generation and Growth in Insulating Materials. Electric Wire & Cable, 2006(6): 5-8.

Journal of Physics: Conference Series **3009** (2025) 012064 doi:10.1088/1742-6596/3009/1/012064

Finite element analysis of front suspension of baja racing car with double forks

Qiang Ding, Jiatong Qin, Biao Jin*

Guangdong University of Science and Technology, Dongguan 523000, China

*Corresponding author's e-mail: 53340450@qq.com

Abstract. In this research, we conducted 3D modelling and failure analysis on the double-arm front suspension of a Baja racing car. Initially, based on automotive suspension design principles, we employed CATIA software to establish the structure of the upper and lower control arms and steering components, while confirming the materials used. Subsequently, we performed strength assessments and modal analysis on the front suspension structure using ANSYS software, conducting finite element analysis under various operational conditions. By analysing stress relief and deformation under different field conditions, our findings indicate that the front suspension exhibits reasonable deformation under bending, flexing, and torsion, aligning with its structural and material attributes. To prevent resonance and structural damage under specific load frequencies during actual racing, we calculated and analyzed the dual-arm structure's capacity. Our results demonstrate that this structure maintains efficient and stable suspension performance in Baja racing cars' extreme off-road environments, catering to diverse field conditions and enhancing overall driving comfort. Furthermore, this study offers valuable insights for future design and research of Baja racing car suspension systems.

1. Introduction

The China Baja Competition is an undergraduate competition focusing on car design and production, comprising both static and dynamic components. As the event evolves, the performance standards for Baja racing vehicles, particularly in terms of driving stability, have been increasingly demanding.

Currently, the domestic and international landscape in car suspension research exhibits the following trends: Domestic efforts in studying car suspensions commenced later and predominantly rely on advanced computer-aided techniques. Huang [1] conducted a finite element analysis on car suspension systems, developing a model capable of addressing nonlinear issues, tailored for elastic suspension kinematics and modal analysis. Yang and Weng [2] employed spatial analytical geometry to investigate suspension motion laws, innovatively integrating ISIGHT and ADAMS software to analyse suspension hard point sensitivity for performance optimization. Their findings demonstrate a reliable and precise optimization method that meets its objectives. Internationally, automotive suspension research evolved in two phases. Initially, researchers utilized graph theory and comparative studies to determine optimal parameter matching. Subsequently, the adoption of multi-body dynamics theory revolutionized automotive research, enhancing the simulation capabilities of complex vehicle systems. Qin [3] pioneeringly validated suspension virtual prototyping through simulations, expediting research and development. Yu [4] devised a rapid method for predicting heavy-vehicle air suspension fatigue life, leveraging modal stress recovery technology for dynamic analysis with reduced simulations. Additionally, Bai and Lei [5] conducted kinematic design studies on dual-arm suspension systems, employing multi-objective, multi-dimensional techniques for link optimization, thereby affirming the method's precision and robustness [6].

Journal of Physics: Conference Series **3009** (2025) 012064 doi:10.1088/1742-6596/3009/1/012064

Currently, research on suspensions primarily focuses on conventional automobile systems, with less attention dedicated to Baja racing cars. Our investigation reveals that as race difficulty increases, vehicle resonance intensifies, compromising driving comfort and impacting drivers' performance. This is particularly evident in 4-hour endurance races, where many drivers struggle to complete the event successfully. Resonance in the suspension structure under specific loads can undermine structural integrity and comfort. Therefore, this paper aims to conduct a comprehensive design, material selection, strength assessment, and modal analysis of the Baja racing car suspension to ensure sustained performance and enhanced vehicle comfort.

2. Material selection

The strength and stiffness of the car should be ensured while reducing the mass of the car as much as possible. From the physical characteristics of the following two commonly used materials, it can be seen that 20 Cr steel is better than 4130 steel in terms of quality, and 4130 steel has high yield strength and high tensile strength, and the overall mechanical properties are better than 20 Cr steel, which also meets the requirements in the actual welding operation. According to the above, the 4130 seamless steel pipe is selected. The specific parameters are shown in Table 1.

Table 1. Material properties.

Material	Density (g/cm^3)	Tensile strength (MPa)	Lower yield strength (MPa)
4130	7.85	930	785
20 Cr	7.82	846	561

3. Establishment of 3D model

Utilizing the CATIA software, 3D solid modeling was conducted based on the specified parameters and dimensions of the front suspension and frame. Initially, the basic geometry was outlined on the sketching workbench, and refined using constraint and dimension tools. Subsequently, 3D solid features such as the upper and lower control arms, steering knuckles, and others were created through stretching, rotating, sweeping, and other modeling techniques. Boolean operations were also applied for solid editing and modification. The final geometric model is illustrated in Figure 1.

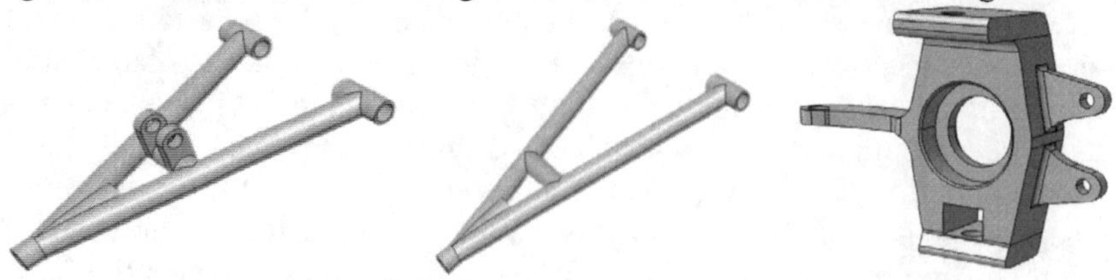

(a) Upper control arm (b) Lower control arm (c) Steering knuckle
Figure 1. Three-dimensional geometric model.

4. Strength Verification and Modal Analysis

4.1. Finite element analysis of front suspension

4.1.1. Finite element analysis of steering knuckle. The peak force on the steering knuckle primarily arises from the maximum impact force experienced by the car upon landing after obstacle clearance. In this instance, it bears a combined force from the upper and lower control arms, along with the wheel rolling bearing, with the maximum force occurring at the juncture of the control arms. Its main pin is then constrained. Finite element analysis using ANSYS, with a force of 2200 N applied in the Y-axis direction (representing full load), reveals a maximum stress of 198.5 MPa and deformation of 0.19 mm. Given that these values are well below the yield strength of 4130 steel (785 MPa), it confirms that the steering knuckle remains intact under these operating conditions, fulfilling the strength design criteria. Thus, the steering knuckle is deemed safe under this scenario [7], as illustrated in Figure 2.

During sudden swerves or high-speed encounters with bumpy roads, significant impact forces are generated [8]. These forces require the steering tie rod to withstand tension and pressure from the steering system for effective wheel maneuvering. The uneven road surface exacerbates wheel vibration and displacement, further stressing the steering tie rod. Additionally, collisions or unforeseen events during driving can subject the steering tie rod to substantial impact forces, potentially reaching maximum loads. Finite element analysis was conducted with the upper and lower kingpins of the steering knuckles constrained, and a load of 2200 N applied in the Z-axis direction. The results showed maximum stress of 109.2 MPa and deformation of 0.18 mm, both below the yield strength of 4130 steel (785 MPa). This confirms that the steering knuckle arm remains intact under these conditions, fulfilling strength requirements and deemed safe. The deformation contour is illustrated in Figure 3.

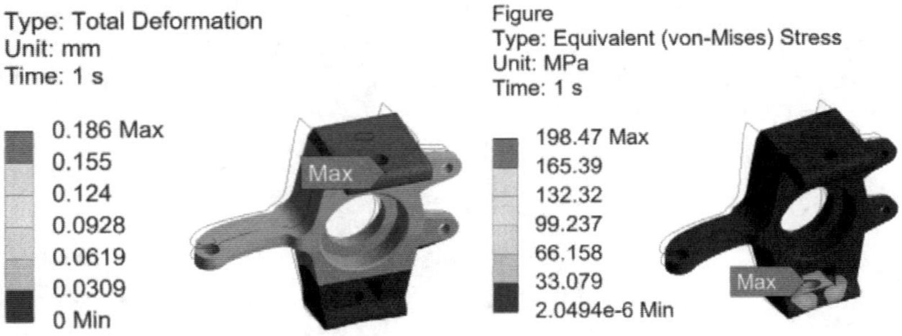

Figure 2. Deformation contour of the steering knuckle.

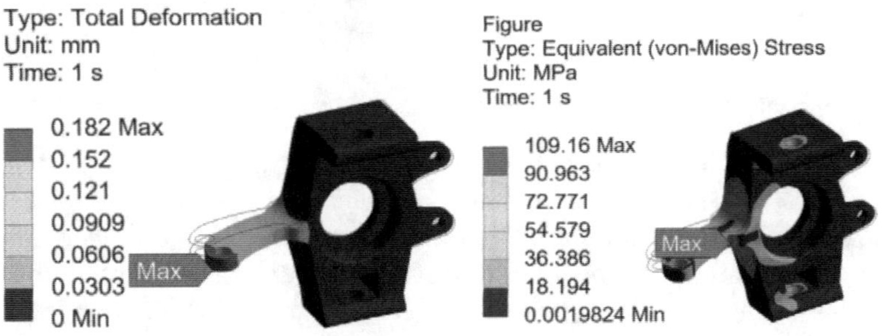

Figure 3. Deformation contour of the steering knuckle arm deformation.

4.1.2. Finite element analysis of upper and lower control arms. During maximum force conditions, the upper control arm experiences force equivalent to half that of the steering knuckles, with the arm-to-frame joint constrained. At the peak impact moment during obstacle clearance, the upper control arm primarily supports forces from the frame, shock absorber, and steering knuckle. Finite element analysis indicates a force of 1100 N along the negative z-axis, yielding a maximum stress of 147.1 MPa and deformation of 1.9 mm, as shown in Figure 4. These values fall below the 785 MPa yield strength of 4130 seamless steel, confirming the upper control arm's structural integrity and compliance with strength criteria under these conditions. Hence, the upper control arm is deemed safe [9].

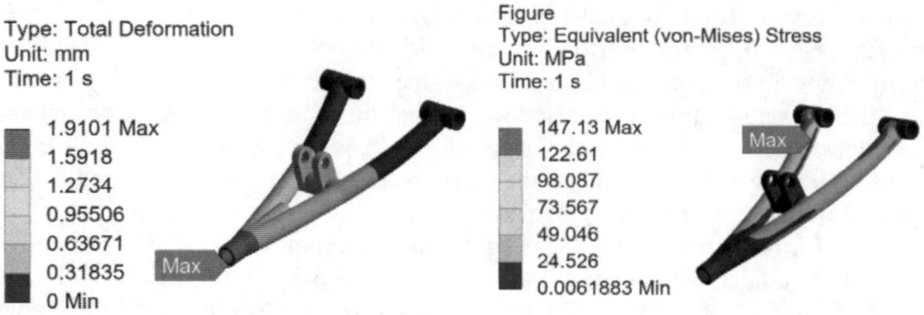

Figure 4. Deformation contour of upper control arm.

Under maximum impact force conditions upon landing after obstacle clearance, the lower control arm experiences force comparable to those of the upper control arm. Primarily, it bears the combined forces from the frame and steering knuckle [10]. Finite element analysis shows a force of 1100 N along the negative z-axis, with a peak stress of 184.2 MPa and deformation of 3.5 mm. These values remain below the 785 MPa yield strength of 4130 seamless steel, confirming the lower control arm's structural integrity and compliance with strength criteria. Thus, the lower control arm is deemed safe under these conditions [11], as shown in Figure 5.

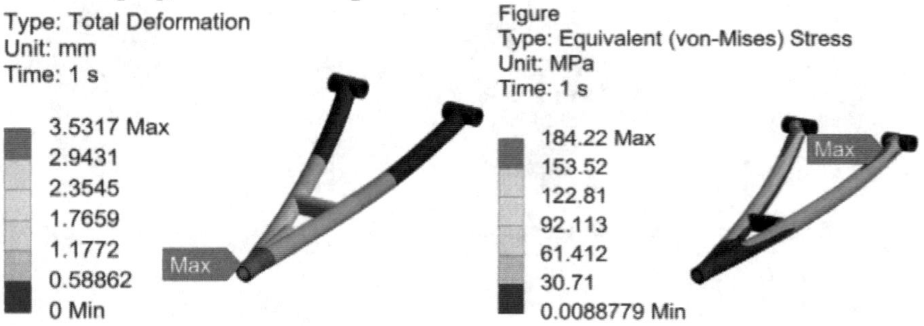

Figure 5. Deformation contour of upper control arm.

4.2. Modal analysis of front suspension

Modal analysis results are typically assessed using analytical or analogical methods; this study employs the analytical approach. Vibration frequencies in a vehicle during operation are primarily induced by road irregularities, engine operation, and power transmission imbalance. The natural vibration frequencies of the Baha racing car were determined through data review and calculation, as tabulated in Table 2.

Table 2. Natural excitation frequency table of Baja racing car.

Item	Exciting frequency (Hz)
Uneven pavement	1~20
Engine running	14.6~31.7
Unbalanced transmission system	≤10.5

Hence, the first-order natural frequencies of the designed front suspension components, including the steering knuckle, and upper and lower control arms, must lie outside the range of the aforementioned excitation frequencies to meet dynamic characteristic requirements.

4.2.1. Modal analysis of steering knuckle. The constructed model underwent modal analysis in ANSYS to develop a finite element representation, characterized by a mesh size of 5 mm and composed of steel grade 4130. The modal assessment of the steering components revealed that the fundamental vibration frequency stands at 4638.1 Hz, primarily involving flexural motion along the Z-axis in a forward and backward direction. The second vibration mode, with a frequency of 5301.1 Hz, exhibits a combination of forward-backward bending and torsional movements within the steering components. The third modal, at 5879.4 Hz, is characterized by vertical oscillation of the steering knuckles. Lastly, the fourth modal, at 6775.9 Hz, demonstrates bending and twisting motions within the steering knuckle. Visual representation of these modals is provided in Figure 6, and Table 3 summarizes the initial four vibration frequencies.

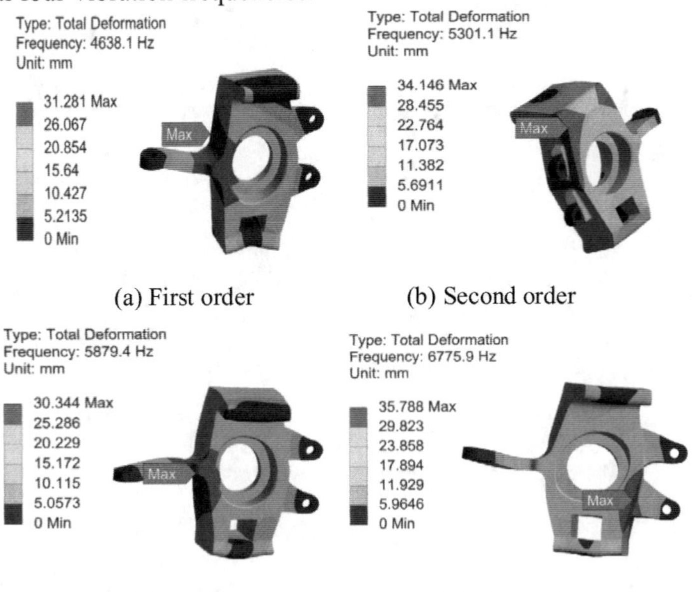

(a) First order (b) Second order

(c) Third order (d) Fourth order

Figure 6. The first to fourth order modal amplitudes of the steering knuckle.

4.2.2. Modal analysis of upper and lower control arm. A finite element model was constructed using ANSYS for modal analysis, incorporating a mesh size of 5 mm and employing 4130 seamless steel material. This analysis was conducted on both the upper and lower control arms. The results indicate that the first-order vibration frequency of the upper control arm is 3473.7 Hz, while the lower control arm exhibits a first-order frequency of 1025 Hz. Specifically, the upper arm undergoes lateral oscillation, whereas the lower arm undergoes vertical oscillation. In the second-order mode, the upper control arm vibrates at 4088.5 Hz and the lower arm at 1088.2 Hz, with the upper arm showing vertical oscillation and the lower arm lateral oscillation. For the third-order mode, the upper and lower control arms vibrate at 4324.5 Hz and 1532.6 Hz respectively, both engaging in overall axial torsional motion. These modal patterns are visually depicted in Figures 7 through 10, and Table 3 summarizes the top four vibration frequencies.

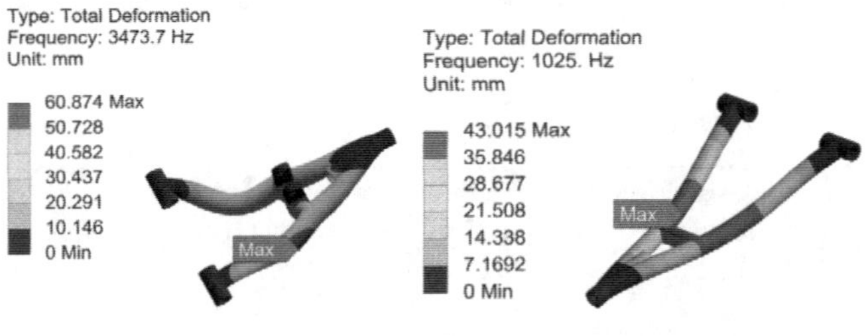

(a) Upper control arm (b) Lower control arm

Figure 7. The first-order modal amplitudes of upper and lower control arm.

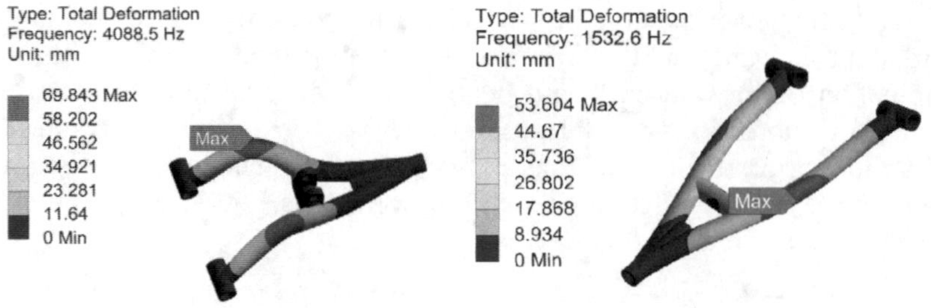

(a) Upper control arm (b) Lower control arm

Figure 8. The second-order modal amplitudes of upper and lower control arm.

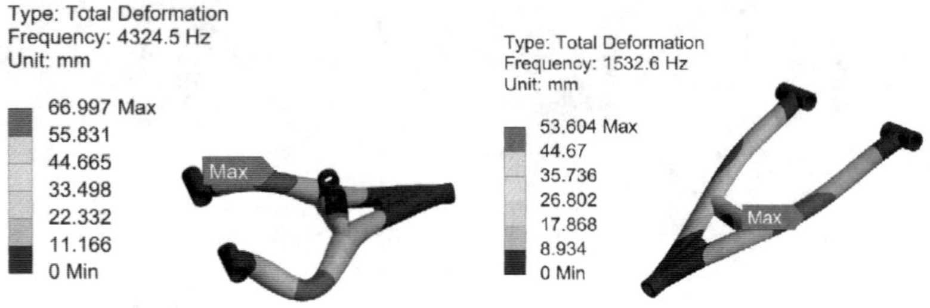

(a) Upper control arm (b) Lower control arm

Figure 9. The third-order modal amplitudes of upper and lower control arm.

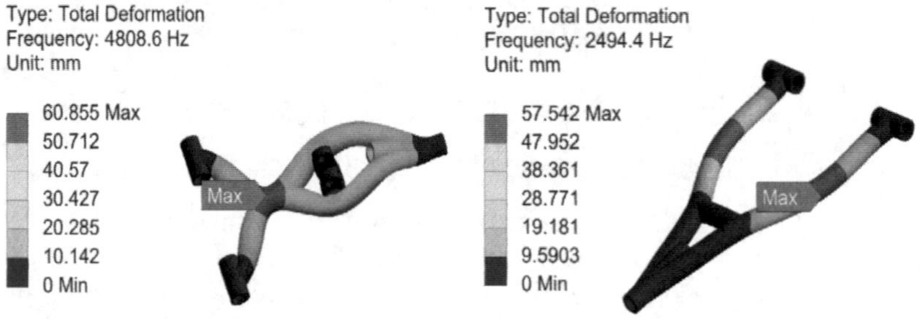

(a) Upper control arm (b) Lower control arm

Figure 10. The fourth-order modal amplitudes of upper and lower control arm.

The simulation outcomes are condensed in Table 3 below. According to Table 3, the steering knuckle's natural frequencies are predominantly concentrated between 4638.1 Hz and 6775.9 Hz, while the upper control arm's frequencies range from 3473.7 Hz to 4808.6 Hz, and the lower control arm's frequencies extend from 1025 Hz to 2494.4 Hz. It is worth noting that these frequencies exceed the Baja racing car's natural vibration frequency by a considerable amount. Therefore, the front suspension's natural frequencies fall within a suitable range, which enhances the vehicle's comfort and stability.

Table 3. The first four vibration modal Frequencies.

Modal order	Knuckle excitation frequency (Hz)	Upper control arm excitation frequency (Hz)	Hem arm excitation frequency (Hz)
1	4638.1	3473.7	1025
2	5301.1	4088.5	1088.2
3	5879.4	4324.5	1532.6
4	6775.9	4808.6	2494.4

5. Conclusion

In this study, CATIA and ANSYS software were utilized to design, assess strength, and perform modal analysis on the double-wishbone front suspension of a Baja racing car. The analysis focused on stress distribution and deformation, revealing a reasonable range under various conditions. The modal analysis confirmed that the suspension's natural frequency exceeded the vehicle's excitation frequency. The objective was to evaluate the suspension's stability in extreme off-road conditions and driving comfort. Results demonstrated reliable suspension performance, meeting operational demands and enhancing overall comfort.

Acknowledgments

This project is funded by the teaching and learning enhancement project team of Guangdong University of Science and Technology (NO. GKJXXZ2023025).

References

[1] Huang X. (2016) Finite element analysis of automobile suspension system and its application. Conference, LNCS, vol. 9999:1–13.
[2] Yang H. Weng J. (2011) Simulation and optimization of double crossbar independent suspension. Nanjing: Mechanical Science and Technology for Aerospace Engineering, 30(10): 1683-1687.
[3] Qin D. (2007) Modeling, Simulation and optimization for handling stability of sport utility vehicle. Wuhan: Huazhong University of Science and Technology.
[4] Yu Z S. (2009) Automobile Theory. Beijing: China Machine Press.
[5] X L B, J Lei. (2019) Internal model-based optimal vibration control for linear vehicle suspension systems with actuator delay. Ferroelectrics, 549(1):195.
[6] Sancibrian R, Garcia P, et al. (2010) Kinematic design of double-wishbone suspension systems using a multiobjective optimization approach. Vehicle System Dynamics, 48(7):793-813.
[7] Yang C. (2021) Research on modeling and optimization design of dual fork arm front suspension system for Baja racing car. Chongqing: Chongqing Jiaotong University.
[8] Vishwakarma A. (2021) Design and finite elements analysis of atv rollcage modeling.
[9] Ning T, Chen W, Chen J. (2021) Energy saving design and optimization of suspension system in Baja competition for college students // IOP Conference Series: Earth and Environmental Science. 706(1): 012023.
[10] Shi L, Zhu C D, Liu X, et al. (2020) Optimum design of the double roll rotary forging machine frame. Mechanical Sciences, 11(1): 101-114.
[11] Da C W, Pedro A, Diego O, et al. (2021) Structural analysis of the Baja SAE vehicle from UERJ Technology Faculty.

Journal of Physics: Conference Series **3009** (2025) 012065 doi:10.1088/1742-6596/3009/1/012065

Online bubble size measurement technology for bulk nanobubbles generation process in directed flow

Zhiwen Ye[1], Jian Qiu*[1], Peng Han[1], Dongmei Liu[1], Li Peng[2], Kaiqing Luo[1], Miao Chen[1], Shaomin Tang[1]

[1] Guangdong Provincial Engineering Research Center for Optoelectronic Instrument, School of Electronic Science and Engineering (School of Microelectronics), South China Normal University, Foshan, 528225, China
[2] School of Physics, South China Normal University, Guangzhou, 510006, China

*Corresponding author's e-mail: qiuj@scnu.edu.cn

Abstract. The current Ethanol-water Exchange method utilizes an offline bubble size measurement technique, which is unable to monitor the process of bubble size stabilization in real time. Therefore, this paper proposes an online measurement technology for the size variation of bubbles with directional flow during the nanobubble displacement generation process. In this context, a sample cell for the Ethanol-water Exchange method capable of online measurement was designed, and a dynamic light scattering (DLS) device with online measurement capabilities was constructed. Regarding bubble size measurement, the DLS measurement technique was refined using the SincModel, achieving accurate online measurements of bubble sizes. The experimental results show that submicron bubbles are generated at the initial stage of substitution and then gradually shrink, reaching a plateau about 35 seconds after the start of substitution for an ideal monodisperse system. We believe that this new measurement method will contribute to a better controlling and deeper exploration of the nature of native nanobubble stability.

1. Introduction

Bulk nanobubbles are gaseous entities at the nanoscale, with diameters typically less than 1 μm. Due to their unique characteristics, bulk nanobubbles have become an increasingly studied subject of interest over the past two decades. Their exceptional stability, high specific surface area, and high mass transfer efficiency have demonstrated significant potential in a variety of applications. In the medical field, it can serve as ultrasound contrast agents [1], and as potential carriers for targeted drug delivery in cancer treatment [2]. Furthermore, due to their outstanding adsorption capacity for organic compounds, their applications have expanded to aquatic environmental remediation [3], wastewater treatment [4], surface cleaning [5], mineral flotation[6], and fuel oil applications.

Monitoring the generation process of nanobubbles can provide insights into the quantity it produces, the changes in particle size, and the stable period. Identifying the optimal displacement duration helps to retain as many nanobubbles as possible, enhancing the controllability and repeatability of the production process. This is significant for optimizing their application effects in various fields, as well as for advancing scientific research and industrial development. In the current research field, the nucleation and stability of bulk nanobubbles [7][8] are usually measured using offline methods. Offline measurement typically involves sampling the sample and then using actual equipment to measure the bubble size and distribution. This method is time-consuming and cannot monitor the process of bubble size changes in real time, whereas online monitoring can solve the aforementioned issues. Furthermore, the Ethanol-water Exchange method causes nanobubbles to superimpose directional motion during the replacement process, leading to significant errors in particle size characterization by traditional DLS instruments.

To address this, we introduced a combination of SincModel and online measurements proposed by Huang et al. [9], to correct the impact of directed flow on particle size measurement results. Concurrently, we have designed and constructed an innovative measurement system that enables real-time monitoring of particle size changes during the nanobubble generation process. We firmly believe that by monitoring the characteristic changes during the generation process of bulk nanobubbles, we can gain a deeper understanding of their stability mechanisms, thereby promoting the practical application of bulk nanobubble technology.

2. Principle

Particle size is a crucial factor in determining the characteristics of bulk nanobubbles. In dynamic light scattering technology, the measured quantity is the time-dependent autocorrelation function of light intensity. For a single particle, the relationship between the autocorrelation function of light intensity and the autocorrelation function of the electric field is shown in Equation (1).

$$G_2(\tau) = B(1 + f^2 \mid g_1(\tau) \mid^2) \tag{1}$$

where $G_2(\tau)$ represents the autocorrelation function of light intensity, $g_1(\tau)$ represents the autocorrelation function of the electric field, τ is the correlation time, B is the baseline, which can be obtained through fitting calculations, and f is the coherence factor. In a polydisperse system, the electric field autocorrelation function is a single exponential, and the normalized autocorrelation function of the scattered light electric field is shown in Equation (2).

$$g_1(\tau) = \int_0^{+\infty} G(\Gamma) \exp(-\Gamma \tau) d\Gamma \tag{2}$$

where $\Gamma = Dq^2$, where q represents the scattering vector, and $G(\Gamma)$ is the distribution function of the decay linewidth. According to the Stokes-Einstein relationship, as shown in Equation (3), the diffusion coefficient D can be correlated with the diameter of the particles under investigation.

$$D = \frac{\mathrm{k}T}{3\pi\eta d} \tag{3}$$

During the process of Ethanol-water Exchange, newly formed bubbles exhibit both Brownian motion and directed motion, which together determine the phase of the scattered light. Due to this phase variation, the traditional light scattering autocorrelation function is superimposed with a Sinc function. Consequently, in the environment of flow dispersion, traditional DLS techniques struggle to accurately measure particle sizes, necessitating the introduction of the SincModel for correction. This model encompasses two measurement methods: the vertical field and the parallel field. This paper primarily utilizes the parallel field for experimentation, as shown in Figure 1. The parallel field refers to the condition where the direction of the solution flow is aligned with the direction of the incident light, and the normalization of its light intensity autocorrelation function follows the Equation (4) indicated.

$$g_2(\tau) = 1 + e^{-2\Gamma\tau} \operatorname{sinc}^2\left(\frac{\sqrt{2}\pi nlv\tau}{\lambda d}\right) \tag{4}$$

where d represents the vertical distance between the scattering point and the detection point, l is the length of the scattering region, v is the velocity of directed motion, and n is the refractive index. Compared to the autocorrelation function $g_2(\tau)=1+e^{-2\Gamma\tau}$ which is solely associated with Brownian motion, it is further multiplied by a Sinc function, hence the designation as the SincModel. $g_2(\tau)$ is the autocorrelation curve of the scattered light intensity obtained from the measured particles. By applying a cumulative inversion algorithm, the decay function Γ can be fitted. In conjunction with Equation (3), this allows for the determination of particle size, polydispersity index, and other information.

Journal of Physics: Conference Series **3009** (2025) 012065 doi:10.1088/1742-6596/3009/1/012065

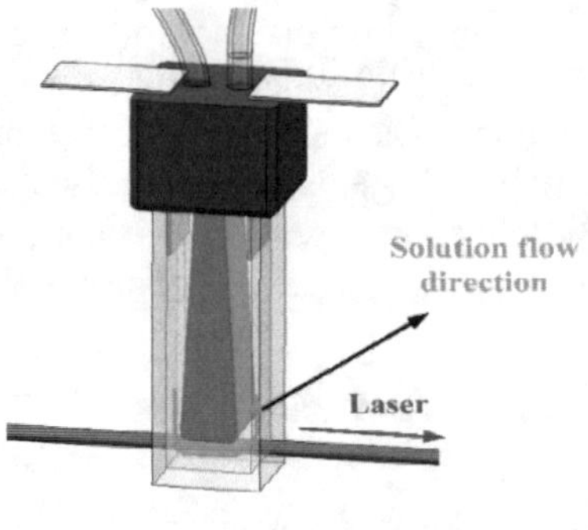

Figure 1. SincModel Parallel Field.

3. Experiment

3.1. Experimental material

In this experiment, the primary materials used were ultrapure water with a conductivity of 18.2 MΩ·cm and a pH of 6.5 at 25°C, as well as anhydrous ethanol (with a mass fraction of C2H6O % ≥ 99.7%, and a density of 0.789 to 0.791 g/mL at 20°C, purchased from Tianjin Zhiyuan). Two cleaned 20 mL glass bottles were used to hold the ultrapure water and waste liquid. Single-use water-based syringe filters (with a pore size of 0.45 μm, purchased from Beijing Bkman Biotechnology Co., Ltd.) were employed to remove potential contaminants during the replacement process.

3.2. Manufacture of bulk nanobubble suspensions

The generation of bulk nanobubbles by the Ethanol-water Exchange method is based on the difference in solubility of gases in the two solutions, where the air is released during the exchange process and a localized supersaturation is formed, resulting in the generation of many nanobubbles. Prior to the experiment, all experimental apparatus were cleaned in an ultrasonic cleaner with anhydrous ethanol and ultrapure water, each for 5 minutes. After cleaning, nitrogen gas was used to blow dry or a blower was used to dry them, to minimize contamination from impurities. Sterile disposable syringes were used to fill the cleaned cuvettes with anhydrous ethanol. A peristaltic pump (model NKCP-C-S04B produced by China Kamoer Company, with a flow rate of 4-14 ml/min) was used to slowly inject ultrapure water into the glass cuvettes. The experimental setup is shown in Figure 2.

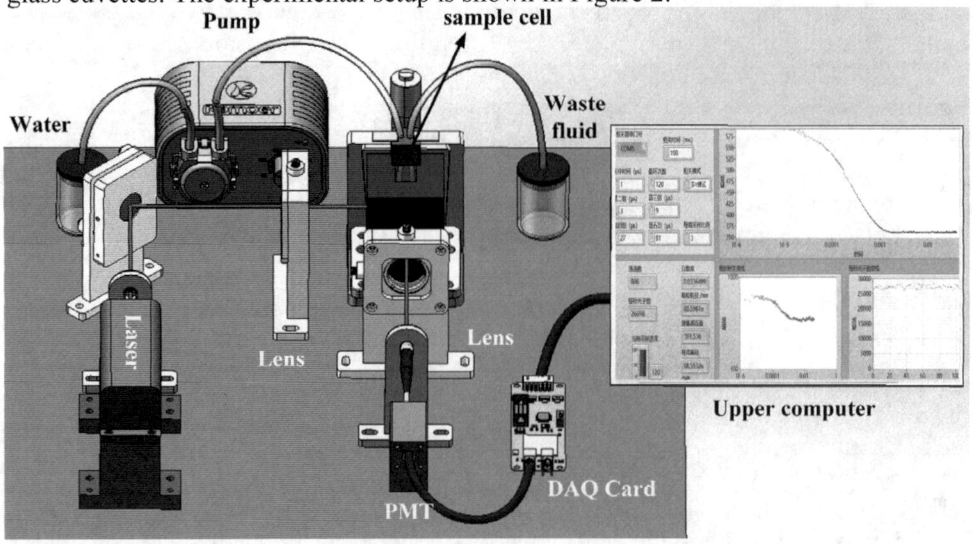

Figure 2. Bulk nanobubbles generation device.

3.3. Validation of SincModel results

The online dynamic light scattering experimental setup was constructed as shown in Figure 2. The wavelength of the laser used is 632.8 nm, with a scattering angle of 90°, and the photon count collection interval (minimum delay time) is 6 microseconds. The measurement target is the 100±3 nm standard particles produced by Haian Hongmeng. A total of three comparative experiments were conducted, with each experiment measured 5 times. The first group used the traditional cumulative inversion algorithm to measure the particle size of standard particles with only Brownian motion. Building upon the first group, the second group of experiments superimposed a directed flow velocity of 1.67 mm/s onto the standard particles. The third group of experiments, based on the second group, employed an accumulated inversion correction algorithm based on the SincModel to calculate the particle size and distribution parameters. All the results are shown in Table 1.

Table 1. Particle size measurements in different scenarios.

Experimental group	1st (nm)	2nd (nm)	3rd (nm)	4th (nm)	5th (nm)	Average particle size (nm)	Relative error
G1	98	96	99	94	97	97	3.2%
G2	61	75	67	68	66	67	32.6%
G3	104	101	97	106	101	102	1.8%

The relative error is derived from a comparison with a standard particle of 100 nm. Our instrument exhibits a relative error of only 3.2% when measuring particles affected solely by Brownian motion, demonstrating high accuracy. Comparing the data from the second and third groups, the traditional inversion algorithm shows a high error rate of 32.6% when measuring particles with directed flow. After applying the SincModel correction, the error is reduced to 1.8%, confirming that the SincModel has a significant corrective effect on the measurement of particle sizes with directed flow.

3.4. Results and discussion

Firstly, a sample cell filled with anhydrous ethanol is placed in the experimental light path and allowed to stand. At time zero, the laser is turned on to monitor the scattered light intensity in real time for 30 seconds, which is the observation period (P1). Then, pure water is injected at a fixed flow rate of 1.67 mm/s to replace the ethanol, and a total of six experiments with different replacement durations (S1-S6) are conducted. The first group has a replacement duration of 10 seconds, increasing by 10-second intervals, so the second group lasts for 20 seconds, and so on. Based on different replacement durations and empirical values for bubble size stabilization, the period between 30 and 75 seconds is defined as the nucleation period (P2), with the stable period (P3) starting after 75 seconds.

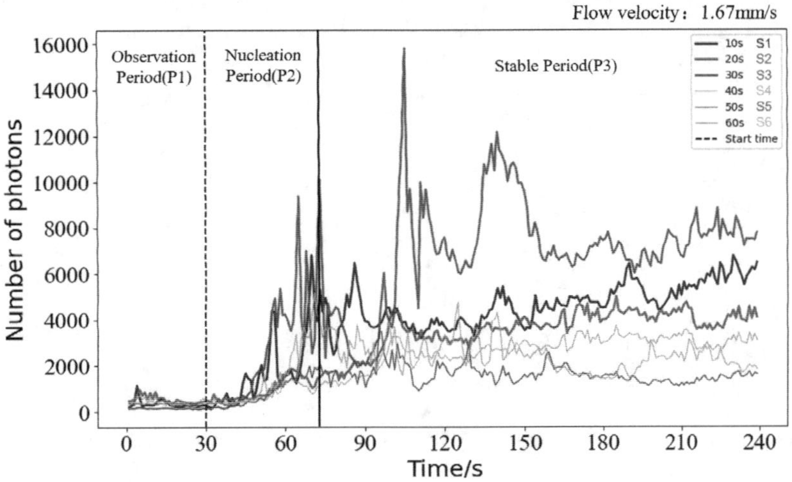

Figure 3. Scattered light intensity variation graph.

As shown in Figure 3, the above six sets of experiments are the curves of scattered light intensity per second obtained from real-time monitoring. During P1, the number of photons per second is lower than 500, indicating that no impurities in the solution interfere with the measurement. At P2, scattered light is generated when a bubble enters the incident beam interval, and the average scattered light intensity rises as the number of bubbles continues to increase. During this period, a peak in the scattered light intensity is evident in all six data sets. At P3, the scattered light intensity gradually flattens out and stabilizes, indicating that the body phase stabilizes after the replacement is completed. It should be noted that in the three sets of experiments from S1 to S3, the scattered light intensity fluctuates greatly between 90 and 150 s in the P3 phase, which may be related to the fact that a small number of unstabilized submicron bubbles still exist in the system.

According to the curve distribution in Figure 3, it can be found that the intensity of the scattered light of S3 is significantly higher than the results of the other five groups of experiments. It can be guessed that the replacement process of the experiment with a time length of 30 seconds is a little more adequate, and the concentration of the bubbles obtained is a little higher. The replacement of S4-S6 with too long a time may appear that the bubbles are discharged along with the solution, resulting in a decrease in the intensity of scattered light [11]. Because the stronger the scattered light intensity is, the higher the signal-to-noise ratio of the correlation curve of DLS will be, the two sets of data (S1 and S3) with the strongest scattered light intensity were selected for comparison in the experiments of particle size monitoring. The data were processed as follows. During the nucleation period (P2) when pure water was injected, the conventional cumulative inversion algorithm was calibrated using SincModel to calculate the average particle size of nanobubbles per second in real time. After the cessation of water injection, during the stabilization period (P3), the traditional cumulative inversion algorithm was used to calculate the average particle size of nanobubbles per second.

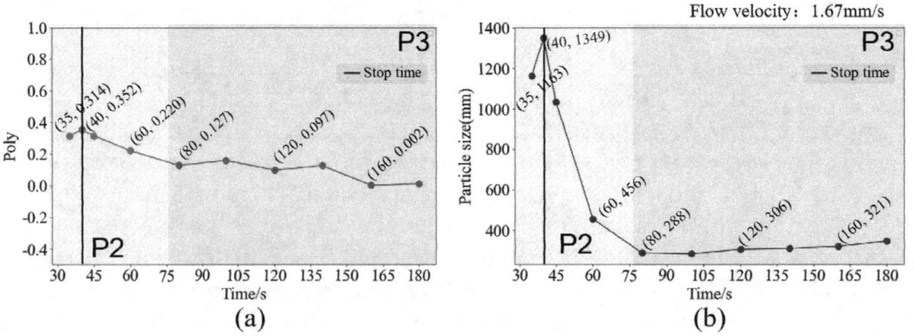

Figure 4. Poly (a) and particle size (b) change with 10 seconds replacement duration.

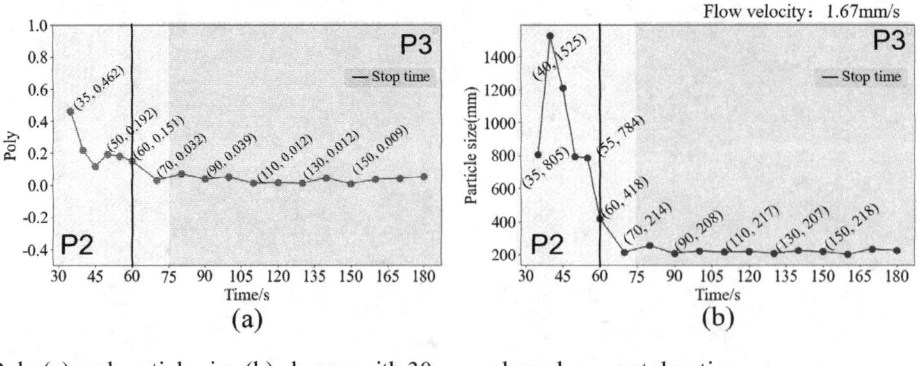

Figure 5. Poly (a) and particle size (b) change with 30 seconds replacement duration.

Figures 4 and 5 show the variation of Poly and bubble size for two sets of experiments, S1 and S3, respectively. First, Polydispersity (Poly) is a key parameter to measure the monodispersity, and when the value of Poly is lower than 0.1, it indicates that the tested sample belongs to a monodisperse system. Second, the black lines in Figures 4 and 5 indicate the moment when the injection of pure water was stopped. It should be noted that the behavior of the bubbles during the water injection process is the result of the combined effect of Brownian motion and liquid flow. Therefore, the bubble size and poly values during this period are the result of applying SincModel corrections to the conventional cumulative inversion algorithm (Equation (3)).

For both sets of Poly values shown in Figure 4(a) and Figure 5(a), the overall Poly values are significantly greater than 0.1 at P2 (yellow area) and gradually decrease from around 0.4. Since the oxygen solubility of ethanol is higher than that of water, air is released and forms a local supersaturation during water replacement of ethanol, nucleating the bubbles [10]. The particle size results shown in Figure 4(b) and Figure 5(b) indicate that during this period, the resulting bubbles are distributed at a submicrometer scale with a wide range of distribution. As the replacement process proceeds, the bubbles will either rupture or shrink [12][13], resulting in an overall smaller bubble particle size. Therefore, it can be seen in the P3 period of Figure 4 and Figure 5 that nanobubbles in the solution become dominant, concentrating in the range of 200-300 nm, and the Poly value gradually decreases and tends to be close to 0.1, indicating that the bubble distribution is becoming more and more consistent and stable. The above trends in online monitoring changes are consistent with the current offline analysis process.

In this paper, DLS online measurement was realized in the proposed 90° scattered light acquisition experimental setup based on a self-developed FPGA data acquisition circuit and 6 us sampling interval strategy. SincModel was introduced to correct the results of the particle size of the oriented flow particles, and the actual variation of the bubble particle size was obtained. By analyzing the result, the optimal substitution duration for generating bulk nanobubbles was obtained to be 30 seconds, and it is known that the Ethanol-water Exchange method requires approximately 35 seconds before the nanobubbles stabilize. These data can be used to determine the optimal substitution time, optimize the production process, reduce the production cost of nanobubbles, and further promote the practical application of nanobubbles, which can be combined with the online measurement of the Zeta potential of nanobubbles to better explore the nature of nanobubble stabilization in the future.

4. Conclusion

This study proposes a method that combines the use of SincModel and online measurement to monitor the changes in light intensity, Polydispersity (Poly), and particle size in real time during the Ethanol-water Exchange method. The trend of these changes is divided into three periods. In the observation period (P1), the photon count per second is less than 500, indicating that the solution is free of impurities. The nucleation period (P2) begins with the replacement, during which the scattered light intensity gradually increases and reaches the first peak. At this time, a Poly value greater than 0.1 indicates that the solution is a polydisperse system, and the particle size results show that microbubbles are predominant in the solution at this stage. In the stable period (P3), the scattered light intensity has become stable. In the S3 experiment, where the scattered light intensity is the strongest, the average Poly value is 0.029, with a variance of 0.018, and the average particle size is 220 nm, with a variance of 13 nm. This indicates that nanobubbles are predominant and have stabilized at this point. According to the experimental results of this paper, it can be known that the optimal replacement duration for the ethanol-water replacement method is found to be 30 seconds, with bubbles tending to stabilize approximately 35 seconds after the start of the replacement. The experimental data can be used to determine the optimal replacement duration, optimize the production process of nanobubbles, and reduce production costs. The introduction of this method may also contribute to exploring the essence of nanobubble stability.

Acknowledgments

This work was supported by funding from the National Natural Science Foundation of China (62375089, 62205110), Guangdong Basic and Applied Basic Research Foundation (2022A1515140139), Natural

Science Foundation of Guang-dong Province (2023A1515011452), and China Postdoctoral Science Foundation Project (2023M741242).

References

[1] Gao, C. Hernandez, H.-X. Yuan, J. Lilly, P. Kota, H. Zhou, H. Wu, A.A. Exner, Ultrasound molecular imaging of ovarian cancer with ca-125 targeted nanobubble contrast agents, Nanomed.: Nanotechnol. Biol. Med. 13(7)(2017) 2159–2168.
[2] S.S. Thakur, Y.-S. Chen, Z.H. Houston, N. Fletcher, N.L. Barnett, K.J. Thurecht, I.D. Ru penthal, H.S. Parekh, Ultrasound-responsive nanobubbles for enhanced intravitreal drug igration: An ex vivo evaluation, Eur. J. Pharmaceut. Biopharmaceut. 136(2019) 102–107.
[3] H. Li, L. Hu, D. Song, F. Lin, Characteristics of micro-nano bubbles and potential application in groundwater bioremediation, Water Environ. Res. 86(9)(2014) 844–851.
[4] A. Azevedo, H. Oliveira, J. Rubio, Bulk nanobubbles in the mineral and environmental areas: Updating research and applications, Adv. Colloid Interface Sci. 271(2019) 101992.
[5] J. Zhu, H. An, M. Alheshibri, L. Liu, P.M. Terpstra, G. Liu, V.S. Craig, Cleaning withbulk nanobubbles, Langmuir 32(43)(2016) 11203–11211.
[6] A. Rosa, J. Rubio, On the role of nanobubbles in particle-bubble adhesion for the flotation of quartz and apatitic minerals, Miner. Eng. 127(2018) 178–184.
[7] Hewage J M J N. Stability of nanobubbles in different salts solutions [J]. Colloids and Surfaces, A. Physicochemical and Engineering Aspects, 2021, 609(1). DOI: 10.1016/j.colsurfa.2020.125669.
[8] Ma X, Li M, Sun C. Effect of ionic environment in aqueous solution on nucleation and stabilization of bulk nanobubbles [J]. Applied Surface Science, 2024, 656. DOI: 10.1016/j.apsusc.2024.159726
[9] Huang G, Xu B, Qiu J, et al. Effect of directional movement on dynamic light scattering [J]. IEEE Photonics Journal, 2021, 13(3): 1-13.
[10] Chen M, Peng L, Qiu J, et al. Monitoring of an Ethanol–Water Exchange Process to Produce Bulk Nanobubbles Based on Dynamic Light Scattering [J]. Langmuir, 2020, 36(34): 10069-10073.
[11] Jie, Qiu, Zhenglei, et al. Formation and Stability of Bulk Nanobubbles Generated by Ethanol-Water Exchange [J]. ChemPhysChem, 2017, 18(10):1345-1350.DOI:10.1002/cphc.201700010.
[12] Meegoda J N, Hewage S A, Batagoda J H. Application of the diffused double layer theory to nanobubbles [J]. Langmuir, 2019, 35(37): 12100-12112.
[13] Tan, B. H.; An, H. J.; Ohl, C. D. How Bulk Nanobubbles Might Survive. Phys. Rev. Lett. 2020, 124, 134503.

Journal of Physics: Conference Series **3009** (2025) 012066 doi:10.1088/1742-6596/3009/1/012066

An integrated scheduling genetic algorithm based on process constraint matrix and family definition coding

Chao Qian[1], **Jianxin Zhang**[1,*]

[1]Inner Mongolia University of Technology, Hohhot, Neimenggu, 010080, China

[*]Corresponding author's e-mail: zhangjianxin@imut.edu.cn

Abstract: To address the complex product-integrated scheduling problem, an improved genetic algorithm based on process constraint matrix and family definition is proposed. The algorithm features a coding method that accurately represents the processing sequence constraints in the product process tree and introduces a simplified method for defining families and sub-processes. On this basis, a hierarchical relationship matrix is derived from the constraint matrix. To resolve the issue of infeasible solutions arising from crossover and mutation operations, a family-based single-parent genetic improvement algorithm is designed. Experimental results validate the effectiveness of the proposed algorithm.

1. Introduction

In the manufacturing field, there is a significant shift in product demand from traditional large-scale production to medium- and small-batch, multi-variety production. This transition has introduced assembly processes during production [1-2], leading to an integrated production scheduling problem (IPSP), which involves both process operations and assembly operations [3-4].

For the integrated scheduling problem of complex products with tight connection constraints, Guo [5] proposed a tight connection integrated scheduling algorithm based on reverse virtual components.Xie [6] considered various scenarios of the IPSP problem, and proposed a dynamic Job-Shop scheduling problem with prioritized process sets from the perspective of dynamic scheduling. Shi [7] introduced a dynamic coding method for process constraint chains but significantly increased computational complexity for crossover and mutation operations. Zhao [8] developed a process-level partitioned coding method, but reducing the computational diversity of the initial population led to a decrease in population diversity.

This paper designs an integrated scheduling genetic algorithm based on process constraint matrix and family definition coding.

2. Problem description

The integrated scheduling problem is a special type of shop scheduling problem for complex product processing, where both process machining and assembly steps occur simultaneously during production. Figure 1 illustrates an example of a product process tree [6]. The product is decomposed into n processes $\{ P_1, P_2, \dots, P_i, \dots, P_n \}$ according to the product structure process tree, the process is divided into the processing process and the assembly process. Some processes have precedence constraints in processing time. All processes are performed on m machines $\{M_1, M_2, \dots, M_k, \dots, M_m\}$. The optimization goal is to minimize the total processing time of the product. At the same time, several constraint conditions must be satisfied.

Based on this optimization objective and the related constraints, this paper establishes the following mathematical description:

$$min\ C_{max} = min\{max(C_i)\} \quad (1)$$

S.t.

$$X_{jk} = \begin{cases} 1, Processes\ O_i\ are\ performed\ on\ machines\ M_k \\ 0, otherwise \end{cases} \quad (2)$$

$$S_i \gg max(S_j + T_{jk'}), X_{jk'} = 1, X_{jk} = 1, o_j \in P_n \quad (3)$$

$$C_{i+1,k} \gg C_{i,k} + \sum_{j=1}^{n} X_{j,k+1} T_{i,k} \quad (4)$$

$$\sum_{k=1}^{m} X_{ik} = 1 \quad (5)$$

where Equation (1) $minC_{max}$ is the scheduling objective, which minimizes the total completion time of the product. Equation (2) X_{jk} is a decision variable, which is 1 if the process O_i is performed on the machine, and 0 otherwise. Equation (3) indicates that a subsequent process O_i can only start after all processes in its predecessor set $\{O_{i1}, O_{i2}, \ldots\}$ are completed. Equation (4) states that at any given time, each machine can only process one operation. Equation (5) indicates that at any given time, each process can only be assigned to one machine for processing.

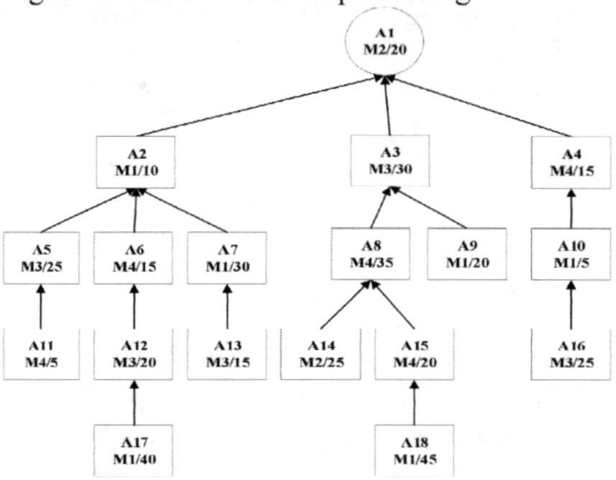

Figure 1. Product Process Tree.

Figure 1 presents a process production diagram with assembly relationships and processing sequence constraints. By appropriately grouping or combining these processes based on their relationships, the entire processing flow can be greatly simplified.

For the convenience of algorithm description, the following definitions are given in this paper.

Definition 1: A set of processes $F_i = \{O_1, O_2, \ldots\}$ that belong to the same operation O_i, where each preceding process is independent of the others, is referred to as a "family ". The subsequent process O_i is referred to as the superior process, while the preceding process $O_n (O_n \in F_i)$ is referred to as the subordinate process of O_i. As shown in Figure 2, it is an example of a family definition.

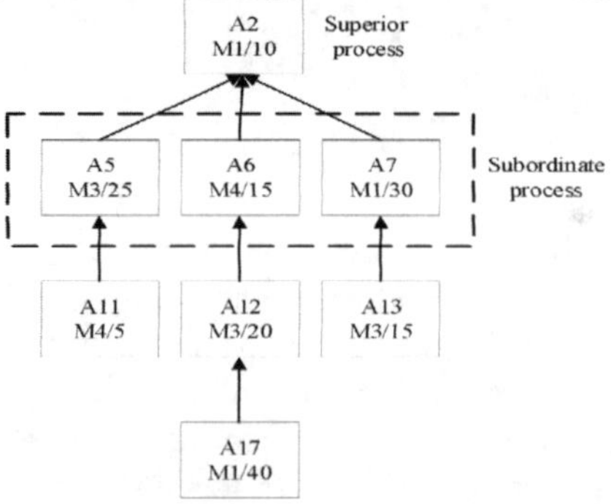

Figure 2. Schematic diagram of the family definition.

Definition 2: Under the same process constraint chain, a new set J_i is formed by the processes of family F_i as a single-element set. The set J_i is called the new process, and the processes in F_i are referred to as the sub-processes of J_i , arranged in the order of their processing sequence.As shown in Figure 3, it is an example diagram of sub-processes.

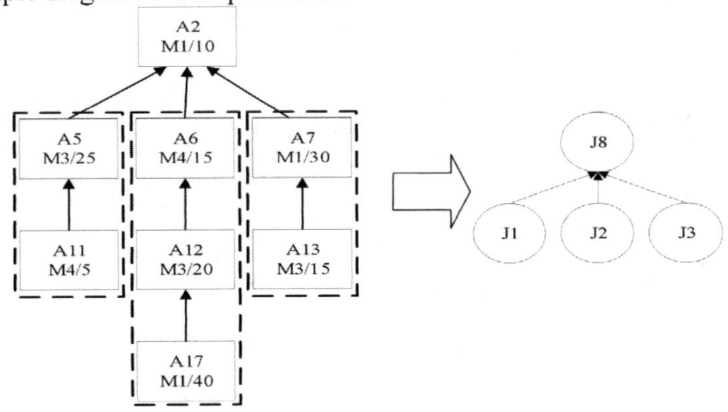

Figure 3. Schematic diagram of sub-processes.

By applying Definition 1, the product processing technology shown in Figure 1 is simplified, reducing the original 18-process problem to an 11-process problem, as shown in Figure 4.

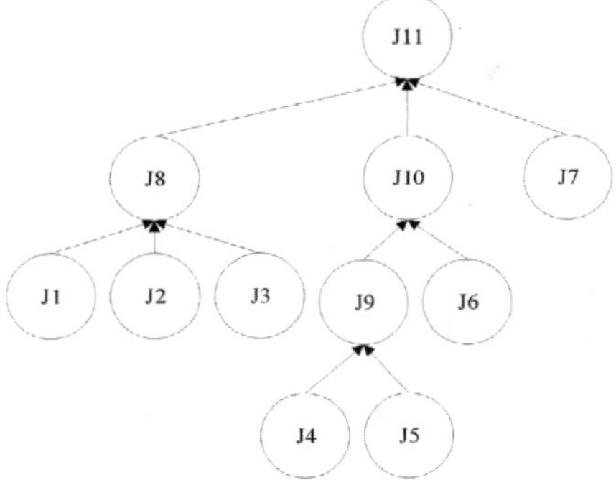

Figure 4. Simplified product process tree.

3. Algorithm Design

3.1. Coding Design

This paper adopts a full-space coding method based on the constraint matrix to address the issues of generating invalid and infeasible solutions.

Based on the process constraint relationships in Figure 4, a constraint matrix can be established A[a_{ij}], where,

$$a_{ij}\begin{cases} a_{ij} = 1, Process\ i\ is\ a\ subordinate\ process\ of\ process\ j. \\ a_{ij} = 0, Otherwise. \end{cases}$$

If the sum of all elements in column j is zero, then process j has no subordinate processes. Otherwise, process j has subordinate processes.

$$A=\begin{bmatrix}0&0&0&0&0&0&0&1&0&0&0\\0&0&0&0&0&0&0&1&0&0&0\\0&0&0&0&0&0&0&1&0&0&0\\0&0&0&0&0&0&0&0&1&0&0\\0&0&0&0&0&0&0&0&1&0&0\\0&0&0&0&0&0&0&0&0&1&0\\0&0&0&0&0&0&0&0&0&0&1\\0&0&0&0&0&0&0&0&0&0&1\\0&0&0&0&0&0&0&0&0&1&0\\0&0&0&0&0&0&0&0&0&0&1\\0&0&0&0&0&0&0&0&0&0&0\end{bmatrix}_{n*n}$$

Definition 3: The matrix representing the hierarchical relationship between processes, derived from the constraint matrix A, is called the Hier matrix.

The first n-1 columns of the Hier matrix represent subordinate processes, and the n-th column represents the superior process.

$$\text{Hier}=\begin{bmatrix}1&2&3&8\\4&5&&9\\6&9&&10\\7&8&10&11\end{bmatrix}_{4*4}$$

The steps of the encoding method are as follows:

Step 1: Construct an empty gene sequence POP and IN, and establish a process count sequence ST. POP is used to store the sequence of chromosomes, representing the completed processing operations. IN is used to store the initially available and selectable operations. ST is used to record the number of operations for each process.

Step 2: Randomly select a process k from IN and store it in POP. At the same time, remove process k from IN, and perform a decrement operation at position k in ST.

Step 3: Check the processed operations in POP. If the process numbers in POP satisfy the first n columns of a certain row in the Hier matrix, proceed to Step 4; otherwise, repeat Step 2.

Step 4: Check the process counts at the positions in ST corresponding to the process numbers in the first n columns. If all the values have been decremented to zero, the process can be added to IN as an initially selectable process and proceed to Step 5. If not, all values are zero, go back to Step 2.

Step 5: If there are still unscheduled process numbers in the gene sequence IN, repeat Step 2 until IN is empty.

3.2. Single-Parent Operator

This paper improves upon the gene inversion method based on the IPSP [9]. The superior process P1 and the superior process P2 are used as gene points to divide the segment. As shown in Figure 5, perform an inversion operation on the gene sequence between these two gene points.

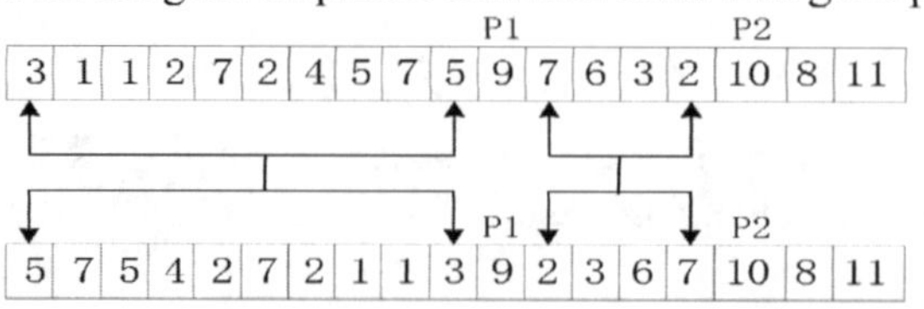

Figure 5. Crossover operation.

3.3. Family Mutation

As shown in Figure 6, the Family mutation occurs randomly between processes that belong to the

Journal of Physics: Conference Series **3009** (2025) 012066 doi:10.1088/1742-6596/3009/1/012066

same superior process and the same family. If the mutated gene sequence contains processes, the order of the processes after mutation is rearranged.

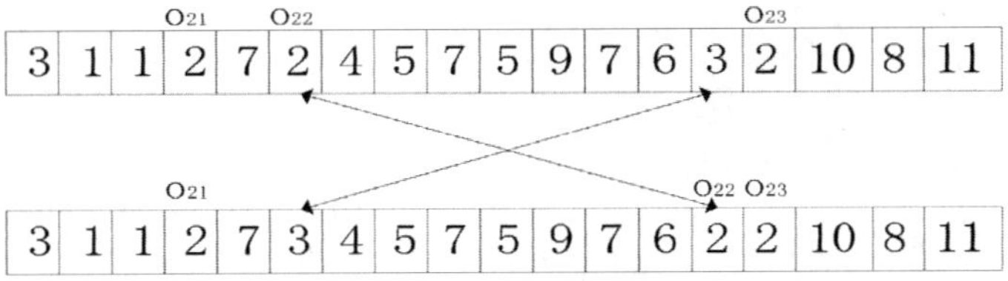

Figure 6. Mutation operation.

4. Algorithm testing

As shown in Figure 7, the simplified product process tree from the literature [10] is obtained based on the simplification method proposed in this paper. A data model is summarized as presented in Table 1.

The algorithm parameters are set as follows: the initial population size is 20, the number of genetic iterations is 100, the crossover probability is 0.8, the mutation probability is 0.05, written using MATLAB 2021 language, and the algorithm is independently run 10 times.

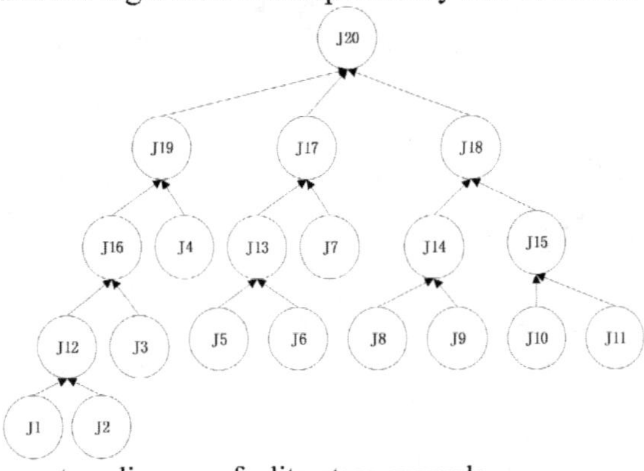

Figure 7. Simplified process tree diagram of a literature example.

Table 1. Data model for comprehensive scheduling problems in literature.

Process	Superior process	Sub-processes	Corresponding node	Processing machine	Processing time
J1		$O_{1,1}$	A20	5	25
J2		$O_{2,1}$	A21	2	15
J3		$O_{3,1}$	A12	4	15
J4		$O_{4,1}$	A6	5	25
		$O_{4,2}$	A13	4	20
J5		$O_{5,1}$	A14	5	20
J6		$O_{6,1}$	A15	1	15
J7		$O_{7,1}$	A8	1	30
J8		$O_{8,1}$	A16	2	35
		$O_{8,2}$	A22	3	25
J9		$O_{9,1}$	A17	4	20
J10		$O_{10,1}$	A18	1	35
J11		$O_{11,1}$	A19	2	15
J12	J1，J2	$O_{12,1}$	A11	4	15
J13	J5，J6	$O_{13,1}$	A7	2	25
J14	J8，J9	$O_{14,1}$	A9	1	20
J15	J10，J11	$O_{15,1}$	A10	3	20

J16	J3, J12	$O_{16,1}$	A5	3	30
J17	J7, J13	$O_{17,1}$	A3	4	25
J18	J14, J15	$O_{18,1}$	A4	5	20
J19	J4, J16	$O_{19,1}$	A2	1	35
J20	J17, J18, J19	$O_{20,1}$	A1	3	15

The algorithm presented in this paper achieves an average makespan of 150 with a runtime of less than 1 second. Figure 8 shows the scheduling results for this data model using two different methods. It demonstrates that as long as the operations on machine M1 and the final operation on machine M3 have zero waiting time during processing, forming a critical path, satisfactory scheduling results can be obtained.

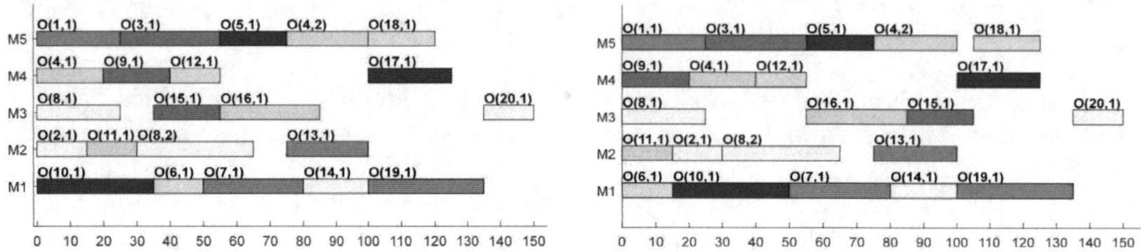

Figure 8. Gantt chart of literature results obtained by the algorithm in this paper.

The comparison of iteration curves between the proposed algorithm and the algorithm in the reference [4] is shown in Figure 9. The IPSP algorithm requires fewer iterations, stabilizing at around 80 iterations, while the algorithm in the reference [4] takes more iterations to stabilize. This indicates that the IPSP algorithm has a faster convergence speed.

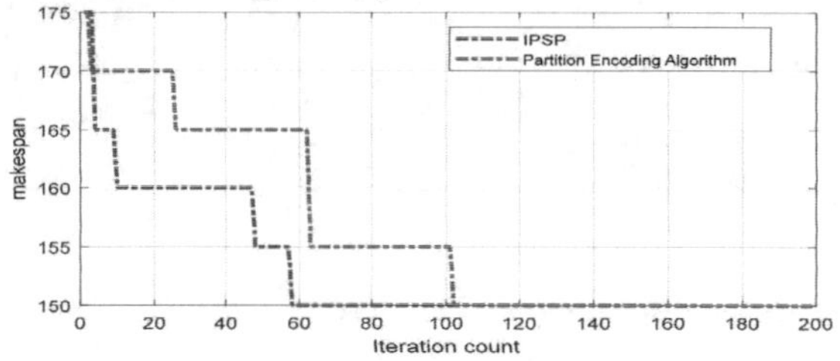

Figure 9. Comparison diagram of iteration curves between two algorithms.

5. Conclusion

This paper addresses the integrated scheduling problem that unifies the machining and assembly processes for complex products, proposing a solution method based on a single-parent genetic algorithm. The designed encoding method, based on the operation constraint matrix, not only satisfies the precedence constraints in the product process tree but also enables population initialization in the full solution space. Test results verify the effectiveness of the proposed single-parent genetic algorithm.

References

[1] Beibei Fan, Wenwei Yang, and Zaifang Zhang(2019) " Solving the two-stage hybrid flow shop scheduling problem based on mutant firefly algorithm, " in J Ambient Intell Human Comput 10, 979–990.

[2] A. Allahverdi, H. Aydilek(2015) " The two-stage assembly flow-shop scheduling problem to minimize total tardiness, " in Journal of Intelligent Manufacturing,26(2),225-237.

[3] M. C. Gomes, A. P. Barbosa-Póvoa(2013) " Reactive scheduling in a make-to-order flexible job shop with re-entrant process and assembly: A mathematical programming approach, " in International Journal of Production Research, 51(17), 5120–5141.

[4] F. Daneshamooz, P. Fattahi, and S. M. H. Hosseini(2022) " Scheduling in a flexible job shop followed by some parallel assembly stations considering lot streaming, " in Engineering Optimization, 54(4), 614–633.
[5] Weifei Guo, Yuchuan Song, Fan Zhou, Qi Lei, and Xiangfei Lv(2020) " Integrated scheduling algorithm of complex product with no-wait constraint based on reversed virtual component " in Computer Integrated Manufacturing Systems, CIMS, 26(12), 3313–3328.
[6] Zhiqiang Xie， Jing Yang(2008) " Dynamic job-shop scheduling algorithm with dynamic set of operation having priority, " in Chinese Journal of Computer,2008,31(3):502-508.
[7] Fei Shi, Shikui Zhao(2017) " Product comprehensive scheduling problems solved by genetic algorithm based on operation constraint chain coding, " in China Mechanical Engineering, 28(20):2483-2492.
[8] Shikui Zhao, Shuiliang Fang(2013) " Operation-based encoding and neighborhood search genetic algorithm for job-shop scheduling optimization, " in Chinese Journal of Mechanical Engineering,49(16):160-169.
[9] Qing Zhang, Xifan Yao(2009) " Research on Job-Shop Problem Based on Hybrid Partheno-Genetic Algorithm, " in Mechanical Design and Manufacturing,7,92-94.
[10] Zhiqiang Xie(2009) " Study on operation scheduling of complex product with constraint among jobs, " in Harbin University of Science and Technology.

Journal of Physics: Conference Series **3009** (2025) 012067 doi:10.1088/1742-6596/3009/1/012067

Flexural stiffness parameters identification using integrated inverse finite element method and pseudo-excitation approach

Tengteng Li[1], Yamei Wang[1], Yang Wu[1], Jian Liu[2], Jie Sun[1], Dongtao Wu[2], Hao Xu[3*]

[1]Suzhou Laboratory, Suzhou 215123, China
[2]Beijing Institute of Aerospace Technology, Beijing 100074, China
[3]School of Materials Science and Engineering, Dalian University of Technology, 116024, Dalian, China

[*]Corresponding author's e-mail: xuhao@dlut.edu.cn

Abstract. By integrating the inverse finite element method (iFEM) with the pseudo-excitation method (PE), the iFEM-PE framework can efficiently reconstruct the deformation of composite structures and accurately identify damage. However, the effectiveness of this approach depends on the prior knowledge of flexural stiffness parameters, which restricts its broader use. This study introduces a new method for identifying flexural stiffness parameters, based on the dynamic equilibrium of laminated elements. Additionally, this technique improves the iFEM-PE framework. Finally, the performance of the proposed method in determining flexural stiffness coefficients is validated using a damaged cantilever laminated panel model.

1. Introduction

Due to their excellent load-bearing capacity and versatile design options, composite thin-walled structures are commonly applied in aerospace industries [1]. However, aircraft are exposed to harsh mechanical and thermal environments during service, which can easily lead to structural damage. At the same time, damage in composites can manifest in various forms, often occurring within the material and making early detection challenging. As damage accumulates, the integrity and safety of the structure are significantly compromised. Consequently, online structural health monitoring (SHM) using advanced sensor networks holds tremendous engineering potential, enhancing safety for both people and the environment while reducing maintenance costs [2].

Recently, vibration-based methods have gained significant use in damage detection, utilizing variations in structural vibration characteristics like natural frequency, mode shape, and the stiffness matrix [3-4]. Among them, the natural frequency approach can only identify the presence of damage but is imprecise in the damage location. On the other hand, approaches based on mode shapes and other factors are frequently heavily reliant on baseline signals. In real engineering applications, getting baseline signals from intact buildings presents major problems in terms of practicality and precision. The pseudo-excitation (PE) method, introduced by Xu et al., has attracted considerable attention due to its distinct advantages, including independence from baseline signals and reference structures [5-6]. Additionally, the PE method identifies damage by analyzing the dynamic equilibrium conditions of structural components on a point-by-point basis, enabling the detection not only of small damage locations but also of their shapes through precise delineation of the damage boundaries. However, the method involves higher-order derivative terms of structural vibration displacement. This presents two challenges: first, it is difficult to achieve high-density dynamic displacement monitoring in engineering applications; second, higher-order derivatives amplify measurement noise, resulting in poor noise immunity.

To overcome this challenge, Li et al. [7] introduced an innovative structural health monitoring (SHM) framework that combines the iFEM [8] with the PE approach. A central element of this framework, the iFEM, is a highly effective shape-sensing algorithm capable of accurately reconstructing vibration displacements with measured strains. The PE is then used to construct a damage index with reconstructed displacements. In comparison to the PE approach, the iFEM-PE method has superior noise immunity and is more suited for engineering applications. Although the iFEM-PE method offers many advantages, the damage identification process necessitates prior mechanical performance data, such as the structure's bending stiffness. However, in practical engineering applications, it is often challenging to obtain the stacking sequence of composite structures, making it impossible to calculate the bending stiffness parameters.

Techniques for identifying composite material parameters mostly use vibration-based approaches and static testing. Static testing is widely used due to its standardized methods, but it provides limited information. To address these limitations, vibration-based methods, such as experimental modal analysis, have been employed. However, modal approaches face challenges at medium and high frequencies due to modal overlap and require specific boundary conditions, which limits their applicability. Furthermore, Ablitzer et al. [9] proposed a method for identifying flexural stiffness characteristics using vibratory measurements of plate-like structures. This strategy does not require previous knowledge, yet it has the same disadvantages as the PE method.

The major goal of this study is to provide a methodology for determining flexural stiffness parameters based on the dynamic equilibrium relationships of laminated parts to enhance the iFEM-PE. The paper's structure is set out as follows. The iQS4 element theory will be given first. Then, a method based on the pseudo-excitation (PE) approach for evaluating flexural stiffness parameters and detecting damage in laminated structures will be discussed. Lastly, through the use of a damaged cantilever laminated panel model, the efficacy of the suggested technique in determining flexural stiffness coefficients and detecting damage will be confirmed.

2. Methodologies

The two primary parts of the iFEM-PE calculation procedure are as follows, as explained in [7]: first, measured dynamic strains are quickly converted into vibration displacements using the iFEM, and then the PE analysis is performed using the reconstructed vibration displacements.

2.1. Inverse Finite Elements (iFEM)

To apply the iFEM, the geometry of the structure being analyzed must first be discretized into several inverse elements, akin to the conventional finite element method. So far, various inverse elements have been developed, including iQS4 and iMIN3. In this work, the iQS4 element is chosen [8] due to its excellent performance in addressing shell structures and membrane issues, particularly because it incorporates drilling rotations.

2.1.1. The iQS4 element. As illustrated in Figure 1, the iQS4 element has a uniform thickness of *h*. Each of the iQS4 element's four nodes has three displacement degrees of freedom (DOFs) and three rotational DOFs. Based on the strain-displacement relation, the analytical strain components within the element are given by:

$$\mathbf{e} = \mathbf{B}_e \mathbf{u}^e \tag{1}$$

$$\boldsymbol{\kappa} = \mathbf{B}_\kappa \mathbf{u}^e \tag{2}$$

$$\boldsymbol{\gamma} = \mathbf{B}_\gamma \mathbf{u}^e \tag{3}$$

where **e**, **k**, and $\boldsymbol{\gamma}$ are membrane, bending and transverse-shear section strains, respectively. $\mathbf{B}_e$, $\mathbf{B}_k$ and $\mathbf{B}_\gamma$ are corresponding strain-displacement matrices that contain derivatives of the shape functionals, as specified in [8].

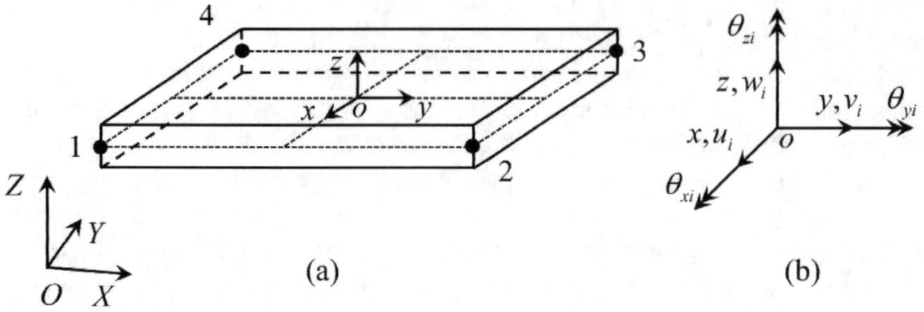

Figure 1. The iQS4 element.

2.1.2. Input data from in-situ strain sensors. The iFEM formulation highlights the significance of discrete in-situ strain data gathered by on-board sensors. To compute the experimental section strains, which correspond to the analytical strains presented in Equations (1) and (2), measurements must be taken simultaneously from both surfaces of the element, as shown in Figure 2. At a given in-plane position, the membrane strains and bending curvatures can be derived using:

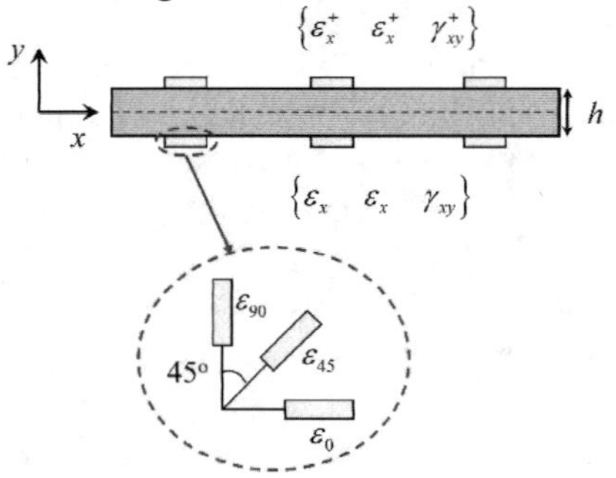

Figure 2. The deployment plan of strain sensors within iQS4.

$$\mathbf{e}_i^{\varepsilon} \equiv \begin{bmatrix} \varepsilon_x^{\varepsilon} \\ \varepsilon_y^{\varepsilon} \\ \gamma_{xy}^{\varepsilon} \end{bmatrix} = \frac{1}{2}\begin{Bmatrix} \varepsilon_x^{+} + \varepsilon_x^{-} \\ \varepsilon_y^{+} + \varepsilon_y^{-} \\ \gamma_{xy}^{+} + \gamma_{xy}^{-} \end{Bmatrix}, \quad \mathbf{k}_i^{\varepsilon} \equiv \begin{bmatrix} \kappa_x^{\varepsilon} \\ \kappa_y^{\varepsilon} \\ \kappa_{xy}^{\varepsilon} \end{bmatrix}_i = \frac{1}{h}\begin{Bmatrix} \varepsilon_x^{+} - \varepsilon_x^{-} \\ \varepsilon_y^{+} - \varepsilon_y^{-} \\ \gamma_{xy}^{+} - \gamma_{xy}^{-} \end{Bmatrix}_i \quad (i = 1, \cdots, n) \tag{4}$$

where the surface strains measured are represented by $\left\{\varepsilon_x^{+} \quad \varepsilon_y^{+} \quad \gamma_{xy}^{+}\right\}_i^{\mathrm{T}}$ and $\left\{\varepsilon_x^{-} \quad \varepsilon_y^{-} \quad \gamma_{xy}^{-}\right\}_i^{\mathrm{T}}$, and the superscript ε signifies the presence of errors in the strain measurement.

2.1.3. Equation of iFEM. Using the derived analytical and experimental strains, the weighted least-square functional, $\mathbf{\Phi}^e(\mathbf{u}^e)$, for an iQS4 element, can then be defined as:

$$\Phi^e(\boldsymbol{u}^e) = w_e \left\|\boldsymbol{e}_i(\boldsymbol{u}^e) - \boldsymbol{e}_i^{\varepsilon}\right\| + w_k \left\|\boldsymbol{\kappa}_i(\boldsymbol{u}^e) - \boldsymbol{\kappa}_i^{\varepsilon}\right\| + w_g \left\|\gamma_i(\boldsymbol{u}^e) - \gamma_i^{\varepsilon}\right\| \tag{5}$$

where w_e, w_k, and w_g are positive weighting constants. When experimental strains are available, the weighting constants is set to one. If not, a tiny number (10^{-4} or 10^{-5}) might be used. Subsequently, minimizing the least-square functional with respect to the nodal DOFs yields a set of equations:

$$\frac{\partial \Phi_e(\mathbf{u}^e)}{\partial \mathbf{u}^e} = \mathbf{k}^e\mathbf{u}^e - \mathbf{f}^e = 0 \Rightarrow \mathbf{k}^e\mathbf{u}^e = \mathbf{f}^e \tag{6}$$

where $\mathbf{k}^e$ represents the stiffness matrix of the element and $\mathbf{f}^e$ is the load vector of the element. The local matrices are then integrated into a global matrix that may be easily solved to yield the displacement field of the target structure.

2.2. *Pseudo-Excitation (PE) approach*

2.2.1. Damage identification. For composite laminates, the damage index (*DI*) is given by:

$$DI = D_{11}w_{,xxxx} + 4D_{16}w_{,xxxy} + 2(D_{12} + 2D_{66})w_{,xxyy} + 4D_{26}w_{,xyyy} + D_{22}w_{,yyyy} - \rho h\omega^2 w \quad (7)$$

where $w = w(x, y)$ is the vibration displacement of the composite laminate at location (x, y); D_{ij} refers to the bending stiffness matrices of the laminate, and ρ, h and ω represent the density, thickness, and angular vibration frequency, respectively. The comma (',') indicates the differentiation of the displacement w with respect to the subscripts.

The *DI* values are zero in regions where equilibrium conditions are satisfied and there is no damage or excitation. In contrast, *DI* shows noticeable singularities in damaged areas. Hence, this *DI* method is effective for detecting damage in laminates.

2.2.2. Identification of the flexural stiffness parameters of laminates. Based on Equation (7), the *DI* in undamaged areas is exactly zero due to the fulfillment of dynamic equilibrium, as explained earlier. Building upon this, a quantitative mapping relationship between the structural vibration displacements and mechanical/material properties (such as flexural stiffness) can also be derived:

$$D_{11}w_{,xxxx} + 4D_{16}w_{,xxxy} + 2(D_{12} + 2D_{66})w_{,xxyy} + 4D_{26}w_{,xyyy} + D_{22}w_{,yyyy} = \rho h\omega^2 w \quad (8)$$

Given that the laminate density ρ is known, Equation (8) can be treated as a linear equation with the unknown bending stiffness components, D_{ij}. It should be noted that D_{12} and D_{66} are coupled which cannot be solved separately. Substituting w_i ($i = 1, \cdots, N$) in the intact region into Equation (8), a set of equations is formed as:

$$\begin{cases} DI_1(D_{11}, D_{16}, D_{12} + 2D_{66}, D_{26}, D_{22}) = \rho h\omega^2 w_1 \\ DI_2(D_{11}, D_{16}, D_{12} + 2D_{66}, D_{26}, D_{22}) = \rho h\omega^2 w_2 \\ \cdots \\ DI_N(D_{11}, D_{16}, D_{12} + 2D_{66}, D_{26}, D_{22}) = \rho h\omega^2 w_N \end{cases} \quad (9)$$

Because the number of Equations (measurement points), denoted as N, is typically much larger than the number of unknowns, D_{ij} can be determined by solving any five of the equations above. It is important to select measurement points located away from regions with damage or excitation to ensure calculation accuracy. Using the identified laminated flexural stiffness parameters, it is possible to further deduce the stiffness matrix and elastic modulus of the laminae that make up the laminate; however, this topic lies beyond the scope of this study, and no additional discussion will be provided.

3. Numerical validation

3.1. *Numerical Simulation*

The accuracy of the improved iFEM-PE technique in identifying bending stiffness and damage is evaluated using a damaged cantilever CFRP laminate, as shown in Figure 3. Since intra-ply and inter-ply damage in composite laminates influences their local stiffness, a reduction in local stiffness is used to represent the damage, simplifying the model without sacrificing generality. The laminate consists of 16 layers of unidirectional lamina stacked in a symmetrical pattern of $[0_2/90_2/0_2/90_2]_s$. Table 1 shows the material parameters for the lamina. The laminate's dimensions are 720 mm wide and 960 mm long, with each lamina having a thickness of 0.15 mm, for a total laminate thickness of 2.4 mm.

A high-fidelity FEM model was initially constructed in ABAQUS® commercial software to simulate strain data, instead of relying on experimental measurements. The model comprised 6912 S4R shell elements, each with a size of 10 × 10 mm^2. Laminate damage was mimicked by halving Young's modulus of the ninth layer (90° layer), and its dimensions are shown in Figure 3. The excitation frequency was set to 200 Hz. In this study, the mode superposition method was employed for the dynamic analysis of the laminate. A critical damping ratio of 5% was applied to each vibration mode.

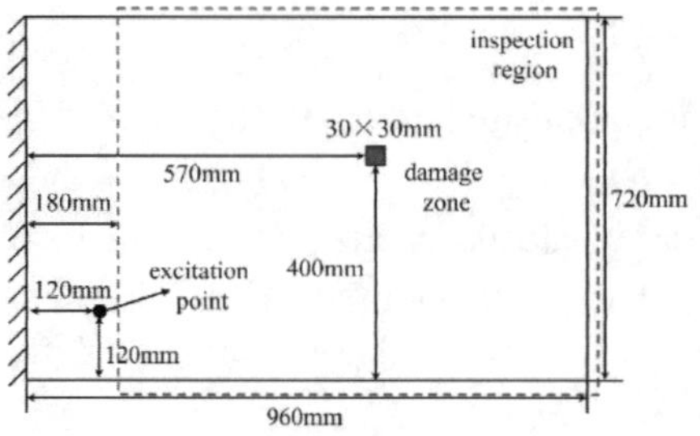

Figure 3. The dimensions of the laminate and damage.

Table 1. Mechanical characteristics of the CFRP lamina.

Young's Modulus [Gpa] (E1/E2/E3)	Shear Modulus [Gpa] (G12/G13/G23)	Poisson's Ratio (v12/v13/v23)	Density [kg/m^3]
157.9/9.6/9.6	5.9/5.9/3.2	0.32/0.32/0.49	1620

3.2. *Reconstruction of Vibration Displacements*

To reconstruct the vibration displacements of the laminate, iFEM analysis was initially conducted using iQS4 elements. The size of the iQS4 elements matches that of the S4R elements used in FEM analysis, simplifying the transmission of strain data. In this study, it is assumed that each iQS4 element is equipped with strain rosettes. Given that the laminate's deformations are primarily due to bending, strains measured on the top surface were used, while the strains on the bottom surface can be computed indirectly.

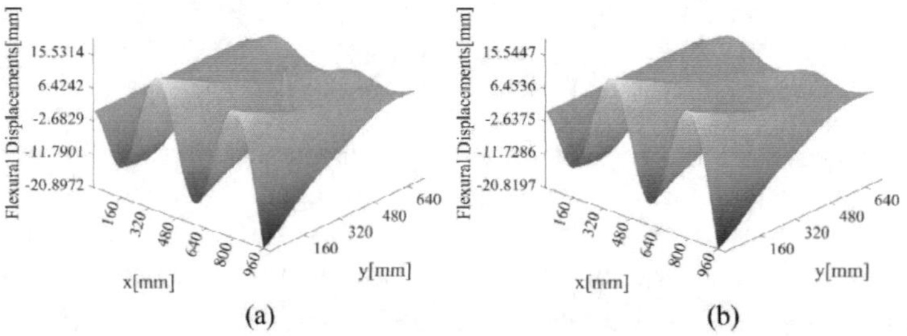

Figure 4. The vibration displacements (a) computed by direct FEM and (d) reconstructed using iFEM.

Figure 4 shows the transverse displacement nephogram from the direct FEM and iFEM analysis. With the exception of the damaged region, the figure shows that the two are essentially consistent, with a large variation in the damaged area. This is because iQS4 elements did not introduce any damage during the iFEM analysis process. The maximum displacement prediction error is only 0.37%. This result confirms the iQS4 element's remarkable performance and provides a guarantee for the subsequent measurement of damage and bending stiffness coefficient.

3.3. *Identification of the flexural stiffness parameters and damage*

The PE method is employed to determine the flexural stiffness parameters of the laminate based on the reconstructed displacements. Subsequently, the PE method, along with the identified stiffness parameters, enables the construction of the damage index (*DI*) as described in Equation (7) through a finite difference scheme, as outlined in [7]. To mitigate the influence of excitation and fixed boundary conditions on the identification of the laminate's flexural stiffness characteristics and damage, an inspection zone is selected, as illustrated in Figure 3.

For orthogonal laminated composite laminates, the flexural stiffness coefficients D_{16} and D_{26} are equal to 0. That is, the number of unknowns to be solved in Equation (9) is decreased to three, implying that only three measurement points need to be picked to obtain the unknown stiffness coefficients. In this study, 10 group points (3 in each group, a total of 30 points) were chosen for parameter identification, and the average of their identification results was utilized as the final identification result, as shown in Table 2. The improved iFEM-PE approach has been shown to be useful in determining the flexural stiffness of laminates, as shown in Table 2. Despite the presence of reconstruction errors and finite differential errors, it maintains good identification accuracy for D_{11} and D_{22}.

Table 2. Identification results of the flexural stiffness parameters.

	D_{11} [10^5Mpa]	D_{22} [10^5Mpa]	D_{12}+2D_{66} [10^5Mpa]
The actual values	1.29	0.65	0.17
The calculation results	1.23	0.54	0.23
Calculation error	4.65%	16.9%	35%

Substituting the identified flexural stiffness coefficients of the laminate and the iFEM reconstructed displacement results into Equation (8) gives the *DI* distribution results as shown in Figure 5. As observed, the DI exhibits sharp fluctuations in the damaged region, enabling the constructed DI to not only identify the damage locations but also capture the precise shapes of the damage. Additionally, noticeable non-zero fluctuations appear in the undamaged areas, primarily due to identification errors in the vibration displacement and flexural stiffness parameters. To further enhance the damage recognition accuracy of the improved iFEM-PE, it may be beneficial to refine the reconstruction accuracy of iFEM through measurement point censoring and to improve the recognition of flexural stiffness parameters by utilizing regularization techniques, among other potential strategies. Nevertheless, even in the absence of these measures, the improved iFEM-PE still retains significant recognition accuracy due to its inclusion of higher-order displacement derivatives.

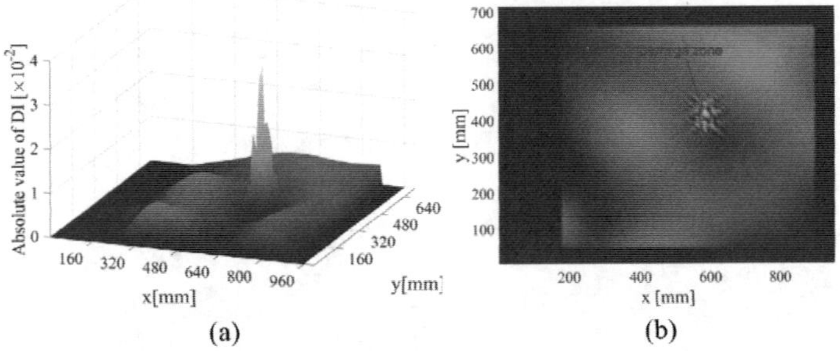

Figure 5. (a) Damage identification results presented in three dimensions and (b) in two dimensions.

4. Conclusion

This study proposes a method for identifying the flexural stiffness coefficients of a laminate, based on the dynamic equilibrium relationships of laminated elements. The accuracy of this identification method was validated using a damaged cantilever CFRP laminate. By comparing the reconstructed results with the corresponding reference values, the following conclusions can be drawn:

(1) The iQS4 element can accurately reconstruct the laminate's vibration displacements with measured strains, providing data for subsequent damage and parameter identification.

(2) Based on the dynamic equilibrium relationship of laminated elements, the improved iFEM-PE method can accurately identify the flexural stiffness of laminates.

(3) Based on the identified flexural stiffness coefficients of the laminate and the iFEM reconstructed displacements, the improved iFEM-PE can identify composite material damage correctly.

Acknowledgments

This work was financially supported by the National Science Foundation of China (No. 12072056)

References

[1] Ren R, Ma X, Yue H, et al. (2024) Stiffness enhancement methods for thin-walled aircraft structures: A review. Thin Wall Struct., 201:111995.
[2] Abdulkarem M, Samsudin K, Rokhani F Z, et al. (2020) Wireless sensor network for structural health monitoring: A contemporary review of technologies, challenges, and future direction. Struct. Health. Monit., 19(3): 693-735.
[3] Kim J T, Ryu Y S, Cho H M, et al. (2003) Damage identification in beam-type structures: frequency-based method vs mode-shape-based method. Eng. Struct., 25(1): 57-67.
[4] Maeck J, Wahab M A, Peeters B, et al. (2000) Damage identification in reinforced concrete structures by dynamic stiffness determination. Eng. Struct., 22(10): 1339-1349.
[5] Xu H, Cheng L, Su Z, Guyader J.L. (2011) Identification of structural damage based on locally perturbed dynamic equilibrium with an application to beam component. J. Sound Vib., 330, 5963–5981
[6] Xu H, Cheng L, Su Z, Guyader, J.L. (2013) Damage visualization based on local dynamic perturbation: Theory and application to characterization of multi-damage in a plane structure. J. Sound Vib., 332, 3438–3462.
[7] Li T, Cao M, Li J, et al. (2021) Structural damage identification based on integrated utilization of inverse finite element method and pseudo-excitation approach. Sensors, 21(2): 606.
[8] Kefal A, Oterkus E, Tessler A, et al. (2016) A quadrilateral inverse-shell element with drilling degrees of freedom for shape sensing and structural health monitoring. Eng Sci Technol, 19(3): 1299-1313.
[9] Tam J H, Ong Z C, Ismail Z, et al. (2017) Identification of material properties of composite materials using nondestructive vibrational evaluation approaches: A review. Mech Adv Mater Struc, 24(12): 971-986.

Journal of Physics: Conference Series **3009** (2025) 012068 doi:10.1088/1742-6596/3009/1/012068

Optimization of mechanical properties of laser welded joints in automotive 304 stainless steel using the response surface methodology

Liqing Liang*

School of Mechanical and Civil Engineering, Jilin Agriculture Science and Technology University Jilin, Jilin, China

* Corresponding author's e-mail: 648346878@qq.com

Abstract: Experimental design methods based on mathematical models are increasingly used to guide practical production. A series of welding process tests were carried out by laser lap welding using 304 stainless steel which is 2 mm thick for automobiles. Based on the response surface method (RSM), a mathematical model of the relationship between laser welding process parameters and the maximum tensile strength of the weld was established. To ensure the formation of the weld and maximize the tensile strength of the weld, the process parameters such as laser power, welding speed, and defocusing amount were optimized and verified by experiments. The results show that with the increase of laser power, welding speed, and defocusing amount, the tensile strength of the weld first increases and then decreases. When the laser power is 2.046 ~ 2.098 kW, the welding speed is 0.965 ~ 1.017 m/min, and the defocus is -0.825 ~ -1.12 mm, the tensile strength of the weld reaches the best. The optimal process parameters obtained by the model were verified through experiments, and the error rate between the experimental results and the model prediction was less than 5%, which verified the accuracy of the model and provided ideas for the subsequent experimental design.

1. Introduction

Experimental design refers to the process of reasonably planning the experimental process, selecting experimental variables, determining experimental samples, and designing experimental steps and data analysis methods according to the purpose and hypothesis of an experimental study. Experimental design methods are important tools in scientific research, helping researchers systematically plan experiments to test hypotheses, explore relationships between variables, and draw reliable conclusions. In the field of laser welding, the application of experimental design methods is also crucial, which helps to optimize welding parameters, improve welding quality, and promote the innovation and development of laser welding technology. 304 stainless steel is widely used in the automotive manufacturing industry due to its excellent workability, high toughness, and strong corrosion resistance [1]. Welding is the most commonly employed processing technique in automotive manufacturing. Stainless steel has excellent corrosion resistance, high-temperature stability, and good mechanical properties. Given the strict requirements for high-temperature resistance and flame-retardant performance of battery packs in the manufacturing of new energy vehicles, the existing process usually uses 0.8 mm thick carbon steel materials to enhance the flame-retardant effect of battery packs. For per 100 kg reduction of car weight, 100 km fuel consumption can save fuel 0.3 ~ 0.5 L and reduce CO_2 emissions 8 ~ 11 g, so car lightweight is an important way to reduce energy consumption and carbon emissions. Body weight will also reduce the performance requirements of the car power system and car chassis to a certain extent.

Among various welding methods, laser welding has become the primary choice in the automotive industry due to its high efficiency, low deformation, and high degree of automation [2-3]. As a modern welding method, laser welding has developed rapidly in recent years, especially for the welding of thin plates. Stainless steel plates have been widely used in automobiles, aerospace, medicine, and other fields. According to the different working modes of the laser, laser welding can be divided into continuous welding and pulse welding, in which pulse laser welding can form a welding spot similar to spot welding, but also can form a continuous weld. The spot diameter of pulsed laser welding is very small, even small to the micron level, and each laser pulse is emitted to form a solder spot on the metal material. The welding of stainless steel has gradually become a research hotspot.

Chen et al. [4] studied the weld forming of 0.2 mm thick 321 stainless steel pulsed laser welding. The results show that the micro pulsed laser butt welding is easy to burn through the weld, and a deep arc pit is formed at the arc retract. When the pulse power fraction is 12% (the average power of the pulse laser is 80 W), the well-formed and defect-free welded joint can be obtained in a wide range of pulse frequency changes. Wu [5] used an Nd: YAG pulsed laser with a wavelength of 1.06 μm to perform laser sealing welding on an ultra-thin stainless steel sheet with a thickness of 0.08 mm. By adjusting the pulse laser frequency, pulse width, welding speed, and defocusing, the ultra-thin stainless steel sheet has been successfully welded by a pulsed laser. The ratio of pulse frequency to welding speed must meet the relationship of 3:1, to get a better overlap rate of welding spots, and then get a better sealing of welding seams. Huo et al. [6] used IPG's quasi-continuous laser YLR-150/1500-QCW-AM to lap weld 316 stainless steel with a thickness of 0.05 mm, but the clamping fixture was not good, which was prone to virtual welding and welding penetration.

Optimizing welding process parameters can ensure good weld formation. However, to further enhance the mechanical properties of the weld while maintaining its quality, it is essential to consider the interactions among various process parameters. Consequently, many researchers have conducted studies on improving the mechanical properties of welds by optimizing welding process parameters.

Luo et al. investigated the effects of process parameters such as laser power, welding speed, welding current, and beam spacing on the weld morphology and tensile strength of 304 stainless steel [7]. Chen optimized laser welding process parameters for 316 stainless steel using tensile strength as the objective through orthogonal experiments, guiding practical production [8]. Chen et al. studied the influence of laser welding process parameters on the microstructure and mechanical properties through single-factor experiments [9]. However, research on the interactive effects of laser welding process parameters on weld tensile strength is still relatively limited. Therefore, in this study, laser lap welding experiments were conducted on 2-mm-thick automotive 304 stainless steel. The interactive effects of laser power, welding speed, and defocusing amount on weld tensile strength were investigated, providing guidance and theoretical reference for non-penetration laser welding of automotive 304 stainless steel.

2. Experimental methods and program design

The laser welding system consisted of a TRUMPF H4006D Nd solid-state laser and a KUKA robotic arm. The experimental setup is shown in Figure 1. The laser beam was vertically irradiated, and 304 stainless steel was selected as the welding material. The test plates, with dimensions of 100 mm × 80 mm × 2 mm, were polished with sandpaper and cleaned with acetone before performing butt welding without a gap. According to preliminary experimental results, good weld formation quality was achieved under the following parameter ranges: laser power of 1.8-2.2 kW, welding speed of 0.8-1.2 m/min, defocusing amount of -2 to 0 mm, and post-weld shielding gas (Ar) at a flow rate of 30 L/min. Based on these ranges, a Box-Behnken design with three factors and three levels was employed for the experimental design. The preparation and testing of tensile specimens were conducted in accordance with GB/T 2651-2008 and GB/T 2652-2008 standards, with a loading rate of 3 mm/min. Tensile tests on the welded specimens were performed by using a universal testing machine manufactured by

Zhongji Machinery. The experimental factors and their coding ranges are listed in Table 1, while the experimental design and tensile strength results of the welds are provided in Table 2.

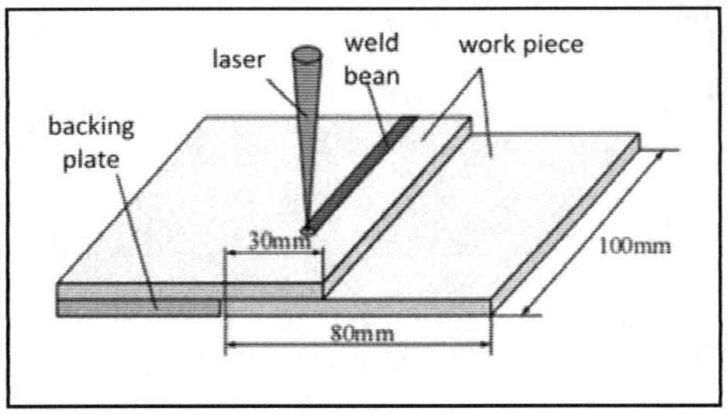

Figure 1. Schematic diagram of the experiment.

Table 1. Process parameter level.

Horizontal coding	-1	0	1
Laser power L/kW	1.8	2	2.2
Welding speed v/(m/min)	0.8	1.0	1.2
Defocusing quantity f/mm	-2	-1	0

Table 2. Trial Protocol and Response.

Serial number	Laser power L/kW	Welding speed $v/m\cdot min^{-1}$	Defocusing quantity f/mm	Maximum tensile force per unit length $TS/kN\cdot mm^{-1}$
1	2	1.2	-2	0.58
2	2	1	-1	0.63
3	2.2	1.2	-1	0.59
4	2	0.8	-2	0.57
5	2.2	1	0	0.60
6	2.2	0.8	-1	0.59
7	2	0.8	0	0.58
8	1.8	1	-2	0.55
9	2.2	1	-2	0.60
10	2	1.2	0	0.58
11	2	1	-1	0.63
12	2	1	-1	0.63
13	2	1	-1	0.64
14	1.8	1	0	0.56
15	1.8	0.8	-1	0.55
16	1.8	1.2	-1	0.55
17	2	1	-1	0.64

3. Results and analysis

3.1 Regression model

The coefficients of different regression models were calculated by using the least squares method, and the comprehensive analysis results of the R^2 values for each regression model are shown in Table 3. As can be seen from Table 3, the optimal regression model for predicting the maximum tensile strength is the quadratic regression model.

Table 3. Regression model R^2 analysis.

Model	Standard deviation	R^2	R^2 calibration value	R^2 projected value	Result
Linear regression	31.26	0.2577	0.0864	-0.0607	
2FI regression	35.63	0.2580	-0.1871	-0.7618	
Square 2 regression	3.30	0.9956	0.9898	0.9447	Adoption
Cubic regression	2.17	0.9989	0.9956		Unintended

3.2 Analysis of variance and regression equations

The analysis of variance (ANOVA) for the regression model of weld tensile strength is presented in Table 4. When the probability P value is less than 0.0001, it indicates that the factor has a highly significant effect on the response. A P value less than 0.05 suggests a significant effect, while a P value greater than 0.05 implies the factor does not significantly affect the response. As shown in Table 5, the P value of the quadratic regression model is less than 0.0001, indicating that the regression model is statistically significant. Among the factors, laser power (L), defocus distance (f), and the squared terms L^2, v^2, and f^2 have a significant impact on the maximum tensile force per unit length. In contrast, welding speed (v) and the interaction terms (L•v), (L•f), and (v•f) have minimal influence on the maximum tensile force per unit length.

Table 4. Analysis of variance of quadratic regression model.

Source	Square sum	Degrees of freedom	Mean square value	F	P	Significance
Model	17, 031.07	9	1, 892.67	174.21	<0.0001	Significant
L	4, 324.50	1	4, 324.50	398.05	<0.0001	
v	6.13	1	6.13	0.5638	0.4772	
f	78.13	1	78.13	7.19	0.0315	
L•v	0.2500	1	0.2500	0.0230	0.8837	
L•f	6.25	1	6.25	0.5753	0.4729	
v•f	0	1	0	0	1	
L^2	4, 896.04	1	4, 896.04	450.65	<0.0001	
v^2	4, 138.20	1	4, 138.20	380.90	<0.0001	
f^2	2, 295.67	1	2, 295.67	211.30	<0.0001	
Residual	76.05	7	10.86			
Lost proposal	57.25	3	19.08	4.06	0.1047	Insignificant
Pure error	18.80	4	4.70			

The multiple nonlinear regression equation for the maximum tensile force per unit length based on the actual factor values is:

$$TS=-3,812.65+3,526.25*L-31.075*f-852.5*L^2-783.75*v^2-23.35*f^2 \quad (1)$$

3.3 Influence of process parameters on the tensile strength of welds

Figure 2 (a) shows the three-dimensional response surface illustrating the interactive effects of laser power and welding speed on the maximum tensile force per unit length. It can be observed that the maximum tensile force per unit length initially increases and then decreases with the increase of laser power and welding speed. The maximum value is achieved when the laser power is approximately 2.1 kW and the welding speed is around 1 m/min. Figure 2 (b) presents the response surface of the interaction between laser power and defocus distance on weld tensile strength. It is evident that the maximum tensile force per unit length increases initially and then decreases as laser power and defocus distance increase. The maximum weld tensile strength is reached when the laser power is around 2.1 kW and the defocus distance is approximately -1 mm. Figure 2 (c) illustrates the response surface of the interaction between welding speed and defocus distance on the maximum tensile force per unit length. Similarly, the tensile force per unit length first increases and then decreases with the increase of welding speed and defocus distance. The maximum value occurs when the welding speed is 1 m/min and the defocus distance is 1 mm. The response surface analysis of various parameters indicates that there exists an optimal combination of process parameters that maximizes the maximum tensile force per unit length.

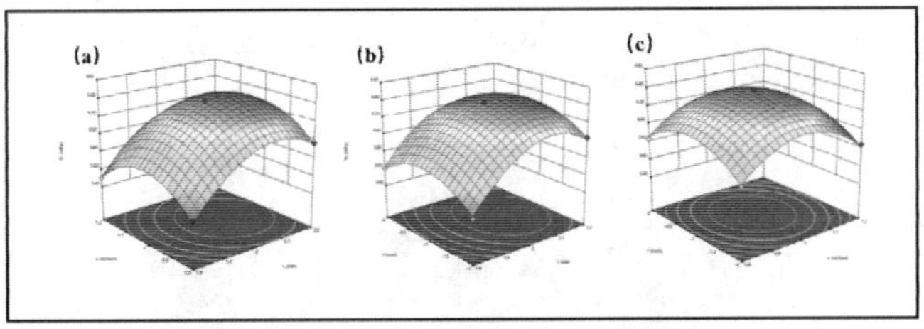

Figure 2. (a) The interactive effect of laser power and maximum tensile force of welding speed; (b) The interactive effect of laser power and defocus amount on the maximum stretching force; (c) The interactive effect of welding speed and defocus amount on the maximum tensile force.

3.4 Parameter optimization and experimental validation

To maximize the maximum tensile force per unit length of the weld, the process parameters were optimized within their respective ranges by setting the laser power, welding speed, and defocus distance as variables. The optimal parameters and the corresponding maximum tensile force per unit length were determined through optimization. When the laser power is in the range of 2.046-2.098 kW, the welding speed is 0.965-1.017 m/min, and the defocus distance is -0.825 to -1.12 mm, the maximum tensile force per unit length reaches 0.64 kN·mm^{-1}, and now, the mechanical performance was the best.

To validate the optimized process parameters, independent welding experiments were conducted under the following conditions: laser power of 2.0 kW, welding speed of 1.0 m/min, and defocus distance of -1 mm. Three tensile tests were performed on the weld specimens, as shown in Table 5. The results showed that the maximum tensile force per unit length ranged between 0.63 and 0.64 kN·mm^{-1}, with a prediction error rate of less than 5%. This confirms the accuracy of the predictive model.

Table 5. Regression model R^2 analysis.

Number	Laser power L/kW	Welding speed v/(m/min)	Defocusing quantity f/mm	Maximum tensile force per unit length TS /kN·mm^{-1}
1	2.0	1.0	-1	0.64
2	2.0	1.0	-1	0.63
3	2.0	1.0	-1	0.64

Figure 3 shows the macroscopic morphology of typical weld joints. The weld zone is still composed of a central equiaxed crystal zone and a marginal columnar crystal zone. The laser welding speed is fast, and the heat input is reduced, increasing the G/R value, which slows down the cooling rate of the metal in the molten pool, and forms a smaller central equiaxed crystal. The average size of the central equiaxed crystal is the smallest, which is about 1.68 times that of the base material.

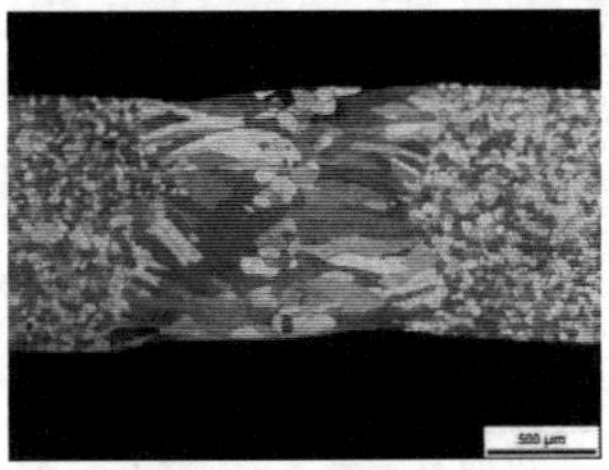

Figure 3. Macro morphology of weld center of a typical sample.

Figure 4 shows the fracture photo at the weld. After magnification, the dimple pattern of ductile fracture can be observed, no expansion crack is generated, and a local cleavage facet exists, which affects the tensile property. At the same time, it is found that the neck shrinkage phenomenon is obvious when observing the macroscopic morphology, which can be judged as a ductile fracture. The size of the dimples varied, with individual dimple sizes varying by tens of times. The second-phase particles or impurities were observed at the bottom of the dimple pits.

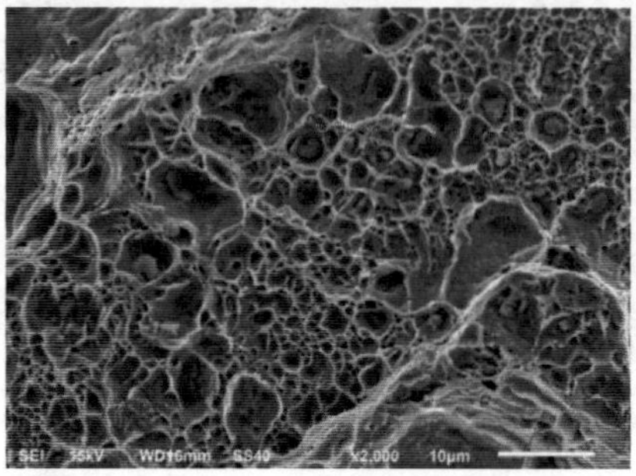

Figure 4. The SEM image of a typical sample weld fracture.

Figure 5 shows the SEM image of the second particle found in fracture analysis and the results of EDS test analysis. The detected particle composition is shown in Figure 5 (b), and the material in the pits is judged to be chromium-rich particles of the second phase, which is closely related to the formation of the dimple. The reason for the formation of the dimple is the cavity aggregation, that is, the formation, expansion, merger, and fracture of the cavity. Most metal materials in the tensile process of the cavity production center are the second phase particles, and because the second phase particles can hinder the displacement, they can play a positive role in improving the tensile strength of the alloy.

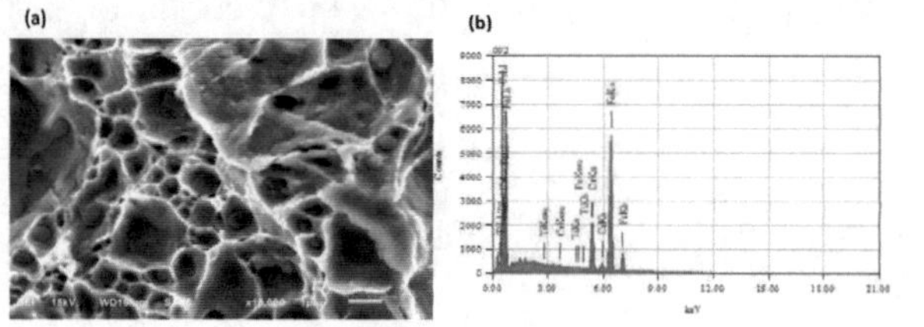

Figure 5. (a) Local analysis of the second phase particle SEM image; (b) EDS analysis at the position of the second phase particles.

4. Conclusions

(1) A mathematical model was established based on the response surface methodology to describe the relationship between laser welding parameters and weld tensile strength for automotive 304 stainless steel. The accuracy of the model was confirmed through analysis of variance and experimental validation. According to the data, the best regression model for predicting the maximum tensile strength is the quadratic regression model. When the P-value of the quadratic regression model is less than 0.0001, it has a significant statistical impact. Through analysis, laser power (L), defocusing distance (f), and square terms L^2, v^2, and f^2 have a significant impact on the maximum tension per unit length. We analyze the influence on the mechanical properties of 304 stainless steel laser welded joints in automobiles: When the laser power is about 2.1 kW and the welding speed is about 1 m/min, the maximum tension per unit length reaches its maximum value; When the laser power is about 2.1 kW and the defocusing distance is about -1 mm, the maximum tensile strength of the weld seam is reached.

(2) With the increase of laser power, welding speed, and defocus distance, the weld tensile strength initially increases and then decreases. The maximum tensile force per unit length of the weld is achieved when the laser power is in the range of 2.046-2.098 kW, the welding speed is 0.965-1.017 m/min, and the defocus distance is -0.825 to -1.12 mm. The accuracy of the prediction model has been confirmed through experiments. It is effective in optimizing the mechanical properties of automotive 304 stainless steel laser welded joints by using response surface methodology by selecting typical macroscopic morphology and SEM images of welds, providing reference value for process optimization and durability strength of automotive 304 laser welded joints.

Acknowledgment

This paper is funded by the Youth Fund Project of Jilin University of Agricultural Science and Technology "Mechanism Analysis of Freeze-Thaw Disease of Channels in Cold Irrigation Area (Project No.: School 20190603)".

References

[1] Alizadeh-Sh M., Pouranvari M., and Marashi S. P. H. Welding metallurgy of stainless steels during resistance spot welding part II – heat affected zone and mechanical performance [J]. Science and Technology of Welding and Joining, 2015, 20 (6): 512-521.
[2] Pouranvari M., Alizadeh-Sh M., and Marashi S. P. H. Welding metallurgy of stainless steels during resistance spot welding Part I: fusion zone [J]. Science and Technology of Welding and Joining, 2015, 20 (6): 502-511.
[3] Chang Yanbo, Guo Liang, and Zhang Qingmao. Research on Laser Lap Welding Process of Automotive Steel/Aluminum Alloy [J]. Applied Laser, 2020, 40 (04): 665-671.
[4] Chen Yuhua, Ke Liming, Xu Shilong, et al. Research on micro-laser welding process of ultra-thin stainless steel sheet [J]. Journal of Materials Engineering, 2008, (11): 28-31.
[5] WU Xiaohong. Ultra-thin stainless steel piece of laser seal welding technology research [J]. Journal of welding technology, The Lancet, 2014 (05): 41-43.
[6] Huo Hongwei, Hu Haijun, Li Zhiguo, et al. Microstructure and properties of laser-welded lap joint of 304 stainless steel sheet [J]. Electric Welding Machine, 2016, 46 (03): 44-47.
[7] Luo Ziyi, Han Shanguo, and Chen Yongcheng. The Influence of Process Parameters on Weld Formation and Tensile Properties in Laser-Arc Hybrid Welding [J]. Materials Reports, 2019, 33 (13): 2, 146-2, 150.
[8] Chen Yinyin. Laser Welding Process for 0.05 mm Thick 316 Stainless Steel [J]. Journal of Netshape Forming Engineering, 2021, 13 (05): 187-192.
[9] Chen Shuqin, Wei Xin, and Zhao Jiekui. Study on Laser Welding Process of 6 mm Thick 304 Austenitic Stainless Steel [J]. Machinery Design & Manufacture, 2021 (09): 55-58.

Investigation of the wetting behavior and evaluation of the warm mixing effects of high viscosity modified asphalt with bio-based warm mixing agent

Qianqiao Zhao[1], Lei Guo[2], Haiyu Wang[3], Zijian Wang[4] and Ruibo Ren[3*]

[1]Qingdao University of Technology, Qingdao, Shandong, 266033, China
[2]Jimo District Housing and Urban-Rural Development Bureau, Qingdao, Shandong, 266200, China
[3]Shandong Jianzhu University, Jinan, Shandong, 250101, China
[4]Jimo District Transportation Bureau, Qingdao, Shandong, 266200, China

[*] Corresponding author's e-mail: rrbgq@sdjzu.edu.cn

Abstract: In this study, the spreading and wetting behavior of bio-based warm-mix high viscosity modified asphalt and its warm-mixing effect were comprehensively investigated by using bio-based warm-mix as a modified material through laboratory preparation and performance testing. It was found that the bio-based warm-mixing agent could significantly improve the construction ease of high-viscosity asphalt, reduce its viscosity, and improve the adhesion performance with aggregates. Through the master curve analysis, BBR test, contact angle measurement, and viscosity reduction effect evaluation, the results showed that the bio-based warm-mixed high viscosity modified asphalt can effectively reduce the construction temperature, reduce energy consumption, and improve the construction ease while maintaining the excellent road performance. In addition, the introduction of a bio-based warm mix enhances the aging resistance and water damage resistance of asphalt. This study provides a theoretical basis and technical support for the warm mix technology of high-viscosity modified asphalt, which promotes the development of road engineering in the direction of environmental protection and sustainability.

1. Introduction

In modern road construction, the performance of asphalt mixture plays a decisive role in the quality and service life of pavement. The utilization rate of plant-mixed hot regeneration technology for old materials in practical engineering applications is generally lower than 30% [1], and warm-mix asphalt technology is an environmentally friendly technology that reduces energy consumption, lowers the construction temperature, and reduces the emission of greenhouse gases [2-5], and has become a research hotspot in the field of road engineering.

High viscosity modified asphalt (HVMA) shows excellent performance under heavy traffic loads and high-temperature environments due to its excellent high-temperature stability and deformation resistance. Li et al. [6] carried out a 60°C dynamic viscosity and dynamic shear rate scanning test on three different viscosities of asphalt and found that high-viscosity asphalt using a dynamic shear rate scanning test is the most appropriate; Li [7] used an SBS modifier, stabilizer and other additives as raw materials to prepare high viscosity modified asphalt, and studied the effect of components on the storage stability of high viscosity asphalt, conventional indicators and microscopic characteristics.

However, the high viscosity characteristics of high-viscosity modified asphalt also cause difficulties in construction and high energy consumption. Bio-based warm mixes, as a new type of environmentally friendly additive, have been extensively studied by scholars [8-14]. In the field of building science, Ferroukhi et al. [15] thermally added organic and non-paraffinic phase change materials (PCM) to cannabis concrete; Borhan et al. [16] investigated the effect of waste cylinder oil on the properties of cold-mix asphalt mixtures; Lu et al. [17] evaluated the effect of five different oil modifiers on the low-temperature cracking resistance of asphalt.

The purpose of this paper is to study the effect of bio-based warm mixing agent on the performance of high viscosity modified asphalt, to study its spreading and wetting behavior, and to provide theoretical basis and technical support for the warm mixing technology of high viscosity modified asphalt, and to promote the development of road engineering in a more environmentally friendly and sustainable direction.

2. Materials and Test Methods

2.1 Materials

Base asphalt is Jingbo 70 # base asphalt, is a good performance, suitable for a variety of road projects, modified asphalt; high viscosity asphalt prepared from the base asphalt, instant high viscosity modifier, and bio-based stabilizers, instant high viscosity modifier and bio-based stabilizers are developed independently of the subject group; due to the self-developed bio-based warm mixing agent and asphalt has a good affinity and is easy to dispersion, only the high viscosity asphalt heating to 180°C ~ 185°C to add bio-based warm mixing agent for stirring 20min, can produce bio-based warm mixing high viscosity asphalt.

2.2 Experimental Methods

2.2.1 Master Curve. Asphalt belongs to viscoelastic materials, and its mechanical properties do not depend on the temperature effect or time effect alone, but on the equivalent effect of time and temperature acting together, i.e., the time-temperature equivalence principle. Following the principle of time-temperature equivalence, the reference temperature is selected, the displacement factor is calculated, and the parameters measured at different temperatures and frequencies are coupled into a smooth curve, i.e., the master curve [18]. The rheological properties of different asphalts were tested using a dynamic shear rheometer, and the master curves were fitted by the Christensen-Anderson-Marasteanu (CAM) model.

2.2.2 Contact Angle Measurement. The tests were conducted by the seated-drop method, and the initial test temperatures of 170°C, 180°C, 190°C, and 200°C were set for the four materials, namely, HVA, WHVA35, WHVA70, and WHVA105, while the initial test temperatures of 90°C, 100°C, 110°C, and 120°C were set for MA, WMA35, WMA70, and WMA105, respectively. The measurement duration of each group of samples was set to 1 minute, and the test interval was accurate to 0.1 seconds to ensure the accuracy and reliability of the data.

2.2.3 Cooling Gradient. The rotary compaction method was used to mold the specimens, and the baseline mixing and molding temperatures of hot-mixed high-viscous asphalt mixtures (binder is HVA) were taken as 185/180°C. The temperature of the hot-mixed high-viscous asphalt mixtures was taken as 185/180°C. According to the different amounts of warm mixing agent contained in the bio-based warm-mixed high viscosity asphalt, the following rules are to determine the temperature reduction range: WHVA35 molding temperature reduction range of 0°C~10°C, WHVA70 molding temperature reduction range of 0°C~20°C, WHVA105 molding temperature reduction range of 0°C~30°C for the test.

3. Results and Discussion

3.1 Preparation and Performance Differences of Bio-Based Warm Mix High Viscosity Modified Asphalt

3.1.1 Preparation of Bio-Based Warm Mix High Viscosity Modified Asphalt. In the process of preparing bio-based warm mix high viscosity asphalt, it needs to heat the high viscosity asphalt to 180°C to 185°C, and then add the bio-based warm mix and stir for 20 minutes to get the desired bio-based warm mix high viscosity asphalt. For the preparation of the finished warm mix matrix asphalt, it is only necessary to adjust the heating temperature to 160°C to 165°C to complete. In the preparation process, the dosage of the bio-based warm mix was set at 0.35%, 0.7%, and 1.05%, respectively. The three prepared bio-based warm mix high viscosity asphalts were named WHVA35, WHVA70, and WHVA105, respectively, while the high viscosity asphalts without the addition of the bio-based warm mixing agent were named HVA. Similarly, the three bio-based warm mix matrix asphalts were named WMA35, WMA70, and WMA105, respectively, while the matrix asphalts without the addition of the bio-based warm mixing agent were named MA.

3.1.2 Test Results of Three Major Indicators

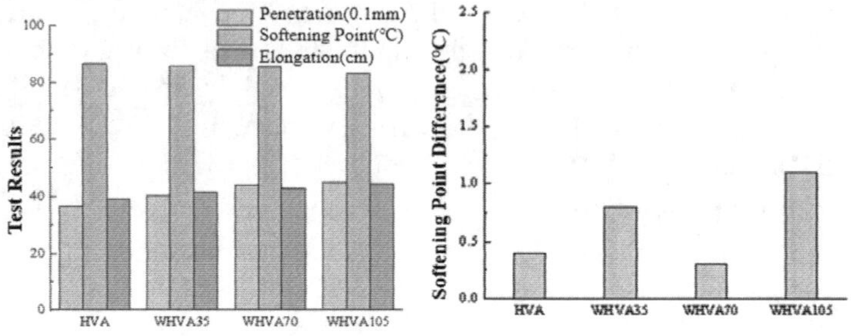

(a) (b)

Figure 1. Warm mix high viscosity asphalt properties.

From Figure 1(a), it can be seen that WHVA35, WHVA70, and WHVA105 all exhibit an increase in needle penetration and an increase in ductility compared to HVA, while the softening point shows a decreasing trend, especially WHVA105, which shows the most significant change in performance. Given that the difference between these three bio-based warm-mix high viscosity asphalts lies only in the amount of warm-mix, it can be inferred that the differences in these performance indexes are mainly attributed to the difference in the amount of warm-mix. As can be seen from Figure 1(b), WHVA70 has the best storage stability, with the smallest segregation softening point difference among the three biobased warm-mix high-viscosity asphalts. This excellent storage stability may be attributed to the specific biobased chain structure in the biobased warm mixes, which are capable of strong interactions with the asphalt molecules, promoting the formation of a more stable network structure.

3.1.3 Workability Analysis

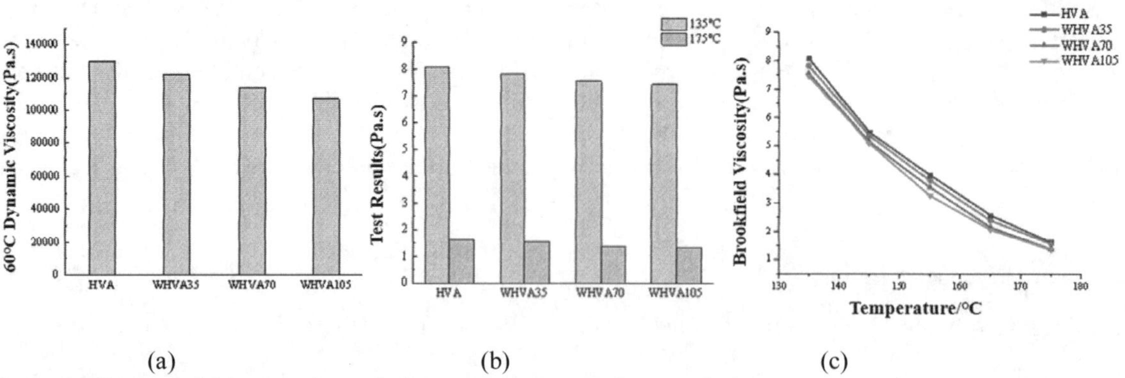

(a) (b) (c)

Figure 2. Warm mix high viscosity asphalt construction ease of use analysis.

From Figures 2(a) and 2(b), it can be seen that the dynamic viscosity at 60°C and the Brinell viscosity at 135°C and 175°C of the three asphalts with the addition of biobased warm-mixing agent decreased compared to the unmodified high viscosity asphalt (HVA), and the decrease in the WHVA105 in particular was more significant. This phenomenon suggests that the introduction of a bio-based warm mixing agent effectively reduces the viscosity of high-viscosity asphalt and thus enhances its fluidity. The Brinell viscosities of bio-based warm mix high viscosity bitumen in Figure 2(c) are generally lower than that of HVA and show the order of WHVA105< WHVA70< WHVA35< HVA. This indicates that there are differences in the effects of different dosages of bio-based warm mix on the performance of high-viscosity asphalt.

3.1.4 Master Curve Results of Complex Modulus

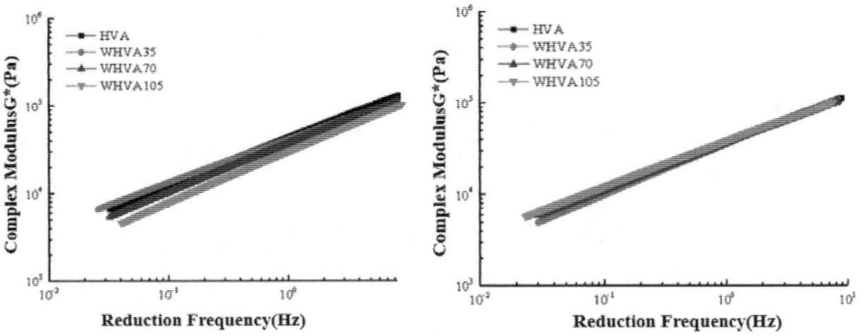

(a) (b)

Figure 3. Master curve of complex modulus of high viscosity asphalt before and after aging.

As can be seen in Figure 3(a), the differences between the complex modulus principal curves of the four high-viscous asphalts are not significant before the TFOT treatment, and the differences between these principal curves gradually decrease with the increase of shear frequency. In the high-temperature and low-frequency region, the complex modulus master curves of the four high-viscous asphalts are arranged in the order of WHVA35 > HVA > WHVA70 > WHVA105, which reveals that WHVA35 shows the best resistance to deformation under the high-temperature and low-frequency loading conditions and that WHVA105 shows the best resistance to deformation under the low-temperature and high-frequency loading conditions. In the low-temperature and high-frequency regions, even though the complex modulus master curves of the four high-viscous asphalts are close to coinciding, the complex modulus master curve of WHVA105 has a slightly lower complex modulus value than the other samples. This slight difference suggests that WHVA105 has superior cracking resistance under low temperatures and high-speed driving conditions. From Figure 3(b), it can be seen that after short-term aging, the complex modulus master curves remain highly consistent in the high-temperature low-frequency region and low-temperature high-frequency region, and there is no obvious morphological change.

3.1.5 BBR Test Results

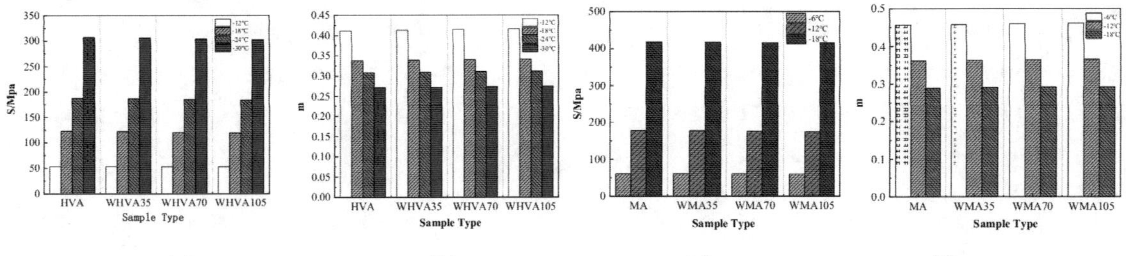

(a) (b) (c) (d)

Figure 4. Evaluation of the low-temperature performance of warm mix high viscosity asphalt.

Based on the analytical results in Figure 4, it can be observed that under the same temperature conditions, the three high-viscosity asphalts, WHVA35, WHVA70, and WHVA105, which are supplemented with biobased warm mixes, do not exhibit significant differences in the values of the modulus of strength, S, and the creep rate, m, when compared to the unmodified high viscosity asphalts (HVAs). However, these asphalt samples fail to meet the specification requirements at a low temperature of -30°C, and therefore, it can be determined that their failure temperature is -30°C. This phenomenon may be related to the fact that the asphalt samples underwent simultaneous aging of the bio-based warm mixes during the aging process, which resulted in the volatilization of the components in the asphalt that maintain the low-temperature cracking resistance properties and thus resulted in the lack of significant changes in the modulus of strength and the creep rate of the high-viscosity asphalt samples.

3.2 Analysis of Wetting Behavior for Bio-Based Warm Mix High Viscosity Modified Asphalt

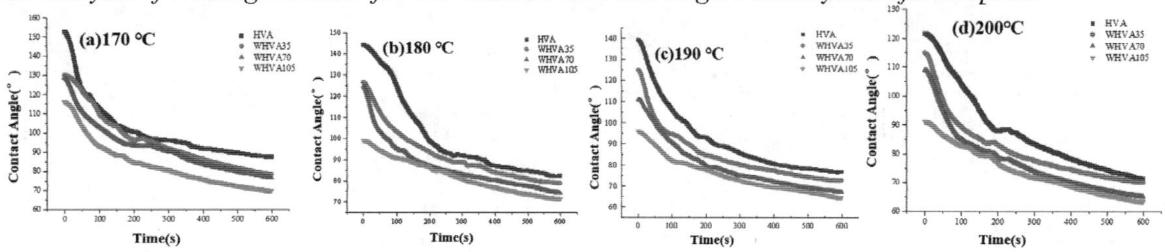

Figure 5. Time-dependent diffusion curves of contact angle of warm-mixed high-viscosity asphalt at different temperatures.

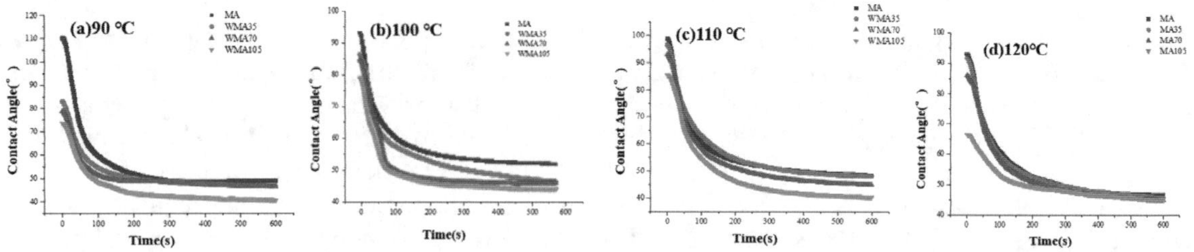

Figure 6. Time-range diffusion curves of contact angle of warm matrix asphalt at different temperatures.

From the analyzed results in Figures 5 to 6, the contact angles of the biobased warm mix high viscosity asphalts are generally smaller compared to the high viscosity asphalts (HVA) in the order of WHVA105, WHVA70, WHVA35, and HVA, with the maximum decrease in the contact angle of about 40°. Similarly, the three bio-based warm mix matrix asphalts showed a trend of smaller contact angles compared to the matrix asphalt (MA) in the order of WMA105, WMA70, WMA35, and MA. Since the difference between the asphalt samples was only in the amount of warm mix blended, it can be deduced that the incorporation of the warm mix significantly enhanced the surface free energy of the asphalts. The enhanced surface activity endowed the bio-based warm mix asphalt with the ability to spread rapidly and adhere firmly to the aggregate surface at lower temperatures.

3.3 Analysis of the Viscosity Reduction Effect of Bio-Based Warm Mix High Viscosity Modified Asphalt Mixture

3.3.1 Heat Dissipation Rate Analysis

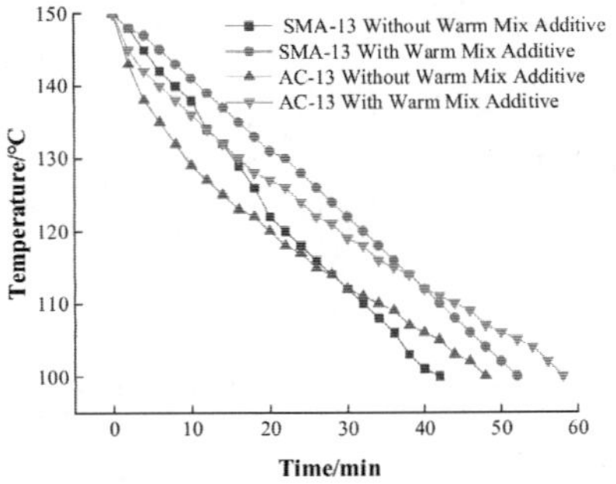

Figure 7. Effect of mineral type on the rate of heat dissipation in asphalt mixtures.

Two kinds of binders, matrix asphalt and SBS-modified asphalt, and mineral composition, AC-13 and SMA-13 two kinds of gradation are used, warm mixing agent dosage is 0% and 0.3%, respectively, mixing temperatures are 150°C, and the impact of asphalt mixtures heat dissipation rate is explored. As can be seen from Figure 7, under the condition of the same warm mix content, the heat dissipation rate of the AC-13 asphalt mixture is slower than that of the SMA-13 asphalt mixture. AC-13 asphalt mixture is a suspended dense structure, with a large degree of densification, a small void ratio, and a small area of heat dissipation; whereas, SMA-13 asphalt mixture is a skeleton dense structure, and after compaction, the void ratio is larger than that of AC-13 asphalt mixture, and the area of heat dissipation is larger than that of AC-13 asphalt mixture. Therefore, the heat dissipation rate of the SMA-13 asphalt mixture is faster than that of the AC-13 asphalt mixture, with a good viscosity reduction effect. A biological warm mix is added to reduce the viscosity of asphalt, reduce the frictional resistance within the mixture, make it easier to transfer heat from the inside of the mixture to the surface, and accelerate the dissipation of heat, thus improving the heat dissipation rate. Through the above analysis, in areas with low temperatures and cold, and small traffic load, priority can be given to the use of AC-13 mineral composition, which can prolong the construction time and increase the construction and ease of construction, as well as delay the aging of asphalt pavement and slow down its water damage.

3.3.2 Performance Evaluation of Bio-Based High Viscosity Asphalt Mixtures

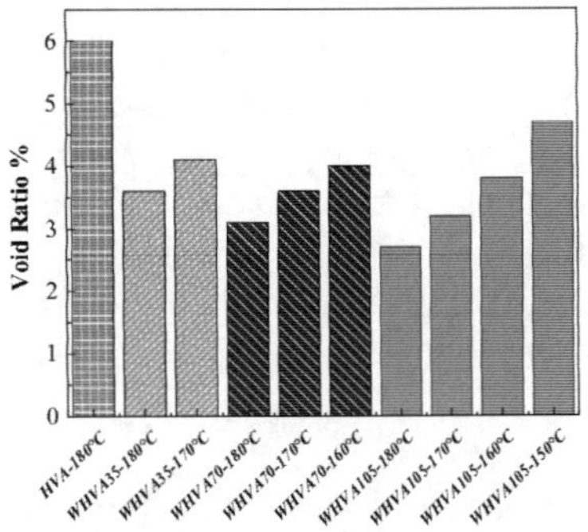

Figure 8. Voidness of asphalt mixtures at different temperatures.

As can be seen from Figure 8, the void ratio of the specimens molded using HVA as a binder was 6.0%. In comparison, when WHVA35, WHVA70, and WHVA105 were used as the mixes for binder molding, all were able to obtain lower void fractions as the temperature reduction was controlled within 10°C, 20°C, and 30°C, respectively.

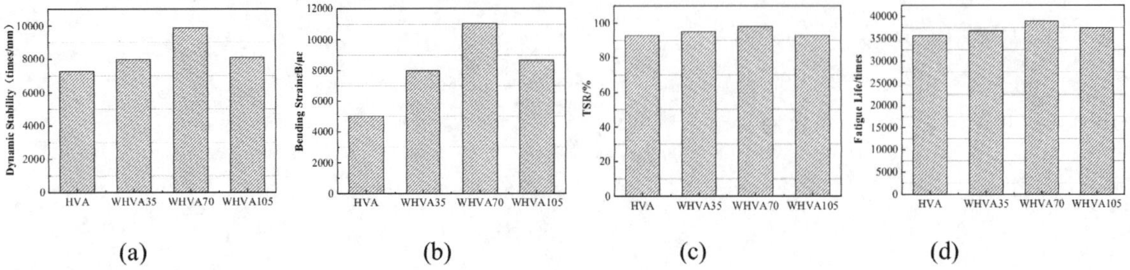

Figure 9. Evaluation of high and low-temperature performance, water stability, and fatigue characteristics of different asphalt mixtures.

In Figure 9(a), rutting test results show that WHVA70 exhibited excellent high-temperature rutting resistance at 60°C, followed by WHVA105 and WHVA35, while HVA was ranked fourth; this result reveals the potential of bio-based warm mixes in improving the high-temperature stability of asphalt mixtures. Low-temperature bending test data in Figure 9(b) shows that the bending and tensile strains of WHVA35, WHVA70, and WHVA105 were improved by 59%, 119%, and 72%, respectively, compared to HVA, with WHVA70 showing a particularly significant performance enhancement, which demonstrates its excellent bending toughness under low-temperature conditions. As can be seen in Figure 9(c), the results of the freeze-thaw splitting test further confirmed the effectiveness of the bio-based warm mixes in improving the water damage resistance of the asphalt mixtures, and all three showed better TSR values of more than 90% compared to HVA, and all of them demonstrated better resistance to water loss, with WHVA70 having the largest TSR value, showing the best water stability performance. In Figure 9(d), fatigue test results of four-point bending trabeculae showed that the fatigue resistance of WHVA35, WHVA70, and WHVA105 was improved by 3%, 9%, and 5%, respectively, when compared to HVA, with WHVA70 showing the most significant fatigue performance improvement.

Comprehensively analyzed above, the bio-based temperature-mixed high-viscosity modified asphalt mixtures showed significant advantages in high-temperature stability, low-temperature bending toughness, water damage resistance, and fatigue resistance, which is conducive to improving the viscosity reduction of asphalt mixtures, and the performance of WHVA70 is the most superior.

4. Conclusion

This paper discusses the preparation and performance differences of bio-based temperature-mixed high-viscosity modified asphalt, analyzes the wetting behavior of temperature-mixed high-viscosity modified asphalt according to the contact angle test results, and analyzes the viscosity-reducing effect of temperature-mixed high-viscosity asphalt through the asphalt mixture heat dissipation rate, void ratio, and rutting test results, and draws the following conclusions:

(1) In terms of macroscopic properties, the asphalt was analyzed by three major indicators: Brinell viscosity, 60°C dynamic viscosity, and segregation. The results show that the bio-based temperature-mixed high-viscosity asphalt exhibits better low-temperature crack resistance and construction ease compared with high-viscosity asphalt.

(2) In terms of microscopic properties, the wetting behavior of bio-based temperature-mixed high-viscosity modified asphalt was analyzed by contact angle test. The results show that compared with high-viscosity asphalt, bio-based warm mix high-viscosity asphalt and aggregate have greater surface energy, which makes it easier to realize the spreading and wetting effect. The asphalt in bio-based warm mix high-viscosity asphalt has better compatibility with the modifier, and there is a chemical reaction between the warm mix agent and other components.

(4) Based on the analysis of heat dissipation rate, the void ratio of the mixture at different temperatures, the viscosity reduction effect of warm-mixed high-viscosity modified asphalt mixtures was analyzed, and it was found that the addition of biobased warm-mixing agent had a better effect on improving the viscosity reduction effect of asphalt mixtures.

(5) The road performance of warm mix high viscosity asphalt mixture was analyzed. The results show that compared with hot-mixed high-viscosity asphalt mixtures, warm-mixed high-viscosity asphalt mixtures still have better high-temperature stability, low-temperature cracking resistance, water damage resistance, and fatigue resistance.

As a new type of environmentally friendly additive, bio-based temperature mixer has an important research value and application prospect by reducing the viscosity of high-viscosity modified asphalt and improving its spreading and wetting behaviors, thus realizing low-temperature construction while maintaining the excellent performance of asphalt mixtures.

References

[1] Li Q, Lu Y, Wang J, et al. (2022) Performance of secondary aging of warm-mix recycled asphalt binder. Journal of Transportation Engineering, 22(4): 102-116.
[2] Editorial Department of China Highway (2013) A review of academic research in Chinese road engineering. China Journal of Highway, 26(3): 1-36.
[3] He H, Wang Z, Sun X, et al. (2014) Development of novel warm-mix modified asphalt and its performance in road applications. Journal of Building Materials, 17(5): 927-93.
[4] Song W, Huang B, and Shu X (2018) Influence of warm-mix asphalt technology and rejuvenator on the performance of asphalt mixtures containing 50% reclaimed asphalt pavement J . Journal of Cleaner Production, 192: 191-198.
[5] Yousefi A, Behnood A, Nowruzi A, et al. (2021) Performance evaluation of asphalt mixtures containing warm mix asphalt (WMA) additives and reclaimed asphalt pavement (RAP) J . Construction and Building Materials, 268: 121200.
[6] Li L, Geng H, and Sun Y (2010) Evaluation methods and indicators for the viscosity of high-viscosity asphalt. Journal of Building Materials, 13(03): 352-356+362.
[7] Li Z (2019) Preparation and research on high-viscosity modified asphalt D. Guangzhou: East China University of Science and Technology.
[8] Li S (2021) Promoting the new energy revolution to achieve carbon neutrality goals. People's Forum: Academic Frontier,(14): 42-51.
[9] Cong X, Liu T, Liang D, et al. (2024) Prospects for the application and recycling of novel bio-based materials. Green Packaging, (03): 21-25. DOI:10.19362/j.cnki.cn10-1400/tb. 03.002.
[10] Chen G, Chen X, Xu J, et al. (2010) Development of environmentally friendly bio-based materials. New Materials Industry, (03): 54-62.
[11] Jiang H, Li X, Gao Q, et al. (2024) Situation analysis of bio-based materials research and development. China Biotechnology Engineering Journal, 44(01): 142-151. DOI:10.13523/j.cb.2307034.
[12] Tan T, Su H, and Yang J (2012) Progress in the industrialization of bio-based materials. Progress in China Materials, 31(02): 1-6.
[13] Wei X and Lin C (2022) Development strategies and suggestions for the bio-based materials industry in China. Chemical Industry, 40(02): 6-11+23.
[14] Bai JY, Lin XF and Yin ZQ (2020) Current status of global bioindustry development and policy implications J. Journal of Biotechnology, 36(08): 28-35.
[15] Ferroukhi, MY, Mesticou Z and Larbi AS (2023) Impact of microencapsulated phase change material on a bio-based building composite, hygrothermal and mechanical behavior. 133925.
[16] Borhan M N, Suja E, Ismail A, et al. (2007) Used Cylinder Oil Modified Cold—Mix Asphalt Concrete J. Journal of Applied Sciences,7(22): 3485-3491.
[17] Lu Q (2016) Research on modifying asphalt with waste oils and its low-temperature performance. Highway Engineering, 41(3).
[18] Liu S (2020) Study on the mechanism of microscopic composition structure of asphalt on low-temperature performance D. Harbin: Harbin Institute of Technology.

Journal of Physics: Conference Series **3009** (2025) 012070 doi:10.1088/1742-6596/3009/1/012070

Study on the strength of steel pedestal for transporting 5-20 t coil steel in 70-t universal gondola

Yanchun Huang and Yanhui Han *

School of Traffic and Transportation, Beijing Jiaotong University, No. 3 Shangyuancun, Haidian District, Beijing, China

E-mail: yhhan@bjtu.edu.cn (* Corresponding author)

Abstract. With the speed increase of China's railways and the continuous increase in the weight and volume of coil steel, the requirements for loading and securement of coil steel have been further improved to ensure transportation safety. China Railway also required the complete elimination of straw products for loading and securement of coil steel, and all coil steel should be loaded and secured by using steel pedestals for transportation. Therefore, it is very necessary to study the special steel pedestal for coil steel. This article analyses the advantages of using a 70-t universal gondola with a horizontal loading of 5-20 tons of steel coils. In response to the transportation needs of 5-20 tons of steel coils, a new type of special steel pedestal is designed. To verify the rationality of the steel pedestal design, a solid model is created, simulating the most extreme loading conditions. Finite element analysis is applied to obtain the stress distribution diagram of the steel pedestal. The simulation results demonstrate that only the edge of the steel pedestal and the contact surface of the steel pedestal exceed the allowable stress of the selected material and can be optimized in the actual design process.

1. Introduction

The steel industry, as an important part of China's heavy industry, plays a pivotal role in the national economy. At present, universal gondola are the main type of vehicles used for railway coil steel transportation in China. Since 2006, all newly manufactured universal gondolas have been upgraded to 70 t, which optimizes the chassis structure, making the central load capacity reach 25 t. Compared to the 60-t and 61-t universal gondola, the 70-t universal gondola is suitable for a wider range of sizes and weights of coil steel. While improving the efficiency of steel enterprises, it also further enhances the capacity of railway coil steel transportation in China.

Loading and securement are an important part of freight transportation. To efficiently and safely complete transportation tasks, the former Ministry of Railways formulated the "The Rules for Railway Freight Loading and Securement" [1], which comprehensively and systematically stipulated the technical conditions for railway cargo loading and securement. The corresponding loading and securement plans have been finalized in Annex 1 "The Scheme for Loading and Reinforcement of Railway Goods" [2] and Annex 6 "The Trial Operation Plan and Trial Operation Materials for Loading and Reinforcement of Railway Goods" [3].

When using a steel pedestal to load and secure coil steel, the steel pedestal is directly rigidly secured onto the gondola, while the coil steel is directly secured onto the steel pedestal. The steel pedestal can be reused, which not only ensures transportation safety, but also has low material costs, short securement time, and high efficiency [4-6]. To reduce the repetitiveness of steel pedestal design work and minimize the consumption of manpower and material resources, simulation methods can be applied to the steel pedestal design process.

North America and Germany researched railway transportation of coil steel early. In 1992, a specialized truck called Colishield was jointly developed for American Steel Company. The main feature of this type of vehicle is the installation of a groove structure inside the vehicle, which increases the range of suitable coil steel [7]. In 2004, Johnstown Corporation in North America developed a new type of coil steel transport vehicle. The coil diameter suitable for this vehicle is 760-2, 030 mm [8]. In Germany, coil steel is mainly transported by using specialized vehicles. There are mainly eleven types of specialized vehicles for transporting rolled steel in Germany [9-10].

China has also conducted specialized research on coil steel pedestals. Zhuzhou Rolling Stock Factory designed a new type of specialized loading and securement device for transporting coil steel. This design is in the form of a "V" groove and can prevent coil steel from rolling [11]. The Institute of Transportation and Economics of China Academy of Railway Sciences designed a split-type steel pedestal for transporting coil steel [12]. Jiang designed a horizontal loading steel bracket [13].

This article intends to study the use of a 70-t universal gondola, coil steel with a weight of 5-20 t is loaded and secured by using a reusable steel pedestal, and SolidWorks is used for simulation analysis.

2. Scheme for horizontal loading of 5-20 tons of coil steel using a dedicated inclined strut for coil steel transportation

2.1 The advantages of using a steel pedestal to transport 5-20 tons of coil steel horizontally

At present, 70-t universal gondolas have become the main type of gondola in China, and the use of 70-t universal gondolas for transporting coil steel is an inevitable development trend.

(1) When the 60-t and 61-t universal gondolas are loaded at both ends and three places in the middle, the weight of the middle cargo shall not exceed 13 t. For coil steel weighing 20 tons, when each vehicle is loaded with 3 pieces, the loading weight in the middle will exceed the requirements of the plan, so only 2 pieces can be loaded, which will cause serious tonnage loss and increase transportation costs. The 70-t universal gondola has a wider range of transportable coil steel due to its central load capacity of 25 t.

(2) The use of a steel pedestal for transporting coil steel simplifies the securement process, and the steel pedestal can also be recycled, reducing transportation costs.

(3) 5-20 tons of coil steel are mostly cold-rolled coil steel, which can only be transported horizontally.

2.2 Scheme for horizontal loading of 5-20 tons of coil steel using a dedicated inclined strut for coil steel transportation

When setting the self-weight of the steel pedestal to 1 t and ideally using a 70-t universal gondola to load and secure 5-20 t of coil steel, the total weight of the coil steel placed on each end of the steel pedestal is at most 34 t. To analyze the most extreme stress condition of the steel pedestal, at least one 20-t coil steel shall be placed on the steel pedestal, the number of stressed inclined struts shall be reduced as far as possible, and the stress on a single inclined strut shall be increased; Only four inclined struts are allowed to be stressed in the scheme of one vehicle with four pieces, so the loading scheme selected in this paper is one vehicle with four pieces. Two pieces of coil steel are loaded in the middle of a single row on the steel pedestal. The coil steel placed near the "V" groove at the end of the vehicle is 20 t, and the coil steel placed in the other "V" groove is 14 t.

Loading 4 pieces in one vehicle means placing a steel pedestal at each end of the vehicle, with 2 pieces of coil steel loaded in a single row in the center on each steel pedestal; The difference in total weight between the two steel pedestals of coil steel should be less than 2 t, and the difference in weight between two pieces of coil steel at symmetrical positions on both sides of the vehicle's horizontal centerline should be less than 1 t. The total weight of each steel pedestal and coil steel should be less than half of the vehicle's marked load capacity, as shown in the loading diagram in Figure 1.

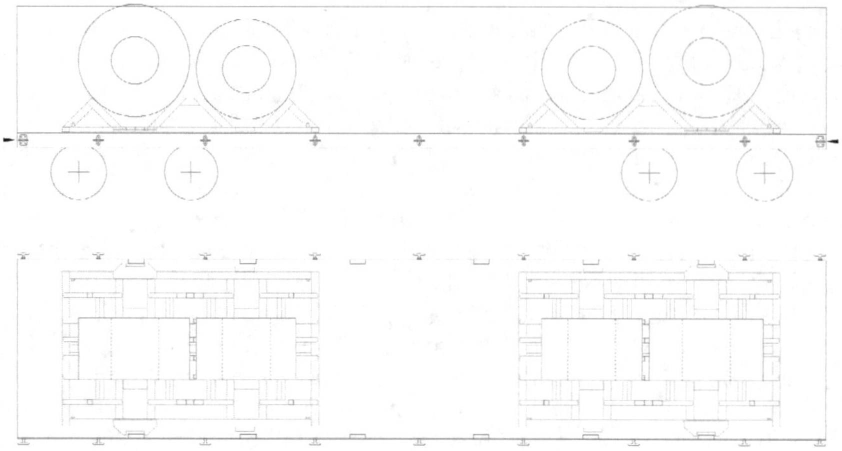

Figure 1. Loading 4 pieces in one vehicle.

After determining the weight of the loaded coil steel, to consider the most extreme loading situation, the coil diameter of the 20-t coil steel is taken as 1, 800 mm, and the contact point between the coil steel and the inclined strut is exactly 50 mm away from the top of the inclined strut. While meeting the safety requirements for loading and securement, the plate width should be minimized to reduce the number of inclined struts contacting the coil steel; The width of the 14-t coil steel plate is 1, 600 mm. When only four inclined struts are stressed, the coil diameter should be reduced as much as possible to increase the vertical distance between the contact point of 14-t and 20-t coil steel and the inclined strut. Referring to the specifications of cold-rolled coil steel produced by various steel plants, the inner diameter of 14-t and 20-t coil steel is taken as 580 mm.

Due to the following relationship between the weight, coil diameter, and plate width of cold-rolled coil steel:

$$3.14 \times \left(\frac{R^2-r^2}{1{,}000{,}000}\right) \times \rho \times \frac{d}{1{,}000} = m \quad (1)$$

where R is the outer radius of coil steel, mm; r is the inner radius of the rolled steel, taken as 290 mm; ρ is the density of rolled steel, taken as 7.9 t/m^3 in this article; d is the plate width of rolled steel, mm;

m is the weight of coil steel, t.

Substituting the above data into the equation, the width of the 20-t steel coil is 1, 107 mm, and the outer diameter of the 14-t steel coil is 1, 320 mm. According to the selected loading scheme, after centering the coil steel on the pedestal, only the bearing surface of the four inclined struts in the middle of the pedestal contacts the surface of the coil steel.

3. Solid modeling and simulation analysis of coil steel pedestal

3.1 Design of steel pedestal for 70-t universal gondola shipping coil steel

To meet the current demand for 5-20 tons of coil steel for the horizontal loading of a 70-t universal gondola, this article designs a new type of coil steel horizontal loading transportation dedicated steel pedestal. The basic principles of steel pedestal design include: ensuring transportation safety, minimizing the weight of the steel pedestal itself, facilitating stacking, minimizing the production cost, protecting the coil steel, adapting to a wide range of coil steel, and facilitating the retrieval and placement of lifting equipment. The coil steel frame adopts a "V"-shaped groove securement structure to effectively fix the coil steel. We set up a blocking structure to clamp the reinforcing seat inside the vehicle, ensuring that the steel pedestal is relatively fixed to the vehicle body.

The inclined strut mainly is as follows:

(1) The chassis is welded from 120 × 80 × 6 mm rectangular tubes. The width of the chassis is 2, 500 mm.

(2) The blocking has a length of 700 mm and a width of 250 mm. The designed length of the blocking is 300 mm and the width is 55 mm, which just fits the securement seat inside the 70-t universal gondola. The entire structure is made of a 12-mm thick steel plate.

(3) It includes a triangular inclined strut and a trapezoidal inclined strut. Inclined struts are the main components of the "V" groove. Steel plate is welded on the inclined strut, with a length of 1, 800 mm and a width of 500 mm; According to the distance between the inclined strut, the transverse connecting beams designed are 330 mm and 220 mm.

3.2 *Force analysis of the loading process of the coil steel frame*

Due to the different force situations of the steel pedestal under static load and impact, force analysis is divided into static and impact states for calculation.

3.2.1 Force analysis under static state. In a stationary or moving state, the force exerted by the coil steel on the steel pedestal is mainly due to the gravity of the coil steel itself. The weight of the coil steel determines the magnitude of the force, and the coil diameter determines the position of the applied force.

Based on the parameters of the coil steel determined in the previous text, the force exerted by 20 tons of coil steel on the steel pedestal is:

$$F_{1(20t)}=F_{2(20t)} = 20 \times cos45° \times 9.8 = 138.593\ (kN) \tag{2}$$

Similarly, the force exerted by 14 tons of coil steel on the steel pedestal is:

$$F_{1(14t)}=F_{2(14t)} = 14 \times cos45° \times 9.8 = 97.015\ (kN) \tag{3}$$

3.2.2 Force analysis under impact state. In the impact state, the main calculation includes the longitudinal inertia force of the coil steel, the frictional force between the bottom surface of the steel pedestal and the vehicle floor, and the horizontal longitudinal force at the stop.

(1) The longitudinal inertia force.

$$T = t_0 Q\ (kN) \tag{4}$$

where Q is the weight of goods, t; t_0 is longitudinal inertia force per ton of cargo, kN/t.

The self-weight of a 70-t universal gondola is 23.8 t, with two coil steel weighing 68 t and two steel pedestals weighing 2 t. Therefore, the total weight is taken as 93.8 t.

The securement between the steel pedestal and the coil steel is rigid, and the longitudinal inertial force per unit mass of round steel is:

$$t_0 = 26.69 - 0.13Q_{gross} = 14.496\ (kN/t) \tag{5}$$

where Q_{gross} is the total weight.

The longitudinal inertial force acting on the single round steel weighing is:

14-t coil steel:

$$T_{14} = t_0 Q = 14.4969 \times 14 = 202.944\ (kN)$$

20-t coil steel:

$$T_{20} = t_0 Q = 14.4969 \times 20 = 289.942\ (kN)$$

The longitudinal inertial force experienced by the coil steel and the entire frame is:

$T_{total} = t_0 Q = 14.4969 \times 35 = 507.36\ (kN)$

(2) The longitudinal friction force.

$$F_{lf} = 9.8\mu Q = 9.8 \times 0.3 \times 35 = 102.9\ (kN) \tag{6}$$

where μ is the friction coefficient between steel and steel.

(3) longitudinal horizontal force.

To prevent longitudinal movement of the coil steel and the entire frame on a horizontal plane, the horizontal longitudinal force that needs to be supported by the clamp is:

$$\Delta T = T - F_{lf} = 507.36 - 102.9 = 404.46\ (kN) \tag{7}$$

When the coil steel is placed on the pedestal, it is tangent to the bearing surface of the inclined strut. Only the middle four inclined struts are in contact with the coil steel. The gravity and longitudinal inertia forces of the coil steel calculated above act on the contact surface set here.

3.3 Solid modeling and simulation analysis of coil steel frame based on SolidWorks

3.3.1 Solid modeling of coil steel frame

(1) Solid modeling of each component

Based on the parameters of each component in the steel pedestal design diagram, we use the component design module to conduct solid modeling of each component. The assembly function module is used to assemble the steel pedestal as a whole based on the connection characteristics between the components, and interference checks are performed on the rationality of the assembly. The final overall model of the steel pedestal is shown in Figure 2.

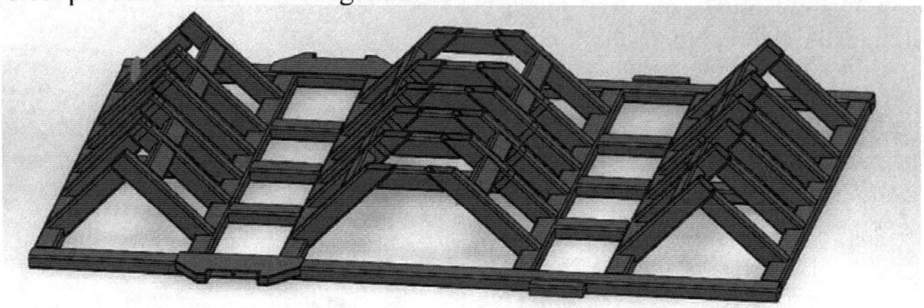

Figure 2. The overall model of the steel pedal.

3.3.2 Simulation analysis of coil steel frame. SolidWorks Simulation plugin is used for simulation analysis.

(1) Creation of a new example

We open the model diagram of the bracket in SolidWorks software, open the Simulation plugin, create a new example for the model, and select the static stress analysis in the conventional simulation.

(2) Application materials

Q450 steel has excellent physical and mechanical properties. Due to its high strength, good processability, and corrosion resistance, it is widely welcomed in the field of mechanical equipment manufacturing. According to the design requirements of the blueprint, steel pedestals are made of Q450 steel material. We define the properties of the material based on the standard parameters of Q450.

(3) Fixtures

In the actual transportation and loading process, there is a certain gap between the stopper and the securement seat. When the vehicle is impacted, there will be a slight relative sliding between the steel pedestal and the vehicle, causing friction on the bottom surface of the steel pedestal. Therefore, the fixture selected in this article is a "roller/sliding rod", and the plane of the added fixture can slide freely on the plane, but cannot move in the vertical direction of the plane. The position for adding the fixture is on the lower surface of the base frame.

(4) Load

The forces exerted on the steel pedestal are divided into the gravity and longitudinal friction force exerted by the coil steel, the friction force between the chassis and the vehicle floor, and the horizontal and longitudinal forces at the stopper. The component force of 20-t coil steel gravity acting on the bearing surface of the inclined strut is 138.593 kN, and the component force of 14-t coil steel gravity acting on the bearing surface of the four middle inclined struts is 97.015 kN. The longitudinal inertia force of 20-t coil steel is 289.92 kN, and that of 14-t coil steel is 202.944 kN, which acts horizontally to the left on the triangular inclined strut. The longitudinal friction force of 102.9 kN acts on the lower surface of the chassis. When the vehicle is impacted, the side surface of the stopper is in close contact with the side surface of the securement seat, generating a horizontal longitudinal force of 404.46 kN, which acts vertically on the side surface of the stopper.

(5) Grid division

When dividing the grid, high-quality second-order solid tetrahedral elements are used as the element type, with a maximum element size of 196.868 mm and a minimum element size of 39.374 mm. The total number of nodes in the model is 315, 370, and the number of elements is 159, 109.

(6) Displacement of the model

The stress cloud is shown in Figure 3. The localized enlarged stress clipping diagram is shown in Figure 4.

Figure 3. Simulated stress cloud map of steel crosswise pedestal.

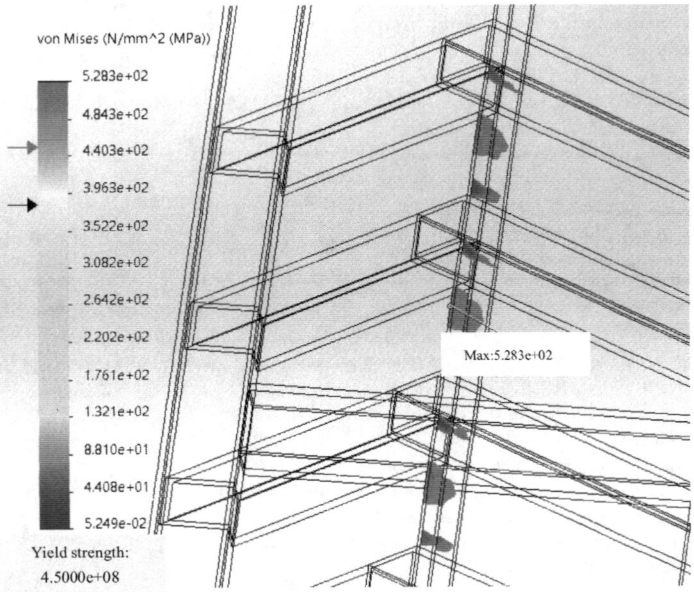

Figure 4. Localized enlarged stress clipping diagram.

From the figure, the maximum stress of the steel pedestal is 528 MPa, which is lower than the maximum stress of the steel pedestal before improvement. The stress is mainly concentrated on the contact surface between the steel plate and the coil steel. Referring to "Material and device for loading and securing freight Part 1: Goods turning rack" (TB/T 3079.1-2016), the allowable stress of Q450 under dynamic conditions is 382 MPa. The cutting function was used to cut the steel pedestal, and the stress distribution of the steel pedestal exceeding the allowable stress was obtained. Local amplification of the steel pedestal revealed that only a small part of the steel plate contact surface and edge, which were subjected to the gravity and longitudinal inertia force of 20 t of rolled steel, exceeded the allowable stress, while the stress in other parts of the steel pedestal was within the allowable stress range. The reason why a very small part of the steel pedestal exceeds the allowable stress may be due to insufficient support from the welded steel plate, resulting in stress concentration at the edge of the steel plate. Local stress concentration points can be optimized in the actual design process and will not be considered in finite element simulation temporarily. Therefore, the steel pedestal design can meet the transportation requirements of round steel.

4. Conclusion

This article proposes a loading and securing scheme for coil steel with a total weight of 5 t-20 t using a 70-t universal gondola and steel pedestal. Using SolidWorks for modeling and simulation, we analyze the obtained stress cloud map, and based on the simulation results, we determine that the structure of the steel pedestal can meet the requirements, and local stress concentration points can be processed in the actual design. Using steel pedestals to transport coil steel can not only improve safety but also reduce waste and make it more economical as steel pedestals can be reused. After production, actual stress testing can be conducted, and the simulation results can be compared with the measurement data to further verify the rationality of the design.

References

[1] China Railway Corporation. Rules for Railway Freight Loading and Securing [M]. Beijing: China Railway Press. 2015.
[2] China Railway Corporation. The Scheme for Loading and Reinforcement of Railway Goods [M]. Beijing: China Railway Press. 2015.
[3] China Railway Corporation. The Trial Operation Plan and Trial Operation Materials for Loading and Reinforcement of Railway Goods [M]. Beijing: China Railway Press. 2015.
[4] Li Jinsong. Practice and Thoughts on Transporting Coil Steel by Using Steel Seat [J]. Railway Freight Transport, 2012 (4): 34-38.
[5] Yue Sengen. Application and Suggestions on Roll Steel Transportation by Using Steel Seat [J]. Shanghai Railway Science & Technology, 2014, (1): 20-21, 38.
[6] Gao Junping. Study on Simulation Analysis and Test of Strength of Rolled-steel Pallet Transported by Railway [D]. Beijing Jiaotong University, 2012.
[7] Translated by Wu Yunxing. Special Trucks for Transporting Rolled Steel [J]. Railway Gazette International, 1992 (2): 117.
[8] Christopher Ytuarte. Development Prospects of Railway Freight Cars in U.S.A [J]. Foreign Rolling Stock, 2004, 41 (6), 11-13.
[9] Wang Dezhan. Thoughts on the use of composite material seat for the transportation of coiled steel [J]. Railway Freight Transport, 2006 (11): 39-40.
[10] Translated by Huang Wancheng. Specialized Truck for Transporting Steel Coils [M]. ETR, 1978, 1/2: 27-32.
[11] China South Locomotive & Rolling Stock Corporation Zhuzhou Rolling Stock Factory. Special Loading and Securement Device for Transporting Coil Steel (CN200710192635.6) [P]. 2007-12-18.
[12] The Institute of Transportation and Economics of China Academy of Railway Sciences. A Split Type Steel Pedestal for Transporting Coil Steel (CN201120019932.2) [P], 2011-08-30.
[13] Jiang Enquan. Horizontal loading steel bracket (CN201120082665.3) [P], 2011-11-08.

Static analysis of taper roller bearings based on finite element method

Jiahua Wu [1]* and Yu Peng [2]

[1] Century College, Beijing University of Posts and Telecommunications, Beijing, China
[2] China United Certification Center (Beijing) Co., Ltd. Beijing, China

*Corresponding author's e-mail: wujiahau@ccbupt.cn

Abstract. With the rapid development of China's social economy, the technology for oil and gas resource development is also continuously upgrading. In this process, various problems have emerged in special wells such as deep wells, high-angle wells, and horizontal wells. This paper mainly utilizes finite element simulation technology to simulate the mechanical performance of a certain drilling bearing under different working conditions, thereby effectively reducing the development cost and production cycle of the bearing. This paper models a certain drilling bearing, then theoretically analyzes its contact stress, and finally sets finite element parameters according to the working conditions and performs finite element analysis.

1. Introduction

In recent years, with the enhancement of national research and development capabilities, exploration and development equipment has also been updated in line with the trends of the times, leading to a qualitative leap in the quality of oil and gas wells [1]. Despite this, in regions with harsh geological environments, such as deep and ultra-deep wells, the quality of well completion is not particularly outstanding. The quality of well completion determines whether oil and gas wells can be applied in production, hence, well completion technology holds a very important position in oil and gas production and extraction [2]. Common well completion technologies currently in use, such as rotating tailpipe completion, are also an indispensable part of the drilling process. The reliability of the rotating tail pipe hoisting frame and its accessories directly affects the quality of well completion. During operation, the tail pipe hoisting frame bears all the gravitational forces, and the greater the drilling depth is, the higher the load it has to bear. The rotating tail pipe hoisting frame not only carries heavy loads but also has a strong load-bearing capacity. When working in mud, it is always in an environment of 120 ~ 150°C, requiring a minimum usage time of 20 hours[3]. As for the bearings currently researched, those suitable for rotating tailpipe completion generally have a short service life and cannot further meet construction requirements [4].

Traditional experiments are time-consuming and costly. Finite element simulation can save the cost of making physical objects and provide results in a short time, thus accelerating the efficiency of designers in product development [5]. Finite element simulation can also simulate various loads and working conditions to quickly obtain the desired results. Therefore, this paper uses finite element analysis for simulation and analysis of bearings.

2. Bearing Modeling

This paper focuses on the study of a rotating tailpipe hanger bearing with a retainer, where the seat ring and the shaft ring are separate, making the structure relatively rigorous. This type of bearing can withstand significant axial load forces, but its ability to bear radial loads is relatively weaker. The bearing is not sensitive to impact loads and has good rigidity. The overall structure is shown in Figure 1, and the structure of a single roller is shown in Figure 2.

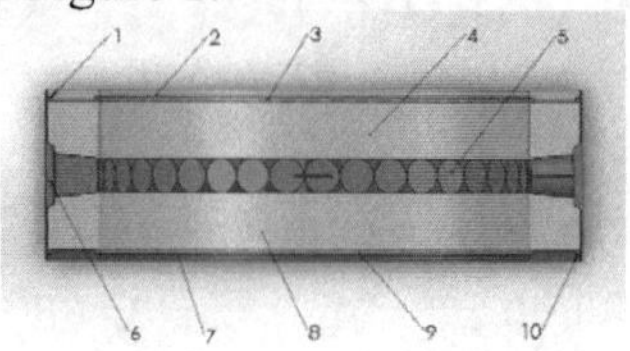

Figure 1. Structure of the rotating tailpipe hanger bearing.

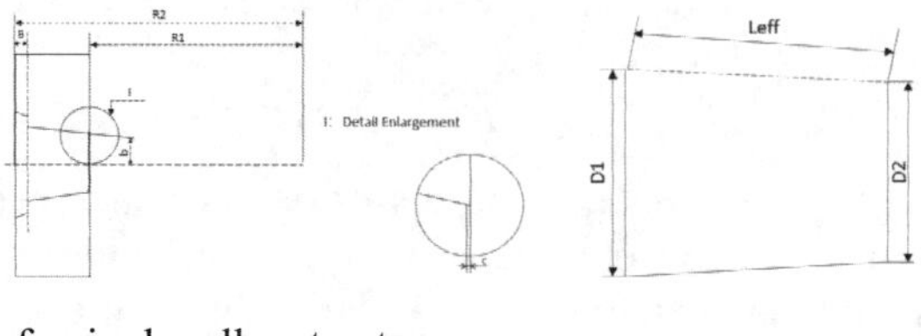

Figure 2. Schematic diagram of a single roller structure.

In Figure 2, R1 is the inner radius of the bearing, R2 is the outer radius of the bearing, H is the height of the bearing, B is the thickness of the bearing retainer wall, c is the safety distance of the roller, D2 is the diameter at the smaller end of the roller, D1 is the diameter at the larger end of the roller, Leff is the effective length of the roller's generatrix, and b is the contact angle of the roller.

Since the rotating tailpipe hanger bearing is a full complement thrust tapered roller bearing, the half inclination angle and contact angle of the rollers are given by Formula (1) and Formula (2), respectively.

$$\beta = \frac{180}{Z} \tag{1}$$

$$\alpha = 90 - \beta = 90 - \frac{180}{Z} \tag{2}$$

The other installation dimensions of the bearing are shown in Table 1 below.

Table 1. External installation dimensions of the rotating tailpipe hanger bearing.

Bearing installation dimension	Value (mm)
Inner radius R1	94
Outer radius R2	117.5
Height H	75
Retainer wall thickness B	4
Roller safety distance c	0.5

The specific dimensions of the bearing rollers are shown in Table 2.

Table 2. Dimensions of the rotating tailpipe hanger bearing rollers.

Bearing structural parameter	Value (mm)
Small end diameter D2	(2R1+2c) tanb
Large end diameter D1	(2R2-2B) sinb
Roller equivalent diameter Dwe	(D1+D2) /2
Roller generatrix length Leff	[2R2cosb- (2R1+2c)]/2cosb-B

The rotating tailpipe hanger is connected to the casing using bearings, located in the middle section between the cone sleeve and the joint. Only when the tailpipe hanger bearing is in a compact state can it accept the full load of the tailpipe, which means there are certain requirements for the dimensions of the casing and the tailpipe. The load on the bearing is directly proportional to the length and the casing is inserted; the greater the load is, the deeper the casing is inserted. The rotating tailpipe hanger bearing is always in a high-load, high-temperature environment and operates in mud, so the bearing's service life must be at least 20 hours [6]. The specific operating conditions of the bearing are shown in Table 3.

Table 3. Specific operating conditions of the rotating tailpipe hanger bearing.

Temperature (°C)	Medium	Sand and gravel particle size/mesh	Mud density (g/cm³)	Speed (r/min)	Life (h)	Load (t)
120-150	Mud	200-400	1.2-1.8	20-25	<20	60-100

The axial load on the bearing during operation is relatively large, and it is always in a harsh working environment, frequently subjected to erosion by drilling fluid. Therefore, the service life of the rotating hanger bearing is generally lower than that of other bearings.

3. Finite Element Analysis

In the early 1880s, the famous German expert Heinrich Rudolf Hertz, in his research on the contact of elastic bodies, proposed Hertz's formula, also known as the Hertz equation. This equation, after being proposed, was mainly used for calculating and studying the interactions between friction, fatigue, or contact friction between two bodies.

4. Contact Stress Calculation

Analyzing the contact relationship between the rollers and the races in the hanger bearing reveals that they have a line contact. During motion, they are subjected to axial load forces, which cause contact to occur between the rollers and the races, resulting in a small deformed area where the contact between the two is now a surface contact. The contact model is shown in Figure 3.The final deformation of the deformed contact surface is related to the curvature that occurs in the line contact between the seat race and the roller, with the contact area increasing as the load increases[7-9].

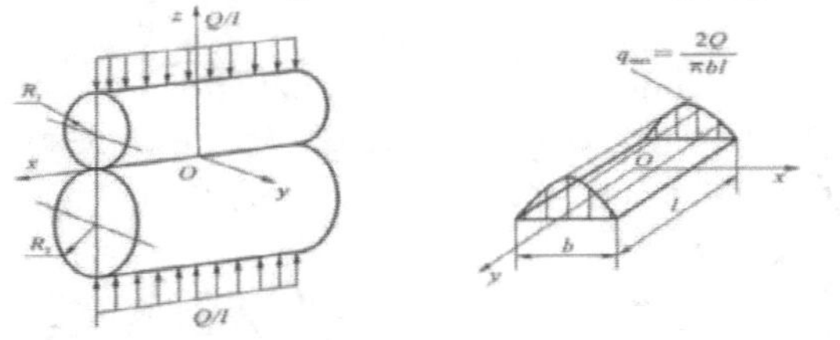

Figure 3. Line contact model of two cylinders of the same length.

It is assumed that the radii of the cylinders in this contact model are R1 and R2, their corresponding Young's moduli are E1 and E2, and their Poisson ratios are v1 and v2. Point contact analysis is used, and to match the contact surface, the distribution of contact stress is assumed to be in the shape of a semi-elliptical cylinder. From the figure, the contact area is rectangular, so the contact stress formula is given by [10]:

$$q(y) = q_{\max}\sqrt{1-\left(\frac{y}{b}\right)^2} \tag{3}$$

By the balance condition of contact stress and external force, using the volume integral of the semi-elliptical cylinder, Formula (4) can be derived:

$$Q = \int_S lq(y)dy = q_{\max} l \int_{-b}^{b} \sqrt{1-\left(\frac{y}{b}\right)^2}\,dy = \frac{\pi q_{\max} bl}{2} \tag{4}$$

Therefore, the maximum contact stress on the contact surface is given by Formula (5):

$$q_{\max} = \frac{2Q}{\pi bl} \tag{5}$$

The maximum contact stress is calculated to be q_{max}=2389.601082 Mpa.

5. Finite Element Preprocessing

In this paper, only the shaft ring, seat ring, and rolling elements are considered for analysis, so only the materials of these three components need to be considered. The shaft ring, seat ring, and rolling elements of the rotating tailpipe hanger bearing analyzed in this paper all use the material ZGCr15, which is commonly used for the raw materials of bearing rollers and bearing races and has good comprehensive performance. The specific material properties are shown in Table 4.

Table 4. Material and property parameters of components.

Component name	Material name	Young's modulus (Gpa)	Poisson's ratio	Density
Shaft ring, seat ring, rolling elements	ZGCr15	206	0.3	7, 830

The main stress contact in this paper occurs at the positions where the rollers contact the seat ring and the shaft ring, so finer meshing at these locations can make the results more accurate. According to the needs, the rollers are divided into three parts, and the final meshing result is shown in Figure 4.

Based on the specific working conditions of the bearing, the bearing is subjected to an axial load of 60 t to 100 t. To increase the safety margin of the bearing, this project takes a higher load of 100 t for analysis. The model selects 1/40 of the entire bearing structure. According to the principle of uniform distribution of load, the axial force Q applied to the analysis model is $Q = 100\ t/40 = 2.5\ t = 2,500\ kg = 2,500 \times 9.8\ N \approx 25,000\ N$, without considering the influence of radial force. Therefore, a vertically downward uniformly distributed axial force $F = 25,000\ N$ is applied to the upper surface of the bearing race; frictionless support is applied to the symmetrical surfaces of the race and the housing sides to set symmetrical boundary conditions; cylindrical support is set for the inner cylindrical surfaces of the race and the housing, allowing movement in the axial direction; the bottom surface of the housing is set with a fixed constraint, allowing movement in the axial direction; the large end surface of the roller is also set with a fixed constraint, allowing movement in the radial direction; the interaction between the roller and the race and housing is achieved through contact algorithms. The specific application of loads and boundary conditions to the model is shown in Figure 5, and the specific experimental data and results are shown in Table 6.

Figure 4. Rotation liner hanger bearing network division.

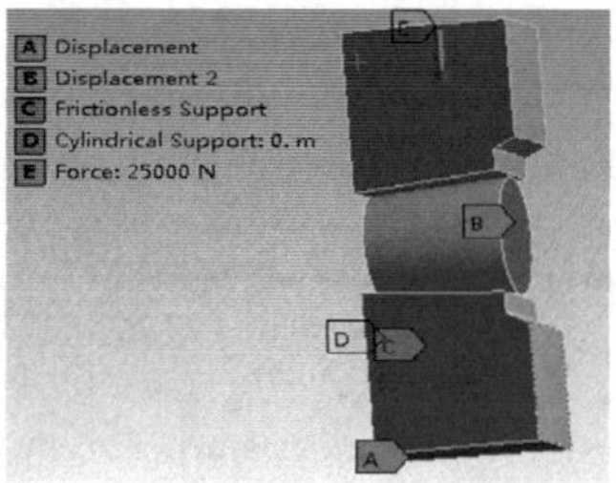

Figure 5. Constraints and loads.

Table 5. Boundary conditions, loads, and constraints.

Constraint position	Constraint nature	Constraint purpose
Underside of the bearing race	Fixed constraint	The seat ring is fixed.
Shaft ring and seat ring on the left- and right-side surfaces	Frictionless constraint	The skewing and rotation of the races are restricted.
The cylindrical surface of the shaft ring and seat ring	Cylindrical constraint	The axial and tangential movements of the races are restricted.
The large end face of rollers	Displacement constraint	The retainer to limit the axial displacement of the rollers is replaced.
The upper surface of the shaft ring	Force load	Axial working load Fa

6. Finite Element Post-Processing

After the loads are set, the contact stress results between the rollers and the shaft ring, and the seat ring are shown in Figure 6 and Figure 7.

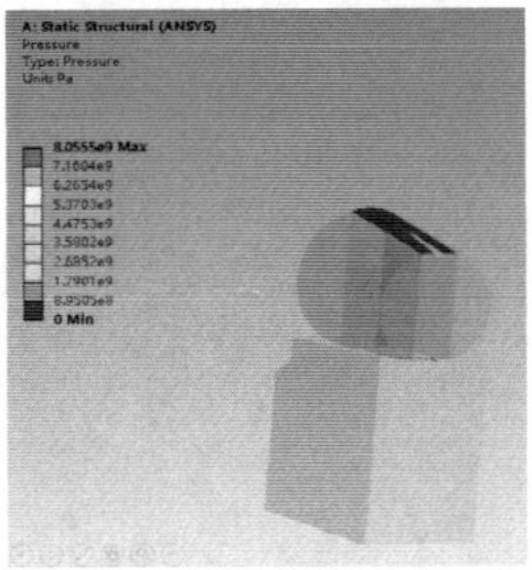

Figure 6. Bearing equivalent stress cloud chart.

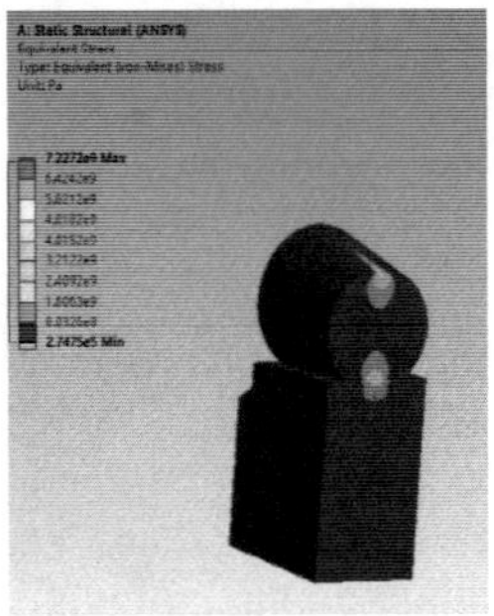

Figure 7. Roller contact stress cloud chart.

By analyzing the equivalent stress distribution cloud chart of the rotating tailpipe hanger bearing and the contact stress distribution chart of the bearing rollers, it can be concluded that when a load of 25, 000 N is applied to the rotating tailpipe hanger bearing, significant contact stress values will occur between the seat ring and shaft ring and the rollers, mainly at the small end face fillet and the large end face fillet of the rollers. Serious stress concentration phenomena will occur. As shown in Figure 6, the maximum equivalent stress value is 7.2272e9 pa.

7. Conclusion

The quality of drilling bearings directly affects the progress of drilling, where the fracture of bearing rollers is the main factor affecting the service life of the bearings. This paper analyzes the working conditions of the bearing during drilling, establishes a 3D model, and performs contact force analysis. The 3D model is then imported into the finite element software for meshing, applying constraints, loads, and other pre-processing. The results of the post-processing analysis show that the maximum equivalent stress when the bearing is under the most severe working conditions is 727 Mpa, which slightly differs from the theoretical calculation value, but this method can greatly reduce experimental costs and is executable. The finite element results indicate that the ends of the rollers experience excessive stress, which is a common type of bearing failure. In subsequent steps, the structure of the bearing rollers should be optimized to reduce the occurrence of stress concentration.

References

[1] Yu Binghao, Zhao Kun, and Zhou Qingqing. Moment Rigidity Analysis of Wheel Hub Bearing Unit Based on ANSYS Workbench [J]. Harbin Bearing, 2023, 44 (04): 25-28+50. DOI: 10.3969/j.issn.1672-4852.2023.04.008.

[2] Wang Jianchong. Finite Element Analysis of Rotating Tailpipe Hanger Bearings Based on ANSYS-Workbench [D]. China University of Geosciences (Beijing), 2014.

[3] Chen Wujun and Ma Kaihua. Development and application of large-sized heavy-duty tailpipe hangers. Petroleum Drilling and Production Technology, 2003, 25 (6): 29-31.

[4] Zhang Jun and Liu Tiansong. Analysis on Load Capacity of Uniform Cross-Section Four-Point Contact Ball Bearings [J]. Modern Machinery, 2023, (05): 29-35. DOI: 10.13667/j.cnki.52-1046/th.2023.05.021.

[5] Luo Liangchuan, Guo Weike, Zhang Huawei, et al. Simulation Study on Contact Characteristics of High-Speed Electric Spindle Bearings [J]. Advanced Technology of Electrical Engineering and Energy, 2024, 53 (08): 215-219. DOI: 10.3969/j.issn.1009-9492.2024.08.046.

[6] Guan Pei. Theoretical research on contact mechanical properties and elastohydrodynamic lubrication of cylindrical roller bits [D]. Beijing: Beijing University of Chemical Technology Library, 2011.

[7] Qiao Xuetao, Sheng Kun, Li Youhua, et al. Fatigue Life and Reliability Analysis of Cycloidal Gears Based on Workbench and nCode [J]. Journal of Mechanical Strength, 2024, 46 (06): 1, 458-1, 464. DOI: 10.16579/j.issn.1001.9669.2024.06.023.

[8] Liang Xianfa, Wang Yi, Tang Baoping, et al. Degradation Model of Rolling Bearings Throughout the Entire Life Cycle Based on Dynamic Modeling [J]. Journal of Vibration and Shock, 2024, 43 (23): 211-219. DOI: 10.13465/j.cnki.jvs.2024.23.024.
[9] Bi Yanchao, Wang Hang, Wang Xinxing, et al. Finite Element Analysis of Surface Cracks in Cylindrical Roller Bearings Based on ANSYS-Workbench [C]. DOI: 10.26914/c.cnkihy.2023.104622.
[10] Zhu Liangliang, Gao Zhu, and Ji Xiaomin. A new method for calculating the life of rolling bearings based on multiple factors [J]. Machinery and Electronics. 2015 (11): 12-16. DOI: 10.3969/j.issn.1001-2257.2015.11.004.

Journal of Physics: Conference Series **3009** (2025) 012072 doi:10.1088/1742-6596/3009/1/012072

A method for detecting anomalies in die forging presses using the pearson correlation coefficient

Ce Han[1*], Hao Zheng[2], Fang He[2]and Tianmin Zhang[3]

[1] Sinomach Sensing Technology Co., Ltd., Shenyang 110043, Liaoning, China
[2]Shenyang Academy Of Instrumentation Science Co., Ltd., Shenyang 110043, Liaoning, China
[3] China National Erzhong Group DeyangWanhang Die Forging Co.,Ltd., Deyang 618000, Sichuan, China

* Corresponding author's e-mail: hance4656@163.com

Abstract. This article explores the correlations among variables during the operation of die forging presses and analyzes the importance of these correlations for anomaly detection in the presses. By dividing the collected data from the die forging presses into training and testing datasets, a correlation matrix analysis is performed using the training dataset to identify groups of variables with strong associations. Subsequently, these selected variable groups are applied to the testing dataset for validation. The results demonstrate that this method can effectively identify specific anomalies in the operation of die forging presses.

1. Introduction

Die forging presses are indispensable key equipment in metal forging processes, utilizing immense pressure to shape metal billets into desired forms[1].

Anomalies in die forging presses not only disrupt production flows and reduce efficiency but can also lead to quality issues such as dimensional inaccuracies and surface defects in the products[2]. Moreover, equipment malfunctions increase the risk of damage, accelerating wear and potentially causing significant harm to the machinery[3]. More critically, such conditions can trigger safety incidents, endangering the personal safety of operators. Additionally, frequent technical failures increase maintenance time and costs[4], thereby affecting the economic benefits of enterprises.

Although there are some existing anomaly warning technologies, they still have certain limitations. Many warning methods rely heavily on the experience of operators, making them susceptible to individual differences and lacking objectivity and consistency. Traditional monitoring techniques often detect problems only after an anomaly has occurred, offering no foresight[5]. Some monitoring systems fail to provide timely warnings[6] because they cannot collect and process data in real-time. Rule-based systems may generate false alarms [7] or miss detections due to improper rule settings. For complex die forging press systems, traditional methods struggle to identify atypical anomaly patterns. Furthermore, current systems lack adequate predictive maintenance capabilities and fail to make full use of historical data to forecast potential future failures.

This article investigates the interrelationships among various monitoring parameters during the operation of die forging presses and analyzes the significance of these relationships for constructing an anomaly warning system. Through correlation matrix analysis, we identified parameter combinations with significant correlations. Using these parameter combinations, an anomaly detection system for die forging presses was developed and validated with a test dataset.

2. The Current State of Research

Ding et al. proposed a method named Latent Correlation-based Anomaly Detection (LCAD) to identify anomalous conditions from data [8]. This approach is based on the assumption that anomalies constitute only a small proportion among a large number of similar entities. Unlike traditional methods that analyze individual monitoring data sequences separately, the LCAD method adopts a new strategy by utilizing a Latent Correlation Probabilistic Model (LCPM) to model the intrinsic relationships between multiple related monitoring data sequences, thereby distinguishing abnormal patterns.

Building upon this foundation, Shisheng Zhong et al. optimized the LCAD method by introducing a Latent Correlation Vector (LCV) to characterize the latent correlations between different parameters [9].By combining LCV with the LCPM, they developed an Anomaly Detection Function (ADF) to assess the state of the equipment. This improvement effectively addresses the issue of increased computational complexity when significant magnitude differences exist between parameters within the same dataset.

To address the dimensionality issues caused by numerous sensors, Pushe Zhao et al., based on Article [9], designed a function for selecting Correlation coefficients that are effective for anomaly detection [10]. This method was applied to compact electric generators with necessary data analysis capabilities. The die forging press has more heterogeneous variables. In order to effectively design the abnormal warning method, we designed a correlation coefficient selector for the die forging press.

3. Correlation Analysis

Correlation analysis is a statistical method used to measure the strength and direction of the linear relationship between two variables. In the context of anomaly warning for die forging presses, analyzing the correlations between different variables can help us understand which variables may have interdependent relationships, thus improving our ability to predict and diagnose anomalies.

The Pearson correlation coefficient is the most commonly used metric for measuring the degree of linear correlation between two variables. It ranges from -1 to +1, where:

- +1 indicates perfect positive correlation, meaning one variable increases as the other increases;
- -1 indicates perfect negative correlation, meaning one variable decreases as the other increases;
- 0 indicates no linear correlation.

The Pearson correlation coefficient can be calculated using the formula (1):

$$\left[r_{xy} = \frac{\sum_{i=1}^{n}(x_i-\bar{x})(y_i-\bar{y})}{\sqrt{\sum_{i=1}^{n}(x_i-\bar{x})^2}\sqrt{\sum_{i=1}^{n}(y_i-\bar{y})^2}} \right] \tag{1}$$

Where:

- $r_{(xy)}$ is the Pearson correlation coefficient between variables $x$and$y$
- x_i and$y_i$are individual sample observations of variables x and y.
- $\bar{x}$ is the mean of the variable x.
- $\bar{y}$ is the mean of the variable y.

Using correlation for anomaly detection involves the following steps: Firstly, we define correlation and establish the metrics for its measurement; in this context, the Pearson correlation coefficient is utilized. Secondly, we calculate the correlations by applying the chosen method to determine the correlation coefficients among different variables. Lastly, we set a threshold based on practical circumstances. A threshold is established to identify combinations of variables that exhibit high linear correlation.

4. Experimental Study

This experiment aims to collect detailed monitor data that can be used to detect anomalies in die forging presses.Before initiating the forging process, meticulous preparation is undertaken to ensure the accuracy and reliability of the data collected. One of the key steps involves setting up the monitoring software associated with the die forging press. This software is configured to start recording the screen display as soon as it is activated, capturing all relevant information displayed on the interface throughout the forging cycle.

Once the setup phase is complete, the die forging press is activated to begin the forging process. The operation of the press is divided into three distinct stages of the forging cycle, each critical to the overall process. The first stage involves the closure of the forging table, where the material to be forged is placed between the dies. Precision during this stage is paramount to ensure the correct alignment and positioning of the material, which directly affects the outcome of the forging process. Following the closure of the forging table, the press initiates the forging action, applying immense pressure to shape the material according to the design of the dies.

Subsequently, the forging table opens, marking the transition into the second stage of the cycle. This stage is equally important as it allows for the removal of the forged product and the preparation of the next piece of material for forging. The final stage involves the cessation of the press operation, signaling the completion of one full cycle. It is at this point that the screen recording of the monitoring system is halted, ensuring that only the relevant data from the forging process is captured.

To enrich the dataset, the recorded content is played back frame by frame, with data being logged once every second. This methodical approach to data collection not only increases the volume of data gathered but also enhances its granularity, allowing for a more comprehensive analysis. The experiment is repeated 'n' times, each iteration contributing to a larger and more diverse dataset. The repetition of the process under controlled conditions helps to identify patterns, anomalies, and areas for improvement in the forging process.

The sensor variables displayed on the press interface, as outlined in Table 1 and shown in Figure 1, play a crucial role in the monitoring and analysis of the forging process. These variables include but are not limited to the force applied by the press, the speed of the forging action, and the position of the forging table. Each variable provides valuable data into different aspects of the forging process, contributing to a holistic detectionof anomaliesin the die forging presses.

Table 1. Monitoring System Variables for a Die Forging Press.

No.	Variable Name
1	Position at W
2	Speed at W
3	Pressure at W
……	……
76	Pilot Valve Pressure

5. Data Analysis and Processing

5.1. Data cleaning and heat map generation

In the realm of data analysis and machine learning, the process of data cleaning plays a pivotal role in ensuring the accuracy and reliability of the results derived from any dataset. In the context of the die forging experiment described earlier, after collecting a substantial amount of data through multiple iterations of the forging process, the next step is to refine this dataset to enhance its utility for further analysis. One of the primary tasks in data cleaning is the identification and removal of variables that do not contribute significantly to the objectives of the study. These unnecessary variables can introduce noise into the analysis, potentially leading to misleading conclusions or suboptimal model performance.

In our case, the initial dataset comprises a series of sensor variables recorded during the forging process, as detailed in Table 1. These variables encompass a wide range of measurements, including the force applied by the press, the speed of the forging action, and the position of the forging table, among others. To streamline the dataset, we must critically evaluate each variable's relevance to the

Journal of Physics: Conference Series **3009** (2025) 012072 doi:10.1088/1742-6596/3009/1/012072

monitoring and optimization goals of the forging process. Variables that exhibit minimal variation or show no clear relationship with the detecting anomalies may be considered for removal. For instance, if a particular sensor consistently records the same value or shows no discernible pattern across different forging cycles, it might be deemed unnecessary for the purposes of our analysis.

During the operation of the die forging press, signal interference, sensor malfunctions, or other external factors may cause some measurement values to be missing. This data loss not only affects real-time monitoring of the equipment but also has adverse effects on subsequent data analysis and fault diagnosis. To address the issues caused by missing data, we directly replace the missing values with 0.

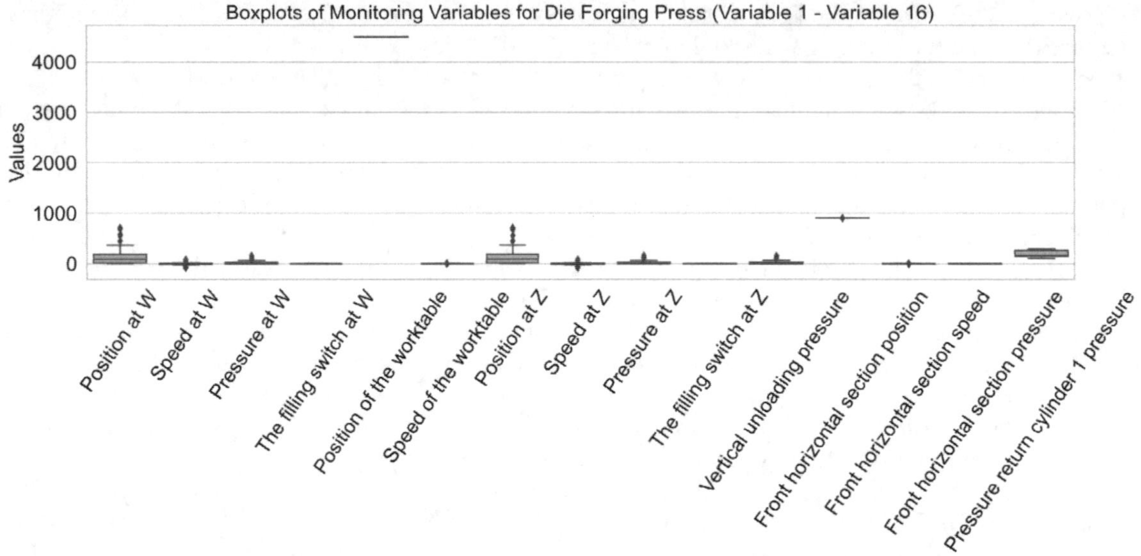

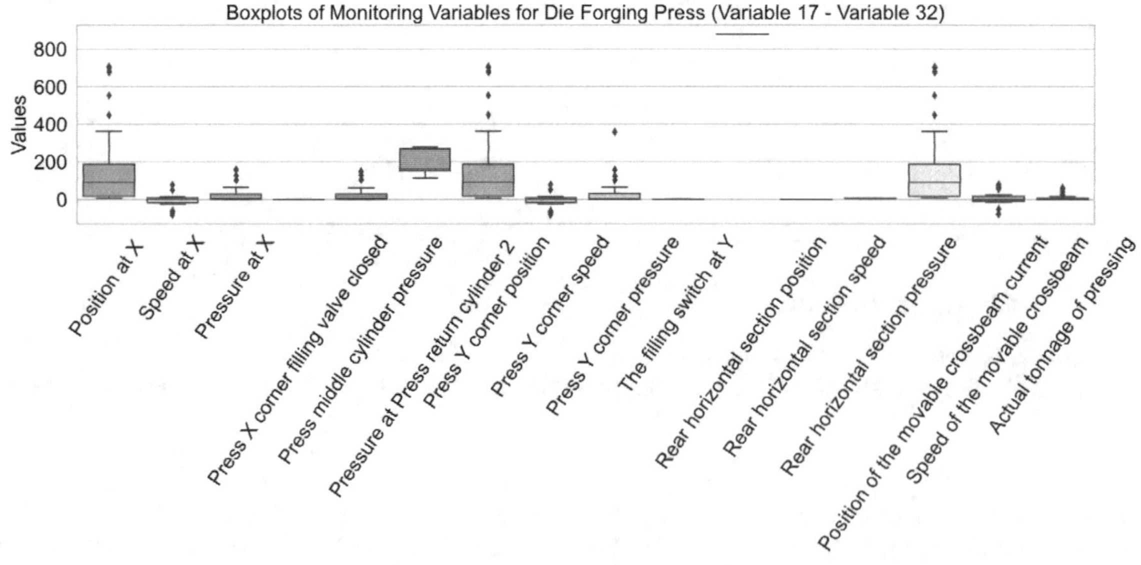

Journal of Physics: Conference Series **3009** (2025) 012072 doi:10.1088/1742-6596/3009/1/012072

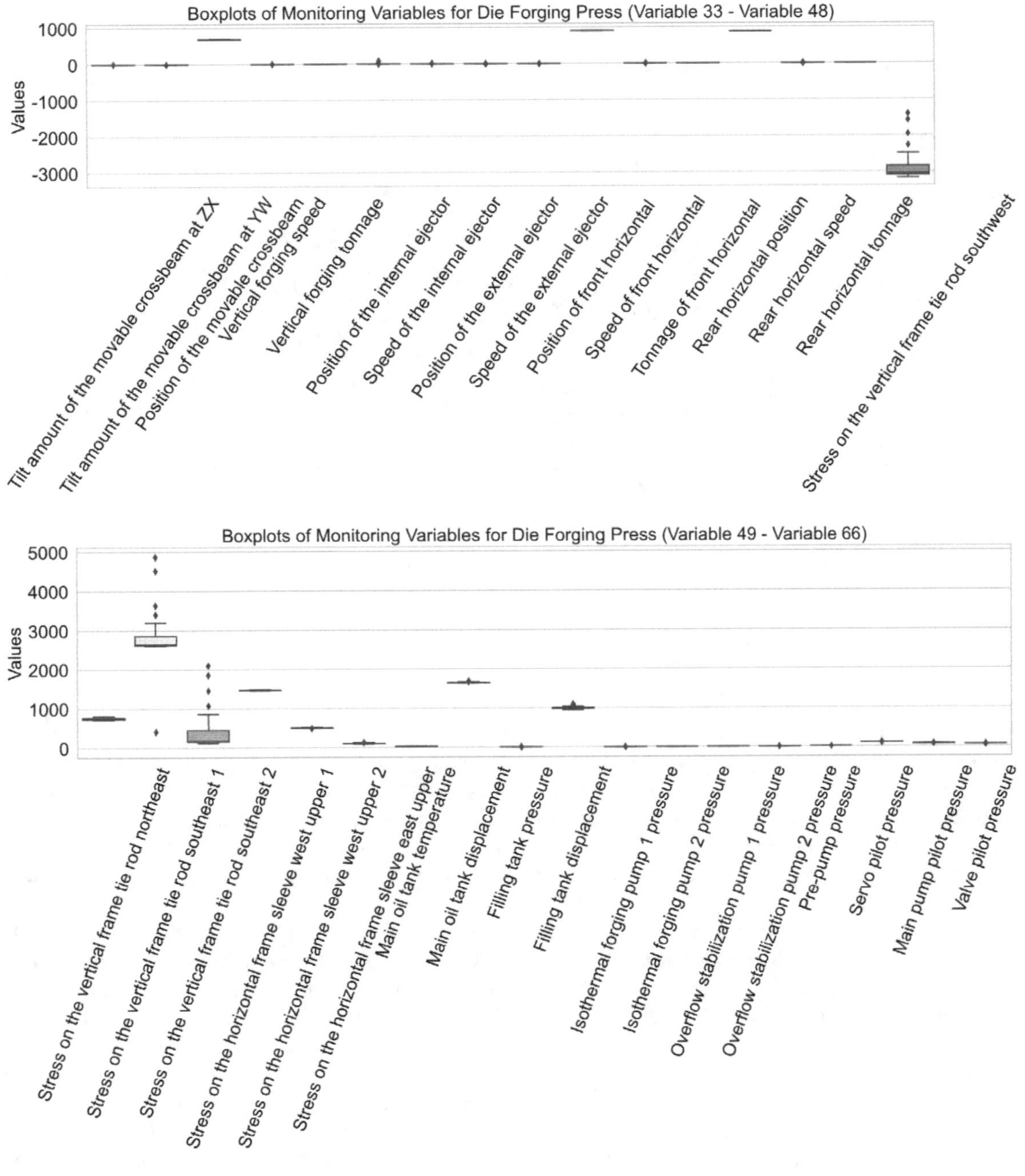

Figure 1. Boxplots of Monitroing Variables for Die Forging Press.

To more clearly present the correlation between various variables, we utilize a Correlation Network Diagram. As shown in Figure 2, we replace the variable names with their corresponding numbers in the diagram, ensuring clarity and readability of the image.

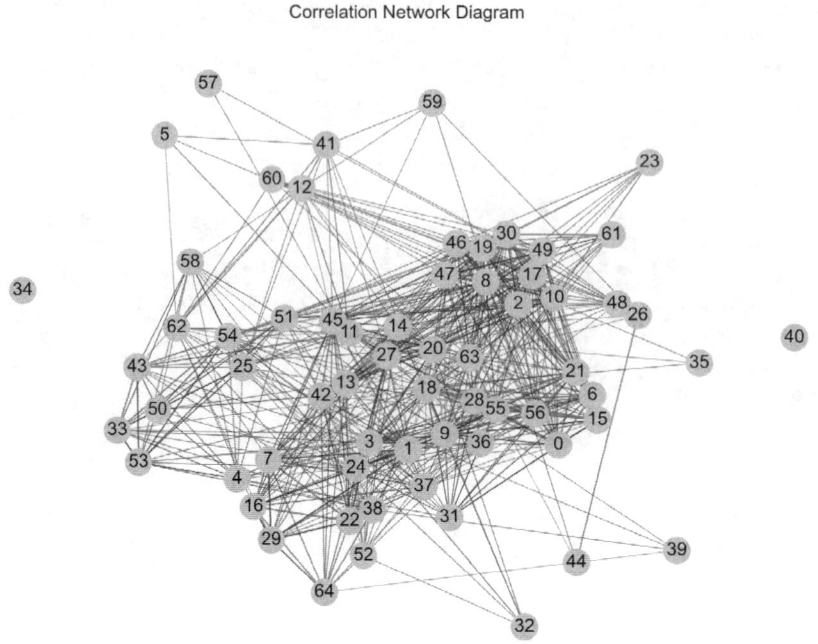

Figure 2. Correlation Network Diagram of Monitoring Variables for Die Forging Press.

After interpolation, the next step in our data analysis pipeline is to explore the relationships between the remaining variables. A powerful tool for this purpose is the computation of a correlation coefficient matrix, which quantifies the strength and direction of the linear relationship between pairs of variables. We will employ Pearson's method to calculate the correlation coefficients, which are particularly suitable for continuous variables and assume a linear relationship between them. Pearson's correlation coefficient ranges from -1 to 1, where values close to 1 indicate a strong positive correlation, values close to -1 suggest a strong negative correlation, and values around 0 imply little to no correlation.

To construct the correlation coefficient matrix, we treat the refined list of variables as both the rows and columns of the matrix. Each cell in the matrix represents the correlation coefficient between the variable corresponding to the row and the variable corresponding to the column. For example, the cell at the intersection of the "Speed at W" row and the "Pressure at W" column would contain the Pearson correlation coefficient between the force applied by the press and the temperature of the material during the forging process.

Once the correlation coefficient matrix is computed, visualizing this information becomes crucial for gaining deeper insights into the data. A heatmap is an effective way to represent the matrix, as it uses color gradients to highlight the strength and nature of the correlations. Deeper colors indicate higher correlations, while lighter colors represent lower correlations.

In this paper, we utilized a visualization library, Seaborn, in Python to create the heatmap, as shown in Figure 3.

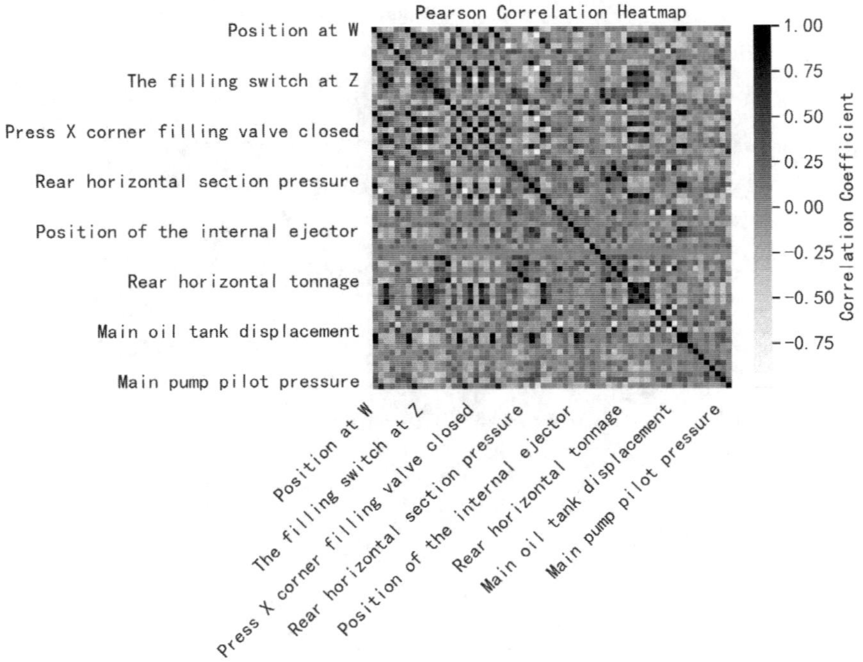

Figure 3. Pearson Correlation Heatmap of Features.

In addition to the Pearson correlation coefficient heatmap, we have also calculated the Spearman correlation and Kendall correlation, with the heatmaps presented in Figure 4 and Figure 5, respectively.

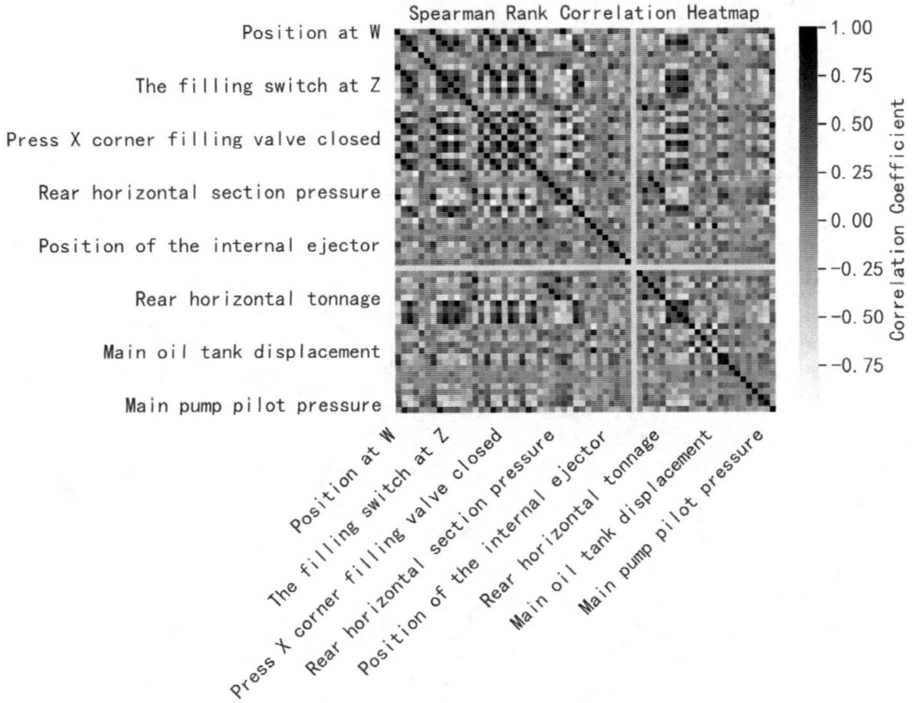

Figure 4. Spearman Correlation Heatmap of Features.

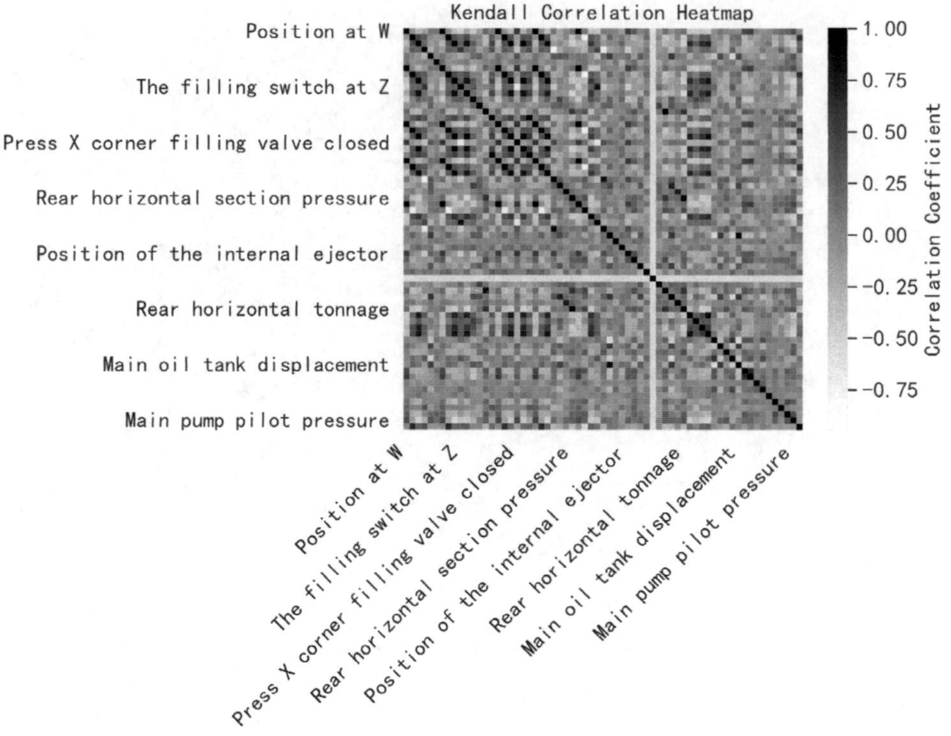

Figure 5. Kendall Correlation Heatmap of Features.

5.2. Related variables filtering

In order to further analyze the correlations between the remaining variables, we select those with a correlation greater than 0.9, as shown in Table 2.

Table 2. Relevant Variables with Correlation Greater Than 0.9.

No.	First variable	Second variable	correlation
0	Position at W	Position at Z	0.998
1	Position at W	Position at X	0.998
2	Position at W	Position at Y	0.998
……		……	
117	Filling tank Position	Filling tank Pressure	0.938

To use correlation coefficients for anomaly detection, it is necessary to ensure that the correlation coefficients remain stable across different operating states of the press. Assuming the press operation includes five steps: start-up, descent, pressing, ascent, and stop, it is essential to maintain consistent correlation coefficients throughout these steps. Since the duration of each step varies during different tasks, the number of recorded data points for each step is inconsistent. If the maximum and minimum values of the correlation coefficients differ significantly across the steps, it leads to drastic changes in the overall correlation coefficients, making it difficult to set an appropriate correlation coefficient threshold.

To study the range of correlation coefficient variations for the aforementioned five steps, we divide the overall dataset into five subsets, labeling the "sequence (total)" as "sequence 1," "sequence 2," "sequence 3," "sequence 4," and "sequence 5." Using these divided datasets, we calculate the correlation coefficients for the variable groups. From the results, we identify 84 groups where the correlation is less than 0.9 in various stages. We select representative groups, as shown in Table 3.

Table 3. Representative Groups with Correlation Less Than 0.9 in Various Stages.

No.	First variable	Second variable	Subset total	Subset 1	Subset 2	Subset 3	Subset4	Subset 5
4	Position at W	Filling tank Pressure	0.950	0.999	0.988	-0.688	-0.694	1.000
5	Position at W	Filling tank Position	0.936	0.995	0.935	0.024	-0.562	0.989
10	Pressure at W	Vertical unloading Pressure	0.997	0.999	0.714	0.972	0.998	0.661
……				……				
117	Filling tank Position	Filling tank Pressure	0.938	0.993	0.893	0.150	0.635	0.992

Further screening was conducted to identify groups with correlations consistently greater than 0.9 across all stages. The resulting groups are shown in Table 4.

Table 4. Correlation Coefficient Table for Highly Correlated Variables in a Die Forging Press.

No.	First variable	Second variable	correlation
0	Position at W	Position at Z	0.998
1	Position at W	Position at X	0.998
2	Position at W	Position at Y	0.998
……		……	
78	Position of the movable crossbeam	Position at Y	1.000

5.3. Design of anomaly detection method

Based on the above analysis, we propose an anomaly detection method based on correlations. First, we collect key monitoring variables of the forging press at fixed time intervals to form a time-series dataset $D_{m \times t}$, where t represents the collection times and m represents the number of variables. Then, we calculate the pairwise correlations to form a correlation matrix CM. Next, we perform the first screening of the correlation matrix CM: elements greater than 0.9 are retained, and elements less than 0.9 are set to 0, forming CM_1.

Following this, we conduct a second screening of the correlation matrix. We divide the dataset into five equal parts and compute the correlation matrix for each part, resulting in five correlation matrixCM_{part1} toCM_{part5}. In CM_{part1} toCM_{part5}, elements greater than 0.9 are set to 1, and elements less than 0.9 are set to 0, forming Mask MatricesMM_1 toMM_5. We can get the anomaly warning Mask Matrix$MM_{threshold}$using the formula (2):

$$CM_{threshold} = CM_1 \times MM_1 \times MM_2 \times MM_3 \times MM_4 \times MM_5 \times Th \quad (2)$$

Where:

Th is the threshold. Th is dynamically adjusted based on the pressure and speed at the four corners of the press. The adjustment method is as shown in formula (3).

$$Th = \begin{cases} Th_{base} & \alpha > Coe \\ 0 & \alpha \leq Coe \end{cases} \quad (3)$$

Where:

Coe is the threshold of dynamic coefficient, Th_{base} is the basic threshold.

The calculation method for α is shown in formula (4).

$$\alpha = 1 - \prod_{i=1}^{4}\left(1 - \frac{P_i}{R_p}\right)\prod_{i=1}^{4}\left(1 - \frac{S_i}{R_s}\right) \quad (4)$$

Where:

P_i is the pressure at W,X, Y, Z;

S_i is the speed at W, X, Y, Z;

R_p is the range of pressure;

R_s is the range of speed.

Elements in $CM_{threshold}$ that are not zero are set to 1, forming the mask correlation matrix MCM.

When we obtain a new time-series dataset D_{test}, we compute its correlation matrix CM_{test1}. We then calculate the Hadamard product of CM_{test1} and MCM to get CM_{test2}. Subtracting $CM_{threshold}$ from CM_{test2} yields CM_{result}. If any element in CM_{result} is negative, it indicates that the corresponding correlation exceeds the threshold, triggering an alarm.

5.4. Comparison of anomaly detection methods

To verify the performance of the algorithm, we designed the following experiment. We collected key monitoring variables of the forging press at fixed time intervals, gathering 40 sets of data, each containing 117 variables, to form the training set. Additionally, we collected four sets of normal data and 12 sets of artificially simulated abnormal data to form the test set. Using the training set, we computed $CM_{threshold}$. Using the test set, we derived CM_{test2} and ultimately CM_{result}. We compared the anomaly detection results with those of two other methods, One-Class SVM and Principal Component Analysis.

One-class SVM is a supervised learning method particularly suitable for scenarios where only normal samples are available. It works by finding an optimal hyperplane that maximizes the separation of the normal samples to one side of the hyperplane while minimizing the distance from the hyperplane to the origin. In this experiment, we trained a One-Class SVM model using the training set data. This model learned the characteristic features of the die forging press during its normal operating state. Subsequently, we input the test set data into the trained One-Class SVM model to compute the decision function value for each test sample. Based on a pre-defined threshold, the test samples can be classified as either normal or abnormal.

Principal Component Analysis (PCA) is a widely used technique for data dimensionality reduction and feature extraction, particularly suitable for handling high-dimensional datasets. In this experiment, we applied PCA to 40 sets of normal operation data from the die forging press. The goal was to extract the main features of the data and reduce its dimensionality to enable more efficient anomaly detection.

Before performing PCA, it is essential to standardize the data. This is because PCA is highly sensitive to the scale of the variables. If some variables have a much larger numerical range than others, they will dominate the principal component analysis, potentially skewing the results. Therefore, we standardized each variable in the training set using Z-score normalization. This involves subtracting the mean of each variable from its values and then dividing it by its standard deviation, ensuring that each variable has a mean of 0 and a standard deviation of 1. This standardization ensures that all variables are analyzed on the same scale.

The standardized data was then used to compute the covariance matrix. The covariance matrix is a square matrix whose elements represent the linear relationships between different variables. After calculating the covariance matrix, the next step is to perform eigenvalue decomposition. The purpose of eigenvalue decomposition is to find a set of orthogonal basis vectors (eigenvectors) that can project the data into a new coordinate system where the variance of the data is maximized.

Following the eigenvalue decomposition, we selected the principal components that explain the majority of the variance in the data. Typically, this involves choosing the components with the largest eigenvalues. Specifically, we determined the number of principal components to retain by calculating the cumulative variance contribution rate. The final step is to project the original data onto the selected principal component space.

In PCA-based anomaly detection, two important monitoring indicators are the T^2 statistic and the Squared Prediction Error (SPE). These metrics are used to assess whether data points deviate from the normal operating state. The T^2 statistic measures the position of a data point in the principal component space, reflecting the degree of deviation from the principal components. The SPE measures the error of a data point in the residual space, reflecting the degree of deviation in non-principal component directions. We set corresponding thresholds based on the training results. If a data point's T^2 value or SPE value exceeds the respective threshold, the data point is flagged as an anomaly.

When comparing the three methods discussed—One-Class SVM, Principal Component Analysis (PCA), and the proposed method—it is evident that the proposed method achieves higher accuracy, as detailed in Table 5.

Table 5. Accuracy Comparison of Different Methods.

Method	Accuracy
correlation coefficient	0.9375
One-class SVM	0.8125
PCA	0.75

6. Conclusions

This study collected monitoring data from a die forging press at a particular forging factory, sorted the data chronologically, and calculated the Pearson correlation coefficients between the data points. Variables with correlation coefficients greater than 0.9 were selected. The filtered data were then divided into five equal groups, and the Pearson correlation coefficients were recalculated. Through a secondary screening process, variables suitable for anomaly detection were identified. These variables were used to perform anomaly detection on the test set. For variables with high correlation, the anomaly detection method based on correlation proved effective in identifying anomalies.

Acknowledgments

Supported by National Key R&D Plan of China (Grant No.2023YFB3210800).

References

[1] Jianjun, L., Xinyun, W. 2024. Die for Forging and Stamping. In: Kuangdi, X. (eds) The ECPH Encyclopedia of Mining and Metallurgy. Springer, Singapore. https://doi.org/10.1007/978-981-99-2086-0_689.

[2] Daniele Farioli,ErturulKaya,AndreaFumagalli,PaoloCattaneoMatteo Strano.2023.A Data-Based Tool Failure Prevention Approach in Progressive Die Stamping.JOURNAL OF MANUFACTURING AND MATERIALS PROCESSING,7(3).https://doi.org/10.3390/jmmp7030092.

[3] Lu, P., Powrie, H. E., Wood, R. J., Harvey, T. J., & Harris, N. R. 2021. Early wear detection and its significance for condition monitoring. Tribology International, 159, 106946.https://doi.org/10.1016/j.triboint.2021.106946.

[4] Arno, R.G., Dowling, N., & Schuerger, R. 2015. Equipment failure characteristics & RCM for optimizing maintenance cost. 2015 IEEE/IAS 51st Industrial & Commercial Power Systems Technical Conference (I&CPS), 1-8.https://doi.org/10.1109/TIA.2015.2483700.

[5] Kashpruk, N., Piskor-Ignatowicz, C., & Baranowski, J. 2023. Time Series Prediction in Industry 4.0: A Comprehensive Review and Prospects for Future Advancements. Applied Sciences, 13(22), 12374.https://doi.org/10.3390/app132212374.

[6] Malekloo, A., Ozer, E., AlHamaydeh, M., & Girolami, M. 2022. Machine learning and structural health monitoring overview with emerging technology and high-dimensional data source highlights. Structural Health Monitoring, 21(4), 1906-1955.https://doi.org/10.1177/14759217211036880.

[7] Mir, A. A. 2024. Adaptive Fraud Detection Systems: Real-Time Learning from Credit Card Transaction Data. Advances in Computer Sciences, 7(1).https://academicpinnacle.com/index.php/acs/article/view/229.

[8]Ding, J., Liu, Y., Zhang, L., & Wang, J. 2014. LCAD: a correlation based abnormal pattern detection approach for large amount of monitor data. In Web Technologies and Applications: 16th Asia-Pacific Web Conference, APWeb 2014, Changsha, China, September 5-7, 2014. Proceedings 16 (pp. 550-558). Springer International Publishing.https://doi.org/10.1007/978-3-319-11116-2_51.

[9] Zhong, S., Luo, H., Lin, L., & Fu, X. 2016. An improved correlation-based anomaly detection approach for condition monitoring data of industrial equipment. In 2016 IEEE International Conference on Prognostics and Health Management (ICPHM) (pp. 1-5). IEEE.https://doi.org/10.1109/ICPHM.2016.7542850.

[10] Zhao, P., Kurihara, M., Tanaka, J., Noda, T., Chikuma, S., & Suzuki, T. 2017. Advanced correlation-based anomaly detection method for predictive maintenance. In 2017 IEEE International Conference on Prognostics and Health Management (ICPHM) (pp. 78-83). IEEE.https://doi.org/10.1109/ICPHM.2017.7998309.

Research on the mechanism of effective remelted layer removal in micro-hole electrical discharge machining with ultrasonic vibration drilling

Jiaqian Chen, Jianyong Liu *, Xueke Luo, Hui Li, and Xinghao Fan

Department of Mechanical Engineering, Beijing Institute of Petrochemical Technology, Beijing 102617, China

*Corresponding author's e-mail: 0020200028@bipt.edu.cn

Abstract: This study addresses the issue of remelted layer removal in the composite machining of micro-holes using electrical discharge machining (EDM) combined with ultrasonic vibration drilling. The effects of machining deviations caused by limited machine tool precision on remelted layer removal and hole roundness are analyzed. Multi-body dynamic error analysis was employed to optimize machining parameters, reduce mechanical errors, and achieve precise control over remelted layer thickness, enabling remelt-free machining. The influence of remelted layer thickness on reaming processes and tool life, particularly for titanium alloys, was investigated, concluding that thinner layers are advantageous for subsequent machining. A multi-physics coupling model was developed by using COMSOL to determine optimal remelted layer thickness and machining parameters. Experimental results demonstrated a 43.90% reduction in electrode wear, a 32.1% reduction in machining time, a 7.1% decrease in hole diameter, and a 25% reduction in remelted layer thickness (from 16.4 μm to 12.3 μm), significantly enhancing machining efficiency.

1. Introduction

Miniaturization, integration, and intelligence are the main directions of current scientific and technological development. Micro-hole Electrical Discharge Machining (EDM) holds irreplaceable advantages in processing materials with high hardness, brittleness, and plasticity. However, achieving low-damage or damage-free surface quality has always been a key concern. Mainstream techniques such as EDM, laser machining, and electro-hydraulic beam machining each have limitations: EDM often produces remelted layers and microcracks; femtosecond laser machining theoretically avoids remelted layers [1] but introduces hollow and multi-layer back-wall structural damage; electro-hydraulic beam machining faces challenges in controlling hole geometry and corrosion from electrolytes. Comp composite machining methods have emerged to meet the high precision and quality demands of film cooling holes, such as remelt-free, heat-affected-zone-free, and crack-free results. EDM's inherent thermal stress deformation necessitates combining other processes to remove remelted layers. Current domestic studies have explored remelted layer removal techniques through composite EDM methods such as electrochemical [2], electrolytic [3], and abrasive flow processing [4], achieving some progress [5] but still struggling with complete layer removal. Over time, extreme conditions of high-temperature and high-pressure environments can lead to remelted layer detachment and microcrack propagation, risking blade failure and engine breakdown.

Micro-drilling, a conventional mechanical process, offers high surface quality, accuracy, and cost-efficiency [6]. Ultrasonic vibration-assisted EDM (UVE), a composite method, integrates ultrasonic vibration with EDM for micro-hole machining. This innovative approach enhances the probability of effective discharges and reduces abnormal discharges, thereby improving material removal rates. Ultrasonic vibration generates cavitation bubbles that remelt debris for effective chip removal and

Journal of Physics: Conference Series **3009** (2025) 012073 doi:10.1088/1742-6596/3009/1/012073

deionization. This composite method, particularly in removing deteriorated layers, demonstrates significant potential and promising application prospects [7].

To address diverse demands for efficient and high-quality micro-hole processing, this study combines EDM and ultrasonic vibration drilling techniques into a composite method, achieving micro-holes free of deteriorated layers for applications in advanced industries. The technique has been extensively validated, and to further improve processing efficiency, a multi-physics plasma channel model is proposed to analyze remelted layer thickness. By integrating multi-body dynamic error analysis, machining errors are reduced, advancing remelted layer evaluation methods and promoting the composite process in high-end applications.

2. Composite machining process solution design

To address the requirements of micro-hole machining with no remelted layer, no microcracks, and no back-wall damage, this study proposes a coaxial coupling scheme combining Electrical Discharge Machining (EDM) and ultrasonic vibration drilling. The ultrasonic vibration drilling system for micro-holes is modularly designed and integrated into the composite machining test platform. Considering the need to process small holes in complex poses and special materials, a high-precision dual-indexing rotary table is configured on the spindle worktable. The architecture of the remelt-free EDM ultrasonic vibration drilling equipment is shown in Figure 1.

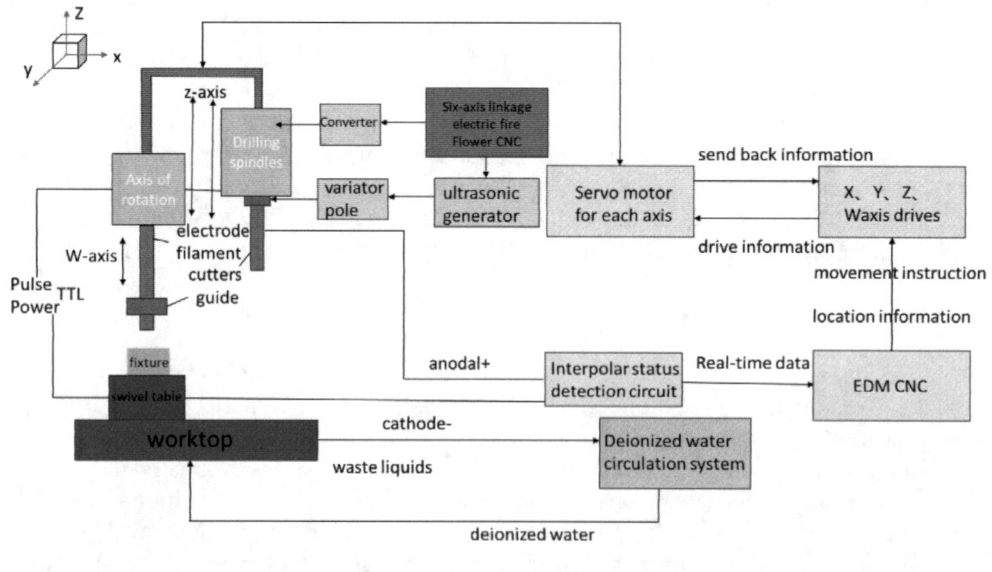

Figure 1. EDM ultrasonic drilling and machining equipment architecture.

Based on the requirements for high surface quality and precise hole geometry in aerospace blade film cooling holes, the rotary motion is achieved by using an indexing rotary table. In addition to the X, Y, and Z linear axes, the machine tool incorporates a dedicated linear W-axis for electrode wire trajectory during film hole machining. This separation from the Z-axis minimizes positioning errors, achieving a repeatable accuracy of ±4 μm. Both the EDM dielectric and drilling coolant use deionized water, with a recirculation system ensuring continuous supply.

The ultrasonic vibration drilling system for micro-holes consists of a drilling spindle, frequency converter, ultrasonic generator, and amplitude transformer. Integrated into the six-axis EDM CNC system, it includes spindle control, ultrasonic parameter tuning, and modularized power systems, enabling seamless automation between EDM micro-hole machining and high-speed micro-hole drilling [8]. Ultrasonic vibration-assisted drilling, a hybrid technique combining traditional drilling and ultrasonic vibration, applies high-frequency vibrations along the feed direction. This creates complex relative motions between the tool and workpiece, achieving periodic zero chip thickness to break chips effectively [9]. The process ensures efficient debris removal, minimizes secondary friction, reduces tool wear, and enhances tool life. This method completely removes remelted layers, improves surface quality, and significantly reduces burr formation, supporting high-performance EDM micro-hole machining.

3. Composite machining error analysis based on multibody dynamics

3.1. Machine topology and low-order body description

The topology of the machine motion system is shown in Figure. 2, which is divided into EDM module and ultrasonic vibratory drilling module.

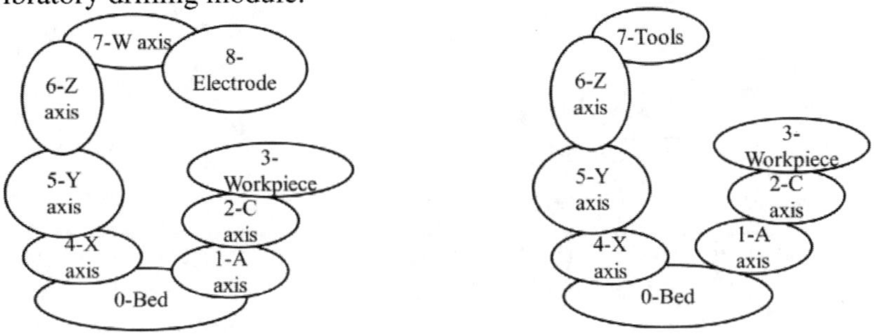

(a) EDM module **(b)** Ultrasonic vibration drilling module

Figure 2. Machine tool motion system topology.

The low order body arrays have to be analyzed separately according to the topology, the low order body arrays for EDM machining are shown in Table 1 and the ultrasonic vibratory drilling low order body arrays are shown in Table 2.

Table 1. Low-order body arrays during EDM machining.

j	1	2	3	4	5	6	7	8
$L(0)$	1	2	3	4	5	6	7	8
$L(1)$	0	1	2	0	4	5	6	7
$L(2)$	0	0	1	0	0	4	5	6
$L(3)$	0	0	0	0	0	0	4	5
$L(4)$	0	0	0	0	0	0	0	4
$L(5)$	0	0	0	0	0	0	0	0

Table 2. Ultrasonic vibratory drilling with low-order body arrays.

	1	2	3	4	5	6	7
$L(0)$	1	2	3	4	5	6	7
$L(1)$	0	1	2	0	4	5	6
$L(2)$	0	0	1	0	0	4	5
$L(3)$	0	0	0	0	0	0	4
$L(4)$	0	0	0	0	0	0	0

3.2. Composite machining error modeling

The primary motion components of the machine tool include three linear axes (X, Y, Z), two rotational axes (A, C), and the W-axis. Errors generated by these components are the main contributors to geometric errors.

Geometric error analysis:

1. Each of the X, Y, Z, A, and C axes has six fundamental errors, while the W-axis has five, totaling 35 error sources.

2. Positional relationships (e.g., perpendicularity or parallelism) among components add nine inter-component posture errors, including perpendicularity errors among XYZ, parallelism between ZW, and perpendicularity errors of the A-axis to YZ and the C-axis to XY.

In total, the machine has 44 geometric errors, as shown in Table 3. During drilling, since the W-axis is not utilized, the geometric errors are reduced to 37.

Table 3. Geometric error terms and their expressions.

Name	Geometric error	Perpendicularity/Parallelism
X	$\delta_{xX}\delta_{yX}\delta_{zX}\varepsilon_{xX}\varepsilon_{yX}\varepsilon_{zX}$	$S_{yX}S_{yZ}S_{xZ}$
Y	$\delta_{xY}\delta_{yY}\delta_{zY}\varepsilon_{xY}\varepsilon_{yY}\varepsilon_{zY}$	1
Z	$\delta_{xZ}\delta_{yZ}\delta_{zZ}\varepsilon_{xZ}\varepsilon_{yZ}\varepsilon_{zZ}$	0
A	$\delta_{xA}\delta_{yA}\delta_{zA}\varepsilon_{xA}\varepsilon_{yA}\varepsilon_{zA}$	$S_{yA}S_{zA}$
C	$\delta_{xC}\delta_{yC}\delta_{zC}\varepsilon_{xC}\varepsilon_{yC}\varepsilon_{zC}$	$S_{xC}S_{yC}$
W	$\delta_{xW}\delta_{yW}\delta_{zW}\varepsilon_{xW}\varepsilon_{yW}$	$\kappa_{xZW}\kappa_{yZW}$

Based on the theory of multi-body kinematics and homogeneous coordinate transformation matrices, the coordinate transformations between various motion components can be represented by using transformation matrices. We set T_{ij} as the ideal inter-body transformation matrix from body i to body j, and ΔT_{ij} represents the error transformation matrix between the same bodies. The relationship can then be expressed as follows:

(1) During EDM machining, ideally, the chi-square coordinate transformation matrix between the electrode and the workpiece is:

$$T_{83} = T_{32}T_{21}T_{10}T_{04}T_{45}T_{56}T_{67}T_{78} \quad (1)$$

The chi-square coordinate transformation matrix between the actual state electrode and the workpiece is:

$$T'_{83} = T_{32}\Delta T_{32}T_{21}\Delta T_{21}T_{10}\Delta T_{10}T_{04}\Delta T_{04}T_{45}\Delta T_{45}T_{56}\Delta T_{56}T_{67}\Delta T_{67}T_{78}\Delta T_{78} \quad (2)$$

(2) During drilling machining, ideally, the chi-square coordinate transformation matrix between the tool and the workpiece is:

$$T_{73} = T_{32}T_{21}T_{10}T_{04}T_{45}T_{56}T_{67} \quad (3)$$

The chi-square coordinate transformation matrix between the actual state tool and the workpiece is:

$$T'_{73} = T_{32}\Delta T_{32}T_{21}\Delta T_{21}T_{10}\Delta T_{10}T_{04}\Delta T_{04}T_{45}\Delta T_{45}T_{56}\Delta T_{56}T_{67}\Delta T_{67} \quad (4)$$

The actual state is:

$$P_{we} = T'_{83}P_e \quad (5)$$

$$P_{wt} = T'_{73}P_t \quad (6)$$

where P_t is the predicted value of the combined model, $P_e$is the coordinates of machining points in the electrode coordinate system, P_{wt} is the coordinates of machining in the workpiece coordinate system during drilling machining, and P_{we} is the coordinates machined in the workpiece coordinate system during EDM machining.

The machining point error in drilling and EDM is:

$$E = P_{we} - P_{wt} = T_{32}\Delta T_{32}T_{21}\Delta T_{21}T_{10}\Delta T_{10}T_{04}\Delta T_{04}T_{45}\Delta T_{45}T_{56}\Delta T_{56}T_{67}\Delta T_{67}T_{78}\Delta T_{78}P_e - T_{32}\Delta T_{32}T_{21}\Delta T_{21}T_{10}\Delta T_{10}T_{04}\Delta T_{04}T_{45}\Delta T_{45}T_{56}\Delta T_{56}T_{67}\Delta T_{67}P_t \quad (7)$$

4. Removal margin control analysis

4.1. Simulation of remelted layer thickness

The formation of the remelted layer involves multiple physical processes, including vaporization, melting, and solidification, and consists of a mixture of the workpiece material, electrode, and decomposed working fluid, which makes it significantly different from the substrate material. The type of workpiece material further determines the microstructure of the remelted layer [10].

Figures 3 (a) and 3 (b) show the heating phase during EDM at 30 μs and 90 μs, where the molten zone rapidly expands. Metallurgical phase transitions are primarily observed in the heat-affected zone, with only a small portion of the material undergoing a complete transition from α+β to β. Most materials remain in the α+β phase, though their volume fractions change. Figures 3 (c) and 3 (d) illustrate the cooling phase at 120 μs and 150 μs, where temperature distribution shows little difference, but the microstructure becomes increasingly complex. When the temperature drops below 1, 298 K, molten material near the boundary of the molten zone resolidifies, reconstructing the martensite structure. Once cooled to room temperature, the organization of the discharge crater surface becomes distinct, with a clear remelted layer boundary. The increased surface hardness is attributed to the formation of high-hardness martensite within the remelted layer [11].

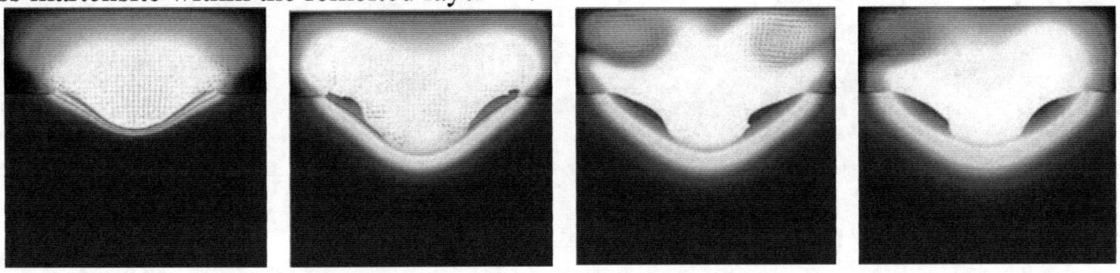

(a) Simulation 30 us (b) Simulation 60 us (c) Simulation 120 us (d) Simulation 150 us

Figure 3. Simulation temperature distribution.

4.2. Optimization of simulation parameters

Theoretically, the remelted layer in Electrical Discharge Machining (EDM) is formed by the continuous accumulation of single-pulse discharges. Due to the randomness of each discharge, the remelted layer thickness on the processed surface tends to be uneven, leading to pits and protrusions that influence surface roughness and machining quality [12]. Previous experimental studies on EDM-drilling composite processing revealed that the efficiency of subsequent reaming processes is closely related to the remelted layer thickness. Given that titanium alloy is a hard-to-machine material, ensuring tool life is critical for reaming processes. Excessive material removal during reaming significantly reduces tool life, making thinner remelted layers under certain processing conditions preferable. This suggests that smaller remelted layer thicknesses from single-pulse discharges benefit subsequent removal processes [13]. In studies aimed at completely removing remelted layers using EDM-ultrasonic drilling composite processing, orthogonal experiments were conducted to analyze the correlation between various machining parameters.

Using gray relational analysis, the optimal parameter combination was evaluated. Since discharge pulse energy significantly impacts remelted layer thickness, parameters such as current and pulse duration, which directly determine energy levels, were optimized. By selecting smaller peak currents and shorter pulse widths under applicable EDM titanium alloy parameters, an optimal parameter combination for straight-hole micro-hole machining was obtained [14]. Based on prior optimal parameter combinations, four key electrical parameters affecting remelted layer thickness were identified: peak current, pulse width, duty cycle, and gap voltage. These parameters were studied to determine their influence on remelted layer formation during single-pulse discharges, with specific selection principles and settings summarized in the table.

Table 4. Experimental condition.

Key electrical parameters	Optimal parameter combination values
Discharge current/A	1.5
Discharge gap/um	6.0
Electrode-workpiece material	Copper-single crystal high-temperature alloys

 doi:10.1088/1742-6596/3009/1/012073

Interpolar medium	Deionized water
Voltage amplitude/V	60
Pulse width/us	20-80

To validate the simulation results of the discharge crater model, pulse discharge experiments were conducted by using an electrical discharge machining (EDM) experimental device. The morphology of the discharge craters was analyzed, and their dimensions were measured. Since EDM operates via tip discharge, the cross-sectional remelted layer thickness at the bottom of the crater corresponds to the surface characteristics of a single-pulse EDM crater, ensuring comparability between experimental and simulation results. The average remelted layer thickness measured in the experiment was 12.4 μm, as shown in Figure 4.

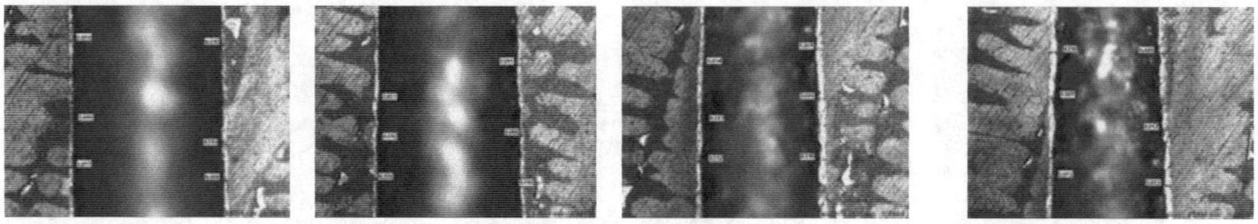

(a) Simulation 30 us (b) Simulation 60 us (c) Simulation 120 us (d) Simulation 150 us

Figure 4. Test measurement of remelted layer thickness.

Experimental data validation:

(1) Based on COMSOL simulation results, the remelted layer thickness before optimization was approximately 16 μm, while after optimization it was reduced to around 10 μm. These results align with the second-order regression fitting formula relating peak current, pulse width, and remelted layer thickness proposed by Zhang [15], expressed as:

$$H = 1.85 + 0.1034A + 0.1102B + 0.0104AB \tag{8}$$

where A is the peak current, B is the pulse width, and H is the remelt layer thickness.

(2) Through orthogonal experiments, a series of parameters were determined, showing significant improvements: electrode wear was reduced by 43.90%, machining time was shortened by 32.1%, hole diameter decreased by 7.1%, and the remelted layer thickness was reduced by 25%, from 16.4 μm to 12.4 μm after optimization.

5. Composite processing test without remelting layer

Based on the optimized parameters, straight-hole composite machining experiments were conducted, with specific parameters listed in Table 5, and the process flow shown in Figure 5. Figures 6 (a)-(c) depict the EDM-formed hole wall surface, revealing pits and remelted material caused by secondary discharges. The wall surface also shows condensed remelted material, forming a deteriorated layer that leads to slight burrs at the hole edge. Figure 7 illustrates the distribution of the remelted layer, where color differences highlight the base material and remelted layer, with the latter unevenly distributed.

Table 5. Straight-hole composite machining parameters.

Processing parameters	Specific values
Pulse width	5.0 u sec
Pulse interval	6.0 u sec
Peak current	1.5 A
Servo reference voltage	60.0 V
Main circuit voltage	120 V
Feed rate	24 u m/r
Spindle speed	12, 000 r/min
EDM bottom hole bore diameter	0.31 mm
Tool size	0.35 mm

(a) EDM drilling (b) Drilling and reaming

Figure 5. Straight hole composite machining process.

(a) (b) (c)

Figure 6. Straight hole EDM small hole picture.

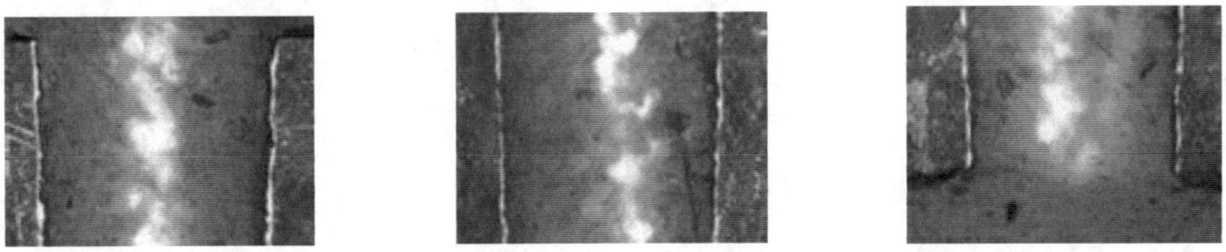

(a) Hole entrance remelting layer (b) Intermediate section remelt layer (c) Hole exit remelt layer

Figure 7. EDM remelted layer metallographic results.

Figures 8 (a)~(c) show the surface morphology of holes processed through composite machining. The material removed during the drilling reaming process effectively eliminated the remelted layer initially attached to the hole wall. The circularity of the micro-holes indicates that machine tool errors did not compromise the hole geometry. As shown in Figure 9, the remelted layer was entirely removed at the inlet, middle, and outlet sections after composite machining, confirming that machine tool errors did not affect the removal process. This validates the effectiveness of EDM-drilling composite machining in achieving remelt-free straight-hole processing.

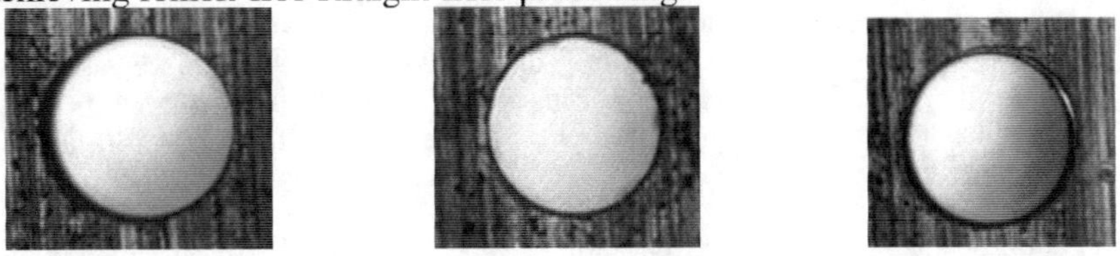

(a) (b) (c)

Figure 8. Straight hole composite machining of small holes picture.

(a) Hole entrance remelting layer (b) Intermediate section remelt layer (c) Hole exit remelt layer

Figure 9. Metallographic inspection of hole wall after composite processing.

6. Conclusions

This study proposes a composite machining method combining Electrical Discharge Machining (EDM) with ultrasonic vibration drilling, effectively addressing the challenge of remelted layer removal in micro-hole processing. Through multi-body dynamic error analysis and parameter optimization, the method significantly reduces mechanical errors, achieving remelt-free machining. Tailored for titanium alloy characteristics, the findings highlight that thinner remelted layers enhance tool life and machining efficiency.

Using a COMSOL multiphysics coupling model, the optimal remelted layer thickness and machining parameters were determined and experimentally validated. Results showed a 43.90% reduction in electrode wear, a 32.1% reduction in machining time, and a 25% decrease in remelted layer thickness from 16.4 μm to 12.3 μm, ensuring no residual remelted layer on hole walls.

This approach improves both machining efficiency and surface quality, demonstrating broad application prospects in advanced manufacturing, and provides theoretical and practical support for further development of micro-hole processing technologies.

Acknowledgments

This work was sponsored by the National Key Research and Development Program of China (Grant No. 2023YFB4605900), Beijing Nova Program (No. 20240484626), and BIPT Zhiyuan Research Fund (No. 2023003).

References

[1] Song Jinbang and Yi Haiyun. Machining technology for cooling film holes in turbine blades of aero-engines [J]. Tool Engineering, 2020, 54 (12): 82-86.
[2] Qiao Wangju. Application of composite machining technology in manufacturing aerospace structural components [J]. Aeronautical Manufacturing Technology, 2009 (12): 40-43.
[3] CORCORAN A., SEXTON L., SEAMAN B., et al. The laser drilling of multilayer aerospace material systems [J]. Journal of Materials Processing Technology, 2002, 123 (1): 100-106.
[4] Shao Junling. Experimental research on composite machining technology for metal micro-holes without remelted layers [D]. Anhui University of Science and Technology, 2016.
[5] Yin Dapeng. Machining technology for cooling film holes in turbine blades of aero-engines [D]. Dalian: Dalian University of Technology, 2013.
[6] Jia Baoxian, Wang Zhenlong, and Zhao Wansheng. Micro-hole machining technology based on special processing [J]. Electrical Machining and Die Mould, 2005 (02): 1-5.
[7] L. Selvarajan, R. Rajavel, K. Venkataramanan, and V. P. Srinivasan. Experimental investigation on surface morphology and a recasting layer of Si_3N_4-TiN composites machined by die-sinking and rotary EDM [J]. Ceramics International, 2022.
[8] Qin Sheng, Mi Sipei, Ming Weiwei, et al. Research on temperature measurement in deep-hole drilling of titanium alloy and machining surface quality [J]. Mechanical Design and Manufacturing, 2021 (07): 155-157+162.
[9] Tian Y., Zou P., Wang X., et al. Study on chip shapes in longitudinal-torsional composite ultrasonic vibration-assisted drilling of Ti-6Al-4V. Int. J. Adv. Manuf. Technol. 128. 2023; 5, 561-5, 571.
[10] Jithin S., Raut A., Bhandarkar V. U., et al. FE Modeling for Single Spark in EDM Considering Plasma Flushing Efficiency [J]. Procedia Manufacturing, 2018, 26, 617-628.
[11] Li En. Thermal-flow coupling simulation and experimental study of single-pulse discharge in electrical discharge machining [D]. Harbin University of Science and Technology, 2021.
[12] Yue Xiaoming. Research on material removal mechanisms and surface layer formation in EDM [D]. Harbin Institute of Technology, 2019.
[13] Luo Xueke, Li Ruobing, Liu Jianyong, et al. Research on composite machining technology for micro-hole EDM-drilling without remelted layers [J]. Aeronautical Manufacturing Technology, 2023, 66 (19): 41-47.
[14] Li Ruobing, Luo Xueke, Liu Jianyong, et al. Optimization of EDM micro-hole processing parameters based on canonical correlation analysis and grey relational analysis [J]. Electrical Machining and Die Mould, 2023 (01): 15-21.
[15] Zhang T. L. Simulation and experimental study of remelted layer forming process on EDM surface[D]. Harbin Institute of Technology, 2023 (in Chinese).

Journal of Physics: Conference Series **3009** (2025) 012074 doi:10.1088/1742-6596/3009/1/012074

A comparative experimental study on fault diagnosis of rolling bearings based on vibration and acoustic emission

Weijie Gao, Leixuan Zhang, Shuyao Ding, Guoan Yang*

School of Mechanical and Electrical Engineering, Beijing University of Chemical Technology, Beijing, China

* Corresponding author e-mail: yangga@mail.buct.edu.cn

Abstract: A comparative experimental study was conducted on the detection performance of vibration and acoustic emission detection methods in the natural degradation of rolling bearings. A fault diagnosis test bench for rolling bearings was designed and accelerated life tests were conducted on rolling bearings under different operating conditions. two methods, vibration and acoustic emission, were used to monitor the tested bearings online simultaneously. the experimental results indicate that for early bearing failures, acoustic emission is more sensitive than vibration. the magnitude of the load on the bearing has a significant impact on the advance of acoustic emission detection. the absolute energy and effective value (RMS) of acoustic emission characteristic parameters can serve as criteria for determining whether a bearing has malfunctioned.

1. Introduction

Rolling bearings are widely used in various types of rotating machinery and have frequent malfunctions [1]. At present, the most commonly used diagnostic method is vibration analysis [2]. However, the vibration signals of early bearing failures are very weak and easily drowned out in background noise, so other means need to be used to detect them in advance.

The high-frequency characteristics of acoustic emission signals have unique advantages for fault detection of rolling bearings. Acoustic emission refers to the transient elastic waves generated by the rapid release of strain energy caused by deformation or damage on the surface or inside of a material [3]. Its application in bearing fault detection has made up for the shortcomings of vibration analysis. Morhain [4] validated the effectiveness of using RMS, absolute energy, amplitude and ringing count for bearing diagnosis. Mba [5] added artificial faults on the inner and outer rings of the bearing and found that the outer ring defect caused an increase in RMS value. However, the same trend was not observed in the inner circle defects. Shiroishi [6] pointed out that in the diagnosis of bearing inner ring faults acoustic emission is not sensitive to inner ring faults, and is even ineffective. Elforjani [7] demonstrated through experiments that there is a clear correlation between the increase in acoustic emission energy and the formation of bearing faults. Yoshioka [8] has demonstrated that through research acoustic emission can detect early bearing failures before vibration response.

This article further develops on the basis of previous research, focusing on the advantages of acoustic emission technology over vibration analysis under natural degradation conditions of rolling bearings.

2. Experimental apparatus

2.1 Test bench structure

The experimental platform was independently designed and constructed by the author (as shown in Figure 1). The test bench consists of a power transformer, a frequency converter, a brushless motor, a transmission device and a loading device. Accelerated life tests of rolling bearings can be conducted under different operating conditions to obtain monitoring data on the entire life cycle of the tested bearings.

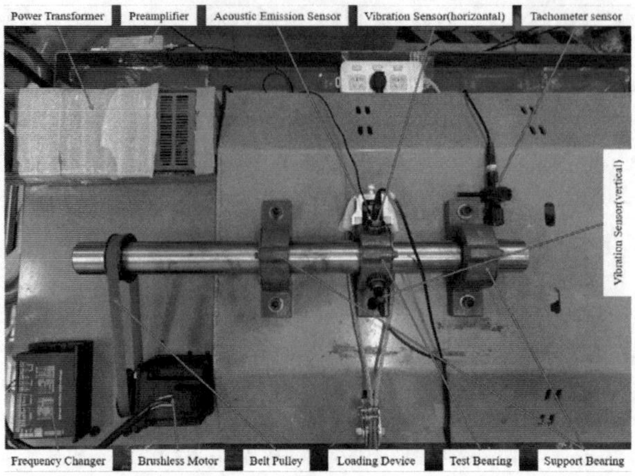

Figure 1. Rolling bearing fault diagnosis test bench.

The test-bearing model is 6710 (as shown in Figure 2), and its main parameters are shown in Table 1.

Figure 2. Testing bearings.

Table 1. Main parameters of test bearing 6710.

Parameter	Value
Rolling element diameter/mm	2.3
Bearing pitch diameter/mm	56
Number of rolling elements	30
Contact angle/°	0
Basic dynamic load rating/kN	2.6

2.2 Test conditions

The speed of this experiment is set at a constant value of 800 r/min. By changing the load on the test bearing, the advance of acoustic emission compared to vibration detection of bearing faults is studied. Five operating conditions were designed for the experiment, as shown in Table 2. The characteristic frequencies and their harmonics of various parts of the bearing are tested, as shown in Table 3.

Table 2. Accelerated Life Test Conditions for Bearings.

Test number	1	2	3	4	5
Radial load/kN	1.2	1.4	1.6	1.8	2

Table 3. Test bearing characteristic frequencies and their harmonics.

Frequency doubling	Frequency/Hz			
	FTF	BSF	BPFO	BPFI
1x frequency	6.4	162.7	191.8	208.2
2x frequency	12.8	325.4	383.6	416.4
3x frequency	19.2	488.1	575.4	624.6
4x frequency	25.6	650.8	767.2	832.8

2.3 Failure threshold

The failure threshold of the tested bearing in this experiment is defined as the complete failure of the bearing when there is a second surge in acoustic emission activity, that is, when the vibration amplitude increases for the first time.

3. Data analysis

The detailed test information for testing bearings is shown in Table 4. It's noted that the sound emission in the inner ring fault did not occur before vibration detection, which is consistent with previous research conclusions. It is speculated that this may be due to the complex signal propagation path and excessive signal attenuation in the inner ring fault.

Table 4. Detailed Test Information for Testing Bearings.

Number	Response time of AE/h	Response time of vibration/h	Fault location	AE advance
1	23.6	23.6	Inner ring, retainer	0
2	7.6	9.4	rolling elements outer ring, retainer	19%
3	4.0	8.2	rolling elements retainer	33%
4	2.2	3.0	rolling elements outer ring retainer	26%
5	4.7	5.4	rolling elements retainer	13%

3.1 AE analysis

The analysis methods for acoustic emission signals mainly include parameter analysis [9] and waveform analysis [10]. Taking condition 5 as an example, this experiment mainly measured the root mean square (RMS) and absolute energy of the signal. In addition, amplitude, ringing count and ASL were also measured.

The mean square voltage and root mean square voltage of a signal V (t) are defined as follows:

$$V_{MS} = \frac{1}{\Delta T}\int_0^{\Delta T} V^2(t)dt \qquad (1)$$

$$RMS = V_{RMS} = \sqrt{V_{MS}} \qquad (2)$$

In the formula, T is the average time, and V (t) is the signal voltage that varies with time.

The absolute energy of a signal is the square of the root mean square voltage divided by the integral of the impedance (10k Ω) over time. The ringing count represents the number of times the threshold signal is crossed, and the amplitude represents the maximum amplitude value of the signal waveform. ASL represents the mean signal level during the sampling time, measured in decibels (dB), and is defined as follows:

$$ASL(dB) = 20\log_{10}(1.4 \times RMS/100) \qquad (3)$$

The unit of RMS in the equation is millivolts (mV).

Journal of Physics: Conference Series **3009** (2025) 012074 doi:10.1088/1742-6596/3009/1/012074

The bearing is in good condition during the initial operation (as shown in Figures 3-4). Starting from 4.7 hours, the appearance of internal defects in the bearing resulted in a significant increase in both absolute energy and RMS amplitude but soon decreased to similar amplitudes to that before. This is because acoustic emission is a real-time dynamic detection technology. As the defect did not further expand, the amplitude returned to normal. Starting from 5.4 hours, new damages continue to appear on the surface of the bearing, and there is an explosive growth in acoustic emission activity, indicating that the bearing has entered the late stage of failure.

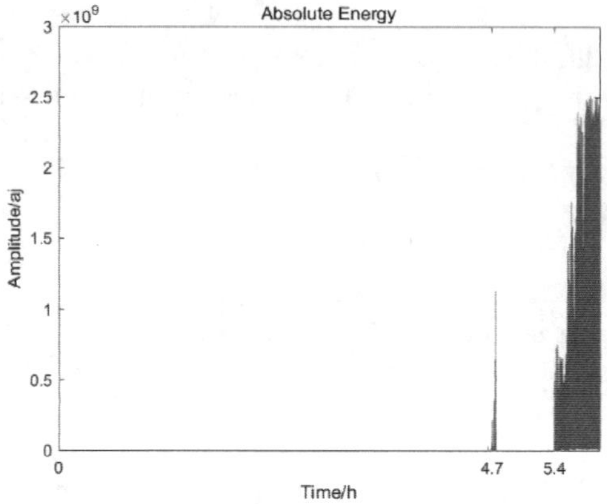

Figure 3. Acoustic emission parameters - absolute energy.

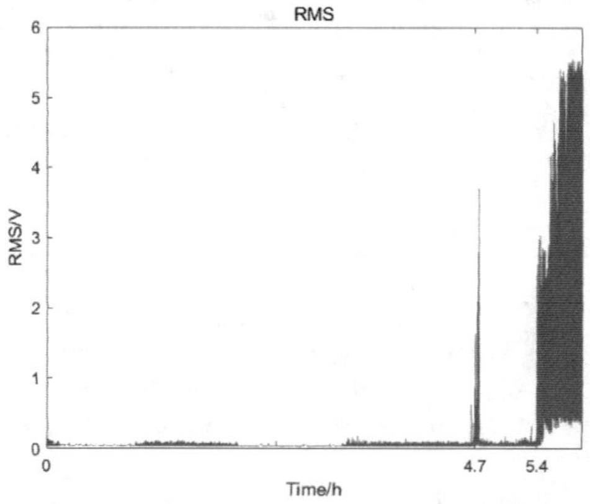

Figure 4. Acoustic emission parameters – RMS.

Figure 5 shows the variation trends of acoustic emission parameter amplitude, ringing count and ASL, which are consistent with absolute energy and RMS. Under different operating conditions, it was found that there was some acoustic emission activity during the initial operation of the test bearings, mainly because the newly replaced test bearings were in the running period.

Journal of Physics: Conference Series **3009** (2025) 012074 doi:10.1088/1742-6596/3009/1/012074

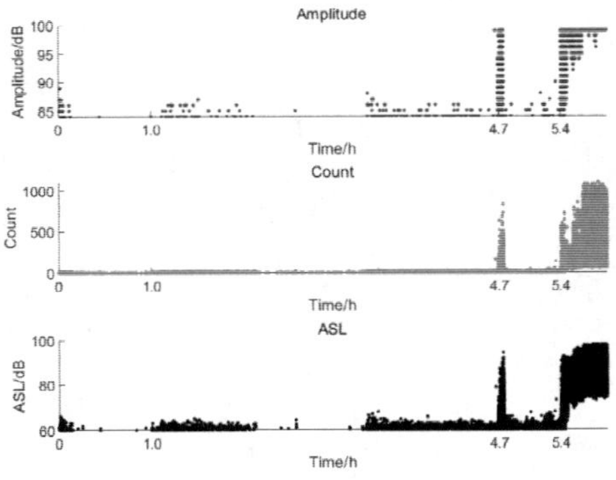

Figure 5. Acoustic emission parameters - amplitude, ringing count and ASL.

The different types of sensors and testing system settings can have a significant impact on acoustic emission parameters. The sound emission waveform contains a large amount of sound source information, which can effectively compensate for the shortcomings of parameter analysis.

As shown in Figures 6-7, when the level of acoustic emission activity increases for the first time after 4.7 hours, periodic impact signals appear in the waveform, and the frequency and its harmonics of the rolling elements are obvious. At this point, defects have already appeared inside the bearing. But the defect did not further expand, the impact signal disappeared, and the frequency amplitude of the rolling element decreased. Starting from 5.4 hours, the development of bearing defects rapidly increased, and acoustic emission activity surged. The amplitude of the collected signal showed a phenomenon of peak shaving, and the amplitude of the second harmonic of the rolling element in the spectrum significantly increased. Indicating that the bearing defect has extended from the subsurface to the surface, resulting in fatigue spalling damage.

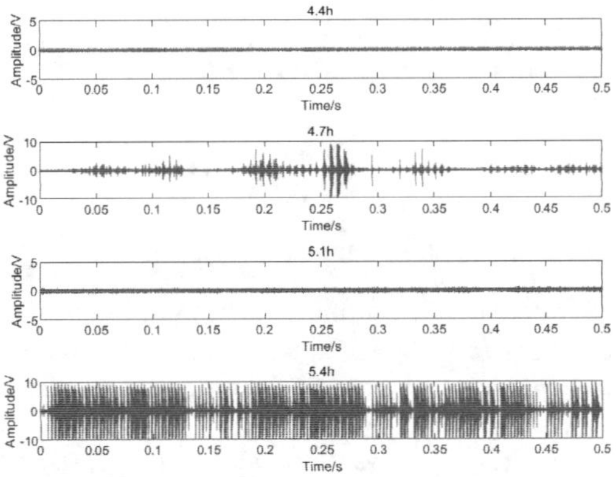

Figure 6. Acoustic emission waveform.

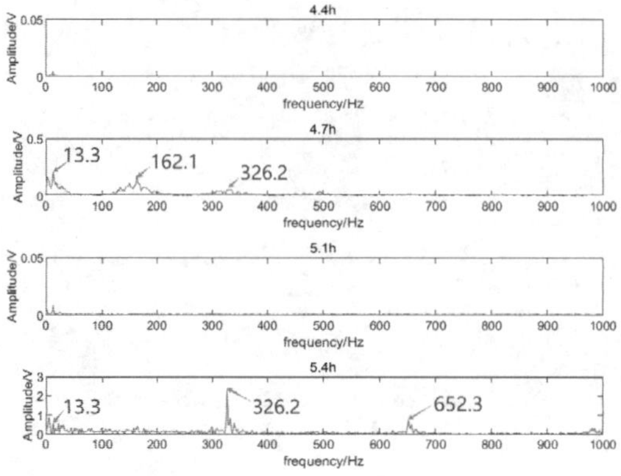

Figure 7. Acoustic emission envelope spectrum.

3.2 Vibration analysis

The vibration signal during the entire life cycle of the test bearing is shown in Figure 8. When the sound emission first detected the occurrence of internal defects in the bearing, the amplitude of the vibration signal did not show any abnormalities. The load on the test bearing is in the horizontal direction, so the vibration signal in the horizontal direction is selected for analysis.

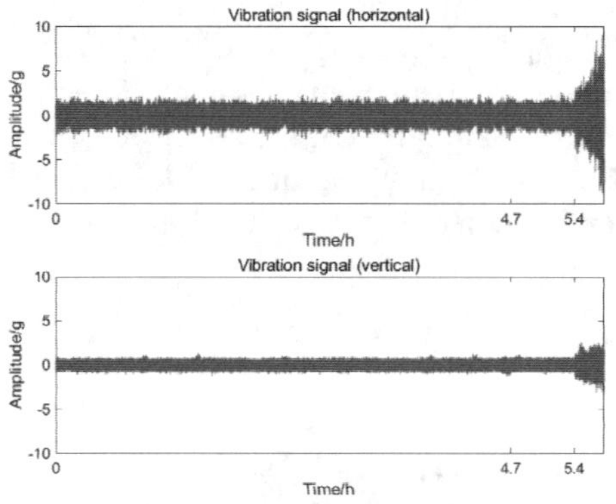

Figure 8. Test the vibration signal of the bearing throughout its entire life cycle.

To verify the superiority of acoustic emission in early fault detection of bearings, vibration signals at the same time were selected for comparison (as shown in Figures 9-10). The occurrence of subsurface damage inside the bearing does not cause a change in the operating state of the bearing. Therefore, the envelope spectrum of the mid-term vibration analysis does not show the frequency of the bearing rolling elements passing through. When the bearing enters the stage of complete failure, surface damage occurs on the rolling elements, causing additional load and friction on the cage, resulting in damage to the cage, which is mainly manifested in the envelope spectrum as the cage passing frequency and its harmonics.

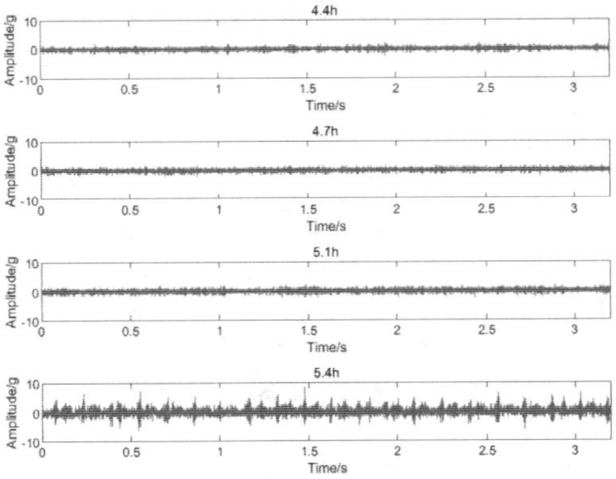

Figure 9. Vibration data waveform.

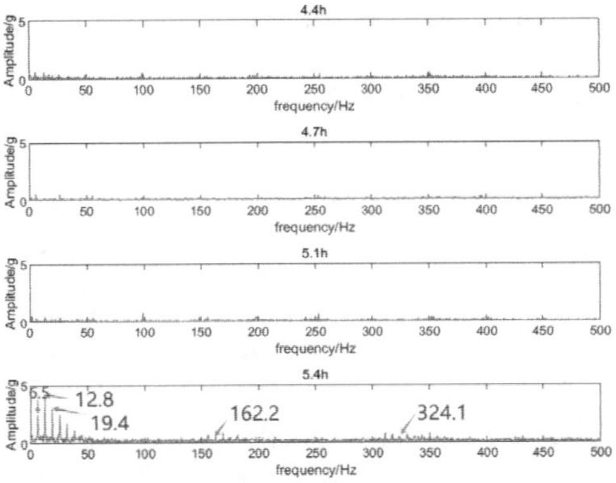

Figure 10. Vibration data envelope spectrum.

From the above analysis, it can be concluded that the advance of fault detection by acoustic emission compared to vibration is about 13% of the total life of the tested bearing, and the fault type is a composite fault of rolling elements and retainer.

3.3 Result Verification

After each experiment, the test bearings were disassembled to verify the test results, as shown in Figure 11.

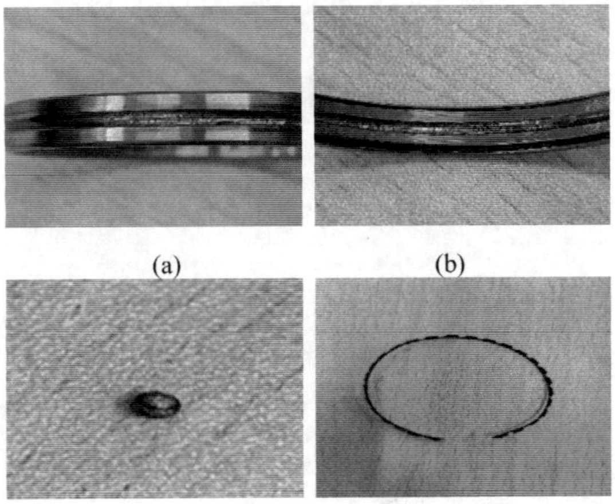

(c) (d)

Figure 11. Test bearing failure (a) Inner ring (b) Outer ring (c) Rolling element (d) Retainer.

4. Conclusion

This article conducts a comparative experimental study on the application of two detection methods, vibration and acoustic emission, in the natural degradation process of rolling bearings. Through the analysis of experimental data, the following conclusions are drawn:

1) Acoustic emission is sensitive to early bearing failures, especially rolling element failures. Acoustic emission effectively compensates for the shortcomings of vibration analysis in early fault detection of bearings.

2) The magnitude of the load on the bearing has a significant impact on the advance of acoustic emission detection. As the load on the bearing increases, this advance tends to decrease.

3) The trend of absolute energy and RMS variation of acoustic emission parameters can serve as a criterion for determining whether early faults occur in bearings. When early bearing failures occur, the absolute energy and RMS amplitude increase exponentially, showing a clear correlation with the fault and not affected by threshold settings, with good robustness.

References

[1] Xie, C. H, Wang X, Xiao, F. H, et al. (2024) Overview of Bearing Fault Modes and Fault Diagnosis Methods. Computer Measurement and Control, 1-11.
[2] Yang, G, Xu, W. Y, Deng, Q, et al. (2024) A Review of Research on Composite Fault Diagnosis of Rolling Bearings Based on Vibration Signals. *Journal of Xihua University* (Natural Science Edition), 43(01): 48-69.
[3] ISO 22096: 2007, Condition Monitoring and Diagnostics of Machines — Acoustic Emission [S].
[4] Morhain A, Mba D. (2003) Bearing Defect Diagnosis and Acoustic Emission. ARCHIVE Proceedings of the Institution of Mechanical Engineers Part J Journal of Engineering Tribology 1994-1996 (vols 208-210), 217(4): 257-272.
[5] Mba D. (2003) Acoustic Emissions and Monitoring Bearing Health. STLE Annual Meeting and Exhibition, Society of Tribologists and Lubrication Engineers (STLE).
[6] Shiroishi J, Li Y, Liang S, et al. (1997) Bearing Condition Diagnostics via Vibration and Acoustic Emission Measurements. Mechanical Systems & Signal Processing, 11(5): 693-705.
[7] Elforjani M, Mba D. (2010) Accelerated Natural Fault Diagnosis in Slow Speed Bearings with Acoustic Emission. Engineering Fracture Mechanics, 77(1): 112-127.
[8] Yoshioka T, Fujiwara T. (1982) A New Acoustic Emission Source Locating System for the Study of Rolling Contact Fatigue. Wear. 81(1): 183-186.
[9] Li, B. (2024) Research on Low Signal-to-noise Ratio Acoustic Emission Signal Processing Technology for Rotating Machinery Bearing Faults. Shenyang University of Technology, 2024. 000023.
[10] Tong, X. Y, Sha, Y. D, Luan, X. C, et al. (2024) Fault Diagnosis Method for Rolling Bearings Based on Acoustic Emission Characteristic Parameters and Waveform Flow Analysis. Science Technology and Engineering, 24(30): 13180-13188.

Study on the initial corrosion behavior of Q235 steel in high-temperature brine

Yalin Chen [1], Jiangyong Guan [1], Zhongxiang Chen[2], Huiwu Fan [1,*], Guanzhong Cao[1], Zhehua Wang[1]

[1] Qingdao Economic and Technological Development Zone Haier Water Heater Co., Ltd, Qingdao, 266000, China
[2]Qingdao Haida Cheng Procurement Service Co., Ltd, Qingdao, 266000, China

[*]Corresponding author's e-mail: fanhuiwu@haier.com (H. Fan).

Abstract: The corrosion behaviors of Q235 steel in brine at 45°C and 60°C were studied in this manuscript. The results showed that the increase in temperature promoted the corrosion. At 45°C, the formation of the rust layer was slow, the adsorption of Cl^- on the rust layer was uniform, the corrosion of metal was relatively uniform, and the average corrosion rate and pitting rate were low. After the formation of a relatively dense rust layer, the corrosion rate decreased rapidly. From then on until the end of the trial, the state of the rust layer and the corrosion process were all relatively stable. Under the condition of 60°C, the formation of the rust layer was faster, the adsorption of Cl^- on the rust layer was not uniform, and the average corrosion rate and pitting rate of metal were faster. After the formation of the rust layer, the corrosion rate also decreased rapidly, but during the period when the rust layer was stable, more serious pitting corrosion occurred. At the end of the trial, parts of the rust layer peeled off, the pitting corrosion was alleviated, but the uniform corrosion rate increased.

1 Introduction

Carbon steel is a common structural material, widely used in pipelines, ships, bridges, and other engineering fields. However, carbon steel will be affected by corrosion when exposed to some specific environments, especially high-salt brine. The Cl^- in brine will accelerate the corrosion rate of the metal, especially for plain carbon steel. It will not only lead to corrosion on the surface of carbon steel but also penetrate into the internal structure, resulting in damage to the structure and performance degradation [1-4]. The ambient temperature is also considered to be one of the important factors affecting corrosion behavior [5-7]. In the high-temperature service environment, the microstructure of carbon steel may change, making it more vulnerable to corrosion, and the anti-corrosion performance will degrade [8, 9]. Therefore, studying the corrosion behavior of carbon steel in brine environments at different temperatures can accumulate corrosion data, obtain the corrosion laws of materials in brine environments, and guide the selection of materials in different scenarios [10-12].

In this study, the corrosion process of carbon steel at 45°C and 60°C was studied. The corrosion mechanism of carbon steel in brine at two different temperatures was systematically explored through a comprehensive analysis of the corrosion rate, microscopic morphology, rust layer composition, polarization curve, and impedance spectrum.

2 Experiment

2.1 Test Materials

The brine was taken from Qarhan Salt Lake in Qinghai Province. The lake is located at 36°50 ' north latitude, 95°06' east longitude, water depth of 1.5~2 meters, elevation of 2680 meters, an average annual temperature of 4.3°C, an average annual rainfall of 38.3 mm, and an average annual evaporation of 3067 mm. The brine is a saturated salt solution with a salt content of about 32.5%. The specific components of the brine are shown in Table 1. The material used in the test was Q235 steel produced in the same batch, and its chemical composition was C of 0.19%, Si of 0.13%, Mn of 0.19%, P of 0.042%, S of less than 0.01%, and Fe of the rest.

Table 1 The Chemical Composition of Brine.

Specific Gravity	°Bé	KCl%	NaCl%	$MgCl_2$%	$CaSO_4$
1.29	32.44	1.48	1.35	28.84	0.1

2.2 Experimental Methods

In accordance with the Chinese Machinery Industry Standard of JB/T 7901-1999, the test method for uniform corrosion of metallic materials in the laboratory [13], the Q235 steel coupons were placed in brine and heated in a constant temperature water bath at 45°C and 60°C respectively, then measured on day 3, day 6, day 9 and day 12. Four parallel samples of 20 mm×20 mm×3 mm were set for every measurement. The coupons were deburred and machined to a surface roughness of 0.8 Ra. Three of the parallel samples were used for weight loss analysis and the rest was used for a microscopic test of the rust layer. The electrochemical sample of 10 mm×10 mm×3 mm was sealed with epoxy resin, with only one square surface exposed. After being fully cured, the detected surface was continuously polished to 1500 mesh with waterproof abrasive paper.

In accordance with the Chinese National Standard of GB/T 16545-2015, removal of corrosion products from corrosion test specimens [14], the coupons for mass loss analysis were pickled after the test, and then cleaned with distilled water, wiped with anhydrous alcohol, dried with hair dryer and weighed at a precision of 0.1 mg. The microtopography and element distribution of the rust layer was analyzed by COXEM EM-30AX scanning electron microscope integrated with energy spectrum. The three-dimensional morphology after pickling was observed by KEYENCE VHX7000 ultra depth of field microscope.

The traditional three-electrode system was adopted for the electrochemical test. The electrochemical sample was the working electrode, the auxiliary electrode was a 15 mm×15 mm platinum electrode, the reference electrode was the saturated calomel electrode, and the electrolyte solution was salt lake brine. The sample was measured by electrochemical impedance spectroscopy and potentiodynamic polarization in turn. The measurement frequency of EIS was from 10^5 Hz to 10^{-2} Hz, the amplitude was 10 mV, and the test results were fitted by ZSimpWin. The potentiodynamic polarization range was -0.4V_{OCP} to +0.6V_{OCP}, the scan frequency was 0.33 mV/S, and the curve was fitted by CView software.

3 Experiment Analysis and Further Discussion

3.1 Corrosion Rate

The corrosion rates during different experimental periods of Q235 steel in brine at 45°C and 60°C were shown in Figure 1. The corrosion rate was calculated according to the weight loss of the coupon, and the result represented the average corrosion rate of the coupon. The result in Figure 1 was the average of 3 parallel samples. The variations of corrosion rate at different experimental temperatures were obviously different, and for the same experimental period, the corrosion rates at 60°C were all faster than those at 45°C. When the experimental temperature was 45°C, the variation of corrosion rate showed a rapid decline and tended to be stable, and the corrosion rate of 0.273 mm/a was the lowest on day 12. When the test temperature was 60°C, the variation of corrosion rate showed an initial decrease which was followed by a subsequent increase, and the corrosion rate for each experimental period was significantly different. The lowest corrosion rate was 0.312 mm/a on day 9. It showed that the temperature had a significant effect on the initial corrosion of Q235 steel in brine, which accelerated corrosion by increasing the electrochemical reaction rate [15]. The dissolved oxygen content in high-concentration brine is low, and the increase in temperature will further reduce the dissolved oxygen. But at least below 60°C, the influence of temperature on the corrosion rate is dominant, and the corrosion was not slowed down by the decrease of dissolved oxygen.

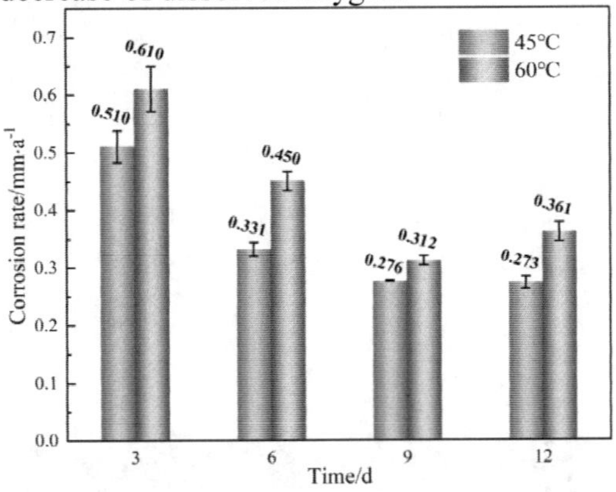

Figure 1 Corrosion Rates of Q235 Steel in 45°C and 60°C Brine.

3.2 Surface Micro-Topography

The microtopographies of the rust layer on coupons for each experimental temperature at different times are shown in Figure 2. Under the two temperature conditions, there were obvious differences in the surface rust layer for the same experimental period. In the 45°C brine, after 3 days of testing, the rust layer was relatively loose and there were a large number of pores, which could be used as the channel for the diffusion of the corrosive medium to the metal matrix [16, 17], which promoted the corrosion. After 6 days of testing, although there were some micro-cracks on the rust layer, the overall rust layer became relatively dense, and this dense rust layer played a certain role in protecting the matrix [18]. On the 9th and 12th day of the test, the rust layer did not change much, the surface was relatively dense with multiple cracks.

In the 60°C brine, after 3 days of the test, the surface of the rust layer has been relatively dense. On the 6th and 9th day, the rust layer became gradually dense, and there were also multiple cracks. On the 12th day of the test, the rust layer showed partial shedding.

The microtopography changes of the rust layer lucidly explained the variations in corrosion rate. At 45°C, after 6 days of testing, the rust layer has become relatively dense, so the corrosion rate decreased rapidly. On the 9th and 12th day, the surface of the rust layer became denser and changed slightly, so the corrosion rate gradually decreased and tended to be stable. At 60°C, due to the rust layer shedding on the 12th day, the corrosion rate rebounded and increased [20].

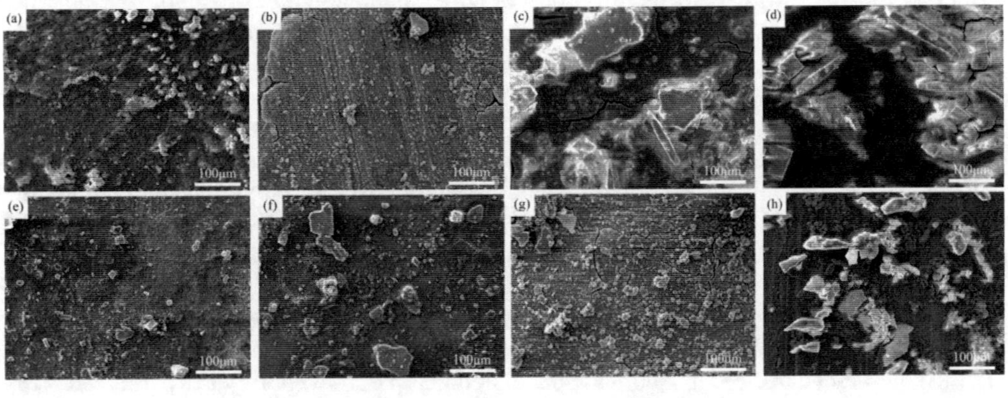

Figure 2 Micro-Topographies of Rust Layer on Coupons for the temperature of 45 °C on day 3 (a), day 6 (b), day 9 (c), day 12 (d), and 60°C on day 3 (e), day 6 (f), day 9 (g), and day 12 (h).

3.3 X-Ray Energy Dispersive Spectrometer of Rust Layer

The EDS test results of the rust layer on coupons which soaked in brine at 45°C and 60°C for 12 days were shown in Figure 3 and Figure 4. After 12 days of testing in brine at 45°C, there were scattered NaCl crystalline salts on the surface of the rust layer [19]. Meanwhile, as the brine contained $MgCl_2$ with a mass percentage of 28.84%, besides iron oxides, $MgCl_2$ was uniformly enriched on the surface of the rust layer, which participated in the growth of the rust layer [21]. Fe was obviously enriched at the crack of the rust layer, where the sample would be preferentially corroded. It could be judged that the crack of the rust layer would directly affect the corrosion rate of the metal.

The element quantitative test was carried out on the three points in Figure 3 respectively. The relative contents of Cl and Na at Point 1 were 46.9% and 36.6% respectively, indicating that this was a NaCl crystal. There were cracks at Point 2, and the Fe content was obviously higher, accompanied by a large amount of $MgCl_2$. Point 3 was the bulge of the oxide on the rust layer. Judging from the element content, this point contained trivalent oxide of iron and was still accompanied by a large amount of $MgCl_2$.

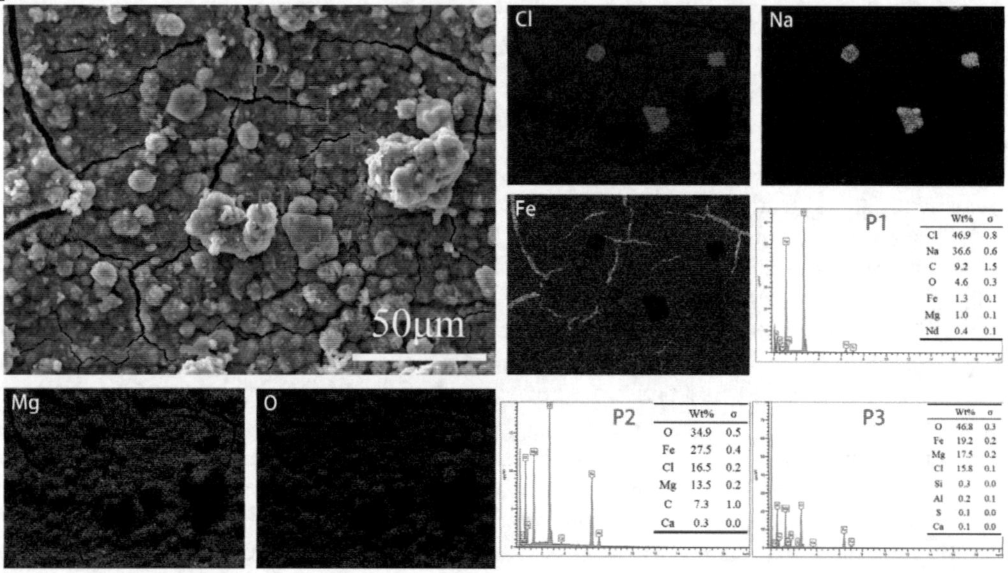

Figure 3 The EDS Results on Coupon in 45°C Brine on Day 12.

After 12 days of testing in brine at 60°C, the main components of the rust layer were the same as those in brine at 45°C, but the distribution characteristics of elements were obviously different. Firstly, the distribution of $MgCl_2$ was characterized by local enrichment, and the distribution of Mg and Cl was similar and complementary to that of Fe. It indicated that $MgCl_2$ crystals grew locally to a certain thickness and covered the iron oxides. Secondly, no obvious phenomenon of Fe enrichment at the crack was found in Figure 4, suggesting that a large amount of $MgCl_2$ crystals might cover the crack and dilute the Fe content at the crack. In short, compared with the detection results in brine at 45°C, it could be concluded that the temperature increase led to the increase of $MgCl_2$ concentration on the rust layer. On the one hand, $MgCl_2$ was acidic when hydrolyzed, destroying the dense oxide layer. On the other hand, a large number of chloride ions might lead to local corrosion of Q235 steel, which will be further explained in the later tests.

The element quantitative test was carried out at the three points in Figure 4 respectively. Point 1 was poor in $MgCl_2$, judging from the content of elements and the point contained a lot of iron oxide. The distribution of all elements at Point 2 was relatively uniform, the content of O was high, and the content of Mg was twice that of Cl, so it was speculated that part of $MgCl_2$ was hydrolyzed. Point 3 was rich in $MgCl_2$, and the iron oxide content was significantly reduced. It was speculated that a large number of $MgCl_2$ crystals were covered on the rust layer, diluting the Fe content.

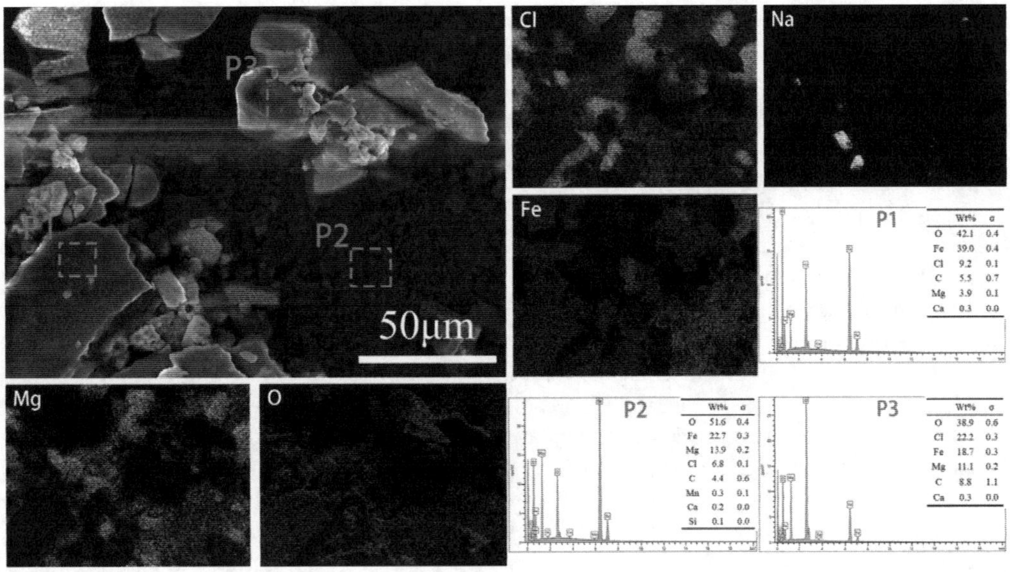

Figure 4 The EDS Results on Coupon in 60°C Brine on Day 12.

3.4 Three-dimensional Corrosion Morphology

The three-dimensional morphologies and local corrosion data of the samples after pickling are shown in Figure 5. It could be seen that after 12 days of testing in brine at 45°C, the maximum depth of the local corrosion pit was about 6.67 μm, and the corrosion was relatively uniform on the whole. While at 60°C, the local corrosion tendency was more obvious, the maximum depth of the corrosion pit reached 17.9 μm, and the number of deep corrosion pits was also more. The results of three-dimensional corrosion morphology confirmed the analysis and speculation of the EDS results as shown in Figure 3 and Figure 4. As shown in Figure 5, both the local corrosion rate and the overall corrosion rate of the sample increased at 60°C, which was inferred to be closely related to $MgCl_2$ enrichment on the rust layer in addition to the influence of temperature [22-24].

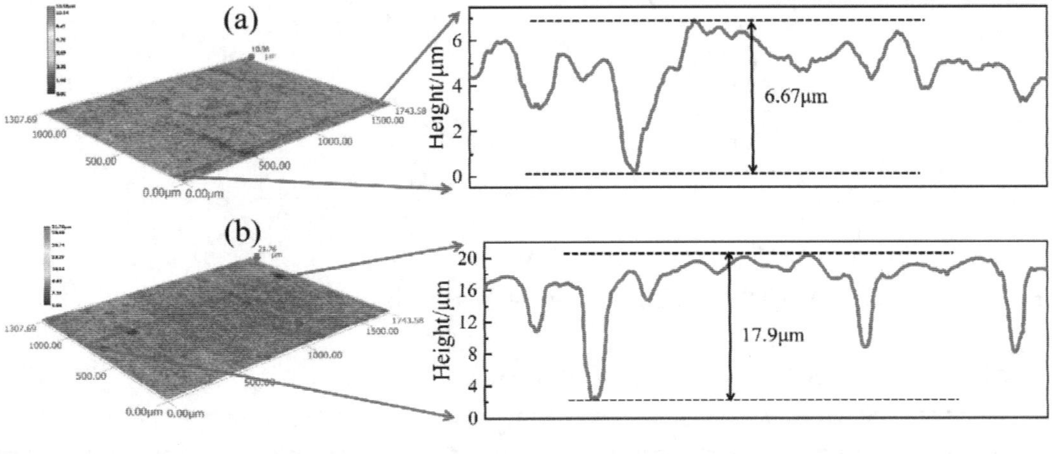

Figure 5 Three-Dimensional Morphology and Maximum Corrosion Pit Depth of Q235 Coupon on Day 12 at The Experimental Temperature of 45°C (a) and 60°C (b).

3.5 Electrochemical Polarization Curve

The polarization curves of Q235 samples in 45°C and 60°C brine are shown in Figure 6, and their fitting parameters are shown in Table 2. The polarization curves were measured under the quasi-steady state and only represented the corrosion state at a certain time, which had some randomness. The fitting results could not be directly compared with the average corrosion rate. In addition, during the testing process, it was found that the self-corrosion potential of the polarization curve differed significantly from the open circuit potential before the start of the test, and there was an obvious current step in the anodic Tafel range for several curves, which may be due to strong polarization changing the original state of the electrode surface [25-27]. On this account, when fitting the curve, linear segments were selected for the anode and cathode regions respectively first, and then the two straights were extrapolated to fit the self-correction potential and corrosion current. However, the fitted parameters may deviate significantly from reality. Here, we will only discuss the approximate trends of the fitted parameters at each experimental temperature. In 45°C brine, the corrosion currents on day 3 and day 6 were significantly higher than those on day 9 and day 12, the overall trend of corrosion current was decreasing. In 60°C brine, the corrosion currents on day 3 and day 6 were significantly higher than those on day 9, and the corrosion current on day 12 was much higher than those on day 3 and day 6. The current trend was first decreasing and then increasing. In the two groups of experiments, the changes in instantaneous corrosion current at different times corresponded to the average corrosion rate of each test period.

On the whole, the polarization curves of each group measured at different times were basically consistent, indicating that the corrosion system of each group was relatively stable. In contrast, the polarization curves measured at 45°C had a large difference, such as the low potential of the polarization curve on day 12. However, combined with the curve fitting, it was found that the data on the 12th day was not much different from those on the other days. On the whole, the four fitting results were close to each other, except for the large cathode Tafel slope on the 6th day. The Tafel range of the polarization curve on the 12th day was basically consistent with other curves, indicating that the corrosion system did not change significantly at that time. In 60°C brine, the curve fitting result on the 9th day was the most special, in which the Tafel slopes of the anode and cathode were both low and the self-corrosion potential was low. It indicated that the corrosion reaction changed significantly around the 6th day. The reason for the change will be further discussed in combination with the electrochemical impedance spectrum.

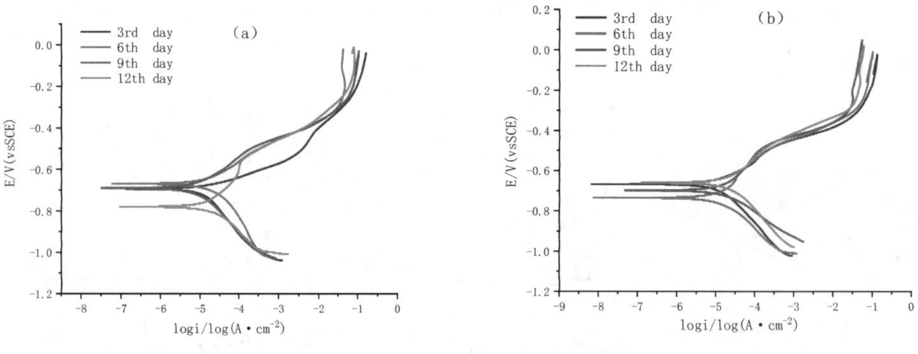

Figure 6 Polarization Curves of Q235 Samples in Brine at the Experimental Temperature of 45°C (a) and 60°C (b).

Table 2 Parameters for Polarization Curve Fitting of Sample.s

	45 °C				60 °C			
	3d	6d	9d	12d	3d	6d	9d	12d
i_{corr} (mA/cm^2)	1.63E^{-2}	2.36E^{-2}	4.11E^{-3}	6.87E^{-3}	6.89E^{-3}	6.39E^{-3}	2.30E^{-3}	1.22E^{-2}
E_0/V (vs SCE)	-0.639	-0.621	-0.630	-0.668	0.622	0.629	0.586	0.580
b_a/(mV)	97.93	88.45	79.13	97.95	67.54	91.49	62.94	88.25
b_c/(mV)	206.52	355.52	190.31	196.81	225.44	200.76	130.89	223.28

3.6 Electrochemical Impedance Spectrum

Figure 7 and Figure 8 respectively showed the electrochemical impedance spectrums of the samples in brine at different temperatures. The Nyquist diagram in Figure 7 was represented by four capacitive loops with close radii. Combined with the phase angle curves of the Bode diagram, which all had only one wave peak, it was determined that every impedance spectrum had one time constant [28]. The corresponding equivalent circuit was shown in Figure 9(a), of which R_s was the solution resistance, and R_p was the polarization resistance. Considering the dispersion effect caused by uneven corrosion on the metal surface, the double layer capacitance on the metal interface was replaced by CPE [28, 29]. It should be noted that the cathode reaction and anode reaction were carried out on the metal surface at the same time, and the impedance spectrum was measured under mixed potential, so the Faraday impedance was expressed by polarization resistance. Although the Nyquist diagram showed that the radius of the capacitive loop on the 3rd day was larger, the R_p values in Table 3 on the 9th and 12th day were slightly larger than those on the 3rd and 6th day, with little difference in the whole. It was also known from the n values, which represented the dispersion coefficient [28], that the non-uniformity of the sample surface continued to increase with the n decreasing. However, the equivalent interface capacitance Q_0 values were close on the 6th, 9th, and 12th day, and significantly greater than that on the 3rd day. It indicated that the rust layer had little influence on the interface capacitance in the middle and late period of the test, it was relatively stable. Combined with the change of R_p, it could be inferred that although the non-uniformity continued superposition with time going, the corrosion reaction was relatively stable in the middle and late period of the test, and the corrosion rate did not fluctuate too much. This is consistent with the weightlessness and microscopic test results of the hanging piece.

In addition, the real part of the impedance in the Nyquist diagram on the 12th day showed an obvious decreasing trend in the low frequency tail. Since the resistance was a specific value and was not changed by the dispersion effect, the variation could only be the influence of other state variables on the Faraday resistance [30], and it was a characteristic of inductive reactance. It required that the effect of the state variable on the Faraday current was the same as that of the potential on the state variable. That was, $\partial X/\partial E<0$ and $\partial I/\partial X<0$, or $\partial X/\partial E>0$ and $\partial I/\partial X>0$, of which X was the state variable, I was the Faraday current, and E was the electrode potential [28]. Since the test environment was fixed and the corrosion system was relatively stable, the state variables of response to Faraday current, such as the thickness of the rust layer and the coverage of the rust layer, could not meet the requirements of the inductive reactance. The inductive element did not conform to the corrosion mechanism. Therefore, the impedance spectrum of the low frequency tail was not discussed further.

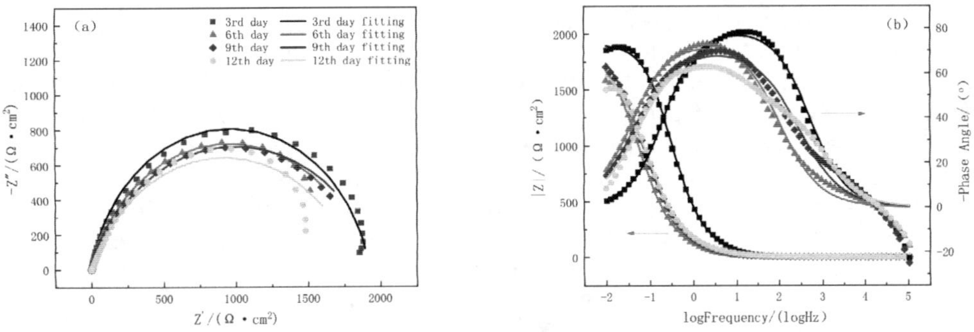

Figure 7 The Nyquist Diagram (a) and Bode Diagram (b) of the Q235 sample in 45°C Brine.

Table 3 Parameters for Impedance Spectrum Fitting of Samples.

	45°C				60°C			
	3d	6d	9d	12d	3d	6d	9d	12d
$R_s/(\Omega\cdot cm^2)$	1.882	2.555	1.971	2.579	1.891	2.287	1.787	2.376

$Q_0/(\Omega^{-1}\cdot cm^{-2}\cdot s^{-n})$	3.98E^{-4}	1.77E^{-3}	1.39E^{-3}	1.36E^{-3}	8.96E^{-4}	7.83E^{-4}	6.86E^{-4}	1.14E^{-3}
n	0.8918	0.8302	0.7882	0.7184	0.8712	0.8402	0.8199	0.7845
$R_p/(\Omega\cdot cm^2)$	1917	1893	1981	2095	2044	1162	800.9	842.2

The Nyquist plot in Figure 8 was significantly different from that in Figure 7. Although there were 4 arcs of capacitive reactance, the 4 radii were different greatly, and the arc radius of capacitive reactance was larger in the early stage of the corrosion. In addition, the low-frequency tail of the Nyquist plot on the 9th day showed a feature similar to Warbur diffusion impedance, but the Bode plot had an insignificant phase angle crest at 0.53 Hz. The solution environment of the test was basically unchanged, and the solute concentration and diffusion rate were relatively stable. Moreover, it could be seen from the previous test results that the corrosion reaction was slow before and after the 9th day, and there should be no diffusion layer with a large concentration gradient and diffusion current. Therefore, the impedance spectrum on the 9th day should result from the superposition of two time constants. In addition to the interface capacitance of the sample, the oxidation rate of Fe at the pitting position where enriched Cl^- was faster than that at other positions, so the Faraday admittance was larger. Therefore, the impedance spectrum was significantly affected by pitting. The equivalent circuit of impedance spectrum on the 9th day was shown in Figure 9(b), of which R_s, CPE1 and R_p were the same as those in Figure 9(a), $R_s^{'}$ was the solution resistance of pitting, CPE2 was the constant phase angle element at the pitting position, and R_{ct} was the charge transfer resistance at the pitting position.

Combined with the previous test results, that was, the self-corrosion potential was more positive and the average corrosion rate was lower, it could be inferred that severe local corrosion occurred at that time. Looking back at the polarization curve on the 9th day, due to the development of pitting, the corrosion current increased rapidly with the increase of potential, so the anode Tafel slope was low. At the same time, due to the formation of a large cathode and a small anode, the relative cathode reaction area became larger, resulting in a decrease in the cathode Tafel slope. The impedance spectrum explained the fitting results of the polarization curve on the 9th day. This correspondence of different electrochemical test results indicated that the judgment of the corrosion condition on the 9th day and the equivalent circuit were reliable.

In Table 3, the change characteristic of fitted data at 60°C was obviously different from that at 45°C. The values of interface capacitance Q_0 on the 3rd, 6th, and 9th day were close, and greater than that on the 12th day. It indicated that the rust layer was deposited faster at 60°C and was relatively stable in the middle and early periods of the test. On the 12th day, due to the rust peeling off, the double layer capacitance on the metal surface was low, so the Q_0 value decreased significantly. This also corresponded to the microtopographies of the corrosion coupons. In addition, it was worth noting that the polarization resistance values at 60°C changed greatly, with a trend of declining rapidly and then tending to be stable. It is generally believed that in a constant corrosion system, the polarization resistance is inversely proportional to the corrosion rate. This seemed to be contrary to the test results of EIS. It should be noted that the impedance spectrum was measured under mixed potential, and the Faraday resistance should be the parallel result of each reaction Faraday resistance. In this test, the anodic reaction was mainly the oxidation reaction of iron, the cathodic reaction was mainly the reduction reaction of oxygen, and the corrosion was controlled by the cathode. This indicated that the Faraday resistance of the cathode is greater than that of the anode, the parallel resistance was greatly affected by the Faraday resistance of the anode.

Therefore, it could be simply understood that R_p mainly represented the resistance of the anode reaction. On the 6th day, the R_p value dropped rapidly, and it was judged that the anodic charge transfer resistance decreased due to pitting. On the 9th day, the pitting further intensified. On the 12th day, the anodic dissolution activity did not increase with the peeling of the rust. The impedance spectrum only showed one time constant, and the pitting degree had been alleviated. Meanwhile, the self-corrosion potential was more positive, and the polarization resistance was still low, so it was judged that pitting was still going on.

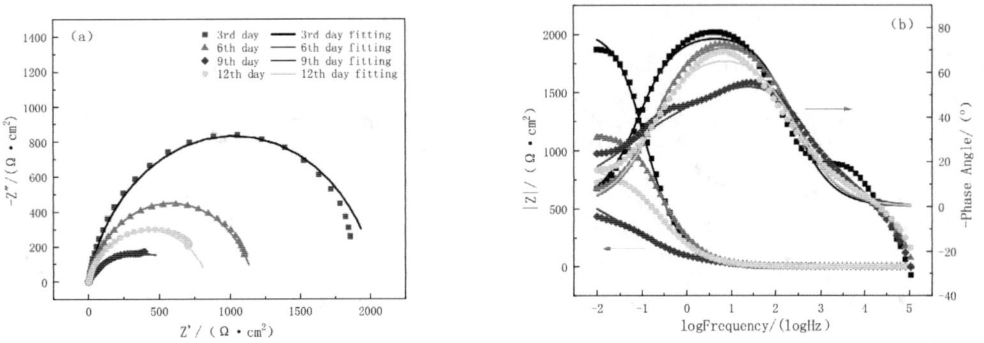

Figure 8 The Nyquist Diagram (a) and Bode Diagram (b) of the Q235 sample in 60°C Brine.

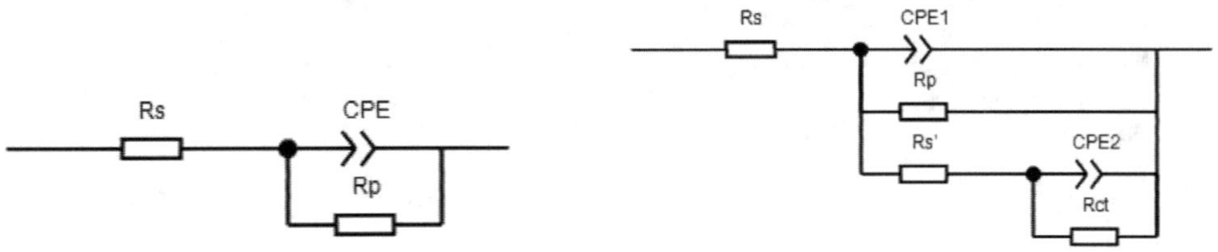

(a) (b)

Figure 9 Equivalent Circuit Diagram.

4 Conclusion

Temperature had a significant effect on the corrosion of Q235 steel in brine [31, 32]. Compared with the corrosion at 45°C, the average corrosion rate and pitting rate of Q235 steel were significantly increased at 60°C, especially when the pitting corrosion was intensified.

At 45°C, the corrosion products were deposited on the metal surface slowly, and the formation time of the dense and stable rust layer was relatively long. When the rust layer formed, the corrosion rate gradually decreased and tended to be stable. During this period, the corrosion development was relatively stable on the whole. Although the corroded surface was not completely uniform, no pitting characteristics were found in the electrochemical test, and the corrosion tended to be uniform overall. The dense rust layer deposited on the metal surface had many cracks. But at least during the test period, the rust layer was relatively stable, and there was no peeling from the metal surface, nor serious local corrosion due to cracks.

At 60°C, the formation of a dense rust layer was faster, and the corrosion rate also decreased rapidly when the rust layer formed. By the 9th day of the test, the rust layer on the metal surface was relatively stable. Although the average corrosion rate decreased, serious pitting corrosion occurred under the rust layer during the period. It was concluded that at high temperatures, the $MgCl_2$ absorbed on the rust layer increased, and the distribution of $MgCl_2$ was not uniform, which led to serious pitting corrosion. At the end of the test, the rust layer peeled off the metal surface, and the pitting corrosion was alleviated, but it still existed.

In conclusion, temperature not only affected the corrosion reaction rate but also affected the formation of a dense rust layer and the distribution of Cl^-, which led to serious pitting corrosion at high temperatures. In addition, the rust layer at high temperature was unstable and would peel off the metal surface, which had an additional effect on corrosion.

Journal of Physics: Conference Series **3009** (2025) 012075 doi:10.1088/1742-6596/3009/1/012075

Reference

[1] Zhang D Q, Hu H P, Gong J, et al. Research on the Corrosion Behavior of Carbon Steel in Bittern[J]. Chemical Engineering & Machinery, 2005(3): 186-189.
[2] Katona R M, Taylor J M, McCready T A, et al. Towards understanding stress corrosion cracking of austenitic stainless steels exposed to realistic sea salt brines[J]. Corrosion Science, 2024, 232: 111992.
[3] Mundhenk N, Huttenloch P, Kohl T, et al. Metal corrosion in geothermal brine environments of the Upper Rhine graben – Laboratory and on-site studies[J]. Geothermics, 2013, 46: 14-21.
[4] Guo K X, Zhao L Y, Zheng W B, et al. Effect of Oilfield Brine on the Corrosion of X70 Pipeline Steel[J]. Welded Pipe and Tube, 2021, 44(7): 49-53+58.
[5] Pan C, Cui Y, Liu L, et al. Effect of Temperature on Corrosion Behavior of Low-Alloy Steel Exposed to a Simulated Marine Atmospheric Environment[J]. Journal of Materials Engineering and Performance, 2020, 29(2): 1400-1409.
[6] Liu S, Li R C, Duan T G, et al. Effects of Temperature and Dissolved Oxygen Concentration on the Corrosion Behavior of 7A09 Aluminum Alloy in Seawater [J]. Structure & Environment Engineering, 2024, 51(5): 44-52.
[7] Huang C, Su X, Song Q, et al. Effects of temperature on acceleration and simulation of indoor corrosion test of Q235 carbon steel[J]. Anti-Corrosion Methods and Materials, 2021, 68(6): 564-570.
[8] Hu Y. Flow Accelerated Corrosion Behavior under Multi-factor Coupling in High Temperature and High Pressure Water Environment[D]. University of Science and Technology Beijing, 2023.
[9] Zhang L Y, Wang X M, Liu Y, et al. Effect of Temperature on the Corrosion Behavior of 1Cr Steel in CO_2 Environment [J]. Materials Protection, 2024, 57(7): 136-141.
[10] Yin Z H, Sun Z C, Ma Z Z, et al. Corrosion Characteristics and Corrosion Prediction Model of Aluminum Alloys in Saturated NaCl Solution[J]. Materials Protection, 2024, 57(4): 90-105.
[11] Wang Q X, LV W S, Yang P, et al. Corrosion of Stainless Steel Shell of Embedded Sensor in Tailings Pond [J]. Journal of Chinese Society for Corrosion and Protection, 2022, 42(2): 331-337.
[12] Han D R, Guo P, Huang G Q. Corrosion Resistance of Metallic Materials in the Brine of Salt Lake[J]. Equipment Environmental Engineering, 2005(3): 70-74.
[13] JB/T 7901-1999, The Test Method for Uniform Corrosion of Metallic Materials in Laboratory[S].
[14] GB/T 16545-2015, Corrosion of Metals and Alloys—Removal of Corrosion Products from Corrosion Test Specimens[S].
[15] ZHANG Zheng-biao. Effect of high temperature on electrochemical corrosion behavior of metallic materials [J]. World Nonferrous Metals, 2020(10): 263-264.
[16] Liu Y, Liu M, Lu X, et al. Effect of temperature and ultraviolet radiation on corrosion behavior of carbon steel in high humidity tropical marine atmosphere[J]. Materials Chemistry and Physics, 2022, 277: 124962.
[17] de la Fuente D, Díaz I, Alcántara J, et al. Corrosion mechanisms of mild steel in chloride-rich atmospheres[J]. Materials and Corrosion, 2016, 67(3): 227-238.
[18] Wang J, Dong C C, WU Y Z. Research on Corrosion Behavior of Carbon Steel in the Brine of ChaerhanSaline Lake[J]. Total Corrosion Control, 2010, 24(9): 45-48+31.
[19] Guo K X, Zhao L Y, Zheng W B, et al. Effect of Oilfield Brine on the Corrosion of X70 Pipeline Steel [J]. Welded Pipe and Tube, 2021, 44(7): 49-53.
[20] Gu H J, Bu M Z, Fu X Q, et al. Corrosion Law Study and Protection Analysis of A3 Steel by Brine Medium in Salt Cave Gas Storage[J]. Welded Pipe and Tube, 2022, 45(10): 12-17.
[21] Wang J, Wang Z Y, Ke W. Corrosion behaviour of weathering steel in diluted Qinghai salt lake water in a laboratory accelerated test that involved cyclic wet/dry conditions[J]. Materials Chemistry and Physics, 2010, 124(2): 952-958.
[22] Yang M J. Research on the risk of pipe corrosion failure during the mining and transportating process of mine salt D]. Kunming University of Science and Technology, 2021.
[23] Huang H. Study on Corrosion Mechanism of Metal Materials in Brine Collecting and Transporting Equipment in Salt Lake[D]. Xi'an University of Technology, 2022.
[24] Gu B S, Gong L, Ji X C, et al. Selective Corrosion of Carbon Steel in the Early Stage in Saline Lake Brine in West China[J]. Journal of Iron and Steel Research, 2009, 21(10): 21-27.
[25] Yang Z Q. Corrosion of coated steel bar in salt spray environment[D]. Lanzhou University of Technology, 2021.
[26] Guo M X. Effect of Salt Deposition and UV Illumination on Corrosion Behavior of Stainless Steel Exposed to Salt Lake Atmosphere[D]. University of Science and Technology of China, 2021.
[27] Ma G F. Corrosion Behavior of Carbon Steel Used in Construction in Salt Water[J]. Foundry Technology, 2014, 35(8): 1663-1665.
[28] Huang G Q, Yan M, Yu C J, et al. Corrosion of Metallic Materials in the Brine of Chaerhan Salt Lake[J]. Corrosion & Protection, 2005(9): 369-372.
[29] Lair V, Antony H, Legrand L, et al. Electrochemical reduction of ferric corrosion products and evaluation of galvanic coupling with iron[J]. Corrosion Science, 2006, 48(8): 2050-2063.
[30] Cao C N, Zhang J Q (2002). An Introduction to Electrochemical Impedance Spectroscopy[M]. Beijing: Science Press.
[31] CUI Lei. Research on Composite of Rust Resistance Suitable for the Salt Lake Area of Inner Mongolia [D]. Inner Mongolia University of Science and Technology, 2020.
[32] Liu Q, Chen B, Li Y. Corrosion Investigation of Reinforced Concrete Under Qinghai Salt Lake Environment By EIS[J]. International Journal of Electrochemical Science, 2016, 11(12): 10238-10245.

Research and optimization of temperature uniformity of high-precision muffle furnace

Yuchen Wang [1], Qianli Ma [2], Tao An [1], and Haisheng Fang [1*]

[1] School of Energy and Power Engineering, Huazhong University of Science and Technology, Wuhan, Hubei, 430074, China
[2] Hubei Feilihua Quartz Glass Co. Ltd., Jingzhou, Hubei 434000, China

[*] Corresponding author's e-mail: hafang@hust.edu.cn

Abstract: Numerical simulation has been widely used to study the temperature field characteristics of muffle sintering, compensating for the limitations of experiments in obtaining temperature data for high-temperature muffle furnaces. We established a numerical model of the transient temperature field of a high-entropy alloys workpiece in a muffle furnace and simulated it. The simulation results are in agreement with the measured results. The muffle furnace structure's influence on the workpiece's temperature uniformity during and after the sintering process is further discussed. The results show that increasing the height of the workpiece can reduce the heat loss and heat absorption of the workpiece during the sintering process. The overall temperature and temperature uniformity of the workpiece were effectively improved by raising the height of the workpiece through the use of shims with a moderate height. The layout of the distance between the heating tubes also affects the temperature uniformity of the workpiece. Reducing the tube spacing causes a small increase in the average temperature of the workpiece, but increasing the tube spacing results in a more uniform temperature field of the workpiece.

1. Introduction

High-entropy alloys are used in aerospace [1], automotive, marine, biomedical, and other fields with high-temperature resistance, corrosion resistance, wear resistance, high strength ductility, and oxidation resistance [2, 3]. High-entropy alloys are manufactured by preparing, de-binding, pre-firing, sintering, and cooling before application [4]. One of the key steps in additive manufacturing of high-entropy alloys is the sintering process. It is difficult to meet the requirements of the thermostatic field inside the muffle furnace, resulting in uneven temperature distribution of the workpiece during the sintering process. This situation leads to porosity, cracks, and deformation of the workpiece [5]. To overcome this challenge, the main concern is the effect of heating time [6, 7] and heating tube [8] on the temperature of the workpiece. It has been shown that the workpiece temperature uniformity can be improved by preserving the heat. The suitable heating rate and diameter of the heating tube can improve the heating efficiency and temperature uniformity of the sintering furnace.

We established a physical model of the process of sintering high-entropy alloys workpiece in a muffle furnace using finite element analysis. The effect of muffle furnace structure on temperature uniformity of high-entropy alloys workpiece was investigated by analyzing workpiece temperature non-uniformity generation and variation. Results show that a change structure can effectively improve oven sinter performance and increase temperature uniformity. This ensures the workpiece performance and increases the muffle furnace's productivity.

2. Simulation models and mathematical methods

2.1. Geometry description

The physical drawing of the muffle furnace is shown in Figure 1(a). To simplify the simulation, the effects of porosity and phase change of the workpiece on the temperature field were neglected in the sintering process. The water-cooled system was simplified to a constant temperature wall interface with a temperature of 74°C. Therefore, a simplified model of the muffle furnace is obtained, as shown in Figure 1(b). The high-entropy alloys workpiece and spacers were placed in the furnace with an initial temperature of 25°C and an inlet argon flow rate of 0.05 m/s. The physical parameters of the heater [8] and the high-entropy alloys [9] workpiece were obtained by reviewing the literature. The geometric model used for the simulation in this paper is half of the full furnace to improve computational efficiency.

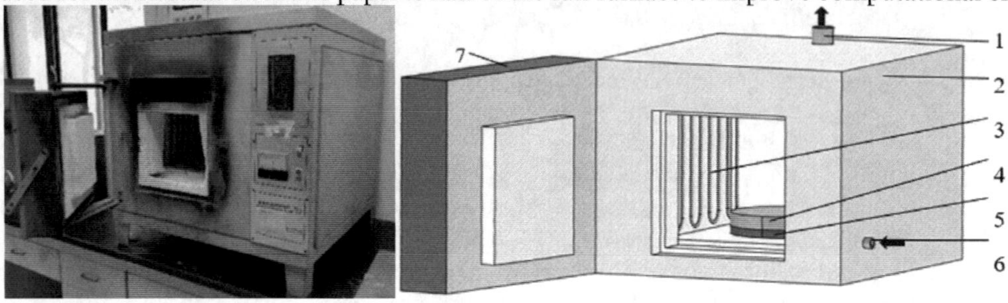

(a) (b)

Figure 1. Muffle furnace (a) physical view, (b) schematic diagram: 1. Air outlet; 2. Furnace body; 3. Heater; 4. High-entropy alloys workpiece; 5. Gasket; 6. Air inlet; 7. Door.

2.2. Governing equation

During the simulated sintering process, the furnace internals and surfaces are assumed to be gray bodies and diffuse reflective interfaces, respectively. Argon is regarded as a transparent substance. The controlling equations mainly include the energy equation and the radiative heat transfer model.

According to Fourier's law, Newton's law of cooling, and Stefan-Boltzmann's law, it can be known that thermal radiation transfers the largest percentage of heat in a high-temperature sintering process, which must compete with conductive and convective heat transfer.

The general form of the energy equation of the high-entropy alloys workpiece may be stated as:

$$\rho C_p \frac{\partial T}{\partial t} + \rho C_p \boldsymbol{u} \cdot \nabla T = \nabla(k \nabla T) + Q - \nabla \cdot \mathbf{q}_R \tag{1}$$

where t is the time, ρ is the density, C_p is the constant-pressure specific heat capacity, $\boldsymbol{u}$ is the velocity vector, ∇ is the Hamiltonian operator, T is the temperature, k is the thermal conductivity, Q is the heat source within the system and $\mathbf{q}_R$ is the radiant heat flux.

Thermal radiation is a method of heat transfer in which energy is transferred between surfaces in the form of electromagnetic waves. The surface-to-surface model is used for simulation in this article. The radiative heat transfer model is as follows:

$$q_{\text{out}} = \varepsilon E_b + \rho q_{\text{in}} \tag{2}$$

where q_{in} and q_{out} denote the inflow and outflow radiative heat fluxes, E_b is the radiative force of the blackbody, ε is the surface emissivity, σ is the Stefan-Boltzmann constant with a value of 5.67×10^{-8} W/(m^2·K^4), and ρ is the reflectance. In muffle sintering furnaces, radiant heat transfer is more complex than heat conduction and heat convection. The main reason is that emissivity and angular coefficients are more difficult to calculate.

2.3. Evaluation index of workpiece temperature uniformity

Here, we propose an index for evaluating the temperature uniformity of a workpiece—the temperature coefficient of variation:

Journal of Physics: Conference Series **3009** (2025) 012076 doi:10.1088/1742-6596/3009/1/012076

$$\gamma = \frac{\sigma}{\overline{T}} \times 100\% = \frac{\sqrt{\frac{1}{N}\sum_{i=1}^{N}\left(T_i - \overline{T}\right)^2}}{\overline{T}} \tag{3}$$

where σ represents the standard deviation of the workpiece, $\overline{T}$ represents the average temperature of the workpiece, and N represents the number of temperature measurement points obtained on the workpiece. When the value of γ is larger, it means that the temperature field uniformity of the workpiece is worse and vice versa.

2.4. Verification model

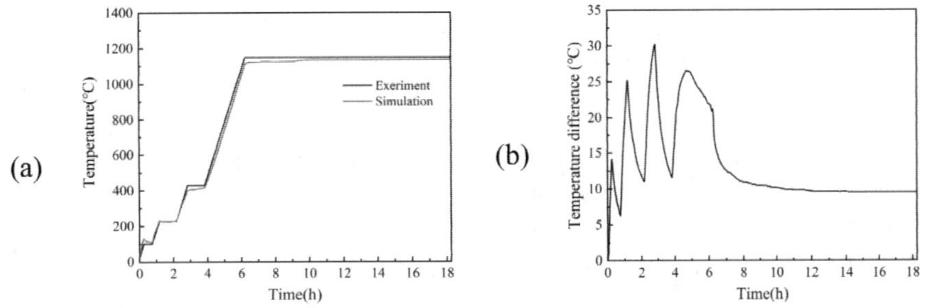

Figure 2. The workpiece (a) overall temperature, (b) temperature difference in the sintering process.

The simulation results and the experimental temperatures [10] were compared under the muffle load condition, which was used to verify the accuracy of the muffle furnace. The temperature variation of the workpiece is shown in the sintering process in Figure 2. Figures 2(a) and (b) demonstrate that the simulated temperatures of the high-entropy alloys workpiece are in general agreement with the experimental temperature profiles. The workpiece temperature difference appears to vary with time. During the warming phase, the furnace chamber is continuously heated by the heating tubes. The workpiece absorbs heat and this leads to an increase in the temperature difference. In the process of heat preservation, the difference between the temperature of the outer layer of the workpiece and the temperature of the furnace chamber is small. Heat is gradually transferred from the outer layer of the workpiece to the center. As the center of the workpiece continues to absorb heat, the center temperature gradually rises, and the temperature difference between the inside and outside decreases. Later in the holding phase, the workpiece as a whole reaches equilibrium. And the workpiece temperature difference tends to equilibrate as shown in Figure 2(b).

3. Results and discussion

3.1. Effect of workpiece height

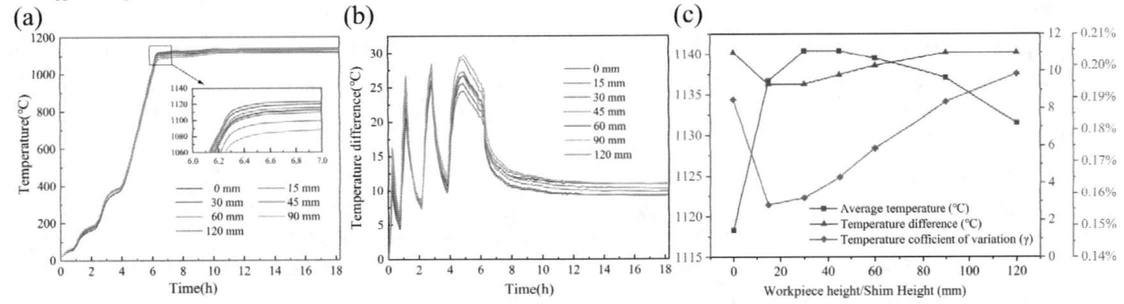

Figure 3. During sintering of the workpiece: (a) Average temperature, (b) Temperature difference; at the end of sintering (c) Temperature, temperature difference, and temperature coefficient of variation under different heights.

The temperature curves and temperature difference curves of the high-entropy alloys workpiece at different heights under the same heating power condition are shown in Figure 3(a) and (b) during the sintering process. Shims were used to increase the height of the workpiece in the study. Figure 3(a) shows that the temperature variation pattern of the workpiece is consistent at different heights in Figure 2(a). The temperature difference curves of the workpiece at different heights demonstrated in Figure 3(b) are highly similar to those in Figure 2(b). That is, the temperature difference fluctuates continuously during the warming-insulation stage and reaches thermodynamic equilibrium in the later part of the insulation stage.

At the end of sintering, Figure 3(c) shows that the temperature of the workpiece increases and then decreases, and the temperature difference and coefficient of variation of the workpiece decreases and then increases, as the height of the workpiece increases. Compared with the temperature coefficient of variation at a workpiece height of 0 mm, the coefficient of variation decreases by 17.42%, 16.27%, 12.82%, 8.06%, -0.35%, and -4.2% in that order. Without shims, the bottom of the workpiece is in direct contact with the muffle furnace insulation. The heat from the workpiece is transferred through the insulation to the outside environment, resulting in heat loss. When the shim is added, it creates a new thermal resistance and reduces heat loss, resulting in a higher workpiece temperature and a smaller temperature difference. However, the heat absorbed by radiation on the surface of the workpiece decreases with increasing height, which leads to a decrease in the angular coefficient. This leads to a gradual contraction of the high-temperature region of the workpiece towards the side regions, a decrease in the temperature of the workpiece, and an increase in the temperature difference, as shown in the complementary diagram in Figure 3(b). Thus, there is a correlation between reduced heat absorption and reduced heat loss, and both are highly correlated with the workpiece. When heated for a longer period, the interior of the furnace chamber gradually tends to thermodynamic equilibrium. At this point, the reduced heat loss becomes the dominant factor shown in Figure 3(a). When the height of the workpiece exceeds 90 mm, the temperature coefficient of variation is larger than that of the working conditions without shims and the uniformity becomes worse.

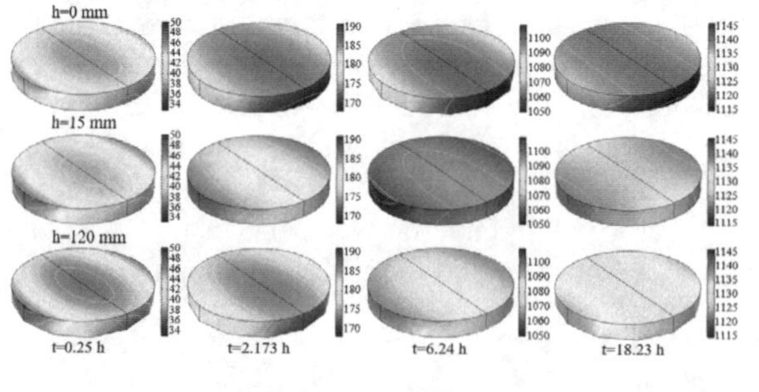

Figure 4. Temperature clouds of workpiece at different heights at different moments in time.

Figure 4 shows the temperature distribution clouds of the high-entropy alloys workpiece in the muffle furnace at four different moments. Firstly, the variation of the temperature field of the workpiece was analyzed in different heating moments. When the sintering starts, the temperature of the heating tube is low, and the main form of heat transfer is natural convection in the furnace chamber. As the heating tube temperature increases, thermal radiation gradually becomes the main form of heat transfer. During the sintering process, the temperature on both sides of the workpiece is always higher than the center temperature, and the temperature on the front side is lower than the temperature on the backside. This is because the heaters are placed on both sides of the furnace chamber and there are symmetrically distributed air inlets on the front side of the chamber. The gas flow at the air inlets leads to a strong convective heat transfer at the front of the workpiece. Then, the variation in temperature distribution of the workpiece is further compared at different heights. At all four moments, the surface temperature of the workpiece at the 0 mm height was less than the other heights. At the beginning of sintering (t=0.25 h), the temperature of the workpiece occupies 4, 5, and 6 temperature intervals, when the height of the workpiece is 0, 15, and 120 mm, respectively. The addition of shims concentrates the high-temperature region to the edge of the workpiece. At the later stage of the holding phase (t=18.23 h), it can be observed that the side surface temperature of the workpiece without added shims is more dispersed. So, Figure 4 again shows the important influence of the appropriate height on the temperature uniformity during the sintering of the workpiece.

3.2. *Effect of distance between heating tubes*

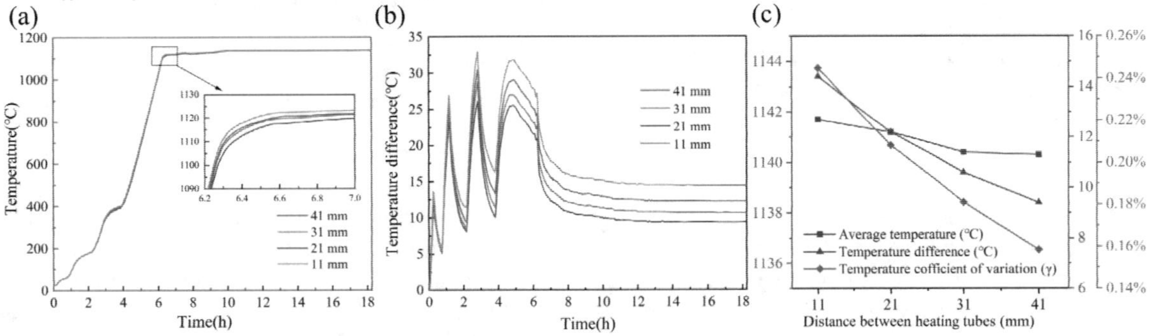

Figure 5. During sintering of the workpiece: (a) Average temperature (b) Temperature difference; at the end of sintering (c) Temperature, temperature difference, and temperature coefficient of variation for different heating tube distances.

Figure 5(a) shows the temperature variation curves of the workpiece made of the high-entropy alloys under the same heating power condition with different distances between the heating tubes. Under different tube spacings, there is no significant difference in the temperature variation curve of the workpiece. Therefore, there is little effect on the overall temperature of the workpiece when the distance between the heating tubes is changed. Figure 5(b) shows that the trend of the temperature difference of the workpiece with time is the same at different inter-tube distances in the sintering process. Figure 5(c) shows the variation of the average temperature, the temperature difference, and the temperature coefficient of variation of the workpiece at the end of the sintering process at different inter-tube distances. The results show that workpiece temperature and temperature difference decreased with increasing tube spacing, but the temperature difference decreased more than the temperature. In comparison with the 11 mm tube distance, the temperature variation coefficients decrease by 15.01%, 26.08%, and 35.29% respectively with an increase in the tube distance.

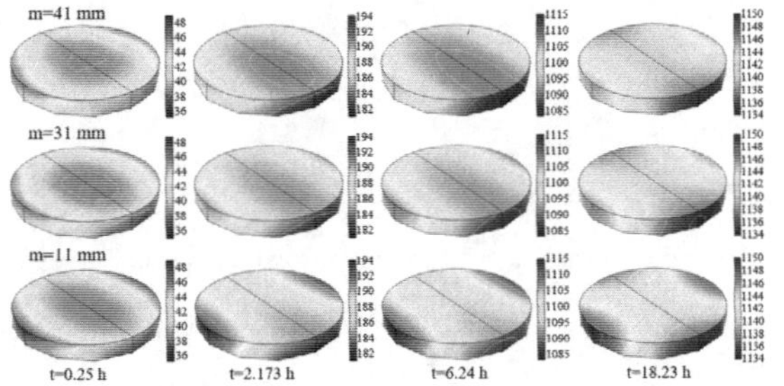

Figure 6. Temperature clouds of the workpiece at different moments under heating with different distances between the heating tubes.

Figure 6 shows the temperature distribution of the workpiece at different times for different inter-tube spacings. The temperature distribution in Figure 6 is similar to Figure 4, with high temperatures at the sides and low temperatures at the front. Further comparisons were made of the temperature distributions of the workpiece at different times. The results show that at each time point, especially near the side of the heater tube, the temperature difference of the workpiece surface with the 11 mm distance between the tubes is greater than that of the other workpieces.

In the sintering process, reducing the distance between the heating tubes results in a closer arrangement of the heating tubes. This leads to a concentration of the heat source and a greater difference in the heat received by the heated surfaces of the workpiece, resulting in a greater temperature gradient in the workpiece. Thus, at the end of the sintering process, the temperature distribution on both sides of the part is in the range of 1136 - 1150°C with a distance between the tubes of 11 mm. However, as the distance between the heating tubes increases, the temperature distribution on both sides of the part is in the range 1138-1144°C. The smaller the temperature gradient of the workpiece is, the better the temperature uniformity achieved. Therefore, the method of increasing the distance between the tubes of the heating tube can effectively reduce the temperature unevenness in the sintering of the workpiece.

4. Conclusion

When studying the effect of workpiece height, a correlation was found between the amount of heat loss and the amount of heat absorption by the workpiece. The temperature of the workpiece increased when the workpiece height was low. This was because the heat loss decreased more than the heat absorption. The maximum temperature of 1140.4°C was obtained for heights of 30 mm and 45 mm. As the height of the workpiece increases, the temperature of the workpiece decreases because the rate of decrease in heat loss is less than the rate of decrease in heat absorption. The temperature uniformity coefficient first decreases and then increases with increasing height. For the workpiece with a shim height of 15 mm, the smallest temperature uniformity coefficient (best temperature uniformity performance) is 0.1564%. The degree of temperature uniformity is comparable to that obtained without shims when the height of the workpiece is increased to 90 mm. When the distance between the heating tubes is studied, it can be seen that as it increases, the heat source becomes more dispersed and the workpiece is heated more uniformly. Increasing the distance between the tubes slightly reduces the average temperature of the workpiece, but the temperature gradient on the surface of the workpiece decreases and the temperature uniformity of the workpiece is improved.

The study results provide a basis for determining the temperature uniformity of high-entropy alloys workpiece and muffle furnace sintering accuracy during muffle furnace modification. Improving the quality of high-entropy alloys workpiece and muffle furnace productivity in industrial production is of great significance. Future work requires further design and parameter optimization of the muffle furnace heating tube structure.

Acknowledgment

This work was supported by the National Natural Science Foundation of China (Grant No. 52276064).

References

[1] Dada M, Popoola P and Mathe N. Recent advances of high entropy alloys for aerospace applications: a review 2023 *World J Eng*, 20(1): 43-74.
[2] Guo J M, Zhou BC and Qiu S. Achieving ultrahigh strength and ductility in high-entropy alloys via dual precipitation 2023 *J Mater Sci Technol*, 166: 67-77.
[3] Majji B G R, Vasudev H and Bansal A. Application of plasma-sprayed AlxCrCoFeNi High-Entropy alloys for high-temperature oxidation resistance 2024 *Mater Lett*, 361.
[4] Kenel C, Casati N P M and Dunand D C. 3D ink-extrusion additive manufacturing of CoCrFeNi high-entropy alloy micro-lattices 2019 *Nat Commun*, 10.
[5] Huang M J, Jiang J F, and Wang Y. Microstructure evolution, phase decomposition and transition of Al0.8Co0.5Cr1.5CuFeNi HEA during high temperature sintering and mechanical properties corresponding to different microstructures 2022 *Mater Design*, 223.
[6] Li B, Su F Y and Li C W. Study on temperature uniformity of workpieces in multi-layer trays inside vacuum sintering furnace for cemented carbide 2024 *Int J Therm Sci*, 196.
[7] Wang Q H, Li J J and Hu S D. Effect of sintering temperature and holding time on microstructures and properties of AlCoCrFeNi2.1 eutectic high entropy alloy by binder jetting additive manufacturing 2024 *Transactions of Nonferrous Metals Society of China* 34(04): 1201-14.

[8] Xiong L, Zhang D C and Song S C. Numerical simulation and system optimization of temperature field in silicon carbide vacuum sintering furnace 2022 *Heat treatment of metal*, 47(06): 259-65.
[9] Chou H P, Chang Y S and Chen S K. Microstructure, thermophysical and electrical properties in AlxCoCrFeNi ($0 \leq x \leq 2$) high-entropy alloys 2009 *Mater Sci Eng B-Adv*, 163(3): 184-9.
[10] Peng S Y, Mooraj S, and Feng R. Additive manufacturing of three-dimensional (3D)-architected CoCrFeNiMn high-entropy alloy with great energy absorption 2021 *Scripta Mater*, 190: 46-51.

Journal of Physics: Conference Series **3009** (2025) 012077 doi:10.1088/1742-6596/3009/1/012077

Characteristics and mechanism analysis of high-temperature damage to metal nozzles used in converter stations

Jun Ji[1,2,*], Dong Hou[1,2], Jingkai Nie[1,2], Xiaoming Wang[3], Changming Shi[4]

[1]China Electric Power Research Institute, Beijing 100192, China
[2]State Key Laboratory of Advanced Power Transmission Technology, Beijing 102209, China
[3]State GRID Shandong Electric Power Research Institute, Jinan, 250003, China
[4]Electric Power Research Institute of State Grid East Inner Mongolia Electric Power Co., Ltd., Inner Mongolia 010020, China

[*]Corresponding author's e-mail: jijun2006@163.com

Abstract. The high-temperature deformation of the fire sprinkler in the ultra-high voltage converter station is a key factor causing fire accidents in ultra-high voltage converter transformers. This paper studies the damage characteristics of the commonly used fire sprinklers for ultra-high voltage converter transformers in China under high temperatures through a series of metal high-temperature experiments. A detailed discussion of the Microstructural evolution of metal materials at high temperatures is provided. The experimental results indicate that stainless steel material has a higher temperature resistance than carbon steel and copper alloy. The microstructure of Q310s stainless steel appears at 400°C and becomes clear at 700°C. After 1100°C, grain aggregation growth and annealing twinning occur. After being kept at 1100°C for 60 minutes, the original grain boundaries basically disappeared. Furthermore, insulation at 1100°C for 90 minutes results in a notable protrusion within the microstructure.

1. Introduction

In recent years, China's ultra-high voltage has been developing toward large capacity [1-2]. The insulation oil reserve of the converter transformer in the 1100 kV HVDC project can reach 130-200 tons. A large amount of insulation oil combustion has caused several serious transformer fire accidents worldwide [3-5]. The water spray system is the most common way to extinguish afire in the converter station, but the mechanical properties of the metal materials represented by the sprinkler head drop sharply in the process of the transformer fire, which seriously affects the spray effect. Power fires are often guarded based on power or hydrocarbon heating curves and can be heated to 1100°C in a short time. When a fire occurs, the fire sprinkler is located at the forefront of the flame, and its reliability in service at high temperatures directly affects the extinguishing effect of the fire sprinkler equipment [6-7]. At present, the domestic standard GB 5135.3-2003"Automatic Sprinkler Fire Extinguishing Systems Part 3: Water Mist Nozzles" tests the high-temperature performance of nozzles based on the standard heating curve, with a detection temperature of 800°C and is not applicable to electric fires. In addition, the strength loss of the nozzle material in different temperature ranges is unclear, which cannot support the design institute's selection of fire sprinklers [8].

The key construction of sprinkler pipelines in converter stations is often made of copper alloys or stainless steel alloys [9]. In recent years, extensive research has been conducted on the high-temperature resistance of metals. Before the maximum firing temperature reaches 700°C, the strength of the steel after firing does not show significant degradation. When the maximum overheating temperature exceeds 700°C, there is a certain loss in the yield strength of the steel, while the method of immersion cooling will cause a certain recovery in the yield strength of the steel [10-11]. In addition, the elastic modulus is basically stable during this process. For the ultimate strength of steel, experiments have shown that even if the maximum overheating temperature reaches 1000°C, the ultimate strength of specimens cooled by immersion method is also improved [12-13]. At the same time, the loss of steel fracture toughness also provides a method for quickly determining the strength of steel after fire, which is to determine the strength of steel at that location through hardness testing [14].

This article focuses on studying the damage performance of key metal materials in the spray pipe network for ultra-high voltage stations at high temperatures. We are trying to obtain the changes in grain boundaries of typical metal materials at high temperatures through microscopic and metallographic analysis, providing data support for the fire protection improvement of converter stations.

2. Study on the damage characteristics of fire sprinklers under high temperature

2.1. Mechanical property test of metal materials at high temperature

The nozzles used for the fire protection pipelines in the converter station are made of materials such as carbon steel, stainless steel, copper alloy, etc. Therefore, the mechanical properties of four materials, including 304, 316, 310s, and HPb59-1, were studied at different temperatures. The experiment was conducted based on the material used for the nozzle to study the high-temperature mechanical properties in the temperature range of 700~1100°C. A high-temperature tensile testing machine was used to test the high-temperature mechanical properties of four materials, namely 304, 310s, 316 stainless steel, and HPb59-1, as shown in Figure 1.

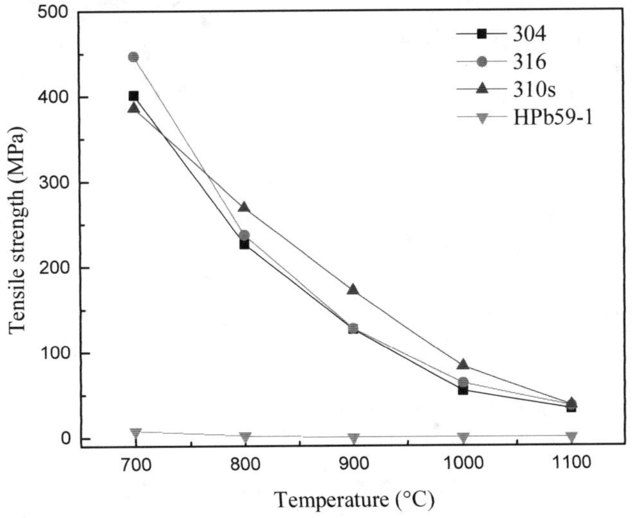

Figure 1. The high-temperature mechanical properties of four materials.

The test results show that the high-temperature mechanical properties of HPb59-1 copper alloy material are poor. The remaining tensile strength is only 2.12 MPa at 800°C, which cannot meet the safety and reliability requirements under hydrocarbon fire. Stainless steel material has relatively good high-temperature mechanical properties, especially 310s stainless steel. At 800 °C, the tensile strength is 268.89 MPa; at 1100°C, the tensile strength is 37.37 MPa.

2.2. Reliability test of fire sprinklers in service at high temperature

According to GB 5135.3-2003 "Automatic Sprinkler Systems Part 3: Water Mist Nozzles", the high-temperature performance testing method for nozzles was selected. Two specifications of stainless steel nozzles, B42.8/90 and B42.8/120, and two specifications of copper chrome plated nozzles, B42.8/90 and B42.8/120, produced by Shou'an Industrial Fire Protection Co., Ltd., were selected as test samples. The test temperature was 1100°C, and the test time was 15 minutes. The comparison of the samples before and after the test is shown in Figure 2.

Journal of Physics: Conference Series **3009** (2025) 012077 doi:10.1088/1742-6596/3009/1/012077

Figure 2. Sample morphology after 15 minutes of the experiment.

Figure 3. Copper chrome plated nozzle melting.

Figure 4. Stainless steel nozzle intact.

After 15 minutes of testing, the copper chrome-plated nozzle has turned into slag, as shown in Figure 2. This indicates that the nozzle of this specification does not have fire resistance at 1100°C because the liquidus temperature of HPb 59-1 copper alloy is only 900°C, and it will melt at 1100°C.

After cooling the two specifications of stainless steel nozzles B42.8/90 and B42.8/120 in water at 15 °C, it was found that the shrapnel of the B42.8/90 stainless steel nozzle had fallen off and could not be used for fire extinguishing, as shown in Figure 3. The stainless steel nozzle of B42.8/120 did not show significant deformation, as shown in Figure 4. This indicates that stainless steel material has higher service reliability compared to HPb59-1 copper alloy material.

3. Microstructure analysis of metal materials

Perform scanning electron microscopy analysis on the tensile fracture surface of 310s stainless steel at 800°C and 1100°C, as shown in Figures 5 and 6. The surface of Q310s material is oxidized at high temperatures, forming a carbonized layer with a thickness of about 100 μ m. The morphology of the middle section is also different from that of stainless steel. The fracture surface is distributed with a small number of dimples, which are sparse and of varying depths. The diameter of the dimples is large, and the walls of the dimples are relatively smooth. There are holes in the centre of the dimples, showing characteristics of ductile fracture.

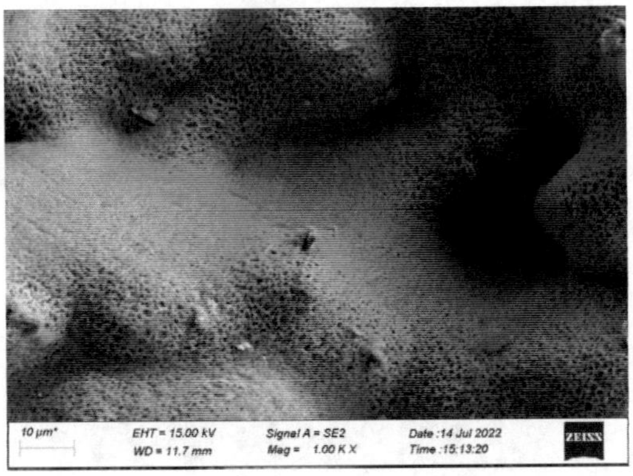

Figure 5. 310s stainless steel SEM (800°C).

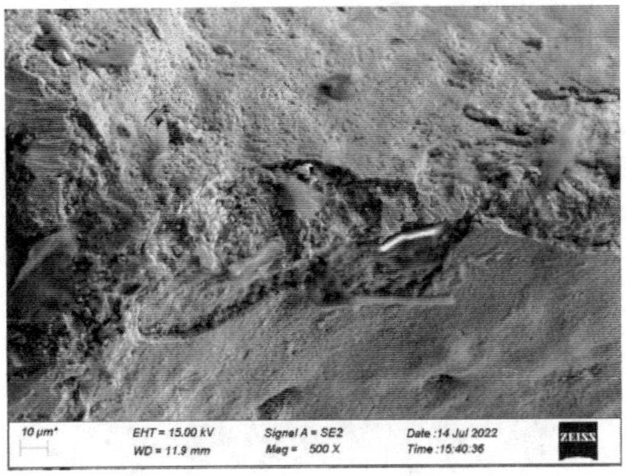

Figure 6. 310s stainless steel SEM (1100°C).

The experiment heats Q310s stainless steel to 400 °C, 700 °C, and 1100 °C, and holds it at 1100 °C for 30 minutes, 60 minutes, and 90 minutes, respectively. Gold phase was observed and shown in Figure 7, Figure 8, Figure 9, Figure 10, Figure 11, and Figure 12. At a high temperature of 400°C, microstructure appeared in the Q310s metallographic examination, and at 700°C, the microstructure became clear. After 1100 °C, grain aggregation growth and annealing twinning occurred. After being kept at 1100°C for 60 minutes, the original grain boundaries basically disappeared. After insulation at 1100°C for 90 minutes, the tissue showed a phenomenon of protrusion.

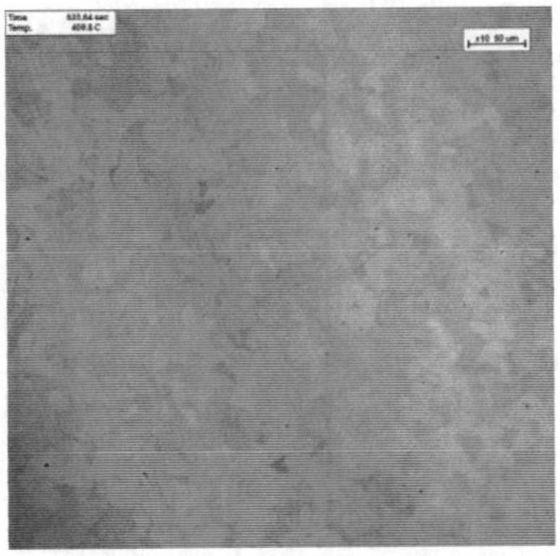

Figure 7. High-temperature metallographic structure of Q310s at 400°C.

Figure 8. High-temperature metallographic structure of Q310s at 700°C.

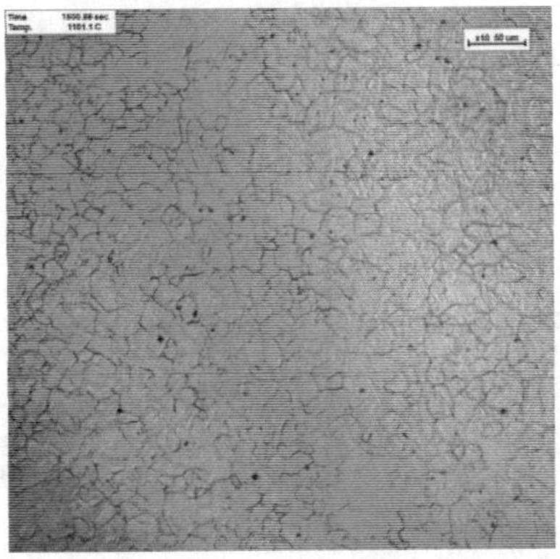

Figure 9. High-temperature metallographic structure of Q310s at 1100°C.

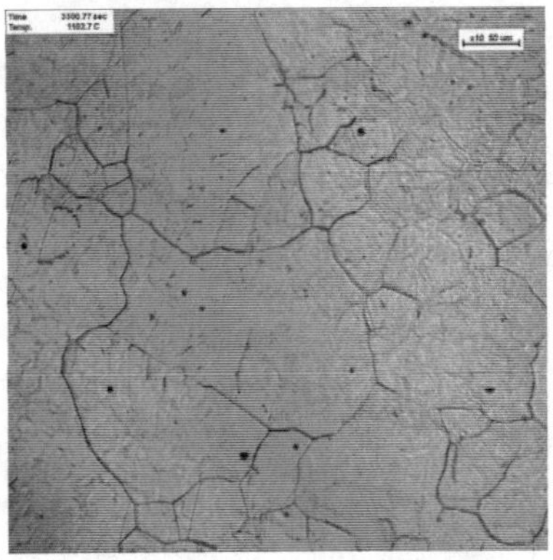

Figure 10. High-temperature metallographic structure of Q310s at 1100°C for 30 minutes.

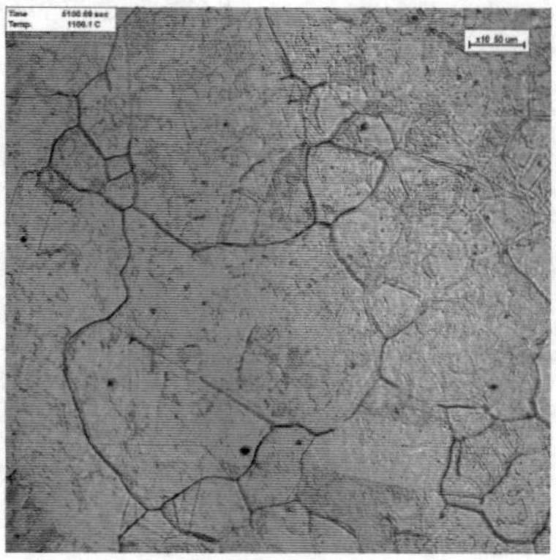

Figure 11. High-temperature metallographic structure of Q310s at 1100°C for 60 minutes.

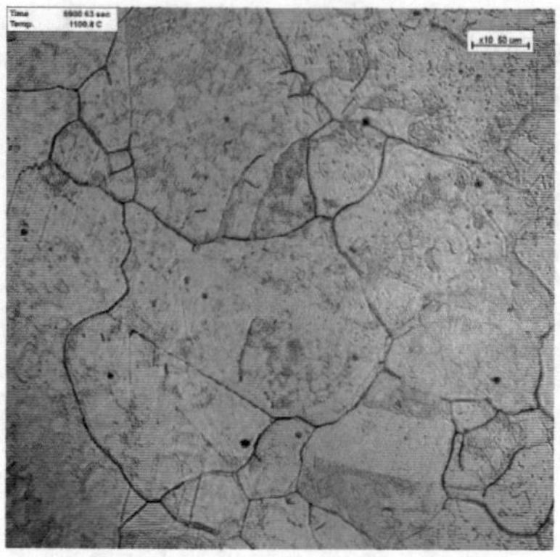

Figure 12. High-temperature metallographic structure of Q310s at 1100°C for 90 minutes.

4. Conclusion

This article investigates the high-temperature resistance and microstructural changes of sprinkler components used in converter station fire protection pipeline networks through high-temperature performance tests of metal materials. This article draws the following conclusions:

(1) The high-temperature mechanical properties of metal materials decrease significantly with increasing temperature. Compared with copper alloy and carbon steel, stainless steel material has relatively better high-temperature mechanical properties. At 800°C, the tensile strength is 268.89 MPa, and at 1100°C, the tensile strength is 37.37 MPa.

(2) Copper alloy nozzles will melt at 1100°C and are not suitable for use as nozzles for fire protection pipelines in converter stations.

(3) Upon heating Q310 to 400°C, the microstructure appeared, and at 700°C, the microstructure became clear. After 10 minutes of insulation at 1100°C, grain aggregation and growth, as well as annealing twinning, occurred. After 30 minutes of insulation at 1100°C, the original grain boundaries basically disappeared, and after 40 minutes of insulation at 1100°C, the microstructure showed a protrusion phenomenon.

Acknowledgements

This work was supported by Science and Technology Projects of State Grid Corporation of China (5500-202216111A-1-1-ZN).

References

[1] Jonsson L. (2008) Fire and Safety Aspects of High Voltage Bushings [M]. CIGRE, Transformer Technology Conference. Sydney, Australia.

[2] Zhang J Q, Huang Y, Zhou Y F, et al. (2023) Study on the temperature of upper space of UHVDC converter transformer fires under the action of sprinklers. J. High Voltage Apparatus, 59 (10): 140-145.

[3] Gou C M, Dong J, Cui D D. (2023) High-temperature rheological behaviour and thermal processing diagram for 34CrNiMo6 steel J. Forging & Stamping Technology, 48 (2): 233-240.

[4] Zhao T, Rong S W, Hao X H, et al. (2022) Effect of Nb-V microalloying on hot deformation characteristics and microstructures of Fe-Mn-Al-C austenitic steel J. Materials Characterization, 183: 111595.

[5] Zhou P W, Song Y R, Jiang H W, et al. (2022) Hot deformation behaviour and processing maps of BG801 bearing steel J. Journal of Materials Research and Technology, 18: 3725-3738.

[6] Hu Y, Wang L H, Ouyang M H, et al. (2023) Hot deformation behaviours and dynamic softening mechanism of 6%Si high-silicon austenitic stainless steel. J. Journal of Materials Research and Technology, 26: 4263-4281.

[7] Zhao J, Wan X L, Ke R, et al. (2023) Effect of grain refinement on mechanical properties of austenitic stainless steels at high temperature. J. Journal of Iron and Steel Research, 35 (1): 88-97.

[8] Liu X S, Nie J K, Chen G H, et al. (2024) Mechanical properties and microstructure evolution of 304 stainless steel under electric fire. J. Metallic Functional Materials, 31(2): 93-99.
[9] Hou D, Nie J K, Han Y, et al. (2024) Thermal deformation behaviour and strain-compensated constitutive equation of Q355B steel. J. Forging & Stamping Technology,49(3): 240-250.
[10] Xing Z Q (2024) Microstructure Regulation and Corrosion Resistance of FeNiCrAl Based High-Entropy Alloys. D. Dalian Jiaotong University.
[11] Balaji V, Anthony X. (2024) Development of high entropy alloys (HEAs): Current trends. J. Heliyon, 10: e26464.
[12] Kumar A, Singh A, Suhane A, et al. (2023) Wear and corrosion behaviour of high entropy alloys. J. Materials Today: Proceedings, In Press.
[13] Gludovatz B, Hohenwarter A, Catoor D, et al. (2014) A fracture-resistant high-entropy alloy for cryogenic applications. J. Science, 345:1153-1158.
[14] Laplanche G, Kostka A, Reinhart C, et al. (2017) Reasons for the superior mechanical properties of medium-entropy CrCoNi compared to high-entropy CrMnFeCoNi. J. Acta Materialia, 128: 292-303.

Journal of Physics: Conference Series **3009** (2025) 012078 doi:10.1088/1742-6596/3009/1/012078

Study on the sound insulation performance of lightweight sandwich panel structures in large cruise ships

Xinrui Peng, Kun Liu *, Shuai Zong and Xinpeng Yan

School of Ship and Ocean Engineering, Jiangsu University of Science and Technology (JUST), Zhenjiang, China

* Corresponding author's e-mail: kunliu@just.edu.cn

Abstract: With the increasing demand for cabin comfort in the cruise industry, noise control has become a key research focus. Traditional sound insulation structures can no longer meet the required soundproofing standards. With their unique advantages, lightweight and high-strength sandwich panels are gaining significant attention in cruise ship design. Therefore, this paper designs a C-type sandwich panel structure based on the principle of equivalent mass substitution and uses ANSYS Workbench to analyze and verify its static performance. A sound insulation theoretical model is also established by using the space-harmonic expansion method and the principle of virtual work, from which the sound insulation expression is derived. The paper also systematically discusses the soundproofing performance of the sandwich panel under different structural parameters. The results show that the theoretical model is in good agreement with the simulation data within a certain frequency range, proving the model's validity. Furthermore, the elastic modulus and thickness of the C-type sandwich panel have a significant impact on the structure's sound insulation performance, providing an important reference for future cabin soundproofing design in cruise ships.

1. Introduction

With the rapid development of China's cruise tourism industry, the issue of cabin comfort has gradually become a focal point of public attention [1]. Noise from sources such as the cruise's power systems and propellers is primarily transmitted to cabins through both the air and the ship's structural frame. If the sound insulation and acoustic radiation performance of the wall panel structures used in the cruise do not meet the required standards, it may not only lead to damage to cruise equipment but also pose health risks to passengers. Therefore, seeking an alternative wall panel structure for cabins to reduce or even eliminate cabin noise is of great importance [2].

Sandwich panels often feature advantages such as light weight, high strength, and high rigidity [3], enabling them to bear external loads while effectively blocking or even isolating sound wave transmission [4]. In the future, these panels are expected to see widespread application in engineering fields such as automotive, aerospace, and marine industries. The types of soundproof sandwich panels are diverse, including honeycomb structures, foam cores, and lattice structures, each demonstrating notable differences in sound insulation performance. Research has shown that honeycomb-core sandwich panels outperform traditional structures in high-frequency noise insulation [5], although they exhibit significant limitations in the mid-and low-frequency ranges. MA and MRM Rejab [6] have pointed out that using porous or closed-cell foam materials as the core in foam sandwich panels enhances the structure's ability to absorb sound waves. However, the strength and durability of foam materials are relatively low, making them unsuitable for environments with high load-bearing requirements.

In contrast, the C-type core sandwich panel consists of two facing plates connected to the C-type stiffener core at the top and bottom. This design not only effectively blocks the direct transmission of sound waves but also dissipates sound energy by increasing the multiple reflection and refraction paths of the sound waves within the structure, creating a cavity-like structure that further helps dissipate sound energy. Li et al. [7] proposed a novel approach in a study, where the shape of the cavity is reshaped to significantly improve sound absorption performance while keeping the total mass and volume of the cavity unchanged. For C-type sandwich panels, the Helmholtz resonance frequency can be adjusted by designing C-type units with different volumes and thicknesses [8], allowing for higher sound transmission loss in specific frequency ranges. Therefore, the C-type sandwich panel isolation units not only enhance the overall stiffness of the structure but also increase the number of sound paths between the panel and the core material, enabling the adjustment of different material parameters in C-type sandwich panels to precisely meet sound insulation requirements in different scenarios.

In recent years, with the increasing demand for sound insulation in buildings, vehicles, and industrial equipment, the sound insulation theory of sandwich panels has become a key area of research. The development of theoretical models for sound insulation provides a foundation for predicting the sound transmission characteristics of sandwich panels across different frequencies and guides their design and optimization [9]. To improve the accuracy of low-frequency sound insulation predictions, researchers have proposed the Mass-Spring-Damper (MSD) model. This model simplifies the sandwich panel as a combination of facing plates and a core, with the core represented as an equivalent system of springs and dampers. This approach effectively simulates the sound insulation characteristics at low frequencies. Xu and Huang [10] developed a theoretical model for calculating the Sound Transmission Loss (STL) of orthogonally rib-reinforced honeycomb double-sandwich structures using the spatial harmonic method and the principle of virtual work. Xin and Lu [11], by incorporating tension, bending moments, and torsion, analyzed the transmission loss of orthogonally ribbed sandwich panels using the spatial harmonic expansion method. They further extended the application of the spatial harmonic expansion method from one-dimensional periodically supported beams to two-dimensional periodic plate structures.

In summary, the C-type sandwich panel, with its unique geometric configuration, can improve sound insulation within a certain frequency range while ensuring structural integrity. However, the research on the sound insulation performance of sandwich panels in domestic cruise ships is inconsistent, particularly in the areas of vibration damping and noise reduction, where the study of vibration and sound transmission mechanisms in sandwich panel structures needs significant improvement. Therefore, this paper combines theoretical modeling with numerical analysis to explore the theoretical foundation and sound insulation characteristics of C-type sandwich panels in depth. This research provides important guidance for the future application and design of more efficient and lightweight sound-insulating sandwich panels in cruise ships.

2. Design of sound-insulating sandwich panel structures

The method of separating noise sources from noise receivers to isolate or reduce noise is referred to as sound insulation. Currently, sound insulation structures in ships are typically single-layer or double-layer panels with air gaps. However, these structures no longer meet the requirements of modern cruise ships. As a structure with favorable acoustic properties, sandwich panels are gradually being applied in marine vessels [12].

2.1 Design of sandwich panel structure

Currently, sandwich panels with semicircular C-shaped stiffeners are recognized as a representative structure among metal sandwich panels. This design not only features a simple structure and ease of fabrication but also offers excellent performance characteristics. Therefore, this paper selects the C-type sandwich panel as the research subject to investigate its acoustic properties.

This study uses the deck of the lounge area in the residential cabins between Deck 1 and Deck 2 of a cruise ship's engine room section (Figure 1) as a prototype. Based on the principle of approximately equal mass, a substitution design is performed for the C-type sandwich panel. Figure 2 shows the form of the original stiffened plate and the C-type sandwich panel after the equal-mass substitution design.

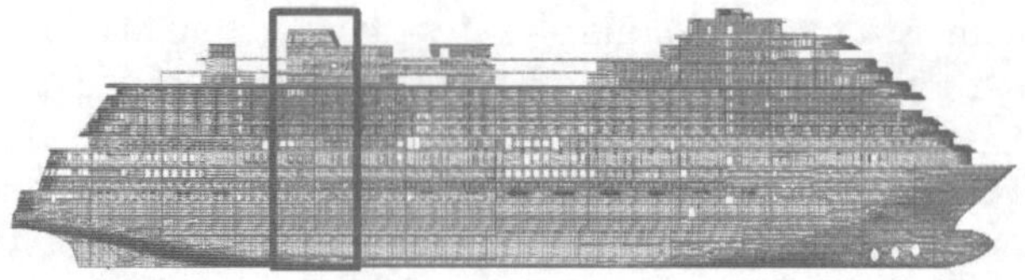

Figure 1. Selection of cabin position.

The panels and stiffeners of both the stiffened plate and the sandwich panel are made of standard steel, with the following material properties: density of 7, 850 kg/m³, Poisson's ratio of 0.3, and elastic modulus of 206 GPa. The dimensional parameters for each structure are as follows: 1. Stiffened Plate: Panel thickness 10 mm and Stiffener dimensions 145 mm×2 mm×13 mm; 2. C-type Sandwich Panel: Panel thickness 5 mm, Panel spacing 100 mm, and Stiffener thickness 5 mm. With this substitution design, the weight of the C-type sandwich panel is 320.46 kg, while the weight of the original stiffened plate is 319.94 kg.

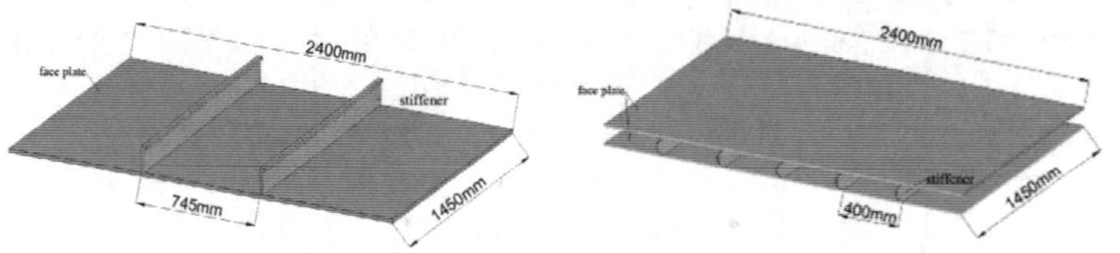

(a) Stiffened plate (b) C-type sandwich panel

Figure 2. Lounge deck is an alternative design form.

2.2 Structural static comparison analysis

To ensure that the strength and stiffness of the C-type sandwich panel, designed based on the principle of approximately equal mass, are not inferior to the original structure, the mechanical performance of the two types of structures under static loading is analyzed. A uniformly distributed pressure of 10, 000 Pa is applied to the top panel of each structure, with boundary conditions set as fixed supports along all four edges. A static analysis was conducted in ANSYS Workbench, resulting in the stress and displacement contour plots for each structure, as shown in Figure 3. The maximum stress and maximum displacement for each structure are summarized in Table 1.

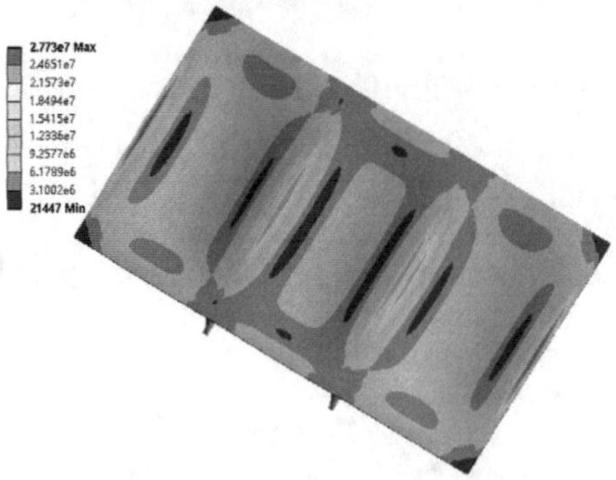

(a) Stress contour plot of stiffened plate

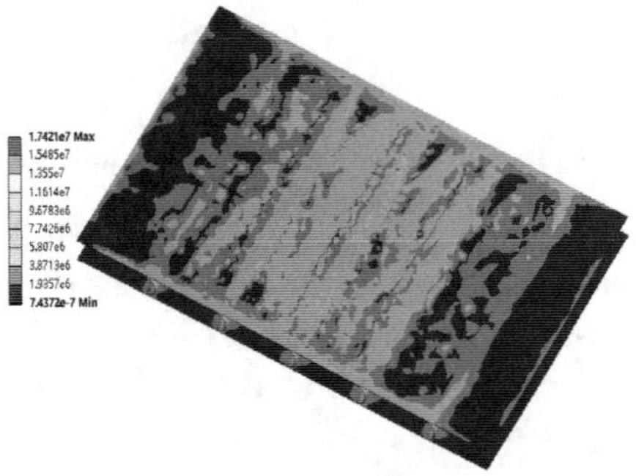

(b) Stress contour plot of C-type sandwich panel

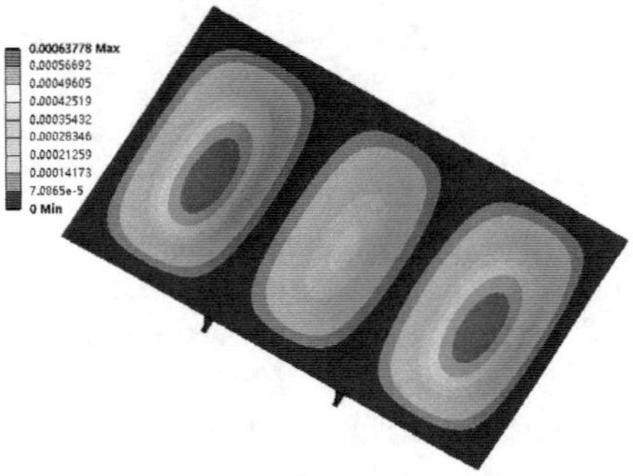

(c) Displacement contour plot of stiffened plate

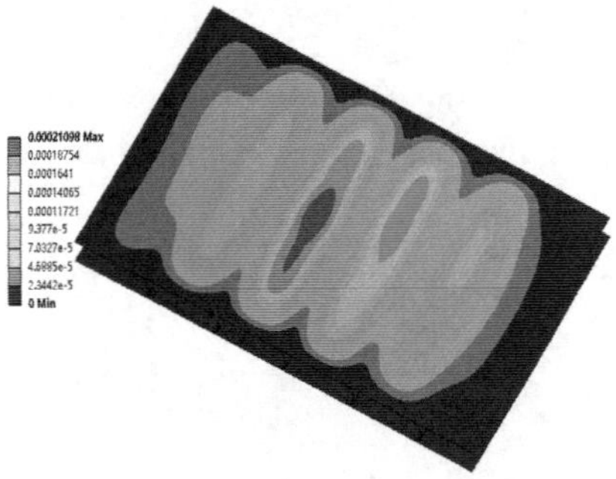

(d) Displacement contour plot of C-type sandwich panel

Figure 3. Cloud diagram of stress and displacement of different structures with equal mass.

Table 1. Structural Static Comparison At Equal Mass.

Structural configuration	P=10, 000 Pa	
	Stress (MPa)	Displacement (mm)
Stiffened plate	27.73	0.637
C-type sandwich panel	17.42	0.211

As observed in Table 1, the maximum stress of the C-type sandwich panel is reduced by 10.31 MPa compared to that of the stiffened plate, and the displacement is decreased by 0.426 mm. This demonstrates that the C-type sandwich panel, designed based on the principle of equal mass substitution, achieves strength and stiffness that are not inferior to those of the original stiffened plate structure.

3. Theoretical model of sound insulation for C-Type sandwich panel

3.1 Theoretical model development

First, it is necessary to understand the sound transmission mechanism in the study of C-type sandwich panels. When the top panel of the sandwich panel structure (as shown in Figure 4) is vibrated by sound waves from the incident region, the vibration is mainly transmitted to the bottom panel of the sandwich structure through two primary pathways: One pathway is through the periodically distributed C-type stiffeners, and the other is through the air gap between the panels. These two mechanisms induce vibration in the bottom panel, which then radiates sound pressure into the transmission region, allowing sound to pass through the sandwich structure.

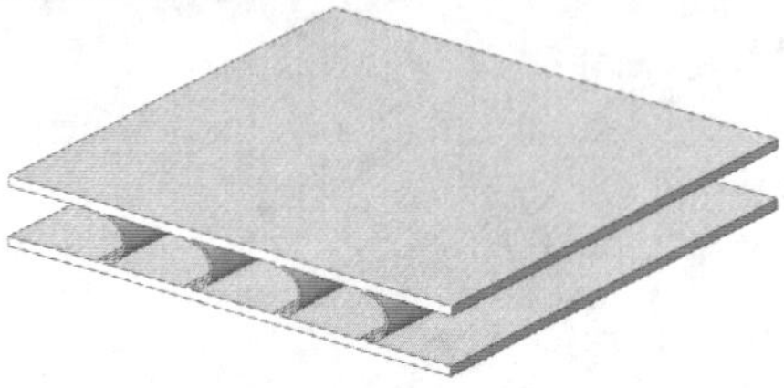

Figure 4. C-type sandwich panel.

Taking the infinitely large periodic C-type sandwich panel structure and its equivalent model as shown in Figure 5, and using the theoretical coordinate system defined by the x-axis and y-axis in the figure, when sound waves with a circular frequency of ω incident from the incident area at an angle θ onto the upper panel of the sandwich structure, part of the sound waves propagate through the sandwich structure to the transmission area, while the remainder is reflected. It is assumed that the elastic modulus of the materials for the upper and lower panels are E_1 and E_2, the Poisson's ratios are ν_1 and ν_2, and the densities are ρ_1 and ρ_2, respectively, while the elastic modulus of the material for the intermediate reinforcements is E_c, the Poisson's ratio is ν_c, and the density is ρ_c. The remaining assumed parameters for the sandwich panel are specified in Figure 5.

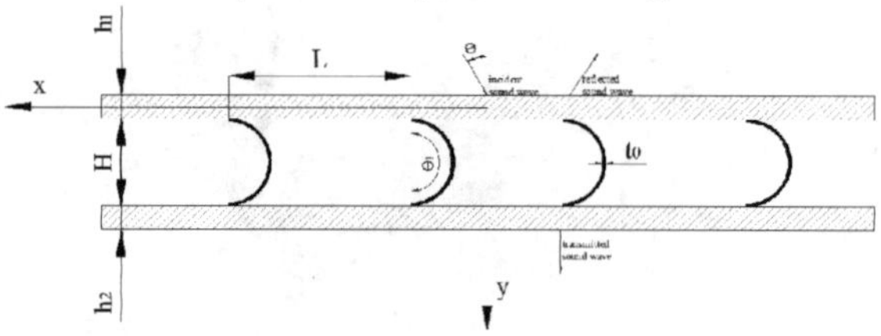

(a) C-type sandwich panel

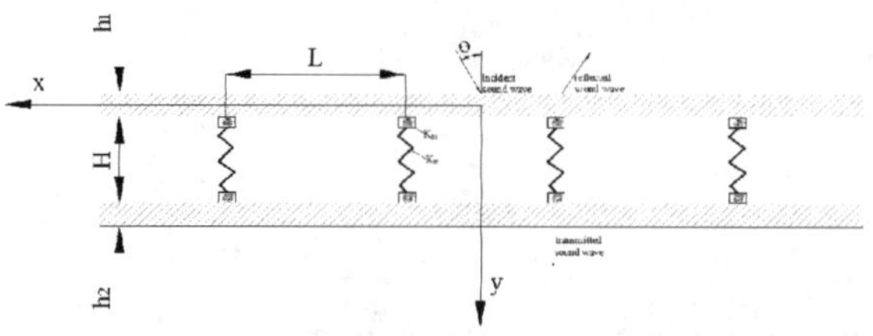

(b) Equivalent structure of C-type sandwich panel

Figure 5. C-type sandwich panel and its equivalent structure.

As shown in Figure 5, the C-type reinforcements in the core layer of the sandwich panel are equivalently replaced by a combination of periodically distributed springs and torsion springs, while the mass of the reinforcements is equivalently replaced by concentrated masses distributed on the upper and lower panels of the structure. The equivalent axial stiffness of the springs is set to K_F, and the torsional stiffness of the equivalent torsion springs is set to K_M. In this paper, the tensile stiffness of the C-type stiffener is derived theoretically and validated through finite element simulation, with the formula provided as Equation (1).

$$K_F = \frac{2EI}{\pi R^3} \tag{1}$$

Since in practical examples, the torsional stiffness of the torsion springs has almost no impact on the calculation results, the torsional stiffness of the torsion springs is not provided here.

Assuming that the two side panels of the sandwich panel are homogeneous, the mass per unit periodic length of the panels are m_1 and m_2, respectively, and the mass of the intermediate C-type reinforcements is 2M. Under these assumptions, the vibration control equation for the equivalent structure of the C-type sandwich panel shown in Figure 5 (b) [13] can be written as:

$$\begin{aligned} &D_1\frac{\partial^4 w_1(x,t)}{\partial x^4} + (m_1 + \frac{M}{L})\frac{\partial^2 w_1(x,t)}{\partial t^2} + K_F[w_1(x,t) - w_2(x,t)] \\ &-K_M\frac{\partial^2}{\partial x^2}[w_1(x,t) - w_2(x,t)] - j\omega\rho_0(\Phi_1 - \Phi_2) = 0 \end{aligned} \tag{2}$$

$$\begin{aligned} &D_2\frac{\partial^4 w_2(x,t)}{\partial x^4} + (m_2 + \frac{M}{L})\frac{\partial^2 w_2(x,t)}{\partial t^2} + K_F[w_2(x,t) - w_1(x,t)] \\ &-K_M\frac{\partial^2}{\partial x^2}[w_2(x,t) - w_1(x,t)] - j\omega\rho_0(\Phi_2 - \Phi_3) = 0 \end{aligned} \tag{3}$$

where $D_i (i = 1,2)$ represents the bending stiffness of the plate, with consideration of the damping loss factor $D_i = \frac{E_i h_i^3 (1 + j\eta_i)}{12(1 - v_i^2)}$, η_i is the material's damping loss factor, $w_i (i = 1,2)$ is the displacement of the upper and lower panels of the sandwich panel, $j = \sqrt{-1}$ and ω are the circular frequencies, ρ_0 is the air density, and $\Phi_i (i = 1,2,3)$ is the velocity potential energy of the sound field.

Since this theory applies to C-type sandwich panels that are periodic and infinitely large structures, when the structure is excited by plane harmonic waves, the sound field around the panel can be regarded, according to Floquet's theorem, as a superposition of a series of spatial harmonics. Therefore, the velocity potential energy of the sound field in the incident area can be expressed as [14]:

$$\Phi_1(x,y,t) = Ie^{-j(k_x x + k_y y - \omega t)} + \sum_{n=-\infty}^{+\infty} \beta_n e^{-j[(k_x + \frac{2n\pi}{L})x - k_{yn} y - \omega t]} \tag{4}$$

where I represents the amplitude of the velocity potential energy of the sound waves in the incident area, β_n is the amplitude of the velocity potential energy of the reflected sound waves in the n th order component, and k_x and k_y are the wavenumbers related to the incident angle θ, which can be expressed as:

$$k_x = k\sin\theta \quad k_y = k\cos\theta \tag{5}$$

The velocity potential energy of the sound field in the cavity area of the middle core layer can be expressed as:

$$\Phi_2 = \sum_{n=-\infty}^{+\infty} \varepsilon_n e^{-j[(k_x + \frac{2n\pi}{L})x - k_{yn} y - \omega t]} + \sum_{n=-\infty}^{+\infty} \zeta_n e^{-j[(k_x + \frac{2n\pi}{L})x - k_{yn} y - \omega t]} \tag{6}$$

where ε_n represents the amplitude of the n th velocity component of the forward wave velocity potential energy in the cavity domain and ζ_n represents the amplitude of the n th velocity component of the backward wave velocity potential energy in the cavity domain.

The velocity potential of the sound field in the transmission region can be expressed as:

$$\Phi_3(x,y,t) = \sum_{n=-\infty}^{+\infty} \xi_n e^{-j[(k_x + \frac{2n\pi}{L})x - k_{yn} y - \omega t]} \tag{7}$$

where ξ_n represents the amplitude of the n th order component of the forward wave velocity potential energy in the transmission area, $k=\omega/c_0$ and c_0 represent the propagation speed of sound waves in air, and k_{yn} is the wavenumber of the harmonic components of the sound wave in the y direction, which can be expressed as:

$$k_{yn}=\sqrt{\left(\frac{w}{c}\right)^2-\left(k_x+\frac{2n\pi}{L}\right)^2} \tag{8}$$

When $\frac{w}{c}<\left|k_x+\frac{2n\pi}{L}\right|$:

$$k_{yn}=j\sqrt{\left(k_x+\frac{2n\pi}{L}\right)^2-\left(\frac{w}{c}\right)^2} \tag{9}$$

Similar to the velocity potential energy of the sound field, the motion of the upper and lower panels of the sandwich panel can be represented by using the method of Mead and Pujara as a superposition of a series of spatial harmonics:

$$W_1(x,t)=\sum_{n=-\infty}^{+\infty}\alpha_{1,n}e^{-j(k_x+\frac{2n\pi}{L})x}e^{j\omega t} \tag{10}$$

$$W_2(x,t)=\sum_{n=-\infty}^{+\infty}\alpha_{2,n}e^{-j(k_x+\frac{2n\pi}{L})x}e^{j\omega t} \tag{11}$$

Since the normal vibration velocity of the top and bottom panels of the sandwich structure is equal to the velocity of the air at the interface between the panel and the air, this condition can be used to derive the continuity condition in Equations (12)-(13).

$$-\frac{\partial \Phi_1}{\partial y}=j\omega W_1,\quad -\frac{\partial \Phi_2}{\partial y}=j\omega W_1,\quad y=0 \tag{12}$$

$$-\frac{\partial \Phi_2}{\partial y}=j\omega W_2,\quad -\frac{\partial \Phi_3}{\partial y}=j\omega W_2,\quad y=H \tag{13}$$

By substituting Equations (4)-(11) into the continuity condition of Equations (12)-(13), we obtain:

$$\beta_0=I-\frac{\omega\alpha_{1,0}}{k_y},\quad \beta_n=-\frac{\omega\alpha_{1,n}}{k_{yn}},n\neq 0 \tag{14}$$

$$\varepsilon_n=\frac{\omega\left(\alpha_{2,n}e^{jk_{yn}H}-\alpha_{1,n}e^{2jk_{yn}H}\right)}{k_{yn}\left(1-e^{2jk_{yn}H}\right)} \tag{15}$$

$$\zeta_n=\frac{\omega\left(\alpha_{2,n}e^{jk_{yn}H}-\alpha_{1,n}\right)}{k_{yn}\left(1-e^{2jk_{yn}H}\right)},\quad \xi_n=\frac{\omega\alpha_{2,n}e^{jk_{yn}H}}{k_{yn}} \tag{16}$$

In Equations (14)-(16), the coefficient $a_{i,n}(i=1,2)$ can be obtained by using the principle of virtual work, and the total virtual work done by the system is the sum of the virtual work done by the upper and lower panels, the translational springs, the torsional springs, and the concentrated masses. Assuming the virtual displacement of the system is [14]:

$$\delta W_i=\delta\alpha_{i,m}e^{-j(k_x+\frac{2n\pi}{L})x} \tag{17}$$

The differential equation controlling the motion of the upper and lower panels can be written as:

$$D_1\frac{\partial^4 W_1}{\partial x^4}+m_{p1}\frac{\partial^2 W_1}{\partial t^2}-j\omega\rho_0\left(\Phi_1-\Phi_2\right)=0 \tag{18}$$

$$D_2\frac{\partial^4 W_2}{\partial x^4}+m_{p2}\frac{\partial^2 W_2}{\partial t^2}-j\omega\rho_0\left(\Phi_2-\Phi_3\right)=0 \tag{19}$$

Therefore, the virtual work done by the upper and lower panels can be represented as:

$$\delta \Pi_{p1} = \int_{x=0}^{L} \left(D_1 \frac{\partial^4 W_1}{\partial x^4} + m_{p1} \frac{\partial^2 W_1}{\partial t^2} - j\omega\rho_0 (\Phi_1 - \Phi_2) \right) \delta W_1^* dx, y = 0 \tag{20}$$

$$\delta \Pi_{p2} = \int_{x=0}^{L} \left(D_2 \frac{\partial^4 W_2}{\partial x^4} + m_{p2} \frac{\partial^2 W_2}{\partial t^2} - j\omega\rho_0 (\Phi_2 - \Phi_3) \right) \delta W_2^* dx, y = H \tag{21}$$

where $m_{pi}(i=1,2)$ represents the mass per unit area of the upper and lower panels, and $\delta W_i^* (i=1,2)$ denotes the complex conjugate of the virtual displacement.

The virtual work done by the translational springs can be represented as:

$$\delta \Pi_{t1} = K_t (W_1(0) - W_2(0)) \delta\alpha_{1,m}^* = K_t \left(\sum_{n=-\infty}^{+\infty} \alpha_{1,n} - \sum_{n=-\infty}^{+\infty} \alpha_{2,n} \right) \delta\alpha_{1,m}^* \tag{22}$$

$$\delta \Pi_{t2} = K_t (W_2(0) - W_1(0)) \delta\alpha_{2,m}^* = K_t \left(\sum_{n=-\infty}^{+\infty} \alpha_{2,n} - \sum_{n=\infty}^{+\infty} \alpha_{1,n} \right) \delta\alpha_{2,m}^* \tag{23}$$

The virtual work done by the torsional springs can be represented as:

$$\begin{aligned} \delta \Pi_{r1} &= K_r (W_1'(0) - W_2'(0)) j \left(k_x + \frac{2m\pi}{L} \right) \delta\alpha_{1,m}^* \\ &= K_r \left[\sum_{n=-\infty}^{+\infty} \alpha_{1,n} \left(k_x + \frac{2n\pi}{L} \right) - \sum_{n=-\infty}^{+\infty} \alpha_{2,n} \left(k_x + \frac{2n\pi}{L} \right) \right] \left(k_x + \frac{2m\pi}{L} \right) \delta\alpha_{1,m}^* \end{aligned} \tag{24}$$

$$\begin{aligned} \delta \Pi_{r2} &= K_r (W_2'(0) - W_1'(0)) j \left(k_x + \frac{2m\pi}{L} \right) \delta\alpha_{2,m}^* \\ &= K_r \left[\sum_{n=-\infty}^{+\infty} \alpha_{2,n} \left(k_x + \frac{2n\pi}{L} \right) - \sum_{n=-\infty}^{+\infty} \alpha_{1,n} \left(k_x + \frac{2n\pi}{L} \right) \right] \left(k_x + \frac{2m\pi}{L} \right) \delta\alpha_{2,m}^* \end{aligned} \tag{25}$$

The virtual work done by the concentrated masses can be represented as:

$$\delta \Pi_{m1} = -\omega^2 M W_1(0) \delta\alpha_{1,m}^* = -\omega^2 M \left[\sum_{n=-\infty}^{+\infty} \alpha_{1,n} \right] \delta\alpha_{1,m}^* \tag{26}$$

$$\delta \Pi_{m2} = -\omega^2 M W_2(0) \delta\alpha_{2,m}^* = -\omega^2 M \left[\sum_{n=-\infty}^{+\infty} \alpha_{2,n} \right] \delta\alpha_{2,m}^* \tag{27}$$

By applying the principle of virtual work to establish the equation:

$$\delta \Pi_{pi} + \delta \Pi_{ti} + \delta \Pi_{ri} + \delta \Pi_{Mi} = 0, (i=1,2) \tag{28}$$

Finally, it can be derived from Equations (24)-(25) as follows:

$$\begin{aligned} &\left[D_1 \left(k_x + \frac{2m\pi}{L} \right)^4 - m_{p1}\omega^2 - \frac{2j\omega^2 \rho_0 e^{2jk_{ym}H}}{k_{ym}\left(1 - e^{2jk_{ym}H}\right)} \right] \alpha_{1,m} + \frac{2j\omega^2 \rho_0 e^{jk_{ym}H}}{k_{ym}\left(1 - e^{2jk_{ym}H}\right)} \alpha_{2,m} \\ &+ \sum_{n=-\infty}^{+\infty} \alpha_{1,n} \left[\frac{K_t - \omega^2 M}{L} + \frac{K_r}{L} \left(k_x + \frac{2n\pi}{L} \right) \left(k_x + \frac{2m\pi}{L} \right) \right] \\ &- \sum_{n=-\infty}^{+\infty} \alpha_{2,n} \left[\frac{K_t}{L} + \frac{K_r}{L} \left(k_x + \frac{2n\pi}{L} \right) \left(k_x + \frac{2m\pi}{L} \right) \right] = \begin{cases} 2j\omega\rho_0 I, m = 0 \\ 0, m \neq 0 \end{cases} \end{aligned} \tag{29}$$

$$\begin{aligned} &\left[D_2 \left(k_x + \frac{2m\pi}{L} \right)^4 - m_{p2}\omega^2 - \frac{2j\omega^2 \rho_0 e^{2jk_{ym}H}}{k_{ym}\left(1 - e^{2jk_{ym}H}\right)} \right] \alpha_{2,m} + \frac{2j\omega^2 \rho_0 e^{jk_{ym}H}}{k_{ym}\left(1 - e^{2jk_{ym}H}\right)} \alpha_{1,m} \\ &+ \sum_{n=-\infty}^{+\infty} \alpha_{2,n} \left[\frac{K_t - \omega^2 M}{L} + \frac{K_r}{L} \left(k_x + \frac{2n\pi}{L} \right) \left(k_x + \frac{2m\pi}{L} \right) \right] \\ &- \sum_{n=-\infty}^{+\infty} \alpha_{1,n} \left[\frac{K_t}{L} + \frac{K_r}{L} \left(k_x + \frac{2n\pi}{L} \right) \left(k_x + \frac{2m\pi}{L} \right) \right] = 0 \end{aligned} \tag{30}$$

By solving Equations (29)-(30), the coefficients $\alpha_{1,n}$ and $\alpha_{2,n}$ can be obtained, and thereafter, the coefficients β_n, ε_n, ς_n, and ξ_n can be determined. Subsequently, the sound wave propagation coefficient can be obtained through Equation (31):

$$\tau(\theta) = \frac{\sum_{n=-\infty}^{+\infty} |\xi_n|^2 \operatorname{Re}(k_{yn})}{|I|^2 k_y} \tag{31}$$

Finally, the sound insulation quantity of the sandwich panel structure can be determined by Equation (32):

$$L_{ST} = -10\log_{10}\tau(\theta) \tag{32}$$

3.2 Convergence analysis

Since the solution of the acoustic-vibration coupling equation is expressed as a sum of a finite number of spatial harmonic terms, it is essential to ensure the convergence of the calculated results. Therefore, prior to analysis, the minimum number of terms required for convergence of the sound transmission loss results of the C-type sandwich panel within the frequency range of interest should be determined. If the absolute difference between two consecutive calculation results is less than a specified tolerance, the results can be considered convergent.

In this section, the C-type sandwich panel formed through equal-mass substitution is taken as the research subject, with the incident sound wave angle set at $\theta = \pi / 4$. As shown in Figure 6, when the number of series terms n is set to 15 or greater, the sound transmission loss of the structure converges. Therefore, setting $n = 15$ ensures both the convergence and accuracy of the calculation results.

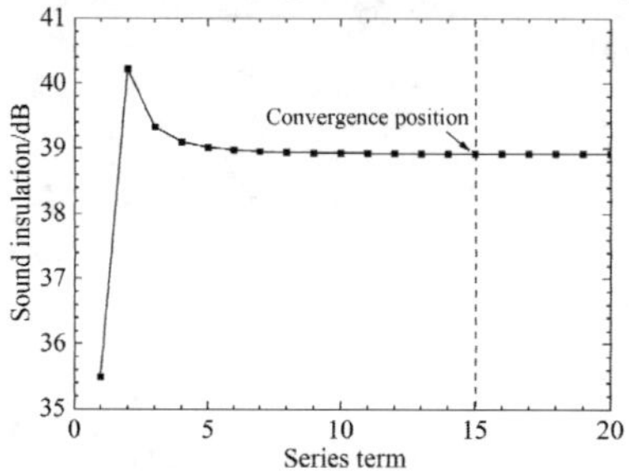

Figure 6. Convergence curve of STL of C foldable sandwich panel.

4. Numerical simulation and validation of the theoretical model

To validate the theoretical model, a C-type sandwich panel structure model is created in the acoustic finite element software LMS Virtual Lab for coupled vibroacoustic analysis (as shown in Figure 7). The maximum calculation frequency in this study is set to 1, 000 Hz. According to the formula $l = c_0 / 6f_{max}$, (where l represents the element length and f_{max} is the upper-frequency limit), the maximum element length is calculated to be 56.6 mm. Therefore, setting the element length to 50 mm in this study ensures the required accuracy for acoustic simulation analysis.

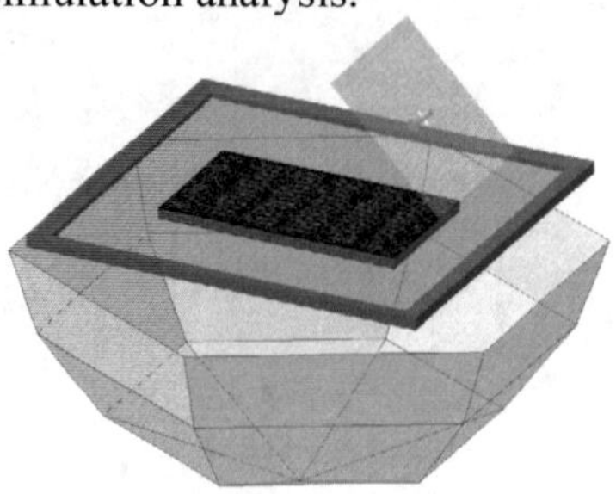

Figure 7. Comparison of theory and simulation of STL.

Figure 8 presents the results of the theoretical model and simulation. From the figure, the overall trends of the theoretical model and simulation match reasonably well within a certain frequency range. This indicates that the theoretical model employed has a certain degree of accuracy.

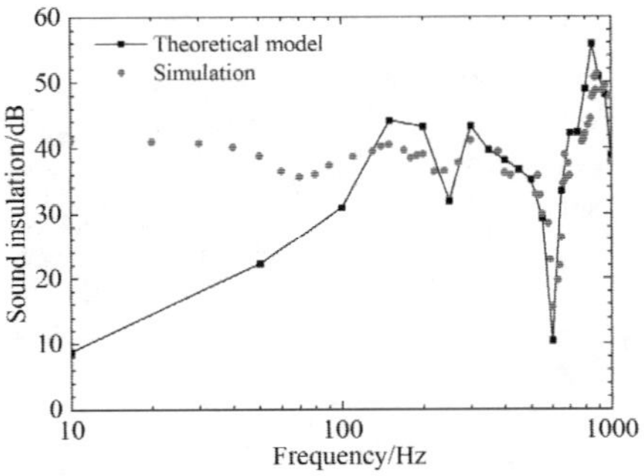

Figure 8. LMS virtual lab simulation diagram.

5. The Effect of system parameters on sound insulation performance

5.1 Panel thickness

Figure 9 presents the transmission loss curves of C-type sandwich panels with different panel thicknesses. To investigate the impact of panel thickness on the sound insulation performance of C-type sandwich panels, other parameters were kept constant while varying the panel thickness, with $h_1 = h_2$ set to 5 mm, 7 mm, and 10 mm. The results demonstrate that at lower frequency ranges, the transmission loss of the sandwich panel structure increases significantly with the panel thickness, and the dips in the transmission loss curves shift to higher frequencies as the panel thickness increases. This occurs because the increase in structural stiffness with thicker panels is faster than the increase in structural mass, leading to a rise in the natural frequencies of the structure. In the frequency range of 600 Hz to 1, 000 Hz, the rate of increase in transmission loss diminishes as the panel thickness increases, indicating that the influence of panel thickness on the sound insulation performance of C-type sandwich panels becomes less significant in this frequency range.

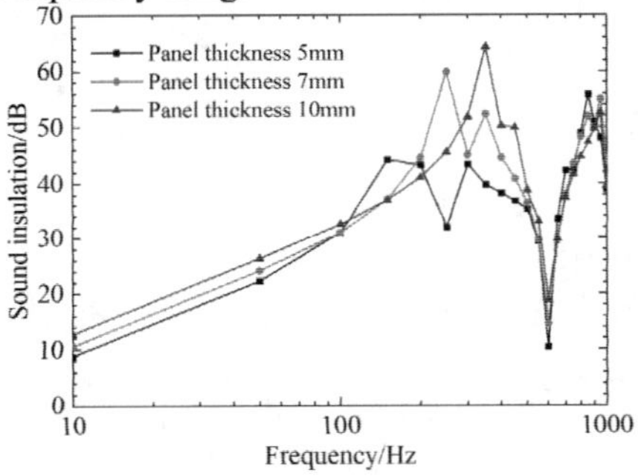

Figure 9. STL curves of the sandwich plate under different panel thicknesses.

5.2 Elastic modulus

Figure 10 presents the transmission loss curves of the sandwich panel structure under different elastic modulus conditions of the panel material. Keeping all other structural parameters constant, the elastic moduli of the steel material $E_1 = E_2$ were set to 103, 000 MPa, 206, 000 MPa, and 412, 000 MPa, respectively. The results indicate that in the frequency range of 10 Hz to 500 Hz, the transmission loss of the sandwich panel structure increases with an increase in the material's elastic modulus. Additionally, the dips in the transmission loss curves shift toward higher frequencies as the elastic modulus increases. This behavior occurs because the overall stiffness of the sandwich panel increases with the elastic modulus. As incident sound waves induce vibrations in the panels and transmit to the radiating-side lower panel, the increased stiffness reduces the vibration displacement of the radiating-side panel. Consequently, the radiated sound power of the structure decreases. Thus, sandwich panels made from materials with higher elastic moduli exhibit superior sound insulation performance.

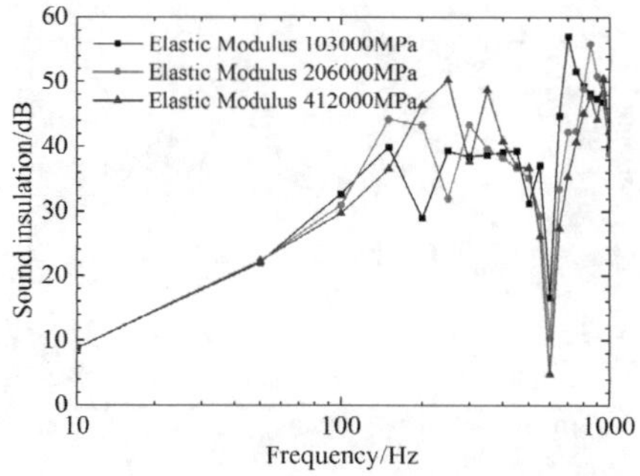

Figure 10. STL curves of the sandwich plate under different elastic modulus.

5.3 Panel spacing

Figure 11 illustrates the sound transmission loss curves of C-type sandwich panel structures under three different panel spacing conditions. To investigate the effect of panel spacing on sound insulation performance, other parameters were kept constant while the panel spacing H was set to 100 mm, 120 mm, and 150 mm, respectively. The results indicate that the STL curves exhibit similar trends. As the panel spacing increases, the sound insulation performance of the C-type sandwich panel structure improves. However, the STL values at the trough points of the curve decrease, while the STL values at the peak points increase. In the frequency range of 300 Hz to 1, 000 Hz, the influence of panel spacing on the sound insulation performance of the sandwich panel structure diminishes.

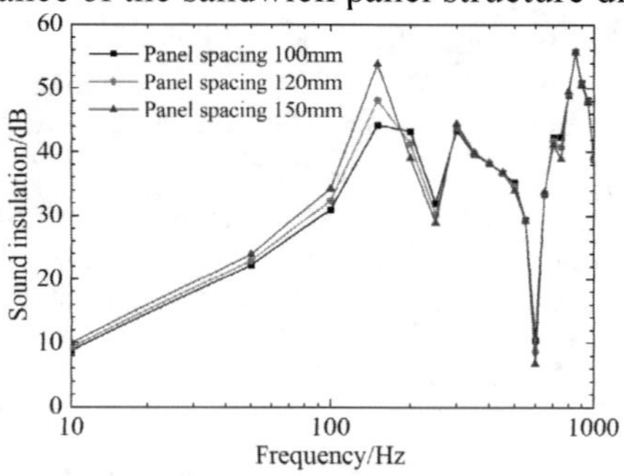

Figure 11. STL curves of the sandwich plate under different panel spacing.

5.4 Core stiffener spacing

Figure 12 illustrates the sound transmission loss curves of the C-type sandwich panel structure under three different stiffener spacing conditions. Keeping all other parameters constant, the stiffener spacing L was set to 0.2 m, 0.3 m, and 0.4 m, respectively. The results indicate that as the stiffener spacing increases, the number of peaks and troughs in the STL curves also increases. Conversely, as the stiffener spacing decreases, the troughs in the STL curves shift toward higher frequencies. This occurs because smaller stiffener spacing enhances the coupling degree of the sandwich panel structure, which shifts the natural frequencies of the structure to higher frequencies. Moreover, the increased overall stiffness of the structure reduces the energy transfer from the incident-side panel to the radiating-side panel, thereby improving the sound insulation performance.

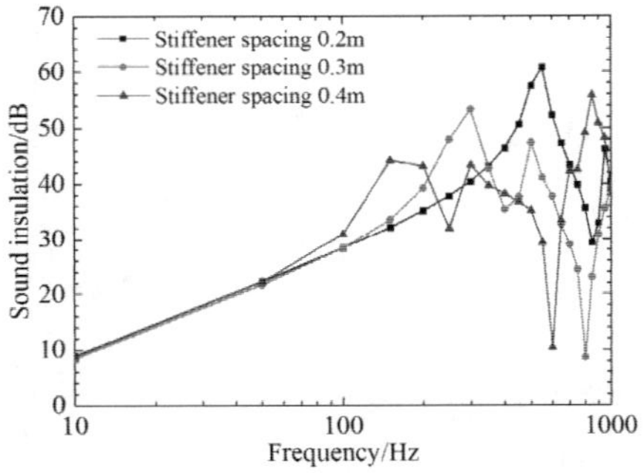

Figure 12. STL curves of the sandwich plate under different panel stiffener spacing.

5.5 Core stiffener thickness

When studying the effect of stiffener thickness on the sound insulation performance of the structure, the stiffness of the stiffeners and other structural parameters were kept constant, while only the thickness t_0 of the C-type stiffeners was varied to 5 mm, 7 mm, and 10 mm, respectively. The corresponding results are shown in Figure 13. In the frequency range of 10 Hz to 600 Hz, stiffener thickness has a significant impact on the sound insulation performance of the structure. As the stiffener thickness increases, the sound transmission loss of the sandwich panel structure also increases, though the improvement is not significant. Additionally, the peaks and troughs of the STL curves shift toward higher frequencies. However, in the frequency range of 600 Hz to 1, 000 Hz, the increase in stiffener thickness does not effectively improve the sound insulation performance.

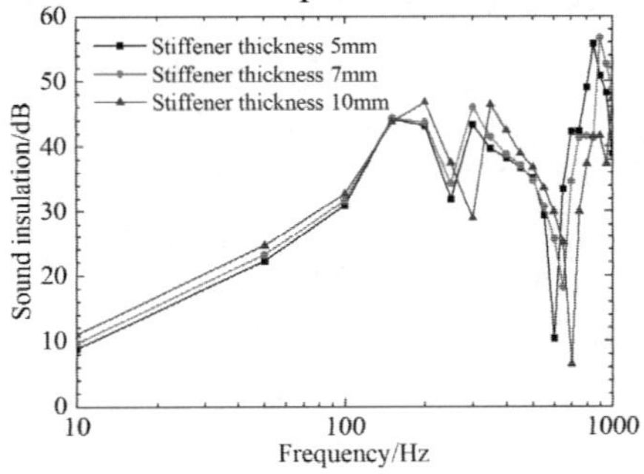

Figure 13. STL curves of the sandwich plate under different panel stiffener thickness.

6. Conclusions

In this paper, the deck of the lounge area in a large cruise ship's engine room section was selected as the research subject. Based on the principle of equal mass, a C-type sandwich panel was designed as a substitute for the traditional stiffened plate, and its mechanical performance under typical conditions was compared and analyzed by using finite element software. On this basis, a vibroacoustic coupling equation for the C-type sandwich panel was established by using acoustic theory, leading to the derivation of the sound transmission loss expression for the panel. The main conclusions are as follows:

(1) Mechanical performance simulations conducted by using finite element software indicate that the C-type sandwich panel offers significant advantages in stiffness and strength while ensuring lightweight characteristics.

(2) By modeling the C-type stiffener as an equivalent system of springs, torsion springs, and concentrated mass, the tensile strength of the C-type stiffener was derived. Finite element calculations confirmed the feasibility of this equivalence, validating its use as an input parameter in the derivation of the sound insulation theory.

(3) Panel spacing and stiffener thickness have a certain degree of improvement on the structure's sound insulation performance in the mid-low frequency range, but the overall trend of impact is not very significant.

(4) Panel thickness, elastic modulus of the material, and core stiffener spacing have a significant impact on the structure's sound insulation performance in the mid-low frequency range. However, their influence on the sound insulation performance in the high-frequency range is not very significant. Therefore, in future optimizations, it is advisable to select appropriate panel thickness, high elastic modulus materials, and smaller stiffener spacing to enhance sound insulation performance. However, when designing, consideration must be given to maintaining control over the weight of the structure.

References

[1] R. Wei, C. Cheng, Y. Yin, and J. Hu. "Discussion on the development status and strategies of China's cruise tourism industry", Special Zone Economy, No. 02, pp. 80-83, 2018.
[2] Z. Li. "Refined prediction study of the internal noise in ship cabins", Dalian University of Technology, 2019.
[3] A. Farrokhabadi, S. Ahmad Taghizadeh, H. Madadi, H. Norouzi, and A. Ataei. "Experimental and numerical analysis of novel multi-layer sandwich panels under three-point bending load", Composite Structures, Vol. 250, p. 112, 631, Oct. 2020.
[4] F. Nentwich and A. Fuchs. "Acoustic behavior of sandwich plates", Journal of Sandwich Structures & Materials, Vol. 17, No. 2, pp. 183-213, Oct. 2014.
[5] F. Nentwich and A. Fuchs. "Acoustic behavior of sandwich plates", Journal of Sandwich Structures & Materials, Vol. 17, No. 2, pp. 183-213, Oct. 2014.
[6] Q. Ma, M. R. M. Rejab, J. P. Siregar, and Z. Guan. "A review of the recent trends on core structures and impact response of sandwich panels", Journal of Composite Materials, Vol. 0, No. 0, pp. 1-43, 2021.
[7] X. Li, X. Yu, Junwei Chua, and W. Zhai. "Harnessing cavity dissipation for enhanced sound absorption in Helmholtz resonance metamaterials", Materials Horizons, Vol. 10, No. 8, pp. 2, 892-2, 903, Jan. 2023.
[8] M. Dannemann et al. "Experimental Study of Advanced Helmholtz Resonator Liners with Increased Acoustic Performance by Utilizing Material Damping Effects", Applied Sciences, Vol. 8, No. 10, pp. 1, 923-1, 923, Oct. 2018.
[9] D. Zhao, G. Squicciarini, and N. Ferguson. "Vibroacoustic response of stiffened thin plates to incident sound", Applied Acoustics, Vol. 172, p. 107, 578, Jan. 2021.
[10] D. Xu and Q. Huang. "Broadband Sound Insulation Performance of Composite Double Sandwich Panels with Periodic Piezoelectric Arrays", 2022 5th World Conference on Mechanical Engineering and Intelligent Manufacturing (WCMEIM), Nov. 2022.
[11] F. Xin and T. Lu. "Analytical and experimental investigation on transmission loss of clamped double panels: Implication of boundary effects", Journal of the Acoustical Society of America, Vol. 125, No. 3, pp. 1, 506-1, 517, Mar. 2009.
[12] G. Palomba, G. Epasto, and V. Crupi. "Lightweight sandwich structures for marine applications: a review", Mechanics of Advanced Materials and Structures, pp. 1-26, Jul. 2021.
[13] J. Wang, T. J. Lu, J. Woodhouse, R. S. Langley, and J. Evans. "Sound transmission through lightweight double-leaf partitions: theoretical modeling", Journal of Sound and Vibration, Vol. 286, No. 4-5, pp. 817-847, Sep. 2005.
[14] J. H. LEE and J. KIM. "Analysis of sound transmission through periodically stiffened panels by space-harmonic expansion method", Journal of Sound and Vibration, Vol. 251, No. 2, pp. 349-366, Mar. 2002.

Journal of Physics: Conference Series **3009** (2025) 012079 doi:10.1088/1742-6596/3009/1/012079

Defect detection based on MobileViTv2 and spatial-channel reconstruction fusion

Rumeng Liang and Tao Wu *

School of Computer Science, Xi'an Polytechnic University, Xi'an, Shaanxi 710048, China

*E-mail: Wutao1976@126.com

Abstract. This research builds an innovative identification mechanism for miniaturized fault features detection on PCB, where we replace the backbone part of the YOLOv8n model with MobileViTv2 to optimize the algorithm's processing efficiency and recognition precision. In terms of spatial-channel reconstruction fusion, we optimize the network structure by integrating an attention-enhanced SCConv block to replace the original C2f component in the header network, effectively reducing spatial and channel redundancy throughout the model, and consequently elevating the model's holistic metrics. Furthermore, we introduce an attention-recalibrated geometric loss to further optimize detection precision and robustness. Benchmark evaluations demonstrate the proposed approach yields a 4.6% improvement in detection accuracy (mAP@50) and a 6.6% improvement in recall rate on the PCB Defect public datasets. Across tests on datasets with different proportions, the detection fidelity across multiple defect categories demonstrates marked enhancement, fully meeting the practical needs of industrial applications.

1. Introduction

As state-of-the-art electronic innovations continue to proliferate, as integrated circuitry substrates, PCBs (printed circuit borads) constitute the foundational elements of modern electronics, and their quality and stability have a decisive impact on the overall performance of electronic systems. However, PCB may encounter various challenges during the production process, such as material defects, and manufacturing process deviations, which can lead to problems such as open circuits, short circuits, and foreign matter residues on the PCB surface. These problems not only reduce the electrical performance of the PCB but also may cause many defective circuit boards to be discarded, causing environmental problems [1]. Therefore, efficient and accurate defect detection of PCB has become a pressing challenge requiring immediate attention. Traditionally, PCB defect detection mainly primarily dependent on labor-intensive manual inspections or conventional detection tools [2], but these approaches are not only substantial temporal overheads and workforce allocation but easily affected by human factors, resulting in missed detection or misjudgment. Propelled by breakthroughs in visual perception algorithms and neural architectures, PCB defect detection strategies based on image processing have gradually emerged [3] and become a research hotspot in the industry.

Deep neural network-driven PCB inspection has emerged as a predominant methodology in contemporary manufacturing quality control systems [4][5]. This technology mainly relies on deep learning target detection methods, including two-stage, single-stage, and Transformer-based (DEtection TRansformer, DETR). The single-stage target detection algorithm calculates the image and generates detection results. It is fast but relatively low in accuracy. The YOLO (You Only Look Once) algorithm family epitomizes this paradigm, establishing benchmark performance in real-time detection tasks. In this paper, we focus on the YOLO-based PCB defect detection algorithm.

Nowadays, most algorithms ignore the optimization of model size when pursuing the model's precision in defect identification, which leads to the increase of network structure depth, the introduction of unnecessary redundancy, and the extension of model detection running time. To address these problems, we integrated the lightweight MobileViTv2 network model based on the YOLOv8 algorithm framework, introduced spatial and channel reconstruction convolution and Focal-EIoU Loss function, and proposed a new algorithm model MobileViTv2-YOLO, which not only markedly augmented the classification fidelity, but also realized the lightweight design of the network structure, effectively reducing the complexity of the model.

The remaining sections of this paper are structured in the following manner: Chapter 2 mainly explains the relevant algorithms in detail; Chapter 3 describes the improved algorithms; Chapter 4 describes the experimental process and results comparison in detail; Chapter 5 summarizes.

2. Algorithms

2.1 MobileViTv2 algorithm

MobileViTv2 [6] is a lightweight visual Transformer [7] model that synthesizes CNN's hierarchical feature abstraction capabilities and Vision Transformer(ViT)to provide low-latency, high-performance visual processing capabilities. MobileViTv2's MobileViT block module combines traditional convolution operations with Transformer Block and passes features to subsequent Transformer Blocks through residual connections, making the network more stable during training. The network model diagram of MobileViT is schematically represented in Figure 1.

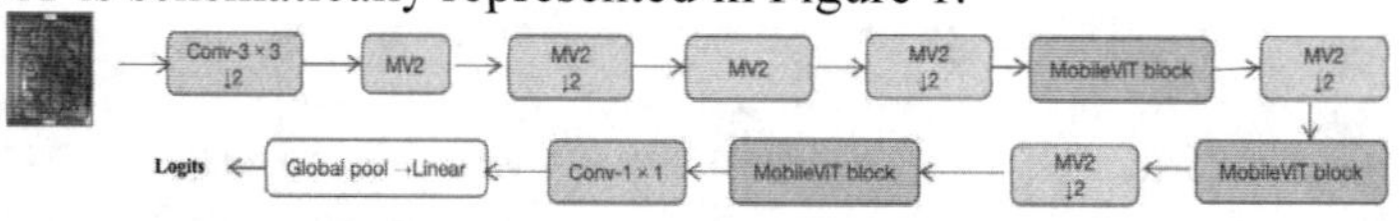

Figure 1. Network model of MobileViT.

As illustrated Figure 1, the network model of MobileViT integrates the Conv layer, the MV2 layer (i.e., MobileNet V2 [8]), and the MobileViT block. Through convolution operations, local spatial relationships in the image can be captured. The MV2 structure (as shown in Figure 2) introduces the inverted residual block in the MobileNetV2 structure. As depicted in Figure 3, the difference between the feature map of this structure and the ordinary residual structure is that its dimensional change shows the characteristics of first increasing and then decreasing. That is, the inverted residual structure increases the number of channels from the beginning, then extracts features through depthwise separable convolution, and finally uses Conv 1×1 for dimensionality reduction. Among them, the MV2 structure marked with a downward arrow represents the case of stride=2, that is, downsampling is required. This architecture optimizes computational efficiency while maintaining high performance.

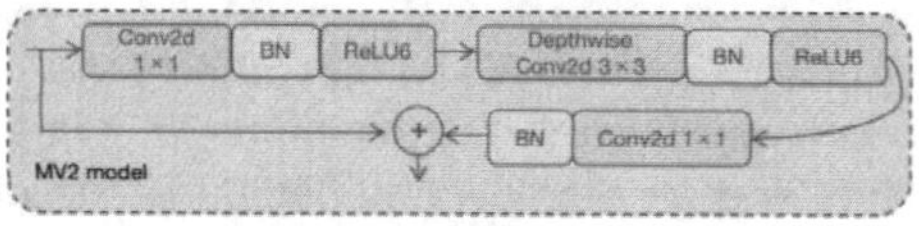

Figure 2. MV2 network structure chart.

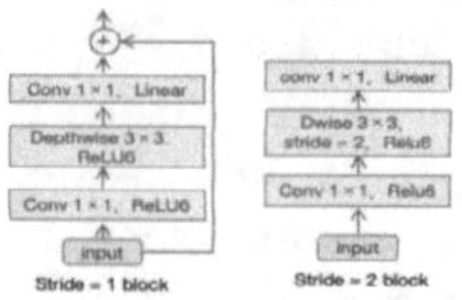

Figure 3. Inverted residual architecture schematic.

The mobileViT block is the core part of the MobileViT network, as shown in Figure 4. The convolution layer first extracts local features from input channel data through a n × n convolution layer, and then performs channel dimension modulation via 1 × 1 convolutional projection; the next is the Unfold operation as presented in Figure 5, which divides the feature tensor obtained by convolution into N non-overlapping patches. For each patch, a Transformer Block is used for learning to capture the correlation within the patch. Next is the Fold operation. We use shortcut connections to splice the output of the Transformer Block with the original input feature map. Finally, these spliced features are fused and processed via an n × n convolutional filter to generate the resultant activation map.

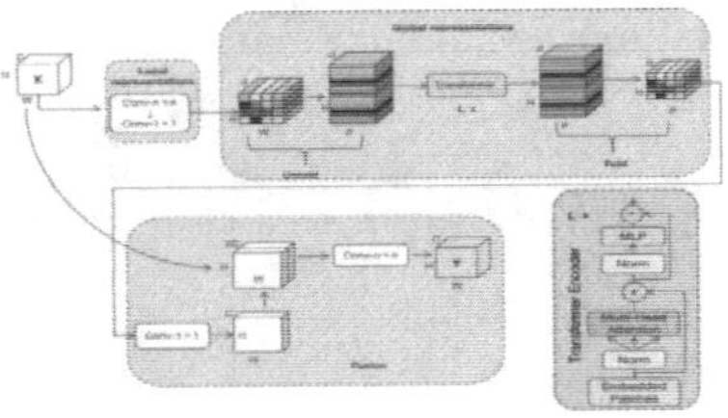

Figure 4. MobileViT block network structure diagram.

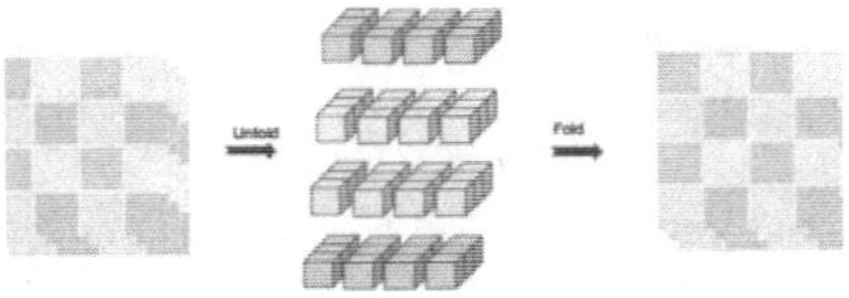

Figure 5. Unfold to fold.

2.2 YOLOv8 algorithm

YOLOv8, hereinafter referred to as v8 [9], is one of the versions in the YOLO series. v8 brings many new features and optimizations, including a new backbone network architecture, Anchor-Free detection mechanism, and innovative loss function. These improvements enable v8 to further accelerate the detection process while maintaining high accuracy. The architectural schematic of v8 is demonstrated in Figure 6. The Backbone part converts image data into features with deep semantics, and the C2f structure improves performance by optimizing gradient flow and simplifying parameters. The Head module includes the detection head and the classification head, which are responsible for accurately predicting bounding boxes and categories, accelerating model convergence, and improving detection accuracy. Therefore, in the detection algorithm, we introduce part of the network architecture of v8 as an important part of the model training process.

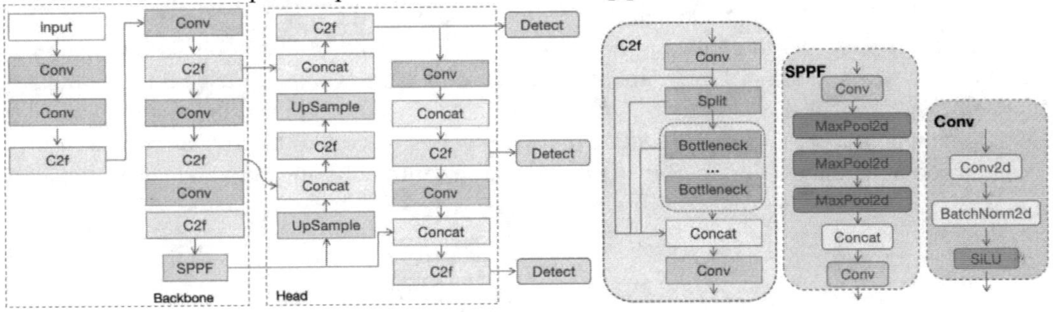

Figure 6. V8 network model diagram.

2.3 SCConv spatial and channel reconstruction convolution

SCConv [10] implements a dual-branch convolutional operation with spatial and channel in order to eliminate spatial and channel correlated features in convolutional neural networks (CNNs). Its core goal is to reduce computing resource usage and maximize the network's task-specific mutual information by refining the feature extraction process. The SCConv module (as shown in Figure 7) is designed with two core components: the spatial refinement unit (SRU) and the channel refinement unit (CRU).

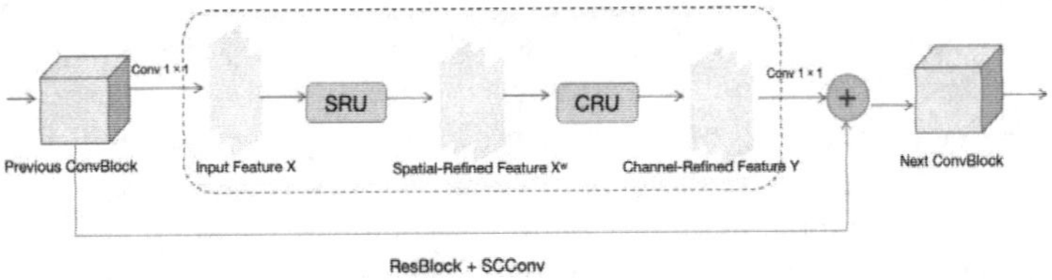

Figure 7. SCConv module diagram.

2.3.1 Spatial reconstruction unit. The Spatial Reconstruction Unit (SRU) usually includes the input level, a group normalization layer, a weight generation layer, a feature separation layer, and a feature reconstruction layer, as shown in Figure 8. First, the input features undergo normalization to minimize scale discrepancies among various feature maps. Next, weights are generated from the normalized feature maps by applying normalization and activation functions (such as Sigmoid in the figure). Then, based on the computed weighting schema, the input features are segmented to form multiple sub-feature sets. Ultimately, these divided feature sets are transformed and reorganized to produce spatially refined feature outputs. SRU reduces the redundancy of features in the spatial dimension through separation and reconstruction methods.

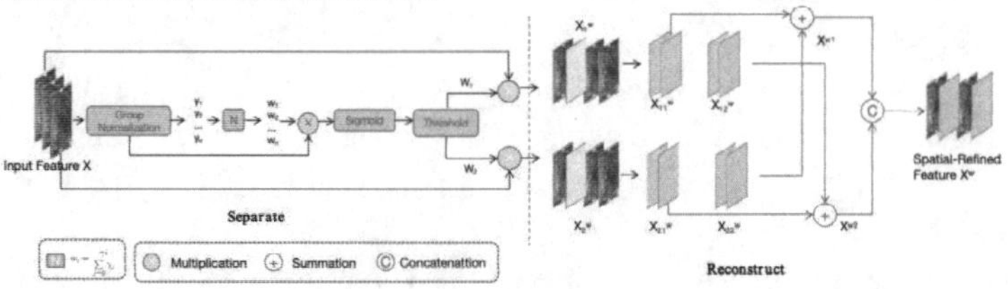

Figure 8. Spatial reconstruction unit structure diagram.

2.3.2 Channel reconstruction unit. The Channel Reconstruction Unit (CRU) aims to minimize the channel information overlap of convolutional neural network features. The working principle of CRU can be summarized into three steps: segmentation, transformation, and fusion, as exhibited in Figure 9.

First, the channels of the input feature map are divided into two parts, usually according to a certain ratio (such as aC and (1-a)C in the figure, where C denotes the overall count of channels and a is a segmentation ratio). The segmented feature map channels go through different processing paths. One part of the channel may be further transformed through global convolution (GWC) and point convolution (PWC) to obtain representative high-level features; the other part of the channel may be compressed and processed through simpler Conv1×1 operations to extract shallow features as a supplement. The two segments of the transformed feature map are weighted fused through pooling and attention mechanisms (such as Figure 9 SoftMax weighted fusion) to form the final channel refined features. The main goal of CRU is to reduce the redundancy of convolutional neural network (CNN) features in the channel dimension to upgrade the overall performance and network's efficiency.

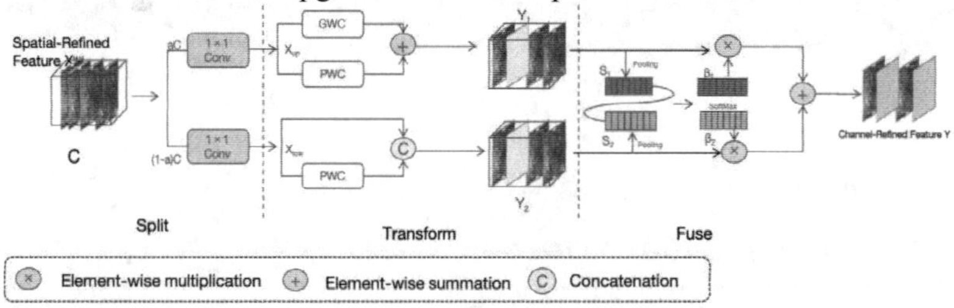

Figure 9. Channel reconstruction unit structure diagram.

2.4 Focal EIOU loss function

The EIoU (Enhanced Intersection over Union) loss function [11] aims to more accurately measure the difference between the predicted box and the true box. The EIoU loss function primarily comprises three components: overlap loss, center distance loss, and width-height loss.

(1) Overlap loss: The Intersection over Union (IoU) loss segment assesses the level of overlap between the predicted box and the ground truth box. In EIoU, the calculation method of IoU loss remains unchanged, but its weight in the overall loss function is adapted to the given situation to balance the relationship with other loss terms.

(2) Center distance loss: The part [12] focuses on disparity between the center of the forecasted box and the center of the actual box. In EIoU, the center point distance loss (as shown by Point b and Point b^{gt} in Figure 10) is usually implemented by measuring the geometric distance or Manhattan distance between the two center points and performing appropriate normalization.

(3) Width-height loss: It refers to the difference in length ratio between the predicted box and the true box. In EIoU, the length ratio loss can be achieved by assessing the disparity in the aspect ratio

and appropriately weighting them.

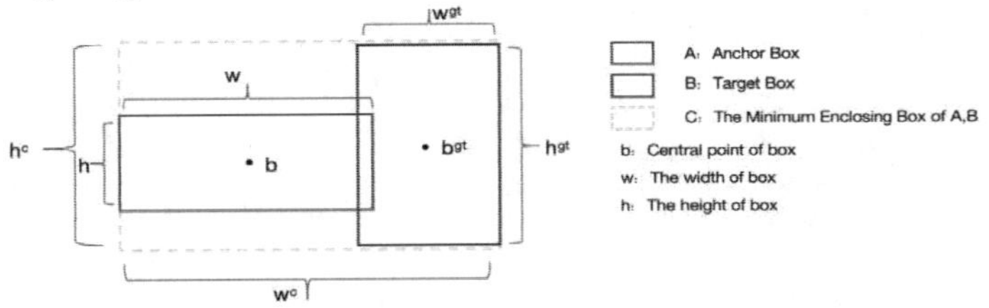

Figure 10. IoU graph.

Among them, the initial two components of the EIoU loss function inherit the design idea of CIOU (Complete Intersection over Union), while the aspect-height ratio loss term accelerates the convergence process by directly minimizing the deviation between the intended box and the anchor point box in the width and height dimensions. The formula for minimizing the width-height difference is given below:

$$L_{EIOU} = L_{IOU} + L_{dis} + L_{asp} = 1 - IOU + \frac{\rho^2(b,b^{gt})}{(w^c)^2+(h^c)^2} + \frac{\rho^2(w,w^{gt})}{(w^c)^2} + \frac{\rho^2(h,h^{gt})}{(h^c)^2} \quad (1)$$

where w^c and h^c represent the width and height of the smallest enclosing rectangle of the forecasted box; L_{EIOU} is the loss measuring overlap between the estimated box and the true box; L_{dis} represents the central distance loss between the forecasted box and the actual box; L_{asp} denotes a substantial loss between the predicted box and the ground truth box.

In the bounding box (BBox) regression task, there is a problem with unbalanced training samples. Low-quality samples often produce excessively large gradients due to their large errors, which has an adverse effect on the training process. To resolve the issue outlined above, a new loss function is introduced, namely, Focal EIOU Loss [13]. This loss function can differentiate between high-quality and low-quality anchor boxes and apply distinct approaches. The specific formula of its penalty term is presented in Formula (2):

$$L_{Focal-EIOU} = IoU^{\gamma} L_{EIoU} \quad (2)$$

where IoU=$|A \cap B|/|A \cup B|$, and γ is used to modulate of deviation dampening. In this new loss function: When the IoU value is higher, it usually means that the predicted box overlaps more with the target box, that is, the regression effect is better. This plays a weighted role, assigning a larger loss weight to the target with a better regression effect.

3. Improved algorithm

The improved network model adds the MobileViT and SCConv attention mechanisms as shown in Figure 11. MobileViT integrates the merits of Transformer, CNN, and loss function. It captures long-distance dependencies in images through the self-attention mechanism of the Transformer and captures local spatial relationships in combination with the convolution operation of CNN so that the model can maintain high performance while reducing computational complexity and memory requirements. SCConv improves the information loss and redundancy problems that may exist in the feature representation paradigm of conventional convolutional architectures by incorporating spatial and channel recalibration modules. It enhances the feature representation capability, allowing the model to more accurately capture key information when analyzing images, thus enhancing the precision and resilience of target detection.

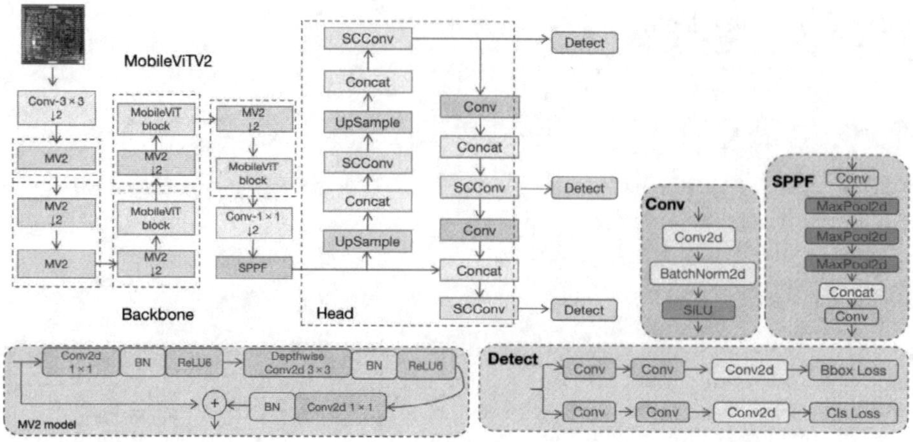

Figure 11. Improved network model.

4. Experiments and analysis

4.1 Experimental configuration environment

In this paper, a Linux server is employed as the platform for training data, a pytorch-based project, and the server is equipped with a relatively good graphics card. The specific experimental hardware parameter configuration is illustrated in Table 1, and the relevant parameters during the experiment are displayed in Table 2.

Table 1. Experimental hardware configuration.

Experimental configuration information	
Python version	3.10.13
Torch version	2.2.1
CUDA version	12.2
Server configuration	NVIDIA GeForce RTX 3090 GPU

Table 2. Experimental parameters.

Relevant parameters of the experiment	
Image size	640×640
Batch size	16
Learning rate	0.01
Momentum size	0.937
Epochs	400

4.2 Datasets setting

This paper uses the defect datasets provided by PKU-Market-PCB [14], which comprises 693 samples, each of which may show six defect categories: leakage, rat bite, open circuit, short circuit, stray, and copper contamination. For ease of reference, we name it the PCB_693 datasets. To improve the framework's capability to learn micro-scale flaw signatures, we use enhancement techniques such as offline cropping and image transformation to expand the datasets to 10, 668 images and named PCB_10668.

To comprehensively assess the generalization efficacy of the experiment on distinct datasets, we selected the DeepPCB datasets [15] as a comparative verification object. In terms of dataset division, the first two datasets are segmented training sets, validation sets and test sets in a ratio of 8:1:1. The DeepPCB datasets use a partition ratio of 7:2:1.

4.3 Comparative analysis

To comprehensively assess the performance gains of the enhanced algorithm, we compared the experimental results of distinct models on the PCB_693 datasets. The model comparison outcomes are presented in Table 3, which presents the specific performance of different models in performance evaluation indicators: recall rate (Recall) and average accuracy (mAP50).

Table 3. Comparison of running results of PCB_693 datasets on different models.

Models	Precision/%	Recall/%	mAP50/%	mAP50-95/%
v8n	94.9	85.3	91.6	47.5
v10n	92.0	87.2	93.9	**50.0**
MobileViTV2+v8	95.5	90.5	95.1	48.8
SCConv+v8	95.3	85.1	92.5	48.2
MobileViTV2+SCConv+v8	**95.8**	88.9	94.7	48.7
Ours	93.3	**91.9**	**96.2**	49.5

Note: The bold font in the table indicates the largest percentage in each column.

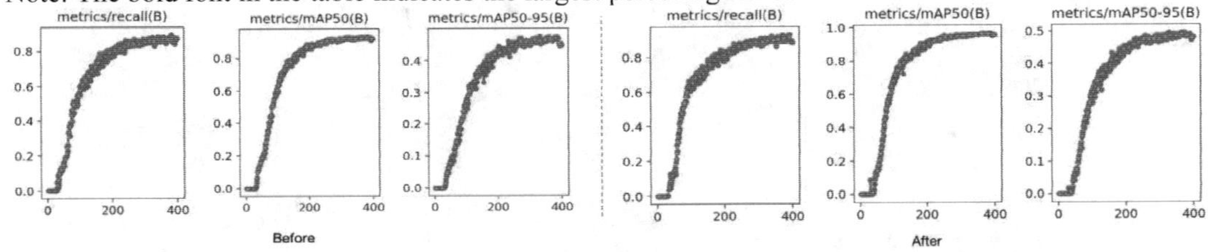

Figure 12. Comparison of recall rate, mAP50, and mAP50-95 before and after model improvement.

We use a chart (as shown in Figure 12) to visually show the comparison of precision and recall. We can see that the mAP50 of the model is 91.6%, and the improved model is close to 96.2%. The mAP50-95 is 47.5%, and the improved model is 49.5%. The recall rate of the model before the improvement was 85.3%, and after the improved model reached 91.9%. It is clear that the accuracy of the enhanced model has significantly increased,mAP50 saw a 4.6% increase, while mAP50-95 rose by 2%. The recall rate increased by 6.6%, which greatly improved the model's false negative and false positive rates. Among them, the recall rate is a core metric for measure the performance of the classification model. In industrial application scenarios, a high recall rate is crucial. It can fully capture defects and greatly reduce missed detection and false detection.

The accuracy of different categories of defects before and after the improvement is shown in Table 4. Before the improvement, we conducted a preliminary evaluation of defect detection in each category. From the data in Table 5, it can be seen that although the model has shown a certain accuracy in multiple categories, there is still a lot of room for improvement. Especially in the key indicator of Recall, the accuracy of the model has not reached the ideal level. After optimizing the model, we evaluated it again. From the data in Table 6, it is evident that the enhanced model has considerably increased the accuracy and recall rate in each category. In particular, the accuracy on mAP50 has reached more than 92%, and the recall rate of missing_hole defects has reached 100%.

Table 4. Recall and precision of the model in each category before and after improvement.

Category	Before		After	
	Recall/%	mAP50/%	Recall/%	mAP50/%
Missing_hole	98.3	98.7	100.0	99.5

Mouse_bite	89.5	90.6	97.3	97.9
Open_circuit	77.8	90.3	85.7	95.4
Short	91.00	95.90	93.5	98.9
Spur	74.3	83.9	87.0	93.5
Spurious_copper	81.0	90.3	87.6	92.0
All	85.3	91.6	91.9	96.2

Before and after the model was improved, we conducted comprehensive verification and testing on a variety of different defect categories. Figure 13 illustrates the experimental results.

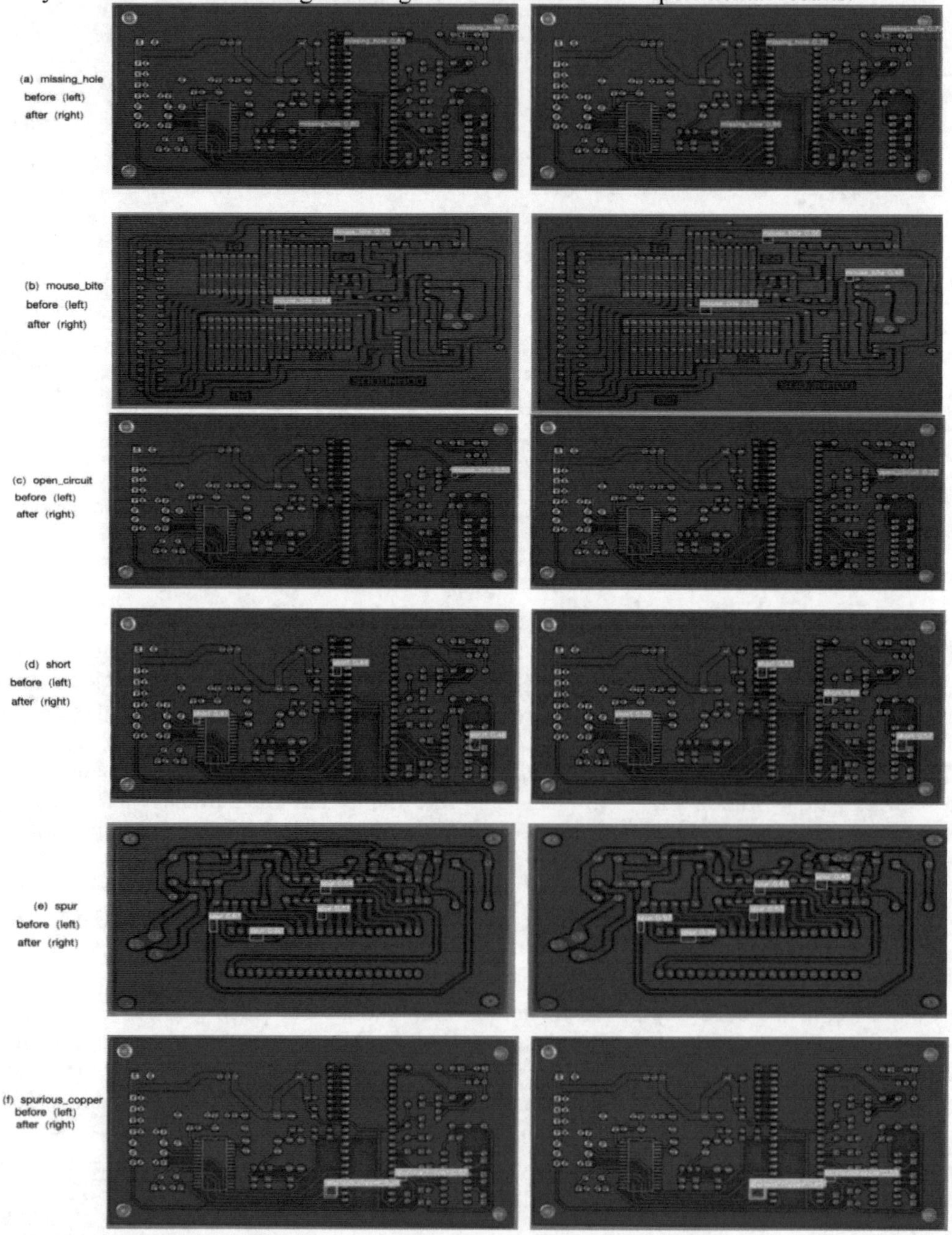

Figure 13. Comparison of detection results of different categories of models.

The above figures (a), (b), (c), (d), (e), and (f) show different types of defects. Leftward each figure illustrates baseline performance metrics before the model was improved, and rightward depicts the running results after the model was improved. In defect categories (b), (d), and (e), the improved model identified 3, 4, and 4 defects respectively, while the original model missed 1 defect respectively. In defect category (c), the improved model correctly identified the open_circuit defect, while the original model misdetected 1 defect. The improved model showed higher accuracy in identifying model defects, while significantly reducing the incidence of false detection and missed detection.

Next, we will conduct a comprehensive verification of the improved algorithm on different datasets. The experimental outcomes demonstrate a significant improvement in Table 5 and Table 6. The improved model in Table 5 has achieved an increase in recall rate; Table 6 further shows that in the partitioning of datasets with different ratios, the improved model not only improves the recall rate by 2.5% but also significantly improves the accuracy by 1.3%. These experimental outcomes fully demonstrate the effectiveness and accuracy of the improved.

Table 5. Comparison models of PCB_10668 datasets.

Models	Precision/%	Recall/%	mAP50/%	mAP50-95/%
v8	97.6	98.5	99.0	66.9
v10	98.2	98.3	99.1	65.0
Improved model	97.2	98.6	98.7	66.9

Table 6. Comparison models of DeepPCB datasets.

Models	Precision/%	Recall/%	mAP50/%	mAP50-95/%
v8n	96.4	92.0	96.9	66.2
v10n	92.8	89.2	95.2	70.2
Improved model	96.1	94.5	98.2	75.7

5. Conclusion

The lightweight MobileViTv2 and SCConv fusion network model constructed in this article has a 2.5% enhancement in recall and a 1.3% improvement in precision on datasets divided into different proportions. On public datasets, the proposed method achieves a 4.6% mAP50 elevation and recall achieves a 6.6% elevation. Despite the growth in parametric quantity, the overall performance can still meet the requirements of industrial detection. Moving forward, we will undertake rigorous investigation into the model's architectural deficiencies, implementing targeted refinements that systematically enhance its cross-domain generalization efficacy.

References

[1] Huang Z., Zhu J., and Ruan J. A novel technology of vacuum low-temperature pyrolysis with NVZI for the hi gh-efficiency debromination of resin particles from waste-printed circuit boards [J]. Resources, Conservation, and Re cycling, 2023, 188: 106, 711. https://doi.org/10.1016/j.resconrec.2022.106711.
[2] Liu Z. and Qu B. Machine vision based online detection of PCB defect [J]. Microprocessors and Microsystems, 2021, 82: 103, 807. https://doi.org/10.1016/j.micpro.2020.103807.
[3] Zhou Y., Yuan M., Zhang J., et al. Review of vision-based defect detection research and its perspectives for printed circuit board [J]. Journal of Manufacturing Systems, 2023, 70: 557-578. https://doi.org/10.1016/j.jmsy.2023.08.019.
[4] Ling Q. and Isa N. A. M. Printed circuit board defect detection methods based on image processing, machine learning, and deep learning: A survey [J]. IEEE Access, 2023, 11: 15, 921-15, 944. Doi: 10.1109/ACCESS.2023.3245093.
[5] Park J. H., Kim Y. S., Seo H., et al. Analysis of training deep learning models for PCB defect detection [J]. Sensors, 2023, 23 (5): 2, 766. https://doi.org/10.3390/s23052766.
[6] Sandler M., Howard A., Zhu M., et al. Mobilenetv2: Inverted residuals and linear bottlenecks [C]// Proceedings of the IEEE conference on computer vision and pattern recognition. 2018: 4, 510-4, 520.

[7] Mehta S. and Rastegari M. Separable self-attention for mobile vision transformers. arXiv 2022 [J]. arXiv preprint arXiv:2206.02680.
[8] Han K., Wang Y., Ch 'en H., et al. A survey on vision transformer [J]. IEEE transactions on pattern analysis and machine intelligence, 2022, 45 (1): 87-110. Doi: 10.1109/TPAMI.2022.3152247.
[9] Jocher Glenn, et al. "YOLOv8 by Ultralytics". GitHub, 1 Jan. 2023. https://ultralytics.com.
[10]Li Jiafeng, et al. SCConv: Spatial and Channel Reconstruction Convolution for Feature Redundancy. 1 June 2023.
[11]Union G. I. O. A metric and a loss for bounding box regression [C]// Proceedings of the IEEE/CVF Conference on Computer Vision and Pattern Recognition (CVPR). 2019: 658-666.
[12]Zheng Z., Wang P., Liu W., et al. Distance-IoU loss: Faster and better learning for bounding box regression [C]// Proceedings of the AAAI conference on artificial intelligence. 2020, 34 (07): 12, 993-13, 000. https://doi.org/10.1609/aaai.v34i07.6999.
[13]Zhang Y. F., Ren W., Zhang Z., et al. Focal and efficient IOU loss for accurate bounding box regression [J]. Neurocomputing, 2022, 506: 146-157. https://doi.org/10.1016/j.neucom.2022.07.042.
[14]DINGR, DAIL, LIG, et al. TDD-net: a tiny detect detection network for printed circuit boards [J]. CAAI Trans actions on Intelligence Technology, 2019, 4 (2): 110-116. https://doi.org/10.1049/trit.2019.0019.
[15]Tang S., He F., Huang X., et al. Online PCB defect detector on a new PCB defect dataset [J]. arXiv preprint arXiv: 1902.06197, 2019. https://doi.org/10.48550/arXiv.1902.06197.

Journal of Physics: Conference Series **3009** (2025) 012080 doi:10.1088/1742-6596/3009/1/012080

Numerical simulation and lifetime study of the high-temperature creep behavior of P91 pipeline with a soft zone

Ge Li [1], Haoliang Wang [2], Qinglin Deng [3], Jiandong Chen [3], Zhibin Qiu [1], Haiyang Li [1], Ze Li [1*]

[1] Huadian Electric Power Research Institute Co., Ltd., Hangzhou, 310030, China;
[2] School of Mechanical Engineering, Dongguan Institute of Technology, Dongguan, 523808, China;
[3] Huadian Sihanoukville Power Generation Co., LTD., Phnom Penh, Cambodia

*Corresponding author's e-mail: ze-li@chder.com

Abstract: In this paper, the local low hardness P91 pipe fittings of a power plant are taken as the research object, and the distribution of the soft area of the pipe fittings is determined by on-site "gridded" hardness testing, and the three-dimensional finite element model of the normal and soft areas of the pipe fittings is constructed. FEA technology was used to numerically simulate the high-temperature creep behaviour of pipe fittings based on the Norton model. Through numerical simulation, it is found that due to the creep relaxation effect of the material, the creep stress of the pipe fitting is redistributed during the creep process at high temperatures and the position of the peak stress shifts from the pipe ID to the OD. The soft zone will have an impact on the creep stress redistribution inside the pipe fitting, the peak stress of the pipe fitting is not in the soft zone, and the stress level in the soft zone is always lower than the normal zone. After numerical simulation, the stress value of the soft zone in the creep steady state is obtained, which can provide data support for safety assessment and life assessment.

1. Introduction

In recent years, in the operation practice and inspection activities of P91 steel, many power plants at home and abroad have found that the hardness of P91 pipe fittings is locally low [1]. As early as 2011, the Electric Power Research Institute (EPRI) first described this phenomenon of low local hardness and defined the low hardness area as the soft zone [2]. The literature research [3~4] shows that when the hardness value of P91 pipe fittings is lower than the standard value, its mechanical properties will decrease significantly, and the remaining life will be significantly shortened, which will bring huge safety hazards to the operation of the unit. Ref. [3] pointed out that the endurance strength of low-hardness P91 specimens is significantly lower than that of normal hardness specimens, and the durable strength of the specimen can only be reached when the hardness value of the specimen is at least 195HV. In Ref. [4], researchers pointed out that the hardness of Gr.91 steel decreases with the increase of creep life fraction (t/tr), and its creep life at 650°C/71 MPa decreases by 50% when the hardness is lower than 190 HV. These research results show that there is a certain correspondence between hardness and the lasting strength and life of P91 steel, and it is necessary to carry out a safety assessment and life assessment of P91 pipe fittings with the soft zone.

2. Research background

Finite Element Analysis Referred to as FEA, it has the advantages and characteristics of convenient and fast operation, saving test time and cost, and rapid accident analysis, and has been widely used in engineering practice. For example, in Ref. [5], the high-temperature creep behaviour of P91 steel was studied by constructing a Graham creep constitutive model, which can accurately simulate the stress state of P91 steel in the creep steady-state stage, which provides a theoretical basis for correctly predicting the life loss of creep of high-temperature components. In Ref. [6], based on the microcavitation damage mechanism and the creep damage equation modelling, the Gurson creep constitutive model of P91 steel was constructed and modified, and a creep damage model based on microcavity damage was deduced to describe the whole creep process, and the finite element simulation results were satisfactory. Based on the theory of precipitate phase coarsening and cavitary nucleation, the creep curve of P91 steel was obtained by creep persistence test, and a unified creep damage constitutive equation based on the micromechanics-hyperbolic sinusoidal model (Sinh) was established, and the creep life of P91 steel was accurately calculated with the help of ABAQUS software [7]. In Ref. [8], based on the theory of damage mechanics, a Kachanov-Rabotnov creep constitutive model of P91 steel at 620°C was established, which provided a theoretical basis for the life prediction of high-temperature components of P91 steel in power plants. It can be seen that many of the above-mentioned research works provide some theoretical support and basis for safety assessment and life assessment through the numerical simulation of the high-temperature creep behaviour of P91 steel, but they still stay at the research level of normal P91 steel, and the research work has not yet been carried out on whether the existence of soft zone has an impact on the high-temperature creep behaviour of 91 steel, what the degree of influence is, and the interaction between stress levels between soft zone and normal zone.

Inspired by this, this paper takes the P91 main steam pipeline with the soft zone in a power plant as a case study and uses ABAQUS software to numerically simulate the high-temperature creep behaviour based on the Norton model to study the influence of the soft zone on the creep stress redistribution of pipe fittings. At the same time, numerical simulation is used to obtain the stress level of the soft zone when it reaches a creep steady state, which provides data support for safety assessment and life assessment.

3. Case Description

3.1. Overview

During the overhaul of a power plant, it was found through metal inspection that the main steam pipeline had problems of low local hardness and abnormal structure, and the pipe material was P91 and the specification was ID229×41 mm.

3.2. On-site testing

According to the requirements of the DL/T 438 standard [9], a PHB-200 magnetic Brinell hardness tester was used to check the lowest position of hardness, and the check result was 163 HBW, that is, it was confirmed that there was a soft area in the main steam pipeline, and metallographic inspection was carried out on the position of the lowest hardness point, and the metallographic structure was found to be abnormal, which was ferrite+Carbide, martensite orientation disappears completely.

To confirm the boundary of the soft area, the hardness testing points are added, the detection and measurement points are arranged in a "gridded" manner, and the spacing between each measurement point is about 10 mm. The specific location of the low hardness measurement point is obtained through the "gridded" hardness test, and the outermost low hardness measurement point of the soft zone is hooked, that is, the boundary of the soft area is displayed. The field test results show that the low value range of hardness is about 135~165 HBHLD, the soft area is distributed in the arc length of the pipe diameter of 10:00~2:00, and the axial length is 200 mm, as shown in Figure 1 below.

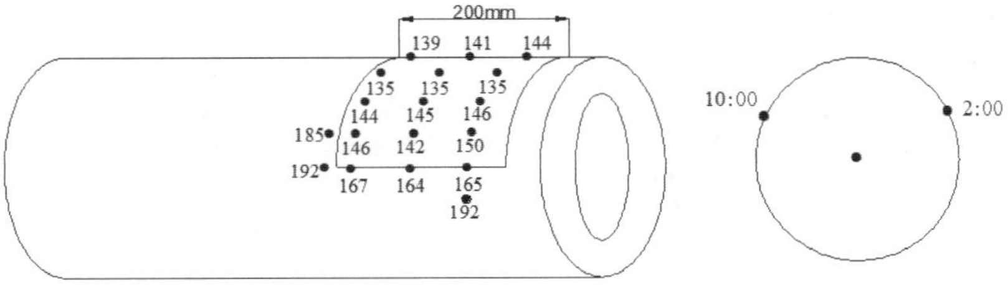

Figure 1. Schematic diagram of "gridded" hardness measurement points and hardness distribution.

4. Numerical simulation of creep behavior

4.1. Choice of creep model

The creep stage II of P91 steel has an approximately stable creep rate. This stable or minimum creep strain rate (dε/dt) varies systematically with the applied stress and temperature. For a creep test at a given temperature, the dependence of the minimum creep strain rate on the applied stress can be described by the well-known Norton's law (power law) [7], as follows:

$$\frac{d_\varepsilon}{d_t} = A\sigma^n$$

This paper uses the Norton model in ABAQUS software (Time Hardening Creep Model) as a creep damage constitutive model to simulate the creep behaviour of P91 steel in the normal and soft regions.

4.2. Establishment of creep law parameters

In this case, for the normal P91 material, the creep law parameters at 600°C and 550°C should be determined first. So that the parameters at 570°C can be obtained by interpolation between the determined parameters. In Ref. [10], scholars show that a large number of creep tests for normal P91 materials at 600°C are consistent with the Watanabe relationship. At the same time, the NIMS (National Institute of Materials Science) of Japan obtained the test results of normal P91 materials through long-term low-stress creep tests at 550°C [11]. The selected Watanabe and NIMS data were used as the interpolation of creep law parameters at 570°C for normal P91 materials in this paper, as shown in Table 1.

Table 1. Creep law parameters of normal 91.

Normal P91	$A(MPa^{-n}Hr)^{-1}$	n	Temperature/°C
Watanabe	4.91E-29	11.10	600
NIMS	1.50E-35	12.90	550

The Norton relationship of soft 91 materials at temperatures below 600°C can be described in the EPRI study, which shows that the Norton relationship between soft 91 materials at 600°C and 550°C is very close to that of T22 materials [2]. Based on the Norton relation of T22 materials, EPRI gave the creep law parameters A and n values of soft 91 materials at 600°C and 550°C [2], as shown in Table 2.

Table 2. Creep law parameters of soft 91.

Soft 91	$A(MPa^{-n}Hr)^{-1}$	n	Temperature/°C
EPRI	7.46E-27	11.18	600
EPRI	4.02E-33	12.96	550

According to Table 1 and Table 2, Norton's creep law parameters A and n at 570°C for normal 91 and soft 91 materials were calculated by interpolation, as shown in Table 3 below.

Journal of Physics: Conference Series **3009** (2025) 012080 doi:10.1088/1742-6596/3009/1/012080

Table 3. Creep law parameters for normal 91 and soft 91 materials at 570°C.

Remark	$A(MPa^{-n}Hr)^{-1}$	$A(MPa^{-n}S)^{-1}$	n	Temperature/°C
Normal 91	8.17E-33	2.27E-36	12.11	570
Soft 91	1.73E-30	4.79E-34	12.18	570

Therefore, the Norton formula used to describe the normal and soft areas of P91 in this case can be expressed as follows:

$$\begin{cases} \dfrac{d_\varepsilon}{d_t} = (2.27E-36)\sigma^{12.11} \\ \dfrac{d_\varepsilon}{d_t} = (4.79E-34)\sigma^{12.18} \end{cases}$$

4.3. Establishment of finite element model

The 3D finite element modelling and meshing were carried out with the help of ABAQUS software, as shown in Figure 2 below. The mesh element type is an eight-node hexahedral element of the ABAQUS standard, besides, the mesh is refined to ensure the accuracy of the local stress values and strain gradients calculated by the simulation and the mesh quality can be verified in the elastic analysis below.

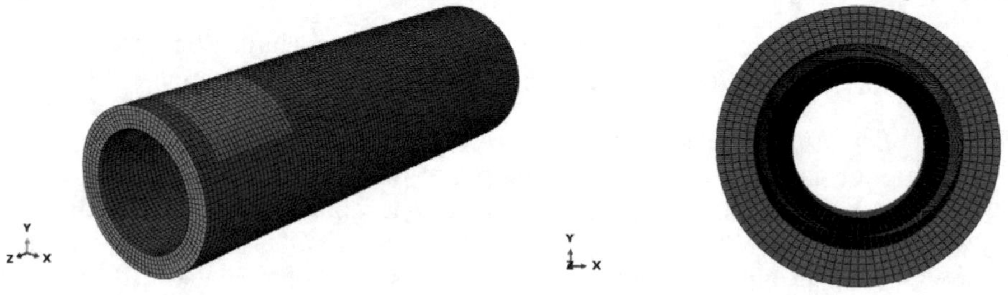

Figure 2. 3D finite element model of pipe fittings.

4.4. Setting of boundary conditions

4.4.1 Loading conditions. In order to ensure the partial safety of the calculation results, the main steam pressure and temperature parameters of the unit under BMCR condition were selected for the loading conditions of the pipe fittings, that is, P=25.50 MPa, t=570°C. The loading process is divided into two steps, the elasticity analysis step and the creep analysis step.

4.4.2 Material Properties. The elastic properties (elastic modulus, Poisson's ratio) of the material have nothing to do with hardness, and the normal 91 material has the same material properties as the soft 91 materials, which can be obtained by interpolation calculation with reference to the GB/T 16507.2-2022 [12]: at 570°C, the elastic modulus of P91 material is $E=1.745\times10^5$ MPa, and the Poisson's ratio $\upsilon=0.3$.

4.5. Numerical simulation of elastic stage

Before evaluating the stress relaxation and redistribution due to creep during the use of the fitting, it is necessary to perform an elastic analysis. The elastic analysis not only provides effective verification for the quality of the FEM, the rationality of the boundary condition setting, and the accuracy of the applied load of the finite element model but also establishes the accurate initial state for creep analysis.

4.5.1 Numerical simulation results. The numerical simulation of the elasticity analysis step of the finite element model was carried out by ABAQUS, and the simulation effect is shown in Figure 3 below.

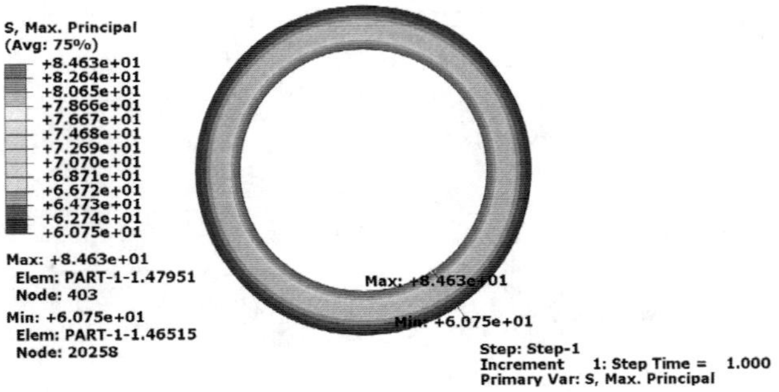

Figure 3. Maximum principal stress distribution.

The simulation results of the elastic phase show that there is no difference in the stress distribution between the normal and soft regions. At the same time, since only the internal pressure is applied to the finite element model, the maximum principal stress is manifested as hoop stress, the peak stress is at the ID of the inner wall of the pipe fitting, $\sigma_{\theta ma}$x=84.63 MPa, and the minimum stress value is at the OD of the outer wall of the pipe fitting, $\sigma_{\theta min}$=60.75 MPa.

4.5.2 Comparison with theoretical solution. As early as the eighteenth century, the French mathematician Gabriel Lamé proposed the Lamé formula, which is widely used in the calculation of the strength of pressure vessels and pipes, by performing stress analysis on thick-walled cylinders [13]. The three-way stress at the outer diameter D of the pipe derived from the Lamé formula is as follows:

$$\begin{cases} \sigma_{\text{周}} = \dfrac{PD_n^{\,2}}{D_w^{\,2} - D_n^{\,2}}\left(1 + \dfrac{D_w^{\,2}}{D^2}\right) \\ \sigma_{\text{轴}} = \dfrac{PD_n^{\,2}}{D_w^{\,2} - D_n^{\,2}} + \sigma_{ZHW} + \sigma_W \\ \sigma_{\text{径}} = \dfrac{PD_n^{\,2}}{D_w^{\,2} - D_n^{\,2}}\left(1 - \dfrac{D_w^{\,2}}{D^2}\right) \end{cases}$$

According to the Lamé formula, the theoretical solutions of the circumferential stress of the outer wall and the inner wall of the pipe fitting under loading conditions can be obtained, respectively: $\sigma_{\theta outer}$= 60.40 MPa, $\sigma_{\theta inner}$ = 85.90 MPa. The results of the finite element analysis of the elastic phase of the pipe fitting under loading conditions are compared with the theoretical solution, as shown in Table 4 below.

Table 4. Comparison of elastic finite element analysis results and theoretical solutions of pipe fittings.

Remark	Theoretical solution/MPa	Finite element solution/MPa	Difference
OD	60.40	60.75	0.58%
ID	85.90	84.63	1.51%

It can be seen that the deviation between the theoretical solution and the finite element solution of the circumferential stress at the ID of the inner wall of the pipe fitting is 1.51%, and the deviation between the theoretical solution and the finite element solution of the circumferential stress at the OD of the outer wall is only 0.58%, which indicates that the established finite element model has high quality and can ensure accurate results in the subsequent finite element analysis.

4.6. Numerical simulation of inelastic stage

The inelastic creep analysis of pipe fittings was carried out to study the stress redistribution of pipe fittings during the creep process. The above elastic analysis stress values are used as the initial stress values for the creep analysis.

4.6.1 Numerical simulation results. The loading time of the finite element model in the inelastic phase is set at 100, 000 hours, which is considered creep stable. After the loading time is over, the stress profile of the pipe after stress redistribution is obtained, in which the stress profile of the maximum principal stress is shown in Figure 4, and the stress profile of Von Mises equivalent stress is shown in Figure 5.

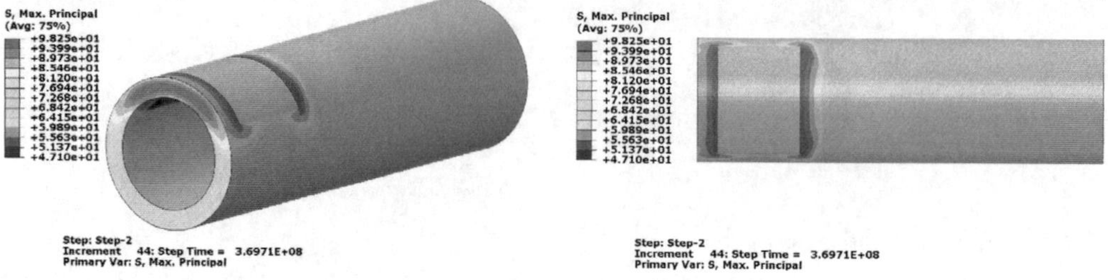

Figure 4. Maximum principal stress distribution after 100,000 hours of creep.

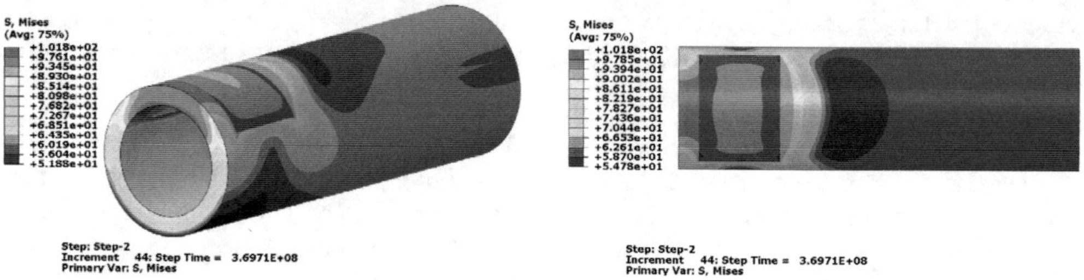

Figure 5. Equivalent stress distribution after 100, 000 hours of creep.

The numerical simulation results show that when the pipe fittings reach creep steady, they have the following characteristics:

1. Due to the creep relaxation of the material, the stress redistribution occurs during the creep of the pipe fitting at high temperatures, and the position of the peak stress shifts from the pipe ID to the OD.

2. When the pipe reaches a creep steady state, the distribution trend of Von Mises equivalent stress is similar to that of the maximum principal stress.

3. The peak stress is not in the soft zone, and the stress level of the soft zone material is always lower than that of the normal material. This is mainly because normal materials surround the soft zone material, the creep behaviour of the soft zone is constrained, the surrounding normal material absorbs the stress part of the soft zone, and the stress of the two regions is in a state of mutual equilibrium [8], which can be called the "binding effect".

4.6.2 Soft zone stress analysis. Studies have shown that the life of soft 91 steel is significantly lower than that of normal 91 steel, and the average life of soft 91 steel is closer to that of P/T22 steel [2]. Therefore, the main focus of this study is the stress levels within the soft zone area. The stress simulation results for the soft zone reaching creep steady state are shown in Figures 6 and 7 below.

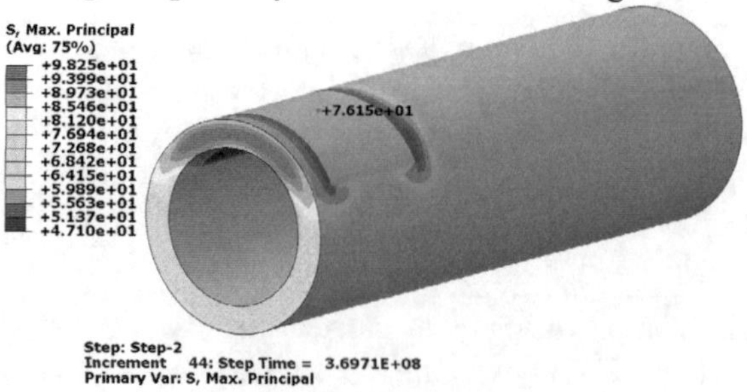

Figure 6. Maximum principal stress distribution after 100,000 hours of creep in the soft zone.

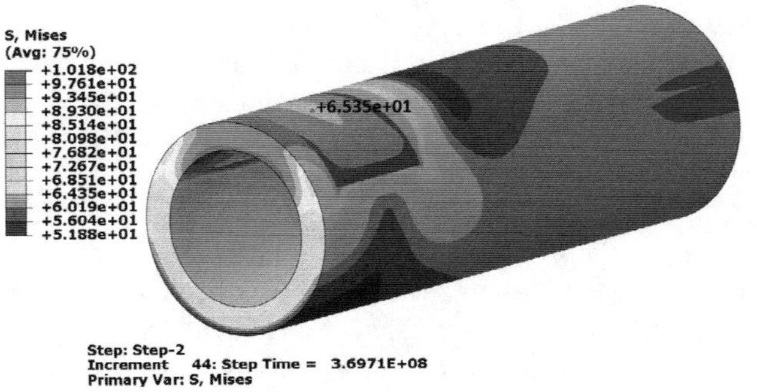

Figure 7. Equivalent stress distribution after 100, 000 hours of creep in the soft zone.

Through numerical simulation, after 100, 000 hours of creep of the pipe fittings, the maximum principal stress and the peak stress of Von Mises equivalent stress in the soft region are located in the center of the soft region, and $\sigma_{\theta softmax}$=76.15 MPa and $\sigma_{esoftmax}$=65.35 MPa. It can be seen from the numerical simulation results that the maximum principal stress in the soft zone is always higher than the equivalent stress of Von Mises, so the peak stress of the maximum principal stress in the soft zone should be used as the damage stress σr in the soft zone when carrying out safety assessment and life assessment, that is, $\sigma_r=\sigma_{\theta softmax}$. Of course, for fittings with soft zones, the application of the maximum principal stress will result in a more stringent safety assessment and a conservative estimate of the life assessment.

5. Safety Assessment

In China, when designing the strength of boiler high-temperature pressure-bearing components, the allowable stress of the material is selected by dividing the average value of the 1×10^5 h endurance strength limit at the design temperature by the safety factor of 1.5 [14~15]. If the 1×10^5 h endurance strength decreases to less than 67% (1/1.5) of the design guarantee value due to the reduction of hardness, the safety margin of 1×10^5 h operation is exhausted, and safety assessment and life assessment are required.

5.1. Safety Assessment Principles

In Ref. [16], researchers pointed out that safety assessment is required before the life prediction of pipe fittings, and the principle of safety assessment is to divide the safety level of pipe fittings into three levels: “low”, “medium” and “high”, with “low” indicating that it is difficult to ensure the safe operation of pipe fittings in one maintenance cycle, “medium” means that pipe fittings need to be evaluated for life, and “high” means that pipe fittings can continue to operate safely. Referring to the above principles, the safety assessment principles for soft zones are formulated, as shown in Table 5 below.

Table 5. Safety assessment principles of P91 soft zone of pipe fittings.

Parameter	Security Level	Suggestion
Hardness≤140 HBW	Low	Replace in time
Creep cracks appear in the tissue	Low	Replace in time
$\sigma_r\geq\sigma_{10^4}$	Low	Replace in time
$\sigma_{10^5}\leq\sigma_r\leq\sigma_{10^4}$	Middle	Life assessment and operation supervision
$\sigma_r\leq\sigma_{10^5}$	High	Continue to operate safely

5.2. Calculation and determination of soft zone endurance strength

At present, the research on low hardness P91 at home and abroad has been very mature, and many scientific research institutions have obtained the curve of lasting strength and creep strength of different hardness levels through a large number of tests, which provides rich data support for the safety assessment and life evaluation of similar low hardness materials. Referring to the relevant literature, the experiment of the literature [17] shows that the lasting strength of P91 with a hardness value of 157 HBW is 85.1 MPa for 105 h and 102.8 MPa for 104 h when extrapolated at 566°C. In Ref. [16], scholars show through experiments that when the hardness range is 150~200 HBW, the lasting strength of P91 extrapolated at 570°C should be 79.7 MPa for 105 h and 103.1 MPa for 104 h.

At the same time, an invention patent applied by Jiangsu State Grid Electrical Science Institute in 2017 - A method for establishing the extrapolation function of the durability curve of low hardness P91 pipe fittings can be used for the extrapolation of the durability curve of 91 steel at 570°C with different hardness [18]. Using the patented technology, it is calculated that the permanent strength of the pipe fitting in this case is 101.22 MPa for 10^4 hours and 81.88 MPa for 10^5 hours at 570°C. The lasting strength values obtained according to the above three methods are shown in Table 6 below.

Table 6. Persistent strength values for different methods.

Serial number	Calculation method	$\sigma^{570^\circ C}_{1\times10^5}$ / MPa	$\sigma^{570^\circ C}_{1\times10^4}$ /MPa
1	Literature [17]	85.1	102.8
2	Literature [16]	79.7	103.1
3	Invention Patent [18]	81.88	101.22

Visible, the calculation results of the above three methods have small deviations, indicating that the three calculation methods are accurate and effective, and the calculation results are within the confidence interval. In this case, a relatively low value of the durable strength should be taken during the safety assessment and remaining life calculation to ensure the safety of the solution results, that is, the durable strength of the soft area of the pipe fitting is 79.7 MPa for 10^5 h and 101.22 MPa for 10^4 h.

5.3. *Evaluation Conclusion*

Through the numerical simulation above, it has been obtained that the damage stress in the soft zone of the pipe fitting in this case $\sigma_r=\sigma_{\theta softmax}$=76.15 MPa. By comparing the sizes of σ_r, $\sigma_{10}{}^4$ and $\sigma_{10}{}^5$ in the soft zone of the pipe fitting, the following relationship is satisfied: $\sigma_r \leq \sigma_{10}{}^5$, according to the safety assessment principle, the safety grade of the pipe fitting is rated as high, and safe operation can continue.

Of course, referring to the research of domestic and foreign scholars on soft P91, under the premise that the pipe fittings with abnormal hardness are not replaced, in order to ensure the safety and reliability of the operation of the unit, the operation supervision should still be strengthened and the life assessment should be carried out.

6. Life Assessment

According to DL/T 940 [19], the remaining life of steam pipelines is generally calculated using the isotherm extrapolation method or the L-M parameter method. When the isothermal extrapolation method is used, the damage stress σ_r in the soft region is substituted into the creep life formula to solve.

In this paper, the isotherm extrapolation method is used to calculate the remaining life of pipe fittings, and the remaining life calculation formula is as follows [20]:

$$\lg\frac{t_r}{10^5}=\frac{\lg\dfrac{\sigma^t{}_{10^5}}{\sigma_r}}{\lg\dfrac{\sigma^t{}_{10^4}}{\sigma^t{}_{10^5}}}$$

$\sigma_r=\sigma_{\theta softmax}$=76.15 MPa, $\sigma_{10}{}^5$=79.7 MPa, $\sigma_{10}{}^4$=101.22 MPa are substituted into the above formula, and the remaining life t_r=155104 h of the soft area 91 material of the pipe fitting in this case is calculated. The unit was put into operation in November 2022 with an operating time of about 6, 100 hours, but due to the local low hardness, the remaining life of the pipe fitting was shortened by nearly 50, 000 hours compared with the design life (calculated as 200, 000 hours).

7. Research Conclusions

1. Due to the creep relaxation effect of the material, the creep stress is redistributed during the creep process of the pipe fitting, and the position of the peak stress shifts from the pipe ID to the OD direction, and the distribution trend of the Von Mises equivalent stress is similar to that of the maximum principal stress when the creep steady state is reached.

2. The soft zone has an effect on the creep stress redistribution inside the fitting, the peak stress is not in the soft zone, and the stress level in the soft zone is always lower than the stress level in the normal zone.

3. The maximum principal stress in the soft zone is always higher than the equivalent stress, and the maximum principal stress in the soft zone and the peak stress of the Von Mises equivalent stress are located in the center of the soft zone, and $\sigma_{\theta softmax}$=76.15 MPa and $\sigma_{esoftmax}$=65.35 MPa.

4. According to the principle of safety assessment, the safety level of the soft area of the pipe fitting is rated as high, and the safe operation can be continued. According to the calculation formula of the isothermal extrapolation method remaining life, the remaining life of the soft zone 91 material of the pipe fitting is about 155104 h and still has a high service life.

5. Compared with the normal zone, the soft zone has a worse creep rupture strength, but the soft zone is also at a relatively lower creep steady-state stress level, which does not necessarily have a short service life and may still have relatively high safety performance.

References:

[1] Li Ge, Li Haiyang, Feng Keyun et al. Evaluation of Local Low Hardness P91 Elbow Creep Life Based on Image Processing Methods[J]. Heat Processing Technology, 2023, 52(16): 21-26.

[2] J. Shingledecker. Effect of Soft-Zone on the Creep Performance of Grade 91 Piping Components[R]. Hillview Avenue Palo Alto: EPRI, 2011.

[3] MASUYAMA F, SHINGLEDECKER J P. Recent status of ASME code on creep strength enhanced ferritic steels[J]. Procedia Engineering, 2013, 55(12): 314-325.

[4] MASUYAMA F et al. Phase transformation and properties of Gr.91 at around critical temperature[C], ASME/JSME Pressure Vessels and Piping Conference. San Diego, 2004: 85-91.

[5] Liu Xue. Numerical Simulation and Life Study of Creep Damage in High-Temperature Components under Multi-axial Stress [D]. North China Electric Power University, 2006.

[6] Ji Runjing. Study on the Full Process of Creep Damage Model of Jirunjing P91 Steel [D]. North China Electric Power University, 2007.

[7] Zhang Yagang. Numerical Study on Creep behavior of P91 Steel at High Temperature [D]. North China Electric Power University,2019.

[8] Wang Jinfeng, Xing Dixiong, et al. Numerical Simulation and Experimental Study on Creep of P91 Steel at High Temperature [J]. Chinese Journal of Construction Machinery, 2020, 18(04): 283-287.

[9] DL/T 438-2023, Guidelines for Technical Supervision of Metals in Thermal Power Plants[[S].

[10] T. Watanabe, et al., Creep Damage Evaluation of 9Cr-1 Mo-V-Nb steel welded joints showing Type IV Fracture, International Journal of Pressure Vessels and Piping; Vol.83, 2006, P63-71.

[11] NIMS, Creep Deformation Properties of 9Cr-1Mo-V-Nb Steel Plates for Boilers and Pressure Vessels, National Institute for Material Science, 1-2-1, Sengen, Tsukuba-shi, Ibaraki, Japan,2008.

[12] GB/T 16507.2-2022, Tubulous Boiler Part 2: Materials[S].

[13] Lv Xueqi. Derivation and applicable conditions of strength calculation formula of pipeline and thin-walled cylinder vessel [J]. Newsletter of Petrochemical Equipment,1977, 1: 38-47.

[14] TSG 11-2020, Technical Specification for Boiler Safety[S].

[15] GB/T 16507.1-2022, Tubulous Boiler Part 1: General Rules[[S].

[16] Yang Chao, Tang Chunpo, et al. Safety Evaluation and Life Prediction of Low hardness P91 pipe Fitting [J]. China Electric Power, 2017, 50(08): 82-86.

[17] Deng Hui. Evaluation of microstructure properties and safety of low hardness P91 steel pipe after long service [J]. Heat treatment of metals, 2021,12(46):209-213.

[18] Yang Chao, Tang Chunpo, et al. A method for establishing extrapolation function of durability curve of low hardness P91 pipe fittings: CN 106769531 A[P]. 2017-03-10.

[19] DL/T 940-2022, Technical Guidelines for Life Assessment of Steam Pipelines in Thermal Power Plants[S].
[20] Li Ge, Li Xiaohui, Feng Keyun et al. Inaccuracy of High-temperature Component Life Evaluation Method Based on Classical Strength Theory [J]. Excellent Proceedings of Zhejiang Electric Power Society 2022, 2022 395-406.

Analysis of the static characteristics of magnetic preloaded small-throat gas thrust bearing rotary table

Yang Zhao [1], Yaping Xue [2*], and Weining Lei [1]

[1] College of Mechanical Engineering, Jiangsu University of Technology, Wujin, Changzhou 213016, Jiangsu, China

[2] College of Mechanical Engineering, Jiangsu University of Technology,Zhongwu Avenue 1801, Changzhou 21300l, China

*Corresponding author: Xue Yaping: E-mail: xyp@jsut.edu.cn

Abstract. To make orifice-compensated aerostatic bearings stiffer, we have added permanent magnets to the design of thrust aerostatic bearings. This study looks at a type of bearing that uses air and magnets to see how changes in magnetic force, air pressure, and hole size affect the bearing's performance. The magnetic preload force is created when permanent magnets interact with magnetic materials, using the no-current permanent magnetic field module in COMSOL. We made a simulation model to study how far the magnetic field reached. How the magnetic forces connect and how the magnetic strength decreases as the distance between objects increases. This study is the first to use permanent magnets in the preload system of aerostatic bearings. It shows that the way magnetic force changes with distance can create a steady magnetic preload force. The study looks at how changes in magnetic preload force, supply pressure, and hole size affect the performance of the bearing. The results indicate that more supply pressure increases the load capacity and stiffness, but adding magnetic preload decreases the load capacity. By changing the magnetic gap, we can control how thick the gas layer is to make sure the aerostatic system can handle the needed strength and load.

1. Introduction

Currently, the application of exactness and ultra-precision machining technologies is no longer limited to a few sectors such as national defense and aerospace. It has expanded into many areas of the national economy, leading to increasing demands for productivity and stability of machine tools. Precision micro-displacement workbench systems, characterized by high care and good equipment stability, play a crucial role in modern advanced industrial production and scientific research fields. It directly affects the levels of precision and ultra-precision machining, precision measurement, and large-scale integrated circuit production [1]. Its various technical indicators are important markers of high-tech development levels in various countries. Precision motion platforms are core components in various precision instruments and precision automation equipment. They not only found extensive applications in the field of precision machining but also held a significant position in the measurement field. The driving technology and rail structure are critical factors for precision platforms that meet high precision and high-speed requirements. Therefore, researchers have developed motion platforms supported by air-floating rails. Static air-bearing platforms are devices that help position and lift objects accurately. Their ability to perform well depends on several factors, and one important factor is how stiff the air floating below them is. Using magnetic preloading technology can make this air floating stiffer, which improves how well the platforms work. This is especially useful for tasks that need very precise positioning and control of vibrations, such as semiconductor making, precision machining, and optical systems.

Journal of Physics: Conference Series **3009** (2025) 012081 doi:10.1088/1742-6596/3009/1/012081

As modern technologies continue to evolve, the demand for high-precision motion platforms has increased significantly. Vainio et al. [2, 3] enhanced the performance of air static bearings by applying cement-based composite materials reinforced with silica, silicon carbide, or carbon microfibers. Energy conservation and emission reduction have always been focal points across various industries. For instance, Su et al. [4] made better performance by lowering low-frequency vibrations using special testing and analysis methods. These methods include looking at how things move in straight lines, studying changes in behavior, and examining different frequency patterns. Ting et al. [5] reduced friction and energy loss by optimizing structural designs. In particular, reducing friction loss and energy, consumption can significantly lower operating costs, which is crucial in the context of current sustainable development. Investigating the impact of magnetic preloading on air-floating platforms is expected to support industrial innovation. This research contributes to the development of new air-floating platform technologies, enhancing industrial automation levels, and creating more opportunities in fields such as manufacturing, semiconductor production, and medical devices. Studying the performance of static air-bearing platforms not only helps address engineering problems but also promotes advancements in engineering science. This research is expected to provide deeper insights and offer more technical support and guidance for engineers and scientists.

Gas-supported systems are generally divided into open and closed types. Closed air-floating bearings, characterized by high stiffness and stability, are widely used in static air-floating rails and turntables. However, they cannot accommodate all application scenarios. Open-air-floating bearing structures have a broader range of application scenarios. For open air-floating structures, new requirements for the stiffness of the air-floating structure have emerged. The study of the static performance of air-floating bearings has become relatively mature. For example, Zhang et al. [6] studied how different design features, like the length, depth, number, and location of pressure grooves. It affects how much weight air-floating bearings can hold and how stiff they are. Luan et al. [7, 8] studied several arrangements of pressure-regulating grooves. They found that these grooves can greatly improve how much weight air-floating bearings can handle and how strong they are, especially when arranged in a mesh pattern.

Additionally, to further explore methods for enhancing the stiffness of air-floating bearings, some researchers have introduced preloading techniques into air-floating bearings. There are three forms of preloading for air-floating bearings: added mass, magnetic preloading, and vacuum preloading. Li et al. [9] conducted electromagnetic preloading by adding magnetic preloading devices, which altered the film thickness and improved the positioning accuracy of the air-floating platform. However, electromagnetic preloading requires an external electromagnetic control system, significantly increasing equipment costs. Zhang et al. [10, 11] suggested a way to calculate how pressure is spread out in VPL static air bearings using a method called FDM. They also studied how vacuum affects the load capacity and stiffness of the bearings. Yao et al. [12, 13] enhanced stiffness through eccentric vacuum preloading.

Compared to traditional vacuum and electromagnetic preloading, both vacuum and electromagnetic preloading methods require external control systems for adjustment, which introduces uncertainties from the control system that significantly affect the platform's stability. Additionally, the vacuum chamber in vacuum preloading occupies part of the air film area, reducing the effective air film area for devices of the same size. Furthermore, the vacuum preloading system exhibits poor impact resistance, as excessive external impact loads may cause instability or failure of the preload. In contrast, permanent magnetic preloading does not rely on external control systems. Instead, it fully leverages the interaction between permanent magnets and ferromagnetic materials to achieve the preloading effect. The permanent magnets and ferromagnetic materials in this method can be positioned outside the air-floating surface, thereby avoiding the occupation of the air-floating area. In summary, compared with traditional preloading methods, permanent magnetic preloading greatly simplifies the structure and enhances the stability of hydrostatic air bearings.

Consequently, this study aims to investigate how the properties of permanent magnets and ferromagnetic materials can provide stable preload forces and analyze their effects on load capacity and stiffness. Permanent magnets offer magnetic preloading as an optimal solution due to their compact and simple structure, impact resistance, and adjustable preload forces. In the study of incorporating permanent magnet preload into the preloading structure of aerostatic bearings, this paper has certain limitations. For instance, it does not delve into the issues of magnetization in different ferromagnetic materials under prolonged operation of the preloading mechanism.

2. Working Principle of the Magnetic Preloaded Hydrostatic Air-Floating Platform

The way the magnetic forces from a permanent magnet work with a magnetic material in its field can be measured by using this equation:

$$F = \frac{\mu_r - 1}{2\mu_r \mu_0} \iint_s B^2 dS \tag{1}$$

where μ_r represents how well a material can transmit magnetic fields; μ_0 is the measure of how easily a magnetic field can pass through a vacuum; S represents the area where the magnetic field interacts with the material (in mm²); B represents the magnetic flux density at the surface of interaction (in Gauss). From Equation (1), it can be observed that the magnetic force is determined by the relative permeability μ_r, the interaction area S, and the strength of the magnetic field at the surface B. When the permanent magnet's cross-sectional area and the ferromagnetic layer's permeability are constant, the strength of the magnetic force mainly depends on the amount of magnetic flux at the surface of the ferromagnetic material.

The working principle of the magnetic preloaded structure in the air-floating rotary table is shown in Figure 1. In Figure 1, the arrows show which way the high-pressure gas from source P_1 is flowing. The gas goes through the small opening and into the tight space between the bearing and the thrust plate, forming a micron-thick air film that supports the thrust plate (along with the connected motor shaft and load). The air within the film is discharged into the atmosphere along the inside and outside ring surfaces of the air-float system, with the pressure dropping to atmospheric pressure P_2. Due to the pressure difference, the air pressure can support a certain amount of weight.

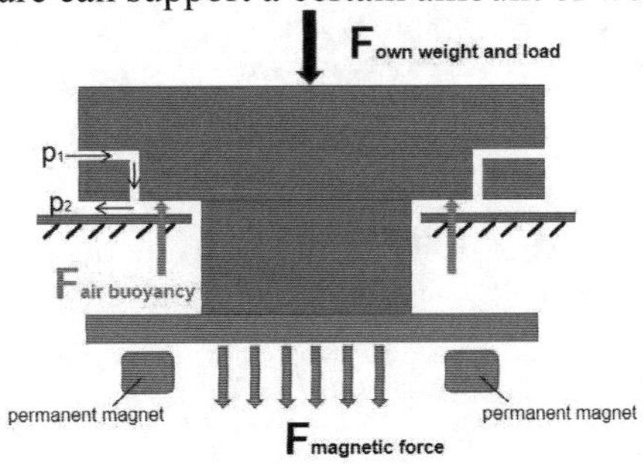

Figure 1. Schematic diagram of magnetic preload structure

The ferromagnetic material is connected to rotating components such as the motor shaft. We apply a downward force by using the magnetic attraction between the permanent magnet and the metal material. By adjusting the magnitude of the preloading force, a balance can be achieved between the two, allowing the air film gap to maintain a stable range. This stabilization is essential for obtaining appropriate stiffness and stability in the bearing.

3. Theoretical Analysis of Orifice Throttling and Permanent Magnets

The static characteristics of the magnetic preloaded small-throat throttling (as shown in Figure 2) static gas bearings can support weight and are stiff. To simplify the calculations, it is assumed that the airflow is radially uniform, and the way pressure is spread out in the space between the bearing parts.

$$\frac{\partial}{\partial x}\left(\frac{\rho h^3}{12\mu}\frac{\partial P}{\partial x}\right)+\frac{\partial}{\partial y}\left(\frac{\rho h^3}{12\mu}\frac{\partial P}{\partial y}\right)=\frac{v_x}{2}\frac{\partial(\rho h)}{\partial x}+\frac{v_y}{2}\frac{\partial(\rho h)}{\partial y}+\frac{\partial(\rho h)}{\partial t} \tag{2}$$

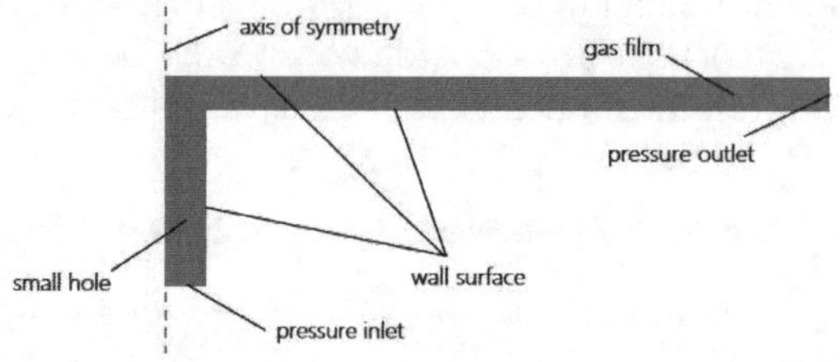

Figure 2. Schematic diagram of orifice structure

where ρ represents the gas density; h represents the thickness of the air film; P indicates film pressure; μ measures how dense or sticky the gas is; V_x and V_y represent the speeds of the bearing in the horizontal and vertical directions, respectively. Based on the gas state equation and the isothermal assumption, Equation (2) can be rewritten as a nonlinear partial differential equation relating pressure P and time t, expressed as follows:

$$\frac{\partial}{\partial x}\left(Ph^3\frac{\partial P}{\partial x}\right)+\frac{\partial}{\partial y}\left(Ph^3\frac{\partial P}{\partial y}\right)=6\mu v_x\frac{\partial(Ph)}{\partial x}+6\mu v_y\frac{\partial(Ph)}{\partial y}+12\mu\frac{\partial(Ph)}{\partial t} \tag{3}$$

When the gas flows rapidly through the throttling device, assuming that the process is adiabatic, the amount of gas that flows into the throttling opening can be described like this:

$$Q = AC_d\psi\frac{P_S}{\sqrt{RT}} \tag{4}$$

where A represents the throttling area; C_d is the flow rate factor of the opening that controls the flow, typically taken as 0.8; P_s denotes the supply pressure; ψ is the flow coefficient for throttling.

The magnetic field in a zero-current state primarily relies on the Maxwell magnetic field model:

$$\nabla\times H = 0 \tag{5}$$

We define the scalar magnetic potential V_m as follows:

$$H = -\nabla V_m \tag{6}$$

From the basic connection between how strong a magnetic field is and the amount of magnetic energy it has, it can be inferred that:

$$B = \mu_0(H + M)$$

$$\nabla\cdot B = 0 \tag{7}$$

where B is magnetic flux density; H is the magnetic field strength; μ_0 is the magnetic permeability of the medium.

4. Permanent Magnet Magnetic Field Analysis and Modeling Method

4.1 Analysis of Strong Magnets and Material Selection

Creating strong magnetic fields is an important challenge in many technological and scientific fields, with the highest magnetic fields obtained directly from permanent sources being extremely limited. Samarium-cobalt and neodymium-iron-boron (NdFeB) magnets are the most common types of rare earth magnets, with residual magnetic flux densities reaching up to 1.48 T [14]. The square hysteresis loops, high coercivity, and the uneven magnetic strengths of rare earth magnets make it easier to design magnetic parts, making each magnet block easy for magnetic fields from other parts to pass through [15, 16]. Additionally, their mechanical properties are stable, making them resistant to demagnetization.

Figure 3. Diagram of the testing platform and apparatus for air bearing performance

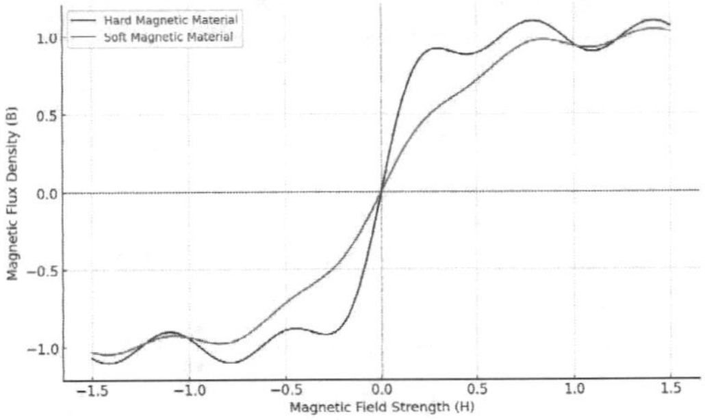

Figure 4. Hysteresis loops of hard and soft materials

The two curves in Figure 4 represent the hysteresis loops of hard magnetic material (NdFeB) and soft magnetic material, illustrating the variation of magnetic flux density with the applied magnetic field intensity during the magnetization and demagnetization processes.

The maximum magnetic flux density (saturation magnetization B_s) of hard magnetic materials is relatively high. After the external magnetic field is removed, the intercept of the curve on the vertical axis is significantly larger, indicating a higher remanence (B_r). This demonstrates that hard magnetic materials can retain strong magnetism even in the absence of an external magnetic field.

In contrast, soft magnetic materials exhibit a lower maximum magnetic flux density, suggesting that they more readily achieve magnetic saturation; however, their saturation magnetization is relatively small. The intercept of the curve on the vertical axis is minimal, with a low remanence (B_r), indicating that the magnetism almost entirely dissipates once the external magnetic field is removed.

The experiment aims to find out how much weight a magnetic preload gas static thrust bearing can hold and how stiff it is. We will also check if the results match with our theoretical predictions to make sure they are correct. A prototype of the magnetic preload static pressure small orifice throttle thrust bearing has been developed in the laboratory. To enhance the accuracy of the tests, the magnetic material is a type of magnet made from neodymium, iron, and boron. It has a round shape with a width of 10 mm and is 8 mm thick, classified as N35. The ferromagnetic material is a high-permeability nickel alloy (1J22). Prior to the experiment, the magnets and ferromagnetic materials (1J22) need to undergo precision machining, and the residual magnetic flux of the magnets and the magnetic permeability of the ferromagnetic materials must be measured to provide essential data for theoretical analysis. During the experiment, the thickness of the gas film and the load applied to the bearing will be measured to calculate its load capacity and stiffness.

4.2 Modeling Method

The static analysis of the magnetic preload orifice-throttling system focuses on the magnetic field distribution of permanent magnets within ferromagnetic materials. The magnet and ferromagnetic material are meshed and imported into COMSOL for simulation analysis (see Figure 5). A spherical region is designated as the air domain, with the outermost spherical surface set to zero magnetic flux potential. Inside the air domain, there is a cylindrical permanent magnet and a rectangular ferromagnetic material, and a parametric scan is performed on the magnet spacing in the X and Y directions.

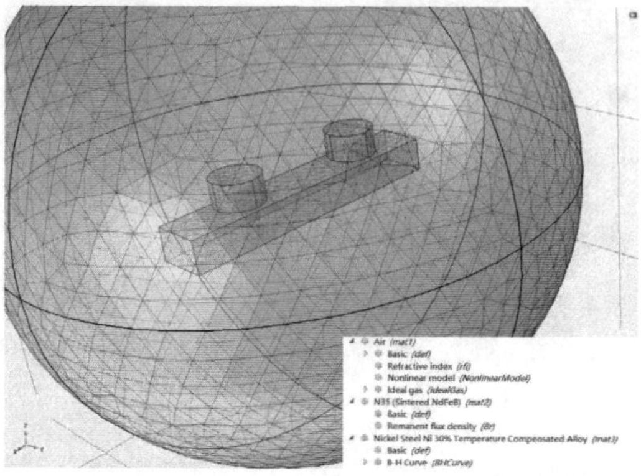

Figure 5. Simulation model of the permanent magnet

The magnetic material is set as NdFeB-N35. With a permeability of 1.05 and a residual magnetic flux density modulus of 1.48 T, the ferromagnetic material is defined as Nickel Steel Ni30% Temperature Compensated Alloy. Simulation results show that the magnetic flux density of the permanent magnet exhibits a fan-shaped distribution within the ferromagnetic material (see Figure 6), which presents a circular distribution. Due to the magnetic field properties of the circular magnet, the flux density is stronger at the edges than in the center. The magnetic force in the ferromagnetic material extends beyond the area relative to the magnet, becoming more concentrated in a toroidal shape. To enhance the stability of the preload force provided by the magnetic preload mechanism, specific dimensions for the ferromagnetic material are required: its length, width, and thickness must exceed those of the magnet to ensure complete saturation of the magnetic flux density within the ferromagnetic material.

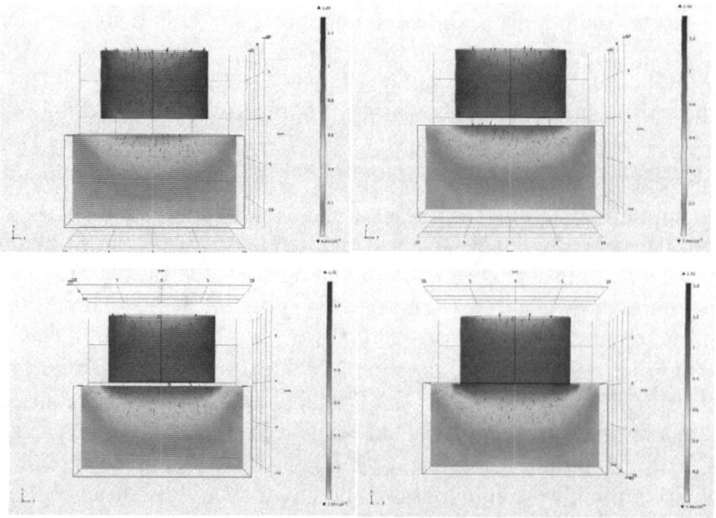

IOP Publishing
doi:10.1088/1742-6596/3009/1/012081

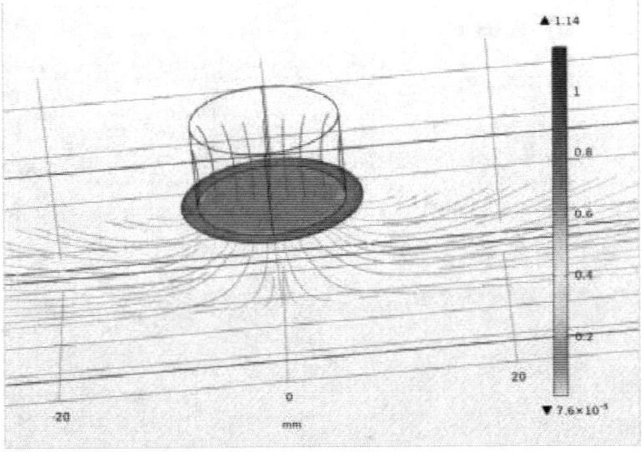

Figure 6. Magnetic field distribution of permanent magnet in ferromagnetic material

The magnetic circuit has significant drawbacks, including severe cross-coupling effects and complex magnetic pole structures [17]. For example, in a Halbach array stacked structure, the complexity of stacking different magnetic poles and their coupling makes it difficult to control the magnetic field distribution. Jiang et al. [18] conducted a simulation analysis to determine the model for the gap and size of the permanent magnet in the Halbach [19, 20] magnetic field configuration. To eliminate the uncertainty of the coupled magnetic field when multiple magnets are aligned in the same direction and to reduce their interactions, we stabilize the distribution of magnetic flux density of the permanent magnets within the ferromagnetic material. Ensuring it does not affect the stability of the preload force in the mechanism. Through simulation analysis, the results indicate that when magnets are aligned in the same direction, the magnetic flux density modulus shifts outward due to mutual repulsion as the magnets come closer (see Figure 7). The magnetic force does not exhibit linear superposition. To avoid magnetic field interactions, the axial spacing between circular magnets should not be less than the diameter of the magnet cross-section, thereby preventing magnetic field coupling.

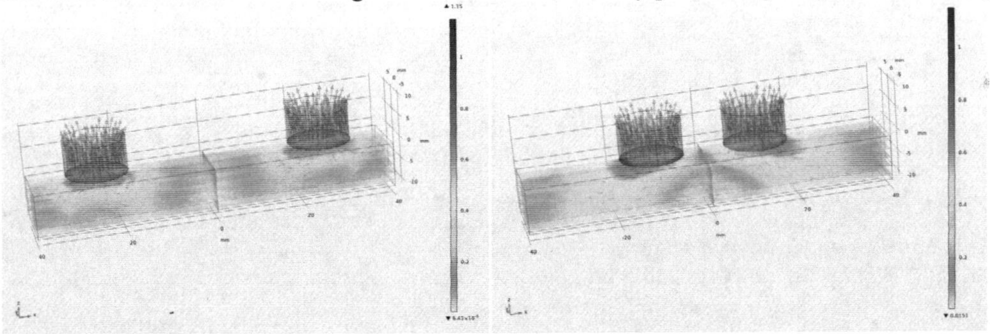

Figure 7. Magnetic coupling

5. Steady-state Characteristics of Attractive Preload

Due to the nonlinear variation in the strength of the attractive force of permanent magnets with distance, the magnetic flux density within the ferromagnetic material is relatively high at small gaps. This results in a strong alignment of magnetic domains within the ferromagnetic material, coupled with magnetization phenomena and the coupling effect with the permanent magnet's magnetic field. Consequently, discrepancies between the experimental and simulation results in magnetic force magnitude occur at both close and distant separations.

The experimental results indicate that the magnetic force varies with changes in the action gap (as shown in Figure 3). The experimental data were fitted, and in both test groups, the theoretical curves showed a high degree of conformity with the experimental curves. Although there is some discrepancy between the theoretical magnetic preload force and the experimental values, the error does not exceed

Journal of Physics: Conference Series **3009** (2025) 012081 doi:10.1088/1742-6596/3009/1/012081

10%. By comparing the required preload force for the bearing, the action gap can be adjusted by using a differential thread mechanism, allowing for micrometer-level control of the gas film thickness.

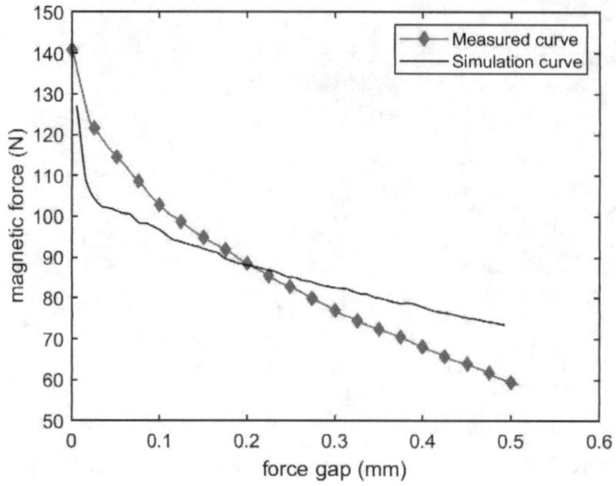

Figure 8. Attenuation curve of magnetic force with varying gap

The parameters for analysis are set as follows: The outer size of the bearing is 150 mm, the inner size is 50 mm, the thickness is 10 mm, and the throttle hole diameter is 0.35 mm. Gas parameters are shown in Table 1.The supply pressures are taken as 0.3, 0.4, 0.5, and 0.6 MPa. The results of the analysis are displayed in Figure 9 .

Table 1. Gas parameter setting

Parameters	Value
Temperature/K	293.15
Density/(kg/m^3)	1.204
Dynamic Viscosity/(Pa • s)	1.81×10^{-5}
Molecular Mass/(kg/mol)	28.97×10^{-3}
Specific Heat Capacity/(J/(kg • k))	1.01×10^{3}
Thermal Conductivity/(W/(m • k))	2.42×10^{-2}

The results show that when the supply pressure goes up, both the load capacity and stiffness go up too. However, the thickness of the gas film at the highest stiffness stays the same. The gas film gap corresponding to the maximum stiffness at different supply pressures is approximately 10 μm.

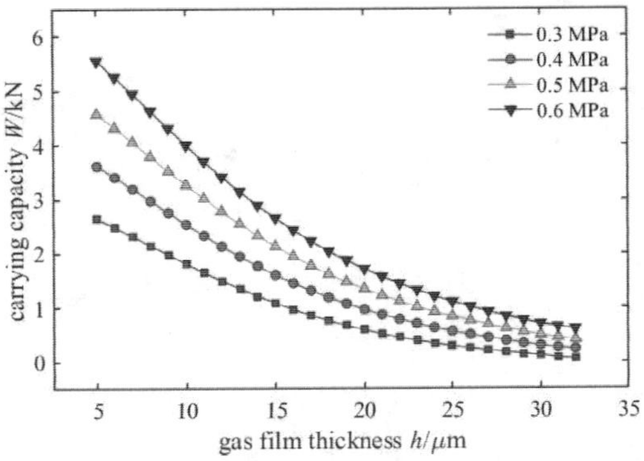

(a) Carrying capacity

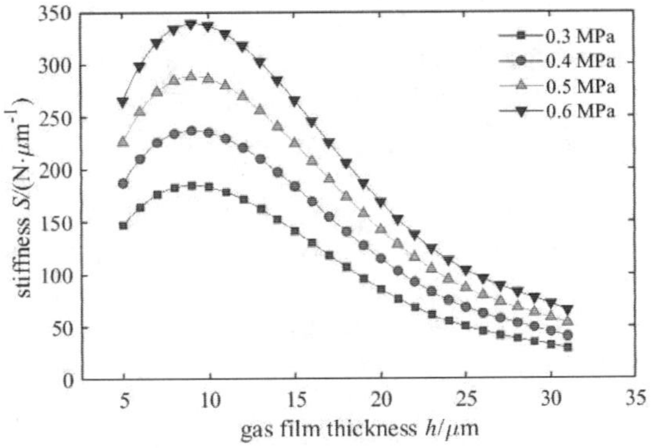

(b) Stiffness

Figure 9. Load capacity curves of gas film under different supply pressures: (a) Carrying capacity; (b) Stiffness.

6. The Effect of Orifice Diameter on Static Characteristics

The supply pressure is set at 0.5 MPa, while the orifice diameter is variable. The static characteristics under different film thicknesses are analyzed, as shown in Figure 10. Figure 10 (a) presents the load capacity curve of the gas film, indicating that, at a constant film thickness, a larger orifice diameter results in greater load capacity. However, the rate of increase in load capacity diminishes as the orifice diameter increases. Figure 10 (b) shows the stiffness curve of the gas film, revealing that, at smaller film thicknesses, a larger orifice diameter corresponds to lower film stiffness; furthermore, a larger orifice diameter leads to a greater film thickness required to achieve maximum stiffness. Under different orifice diameters, the film thickness that gives the strongest stiffness is between 6 and 12 μm.

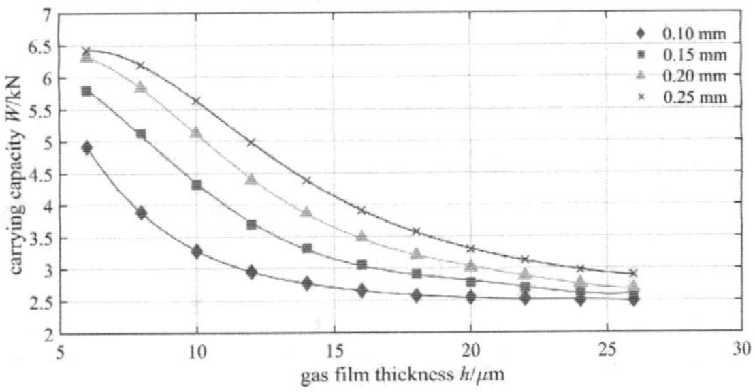

(a) Load capacity curves of air films at different orifice diameters

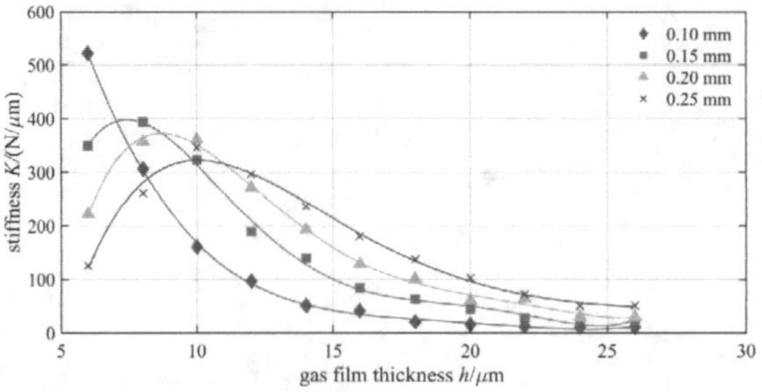

(b) Stiffness curves of air films at different orifice diameters

Figure 10. The effect of different orifice diameters on the static characteristics of the gas film

7. Conclusion

This paper analyzes the distribution of the attractive field of lasting lodestones within ferromagnetic materials, as well as the seductive field coupling between two permanent magnets. Based on the variation laws of the magnetic force in the magnetic preloading system, a preloading system has been established, which enables precise control of the air film gap in the static pressure air-bearing mechanism using the attractive force of the permanent magnets. After applying the magnetic preloading arrangement to the static pressure air-bearing mechanism, the study investigated the variations in air film load ability and stiffness concerning the air film gap, as well as the effects of the orifice diameter on stiffness and load capacity. The following four conclusions were drawn from the above study: (1) The factors influencing the magnitude of the magnetic force of permanent magnets are numerous, especially when the operating gap is small. When multiple magnets interact, magnetic field coupling is a major factor. When there is no magnetic field affecting it, the weight that the air bearing can hold is equal to the lift created near the opening minus the total push from the magnets. (2) Simulation results indicate that when the magnetic force acting gap is fixed, magnetic field coupling between cylindrical permanent magnets has a significant effect on the magnitude of the magnetic force. When the spacing exceeds the diameter of the cylindrical magnets, the effect of the magnetic field coupling on the magnetic force is reduced, resulting in a more stable variation in magnetic force with gap changes. (3) Based on experimental and simulation results, the air film thickness can be adjusted by fine-tuning the operating gap to change the magnitude of the magnetic force. When the air film width is fixed, the supply pressure is positively correlated with both load capacity and stiffness. As the air film thickness decreases, the effect of supply pressure on load capacity and stiffness increases. (4) When magnetic preloading is applied to adjust the air film thickness, increasing the orifice diameter enhances the bearing's load capacity. Conversely, with a thinner air film, the stiffness of the bearing decreases as the orifice diameter increases. To achieve maximum stiffness, a larger orifice diameter demands a greater air film thickness. (5) In all cases analyzed, once the air film thickness surpasses 25 μm, the stiffness of the bearing sharply decreases, ultimately approaching zero.

Acknowledgment

This work was supported by the National Natural Science Foundation of China (No. 51975264).

References

[1] Wang X., Li X., Ma X., Li W., and Yang S. Advance on surface finishing technology of precision bearing cylindrical rollers. The International Journal of Advanced Manufacturing Technology. 2024; 131 (5): 2, 341-63.
[2] Vainio V., Miettinen M., Majuri J., Theska R., and Viitala R. Manufacturing and static performance of porous aerostatic bearings. Precision Engineering. 2023; 84: 177-90.

[3] Jamwal K. S., Singh A. K., Arora K., and Paswan S. K. Performance Analysis of a Designed Aerostatic Bearing with Effect of Surface Roughness. Iranian Journal of Science and Technology, Transactions of Mechanical Engineering. 2024; 48 (3): 1, 363-81.
[4] Su Z., Zhang J., Cai Y., and Han D. Experimental Research on Dynamic Characteristics of a Multi-Disc Rotor System Supported by Aerostatic Bearings. Lubricants [Internet]. 2024; 12 (5).
[5] Ting G. C., Lee Y. H., and Huang K. Y. Development and experimental verification of radial-axial integrated cup-shaped aerostatic bearing for the miniature turbine of dental handpiece. Tribology International. 2022; 174: 107, 740.
[6] Zhang H., Guan X., Wang T., Robin A. U. Z., and Mu C. H. Influence of Pressure-Equalizing Groove on Static Load Performance of Aerostatic Guideway. Journal of Applied Fluid Mechanics. 2023; 16 (5): 992-1, 004.
[7] Luan W., Liu Y., Wang Y., and Xu F. Effect of herringbone groove structure parameters on the static performance of gas foil herringbone groove thrust bearings. Tribology International. 2023; 177: 107, 979.
[8] Wu Y., Li C., Du J., Liu T., and Zuo X. Pneumatic hammer characteristics of the aerostatic thrust bearing with central orifice and pressure-equalizing groove. Nonlinear Dynamics. 2023; 111 (3): 2, 161-82.
[9] Li Z., Yang K., and Wang X. Electromagnetic and Flow Field Analysis of Multi-DOF Motor with Air Bearing Structure. IEEE Transactions on Applied Superconductivity. 2021; 31 (8): 1-4.
[10] Zhang S., Yang D., Li G., Cheng Y., Li J., et al. Carrying Capacity of Spherical Hydrostatic Bearings Including Dynamic Pressure. Lubricants [Internet]. 2024; 12 (10).
[11] Li Z., Liu Y., Liu S. B., Xu Q. Q., and Sun H. X. Structural Design and Optimization Analysis of Multi-degree-of-freedom Motor Using Gas Bearing. Journal of Electrical Engineering & Technology. 2024.
[12] Yao J., Yin Z., Chai N., Meng S., and Li Y. Design of an Eccentric Vacuum Preloaded Aerostatic Bearing to Improve the Lateral Stiffness of Guideway. Tribology Transactions. 2022; 65 (2): 321-36.
[13] Cui H., Wang Y., Zhang M., Wang W., and Zhao C. A fractal method to calculate the permeability for compressible gas flow through a porous restrictor in aerostatic bearings. International Journal of Heat and Mass Transfer. 2018; 121: 437-52.
[14] Przybył A., Gębara P., Gozdur R., and Chwastek K. Modeling of Magnetic Properties of Rare-Earth Hard Magnets. Energies [Internet]. 2022; 15 (21).
[15] Zhou Y., Ji P., Li M., Zhuge L., and Wu X. Influence of magnetic field on power deposition in high magnetic field helicon experiment. Chinese Physics B. 2023; 32 (2): 025205.
[16] Su Z., Wei R., Du Y., Fan W., and Chen J. Numerical Simulation of Magnetic Field and Flow Field of Slab under Composite Magnetic Field. Metals [Internet]. 2023; 13 (7).
[17] Yang G., Liu W., Zhou Y., and Lin M. Variable Magnetic Characteristic Analysis of Series Hybrid Permanent Magnet Magnetic Circuit Using Low Coercive Force Permanent Magnet. IEEE Transactions on Applied Superconductivity. 2021; 31 (8): 1-4.
[18] Jiang Y., Deng Y., Zhu P., Yang M., and Zhou F. Optimization on Size of Halbach Array Permanent Magnets for Magnetic Levitation System for Permanent Magnet Maglev Train. IEEE Access. 2021; 9: 44, 989-5, 000.
[19] Kim D. H. and Lee C. M. A study on the development of a new conceptual automatic variable preload system for a spindle bearing. The International Journal of Advanced Manufacturing Technology. 2013; 65 (5): 817-24.
[20] He H., Jing H., Gu N., Liu Z., Gao M., et al. A new radial quadrupole Nd2Fe14B Halbach ring for application in motor. Journal of Magnetism and Magnetic Materials. 2024; 589: 171, 577.

Study on the crack resistance of high-viscosity asphalt mixture

Shunli Fu[1], Fangzhan Hou[2], Mingzhi Lu[3], Luhai Wang[4], Hao Zhou[5]*

[1]QingdaoTransportation Science Institute, Qingdao, 266000, China
[2]Jiaozhou Transportation Bureau, Qingdao, 266300, China
[3]Qingdao Jimo District Urban Development Investment Co., Qingdao, Ltd., 266200, China
[4]Qingdao Jimo District Transportation Bureau, Qingdao, 266200, China
[5]Shandong Jianzhu University, School of Transportation Engineering, Jinan, 250101, China

*Corresponding author's e-mail: 523475585@qq.com

Abstract. The bridge deck paving is subjected to harsh environments, and low-temperature cracking is a common disease of bridge deck paving asphalt mixtures. This paper determined the three gradations of SMA-10, SMA-13, and SMA-20, and used high-viscosity and SBS-modified asphalt to prepare specimens. This study systematically evaluates the effects of nominal maximum aggregate size (NMPZ) and asphalt type on the flexural performance and low-temperature cracking behavior of asphalt mixtures. Key results show that SMA-10 specimens exhibit 1.14-fold and 1.86-fold higher maximum bending strain compared to SMA-13 and SMA-20 mixtures, respectively, with consistent strain trends observed across different gradations. The high-viscosity asphalt bending strain is approximately 2.2 times that of the SBS-modified asphalt mixture, which demonstrates significantly higher crack resistance ability. At -5C°, -10C° and -15°C, the fracture energy of the high-viscosity modified asphalt mixture is 1.53 times, 1.75 times, and 2.05 times that of the SBS-modified asphalt mixture, the high-viscosity modified asphalt mixture exhibits greater crack resistance at low temperatures. At -15C°, the crack resistance index of high-viscosity asphalt mixtures is still 9.1 times that of SBS asphalt mixtures.

1. Introduction

Bridge deck pavement [1-2] is not only the appearance of the bridge structure, but also the key component to guarantee the structural integrity of the bridge, prolong service life, improve service performance, and reduce maintenance cost, and its importance can not be ignored. Therefore, in the process of design, construction, and maintenance, we should pay enough attention to the bridge deck pavement to ensure that it can play its due role during the service period. The cement concrete bridge deck above the steel box girder is is affected by external conditions such as vehicle load and temperature, as well as the deformation of the steel beam structure [3]. The stress and deformation are more complicated. The pavement needs to have a certain bearing capacity and uniform force transmission characteristics, which can disperse the vehicle load over a larger area, reduce local stress concentration, and prevent the bridge deck from early fatigue failure due to excessive local stress. At the same time, the pavement can also absorb some impact energy through its own deformation, thus reducing the impact load on the bridge deck. Therefore, the performance requirements for the pavement are extremely demanding.

To develop materials suitable for bridge deck paving, researchers have conducted continuous studies and have identified the following three types of bridge deck paving materials: pouring asphalt concrete [4], epoxy-modified asphalt mixture [5-6], and stone matrix asphalt [7]. Pouring asphalt concrete shows obvious advantages on specific occasions, such as steel bridge deck pavement, due to its unique construction method and excellent performance. However, it also has limitations such as high construction conditions, high cost, and difficult maintenance [8]. Epoxy asphalt concrete is well-known for its superior road performance, however, its material properties are highly dependent on molding temperature and time, making precise control challenging during construction. Being a thermosetting material, repairing damaged epoxy asphalt concrete is not only extremely difficult but also often results in less-than-ideal outcomes. In contrast, SMA mixture has obvious advantages such as rutting resistance, sliding resistance, durability, and noise reduction, but SMA mixture designed by ordinary modified asphalt is difficult to adapt to the complex working environment of steel-concrete composite beam bridge pavement. Measuring the deformation ability of materials is an important means of studying the performance of bridge deck paving materials. The most conventional method is to use the built-in sensors of low-temperature bending(LTB) equipment. Some researchers also test deformation [9] using image recognition [10-11] methods.

Therefore, this paper uses high-viscosity asphalt(HVA) [12-13] as cementing material to prepare the SMA asphalt mixture as bridge deck pavement material. The mix proportion of high-viscosity SMA asphalt mixture is designed, and the flexural and tensile properties are systematically studied, which provides a basis for the use environment and conditions of high-viscosity SMA asphalt mixture pavement materials.

2. Materials and Experimental Method

2.1 Materials

2.1.1 Asphalt. In this study, SMA modified asphalt (SBSA) and self-developed HVA were selected as experimental materials. When preparing high-viscosity asphalt, the technological process of "first stirring-then shearing-then stirring" is followed. The operation involves heating the 70# base asphalt to a temperature range of 180°C to 185°C. The high-viscosity modifier is then added in three equal portions over a 30-minute period while continuously stirring. After the final addition, it is switched to shear mode and the shear rate is set to 4000 rpm. Shearing is continued at 180°C for 1 hour. After confirming that the modifier was completely dispersed, it was prepared at a temperature of 180-185°C using a stirring mode, adding stabilizers in three batches within 30 minutes, and allowed to develop stably for 3.5 hours for comparative study. The related indexes of asphalt can be found in Table 1.

Table 1 Key Indicators of Asphalt.

Test Item	HVA	SBSA
Penetration (25℃, 100 g, 5 s)/0.1 mm.	45.4	46.8
Ductility (5 cm/min)/cm	34.7	32.0
Softening point (R&B)/℃	108.5	75.4
Viscosity and toughness (30℃)/n m	31.3	24.8
Toughness (30℃)/n m	24.5	18.2
Brinell viscosity at 135℃/pa s	12.325	1.882
Brinell viscosity at 175℃/pa s	2.86	0.451
Elastic recovery (25℃)/%	99.3	92.1

2.1.2 Mineral Aggregate Grading Design. Using the volume design method, the asphalt mixture gradation in this paper has been carefully designed to ensure it meets the technical requirements of bridge deck pavement. Specific information on the synthetic gradation of nominal maximum particle sizes (NMPZ) has been listed in Table 2.

Table 2 Gradation of SMA

Mesh Size (mm)	Percentage of Passing Through Each Sieve Hole (%)											
	26.5	19	16	13.2	9.5	4.75	2.36	1.18	0.6	0.3	0.15	0.075
SMA-10	100	100	100	100	100	60	32	26	22	18	16	13

SMA-13	100	100	99.9	95.1	74.0	30.5	23.0	19.6	17.3	15.1	13.1	11.0
SMA-20	100	94.8	83.4	71.2	54	23.7	18.9	16.8	15.4	14	12.4	10.5

2.1.3 Determination of Optimum Asphalt Dosage. Using different amounts of high-viscosity asphalt, Marshall tests were conducted on three different types of asphalt mixture. Relevant test results are shown in Table 3.

Table 3 Basic Parameters of SMA

Mixture Type	SMA-10	SMA-13	SMA-20
Asphalt consumption (%)	6.4	6.3	5.8
Gross bulk density (g/cm^3)	2.46	2.461	2.422
Void ratio VV/%	3.5	4	4
Void ratio of mineral aggregate VMA/%	17	17.4	16.6
Asphalt saturation VFA/%	79.4	76.8	75.9
Stability /kN	19.2	23.4	18.1
Flow value /mm	6.13	5.09	5.02
Asphalt leakage loss/%	0.03	0.02	0.03
Mixture scattering loss/%	3.4	2.8	3.6

2.2 Experimental Method

2.2.1 LTB Test. The LTB test can effectively reflect asphalt mixture deformation ability under low-temperature conditions. In this paper, according to the test procedure, the rutting slab specimens are cut into standard beam specimens. They are kept at -10C° for 5 hours, and a concentrated load is applied to the mid-span of the specimens until they break using the UTM-100. The flexural tensile strength, flexural strain, and flexural stiffness modulus are used as evaluation indicators for the low-temperature deformation resistance of asphalt mixes.

2.2.2 Indirect Tensile Asphalt Cracking Test(IDEAL-CT). The IDEAL-CT was carried out according to ASTM D8225-2019 [14]. When conducting the IDEAL-CT test, the special fixture for the splitting test is selected, and the width of the clamping strip is set to 2 cm. The experiment was carried out on an MTS material testing system, and the loading direction was along the diameter of the specimen. Before the test, the specimen had to be preheated in a 25°C thermostat for 2 hours. The test termination condition is that the loading force drops below 0.1 KN. The force-displacement profile recorded by the MTS testing machine is demonstrated in Figure 1.

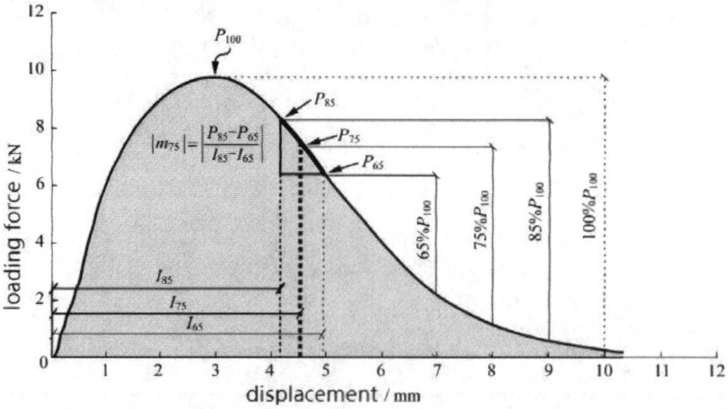

Figure 1 Schematic diagram of ideal-CT test.

Figure 1 is the stress-strain relationship was quantified through experimental measurements, generating characteristic curves that delineate material response under progressive loading conditions, and reflects the stress and displacement changes of SGC specimens during loading. The state of the specimen is different in different stages of curve. In the rising stage, the stress on the specimen increases continuously until the load reaches the maximum value, and the crack of the specimen expands continuously in the falling stage.

$$G_f = \frac{\int Pdl}{Dt} \times 10^6 \quad (1)$$

where: G_f— the fracture energy (j m -2 ;
P— the load (KN);
D— the displacement (mm);

$$|m_{75}| = \left|\frac{P_{85} - P_{65}}{l_{85} - l_{65}}\right| \quad (2)$$

where: $|m_{75}|$—crack propagation speed (KN·mm -1);
P_{85}, P_{65}—85% peak load (KN), and 65% peak load (KN) respectively;
l_{85}, l_{65}—85% peak displacement (mm) and 65% peak displacement (mm) respectively;

$$I_{CT} = \frac{t}{62}\frac{l_{75}}{D}\frac{G_f}{|m_{75}|} \quad (3)$$

where l_{75}—Displacement at 75% peak displacement (mm)

3. Results and Discussion

3.1 Effect of NMPZ on LTB Performance

To compare the influence of the NMPZ of asphalt mixture on LTCR(LTCR), SMA-10, SMA-13, and SMA-20 were utilized. SBSA and HVA were also included for comparison, with the impact of the NMPZ of asphalt mixture on LTCR [15] illustrated in Figure 2

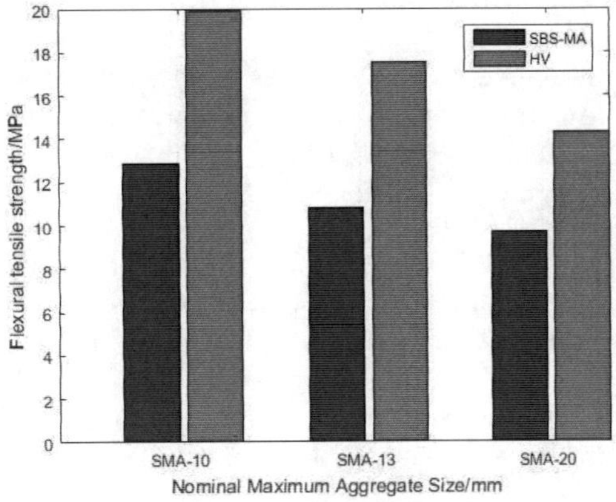

(a) Flexural tensile strength

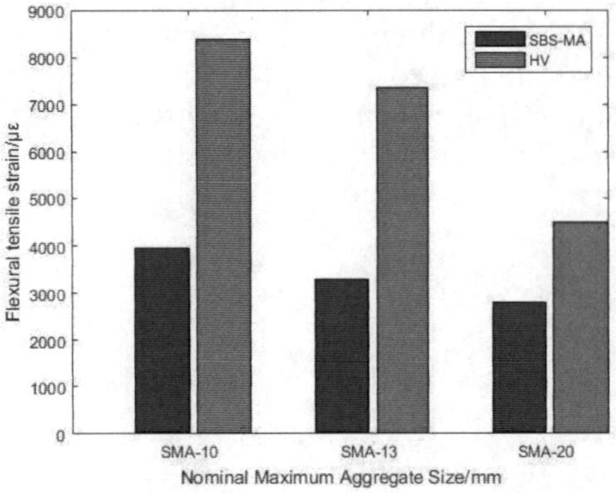

(b) Maximum bending tensile strain

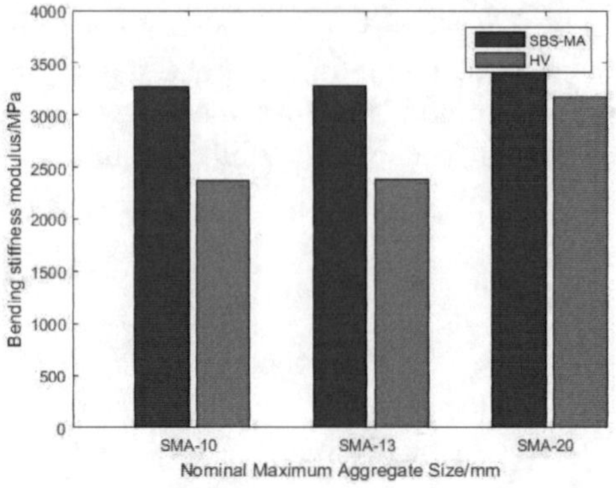

(c) Bending stiffness modulus

Figure 2 LTB results of asphalt mixture with different NMPZ.

Experimental data demonstrate that increased strain levels in trabecular specimens under combined bending and tensile loading conditions are indicative of greater mixture flexibility, with concurrently lower stiffness modulus values reflecting enhanced deformation resistance[16]. The low-temperature evaluation index of modified asphalt mixtures should be above 2500. As can be seen from Figure 2, SBS and HVA SMA mixtures with different NMPZ all meet the requirements. The high-viscosity SMA-10 asphalt mixture exhibited a flexural strength of 19.91 MPa, accompanied by the highest flexural strain and lowest stiffness modulus. For a given asphalt type, increasing the nominal maximum aggregate size (NMPZ) of SMA mixtures led to a gradual reduction in both flexural strength and strain, while the flexural modulus increased correspondingly.

The binder used in the SMA-20 mixture is high-viscosity asphalt. The maximum bending strain of the SMA-20 is 4500 με, while that of the SMA-10 is 8390 με. This is 1.14 times to that of the SMA-13 mixture and 1.86 times to that of the SMA-20 mixture. The performance differentials in key flexural metrics (tensile strength, strain, and stiffness modulus) exhibit reduced magnitude between SMA-10 and SMA-13 mixtures, whereas SMA-13 and SMA-20 formulations demonstrate substantially greater variation.This difference can be attributed to the larger particle size of the SMA-20 graded aggregate, which makes it challenging to form a dense structure and results in poorer low-temperature deformation resistance. It is concluded that SMA-10 and SMA-13 asphalt mixtures have better crack resistance at low temperatures, which is more in line with the requirements for deck pavement of extra-large steel-concrete composite bridges.

3.2 Effect of Asphalt Types on LTB Performance

Small beam samples of SMA-13 asphalt mixture were made with SBSA, 70 # matrix asphalt, 90 # matrix asphalt, rubber-modified materials, and asphalt with 8%, 10%, and 12% high viscosity modifiers. The LTB test was then carried out, as shown in Figure 3. Previous research in the laboratory has shown that the low-temperature performance(LTP) of HVA increases with the increase in modifier content. The HVA with 12% modifier content was found to be the best content selected, and this is the HVA used in this paper.

IOP Publishing
doi:10.1088/1742-6596/3009/1/012082

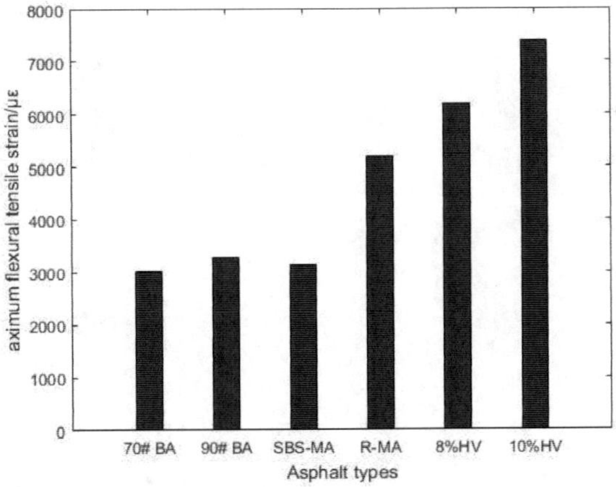

Figure 3 LTB tensile strain of different asphalt type.

Figure 3 illustrates that the 70# matrix asphalt mixture exhibits lower strength in the bending strain test and has poor resistance to low temperatures. In contrast, the HVA mixture demonstrates significantly higher bending strain compared to other asphalt mixtures. Specifically, its bending strain is approximately 2.2 times greater than that of the SBSA mixture and 3.2 times greater than the mixture utilizing 70# matrix asphalt. The 12% is the optimal dosage of a high-viscosity modifier, 12% HVA bending strain is 1.2 times that of a 10% dosage. In comparison to 90# matrix asphalt and asphalt rubber, SBSA exhibits marginally higher bending and tensile strain, with the SBS asphalt mixture demonstrating robust low-temperature crack resistance (LTCR). Increasing the dosage of the high-viscosity modifier leads to a notable upward trend in the bending and tensile strain of high-viscosity asphalt (HVA) mixtures, indicating superior LTP and enhanced LTCR due to modifier incorporation.

Comparing the LTB and tensile strains of different asphalt mixtures reveals that the mixture's LTB is greatly limited by the asphalt materials used, even when the gradation remains consistent.

3.3 Experimental Study on Indirect Tensile Cracking (IDEAL-CT)

IDEAL-CT test was used to study the influence of temperature on low-temperature cracking of HVA mixture. SMA-13 asphalt mixture type was selected and SBS modified asphalt was used for comparison. IDEAL-CT test was carried out at -5°C, -10°C and -15°C. The comparison of evaluation index results of the asphalt mixture IDEAL-CT is depicted as Figure 4.

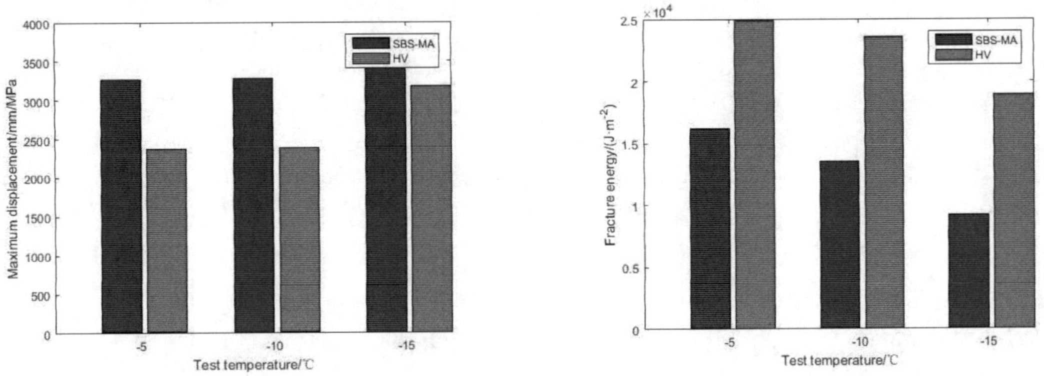

(a) Maximum displacement

(b) Fracture energy

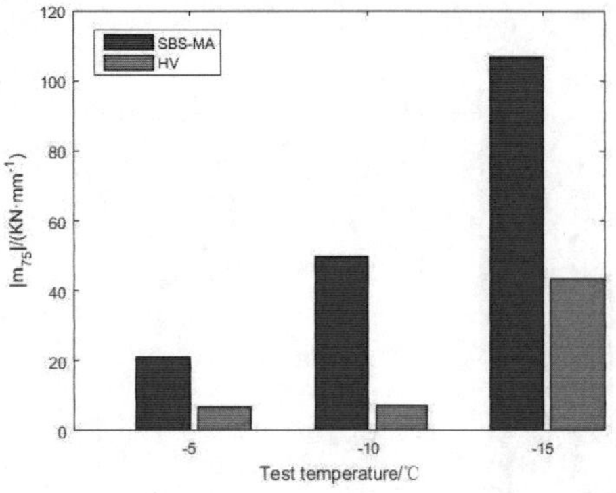

(c) $|m_{75}|$

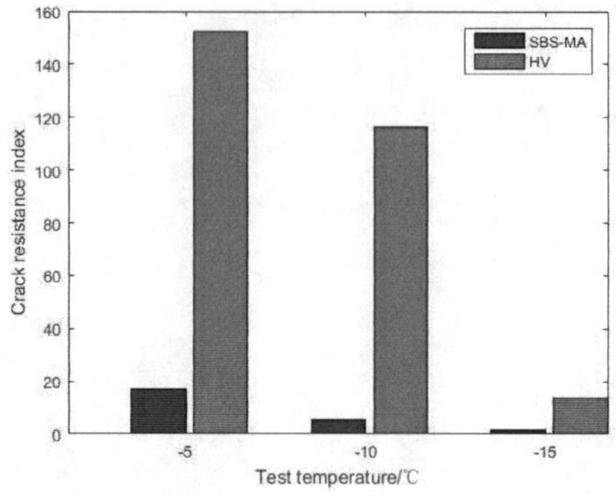

(d) Crack resistance indexI_{CT}

Figure 4 Comparison of evaluation index results of asphalt mixture IDEAL-CT test.

Figure 4(a) illustrates that the IDEAL-CT test results indicate a higher maximum displacement for the HVA mixture compared to the SBSA mixture at a constant test temperature. Specifically, at -5°C, -10°C, and -15°C, the maximum displacement of the HVA mixture is 37.7%, 65.7%, and 74.4% higher, respectively. This suggests that the HVA mixture exhibits greater resistance to low-temperature deformation, with a more pronounced increase in maximum displacement as the temperature decreases. For the same type of asphalt mixture, as the temperature decreases, the maximum displacement of the asphalt mixture also decreases gradually. Specifically, the maximum displacement of the SBSA mixture decreases by 36.9% and 56.2% when the temperature drops from -5°C to -10°C and from -10°C to -15°C, respectively. Similarly, the maximum displacement of the HVA mixture decreases by 24.1% and 53.9% in the same temperature ranges.

Figure 4(b) shows that the fracture energy of the HVA mixture is significantly higher than that of the SBSA mixture at the same test temperature. At -5°C, -10°C, and -15°C, the fracture energy of HVA mixture is 1.53 times, 1.75 times, and 2.05 times to that of SBSA mixture, respectively. Lower temperatures significantly enhance the LTCR performance advantage of HVA mixtures. When the type of asphalt mixture is unchanged, the fracture energy of the mixture increases with the decrease in test temperature. The fracture energy of the SBS asphalt mixture decreases by 2714 $J\cdot m^{-2}$ and 4284.2 $J\cdot m^{-2}$ when the temperature decreases from -5°C to -10°C and from -10°C to -15°C, respectively. Similarly, the fracture energy of HVA mixture decreases by 1295.4 $J\cdot m^{-2}$ and 4688 $J\cdot m^{-2}$ in the same temperature ranges. It is evident that the HVA mixture exhibits greater crack resistance at low temperatures based on the analysis of fracture energy.

Figure 4(c) shows that, during the same type of asphalt mixtures, the $|m_{75}|$ value tends to increase gradually as the test temperature decreases. The SBS asphalt mixture $|m_{75}|$ value increases by 28.5 $KN\cdot mm^{-1}$ from -5C° to -10C°, and by 57.1$KN\cdot mm^{-1}$ from -10C° to -15C°. The HVA mixtures $|m_{75}|$ value increases by 0.4 $KN\cdot mm^{-1}$ and 36.3 $KN\cdot mm^{-1}$. The increase in SBSA mixtures is greater than that of HVA mixtures. Under the same test temperature, the fracture energy of HVA mixtures is significantly lower than that of SBSA mixtures. This indicates that after the occurrence of cracking, the cracks propagate speed of HVA mixture is slower.

Figure 4(d) shows that, under the same test temperature, the crack resistance index of HVA mixtures is much greater than that of SBSA mixtures. For example, at -15 C°, the crack resistance index of HVA mixtures is still 9.1 times that of SBS asphalt mixtures, indicating a significant advantage in crack resistance in low-temperature environments.

4. Conclusion

(1) This paper investigates the flexural and tensile properties of asphalt mixture, as well as its crack

resistance through a LTB test.

(2) The maximum bending strain of the SMA-10 mixture is 8390 με, which is 1.14 times and 1.86 times the strain of the SMA-20 mixture, and the maximum strain has similar patterns. This difference can be attributed to the larger particle size, which makes it challenging to form a dense structure. It could be concluded that SMA-10 and SMA-13 asphalt mixtures have better crack resistance at low temperatures.

(3) The HVA bending strain is approximately 2.2 times and 3.2 times that of the SBS-modified and 70# matrix asphalt mixture, which demonstrates significantly higher crack resistance ability. The 12% is the optimal dosage of a high-viscosity modifier, 12% HVA bending strain is 1.2 times that of a 10% dosage.

(4) When the temperature drops from -5°C to -10°C and -15°C, The maximum displacement of the SBS asphalt mixture decreases by 36.9% and 56.2%, while that of HVA mixture decreases by 24.1% and 53.9%.

At -5°C, -10°C and -15°C, the fracture energy of the HVA mixture is 1.53 times, 1.75 times, and 2.05 times that of SBSA mixture, and the HVA mixture exhibits greater crack resistance at low temperatures.

The $|m_{75}|$ value growth trend of HVA mixture is slight, which means that the cracks propagate speed is slower.

At -15 ℃, the crack resistance index of HVA mixtures is still 9.1 times that of SBS asphalt mixtures.

References

[1] Zhang Xia. Research on the Multi-scale Dynamic Behavior and Damage Evolution of Asphalt Bridge Deck Pavement under the Vehicle-bridge Coupling Effect [D]. Shijiazhuang Tiedao University, 2022.

[2] Wei H, Li L W. Application Research of Asphalt concrete composite structure in the bridge deck pavement [J]. 2015.DOI:10.2991/icmra-15.2015.251.

[3] Zhou Yan, Li Peilin. Fatigue Test Research on Composite Beams of Double-layer SMA Steel Bridge Deck Pavement [J]. Highways & Automotive Applications, 2013, (3): 181-185.

[4] Wan Taotao. Optimal Design of UK Cast-in-place Bridge Deck Pavement Mixture MA [J]. Journal of China & Foreign Highway, 2012, 32(6): 5. DOI: 10.3969/j.issn.1671-2579.2012.06.066.

[5] Xie Hongfeng. Application of High-performance Epoxy Resin Road Materials in Bridge Deck Pavement in China [C]. Proceedings of the National Highway Maintenance New Materials Application Technology Conference. 2015.

[6] Stenko M, Chawalwala A. Thin Polysulfide Epoxy Bridge Deck Overlays [J]. Transportation Research Record Journal of the Transportation Research Board, 2001, 1749:64-67.DOI:10.3141/1749-10.

[7] Dai Jingling. Comprehensive Evaluation and Scheme Optimization of Bridge Deck Waterproof Pavement System [D]. Xiangtan: Xiangtan University, 2020.

[8] Liao M, Luo R. Influencing factors and mechanisms of blistering in epoxy asphalt mixtures for steel deck pavements [J]. Construction and Building Materials, 2022, 316: 126036.

[9] Deep learning-augmented micro defect detection of plastic straw by machine vision.Micro 2023,3(2),484-495.

[10] Diao Xingjian, Chunhui Zhang. et.al. Learning Musical Representations for Music Performance Question Answering. In Findings of the Association for Computational Linguistics: EMNLP 2024, pp. 2803-2813. 2024.

[11] Diao Xingjian, Ming Cheng, Wayner Barrios, and SouYoung Jin. "Ft2tf: First-person statement text-to-talking face generation." arXiv preprint arXiv:2312.05430 (2023).

[12] Geng, Li-tao, et al. "Performance research of high-viscosity asphalt mixture as deck-paving materials for steel bridges." *Road Materials and Pavement Design* 18.1 (2017): 208-220.

[13] Wang, Jianwei, et al. "Application of High-Viscosity Modified Asphalt Mixture in Curved Bridge Pavement." *Sustainability* 15.4 (2023): 3411.

[14] American Society of Testing Materials. Standard test method for determination of cracking tolerance index of asphalt mixture using the indirect tensile cracking test at intermediate temperature: D8225—2019[S]. West Conshohocken: ASTM International, 2019.

[15] Xu Haijun, Tian Ping. Research on Low-temperature Anti-cracking Performance of Asphalt Mixture for Bridge Deck Pavement [J]. Technology of Highway and Transport, 2008(4): 2. DOI: 10.3969/j.issn.1009-6477.2008.04.022.

[16] Shi Yong. Research on the Performance of High-viscosity and High-elasticity Modified Asphalt and Mixtures for Bridge Deck Pavement Disease Repair [D]. Chongqing: Chongqing Jiaotong University, 2021.

Aerodynamic performance of blade based on owl trailing-edge serrated design

Ruoshui Zha*, Yi Han, Xinru Du, Li Li, Wei Liu and Dong Xu

Power China Northwest Engineering Corporation Limited, Xi'an, Shaanxi Province, China

* Corresponding author: 395473241@qq.com

Abstract—As the noise reduction induced by owl trailing-edge serrated design of wind turbine blade during operation has gradually become the industrial consensus, the corresponding effect of the aerodynamic performance on blade has still not been investigated systematically. In the present study, two types of bionic blades (B-A and B-B) with different serration lengths have been designed based on the NREL phase VI blade (W-B). The flow control mechanism and aerodynamic performance of blade under different incoming wind speeds have been studied through the numerical simulation approach. The findings indicate that, with the wind turbine's rotation speed held constant at 72 r/min, the pressure difference between the suction and pressure surfaces of the wind blades rises progressively as the incoming wind speed increases from 7m/s to 13m/s. This increase in pressure difference, in turn, leads to a corresponding rise in the torque exerted on blade. As the incoming wind speed is faster than 13m/s, the velocity gradient at the base of the bionic blade is higher than that of W-B, which increases the pressure difference of blade and promotes the aerodynamic performance of blade. The larger blade training-edge serrations size, the more obvious aerodynamic performance improvement of blade.

1. Introduction

As a vital element of wind turbine, blade optimization is increasingly drawing the focus of many researchers. It has been studied that owl trailing-edge serrated design of wind turbine blade can alleviate the aero-acoustic noise during the operation, due to its simple installation and with no additional mechanical drive required. However, the effect of this bionic design on the aerodynamic performance of the wind blade is still under exploration. An extensive experimental investigation has been conducted on wind turbine blades featuring symmetric NACA0012 airfoils and cambered NACA 65(12)-10 airfoils, incorporating a range of trailing-edge serration designs [1]. Measurements of the aerodynamic forces revealed that the serrations on the trailing-edge of cambered airfoils can significantly reduce the lift coefficient at low angles of attack, yet these serrations do not notably alter the stall behavior of the airfoils. A study of the aerodynamic performance of a NACA 643418 airfoil with trailing-edge serrations was conducted in both numerical and experimental approaches [2]. The apparent chord length was extended resulting from trailing-edge serrations, which triggered the incensement of the lift coefficient. The reduction in boundary layer thickness at the trailing-edge led to a decrease in the drag coefficient.Aerodynamic performance of NACA 65-010 blade with the inclined serrated trailing-edge was studied in the range of Re = $1\times10^5 \sim 2.8\times10^6$[3]. Lift-drag ratio for blades with an inclined trailing edge is much higher than that of blades with a blunt trailing edge at low Re. Conversely, at high Re, lift-drag ratio of the inclined trailing-edge blade is much lower than that of the blunt trailing-edge blade. The impact of adding trailing-edge serrations on the aerodynamic performance of a Wells turbine, which is composed of eight NACA 0015 blades, was studied through numerical simulation [4]. An increase in peak efficiency of 1.51% is noted for partial-span trailing-edge serrations, while full-span serrations result in a 1.22% improvement. The wind tunnel tests of lift, drag and surface pressure distribution of the airfoil were carried out to verify the influence of varying trailing-edge serrations on the aerodynamic performance [5]. The power coefficient calculated by using large-scale wind turbine design and analysis software BLADED for the wind blade with trailing-edge serrations increased by 1% ~3% in comparison with that of the normal blade. The impact of trailing-edge serrations on wind turbine aerodynamics was evaluated with an optimal design [6]. The serration shape was optimized using a statistical-numerical method to achieve significant aerodynamic enhancements. Given that the wavelength of trailing-edge serration corresponds to 5% of the tip chord length, the amplitude is 20% of the chord length, and the thickness of the serration plate is 1 mm, which is capable of enhancing the annual power generation by up to 12%. While research has been conducted on how trailing-edge serrations affect the aerodynamic performance of wind turbine blades, a systematic investigation is still needed to understand the wake distribution near the trailing edge and the pressure distribution along blade at various incoming wind speeds. The resulting change in the micro-scale flow field and the lift and drag forces could induce variations in blade aerodynamic characteristics and power of wind turbine.

In this study, the aerodynamic performance of the widely used owl trailing-edge serrated design centered around the classic NERL Phase VI wind turbine blade with S809 airfoil is investigated under different incoming wind speeds and serration lengths [7]. The computational fluid dynamic (CFD) technique using the RANS approach with the SST k-ω turbulence model is applied to simulate the mean flow field, turbulence kinetic energy (TKE) under the influence of owl trailing-edge serration. In addition, the distribution curves of blade torque and net power under different wind speeds are described. The results indicate owl trailing-edge serration shows an obvious increase in the aerodynamic performance when the wind speed exceeds 13m/s under the fixed rotation speed (72 rpm), in comparison with blade without trailing-edge serrated design. This study provides technical guidance for the efficient and low-noise operation of wind turbine blades, which might help promote the widespread use of owl trailing-edge serration during the development stage of wind turbine blades.

2. Blade design with owl trailing-edge serration

The prototype blade is chosen from the NREL Phase VI wind turbine, which consists of the S809 airfoil sections, the shape transition and the connection part to the hub [8]. The length from the tip of blade to the center of rotation is 5.029 m [9-13]. The rotation speed in general is set up to be 72 rpm. This paper focuses on the simulation of a blade with a conventional profile. The specific airfoil profiles, along with their respective chord lengths and twist angles at various locations across the span of blade, were thoroughly discussed in the preceding publication. Hence, this paper will not repeat it [14]. Two types of bionic serrations are designed in this study. One sawtooth amplitude is 0.1 c, forming the bionic blade A. The other sawtooth amplitude is 0.2 c, forming the bionic blade B. The amplitude-to-wavelength ratio of the sawtooth is 2.0 for each type. Both owl trailing-edge serrations are attached at the trailing-edge of wind turbine blades, along the spanwise direction from the location of 47% of blade radius (R) to 100% R, as illustrated in Figure 1.

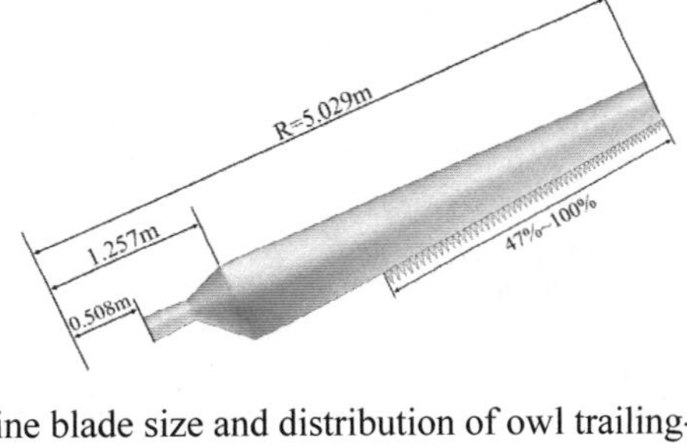

(a) Wind turbine blade size and distribution of owl trailing-edge serration

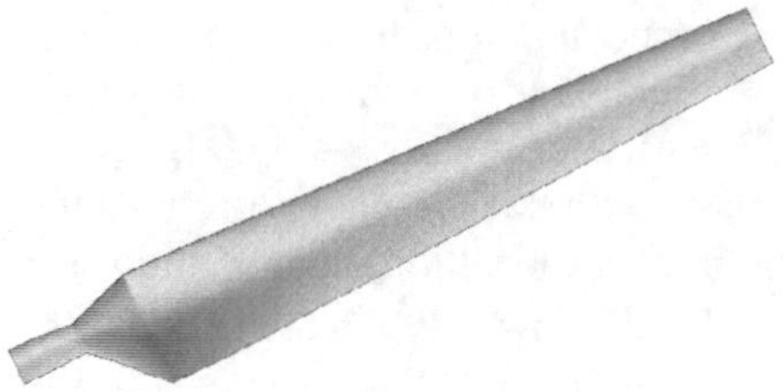

(b) Wind turbine blade_W-B

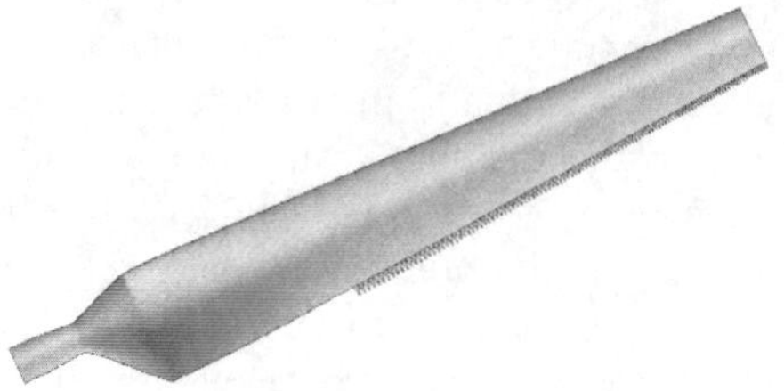

(c) Bionic blade A_B-A

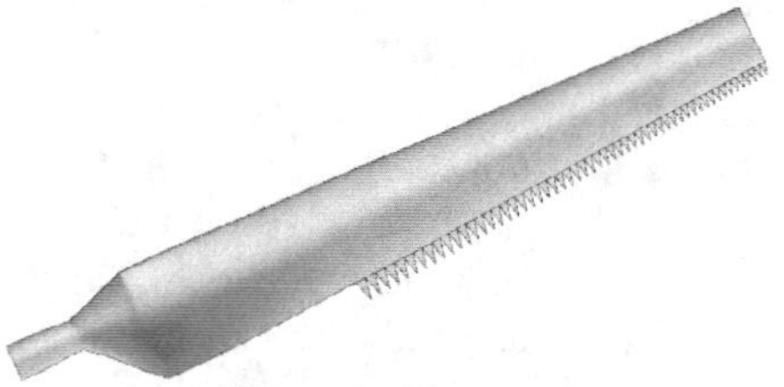

(d) Bionic blade B_B-B

Figure 1. Schematic diagrams for three-dimensional (3D) models of bionic blades.

3. Numerical method

3.1 Flow solver

The flow field is simulated by numerically solving the RANS Equations for incompressible airflow with the two-equation SST *k*-ω within the ANSYS Fluent software. The pressure-velocity coupling is handled using the PISO algorithm [9]. The TKE transport equation of SST *k*-ω is presented in Equation (1):

$$\frac{\partial}{\partial t}(\rho k)+\frac{\partial}{\partial x_j}\left[\rho u_j k-(\mu+\sigma_k\mu_t)\frac{\partial k}{\partial x_j}\right]=\tau_{tij}S_{ij}-\beta^*\rho\omega k \tag{1}$$

ω represents the specific turbulent dissipation rate:

$$\begin{aligned}&\frac{\partial}{\partial t}(\rho\omega)+\frac{\partial}{\partial x_j}\left[\rho u_j k-(\mu+\sigma_\omega\mu_t)\frac{\partial\omega}{\partial x_j}\right]=P_\omega-\beta\rho\omega^2\\&+2(1-\mathrm{F}_1)\frac{\rho\sigma_{\omega 2}}{\omega}\frac{\partial k\partial w}{\partial x_j\partial x_j}\end{aligned} \tag{2}$$

The eddy current viscosity model of Reynolds stress is shown in Equation (3):

$$\frac{\partial}{\partial t}(\rho k)+\frac{\partial}{\partial x_j}\left[\rho u_j k-(\mu+\sigma_k\mu_t)\frac{\partial k}{\partial x_j}\right]=\tau_{tij}S_{ij}-\beta^*\rho\omega k \tag{3}$$

In Equation (3), S_{ij} denotes the average deformation rate component and δ_{ij} is the Kronecker delta.

3.2 Computational domain and mesh generation

Due to the symmetrical feature of blade rotation, the semi-cylinder flow field is modelled considering the computational resource. Blade is positioned at a distance of 2R away from the inlet along streamwise direction. The flow field extends from the location of blade root to 4R along the spanwise direction. The downstream region behind the rotor blade is set up to 4R till the outlet. The simulated volume flow field is $\frac{1}{2}\times\pi(4R)^2\times(2R+5R)$ in total [15]. In computational domain, velocity inlet and pressure outlet are positioned at locations that are twice and five times the radius of blade, respectively, upstream and downstream. Additionally, the radius of the rotational area is set to four times the radius of blade. Hybrid grid approach is used such that triangular prismatic unstructured meshes are used in viscous regions near blade, and the tetrahedral structured meshes are used in other flow regions. The mesh refinement is performed near blade and the sawtooth regions to simulate details of boundary layer flow. Y^+ of blade surface area is less than 1 [16-17]. The mesh generation is conducted by using the ICEM embedded in the ANSYS Fluent, as shown in Figure 2.

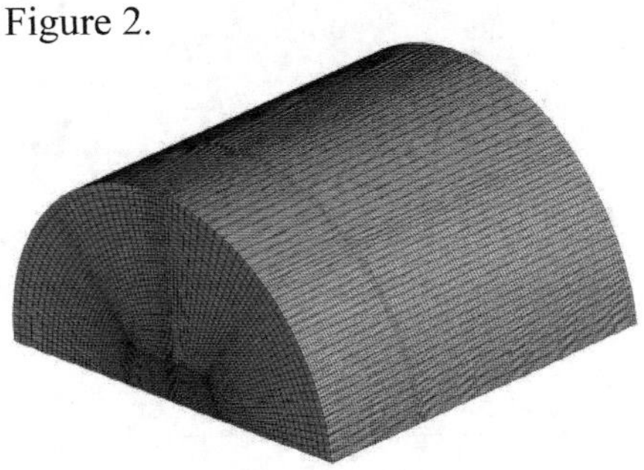

(a) Mesh of whole domain

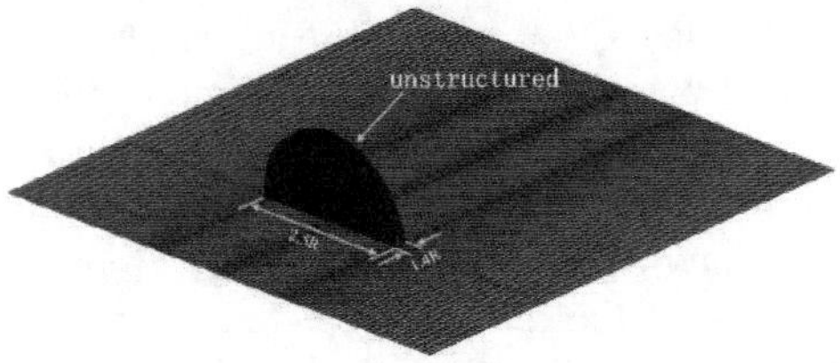

(b) Mesh of internal domain

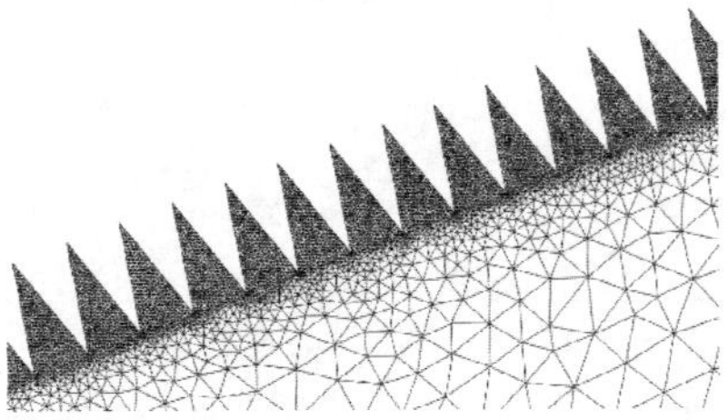

(c) Mesh of serration

Figure 2. Mesh generation for three computational domains.

3.3 Boundary conditions

As shown in Figure 3, in order to economize computational resources and achieve computational accuracy, only symmetry plane is treated as a rotating periodic boundary. Aim is to allow full development of blade wake and to reduce the influence of reflux. In this study, the calculations are performed using ANSYS Fluent, which adopts a steady-state solver with the SST k-ω. To maintain the precision and robustness of the computational analysis, the numerical simulation methodology is presented in Table 1. The rotational speed is 72 rpm, the inner and outer drainage basin and the blade are solid walls.

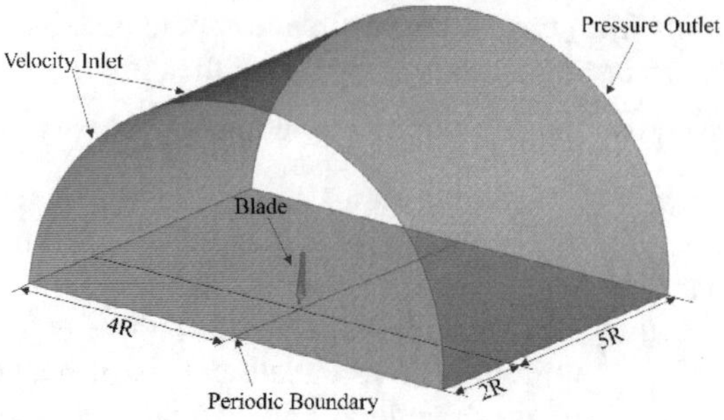

Figure 3. Boundary conditions for the computational domain.

Table 1 Simulation configuration details.

Coupling Scheme	PISO pressure-velocity
Gradient format	Green-Gauss Node Based gradient
pressure space discretization	second-order format,
momentum space discretization	second-order Upwind
transient Formulation	second-order Implicit

3.4 Validation of mesh independent

The grid independence is utilized to ascertain the suitable grid density. The verified analysis is presented in Table 2 by examining the torque of the wind speed at 10m/s for three different grid numbers. As depicted in Figure 4, with the increase in the number of grids, the variation in blade torque gradually stabilizes. Moreover, the maximum discrepancy in torque among the three types of blades is merely 0.23%. Therefore, the grid of 7.28×10^6, 8.37×10^6 and 8.52×10^6 are utilized respectively as the final grid number for the present simulations of the W-B, B-A, B-B. At the same moment, pressure coefficient distribution on different cross-sections of blade at 10m/s and 15m/s is plotted and compared with the previous experiment. Figure 5 presents the outcomes of the validation test for the current numerical simulation. Despite tiny differences between the numerical data obtained here and the existing experimental data, the numerical outcomes are in good agreement with the reference experimental results, thereby validating the efficiency and accuracy of the numerical approach utilized in this paper. [18-20].

Table 2 Results of grid-independence study.

W-B	Grid number/10^6	2.49	5.37	7.28
	Torque (Nm)	1357.732	1380.834	1384.108
B-A	Grid number/10^6	2.98	6.75	8.73
	Torque (Nm)	1295.782	1312.422	1314.146
B-B	Grid number/10^6	2.86	6.48	8.52
	Torque (Nm)	1184.392	1214.734	1219.408

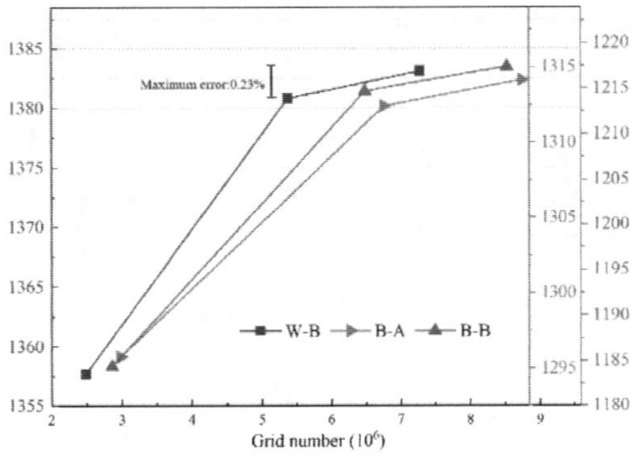

Figure 4. Error of grid-indendence.

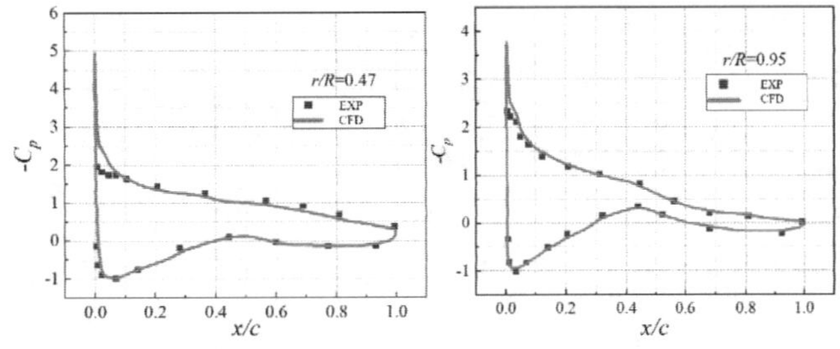

(a) U_∞ =10m/s

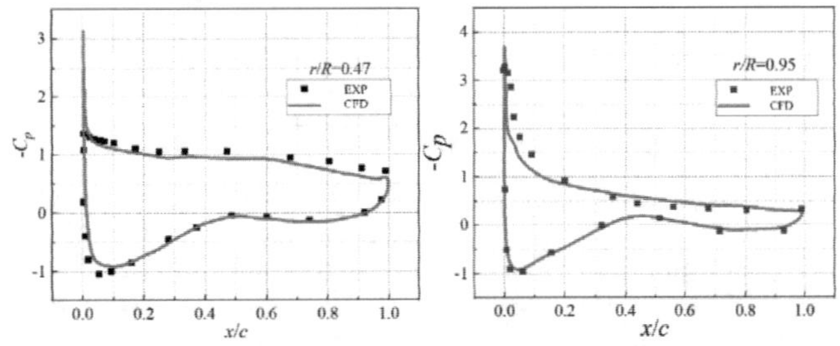

(b) U_∞ =15m/s

Figure 5. Validation test of the numerical simulation.

4. Numerical Result And Analysis

Firstly, the velocity distributions of the serration root in different sections of B-A and B-B are shown in Figure 6, in which W-B extracts the velocity distributions at the same position as the bionic blade. As shown in Figure 6, the flow velocity distributions of three blades at r/R=0.55, r/R=0.7, r/R = 0.8, r/R = 0.9 sections are shown respectively. It is discernible that the flow velocity distribution of different sections of the three types of blades in the upstream of blade is very similar, but there are obvious differences in the downstream of blade. From Figure 6(a), It is revealed that at 7m/s, B-A and B-B have a little speed loss at r/R = 0.55. The flow velocity at r/R=0.7 is larger, and the flow near the tip of the leaf is more stable. Figure 6(b) demonstrates that under a wind speed condition of 10 m/s, serrations significantly affect flow speed downstream of blade. The velocity is first promoted and then suppressed with the flow direction, while the opposite is true when r/R=0.55. When wind speed gradually increases to 13m/s, 15m/s and 20m/s, the downstream flow velocity of the three blades gradually tends to be the same, and there is a partial speed loss and delay attenuation phenomenon at the trailing-edge, as indicated in Figure 6(c, d, e). And at 20m/s, this phenomenon is most obvious. It can be inferred that at 7m/s and 10m/s, trailing-edge serration produce different aerodynamic properties for blades due to the different velocity gradients of different sections. High wind speed will intensify the flow separation at the trailing-edge of blade and delay flow separation position. In addition, it is evident that at a high wind speed, the velocity gradient at the trailing-edge of B-A and B-B is higher than that of the W-B, thus increasing the pressure difference of blade and improving the aerodynamic performance of blade.

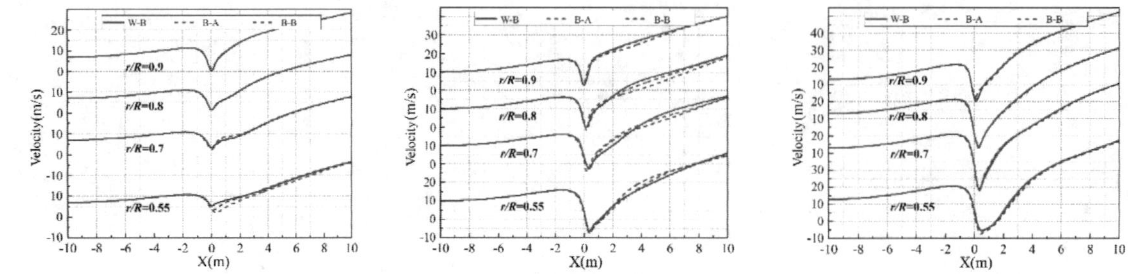

(a) U_∞ =7m/s (b) U_∞ =10m/s (c) U_∞ =13m/s

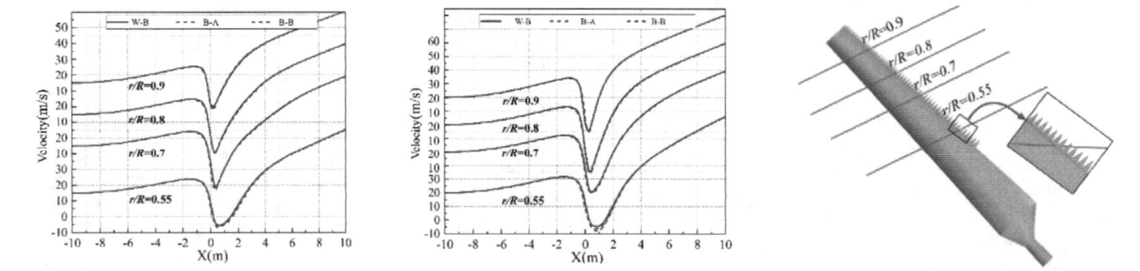

(d) U_∞ =15m/s (e) U_∞ =20m/s (f) Training-edge of cross-sectional

Figure 6. Flow speeds at different cross-sectional along blade spanwise directions for a given wind speed.

Figures 7-11 show the TKE at 55%-90% of cross-sectional along blade spanwise direction to reveal the flow mechanism under different wind speeds is compared for W-B, B-A, and B-B. Due to the significant difference in the distribution of TKE of each blade section under different wind speeds, different scale levels are adopted. With at 7m/s, the TKE at the trailing-edge of blade is weak and the range is smaller, so aerodynamic performance difference of the three blades is small at this wind speed. At 10m/s, the TKE of the suction surface in front of the spanwise direction for B-A and B-B is intensified, which makes the aerodynamic performance of the bionic blade low. As the wind speed rises to 13m/s, compared with W-B, B-A, and B-B effectively alleviate the TKE at the trailing-edge of blade. When wind speed continues to rise to 15m/s and 20m/s, the TKE of B-A and B-B significantly decreases the trailing-edge and suction surface of blade, and longer the length of serrations, the more obvious this phenomenon. This means that at a high wind speed, compared with W-B, the airflow can be better attached to the surface of B-A and B-B, and serrations effectively weaken the TKE of blade, thus improving the aerodynamic performance of blade. The weakening effect on TKE becomes more pronounced with higher wind speeds and longer serrations lengths.

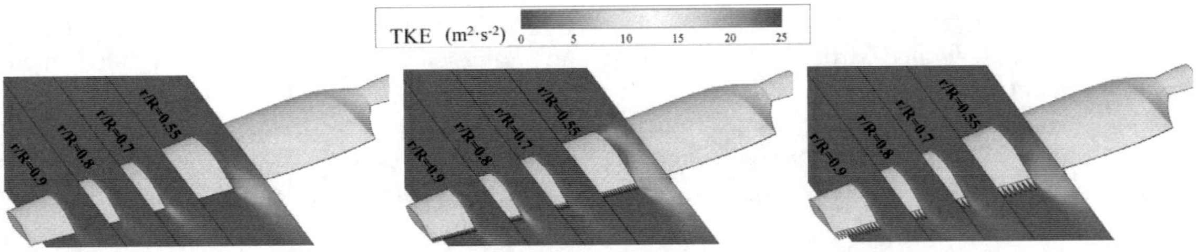

Figure 7. When U_∞ =7m/s, TKE variations captured at multiple spanwise positions.

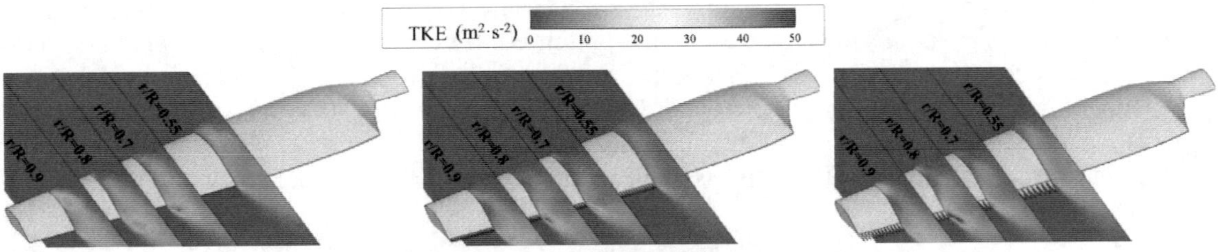

Figure 8. When U_∞ =10m/s, TKE variations captured at multiple spanwise positions.

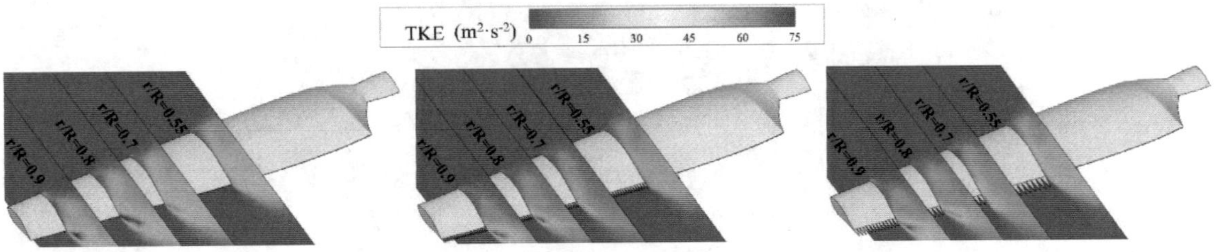

Figure 9. When U_∞ =13m/s, TKE variations captured at multiple spanwise positions.

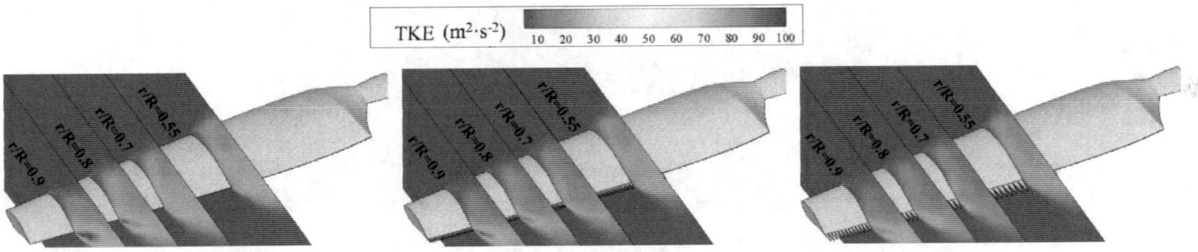

Figure 10. When U_∞ =15m/s, TKE variations captured at multiple spanwise positions.

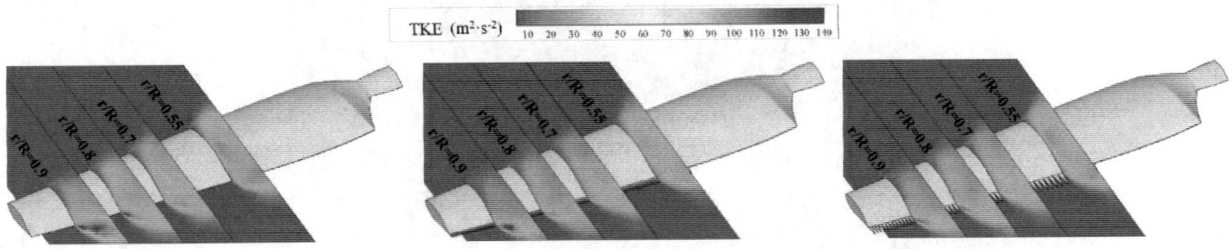

Figure 11. When U_∞ = 20m/s, TKE variations captured at multiple spanwise positions.

From Figure 11, it can be found that the serration structure can better decrease the intensity and range of TKE upon reaching a wind velocity of 20m/s, and it is most significant for 90% of blade spanwise direction. Therefore, Figures 12 and 13 show the vortices distribution and pressure coefficient distribution for 90% of blade spanwise direction of the three types of blades upon achieving a wind speed magnitude of 20m/s, respectively. Refer to Figure 12, due to the asymmetry and angle of attack of blade, airflow generates separated vortices at leading-edges and trailing-edges of blade. On suction side of blade, airflow separation initiates in the rear section, with a marked increase in the separation intensity of bubbles in the tail region. As distance from the trailing edge grows, the wake vortex structure expands rapidly. Meanwhile, the vortex structure near the sawtooth progressively fragments the large-scale vortices into numerous small-scale spiral vortices, driven by the presence of tooth tips.Because of serration structure, generation of leading-edge separation vortices is effectively suppressed while the shedding of training-edge wake vortices is significantly controlled. Also as shown in Figure 13, at the front of suction surface of blade, the pressure coefficients B-A and B-B are higher compared to W-B, and at the middle and rear of the pressure surfaces, the pressure coefficients of B-A and B-B are lower compared to B-B. This points to an increase in the pressure difference between the pressure and suction surfaces, thus enhancing the lift of blade and the torque of blade rotation. This aligns with serrations depicted in Figure 12, which has the effect of diminishing the intensity of vortices. Also as shown in Figures 12 and 13, the longer the serration, the more significant the effect of suppressing the generation of shedding vortices and increasing the pressure difference between the pressure surface of blade and the suction surface.

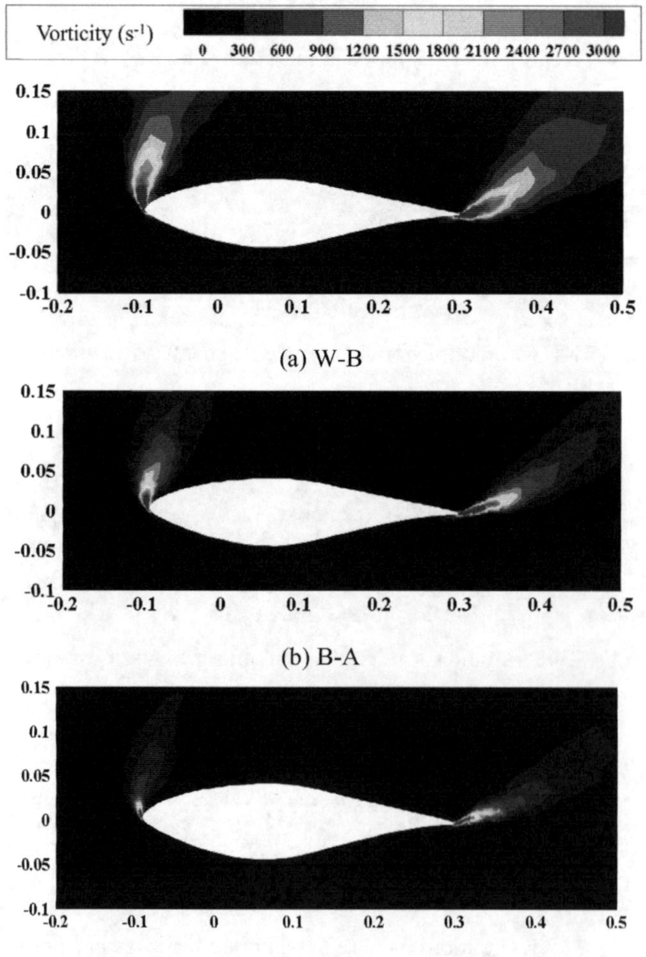

(a) W-B

(b) B-A

(c) B-B

Figure 12. Vortices distributions for three kinds of blades at r/R=0.9 spanwise direction when U_∞ =20m/s.

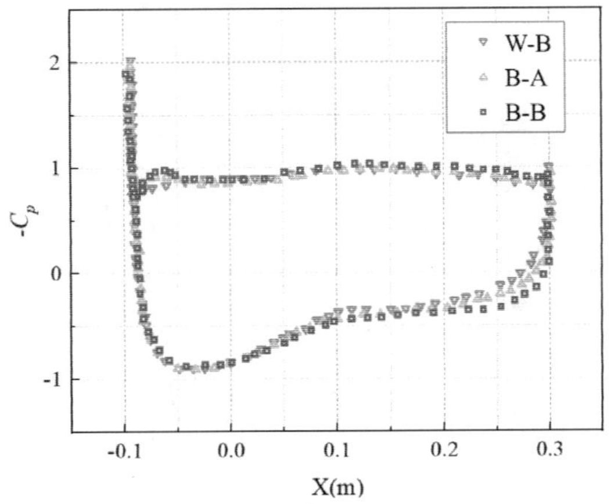

Figure 13. Pressure distributions for three kinds of blades at r/R=0.9 spanwise direction when U_∞ =20m/s.

Finally, torque and net power variation trends of W-B, B-A, and B-B under wind speeds of 7m/s, 10m/s, 13m/s, 15m/s, and 20m/s are used to evaluate the influence of different serrated blades on aerodynamic performance, as shown in Figure 14. The three blades are similar to that of the torque, first increasing with the increase of wind and then decreasing, showing a unimodal feature, but the distribution position of unimodal is different due to the existence of trailing-edge serration. At 7m/s, serration on trailing-edge of blades Exerts negligible influence on the aerodynamic performance characteristics of the wind turbine. However, at 10m/s the serration will reduce the torque of blade. When wind speed reaches 13m/s, serration can increase the torque of blade, and it can be found that B-A has a more significant effect on the torque. As wind speed continues to rise, torque and power of blade start to decline. This could be attributed to the significant TKE generated at blade tips under high wind speeds.Torque and power distributions of W-B, B-A, and B-B are consistent as the speed distribution increases under the same conditions. Upon achieving a wind speed magnitude of 13m/s, 15m/s, and 20m/s, power of blade gradually increases with increasing size of serration, so that it has better aerodynamic performance.

In summary, at 7m/s, changes in the length of serrations have minimal impact on aerodynamic behavio of wind turbines. However, as the serration length decreases, both the torque and power of the wind turbine reduce at 10m/s. When the wind speed increases to 13m/s, the increase in torque and power for the B-A and B-B configurations is the most substantial compared to the W-B configuration.

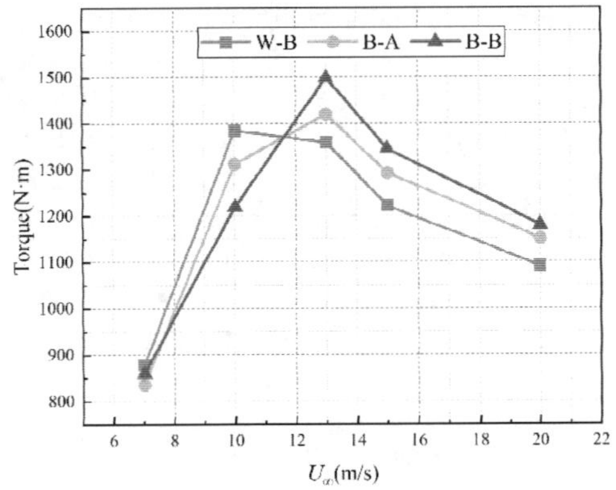

(a) Comparison of torque

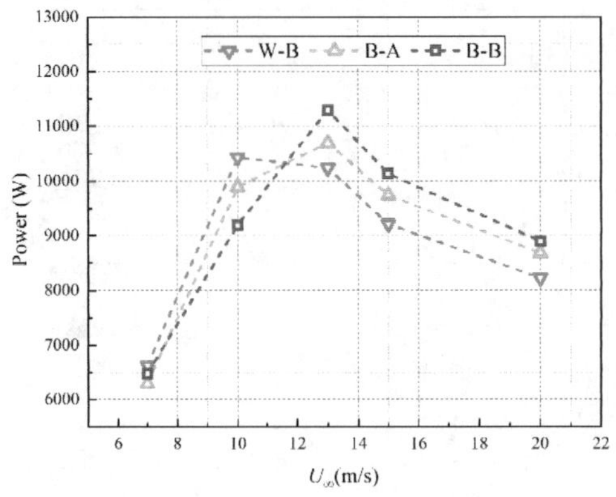

(b) Comparison of net power

Figure 14. Comparison of the torque and net power of three kinds of blades varies with wind speed.

5. Conclusions

In this study, inspired by the serration of the trailing-edge of an owl wing, bionic blade A (B-A) and bionic blade B (B-B) are constructed by installing different sizes of vortices of serration on the trailing-edge of blade. A numerical simulation is conducted to investigate the flow state of different sections of blades under different wind speeds. The effects of different blade lengths on the aerodynamic performance of blades are systematically studied. Based on the findings of this study, the following conclusions can be deduced:

(1) As wind speed gradually increases to 13m/s, 15m/s and 20m/s, the velocity gradient at the root of B-A and B-B is higher than that of W-B, thus increasing the pressure difference of blade and improving the aerodynamic performance of blade. At the same time, with the increase of blade extension distance, the lowest points of B-A and B-B velocity gradually move forward, and the velocity gradient decreases. This phenomenon is most obvious in blade tip r/R = 0.9 and B-B blade. With a further increase in wind speed, b-A and b-B significantly reduce the TKE of blade trailing-edge and suction surface, and this phenomenon is most obvious at r/R = 0.9 cross-sectional.

(2) The current numerical outcomes indicate that Upon reaching a wind velocity of 20m/s, serration at r/R=0.9 section can effectively inhibit the generation of leading-edge separation vortex, and significantly control the loss of wake vortex at the trailing-edge. At the same time, serration in this section has a larger pressure difference, so the lift force of blade increases, and the torque of blade increases. These phenomena become more pronounced as the size of the serrations increases.

(3) Torque and net power distribution of W-B, B-A, and B-B are consistent with speed, and the torque and net distribution curves of each blade are different under different wind speeds. When wind speed is 7m/s, the torque and net power of the bionic blade are not obvious compared with W-B. At a wind speed of 10 m/s, the bionic blade has a negative effect on the torque and net power of wind turbine. When wind speed is 13m/s, 15m/s and 20m/s, the torque and net power of blade gradually increase with the increase in the serration size, thus enhancing the aerodynamic performance of blade.

The bionic structure used in this paper has been widely applied to low-noise blades. In general, the bionic blade has better aerodynamic performance under high wind speeds, so it has certain reference significance for designing an efficient and low-noise wind turbine blade under certain working conditions.

Acknowledgment

This work was supported by the Innovation Capability Support Program of Shaanxi (Program No. 2024ZC-KJXX-103) and the Innovation Capability Support Program of Shaanxi (Program No. 2023-CX-TD-30).

References

[1] Liu, X., Kamliya Jawahar, H., Azarpeyvand, M. and Theunissen, R., 2017. Aerodynamic performance and wake development of airfoils with serrated trailing edges. AIAA Journal, 55(11), pp.3669-3680.
[2] Llorente, E. and Ragni, D., 2019. Trailing edge serrations effects on the aerodynamic performance of a NACA 643418. Wind Energy, 22(3), pp.392-406.
[3] Leng, L., Yang, A, L. and Chen, E, Y 2020. Influence of Reynolds number on aerodynamic and noise characteristics of blades with inclined sawtooth trailing edge. Journal of Engineering for Thermal Energy and Power, 35(07), pp.177-184+222.
[4] Alkhalifa, A.S., Uddin, M.N. and Atkinson, M., 2022. Aerodynamic Performance Analysis of Trailing Edge Serrations on a Wells Turbine. Energies, 15(23), p.9075.
[5] Zeng M W., Yang J W., Sun Z Y. and Wu X B., 2022. effect of sawtooth trailing edge on aerodynamic performance of wind turbine. Dongfang Electric Review, 36(01), pp.50-55.
[6] Qaissi, K., Elsayed, O., Faqir, M. and Essadiqi, E., 2023. Aerodynamic Optimization of Trailing-Edge-Serrations for a Wind Turbine Blade Using Taguchi Modified Additive Model. Energies, 16(3), p.1099.
[7] Li, D., Liu, X., Hu, F. and Wang, L., 2019. Effect of trailing-edge serrations on noise reduction in a coupled bionic aerofoil inspired by barn owls. Bioinspiration & biomimetics, 15(1), p.016009.
[8] Hand, M.M., Simms, D.A., Fingersh, L.J., Jager, D.W., Cotrell, J.R., Schreck, S. and Larwood, S.M., 2001. Unsteady aerodynamics experiment phase VI: wind tunnel test configurations and available data campaigns (No. NREL/TP-500-29955). National Renewable Energy Lab. (NREL), Golden, CO (the United States).
[9] Jonkman, J.M., 2003. Modeling of the UAE Wind Turbine for Refinement of FAST {_} AD (No. NREL/TP-500-34755). National Renewable Energy Lab. (NREL), Golden, CO (the United States).
[10] Hand, M.M., Simms, D.A., Fingersh, L.J., Jager, D.W., Cotrell, J.R., Schreck, S. and Larwood, S.M., 2001. Unsteady aerodynamics experiment phase VI: wind tunnel test configurations and available data campaigns (No. NREL/TP-500-29955). National Renewable Energy Lab. (NREL), Golden, CO (the United States).
[11] Giguere, P. and Selig, M.S., 1999. Design of a tapered and twisted blade for the NREL combined experiment rotor (No. NREL/SR-500-26173). National Renewable Energy Lab. (NREL), Golden, CO (the United States).
[12] Simms, D., Schreck, S., Hand, M. and Fingersh, L.J., 2001. NREL unsteady aerodynamics experiment in the NASA-Ames wind tunnel: a comparison of predictions to measurements (No. NREL/TP-500-29494). National Renewable Energy Lab. (NREL), Golden, CO (the United States).
[13] Sørensen, N.N., Michelsen, J.A. and Schreck, S., 2002. Navier–Stokes predictions of the NREL phase VI rotor in the NASA Ames 80 ft× 120 ft wind tunnel. Wind Energy: An International Journal for Progress and Applications in Wind Power Conversion Technology, 5(2-3), pp.151-169.
[14] Wang, Y., Li, G., Shen, S., Huang, D. and Zheng, Z., 2018. Investigation of aerodynamic performance of horizontal axis wind turbine by setting micro-cylinder in front of the blade leading edge. Energy, 143, pp.1107-1124.
[15] Jiang, R, F., Zhao, Z, Z., Feng, J, X., et al., 2021. Numerical Study on Flow Control of Wind Turbine Blade by Vortex Generators. Journal of Engineering Thermophysics, 42(12), pp.3170-3177.
[16] Kim, H.J., Lee, S. and Fujisawa, N., 2006. Computation of unsteady flow and aerodynamic noise of NACA0018 airfoil using large-eddy simulation. International Journal of Heat and Fluid Flow, 27(2), pp.229-242.
[17] Zhu, L., Wang, P. and Guo, Z., 2017. Performance evaluation of the general characteristics based off-lattice Boltzmann scheme and DUGKS for low-speed continuum flows. Journal of Computational Physics, 333, pp.227-246.
[18] Sørensen, N.N., Michelsen, J.A. and Schreck, S., 2002. Navier–Stokes predictions of the NREL phase VI rotor in the NASA Ames 80 ft× 120 ft wind tunnel. Wind Energy: An International Journal for Progress and Applications in Wind Power Conversion Technology, 5(2-3), pp.151-169.
[19] Yelmule, M.M. and Vsj, E.A., 2013. CFD predictions of NREL phase VI rotor experiments in NASA/AMES wind tunnel. International journal of renewable energy research, 3(2), pp.261-269.s
[20] Potsdam, M. and Mavriplis, D., 2009, January. Unstructured mesh CFD aerodynamic analysis of the NREL Phase VI rotor. In 47th AIAA Aerospace Sciences Meeting Including The New Horizons Forum and Aerospace Exposition (p. 1221).

Journal of Physics: Conference Series **3009** (2025) 012084 doi:10.1088/1742-6596/3009/1/012084

Modal analysis and resonance characteristics research of AGV forklift

Yafeng Jiang [1], Shuaimei Lian [2*]

[1] School of Intelligent Manufacturing and Information Engineering, Sichuan Technology & Business College, Chengdu 611830, China
[2] School of Intelligent Manufacturing, Chengdu Technological University, Chengdu, China

*Corresponding author's e-mail: 317760287@qq.com here

Abstract. AGV forklift truck is a very widely used intelligent warehousing logistics equipment; as a logistics handling tool with a very high degree of intelligence, its stability in the working process has a very high requirement. For the study of dynamic stability, the most common method is to conduct a modal analysis of the structure and analyze the natural vibration frequency characteristics of the structure so as to judge its stability. From the perspective of the dynamic stability of the whole vehicle, Ansys Workbench software was used to analyze and study the mode of a special AGV forklift truck in a crankshaft production workshop, extract the first four natural frequencies and the first four vibration modes, analyze the deformation characteristics of the entire AGV forklift truck in the working process, and predict the risk of cargo falling. Further, the excitation frequency from the ground is calculated, and through the comparison and analysis with the first four-order natural frequency of the vehicle, the conclusion that the AGV forklift will not resonate in normal operation is obtained, which provides a reference for the dynamic stability analysis of AGV forklift truck.

1. Introduction

AGV forklift is a common intelligent warehousing and logistics equipment that is widely used in factory workshop material handling, warehouse palletizing, and other links. It is a key component of the factory's intelligent logistics with a high degree of intelligence, strong carrying capacity, flexible movement, and other characteristics [1]. Stability is the basis of the normal operation of the AGV forklift. As an intelligent and unmanned logistics handling tool, it puts higher stability requirements [2]. Stability can be divided into horizontal and vertical, static, and dynamic [3]. In general engineering design, designers pay more attention to static stability and do not pay enough attention to dynamic stability. As a result, the dynamic characteristics of the designed equipment are poor. Modal analysis is the most basic linear dynamic analysis method used to analyze structures' natural frequency characteristics, such as natural frequencies, vibration modes, and so on. Modal analysis of the equipment structure can effectively prevent the structure from not resonating or vibrating at a fixed frequency.

In the actual working process of the AGV forklift, when the natural frequency of a component is the same or similar to the excitation frequency from the outside, the mechanism will resonate, resulting in a sharp increase in the dynamic load of the mechanism, resulting in structural breakage or greater deformation, and affecting the normal working performance of the mechanism[4,5]. At present, relevant researchers have done a lot of research work around the dynamic analysis of AGV forklift. For example, Du [6], Li [7], and Sun [8] carried out a modal analysis on the key parts of the AGV forklift, such as the door frame, fork, body, and so on, and put forward some improvement measures to reduce the vibration. Chu [9] designed a kind of omnidirectional AGV forklift suitable for a narrow logistics warehouse, and completed the modal analysis and harmonious response analysis of key components. Li and Cai [10] designed a new type of AGV forklift structure suitable for material transportation in the intelligent automatic production line. The modal analysis, dynamic simulation test, and analysis of the AGV forklift truck's stability were carried out using ABAQUS and ADAMS software. The weak position of the structure was obtained, and the improvement strategy was put forward. Lemerle P et al. [11] established the multi-body dynamics model of the forklift, studied its lateral stability during turning, and obtained the relationship between forklift stability and related parameters. Sun [12] and Zhu [13] completed the structural design of the AGV forklift according to different application scenarios, carried out modal analysis from two aspects of key components and the whole vehicle, and obtained the first six vibration modes and vibration frequencies, which verified the overall structural stability of the AGV forklift. Hu [2] carried out the modal analysis of the whole vehicle according to the three working conditions of the loom shaft loader, the empty shaft, and the full shaft, obtained the changing trend of the first six vibration modes and natural frequency of the robot under the three working conditions and verified its reliability.

A whole vehicle is a rigid system composed of multiple parts fixed together through different connection methods. The stiffness and constraints of these interconnected parts affect each other, which leads to differences between the modal characteristics of the whole vehicle and those of individual parts [12-13]. Therefore, it is very necessary to analyze the modal characteristics of the whole vehicle. Based on the existing research, this paper uses Ansys Workbench software to analyze and study the whole vehicle mode of a special AGV forklift truck in a crankshaft production workshop.

2. Modal analysis of forklift AGV whole vehicle

2.1. Model processing

As shown in Figure 1, the structure of the AGV forklift truck is mainly composed of the following parts: working device (including fork, slide frame, door frame, sprocket, chain, hydraulic cylinder, etc.), frame, automatic battery changing mechanism, shell, etc. The vehicle is composed of multiple components. Suppose all the models are imported into the modal analysis. In that case, the calculation amount will be increased, and some models will have an impact on the whole vehicle mode due to modeling errors. Therefore, it is necessary to simplify the entire vehicle model. Common simplification methods include removing unnecessary components, equivalent treatment of standard parts, retention of key components and features, consideration of interactions between components, and reasonable mesh partitioning and adjustment. Specifically, removing some components that do not have a significant impact on the analysis results, such as electrical components, and using the derivative function of Inventor software to derive components such as frames, door frames, and forks into multi-body parts, can reduce the number of contacts in the model. They can also simplify the driving unit as a solid steel structure, do not consider hydraulic cylinders and chains, and simplify the battery as an external load. In addition, the accuracy of the definition of the connection method between each component will also affect the analysis results. In actual conditions, the drive unit housing is connected to the frame through bolts, and the door frame is welded to the frame, so a bonded connection relationship can be added. The fork is hooked to the fork frame, and there is no relative displacement during normal operation, but it can actually move. Therefore, a No Separation connection relationship can be added. Reasonable model simplification can improve computational efficiency, but in simplifying the model, attention should be paid to retaining key factors that significantly impact the modal analysis results, such as the stiffness and mass distribution of the structure. Improper simplification may lead to distortion of the analysis results.

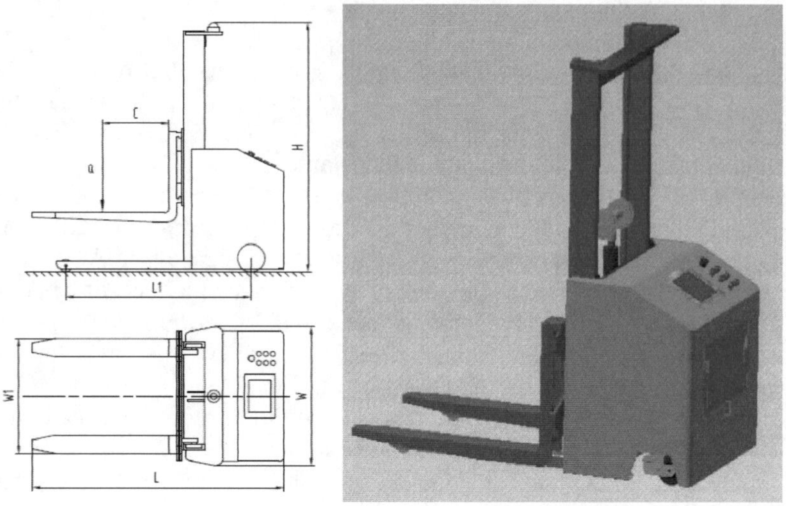

Figure 1. The structure and layout of the AGV forklift schematic diagram.

2.2. Modal analysis and result analysis

The most significant influence on structural dynamic characteristics is low-order vibration, which has practical engineering significance. Therefore, the first four order natural frequencies and AGV forklift truck vibration modes are extracted here for analysis and evaluation. According to the steps of modal analysis, the corresponding boundary and constraint conditions are set. The vehicle finite element model is calculated and solved, and the first four natural frequencies and vibration modes are extracted for analysis. Table 1 shows the natural frequencies of the first four orders of the vehicle, and Figure 2 to Figure 5 shows the first four vibration modes of the vehicle.

Table 1 The first four order natural frequency.

Order	*1*	*2*	*3*	*4*
Natural frequency (HZ)	22.907	24.443	36.113	49.197

As can be seen from Figure 2, the first-order vibration mode of the whole vehicle is mainly the overall bending of the gantry, which shows that the gantry swings back and forth in the Z-axis direction, the maximum swing is at the top of the gantry, and the deformation is about 4.87mm. As can be seen from Figure 3, there is a large bending deformation of the second-order vibration mode gantry and fork of the whole vehicle, in which the gantry swings back and forth in the X-axis direction. The maximum swing is also at the top of the gantry, the deformation is about 4.17mm, and the fork swings up and down in the Y-axis direction. The maximum swing appears at the front end of the fork, which has an impact on the stable state of the goods in the process of handling.

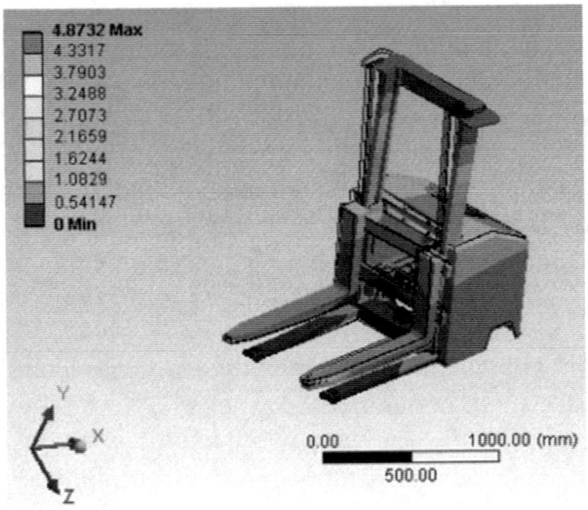

Figure 2. First-order vibration mode.

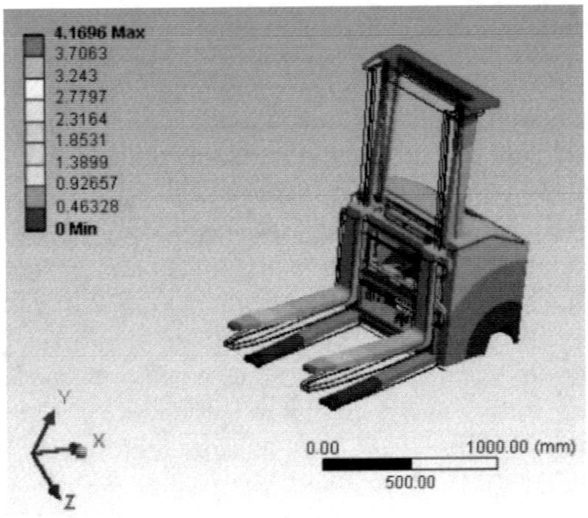

Figure 3. Second-order vibration mode.

As can be seen from Figure 4, the third-order vibration mode of the whole vehicle is mainly the overall torsion of the gantry and the bending of the fork, in which the gantry XZ plane twists and turns, the maximum torsion deformation occurs on the top of the gantry, the fork swings back and forth in the Y-axis direction, accompanied by a slight torsion. The maximum deformation appears in the front end of the fork, and the amount of deformation is about 6.95mm, which will also affect the stability of the goods in the handling process. As can be seen from Figure 5, the fourth vibration mode of the whole vehicle is that both the gantry and the fork are bent. The fork is the main deformation, in which the gantry swings back and forth in the X-axis direction. The maximum swing is also at the top of the gantry, the fork swings up and down in the direction of the Y-axis, the maximum swing appears at the front end of the fork, and the deformation is about 6.87mm, which will also affect the steady state of the goods in the process of handling.

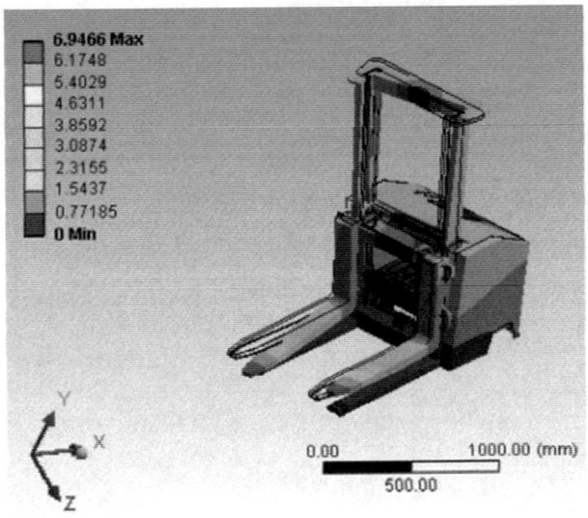

Figure 4. Third-order vibration mode.

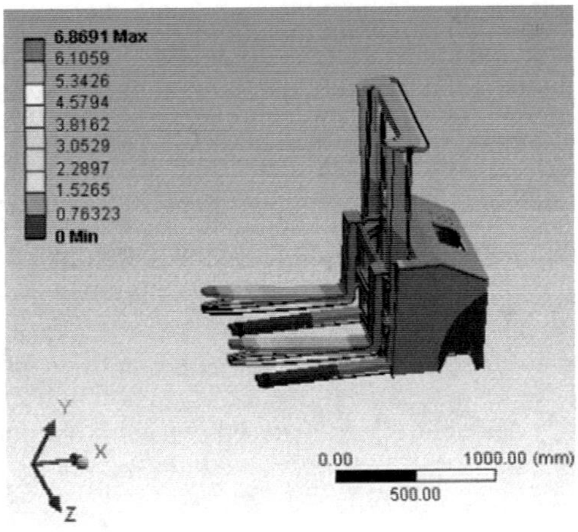

Figure 5. Fourth-order vibration mod.

To sum up, through the analysis of the first four vibration modes of the whole vehicle, the second, third, and fourth vibration modes of the whole vehicle have a large deformation in the front Y direction of the fork, and the maximum deformation is about 6.95mm. There is a risk that the goods will slip forward during the driving process of the AGV forklift. Therefore, the rigidity of the fork can be further improved, and its deformation can be reduced.

3. Resonance analysis

The vibration excitation factors of AGV forklift mainly come from the engine and road [13,14]. As for the research object of this paper, one is the excitation caused by the driving motor and hydraulic system motor. Because the motor runs smoothly at a constant speed, it is not discussed here. Second, in the process of driving, the wheel excitation caused by the road roughness is transmitted to the various parts of the AGV forklift through the supporting parts, which is also the main cause of noise and vibration.

In the driving process of the AGV forklift, the uneven road surface or unique form will stimulate the forklift, resulting in bending or torsional vibration of the parts of the AGV forklift. These excitation frequency devices have a fixed range of values for different walking speeds and road surface conditions. If the natural frequency range of the main components or the AGV forklift's whole vehicle overlaps, it is likely to cause resonance. According to the relevant provisions of the national standard GB/T 7031-1986 "expression method of vehicle vibration input road surface flatness", the excitation frequency produced by the road surface can be calculated according to the following formula:

$$f = \frac{v}{3.6\lambda}$$

Formula: v—Forklift driving speed, unit: m/s;

λ—road wavelength, unit: m; the λ value of the general flat road is 1 ~ 6.3m, and the minimum value of the poor road, such as the gravel road, is 0.32m.

The driving speed of the AGV forklift designed in this paper is 30-50m /min, the use place is inside the workshop, and the road surface has special requirements. Therefore, the flat pavement is selected for analysis, and the corresponding frequency range of pavement excitation can be obtained by calculating the formula.

$$f_{\max} = \frac{50}{3.6\times 1} = 13.88Hz, f_{\min} = \frac{50}{3.6\times 6.3} = 1.32Hz$$

After sorting out the calculation conditions and results, the pavement wavelength and excitation frequency shown in Table 2 are obtained.

Table 2 The road excitation frequency table.

Road condition	*Road wavelength λ (m)*	*Road excitation frequency (Hz)*
Flat pavement	1.0—6.3	1.32—13.88

It can be seen from the calculation results that the excitation frequency of the road is between 1.32-13.88Hz, so the natural frequency of the AGV forklift cannot overlap with this range so as to avoid resonance caused by the road excitation during the driving process of the equipment. The vehicle's natural frequency and the ground excitation frequency are summarized, as shown in Table 3.

Table 3 Comparison table of vehicle natural frequency and road excitation.

Natural frequency (Hz)	*Rank 1*	22.907
	Rank 2	24.443
	Rank 3	36.113
	Rank 4	49.197
Road excitation frequency (Hz)		1.32—13.88

As can be seen from Table 3, the natural frequency distribution of AGV forklift is between 22 and 50Hz, and the maximum frequency of road excitation is 13.88Hz. Therefore, road excitation will not cause resonance of the AGV forklift under normal operation.

4. Conclusion

Ansys Workbench was applied to the modal analysis of the AGV forklift, and the first four natural frequencies and the first four vibration modes of the vehicle were extracted. The first and second vibration modes were mainly bending deformation, and the third and fourth vibration modes were accompanied by torsional deformation in addition to bending deformation, which predicted the risk of cargo falling during the operation of the AGV forklift. The excitation frequency from the ground is analyzed and calculated, and compared with the natural frequency of the AGV forklift, the conclusion that AGV will not resonate under normal operating conditions is obtained, which provides reference data for product design.

Acknowledgement

Fund Project: Sichuan Technology&Business College campus level project funding; Project Number: 2024NC05.

References

[1] Eui-Jung Jung, Jae Yeon Choi, Sung Ho Hong, "Localization for an unmanned forklift in arefrigerated warehouse", 2016 13th International Conference on Ubiquitous Robots and Ambient Intelligence (URAI);
[2] Hu Ruiqing. Structural Optimization and Operational Stability Analysis of a Sliding Robot with Spinning Axis [D]. Donghua University, 2023;
[3] Beom K J, Woonchul S, Hak P J. Stability Analysis of Counter-balanced Forklift Trucks [J]. Journal of the Korean Society of Safety, 2015,30(2):1-8;
[4] Zhang Guangjie Research on AGV Structure Design and Path Planning Algorithm [D]. Jiangnan University,2023r;
[5] Cai Wenwei. Research on Forklift AGV Structure Design and Path Tracking Control [D]. South China University of Technology,2021;
[6] Du Xingjie, Wang Fei. Structural analysis of forklift AGV gantry system based on finite element method [J]. Science and Technology Wind, 2024, (10): 56-59.
[7] Li Kaiyan, Liu Qiang. Finite element analysis of forklift AGV fork structure for heavy load [J]. Science and Technology Wind, 2024, (11): 60-63.
[8] Sun Linlin. Research on Structure Design and Motion Control of Forklift AGV [D]. Shanghai University, 2022;
[9] Chu Weijie. Structural Optimization Design of AGV Forklift Truck [D]. Southeast University,2020;
[10] Li Yang, CAI Wenwei, LI Weiguang. Structure Design and Simulation Analysis of Forklift AGV [J]. Mechanical and Electrical Engineering Technology,2021,50(12):16-22;
[11] Lemerle P, Höppner O, Rebelle J. Dynamic stability of forklift trucks in cornering situations: parametrical analysis using a driving simulator [J]. Vehicle system dynamics, 2011, 49(10): 1673-1693;
[12] Sun Binbin. Study on Structural Design and Stability of Forklift AGV [D]. Henan University of Technology, 2017;
[13] Zhu Tianbao. Research on Structure and Stability of AGV car for Intelligent Storage [D]. Shanghai University of Engineering Science, 2020;
[14] Zhu Heming. Simulation study on mechanical properties and structural optimization design of forklift gantry [D]. Zhejiang University, 2021.

Journal of Physics: Conference Series **3009** (2025) 012085 doi:10.1088/1742-6596/3009/1/012085

Numerical simulation of defect detection in curved pipes using L (0,2) mode ultrasonic guided waves

Jingchi Yan [1], Liqing Yang *[1], Zhonghu Li [1], Xinyu Zhang [2], and Jinming Wang [1]

[1] School of Automation and Electrical Engineering, Inner Mongolia University of Science and Technology, Baotou 014010, Inner Mongolia, China
[2] School of Mechanical Engineering, Inner Mongolia University of Science and Technology, Baotou 014010, Inner Mongolia, China

*Corresponding author's e-mail: nmbtylq@163.com

Abstract. This research explores the potential of ultrasonic-guided waves for defect detection in bent pipes by utilizing COMSOL to create a numerical model of a defective bent pipe. Comparative analysis of different excitation frequencies identified the optimal detection frequency as 70 kHz. For a single-defect bent pipe, the influence of defect size and type on echo signals was studied, enabling axial defect localization through the echo method and assessment of defect severity based on echo amplitude. Furthermore, dual-defect simulations analyzed how defect size and location in the straight pipe section affect defect detection in the bend. Results demonstrate that the proposed simulation model effectively simulates guided wave propagation in bent pipes, providing a theoretical foundation for defect detection in practical engineering applications.

1. Introduction

Ultrasonic-guided waves are characterized by their long-range propagation, high speed, and efficiency. Numerous studies have been carried out on straight pipes [1][2], while reflections and transmissions occur in bends that are common in piping systems and bring complexity to wave propagation. This makes it challenging to identify defect signals in the complex echoes. Researchers worldwide have studied the propagation mechanism of ultrasonic-guided waves in curved pipes [3][4][5]. DEMMA et al. [6] explored the effects of varying bend radii on guided waves through numerical simulations and experiments. NISHINO et al. [7] established a relationship between bend angle and mode conversion efficiency. ZHU et al. [8] investigated the energy flow density at the elbow and used L (0,1) modes to detect the elbow defects. HE et al. [9] verified the effectiveness of guided wave detection for elbow defects using finite element simulation and experiments.

This paper analyses the feasibility of detecting defects in L (0,2) bent pipes, selecting the appropriate detection frequency through frequency comparison tests, using the echo method for defect localization, and researching the effects of defect size, type, and location on the detection sensitivity. This provides insights for multi-defect bent pipe detection.

2. Ultrasonic-guided wave detection theory

2.1 Dispersion characteristics and modal transitions

Due to the effects of anisotropy, discontinuity, and defects of the medium, the ultrasonic guided wave will be dispersed when it propagates in the pipe [10], i.e.

$$\left(\begin{bmatrix} 0 & K_1-\omega^2 M \\ K_1-\omega^2 M & i\left(K_2-K_2^T\right) \end{bmatrix} - k\begin{bmatrix} K_1-\omega^2 M & 0 \\ 0 & K_3 \end{bmatrix}\right)\begin{bmatrix} U_1 \\ kU_1 \end{bmatrix} = \begin{bmatrix} 0 \\ 0 \end{bmatrix} \tag{1}$$

Journal of Physics: Conference Series **3009** (2025) 012085 doi:10.1088/1742-6596/3009/1/012085

The dispersion equation is utilized to derive the dispersion curve in a particular pipe to guide in selecting the appropriate mode and frequency. Figure 1 shows the dispersion curve for a steel pipe with dimensions of 108 mm × 4 mm (OD × wall thickness).

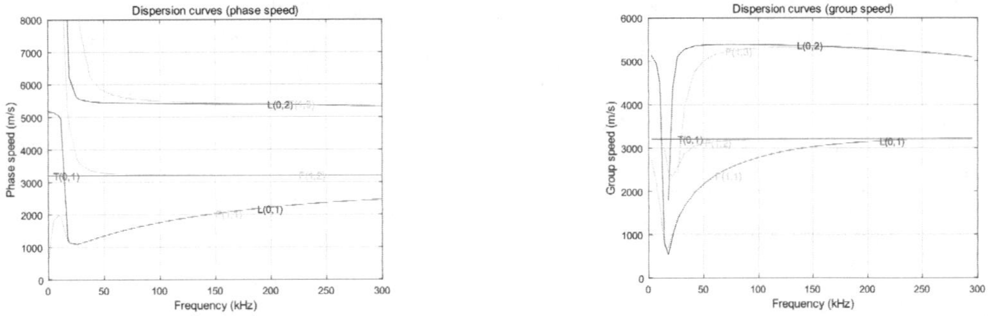

(a) Phase speed (b) Group speed

Figure 1. 108 mm × 4 mm (outer diameter × wall thickness) dispersion curve.

There are three modes of ultrasonic guided waves propagating in the pipe, namely axisymmetric L and T modes and non-axisymmetric F modes. When the guided wave passes through elbows or defects in the pipe, reflection and transmission will occur, and then the specified modes will be converted into several different modes, i.e., the phenomenon of mode conversion.

2.2 Ultrasonic guided wave detection principle of pipe bend defects

The excitation transducer generates ultrasonic-guided waves in the pipeline, which are reflected and mode-shifted when they encounter defects and discontinuities in the propagation process. The receiver transducer receives the echo signal and analyses it to determine whether there are defects in the pipe and where they are located.

$$C = \frac{2L - \Delta L}{t_3 - t_1} \quad (2)$$

where ΔL is the distance between the excitation end and the receiving end, L is the total length of the pipe, d is the defect location, t_1 is the excitation wave crest time, and t_2 is the defect wave crest time. The defect location d can be expressed as:

$$d = C \times (t_2 - t_1) \div 2 \quad (3)$$

3. Finite element model establishment and parameter setting

3.1 Simulation modeling

The pipe consists of two sections of straight pipe with a length of 1, 000 mm and an elbow with a bending angle of 90°. A 108 mm × 4 mm (outer diameter × wall thickness) pipe has a bend radius of 305 mm. The material of the pipe adopts Q235 steel, with Young's modulus of $E = 210GPa$, a Poisson's ratio of $\mu = 0.3$, and a density of $\rho = 7859kg/m^3$. Grid division, to simulate the real guided wave propagation process, here takes 1/12; the time step, affecting the accuracy of the calculation results, is 1 μs. We set two common defects, penetrating hole defects and crack defects. For the hole defects, the damage degree is changed by the radius of the hole; crack defects are set up with an axial width of 5 mm, circumferential length of 1/10, and 1/5 of the circumference of the cylindrical surface of the pipeline respectively, and two different lengths of penetrating crack defects are set, too. Figure 2 is a schematic diagram of the pipeline defects.

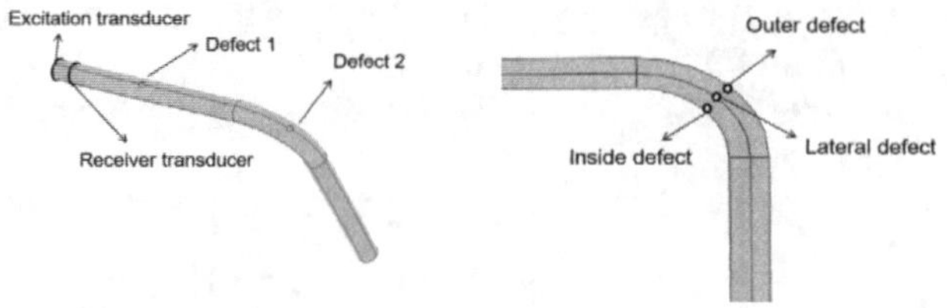

Figure 2. Schematic diagram of defect setting.

3.2 Excitation signal parameter settings

Sixteen signal excitation points are uniformly distributed circumferentially to the pipe to apply axially specified displacements to simulate L (0,2) modal-guided waves. The other end is set as a fixed boundary. Signal-receiving points are set at 100 mm from the excitation points, and the setup is uniformly arranged in the circumferential direction. The signal modulated by the Hanning window is as follows:

$$f(t)=0.5\left(1-\cos\frac{2\pi ft}{n}\right)\sin\left(2\pi ft\right),0<t<n/f \tag{4}$$

where n =5; f is the center frequency.

4. Numerical simulation of bend defect detection

4.1 Excitation frequency selection

The dispersion curve in Figure 1 shows that the L (0,2) mode-guided wave exhibits dispersion in certain frequency bands, which can interfere with ultrasonic-guided wave detection. Therefore, selecting an appropriate excitation frequency is essential for effective pipeline damage detection. The L (0,2) mode has less dispersion in the range of 30 kHz~80 kHz. An excitation signal from 30 to 80 kHz with a step of 10 kHz and a period of 5 was used for comparison to select the best detection frequency. The receiver transducer is set at 200 mm. The data was extracted by using MATLAB and plotted in the time domain, as shown in Figure 3.

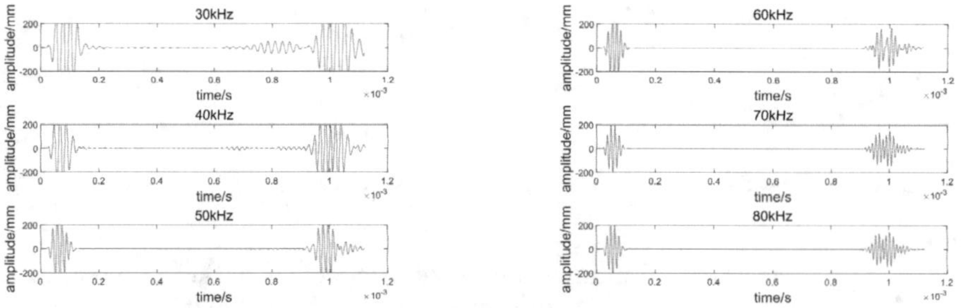

Figure 3. Time-domain plots of damage-free bends at different frequencies.

From the time domain in Figure 3, when the excitation frequency is a low-frequency signal, there is a mode conversion phenomenon that generates clutter, which affects the detection effect, but it will gradually decrease with the increase of frequency. Higher excitation frequency end echo signal will appear a lot of overlap, which is not conducive to signal detection.

4.2 Numerical simulation of defect detection in single-defect bends

(1) Detection of round hole defects at elbows

Penetrating circular hole defects with radii of 5 mm, 10 mm, and 15 mm are set up at the elbow, and Figure 4 gives the echo signal maps of circular hole defects of different sizes at the elbow with the excitation signal calculated by using COMSOL at a frequency of 70 kHz.

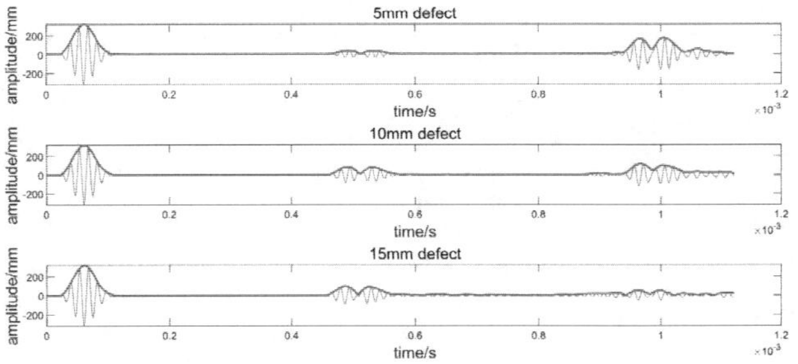

Figure 4. Echo signal of the defective round hole at the elbow.

As shown in the defect echo signal curve in Figure 4, a clear defect echo appears after the excitation signal. As the defect size increases, the defect echo amplitude increases, while the end-face echo amplitude decreases. This is because the guided wave reflects more energy when encountering defects, and after passing through the defects, energy attenuation accelerates, leading to a decrease in the end-face echo.

(2) Circumferential defects at elbows

Numerical simulations were conducted at a frequency of 70 kHz. The extracted simulated data was used to plot the time-domain signal, and the envelope was generated via the Hilbert transform. Figure 5 illustrates the defect echo signal curves for various circumferential angles.

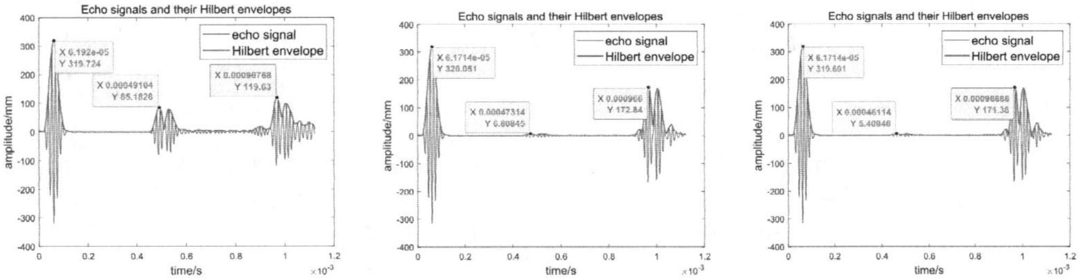

(a) Outer side (b) Sidebar (c) Inner side

Figure 5. Defect echo signal curve at different circumferential angles.

In Figure 5 (a), three wave packets are observed: the first corresponds to the excitation signal, the second to the defect echo signal, and the third to the end-face echo signal. The peak times for these signals are ($t_1 = 61.92$ μs), ($t_2 = 490$ μs), and ($t_3 = 967$ μs), respectively. Using Equation (2), the calculated actual guided wave speed is 5, 361.24 m/s and the theoretical speed is 5, 472 m/s. By substituting the guided wave speed into Equation (3), the defect position is calculated as 1, 147.5 mm, while the actual defect position is 1, 178.8 mm, yielding an error of 2.65%. This demonstrates accurate axial defect positioning and confirms the model's validity and reliability.

Figure 5 demonstrates that defect echo amplitude is highest on the outer side of the elbow compared to the lateral and medial sides, indicating guided wave energy concentration and higher detection sensitivity on the outer side. The end-face echo amplitude decreases as defect location changes, with outer defects showing significantly smaller end-face echo amplitudes than lateral and medial defects.

4.3 Numerical simulation of double defective bends

In practical engineering applications, there will be more than one defect existing at the same time. This section sets up double defects, by changing the defect size and location of the straight section.

(1) Influence of defect size of round hole in straight pipe section on defects at elbows

Based on the undamaged bends, penetrating circular hole defects with radii of 5 mm, 10 mm, and 15 mm are set up in the straight pipe section at 500 mm from the excitation transducer, and penetrating circular hole defects of 10 mm are always set up at the bends. The defect signal amplitude is extracted. Table 1 shows the defect echo signal amplitude.

Table 1. Defect echo signal amplitude.

	5 mm defect	5 mm defect	5 mm defect
Defect 1 amplitude	7.66036	14.4959	19.2234
Defect 2 amplitude	79.757	73.2902	74.7618
Defect 2 reflection coefficient	0.262	0.246	0.24

The ratio of the peak value of the wave peak of Defect 2 to the peak value of the excitation wave peak is taken as the reflection coefficient. As Defect 1 increases, the reflection coefficient of Defect 2 decreases, and the degree of damage of Defect 1 can be judged from this.

(2) The effect of the location of round hole defects on the elbow at the defective

Circular hole defects are set 500 mm, 600 mm, 700 mm, and 800 mm away from the excitation transducer, and a 10 mm penetrating circular hole defect is set at the elbow. The defect echo amplitude is shown in Table 2.

Table 2. Effect of location of defects in straight sections on defects at elbows.

	500 mm	600 mm	700 mm	800 mm
Defect 1 amplitude	16.0051	16.219	16.255	15.7064
Defect 2 amplitude	78.747	80.6019	80.26	75.399

From Table 2, straight pipe section defect 1 and the elbow at the defect 2 distance for the defect 2 amplitude do not have a greater impact but have a greater impact for the end face echo amplitude. With Defect 1 close to Defect 2, the end face echo amplitude gradually decreased.

5. Conclusion

This paper establishes a defect detection model for pipe bends by using finite elements, identifying 30-80 kHz as the low-frequency scattering range and 70 kHz as the optimal detection frequency. Simulations on single-defect elbows show that defect size and type significantly impact the defective echo and end-face echo amplitudes, enabling precise defect localization and damage assessment via echo variations. For double-defect pipes, larger straight-section defects reduce the elbow defect reflection coefficient, while the distance between straight-section and elbow defects primarily affects end-face echo amplitude with minimal impact on elbow defect echoes.

References

[1] LI Z. H., LI J., WANG J. M., et al. Numerical Simulation of Circumferential Positioning Defects in Piping from Ultrasonic Guiding Wave Detection [J]. PROCESS EQUIPMENT & PIPING, 2022, 59 (01): 81-87.
[2] LIU D. N., LI F. C., and SU W. S. Ultrasonic Guided Wave Propagation Characteristic and Damage Localization in Liquid Filled Pipes [J]. NOISE AND VIBRATION CONTROL, 2023, 43 (04): 135-141.
[3] CHEN L., XU J. C., WU J., et al. Numerical Simulation of Ultrasonic Conductive Waves on Defective Elbows [J]. Machine Building & Automation, 2021, 50 (06): 64-67.
[4] WANG C., ZHAO K., and LIU J. Y. Influence of elbow defect size and position on ultrasonic propagation characteristics [J]. JOURNAL OF APPLIED ACOUSTICS, 2023, 42 (04): 710-721.
[5] GAO L. Y., WANG J. J., and WU X. J. Defect detection experiment of a right-angle bend based on open magnetostrictive guided wave sensor [J]. Corrosion & Protection in Petrochemical Industry, 2023, 40 (4): 43-46.
[6] Demma A., Cawley P., and M. J. S. Mode Conversion of Longitudinal and Torsional Guided Modes due to Pipe Bends [J]. Review of Progress in Quantitative Nondestructive Evaluation, 2001, 20: 172-179.

[7] NISHINO H., YOSHIDA K., CHO H., et al. Propagation phenomena of wideband guided waves in a bent pipe [J]. Ultrasonics, 2006, (44): 1, 139-1, 143.
[8] ZHU L. X., ZHANG H. B., WANG Y. M., et al. Defect detection in pipe elbow using L (0,1) mode guided wave [J]. JOURNAL OF NAVAL UNIVERSITY OF ENGINEERING, 2022, 34 (02): 62-68.
[9] HE C. F., LIU Z. H., and SUN Y. X. Experimental Study on Defect Detection in Curved Pipes Using Ultrasonic Guided Waves Technique [J]. China Mechanical Engineering, 2005, (18): 1, 662-1, 665.
[10] ZHU L. X., WANG Y. M., and ZHAO S. Influence of Geometric Parameters on Propagation Characteristic of Guided Waves in Curved pipe [J]. JOURNAL OF NAVAL UNIVERSITY OF ENGINEERING, 2021, 33 (02): 57-63.

Journal of Physics: Conference Series **3009** (2025) 012086 doi:10.1088/1742-6596/3009/1/012086

Failure analysis of cracking of 304 stainless steel pipeline at the outlet of heat exchanger in ethylene plant

Xiaowei Li[1, 2], Xiaoqiang Hong [3], Lei Cheng[1, 2] and Chenyang Du[1, 2,*]

[1] China Special Equipment Inspection and Research Institute, Beijing 100029, China.
[2] Technology Innovation Center of Risk Prevention and Control of Refining and Chemical Equipment for State Market Regulation, Beijing 100029, China.
[3] ZIBO QIXIANG TENGDA CHEMICAL CO., LTD., Zibo City, Shandong Province 255400, China.

*Correspondence: Chenyang Du, chenyangdu@126.com

Abstract: The 304 stainless steel pipeline at the outlet of the ethylene heat exchanger in a petrochemical plant cracked and failed in use. Failure analysis was carried out to determine the reason for the failure and ensure the safe, stable, and continuous operation of the plant. Through macroscopic examination, chemical composition analysis, metallographic analysis, hardness test, scanning electron microscope analysis, energy spectrum analysis, etc., this paper finds that the main reason for cracking of stainless steel pipes is material sensitization-intergranular corrosion and chloride stress corrosion cracking failure occurs due to the invasion of chloride ions in superimposed media. It gives some suggestions to avoid similar problems in the future. Hopefully, this fault analysis can provide a reference for the failure of similar stainless steel-related equipment in the future.

1 Introduction

Stainless steel has high resistance to rust and corrosion, so it has a wide range of applications in various industries [1-3], and in petrochemical enterprises, it is mainly used in the environment with corrosive media. Stainless steel's excellent rust resistance and corrosion resistance is mainly due to its composition of the Cr element, which will form a layer of chromium-rich oxide film on the surface of the stainless steel, which is also a layer of corrosion media to prevent erosion of its internal passivation film [4].

According to its composition structure, stainless steel can be divided into austenitic stainless steel, ferritic stainless steel, martensitic stainless steel, and duplex steel [5]. Among them, austenitic stainless steel is mainly composed of chromium and nickel, and its structure is austenite at room temperature. Increasing chromium content increases the hardness of austenitic stainless steel, and the plasticity and toughness of austenitic stainless steel can be improved by adding nickel to make it a face-centered structure. However, austenitic stainless steel is prone to intergranular corrosion [6, 7], and measures such as ultra-low carbon content, increasing carbide-forming elements (such as titanium), and solid solution treatment can be adopted to reduce the possibility of grain boundary corrosion.

The stainless steel pipeline at the outlet of the heat exchanger of an ethylene plant in a petrochemical enterprise cracked and failed in use. The pipeline was put into use in July 2007, and it was found to be cracked and failed during inspection in July 2024. The medium inside the pipeline is cracked gas (containing C1-C9 and H2, with water content of about 20% and S content of about 0.03%), working temperature of 350°C, operating pressure of 0.3 MPa, material of 30408, and pipeline diameter of Φ610 mm and nominal wall thickness of 7.0 mm.

2 Experimental methods and apparatus

This failure analysis program carries out macro-inspection, material chemical composition analysis, metallographic analysis, hardness testing, scanning electron microscopy analysis, and energy spectrum analysis respectively.

SPECTROMAXx photoelectric direct reading spectrometer was used for chemical composition analysis. AXIO Imager A2m metallographic microscope was used for metallographic analysis. Hardness analysis was performed using a FALCON 600 Vickers hardness tester. Scanning Electron Microscope (SEM) instrument was EVO 10 / AZteLiveOne30 environmental scanning electron microscope. Energy Spectrometer (ESI) analysis was carried out by Ultim Max 65 energy spectrometer.

3 Experimental result

3.1 Macroscopic examination

Figure. 1 shows the overall shape of the crack in the cracked pipe. The pink part is the location of the crack, which is the full picture of the crack after penetration testing. Cracks are located in the elbow area of the pipe, expanding radially along the pipe, and the length of the crack is about 110 mm. The crack is not straight and obvious meandering and bifurcation can be observed. Figure. 2 is a partial enlargement of the crack. The figure can be seen on the crack's intermittent yellow-brown pitting corrosion traces, while the crack can also be seen in the vicinity of the area of scattered pitting corrosion traces.

Figure. 1 Overall macroscopic morphology of cracks.

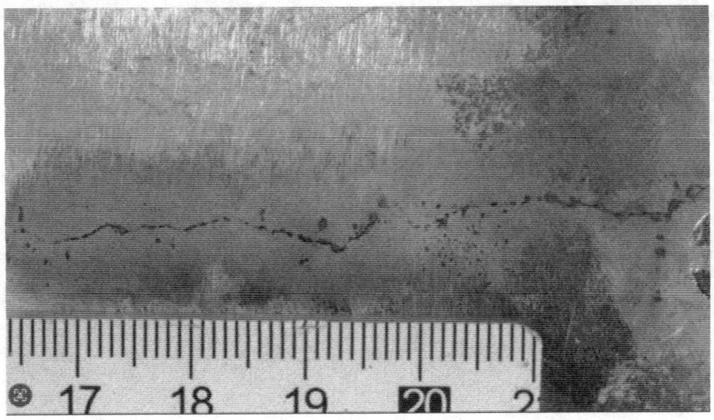

Figure. 2 Local morphology of cracks.

Journal of Physics: Conference Series **3009** (2025) 012086 doi:10.1088/1742-6596/3009/1/012086

3.2 Chemical composition of materials

We analyzed the chemical composition of cracked pipeline samples to determine whether the material composition of the pipeline meets the standard requirements. The analysis results are shown in Table 1 below. According to the results of material composition, the chemical composition analysis results of pipeline body material and weld material meet the requirements of GB/T 12771-2019 welded stainless steel pipes for liquid delivery, but the contents of Cr and Ni in base metal are lower than those in the weld.

Table. 1 Analysis results of the chemical composition of the cracked pipeline.

Chemical composition	C	Si	Mn	P	S	Cr	Ni
GB/T 12771-2019	≤0.08	≤1.00	≤2.00	≤0.040	≤0.030	18.00-20.00	8.00-11.00
Parent metal	0.07	0.47	1.15	0.011	0.01	18.27	8.26
Welding seam	0.06	0.60	1.89	0.035	0.001	19.96	9.35

3.3 Metallographic analysis

The cracked pipeline was sampled and analyzed by microscope. Figure. 3 shows the metallographic structure at the crack. Figure. 3(a) displays the metallographic structure of cracks on the inner wall of the pipeline. The figure shows that the main crack is wide and there are many secondary cracks in the branches of the main crack. There are slight corrosion marks on the inner wall of the pipeline, and obvious grain shedding around the crack. Figure. 3(b) shows the metallographic structure of the crack near the inner wall of the pipeline. From this figure, it can be seen that the overall shape of the crack is dendritic and there are a lot of crystal grains falling off around the crack. Crack propagation along the grain. Figure. 3(c) shows the metallographic structure of cracks at the weld on the inner wall of the pipeline. From the figure, it can be seen that the phenomena of cracking and grain shedding are better than those of the base material, but dendritic cracking and sporadic grain shedding can still be seen. Figure. 3(d) shows the metallographic structure and morphology of cracks at the base metal of the outer wall of the pipeline. The morphology is similar to that of the inner wall, and it is also dendritic cracking plus grain shedding. The difference is that the outer wall is flat and there is no obvious corrosion trace.

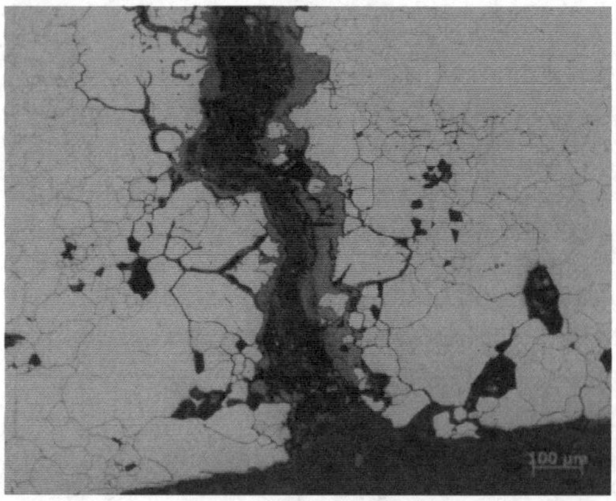

(a) Metallographic of parent metal crack at the inner wall

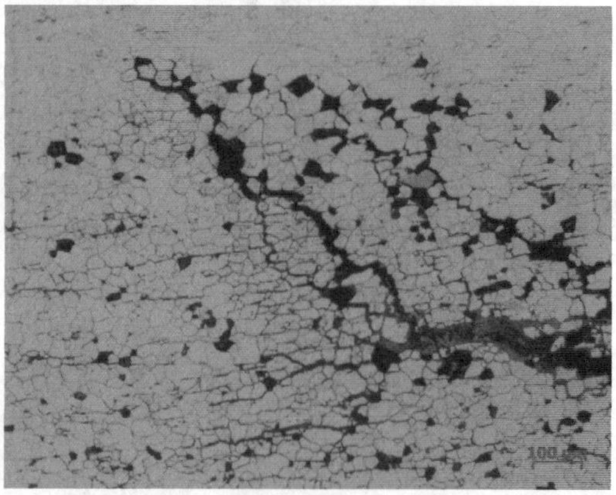

(b) Metallographic of parent metal crack near the inner wall

(c) Metallographic of inner wall weld crack

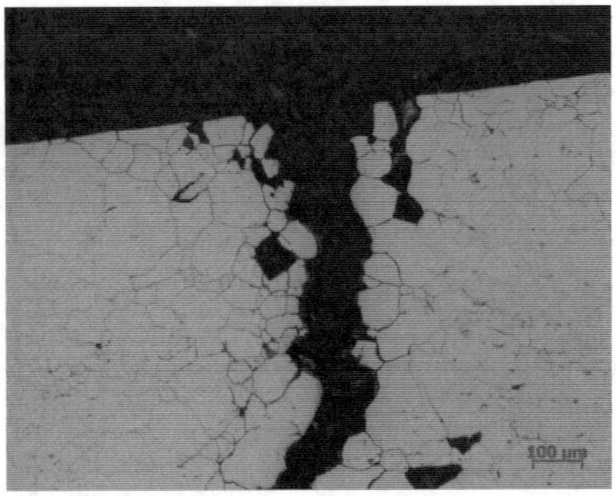

(d) Crack morphology of outer wall parent material

Figure. 3 Metallographic structure and morphology of cracks in cracked pipeline.

3.4 Hardness analysis

Table 2 shows the results of the hardness analysis of the cracked pipeline. It can be seen that the average hardness of the pipeline substrate is about 199 HV, and the average hardness of the weld is about 203 HV. According to GB/T 713.7-2023 "Pressurized Equipment Steel Plates and Strips Part 7: Stainless Steels and Heat-Resistant Steels," the requirements, the austenitic stainless steel solid solution processed hardness value at room temperature shall not be greater than 210 HV. The measured hardness value of the base material and weld seam meets the standard requirements.

Table 2. Pipeline hardness analysis results.

Position	Measuring point	Hardness(HV)
	1	203
Left parent metal	2	196
	3	192
	4	200
Welding seam	5	192
	6	203
	7	195
Right parent metal	8	205
	9	205

3.5 SEM analysis

The fracture was analyzed by SEM, as shown in Figure. 4. No plastic deformation is found in the fracture surface, and it is a brittle fracture surface. The phenomenon of grain cracking and shedding can be observed on the fracture surface, showing obvious intergranular cracking characteristics.

Figure. 4 Fracture surface morphology.

3.6 Energy spectrum analysis

The energy spectrum analysis of the fracture surface is carried out, and the analysis results are shown in Figure. 5 and Table 3. Table 3 shows that the fracture surface is mainly carbon, oxygen, iron, and other elements, and chlorine is found in the energy spectrum.

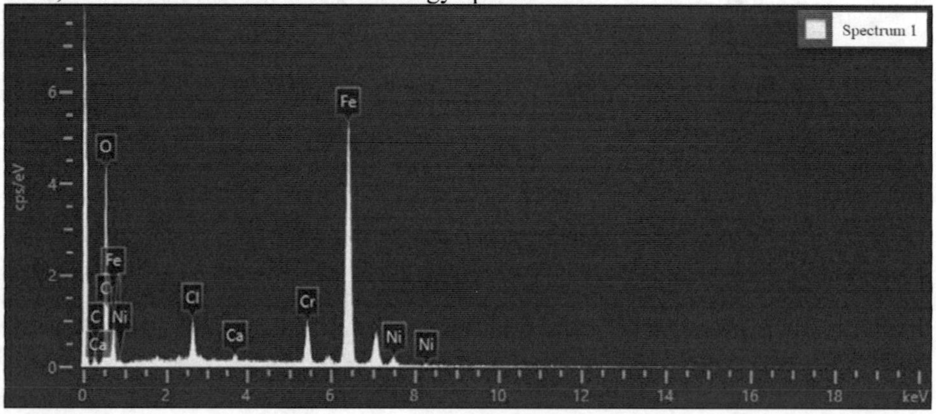

Figure. 5 Element distribution diagram of energy spectrum analysis.

Table. 3 Energy spectrum analysis results.

Element	C	O	Cl	Ca	Cr	Ni	Fe
wt, [%]	13.52	24.90	2.08	0.47	5.09	2.55	51.40

4 Analysis and discussion on failure reasons

Through the macroscopic observation of the cracked pipeline through penetrant inspection, it can be seen that the crack is tortuous, not straight, and bifurcated. At the same time, scattered dark pitting marks can be seen on the crack and the surrounding parent material, indicating that the stainless steel pipeline has been corroded. The results of the chemical composition analysis of the weld seam and base material of the cracked pipeline show that the chemical composition meets the requirements of GB/T 12771-2019 "Welded Stainless Steel Pipe for Liquid Transmission".

Journal of Physics: Conference Series **3009** (2025) 012086 doi:10.1088/1742-6596/3009/1/012086

From the metallographic analysis results of the failed pipeline, it can be seen that:

a) In the metallographic analysis of the inner wall and outer wall of the pipeline corresponding to the crack position, the phenomenon of grain shedding was found;

b) The crack has obvious dendritic bifurcation;

c) Cracking is intergranular cracking;

d) The grain boundary around the crack becomes wider and reticular, with obvious intergranular corrosion characteristics.

The results of the hardness test carried out on the base material and welds of the cracked pipe show that the hardness of the base material and welds comply with the relevant requirements of GB/T 713.7-2023 “Steel Plates and Strips for Pressurized Equipment Part 7: Stainless Steel and Heat-Resistant Steel”. Scanning electron microscopy (SEM) showed that there was no plastic deformation in the fracture surface obvious brittle fracture characteristics could be seen and the phenomenon of grain separation and shedding was also obvious. Finally, the results of energy spectrum analysis of pipeline cracking position show that in addition to the common elements such as C, O, Fe, Cr, and Ni, Cl was also found. Combined with the metallographic and macroscopic dendritic cracking characteristics, it can be judged that the cracking of the stainless steel pipeline was caused by chloride ion stress corrosion cracking [8, 9].

From the metallographic analysis, it can be inferred that intergranular corrosion occurred in the cracked pipeline, which is mainly due to the chemical reaction between chromium and carbon in austenitic stainless steel, resulting in poor chromium at grain boundaries and a decrease in corrosion resistance at grain boundaries. After corrosion, the mechanical properties of the material deteriorate and become brittle [10-12]. At the same time, the lack of chromium in grain boundaries will not only reduce the corrosion resistance but also increase the sensitivity of materials to stress corrosion cracking [13, 14]. Under the action of chloride ion, stress corrosion cracking will start from the chromium-poor area of the grain boundary, and the stress corrosion cracking will show obvious intergranular and brittle fracture characteristics due to the widening of grain boundary corrosion.

This failure is caused by the intergranular corrosion of materials, which leads to the decline of corrosion cracking resistance, and chloride ion invasion leads to chloride ion stress corrosion cracking.

5 Conclusion

The cracking failure of the 304 stainless steel pipeline at the outlet of the heat exchanger in the ethylene plant is mainly caused by the intergranular corrosion of the pipeline material. This can lead to the decline of its corrosion and cracking resistance, and chloride ion stress corrosion cracking is caused by chloride ion invasion in the medium. It is recommended that the factory staff in the stainless steel equipment and pipeline into the factory strengthen the sampling and inspection work, and increase sensitization - intergranular corrosion sampling and inspection, to prevent similar accidents from occurring again.

References

[1] Baddoo, N. R. (2008). Stainless steel in construction: A review of research, applications, challenges and opportunities. Journal of Constructional Steel Research, 64(11), 1199-1206.

[2] Gardner, L. (2019). Stability and design of stainless steel structures–Review and outlook. Thin-Walled Structures, 141, 208-216.

[3] Lo, K. H., Shek, C. H., & Lai, J. K. L. (2009). Recent developments in stainless steel. Materials Science and Engineering: R: Reports, 65(4-6), 39-104.

[4] Cunat, P. J. (2004). Alloying elements in stainless steel and other chromium-containing alloys. Euro Inox, 2004, 1-24.

[5] Kelly, J. (2006). Stainless steels. Materials and Mechanical Design, 39.

[6] Ettefagh, A. H., Guo, S., & Raush, J. (2021). Corrosion performance of additively manufactured stainless steel parts: A review. Additive manufacturing, 37, 101689.

[7] Du, C., Song, C., Liu, C., Li, X., Xin, Y., & Liu, B. (2023, October). Butyl rubber final cooler heat exchanger tube cracking failure analysis. In Ninth International Conference on Mechanical Engineering, Materials, and Automation Technology (MMEAT 2023) (Vol. 12801, pp. 89-95). SPIE.

[8] Rhodes, P. R. (1969). Mechanism of chloride stress corrosion cracking of austenitic stainless steels. Corrosion, 25(11), 462-472.
[9] Yazdanpanah, A., Pezzato, L., & Dabalà, M. (2022). Stress corrosion cracking of AISI 304 under chromium variation within the standard limits: Failure analysis implementing microcapillary method. Engineering Failure Analysis, 142, 106797.
[10] Qin, J., Zhao, W., Yan, Z., Wang, R., Yu, Z., Shi, Z., & Wang, B. (2024). Intergranular corrosion and stress corrosion cracking properties evaluation and mechanism study of multi-microalloyed 2519 Al alloy. Corrosion Science, 236, 112288.
[11] Gunawan, G., & Arifin, A. (2021). Intergranular Corrosion and Ductile-Brittle Transition Behaviour in Martensitic Stainless Steel. Indonesian Journal of Engineering and Science, 2(3), 029-039.
[12] Shit, G., Mariappan, K., & Ningshen, S. (2023). Improvement of sensitization and intergranular corrosion of AISI type 304L stainless steel through thermo-mechanical treatment. Corrosion Science, 213, 110975.
[13] Srinivasan, N. (2021). Sensitization of austenitic stainless steels: current developments, trends, and future directions. Metallography, Microstructure, and Analysis, 10(2), 133-147.
[14] Abou-Elazm, A., Abdel-Karim, R., Elmahallawi, I., & Rashad, R. (2009). Correlation between the degree of sensitization and stress corrosion cracking susceptibility of type 304H stainless steel. Corrosion Science, 51(2), 203-208.

Journal of Physics: Conference Series **3009** (2025) 012087 doi:10.1088/1742-6596/3009/1/012087

Study on the constant volume combustion performance of single-based propellant by considering heat dissipation correction

Guangpei Sun [1, 2], **Ling Chen** [1, 2*] **and Weidong He** [1, 2*]

[1] School of Chemistry and Chemical Engineering, Nanjing University of Science and Technology, Nanjing 210094, China
[2] Key Laboratory of Special Energy Materials, Ministry of Education, Nanjing University of Science and Technology, Nanjing 210094, China

* Corresponding author's e-mail: l_chen_zf_139@njust.edu.cn, hewedong@njust.edu.cn

Abstract: Aiming at the problem of the difference between the combustion performance of propellant obtained according to the current processing method and the actual situation, a one-dimensional cylindrical wall axisymmetric unsteady heat transfer model for the burning of the propellant in the closed vessel has been established. By utilizing the original pressure-time curve obtained from experiments, the heat dissipation and the corresponding pressure drop were calculated and the original curve was corrected during the combustion process. The corrected curves were used to investigate the impact of heat dissipation corrections on the combustion performance of single-based propellants under various loading densities and test temperatures. Compared with the result without heat dissipation correction, the maximum pressure and the maximum burning rate of propellant (0.12 g·cm^{-3}, 20°C) were increased by 7.08% and 6.40% respectively after considering the heat dissipation correction.

1. Introduction

Currently, the constant-volume combustion performance of propellant is obtained by the *p-t* curve from closed vessel experiments through a series of mathematical treatments [1-3]. The method for obtaining relationships such as *u*-p and *Γ-Ψ* from the combustion *p-t* curve is as follows: First, the *p-Ψ* curve is obtained from the *p-t* curve, then the *Ψ-Z* and *p-Z* curves are derived using the shape function formula. Finally, the *u-p* and *Γ-Ψ* curves are obtained [4-7]. However, convective heat transfer and radiative heat transfer between the combustion gas and the wall consume a certain amount of heat, leading to a pressure from the experimental test that is lower than the pressure without heat dissipation. Therefore, the *p-Ψ* relationship obtained during the processing is not accurate, which in turn leads to errors in the *u-p*, *Γ-Ψ* curves, and others [8].

In the study, a one-dimensional cylindrical wall axisymmetric unsteady heat transfer model for single-based propellant combustion in a closed vessel was established. By calculating the convective heat transfer coefficient using the Nusselt number, heat dissipation during the combustion was determined. Additionally, the original *p-t* curves of the single-based propellant at different densities and different test temperatures were processed. Differences in combustion performance before and after correction for heat dissipation were compared.

2. Closed Vessel Propellant Combustion Heat Transfer Model

2.1. Basic Assumptions of the Heat Transfer Process

Figure 1 shows a cross-sectional view of the closed vessel, typically a short cylindrical body with thick walls, and the propellant loading density ranges from 0.1 g cm^{-3} to 0.3 g cm^{-3} [9]. The basic assumptions for the heat transfer process are as follows:

1. The single-based propellant gas is uniformly distributed within the closed vessel.

2. The generated gas undergoes natural convection and radiative heat transfer with the wall in a confined space within the closed vessel.

3. The effects of heat dissipation and gas expansion work on gas composition and temperature are neglected. The gas temperature within the vessel is assumed to be constant (explosion temperature T_v).

4. The initial temperature of the closed vessel's wall is set to room temperature.

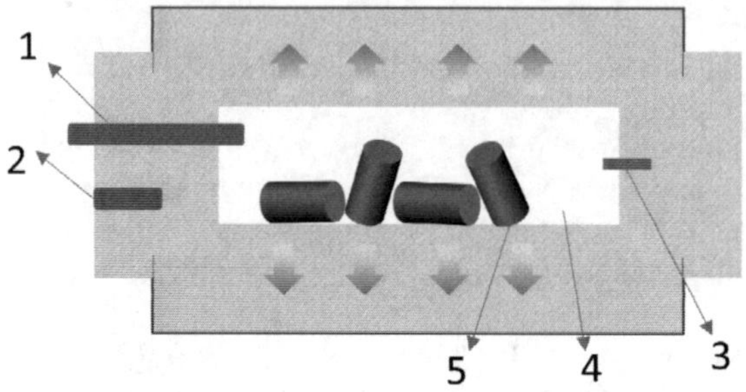

1. Ignition device; 2. Vent hole; 3. Pressure sensing device; 4. Combustion chamber; 5. Propellant.

Figure 1. The sectional diagram of a closed vessel.

2.2. Heat Transfer Between Propellant Gas and Wall

The combustion process of the single-based propellant is a typical solid-gas two-phase coupled reaction process. The solid components are first decomposed to produce gaseous products. Then complex chemical reactions occur in the gas phase. For example, the single-based propellants first generate intermediate gaseous species, such as NO_2, CO, and H_2, which subsequently undergo gas-phase reactions involving free radicals OH, H, and O. These reactions release significant thermal energy, establishing a high-temperature zone that promotes thermal radiation and convective heat transfer.

$$C_6H_7O_2(ONO_2)_3 \rightarrow CO_2 + H_2O + NO_2 + CO + C_xH_y \quad (1)$$

$$CO_2 + OH \rightleftharpoons CO + H_2O \quad (2)$$

$$H_2 \rightleftharpoons 2H \quad (3)$$

$$CO_2 \rightleftharpoons CO + O \quad (4)$$

$$H_2O \rightleftharpoons OH + H_2/2 \quad (5)$$

$$CO_2 \rightleftharpoons CO + O_2/2 \quad (6)$$

$$H_2O + N_2/2 \rightleftharpoons H_2 + NO \quad (7)$$

$$k = Ce^{\frac{-E_a}{RT}} \quad (8)$$

$$r = kA^mB^n \quad (9)$$

E_a is the activation energy, C is the pre-exponential factor, T is the temperature, R is the gas constant, k is the reaction rate constant, and A and B are the molar concentrations of the reactants.

Figure 2 shows the pressure change and the rate of pressure change of the gas in the closed vessel. In the initial stage of combustion, partial burning occurs on the surface of the propellant, with the flame propagating along its surface. At this stage, combustion primarily involves an increase in the single-based propellant's temperature and sublimation, accompanied by some chemical reactions. As combustion progresses, the intermediate products of combustion continue to react. Due to the rapid and high-energy nature of combustion, the temperature of the chamber gases quickly rises to the explosion temperature. Furthermore, as the combustion products increase, the chamber pressure rises accordingly. The rise in chamber temperature and pressure further accelerates the reaction rate, resulting in a faster rate of pressure increase within the chamber. During this stage, a substantial amount of heat and energy is released, making it the most intense phase of combustion. Simultaneously, the heat flux density between the combustion gases and the chamber walls increases significantly. At the end of combustion, the mass of combustion gases no longer increases, and the rate of pressure rises rapidly approaches zero.

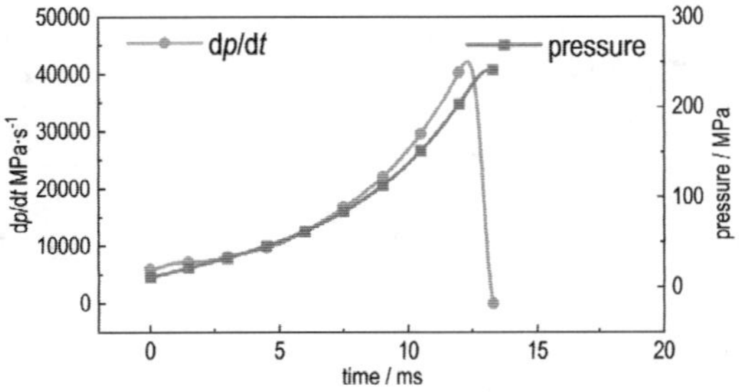

Figure 2. The pressure change and rate of pressure change in the closed vessel.

During the combustion process of single-based propellants in a closed vessel, the heat transfer phenomena are primarily governed by two mechanisms: convective heat transfer induced by the flow of high-temperature combustion gases and radiative heat transfer from the high-temperature gas-phase products and flame.

The high-temperature combustion gases generated during the burning of propellants flow rapidly within the confined space of the closed vessel, as the combustion progresses, the high-pressure gas flows in a highly turbulent manner, enhancing the heat transfer between the gas and the walls. This flow creates forced convective heat transfer between the gas and the vessel's inner wall. Convection heat transfer is the result of the joint action of two heat transfer methods, heat conduction and heat convection. Gas internal heat exchange occurs through heat convection, where the heat transfer between the gas and the wall involves heat conduction. According to molecular kinetic theory, when the temperature and pressure are higher, the molecules interact more intensely with the wall, resulting in a greater convection heat transfer coefficient. The rate of convective heat transfer can be expressed as [10]:

$$q_h = h(T_g - T_0) \tag{10}$$

Radiative heat transfer plays a significant role in high-temperature environments like combustion chambers. The flame and the combustion gases, primarily composed of CO_2, H_2O, and soot particles, emit thermal radiation due to their elevated temperatures. Unlike convective heat transfer, radiative heat transfer does not require a medium and can directly transfer energy to the vessel's walls even in the absence of gas flow. This mechanism is particularly significant in the initial stages of combustion, where the gas velocity is low, but the temperature is extremely high. The radiative heat flux can be described by the Stefan-Boltzmann law [10]:

$$q_r = \varepsilon\sigma(T_g^4 - T_0^4) \tag{11}$$

$$Q = \int q\, dt = \int q_h + q_r\, dt \tag{12}$$

h is the convective heat transfer coefficient between the gas and the wall at the inner wall of the closed vessel, W·m^{-2}·K^{-1}; q_h is the convective heat flux density at the inner wall; q_r is the radiative heat flux density; q is the total heat flux density between the gas and the inner wall, W·m^{-2}; Q is the amount of heat dissipation during the propellant combustion process, J; ϵ is the emissivity; σ is the Stefan-Boltzmann constant; T_g is the gas temperature; T_0 is the inner wall temperature, K.

$$\frac{\partial T}{\partial \tau} = \alpha\left(\frac{\partial^2 T}{\partial r^2} + \frac{1}{r}\frac{\partial T}{\partial r}\right), r_1 < r < r_2, \tau > 0 \tag{13}$$

$$-\lambda\frac{\partial T}{\partial r} = q, r = r_1, \tau > 0 \tag{14}$$

$$T = T_{atom}, r_1 < r < r_2, \tau = 0 \tag{15}$$

$$T_0 = T_1 + q \bullet \frac{dr}{\lambda} \tag{16}$$

$$T_i^{n+1} = \alpha \bullet dt \bullet \frac{T_{i+1} - 2T_i + T_{i-1}}{dr^2} + \frac{T_{i+1} - T_{i-1}}{2rdr} \tag{17}$$

τ is time; r_1 is the inner radius of the closed vessel; r_2 is the outer radius of the closed vessel, m; α is the thermal diffusivity of the container material, m^2·s^{-1}; λ is the thermal conductivity of the closed vessel, W·m^{-1}·K^{-1}; T_{-1} is the outer wall temperature; T_{atom} is room temperature, K; T_n^i refers to a point along the cylindrical wall thickness, divided into 100 equal segments, where i represents a point, and n refers to a specific moment during combustion.

2.3. Calculation of Convective Heat Transfer Coefficient

The convective heat transfer coefficient is calculated with the Nusselt number:

$$Ra = \frac{g \bullet \beta \bullet (T_g - T) \bullet d_1^3 \bullet \rho^2 \bullet c}{\mu \bullet k} \tag{18}$$

$$Nu = C \bullet Ra^n \tag{19}$$

$$h = \frac{Nu \bullet k}{d_1} \tag{20}$$

$$q_h = \frac{Nu \bullet k}{d_1}\left(T_g - T_0\right) \tag{21}$$

In Equations (18)-(21), d_1 is the inner surface diameter of the closed vessel, m; Nu is the Nusselt number; Ra is the Rayleigh number, dimensionless constants; g is the gravitational constant, N·kg^{-1}; β is the thermal expansion coefficient, K^{-1}; k is the thermal conductivity of the gas, W·m^{-1}·K^{-1}; μ is the dynamic viscosity, N·s·m^{-2}; ρ is the gas density, kg·m^{-3}; c is the specific heat capacity of the gas, J·kg^{-1}·K^{-1}.

3. Accuracy Verification of the Heat Dissipation Model

The closed vessel had a chamber volume of 112 cm³, a wall surface area of 139 cm², an ignition pressure of 10.98 MPa, and a sample consisting of the single-based propellant with a length of 40 mm, outer diameter of 4.98 mm, and inner diameter of 1.90 mm. Table 1 shows the propellant loading densities and test temperatures.

Table 1. The loading density and test temperature of the single-based propellant.

Number	Loading density / g.cm^{-3}	Temperature / ℃
1#	0.12	20
2#	0.20	20
3#	0.20	50
4#	0.20	-40

3.1. Heat Flux Density and Heat Dissipation

Analysis of Figure 3 showed that as combustion proceeded, the heat flux density in the closed environment gradually increased. The rate of heat dissipation accelerated. This was mainly because the pressure increased during combustion, leading to an increase in gas density, which in turn improved the convective heat transfer coefficient. From Table 2, it can be observed that the heat flux density of 2# was significantly higher than that of 1#. This was mainly because, under high loading density conditions, the propellant mass was greater, resulting in more heat being released during combustion. The higher chamber pressure further accelerated the reaction, increasing the chamber pressure even more, which enhanced the convective heat transfer coefficient and consequently increased the heat flux density. For propellants with the same loading density (2#, 3#, and 4#), the maximum heat flux densities were similar. However, under the low-temperature test conditions for 4#, the combustion time of the propellant was longer, leading to higher heat dissipation.

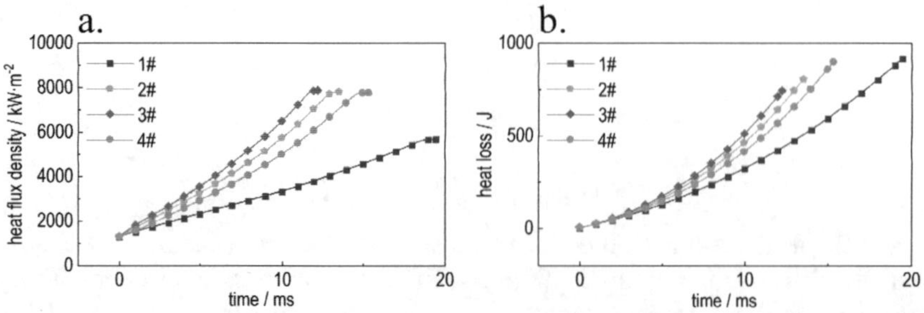

Figure 3. The relationship between time and heat flux density, heat dissipation during the burning process.

Table 2. The maximum heat flux density and maximum heat dissipation during the burning process.

Number	Heat flux density / kW·m^{-2}	Heat loss / J
1#	5675	914.6
2#	7798	805.9
3#	7865	743.3
4#	7772	899.4

3.2. Reliability Verification of the Heat Dissipation Correction Method

The original data obtained from the tests did not consider the effect of heat dissipation on pressure. The corrected pressure after heat dissipation was:

$$\Delta p = \frac{Q}{W_V} \tag{22}$$

$$p = p_{tested} + \Delta p \tag{23}$$

w was the mass of single-based propellants, g; W_V is the volume of the combustion chamber, cm^3; f is the propellant force, J·g^{-1}; p_{tested} is the measured pressure in the closed vessel; Δp is the pressure loss, MPa.

Figure 4 shows the relationship between the original pressure of the propellant and the corrected pressure after heat dissipation. The curves labeled 1#modified, 2#modified, etc., represented the pressure curves after heat dissipation correction. It can be seen that at the early stage of combustion, the pressure before and after heat dissipation correction was nearly the same. However, as combustion progressed, the difference between the original pressure and the corrected pressure increased. This was mainly because as combustion progressed, the heat dissipation increased, leading to a greater pressure loss.

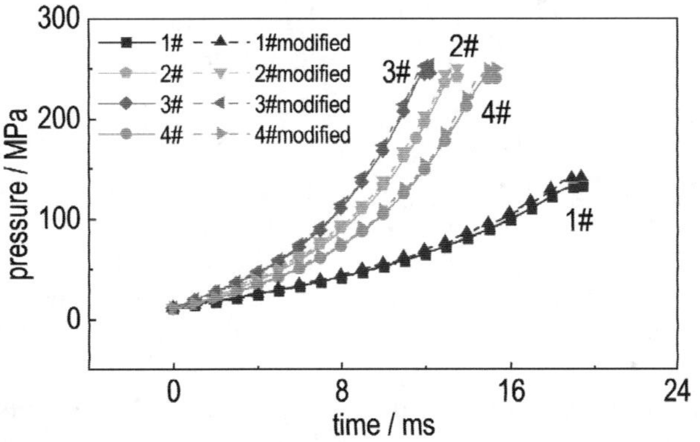

Figure 4. The p-t curve of single-based propellant before and after correction for heat dissipation.

When the propellant burned in the closed vessel, the pressure drops caused by heat dissipation were typically calculated using the Crow-Grimshaw formula. The calculation expression was as follows:

$$\frac{\Delta p}{p_m} = \frac{4.52 \bullet 10^{-2} \bullet S \bullet \sqrt{t_k}}{w} \tag{24}$$

t_k is the burning time, s; p_m is the maximum pressure, MPa; S is the inner surface area of the closed vessel, cm^2.

Table 3 provides a comparison of the maximum pressures obtained using the Crow-Grimshaw method and the pressure corrected using the model in this study. p_{tested} was the maximum measured pressure in the closed vessel, $p_{modified}$ was the maximum pressure after correcting for heat dissipation using the method in this study, p_{crow} was the maximum pressure corrected using the Crow-Grimshaw formula, and Δp was the difference between $p_{modified}$ and p_{crow}. Analysis of Table 3 showed that the corrected maximum pressure was close to the maximum pressure obtained by using the Crow-Grimshaw method, indicating that the one-dimensional cylindrical wall axisymmetric unsteady heat transfer model established in this study provided a reasonable and reliable correction for the pressure loss during the combustion process of the single-based propellant.

Journal of Physics: Conference Series **3009** (2025) 012087 doi:10.1088/1742-6596/3009/1/012087

Table 3. The maximum pressure of different modified methods and the maximum measured pressure.

Number	p_{tested}/ MPa	p_{modified}/ MPa	p_{crow} / MPa	Δp / MPa
1#	131.4	140.7	141.0	-0.3
2#	241.3	250.2	250.2	0.0
3#	245.4	253.6	254.0	-0.4
4#	240.0	249.9	249.3	0.6

4. Propellant Combustion Performance Processing Results

The original pressure-time (*p-t*) curves obtained from the closed vessel experiments were corrected for pressure loss caused by heat dissipation. The corrected *p-t* curves were used to determine the burning rate, gas release intensity, and other parameters of the propellant combustion process. These values were compared with the parameters obtained directly from the original *p-t* curves. The differences were analyzed.

4.1. Burning Rate

The burning rate of propellants was typically calculated using the pressure measured in the closed vessel. Since the ψ value calculated from the measured pressure deviated from the actual value, the burning mass percentage corrected for heat dissipation under the conditions of measured pressure was used to replace the burning mass percentage calculated from the measured pressure when calculating the burning rate-pressure curve, in order to better reflect the real situation. The calculation expressions were as follows:

$$u = \frac{de}{dt} = \frac{e_1}{\sigma\chi}\frac{d\psi}{dt} \tag{25}$$

$$\psi = \frac{\frac{1}{\Delta}-\frac{1}{\delta}}{\frac{f}{P_\psi - P_B}+a-\frac{1}{\delta}} \tag{26}$$

u was the burning rate, cm·s^{-1}; χ and σ were shape functions; e_1 was the propellant arc thickness, cm; Δ was the loading density, δ was the propellant density, g·cm^{-3}; P_B was the ignition pressure, p_ψ was the pressure at a relative burning mass of ψ, MPa; a was the remaining volume, cm^3·g^{-1}.

After correcting heat dissipation, the gas pressure increased, and the ψ value calculated from the corrected pressure curve increased. As a result, the burning rate calculated after considering heat dissipation also increased.

Figure 5 shows the relationship between the burning rate and pressure for single-based propellant before and after heat dissipation correction. P_{umax} is the pressure at which the burning rate reaches its maximum value. After considering heat dissipation, as the combustion progressed, the absolute value of the difference in burning rate before and after heat dissipation correction gradually increased, while the relative difference in burning rate gradually decreased. This was primarily because, in the initial stages of combustion, the propellant's burning rate was slight, and heat dissipation had a larger effect on the burning rate variation. As combustion progressed and the burning rate increased, the impact of heat dissipation on the rate of change in the burning rate diminished.

From Table 4, compared with the original data without correcting for heat dissipation, the maximum burning rate of propellant (1#) increased by 6.40%, which was significantly higher than the 2.56% increase in the maximum burning rate of propellant (2#), indicating that the heat dissipation had a more significant effect on the burning rate curve for propellants with low loading density. Under low-temperature test conditions, the maximum burning rate of propellant (4#) increased by 2.91%, which was slightly higher than the increase of 2.40%. For propellant under high-temperature test conditions (3#), due to the shorter combustion time and lower heat dissipation under higher loading density and higher test temperature conditions, the impact on the pressure variation of the propellant was relatively small. As a result, the changes in the *u-p* curve calculated from the *p-t* curve after correcting for heat dissipation were minimal, leading to smaller differences in the burning rate before and after the heat dissipation correction. The difference in temperature had a relatively small impact on the burning rate results, while the difference in loading density had a greater impact on the burning rate results.

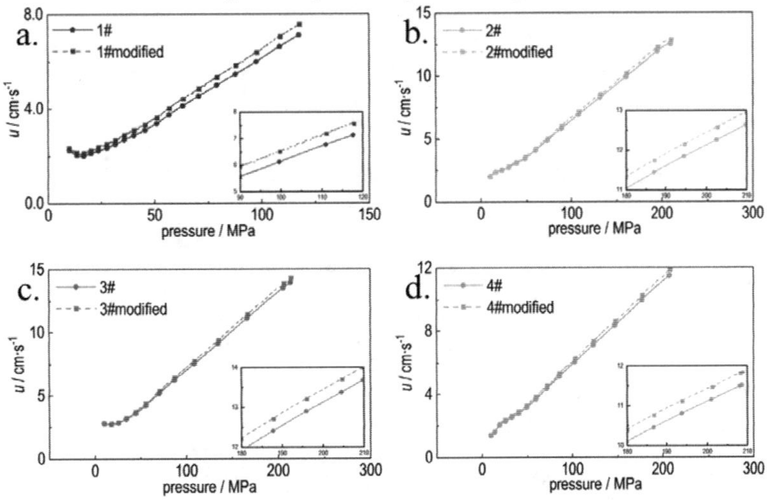

Figure 5. The *u-p* curve of single-based propellant before and after correction for heat dissipation.

Table 4. The burning rate of single-based propellant before and after correction for heat dissipation.

Number	50MPa			100MPa			P_{umax}		
	u / cm·s^{-1}	$u_{modified}$ / cm·s^{-1}	Δu	u / cm·s^{-1}	$u_{modified}$ / cm·s^{-1}	Δu	u / cm·s^{-1}	$u_{modified}$ / cm·s^{-1}	Δu
1#	3.28	3.54	7.93%	6.10	6.50	6.56%	7.03	7.48	6.40%
2#	3.43	3.59	4.66%	6.36	6.59	3.62%	12.49	12.81	2.56%
3#	3.79	3.94	3.96%	6.93	7.15	3.17%	13.77	14.10	2.40%
4#	3.11	3.26	4.82%	5.79	6.02	3.97%	11.34	11.67	2.91%

4.2. Gas Generation Intensity

The gas release intensity of the propellant was typically calculated from the measured pressure in the closed vessel experiment. The formula for calculating gas release intensity was:

$$\Gamma = \frac{1}{p} \bullet \frac{d\psi}{dt} \tag{27}$$

Γ was the gas release intensity, MPa^{-1}·s^{-1}. The new Γ–ψ curve was generated based on the corrected burning mass percentage ψ and measured pressure.

From Figure 6, it can be seen that in the early stages of combustion, the gas generation intensity after considering heat dissipation was higher than the original gas generation intensity. As combustion continued, the difference between the gas generation intensity after considering heat dissipation and the original gas generation intensity gradually decreased. By the end of the combustion, the two curves nearly coincided.

Table 5 showed that the relative difference in gas release intensity before and after heat dissipation correction for 1# and 2# were 7.56% and 4.44%, respectively. For 3# and 4#, the relative difference in gas release intensity before and after heat dissipation correction was 4.01% and 4.81%, respectively. Heat dissipation had a more significant impact on the gas generation intensity of the propellant under low loading density and low-temperature test conditions. Consistent with the analysis of the burning rate curve results, under higher loading density and higher test temperature conditions, the combustion time was shorter, and the heat dissipation was lower. Therefore, the impact on the pressure variation of the propellant was relatively small, resulting in minimal changes in the gas generation intensity calculated from the p–t curve after correcting for heat dissipation. The difference in loading density had a greater impact on the gas generation intensity results.

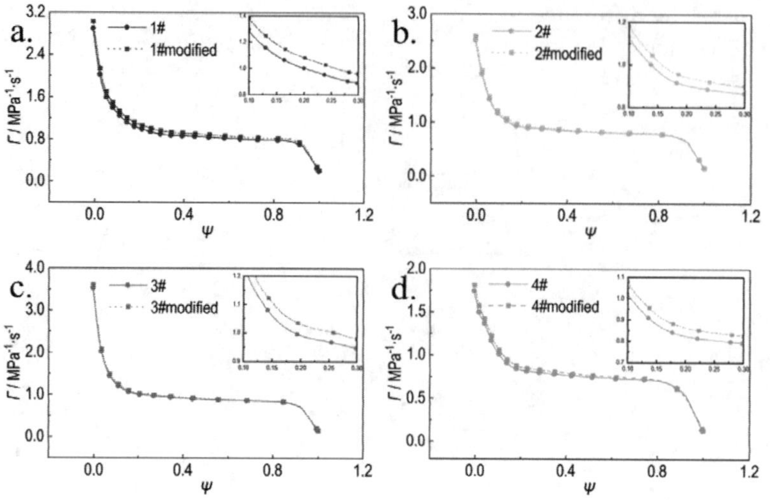

Figure 6. The Γ-Ψ curve of single-based propellant before and after correction for heat dissipation.

Table 5. The gas generation intensity of single-based propellant before and after correction for heat dissipation.

Number	Γ / $MPa^{-1}\cdot s^{-1}$	$\Gamma_{modified}$ / $MPa^{-1}\cdot s^{-1}$	$\Delta\Gamma$
1#	1.244	1.338	7.56%
2#	1.125	1.175	4.44%
3#	1.223	1.272	4.01%
4#	0.997	1.045	4.81%

5. Conclusion

To address the heat dissipation phenomenon during propellant combustion in the closed vessel, a one-dimensional cylindrical wall axisymmetric unsteady heat transfer model was established. The effect of heat dissipation on the constant-volume combustion performance of the single-based propellant was investigated. The conclusions were as follows:

1. A cylindrical wall axisymmetric heat transfer model was established using the one-dimensional unsteady heat conduction equation. Using this model, the pressure drop during the entire combustion process caused by heat dissipation in the closed vessel experiment was corrected. The corrected maximum pressure was close to the maximum pressure obtained using the Crow-Grimshaw correction method, verifying the reliability of the model.

2. Due to heat dissipation, the measured pressure was lower, which led to a lower ψ. As a result, the u and Γ calculated from the measured pressure were underestimated. The maximum pressure of propellant (1#) increased by 7.08%, the maximum burning rate increased by 6.40%, and the gas generation intensity increased by 7.56%.

3. The differences in loading density and test temperature had a certain impact on the combustion performance before and after heat dissipation correction. Under high loading density and high-temperature test conditions, the combustion time of the propellant was shorter, and the heat dissipation was lower, resulting in smaller changes in the p-t curve and ψ after correcting for heat dissipation, which in turn had a smaller effect on the combustion performance. Among them, the variation in temperature had a relatively smaller impact on the results.

Acknowledgment

This work thanks to the support of the Jiangsu Funding Program for Excellent Postdoctoral Talent (2023ZB472), China Postdoctoral Science Foundation (2023TQ0158), the National Funding

Postdoctoral Researcher Program (GZC20233496), and Young Elite Scientists Sponsorship Program by CAST (Grant No. 2024QNRCxxx).

References

[1] Chen, L., Sun, A. N., Zhang, J. W., Meng, D. R., Han, P. C., Chen, F. Y., Nan, F. Q., Du P., He, W. D. (2024) Carbon nanofibers introduce nitroguanidine propulsion to improve mechanical properties, and the study of thermal decomposition, combustion behavior and molecular dynamics. Chemical Engineering Journal, 496: 153932. https://doi.org/10.1016/j.cej.2024.153932.
[2] Yue, C. H., Liu, B., Li, M. M., Wang, F., Wei, L., Wang, Q. L. (2024) Progress in the synthesis of DIANP and its applications in propellants. Chinese Journal of Energetic Materials, 32(8): 871-884. https://doi.org/10.11943/CJEM2024173.
[3] Wu, R., Xiao, Y. J., Li, Q., Yuan, M. B., Liu, B., Zhang, Y. C. (2024) Thermal erosion characteristics and variation law of triple base propellant. Chinese Journal of Energetic Materials, 32(8): 818-823. https://doi.org/10.11943/CJEM2024033.
[4] Chen, L., Li, Q., Zhao, L. Y., Nan, F. Q., Liu J., Wang, X. J., Chen, F. Y., Shao, Z. Q., He, W. D. (2022) Enhancement strategy of mechanical property by constructing of energetic RDX@CNFs composites in propellants, and investigation on its combustion and sensitivity behavior. Combustion and Flame, 244: 112249. https://doi.org/10.1016/j.combustflame.2022.112249.
[5] Chen, L., Sun, A. N., Meng, D. R., Wang, B. B., Chen, F. Y., Nan, F. Q., Du P., He, W. D. (2024) Modified mechanical strength, thermal decomposition, and combustion characteristics of nitroguanidine propellant with graphene nanosheets as reinforcement[J]. Industrial & Engineering Chemistry Research, 63(13): 5527-5541. https://doi.org/10.1021/acs.iecr.3c04467.
[6] Zhao, Y. H., Xiao, Z. G., Yan, W. R., Yan, G. H., Liang, L. (2020) Interior ballistic performance prediction of gun propellant charge based on closed-bomb test. Chinese Journal of Explosives & Propellants, 43(1): 81-84+89. https://doi.org/10.14077/j.issn.1007-7812.201901010.
[7] Zhu, W. F., Wang, Y. W., Liu, W., Liu, Q., Wang, H. (2023) Research on the internal relationship between dynamic and static combustion laws of propellant for a large-caliber artillery. Journal of Physics: Conference Series, 2478: 112022. https://doi.org/10.1088/1742-6596/2478/11/112022.
[8] Wang, Y. W., Zhu, W F., Di, J. W., Hu, X. H. (2018) Study on the analysis method on ballistic performance of deterred propellant with large web size in large caliber artillery. Defence Technology, 14(5): 522-526. https://doi.org/10.1016/j.dt.2018.07.027.
[9] Christopher, H., Nicholaus, P. (2024) A numerical determination of complex solid gun propellant burn rates through closed bomb simulation. Propellants Explosives Pyrotechnics, 49(6): 4125-4339. https://doi.org/10.1002/prep.202300258.
[10] Giovanni Maria, C., Cardone, G. (2010) Infrared thermography for convective heat transfer measurements. Experiments in fluids, 49: 1187-1218. https://doi.org/10.1007/s00348-010-0912-2.

Journal of Physics: Conference Series **3009** (2025) 012088 doi:10.1088/1742-6596/3009/1/012088

A Simulation study on the service deformation characteristics of ht300 hydrostatic linear guides considering residual stress distribution at contact surfaces

Xiaozhi Yang, Xiao Cao, Huizhen Wang, Xifeng Wang*, Yuewen Zhai, Weidong Zhu and Zibo Zhang

No. 18 Xueqing Road, Haidian District, Beijing, China, China Academy of Machinery Beijing Research Institute of Mechanical & Electrical Technology Co., Ltd.

*Corresponding author's e-mail: magus0526@163.com

Abstract. This study simulates the deformation characteristics of HT300 hydraulic linear guides under varying residual stress distributions, focusing on the effects of residual stress and temperature fields on precision and deformation. Using ANSYS, a finite element model is analysed over 2100 seconds. Results show that slider position minimally affects temperature and displacement but significantly impacts stress and creep strain. Temperature rise leads to thermal expansion outweighing creep subsidence. Residual stress is positively correlated with creep strain, with creep partially reducing high stress. The study emphasizes controlling temperature and minimizing residual stress to enhance guide precision, recommending static models for efficiency. This research supports improving the precision and stability of domestic CNC machine tools.

1. Introduction

In the context of the transformation and upgrading of the manufacturing industry, CNC coordinate boring machines in China are advancing towards high-end and high-reliability development, which imposes higher demands on the design and manufacturing of their key components. Residual stress, as an internal stress present during the manufacturing process, gradually releases during the service period of components, leading to changes in the geometric accuracy of key components and becoming one of the critical factors that reduce the precision reliability of machine tools [1]. However, current analyses of residual stress in machine tools primarily serve as evaluation metrics during the aging process, and the relationship between residual stress in key components and the precision reliability of machine tools remains unclear. The standards for residual stress elimination are difficult to quantify, posing a longstanding challenge for machine tool enterprises.

Residual stress is defined as the stress that remains self-balanced within an object after the external load has been removed [2–7]. During the casting process of large foundational components of machine tools, the complexity of the structure causes uneven cooling shrinkage rates in different areas as the casting solidifies and cools, resulting in the generation of casting residual stress. In the machining process, the tool's cutting action disrupts the original equilibrium state of the workpiece's residual stress, causing deformation of the workpiece (Figure 1). Simultaneously, plastic deformation occurs at the contact area between the workpiece and the tool, generating significant heat and introducing new machining residual stress. The guideways of CNC machine tools, as critical moving components, directly affect the machining accuracy and stability of the machine tools. Sliding guideways, with their excellent rigidity, vibration absorption, and damping properties, possess irreplaceable advantages in scenarios requiring high rigidity, high damping characteristics, and high metal removal rates. Therefore, by considering the distribution of residual stress after workpiece machining during the design phase and analysing the impact of residual stress layers on creep during use, we can improve the flatness of the sliding guideway contact surfaces.

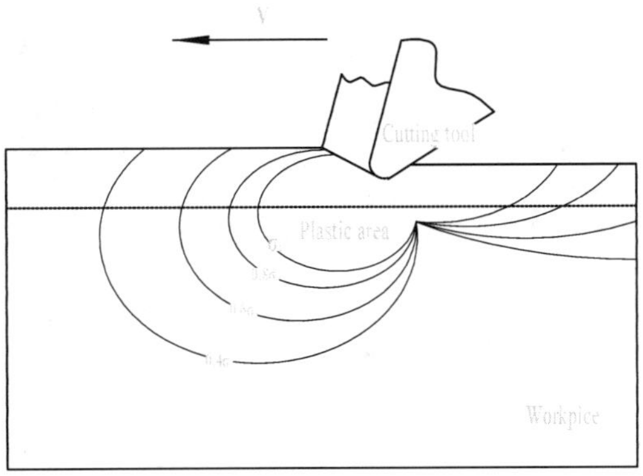

Figure 1 Distribution of residual stress layers in machining.

In machine tools and other assembled structures, the contact surfaces between parts, components, and assemblies form "mechanical interfaces." From an assembly perspective, to achieve proper connections between these contact surfaces, they can be classified into interference fits, transition fits, and clearance fits. From a motion perspective, these interfaces can be categorized as fixed interfaces, semi-fixed interfaces, and moving interfaces. The presence of numerous interfaces causes discontinuities in mechanical structures at the joints, leading to abrupt changes in stiffness and damping. To address this, scholars have employed the finite element method, taking into account the contact stiffness of the interfaces and the stiffness of the surrounding components, to analyse the deformation and pressure distribution of mechanical structures, thereby enhancing computational accuracy [8]. Due to its high computational precision and strong adaptability, the finite element method has become a powerful tool for analysing complex mechanical structures. By establishing precise finite element models, the impact of residual stress and mechanical interfaces on machine tool performance can be studied in depth, providing a theoretical basis for the design and optimization of key components.

In summary, investigating the influence of the distribution and magnitude of residual stress in key components of machine tools on the operational accuracy of guideways is of significant theoretical and practical importance. By employing advanced finite element analysis techniques, the relationship between residual stress and machine tool performance can be explored in depth, supporting the formulation of scientific standards for residual stress elimination, thereby enhancing the overall performance and competitiveness of domestic CNC machine tools.

2. Description of the physical models

Figure 2 illustrates a schematic of the linear guideway with a residual stress layer distribution used in this study. The entire linear guideway has a total length of 430 mm and consists of three parts: the guideway, the slider, and the pressure plate. The material of the guideway is HT300, while the pressure plate and slider are made of QT500. The guideway undergoes casting, followed by milling and grinding of the assembly surfaces. During operation, the contact surface between the guideway and the slider is uniformly covered with lubricating oil, and oil storage grooves are distributed on both sides of the guideway and slider to store excess lubricating oil. Stoppers are installed on the upper and left sides of the guideway to restrict the movement of the slider. To accurately reflect the time-varying characteristics of the interfaces, five layers of residual stress are uniformly distributed on the machined surface of the guideway. The thickness of each stress layer is 0.2 mm, and the stress values decrease in a gradient from the interface to the interior of the model. The slider contacts the guideway and pressure plate on four sides, with a height and width of 50 mm and 100 mm, respectively.

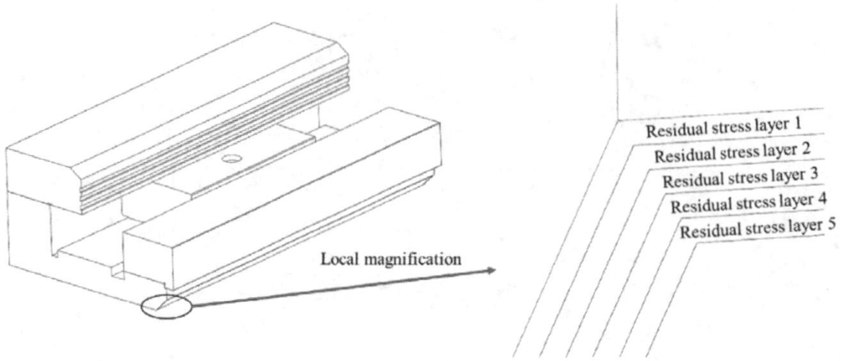

Figure 2 Residual stress layer settings.

In this study, different residual stress values and slider positions are assigned to the model to investigate the sensitivity of guideway accuracy to various design parameters. For all groups in this study, the residual stress layers of the guideway are distributed with the same thickness on the surface. The main parameters involved include the variation of the slider's position on the guide rail (Group A) and the distribution of maximum residual stress layer values (Group B). Under the condition where the residual stress layer values remain unchanged, the sensitivity of the guide rail surface deformation to the slider position is verified by placing the slider at three positions along the guide rail: at 0, 1/4, and 1/2 of its length. These are referred to as Case A1, Case A2, and Case A3, respectively, in the following text. The maximum stress layer values range from 100 to 300 MPa, corresponding to different machining processes, and these cases are referred to as Cases B1 through B4. As friction between the slider and the contact surface during operation leads to changes in the lubricating oil temperature, the temperature between the guideway and slider is set as a constant to validate the sensitivity of temperature rise on guideway accuracy, referred to as Case C in the text.

- Case A1: The front surface of the slider is flush with the front surface of the guideway. The residual stress layers decrease from the outside to the inside with values of 200, 160, 120, 80, and 40 MPa, respectively.
- Case A2: The symmetry plane of the slider is located at the front 1/4 of the guideway length, with all other dimensions and arrangements identical to Case A1.
- Case A3: The symmetry plane of the slider is located at the front 1/2 of the guideway length, with all other dimensions and arrangements identical to Case A1.
- Case B1: The residual stress layers decrease from the outside to the inside with values of 100, 80, 60, 40, and 20 MPa, respectively, with all other dimensions and arrangements identical to Case A3.
- Case B2: The residual stress layers decrease from the outside to the inside with values of 160, 128, 96, 64, and 32 MPa, respectively, with all other dimensions and arrangements identical to Case A3.
- Case B3: The residual stress layers decrease from the outside to the inside with values of 180, 144, 108, 72, and 36 MPa, respectively, with all other dimensions and arrangements identical to Case A3.
- Case B4: The residual stress layers decrease from the outside to the inside with values of 300, 240, 180, 120, and 60 MPa, respectively, with all other dimensions and arrangements identical to Case A33.
- Case C: No temperature load is applied, and the overall temperature is maintained consistent with the ambient temperature, with all other dimensions and arrangements identical to Case A3.

Journal of Physics: Conference Series **3009** (2025) 012088 doi:10.1088/1742-6596/3009/1/012088

3. Computational method and procedure

3.1. Overview

In this study, a creep model is constructed to investigate the time-varying characteristics of guideway surface deformation over time, considering the effects of temperature and residual stress[9–11]. The construction of the creep model is based on sampling the study object and conducting two sets of tensile creep experiments at room temperature. The model is then fitted using a simplified classical strain hardening formula based on the experimental results. Since the entire analysis process involves temperature field analysis, structural analysis, and creep analysis, the transient temperature analysis module and the static structural analysis module of ANSYS Workbench are used in a combined simulation to determine the sensitivity of guideway deformation to various influencing variables.

3.2. Boundary conditions

In this study, the contact surfaces of a linear sliding guideway with varying residual stress and slider position distributions are considered. The constraint relationship between the slider and the surrounding contact surfaces is assumed to be frictional contact, with a friction coefficient of 0.15. The total thickness of the residual stress layer is 1 mm, with a gradient distribution decreasing in 0.2 mm increments, and it is assumed that the residual stress distribution within each layer is uniform. During the calculation process, the contact relationship between the residual stress layers and the inner surface of the guideway is considered "bonded" (nodes at the connection maintain consistent force or displacement).

The temperature data of the oil film over time is obtained by measuring the lubricating oil temperature in the nozzle during the experiment. It is assumed that the temperature distribution of the oil film is uniform, and the oil film layer is extremely thin and incompressible, allowing the volume of the oil film to be neglected. Oil can be stored in the grooves of the guideway and slider, which heats the contact surface. A temperature load measured over 2100 seconds is applied, and a creep analysis considering temperature effects is conducted based on this data. The initial temperature of the entire model is assumed to be 11°C (the constant factory temperature). The constraint relationship between the guideway and the pressure plate is bonded to simplify the modelling of bolts. The surface of the slider is subjected to a pressure load of 600 N to simulate the pressure load exerted on the slider during lathe processing.

3.3. Creep model establishment

Creep refers to the continuous deformation of a material under constant stress over time. Creep models are used to describe the deformation behaviour of materials under long-term loads, particularly for materials subjected to stress for extended periods. In this study, the material of the guideway, HT300, requires consideration of creep effects. Therefore, creep tests were conducted at room temperature (30°C) under tensile stresses of 100 MPa and 200 MPa to obtain the corresponding results. Based on actual working conditions, the classical creep equation was simplified and fitted to obtain the relevant creep parameters. The geometry of the test specimen for the creep tests is shown in Figure 3.

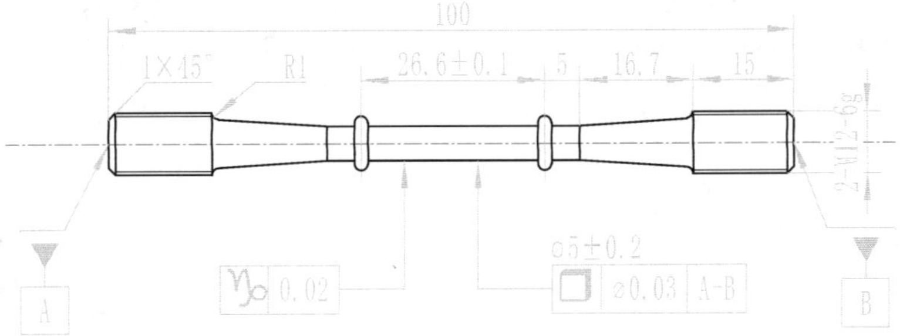

Figure 3 Creep test tensile specimens.

Based on the results obtained from the creep tests, as shown in Figure 4, the creep curve is typically divided into three stages:

- Primary Stage: The strain in the material increases rapidly with time.
- Secondary Stage: The strain increases steadily and stabilizes over time.
- Tertiary Stage: The strain once again increases rapidly.

From the perspective of strain rate analysis: In the primary stage, the strain rate decreases gradually over time. In the secondary stage, the strain rate remains constant. In the tertiary stage, the strain rate increases significantly again. In this experiment, due to the limited creep time, only the phenomena of the first two stages can be observed. Additionally, since the guideway operates at room temperature, the influence of temperature on creep does not need to be considered. This also indicates that after the primary stage, the slope of the secondary stage is approximately constant. To simplify the analysis, the classical strain hardening formula (Equation 1) has been simplified (Equation 2). In the secondary stage, it is assumed that the strain rate is only related to stress.

$$\dot{\varepsilon} = A\sigma^n \varepsilon^m e^{-\frac{C}{T}} \tag{1}$$

$$\dot{\varepsilon} = A\sigma^n \tag{2}$$

In the given equation, represents the strain rate, denotes the stress, is the strain, and signifies the temperature. The parameters A, n, m, and C are coefficients.

Through analysis, it was determined that under a stress of 100 MPa, the strain rate in the secondary stage of creep is 1.5e-8 (1/s), whereas the strain rate is 4.5e-4(1/s) under a stress of 200 MPa. Calculations yielded a constant 4.5e-4(1/s) and n = 14.88. Consequently, the creep equation for the HT300 material is formulated as follows:

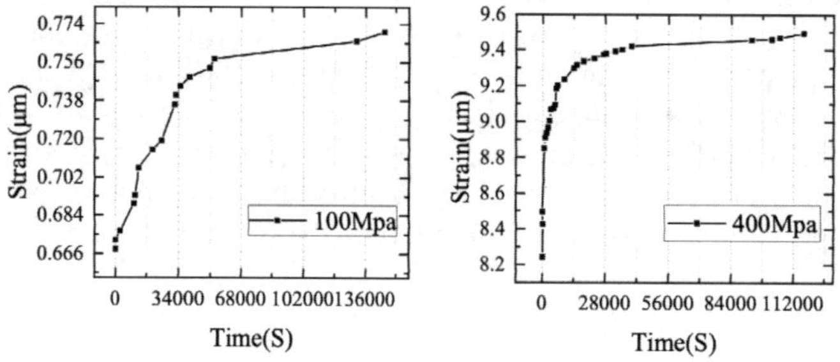

Figure 4 Creep test strain curves.

3.4. Grid meshing

Generally, hexahedral meshes offer higher mesh accuracy compared to tetrahedral meshes. However, hexahedral meshes impose stricter requirements on the geometry of the target. Therefore, in this study, a combination of hexahedral, pentahedral, and tetrahedral elements was employed to mesh the entire model, using linear elements with a size of 3 mm. Figure 5 illustrates the computational mesh around the guide rail slider. As depicted in the figure, the mesh in the residual stress distribution layer is denser compared to the mesh within the guide rail. This is due to the relatively small thickness of the residual stress layer compared to the thickness of the guide rail. Consequently, to maintain a balance between computational efficiency and prediction accuracy, a finite element model comprising 1.12 million elements and 0.4 million nodes was generated.

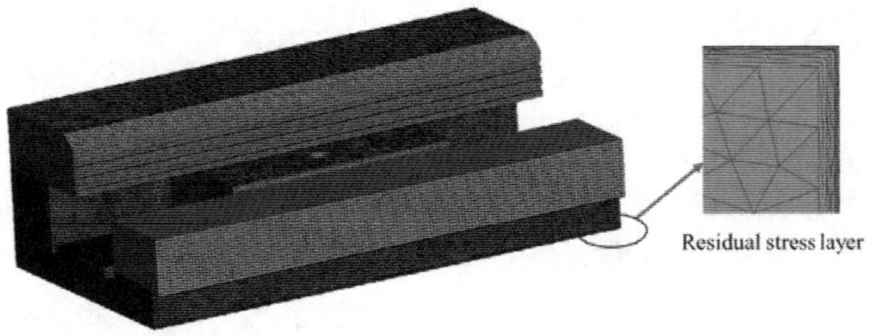

Figure 5 Meshing results.

4. Computational results and discussion

4.1. Parameter definition

The variation in the displacement of the guide rail is the primary focus of this study. To describe the analysis results of each model, the investigation should be conducted across three analysis steps. Analysis Step 1 involves the application of residual stress, representing the state of the model at the initial working moment. Analysis Step 2 pertains to the application of working loads, while Analysis Step 3 involves the application of thermal loads and creep, both of which are relative to the results of Analysis Step 1. Since residual stress is generated during the forming stage of the guide rail, Analysis Step 1 serves as the initial state. The study primarily focuses on the changes in parameters such as surface dimensions, stress, and strain of the guide rail during Analysis Steps 2 and 3.

In terms of displacement, the analysis focuses on the maximum deformation, displacement fluctuation values of the guide rail interface (upper surface of the guide rail) in the Y direction (gravity direction), and the displacement fluctuation values in the main region. The Y-component displacement of the upper surface of the guide rail primarily reflects the overall deformation trend of the guide rail surface, while the fluctuation values characterize surface roughness and precision. The displacement fluctuation values in the main region are described by selecting three non-adjacent nodes and measuring the actual displacement differences in the Y direction (Figure 6).

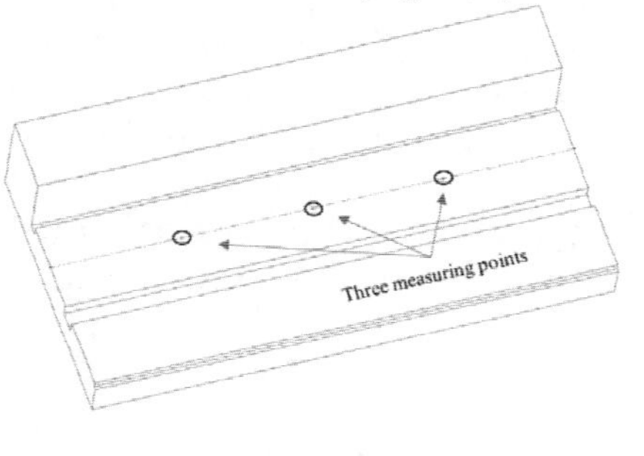

Figure 6 Nodes location selection.

Thermal load serves as the boundary condition for creep analysis, so this study only considers the overall temperature distribution and the temperature distribution on the upper surface of the guide rail. For stress components, it is sufficient to directly consider the maximum value of the von Mises equivalent stress. Regarding strain components, in addition to the equivalent total strain, it is necessary to consider thermal strain, creep strain, equivalent elastic strain, and the results of equivalent total strain. The equivalent total strain is the sum of the creep strain and equivalent elastic strain, excluding thermal strain.

4.2. Temperature sensitivity analysis

Figure 7. illustrates the temperature distribution contours on the surface of all linear guide rails after applying the actual temperature field. Overall, the maximum temperature on the interface (upper surface of the guide rail) for Case A1 to Case A3 consistently remains at 20.8°C, while the minimum

Journal of Physics: Conference Series **3009** (2025) 012088 doi:10.1088/1742-6596/3009/1/012088

temperature stays at 20.3°C and 20.4°C. For the overall structure, the temperature increase around the slider is relatively higher, with the maximum and minimum temperature difference initially increasing and then stabilizing after 800 seconds. For the upper surface of the guide rail, the maximum and minimum temperature difference stabilizes after 100 seconds, maintaining a temperature difference range of 0.5°C. Therefore, using the current temperature boundary conditions, the guide rail interface achieves a relatively uniform temperature distribution. The temperature results are presented in Table 1.

Table 1. Temperature results.

	Case A1		Case A2		Case A3	
	Min/℃	Max/℃	Min/℃	Max/℃	Min/℃	Max/℃
Monolithic model	16.6	20.9	17.1	20.9	17.9	20.9
The upper surface of the rail	20.3	20.8	20.3	20.8	20.4	20.8

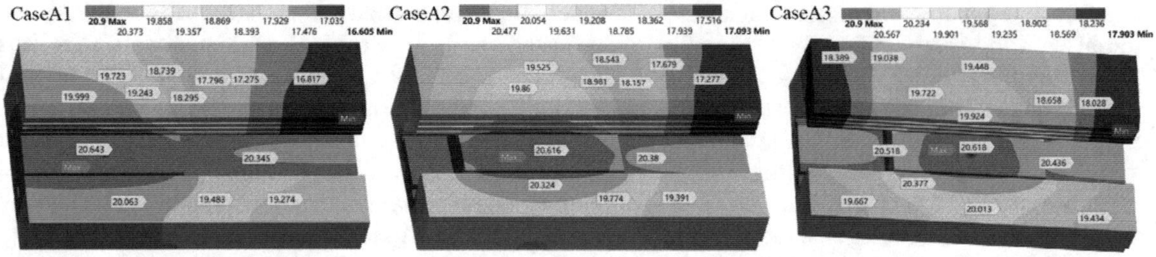

Figure 7. The temperature distribution of different slide positions after operation.

In both Case A3 and Case C, where the relative position of the slider is 1/2 and the residual stress is 200 MPa, the study investigates the impact of considering temperature field boundary conditions on creep results. To more accurately describe the deformation fluctuation values of the main region during temperature rise, the actual displacement differences in the y-direction are measured using three distantly spaced nodes on the guide rail interface. Figure 8 and Table 2. show that when the temperature field is considered, the displacement and fluctuation values of the guide rail interface increase more than tenfold compared to the results without considering the temperature field. However, the displacement fluctuation value in the interface region remains below 10 nm. Therefore, controlling the temperature rise during operation has a significant impact on the precision of the guide rail.

Table 2. Temperature results.

	True Displacement in the Y Direction of the Rail Joint Surface /mm		
	Maximum Value	Fluctuation Value	Fluctuation Value of the Main Region
Case A3	0.0057	0.0032	0.00033
Case C	0.0002	0.0003	5e-6

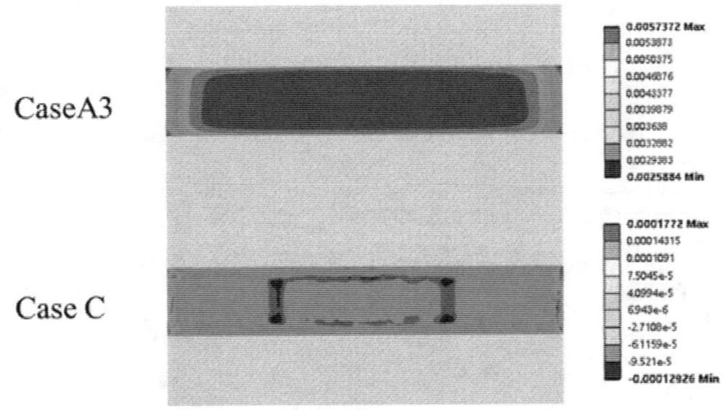

Figure 8 The true displacement of the rail bonding surface.

Regarding stress, the equivalent stress distribution contour plot after creep on the upper surface of the guide rail and the curve of the maximum value over time are shown in Figure 9 and Figure 10. It can be observed that after 2100 seconds, considering the temperature field, the stress value for Case C is 106.6 MPa, while for Case A3 it is 103.7 MPa, a reduction of 2.9 MPa or approximately 2.7%. This occurs because the gravity and working load pressure on the guide rail-slider system are both directed downward, and the bottom of the guide rail is fixed. During the application of thermal load, the temperature in the main region exceeds the initial temperature by 11°C, resulting in thermal strain directed upward, which is opposite to the strain direction caused by gravity and working load pressure. These opposing strains can partially offset each other. Throughout the creep analysis process, the maximum equivalent stress in Case A3 remains consistently lower than in Case C.

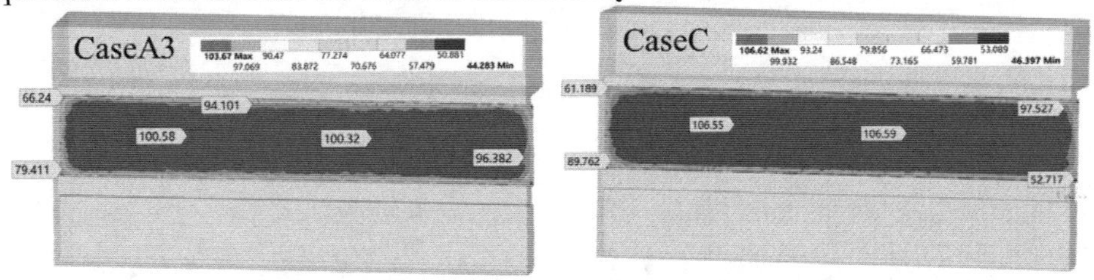

Figure 9. Equivalent stress distribution contour.

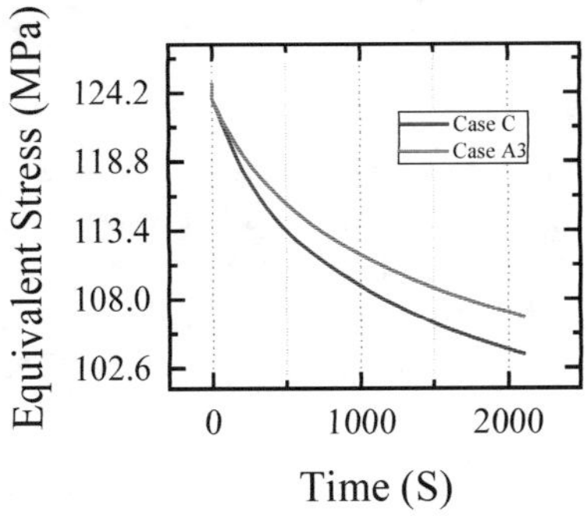

Figure 10 Equivalent stress curve.

Regarding strain, the creep strain contour plot and the creep strain curve on the upper surface of the guide rail after creep are shown in Figure 11 and Figure 12. It can be observed that when considering the temperature field, the creep results for Case A3 are lower than those for Case C. After 2100 seconds, the creep strain value decreases from 204 με to 154 με, a reduction of approximately 24.5%. This is because the increase in temperature leads to a reduction in stress on the upper surface of the guide rail. According to the creep formula, this results in a lower creep strain rate, which, through the effect of time integration, leads to significant changes in creep strain. Consequently, as the operating duration increases, the difference in the creep curves between considering and not considering the temperature field becomes more pronounced.

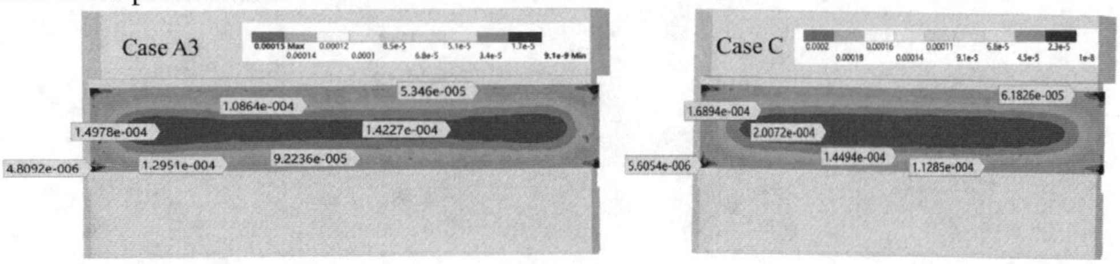

Figure 11. Creep strain contour map.

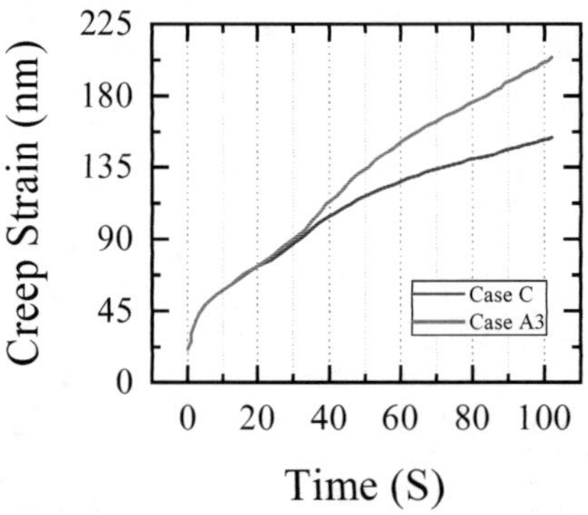

Figure 12. Creep strain on the rail surface.

Based on the foregoing analysis, it is evident that the temperature field has a significant impact on the deformation behaviour of the guide rail structure. The following conclusions can be drawn:

- To reduce creep strain, the temperature should be increased;
- To reduce equivalent stress, the temperature should be increased;
- To reduce surface displacement and displacement fluctuations, the temperature should be decreased.

4.3. Slider position sensitivity analysis

Figure 13 presents the time-dependent curves of the maximum displacement and displacement fluctuation values for Cases A1 to A3. The results indicate that different slider positions have minimal effect on the maximum displacement of the rail's upper surface. Although there are slight differences in the displacement fluctuation values, the magnitude of variation is less than 0.001 mm. Consequently, changes in slider position have no significant impact on the displacement results induced by creep.

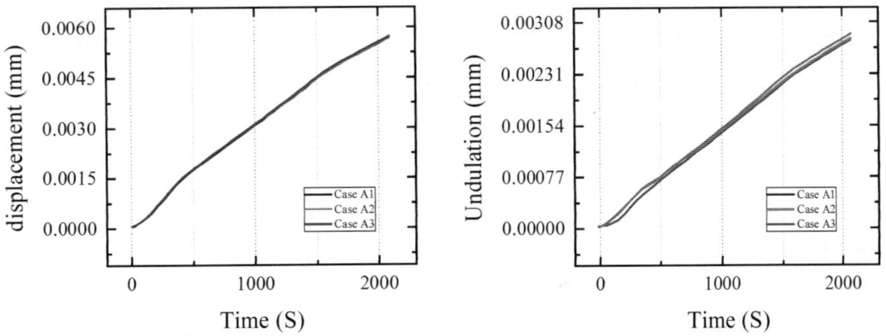

Figure 13. The displacement value and fluctuation value of the y direction of the combined surface of different slider positions.

In the context of stress analysis, the maximum equivalent stress value on the rail's upper surface post-creep was uniformly measured at 103.7 MPa, indicating that the equivalent stress is identical across various slider positions. Figure 14 presents the temporal variation of the maximum equivalent stress on the rail's upper surface, where the three curves are completely superimposed and any specific differences are indistinguishable. This observation suggests that the influence of slider position on the equivalent stress is negligible.

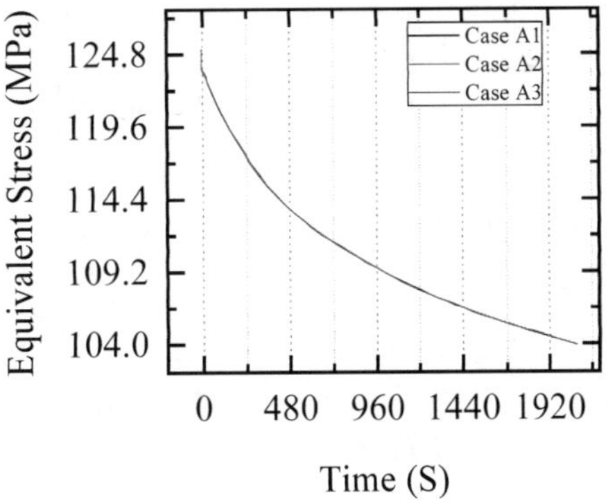

Figure 14. Equivalent stress on the rail surface at different slider positions.

Figure 15 presents the time-dependent curves of thermal strain and creep strain, where the curves for all three slider positions overlap entirely. This overlap indicates that the slider position has a negligible effect on the strain at the rail interface (the upper surface of the rail). Specifically, the maximum variation in creep strain is 154 $\mu\varepsilon$, corresponding to a change rate of only 0.7%.

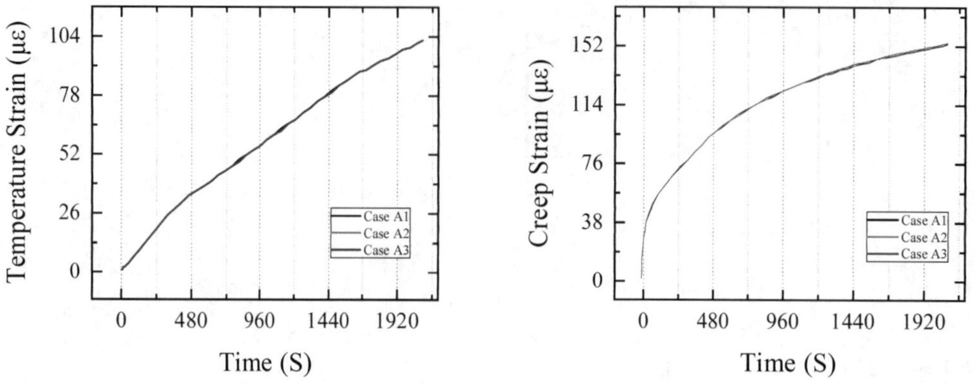

Figure 15. The displacement value and fluctuation value of the y direction of the combined surface of different slider positions.

By comparing the temperature, displacement, stress, and strain results of models with different slider positions, it can be determined that the slider position has no significant impact on these parameters, with variation rates all below 1%. Therefore, the slider position does not affect the creep results of the rail structure. This indicates that in dynamic simulations of the rail-slider mechanism, a static model with a fixed slider position can be employed as a simplified model. This approach not only ensures the accuracy of the results but also enhances simulation efficiency.

4.4. Creep strain sensitivity analysis

To ensure that the slide rail can maintain high-precision operation over an extended period, it is necessary to conduct a creep analysis of the structure. This paper investigates the displacement changes at the interface of slide rails with different residual stresses (due to varying machining processes) after considering creep strain. We examine the temporal distribution patterns of the residual stress layers and quantify the creep strain values under different residual stresses. Through this study, we derive the variation law of rail precision over time within the model, as well as the trend in maximum equivalent stress values at the interface after a certain period.

Figure 16 depicts the curve of the Y-direction displacement at the rail interface over time. Table 3 shows the displacement of the guide rail bonding surface in the Y direction caused by creep and temperature respectively after 2100 s. The results show that as time increases, the displacement in the Y-direction at the rail interface gradually increases. This is due to the rail's expansion caused by the temperature rise, where the amount of expansion offsets the surface subsidence phenomenon brought about by creep. However, under different residual stress levels, the maximum displacement at the rail interface remains essentially the same, indicating that the residual stress level is not the main factor affecting the deformation of the rail surface.

Table 3. Y-direction displacement of guide rail joint surface caused by creep and temperature.

	Case A3			Case C		
	Point 1	Point 2	Point 3	Point 1	Point 2	Point 3
Y-direction displacement	5.5e-3	5.3e-3	5.4e-3	-2.4e-6	-2.9e-6	-2.5e-6

Journal of Physics: Conference Series **3009** (2025) 012088 doi:10.1088/1742-6596/3009/1/012088

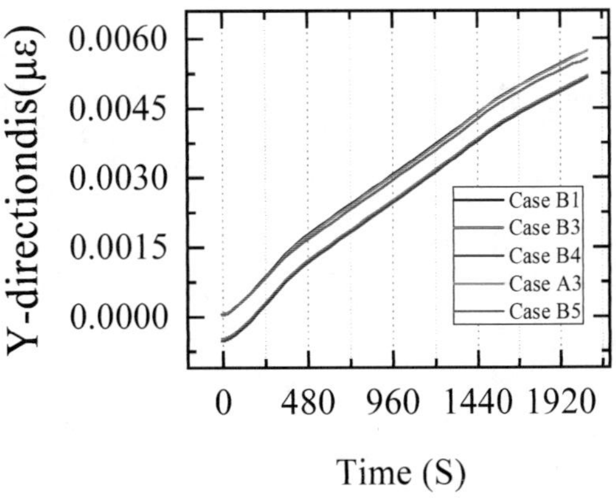

Figure 16 Displacement in Y-direction of guide rail joint surface with different residual stress.

In terms of stress, the maximum equivalent stress on the upper surface of the rail after creep was studied. The results indicate that different residual stress levels lead to significant differences in the final equivalent stress (Figure 17). Specifically, for residual stress levels ranging from 180 MPa to 300 MPa, the maximum stresses after creep analysis are 105.7 MPa, 103.7 MPa, and 107.6 MPa, respectively, demonstrating that higher residual stress levels eventually tend to stabilize. By comparing the curves of the maximum equivalent stress on the upper surface of the rail over time, the influence of different residual stress levels on the rate of stress change can be observed: the higher the residual stress level, the higher the initial stress, and the faster the rate of stress decay. However, when the residual stress level exceeds a certain threshold (such as 180 MPa in this study), further increases in residual stress have an insignificant effect on the final stress state.

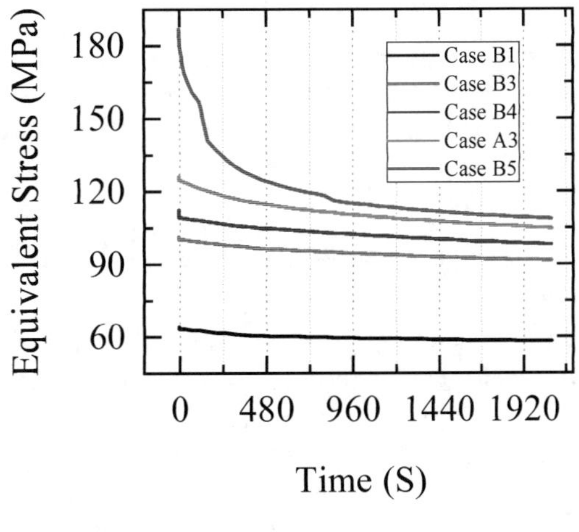

Figure 17 Equivalent stress of guide rail joint surface under different residual stress.

Regarding the strain, we studied the maximum values of thermal strain and creep strain on the upper surface of the guide rail after creep. The results show that under different residual stress levels, the thermal strain is completely consistent, with no differences (Table 4). The maximum variation in creep strain is 867 με, with a variation rate of 0.7%, which can also be neglected. Further comparing the curves of creep strain over time under different residual stress levels (Figure 18), due to the creep calculation formula of the guide rail material HT300: $\dot{\varepsilon} = 2.6 \times 10^{-38} \sigma^{14.88}$, under the influence of the temperature field, the stress on the upper surface of the guide rail decreases. This leads to a change in the creep strain rate, and through the effect of time integration, the creep strain undergoes significant changes.

Table 4. Y-direction displacement of guide rail joint surface caused by creep and temperature.

	Temperature strain （）	Creep strain （）
Case A3	102	154
Case B1	102	0.4
Case B2	102	41.3
Case B3	102	87.4
Case B4	102	867

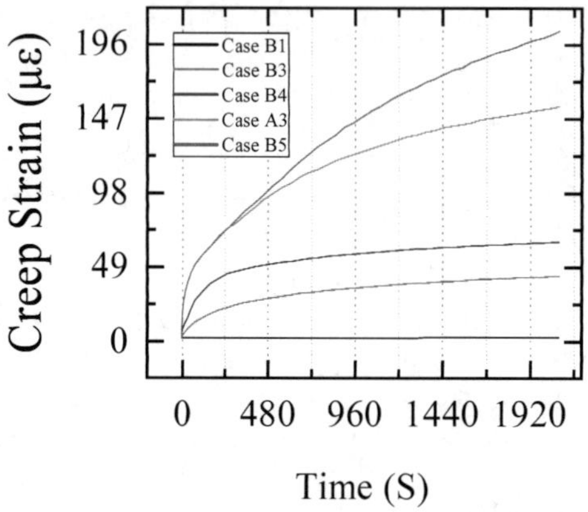

Figure 18 Creep strain of guide rail joint under different residual stresses.

By comparing the temperature results, displacement results, stress results, and strain results under different residual stress levels, it can be determined that the residual stress level has a negligible effect on the temperature and displacement results of the guide rail structure. However, it has a significant impact on the stress results and creep strain results. The higher the residual stress level, the greater the stress and the greater the creep strain.

5. Summary and conclusions

This study conducted a comprehensive analysis of a specific rail structure under various conditions, including different slider positions, levels of residual stress, and the influence of a temperature field. By simulating the creep process of eight models over 2100 seconds using ANSYS software, we examined the effects on temperature, displacement, stress, and strain at critical locations on the rail surface. Our findings reveal that:

- Minimal Displacement Fluctuation: The fluctuation in Y-direction displacement across all models is approximately 0.003 mm, with the main area exhibiting fluctuations less than 0.001 mm, indicating satisfactory operational performance.
- Negligible Impact of Slider Position: The slider position has a negligible effect on temperature and displacement results but significantly influences stress and creep strain. This suggests that a static model with a fixed slider position can be used to simplify simulations without compromising accuracy.
- Temperature Thermal Effect: Due to the temperature rise during operation (from 11°C to 20°C), the overall displacement of the guide rail surface and the surface fluctuation values can

increase by more than tenfold. Therefore, timely reduction of lubricant and environmental temperatures is an important measure to ensure the accuracy of the guide rail.

- Surface Residual Stress: Residual stress introduced by manufacturing processes is positively correlated with creep strain during the operation of the guide rail. The creep process effectively reduces residual stress; after 2100 seconds, initial stresses of 200 MPa and 300 MPa decreased to 100 MPa.

In summary, this paper constructs a creep model to experimentally study and summarize the impact of residual stress and thermal effects on the surface accuracy of the guide rail. It also demonstrates that the slider position has little impact on the calculation results, providing a practical method to simplify modelling and improve simulation efficiency. Based on this work, future research can continue in directions such as using forged components processed after tempering or aging treatment, reducing the working temperature of the guide rail surface, and extending the simulation time to observe changes in guide rail surface accuracy when the creep effect dominates.

References

[1] Anonymous. Linear slides - an essential ingredient for automated Production [J]. Assembly Automation, 1998, 18(2).

[2] Spence T, Makhlouf M M. Modelling Creep Induced by Machining Residual Stresses in Aluminium Alloys [J]. Materials Science Forum, 2013, 765: 585–589.

[3] Jiang X, Kong X, He S, et al. Modeling the superposition of residual stresses induced by cutting force and heat during the milling of Thin-walled Parts [J]. Journal of Manufacturing Processes, 2021, 68: 356–370.

[4] Ding W, Zhang L, Li Z, et al. Review on Grinding-induced residual stresses in metallic Materials [J]. The International Journal of Advanced Manufacturing Technology, 2016, 88(9–12): 2939–2968.

[5] Gong P, Zhang Y, Wang C, et al. Residual stress generation in grinding: Mechanism and Modeling [J]. Journal of Materials Processing Technology, 2024, 324: 118262.

[6] Su J-C, Young K A, Srivatsa S, et al. Predictive modeling of machining residual stresses considering tool edge Effects [J]. Production Engineering, 2013, 7(4): 391–400.

[7] Naghdi Sedeh M R, Ghaei A. The effects of machining residual stresses on springback in deformation machining bending Mode [J]. The International Journal of Advanced Manufacturing Technology, 2021, 114(3–4): 1087–1098.

[8] Jiang H, Man Z, Guo Z, et al. Determination of residual stress field in laser cladding using finite element updating method driven by contour Deformation [J]. Optics & Laser Technology, 2024, 179: 111371.

[9] Taherinia A, Eslami A, Golozar M A, et al. High-Temperature Creep Analysis of Carbon Steel A516-Gr70 Used in Thin-Walled Pressure Vessels Under Different Loads at Constant Temperature [J]. Arabian Journal for Science and Engineering, 2024, 49(11): 15667–15677.

[10] Yao H-T, Xuan F-Z, Wang Z, et al. A review of creep analysis and design under Multi-axial stress States [J]. Nuclear Engineering and Design, 2007, 237(18): 1969–1986.

[11] Sattar M, Othman A R, Kamaruddin S, et al. Limitations on the computational analysis of creep failure models: A Review [J]. Engineering Failure Analysis, 2022, 134: 105968.

Journal of Physics: Conference Series **3009** (2025) 012089 doi:10.1088/1742-6596/3009/1/012089

Research on thermal cycling reliability of GaN chip based on micro/nanostructure-assisted nano-silver sintering interconnection

Lei Wu[1,2,*]

[1] The 10th Research Institute of China Electronics Technology Corporation, Chengdu 610036, China
[2] Sichuan Engineering Technology Research Center for Electricity Interconnect of Electronic Information Equipment, Chengdu 610036, China

[*]Corresponding author's e-mail: 18140049281@163.com

Abstract. The high-reliability interconnection of generation semiconductors represented by GaN is of significance for promoting the development of aerospace and 5G/6G communication. Herein, the micro/nanostructure arrays were introduced into the interconnection between the GaN chip and the Al2O3 substrate to reduce the dependence on harsh processing environments. A simulation model of micro/nanostructure-assisted nano-silver sintered GaN chip interconnection was established, and the deformation, stress, and strain behaviours of the entire interconnection structure and nano-silver sintering layer were further analyzed and revealed. The failure lifetime of the nano-silver sintering layer was calculated using a modified Coffin-Manson equation. The results demonstrated that micro/nanostructures within interconnection structures can ensure thermal cycle reliability while reducing the dependence on harsh processing environments.

1. Introduction

With the rapid development of new energy vehicles, aerospace, and 5G/6G communication, third-generation semiconductors represented by GaN have attracted much attention due to their high electron mobility, high thermal conductivity, and high-temperature resistance [1-2]. To fully ensure these devices' high efficiency, reliability, and stability during the application, interconnection materials between the chip and substrate are required to exhibit excellent electrical, thermal, and mechanical properties [3]. However, due to its unsatisfactory remelting temperature, thermal conductivity, and high-temperature reliability, the common lead-based solder in the electronic packaging industry cannot meet the strict requirements of third-generation semiconductor devices. Particle sintering technology can interconnect metal nanoparticles to form block joints at specific temperature and pressure environments, and the joints can still serve with high reliability at high temperatures. Therefore, it is considered an effective packaging and interconnection method for third-generation semiconductor devices.

Silver nanoparticles have become a commonly used interconnect material in packaging third-generation semiconductor devices because of their high melting point, low resistivity, and good mechanical properties [4-5]. It is worth noting that the sintering rate between silver nanoparticles and the chip substrate is considerably slower than the self-sintering rate of silver nanoparticles because of their differences in chemical and physical properties, which hampers the efficient formation of strong interconnections. Correspondingly, increasing the sintering temperature, prolonging the processing time, and applying higher pressures are required to achieve satisfactory interconnection quality [6-7]. To avoid the drawbacks associated with such harsh processing conditions, nanostructure-assisted nano-silver sintering interconnection has been put forward to form nanoscale contact surfaces between chip back, nano-silver sintering layer, and substrate pad, which can provide a compatible chemical potential, thus reducing the dependence on harsh processing environments. Nevertheless, introducing nanostructures can give rise to stress concentration within the interconnection structure. During thermal cycling, which chips usually experience in their actual working environment, these stress concentrations can lead to crack initiation and propagation and eventually result in interconnection failure. Hence, a comprehensive understanding of the thermal cycling reliability of nanostructure-assisted nano-silver sintering interconnection is of utmost importance.

In this paper, micro/nanostructure arrays were introduced into the backside of GaN chip and copper pads on Al2O3 substrate to reduce the dependence on harsh interconnection environments. To demonstrate the thermal cycling reliability, a simulation model of micro/nanostructure-assisted nano-silver sintered GaN chip interconnection was established in ANSYS software, and the stress and strain behaviors of the interconnection structures under thermal cycle conditions were analyzed. The thermal fatigue life of the interconnection structures was predicted based on the Coffin-Manson equation to further verify the service reliability. This work not only provides theoretical support for the optimization of GaN chip interconnection technology but also has important guiding significance for promoting the application of GaN devices in harsh environments.

2. Simulation methods

In this study, a simulation model for evaluating the thermal cycle reliability of GaN chips relying on nanostructure-assisted nano-silver sintering interconnection was established using ANSYS workbench software. The constructed simulation model consisted of a micro/nanostructured GaN chip on the backside, an Al_2O_3 substrate with micro/nanostructured copper cladding, and a nano-silver sintering layer sandwiched in between. A coupled transient thermal and transient structural simulation model was adopted to accurately capture the complex physical phenomena. A thermal cycling load was applied to the simulation model to mimic the actual working conditions that chip interconnection may encounter in real applications. Considering the practical installation scenario where the Al_2O_3 substrate is firmly affixed to the base, the Al_2O_3 substrate is fixed on the base, a fixed constraint is applied to the center of the bottom surface, and the fixed displacement in the vertical direction is applied to the bottom surface of the substrate to mimic the actual fixation condition. Anand model was utilized to describe the deformation behavior of the nano-silver sintering layer. Regarding the material parameters, those of the DBC substrate, GaN chip, and copper pads have been reported in other relevant literature [8].

3. Results and discussion

Figure 1 illustrates the simulation model for evaluating the thermal cycle reliability of GaN chip interconnection. As shown in Figure 1a, the constructed simulation model includes a GaN chip, nano-silver layer, copper cladding, and Al_2O_3 substrate. Figure 1b zooms in on the details of the uniform micro/nanostructure arrays on the GaN chip and copper cladding. These precisely patterned arrays can be successfully fabricated through advanced femtosecond laser processing techniques. The dimensions of a single micro/nanostructure are precisely defined, with a height of 80 μm and a width of 50 μm, respectively. According to the JEDEC-A104F standard [9], the temperature cycling regime was meticulously set. The temperature range spanned from -55 to 150°C, and a total of 4 cycles were implemented. Both heating and cooling temperatures were set as 16 minutes, while the holding periods at both the high and low temperatures were maintained at 15 minutes, as depicted in Figure 1c

Journal of Physics: Conference Series **3009** (2025) 012089 doi:10.1088/1742-6596/3009/1/012089

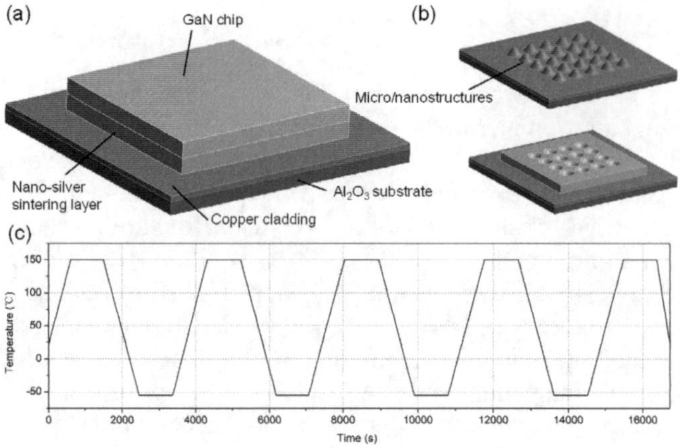

Figure 1 The simulation model for evaluating thermal cycle reliability of GaN chip interconnection. (a) The interconnection structure; (b) The details of the micro/nanostructured GaN chip, Al2O3 substrate with micro/nanostructured copper cladding, and the nano-silver sintering layer; (c) Temperature cycle curve.

Due to the differences in the thermal expansion coefficient among component materials of the interconnection structure, each component's expansion and contraction degrees are not identical during temperature cycles, generating thermal stress within the interconnection structure. The nano-silver sintering layer exhibits viscoplastic mechanical properties. The plastic deformation could occur after loading thermal cycling, thereby leading to thermal fatigue failure. Figure 2 displays the simulation results regarding the deformation behavior of the interconnection structure during temperature cycles. When the temperature increases from -55 to 150°C, the maximum deformation of the entire interconnection structure (68 μm) occurred at the corners of the Al_2O_3 substrate, as shown in Figure 2a. This can be attributed that each component expanded during the heating process. The corners of the Al_2O_3 substrate had less constraint from adjacent parts and thus can deform more freely to relieve thermal stress, thereby resulting in the generation of larger deformation. The deformation distribution cloud of the nano-silver sintering layer was further displayed in Figure 2b. The maximum deformation (55 μm) was also observed at the corners of the nano-silver sintering layer, and it was slightly less than that of the Al_2O_3 substrate. Such results demonstrated that the micro/nanostructures on the Al_2O_3 substrate and GaN chip exhibited little effect on the deformation distribution of the nano-silver sintering layer while increasing the interfacial shear strength.

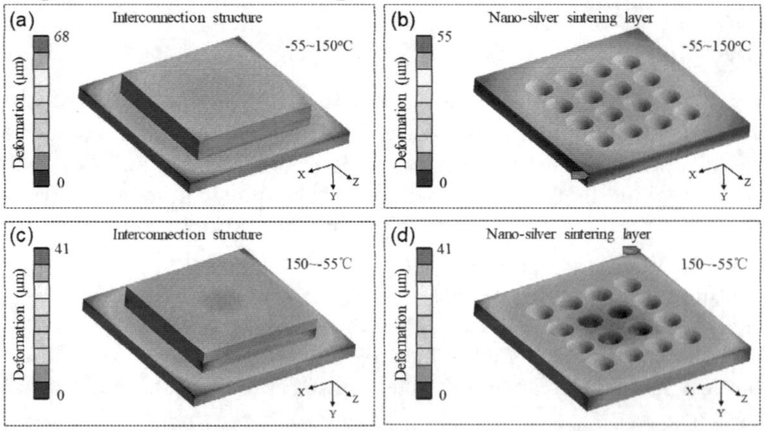

Figure 2 Distribution cloud maps of the deformation. (a) & (b) The distribution cloud of interconnection structure and nano-silver sintering layer when the temperature rises from -55 to 150°C; (c) & (d) The distribution cloud of interconnection structure and nano-silver sintering layer when the temperature decreases from 150 to -55°C.

When the temperature decreased from 150 to -55°C, the maximum deformation of the entire interconnection structure and nano-silver sintering layer still occurred at corresponding corners, as shown in Figs. 2c and 2d. Moreover, the maximum deformation values in the cooling process are lower than during the heating process. It is also noted that the middle part of the nano-silver sintering layer presented extremely small deformation. Further analysis deduced that residual stress was generated during the cooling process, and each component in the interconnection structure was subjected to both residual stress and mechanical constraints. On the other hand, the nano-silver particles within the sintering layer might attempt to rearrange to relieve thermal stress. Still, the viscoplastic properties and the existing deformation state limit the rearrangement. The middle part of the nano-silver sintering layer still exhibited a small deformation because it is less affected by the boundary-induced stress changes. In comparison, the boundaries are constantly influenced by the interaction with other materials, and contraction differences among different components need to be accommodated. Yet, the overall deformation was constrained by the combined effects of residual stress and material properties, resulting in smaller deformation compared to the heating process.

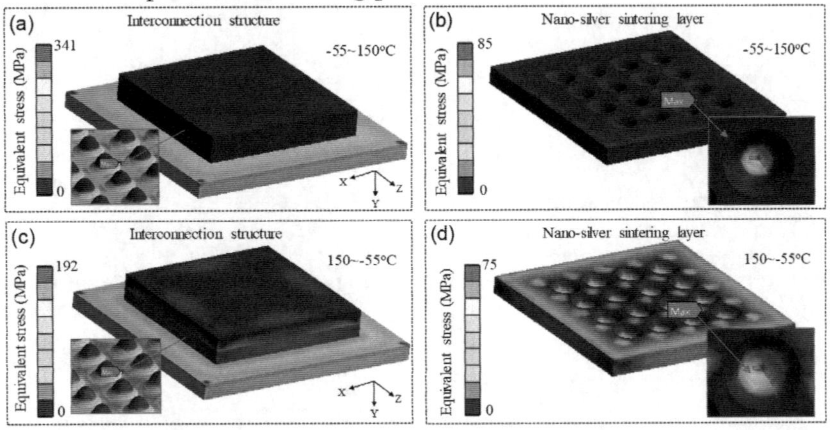

Figure 3 Distribution cloud maps of equivalent stress. (a) & (b) The distribution cloud of interconnection structure and nano-silver sintering layer when the temperature rises from -55 to 150°C; (c) & (d) The distribution cloud of interconnection structure and nano-silver sintering layer when the temperature decreases from 150 to -55°C.

Figure 3 displays the simulation results regarding the equivalent stress of interconnection structure during temperature cycles. Whether during the heating or cooling process, the maximum equivalent stress occurred near the micro/nanostructures of copper pads on the Al_2O_3 substrate surface (Figs. 3a and 3c), and the maximum equivalent stress during heating process (341 MPa) was obviously higher than that during cooling process (192 MPa). Figs. 2b and 2d further display distribution cloud maps regarding the equivalent stress of the nano-silver sintering layer during the heating and cooling process, respectively. The maximum equivalent stress occurred inside the concave structure. Correspondingly, structural defects, such as cracks, were generated in this area more easily, which were hazardous points that led to the failure of the entire interconnection structure. It was also observed that the maximum equivalent stress during the heating process (85 MPa) was higher than during the cooling process (75 MPa).

Figures. 4a and 4b show the curve of equivalent stress and equivalent plastic strain as a function of applied time during temperature cycling of hazardous points in the nano-silver sintering layer, respectively. The results indicated that both exhibited periodic changes under the loading of temperature cycles. During the high-temperature maintenance stage (150°C), equivalent stress significantly decreased, while equivalent plastic strain remarkably increased (-55°C). During the low-temperature maintenance stage, equivalent stress significantly decreased, but equivalent plastic strain slowly decreased. During the transition stage from -55 to 150°C, equivalent stress dropped sharply, while equivalent plastic strain remained basically stable. During the transition stage from 150 to -55°C,

Journal of Physics: Conference Series **3009** (2025) 012089 doi:10.1088/1742-6596/3009/1/012089

equivalent stress significantly increased, while equivalent plastic strain drooped remarkably. It was also noted that during the temperature cycling process, the maximum and minimum values of equivalent stress basically kept stable (Figure 4a). In comparison, both maximum and minimum values of equivalent plastic strain gradually increased, and the difference between the two basically remained unchanged. Therefore, during the temperature cycling process, plastic stains could continue to accumulate until failure occurred at the hazardous point.

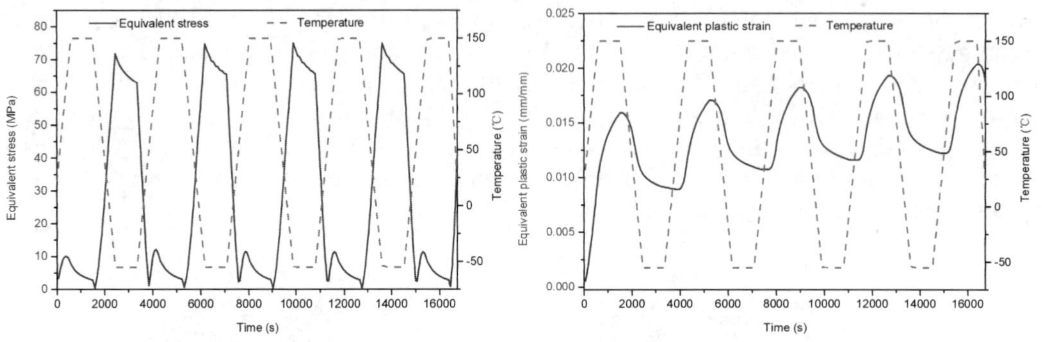

(a) (b)

Figure 4. The curve of (a) equivalent stress and (b) equivalent plastic strain as a function of applied time during temperature cycling of hazardous points in the nano-silver sintering layer.

The failure lifetime (N_f) of the nano-silver sintering layer can be evaluated using a modified Coffin-Manson equation.

$$N_f = \frac{1}{2}\left(\frac{\sqrt{3}\Delta\varepsilon}{2\varepsilon_f}\right)^{\frac{1}{C}}$$

Here, $\Delta\varepsilon$ represents the range of equivalent plastic strain, which can be obtained from Figure 4b, and ε_f is the fatigue toughness coefficient, often taken as 0.325 [10]. The C refers to the fatigue toughness index, which can be calculated using the following equation:

$$C = -0.442 - 6 \times 10^{-4} T_m + 1.74 \times 10^{-4} \ln(1 + f)$$

where T_m represents the arithmetic mean of the highest and lowest temperatures during the cycles. The f refers to the cycle number within a day. Based on the above equation, the N_f of the nano-silver sintering layer was calculated as 4998. Such results demonstrated that micro/nanostructures within interconnection structures can ensure thermal cycle reliability while reducing the dependence on harsh processing environments.

4. Conclusion

In this paper, a simulation model of micro/nanostructure-assisted nano-silver sintered GaN chip interconnection was constructed to demonstrate its thermal cycling reliability. The simulation results indicated that the maximum deformation of the entire interconnection structure and nano-silver sintering layer still occurred at the corners, and the maximum deformation values in the cooling process were lower than those during the heating process. Both equivalent stress and equivalent plastic strain exhibited periodic variations with applied time under the loading of temperature cycles. The failure lifetime of the nano-silver sintering layer was calculated as 4998 based on the modified Coffin-Manson equation, demonstrating a high thermal cycle reliability. This work not only provides theoretical support for the optimization of GaN chip interconnection technology but also has important guiding significance for promoting the application of GaN devices in harsh environments.

References

[1] Ahn H B, Ji H G, Choi Y, Lee S, Kang D M and J Han (2022) 25–31 GHz GaN-based LNA MMIC employing hybrid-matching topology for 5G base station applications. IEEE Microw. Wirel. Tech., 33: 47-50.
[2] Zhong S, Chen T, Chen Y and Mei B (2024) L-Band High Performance Broadband GaN Power Amplifier for Aerospace Application//2024 International Conference on Microwave and Millimeter Wave Technology (ICMMT). IEEE, 1: 1-3.
[3] Kozak J P, Zhang R, Porter M, Song Q, Liu J, Wang B, Wang R, Saito W and Zhang. Y (2023) Stability, reliability, and robustness of GaN power devices: A review. IEEE T. Power Electr., 38: 8442-8471.
[4] Meng F, Peng J, Huang J, Ping B, Lu W, Zhou J and Li P (2022) Computational analysis model of intense pulsed sintering of silver nanoparticles. Addit. Manuf., 51: 102594.
[5] Lv W, Hu J, Liu J, Xiong C and F Zhu (2023) Porosity effect on the mechanical properties of nano-silver solder. Nanotechnology, 34: 165701.
[6] Zhou W, Zheng Z, Wang C, Wang Z and An R (2017) One-step fabrication of 3D nanohierarchical nickel nanomace array to sinter with silver NPs and the interfacial analysis. ACS Appl. Mater. Interfaces, 9: 4798-4807.
[7] Fang Z Z and Wang H (2008) Densification and grain growth during sintering of nanosized particles. Int. Mater. Rev., 53: 326-352.
[8] Yan L, Liu P, Xu P, Tan L and Zhang Z (2023) Reliability Analysis of Flip-Chip Packaging GaN Chip with Nano-Silver Solder BUMP. Micromachines, 14: 1245.
[9] Yang B X, Fei J B, Liang S B, Zhou M B, Hu W L, Huang H J and Zhang X P (2021) 3D Finite Element Simulation Study of Chip Stacking Structure Considering Different Numbers of Stacked Dies and the Effects of Underfill and Intermetallic Compound Layer of Micro-joints//2021 IEEE 71st Electronic Components and Technology Conference (ECTC). IEEE, 1405-1411.
[10] Yang H (2021) Influencing factors of fatigue life of nano-silver paste in chip interconnection. J. Electron. Mater., 50:224-232.

Journal of Physics: Conference Series **3009** (2025) 012090 doi:10.1088/1742-6596/3009/1/012090

Simulation of multi-motor synchronous control for shield main drive based on modelica

Peng Zhou, Pengcheng Sun* and Xin Liang

Aviation Maintenance NCO School, Air Force Engineering University, Xinyang 464000, China

*Corresponding author's e-mail: 316536985@qq.com

Abstract. The main drive system of a shield tunneling machine is typically powered by multiple motors working in unison. Any desynchronization among these motors can lead to significant damage to the entire transmission system. The complexity of synchronization control structures necessitates an advanced approach. Under the premise of a circular coupling control strategy, this study combines fuzzy control with PID control to design a fuzzy adaptive PID control algorithm. This algorithm is integrated with the multi-motor circular coupling control strategy and modeled by using Modelica, a multi-domain simulation language, for analysis. Comparative results against traditional PID control methods indicate that the implementation of fuzzy adaptive PID circular coupling synchronization significantly enhances the synchronous performance of each motor. Furthermore, it effectively improves the stability and dynamic performance of the main drive system, increases control accuracy, accelerates system response speed, and demonstrates robust characteristics.

1. Introduction

The main drive system, as a crucial component of the shield tunneling machine, is responsible for providing power to the cutter head for soil excavation. The stability and reliability of the control system play an essential role in this process [1,2]. Given that shield machines operate in complex and diverse geological strata with poor stability, sudden changes in cutter head torque or uneven distribution of torque among motor drive shafts can easily occur. In severe cases, this may lead to axle breakage incidents [3, 4]. Therefore, it is imperative to study the stability of motor control methods for the main drive system of shield machines and assess the efficiency and feasibility of multi-motor synchronous control.

An improved master-slave multi-motor synchronous control strategy is proposed, concluding that this enhanced approach is more suitable for multi-motor synchronous control under non-homogeneous geological conditions encountered by shield machines [5]. Simulation analysis using Simulink software is conducted and fuzzy PID control exhibits excellent performance under parallel synchronous control strategies [6]. A controller employing an improved master-slave control strategy combined with vector frequency conversion methods is designed, resulting in a dual closed-loop speed-torque control system [7].

These studies primarily focus on relatively simple PI controls or uncoupled multi-motor controls, which do not yield ideal results. Since fuzzy control does not require establishing precise mathematical models and is well-suited for nonlinear complex systems, this paper first conducts a comparative analysis of various control strategies' advantages and disadvantages. A specific multi-motor synchronous control strategy will be selected based on these findings; subsequently, leveraging the strengths of fuzzy logic alongside PID methods allows us to design a multi-motor fuzzy adaptive PID controller based on circular coupling synchronization using Modelica language [8]. This proposed method features a straightforward structure that is easy to implement; simulation results indicate rapid response times and high precision in controlling speed while effectively suppressing disturbances caused by load fluctuations.

2. Research on fuzzy PID control of single motor

2.1 System composition

The main drive system of the shield has two driving modes: hydraulic motor drive and motor drive. In the motor drive mode, it is mainly composed of a three-phase asynchronous motor, frequency converter, reducer, big gear ring, main bearing, and lubricating seal device. As shown in Figure 1, the big gear ring is fixed with the flange, and the motor output power is transmitted to the big gear ring through the reducer, thus driving the knife to rotate.

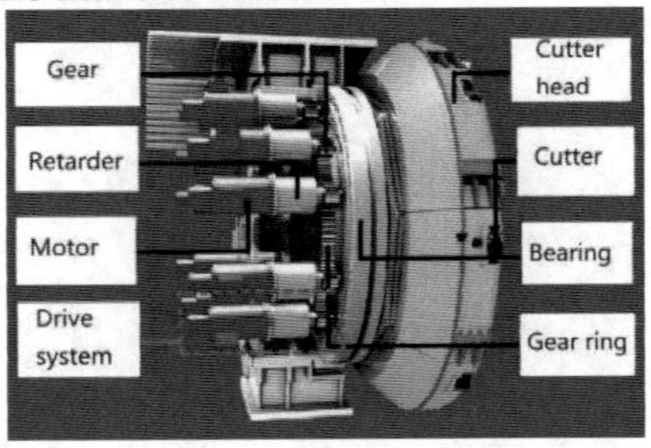

Figure 1. 3D diagram of the main drive system of the shield machine.

2.2 Mathematical model of three-phase asynchronous motor based on vector control

To simplify the mathematical model equation of the motor, the coordinate transformation theory is applied to convert the three-phase stationary ABC coordinate system of the stator of the asynchronous motor into a two-phase synchronous rotating d-q coordinate system to realize the decoupling of the mathematical model of the motor [9]. The mathematical model in the two-phase rotation d-q coordinate system can be expressed by Equations (1), (2), (3), and (4).

$$\begin{bmatrix} u_{sd} \\ u_{sq} \\ u_{rd} \\ u_{rq} \end{bmatrix} = \begin{bmatrix} R_S + L_s p & -\omega_{dqs} L_s & L_m p & -\omega_{dqs} L_m \\ \omega_{dqs} L_s & R_s + L_s p & \omega_{dqs} L_m & L_m p \\ L_m p & -\omega_{dqr} L_m & R_r + L_r p & -\omega_{dqr} L_r \\ \omega_{dqr} L_m & L_m p & \omega_{dqr} L_r & R_r + L_r p \end{bmatrix} \begin{bmatrix} i_{sd} \\ i_{sq} \\ i_{rd} \\ i_{rq} \end{bmatrix} \tag{1}$$

$$\begin{bmatrix} \Psi_{sd} \\ \Psi_{sq} \\ \Psi_{rd} \\ \Psi_{rq} \end{bmatrix} = \begin{bmatrix} L_s & 0 & L_m & 0 \\ 0 & L_S & 0 & L_m \\ L_m & 0 & L_r & 0 \\ 0 & L_m & 0 & L_r \end{bmatrix} \begin{bmatrix} i_{sd} \\ i_{sq} \\ i_{rd} \\ i_{rq} \end{bmatrix} \tag{2}$$

$$T_e = n_p L_m (i_{sq} i_{rd} - i_{sd} i_{rq}) \tag{3}$$

$$T_e = T_L + \frac{J}{n_p} \frac{d\omega}{dt} \tag{4}$$

$$\omega = \omega_{dqs} - \omega_{dqr} \tag{5}$$

2.3 Fuzzy PID controller design

Fuzzy control rules evolved from the empirical knowledge of engineers and technicians and do not require accurate mathematical models to control objects, which has strong robustness [10]. In this paper, a fuzzy adaptive PID control algorithm is adopted to change the PID control parameters according to the needs of the system in real time, which resolves the shortcoming that the traditional PID control parameters cannot be flexibly changed according to the changes of the system. The speed error e(t) and the error change rate ec(t) are used as the input of the fuzzy controller and the PID controller. Fuzzy variables E and EC are obtained. We input into the fuzzy control rule module for fuzzy operation, output fuzzy control quantity U, and finally get three output variables ΔK_P, ΔK_i, and ΔK_d of the fuzzy controller after clear processing in the defuzzy module. They are used as the parameter adjustment amount of the PID controller, and the three parameters of the PID controller are adjusted online by Equations (6), (7), and (8).

$$K_p = \Delta K_p + K_{p0} \tag{6}$$

$$K_i = \Delta K_i + K_{i0} \tag{7}$$

$$K_d = \Delta K_d + K_{d0} \tag{8}$$

To improve the steady-state accuracy of the system, the fuzzy set of error E is selected as 8 elements, and the discourse domain of the control quantity is discretized to 15 points. The fuzzy set of E is {NB, NM, NS, NO, PO, PS, PM, PB}, the error rate of EC and the amount of U fuzzy sets are shown for {NB, NM, NS, O, PS, PM, PB}, E and EC theory domain are {-6, -5,- 4, -3, -2, -1, 0, 1, 2, 3, 4, 5, 6}, and U theory domain is {-7, -6, -5, -4, -3, -2, -1, 0, 1, 2, 3, 4, 5, 6, 7}.

Membership functions of input and output variables are selected as triangular membership functions, which have less work and can meet the requirements. The membership function of E and EC is shown in Figure 2 and Figure 3, and the membership function of control quantity U is shown in Figure 4. The slope of the trigonometric membership function can be adjusted in practical application.

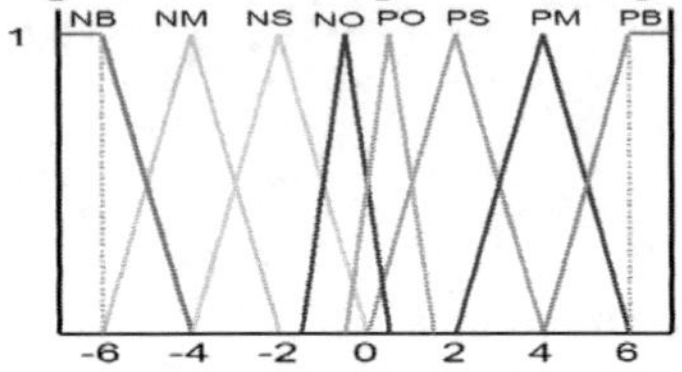

Figure 2. Membership function of E.

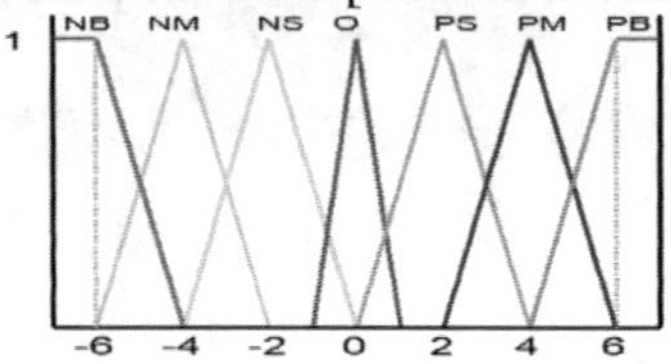

Figure 3. Membership function of EC.

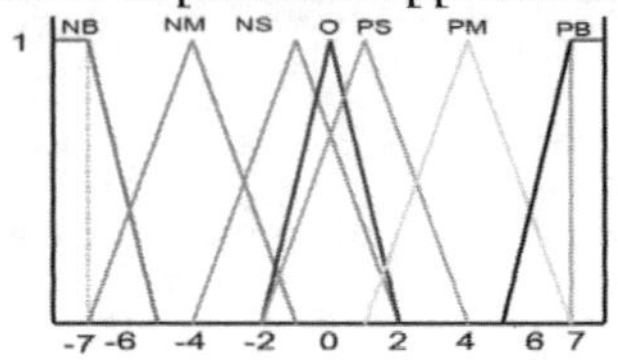

Figure 4. Membership function of U.

2.4 Develop fuzzy PID control rules

The formulation of control rules is the core step of fuzzy controller design, which is derived from the operator's experience and expert knowledge and can be realized by fuzzy conditional statements. In this paper, the Mamdani fuzzy inference algorithm, which is easy to implement, is adopted. According to the principle of the Mamdani minimax fuzzy inference method, 56 rules can be deduced from 8 linguistic variables of velocity error E and 7 linguistic variables of error rate of change EC. Each rule is derived from experience and many tests. Tables 1-3 list the fuzzy control rules.

Table 1. ΔK_p control rule.

U_p		NB	NM	NS	N0 (E)	PO	PS	PM	PB
(EC)	NB	PB	PB	PM	PM	PM	PS	PS	O
	NM	PB	PB	PM	PM	PM	PS	O	O
	NS	PM	PM	PS	PS	PS	O	NS	NM
	O	PM	PS	PS	0	O	NS	NM	NM
	PS	PS	PS	O	NS	NS	NS	NM	NM
	PM	O	O	NS	NM	NM	NM	NM	NB
	PB	O	O	NS	NM	NM	NM	NM	NB

Table 2. ΔK_i control rule.

U_i		NB	NM	NS	N0 (E)	PO	PS	PM	PB
(EC)	NB	NB	NB	NB	NM	NM	NM	O	O
	NM	NB	NM	NM	NM	NM	NS	O	O

NS	NM	NM	NS	NS	NS	O	PS	PM
O	NM	NS	NS	0	O	PS	PS	PM
PS	NS	NS	O	PS	PS	PS	PM	PM
PM	O	O	PS	PM	PM	PM	PB	PB
PB	O	O	PS	PM	PM	PB	PB	PB

Table 3. $\triangle K_d$ control rule.

U_d		NB	NM	NS (E)	N0	PO	PS	PM	PB
(EC)	NB	PS	PS	O	O	O	O	PB	PB
	NM	NS	NS	NS	NS	NS	O	PS	PM
	NS	NB	NB	NM	NS	NS	O	PS	PM
	O	NB	NM	NM	NS	NS	O	PS	PM
	PS	NB	NM	NS	NS	NS	O	PS	PS
	PM	NM	NS	NS	NS	NS	O	PS	PS
	PB	PS	O	O	O	O	O	PB	PB

3. Simulation analysis

A single motor control system model was built on MWorks, a multi-field simulation platform, based on Modelica, as shown in Figure 5. It mainly consists of the control module, rectifier module, inverter module, and motor module. The speed controller is the encapsulated fuzzy adaptive PID control model, which can modify the parameters in the model according to the actual situation. The motor is a three-phase AC asynchronous motor, with a rated voltage of 400 V, a rated frequency of 50 Hz, a pole number of 4, a rated current of 415.8 A, and a rotor moment of inertia 35 kg·m^2. The Diode Bridge module can convert three-phase AC into DC, and the inverter module in the motor package module can convert DC into AC with adjustable frequency to achieve accurate control of the motor.

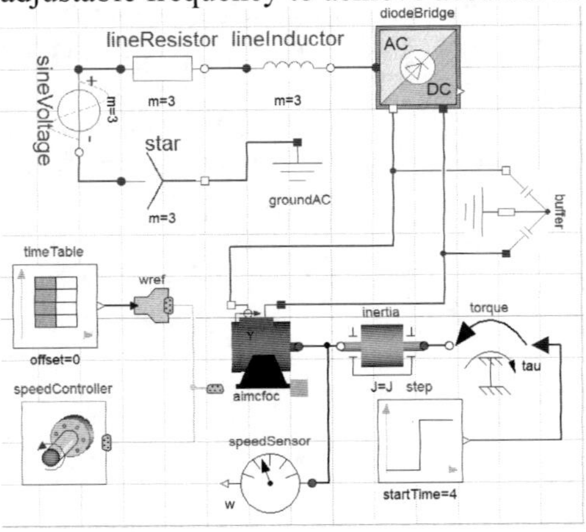

Figure 5. Modelica model of single motor fuzzy PID control system.

3.1 Single motor starting and load disturbance

To verify the effectiveness of the proposed fuzzy adaptive PID control method, it is compared with the traditional PID control method under start-up and load disturbance. In the start-up stage, as shown in Figure 6, the motor under the fuzzy PID control mode does not fluctuate after reaching the set value near 2.0 s and always runs at the set value. In the traditional PID control mode, the motor reaches the set point near 2.25 s, the response is delayed and there is a speed fluctuation phenomenon, and it becomes stable until about 2.7 s. Due to the harsh working environment of the shield machine, uneven ground distribution is very common, resulting in uneven load distribution of each motor of the main drive, which can be simulated by applying sudden load. When the sudden load is applied at 4.0 s, the load change rate is 120%, and the motor load disturbance curve is shown in Figure 7. The fuzzy adaptive PID control mode fluctuates more easily and becomes stable quickly. Under the traditional PID control mode, the load disturbance is severe, the maximum is 10 r/min, and the fluctuation time is long, and it becomes stable after 4.18 s.

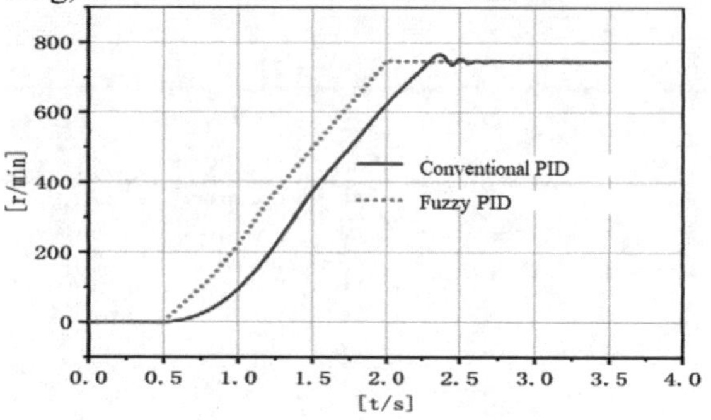

Figure 6. Comparison of speed fluctuations in the start-up phase.

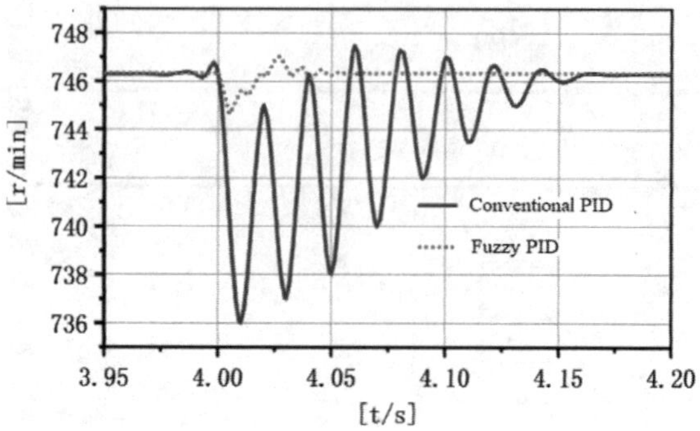

Figure 7. Comparison of velocity fluctuations under sudden load change.

3.2 Multi-motor starting and load disturbance

To realize the simulation of multi-motor synchronous control, the fuzzy PID control system model of a single motor is connected based on the loop coupling control strategy diagram, and the same reference speed is used for the simulation analysis of each motor. In the start-up stage, due to the uneven distribution of strata, the load on each motor is not the same, so the motor is treated differently and different loads are applied. There are inevitably differences in the manufacturing process of the motor, and the parameters of the four motors can be fine-tuned to make them different. To distinguish more clearly, we take the local period to enlarge. As shown in Figure 8, the starting speed response of the four motors is slightly different, which is small and negligible overall. At 4.0 s, a sudden load is applied to four motors, and the load change rate is 120%. When the load changes, the speed response curve of each motor is shown in Figure 9. The speed of each motor fluctuates slightly and quickly and becomes stable at 746 r/min after 4.06 s, indicating that fuzzy adaptive PID control has obvious advantages under the ring coupling control strategy. The steady-state performance and dynamic performance are improved.

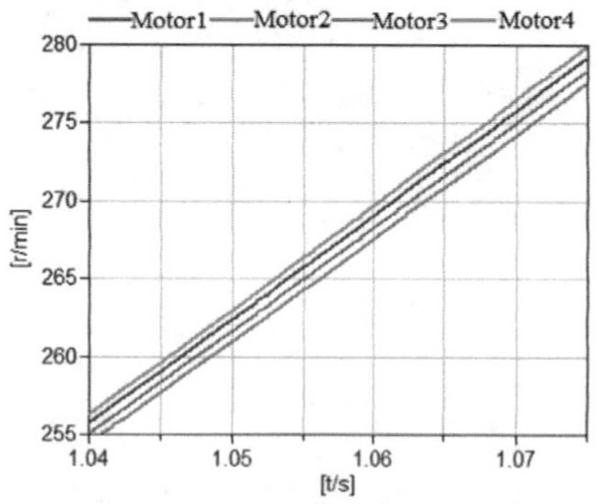

Figure 8. Local velocity response curve during startup.

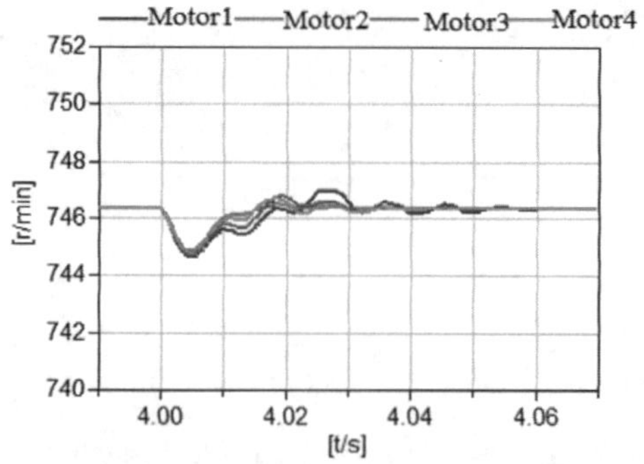

Figure 9. The velocity response curve of each motor under sudden load change.

3.3 Analysis of simulation results

It can be seen from the above simulation results that when a single motor is started, the response of fuzzy adaptive PID control is faster than that of traditional PID control, and the speed fluctuation is small enough to be ignored. When the load changes abruptly, the advantage of fuzzy adaptive PID control is more obvious, the speed fluctuation is small and can be quickly stabilized, while the traditional PID control speed fluctuation is obvious and the steady-state performance is poor.

According to the speed response curve of the multi-motor starting process under the ring coupling control strategy, the increase of motor number does not affect the overall control performance, and the starting performance is the same as that of single motor starting. When the load changes abruptly, the speed of each motor fluctuates to a certain extent but quickly recovers to a stable state, and the fluctuation trend of each motor is the same, which reflects the effectiveness of the synchronous control strategy. In summary, it can be concluded that the motor fuzzy adaptive PID control has good robustness, accelerates the system response speed, improves the control accuracy, and greatly improves the motor synchronization performance under the ring coupling control strategy.

4. Conclusion

In this paper, a fuzzy adaptive PID controller is designed based on Modelica language and applied to the synchronous control system of shield main drive multi-motors. Then, according to the ring coupling control strategy of multi-motors, modeling, and simulation analysis are carried out. The simulation results show that compared with the traditional PID control mode, under this control strategy, the steady-state performance and dynamic performance of the system are greatly improved, the response is faster, the robustness is good, the multi-motor following performance is excellent, and the synchronization error is significantly reduced. After the technology is mature, it can be used in the actual construction control process of the shield, which has important significance for high-end equipment manufacturing to a higher level.

Reference

[1] Wu H. (2020). Electromechanical coupling modeling and dynamic design of TBM main drive system. Dalian University of Technology, Dalian. https://doi.org/10.26991/d.cnki.gdllu. 2020.004080.
[2] Sun C., Gong G., Yang H., and Wang F. (2019). Fuzzy Sliding Mode Control for Synchronization of Multiple Induction Motors. Transactions of the Institute of Measurement and Control, 00 (0), 1-12. https://doi.org/10.1177/0142331218817100.
[3] Feng T., Hu J., and Li Y. (2024). Research on Real-time Prediction of Shield Cutterhead Torque Based on TCN-LSTM. Modern Tunnelling Technology, 61 (5): 120-128+166. https://doi.org/10.13807/ j.cnki.mtt.2024.05.013.
[4] Li F., Xiao L., and Sun C. (2021). Ring Coupling Synchronous Control of Multi-motor for Monorail Crane Based on PID Controller. Coal Mine Machinery, 42 (07): 191-193. https://doi.org/10.13436/ j.mkjx. 202107060.
[5] Ye Y., Peng F., and Huang Y. (2021). Overview of Multi-Motor Synchronous Motion Control Technology. Transactions of China Electrotechnical Society, 36 (14): 2, 922-2, 935. https://doi.org/10.19595/j.cnki. 1000-6753.tces.200343.
[6] Ren Y., Zhang H., and Zhou Z. (2020). Study on adaptive coupling synchronous control of cutter head driving motors of tunnel boring machine. Tunnel Construction, 40 (12): 1, 800.
https://kns.cnki.net/kcms2/article/abstract?v=TI3o-D_a8xm0W5IrgyAAa3D_FmajVtKv2p-W5-l1NDE-_4vIE2LCwCx_Bxcud-UXRHVE0NWZM9nbMNfR6J6CZKXnaxOIqr8xkNsjXLR3QxaM_VTLxDRnpG9jvp2CDI0aeZiBso3geOMZWGddn8Gw1YQUC5srO102R8NxgIKb7F5mKfOQyWYR69fH3LWY7hMahUuP6U4C2zs=&uniplatform=NZKPT&language=CHS.
[7] Liu J. (2022). Research on Synchronous Control Strategy of Networked Multi-Motor System. University of Jinan, Jinan. https://doi.org/10.27166/d.cnki.gsdcc.2022.000850.
[8] Qin D., Yang J., and Wang Y. (2017). Research on Modeling and Simulation of Pure Electric Bus Power System Based on Modelica. Mechanical Design and Manufacture, (06): 179-182. https://doi.org/ 10.19356/j.cnki.1001-3997.2017.06.047.
[9] Liu J. and Zhang C. (2024). Electric Motors and Drive (3rd Edition). Tsinghua University Press, Beijing.
[10] Wang B. and Li P. (2013). Fuzzy PID Control and Simulation of CNC Servo System Based on Modelica. Journal of System Simulation, 25 (12): 2, 924-2, 929. https://doi.org/10.16182/j.cnki.joss. 2013.12.018.

Metal rough surface leakage model based on porous percolation theory

*** Ying Lin[1*], Wanqin Zhu[1], Xinhai Yu[2], Shijian Chen[3] and Yongning Guo[3]**

[1]School of Mechanical Engineering, Shanghai Institute of Technology, Shanghai, China
[2]School of Mechanical Engineering, East China University of Science and Technology, Shanghai, China
[3]CNNC SUFA Technology Industry Co., Ltd. Suzhou, China

[*] Corresponding author: liny@sit.edu.cn

Abstract: Metal seals have been widely used in many industries. The micro-geometry of rough surfaces is an important factor affecting the leakage characteristics of metal static seals. Aiming at the surface structure of the metal plane seal, the contact analysis of the rough surface was carried out. Based on the porous percolation theory, the porosity and permeability of the sealing interface were calculated, and the leakage model of the metal plane seal was established. It was found that when the sealing specific pressure increases, the contact area of the asperity at the sealing interface expands, and the pores gradually close. When the sealing specific pressure reaches the critical value, the pores are completely closed and the seepage stops. Finally, the effectiveness of the leakage model was verified by the experimental study on the leakage characteristics of the toroidal metal seal. The model is helpful to explore the leakage mechanism and law of valve sealing structure and guide the practical engineering application.

1. Introduction

Metal static seals have been extensively utilized in aviation, navigation, nuclear energy, and various industrial sectors. Seal failure can result in environmental contamination and energy wastage due to the escape of sealing media, potentially leading to severe reliability issues [1-2]. As a result, the analysis of sealing mechanisms and the prediction of leakage in metal static seals have consistently garnered research focus.

To address the problem of static seal leakage, researchers investigated the failure mechanism at macroscopic and microscopic levels. They developed a mathematical model to calculate the leakage rate of the seal pair under specific working conditions. Ji et al. [3] studied the contact mechanics of mechanical seal interface by Hertz theory, established the leakage channel model of seal interface by percolation theory, and revealed the variation of porosity with parameters such as seal surface morphology and end load. Shi et al. [4, 5] proposed a static seal leakage model based on percolation theory and grid model, revealing a significant increase in leakage probability near the percolation threshold of the sealing pair. Furthermore, the sealing performance of the sealing pair was found to be significantly influenced by surface roughness and contact area.

At present, the research of many disciplines has established a method for quantitatively predicting the leakage rate of sealing structures through the seepage theory of porous media, which is of great significance to the study of the leakage rate of metal seals. Xin and Du [6] used the theoretical prediction method of the leakage rate of the typical permeable static sealing structure of the closed structure. Based on porous media seepage mechanics, the effects of rough surface morphology, material mechanical properties, and sealing load factors on leakage rate were studied. Tian et al. [7] analyzed the rough surface of the metal seal structure of the safety valve and deduced the calculation equation of the leakage rate when the medium was water based on Darcy's law. Cui et al. [8] established a leakage rate model of a double rough surface annular sealing interface. The effects of surface texture direction, rough surface fluctuation frequency, and material properties on the leakage characteristics of metal static seals were analyzed. The research shows that the sealing performance could be effectively improved by optimizing the design of sealing surface roughness and controlling the direction and fluctuation frequency of surface texture. In short, the percolation theory was applied to study the seepage behavior of fluid between rough surfaces. However, the existing research has insufficient consideration of pore connectivity and actual surface complexity, and the prediction accuracy and applicability of working conditions are weak.

In this paper, the leakage prediction theory of annular metal static seals was proposed. The leakage model was constructed based on the contact analysis and percolation model of a double rough surface. The relationship between sealing specific pressure and microscopic parameters was studied by the finite element method. The pore connectivity was analyzed and the leakage rate was obtained by porous model. Finally, the validity of the model was verified by experiments.

2. Leakage model

Figure 1 shows the process of calculating the leakage of the annular metal seal. Figure 1(a) is the whole annular sealing surface, and the local region was selected, such as Figure 1(c), to study the contact of micro seal. The W-M function was used to reconstruct the sealing surface in the local region, such as Figure 1(b), and a three-dimensional contact model was established. The pore state of the sealing surface before and after contact was studied, and a three-dimensional annular sealing leakage calculation model was established.

2.1 Characterization of 3D rough surface geometry

In the study of contact, the contact surface is generally regarded as a smooth surface. In practice, the surface roughness will affect the contact state. The surface roughness of the sealing surface was simulated and analyzed by using the bivariate W-M (Weierstrass-Mandelbrot) function and the measured roughness of the sealing surface by the three-dimensional topography instrument. Li et al. [9] studied the surface roughness and found that the surface image of the sealing surface was adaptive, which verified the feasibility of the W-M function. The three-dimensional fractal W-M function in Cartesian coordinates is capable of generating a rough surface, and its mathematical expression is given as follows [10]:

$$z(x,y) = L\left(\frac{G}{L}\right)^{D-2}\left(\frac{\ln\gamma}{M}\right)^{1/2}\sum_{m=1}^{M}\sum_{n=m_1}^{n_{\max}}\gamma^{(D-3)n} \times\left\{\cos\phi_a - \cos\left[\frac{2\pi\gamma^n(x^2+y^2)^{1/2}}{L}\cos\left(\arctan\left(\frac{y}{x}\right)-\frac{\pi m}{M}\right)+\phi_a\right]\right\} \tag{1}$$

where z(x, y) is the height of the point on the rough surface (x, y); L is the sampling length; D is the fractal dimension of the three-dimensional rough surface ($2<D<3$), which describes the irregularity of the surface height on all scales and is determined by the rough surface parameters; γ is the spatial frequency of the random contour; G is the scale coefficient of the magnitude of z (x, y); M is the overlap number of surface folds; $n_{\max}$ is the upper limit of n, $n_{\max} = \text{int}[1\log(L/Ls)/\log\gamma]$, Ls is the highest resolution, that is, the cutoff length; ϕ_a is a random phase with the value range of $[0, 2\pi]$, which is generated by a computer.

2.2 Finite element analysis model

Microscopically, the metal sealing surface is a rough surface composed of many asperities. The contact of the rough surface will begin from the contact of the local asperities. With the increase of the contact area, the uncontacted area is divided into some connected or disconnected pores. Therefore, it is necessary to analyze the characteristics of asperity contact mechanics. Based on the local region sealing surface reconstructed by the W-M function, the finite element ABAQUS software was used to calculate the contact of the local region. The contact model of the metal sealing surface was established, as shown in Figure 1(c), to obtain the contact state between the two sealing surfaces during the contact process and extract the contact data.

In the contact process, the lower surface of the model was fixed, the upper surface was applied to the downward displacement, and there was no friction and no slip between the two surfaces. In Figure 1(c), when the contact stress was 0, the asperity of the rough surface contacts did not deform. The contact state of the sealing interface at this time was defined as the initial state and the height of the sealing interface h0 was defined as the initial sealing gap height.

$$h_0 = \max(Z^1_{i,j} + Z^2_{i,j}), i, j = 1, 2, \cdots, n \tag{2}$$

In Figure 1(e), under the action of contact stress P_B, the sealing interface deformed and produced compression ω. At this time, the height h of the sealing interface changes from h_0 to h_0-ω. By extracting the average contact stress of the two contact surfaces and the average sealing interface height at each time step, the relationship between contact stress and compression was obtained.

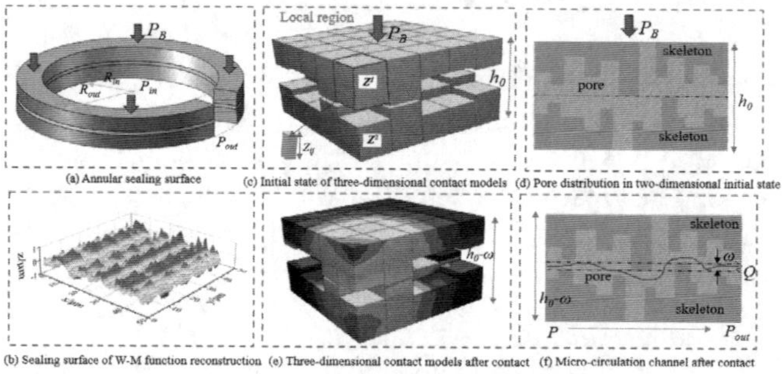

Figure 1. Sealing leakage calculation process (a) Annular sealing surface (b) Sealing surface of the W-M function reconstruction (c) Initial state of three-dimensional contact models (d) Pore distribution in two-dimensional initial state (e) Three-dimensional contact models after contact (f) Micro-circulation channel after contact.

2.3 Leakage model of annular metal static seal

It can be seen from Figure 1(d) that the sealing interface is composed of pores and skeletons. The prerequisite for the leakage of the sealing interface is that the pores between the sealing interfaces are interconnected to form a leakage channel through the interface. Under the action of contact stress P_B, the sealing interface produces a compression amount ω, forming a leakage channel through the interface, as shown in Figure 1(f). The fluid flows along the left boundary of the sealing interface to the right boundary driven by the pressure difference.

In this paper, a three-dimensional percolation grid model was proposed to simulate the local region sealing interface, as shown in Figure 1(c). The seepage between pores was calculated in a porous manner. To calculate the porosity, only the volume of the skeleton and the pores in the closed space can be calculated, which are composed of the highest and lowest peaks and valleys of the surface. It should be pointed out that the positions of the highest and lowest peaks can be determined by the sealing surface generated by the W-M function.

According to percolation theory, the initial porosity of the sealing interface $\phi(0)$ is:

$$\begin{aligned}\phi(0) &= 1 - \frac{V_{solid}(0)}{V_0(0)} \\ &= 1 - \frac{xy\sum_{i=1}^{n}\sum_{j=1}^{n}(Z^1_{i,j} + Z^2_{i,j})}{XYh_0}\end{aligned} \tag{3}$$

where $V_{solid}(0)$ is the solid volume of the sealed interface in the initial state, and $V_0(0)$ is the volume of the sealed interface.

Under the action of contact stress, the upper and lower sealing surfaces were in contact, and the compression amount ω was generated. The volume on the sealing interface changes from V_0=X×Y×h_0 to V_0', and the porosity of the sealing interface after deformation is:

$$\phi(\omega)=\frac{V_0^{'}-V_{孔}^{'}}{V_0^{'}}=\frac{h_0\phi(0)-\omega}{h_0-\omega} \tag{4}$$

When the sealing surface has a highly uniform material distribution and processing technology, the porosity of the whole sealing surface can be approximately estimated by the local region porosity calculated. The whole sealing interface is shown in Figure 1(a). The light green area in the figure is the sealing interface formed by the mutual extrusion of the upper and lower annular rough surfaces. The pressure difference between the inside and outside of the sealing interface produces a pressure difference in the porous area, which drives the fluid to flow through the sealing interface. To calculate the seepage caused by pressure drop, Darcy's law was used to calculate the seepage velocity of porous media, which shows that the seepage velocity is proportional to pressure gradient and permeability, and inversely proportional to fluid viscosity, as shown in Equation (5). The flow of single-phase Newtonian fluid in porous media shows that the permeability is only related to the geometric structure of porous media and has nothing to do with the properties of the fluid.

$$G(\omega)=\frac{K(\omega)}{\mu}(\nabla P) \tag{5}$$

$$K(\omega)=\frac{\phi(\omega)^3}{c(1-\phi(\omega))^2 s^2} \tag{6}$$

where G is the seepage velocity; K is permeability; μ is fluid viscosity; ∇P is the fluid pressure gradient; c and s are the Kozeny-Carman constant and the specific surface area of the solid phase, respectively.

Furthermore, the leakage model considering the microscopic surface morphology and macroscopic structure size of the sealing interface could be obtained, as shown in Equation (7).

$$Q=2\pi rhG(\omega)=-2\pi h\frac{K(\omega)}{\mu}\frac{P_{in}}{\ln\frac{R_{out}}{R_{in}}} \tag{7}$$

where R_{in} is the inner diameter of the sealing surface and R_{out} is the outer diameter.

3. Results and Discussion

3.1 Effect of surface topography on initial porosity

It can be seen from the above content that the establishment of the leakage model depends on the initial porosity, so the initial porosity of the local region needs to be studied in detail. The sealing surface was measured by using the Alicona Infinite Focus G4, a high-precision 3D surface measurement system employing optical profilometry, designed for detailed surface and microstructural analysis. The measured and W-M reconstructed local region surfaces were both used to calculate the porosity under four groups of roughness. The initial porosity data are presented in Figure 2. It can be seen that the initial porosity increases as the surface roughness increases, which is because there are many random and irregular asperities and gaps on the sealing surface. These irregular structures form more voids on the surface with larger roughness, increasing initial porosity. Error bars were derived by performing multiple independent measurements for each data point and calculating the standard deviation and standard error. Through the error bar, the reliability and uncertainty of the experimental data are accurately evaluated. The analysis of the error bars revealed a maximum measurement error of 14.7% and a minimum of 3.4%, indicating that the data fluctuations fall within a reasonable range, with the measurement method being both effective and precise. Overall, the error bars validate the reliability of the experimental data and support the accuracy of the W-M model in describing the relationship between surface roughness and porosity.

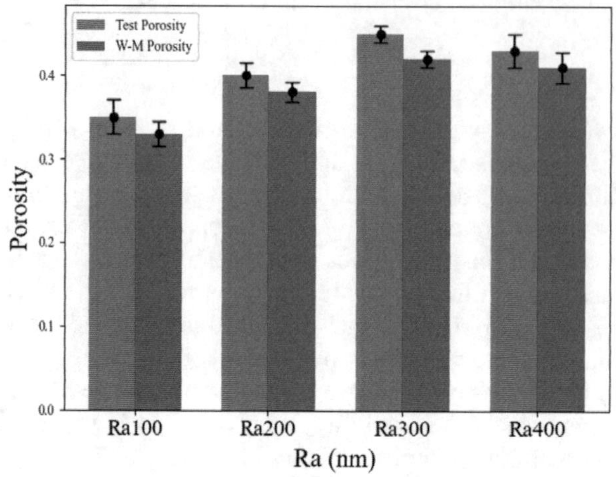

Figure 2. Comparison between the theoretical values and the reconstructed initial porosity.

3.2 Finite element analysis results

A pair of rough surfaces were constructed by the W-M function, and the contact analysis was conducted. Stellite 6 (STL6) was utilized as the material of surfaces to ensure the sealing performance, and its performance parameters were measured by a tensile machine. The results are shown in Table 1.

Table 1 Performance parameters of STL6 material.

Material property parameters	Measured value
Density g/cm^3	8.1
Young modulus GPa	202.11
Poisson's ratio	0.21
Yield limit MPa	592
Tensile limit MPa	859

The sealing surface with a roughness of 200 nm was selected for contact analysis. The contact results of the local region are shown in Figure 3. Figure 3 shows in detail the stress distribution and deformation of the sealing interface under different loading conditions. During the contact process, the distribution of the contact stress between the asperities will also change as the external sealing-specific pressure increases. When the contact stress is low, the contact points on the sealing surface are sparse. At this time, the sealing-specific pressure is mainly concentrated on a few asperities. When the sealing-specific pressure increases, more asperities begin to participate in the contact, the number of contact points increases, the contact area gradually expands, and the sealing interface area produces obvious contact deformation and the deformation is ω.

In the initial stage, due to the small sealing-specific pressure, only a few asperities are involved in the contact, the pore distribution is sparse and not completely closed, and seepage is easy to occur. As the sealing-specific pressure increases, the pores are further compressed or even closed. When the contact stress reaches a certain critical value, the sealing interface forms a penetrating contact area, and the pores are completely closed and no longer have seepage channels. This provides a theoretical basis for further study of the sealing performance and formation of leakage paths.

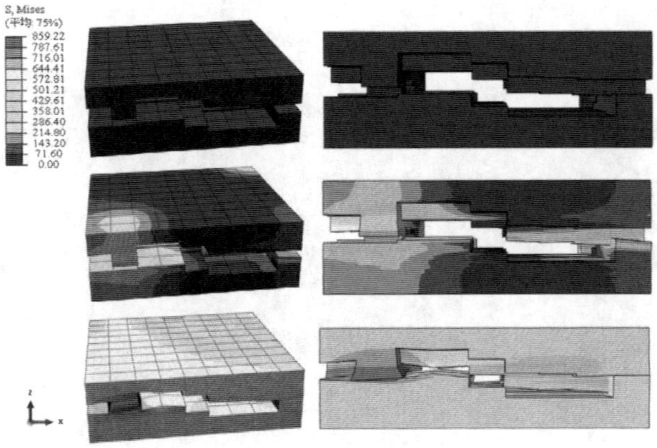

Figure 3. Finite element contact calculation results.

Figure 4 shows the relationship between the sealing-specific pressure PB extracted by the finite element contact calculation, the sealing gap h, and compression amount ω, reflecting the trend of the sealing gap closure and compression amount of the sealing interface during the loading process. With the increase of the sealing-specific pressure, the sealing gap decreases nonlinearly, and the decrease is more obvious in the initial stage and then tends to be gentle. At the same time, the amount of compression increases with the increase of sealing specific pressure, the initial stage increases rapidly, and the subsequent changes tend to be stable. The curves in the figure clearly show the direct effect of contact stress on the deformation of the sealing interface, and the results reveal the key role of contact stress in sealing performance.

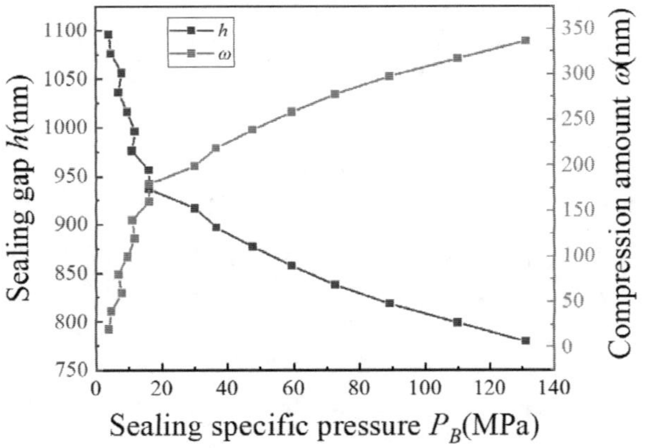

Figure 4. The relationship between sealing specific pressure and sealing gap and compression amount.

As the sealing pressure increases, the gap height decreases, some areas contact to play a sealing effect, and the uncontacted area is divided into some connected or disconnected holes, thus forming a leakage path. To further analyze the sealing surface conditions of different gap heights under the same sealing ratio, the pore distribution maps of the model heights *H* of the same sealing surface and different sections were drawn, as shown in Figure 5. The model height *H* refers to the vertical distance from the lower surface to the upper surface of the local region model. In Figure 5 (a), the black cluster was the skeleton area and the white cluster was the pore. From the diagram, when the sealing height H was 400 μm, there was no connected pore, and there would be no leakage at this time. At the same time step, as the height of the sealing interface section increases, as shown in Figures 5(b) and 5(c), the more connected pores there are between the sealing interfaces, the more connected channels there will be, resulting in enhanced leakage characteristics of the sealing interface.

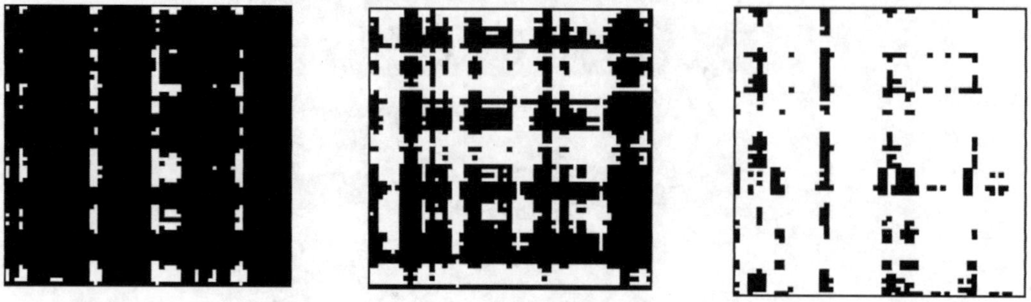

(a) ω = 240 nm, H = 400 nm (b) ω = 240 nm, H = 800 nm (c) ω = 240 nm, H = 1200 nm

Figure 5. Effect of model height on pore distribution.

According to the relationship between the sealing-specific pressure and the compression amount in Figure 6, the porosity of the sealing interface after loading under different contact pressures was calculated by Equation (4). The relationship between porosity and sealing-specific pressure is shown in Figure 6. It can be seen that with the increase of sealing specific pressure, the porosity shows a decreasing trend. In the initial stage, the increase of the sealing specific pressure makes the pores of the sealing material quickly compressed and filled, and the porosity decreases rapidly. When the sealing-specific pressure reaches a certain value, the decrease rate of porosity gradually slows down and eventually tends to be stable. At this time, when the porosity reaches the critical value ϕc and the porosity is less than the percolation threshold, the sealing interface will not form a leakage channel through the interface, and seepage will not occur.

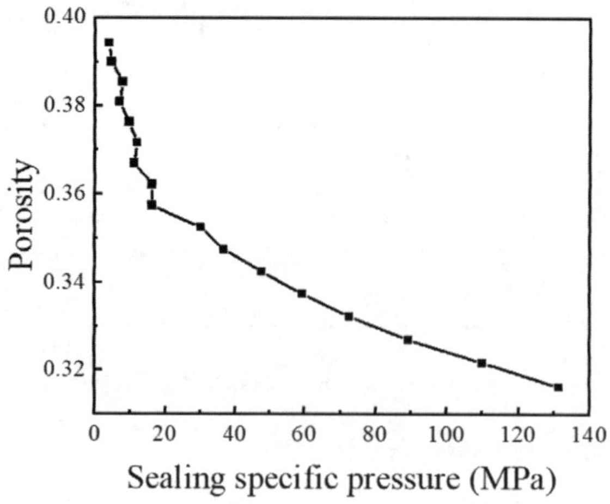

Figure 6. Sealing specific pressure vs. porosity.

3.3 Metal seal test and model verification

To validate the leakage model, a test device was built by using the test schematic diagram shown in Figure 7, and the leakage rate of the annular metal static seal under different working conditions can be obtained. The valve with a nominal diameter of 100 mm was selected to determine the sealing surface size. Two seals are placed in the test valve to contact to form an annular metal static seal. The high-pressure helium cylinder is connected by a high-pressure tee in the pipeline and a pressure-reducing valve that controls the pressure of the medium to provide pressure fluid for the sealing system. The medium pressure PN is measured by a high-precision pressure sensor. A weighing sensor is arranged on the valve disc to provide a contact load F for the sealing interface through the valve stem. The leakage rate data is obtained by connecting the flowmeter behind the test valve.

The sealing-specific pressure under the current working condition can be obtained by controlling the load and medium pressure, and the relationship is shown in Equation (8). At the same time, it is ensured that the calculated sealing specific pressure is lower than the minimum specific pressure required for sealing, to ensure the reliability and feasibility of the test results.

$$P_B = \frac{F - P_N \times (R_{in} + R_{out})^2 \times \pi}{(R_{out}{}^2 - R_{in}{}^2) \times \pi} \tag{8}$$

where R_{in} =48 mm, R_{out} = 52.5 mm.

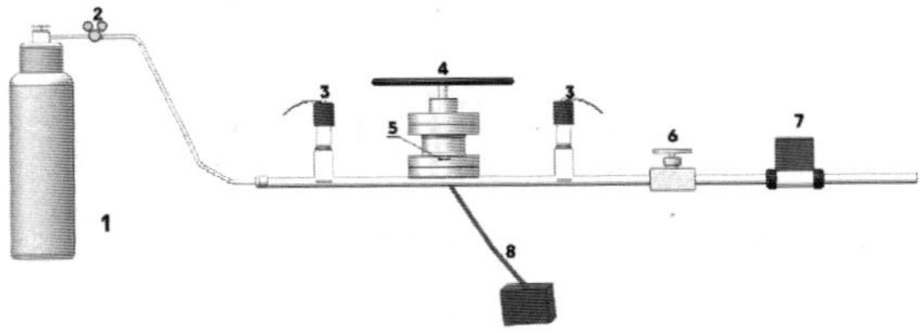

1) Helium bottle 2) Pressure reducing valve 3) Pressure transmitter 4) Test valve 5) Load cell 6) Isolation valve 7) Flowmeter 8) Temperature display instrument

Figure 7. Schematic diagram of the test platform.

The contact stress extracted by contact calculation was matched with the sealing-specific pressure in the experiment, and the results of the theoretical calculation leakage rate and experimental leakage rate were compared. We compared the porous percolation with the conventional Computational Fluid Dynamics (CFD) method, and the results are shown in Figure 8. The calculation results of the two methods are in good agreement with the experimental data, and the accuracy can meet the requirements. CFD uses a more complex multi-physical field coupling model, which can provide higher calculation accuracy, but its calculation process is cumbersome and requires higher computing resources and time. In contrast, the porous percolation method simplifies the calculation process by simplifying assumptions, and can significantly improve the calculation efficiency while ensuring reasonable accuracy. It is verified that the porous percolation method has good calculation accuracy under certain conditions, which proves its feasibility as a theoretical calculation model.

3.4 Effect of surface topography on leakage rate

Taking the test valve sealing surface with inner diameter R_{in} = 48 mm and outer diameter R_{out} = 52.5 mm as an example, the sealing performance of the sealing surface with roughness Ra400, Ra200, and Ra100 was studied under different medium pressures with sealing load of 80 kN. As shown in Figure 9, when the sealing surface structure and sealing load are determined, the leakage rate gradually increases with the increase of medium pressure, and the leakage rate of the sealing surface with large roughness is larger. This is because the increase of medium pressure leads to the decrease of sealing specific pressure, and the irregularity of the sealing surface leads to incomplete contact, which makes the leakage channel larger. The leakage rate is influenced by both surface roughness and sealing-specific pressure, highlighting that medium pressure is the dominant factor determining sealing performance.

For practical applications, this means that in a high-pressure environment, the sealing performance is mainly dominated by the medium pressure, and the optimization effect of roughness is limited. Therefore, in the design of the high-pressure sealing system, the requirements for surface roughness can be appropriately reduced, focusing on improving the medium pressure and optimizing the sealing-specific pressure to achieve a higher sealing effect and system reliability.

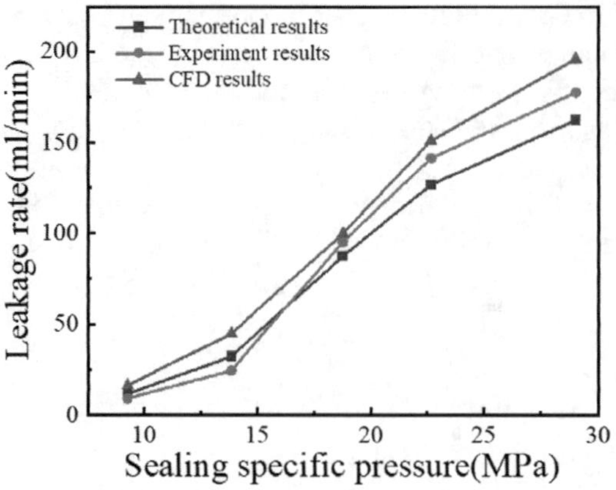

Figure 8. Comparison of leakage rate test results with calculated results.

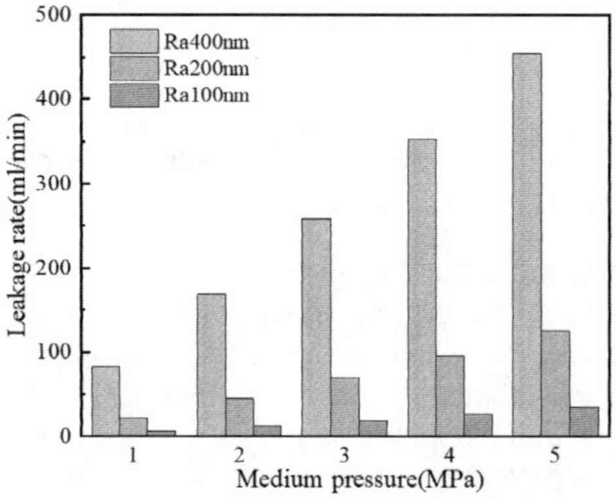

Figure 9. The relationship between surface roughness and leakage rate.

4. Conclusions

In this paper, based on the porous percolation theory, the leakage characteristics of metal plane seals were studied, and the corresponding leakage model was established. By analyzing the relationship between sealing surface roughness, sealing specific pressure and porosity, and permeability, the key role of contact stress in gap closure and interface deformation was revealed. The research shows that with the increase of sealing specific pressure, the pores gradually close and the seepage stops.

The leakage characteristics of the toroidal metal seal were experimentally studied by building a test platform. The results show that the model significantly improves the computational efficiency while ensuring the accuracy. The research on the relationship between different roughness and leakage rates shows that under high medium pressure, the influence of roughness on leakage rate is weakened, and the sealing performance is mainly dominated by medium pressure. Therefore, in the design of high-pressure seals, the roughness requirement can be appropriately reduced, and the sealing-specific pressure can be optimized to improve the sealing effect and system reliability.

References

[1] Wang, J., Zhang, T., Zhou, D., Zhang, W., Wang, T., Xu, G. (2024). Experimental Study of the Factors Influencing the Sealing Performance of Metal-to-Metal Sealing Cut-Off Valves. Lubrication Engineering, 1-8.

[2] Yu, L., Qian, J., Jin, Z. (2021). Research progress on sealing performance of valve Lubrication Engineering, 46(9). 134-142.

[3] Ji, Z., Sun, J., Lu, J., Ma, C., & Yu, Q. (2017). Predicting method for static leakage of contacting mechanical seals interface based on percolation theory. Tribology, 37(6), 734-742.
[4] Shi, J., & Liu, J. (2016). Study on performance of Krylov subspace methods at solving large-scale contact problems. The International Journal of Advanced Manufacturing Technology, 84(1-4), 435-444.
[5] Shi, J., Liu, J., Yang, Z., Zhang, M., Wang, L. (2016). Study on the evolution characteristics of static sealing interface state based on grid leakage model. Journal of Mechanical Engineering, 52(7), 104-113.
[6] Xin, L., & Du, Z. (2022). Study on leakage model of typical penetration of closed structures based on porous media seepage theory. Mathematical Problems in Engineering, Article 6901534, 15 pages.
[7] Tian, X., Zhang, D., Yang, Y., Tang, Y., Tan, S., Xie, T. (2022). Numerical Analysis of Leakage Characteristics of Valve Sealing Structure. Nuclear Power Engineering, 43(3): 151-157.
[8] Cui, Y., Yu, Y., Zhang, H., Zhong, J. (2022). Leakage model of metal static seal with two rough surfaces based on the theory of porous medium. Journal of Mechanical Engineering, 58(21), 215-224.
[9] Li, J., Tang, Y., Li, Z., Ding, X., Li, Z. (2017). Study on the optical performance of thin-film light-emitting diodes using fractal micro-roughness surface model. Appl. Surf. Sci, 410, 60-69.
[10] Deng, K., Liu, Z., Deng, J., Wang, Y., & Li, X. (2017). Variation law of surface profile morphology under W-M function model. Mechanical Design and Manufacturing, (1), 47-5.

Structural analysis of basalt fiber pultruded profiles for photovoltaic support in inner Mongolia desert region

Ruifeng Chen[1], Li Qin[1], Zhaoyang You[2*], Peng Hou[3], Jiuzhu Hao[3], Danhui Niu[4]

[1]New Energy |Business Department, Guangdong Power Grid Energy Development Co., Ltd., Guangzhou, China
[2]School of Electrical Engineering, Guangzhou City University of Technology, Guangzhou, China
[3]Engineering Department, Inner Mongolia Electric Power Survey & Design Institute Co., Ltd. Inner Mongolia Autonomous Region, China
[4] New Energy Development Department, Inner Mongolia Ulanbuh Energy Co., Ltd., Inner Mongolia Autonomous Region, China

[*]Corresponding author's e-mail: 1668394121@qq.com

Abstract: Driven by the 'dual-carbon' goal, photovoltaic (PV) power generation, as a key part of clean energy, is developing rapidly. However, the construction of PV power plants in desert regions suffers from extreme temperature differences, strong winds, and difficulty in construction, while traditional photovoltaic support materials, such as stainless steel and aluminum alloys, come up against various challenges. In this study, a new basalt fiber reinforced profile support (BFRPS) is proposed. Compared with traditional support, BFS shows lighter specific gravity, stronger UV resistance, and excellent corrosion resistance, which effectively solves the stability and durability problems of the support in the desert, making it an ideal alternative to traditional photovoltaic support (PVS). This study verifies the reliability and stability of BFRPS in such harsh environments through structural analysis and research, providing a new solution for the application of PV power plants in such areas.

1. Introduction

Since President Xi Jinping proposed carbon peaking and carbon neutrality targets in 2020 [1], the "dual-carbon" strategy has been at the center of China's energy transition. Northwest China, with its abundant light resources [2], has become a key area for solar energy development. In 2023, China's National Energy Administration further encouraged large-scale PV power generation base projects focusing on deserts, the Gobi deserts regions [3]. However, the extreme temperature differences and strong wind environments in these regions pose a serious challenge to PVS. Currently, stainless steel and aluminum alloys are the mainstream materials for PVS, but they have significant limitations in desert environments: stainless steel supports have high specific gravity and poor corrosion resistance, while aluminum alloy supports are corrosion-resistant but have insufficient strength and wind resistance.

In modern materials science, fiber-reinforced resin composites have attracted attention for their lightweight, high strength, and excellent corrosion resistance, showing great potential, especially in extreme environments. Studies by Wang Bunjian et al. have shown that basalt fibers (BF) are superior to glass fibers in terms of corrosion, high temperature, UV, and fatigue resistance [4]. In addition, profiles of BF and glass fibers showed feasibility in the application study of PVS by Zhang Wei et al. [5]. and the design parameters of BFRPS have been elaborated by Wang Wei et al. [6].

This study explores the application of BFRPS design in desert-type areas, based on a 1 million kW PV generation power project in the Ulanbuh Desert region of Inner Mongolia. And compares the performance of BFRPS with E-glass fiber reinforced profile supports (E-GFRPS) in the desert environment of Inner Mongolia through simulation analysis. The results show that BFRPS have greater strength margins and less deformation occurs under the same load than E-GFRPS. In addition, compared with stainless steel support (SSS), BFRPS offers lower production costs and higher economic benefits, while the production and use process is more environmentally friendly. Therefore, BFRPS are technically advantageous and economically viable for PVS applications, supporting the sustainable development of PV power generation in harsh climatic environments.

This study covers the physical and chemical properties of the support material, long-term performance evaluation, support structure design, and simulation analysis, aiming to provide new solutions and references for the PVS design in Inner Mongolia and the wider region.

2. BF And its Composite

2.1 Introduction of BF

BF stands out from other fibrous materials because of their excellent physical and mechanical properties and stable chemical characteristics. As shown in Table 1, In neutral, alkaline, and acidic corrosive environments, the mass loss rate of BF is significantly lower than that of E-glass fiber (E-GS), especially in acidic environments, where it is only 5.7 % of that of E-GF demonstrating superior corrosion resistance.

Table 1. Mass loss rate of BF and E-GF after boiling for three h in different mediums [7].

Medium Type	BF	E-GF
H2O (Water)	1.6	6.2
2N NaOH (Alkali)	2.8	6.0
2N HCl (Acid)	2.2	38.9

In addition, Table 2 shows that the material properties of BF are better than aramid but lower than carbon fiber, with higher tensile strength and lower elongation at break than E-GF, which is an ideal reinforcing material for high-performance composite pultruded profiles.

Table 2. Comparison of physical and mechanical properties of BFs with other fiber materials [8].

Performance	BF	E-GF	Aramid	Carbon fiber
Tensile strength (MPa)	3800-4840	3100-3800	2900-3400	3500-6000
Modulus of elasticity (GPa)	79.3-93.1	72.5-75.5	70-140	230-600
Elongation at break (%)	3.1	4.7	2.8-3.6	1.5-2.0
Density (g/cm3)	2.6-2.8	2.54	1.45	1.78
Operating temperature (C)	-969	-510	250	500
Thermal conductivity (w/m·k)	0.031-0.038	0.034-0.040	0.04-0.13	5-185

2.2 BF Composite

BFRP usually refers to a new type of structural material with BF and its products (cloth, tape, felt, yarn, etc.) as reinforcement and reinforced resin as a matrix.

In terms of the material selection of PVS, the Standard GB/T 31539-2015 [9] must be followed to ensure the reliability and safety of the material. According to the test results of the National Quality Inspection and Testing Centre for Glass Reinforced Plastic Products in Table 3, as compared with the standard values, BFRP fully satisfies the requirements of this standard.

Table 3. Comparison between the test results of the mechanical properties of BFRP and the standard values.

Testing Items	Test Results	Standard Value
Longitudinal Tensile Strength (Mpa)	1059	400
Longitudinal Tensile Modulus of Elasticity (GPA)	50.1	30
Transverse Tensile Modulus of Elasticity (GPA)	9.99	7
Longitudinal Compression Strength (Mpa)	377	300
Transverse Compression Strength (Mpa)	87.9	70
Longitudinal Bending Strength (Mpa)	1303	200
Laminated Shear Strength (Mpa)	65.5	28
Bending Modulus of Elasticity (Longitudinal) (GPA)	38.6	0.017

Besides, considering the strong ultraviolet radiation in the desert area, the ultraviolet radiation resistance of the material is especially critical. The test results in Table 4 show that the performance of BF is improved slightly after UV irradiation instead, which indicates that it is almost unaffected by sunlight.

Table 4. Test Data of BFRP under Ultraviolet Radiation.

Testing Items	The Original	UV 1000h	UV 2000h	UV 3000h
Tensile Strength (Mpa)	923	986	1050	979
Tensile Elastic Modulus (GPa)	43.5	42.3	39.4	41.2
Tensile Failure Strain (%)	2.37	2.57	3.05	2.63
Compressive Strength (Mpa)	862	829	843	645
Compressive Elastic Modulus (GPA)	54.8	44.3	45.7	43.3
Bending Strength ((Mpa)	933	1015	1135	1030
Bending Elastic Modulus (GPA)	25.6	25.4	26.8	26.1
Full Cross-section Compressive Performance (Mpa)	196	177	246	206

With its enhanced performance, BFRPs have been widely used in aerospace, defense, and military industries and other advanced fields. Meanwhile, this profile also shows great application potential in construction engineering, petrochemical, automobile, and shipbuilding industries. Especially BFRPs not only meet the physical and mechanical property requirements needed but also show significant advantages in terms of UV resistance and economy, making them an ideal material for PVS in harsh desert-type environments.

3. Overview of the project environment and support design requirements

The site area of the Pilot Project 1 million kW PV power generation project is located in the Ulanbuh Desert region of Bayannur City, Inner Mongolia Autonomous Region, covering an area of about 153,000

IOP Publishing
doi:10.1088/1742-6596/3009/1/012092

m², with an elevation of about 1,025 m. The environmental parameters in this region are shown in the table 5 below.

Table 5. Environment parameters of the Ulanbuh Desert.

Environment Parameters	Value
Average temperature (°C)	8.4
Limit temperature (°C)	40.1 （max) / -34.2 （min）
Average wind speed (m/s)	2.8
Maximum wind speed (m/s)	28
Average annual rainfall (mm)	144.3
Average annual snowfall (mm)	120

The design life of the PVS in this project is set at 25 years, and the structural design strictly follows the national and industry standards such as GB 50797-2012 [10], GB 50794-2012 [11], GB 50009-2012 [12], NB/T10115-2018 [13] and other national and industry standards to ensure that the support performance reaches M30 grade.

4. Analysis of BF Bracket Structures AND SIMULATION analysis

4.1 BFRPS structure

In the structural design of BFRPS, the upper part adopts fixed structural support, while the lower part adopts PHC pipe pile strut; the actual structure of the support is shown in Figure 1.

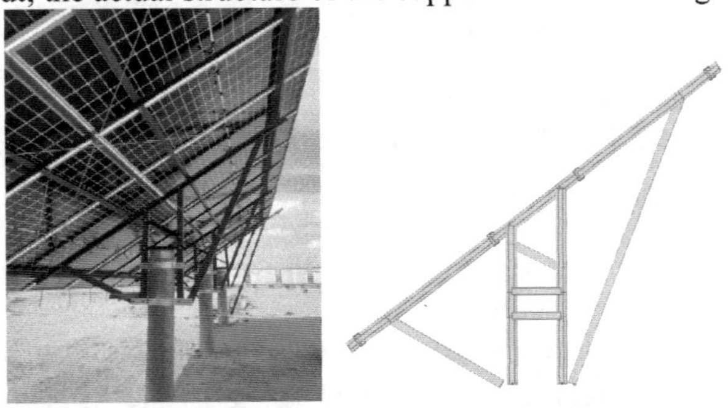

Figure 1. Structure of BFRPS (The left is the actual photo, and the right is the engineering design drawing.)

Considering the characteristics of high wind speed and high wind pressure in the Ulanbuh Desert area, structures such as diagonal bracing are added in the standard bracket structural design to enhance the stability of the bracket system, effectively resist the deformation caused by the wind load, and ensure the long-term and stable operation of PV power plants in the desert environment. The overall model and constraint settings are shown in Figure 2.

4.2 3D modeling of BFRPS structures

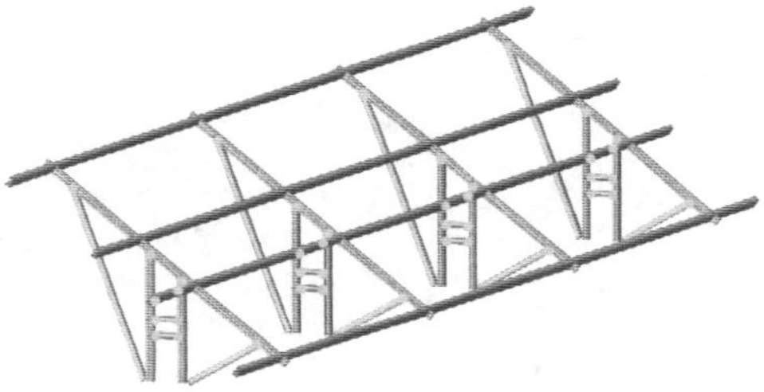

Figure 2. Overview of BFRPS 3D model(green dots in the figure represent part of boundary condition settings).

4.3 Load setting

To assess the mechanical performance of BFRPS under extreme climatic conditions, this study uses a comprehensive load case for simulation and analysis, i.e., the superimposed effects of permanent load, wind load, and snow load are considered to simulate the most unfavorable conditions that may be encountered by the supports in the actual use.

Among them, the permanent load mainly comes from the gravity of the PV module, and its line load can be calculated by equations (1) and (2).

$$P = G / A \tag{1}$$

$$P_L = \frac{1}{2} PL \tag{2}$$

According to the regulations, taking the maximum data from the local meteorological office over 50 years, the basic wind pressure ω_0 is 0.5kN/m^2 and the basic snow pressure s_0 is $0.25kN/m^2$. Wind loads are divided into positive and negative wind loads. Positive wind load means that the wind load acts on the positive surface of the PV module, and negative wind load is the opposite. According to equations (3)-(6), the wind load and snow line load can be found, respectively.

$$\omega_k = \beta_z \mu_s \mu_z \omega_0 \tag{3}$$

$$\omega_L = \frac{1}{2} \omega_k L \sin \alpha \tag{4}$$

$$s_k = \mu_r s_0 \tag{5}$$

$$S_L = \frac{1}{2} S_k L \cos \alpha \tag{6}$$

The names and values of each parameter in the above equation are listed in the table 6 below.

Table 6. List of wind load and snow load calculation parameters.

Name of parameters	Value
β_z Wind vibration coefficient at height z	1.2
μ_s Wind load overall body type coefficient	(Positive) 1.18 (Negative) 1.48
μ_z Wind pressure height change coefficient	1.09
L Span of the PVS (m)	2.278
α Tilt angle of PVS (°)	36
μ_r Roof snow distribution	0.56

4.4 Simulation analysis

According to the load calculation results, the loads are applied to the finite element model, mainly considering two typical loading conditions: 1) permanent load, positive wind load, and snow load superimposed; 2) permanent load superimposed with negative wind load.

Because of the high cost of carbon fiber, which is rarely used in the field of PVS, E-GFRP, with wider use and better performance, is chosen for comparative analysis. The BFRPS and E-GFRPS are simulated, respectively, and Figures 3-6 show the simulation results of BFRPS and E-GFRPS for condition 1, respectively.

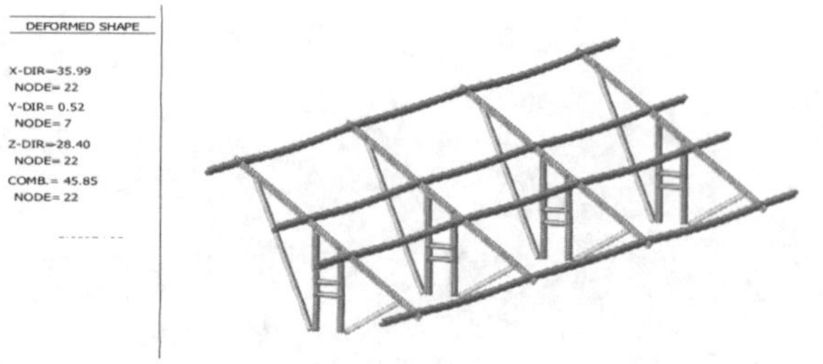

Figure 3 Deformation distribution Cloud map of BFRPS.

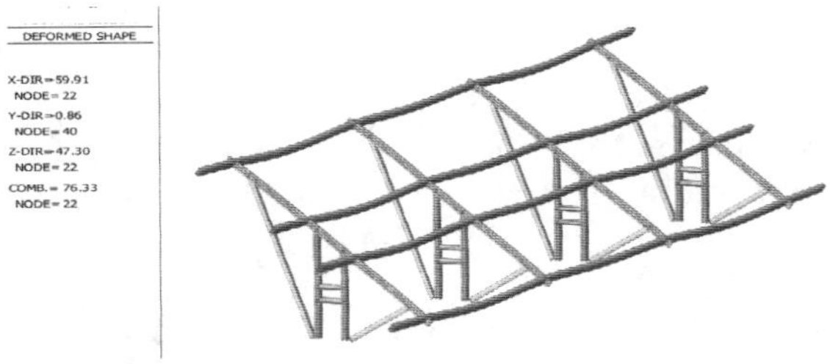

Figure 4 Deformation distribution Cloud map of E-GRPS.

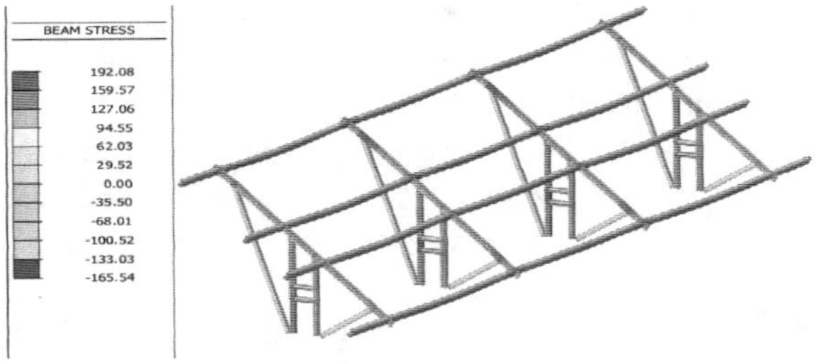

Figure 5 Stress distribution of BFRPS.

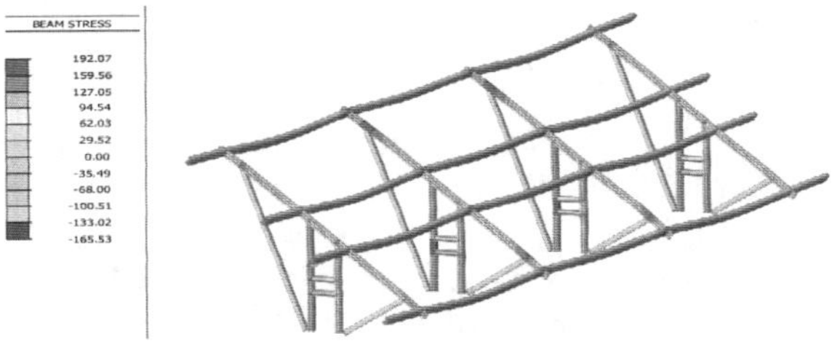

Figure 6 Stress distribution of E-GRPS.

The results show that under the same load, the stress results of BFRPS and E-GFRP are similar because of their similar density, but the performance of BFRPS is better than that of E-GFRP, so the deformation displacement is about 40 % smaller than that of E-GFRP.

Table 7. Part of Simulation Result of BFRPS and E-GFRPS.

	Result	BFRP	E-GFRP
Condition 1	Maximum Deformation (mm)	45.85	76.33
	Maximum Absolute Stress (Mpa)	192.08	192.07
Condition 2	Maximum Deformation (mm)	51.96	86.48
	Maximum Absolute Stress (Mpa)	199.19	199.2

According to the simulation results presented in Table 7, the maximum stress occurs at the columns and diagonal beams, and its value is much lower than the ultimate strength of the two materials, which indicates that the structure meets the relevant requirements in terms of stress. In addition, the maximum displacement is located in the top purlin, and the span of this purlin is about 15 meters. According to the specification, the maximum displacement of the PV must be less than 1/250 of the span, i.e., less than 60 mm. After comparison, the displacement of BFRPS is within the specified range and complies with the requirement, while the deflection of E-GFRPS is too large, which exceeds the standard limit and does not comply with the requirement.

Overall, BFRP can be used as a PVS material in desert areas such as Ulaanbaatar.

5. Engineering economics

The cost-benefit comparison of BFRPS with stainless steel mounts is particularly important in the economic analysis of photovoltaic mounts in the desert region of Inner Mongolia.

In terms of direct costs, the price of BFRPS is about 15,000 RMB/ton, while the unit price of stainless steel is about 6,000 RMB/ton. However, considering that the density of BFRP (about 2.5g/cm3) is only 31.3 % of that of stainless steel (about 7.98 g/cm3), the material cost of BFRPS is approximately 21.8 % lower than that of stainless steel when making the same volume of PVS. In addition, the price of BFRP raw materials (about 650 RMB/ton) is also lower tha190Mpan the price of stainless-steel raw materials (about 700 RMB/ton), it can be foreseen that in the future, with the expansion of BFRP production scale, the production cost is reduced, the cost of BFRPS will be further reduced.

Indirect costs include labor costs, transport costs, and maintenance costs. Due to the lightweight of BFRP, it requires less labor during installation, which reduces labor costs. Also, due to its low density, BFRPS can be transported more efficiently than stainless steel, an advantage that is particularly evident over the vast transport distances in desert areas. In terms of maintenance costs, the superior corrosion and UV resistance of BFRPS means that only a little additional maintenance is required, even in extreme climatic conditions such as the desert.

Considering the material, labor, transport, and maintenance costs, BFRPS are significantly better than stainless steel supports in terms of economy. Therefore, BFRPS are not only technically superior but also economically viable.

6. Conclusion

Among the common fiber reinforcing materials, BF has a small specific gravity, second only to carbon fiber in terms of mechanical properties, and has good chemical stability, which makes it suitable for PVS applications in desert-type environments.

The mechanical properties of BFRP meet the specification requirements and are UV resistant.

Simulation analysis results show that the strength of BFRPS can be used in the Ulanbuh desert region environment. By comparing the simulation results with E-GFRPS it proves that the deformation displacement of BFRPS is smaller than E-GFRPS under the same load.

For the same dimensions, a set of BFRPS is only 31.3 % of the mass of SSS, and the material cost is reduced by about 21.8 %. In addition, because of its lighter mass, BFRPS also has lower indirect costs, including labor costs, transport costs, and maintenance costs, than SSS, which means BFRPS is an ideal alternative to stainless steel braces.

Reference

[1] Li Mingdong, LI Jingwen... Development status and prospect of distributed photovoltaic power generation in China under the goal of "dual carbon" [J]. Solar Energy, 2023, (05): 5-10.
[2] Environmental Planning Institute, Ministry of Ecology and Environment. Evaluation of wind and solar power generation Potential in China (2024) [R], 2024.
[3] Guidance on energy work in 2023 [J]. Power Station Auxiliary Machinery, 2023, 44(02): 29.
[4] Wang Bang-Jian, LIU Jian-Xun, KUANG Wikstroemia.. The development trend of basalt fiber reinforced composite photovoltaic scaffolds [C]// The 4th National Conference on Solar Cell Materials and Devices, 2024:68-69.
[5] Zhang Wei, JIANG Houwen... Application research of a basalt fiber composite photovoltaic scaffold [J]. Glass Fiber, 2019, (02): 18-21.
[6] Wang Wei, Yang Jianlin, Wei Zhiqiang, et al. Design and application of fiber-reinforced composites for photovoltaic scaffolds [J]. Solar Energy, 2023, (11): 95-100.
[7] Xie Ergai 1 Li Zhongying 2 (1.Shengzheng International Technology Innovation Research Institute) (2.Harbin Institute of Technology). Application Prospect of Basalt Fiber[J]. Fiber Composites, 2003(03): 17-20.
[8] Song Ping, Gao Huan, Wang Ling, et al. Basic characteristics and application prospect of basalt fiber [J]. Mineral Resources Conservation and Utilization, 2022, (04): 173-178
[9] General Administration of Quality Supervision, Inspection and Quarantine of the People's Republic of China, Standardization Administration of China. Pultrusion profiles of fiber-reinforced composites for structural purposes. [S], 2015.

Journal of Physics: Conference Series **3009** (2025) 012092 doi:10.1088/1742-6596/3009/1/012092

[10] Ministry of Housing and Urban-Rural Development of the People's Republic of China, General Administration of Quality Supervision, Inspection and Quarantine of the People's Republic of China. Design specification for photovoltaic power stations. [S],2012.
[11] Ministry of Housing and Urban-Rural Development of the People's Republic of China, General Administration of Quality Supervision, Inspection and Quarantine of the People's Republic of China. Construction code for photovoltaic power stations. [S], 2012
[12] Code for Load on building structures. [S], 2012.
[13] National Energy Administration. Code for design of photovoltaic support structures. [S], 2018.

AUTHOR INDEX